Dopamine Handbook

Dopamine Handbook

Edited by

LESLIE L. IVERSEN, PhD, FRS
Professor of Pharmacology
University of Oxford
Oxford, UK

SUSAN D. IVERSEN, PhD
Department of Experimental Psychology
University of Oxford
Oxford, UK

STEPHEN B. DUNNETT, PhD
School of Biosciences
Cardiff University
Cardiff, UK

ANDERS BJÖRKLUND, MD, PhD
Wallenberg Neuroscience Center
Division of Neurobiology
Lund University
Lund, Sweden

OXFORD
UNIVERSITY PRESS
2010

OXFORD
UNIVERSITY PRESS

Oxford University Press, Inc., publishes works that further
Oxford University's objective of excellence
in research, scholarship, and education.

Oxford   New York
Auckland   Cape Town   Dar es Salaam   Hong Kong   Karachi
Kuala Lumpur   Madrid   Melbourne   Mexico City   Nairobi
New Delhi   Shanghai   Taipei   Toronto

With offices in
Argentina   Austria   Brazil   Chile   Czech Republic   France   Greece
Guatemala   Hungary   Italy   Japan   Poland   Portugal   Singapore
South Korea   Switzerland   Thailand   Turkey   Ukraine   Vietnam

Copyright © 2010 by Oxford University Press, Inc.

Published by Oxford University Press, Inc.
198 Madison Avenue, New York, New York 10016

www.oup.com

Oxford is a registered trademark of Oxford University Press.

All rights reserved. No part of this publication may be reproduced,
stored in a retrieval system, or transmitted, in any form or by any means,
electronic, mechanical, photocopying, recording, or otherwise,
without the prior permission of Oxford University Press.

Library of Congress Cataloging-in-Publication Data
Dopamine handbook / [edited by] Leslie L. Iversen ... [et al.].
p. ; cm.
Includes bibliographical references and index.
ISBN: 978-0-19-537303-5
1. Dopaminergic neurons—Handbooks, manuals, etc.   2. Dopaminergic mechanisms—Handbooks, manuals, etc.
3. Dopamine—Handbooks, manuals, etc.   I. Iversen, Leslie L.
[DNLM: 1. Dopamine—physiology—Handbooks.   2. Brain—physiology—Handbooks.
3. Dopamine Agents—Handbooks.   4. Receptors, Dopamine—Handbooks. WK 39 D692 2010]
QP364.7.D666 2010
612.8′042—dc22
2009011010

9 8 7 6 5 4 3 2 1

Printed in the United States of America
on acid-free paper

Preface

The discovery of dopamine as a neurotransmitter in the brain by Arvid Carlsson and colleagues in 1957 proved to be a seminal event in the development of modern neuroscience. Research on dopaminergic mechanisms in the past 50 years has been extremely productive in providing insights into such fundamental aspects of brain function as motor control, cognition, addiction, and reward. More than any other field of neurotransmitter research, dopamine research has provided important links between basic research and clinical practice—with the discovery of the first effective treatments for Parkinson's disease, schizophrenia, and attention deficit hyperactivity disorder (ADHD) based on dopaminergic mechanisms.

The *Dopamine Handbook* arose as an outcome of the symposium "Fifty Years of Dopamine Research" organized by Anders Björklund and colleagues in Goteborg, Sweden, in 2007 to celebrate the 50th anniversary of Carlsson's discovery. Although the proceedings gave rise to a series of excellent short articles in a special edition of *Trends in Neuroscience* (May 2007: Vol 30, No 5, pp. 185–250), we felt that the subject deserved a more detailed review. This culminated in the present handbook, with comprehensive reviews of all major topics in the dopaminergic field written by international experts.

We are very grateful to the many contributors for undertaking the onerous task of writing lengthy reviews, and to the editors, Craig Panner and David D'Addona, at Oxford University Press for their patient overseeing of the project.

Leslie Iversen
Susan Iversen
Stephen Dunnett
Anders Björklund

Contents

Contributors xi

1 Overview: A personal view of the dopamine neuron in historical perspective 3
   Floyd E. Bloom

2 Neuroanatomy 9

   2.1. Functional Neuroanatomy of Dopamine in the Striatum 11
        Charles R. Gerfen

   2.2. Functional Implications of Dopamine D2 Receptor Localization in Relation to
        Glutamate Neurons 22
        Susan R. Sesack

   2.3. Convergence of Limbic, Cognitive, and Motor Cortico-Striatal Circuits with Dopamine
        Pathways in Primate Brain 38
        Suzanne N. Haber

   2.4. The Relationship between Dopaminergic Axons and Glutamatergic Synapses in the Striatum:
        Structural Considerations 49
        Jonathan Moss and *J. Paul Bolam*

3 Molecular pharmacology 61

   3.1. Molecular Pharmacology of the Dopamine Receptors 63
        *Michele L. Rankin, Lisa A. Hazelwood, R. Benjamin Free, Yoon Namkung,
        Elizabeth B. Rex, Rebecca A. Roof,* and *David R. Sibley*

   3.2. Role of Dopamine Transporters in Neuronal Homeostasis 88
        Marc G. Caron and *Raul R. Gainetdinov*

   3.3. Intracellular Dopamine Signaling 100
        Gilberto Fisone

   3.4. Ion Channels and Regulation of Dopamine Neuron Activity 118
        Birgit Liss and *Jochen Roeper*

4 Genes in development 139

   4.1. Genetic Control of Meso-diencephalic Dopaminergic Neuron Development in Rodents 141
        Wolfgang Wurst and *Nilima Prakash*

   4.2. Factors Shaping Later Stages of Dopamine Neuron Development 160
        Robert E. Burke

   4.3. Postnatal Maturation of Dopamine Actions in the Prefrontal Cortex 177
        Patricio O'Donnell and *Kuei Y. Tseng*

viii   CONTENTS

    4.4.   Genetic Dissection of Dopamine-Mediated Prefrontal-Striatal Mechanisms and Its Relationship to Schizophrenia   187
          Hao-Yang Tan and *Daniel R. Weinberger*

5   **Dopamine in prefrontal cortex and cognition**   201

    5.1.   From Behavior to Cognition: Functions of Mesostriatal, Mesolimbic, and Mesocortical Dopamine Systems   203
          Trevor W. Robbins

    5.2.   Contributions of Mesocorticolimbic Dopamine to Cognition and Executive Function   215
          Stan B. Floresco

    5.3.   Dopamine's Influence on Prefrontal Cortical Cognition: Actions and Circuits in Behaving Primates   230
          Amy F.T. Arnsten, Susheel Vijayraghavan, Min Wang, Nao J. Gamo, and *Constantinos D. Paspalas*

    5.4.   Dopaminergic Modulation of Flexible Cognitive Control in Humans   249
          Roshan Cools and *Mark D'Esposito*

    5.5.   Neurocomputational Analysis of Dopamine Function   261
          Daniel Durstewitz

6   **Striatum and midbrain—motor and motivational functions**   277

    6.1.   Dopamine and Motor Function in Rat and Mouse Models of Parkinson's Disease   279
          Timothy Schallert and *Sheila M. Fleming*

    6.2.   Involvement of Nucleus Accumbens Dopamine in Behavioral Activation and Effort-Related Functions   286
          John D. Salamone

    6.3.   Functional Heterogeneity in Striatal Subregions and Neurotransmitter Systems: Implications for Understanding the Neural Substrates Underlying Appetitive Motivation and Learning   301
          Brian A. Baldo and *Matthew E. Andrzejewski*

    6.4.   Behavioral Functions of Dopamine Neurons   316
          Philippe N. Tobler

7   **Plasticity of forebrain dopamine systems**   331

    7.1.   Dynamic Templates for Neuroplasticity in the Striatum   333
          Ann M. Graybiel

    7.2.   Dopamine and Synaptic Plasticity in Mesolimbic Circuits   339
          F. Woodward Hopf, Antonello Bonci, and *Robert C. Malenka*

    7.3.   Dopaminergic Modulation of Striatal Glutamatergic Signaling in Health and Parkinson's Disease   349
          D. James Surmeier, Michelle Day, Tracy S. Gertler, C. Savio Chan, and *Weixing Shen*

8   **Dopamine mechanisms in addiction**   369

    8.1.   The Role of Dopamine in the Motivational Vulnerability to Addiction   371
          George F. Koob and *Michel Le Moal*

    8.2.   Dopaminergic Mechanisms in Drug-Seeking Habits and the Vulnerability to Drug Addiction   389
          Barry J. Everitt, David Belin, Jeffrey W. Dalley, and *Trevor W. Robbins*

    8.3.   Imaging Dopamine's Role in Drug Abuse and Addiction   407
          Nora D. Volkow, Joanna S. Fowler, Gene-Jack Wang, Frank Telang, and *Ruben Baler*

9 Parkinson's disease 419

- 9.1. Exploring the Myths about Parkinson's Disease 421
  Yves Agid and *Andreas Hartmann*

- 9.2. Pathophysiology of L-DOPA-Induced Dyskinesia in Parkinson's Disease 434
  M. Angela Cenci

- 9.3. Progression of Parkinson's Disease Revealed by Imaging Studies 445
  David J. Brooks

- 9.4. Transplantation of Dopamine Neurons: Extent and Mechanisms of Functional Recovery in Rodent Models of Parkinson's Disease 454
  Stephen B. Dunnett and *Anders Björklund*

- 9.5. Clinical Experiences with Dopamine Neuron Replacement in Parkinson's Disease: What Is the Future? 478
  Olle Lindvall

- 9.6. Novel Gene-Based Therapeutics Targeting the Dopaminergic System in Parkinson's Disease 489
  Deniz Kirik, Tomas Björklund, Shilpa Ramaswamy, and *Jeffrey H. Kordower*

- 9.7. Neuroprotective Strategies in Parkinson's Disease 498
  C. Warren Olanow

10 Schizophrenia and other psychiatric illnesses 509

- 10.1. Dopamine Dysfunction in Schizophrenia 511
  Anissa Abi-Dargham, Mark Slifstein, Larry Kegeles, and *Marc Laruelle*

- 10.2. Neuropharmacological Profiles of Antipsychotic Drugs 520
  Bryan L. Roth and *Sarah C. Rogan*

- 10.3. How Antipsychotics Work: Linking Receptors to Response 540
  Nathalie Ginovart and *Shitij Kapur*

- 10.4. Dopamine Dysfunction in Schizophrenia: From Genetic Susceptibility to Cognitive Impairment 558
  Heike Tost, Shabnam Hakimi and *Andreas Meyer-Lindenberg*

- 10.5. The Role of Dopamine in the Pathophysiology and Treatment of Major Depressive Disorder 572
  Boadie W. Dunlop and *Charles B. Nemeroff*

- 10.6. Dopamine Modulation of Forebrain Pathways and the Pathophysiology of Psychiatric Disorders 590
  Anthony A. Grace

Index 599

Contributors

Anissa Abi-Dargham, MD
Department of Psychiatry
Columbia University and New York State
 Psychiatric Institute
New York, NY

Yves Agid, MD, PhD
Centre d'Investigation Clinique
Hôpital de la Pitié-Salpêtrière
Paris, France

Matthew E. Andrzejewski, PhD
Waisman Center
University of Wisconsin-Madison
Madison, WI

Amy F.T. Arnsten, PhD
Department of Neurobiology
Yale Medical School
New Haven, CT

Brian A. Baldo, PhD
Department of Psychiatry
University of Wisconsin-Madison
School of Medicine and Public Health
Madison, WI

Ruben Baler, PhD
Office of Science Policy and Communications
National Institute on Drug Abuse
National Institutes of Health
Bethesda, MD

David Belin, PhD
Pôle Biologie Santé
CNRS UMR 6187 & Université de Poitiers
40, avenue du Recteur Pineau
86022 POITIERS cedex – FRANCE

Anders Björklund, MD, PhD
Wallenberg Neuroscience Center
Division of Neurobiology
Lund University
Lund, Sweden

Tomas Björklund, M. Med
Department of Experimental Medical Science
Lund University
Lund, Sweden

Floyd E. Bloom, MD
Molecular and Integrative Neuroscience
 Department
The Scripps Research Institute
La Jolla, CA

J. Paul Bolam, PhD
Medical Research Council, Anatomical
 Neuropharmacology Unit
Department of Pharmacology
University of Oxford
Oxford, UK

Antonello Bonci, MD
Ernest Gallo Clinic and Research Center
Department of Neurology
Wheeler Center for the Neurobiology of Addiction
Program in Neuroscience, University
 of California, San Francisco
San Francisco, CA

David J. Brooks, MD, DSc, FRCP, FMedSci
Division of Neurosciences and Mental Health
Imperial College
London, UK

Robert E. Burke, MD
Department of Neurology
The College of Physicians and Surgeons
Columbia University
New York, NY

Marc G. Caron, PhD
Department of Cell Biology
Duke University Medical Center
Durham, NC

M. Angela Cenci, MD, PhD
Department of Experimental Medical Science
Lund University
Lund, Sweden

C. Savio Chan, PhD
Department of Physiology
Feinberg School of Medicine
Northwestern University
Chicago, IL

Roshan Cools, PhD
Donders Institute for Brain, Cognition and Behavior
Radboud University Nijmegen Medical Centre
Centre for Cognitive Neuroimaging
Nijmegen, The Netherlands

Jeffrey W. Dalley, PhD
Department of Psychiatry
Behavioral and Clinical Neuroscience Institute
University of Cambridge
Cambridge, UK

Michelle Day, PhD
Department of Physiology
Feinberg School of Medicine
Northwestern University
Chicago, IL

Mark D'Esposito, MD
Helen Willis Neuroscience Institute
University of California, Berkeley
Berkeley, CA

Boadie W. Dunlop, MD
Department of Psychiatry and Behavioral Sciences
Emory University School of Medicine
Atlanta, GA

Stephen B. Dunnett, PhD
School of Biosciences
Cardiff University
Cardiff, UK

Daniel Durstewitz, PhD
Computational Neuroscience Group
Central Institute of Mental Health
University of Heidelberg
Mannheim, Germany

Barry J. Everitt, ScD
Department of Experimental Psychology
Behavioral and Clinical Neuroscience Institute
University of Cambridge
Cambridge, UK

Gilberto Fisone, PhD
Department of Neuroscience
Karolinska Institutet
Stockholm, Sweden

Sheila M. Fleming, PhD
Assistant Professor
Departments of Psychology and Neurology
University of Cincinnati
Cincinnati, OH 45221

Stan B. Floresco, PhD
Department of Psychology
Brain Research Centre
University of British Columbia
Vancouver, B.C. Canada

Joanna S. Fowler, PhD
National Institute on Drug Abuse
National Institutes of Health
Bethesda, MD

R. Benjamin Free, PhD
Molecular Neuropharmacology Section
National Institute of Neurological Disorders and Stroke
National Institutes of Health
Bethesda, MD

Raul R. Gainetdinov, MD, PhD
Department of Neuroscience and Brain Technologies
Italian Institute of Technology
Genova, Italy

Nao J. Gamo
Department of Neurobiology
Yale Medical School
New Haven, CT

Charles R. Gerfen, PhD
Laboratory of Systems Neuroscience
National Institute of Mental Health
National Institutes of Health
Bethesda, MD

Tracy S. Gertler, BS
Department of Physiology
Feinberg School of Medicine
Northwestern University
Chicago, IL

Nathalie Ginovart, PhD
Neuroimaging Unit
Department of Psychiatry
University of Geneva
Geneva, Switzerland

Anthony A. Grace, PhD
Departments of Neuroscience, Psychiatry, and
  Psychology
University of Pittsburgh
Pittsburgh, PA

Ann M. Graybiel, PhD
Department of Brain and Cognitive Sciences
McGovern Institute for Brain Research
Massachusetts Institute of Technology
Cambridge, MA

Suzanne N. Haber, PhD
Department of Pharmacology and Physiology
University of Rochester School of Medicine
Rochester, NY

Shabnam Hakimi, BA
Clinical Brain Disorders Branch
National Institute of Mental Health
National Institutes of Health
Bethesda, MD

Andreas Hartmann, MD, PhD
Centre d'Investigation Clinique
Hôpital de la Pitié-Salpêtrière
Paris, France

Lisa A. Hazelwood, PhD
Molecular Neuropharmacology Section
National Institute of Neurological Disorders
  and Stroke
National Institutes of Health
Bethesda, MD

F. Woodward Hopf, PhD
Ernest Gallo Clinic and Research Center
Department of Neurology
Wheeler Center for the Neurobiology of
  Addiction
University of California, San Francisco
San Francisco, CA

Shitij Kapur, MD, PhD, FRCPC
Institute of Psychiatry
King's College London
London, UK

Larry Kegeles, MD
Departments of Psychiatry and Radiology
Columbia University
New York State Psychiatric Institute
New York, NY

Deniz Kirik, MD, PhD
Department of Experimental Medical Science
Lund University
Lund, Sweden

George F. Koob, PhD
Committee of the Neurobiology of Addictive
  Disorders
The Scripps Research Institute
La Jolla, CA

Jeffrey H. Kordower, PhD
Department of Neurological Sciences
Rush University Medical Center
Chicago, IL

Marc Laruelle, MD
Schizophrenia and Cognitive Disorder Discovery
  Performance Unit
Neurosciences Center of Excellence in Drug
  Discovery
GlaxoSmithKline
Harlow, UK

Michel Le Moal, MD, PhD
Neurocentre Magendie
Institut National de la Santé et de la Recherche
  Médicale, Unite 862
Institut François Magendie
Université Victor Ségalen Bordeaux 2
Bordeaux, France

Olle Lindvall, MD, PhD
Laboratory of Neurogenesis and Cell Therapy
Section of Restorative Neurology
Wallenberg Neuroscience Center
Lund University Hospital
Lund, Sweden

Birgit Liss, PhD
Section Molecular Neurophysiology
Department of General Physiology
University of Ulm
Ulm, Germany

Robert C. Malenka, MD, PhD
Nancy Pritzker Laboratory
Department of Psychiatry and Behavioral Sciences
co-Director, Stanford Institute for Neuro-Innovation and Translational Neurosciences
Stanford University School of Medicine
Palo Alto, CA

Andreas Meyer-Lindenberg, MD, PhD
Department of Psychiatry and Psychotherapy
Central Institute of Mental Health
University of Heidelberg
Mannheim, Germany

Jonathan Moss
Medical Research Council, Anatomical Neuropharmacology Unit
Department of Pharmacology
University of Oxford
Oxford, UK

Yoon Namkung, PhD
Molecular Neuropharmacology Section
National Institute of Neurological Disorders and Stroke
National Institutes of Health
Bethesda, MD

Charles B. Nemeroff, MD, PhD
Department of Psychiatry and Behavioral Sciences
Emory University School of Medicine
Atlanta, GA

Patricio O'Donnell, MD, PhD
Department of Anatomy and Neurobiology
University of Maryland School of Medicine
Baltimore, MD

C. Warren Olanow, MD, FRCPC
Department of Neurology and Neuroscience
Mount Sinai School of Medicine
New York, NY

Constantinos D. Paspalas, PhD
Division of Neuroanatomy
University of Crete School of Medicine
Heraklion, Greece

Nilima Prakash, PhD
Institute of Developmental Genetics
Helmholtz Center Munich
German Research Center for Environmental Health
Technical University Munich
Munich/Neuherberg, Germany

Shilpa Ramaswamy
Department of Neurological Sciences
Rush University Medical Center
Chicago, IL

Michele L. Rankin, PhD
Molecular Neuropharmacology Section
National Institute of Neurological Disorders and Stroke
National Institutes of Health
Bethesda, MD

Elizabeth B. Rex, PhD
Molecular Neuropharmacology Section
National Institute of Neurological Disorders and Stroke
National Institutes of Health
Bethesda, MD

Trevor W. Robbins, PhD
Department of Experimental Psychology and Behavioral and Clinical Neuroscience Institute
University of Cambridge
Cambridge, UK

Jochen Roeper, MD, PhD
Institute of Neurophysiology
Neuroscience Center
Goethe-University Frankfurt
Frankfurt, Germany

Sarah C. Rogan
Department of Pharmacology
University of North Carolina, Chapel Hill
Chapel Hill, NC

Rebecca A. Roof, PhD
Molecular Neuropharmacology Section
National Institute of Neurological Disorders and Stroke
National Institutes of Health
Bethesda, MD

Bryan L. Roth, MD, PhD
Departments of Pharmacology, Medicinal Chemistry, and Psychiatry
Linberger Comprehensive Cancer Center and Neuroscience Center
University of North Carolina, Chapel Hill
Chapel Hill, NC

John D. Salamone, PhD
Division of Behavioral Neuroscience
Department of Psychology
University of Connecticut
Storrs, CT

Timothy Schallert, PhD
Department of Psychology
University of Texas at Austin
Austin, TX

Susan R. Sesack, PhD
Departments of Neuroscience and Psychiatry
University of Pittsburgh
Pittsburgh, PA

Weixing Shen, PhD
Department of Physiology
Feinberg School of Medicine
Northwestern University
Chicago, IL

David R. Sibley, PhD
Molecular Neuropharmacology Section
National Institute of Neurological Disorders and Stroke
National Institutes of Health
Bethesda, MD

Mark Slifstein, PhD
Departments of Psychiatry and Radiology
Columbia University
New York State Psychiatric Institute
New York, NY

D. James Surmeier, PhD
Department of Physiology
Feinberg School of Medicine
Northwestern University
Chicago, IL

Hao-Yang Tan, MBBS, MMed, FRCPC
Clinical Brain Disorders Branch
National Institute of Mental Health
National Institutes of Health
Bethesda, MD

Frank Telang, MD
Brookhaven National Laboratory
Upton, NY

Philippe N. Tobler, PhD
Department of Physiology, Development, and Neuroscience
University of Cambridge
Cambridge, UK

Heike Tost, MD, PhD
Clinical Brain Disorders Branch
National Institute of Mental Health
National Institutes of Health
Bethesda, MD

Kuei Y. Tseng, MD, PhD
Department of Cellular and Molecular Pharmacology
Rosalind Franklin University of Medicine and Science
The Chicago Medical School
Chicago, IL

Susheel Vijayraghavan, PhD
Laboratory of Neuropsychology
National Institute of Mental Health
National Institutes of Health
Bethesda, MD

Nora D. Volkow, MD
Director, National Institute on Drug Abuse
National Institutes of Health
Bethesda, MD

Gene-Jack Wang, MD
Brookhaven National Laboratory
Upton, NY

Min Wang, PhD
Department of Neurobiology
Yale Medical School
New Haven, CT

Daniel R. Weinberger, MD
Clinical Brain Disorders Branch
National Institute of Mental Health
National Institutes of Health
Bethesda, MD

Wolfgang Wurst, PhD
Institute of Developmental
 Genetics
Helmholtz Center Munich
German Research Center for Environmental
 Health
Technical University Munich
Max Planck Institute of
 Psychiatry
Munich, Germany

Dopamine Handbook

1 | Overview: A personal view of the dopamine neuron in historical perspective

FLOYD E. BLOOM

INTRODUCTION

My goals for this overview perspective are to portray what I consider to have been the major discoveries in dopamine (DA) research in the central nervous system from the anatomical, synaptic, and neurohistochemical perspectives, in keeping with my own didactic hypothesis that "the gains in brain are mainly in the stain."[1-5] This handbook was developed in the postcelebratory interval following the semicentennial of the discoveries that led to the recognition of DA as a neurotransmitter in its own right, independent of the other central catecholamines, norepinephrine and epinephrine. My historical perspective began with the first comprehensive review of the structure and function of the central DA neuronal systems done after the first 20 years of DA research.[6]

What stands out now, even more profoundly than it did in 1978, was how primitive our concepts were then of the range of regulatory functions that could be carried out by the systems of neurons characterized primarily on the basis of employing DA as their primary neurotransmitter (recall that 30 years ago, there was no thought that neurons might use more than one substance to transmit their interneuronal signals). Indeed, it was the ability to localize—neurocytologically—the neurotransmitters of the monoamine families that permitted what now can be seen as the progressive analysis of the synthesis, storage, release, conservation, and catabolism that has in consequence pioneered the functional conceptualizations of synaptic transmitter metabolism, not to mention the modes of action of most psychotropic drugs. A large proportion of those early studies were done biochemically on samples of brain tissues dissected without regard to the characterizations of the neuronal circuits that made, stored, and released DA, and partly as a result, our understanding of the structure and function of the DA-containing cells and their circuits in the central nervous system emerged much more slowly.

HOW DO WE 'SEE' DA?

Research into brain DA systems was accelerated because of the role attributed to this transmitter's being deficient in Parkinson's disease (see Agid and Hartmann, Chapter 9.1, this volume), acting excessively in schizophrenia (see Abi-Dargham et al., Chapter 10.1, this volume) and mediating the internal reward systems of the brain (see Koob and Le Moal, Chapter 8.1, this volume). Thus, it was recognized early in the course of this work that DA neuron systems are more complex in their anatomy, more diverse in localization and apparent function, and more numerous, both in terms of definable functional systems (i.e., motor, sensory and reward) and in numbers of neurons, than the other central catecholamine systems. The initial results from formaldehyde-induced fluorescence microscopy indicated that the DA neuron systems were principally located in the upper mesencephalon and diencephalon.[7-10] These systems were not immediately accepted by the professional neuroanatomists of the time because the DA fiber systems were fine, unmyelinated, and largely invisible to the degenerative tract tracing methods then in vogue and because the freeze-dry process that was required to drive the histochemical coupling with formaldehyde fragmented the brains. These problems were later overcome with the development of the vibratome-glyoxylic methods[11,12] and by other molecular probes of the enzymes, receptors, and transporters (see below).

In its early stages, the vast majority of DA research focus was on the major projection from the substantia nigra, the pars compacta to the neostriatum, and, somewhat later, to the projections between the ventral tegmental area and the olfactory tubercle and ventral striatum.[6] With the advent of new, powerful neuroanatomical methods including optical[13,14] and ultrastructural immunocytochemistry for synthetic enzymes,[15] receptors, and transporters (see Sibley et al., Chapter 3.1, and Caron and Gainetdinov, Chapter 3.2, this volume), there has been a very rapid, marked increase in our understanding of the extent and organization of DA neuron systems. This understanding prominently included the detection of bone fide synaptic specializations when the ultrastructurally defined DA terminals in cortex and other forebrain target areas were analyzed with serial section reconstructions.[16,17] Subsequent work has demonstrated DA neurons within the retina,[18] within the olfactory bulb,[19,20] and in projections to the hair cells in the cochlea.[21,22] Not explicitly included in this handbook are the roles of hypothalamic dopaminergic systems in neuroendocrine regulation (see [23] for a recent review).

The surprise finding of the early work was the demonstration of a small but concentrated projection of DA fibers to the rat prefrontal cortex.[24-26] As the compelling involvement of DA in human brain diseases became evident, repetition of the DA molecular mapping tools to the human and nonhuman primate cortex revealed a major, regionally and laminarly selective distribution of DA fibers in the primate medial prefrontal cortex[27,28] that was far more prominent than that observed in the rodent's limited frontal cortical area. Comparisons of the nigrostriatal synaptic arrangements of receptors and transporters with those of the cortical DA terminal fields revealed a striking difference: in striatum, based on the ultrastructural localization of the DA transporter, reuptake seems to occur close to the presumptive synaptic sites bearing the postsynaptic receptors, while in cortex, the transporters are quite remote from presumptive response sites, raising the reality of the popular theory of localized *volume transmission* (see Sibley et al., Chapter 3.1, and Caron and Gainetdinov, Chapter 3.2, this volume; but also see below and [29]). In addition, when DA neurons projecting to cortical target areas were characterized physiologically and neurochemically, it was observed that the cortically projecting neurons exhibited more burst firing than those projecting to the striatum, that their turnover of DA was faster and critically dependent on extracellular tyrosine, that these neuronal somata were less responsive to DA agonists and antagonists, and that in contrast to those projecting to striatum, they showed no tolerance to antipsychotic drugs.[30-32]

Those observations set the stages for the stunning functional depictions of the cellular effects of DA within the primate cortical circuitry that have been linked to cognition (see Robbins, Chapter 5.1; Floresco, Chapter 5.2; and Arnsten et al., Chapter 5.3, this volume) and provided hypotheses of the sequences of pathophysiology in schizophrenia[33-35] (also see Meyer-Lindenberg et al., Chapter 10.4, this volume). Thus, were one forced to pick one thread that has empowered the dynamic thrust of DA research and its relevance for human disease, one need go no further than the ability to localize DA neurons and their circuits.

While this handbook provides a remarkably complete analysis of the many active facets of DA research today, I want to concentrate my analysis of the question I first asked in my brief neurophysiological studies of DA, namely, what does DA do?

It is important to note that the relatively slower rate of progress in elucidating the physiological properties of the DA neuron resulted in part from the fact that methods for the analysis of any catecholamine-containing pathway required techniques not previously available because at the time there were no other neurochemically defined neuronal pathways to be studied, and in part from the fact that once these techniques were available and in widespread use, the body of physiological data that was produced indicated that catecholamine neuron systems had properties that differed dramatically from those of "classical" central systems[6]—by refusing to align with the prototypical bimodal characterization of being either an excitatory or an inhibitory neurotransmitter. The current status of these issues is comprehensively addressed in this handbook (see Gerfen, Chapter 2.1; Sesack Chapter 2.2; Caron and Gainetdinov, Chapter 3.2; Malenka et al., Chapter 7.2; and Surmeier et al., Chapter 7.3, this volume). Thus, my overview of the progress here will be highly selective in order to address two issues: (1) When do DA neurons fire? and (2) What does DA do to the postsynaptic targets of DA neurons?

WHEN DO DA NEURONS FIRE?

The function of a DA neuron, like that of any other neuron, is established by its innate electrophysiological properties and by their synaptic input, which determines the activity of the system. Dopamine neurons in rats were initially localized post hoc by marking the recording sites using any of several methods and examining the recording sites cytologically after the experiment (see[36] and Grace, Chapter 10.6, this volume). However, they were soon found to have a unique and reliable electrophysiological signal that allowed research to progress more rapidly, first in anesthetized animals and somewhat later in awake, behaving animals.[37] Their signature activity consisted of polyphasic action potentials, initially positive or negative waveforms, followed by a prolonged positive component with a relatively long duration (1.8–3.6 ms) and an irregular firing at low baseline frequencies (0.5–8.5 spikes/s). In several experiments, it was possible to confirm the changes in activity in striatum and nucleus accumbens by microdialysis of DA.[38] These early studies confirmed that DA neuron firing was slow in its basal state but shifted to burst firing when significant amounts of extracellular DA were detected.

When studies of DA neuron function were performed in the nonhuman primate model, the activity of the neurons was also characterized by their electrophysiological signature, which was similar to that previously observed in the rat.[39] This transition has proven to be very fruitful in that the greater cognitive skills of the primate permitted much more complex behavioral paradigms beyond drug self-administration and revealed that DA neuronal firing was not simply a signal that a reward to a behavioral response had been generated, but instead suggested that the fluctuating output of the primate DA neuron apparently signals changes or errors in the predictions of future salient and rewarding events. This has led to the emergence of quantitative theories of adaptive optimizing control.[40,41]

An interesting question is how DA, released from axon collaterals and dendrites, regulates the activity of the DA neuron. Using whole neuron recordings in slices from mouse substantia nigra and the ventrotegmental area, Beckstead and Williams[42] observed an inhibitory postsynaptic current (IPSC) that was elicited by localized electrical stimulation of nearby DA neurons This IPSC was tetrodotoxin sensitive, calcium dependent, and blocked by a D2 receptor antagonist. Inhibition of monoamine transporters prolonged the IPSC, indicating that the time course of DA neurotransmission is tightly regulated by reuptake. Changing the stimulus intensity altered the amplitude but not the time course of the IPSC, whose onset was faster than could be reproduced with iontophoresis. The results indicate a rapid rise in local DA concentration at the D2 receptors, suggesting that the DA that is released by a train of action potentials acts in a localized somatodendritic area, observations the authors conclude are incompatible with volume transmission.

WHAT DOES DA DO TO THE POSTSYNAPTIC TARGETS OF DA NEURONS?

My singular adventure into the terminal fields of the nigra neurons focused on a question frequently raised at the beginning of cellular neuropharmacology, namely, "What does DA do?" In the parlance of the times, this question really meant "Is DA an excitatory or an inhibitory transmitter?" since those were the only possibilities then thought to exist. The caudate nucleus had been shown to be the region of the brain that had the highest content of DA, and we elected to probe for its effects on the spontaneous activity of caudate neurons in the cat because the caudate nucleus is impossible to miss with stereotaxic micromanipulators. The challenge of these experiments became immediately obvious when we could find virtually no continuously active spontaneous neurons in intact but anesthetized cats, and so before we could ask what DA did, we needed more active neurons on which to test it. In order to avoid the effects of general anesthetics, we resorted to a surgical method of forebrain isolation, which, unbeknown to us then, also severed the nigrostriatal pathway and greatly improved spontaneous activity. With these slowly discharging striatal units, we observed that acetylcholine or glutamate enhanced activity, while DA not only slowed the discharge, it did so for a prolonged period after its iontophoretic application, during which time excitatory responses to acetylcholine gradually returned.[1] This was our starting point in confirming that DA did indeed act differently than the effects of other transmitters known at that time.

As this handbook makes abundantly clear, the neuropharmacology of DA cannot be considered meaningful without characterization of the receptor subtypes under study, which sets the stage for the intracellular transductive pathways through which the transmitter is acting, and the range of pharmacological tools with which to simulate or antagonize the effects of DA (see Sibley et al., Chapter 3.1; and Fisone, Chapter 3.3, this volume).

However, as with the conceptualizations of when DA neurons fire and what that firing connotes to the individual whose behavior is under observation, major steps forward in elucidating the effects of DA on its target neurons in the nonhuman primate prefrontal cortex (PFC) also resulted from studies recording neuron activity and behavior. The meticulous experimental work of Sawaguchi and Goldman-Rakic,[43,44] first demonstrated with a delayed oculomotor response task, showed that when the D1 receptor antagonists SCH23390 and SCH39166 were injected into the PFC, rhesus monkeys increased their errors and increased performance latency in a task that required memory-guided saccades, in a dose-dependent manner that was proportional to the duration of the delay period, but the D1 antagonists had no effect on performance in a control task requiring visually guided saccades. Based on a variety of other evidence including aging, disease, prior drug treatments, and experience, Sawaguchi and Goldman-Rakic proposed that when cortical DA was reduced below normal levels of function, a D1 agonist would improve function, while when cortical DA function was excessive, as hypothesized in schizophrenia, a D1 antagonist would improve function.

When the Goldman-Rakic group turned their efforts to the functional effects of the D2 receptor in PFC using similar methods with behavioral tasks employing a sequence of phasic and tonic activations linked to a train of sensory, mnemonic, and response-related events, they observed that the DA D2 receptor selectively modulated the neural activities associated with memory-guided saccades in oculomotor delayed-response tasks but had little or no effect on the persistent mnemonic-related activity regulated by the D1 receptors[45] (also see Arnsten et al., Chapter 5.3, this volume).

Still, for some observers, the question of what DA does remains, and for those for whom this query remains unanswered, recent observations on neuronal actions in the mouse caudate nucleus should help settle the issue. Not incidentally, these recent experiments were empowered by observations derived from neuronal localizations through stains. The vast majority of neurons in the striatum are medium spiny neurons, so named for the dendritic spines that characterize their morphology. Their organization was termed a *complex mosaic* organization (see Gerfen, Chapter 2.1, this volume). Through combinations of tract tracing and experimental lesions, it was recognized that the output circuitry of the caudate consists of two principal pathways: a striatonigral projection to the substantia nigra and the entopeduncular nucleus (referred to as the *external pathway*) and a striatopallidal projection to the globus pallidus (referred to as the *internal pathway*). Although all medium spiny neurons contain GABA, their associated neuropeptides and DA receptor subtype differ, depending on which output pathway they are in:

by immunocytochemistry and in situ hybridization, the striatonigral neurons express the neuropeptides substance P, and dynorphin, as well as the D1 receptor, while the striatopallidal neurons express proenkephalin and the D2 receptor.

By developing transgenic mice in which the expression of D1 or D2 receptors was molecularly reported by the coexpression of green fluorescent protein, it was possible for Surmeier and colleagues[46] to investigate two forms of striatal synaptic plasticity: long-term potentiation (LTP) and long-term depression (LTD). By controlling the sequences of presynaptic and postsynaptic activity in striatal slices with small microelectrodes that stimulated either afferent fibers close to a neuron or the neuron itself, when presynaptic activity precedes postsynaptic activity, LTP is produced, and when the sequence is reversed, LTD is produced. D2-expressing neurons were capable of reacting with either LTP or LTD according to the stimulation sequences, and here D2, but not D1, antagonists could block both potentiations. In D1-expressing spiny neurons, an LTP mediated by *N*-methyl-d-aspartic acid (NMDA) receptors was observed, but LTD was difficult to produce unless the D1 receptors were blocked with SCH23390; this LTD was, in turn, antagonized by blockade of the metabotropic Glutamate (GLU) type 5 receptor (mGluR5). These observations indicate that DA "is critical for the induction of the plasticity" acting in concert with GLU, adenosine, and activity in the external world. In conditions in which there are few if any behaviorally interesting stimuli, DA neurons fire slowly to keep high-affinity D2 receptors activated but not low-affinity D1 receptors; the latter are engaged when DA neurons fire in bursts to raise DA levels. Thus, the direction of the plasticity shaped by the same transmitter under different conditions will have distinct but consistent effects. These sorts of modulatory effects help establish in my mind, if not for others, the dynamic synaptic vocabulary for monoamine neurons that I have long envisioned.[47]

REFERENCES

1. Bloom FE, Costa E, Salmoiraghi GC. Anesthesia and the responsiveness of individual neurons of the caudate nucleus of the cat to acetylcholine, norepinephrine and dopamine administered by microelectrophoresis. *J Pharmacol Exp Ther.* 1965;150:244–252.
2. Bloom FE, Algeri S, Groppetti A, Revuelta A, Costa E. Lesions of central norepinephrine terminals with 6-OH-dopamine: biochemistry and fine structure. *Science.* 1969;166:1284–1286.
3. Bloom FE, Costa E. The effects of drugs on serotonergic nerve terminals. *Adv Cytopharmacol.* 1971;1:379–395.

4. Bloom FE, Costa E, Salmoiraghi GC. Analysis of individual rabbit olfactory bulb neuron responses to the microelectrophoresis of acetylcholine, norepinephrine and serotonin synergists and antagonists. *J Pharmacol Exp Ther*. 1965;146:16–23.
5. Bloom FE, Von B, Oliver AP, Costa E, Salmoiraghi GC. Microelectrophoretic studies of adrenergic mechanisms of rabbit olfactory neurons. *Life Sci*. 1964;3:131–136.
6. Moore RY, Bloom FE. Central catecholamine neuron systems: anatomy and physiology of the dopamine systems. *Annu Rev Neurosci*. 1978;1:129–169.
7. Dahlström A, Fuxe K. A method for the demonstration of monoamine-containing nerve fibres in the central nervous system. *Acta Physiol Scand*. 1964;60:293–294.
8. Dahlström A, Fuxe K. Evidence for the existence of monoamine-containing neurons in the central nervous system. I. Demonstration of monoamines in the cell bodies of brain stem neurons. *Acta Physiol Scand Suppl*. 1964;232(suppl):231–255.
9. Ungerstedt U. On the anatomy, pharmacology, and function of the nigro-striatal dopamine system. Thesis: Stockholm: The Karolinska Institute; 1971:122.
10. Fuxe K, Hoekfelt T, Nilsson O. Observations on the cellular localization of dopamine in the caudate nucleus of the rat. *Z Zellforsch Mikrosk Anat*. 1964;63:701–706.
11. Axelsson S, Bjorklund A, Falck B, Lindvall O, Svensson LA. Glyoxylic acid condensation: a new fluorescence method for the histochemical demonstration of biogenic monoamines. *Acta Physiol Scand*. 1973;87:57–62.
12. Lindvall O, Bjorklund A, Hokfelt T, Ljungdahl A. Application of the glyoxylic acid method to vibratome sections for the improved visualization of central catecholamine neurons. *Histochemie*. 1973;35:31–38.
13. Hökfelt T, Johansson O, Fuxe K, Goldstein M, Park D. Immunohistochemical studies on the localization and distribution of monoamine neuron systems in the rat brain. I. Tyrosine hydroxylase in the mes- and diencephalon. *Med Biol*. 1976;54:427–453.
14. Hökfelt T, Johansson O, Fuxe K, Goldstein M, Park D. Immunohistochemical studies on the localization and distribution of monoamine neuron systems in the rat brain II. Tyrosine hydroxylase in the telencephalon. *Med Biol*. 1977;55:21–40.
15. Pickel VM, Joh TH, Reis DJ. Monoamine-synthesizing enzymes in central dopaminergic, noradrenergic and serotonergic neurons. Immunocytochemical localization by light and electron microscopy. *J Histochem Cytochem*. 1976;24:792–306.
16. Smiley JF, Goldman-Rakic PS. Silver-enhanced diaminobenzidine-sulfide (SEDS): a technique for high-resolution immunoelectron microscopy demonstrated with monoamine immunoreactivity in monkey cerebral cortex and caudate. *J Histochem Cytochem*. 1993;41:1393–1404.
17. Smiley JF, Goldman-Rakic PS. Heterogeneous targets of dopamine synapses in monkey prefrontal cortex demonstrated by serial section electron microscopy: a laminar analysis using the silver-enhanced diaminobenzidine sulfide (SEDS) immunolabeling technique. *Cereb Cortex*. 1993;3:223–238.
18. Dowling JE. Retinal neuromodulation: the role of dopamine. *Vis Neurosci*. 1991;7:87–97.
19. Ljungdahl A, Hokfelt T, Halasz N, Johansson O, Goldstein M. Olfactory bulb dopamine neurons – the A15 catecholamine cell group. *Acta Physiol Scand Suppl*. 1977;452:31–34.
20. Halasz N, Hokfelt T, Ljungdahl A, Johansson O, Goldstein M. Dopamine neurons in the olfactory bulb. *Adv Biochem Psychopharmacol*. 1977;16:169–177.
21. Cransac H, Peyrin L, Cottet-Emard JM, Farhat F, Pequignot JM, Reber A. Aging effects on monoamines in rat medial vestibular and cochlear nuclei. *Hear Res*. 1996;100:150–156.
22. Darrow KN, Simons EJ, Dodds L, Liberman MC. Dopaminergic innervation of the mouse inner ear: evidence for a separate cytochemical group of cochlear efferent fibers. *J Comp Neuro.l* 2006;498:403–414.
23. Pivonello R, Ferone D, Lombardi G, Colao A, Lamberts SW, Hofland LJ. Novel insights in dopamine receptor physiology. *Eur J Endocrinol*. 2007;156(suppl 1):S13-S21.
24. Berger B, Thierry AM, Tassin JP, Moyne MA. Dopaminergic innervation of the rat prefrontal cortex: a fluorescence histochemical study. *Brain Res*. 1976;106:133–145.
25. Berger B, Tassin JP, Blanc G, Moyne MA, Thierry AM. Histochemical confirmation for dopaminergic innervation of the rat cerebral cortex after destruction of the noradrenergic ascending pathways. *Brain Res*. 1974;81:332–337.
26. Thierry AM, Blanc G, Sobel A, Stinus L, Glowinski J. Dopaminergic terminals in the rat cortex. *Science*. 1973;182:498–500.
27. Lewis DA. The catecholaminergic innervation of primate prefrontal cortex. *J Neural Transm Suppl*. 1992;36:179–200.
28. Lewis DA, Sesack SR. Dopamine systems in the primate brain. In: Bloom FE, Hokfelt T, eds. *The Primate Nervous System, Part I*. Amsterdam: Elsevier; 1997:187–216.
29. Beckstead MJ, Grandy DK, Wickman K, Williams JT. Vesicular dopamine release elicits an inhibitory postsynaptic current in midbrain dopamine neurons. *Neuron*. 2004;42:939–946.
30. Chau DT, Roth RM, Green AI. The neural circuitry of reward and its relevance to psychiatric disorders. *Curr Psychiatry Rep*. 2004;6: 391–399.
31. Tam SY, Roth RH. Mesoprefrontal dopaminergic neurons: can tyrosine availability influence their functions? *Biochem Pharmacol*. 1997;53:441–453.
32. Horger BA, Roth RH. The role of mesoprefrontal dopamine neurons in stress. *Crit Rev Neurobiol*. 1996;10:395–418.
33. Lewis DA, Gonzalez-Burgos G. Pathophysiologically based treatment interventions in schizophrenia. *Nat Med*. 2006;12:1016–1022.
34. Goldman-Rakic PS. The physiological approach: functional architecture of working memory and disordered cognition in schizophrenia. *Biol Psychiatry*. 1999;46:650–661.
35. Goldman-Rakic PS, Castner SA, Svensson TH, Siever LJ, Williams GV. Targeting the dopamine D1 receptor in schizophrenia: insights for cognitive dysfunction. *Psychopharmacology (Berl)*. 2004;174:3–16.
36. Grace AA. Dopamine. In: Davis KL, Charney D, Coyle JT, Nemeroff C, eds. *Neuropsychopharmacology—The Fifth Generation of Progress*. Philadelphia, PA: Lippincott-Williams & Wilkins; 2002:119–132.
37. Diana M, Garcia-Munoz M, Richards J, Freed CR. Electrophysiological analysis of dopamine cells from the substantia nigra pars compacta of circling rats. *Exp Brain Res*. 1989;74:625–630.
38. Hurd YL, Weiss F, Koob GF, And NE, Ungerstedt U. Cocaine reinforcement and extracellular dopamine overflow in rat nucleus accumbens: an in vivo microdialysis study. *Brain Res*. 1989;498:199–203.
39. Schultz W, Romo R. Dopamine neurons of the monkey midbrain: contingencies of responses to stimuli eliciting immediate behavioral reactions. *J Neurophysiol*. 1990;63:607–624.
40. Kobayashi S, Schultz W. Influence of reward delays on responses of dopamine neurons. *J Neurosci*. 2008;28:7837–7846.

41. Fiorillo CD, Newsome WT, Schultz W. The temporal precision of reward prediction in dopamine neurons. *Nat Neurosci.* 2008;11:966–973.
42. Beckstead MJ, Williams JT. Long-term depression of a dopamine IPSC. *J Neurosci.* 2007;27:2074–2080.
43. Sawaguchi T, Goldman-Rakic PS. D1 dopamine receptors in prefrontal cortex: involvement in working memory. *Science.* 1991;251:947–950.
44. Sawaguchi T, Goldman-Rakic PS. The role of D1-dopamine receptor in working memory: local injections of dopamine antagonists into the prefrontal cortex of rhesus monkeys performing an oculomotor delayed-response task. *J Neurophysiol.* 1994;71:515–528.
45. Wang M, Vijayraghavan S, Goldman-Rakic PS. Selective D2 receptor actions on the functional circuitry of working memory. *Science.* 2004;303:853–856.
46. Shen W, Flajolet M, Greengard P, Surmeier DJ. Dichotomous dopaminergic control of striatal synaptic plasticity. *Science.* 2008;321:848–851.
47. Bloom FE. Dynamic synaptic communication: finding the vocabulary. *Brain Res.* 1973;62:299–305.

2 | Neuroanatomy

2.1 Functional Neuroanatomy of Dopamine in the Striatum

CHARLES R. GERFEN

From the perspective of a neuroanatomist, a major significance of the discovery of dopamine as a neurotransmitter was the subsequent development of the fluorescent catecholamine histochemical methods, which established the ability to visualize neuroanatomical circuits on the basis of their neurotransmitter.[1,2] In the early 1960s, Swedish researchers had the insight to exploit the ability to convert catecholamines into fluorescent molecules by a condensation reaction with formaldehyde to develop a histochemical method that revealed catecholamine-containing neurons and their axons.[3,4] Axonal tracing studies identified additional details of the organization of the nigrostriatal dopamine system and other neuroanatomical circuits of the basal ganglia. Further, the introduction of immunohistochemical methods in the late 1970s and molecular techniques employing in situ hybridization histochemical localization of messenger RNAs in the late 1980s generalized the ability to map the neuroanatomical organization of neurochemically characterized circuits. These latter techniques have provided considerable understanding of the functional circuits of the basal ganglia, built upon the pioneering work using the catecholamine histofluorescence techniques.

THE MESOSTRIATAL DOPAMINE SYSTEM

The most prominent of the dopamine systems revealed by the catecholamine histochemical techniques is the group of dopamine neurons in the midbrain, which provide a dense axonal projection to the striatum and the nucleus accumbens.[3,4] Three cell groups were originally identified based on their location: the A10, A9, and A8 dopamine cell groups located respectively in the ventral tegmental area, substantia nigra pars compacta, and retrorubral area. For the most part, there are no clear boundaries between these groups; they form a single continuous system whose axonal projections provide a dense input to all parts of the striatum, including the nucleus accumbens. The projection of the mesostriatal dopamine system is topographically organized such that medially located neurons project ventrally to the nucleus accumbens and ventral striatum, and projections to the dorsal striatum originate from more laterally positioned neurons in the substantia nigra pars compacta and the retrorubral area. While the most prominent projection of the mesostriatal dopamine neurons is to the striatum, some neurons in this complex also provide inputs to other forebrain areas including the cerebral cortex.

Dopamine axons are densely and rather homogeneously distributed in the striatum. In turn, the neurons of the striatum itself are distributed homogeneously, without any obvious cytoarchitectural features. There is, for example, no clear cytoarchitectural feature that separates the nucleus accumbens from the ventral parts of the striatum. However, this homogeneity in both the dopamine input system and the distribution of striatal neurons masks a number of underlying neuroanatomical circuits that are key to understanding the functional organization of the basal ganglia.

STRIATAL PATCH-MATRIX COMPARTMENTS

Indications that there are compartments within the striatum came from studies that observed islands or patches of dopamine innervation distributed within the neuropil of the striatum during early postnatal development that give way to a homogeneous distribution as development progresses.[5] A number of neurochemical markers were found to coincide with these patches, including staining for acetylcholinesterase[6] and opiate receptor binding.[7] The striatal neurons that are the target of this early dopamine input also develop first within the striatum, with later-born striatal neurons filling in the surrounding matrix regions of the striatum.[8] These two developmental compartments, the early-developing patches or islands and the later-developing matrix, give rise to the adult patch (or striosome)[6] and matrix compartments of the striatum. As the adult striatum appears homogeneous, neurochemical markers are required to reveal them; notably, among others, calbindin marks the matrix compartment[9] and mu opiate receptors mark the patch compartment.[7]

Distinct subsets of dopamine neurons differentially target the striatal patch and matrix compartments[10,11] (Fig. 2.2.1). While the mesostriatal dopamine neurons in the ventral tegmental area, susbstantia nigra pars compacta, and retrorubral area appear as a continuous grouping of neurons, within the substantia nigra pars compacta there are two parts that form a dorsal and a ventral tier of neurons. Dorsal tier pars compacta dopamine neurons are continuous with ventral tegmental dopamine neurons and extend their dendrites in the plane of the pars compacta medial and laterally. Ventral tier dopamine neurons are organized in two parts. One part is immediately ventral to the dorsal tier in the pars compacta and extend its dendrites ventrally into the substantia nigra pars reticulata. The second part consists of dopamine neurons that are grouped within the substantia nigra pars reticulata itself. Axonal tracing studies demonstrated that dopamine neuron projections from the ventral tegmental area, dorsal tier of the substantia nigra pars compacta, and retrorubral area provide input to the striatal matrix compartment, whereas projections from the two groups of ventral tier dopamine neurons of the substantia nigra provide input to the striatal patch compartment.[10] Moreover, matrix projecting neurons coexpress the calcium binding protein, calbindin, which provides a neurochemical marker for these dopamine neurons.[9,11] To confirm this organization, we took advantage of the differential development of the patch- and matrix-directed dopamine systems. Injecting the neurotoxin 6-hydroxydopamine into the striatum on the day of birth resulted in the selective degeneration of the ventral tier dopamine neurons and the dopamine input to the patch compartment.[11] As adults, the calbindin-expressing dopamine neurons in the ventral tegmental area, dorsal tier of the pars compacta, and retrorubral area survived, as did the dopamine input to the striatal matrix compartment. These studies in the rat demonstrate distinct sets of mesostriatal dopamine neurons that differerentially target the striatal patch and matrix compartments, which has also been demonstrated in the primate.[9,12] A recent study by Matsuda et al.[13] adds important details concerning the compartmental organization of the nigrostriatal dopamine system. Using a method that labels the full axonal arborization of single neurons, these authors show that individual dopamine neurons in both the dorsal and ventral tiers distribute axons to both striatal patch and matrix compartments, although each neuron's arborization tend to favor one or the other.

INPUT–OUTPUT ORGANIZATION OF THE STRIATAL PATCH AND MATRIX COMPARTMENTS

In addition to dopaminergic inputs, other input and output connections of the striatum are organized relative to the patch-matrix compartments (Fig. 2.1.2). The major neuron type in the striatum is the medium spiny neuron, which constitutes up to 90% of the striatal neuron population. These neurons provide the axonal output of the striatum, with different populations of these neurons targeting different basal ganglia nuclei. Although the distribution of these neurons is homogeneous and does not reveal striatal compartments, the dendrites of the neurons in patch and matrix compartments remain confined within their respective compartments.[14,15] Axonal tracing studies demonstrate that projections from the striatal patch compartment provide input directed principally to the ventral tier dopamine neurons in the substantia nigra, whereas the striatal matrix neurons project to the globus pallidus, entopeduncular nucleus, and substantia nigra pars reticulate.[14] Thus, the striatal output of the patch compartment is directed principally at the same ventral tier dopamine neurons that provide input to this compartment. In this regard, the recent finding that ventral tier dopamine neurons provide dopamine input to both patch and matrix compartments[13] is important. This finding suggests that the striatal patch output is not part of a closed loop with the dopamine neurons providing patch input, but rather affects dopamine feedback to both compartments. The target of the output of striatal matrix neurons is directed to components of the basal ganglia that provide the output of this system. In particular, the entopeduncular nucleus and substantia nigra pars reticulata are composed of GABAergic neurons, which project to the thalamus and superior colliculus and other midbrain systems connected with motor control. Thus, the output of neurons in the striatal patch and matrix respectively target dopamine feedback to the striatum and basal ganglia output systems.

Dopamine input to striatal medium spiny neurons is directed principally to dendritic shafts and spine necks and likely functions to modulate excitatory input that is directed to the dendritic spines. There are two main

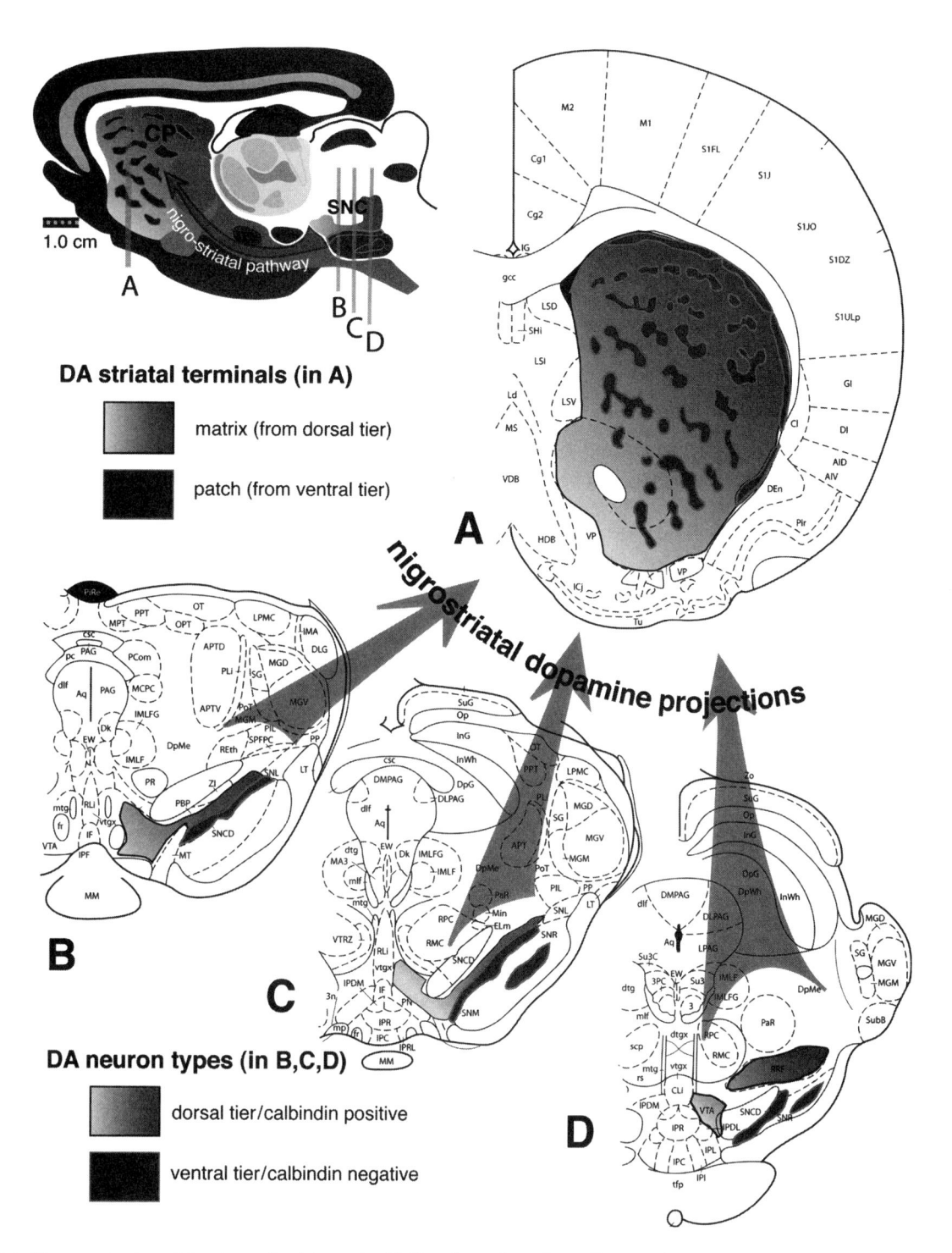

FIGURE 2.1.1. The organization of the nigrostriatal dopamine (DA) pathway from the midbrain to the striatum (sagittal diagram at upper right) is diagrammed to show the organization of this system to the striatal patch and matrix compartments. Coronal sections at three levels through the striatum (A) are depicted to show the innervation of the patch and matrix compartments from different subsets of midbrain DA neurons from three levels (B, C, D). Neurons providing inputs to the striatal matrix compartment (white in B, C, D) are located in the ventral tegmental area (VTA, A10 DA cell group), in the dorsal tier of the substantia nigra pars compacta (in B,C: SNCD, A9), and in the retrorubral area (in D: RRF, A8 DA cell group). Neurons providing input to the striatal patch compartment are located in the ventral tier of the substantia nigra pars compacta (in B, C, D: DA neurons in dark gray areas) and project from A9 DA cells located in the substantia nigra pars reticulata (in C and D). There is a general topography in that medially located cells project to the ventral striatum and laterally located cells project to the dorsal striatum. Neurons at each rostral-caudal level in the midbrain project rather extensively throughout the rostral-caudal extent of the striatum. (See Color Plate 2.1.1.)

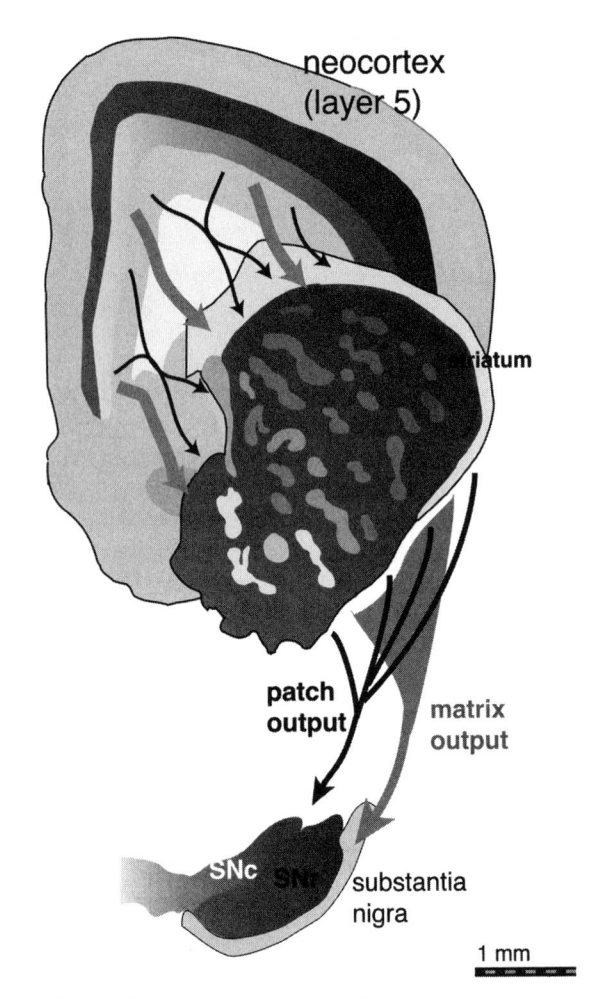

FIGURE 2.1.2. Organization of the striatal patch-matrix compartments provides parallel pathways from the cerebral cortex through the striatum that provide differential input to the dopamine and GABA neurons in the substantia nigra. Deep layer 5 corticostriatal neurons provide selective inputs to the striatal patch compartment, whose neurons provide inputs targeting dopamine neurons in the substantia nigra pars compacta. Superficial layer 5 corticostriatal neurons provide inputs to the striatal matrix compartment, whose neurons project to the substantia nigra pars reticulata, which contains the GABAergic output neurons of the basal ganglia. This organization arises from most neocortical areas, although there is a gradient such that those areas closer to the allocortex provide greater input to the patch compartment, whereas primary sensorimotor areas provide greater input to the matrix compartment. (See Color Plate 2.1.2.)

sources of glutamatergic excitatory input, the cerebral cortex and thalamus, and some aspects of each of these are organized relative to patch and matrix compartments. For the thalamus, parts of the intralaminar thalamic nuclei differentially target the patch matrix compartments, with projections of the parafascicular and centromedian nuclei directed to the matrix and projections of the paraventricular nucleus directed to the patch compartment.[7,16,17]

Corticostriatal projections arise from pyramidal neurons in layer 5. Initial studies examining the compartmental targets of corticostriatal projections suggested that different cortical areas projected selectively to either patch or matrix. Limbic cortical areas were shown to provide input to the patch compartment, whereas neocortical somatosensory and motor cortical area projections targeted the matrix.[18,19] However, more detailed analysis of corticostriatal projections demonstrated that most cortical areas provided inputs to both compartments, but that neurons in different sublayers of layer 5 differentially project to the patch and matrix compartments.[20] For each specific cortical area, neurons with patch-directed inputs are located in deep layer 5, whereas those with matrix-directed inputs are located in superficial layer 5. Corticostriatal projections are topographically organized such that motor cortical areas project to the dorsolateral striatum and prelimbic and infralimbic areas project to the medial and ventral striatum. Importantly, for each cortical area, both patch- and matrix-directed projections target the topographic region within the striatum such that from a given cortical area, its projections to the matrix surround the patches to which it also projects. While this pattern of organization of the corticostriatal projections is apparent in most cortical areas, the relative contribution of inputs to the patch and matrix compartments varies among cortical areas. Neocortical areas, such as motor, supplementary motor, and somatosensory cortices, provide greater inputs to the matrix compartment, whereas allocortical and periallocortical areas such as the prelimbic and infralimbic cortical areas provide greater inputs to the patch compartment. This transition of a predominance of patch-directed inputs from limbic-related cortical areas to matrix-directed inputs from neocortical areas is likely responsible for the earlier findings suggesting that different cortical areas provide inputs only to one compartment. The major significance of the organization of corticostriatal projections is that the striatal patch and matrix compartments are related to the laminar organization of the cerebral cortex rather than to tangential or columnar features of its organization.[20]

D1 AND D2 DOPAMINE RECEPTORS IN DIRECT AND INDIRECT STRIATAL PROJECTIONS

A major discovery concerning the function of dopamine in the basal ganglia was the demonstration that D1 and D2 dopamine receptors are segregated in the direct and indirect striatal projection neurons[21] (Fig. 2.1.3). Striatal medium spiny neurons, which constitute up to 90% of the neuron population of the striatum and nucleus accumbens, are composed of two major subtypes based on their axonal projections. One subtype projects axons through the globus pallidus, making some contacts, but extends axons to terminate in the internal segment of the globus pallidus (or entopeduncular nucleus) and substantia nigra. These nuclei constitute the major output system of the basal ganglia such that the striatal neurons that project to them directly are considered to provide the "direct" striatal projection pathway. The other subtype of striatal projection neuron extends its axon only to the globus pallidus. Neurons in this nucleus provide inputs to the internal segment of the globus pallidus and substantia nigra and to the subthalamic nucleus, which in turn projects to these basal ganglia output nuclei. Thus, striatal neurons that project only to the globus pallidus are connected through multiple synaptic connections to the output of the basal ganglia and are considered to provide the "indirect" striatal projection pathway. Neurons giving rise to the direct and indirect pathway are approximately equal in number and are intermingled with one another in both the patch and matrix compartments.[22]

The functional significance of the striatal direct and indirect pathways was established by the observation that following dopamine depletion in the striatum, there are differential changes in GABA receptor binding in the globus pallidus and substantia nigra[23] and in the expression of peptides expressed by striatal direct and indirect pathway neurons.[24] These findings led to the hallmark theory that clinical movement disorders such as Parkinson's disease result from an imbalance in the output activity of the striatal direct and indirect pathways.[25,26] This theory suggested that akinesia, which characterizes Parkinson's disease, is a consequence of increased functional activity in the indirect striatal pathway.

The underlying mechanism responsible for dopamine-mediated differential changes in the functional activity of the striatal direct and indirect pathways was shown to be that D1- and D2-dopamine receptors are respectively segregated in the neurons giving rise to these projections.[21] Two lines of evidence were provided in this study. The first line of evidence was provided by neuroanatomical studies. In situ hybridization histochemical localization of the mRNAs encoding D1-dopamine receptors demonstrated the selective expression of these receptors in neurons that project to the substantia nigra and coexpress the peptides dynorphin and substance P, markers of the direct striatal pathway. On the other hand, D2 mRNA was shown to be expressed selectively in neurons that project to the globus pallidus and coexpress the peptide enkephalin, a marker of indirect striatal pathway neurons. The second line of evidence was provided by functional studies. Following dopamine depletion of the nigrostriatal pathway, enkephalin expression increases in indirect pathway neurons, whereas substance P and dynorphin expression decreases in direct pathway neurons. These dopamine lesion–induced changes in gene expression were demonstrated to be selectively reversed in indirect pathway neurons with D2 receptor agonist treatment and in direct pathway neurons with D1 receptor agonist treatment. This finding was somewhat controversial, as some investigators maintained that D1 and D2 dopamine receptors are coexpressed in most striatal medium spiny neurons.[23] However, the segregation of D1 and D2 dopamine receptors in direct and indirect striatal pathway neurons has been confirmed by numerous

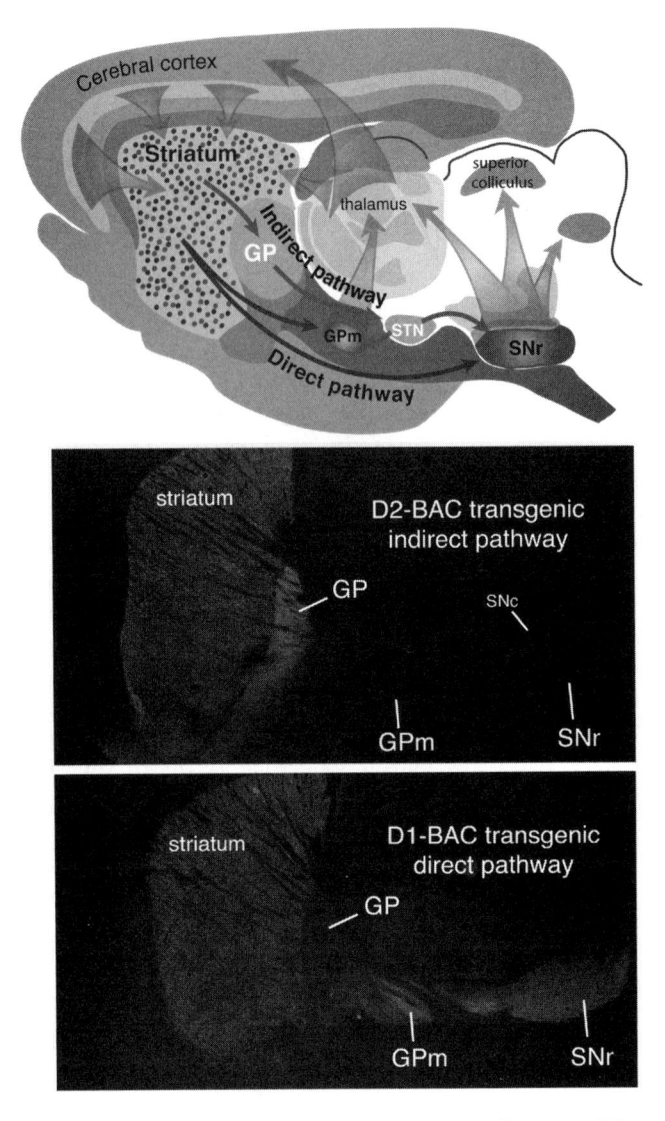

FIGURE 2.1.3. Circuitry involved in Parkinson's disease. Upper diagram: Imbalances in the function of direct and indirect pathways of the basal ganglia in Parkinson's disease, shown in a sagittal brain section of the mouse. The cerebral cortex and thalamus provide excitatory inputs to the striatum, the main input nucleus of the basal ganglia. The output of the basal ganglia originates from the medial globus pallidus (GPm) and the substantia nigra pars reticulata (SNr) and is directed primarily to thalamic nuclei, which project to frontal areas of the cerebral cortex. The direct pathway originates from striatal projection neurons whose axons extend directly to the GPm and SNr output nuclei. The indirect pathway originates from striatopallidal neurons whose axons terminate within the globus pallidus (GP). Neurons in the GP, in turn, project to the subthalamic nucleus (STN), which projects to the GPm and SNr. Thus, striatopallidal neurons are connected indirectly, through the GP and STN, with the output of the basal ganglia. Lower images: D1 and D2 dopamine neurons are segregated to direct- and indirect-pathway neurons, respectively. Sagittal sections from BAC transgenic mice in which these receptors are labeled with enhanced green fluorescence protein (EGFP) show labeling of the neuron cell bodies in the striatum as well as their axonal projections. D2-BAC transgenic mice show labeling of the indirect-pathway neurons (these axon projections terminate in the GP), whereas D1-BAC mice show labeling of the direct pathway, as seen by labeling of axon terminals in the GPm and SNr.[30,31] (See Color Plate 2.1.3.)

studies such that there is now a consensus in the field upholding the original finding.[27–31]

The demonstration of segregation of D1 and D2 dopamine receptors, respectively, in direct and indirect pathway neurons[21] provided the basis for understanding the functional changes in movement disorders such as Parkinson's disease.[25,26] The central tenet of the theory of movement disorders is that they result from imbalanced activity in the direct and indirect striatal pathways. In Parkinson's disease, which is marked by akinesia, the theory suggests that there is increased activity in the indirect pathway. Neurons of this pathway express the D2 dopamine receptor, which is coupled to the inhibitory G protein, Gi. In the normal animal, dopamine binding to the D2 receptors provides an inhibitory function. On the other hand, the D1 receptor expressed on direct pathway neurons is coupled to stimulatory G proteins, Gs and Golf. Consequently, in Parkinson's disease, the loss of dopamine input to the striatum has opposite effects on the direct and indirect pathways, with increased function in the indirect pathway and decreased function in the direct pathway. Surgical therapies developed to reverse this imbalance by interfering with altered function in the indirect pathway proven to have considerable clinical benefit.[32,33]

L-DOPA-INDUCED DYSKINESIA IN PARKINSON'S DISEASE

Treatment of Parkinson's disease with L-DOPA[34] was a direct result of the discovery of dopamine and its depletion in the disease and remains the primary therapeutic treatment. While it is a very effective therapy, long-term treatment invariably leads to the development of dyskinesias.[35] We have proposed that L-DOPA-induced dyskinesia in the treatment of Parkinson's disease results from an aberrant switch in the linkage of the D1 dopamine receptor to signal transduction systems that activate the protein kinase, extracellular signal-regulated protein kinase (ERK1/2).[36] As discussed, dopamine depletion of the striatum results in opposite effects on the function of D2-indirect and D1-direct pathway striatal neurons evidenced by changes in gene expression.[21] While either L-DOPA or selective D2 and D1 receptor agonist treatments reverse some of the gene expression changes, the response of D1-receptor-bearing direct pathway neurons is supersensitive to these treatments, as demonstrated by the induction of a large number of so-called immediate early genes.[37]

The supersensitive response of striatal neurons following lesioning of the nigrostriatal dopamine system was first described by Ungerstedt,[38] who observed that animals with unilateral lesions exhibited a robust rotation contralateral to the lesioned side in response to direct and indirect dopamine agonist treatment. This experimental paradigm remains the standard animal model for the study of Parkinson's disease. The reasonable explanation of why these animals display contralateral rotation following dopamine receptor agonist treatment was that striatal neurons compensated for the loss of dopamine by increasing their expression of dopamine receptors in order to increase their response to decreased levels of neurotransmitter. Thus, striatal neurons in the lesioned striatum would produce a supersensitive response relative to the dopamine-intact striatum, which resulted in the behavioral rotation. However, our studies demonstrating the segregation of D1 and D2 dopamine receptors on direct and indirect striatal neurons provide a different model.[21] Following dopamine lesioning, there is an increase in D2 dopamine receptor expression in indirect pathway neurons and a decrease in D1 receptor expression in direct pathway neurons. Rather than reflecting a compensatory response of striatal neurons to decreased dopamine input, these changes in receptor expression reflect the simple consequence of the loss of dopamine function on these neurons. Thus, in indirect pathway neurons, the absence of dopamine acting on D2 dopamine receptors, coupled to the inhibitory G protein, Gi, results in increased gene expression in these neurons, including the D2 dopamine receptor. On the other hand, in direct pathway neurons, the absence of dopamine acting on D1 dopamine receptors, coupled to the stimulatory G proteins, Gs and Golf, results in decreased gene expression, including the D1 dopamine receptor.

In addition to the behavioral rotational response in the unilateral dopamine lesion paradigm, a cellular response was shown by the demonstration of the induction of immediate early genes (IEGs), such as c-fos, in striatal neurons in response to dopamine agonists.[39,40] Significantly, the IEG response to L-DOPA or dopamine agonists such as apomorphine was found to occur exclusively in D1-expressing direct pathway striatal neurons. This IEG response provides a cellular measure of receptor supersensitivity. What is most interesting about this IEG response in D1 receptor–expressing neurons is that it occurs upon the first treatment with the dopamine receptor agonist, when the level of D1 receptor expression is decreased compared with that of neurons in the dopamine-intact striatum.[37] This finding suggested that in the dopamine-lesioned striatum, the supersensitive response of D1 dopamine receptor–expressing neurons is not a consequence of increased D1 dopamine receptor expression, but rather is due to a change in the coupling of this receptor to signal transduction systems.

Psychostimulants, such as cocaine and amphetamine, produce robust induction of IEGs in the normal dopamine-innervated striatum.[41] This raises the question as to whether the D1 dopamine receptor–mediated supersensitive induction of IEGs in the dopamine-depleted striatum is due to an amplification of the normal D1 receptor coupling to signal transduction. However, psychostimulant striatal IEG induction differs in important ways from the D1 response in the dopamine-depleted striatum. First, whereas the psychostimulant response is dependent on glutamate N-methyl-D-aspartate (NMDA) receptor activation,[42] the D1-mediated IEG induction in the dopamine-depleted striatum occurs independently of NMDA receptor function.[43] Second, repeated psychostimulant treatment produces an attenuated striatal IEG response[44]; the response in the dopamine-depleted striatum increases with extended dopamine-receptor agonist treatment.[45]

Using pharmacologic treatment paradigms to compare D1 receptor–mediated signaling in the dopamine-intact and -lesioned striatum, activation of ERK1/2 was demonstrated to occur exclusively in the dopamine-depleted striatum.[36] In this study, pharmacologic treatments with high doses of D1 receptor agonists, or combined D1 and D2 dopamine agonists, produced induction of IEGs in the dopamine-intact striatum at levels comparable to those produced in the dopamine-depleted striatum. However, activation of ERK1/2 occurred only in the dopamine-depleted striatum. In the dopamine-intact striatum, dopamine-agonist treatment activation of ERK1/2 was limited to the nucleus accumbens. This finding suggests that the depletion of dopamine in the striatum produces an aberrant coupling of the D1 receptor with activation of ERK1/2.

Psychostimulant treatments activate ERK1/2 in the nucleus accumbens and in a small percentage of neurons in the dorsal striatum, which was proposed to be dependent on dopamine- and cyclic adenosine monophosphate (cAMP)–regulated phosphoprotein, 32 kDa (DARPP32).[46] However, in transgenic mice with either the D1 dopamine receptor or DARPP32 knocked out, psychostimulant activation of ERK1/2 occurs in the dorsal striatum but is reduced in the nucleus accumbens.[37] Thus, psychostimulant activation of ERK1/2 in the dorsal striatum does not appear to involve the D1 dopamine receptor. Moreover, it is important to note that psychostimulant activation occurs in a small percentage of dorsal striatal neurons (approximately 10%); in the dopamine-depleted striatum, D1 dopamine receptor activation of ERK1/2 occurs in nearly all D1 receptor–expressing dorsal striatal neurons. On the other hand, in mice with the D1 dopamine receptor knocked out, L-DOPA treatment did not produce activation of ERK1/2 in the dopamine-depleted striatum. However, in mice with DARPP-32 knocked out, L-DOPA treatment produced robust activation of ERK1/2 in the dopamine-depleted striatum similar to that in wild-type mice.[47]

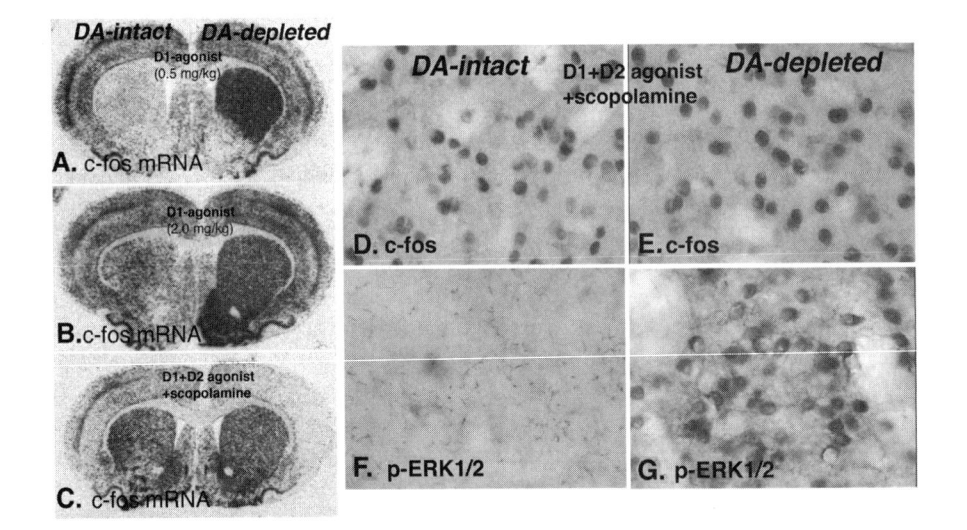

FIGURE 2.1.4. Demonstration of distinct mechanisms of D1 dopamine receptor–mediated gene regulation in the dopamine (DA)-intact and-depleted striatum, using the full D1 agonist SKF81297 alone or combined with other drugs. (A–C) In situ hybridization histochemical localization of mRNA encoding c-fos 45 min after different drug combinations: A, SKF81297 (0.5 mg/kg); B, SKF81297 (2.0 mg/kg); C, SKF81297 (2.0 mg/kg) combined with the D2 DA receptor agonist (1 mg/kg) and scopolamine. The low dose of agonist alone (A) demonstrates the supersensitive response by the selective induction of c-fos in the DA-depleted striatum. Bilateral induction of c-fos IEG in both the DA-intact and -depleted striatum follows treatment with a high dose of the full D1 agonist alone (B) or in combination with other drugs (C). However, when animals receiving any of these treatments are killed at 15 min, p-ERK1/2- immunoreactive neurons are evident only in the DA-depleted striatum and not in the DA-intact striatum (data not shown). The treatment combining the full D1 agonist with both the D2 agonist and scopolamine produces a robust c-fos IEG response in both the DA-intact (D) and DA-depleted striatum (E). This treatment also results in persistent p-ERK1/2 (H) in the DA-depleted striatum but does not activate p-ERK1/2 (G) in neurons in the DA-intact striatum. These results demonstrate that, although D1-DA receptor-mediated induction of the IEG c-fos occurs in both the DA-intact and -depleted striatum, activation of ERK1/2 occurs only in the DA-depleted striatum.[36]

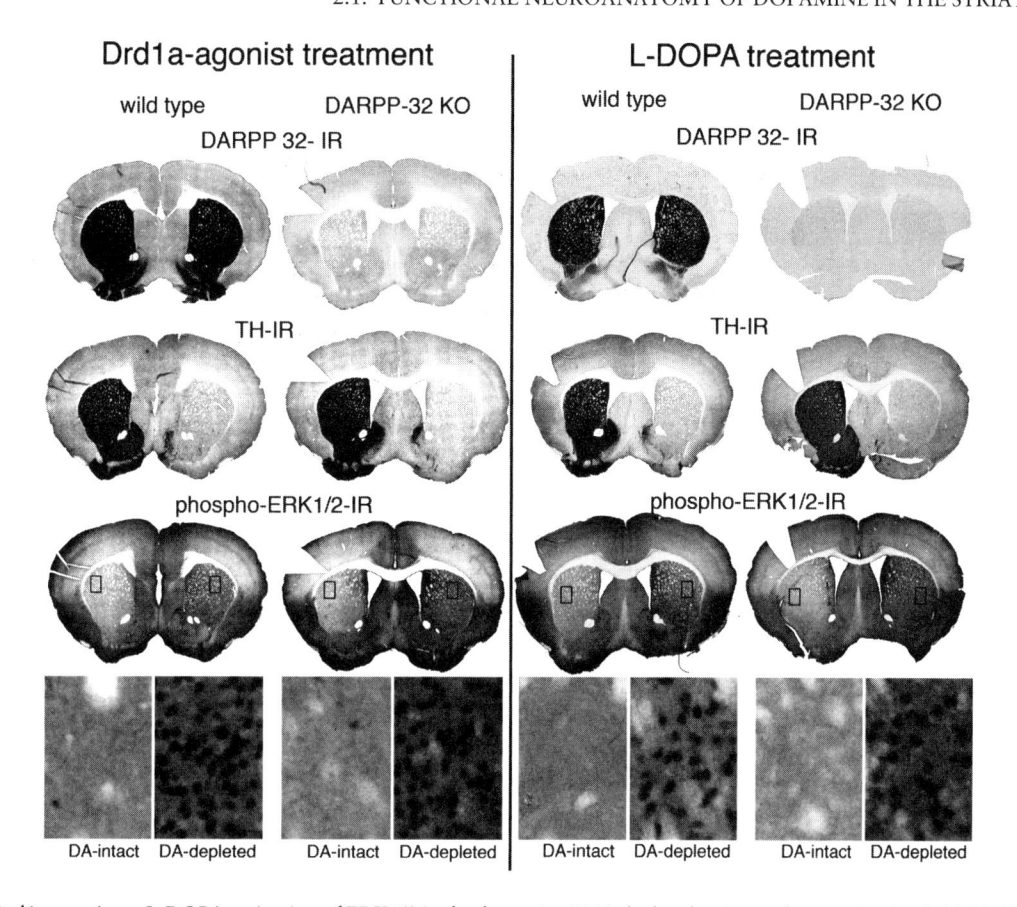

FIGURE 2.1.5. Drd1a-agonist or L-DOPA activation of ERK1/2 in the dopamine (DA)-depleted striatum does not involve DARPP-32. Comparison of coronal brain sections at the level of the rostral striatum from wild-type and DARPP-32 knockout (KO) mice, with unilateral lesions of the nigrostriatal DA system and treated with a drd1a agonist (SKF81298, 5 mg/kg, 1 day) or L-DOPA (20 mg/kg with 12 mg/kg benserazide, 10 days). DARPP-32 immunoreactivity (IR) labels neurons in the striatum in wild-type mice, which are unlabeled in DARPP-32 KO mice. The unilateral lesion of the nigrostriatal DA pathway in these animals is shown by the absence of tyrosine hydroxylase (TH)-IR in the axonal terminals in the right striatum. Activation of ERK1/2 in response to either drd1a-agonist treatment (left-side figures) or L-DOPA treatment (right-side figures) is demonstrated by phospho-ERK1/2-IR throughout the DA-depleted striatum. High-power images from the dorsolateral striatum (inset boxes, 100 um wide) show few to no phospho-ERK1/2 IR neurons in the DA-intact striatum. In contrast, there are numerous phospho-ERK1/2 IR neurons in the DA-depleted striatum in both the wild-type and DARPP-32 KO animals.[47]

Together these studies suggest that following dopamine depletion, there is an aberrant coupling of the D1 dopamine receptor to activation of ERK1/2 that does not involve DARPP32. In addition, these studies point to distinct regional differences between the nucleus accumbens and dorsal striatum in the coupling of the D1 receptor with signal transduction systems. Based on these findings, we suggested that following dopamine depletion in the striatum, repeated activation of the aberrant coupling of the D1 dopamine receptor by L-DOPA is responsible for the development of dyskinesias.[36] Subsequent studies by a number of groups have demonstrated in Parkinson's disease animal models strong correlations between L-DOPA induction of ERK1/2 and dyskinesias.[48-50]

REFERENCES

1. Carlsson A, Falck B, Hillarp NA. Cellular localization of brain monoamines. *Acta Physiol Scand Suppl.* 1962;56:1–28.
2. Dahlstroem A, Fuxe K. A method for the demonstration of monoamine-containing nerve fibers in the central nervous system. *Acta Physiol Scand.* 1964;60:293–294.
3. Anden NE, Carlsson A, Dahlstroem A, Fuxe K, Hillarp NA, Larsson K. Demonstration and mapping out of nigro-neostriatal dopamine neurons *Life Sci.* 1964;3:523–530.
4. Dahlstroem A, Fuxe K. Evidence for the existence of monoamine containing neurons in the central nervous system. I. Demonstration of monoamines in the cell bodies of brain stem neurons. *Acta Physiol Scand.* 1964(suppl);232:1–55.
5. Tennyson VM, Barrett RE, Cohen G, Côté L, Heikkila R, Mytilineou C. Correlation of anatomical and biochemical development of the rabbit neostriatum. *Prog Brain Res.* 1973;40:203–217.

6. Graybiel AM, Ragsdale CW Jr. Histochemically distinct compartments in the striatum of human, monkeys, and cat demonstrated by acetylthiocholinesterase staining. *Proc Natl Acad Sci USA.* 1978;75:5723–5726.
7. Herkenham M, Pert CB. Mosaic distribution of opiate receptors, parafascicular projections and acetylcholinesterase in rat striatum. *Nature.* 1981;291:415–418.
8. van der Kooy D, Fishell G. Neuronal birthdate underlies the development of striatal compartments. *Brain Res.* 1987;401:155–161.
9. Gerfen CR, Baimbridge KG, Miller JJ. The neostriatal mosaic: compartmental distribution of calcium-binding protein and parvalbumin in the basal ganglia of the rat and monkey. *Proc Natl Acad Sci USA.* 1985;82:8780–8784.
10. Gerfen CR, Herkenham M, Thibault J. The neostriatal mosaic: II. Patch- and matrix-directed mesostriatal dopaminergic and non-dopaminergic systems. *J Neurosci.* 1987;7:3915–3934.
11. Gerfen CR, Baimbridge KG, Thibault J. The neostriatal mosaic: III. Biochemical and developmental dissociation of patch-matrix mesostriatal systems. *J Neurosci.* 1987;7:3935–3944.
12. Jimenez-Castellanos J, Graybiel AM. Subdivisions of the dopamine containing A8-A9-A10 complex identified by their differential mesostriatal innervation of striosomes and extrastriosomal matrix. *Neuroscience.* 1987;23:223–242.
13. Matsuda W, Furuta T, Nakamura KC, Hioki H, Fujiyama F, Arai R, Kaneko T. Single nigrostriatal dopaminergic neurons form widely spread and highly dense axonal arborizations in the neostriatum. *J Neurosci.* 2009;29:444–453.
14. Gerfen CR. The neostriatal mosaic. I. Compartmental organization of projections from the striatum to the substantia nigra in the rat. *J Comp Neurol.* 1985;236:454–476.
15. Bolam JP, Izzo PN, Graybiel AM. Cellular substrate of the histochemically defined striosome/matrix system of the caudate nucleus: a combined Golgi and immunocytochemical study in cat and ferret. *Neuroscience.* 1988;24:853–875.
16. Gerfen CR, Staines WA, Arbuthnott GW, Fibiger HC. Crossed connections of the substantia nigra in the rat. *J Comp Neurol.* 1982;207:283–303.
17. Berendse HW, Voorn P, te Kortschot A, Groenewegen HJ. Nuclear origin of thalamic afferents of the ventral striatum determines their relation to patch/matrix configurations in enkephalin-immunoreactivity in the rat. *J Chem Neuroanat.* 1988;1:3–10.
18. Gerfen CR. The neostriatal mosaic: compartmentalization of corticostriatal input and striatonigral output systems. *Nature.* 1984;311:461–464.
19. Donoghue JP, Herkenham M. Neostriatal projections from individual cortical fields conform to histochemically distinct striatal compartments in the rat. *Brain Res.* 1986;365:397–403.
20. Gerfen CR. The neostriatal mosaic: striatal patch-matrix organization is related to cortical lamination. *Science.* 1989;246:385–388.
21. Gerfen CR, Engber TM, Mahan LC, Susel Z, Chase TN, Monsma FJ Jr, Sibley DR. D1 and D2 dopamine receptor-regulated gene expression of striatonigral and striatopallidal neurons. *Science.* 1990;250:1429–1432.
22. Gerfen CR, Young WS 3rd. Distribution of striatonigral and striatopallidal peptidergic neurons in both patch and matrix compartments: an in situ hybridization histochemistry and fluorescent retrograde tracing study. *Brain Res.* 1988;460:161–167.
23. Pan HS, Penney JB, Young AB. Gamma-aminobutyric acid and benzodiazepine receptor changes induced by unilateral 6-hydroxydopamine lesions of the medial forebrain bundle. *J Neurochem.* 1985;45:1396–1404.
24. Young WS 3rd, Bonner TI, Brann MR. Mesencephalic dopamine neurons regulate the expression of neuropeptide mRNAs in the rat forebrain. *Proc Natl Acad Sci USA.* 1986;83:9827–9831.
25. Albin RL, Young AB, Penney JB. The functional anatomy of basal ganglia disorders. *Trends Neurosci.* 1989;12:366–375.
26. DeLong MR. Primate models of movement disorders of basal ganglia origin. *Trends Neurosci.* 1990;13:281–285.
27. LeMoine C, Normand E, Guitteny AF, Fouque B, Teoule R, Bloch B. Dopamine receptor gene expression by enkephalin neurons in rat forebrain. *Proc Natl Acad Sci USA.* 1990;87:230–234.
28. LeMoine C, Bloch B. D1 and D2 dopamine receptor gene expression in the rat striatum: sensitive cRNA probes demonstrate prominent segregation of D1 and D2 mRNAs in distinct neuronal populations of the dorsal and ventral striatum. *J Comp Neurol.* 1995;355:418–426.
29. Hersch SM, Ciliax BJ, Gutekunst CA, Rees HD, Heilman CJ, Yung KK, Bolam JP, Ince E, Yi H, Levey AI. Electron microscope analysis of D1 and D2 dopamine receptor proteins in the dorsal striatum and their synaptic relationships with motor corticostriatal afferents. *J Neurosci.* 1995;15:5222–5237.
30. Gong S, Zheng C, Doughty ML, Losos K, Didkovsky N, Schambra UB, Nowak NJ, Joyner A, Leblanc G, Hatten ME, Heintz N. A gene expression atlas of the central nervous system based on bacterial artificial chromosomes. *Nature.* 2003;425:917–925.
31. Gong S, Doughty M, Harbaugh CR, Cummins A, Hatten ME, Heintz N, Gerfen CR. Targeting Cre recombinase to specific neuron populations with bacterial artificial chromosome constructs. *J Neurosci.* 2007;27:9817–9823.
32. Bakay RA, DeLong MR, Vitek JL. Posteroventral pallidotomy for Parkinson's disease. *J Neurosurg.* 1992;77:487–488.
33. Lozano AM, Lang AE, Hutchison WD, Dostrovsky JO. New developments in understanding the etiology of Parkinson's disease and in its treatment. *Curr Opin Neurobiol.* 1998;8:783–790.
34. Birkmayer W, Hornykiewicz O. The L-dihydroxyphenylalanine L-DOPA) effect in Parkinson's syndrome in man: on the pathogenesis and treatment of Parkinson akinesis. *Arch Psychiatr Nervenkr Z Gesamte Neurol Psychiatr.* 1962;203:560–574.
35. Bergmann KJ, Mendoza MR, Yahr MD. Parkinson's disease and long-term levodopa therapy. *Adv Neurol.* 1987;45:463–467.
36. Gerfen CR, Miyachi S, Paletzi R, Brown P. D1 dopamine receptor supersensitivity in the dopamine-depleted striatum results from a switch in the regulation of ERK1/2MAP kinase. *J Neurosci.* 2002;22:5042–5054.
37. Berke JD, Paletzki RF, Aronson GJ, Hyman SE, Gerfen CR. A complex program of striatal gene expression induced by dopaminergic stimulation. *J Neurosci.* 1998;18:5301–5310.
38. Ungerstedt U. Postsynaptic supersensitivity after 6-hydroxydopamine induced degeneration of the nigro-striatal dopamine system. *Acta Physiol Scand Suppl.* 1971;367:69–93.
39. Robertson GS, Herrera DG, Dragunow M, Robertson HA. L-dopa activates c-*fos* in the striatum ipsilateral to a 6-hydroxydopamine lesion of the substantia nigra. *Eur J Pharmacol.* 1989;159:99–100.
40. Robertson GS, Vincent SR, Fibiger HC. Striatonigral projection neurons contain D1 dopamine receptor-activated c-*fos*. *Brain Res.* 1990;523:288–290.
41. Graybiel AM, Moratalla R, Robertson HA. Amphetamine and cocaine induce drug-specific activation of the c-*fos* gene in striosome-matrix compartments and limbic subdivisions of the striatum. *Proc Natl Acad Sci USA.* 1990;87:6912–6916.
42. Konradi C, Leveque JC, Hyman SEJ. Amphetamine and dopamine-induced immediate early gene expression in striatal neurons

depends on postsynaptic NMDA receptors and calcium. *Neuroscience.* 1996;16:4231–4239.
43. Keefe KA, Gerfen CR. D1 dopamine receptor-mediated induction of zif268 and c-*fos* in the dopamine-depleted striatum: differential regulation and independence from NMDA receptors. *J Comp Neurol.* 1996;367:165–176.
44. Steiner H, Gerfen CR. Cocaine-induced c-*fos* messenger RNA is inversely related to dynorphin expression in striatum. *J Neurosci.* 1993;13:5066–5081.
45. Steiner H, Gerfen CR. Dynorphin regulates D1 dopamine receptor-mediated responses in the striatum: relative contributions of pre- and postsynaptic mechanisms in dorsal and ventral striatum demonstrated by altered immediate-early gene induction. *J Comp Neurol.* 1996;376:530–541.
46. Valjent E, Pascoli V, Svenningsson P, Paul S, Enslen H, Corvol JC, Stipanovich A, Caboche J, Lombroso PJ, Nairn AC, Greengard P, Hervé D, Girault JA. Regulation of a protein phosphatase cascade allows convergent dopamine and glutamate signals to activate ERK in the striatum. *Proc Natl Acad Sci USA.* 2005;102:491–496.
47. Gerfen CR, Paletzki R, Worley P. Differences between dorsal and ventral striatum in drd1a dopamine receptor coupling of dopamine- and cAMP-regulated phosphoprotein-32 to activation of extracellular signal-regulated kinase. *J Neurosci.* 2008; 28:7113–7120.
48. Aubert I, Guigoni C, Håkansson K, Li Q, Dovero S, Barthe N, Bioulac BH, Gross CE, Fisone G, Bloch B, Bezard E. Increased D1 dopamine receptor signaling in levodopa-induced dyskinesia. *Ann Neurol.* 2005;57:17–26.
49. Santini E, Valjent E, Usiello A, Carta M, Borgkvist A, Girault JA, Hervé D, Greengard P, Fisone G. Critical involvement of cAMP/DARPP-32 and extracellular signal-regulated protein kinase signaling in L-DOPA-induced dyskinesia. *J Neurosci.* 2007; 7:6995–7005.
50. Nadjar A, Gerfen CR, Bezard E. Priming for l-dopa-induced dyskinesia in Parkinson's disease: a feature inherent to the treatment or the disease? *Prog Neurobiol.* 2009;87:1–9.

2.2 Functional Implications of Dopamine D2 Receptor Localization in Relation to Glutamate Neurons

SUSAN R. SESACK

INTRODUCTION

Dopamine (DA) cell clusters project to large parts of the neural axis, where they provide a critical modulation of diverse functions. Midbrain DA neurons and their projections to forebrain targets have been the subject of extensive research. These cells reside mainly in the substantia nigra pars compacta (SNc) and ventral tegmental area (VTA) and project to widespread areas of the cortex, basal ganglia, limbic forebrain, and thalamus. They sustain tonic rates of firing due to pacemaker potentials and show additional bursts and pauses of activity that reflect their synaptic inputs and signal future expectancy and shifts in attentional resources.[1,2] The release of DA in target regions modulates cell excitability in a manner that contributes to motor control, goal-directed behavior, and cognitive function. Malfunctions in midbrain DA neurons and their activity patterns have been implicated in several disease states, including Parkinson's disease, attention deficit hyperactivity disorder, substance abuse, and schizophrenia.[1,3]

Among the critical functions of the midbrain DA system is the modulation of glutamate (Glut) transmission at multiple sites where these two transmitter systems interact[4,5] (see also Chapter 2.4 by Bolam and Moss and Chapter 7.3 by Surmeier et al. in this volume). As illustrated in Figure 2.2.1, the anatomical connectivity between DA and Glut neurons occurs at three major levels: the dendrites of DA neurons are innervated by Glut axons originating from cortical and subcortical regions (level A); DA axons target the dendrites of Glut neurons primarily in the cortex, hippocampus, amygdala, and thalamus (level B); DA and Glut axons converge onto common target neurons, where they synapse in close proximity (level C). The last configuration occurs at both cortical and basal ganglia sites and facilitates additional presynaptic interactions between DA and Glut. Other, less direct interactions between DA and Glut involve intermediary cell types. In regulating Glut transmission, DA utilizes multiple receptors in two major classes: D1, consisting of D1 and D5 subtypes, and D2, which includes D2, D3, and D4 subtypes. These receptors are extensively expressed at extrasynaptic portions of the plasma membrane, consistent with evidence that DA communicates via volume transmission, in addition to standard synaptic communication.[6-8] Consequently, a complete picture of sites where DA and Glut interact can only be achieved by consideration of DA receptor distribution.

This review will focus on DA receptors of the D2 class and their spatial and functional relationships with Glut neurons within the circuitry that comprises midbrain DA neurons and their ascending projections to forebrain targets, especially the cerebral cortex and basal ganglia. Interest in D2 receptors has been fueled primarily by their correlation to antipsychotic drug efficacy and their role as autoreceptors. Given the overall similar pharmacology and functions of the D2 receptor class,[9] a brief consideration of D3 and D4 receptor subtypes will also be provided. DA also modulates Glut transmission via D1 and D5 receptors, but this subject is beyond the scope of the chapter. The reader is referred to several important papers on the subcellular localization of D1 and D5 receptors in relation to Glut neurons and synapses.[10-19]

This review will also focus mainly on animal and postmortem human studies. For a review of DA receptor localization in the living human brain and its relationship to Glut function, please see work by Abi-Dargham and Laruelle.[20] For in-depth considerations of DA cell anatomy and projections, the reader is referred to comprehensive reviews,[21-24] including chapters in this

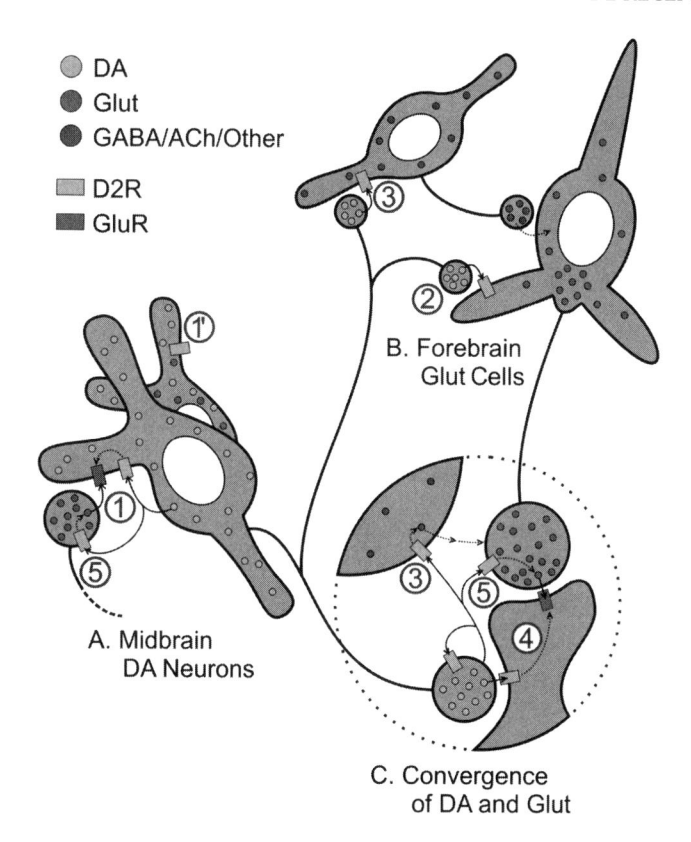

FIGURE 2.2.1. Dopamine (DA) acting on D2 receptors (D2Rs) modulates glutamate (Glut) transmission at multiple levels. Solid lines indicate binding of transmitters to their respective receptors; dashed lines indicate secondary actions subsequent to transmitter binding. (A1) Midbrain DA neurons express D2 autoreceptors and receive synaptic Glut afferents. Glut-mediated depolarization would release DA from dendrites to act on D2Rs, which would counteract further excitatory influence from Glut. D2 receptors might also directly modulate Glut transmission at membrane sites where both D2Rs and Glut receptors (GluRs) are distributed in close proximity. (A1′) A subpopulation of DA neurons coexpresses Glut; these neurons may also contain D2Rs that could modulate Glut release from these neurons. (B2) Dopamine axons directly innervate Glut cells in several target areas, most notably the cortex, amygdala, hippocampus, and thalamus. Modulatory actions on D2Rs at these sites can directly alter the excitability of Glut neurons. (B3, C3) D2 receptors can also influence Glut transmission indirectly via actions on local circuit neurons (e.g., GABA cells in the cortex and cholinergic cells in the striatum). (C4) Dopamine and Glut axons often form closely convergent synapses on target neurons throughout the forebrain, especially the cortex and basal ganglia. Postsynaptic cells expressing D2Rs bind DA following synaptic release or via volume transmission. Activation of D2Rs subsequently modulates Glut transmission, primarily via regulatory actions on AMPA receptors. (A5, C5) Many Glut axon terminals express D2Rs, allowing DA to modulate Glut release presynaptically. (See Color Plate 2.2.1.)

volume by Gerfen (2.1), Haber (2.3), and Moss and Bolam (2.4). More complete consideration of DA receptor subtypes, distribution, and signaling is provided in the chapters by Rankin et al. (3.1) and Fisone (3.3). Many other chapters in this volume cover the physiology and functional importance of midbrain DA systems.

METHODOLOGICAL APPROACHES FOR RECEPTOR LOCALIZATION

In considering the anatomical evidence for relationships between D2 receptors and Glut neurons, it is important to keep in mind the advantages and limitations of the various methods used for these studies. Indeed, the technological state of the art is typically the limiting factor in determining exact receptor distribution, particularly at the cellular and subcellular levels. A full consideration of methodological limitations is beyond the scope of this review and will only be discussed briefly. The reader is encouraged to consult more in-depth reviews of methodology for receptor autoradiography,[25,26] in situ hybridization,[27,28] and immunocytochemistry.[25,29,30]

Receptor Autoradiography

Receptor autoradiography localizes receptors by the binding of radioactive ligands and has the advantage of revealing receptor distribution in brain regions at high sensitivity regardless of whether or not the receptor is synthesized there. It also can be quantified for determination of changes in receptor density with disease and/or environmental manipulations. However, its usefulness depends on the availability of selective ligands with high affinity and high specific activity, and such ligands are not always available. For example, DA D1 and D5 receptor subtypes have a pharmacology that is too similar to allow their distinction using this approach. Many ligands also bind multiple receptor types, requiring blocking and subtraction methods to distinguish them. Finally, receptor autoradiography allows regional but not cellular or subcellular localization. Selective lesions may help to reveal whether receptors are pre- versus postsynaptic, but interpretation of such experiments is often complicated by compensatory alterations in receptor expression.

In Situ Hybridization

In situ hybridization identifies the presence of mRNA for a given protein within cells and is therefore essential for determining the capacity of any cell to express a particular receptor. Hybridization probes must be specific for the mRNA species of interest and applied under conditions that prevent nonspecific labeling. Antisense strands tagged with radioisotopes provide a high-sensitivity signal that can be quantified for documenting changes in mRNA levels under natural or experimental conditions. However, radioactive decay reduces the stability of the probes and may necessitate long exposure times for detecting some mRNA species. Digoxigenin or fluorophore tags allow for dual in situ hybridization approaches as well as for combinations of in situ hybridization with immunocytochemistry and with tract tracing. However, these can be less useful for quantitative estimates of mRNA levels. Studies using real-time reverse transcriptase polymerase chain reaction (RT-PCR) indicate that standard in situ hybridization methods are sometimes unable to reveal low levels of receptor mRNA.[31] Nevertheless, the results of single-cell RT-PCR experiments also need to be interpreted with caution because amplification may overestimate the actual copy number. In situ hybridization mainly labels cell soma and proximal dendrites where mRNA is most abundant and provides little information regarding the ultimate compartmental destination of receptor proteins following synthesis. Still, in situ hybridization has been used to localize mRNA in distal neuronal compartments,[32] although such methodology has not yet been applied to the study of DA receptors.

Immunocytochemistry

Immunocytochemistry utilizes antibodies directed against various epitopes of receptor proteins and can detect receptors wherever they are expressed within regions and within cells. The validity of the results is highly dependent on the specificity of antibodies for discrete antigens that are unique to the receptor of interest and are not cross-reacting with other proteins.[33] Even when specific reagents are available, different antibodies can suggest different distributions for the same receptor. For reasons that are unclear, some antibodies may preferentially label pre- versus postsynaptic receptors or glial versus neuronal receptors. Hence, combined results using multiple specific antibodies may sometimes be necessary for a complete understanding of receptor distribution. Immunoperoxidase and immunofluorescence methods are ideal for light microscopic studies and are employed with high sensitivity for delineating receptor distribution within cells and neuropil. Efforts to quantify changes in receptor levels following various treatments can be hampered by the amplification methods used to achieve high sensitivity, as well as by variation in tissue fixation, immunoreagent batches, and antibody penetration. Nevertheless, computer-assisted image analysis can improve quantification in many cases.[25,34]

Immunocytochemistry also provides the best approach for subcellular localization of receptor proteins when their exact distribution within neuronal compartments is investigated. It should be noted, however, that when receptors are predominantly trafficked into distal processes, their levels within cell soma can sometimes fall below the detection limits of light microscopy. This has often been the case for DA receptors, making it difficult to apply dual immunolabeling or tract tracing to the phenotypic identification of cells that express these receptors. The use of confocal microscopy can improve detection of low DA receptor levels in soma,[35,36] and electron microscopy also permits detection of sparse immunoreactivity in perikarya as well as distal processes.[15,37,38]

Two additional methodological issues affect the interpretation of subcellular localization studies of DA receptors. First, there is often a trade-off between sensitivity and spatial resolution. Specifically, immunoperoxidase methods permit only a relative illustration of subcellular distribution because the reaction product is diffusible. Nevertheless, these approaches typically employ signal amplification, giving them the sensitivity to detect receptors present at low levels. Immunogold techniques use a nondiffusible marker and so have better spatial resolution for more precise subcellular localization, although this is often achieved at the expense of sensitivity.[30] Second, pre-embedding immunogold methods tend to bias detection toward extrasynaptic receptors, in part because of limited antibody penetration of the protein complex at synaptic junctions. Postembedding techniques involve application of antibodies after ultrathin sectioning and therefore afford better penetration of synapses. Nevertheless, the processes required for plastic embedding (especially lipid fixation) often have the effect of destroying antigenicity. Such limitations can sometimes be overcome, as they were for an elegant series of studies on Glut receptor localization by Somogyi and colleagues.[39] Unfortunately, the precise distribution of DA receptors with respect to synapses has yet to be accomplished with similar methodology. Hence, it is important to keep in mind while reading this review that the exact spatial location of D2 receptors with respect to Glut or other synapses is not yet known, although some speculative observations on this subject are occasionally offered.

D2 RECEPTOR DISTRIBUTION AND FUNCTION

Midbrain DA Neurons

D2 autoreceptors on DA neurons

D2 autoreceptors expressed by midbrain DA neurons are activated by dendritically released DA and moderate cell activity in the face of phasic excitation[40] driven by Glut afferents from multiple sources.[41] In addition to counteracting excitatory drive, D2 receptors occur in close proximity to the synapses formed by Glut axons and can therefore modulate transmission via actions on Glut receptors (Fig. 2.2.1, site 1).

Numerous anatomical studies document the expression of D2 receptors in the ventral midbrain.[9,13,37,38,42–46] D2 receptor protein has been specifically localized to DA cells,[13,38] with occasional non-DA neurons also appearing to express this receptor.[38,47] One study suggests that D2 receptors are not uniformly expressed by all populations of DA cells, being notably lower in the parabrachial than the paranigral division of the VTA.[47] Although retrograde tract tracing was not employed in this study, projections to the prefrontal cortex (PFC) originate more commonly from the parabrachial subdivision, making these observations consistent with physiological and neurochemical reports that mesocortical DA neurons are less sensitive to autoreceptor control than mesolimbic or nigrostriatal cells.[48]

For many years, midbrain DA neurons were thought to receive Glut afferents from a limited number of sources, most prominently the PFC and brainstem mesopontine tegmentum. More recently, the seminal work of Geisler and colleagues has revealed numerous sources of Glut afferents to the VTA that, in addition to the cortex, arise from widespread subcortical structures including the lateral hypothalamic and lateral preoptic areas, central gray, raphe nuclei, brainstem tegmentum, and reticular formation.[41] Additional Glut afferents to the SNc originate from the subthalamic nucleus.[49] Our own ultrastructural studies have verified that the majority of Glut axons in the VTA originate from probable subcortical as opposed to cortical sites.[50] Collectively, DA neurons receive relatively abundant synaptic input from these various Glut afferents, although SNc DA cells exhibit synaptic input more commonly from GABA axons.[51] Dopamine cells also express ionotropic and metabotropic Glut receptors, in some cases near sites of presumed Glut synapses.[52–54]

Within midbrain DA and non-DA cells, the majority of D2 receptor immunoreactivity has been localized to the cytoplasm along the surface of smooth endoplasmic reticulum as well as to the plasma membrane at extrasynaptic sites.[13,38,47] The D2 receptor has occasionally been detected near symmetric or asymmetric synapses when the subcellular distribution was examined using immunoperoxidase methods. Immunogold localization has verified the cytoplasmic and endosomal labeling as well as the distribution of D2 receptors to extrasynaptic portions of the plasmalemma. However, punctate gold labeling near asymmetric synapses has not yet been observed,[13] suggesting that autoreceptors may not directly modulate Glut transmission in the SNc or VTA. Moreover, there are as yet no studies examining the proximity of D2 receptors to *N*-methyl-D-aspartate (NMDA), AMPA, or metabotropic Glut receptors in DA neurons. Nevertheless, some electrophysiological studies do report modulatory interactions. Specifically, long-term depression (LTD) in VTA DA neurons can be completely blocked by amphetamine via a D2-receptor-dependent mechanism.[55,56] Such an effect may consequently result in the expression of long-term potentiation (LTP) and excessive Glut transmission with chronic amphetamine abuse.[55] Moreover, it has been conjectured that blocking D2 receptors with antipsychotic drugs may contribute to the treatment of schizophrenia by facilitating the development of LTD.[55] The exact mechanism whereby D2 receptors alter Glut plasticity in DA cells remains to be determined.

Possible D2 autoreceptors on DA neurons that colocalize Glut

Dopamine and Glut also reportedly intersect *within* DA neurons in that some DA cells have been reported to colocalize Glut. An in-depth evaluation of this subject is beyond the scope of this chapter; nevertheless, a few comments are in order. Early reports of extensive colocalization of Glut in most DA neurons were based on indirect measures or on analysis of cultured cells.[57–59] In addition, many indications of a Glut phenotype have been complicated by the presence of this substance for metabolic purposes or as a precursor for GABA. More recently, several laboratories have reported a substantial population of VTA (but not SNc) neurons expressing mRNA for the vesicular glutamate transporter type 2 (VGlut2), an accepted marker of Glut cells.[7,60–62] Most midbrain Glut neurons appear to be a separate population from DA cells, although there is some colocalization. Standard in situ hybridization estimates in adult animals range from 2% or less[7,61,62] to as high as 20%–50% in certain midline divisions[60] (but see [61,62]). Single-cell RT-PCR experiments suggest that 25% of DA neurons may express a low abundance of VGlut2 mRNA.[7] This coexpression is developmentally regulated and largely suppressed in adulthood, although it might be reinduced under pathological conditions.[7] In any case, Glut appears to be colocalized only in a subset of DA neurons.

Regarding the targets of VTA cells containing both DA and Glut, at least some of these neurons are reported to project to the core of the NAc[7] but not to the dorsal striatum.[63] VGlut2-containing VTA cells also project to the PFC,[64] but it has not yet been determined whether any of these neurons coexpress DA. A preliminary report indicates that Glut VTA cells have local collaterals that innervate DA neurons.[65] Hence, Glut afferents to DA cells derive from intrinsic as well as extrinsic sources. It is unclear how the combined expression of DA and Glut interacts functionally within target regions, although some theoretical models have been generated.[7,66,67] In future studies, it will be important to determine the extent to which D2 autoreceptors at soma (Fig. 2.2.1, site 1′) or nerve terminals are involved in regulating Glut corelease. Such an action has been suggested from cultured cell studies.[58]

Forebrain Glut Cells

D2 receptors on Glut neurons

Dopamine cells directly innervate Glut neurons within the cerebral cortex, hippocampus, amygdala, and thalamus[68–71] (Fig. 2.2.1, site 2). The cortical DA projection is densest in frontal areas, including various divisions of the medial PFC, orbital cortex, and cingulate cortex. A projection to the motor cortex is prominent in primates but only modest in the rat.[23] Dopamine axons also innervate Glut cells in the subthalamic nucleus,[72,73] but this structure expresses mainly D5 receptors[74] and will not be considered further here.

D2 receptors as measured by receptor autoradiography or in situ hybridization are expressed in the cerebral cortex at levels noticeably lower than in the basal ganglia and are also lower than the dominant D1 subtype.[9,42,45,75–78] D2 receptor binding and/or mRNA have also been described in the amygdala, hippocampus, and thalamus.[9,79,80] In multiple cortical regions, and especially in the PFC and cingulate cortex, D2 mRNA is expressed most heavily in the pyramidal cell layers 2–3 and 5–6,[75–78] and specifically in neurons projecting to the striatum and to other cortical regions.[75] The results of immunohistochemical localization studies are relatively well matched to the distribution of D2 receptor mRNA.[15,37,43,46,80–83] In addition, antibodies have identified D2 receptors within astrocytic processes, where they reportedly increase calcium levels.[82,84] Whether this ultimately provides a mechanism for regulation of Glut transmission[85] has not yet been determined.

A few ultrastructural studies using immunogold methods have described the subcellular distribution of D2 receptor immunoreactivity in the cortex, which is found most commonly in the cytoplasm associated with endosomes, smooth endoplasmic reticulum, or Golgi apparatus, and along the plasma membrane at nonsynaptic sites.[15,82,84,86–88] D2 receptors are also frequently associated with clathrin-coated vesicles, suggesting that these organelles form an important component of recycling for this receptor.[87] To date, none of these studies has localized D2 receptors by immunogold in close proximity to asymmetric, presumed Glut synapses or to identified Glut receptors on the dendrites of cortical cells. Nevertheless, considerable functional interactions between D2 and Glut receptors have been documented. Many of these interactions exemplify DA regulation of Glut transmission via convergence onto common targets (see the section "Functional Implications of DA and Glut Convergence" below). However, as the target neurons are themselves glutamatergic, the interactions will be discussed here.

Electrophysiological evidence, particularly from the PFC, suggests that DA acting on D2 receptors decreases the excitability of cortical pyramidal cells and suppresses Glut synaptic responses, especially those of AMPA receptors[89–92] (see also [93]). In other cortical regions, the direction of D2 modulation of Glut transmission varies from attenuating to facilitating.[81,94–96] These variable responses appear to depend on multiple factors, including target region, cortical layer, the activity state of neurons, and timing issues relative to short-term plasticity.[5,97]

Within the PFC, the actions of D2 receptors on Glut cells are often opposite those of D1 receptors, which typically increase cell excitability and Glut transmission.[5,90,91,93,98,99] D2 receptors also alter plasticity at pyramidal neurons by favoring LTD of Glut transmission[100]; D1 receptors again have opposing actions by facilitating LTP.[93,99,101] Collectively, the marked differences in D2 versus D1 receptor actions in the PFC have important implications for the working memory functions of this region.[5]

D2 receptors on secondary neurons

The activity patterns of cortical Glut pyramidal cells are strongly influenced by GABA local circuit neurons,[102] which are also innervated by DA axons[103,104] and express D2 receptors.[46,84] Hence, non-Glut neurons constitute another site at which DA can influence Glut transmission, albeit indirectly (Fig. 2.2.1, site B3). As described above, the dominant action of D2 receptors on cortical Glut neurons is suppression of excitability and synaptic responses. This suppressive effect is likely to be accomplished, at least in part, by increasing GABAergic drive onto these cells.[92,105-107] Interestingly, the strength and duration of this effect increase in postpubertal animals.[92,107] Secondary neurons in other target areas also mediate some of the indirect actions of D2 receptors on Glut transmission,[5] although many of these are mediated at the presynaptic level (see below).

Convergence of DA and Glut

D2 receptors on the postsynaptic targets of DA and Glut inputs

Throughout their target fields, DA axons synapse onto neurons that also receive convergent Glut afferents, providing one of the main sites where these two neuroregulators interact (Fig. 2.2.1, site 4). Within the cortex, these convergence sites include mainly the spines of Glut pyramidal neurons,[68] although DA synapses onto local circuit neurons also occur in close proximity to Glut contacts.[103] Within the basal ganglia, convergence sites for DA and Glut axons are primarily the spines and distal dendrites of GABA projection neurons.[72,108,109] For a more thorough description of the anatomy of DA and Glut convergence, the reader should consult Chapter 2.4 by Moss and Bolam in this volume. Functional sites where DA and Glut can interact via postsynaptic receptors range from highly compartmentalized in the case of spine convergence[110] to more diffuse in instances where DA receptors alter overall cell excitability regardless of the focal sites where Glut is released and initiates depolarization.[111,112] Evidence for D2 receptor expression in cortical structures was previously considered in the section "Forebrain Glut Cells," so this section will focus on D2 receptor distribution within basal ganglia regions.

Cellular localization studies. The presence of DA D2 receptors throughout the dorsal to ventral aspects of the striatopallidal complex has been amply demonstrated by receptor autoradiography, in situ hybridization, and immunohistochemical methods.[8,9,13,37,42-46,76,113-115] D2 receptors are expressed by cholinergic and neurotensin interneurons[116-118] and GABAergic medium spiny neurons.[13,114,117,119-121] The D2 receptor has also occasionally been identified within glial processes in the striatum and pallidum.[38,115]

Within the striatum, the D2 receptor is localized mainly to cells that express enkephalin and project to the external globus pallidus (i.e., the indirect striatal output pathway).[36,117,119,120,122,123] Most of these neurons do not express detectable levels of D1 receptor and therefore do not appear to contribute substantially to the direct output pathway projecting to the substantia nigra.[12,120,124] Nevertheless, recent studies indicate a greater degree of collateralization in both direct and indirect striatal output channels than was previously appreciated.[125] Moreover, mRNA amplification indicates that nearly half of striatal medium spiny neurons have the capacity to express a combination of DA receptor subtypes when one includes the extended D1–D5 and D2–D3–D4 families.[31] This finding does not necessarily mean that protein for multiple receptor types is expressed at functional levels. Indeed, early quantification revealed little evidence for receptor coexpression within striatal spines.[12] Nevertheless, confocal microscopic studies suggest a greater extent of colocalization within soma, and physiological evidence indicates that many striatal neurons show comparable responses to D1- and D2-selective agonists.[31,35,122,123]

Such conflicting reports have raised questions regarding the sensitivity and accuracy of various methodologies for estimating receptor expression. Some resolution of this controversy appears to come from BAC transgenic mouse strains in which expression of enhanced green fluorescent protein (EGFP) is selectively linked to the promoter for either D1 or D2 receptors. In these animals, the D1 and D2 subtypes are largely segregated in separate medium spiny cell populations.[126-128] Moreover, experiments in BAC transgenic mice have begun to reveal some of the complex mechanisms whereby D1- or D2-selective agonists can induce comparable physiological actions in direct and indirect pathway cells. For example, D2 receptors on cholinergic interneurons reduce acetylcholine release, which triggers a cascade of events that ultimately attenuates Glut release onto principal cells, regardless of whether they express D1 or D2 receptors[128] (Fig. 2.2.1, site C3). Such a mechanism serves as a reminder that indirect actions of D2 receptors can occur even when the Glut neuronal elements do not express these receptors or lie in close proximity to the site of DA release.

Subcellular localization studies. Within spiny striatal neurons, some studies have noted that immunoreactivity for the D2 receptor is more intense in distal portions of the dendritic tree than in soma and proximal dendrites.[12,37,38,129] This distribution is consistent with physiological studies emphasizing the dominant functionality of distal DA receptors[96,130,131] and with the location of Glut synapses and receptors in dendrites.[132,133] The subcellular distribution of the D2 receptor is similar to that reported in the cortex (see above), namely, being cytoplasmic in association with organelles (e.g., endosomes and Golgi apparatus) and plasmalemmal at nonsynaptic locations.[13,86,88,115,134] The proportion of membrane to cytoplasmic receptor appears to increase from the soma outward into the dendritic tree.[134]

In addition to extrasynaptic sites, D2 receptor immunolabeling occurs near asymmetric and symmetric synapses,[12,14,37,38,135] and this distribution has been confirmed by discrete immunogold methods.[13] D2 receptors near asymmetric synapses provide potential anatomical substrates for modulation of Glut transmission at these sites. This interpretation is supported by observations of D2 receptor immunolabeling within spines that receive synapses from axons labeled for Glut[14] or tracer transported anterogradely from the motor cortex[12] or the PFC.[135] Receptors associated with symmetric synapses might be postsynaptic to DA axons, although this has not yet been demonstrated directly. Some supportive evidence for D2 postsynaptic receptors does exist (Sesack, unpublished observations), and functional studies have speculated that these are the logical subtype to receive synaptic DA signals.[136] D2 receptors also occur within dendritic structures receiving symmetric synaptic input from GABA-labeled terminals.[14,121]

Functional implications of DA and Glut convergence. Sites where DA and Glut axons synapse onto the same spines place DA in a position to modulate postsynaptic responses to highly specific sources of Glut drive. This "triadic" configuration also places DA in a favorable position for reaching extrasynaptic heteroreceptors on Glut nerve terminals (see below). Spines are now understood to be biochemically rather than electrically isolated compartments. Diffusion of large signaling molecules is relatively restricted, allowing individual spines with specific synaptic inputs to undergo selective alterations in activity and plasticity.[110] Moreover, evidence suggests that the movement of membrane-bound receptors (at least for Glut) is tightly regulated within and between spines in an activity-dependent manner.[137] How DA receptors are trafficked in accordance with such phenomena or themselves contribute to the movement of Glut receptors is only beginning to be explored.[93,110,138]

The interesting speculation has been put forward that spines with Glut synapses but no converging DA synapse (i.e., dyads versus triads) may be the main detectors of increased DA levels, assuming that they express DA receptors.[110] In this case, DA modulation of Glut transmission would occur only when extracellular levels of DA rise, either from increased tonic activity or short-term phasic bursts. Moreover, depending on the distance from the focal source of DA release, each spine could be set to different levels of activity and plasticity. Hence, spines that express D2 receptors do not necessarily need to receive DA from a synaptic release event. In this regard, it is interesting to note evidence suggesting that DA axons form a regular lattice in the striatum that maintains relatively constant spacing (~1 μm) from thalamostriatal and corticostriatal Glut axons[112] (see also Chapter 2.4 by Moss and Bolam in this volume). Moreover, it has been estimated that DA release from synaptic or extrasynaptic sites can diffuse 7–8 μm and still be in sufficient concentration to stimulate D2 receptors.[8]

Acute D2 modulation of Glut transmission is believed to occur primarily through regulation of AMPA receptors, which are located in distal dendrites and spines in striatal neurons.[133] Specifically, D2-induced reduction of membrane AMPA receptors decreases the excitability of neurons.[4,130,131] D2 receptors have additional actions that further reduce excitability and the response to synaptic Glut, including indirect effects mediated by reducing acetylcholine release from cholinergic interneurons.[131] As was found in the cortex, the actions of striatal D2 receptors are typically opposite the actions of D1 receptors that facilitate Glut transmission.[4,131] Interestingly, D1 and D2 receptor–expressing cell populations also appear to receive Glut input from different cortical sources.[139] It is likely that D2-mediated reduction of the corticostriatal Glut drive serves to suppress the selection of inappropriate motor programs elicited by uncoordinated cortical events.[131] However, striatal neurons expressing D2 receptors have recently been shown to be more excitable than those expressing D1 receptors and to more faithfully represent the corticostriatal drive.[140] These characteristics appear to be independent of D2 and AMPA receptors, and how they contribute to adaptive motor control remains to be established.

Structural plasticity regulated by D2 receptors. Long-term structural plasticity at the spines of striatal medium spiny neurons represents an important functional event that reflects DA interactions with Glut via D2 receptors. Many studies document the importance of DA receptors for synaptic plasticity, with D2 receptors favoring LTD and D1 receptors facilitating LTP[131,138,141] (see also Chapter 7.3 by Surmeier et al. in this volume). That DA is at least partly responsible for structural maintenance in the striatum is evidenced by reports of 27% spine loss in postmortem Parkinson's brain[142] and 15%–30% reduction in spines and axo-spinous synapses in animal models with DA denervation.[143,144] The degree of spine loss is tightly correlated to the reduction in DA fibers,[144] and lesion-induced spine loss is also noted in the nucleus accumbens (NAc) and PFC.[145] Disruptions of other aspects of dendritic morphology have also been reported following DA depletion in animal models and in Parkinson's disease (for a review, see [73]).

A recent collaborative study indicates that loss of axo-spinous synapses induced by DA denervation actually reflects a higher-magnitude reduction (~35%–50%) in the population of neurons expressing the D2 receptor combined with no change in D1 receptor–expressing cells.[126] Both the spines themselves and the Glut synapses innervating the heads of spines are lost, and postsynaptic excitatory potentials are also substantially reduced. These findings suggest that D2 receptors stabilize both the pre- and postsynaptic elements in corticostriatal or thalamostriatal axo-spinous synapses on striatopallidal neurons. Reduced spine density has also been observed with reserpine treatment that merely depletes DA,[126] suggesting that it is the loss of DA itself that destabilizes spines and not the physical removal of the nerve terminal. This is consistent with the well documented ability of DA to communicate via extrasynaptic receptors.[6]

The exact mechanism linking D2 receptors to spine stabilization has not yet been fully elucidated. D2 receptor activation can produce antioxidant effects and protect against Glut cytotoxicity.[146,147] However, the physiological studies of Surmeier indicate that D2 receptor–mediated spine stabilization depends at least in part on selective inhibition of L-type calcium channels with a Cav1.3α1 subunit.[126] This channel is expressed in spines and linked to Glut synapses by scaffolding proteins. Loss of D2 inhibitory influence reduces spines and Glut axo-spinous synapses but leads conversely to compensatory increases in Glut-driven activity in striatopallidal cells that may contribute to the symptoms of Parkinson's disease. This mechanism might be specific for the striatum, as a similar loss of spines in the PFC following DA lesioning appears not to be linked to D2 receptors.[145,148] The selective spine stabilization by D2 receptors in the striatum may also be specific to rodents, given that DA denervation in the primate basal ganglia produces spine loss that is not selective for D2-expressing cells.[144] However, it should be noted that the latter study did not employ unbiased stereological measurements.

Despite reducing the density of excitatory axo-spinous synapses overall, selective DA lesions actually increase the number of complex synapses with perforations,[143] a morphological feature thought to reflect enhanced synaptic efficacy.[149,150] This effect clearly involves synapses formed by Glut axons and is mimicked by chronic treatment with D2 receptor antagonists, suggesting a role in the neurological side effects produced by antipsychotic drugs.[151] Nevertheless, the exact contribution of D2 receptors to this phenomenon is unclear.[152] Changes in spine density in various forebrain regions also accompany behavioral sensitization to chronic psychostimulants that enhance DA levels.[153] However, a clear correlation to D2 receptor activation also has not yet been demonstrated.[127]

D2 heteroreceptors on Glut nerve terminals

Glut neurons that express D2 receptors have the potential to transport these receptors into axons, where they can be inserted into the presynaptic membrane (Fig. 2.2.1, sites A5, C5). Although DA and Glut axons do not exhibit axo-axonic synapses, DA can reach presynaptic heteroreceptors via diffusion over distances of several microns.[8,112] Moreover, the ability of DA to act on these receptors is likely to be facilitated by the frequent close apposition of DA and Glut axons and their synaptic convergence onto common distal dendrites, as observed in the cortex and striatum (see above). Although presynaptic D1 receptors are sometimes described on Glut nerve terminals,[16] it is the D2 class that is the most common presynaptic receptor type.[154]

Forebrain. Presynaptic D2 receptors have been reported in numerous brain regions, including the striatal complex, ventral pallidum, amygdala, and cerebral cortex.[12,13,23,37,82,83,87,88,114–116,121,129,135,139] Within axons, D2 receptor immunolabeling is distributed to both cytoplasmic and plasmalemmal sites.[82,87,135] D2 receptor–bearing varicosities forming asymmetric synapses presumably represent Glut nerve terminals. Although the presence of Glut has not been directly demonstrated in these profiles, some have been shown to originate from the PFC.[83,135] These observations are consistent with physiological demonstrations of presynaptic inhibition of Glut transmission via D2 receptors on axons from the PFC and other Glut sources.[154–158] Other presynaptic D2 receptors represent autoreceptors on DA nerve terminals[38,115] or heteroreceptors regulating the release of other transmitters like GABA.[121]

AMPA receptors clearly serve as presynaptic autoreceptors in striatal structures.[63] Hence, the functional interactions reported between D2 and AMPA receptors in the soma and dendrites of cortical and striatal cells[4,130,131] (see above) may also play out within axon terminals. However, to date, no study has endeavored to localize both D2 heteroreceptors and AMPA autoreceptors within the same axon terminals. In the striatum, the extent of D2 inhibition of Glut release increases with the firing frequency of cortical afferents and is selective for terminals with low release probability.[154] In this way, DA acts as a low-pass filter to favor corticostriatal transmission via the most active synaptic connections. Although the ability of D2 receptors to reduce Glut release is likely to occur through presynaptic receptors, studies also suggest that this effect may involve a postsynaptic D2 receptor action and retrograde signaling via endocannabinoids[159] (see also [128]).

Midbrain. Within the VTA, D2 receptor labeling has also been reported in axons, most of which do not form synapses in single sections.[38,47] Some axon varicosities containing weak immunolabeling for D2 receptor and forming asymmetric synapses have been described.[47] This observation is consistent with one electrophysiological study reporting presynaptic inhibition of Glut transmission by D2 receptors in the VTA.[160] The latter authors have speculated that dendritically released DA might act on these receptors in order to reduce the Glut drive associated with burst firing.

D3 RECEPTORS IN RELATION TO GLUTAMATE

The D3 receptor subtype has garnered extensive interest in neuropsychopharmacology research because of its relatively restricted expression to the ventral parts of the striatal complex and to the cortex. This distribution opens up the possibility that novel D3-selective antipsychotics might be free of the adverse effects mediated by D2 receptors in dorsal striatal regions.[161–163] The D3 receptor subtype has also been suggested as a potential therapeutic target for the treatment of drug addiction.[164]

D3 receptors have been localized by receptor autoradiography, in situ hybridization, and immunocytochemistry. Despite some discrepancies among the three approaches, there is good agreement that the most robust D3 expression is in the ventral striatal complex, including the NAc shell, olfactory tubercle, and islands of Calleja.[9,42,46,124,161,163–169] Lower levels of D3 receptors have also been reported in the caudate putamen, ventral pallidum, hippocampus, amygdala, and midbrain. In striatal neurons, the D3 receptor is at least partially colocalized with D1 and D2 receptors and is most abundant in neurons that coexpress substance P and project to the substantia nigra.[31,124,165,166] Many investigators report the presence of D3 receptors in the cerebral cortex, most notably the parietal region and PFC.[46,163,165,167,168] D3 receptors localized to layer 3–5 cells may be present in Glut pyramidal neurons, although many may be in nonpyramidal cells.[46]

A fair amount of controversy surrounds the issue of whether midbrain D3 receptors represent actual autoreceptors. Estimates range from weak expression in small populations of DA cells to nearly complete expression in all DA neurons.[166,168] Immunoreactivity also occurs within non-DA cells and in the neuropil of the substantia nigra, which may correspond in part to presynaptic receptors on striatonigral axons.[165,168,169] Selective lesions of DA neurons reduce D3 receptor binding in the NAc, but they also reduce mRNA for this receptor, suggesting that an anterograde factor in DA axons may be necessary for sustaining postsynaptic D3 receptors.[162,169] Further challenges to the establishment of D3 autoreceptors come from transgenic mouse models, where knockouts of D3 receptors produce little change in autoreceptor functions, whereas D2 knockouts eliminate evidence of autoreceptor activity.[40,170] Most recently, it has been suggested that D3 autoreceptors control only basal levels of DA and not release or firing activity.[171]

Compared to the other D2 family receptors, much less is known about the subcellular localization of the D3 subtype. Ultrastructural studies are especially lacking, perhaps due to concerns regarding antibody specificity.[46,165,168] Where it has been estimated from light microscopy, the D3 receptor has been described as occurring in the cytoplasm and at extrasynaptic portions of the plasma membrane.[168] Given the limited subcellular information, one can only speculate regarding potential sites where D3 receptors might interact with Glut transmission: (1) as potential autoreceptors on some midbrain DA neurons that receive Glut input (or colocalize Glut), (2) as postsynaptic receptors on cortical pyramidal or non-pyramidal cells, and (3) as postsynaptic receptors on ventral striatal neurons that occur in proximity to Glut inputs. Clearly, there is a need for more detailed studies of D3 receptor localization before more specific relationships to Glut transmission can be established.

D4 RECEPTORS IN RELATION TO GLUTAMATE

There is considerable interest in the localization of the DA D4 receptor because of its higher affinity for the atypical antipsychotic drug clozapine and the potential contribution of this receptor subtype to the understanding of schizophrenia pathophysiology and treatment.[172,173] D4 receptors have also been linked to attention deficit hyperactivity disorder and the regulation of novelty seeking.[174]

The DA D4 receptor is expressed predominantly in cortical regions, suggesting a particular role in the modulation of cognitive functions. However, the various approaches for detecting D4 receptors have produced somewhat conflicting results. By in situ hybridization or immunocytochemistry, the density of D4 receptor expression is as follows: densest in motor, sensory, and cingulate cortices; intermediate in temporal, retrosplenial, and granular association areas; and lowest in the PFC and cortical regions near the rhinal sulcus. D4 receptors have also been observed in the hippocampus and amygdala.[9,46,76,175–179] D4 receptor mRNA is reportedly concentrated in deep cortical layers, especially layer 5.[76–78] Immunoreactivity for the D4 receptor protein is densest in layers 2–5, including a modest expression in Glut pyramidal cells and dense expression in GABA local circuit neurons.[46,173,176,179] GABA neurons in the striatum and pallidum also appear to express measurable levels of D4 receptor mRNA or protein, although these have been difficult to detect in some studies.[31,46,173,176,178,180]

Recently, BAC transgenic mice have been used to selectively express EGFP in cells transcribing the D4 receptor gene.[181] The densest expression was reported in deep layers of the PFC, including prelimbic, cingulate, orbital, and agranular insular areas. D4 receptors were also observed in motor and piriform cortices, the anterior olfactory nucleus, ventral pallidum, and brainstem parabrachial nucleus. No signal was detected in the striatum or NAc, amygdala, hippocampus, thalamus, or midbrain. It is not yet known whether the discrepancies in D4 receptor expression between BAC transgenic mice and other anatomical localization methods are due to sensitivity issues, problems with the specificity of some D4 receptor antibodies, or species differences.

At the subcellular level in both the cortex and striatum, immunoreactivity for the D4 receptor has been reported along nonsynaptic portions of the plasma membrane throughout soma, dendrites, and spines; receptor localization to the smooth endoplasmic reticulum and other cytoplasmic organelles has also been described.[173,180] The dominant nonsynaptic distribution is consistent with observations that D4 receptors occur at considerable distances from DA axons[179] (but see [181]). D4 receptor immunoreactivity has specifically been localized in proximity to Glut synapses in NAc spines by both immunoperoxidase and immunogold methods,[180] suggesting that the D4 receptor may directly modulate Glut transmission at these sites. As yet, there are no physiological studies to support this, although there is evidence for D4 receptor-mediated attenuation of Glut transmission in the amygdala and PFC.[182–184] D4 receptors may also reduce transmission through GABA-A receptors in the PFC.[185] The anatomical substrates for these interactions have not yet been explored.

The majority of D4 receptors in the striatal complex appear to be presynaptic.[180] These D4 immunoreactive axons either fail to form synapses in single sections, precluding their identification, or make asymmetric contacts characteristic of Glut connections. The source of these latter axons is not known but is most likely the cerebral cortex.[177,180] The thalamus is another potential source, but thalamic D4 receptors are primarily expressed by intrinsic reticular cells[173] and not projection neurons. It should be noted that some cross-reaction of the D4 antibody with the D2 receptor could not be ruled out in the NAc ultrastructural study.[180] This is important to consider given that much of the distribution reported for D4 receptors has also been observed in D2 localization studies. Importantly, Glut responses seem to be unaltered in the striatum of mice with transgenic deletion of D4 receptors, whereas D2 receptor knockout mice do show substantial changes in excitatory synaptic activity.[157] This suggests that D2 receptors may be the dominant presynaptic controller of Glut transmission, at least in the striatum. In the cortex, D4 receptors do have presynaptic actions on Glut axons,[186] consistent with the greater density of D4 receptors there compared to the striatum.

In summary, the main sites of probable DA modulation of Glut transmission via the D4 receptor are (1) directly onto Glut pyramidal neurons of PFC and other cortical regions, (2) indirect regulation of pyramidal neurons via actions on GABA local circuit neurons, (3) modulation of Glut transmission in NAc spines, and (4) presynaptic actions on Glut axons in the NAc.

REFERENCES

1. Schultz W. Multiple dopamine functions at different time courses. *Annu Rev Neurosci.* 2007;30:259–288.
2. Redgrave P, Gurney K, Reynolds J. What is reinforced by phasic dopamine signals? *Brain Res Rev.* 2008;58(2):322–339.
3. Liss B, Roeper J. Individual dopamine midbrain neurons: functional diversity and flexibility in health and disease. *Brain Res Rev.* 2008;58(2):314–321.
4. Cepeda C, Buchwald NA, Levine MS. Neuromodulatory actions of dopamine in the neostriatum are dependent on the excitatory amino acid receptor subtypes activated. *Proc Natl Acad Sci USA.* 1993;90(20):9576–9580.
5. Seamans JK, Yang CR. The principal features and mechanisms of dopamine modulation in the prefrontal cortex. *Prog Neurobiol.* 2004;74(1):1–58.
6. Zoli M, Torri C, Ferrari R, et al. The emergence of the volume transmission concept. *Brain Res Rev.* 1998;26(2-3):136–147.
7. Descarries L, Berube-Carriere N, Riad M, Bo GD, Mendez JA, Trudeau LE. Glutamate in dopamine neurons: synaptic versus diffuse transmission. *Brain Res Rev.* 2008;58(2):290–302.
8. Rice ME, Cragg SJ. Dopamine spillover after quantal release: rethinking dopamine transmission in the nigrostriatal pathway. *Brain Res Rev.* 2008;58(2):303–313.
9. Missale C, Nash SR, Robinson SW, Jaber M, Caron MG. Dopamine receptors: from structure to function. *Physiol Rev.* 1998;78(1):189–225.
10. Smiley JF, Levey AI, Ciliax BJ, Goldman-Rakic PS. D_1 dopamine receptor immunoreactivity in human and monkey cerebral cortex: predominant and extrasynaptic localization in dendritic spines. *Proc Natl Acad Sci USA.* 1994;91(12):5720–5724.
11. Bergson C, Mrzljak L, Smiley JF, Pappy M, Levenson R, Goldman-Rakic PS. Regional, cellular, and subcellular variations in the distribution of D_1 and D_5 receptors in primate brain. *J Neurosci.* 1995;15(12):7821–7836.
12. Hersch SM, Ciliax BJ, Gutekunst CA, et al. Electron microscopic analysis of D1 and D2 dopamine receptor proteins in the dorsal striatum and their synaptic relationships with motor corticostriatal afferents. *J Neurosci.* 1995;15(7 pt 2):5222–5237.
13. Yung KK, Bolam JP, Smith AD, Hersch SM, Ciliax BJ, Levey AI. Immunocytochemical localization of D1 and D2 dopamine receptors in the basal ganglia of the rat: light and electron microscopy. *Neuroscience.* 1995;65(3):709–730.
14. Yung KK, Bolam JP. Localization of dopamine D1 and D2 receptors in the rat neostriatum: synaptic interaction with glutamate- and GABA-containing axonal terminals. *Synapse.* 2000;38(4):413–420.
15. Paspalas CD, Goldman-Rakic PS. Microdomains for dopamine volume neurotransmission in primate prefrontal cortex. *J Neurosci.* 2004;24(23):5292–5300.
16. Paspalas CD, Goldman-Rakic PS. Presynaptic D1 dopamine receptors in primate prefrontal cortex: target-specific expression in the glutamatergic synapse. *J Neurosci.* 2005;25(5):1260–1267.
17. Hara Y, Pickel VM. Overlapping intracellular and differential synaptic distributions of dopamine D1 and glutamate N-methyl-D-aspartate receptors in rat nucleus accumbens. *J Comp Neurol.* 2005;492(4):442–455.
18. Pickel VM, Colago EE, Mania I, Molosh AI, Rainnie DG. Dopamine D1 receptors co-distribute with N-methyl-D-aspartic acid type-1 subunits and modulate synaptically-evoked N-methyl-D-aspartic acid currents in rat basolateral amygdala. *Neuroscience.* 2006;142(3):671–690.
19. Bordelon-Glausier JR, Khan ZU, Muly EC. Quantification of D1 and D5 dopamine receptor localization in layers I, III, and V of *Macaca mulatta* prefrontal cortical area 9: coexpression in dendritic spines and axon terminals. *J Comp Neurol.* 2008;508(6):893–905.
20. Abi-Dargham A, Laruelle M. Mechanisms of action of second generation antipsychotic drugs in schizophrenia: insights from brain imaging studies. *Eur Psychiatry.* 2005;20(1):15–27.
21. Oades RD, Halliday GM. Ventral tegmental (A10) system: neurobiology. 1. Anatomy and connectivity. *Brain Res Rev.* 1987;12(2):117–165.
22. Fallon JH, Loughlin SE. Substantia nigra. In: Paxinos G, ed. *The Rat Nervous System.* 2nd ed. San Diego, CA: Academic Press; 1995:215–237.
23. Lewis DA, Sesack SR. Dopamine systems in the primate brain. In: Bloom FE, Björklund A, Hökfelt T, eds. *Handbook of Chemical Neuroanatomy, The Primate Nervous System, Part I.* Vol 13. Amsterdam: Elsevier Science Publishers; 1997:261–373.
24. Björklund A, Dunnett SB. Dopamine neuron systems in the brain: an update. *Trends Neurosci.* 2007;30(5):194–202.
25. Peretti-Renucci R, Feuerstein C, Manier M, et al. Quantitative image analysis with densitometry for immunohistochemistry and autoradiography of receptor binding sites–methodological considerations. *J Neurosci Res.* 1991;28(4):583–600.
26. Maggio JE, Mantyh PW. Autoradiography of reversible ligands. In: Ariano MA, ed. *Receptor Localization: Laboratory Methods and Procedures* New York, NY: Wiley-Liss; 1998:17–30.
27. Chesselet MF. Localization of mRNAs encoding receptors with *in situ* hybridization histochemistry. In: Ariano MA, ed. *Receptor Localization: Laboratory Methods and Procedures.* New York, NY: Wiley-Liss; 1998:140–159.
28. Stornetta RL, Guyenet PG. Nonradioactive in situ hybridization in combination with tract-tracing. In: Zaborszky L, Wouterlood FG, Lanciego JL, eds. *Neuroanatomical Tract-Tracing 3: Molecules, Neurons, Systems.* New York, NY: Springer; 2006:237–262.
29. Mathiisen TM, Nagelhus EA, Jouleh B, et al. Postembedding immunogold cytochemistry of membrane molecules and amino acid transmitters in the central nervous system. In: Zaborszky L, Wouterlood FG, Lanciego JL, eds. *Neuroanatomical Tract-Tracing 3: Molecules, Neurons, Systems.* New York, NY: Springer; 2006:72–108.
30. Sesack SR, Miner LAH, Omelchenko N. Pre-embedding immunoelectron microscopy: applications for studies of the nervous system. In: Zaborszky L, Wouterlood FG, Lanciego JL, eds. *Neuroanatomical Tract-Tracing 3: Molecules, Neurons, Systems.* New York, NY: Springer; 2006:6–71.
31. Surmeier DJ, Song W-J, Yan Z. Coordinated expression of dopamine receptors in neostriatal medium spiny neurons. *J Neurosci.* 1996;16(20):6579–6591.
32. Eberwine J, Belt B, Kacharmina JE, Miyashiro K. Analysis of subcellularly localized mRNAs using in situ hybridization, mRNA amplification, and expression profiling. *Neurochem Res.* 2002;27(10):1065–1077.
33. Saper CB. Magic peptides, magic antibodies: guidelines for appropriate controls for immunohistochemistry. *J Comp Neurol.* 2003;465(2):161–163.
34. Brey EM, Lalani Z, Johnston C, et al. Automated selection of DAB-labeled tissue for immunohistochemical quantification. *J Histochem Cytochem.* 2003;51(5):575–584.
35. Aizman O, Brismar H, Uhlen P, et al. Anatomical and physiological evidence for D1 and D2 dopamine receptor colocalization in neostriatal neurons. *Nature Neurosci.* 2000;3(3):226–230.

36. Deng YP, Lei WL, Reiner A. Differential perikaryal localization in rats of D1 and D2 dopamine receptors on striatal projection neuron types identified by retrograde labeling. *J Chem. Neuroanat.* 2006;32(2-4):101–116.
37. Levey A, Hersch S, Rye D, et al. Localization of D_1 and D_2 dopamine receptors in brain with subtype-specific antibodies. *Proc Natl Acad Sci USA.* 1993;90(19):8861–8865.
38. Sesack SR, Aoki C, Pickel VM. Ultrastructural localization of D2 receptor-like immunoreactivity in midbrain dopamine neurons and their striatal targets. *J Neurosci.* 1994;14(1):88–106.
39. Somogyi P, Tamás G, Lujan R, Buhl EH. Salient features of synaptic organisation in the cerebral cortex. *Brain Res Rev.* 1998;26(2-3):113–135.
40. Mercuri NB, Saiardi A, Bonci A, et al. Loss of autoreceptor function in dopaminergic neurons from dopamine D2 receptor deficient mice. *Neuroscience.* 1997;79(2):323–327.
41. Geisler S, Derst C, Veh RW, Zahm DS. Glutamatergic afferents of the ventral tegmental area in the rat. *J Neurosci.* 2007;27(21):5730–5743.
42. Bouthenet ML, Souil E, Martres MP, Sokoloff P, Giros B, Schwartz JC. Localization of dopamine D3 receptor mRNA in the rat brain using in situ hybridization histochemistry: comparison with dopamine D2 receptor mRNA. *Brain Res.* 1991;564(2):203–219.
43. Ariano MA, Fisher RS, Smyk-Randall E, Sibley DR, Levine MS. D_2 dopamine receptor distribution in the rodent CNS using anti-peptide antibodies. *Brain Res.* 1993;609(1-2):71–80.
44. Meador-Woodruff JH, Damask SP, Watson SJ Jr. Differential expression of autoreceptors in the ascending dopamine systems of the human brain. *Proc Natl Acad Sci USA.* 1994;91(17):8297–8301.
45. Choi WS, Machida CA, Ronnekleiv OK. Distribution of dopamine D1, D2, and D5 receptor mRNAs in the monkey brain: ribonuclease protection assay analysis. *Mol Brain Res.* 1995;31(1-2):86–94.
46. Khan ZU, Gutierrez A, Martin R, Penafiel A, Rivera A, De La Calle A. Differential regional and cellular distribution of dopamine D2-like receptors: an immunocytochemical study of subtype-specific antibodies in rat and human brain. *J Comp Neurol.* 1998;402(3):353–371.
47. Pickel VM, Chan J, Nirenberg MJ. Region-specific targeting of dopamine D2-receptors and somatodendritic vesicular monoamine transporter 2 (VMAT2) within ventral tegmental area subdivisions. *Synapse.* 2002;45(2):113–124.
48. Tzschentke TM. Pharmacology and behavioral pharmacology of the mesocortical dopamine system. *Prog Neurobiol.* 2001;63(3):241–320.
49. Kita H, Kitai ST. Efferent projections of the subthalamic nucleus in the rat: light and electron microscopic analysis with the PHA-L method. *J Comp Neurol.* 1987;260(3):435–452.
50. Omelchenko N, Sesack SR. Glutamate synaptic inputs to ventral tegmental area neurons in the rat derive primarily from subcortical sources. *Neuroscience.* 2007;146(3):1259–1274.
51. Bolam JP, Smith Y. The GABA and substance P input to dopaminergic neurones in the substantia nigra of the rat. *Brain Res.* 1990;529(1-2):57–78.
52. Kosinski CM, Standaert DG, Testa CM, Penney JB Jr, Young AB. Expression of metabotropic glutamate receptor 1 isoforms in the substantia nigra pars compacta of the rat. *Neuroscience.* 1998;86(3):783–798.
53. Chatha BT, Bernard V, Streit P, Bolam JP. Synaptic localization of ionotropic glutamate receptors in the rat substantia nigra. *Neuroscience.* 2000;101(4):1037–1051.
54. Rodriguez JJ, Doherty MD, Pickel VM. N-methyl-D-aspartate (NMDA) receptors in the ventral tegmental area: subcellular distribution and colocalization with 5-hydroxytryptamine(2A) receptors. *J Neurosci Res.* 2000;60(2):202–211.
55. Jones S, Kornblum JL, Kauer JA. Amphetamine blocks long-term synaptic depression in the ventral tegmental area. *J Neurosci.* 2000;20(15):5575–5580.
56. Thomas MJ, Malenka RC, Bonci A. Modulation of long-term depression by dopamine in the mesolimbic system. *J Neurosci.* 2000;20(15):5581–5586.
57. Sulzer D, Joyce MP, Lin L, et al. Dopamine neurons make glutamatergic synapses *in vitro*. *J Neurosci.* 1998;18(12):4588–4602.
58. Congar P, Bergevin A, Trudeau LE. D2 receptors inhibit the secretory process downstream from calcium influx in dopaminergic neurons: implication of K+ channels. *J Neurophysiol.* 2002;87(2):1046–1056.
59. Lavin A, Nogueira L, Lapish CC, Wightman RM, Phillips PE, Seamans JK. Mesocortical dopamine neurons operate in distinct temporal domains using multimodal signaling. *J Neurosci.* 2005;25(20):5013–5023.
60. Kawano M, Kawasaki A, Sakata-Haga H, et al. Particular subpopulations of midbrain and hypothalamic dopamine neurons express vesicular glutamate transporter 2 in the rat brain. *J Comp Neurol.* 2006;498(5):581–592.
61. Yamaguchi T, Sheen W, Morales M. Glutamatergic neurons are present in the rat ventral tegmental area. *Eur J Neurosci.* 2007;25(1):106–118.
62. Nair-Roberts RG, Chatelain-Badie SD, Benson E, White-Cooper H, Bolam JP, Ungless MA. Stereological estimates of dopaminergic, GABAergic and glutamatergic neurons in the ventral tegmental area, substantia nigra and retrorubral field in the rat. *Neuroscience.* 2008;152(4):1024–1031.
63. Fujiyama F, Kuramoto E, Okamoto K, et al. Presynaptic localization of an AMPA-type glutamate receptor in corticostriatal and thalamostriatal axon terminals. *Eur J Neurosci.* 2004;20(12):3322–3330.
64. Hur EE, Zaborszky L. Vglut2 afferents to the medial prefrontal and primary somatosensory cortices: a combined retrograde tracing in situ hybridization. *J Comp Neurol.* 2005;483(3):351–373.
65. Dobi A, Morales M. Dopaminergic neurons in the rat ventral tegmental area (VTA) receive glutamatergic inputs from local glutamatergic neurons. *Soc Neurosci Abstr.* 2007:916–918.
66. Trudeau LE. Glutamate co-transmission as an emerging concept in monoamine neuron function. *J Psychiatry Neurosci.* 2004;29(4):296–310.
67. Lapish CC, Kroener S, Durstewitz D, Lavin A, Seamans JK. The ability of the mesocortical dopamine system to operate in distinct temporal modes. *Psychopharmacology.* 2007;191(3):609–625.
68. Goldman-Rakic PS, Leranth C, Williams SM, Mons N, Geffard M. Dopamine synaptic complex with pyramidal neurons in primate cerebral cortex. *Proc Natl Acad Sci USA.* 1989;86(22):9015–9019.
69. Gasbarri A, Sulli A, Packard MG. The dopaminergic mesencephalic projections to the hippocampal formation in the rat. *Prog Neuro-Psychopharm Biol Psychiatr.* 1997;21(1):1–22.
70. Melchitzky DS, Erickson SL, Lewis DA. Dopamine innervation of the monkey mediodorsal thalamus: location of projection neurons and ultrastructural characteristics of axon terminals. *Neuroscience.* 2006;143(4):1021–1030.
71. Muller JF, Mascagni F, McDonald AJ. Dopaminergic innervation of pyramidal cells in the rat basolateral amygdala. *Brain Struct Funct.* 2009;213(3):275–288.

72. Smith Y, Kieval JZ. Anatomy of the dopamine system in the basal ganglia. *Tr Neurosci.* 2000;23(suppl 10):S28–S33.
73. Smith Y, Villalba R. Striatal and extrastriatal dopamine in the basal ganglia: an overview of its anatomical organization in normal and Parkinsonian brains. 2008;23:S534–S547.
74. Svenningsson P, Le Moine C. Dopamine D1/5 receptor stimulation induces c-*fos* expression in the subthalamic nucleus: possible involvement of local D5 receptors. *Eur J Neurosci.* 2002;15(1):133–142.
75. Gaspar P, Bloch B, Le Moine C. D1 and D2 receptor gene expression in the rat frontal cortex: cellular localization in different classes of efferent neurons. *Eur J Neurosci.* 1995;7(5):1050–1063.
76. Meador-Woodruff JH, Damask SP, Wang JC, Haroutunian V, Davis KL, Watson SJ. Dopamine receptor mRNA expression in human striatum and neocortex. *Neuropsychopharmacology.* 1996;15(1):17–29.
77. Lidow MS, Wang F, Cao Y, Goldman-Rakic PS. Layer V neurons bear the majority of mRNAs encoding the five distinct dopamine receptor subtypes in the primate prefrontal cortex. *Synapse.* 1998;28(1):10–20.
78. de Almeida J, Palacios JM, Mengod G. Distribution of 5-HT and DA receptors in primate prefrontal cortex: implications for pathophysiology and treatment. *Prog Brain Res.* 2008;172:101–115.
79. Eliava M, Yilmazer-Hanke D, Asan E. Interrelations between monoaminergic afferents and corticotropin-releasing factor-immunoreactive neurons in the rat central amygdaloid nucleus: ultrastructural evidence for dopaminergic control of amygdaloid stress systems. *Histochem Cell Biol.* 2003;120(3):183–197.
80. Rieck RW, Ansari MS, Whetsell WO Jr, Deutch AY, Kessler RM. Distribution of dopamine D2-like receptors in the human thalamus: autoradiographic and PET studies. *Neuropsychopharmacology.* 2004;29(2):362–372.
81. Awenowicz PW, Porter LL. Local application of dopamine inhibits pyramidal tract neuron activity in the rodent motor cortex. *J Neurophysiol.* 2002;88(6):3439–3451.
82. Negyessy L, Goldman-Rakic PS. Subcellular localization of the dopamine D2 receptor and coexistence with the calcium-binding protein neuronal calcium sensor-1 in the primate prefrontal cortex. *J Comp Neurol.* 2005;488(4):464–475.
83. Pinto A, Sesack SR. Ultrastructural analysis of prefrontal cortical inputs to the rat amygdala: spatial relationships to presumed dopamine axons and D1 and D2 receptors. *Brain Struct Funct.* 2008;213(1-2):159–175.
84. Khan ZU, Koulen P, Rubinstein M, Grandy DK, Goldman-Rakic PS. An astroglia-linked dopamine D2-receptor action in prefrontal cortex. *Proc Natl Acad Sci USA.* 2001;98(4):1964–1969.
85. Agulhon C, Petravicz J, McMullen AB, et al. What is the role of astrocyte calcium in neurophysiology? *Neuron.* 2008;59(6):932–946.
86. Kabbani N, Negyessy L, Lin R, Goldman-Rakic P, Levenson R. Interaction with neuronal calcium sensor NCS-1 mediates desensitization of the D2 dopamine receptor. *J Neurosci.* 2002;22(19):8476–8486.
87. Paspalas CD, Rakic P, Goldman-Rakic PS. Internalization of D2 dopamine receptors is clathrin-dependent and select to dendro-axonic appositions in primate prefrontal cortex. *Eur J Neurosci.* 2006;24(5):1395–1403.
88. Pickel VM, Chan J, Kearn CS, Mackie K. Targeting dopamine D2 and cannabinoid-1 (CB1) receptors in rat nucleus accumbens. *J Comp Neurol.* 2006;495(3):299–313.
89. Gulledge AT, Jaffe DB. Dopamine decreases the excitability of layer V pyramidal cells in the rat prefrontal cortex. *J Neurosci.* 1998;18(21):9139–9151.
90. Zheng P, Zhang XX, Bunney BS, Shi WX. Opposite modulation of cortical N-methyl-D-aspartate receptor-mediated responses by low and high concentrations of dopamine. *Neuroscience.* 1999;91(2):527–535.
91. Tseng KY, O'Donnell P. Dopamine-glutamate interactions controlling prefrontal cortical pyramidal cell excitability involve multiple signaling mechanisms. *J Neurosci.* 2004;24(22):5131–5139.
92. Tseng KY, O'Donnell P. D2 dopamine receptors recruit a GABA component for their attenuation of excitatory synaptic transmission in the adult rat prefrontal cortex. *Synapse.* 2007;61(10):843–850.
93. Sun X, Zhao Y, Wolf ME. Dopamine receptor stimulation modulates AMPA receptor synaptic insertion in prefrontal cortex neurons. *J Neurosci.* 2005;25(32):7342–7351.
94. Kotecha SA, Oak JN, Jackson MF, et al. A D2 class dopamine receptor transactivates a receptor tyrosine kinase to inhibit NMDA receptor transmission. *Neuron.* 2002;35(6):1111–1122.
95. Kroner S, Rosenkranz JA, Grace AA, Barrionuevo G. Dopamine modulates excitability of basolateral amygdala neurons in vitro. *J Neurophysiol.* 2005;93(3):1598–1610.
96. Rosenkranz JA, Johnston D. State-dependent modulation of amygdala inputs by dopamine-induced enhancement of sodium currents in layer V entorhinal cortex. *J Neurosci.* 2007;27(26):7054–7069.
97. O'Donnell P. Dopamine gating of forebrain neural ensembles. *Eur J Neurosci.* 2003;17(3):429–435.
98. Tseng KY, O'Donnell P. Post-pubertal emergence of prefrontal cortical up states induced by D1-NMDA co-activation. *Cereb Cortex.* 2005;15(1):49–57.
99. Chen L, Bohanick JD, Nishihara M, Seamans JK, Yang CR. Dopamine D1/5 receptor-mediated long-term potentiation of intrinsic excitability in rat prefrontal cortical neurons: Ca2+-dependent intracellular signaling. *J Neurophysiol.* 2007;97(3):2448–2464.
100. Otani S, Blond O, Desce JM, Crepel F. Dopamine facilitates long-term depression of glutamatergic transmission in rat prefrontal cortex. *Neuroscience.* 1998;85(3):669–676.
101. Gurden H, Takita M, Jay TM. Essential role of D1 but not D2 receptors in the NMDA receptor-dependent long-term potentiation at hippocampal-prefrontal cortex synapses in vivo. *J Neurosci.* 2000;20:RC106(22):1–5.
102. Gonzalez-Burgos G, Lewis DA. GABA neurons and the mechanisms of network oscillations: implications for understanding cortical dysfunction in schizophrenia. *Schizophr Bull.* 2008;34(5):944–961.
103. Sesack SR, Snyder CL, Lewis DA. Axon terminals immunolabeled for dopamine or tyrosine hydroxylase synapse on GABA-immunoreactive dendrites in rat and monkey cortex. *J Comp Neurol.* 1995;363(2):264–280.
104. Pinard CR, Muller JF, Mascagni F, McDonald AJ. Dopaminergic innervation of interneurons in the rat basolateral amygdala. *Neuroscience.* 2008;157(4):850–863.
105. Gulledge AT, Jaffe DB. Multiple effects of dopamine on layer V pyramidal cell excitability in rat prefrontal cortex. *J Neurophysiol.* 2001;86(2):586–595.
106. Tseng KY, Mallet N, Toreson KL, Le Moine C, Gonon F, O'Donnell P. Excitatory response of prefrontal cortical fast-spiking interneurons to ventral tegmental area stimulation in vivo. *Synapse.* 2006;59(7):412–417.

107. Tseng KY, O'Donnell P. Dopamine modulation of prefrontal cortical interneurons changes during adolescence. *Cereb Cortex*. 2007;17(5):1235–1240.
108. Totterdell S, Smith AD. Convergence of hippocampal and dopaminergic input onto identified neurons in the nucleus accumbens of the rat. *J Chem Neuroanat*. 1989;2(5):285–298.
109. Sesack SR, Pickel VM. Prefrontal cortical efferents in the rat synapse on unlabeled neuronal targets of catecholamine terminals in the nucleus accumbens septi and on dopamine neurons in the ventral tegmental area. *J Comp Neurol*. 1992;320(2):145–160.
110. Yao WD, Spealman RD, Zhang J. Dopaminergic signaling in dendritic spines. *Biochem Pharmacol*. 2008;75(11):2055–2069.
111. Henze DA, Gonzalez-Burgos GR, Urban NN, Lewis DA, Barrionuevo G. Dopamine increases excitability of pyramidal neurons in primate prefrontal cortex. *J Neurophysiol*. 2000;84(6):2799–2809.
112. Moss J, Bolam JP. A dopaminergic axon lattice in the striatum and its relationship with cortical and thalamic terminals. *J Neurosci*. 2008;28(44):11221–11230.
113. Boundy VA, Luedtke RR, Artymyshyn RP, Filtz TM, Molinoff PB. Development of polyclonal anti-D2 dopamine receptor antibodies using sequence-specific peptides. *Mol Pharmacol*. 1993;43(5):666–676.
114. Fisher RS, Levine MS, Sibley DR, Ariano MA. D2 dopamine receptor protein localization: golgi impregnation-gold toned and ultrastructural analysis of the rat neostriatum. *J Neurosci Res*. 1994;38(5):551–564.
115. Mengual E, Pickel VM. Ultrastructural immunocytochemical localization of the dopamine D2 receptor and tyrosine hydroxylase in the rat ventral pallidum. *Synapse*. 2002;43(3):151–162.
116. Delle Donne KT, Sesack SR, Pickel VM. Ultrastructural immunocytochemical localization of neurotensin and the dopamine D_2 receptor in the rat nucleus accumbens. *J Comp Neurol*. 1996;371(4):552–566.
117. Aubert I, Ghorayeb I, Normand E, Bloch B. Phenotypical characterization of the neurons expressing the D1 and D2 dopamine receptors in the monkey striatum. *J Comp Neurol*. 2000;418(1):22–32.
118. Alcantara AA, Chen V, Herring BE, Mendenhall JM, Berlanga ML. Localization of dopamine D2 receptors on cholinergic interneurons of the dorsal striatum and nucleus accumbens of the rat. *Brain Res*. 2003;986(1–2):22–29.
119. Gerfen CR, Engber TM, Mahan LC, et al. D1 and D2 dopamine receptor regulated gene expression of striatonigral and striatopallidal neurons. *Science*. 1990;250(4986):1429–1432.
120. Le Moine C, Bloch B. D1 and D2 dopamine receptor gene expression in the rat striatum: sensitive cRNA probes demonstrate prominent segregation of D1 and D2 mRNAs in distinct neuronal populations of the dorsal and ventral striatum. *J Comp Neurol*. 1995;355(3):418–426.
121. Delle Donne KT, Sesack SR, Pickel VM. Ultrastructural immunocytochemical localization of the dopamine D_2 receptor within GABAergic neurons of the rat striatum. *Brain Res*. 1997;746(1-2):239–255.
122. Surmeier DJ, Eberwine J, Wilson CJ, Cao Y, Stefani A, Kitai ST. Dopamine receptor subtypes colocalize in rat striatonigral neurons. *Proc Natl Acad Sci USA*. 1992;89(21):10178–10182.
123. Surmeier DJ, Reiner A, Levine MS, Ariano MA. Are neostriatal dopamine receptors co-localized? *Trends Neurosci*. 1993;16(8):299–305.
124. Le Moine C, Bloch B. Expression of the D3 dopamine receptor in peptidergic neurons of the nucleus accumbens: comparison with the D1 and D2 dopamine receptors. *Neuroscience*. 1996;73(1):131–143.
125. Parent A, Sato F, Wu Y, Gauthier J, Levesque M, Parent M. Organization of the basal ganglia: the importance of axonal collateralization. *Trends Neurosci*. 2000;23(suppl 10):S20–S27.
126. Day M, Wang Z, Ding J, et al. Selective elimination of glutamatergic synapses on striatopallidal neurons in Parkinson disease models. *Nat Neurosci*. 2006;9(2):251–259.
127. Lee KW, Kim Y, Kim AM, Helmin K, Nairn AC, Greengard P. Cocaine-induced dendritic spine formation in D1 and D2 dopamine receptor-containing medium spiny neurons in nucleus accumbens. *Proc Natl Acad Sci USA*. 2006;103(9):3399–3404.
128. Wang Z, Kai L, Day M, et al. Dopaminergic control of corticostriatal long-term synaptic depression in medium spiny neurons is mediated by cholinergic interneurons. *Neuron*. 2006;50(3):443–452.
129. Delle Donne KT, Chan J, Boudin H, Pelaprat D, Rostene W, Pickel VM. Electron microscopic dual labeling of high-affinity neurotensin and dopamine D2 receptors in the rat nucleus accumbens shell. *Synapse*. 2004;52(3):176–187.
130. Hernández-Echeagaray E, Starling AJ, Cepeda C, Levine MS. Modulation of AMPA currents by D2 dopamine receptors in striatal medium-sized spiny neurons: are dendrites necessary? *Eur J Neurosci*. 2004;19(9):2455–2463.
131. Surmeier DJ, Ding J, Day M, Wang Z, Shen W. D1 and D2 dopamine-receptor modulation of striatal glutamatergic signaling in striatal medium spiny neurons. *Trends Neurosci*. 2007;30(5):228–235.
132. Smith AD, Bolam JP. The neural network of the basal ganglia as revealed by the study of synaptic connections of identified neurones. *Trends Neurosci*. 1990;13(7):259–265.
133. Bernard V, Somogyi P, Bolam JP. Cellular, subcellular, and subsynaptic distribution of AMPA-type glutamate receptor subunits in the neostriatum of the rat. *J Neurosci*. 1997;17(2):819–833.
134. Guigoni C, Doudnikoff E, Li Q, Bloch B, Bezard E. Altered D(1) dopamine receptor trafficking in parkinsonian and dyskinetic non-human primates. *Neurobiol Dis*. 2007;26(2):452–463.
135. Wang H, Pickel VM. Dopamine D2 receptors are present in prefrontal cortical afferents and their targets in patches of the rat caudate-putamen nucleus. *J Comp Neurol*. 2002;442(4):392–404.
136. Seamans JK, Gorelova NA, Yang CR. Contributions of voltage-gated Ca^{2+} channels in the proximal versus distal dendrites to synaptic integration in prefrontal cortical neurons. *J Neurosci*. 1997;17(15):5936–5948.
137. Ehlers MD, Heine M, Groc L, Lee MC, Choquet D. Diffusional trapping of GluR1 AMPA receptors by input-specific synaptic activity. *Neuron*. 2007;54(3):447–460.
138. Wolf ME, Mangiavacchi S, Sun X. Mechanisms by which dopamine receptors may influence synaptic plasticity. *Ann NY Acad Sci*. 2003;1003:241–249.
139. Lei W, Jiao Y, Del Mar N, Reiner A. Evidence for differential cortical input to direct pathway versus indirect pathway striatal projection neurons in rats. *J Neurosci*. 2004;24(38):8289–8299.
140. Cepeda C, Andre VM, Yamazaki I, Wu N, Kleiman-Weiner M, Levine MS. Differential electrophysiological properties of dopamine D1 and D2 receptor-containing striatal medium-sized spiny neurons. *Eur J Neurosci*. 2008;27(3):671–682.

141. Shen W, Flajolet M, Greengard P, Surmeier DJ. Dichotomous dopaminergic control of striatal synaptic plasticity. *Science.* 2008;321(5890):848–851.
142. Stephens B, Mueller AJ, Shering AF, et al. Evidence of a breakdown of corticostriatal connections in Parkinson's disease. *Neuroscience.* 2005;132(3):741–754.
143. Ingham CA, Hood SH, Taggart P, Arbuthnott GW. Plasticity of synapses in the rat neostriatum after unilateral lesion of the nigrostriatal dopaminergic pathway. *J Neurosci.* 1998;18(12): 4732–4743.
144. Villalba RM, Lee H, Smith Y. Dopaminergic denervation and spine loss in the striatum of MPTP-treated monkeys. *Exp Neurol.* 2008;215(2):220–227.
145. Solis O, Limon DI, Flores-Hernandez J, Flores G. Alterations in dendritic morphology of the prefrontal cortical and striatum neurons in the unilateral 6-OHDA-rat model of Parkinson's disease. *Synapse.* 2007;61(6):450–458.
146. Smythies J. Redox mechanisms at the glutamate synapse and their significance: a review. *Eur J Pharmacol.* 1999;370(1):1–7.
147. Kihara T, Shimohama S, Sawada H, et al. Protective effect of dopamine D2 agonists in cortical neurons via the phosphatidylinositol 3 kinase cascade. *J Neurosci Res.* 2002;70(3):274–282.
148. Wang HD, Deutch AY. Dopamine depletion of the prefrontal cortex induces dendritic spine loss: reversal by atypical antipsychotic drug treatment. *Neuropsychopharmacology.* 2008;33(6): 1276–1286.
149. Pierce JP, Lewin GR. An ultrastructural size principle. *Neuroscience.* 1994;58(3):441–446.
150. Sorra KE, Fiala JC, Harris KM. Critical assessment of the involvement of perforations, spinules, and spine branching in hippocampal synapse formation. *J Comp Neurol.* 1998;398(2): 225–240.
151. Meshul CK, Stallbaumer RK, Taylor B, Janowsky A. Haloperidol-induced morphological changes in striatum are associated with glutamate synapses. *Brain Res.* 1994; 648(2): 181–195.
152. Meshul CK, Janowsky A, Casey DE, Stallbaumer RK, Taylor B. Coadministration of haloperidol and SCH-23390 prevents the increase in "perforated" synapses due to either drug alone. *Neuropsychopharmacology.* 1992;7(4):285–293.
153. Robinson TE, Kolb B. Structural plasticity associated with exposure to drugs of abuse. *Neuropharmacology.* 2004;47(suppl 1): 33–46.
154. Bamford NS, Zhang H, Schmitz Y, et al. Heterosynaptic dopamine neurotransmission selects sets of corticostriatal terminals. *Neuron.* 2004;42(4):653–663.
155. Yang CR, Mogenson GJ. Dopamine enhances terminal excitability of hippocampal-accumbens neurons via D2 receptor: role of dopamine in presynaptic inhibition. *J Neurosci.* 1986;6(8):2470–2478.
156. O'Donnell P, Grace AA. Tonic D_2-mediated attenuation of cortical excitation in nucleus accumbens neurons recorded in vitro. *Brain Res.* 1994;634(1):105–112.
157. Cepeda C, Hurst RS, Altemus KL, et al. Facilitated glutamatergic transmission in the striatum of D2 dopamine receptor-deficient mice. *J Neurophysiol.* 2001;85(2):659–670.
158. Rosenkranz JA, Grace AA. Dopamine attenuates prefrontal cortical suppression of sensory inputs to the basolateral amygdala of rats. *J Neurosci.* 2001;21(11):4090–4103.
159. Yin HH, Lovinger DM. Frequency-specific and D2 receptor-mediated inhibition of glutamate release by retrograde endocannabinoid signaling. *Proc Natl Acad Sci USA.* 2006; 103(21):8251–8256.
160. Koga E, Momiyama T. Presynaptic dopamine D2-like receptors inhibit excitatory transmission onto rat ventral tegmental dopaminergic neurones. *J Physiol.* 2000;523(pt 1):163–173.
161. Sokoloff P, Giros B, Martres MP, Bouthenet ML, Schwartz JC. Molecular cloning and characterization of a novel dopamine receptor (D3) as a target for neuroleptics. *Nature.* 1990;347(6289):146–151.
162. Lévesque D, Martres MP, Diaz J, et al. A paradoxical regulation of the dopamine D3 receptor expression suggests the involvement of an anterograde factor from dopamine neurons. *Proc Natl Acad Sci USA.* 1995;92(5):1719–1723.
163. Levant B. The D3 dopamine receptor: neurobiology and potential clinical relevance. *Pharmacol Rev.* 1997;49(3):231–252.
164. Heidbreder CA, Gardner EL, Xi ZX, et al. The role of central dopamine D3 receptors in drug addiction: a review of pharmacological evidence. *Brain Res Rev.* 2005;49(1):77–105.
165. Ariano AA, Sibley DR. Dopamine receptor distribution in the rat CNS: elucidation using anti-peptide antisera directed against D_{1A} and D_3 subtypes. *Brain Res.* 1994;649(1-2):95–110.
166. Diaz J, Levesque D, Lammers CH, et al. Phenotypical characterization of neurons expressing the dopamine D3 receptor in the rat brain. *Neuroscience.* 1995;65(3):731–745.
167. Suzuki M, Hurd YL, Sokoloff P, Schwartz JC, Sedvall G. D3 dopamine receptor mRNA is widely expressed in the human brain. *Brain Res.* 1998;779(1-2):58–74.
168. Diaz J, Pilon C, Le Foll B, et al. Dopamine D3 receptors expressed by all mesencephalic dopamine neurons. *J Neurosci.* 2000;20(23):8677–8684.
169. Stanwood GD, Artymyshyn RP, Kung MP, Kung HF, Lucki I, McGonigle P. Quantitative autoradiographic mapping of rat brain dopamine D3 binding with [(125)I]7-OH-PIPAT: evidence for the presence of D3 receptors on dopaminergic and nondopaminergic cell bodies and terminals. *J Pharmacol Exp Ther.* 2000;295(3):1223–1231.
170. Koeltzow TE, Xu M, Cooper DC, et al. Alterations in dopamine release but not dopamine autoreceptor function in dopamine D3 receptor mutant mice. *J Neurosci.* 1998;18(6):2231–2238.
171. Le Foll B, Diaz J, Sokoloff P. Neuroadaptations to hyperdopaminergia in dopamine D3 receptor-deficient mice. *Life Sci.* 2005;76(11):1281–1296.
172. Van Tol HH, Bunzow JR, Guan HC, et al. Cloning of the gene for a human dopamine D4 receptor with high affinity for the antipsychotic clozapine. *Nature.* 1991;350(6319): 610–614.
173. Mrzljak L, Bergson C, Pappy M, Huff R, Levenson R, Goldman-Rakic PS. Localization of dopamine D4 receptors in GABAergic neurons of the primate brain. *Nature.* 1996;381(6579): 245–248.
174. Avale ME, Falzone TL, Gelman DM, Low MJ, Grandy DK, Rubinstein M. The dopamine D4 receptor is essential for hyperactivity and impaired behavioral inhibition in a mouse model of attention deficit/hyperactivity disorder. *Mol Psychiatry.* 2004;9(7):718–726.
175. Ariano MA, Wang J, Noblett KL, Larson ER, Sibley DR. Cellular distribution of the rat D4 dopamine receptor protein in the CNS using anti-receptor antisera. *Brain Res.* 1997; 752(1-2):26–34.
176. Wędzony K, Chocyk A, Maćkowiak M, Fijał K, Czyrak A. Cortical localization of dopamine D4 receptors in the rat brain—immunocytochemical study. *J Physiol Pharmacol.* 2000;51(2):205–221.
177. Berger MA, Defagot MC, Villar MJ, Antonelli MC. D4 dopamine and metabotropic glutamate receptors in cerebral cortex

and striatum in rat brain. *Neurochem Res.* 2001;26(4): 345–352.
178. Rivera A, Cuellar B, Giron FJ, Grandy DK, de la Calle A, Moratalla R. Dopamine D4 receptors are heterogeneously distributed in the striosomes/matrix compartments of the striatum. *J Neurochem.* 2002;80(2):219–229.
179. Rivera A, Peñafiel A, Megías M, et al. Cellular localization and distribution of dopamine D(4) receptors in the rat cerebral cortex and their relationship with the cortical dopaminergic and noradrenergic nerve terminal networks. *Neuroscience.* 2008;155(3):997–1010.
180. Svingos AL, Periasamy S, Pickel VM. Presynaptic dopamine D4 receptor localization in the rat nucleus accumbens shell. *Synapse.* 2000;36(3):222–232.
181. Noain D, Avale ME, Wedemeyer C, Calvo D, Peper M, Rubinstein M. Identification of brain neurons expressing the dopamine D4 receptor gene using BAC transgenic mice. *Eur J Neurosci.* 2006;24(9):2429–2438.
182. Martina M, Bergeron R. D1 and D4 dopaminergic receptor interplay mediates coincident G protein-independent and dependent regulation of glutamate NMDA receptors in the lateral amygdala. *J Neurochem.* 2008; 106(6):2421–2435.
183. Wang X, Zhong P, Gu Z, Yan Z. Regulation of NMDA receptors by dopamine D4 signaling in prefrontal cortex. *J Neurosci.* 2003;23(30):9852–9861.
184. Onn SP, Wang XB, Lin M, Grace AA. Dopamine D1 and D4 receptor subtypes differentially modulate recurrent excitatory synapses in prefrontal cortical pyramidal neurons. *Neuropsychopharmacology.* 2006;31(2): 318–338.
185. Wang X, Zhong P, Yan Z. Dopamine D4 receptors modulate GABAergic signaling in pyramidal neurons of prefrontal cortex. *J Neurosci.* 2002;22(21):9185–9193.
186. Rubinstein M, Cepeda C, Hurst RS, et al. Dopamine D4 receptor-deficient mice display cortical hyperexcitability. *J Neurosci.* 2001;21(11):3756–3763.

2.3 Convergence of Limbic, Cognitive, and Motor Cortico-Striatal Circuits with Dopamine Pathways in Primate Brain

SUZANNE N. HABER

INTRODUCTION

Dopamine plays a central role in a wide variety of behaviors including reward, cognition, and motor control. Subpopulations of dopamine neurons have been associated with these different functions: the mesolimbic, mesocortical, and nigrostriatal pathways, respectively. Recently, all dopamine cell groups have been associated with the development of reward-based learning, leading to goal-directed behaviors. These behaviors require a complex interface between motivational drive, cognition, and action planning.[1,2]

The dopamine cells are an integral part of the basal ganglia. They send a massive output to the striatum, the main input structure of the basal ganglia. Moreover, this is a bidirectional pathway, with the dopamine cells receiving a major input from the striatum. Historically, the basal ganglia is best known for their motor functions, in large part because of the association between its neuropathology and neurodegenerative disorders affecting motor control. In particular, the degeneration of the substantia nigra pars compacta, is clearly linked to Parkinson's disease. While a role for the basal ganglia in the control of movement is clear, our concept of basal ganglia function has dramatically changed in the past 20 years. It is now recognized to mediate the full range of behaviors leading to the development and execution of action plans, including the emotions, motivation, and cognition that drive them. Regions within each of the basal ganglia nuclei have been identified as serving these functions. The ventral striatum plays a key role in reward and reinforcement, central striatal areas are involved in executive functions, and dorsolateral regions are associated with sensorimotor control. Likewise, within the midbrain, the ventral tegmental area (VTA) is most closely linked to reward and reinforcement, while the substantia nigra, pars compacta is linked to sensorimotor function. Thus, this set of subcortical nuclei works in tandem with cortex (particularly frontal cortex) via complex cortico-basal ganglia networks that are fundamentally linked to incentive-based learning and the development and execution of goal-directed behaviors.

The basal ganglia are traditionally considered to process this information in parallel and segregated functional streams consisting of reward (limbic), associative (cognitive), and motor control circuits.[3] Moreover, microcircuits within each region are thought to mediate different aspects of each function.[4] However, while frontal cortex is indeed divided based on specific functions, expressed behaviors are the result of a combination of complex information processing that involves all of the frontal cortex. Indeed, appropriate responses to environmental stimuli require continual updating, learning to adjust behaviors according to new data. This necessitates coordination between limbic, cognitive, and motor systems to form smoothly executed, goal-directed behaviors. Parallel processing of functional information through different basal ganglia circuits does not address how information flows between circuits, thereby developing new behaviors (or actions) or adapting to those previously learned. While the anatomical pathways appear to be generally topographic from cortex through basal ganglia circuits, and while there are some physiological correlates to the functional domains of the striatum, a large body of growing evidence supports a duel processing system in which not only is information processed in parallel, but also integration occurs between functional circuits.[5-9] This chapter will first briefly review the basic circuitry that underlies parallel processing; second, the anatomical basis for integration across different corticobasal ganglia circuits, with a particular emphasis on dopamine; and finally, functional support for integrative processes. While the focus is on primate studies, key rodent experiments are also highlighted when primate data are unavailable.

PARALLEL PROCESSING

Functional Organization of Frontal Cortex

Frontal cortex is organized in a hierarchical manner and can be divided into functional regions[10]: the orbital (OFC) and anterior cingulate (ACC) prefrontal cortices are involved in emotions and motivation, the dorsal prefrontal cortex (DPFC) is involved in higher cognitive processes or executive functions, and the premotor and motor areas are involved in motor planning and the execution of those plans. The ACC is divided into ventral, or subgenual, ACC and dorsal ACC (dACC) areas. Medial orbital area 14 and the subgenual ACC cortex are collectively referred to as the *ventral medial prefrontal cortex* (vmPFC) and are particularly important in the expression of emotion.[11,12] The OFC is involved in the development of reward-based learning, aversive, and goal-directed behaviors.[13–17] This area receives input from multimodal sensory regions and is closely linked to the vmPFC.[18,19] Lesions of the OFC, vmPFC, and dACC areas result in an inability to initiate and carry out goal-directed behaviors, and lead to socially inappropriate and impulsive behaviors.[20–23] The DPFC is involved in working memory, set shifting, and strategic planning, often referred to as *executive functions*.[24–28] Motor cortices are the most clearly defined areas of the frontal cortex. Caudal motor areas are highly microexcitable, closely timed to the execution of movement, and send a direct descending projection to spinal motor nuclei. Rostral motor areas are involved in sequence generation and motor learning. They are less microexcitable than the caudal motor areas but more so than the prefrontal cortex (PFC).[29,30] Each of these frontal areas projects to specific striatal regions. However, in addition to the well-described topographic organization, they also follow non-topographic rules.

Functional Projections Through the Basal Ganglia

Together, the frontal regions that mediate reward, motivation, and affect regulation project primarily to the rostral striatum, including the n. accumbens, the medial caudate n., and the medial and ventral rostral putamen, collectively referred to as the *ventral striatum*. While the ventral striatum is similar to the dorsal striatum in most respects, it also has some unique features. The ventral striatum contains a subterritory, called the *shell,* which has been shown in rodents to play a particularly important role in the circuitry underlying goal-directed behaviors, behavioral sensitization, and changes in affective states.[31,32] Moreover, the ventral striatum alone receives a dense projection from the amygdala and from the hippocampus.[33,34] The hippocampal projection is mostly limited to the shell, while the amygdala projects throughout a wider ventral striatal area. There are no clear histochemical boundaries between the ventral and dorsal striatum. Thus, the best way to define the ventral striatum is by its afferent projections, primarily the vmPFC, OFC, dACC, and the medial temporal lobe, particularly the amygdala.[35] The vmPFC projects to the ventral medial striatum, including the shell, and extends along the medial edge of the dorsal ventral caudate n.[9] The shell receives the densest innervation from medial areas 25, and 32 and from agranular insular cortex. The lateral orbital regions project to the central and lateral parts of the ventral striatum and extend into the central rostral caudate n. The dACC terminates in a wide medial striatal region, overlapping with inputs from both the OFC and vmPFC. Consistent with these inputs to the ventral striatum, physiological and imaging studies demonstrate the important role of this ventral striatal region in reward-based learning and motivation.[36–38]

The DPFC projects to the head of the caudate n. and to the putamen rostral to the anterior commissure. Caudal to the commissure, this projection is confined to the medial, central portion of the head of the caudate n., with few terminals in the central and caudal putamen.[9,39] Different parts of the DPFC project with complex topography to different parts of the rostral caudate and putamen.[40] Physiological, imaging, and lesion studies support the idea that these areas are involved in working memory and strategic planning processes, working together with the DPFC in mediating this function.[41–43] Rostral premotor areas terminate in both the caudate and putamen, bridging the two with a continuous projection. Projections from caudal motor areas terminate almost entirely in the dorsolateral putamen, caudal to the anterior commissure. Few terminals are found rostral to the anterior commissure. Both caudal and rostral motor areas occupy much of the putamen caudal to the anterior commissure, a region that also receives overlapping projections from somatosensory cortex, resulting in a somatotopically organized sensorimotor area.[44–46] In summary, projections from frontal cortex form a functional gradient of inputs from the ventromedial sector through the dorsolateral striatum, with the medial and orbital PFCs terminating in the ventromedial part and the motor cortex terminating in the dorsolateral region. Like corticostriatal projection, thalamstriatal projections are organized in a general topographical manner such that interconnected and functionally associated thalamic and cortical regions terminate in the same general striatal region.[47]

The striatal projection to the pallidal complex and substantia nigra pars reticulata are also generally topographically organized, thus maintaining the functional organization of the striatum in these output nuclei.[4,48–51] The ventral striatum terminates in the ventral pallidum and in the dorsal part of the midbrain. Terminals from the central striatum terminate more centrally in both the pallidum and the pars reticulata, while those from the sensorimotor areas of the striatum innervate the ventrolateral part of each pallidal segment and the ventrolateral substantia nigra.. Finally, the pallidum and pars reticulata project to the different basal ganglia output nuclei of the thalamus, the mediodorsal, and the ventral anterior and ventral lateral cell groups. The thalamic-cortical pathway is the last link in the circuit, and the outputs from the mediodorsal, ventral anterior, and ventral lateral thalamic n. are connected respectively to the collective limbic areas, the associative control areas, and the motor control areas.[4,52–55] Thus, the organization of connections through the cortico-basal ganglia cortical network preserves a general functional topography within each structure, from the cortex through the striatum, from the striatum to the pallidum/pars reticulata, from these output structures to the thalamus, and finally, back to the cortex (Fig. 2.3.1).

This organization has led to the concept that each functionally identified cortical region drives (and is driven by) a specific basal ganglia loop or circuit, leading, in turn, to the idea of parallel processing of cortical information through segregated basal ganglia circuits.[3] This concept focuses on the role of the basal ganglia in the selection and implementation of an appropriate motor response while inhibiting unwanted ones.[56] The model assumes, however, that the behavior has been learned and that the role of the basal ganglia is to carry out a coordinated action. We now know that the cortico-basal ganglia network is critical in mediating the learning process to adapt and to accommodate past experiences to modify behavioral responses.[41,57–60] This requires some communication across circuits.

INTEGRATIVE PATHWAYS

Growing evidence has identified possible anatomical substrates through which transfer of information can occur across functional domains.[5–9] Integration between different aspects of reward processing, as well as interaction with cognitive and motor control regions, likely occur at several stations throughout the system. For example, as indicated above, while there is a general topographic organization to the dense (or focal)

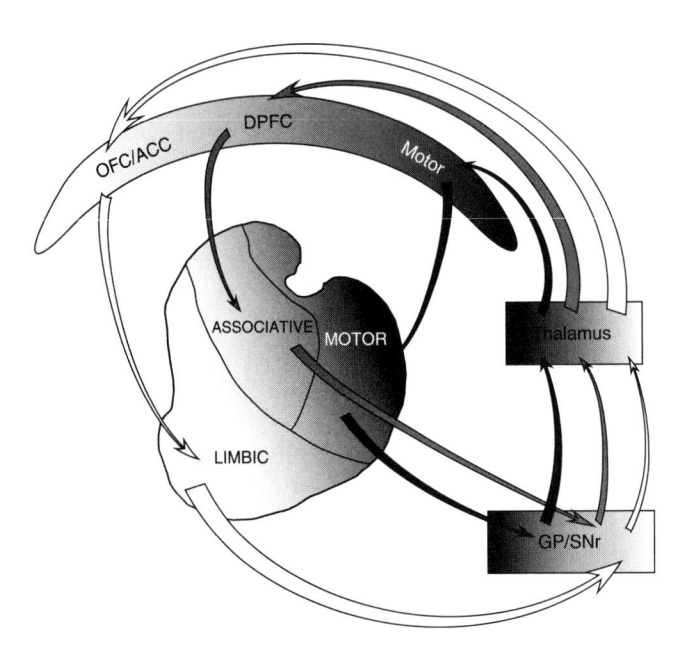

FIGURE 2.3.1. Schematic illustrating parallel circuits through corticobasal ganglia pathways. Corresponding shaded striatal and cortical areas demonstrate topographic projections; white, limbic circuit; light gray, associative circuit; dark gray, motor control circuit.

corticostriatal terminal fields, this projection system also has non-topographic rules. Focal projections from different functional cortical areas also converge in specific striatal areas. These areas of convergence create nodal points of integration embedded within a generally parallel system. Such an arrangement may set the stage for a differential impact on midbrain dopamine cells during learning. In this chapter, we emphasize the role of dopamine in this transfer through its anatomical relationships to the corticostriatal network. First, however, we review the non-topographical aspects of the corticostriatal projection system.

Corticostriatal Projections

The ventral striatum, the area that receives input from the vmPFC, dACC, and OFC, is concentrated in the rostral striatum. Collectively, the terminal fields from these cortical areas occupy approximately 22% of the striatum. As noted above, terminal fields from these cortical areas are concentrated in different striatal regions. However, they also converge extensively.[9]

The areas in which convergence occurs may be particularly critical for the coordination of different aspects of affect regulation (Fig. 2.3.2a-b). Projection fields from the DPFC terminate from the rostral pole and continue throughout much of the body of the caudate n. and medial putamen. However, while focal corticostriatal projections from different limbic and cognitive regions generally occupy separate positions in the striatum, they also converge at specific locations, primarily at the rostral levels (Fig. 2.3.2b). Here, terminals from the DPFC partially converge with those from both the dACC and OFC. In fact, projections from all PFC areas occupy a central region, with each cortical projection extending into nonoverlapping zones.[9] Convergence is less prominent caudally, with almost complete separation of the dense terminals from the DPFC and dACC/OFC/vmPFC just rostral to the anterior commissure. This pattern of PFC projection fields implies a central role, particularly for rostral striatal subregions, in synchronizing different aspects of reward and learning for long-term strategic planning and habit formation.

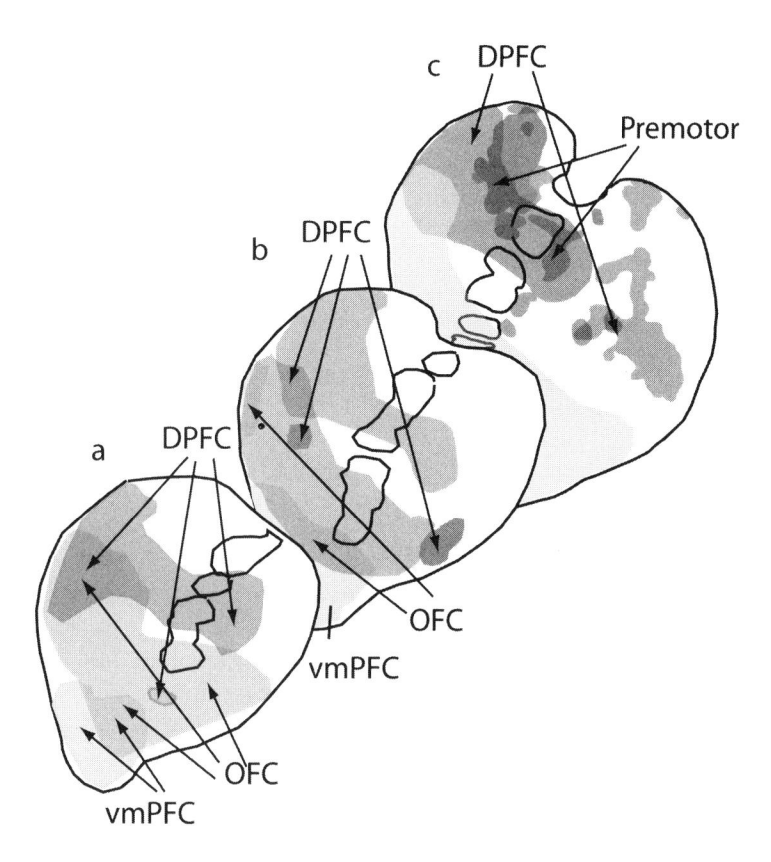

FIGURE 2.3.2. Schematics demonstrating convergence of corticostriatal focal projections from different limbic, associative, and motor areas: (a,b) convergence between projections from different prefrontal regions; (c) convergence between prefrontal regions and motor control areas. DPFC, dorsal prefrontal cortex; OFC, orbital prefrontal cortex; vmPFC, ventral medial prefrontal cortex. DPFC=dorsal prefrontal cortex; OFC=orbital prefrontal cortex; vmPFC=ventral, medial prefrontal cortex.

Just rostral to and at the level of the anterior commissure, convergence occurs between terminals from the DPFC and premotor regions.[61] Interestingly, at more anterior levels, these projections remain relatively segregated. This is a place of prominent convergence between focal projections from the[41] DPFC and those from the OFC/ACC/vmPFC. Figure 2.3.2c is a schematic of a coronal section just anterior to the anterior commissure demonstrating the relatively little convergence with limbic input but an interface with rostral motor control areas. Importantly, there are few convergent terminals between afferent projections from limbic and motor control regions. Projections from DPFC are therefore in a pivotal position in the striatum, converging at one level with inputs from areas associated with motivation and reward and, at a more posterior level, with those from cortical areas associated with action planning. Convergence between terminals from limbic and cognitive areas rostrally, and from cognitive and premotor motor regions more caudally, provides a possible neural substrate for executive control over the development of incentive-based actions. Taken together, the frontostriatal network therefore constitutes a dual system comprising both clearly segregated connections and subregions that contain convergent pathways derived from functionally discrete cortical areas. This dual projection system is further supported using probabilistic tractography methods in humans.[8] It provides an anatomical substrate for a recent finding in rodents in which cross-encoding cortical information influenced the future firing of medium spiny neurons.[62] The nodal points of convergence from different cortical regions may therefore constitute zones for dynamic restructuring of neural ensembles fundamental to learning. These subregions are in a position to send a more functionally integrated input to the dopamine cells compared to the majority of the striatum. Moreover, dopamine input to these subregions is likely to have a different impact on information flow through the basal ganglia circuits.

The Midbrain Dopamine System

The striatal incentive-related learning process is thought to originate, in part, from the midbrain dopamine cells, which signal reward prediction error or reward saliency.[1,63,64] However, the latency between the presentation of the stimuli and the activity of the dopamine cells is too short to reflect the higher cortical processing necessary for linking a stimulus with its rewarding properties.[65] This is consistent with the fact that in behavioral studies, animals have already been trained to link the response to the reward. Indeed, in studies that associate reward prediction error with the dopamine neurons, animals are overtrained to recognize the reward. In other words, the animals first must learn to associate the brown, wrinkled, sticky substance with the sweet taste of a raisin. A critical issue, therefore, is, how do the dopamine cells receive information consolidating this association? The largest forebrain input to the dopamine neurons is from the striatum, and the largest input to the striatum is from cortex. Collectively, the PFC inputs to the striatum are in a position to modulate the striatal response to different aspects of reward saliency and value. Thus, although the short-latency burst firing activity of dopamine that signals the immediate reinforcement is likely to be triggered from brainstem nuclei, the cortico-striato-midbrain pathway is in a position to "train" dopamine cells to distinguish rewards in order to calculate error prediction. The nodal points of convergence between different functional cortical pathways within the striatum may be critical in this initial role and play a particularly important role in the temporal training of dopamine cells, placing these cells in a position to respond with a short-latency signal derived from incoming sensory systems.[65] Over time, the fast burst firing activity of the dopamine cells is quickly activated by the brainstem as incoming stimuli are perceived. This, in turn, impacts the striatum and can influence progressively more dorsal regions during learning and the development of habit formation.[66–68]

The organization of dopamine neurons

Anatomically, the midbrain dopamine neurons are not clearly defined within the mesolimbic, mesocortical, and nigrostriatal categories. In rodents, the midbrain dopamine neurons are generally divided into the substantia nigra pars compacta (SNc), the VTA, and the retrorubral cell groups.[69] In primates, the SNc is further divided into three groups: a dorsal group (the α group); a main densocellular region (the β group); and a ventral group (the γ group), or cell columns.[70,71] The dorsal group is oriented horizontally and extends dorsolaterally, circumventing the ventral and lateral superior cerebellar peduncle and the red nucleus. These cells merge with the immediately adjacent dopamine cell groups of the VTA to form a continuous mediodorsal band of cells. Calbindin, a calcium binding protein (CaBP), marks both the VTA and the dorsal SNc. In contrast, the ventral cell groups (the densocellular group and the cell columns) are calbindin negative and, unlike the dorsal tier, have high expression levels for the dopamine

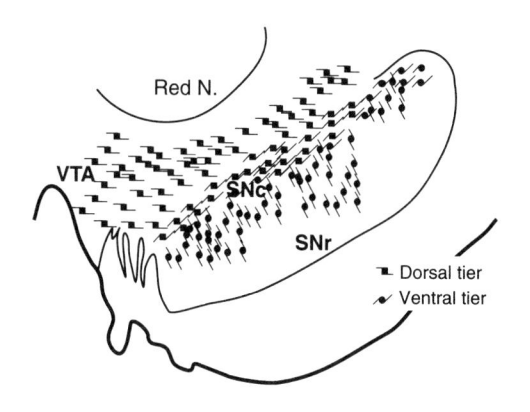

FIGURE 2.3.3. Schematic illustrating the organization of the midbrain dopamine neurons into the dorsal and ventral tiers. SNc, substantia nigra pars compacta; SNr, substantia nigra pars reticulata; VTA, ventral tegmental area.

transporter and for the D2 receptor mRNAs.[71,72] Thus, the midbrain cells are divided into a dorsal tier that includes the VTA and the dorsal SNc and a ventral tier that includes the densocellular group and cell columns (Fig. 2.3.3).

Afferent projections

Input to the midbrain dopamine neurons is primarily from the striatum, from both the external segment of the globus pallidus and the ventral pallidum, and from the brainstem (for review, see[73]). Descending projections from the central nucleus of the amygdala also terminate in a wide mediolateral region but are limited primarily to the dorsal tier cells. In addition, there are projections to the dorsal tier from the bed nucleus of the stria terminalis and from the sublenticular substantia innominata that travel together with those from the amygdala.[74] While the dopamine neurons receive input from these several sources, perhaps the most massive projection is from the striatum. Striatal projections terminate on both the dorsal and ventral tiers in addition to the pars reticulata. This afferent projection is organized with an inverse ventral/dorsal topography. The ventral striatum projects widely to the dorsal tier and much of the dorsal part of the densocellular pars compacta cells. This ventral striatal terminal field extends laterally to include a large mediolateral region. Descending projections from the extended amygdala also terminate in a wide mediolateral region but primarily in the dorsal tier. Therefore, the dorsal tier receives a massive limbic input through an indirect projection from the OFC/dACC/vmPFC (via the striatum) and a direct projection from the extended amygdala.

The central striatum, which receives input from the DPFC, projects extensively to the central and ventral parts of the densocellular region, extending into the cell columns and surrounding pars reticulata. Finally, the dorsolateral striatal projection is concentrated in the ventral and lateral parts of the substantia nigra. Unlike the widespread terminal fields of the ventral and central striatum, the distribution of efferent fibers from the dorsolateral striatum is more restricted and terminates primarily in the pars reticulata. However, their terminal fields do project to the cell columns of dopamine neurons that penetrate deep into the pars reticulata. Thus, in addition to the inverse dorsoventral topographic organization to the striatonigral projection, there is an important difference in the extent of the projection fields from the functional striatal domains. Projections from regions receiving PFC inputs have wide projection fields throughout the midbrain dopamine cells, while those from motor control areas have a relatively limited projection field (Fig. 2.3.4a).

Efferent projections

Like the descending striatonigral pathway, the ascending nigrostriatal projection exhibits an inverse dorsoventral topographic arrangement. Here, there is

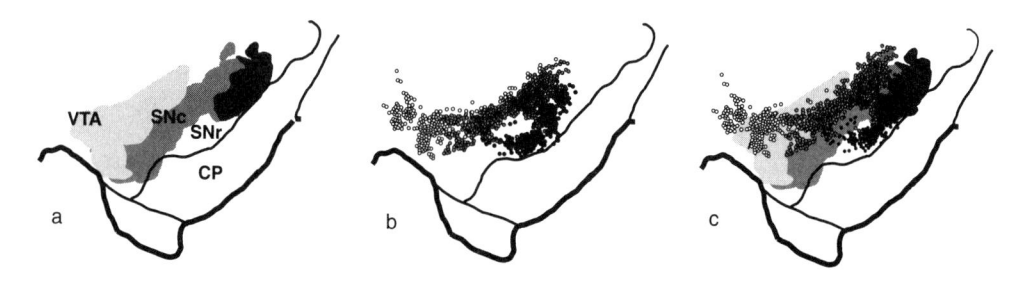

FIGURE 2.3.4. Schematic of the substantia nigra showing the combined distribution of striatonigral terminal fields (a, c) and nigrostriatal cells (b, c) associated with different functional regions of the striatum. Light gray, inputs and outputs from the limbic striatum; medium gray, inputs and outputs from the associative striatum; dark gray, inputs and outputs from the motor striatum. CP, cerebral peduncles; SNc, substantia nigra pars compacta; SNr, substantia nigra pars reticulata; VTA, ventral tegmental area.

also a mediolateral topographic organization. Thus, the dorsal and medial dopamine cells project to the ventral and medial parts of the striatum, while the ventral and lateral cells project to the dorsal and lateral parts of the striatum.[5,48,75,76] Moreover, as with the striatonigral projection, the proportional distribution of cells that project to different functional domains of the striatum differs (Fig. 2.3.4b). The shell region of the ventral striatum receives the most limited midbrain input, primarily derived from the VTA. The rest of the ventral striatum receives input primarily from the dorsal tier, including the retrorubral cell group, and from the medial and dorsal regions of the densocellular group. The central part of the striatum, which receives input from the DPFC, also receives input from the central part of the densocellular region of the dopamine cells. In contrast, the dorsolateral part of the striatum receives input from a wide range of dopamine cells derived from the ventral tier, including both the densocellular and cell columns groups (Fig. 2.3.4b).

When considered separately, each limb of the system creates a loose topographic organization. The VTA and medial substantia nigra are associated with the limbic regions, the central substantia nigra with associative regions, and the lateral and ventral substantia nigra are related to the motor control striatal regions (Fig. 2.3.4c). However, the fact that the descending and ascending limb of each functional striatonigral and nigrostriatal pathways differs in its proportional projections significantly alters the relationship of different functional striatal areas with the midbrain. The ventral striatum receives a relatively limited midbrain input but projects to a large region, which includes dorsal and ventral tiers and the dorsal pars reticulata. In contrast, the dorsolateral striatum receives input from a wide range of dopamine cells but projects to a limited region (Fig. 2.3.4c).[5]

The striato-nigro-striatal projection system

The proportional differences between inputs and outputs of the dopamine neurons, coupled with their topography, result in complex interweaving of functional pathways. For each striatal region, the afferent and efferent striato-nigro-striatal projection system contains three components in the midbrain. There is a reciprocal connection that is flanked by two nonreciprocal connections. The reciprocal component contains cells that project to a specific striatal area. These cells are embedded within terminals from that same striatal area. Dorsal to this region lies a group of cells that project to the same striatal region but do not lie within its reciprocal terminal field. In other words, these cells receive a striatal projection from a region to which they do not project.

Finally, ventral to the reciprocal component are efferent terminals. However, there are no cells embedded in these terminals that project to that same specific striatal region. The cells that are located in this terminal field project to a different striatal area. These three components for each striato-nigro-striatal projection system occupy different positions within the midbrain. The ventral striatum system lies dorsomedially, the dorsolateral striatum system lies ventrolaterally, and the central striatal system is positioned between the two. Moreover, as indicated above, each functional region differs in its proportional projections that significantly alter their relationship to each other. The ventral striatum receives a limited midbrain input but projects to a large region. In contrast, the dorsolateral striatum receives a wide input but projects to a limited region. In other words, the ventral striatum influences a wide range of dopamine neurons but is itself influenced by a relatively limited group of dopamine cells. On the other hand, the dorsolateral striatum influences a limited midbrain region but is affected by a relatively large midbrain region.

Thus, the size and position of the afferent and efferent connections for each system, together with the arrangement into three components, allow information from the limbic system to reach the motor system through a series of connections[5] (see Fig. 2.3.5). The ventral striatum receives input from limbic regions and projects to the dorsal tier. The dorsal tier projects back to the ventral striatum. However, the ventral striatum efferent projection to the midbrain extends beyond the tight ventral striatal/dorsal tier/ventral striatal circuit, terminating lateral and ventral to the dorsal tier. This area of terminal projection does not project back to the ventral striatum. Rather, cells in this region project more dorsally, into the striatal area that receives input from the DPFC. Through this connection, the same cortical information that influences the dorsal tier through the ventral striatum also modulates the densocellular region that projects to the central striatum. This central striatal region is reciprocally connected to the densocellular region. But it also projects to the ventral densocellular area and into the cell columns. Thus, projections from the DPFC, via the striatum, are in a position to influence cells that project to motor control areas of the striatum. The dorsolateral striatum is reciprocally connected to the ventral densocellular region and cell columns. The confined distribution of efferent dorsolateral striatal fibers limits the influence of the motor striatum to a relatively small region involving the cell columns and the pars reticulata. Taken together, the interface between different striatal

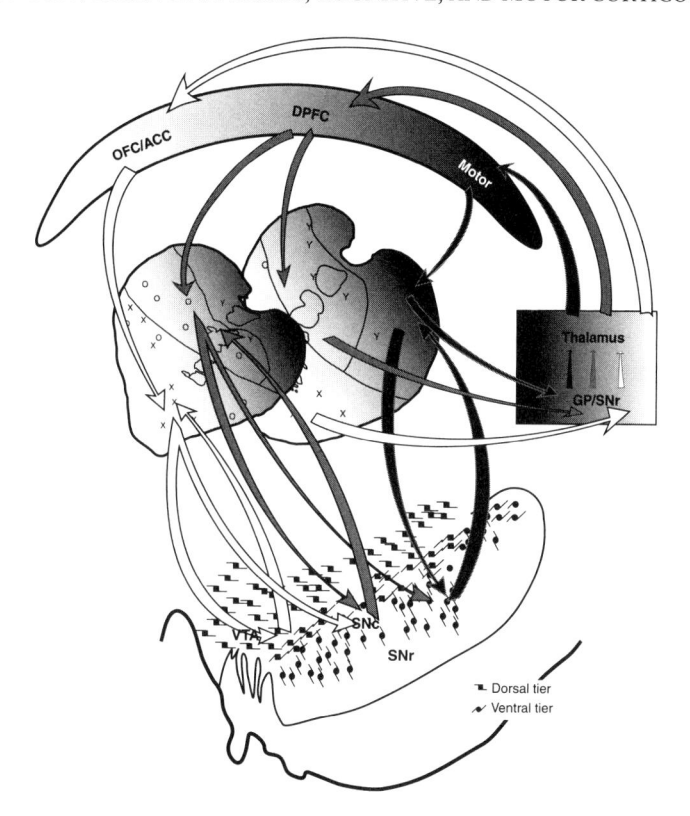

FIGURE 2.3.5. Schematic illustrating both dual parallel and integrative processing through corticobasal ganglia pathways. Corresponding shaded striatal and cortical areas demonstrate topographic of projections. X marks substriatal regions where convergence between terminals from limbic cortical areas occurs; O, convergence between terminals from limbic and cognitive cortical areas; Y, convergence between terminals from cognitive and motor control cortical areas occurs. Arrows connecting the striatum and substantia nigra illustrate how the ventral striatum can influence the dorsal striatum through the midbrain dopamine cells. The connections between integrated areas also enter the parallel processing system, back to cortex, as indicated by the arrows connecting the striatum via the pallidum and thalamus. DPFC, dorsolateral prefrontal cortex; GP/SNr, globus pallidus/substantia nigra pars reticulata; OFC/ACC, orbital prefrontal/anterior cingulate cortex; SNc, substantia nigra pars compacta; SNr, substantia nigra pars reticulata; VTA, ventral tegmental area.

regions via the midbrain dopamine cells is organized in an ascending spiral interconnecting different functional regions of the striatum. This creates a feedforward organization (Fig. 2.3.5). Through this spiral of inputs and outputs between the striatum and midbrain dopamine neurons, information can be channeled from the shell and ventral striatum, through the central striatum, and to the dorsolateral striatum. In this way, information can flow from limbic to cognitive to motor circuits, providing a mechanism by which motivation and cognition can influence motor decision-making processes and appropriate responses to environmental cues.

Functional Considerations

A key component in developing appropriate goal-directed behaviors is the ability to first correctly evaluate different aspects of reward, including value versus risk and predictability, and inhibit maladaptive choices, based on previous experience. These calculations rely on integration of different aspects of reward processing and cognition to develop and execute appropriate action plans. While parallel networks that mediate different functions are critical to maintaining coordinated behaviors, cross-talk between functional circuits during learning is critical. Indeed, reward and associative functions are not clearly and completely separated within the striatum. Consistent with human imaging studies, reward-responsive neurons are not restricted to the ventral striatum, but rather are found throughout the striatum. Moreover, cells responding in working memory tasks are often found also in the ventral striatum.[37,43,77–79]

As described above, embedded within limbic, associative, and motor control striatal territories are subregions containing convergent terminals between different reward-processing cortical areas, between these projections and those from the DPFC, and between the DPFC and rostral motor control areas.

Given that a single corticostriatal axon can innervate 14% of the striatum[80] and that terminals from different cortical areas synapse on a subpopulation of interneurons that are important for integrating information across functions,[81] these nodes of converging terminals may represent "hot spots" that may be particularly sensitive to synchronizing information across functional areas to impact on long-term strategic planning, and habit formation.[62] Indeed, cells in the dorsal striatum are progressively recruited during different types of learning, from simple motor tasks to drug self-administration.[41,66,82,83] Convergent fibers from cortex within the ventral striatum, taken together with hippocampal and amygdalo-striatal projections, place the ventral striatum in a key entry port for processing emotional and motivational information that, in turn, drives basal ganglia action output. The ventral, reward-based striatal region and the associative, central striatal region can impact on motor output circuits, not only through convergent terminal fields within the striatum, but also through the striato-nigro-striatal pathways. One can hypothesize that initially the nodal points of interface between the reward and associative circuits, for example, send a coordinated signal to dopamine cells. This pathway is in a pivotal position for temporal training dopamine cells. In turn, these nodal points may be further reinforced through the burst firing activity of the nigrostriatal pathway, thus transferring that impact back to the striatum (Fig. 2.3.5). Moreover, since the midbrain dopamine neurons project to a wider dorsal striatal region, information is transferred to other functional regions during learning and habit formation.[66,67] This signal then enters the parallel system and, via the pallidum and thalamus, impacts on frontal cortex (Fig. 2.3.5). Indeed, when the striato-nigro-striatal circuit is interrupted, information transfer from Pavlovian to instrumental learning does not take place.[84].

Parallel circuits and integrative circuits must work together, allowing the coordinated behaviors to be maintained and focused (via parallel networks), but also to be modified and changed according to the appropriate external and internal stimuli (via integrative networks) (Fig. 2.3.5). Both the ability to maintain focus in the execution of specific behaviors and the ability to adapt appropriately to external and internal cues are key deficits in basal ganglia diseases that affect these aspects of motor control, cognition, and motivation. Within each interconnected corticobasal ganglia loop, there are subregions that cross functional domains. Their locations (within the striatum or midbrain) are likely to impact differentially on how the dopamine neurons mediate learning and the development of action plans. Dopamine neurons are in a position, therefore, not only to impact on the striatum during learning, but also to be modulated by it during the development of learning and habit formation.[66,82,84]

REFERENCES

1. Schultz W. Getting formal with dopamine and reward. *Neuron*. 2002;36:241–263.
2. Matsumoto N, Hanakawa T, Maki S, Graybiel AM, Kimura M. Nigrostriatal dopamine system in learning to perform sequential motor tasks in a predictive manner. *J Neurophysiol*. 1999;82:978–998.
3. Alexander GE, Crutcher MD. Functional architecture of basal ganglia circuits: neural substrates of parallel processing. *Trends Neurosci*. 1990;13:266–271.
4. Middleton FA, Strick PL. Basal-ganglia 'projections' to the prefrontal cortex of the primate. *Cereb Cortex*. 2002;12:926–935.
5. Haber SN, Fudge JL, McFarland NR. Striatonigrostriatal pathways in primates form an ascending spiral from the shell to the dorsolateral striatum. *J Neurosci*. 2000;20:2369–2382.
6. McFarland NR, Haber SN. Thalamic relay nuclei of the basal ganglia form both reciprocal and nonreciprocal cortical connections, linking multiple frontal cortical areas. *J Neurosci*. 2002;22:8117–8132.
7. Percheron G, Filion M. Parallel processing in the basal ganglia: up to a point. *Trends Neurosci*. 1991;14:55–59.
8. Draganski B, Kherif F, Kloppel S, Cook PA, Alexander DC, Parker GJ, Deichmann R, Ashburner J, Frackowiak RS. Evidence for segregated and integrative connectivity patterns in the human basal ganglia. *J Neurosci*. 2008;28:7143–7152.
9. Haber SN, Kim KS, Mailly P, Calzavara R. Reward-related cortical inputs define a large striatal region in primates that interface with associative cortical inputs, providing a substrate for incentive-based learning. *J Neurosci*. 2006;26:8368–8376.
10. Fuster JM. The prefrontal cortex–an update: time is of the essence. *Neuron*. 2001;30:319–333.
11. Mayberg HS, Liotti M, Brannan SK, McGinnis S, Mahurin RK, Jerabek PA, Silva JA, Tekell JL, Martin CC, Lancaster JL, Fox PT. Reciprocal limbic-cortical function and negative mood: converging PET findings in depression and normal sadness. *Am J Psychiatry*. 1999;156:675–682.
12. Milad MR, Quinn BT, Pitman RK, Orr SP, Fischl B, Rauch SL. Thickness of ventromedial prefrontal cortex in humans is correlated with extinction memory. *Proc Natl Acad Sci USA*. 2005;102:10706–10711.
13. Hikosaka K, Watanabe M. Delay activity of orbital and lateral prefrontal neurons of the monkey varying with different rewards. *Cereb Cortex*. 2000;10:263–271.
14. Schultz W, Tremblay L, Hollerman JR. Reward processing in primate orbitofrontal cortex and basal ganglia. *Cereb Cortex*. 2000;10:272–284.
15. Padoa-Schioppa C, Assad JA. Neurons in the orbitofrontal cortex encode economic value. *Nature*. 2006;441:223–226.
16. Kringelbach ML, Rolls ET. The functional neuroanatomy of the human orbitofrontal cortex: evidence from neuroimaging and neuropsychology. *Prog Neurobiol*. 2004;72:341–372.
17. O'Doherty J, Kringelbach ML, Rolls ET, Hornak J, Andrews C. Abstract reward and punishment representations in the human orbitofrontal cortex. *Nat Neurosci*. 2001;4:95–102.

18. Barbas H. Architecture and cortical connections of the prefrontal cortex in the rhesus monkey. In: Chauvel P, Delgado-Escueta AV, eds. *Advances in Neurology.* New York, NY: Raven Press; 1992:91–115.
19. Price JL, Carmichael ST, Drevets WC. Networks related to the orbital and medial prefrontal cortex; a substrate for emotional behavior? *Prog Brain Res.* 1996;107:523–536.
20. Butter CM, Snyder DR. Alterations in aversive and aggressive behaviors following orbital frontal lesions in rhesus monkeys. *Acta Neurobiol Exp.* 1972;32:525–565.
21. Fuster JM. Lesion studies. In: *The Prefrontal Cortex: Anatomy, Physiology, and Neuropsychology of the Frontal Lobe.* 2nd ed. New York, NY: Raven Press; 1989:51–82.
22. Milad MR, Rauch SL. The role of the orbitofrontal cortex in anxiety disorders. *Ann NY Acad Sci.* 2007;1121:546–561.
23. O'Doherty J, Critchley H, Deichmann R, Dolan RJ. Dissociating valence of outcome from behavioral control in human orbital and ventral prefrontal cortices. *J Neurosci.* 2003;23:7931–7939.
24. Smith EE, Jonides J. Working memory: a view from neuroimaging. *Cogn Psychol.* 1997;33:5–42.
25. Passingham D, Sakai K. The prefrontal cortex and working memory: physiology and brain imaging. *Curr Opin Neurobiol.* 2004;14:163–168.
26. Blumenfeld RS, Ranganath C. Dorsolateral prefrontal cortex promotes long-term memory formation through its role in working memory organization. *J Neurosci.* 2006;26: 916–925.
27. Fuster JM. Prefrontal neurons in networks of executive memory. *Brain Res Bull.* 2000;52:331–336.
28. Goldman-Rakic PS. The prefrontal landscape: implications of functional architecture for understanding human mentation and the central executive. *Philos Trans R Soc Lond-Series B: Biol Sci.* 1996;351:1445–1453.
29. Mushiake H, Inase M, Tanji J. Neuronal activity in the primate premotor, supplementary, and precentral motor cortex during visually guided and internally determined sequential movements. *J Neurophysiol.* 1991;66:705–718.
30. Tanji J, Mushiake H. Comparison of neuronal activity in the supplementary motor area and primary motor cortex. *Brain Res Cogn Brain Res.* 1996;3:143–150.
31. Ito R, Robbins TW, Everitt BJ. Differential control over cocaine-seeking behavior by nucleus accumbens core and shell. *Nat Neurosci.* 2004;7:389–397.
32. Carlezon WA, Wise RA. Rewarding actions of phencyclidine and related drugs in nucleus accumbens shell and frontal cortex. *J Neurosci.* 1996;16:3112–3122.
33. Fudge JL, Kunishio K, Walsh C, Richard D, Haber SN. Amygdaloid projections to ventromedial striatal subterritories in the primate. *Neuroscience.* 2002;110:257–275.
34. Friedman DP, Aggleton JP, Saunders RC. Comparison of hippocampal, amygdala, and perirhinal projections to the nucleus accumbens: combined anterograde and retrograde tracing study in the macaque brain. *J Comp Neurol.* 2002; 450:345–365.
35. Haber SN, McFarland NR. The concept of the ventral striatum in nonhuman primates. In: McGinty JF, ed. *Advancing from the Ventral Striatum to the Extended Amygdala.* New York, NY: New York Academy of Sciences; 1999:33–48.
36. Tremblay L, Hollerman JR, Schultz W. Modifications of reward expectation-related neuronal activity during learning in primate striatum. *J Neurophysiol.* 1998;80:964–977.
37. Hassani OK, Cromwell HC, Schultz W. Influence of expectation of different rewards on behavior-related neuronal activity in the striatum. *J Neurophysiol.* 2001;85:2477–2489.
38. Knutson B, Adams CM, Fong GW, Hommer D. Anticipation of increasing monetary reward selectively recruits nucleus accumbens. *J Neurosci.* 2001;21:RC159.
39. Selemon LD, Goldman-Rakic PS. Longitudinal topography and interdigitation of corticostriatal projections in the rhesus monkey. *J Neurosci.* 1985;5:776–794.
40. Calzavara R, Mailly P, Haber SN. Relationship between the corticostriatal terminals from areas 9 and 46, and those from area 8A, dorsal and rostral premotor cortex and area 24c: an anatomical substrate for cognition to action. *Eur J Neurosci* 2007; 26:2005–2024.
41. Pasupathy A, Miller EK. Different time courses of learning-related activity in the prefrontal cortex and striatum. *Nature.* 2005;433:873–876.
42. Battig K, Rosvold HE, Mishkin M. Comparison of the effect of frontal and caudate lesions on delayed response and alternation in monkeys. *J Comp Physiol Psychol.* 1960;53: 400–404.
43. Levy R, Friedman HR, Davachi L, Goldman-Rakic PS. Differential activation of the caudate nucleus in primates performing spatial and nonspatial working memory tasks. *J Neurosci.* 1997;17: 3870–3882.
44. Flaherty AW, Graybiel AM. Input-output organization of the sensorimotor striatum in the squirrel monkey. *J Neurosci.* 1994;14:599–610.
45. Aldridge JW, Anderson RJ, Murphy JT. Sensory-motor processing in the caudate nucleus and globus pallidus: a single-unit study in behaving primates. *Can J Physiol Pharmacol.* 1980;58:1192–1201.
46. Kimura M. The role of primate putamen neurons in the association of sensory stimulus with movement. *Neurosci Res.* 1986;3:436–443.
47. McFarland NR, Haber SN. Convergent inputs from thalamic motor nuclei and frontal cortical areas to the dorsal striatum in the primate. *J Neurosci.* 2000;20:3798–3813.
48. Hedreen JC, DeLong MR. Organization of striatopallidal, striatonigral, and nigrostriatal projections in the macaque. *J Comp Neurol.* 1991;304:569–595.
49. Haber SN, Lynd E, Klein C, Groenewegen HJ. Topographic organization of the ventral striatal efferent projections in the rhesus monkey: an anterograde tracing study. *J Comp Neurol.* 1990;293:282–298.
50. Selemon LD, Goldman-Rakic PS. Topographic intermingling of striatonigral and striatopallidal neurons in the rhesus monkey. *J Comp Neurol.* 1990;297:359–376.
51. Lynd-Balta E, Haber SN. Primate striatonigral projections: a comparison of the sensorimotor-related striatum and the ventral striatum. *J Comp Neurol.* 1994;345:562–578.
52. Strick PL. Anatomical analysis of ventrolateral thalamic input to primate motor cortex. *J Neurophysiol.* 1976;39:1020–1031.
53. Ilinsky IA, Jouandet ML, Goldman-Rakic PS. Organization of the nigrothalamocortical system in the rhesus monkey. *J Comp Neurol.* 1985;236:315–330.
54. Kuo J, Carpenter MB. Organization of pallidothalamic projections in the rhesus monkey. *J Comp Neurol.* 1973;151:201–236.
55. McFarland NR, Haber SN. Thalamic connections with cortex from the basal ganglia relay nuclei provide a mechanism for integration across multiple cortical areas. *J Neurosci.* 2002; 22: 8117–8132.
56. Mink JW. The basal ganglia: focused selection and inhibition of competing motor programs. *Prog Neurobiol.* 1996;50: 381–425.
57. Wise SP, Murray EA, Gerfen CR. The frontal cortex-basal ganglia system in primates. *Crit Rev Neurobiol.* 1996;10:317–356.

58. Cools R, Clark L, Robbins TW. Differential responses in human striatum and prefrontal cortex to changes in object and rule relevance. *J Neurosci.* 2004;24:1129–1135.
59. Muhammad R, Wallis JD, Miller EK. A comparison of abstract rules in the prefrontal cortex, premotor cortex, inferior temporal cortex, and striatum. *J Cogn Neurosci.* 2006;18:974–989.
60. Hikosaka O, Miyashita K, Miyachi S, Sakai K, Lu X. Differential roles of the frontal cortex, basal ganglia, and cerebellum in visuomotor sequence learning. *Neurobiol Learning Memory.* 1998;70:137–149.
61. Calzavara R, Mailly P, Haber SN. Relationship between the corticostriatal terminals from areas 9 and 46, and those from area 8A, dorsal and rostral premotor cortex and area 24c: an anatomical substrate for cognition to action. *Eur J Neurosci.* 2007;26:2005–2024.
62. Kasanetz F, Riquelme LA, Della-Maggiore V, O'Donnell P, Murer MG. Functional integration across a gradient of corticostriatal channels controls UP state transitions in the dorsal striatum. *Proc Natl Acad Sci USA.* 2008;105:8124–8129.
63. Satoh T, Nakai S, Sato T, Kimura M. Correlated coding of motivation and outcome of decision by dopamine neurons. *J Neurosci.* 2003;23:9913–9923.
64. Pagnoni G, Zink CF, Montague PR, Berns GS. Activity in human ventral striatum locked to errors of reward prediction. *Nat Neurosci.* 2002;5:97–98.
65. Redgrave P, Gurney K. The short-latency dopamine signal: a role in discovering novel actions? *Nat Rev Neurosci.* 2006;7:967–975.
66. Volkow ND, Wang GJ, Telang F, Fowler JS, Logan J, Childress AR, Jayne M, Ma Y, Wong C. Cocaine cues and dopamine in dorsal striatum: mechanism of craving in cocaine addiction. *J Neurosci.* 2006;26:6583–6588.
67. Porrino LJ, Smith HR, Nader MA, Beveridge TJ. The effects of cocaine: a shifting target over the course of addiction. *Prog Neuropsychopharmacol Biol Psychiatry.* 2007;31: 1593–1600.
68. Everitt BJ, Robbins TW. Neural systems of reinforcement for drug addiction: from actions to habits to compulsion. *Nat Neurosci.* 2005;8:1481–1489.
69. Hokfelt T, Martensson R, Bjorklund A, Kleinau S, Goldstein M. Distributional maps of tyrosine-hydroxylase immunoreactive neurons in the rat brain. In: Bjorklund A, Hokfelt T, eds. *Handbook of Chemical Neuroanatomy, Vol. II: Classical Neurotransmitters in the CNS, Part I.* Amsterdam: Elsevier; 1984:277–379.
70. Olszewski J, Baxter D. Cytoarchitecture of the *Human Brain Stem.* 2nd ed. Basel: S. Karger; 1982.
71. Haber SN, Ryoo H, Cox C, Lu W. Subsets of midbrain dopaminergic neurons in monkeys are distinguished by different levels of mRNA for the dopamine transporter: Comparison with the mRNA for the D2 receptor, tyrosine hydroxylase and calbindin immunoreactivity. *J Comp Neurol.* 1995;362:400–410.
72. Lavoie B, Parent A. Dopaminergic neurons expressing calbindin in normal and parkinsonian monkeys. *Neuroreport.* 1991;2(10):601–604.
73. Haber SN, Gdowski MJ. The basal ganglia. In: Paxinos G, Mai JK, eds. *The Human Nervous System.* 2nd ed. New York, NY: Elsevier Press; 2004:677–738.
74. Fudge JL, Haber SN. Bed nucleus of the stria terminalis and extended amygdala inputs to dopamine subpopulations in primates. *Neuroscience.* 2001;104:807–827.
75. Lynd-Balta E, Haber SN. The organization of midbrain projections to the striatum in the primate: sensorimotor-related striatum versus ventral striatum. *Neuroscience.* 1994;59: 625–640.
76. Parent A, Mackey A, De Bellefeuille L. The subcortical afferents to caudate nucleus and putamen in primate: a fluorescence retrograde double labeling study. *Neuroscience.* 1983;10(4): 1137–1150.
77. Watanabe K, Lauwereyns J, Hikosaka O. Neural correlates of rewarded and unrewarded eye movements in the primate caudate nucleus. *J Neurosci.* 2003;23:10052–10057.
78. Takikawa Y, Kawagoe R, Hikosaka O. Reward-dependent spatial selectivity of anticipatory activity in monkey caudate neurons. *J Neurophysiol.* 2002; 87:508–515.
79. Tanaka SC, Doya K, Okada G, Ueda K, Okamoto Y, Yamawaki S. Prediction of immediate and future rewards differentially recruits cortico-basal ganglia loops. *Nat Neurosci.* 2004;7:887–893.
80. Zheng T, Wilson CJ. Corticostriatal combinatorics: the implications of corticostriatal axonal arborizations. *J Neurophysiol.* 2002;87:1007–1017.
81. Mallet N, Le Moine C, Charpier S, Gonon F. Feedforward inhibition of projection neurons by fast-spiking GABA interneurons in the rat striatum in vivo. *J Neurosci.* 2005;25: 3857–3869.
82. Porrino LJ, Lyons D, Smith HR, Daunais JB, Nader MA. Cocaine self-administration produces a progressive involvement of limbic, association, and sensorimotor striatal domains. *J Neurosci.* 2004; 24:3554–3562.
83. Lehericy S, Benali H, Van de Moortele PF, Pelegrini-Issac M, Waechter T, Ugurbil K, Doyon J. Distinct basal ganglia territories are engaged in early and advanced motor sequence learning. *Proc Natl Acad Sci USA.* 2005;102:12566–12571.
84. Belin D, Everitt BJ. Cocaine seeking habits depend upon dopamine-dependent serial connectivity linking the ventral with the dorsal striatum. *Neuron.* 2008;57:432–441.

2.4 The Relationship between Dopaminergic Axons and Glutamatergic Synapses in the Striatum: Structural Considerations

JONATHAN MOSS AND J. PAUL BOLAM

INTRODUCTION

The basal ganglia are a group of highly interconnected nuclei involved in a variety of functions including movement and cognition. The dorsal component, which is primarily related to motor and associative functions, consists of the striatum, external segment of the globus pallidus, subthalamic nucleus, internal segment of the globus pallidus, and substantia nigra pars reticulata. The last two structures form the output nuclei of the basal ganglia (Fig, 2.4.1). The major inputs to the basal ganglia arise in the cerebral cortex and thalamus and are carried by the corticostriatal and thalamostriatal pathways. This information is processed in the striatum and transmitted by various routes to the output nuclei. The basal ganglia then influence behavior by these structures projecting to thalamus and thence to the cortex or to other subcortical structures involved in movement. Overlying this feedforward system of the basal ganglia is feedback from dopamine neurons in the substantia nigra pars compacta (SNc). These neurons massively innervate the striatum and also provide innervation of other regions of the basal ganglia, albeit at a much lower density. At the level of the striatum, the principal role of the dopaminergic innervation is to modulate the flow of cortical and thalamic information through the basal ganglia. The objective of this brief review is to summarize data relating principally to the anatomical substrate of the interaction between both glutamatergic corticostriatal synapses and thalamostriatal synapses with dopaminergic axons and terminals in the striatum.

GENERAL ASPECTS OF DOPAMINERGIC INNERVATION OF THE STRIATUM

Dopamine neurons of the SNc account for a remarkably small number of neurons. It is estimated that there are 7000–8000 neurons[1,2] in each SNc of the rat, with a total of 12,000 in the entire SN and about another 20,000 dopamine neurons in the ventral tegmental area (VTA).[2] The remarkable nature of these neurons lies not only in their numbers but also in their innervation of the forebrain. It is well known that the density and distribution of markers of dopaminergic neurons and transmission in the striatum are the highest in the brain,[3,4] but although many studies have examined the somatodendritic properties and locations of individual dopamine neurons filled in vivo, it was not until very recently that the axonal field of individual dopamine neurons was revealed.[5] Analysis of individual dopamine neurons (revealed by infection with a viral vector expressing membrane-targeted green fluorescent protein) in the rat brain showed that on average the total length of the axon in the striatum is in the region of 47 cm and the arborization can extend to occupy up to 5.7% of the volume of the striatum.[5] The remarkable length of the axons of individual neurons is reflected in the estimates of the number of dopaminergic synapses formed in the striatum. Based on the number of neurons in the SNc and the striatum[1,2] and the known synaptic organization of the dopaminergic nigrostriatal projection,[6] we estimate that an individual dopaminergic neuron gives rise to between 170,00 and 408,000 synapses in the striatum (Table 2.4.1). This figure is close to the estimate of Wickens and Arbuthnott,[7] who, using a completely different approach and set of assumptions based on densities of synapses and neurons, concluded that individual dopamine neurons give rise to about 370,000 synapses in the striatum.[7] Furthermore, these figures are close to the estimates of Anden et al.[3] of a total axon length of 30 cm and 250,000 varicosities per SNc dopaminergic neuron, based on the analysis of tissue stained by the histofluorescence method[3] (cited by, and figures recalculated by, Björklund and Lindvall[4]). To put these figures in perspective, data from rats have indicated that neurons of the external globus pallidus give rise to approximately 2000 synapses,[8,9] striatal spiny neurons probably give rise

49

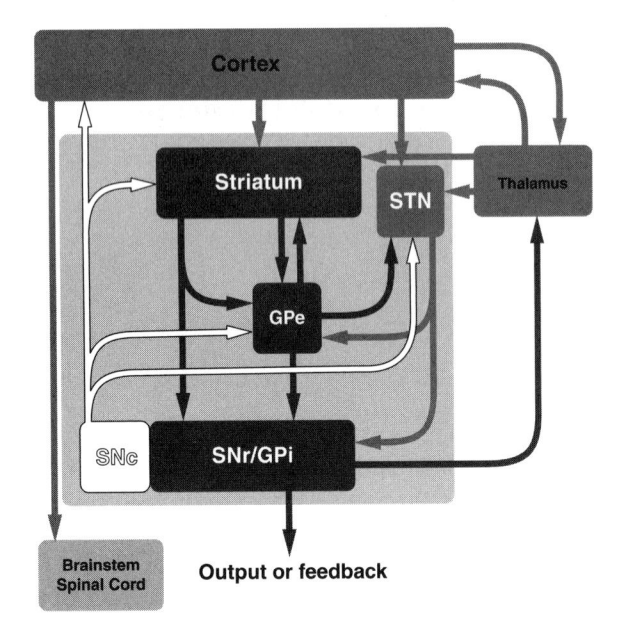

FIGURE 2.4.1. Simplified block diagram of the basal ganglia showing the principal connections of dopamine neurons. The nuclei of the basal ganglia (included in the light gray box) consist of the striatum, the external segment of the globus pallidus (GPe), the subthalamic nucleus (STN), the substantia nigra pars reticulata and the internal segment of the globus pallidus (SNr/GPi), and the substantia nigra pars compacta (SNc). The two major inputs to the basal ganglia are from the cortex and the thalamus (mainly the intralaminar nuclei). The SNr and GPi constitute the output nuclei of the basal ganglia projecting to the thalamus and thence back to the cortex or to other subcortical structures. Dopamine neurons of the SNc provide massive feedback innervation of the striatum but also of other regions of the basal ganglia plus the prefrontal cortex. Dopamine neurons may also modulate neurons of the SNr by the dendritic release of dopamine.

to about 300 synapses,[10,11] and fast-spiking interneurons in the striatum give rise to about 5000 synapses.[12,13] It should be noted that the figures for dopamine neurons are also probably underestimates, as individual dopamine neurons innervate multiple regions of the basal ganglia, where they form synapses and release dopamine (see, for instance,[5,14,15]). Whatever the precise figures for dopamine neurons are, they are remarkable neurons when compared to classical central nervous system (CNS) neurons. Such a large axonal arborization raises questions about the control of the activity of individual boutons, which may be as far as several tens of millimeters from the site of initiation of the axon potential at the axon hillock or proximal dendrite. Do all axon potentials invade all of the extensive and tortuous branches of the axonal arbor? Does such a large axonal arbor, which requires supply and support, render the neuron particularly susceptible to stressors that lead to cell death? It is interesting to note that we estimate, using similar methods and assumptions as described in Table 1, that dopaminergic neurons in the VTA, which are less susceptible to dying in Parkinson's disease, give rise to far less synapses (in the range of 12,000–30,000 per neuron).

SYNAPTIC ORGANIZATION OF THE DOPAMINE INNERVATION OF THE STRIATUM

At the level of the striatum, axons of dopaminergic neurons give rise to small vesicle-containing varicosities that form small, mainly symmetrical (Gray's Type 2) synapses (Fig. 2.4.2). The synapses are often difficult to visualize because of the small size of the specialization and the fact that their integrity is easily lost in suboptimally fixed tissue. Nonsynaptic segments of dopaminergic axons may also contain vesicles, and synapses may be formed by nonvaricose segments of the axons.[16,17] Most studies in the rat agree that one of the principal synaptic targets of dopaminergic terminals in the striatum are dendritic spines (51%–65% of synapses formed by dopaminergic terminals are with spines).[16,18–20] The remaining synapses are with dendritic shafts (30%–46%) and perikarya (2%–6%). One study in the rat, however, identified a smaller proportion in contact with spines (30%) and a correspondingly higher proportion in contact with dendritic shafts (67%).[21] Interestingly, there is very little difference in the distribution of synaptic targets between the patch/striosome and matrix compartments of the striatum.[19] In primates (squirrel monkey), it appears that there is greater heterogeneity in the type of synaptic specialization (i.e., a greater proportion form asymmetrical synapses), and only 22.5% of synapses were identified to be in contact with spines and 72% in contact with dendritic shafts.[22] It is commonly the case that the synaptic contacts with dendritic spines are associated with the necks of spines that are also in synaptic contact with a terminal forming an asymmetrical, presumably excitatory, synaptic contact (Gray's Type 1),[16,22] and indeed, some have been shown to be derived from the cortex[22,23] (see below and Fig. 2.4.3A).

SYNAPTIC ORGANIZATION OF THE CORTICAL INNERVATION OF THE STRIATUM

The corticostriatal projection is both bilateral and topographical in nature, and several organizational principles have been described that contribute to a complex

TABLE 2.4.1. *Number of Synapses Formed by a Single Dopamine Neuron in the Striatum*

1. Average number of dendritic spines on one MSN	6250–15,000
2. Percentage of axo-spinous synapses that involve cortical and thalamic terminals	64.8%
3. Average number of dendritic spines postsynaptic to cortical or thalamic terminals on one MSN (1) x (2)	4050–9720
4. Percentage of these dendritic spines in synaptic contact with a dopamine terminal	6.666%
5. Number of these dendritic spines on one MSN in synaptic contact with a dopamine terminal (3) x (4)	270–648
6. Percentage of dopamine terminals that contact dendritic spines (not shafts or cell bodies)	61.3%
7. Multiplying factor to incorporate synapses with dendritic shafts and cell bodies; reciprocal of 0.613 (6)	1.632
8. Number of dopamine terminals forming synapses with one MSN (5) x (7)	441–1058
9. Number of MSNs in the striatum (one hemisphere)	2,780,000
10. Number of dopamine terminals forming synapses with all MSNs in one hemisphere (8) x (9)	1,224,979,200–2,939,950,080
11. Number of dopamine neurons in the substantia nigra pars compacta (one hemisphere)	7200
12. Number of symmetrical synapses formed by one dopamine neuron in the rat striatum (10)/(11)	170,136–408,326

Note that to arrive at the number of dopaminergic synapses in contact with an individual MSN, we have used quantitative data from spines that are postsynaptic to cortical or thalamic terminals. This will introduce error because dopaminergic terminals may contact other spines and for the reason indicated below.
Value in 1 is the range indicated by Kincaid et al.[45]
Value in 2 is from Lacey et al.[37] This figure represents the percentage of axo-spinous synapses involving VGluT1-positive (cortical) and VGluT2-positive (thalamic) terminals. It is likely to be an underestimate due to false-negative labeling of terminals.
Values in 4 and 6 are from Moss and Bolam.[6]
Values in 9 and 11 are from Oorschot.[1] The figure of 7200 dopamine neurons may be an underestimate of the true number of nigro-striatal dopaminergic neurons, as dopamine neurons located in the pars reticulata also project to the striatum.

and heterogeneous projection. On the basis of single cell filling, several classes of corticostriatal neurons have been described that differ in their cortico-cortical and cortico-fugal projections as well as in their pattern of innervation of the striatum (see [24]). The corticostriatal projection is also heterogeneous with respect to the patch/striosome and matrix subdivisions of the striatum.[24–32] Limbic cortical areas show selectivity for the innervation of the patch/striosome, whereas other areas show selectivity for the matrix. Furthermore, neurons in deep layer 5 and layer 6 of the cortex selectively innervate the patch/striosome, whereas neurons located in upper layer 5 and layer 3 selectively innervate the matrix. The modular organization of the corticostriatal termination within the matrix, referred to as *matrisomes*, represents a further level of heterogeneity.[27,32–35]

Terminals in the striatum that are derived from the cortex form asymmetrical (Gray's Type 1) synaptic specializations (Fig. 2.4.3A; for references see [36]). Ultrastructural analysis of corticostriatal terminals labeled by anterograde degeneration, anterograde tracing, or immunolabeling for vesicular glutamate transporter type 1 (VGluT1; see below)[36–39] reveals that a high proportion (>95%) make synaptic contact with dendritic spines. Since medium-sized spiny projection neurons (MSNs) account for the majority of spines in the striatum, this observation suggests that they are likely to be the major targets of the cortical projection, and this is supported by direct analysis of MSNs.[40–42] Furthermore, spiny neurons giving rise to both the direct and indirect pathways of information flow through the basal ganglia receive input from the cortex.[43] Corticostriatal terminals originating from the

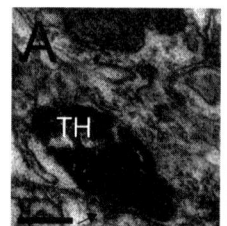

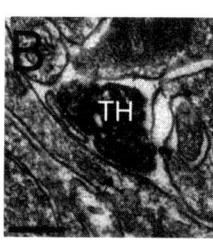

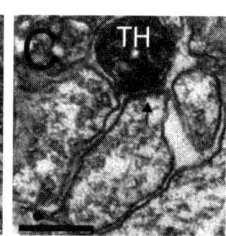

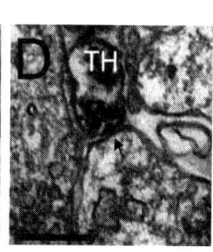

FIGURE 2.4.2. Dopaminergic neurons give rise to small, symmetrical synapses in the striatum. TH-positive terminals (TH) making symmetrical synaptic contact (arrows) with a dendritic shaft in (A) and dendritic spines in (B–D). Scale bars: 200 nm.

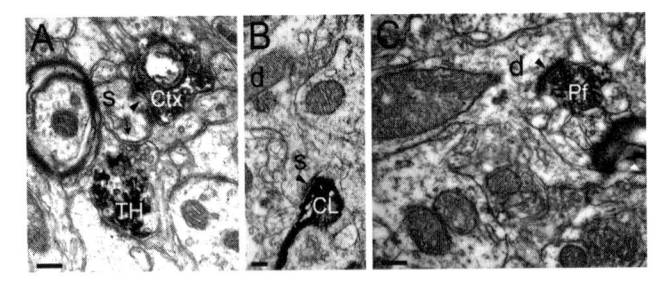

FIGURE 2.4.3. Synaptic targets of cortical and thalamic terminals in the striatum. (A) An axon terminal in the monkey putamen, anterogradely labeled from the cortex (Ctx), makes asymmetrical synaptic contact (arrowhead) with a dendritic spine (s). A TH-positive axon terminal (TH) makes symmetrical synaptic contact (small arrow) with the same spine. Dendritic spines are the main synaptic target of corticostriatal terminals. (B) An axon terminal in the rat striatum derived from a neuron in the central lateral nucleus of the thalamus (CL) makes asymmetrical synaptic contact (arrowhead) with a dendritic spine (s) that can be seen to arise from a dendritic shaft (d). Most terminals derived from the CL make asymmetrical axospinous synapses. (C) An axon terminal in the rat striatum derived from a neuron in the parafascicular nucleus of the thalamus (Pf) makes asymmetrical synaptic contact with a dendritic shaft (d). About 63%–89% of terminals from the parafascicular nucleus make synaptic contact with dendritic shafts and the remainder with dendritic spines. The axon terminals in (A) and (C) were derived from neurons that were recorded and juxtacellularly labeled in vivo. Scale bars: 200 nm. *Source*: Data in (A) derived and modified from Smith et al.[22] Data in (B) and (C) derived and modified from Lacey et al.[65]

contralateral cortex form contact more frequently with dendritic spines,[43] and the two broad classes of corticostriatal neurons,[24] that is, those that project preferentially within the telencephalon and corticopyramidal neurons that give rise to a collateral to the striatum, have different patterns of innervation of the striatum.[44] The former give rise to relatively small axonal boutons that preferentially innervate the spines of spiny neurons giving rise to the direct pathway, whereas the latter give rise to relatively large boutons that preferentially innervate the spines of spiny neurons giving rise to the indirect pathway. Quantitative analysis of the pattern of innervation of the striatum by individual cortical axons suggests that individual spiny neurons are likely to receive only about four synapses from an individual corticostriatal axon[45,46]; there is thus a high degree of divergence of cortical axons in the striatum and a high degree of convergence at the single striatal cell level.

Striatal interneurons that express the calcium-binding protein, parvalbumin, often referred to as *fast-spiking interneurons*,[13,47–50] are prominently innervated by cortical axons,[51,52] but the pattern of innervation is different from that of cortical input to MSNs. Individual corticostriatal axons (which may arise from functionally diverse regions of the cortex) make multiple synaptic contacts with individual parvalbumin-positive GABAergic interneurons.[53] The population of striatal GABAergic interneurons that express somatostatin and neuropeptide Y immunoreactivity and nitric oxide synthase have also been shown to receive synaptic input to their dendrites from corticostriatal terminals.[54] Although electrophysiological analyses indicate that the cholinergic neurons readily respond to cortical stimulation,[55,56] analysis of choline acetyltransferase (ChAT)–immunostained structures in striatal tissue containing terminals anterogradely labeled from frontal cortex has failed to identify a cortical input.[52] This situation is similar in the nucleus accumbens with respect to the hippocampal input.[57] These findings indicate that cholinergic neurons receive little, if any, synaptic input from the frontal cortex in their proximal regions, that is, the regions of the neurons that were labeled by ChAT immunocytochemistry. It is, of course, possible that other cortical regions make synaptic contact with spiny neurons or indeed that the cortical input occurs in the most distal regions of the dendritic tree that were not immunostained (but see reference Thomas et al.[57a]).

SYNAPTIC ORGANIZATION OF THALAMIC INNERVATION OF THE STRIATUM

The thalamostriatal projection originates mainly from the intralaminar thalamic nuclei, although minor projections arise in other thalamic nuclei (see [58,59]). Similar to corticostriatal projections, the thalamostriatal projections are topographically organized and show heterogeneities in relation to the morphology of the projecting axons, their distribution with respect to the patch/striosome-matrix organization of the striatum, and the distribution of postsynaptic targets in the patches/striosomes and matrix.[38,60–63] Single-cell labeling and tracing studies have identified at least two axonal morphologies of the thalamostriatal projection.[61,64–66]

Ultrastructural analyses of the thalamostriatal system have shown that terminals derived from the thalamus are similar in morphology to cortical terminals, they form asymmetrical synaptic specializations, they are packed with vesicles and they usually contain one or two mitochondria (Fig. 2.4.3B,C).[22,38,57,62,65–71] When considering the projection as a whole, by the analysis of terminals immunolabeled for VGluT2 (see below), it is apparent that the principal targets of the thalamostriatal projections, like those of the corticostriatal projections, are dendritic spines of MSNs. Approximately 60%–75% of VGluT2-positive terminals making synaptic contact with the spines and about 25%–40% making contact with dendritic shafts have been reported.[6,37–39] However, analysis of projections from individual different subnulcei of the thalamus by anterograde labeling[38,66] or by juxtacellular labeling of individual thalamostriatal neurons[65] has revealed that the proportion of terminals contacting spines or dendrites is related to the nucleus from which the projection originates. Hence, boutons derived from the parafascicular nucleus terminate primarily on dendritic shafts,[65,66,71,72] but the precise ratio of spines to shafts varies among individual neurons.[65] Other nuclei giving rise to thalamostriatal projections principally target dendritic spines,[38,65,66,68] and it has been proposed that neurons in the centromedian nucleus of the thalamus, at least, preferentially target MSNs that give rise to the direct pathway.[73] Variability also exists in the ratio of dendritic spines to shafts when considering the patch/striosomes and matrix subcompartments of the striatum.[38,39,62]

In addition to MSNs, thalamostriatal neurons innervate interneurons. Cholinergic interneurons, which possess large perikarya and long, essentially spine-free dendrites, receive asymmetrical synaptic input from terminals derived from the parafascicular nucleus of the thalamus.[67,74] A similar arrangement exists in the nucleus accumbens.[57] Parvalbumin-expressing and neuropeptide Y–expressing, but not calretinin-expressing, GABA interneurons have been shown to receive thalamic input in rat and monkey[74,75] (but see [76]).

INTERACTIONS BETWEEN DOPAMINE AND GLUTAMATE IN STRIATUM

Central to our understanding of basal ganglia function is the concept that the role of dopamine is to modulate transmission at glutamatergic synapses within the striatum. This has traditionally been considered to be a modulatory effect on corticostriatal synapses, but it is also likely to be an effect on thalamostriatal synapses (see below).[77–80] This modulatory effect, or interaction between dopaminergic transmission and glutamatergic transmission, takes many forms. Pharmacological analyses have shown that dopamine can directly influence the release of glutamate at glutamatergic synapses within the striatum. Plasticity of corticostriatal synapses in the form of long-term potentiation and long-term depression is dependent on many factors, including dopamine acting upon the D1 or D2 subtypes of dopamine receptors.[78–85] Plasticity of thalamostriatal synapses may also be dependent on released dopamine.[72,86,87] Furthermore, the loss of dopamine innervation of the striatum leads not only to a loss of spines but also to a loss of excitatory synapses.[88–90]

The anatomical substrates of such interactions have long been considered to be the convergent input of excitatory synapses at the head of dendritic spines of MSNs and dopaminergic synapses at the neck of the spines (see Fig. 2.4.3A).[16,22,23,68,91] This synaptic relationship has been identified for corticostriatal synapses in the dorsal striatum[22,23] and thalamostriatal synapses (derived from the paraventricular nucleus of the thalamus) in the ventral striatum,[92] although evidence from tract tracing studies is lacking in the dorsal striatum[22] (but see below). This triadic arrangement of excitatory input at the head of the dendritic spine and dopamine input at the neck has also been identified in other regions of the brain, including cortex[93] and amygdala.[94] Thus, dopamine acting upon receptors at the necks of spines influences the intracellular signaling pathways initiated by the activation of glutamate receptors at the heads of spines, thereby modulating the responsiveness of the postsynaptic spine to the released glutamate.

QUANTITATIVE ANALYSIS OF THE DOPAMINERGIC INNERVATION OF THE STRIATUM REVEALS THE PRINCIPLES OF INNERVATION

Quantitative analysis of possible sites of interaction between dopaminergic and glutamatergic synapses based on anterograde labeling studies or combined anterograde labeling and immunocytochemical studies is limited by the problems of false-negative labeling of terminals. Anterograde tracing will label only a small proportion of the population of terminals in a pathway. These problems, in relation to the quantitative analysis of the corticostriatal and thalamostriatal pathways, have been overcome by the discovery (that Na$^+$-dependent inorganic phosphate transporters that act as VGluTs)[95–98] selectively label corticostriatal and thalamostriatal terminals, respectively[37–39,72,99] (but see [100,101]). Immunocytochemical analyses using antibodies against these transporters have enabled large parts, if not the whole, of these projections to be studied, and have found that the percentage of axon terminals in the striatum that are derived from the thalamus (25% of asymmetrical synapses) is of the same order of magnitude as that from the cortex (35% of asymmetrical synapses).[37]

We have taken advantage of these markers to define quantitatively the spatial relationship between corticostriatal terminals and dopaminergic axons, as well as that between thalamostriatal terminals and dopaminergic axons.[6] Quantitative electron microscopic analysis was performed on sections of rat striatum immunolabeled to reveal tyrosine hydroxylase, as a marker dopaminergic axons, and either VGluT1 or VGluT2 as markers of corticostriatal and thalamostriatal terminals, respectively (Fig. 2.4.4). The essential findings of the study were as follows:

- The majority of cortical terminals made synaptic contact with dendritic spines (96%), and 20% of the postsynaptic spines were apposed by a dopaminergic axon. In 9% of the cases, the postsynaptic structure received synaptic input from the dopaminergic axon.
- Like the cortical terminals, the majority of thalamic terminals made synaptic contact with dendritic spines (71%) and, similar to the cortical terminals, 27% of the structures postsynaptic to the thalamic terminals (spines and dendrites) were also apposed by a dopaminergic axon. In 9% of the cases, the dopaminergic axon formed a synapse with the structure postsynaptic to the thalamic terminal.
- Randomly selected cellular profiles within the striatum, when corrected for the length of their perimeter within the electron micrographs, have a probability of being apposed by, or in synaptic contact with, a dopaminergic axon similar to that of the spines and dendrites postsynaptic to cortical and thalamic terminals.
- Similarly, glutamatergic synaptic terminals from the cortex or thalamus, when corrected for their size, have a probability of being apposed by a dopaminergic axon similar to that of the structures postsynaptic to them.

These results demonstrate that a proportion of those spines and dendrites postsynaptic to cortical terminals receive synaptic input from dopaminergic terminals, and that this is also the case for structures postsynaptic to thalamic terminals (Fig. 2.4.5). There are several important implications of these observations:

- The anatomical substrate for the interaction of dopamine and glutamate applies equally to corticostriatal and thalamostriatal synapses.
- The plasticity of thalamostriatal synapses is therefore likely to have the same degree of dependency upon dopamine as the plasticity of corticostriatal synapses.
- Since the frequency of the spatial relationship between an excitatory synapse and a dopaminergic synapse was the same for randomly selected cellular profiles and a dopaminergic synapse, the relationship between dopaminergic synapses and glutamatergic synapses is unlikely to be a selective

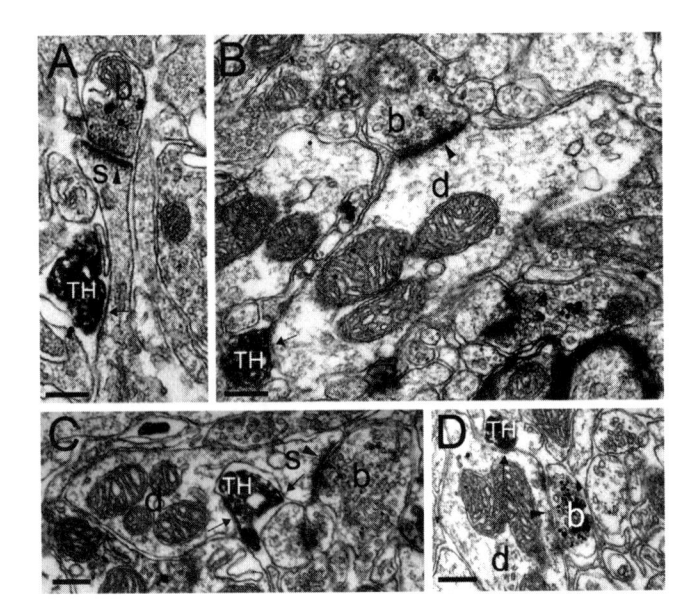

FIGURE 2.4.4. The spatial relationship between excitatory synapses and dopaminergic synapses in the striatum. Corticostriatal terminals were revealed by immunogold labeling for VGluT1, thalamostriatal terminals by immunogold labeling for VGluT2, and dopaminergic axons by immunoperoxidase labeling for tyrosine hydroxylase (TH). (A) A VGluT1-positive bouton (b; corticostriatal) makes asymmetrical synaptic contact (arrowhead) with the head of a long, thin spine (s). A TH-positive terminal (TH; dopaminergic) makes symmetrical synaptic contact (arrow) with the neck of the same spine. (B) A VGluT1-positive bouton (b; corticostriatal) makes asymmetrical synaptic contact with a dendritic shaft (d) that is apposed (arrow) by a TH-positive axon (TH; dopaminergic). Note the additional corticostriatal terminal forming a synapse with a spine at the bottom right of this micrograph. (C) A VGluT2-positive bouton (b; thalamostriatal) makes asymmetrical synaptic contact (arrowhead) with a spine (s) that arises from a dendritic shaft (d). A TH-positive bouton (TH; dopaminergic) makes symmetrical synaptic contact (arrows) with both the spine (s) and the dendritic shaft (d). (D) A VGluT2-positive bouton (b; thalamostriatal) makes asymmetrical synaptic contact (arrowhead) with a dendritic shaft (d) that is in symmetrical synaptic contact (arrow) with a TH-positive terminal (TH; dopaminergic). Scale bars: 200 nm. *Source:* Data modified from Moss and Bolam.[6]

or targeted phenomenon. The chance of a striatal structure being apposed by, or in synaptic contact with, a dopaminergic axon seems to be solely dependent on the size of the structure. The spatial relationship between dopaminergic axons and glutamatergic synapses simply relates to the size of the presynaptic and postsynaptic structures, not to their phenotype.

- If the role of dopamine in the modulation of glutamatergic transmission is so critical to our understanding of the function of the striatum and the basal ganglia in general, then the question arises as to why such a small proportion (9%) of the structures postsynaptic to glutamatergic synapses also receive synaptic input from a dopaminergic terminal. Are the other glutamatergic synapses in the striatum not modulated by dopamine?

One possible explanation is that, in addition to synaptic transmission, dopaminergic transmission may also occur by *volume transmission* as a consequence of spillover of synaptically released dopamine or the release of dopamine at nonsynaptic sites.[102-106] The "sphere of influence" of released dopamine is likely to depend on many factors, including quantal size and the density and distribution of dopamine transporters and receptors. It has been proposed that the sphere of influence of dopamine spillover in a concentration sufficient to stimulate dopamine receptors has a radius of 2-8 μm.[103] A precise synaptic relationship between a glutamatergic terminal and a dopaminergic terminal may thus not be necessary for released dopamine to modulate the strength of a glutamatergic synapse. In order to address this and to see, on average, how close dopaminergic structures are to glutamatergic synapses, we examined the proximity of dopaminergic axons and synapses to glutamatergic synapses.[6] Using the same data set described above, we found that every glutamatergic synapse is within 0.5 μm of a dopaminergic axon and within about 1 μm of a dopaminergic *synapse* (Table 2.4.2). In view of the estimates of the distance that dopamine may diffuse from the synapse, these findings suggest that every structure, including glutamatergic synapses, will be within overlapping spheres of influence of synaptically released dopamine. Thus, all glutamatergic synapses are likely to be within reach of a concentration of dopamine high enough to stimulate both high- and low-affinity receptors.[103] Efficacy of transmission will thus depend on the density and distribution of extrasynaptic dopamine receptors[43,107-109] and, of course, will have different temporal characteristics to synaptic transmission.

It should be noted that, as one would predict from the analysis described above, all randomly selected cellular profiles within the striatum are also located within about 0.5 μm of a dopaminergic axon and within about 1 μm of a dopaminergic synapse (Table 2.4.2). This reinforces the idea that there is no selectivity in the dopaminergic nigrostriatal pathway; rather, the potential for functional connectivity is dependent on the size of the target structure, the density of the projection, and the particular axon involved.[5] *Functional* connectivity thus depends almost entirely on the density, distribution, and location of dopamine receptors.

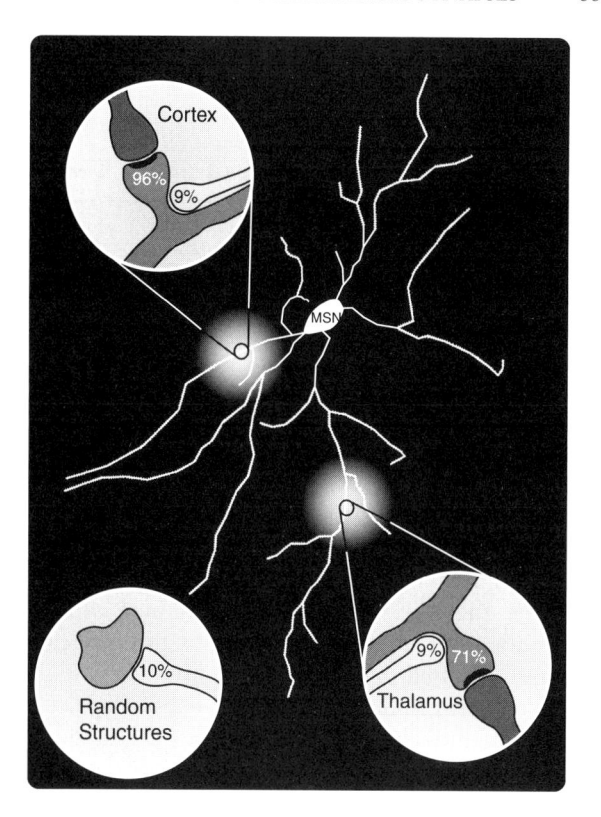

FIGURE 2.4.5. Summary diagram of the convergence of glutamatergic and dopaminergic signals in the striatum and its nonselective nature. Cortical and thalamic afferents to the striatum (red) make asymmetrical synaptic contact with dendritic structures (blue) of a medium-sized spiny projection neuron (MSN, white). The majority of these contacts are with dendritic spines (cortical, 96%; thalamic, 71%), of which 9% receive a second input from a dopaminergic axon from the substantia nigra pars compacta (yellow). This, however, is no different from the proportions of random striatal structures (green) contacted by dopaminergic axons (10%), which demonstrates the nonselective nature of the relationship. In addition, dopamine (yellow clouds) spill over from the synapse and diffuse in concentrations capable of activating dopamine receptors for up to 8 μm. (See Color Plate 2.4.5.)

CONCLUDING COMMENTS

Dopamine neurons are remarkable in their complexity: a small population of neurons gives rise to a phenomenally dense innervation of the striatum, and individual neurons have vast axonal arbors that give rise to hundreds of thousands of synapses. The organization of what is central to basal ganglia function (i.e., the interaction between dopamine and glutamate) is such that striatal neurons are embedded in a dense network of dopamine axons and every structure has a similar probability of being apposed by, or in synaptic contact with, a dopaminergic axon. Furthermore, every structure in the striatum is within overlapping spheres of

TABLE 2.4.2. *The Proximity of Cortical Synapses, Thalamic Synapses, and Random Points in the Striatum to Dopaminergic Axons and Synapses*

	Proximity to Dopaminergic Axons		Proximity to Dopaminergic Synapses	
	Proportion within 0.5 μm	Average distance between	Proportion within 0.5 μm	Average distance between
Cortical Synapses	108%	0.49 μm	20%	0.85 μm
Thalamic Synapses	104%	0.49 μm	10%	1.08 μm
Random Structures	96%	0.51 μm	11%	1.04 μm

Source: Data derived from the quantitative analysis of Moss and Bolam.[6]

influence of synaptically released dopamine that may spill over and diffuse from the synapse. These structural characteristics thus underlie the phasic actions of dopamine at synapses, presumably in response to bursts of activity of dopamine neurons. They also underlie the tonic effects of dopamine, which are likely to occur as a consequence of tonic release at synapses, as well as the diffuse spillover of dopamine from synapses and possibly nonsynaptic sites. Given these structural characteristics, the critical factor in the expression of tonic dopamine function is not the specific location of dopamine synapses, but rather the distribution and density of dopamine receptors that, like most metabotropic receptors, are located at both synaptic and extrasynaptic sites.[43,107–109]

ACKNOWLEDGMENTS

The authors' work described in this chapter was supported by the Medical Research Council, The European community (FP7 project number 201716) and The Parkinson's Disease Society (UK). JM was in receipt of a Medical Research Council studentship.

REFERENCES

1. Oorschot DE. Total number of neurons in the neostriatal, pallidal, subthalamic, and substantia nigral nuclei of the rat basal ganglia: a stereological study using the Cavalieri and optical disector methods. *J Comp Neurol.* 1996;366(4):580–599.
2. Nair-Roberts RG, Chatelain-Badie SD, Benson E, White-Cooper H, Bolam JP, Ungless MA. Stereological estimates of dopaminergic, GABAergic and glutamatergic neurons in the ventral tegmental area, substantia nigra and retrorubral field in the rat. *Neuroscience.* 2008;152(4):1024–1031.
3. Anden NE, Fuxe K, Hamberger B, Hökfelt T. A quantitative study on the nigro-neostriatal dopamine neuron system in the rat. *Acta Physiol Scand.* 1966;67(3):306–312.
4. Björklund A, Lindvall O. Dopamine-containing systems in the CNS. In: Björklund A, Hökfelt T, eds. *Handbook of Chemical Neuroanatomy.* Amsterdam: Elsevier; 1984:2:55–122.
5. Matsuda W, Furuta T, Nakamura KC, et al. Single nigrostriatal dopaminergic neurons form widely spread and highly dense axonal arborizations in the neostriatum. *J Neurosci.* 2009;29(2):444–453.
6. Moss J, Bolam JP. A dopaminergic axon lattice in the striatum and its relationship with cortical and thalamic terminals. *J Neurosci.* 2008;28(44):11221–11230.
7. Wickens J, Arbuthnott GW. Structural and functional interactions in the striatum at the receptor level. In: Dunnett SB, Bentivoglio M, Björklund A, Hökfelt T, eds. *Handbook of Chemical Neuroanatomy.* Amsterdam: Elsevier; 2004:21: 199–236.
8. Bevan MD, Clarke NP, Bolam JP. Synaptic integration of functionally diverse pallidal information in the entopeduncular nucleus and subthalamic nucleus in the rat. *J Neurosci.* 1997;17(1):308–324.
9. Kita H, Kitai ST. The morphology of globus pallidus projection neurons in the rat: an intracellular staining study. *Brain Res.* 1994;636:308–319.
10. Kawaguchi Y, Wilson CJ, Emson PC. Projection subtypes of rat neostriatal matrix cells revealed by intracellular injection of biocytin. *J. Neurosci.* 1990;10:3421–3438.
11. Wu Y, Richard S, Parent A. The organization of the striatal output system: a single-cell juxtacellular labeling study in the rat. *Neurosci Res.* 2000;38(1):49–62.
12. Tepper JM, Bolam JP. Functional diversity and specificity of neostriatal interneurons. *Curr Opin Neurobiol.* 2004;14(6): 685–692.
13. Koos T, Tepper JM. Inhibitory control of neostriatal projection neurons by GABAergic interneurons. *Nat Neurosci.* 1999;2:467–472.
14. Prensa L, Parent A. The nigrostriatal pathway in the rat: a single-axon study of the relationship between dorsal and ventral tier nigral neurons and the striosome/matrix striatal compartments. *J Neurosci.* 2001;21:7247–7260.
15. Cragg SJ, Baufreton J, Xue Y, Bolam JP, Bevan MD. Synaptic release of dopamine in the subthalamic nucleus. *Eur J Neurosci.* 2004;20(7):1788–1802.
16. Freund TF, Powell J, Smith AD. Tyrosine hydroxylase-immunoreactive boutons in synaptic contact with identified striatonigral neurons, with particular reference to dendritic spines. *Neuroscience.* 1984;13:1189–1215.
17. Pickel VM, Beckley SC, Joh TH, Reis DJ. Ultrastructural immunocytochemical localization of tyrosine hydroxylase in the neostriatum. *Brain Res.* 1981;225:373–385.

18. Groves PM, Linder JC, Young SJ. 5-Hydroxydopamine-labeled dopaminergic axons: three-dimensional reconstructions of axons, synapses and postsynaptic targets in rat neostriatum. *Neuroscience.* 1994;58:593–604.
19. Hanley JJ, Bolam JP. Synaptology of the nigrostriatal projection in relation to the compartmental organization of the neostriatum in the rat. *Neuroscience.* 1997;81(2):353–370.
20. Zahm DS. An electron microscopic morphometric comparison of tyrosine hydroxylase immunoreactive in the neostriatum and the nucleus accumbens core and shell. *Brain Res.* 1992;575:341–346.
21. Descarries L, Watkins KC, Garcia S, Bosler O, Doucet G. Dual character, asynaptic and synaptic, of the dopamine innervation in adult rat neostriatum: a quantitative autoradiographic and immunocytochemical analysis. *J Comp Neurol.* 1996;375(2): 167–186.
22. Smith Y, Bennett BD, Bolam JP, Parent A, Sadikot AF. Synaptic relationships between dopaminergic afferents and cortical or thalamic input in the sensorimotor territory of the striatum in monkey. *J Comp Neurol.* 1994;344:1–19.
23. Bouyer JJ, Park DH, Joh TH, Pickel VM. Chemical and structural analysis of the relation between cortical inputs and tyrosine hydroxylase–containing terminals in rat neostriatum. *Brain Res.* 1984;302:267–275.
24. Gerfen CR, Wilson CJ. The basal ganglia. In: Swanson LW, Björklund A, Hökfelt T, eds. *Handbook of Chemical Neuroanatomy.* Amsterdam: Elsevier; 1996:12:371–468.
25. Ragsdale CW, Graybiel AM. The fronto-striatal projection in the cat and monkey and its relationship to inhomogeneities established by acetyl-cholinesterase histochemistry. *Brain Res.* 1981;208:259–266.
26. Donoghue JP, Herkenham M. Neostriatal projections from individual cortical fields conform to histochemically distinct striatal compartments in the rat. *Brain Res.* 1986;397:397–403.
27. Malach R, Graybiel AM. Mosaic architecture of the somatic sensory-recipient sector of the cat's striatum. *J Neurosci.* 1986;6:3436–3458.
28. Gerfen CR. The neostriatal mosaic: compartmentalization of corticostriatal input and striatonigral output systems. *Nature.* 1984;311:461–464.
29. Gerfen CR. The neostriatal mosaic: striatal patch-matrix organization is related to cortical lamination. *Science.* 1989;246:385–388.
30. Berendse HW, Galis-de-Graaf Y, Groenewegen HJ. Topographical organization and relationship with ventral striatal compartments of prefronto corticostriatal projections in the rat. *J Comp Neurol.* 1992;316:314–347.
31. Kincaid AE, Wilson CJ. Corticostriatal innervation of the patch and matrix in the rat neostriatum. *J Comp Neurol.* 1996; 374(4):578–592.
32. Graybiel AM. Network-level neuroplasticity in cortico-basal ganglia pathways. *Parkinsonism Relat Disord.* 2004; 10(5):293–296.
33. Flaherty AW, Graybiel AM. Input-output organization of the sensorimotor striatum in the squirrel monkey. *J Neurosci.* 1994;14:599–610.
34. Parthasarathy HB, Graybiel AM. Cortically driven immediate-early gene expression reflects modular influence of sensorimotor cortex on identified striatal neurons in the squirrel monkey. *J. Neurosci.* 1997;17(7):2477–2491.
35. Parthasarathy HB, Schall JD, Graybiel AM. Distributed but convergent ordering of corticostriatal projections: analysis of the frontal eye field and the supplementary eye field in the macaque monkey. *J. Neurosci.* 1992;12:4468–4488.
36. Smith Y, Bevan MD, Shink E, Bolam JP. Microcircuitry of the direct and indirect pathways of the basal ganglia. *Neuroscience.* 1998;86(2):353–387.
37. Lacey CJ, Boyes J, Gerlach O, Chen L, Magill PJ, Bolam JP. GABA(B) receptors at glutamatergic synapses in the rat striatum. *Neuroscience.* 2005;136(4):1083–1095.
38. Raju DV, Shah DJ, Wright TM, Hall RA, Smith Y. Differential synaptology of vGluT2-containing thalamostriatal afferents between the patch and matrix compartments in rats. *J Comp Neurol.* 2006;499(2):231–243.
39. Fujiyama F, Unzai T, Nakamura K, Nomura S, Kaneko T. Difference in organization of corticostriatal and thalamostriatal synapses between patch and matrix compartments of rat neostriatum. *Eur J Neurosci.* 2006;24(10):2813–2824.
40. Frotscher M, Rinne U, Hassler R, Wagner A. Termination of cortical afferents on identified neurons in the caudate nucleus of the cat. A combined Golgi-EM degeneration study. *Exp Brain Res.* 1981;41:329–337.
41. Somogyi P, Bolam JP, Smith AD. Monosynaptic cortical input and local axon collaterals of identified striatonigral neurons. A light and electron microscopic study using the Golgi-peroxidase transport-degeneration procedure. *J Comp Neurol.* 1981;195:567–584.
42. Kemp JM, Powell TPS. The structure of the caudate nucleus of the cat: light and electron microscopy. *Philos Trans R Soc Lond.* 1971;B 262:383–401.
43. Hersch SM, Ciliax BJ, Gutekunst CA, et al. Electron microscopic analysis of D1 and D2 dopamine receptor proteins in the dorsal striatum and their synaptic relationships with motor corticostriatal afferents. *J Neurosci.* 1995;15:5222–5237.
44. Lei W, Jiao Y, Del Mar N, Reiner A. Evidence for differential cortical input to direct pathway versus indirect pathway striatal projection neurons in rats. *J Neurosci.* 2004;24(38):8289–8299.
45. Kincaid AE, Zheng T, Wilson CJ. Connectivity and convergence of single corticostriatal axons. *J Neurosci.* 1998;18(12):4722–4731.
46. Zheng T, Wilson CJ. Corticostriatal combinatorics: the implications of corticostriatal axonl arborizations. *J Neurophysiol.* 2002;87:1007–1017.
47. Cowan RL, Wilson CJ, Emson PC, Heizmann CW. Parvalbumin-containing GABAergic interneurons in the rat neostriatum. *J Comp Neurol.* 1990;302:197–205.
48. Kita H, Kosaka T, Heizmann CW. Parvalbumin-immunoreactive neurons in the rat neostriatum: a light and electron microscopic study. *Brain Res.* 1990;536:1–15.
49. Kawaguchi Y. Physiological, morphological, and histochemical characterization of three classes of interneurons in rat neostriatum. *J Neurosci.* 1993;13:4908–4923.
50. Tepper JM, Koos T, Wilson CJ. GABAergic microcircuits in the neostriatum. *Trends Neurosci.* 2004;27(11):662–669.
51. Bennett BD, Bolam JP. Synaptic input and output of parvalbumin-immunoreactive neurones in the neostriatum of the rat. *Neuroscience.* 1994; 62:707–719.
52. Lapper SR, Smith Y, Sadikot AF, Parent A, Bolam JP. Cortical input to parvalbumin-immunoreactive neurones in the putamen of the squirrel monkey. *Brain Res.* 1992;580: 215–224.
53. Ramanathan S, Hanley JJ, Deniau J-M, Bolam JP. Synaptic convergence of motor and somatosensory cortical afferents onto GABAergic interneurons in the rat striatum. *J Neurosci.* 2002; 22:8158–8169.

54. Vuillet J, Kerkerian L, Kachidian P, Bosler O, Nioeullon A. Ultrastructural correlates of functional relationships between nigral dopaminergic or cortical afferent fibres and neuropeptide Y–containing neurons in the rat striatum. *Neurosci Lett.* 1989;100:99–104.
55. Wilson CJ, Chang HT, Kitai ST. Firing patterns and synaptic potentials of identified giant aspiny interneurons in the rat neostriatum. *J Neurosci.* 1990;10(2):508–519.
56. Reynolds JN, Wickens JR. The corticostriatal input to giant aspiny interneurons in the rat: a candidate pathway for synchronising the response to reward-related cues. *Brain Res.* 2004;1011(1):115–128.
57. Meredith GE, Wouterlood FG. Hippocampal and midline thalamic fibres and terminals in relation to the choline acetyltransferase–immunoreactive neurons in nucleus accumbens of the rat: a light and electron microscopic study. *J Comp Neurol.* 1990;296:204–221.
57a. Thomas TM, Smith Y, Levey AI, Hersch SM. Cortical inputs to m2-immunoreactive striatal interneurons in rat and monkey. *Synapse.* 2000; 37(4):252–261.
58. Smith Y, Raju DV, Pare JF, Sidibé M. The thalamostriatal system: a highly specific network of the basal ganglia circuitry. *Trends Neurosci.* 2004;27(9):520–527.
59. Groenewegen HJ, Berendse HW. The specificity of the 'nonspecific' midline and intralaminar thalamic nuclei. *Trends Neurosci.* 1994;17(2):52–57.
60. Herkenham M, Pert CB. Mosaic distribution of opiate receptors, parafascicular projections and acetylcholinesterase in rat striatum. *Nature.* 1981;291:415–418.
61. Deschênes M, Bourassa J, Doan VD, Parent A. A single-cell study of the axonal projections arising from the posterior intralaminar thalamic nuclei in the rat. *Eur J Neurosci.* 1996;8(2):329–343.
62. Sadikot AF, Parent A, Smith Y, Bolam JP. Efferent connections of the centromedian and parafascicular nuclei in the squirrel monkey. A light and electron microscopic study of the thalamostriatal projection in relation to striatal heterogeneity. *J Comp Neurol.* 1992;320:228–242.
63. Ragsdale CW, Graybiel AM. Compartmental organization of the thalamostriatal connection in the cat. *J Comp Neurol.* 1991;311:134–167.
64. Deschênes M, Bourassa J, Parent A. Two different types of thalamic fibers innervate the rat striatum. *Brain Res.* 1995;701(1–2):288–292.
65. Lacey CJ, Bolam JP, Magill PJ. Novel and distinct operational principles of intralaminar thalamic neurons and their striatal projections. *J Neurosci.* 2007;27(16):4374–4384.
66. Xu ZC, Wilson CJ, Emson PC. Restoration of thalamostriatal projections in rat neostriatal grafts: An electron microscopic analysis. *J Comp Neurol.* 1991;303:22–34.
67. Lapper SR, Bolam JP. Input from the frontal cortex and the parafascicular nucleus to cholinergic interneurones in the dorsal striatum of the rat. *Neuroscience.* 1992;51:533–545.
68. Kemp JM, Powell TPS. The site of termination of afferent fibres in the caudate nucleus. *Philos Trans R Soc Lond.* 1971;B 262:413–427.
69. Sidibé M, Smith Y. Differential synaptic innervation of striatofugal neurones projecting to the internal or external segments of the globus pallidus by thalamic afferents in the squirrel monkey. *J Comp Neurol.* 1996;365(3):445–465.
70. Chung JW, Hassler R, Wagner A. Degeneration of two of nine types of synapses in the putamen after center median coagulation in the cat. *Exp Brain Res.* 1977;28:345–361.
71. Dubé L, Smith AD, Bolam JP. Identification of synaptic terminals of thalamic or cortical origin in contact with distinct medium size spiny neurons in the rat neostriatum. *J Comp Neurol.* 1988; 267:455–471.
72. Raju DV, Ahern TH, Shah DJ, et al. Differential synaptic plasticity of the corticostriatal and thalamostriatal systems in an MPTP-treated monkey model of parkinsonism. *Eur J Neurosci.* 2008;27(7):1647–1658.
73. Sidibé M, Smith Y. Differential synaptic innervation of striatofugal neurones projecting to the internal or external segments of the globus pallidus by thalamic afferents in the squirrel monkey. *J Comp Neurol.* 1996;365(3):445–465.
74. Sidibé M, Smith Y. Thalamic inputs to striatal interneurons in monkeys: synaptic organization and co-localization of calcium binding proteins. *Neuroscience.* 1999;89:1189–1208.
75. Rudkin TM, Sadikot AF. Thalamic input to parvalbumin-immunoreactive GABAergic interneurons: organization in normal striatum and effect of neonatal decortication. *Neuroscience.* 1999;88(4):1165–1175.
76. Kachidian P, Vuillet J, Nieoullon A, Lafaille G, Goff LK-L. Striatal neuropeptide Y neurones are not a target for thalamic afferent fibres. *Neuroreport.* 1996;7(10):1665–1669.
77. Bolam JP, Bergman H, Graybiel A, et al. Molecules, microcircuits and motivated behaviour: microcircuits in the striatum. In: Grillner S, Graybiel A, eds. *Microcircuits: The Interface between Neurons and Global Brain Function.* Dahlem Workshop Report 93. Cambridge, MA: MIT Press; 2006:165–190.
78. Surmeier DJ, Ding J, Day M, Wang Z, Shen W. D1 and D2 dopamine-receptor modulation of striatal glutamatergic signaling in striatal medium spiny neurons. *Trends Neurosci.* 2007;30(5):228–235.
79. Calabresi P, Picconi B, Tozzi A, Di Filippo M. Dopamine-mediated regulation of corticostriatal synaptic plasticity. *Trends Neurosci.* 2007;30(5):211–219.
80. Shen W, Flajolet M, Greengard P, Surmeier DJ. Dichotomous dopaminergic control of striatal synaptic plasticity. *Science.* 2008;321(5890):848–851.
81. Reynolds JN, Hyland BI, Wickens JR. A cellular mechanism of reward-related learning. *Nature.* 2001;413(6851):67–70.
82. Reynolds JN, Wickens JR. Substantia nigra dopamine regulates synaptic plasticity and membrane potential fluctuations in the rat neostriatum in vivo. *Neuroscience.* 2000;99(2):199–203.
83. Reynolds JN, Wickens JR. Dopamine-dependent plasticity of corticostriatal synapses. *Neural Netw.* 2002;15(4–6):507–521.
84. Wickens JR, Budd CS, Hyland BI, Arbuthnott GW. Striatal contributions to reward and decision making: making sense of regional variations in a reiterated processing matrix. *Ann NY Acad Sci.* 2007;1104:192–212.
85. Wickens JR, Reynolds JN, Hyland BI. Neural mechanisms of reward-related motor learning. *Curr Opin Neurobiol.* 2003;13(6):685–690.
86. Smeal RM, Gaspar RC, Keefe KA, Wilcox KS. A rat brain slice preparation for characterizing both thalamostriatal and corticostriatal afferents. *J Neurosci Methods.* 2007;159(2): 224–235.
87. Ding J, Peterson JD, Surmeier DJ. Corticostriatal and thalamostriatal synapses have distinctive properties. *J Neurosci.* 2008; 28(25):6483–6492.
88. Ingham CA, Hood SH, Taggart P, Arbuthnott GW. Plasticity of synapses in the rat neostriatum after unilateral lesion of the nigrostriatal dopaminergic pathway. *J Neurosci.* 1998; 18:4732–4743.

89. Day M, Wang Z, Ding J, et al. Selective elimination of glutamatergic synapses on striatopallidal neurons in Parkinson disease models. *Nat Neurosci.* 2006;9(2):251–259.
90. Ingham CA, Hood SH, Arbuthnott GW. Spine density on neostriatal neurones changes with 6-hydroxydopamine lesions and with age. *Brain Res.* 1989;503:334–338.
91. Smith AD, Bolam JP. The neural network of the basal ganglia as revealed by the study of synaptic connections of identified neurones. *Trends Neurosci.* 1990;13:259–265.
92. Pinto A, Jankowski M, Sesack SR. Projections from the paraventricular nucleus of the thalamus to the rat prefrontal cortex and nucleus accumbens shell: ultrastructural characteristics and spatial relationships with dopamine afferents. *J Comp Neurol.* 2003;459(2):142–155.
93. Goldman-Rakic PS, Leranth C, Williams SM, Mons N, Geffard M. Dopamine synaptic complex with pyramidal neurons in primate cerebral cortex. *Proc Natl Acad Sci USA.* 1989;86:9015–9019.
94. Pinto A, Sesack SR. Ultrastructural analysis of prefrontal cortical inputs to the rat amygdala: spatial relationships to presumed dopamine axons and D1 and D2 receptors. *Brain Struct Funct.* 2008;213(1–2):159–175.
95. Ni B, Wu X, Yan GM, Wang J, Paul SM. Regional expression and cellular localization of the Na(+)-dependent inorganic phosphate cotransporter of rat brain. *J Neurosci.* 1995;15(8): 5789–5799.
96. Otis TS, Kavanaugh MP. Isolation of current components and partial reaction cycles in the glial glutamate transporter EAAT2. *J Neurosci.* 2000;20(8):2749–2757.
97. Bellocchio EE, Reimer RJ, Fremeau RT Jr, Edwards RH. Uptake of glutamate into synaptic vesicles by an inorganic phosphate transporter. *Science.* 2000;289(5481):957–960.
98. Fremeau RT Jr, Troyer MD, Pahner I, et al. The expression of vesicular glutamate transporters defines two classes of excitatory synapse. *Neuron.* 2001;31(2):247–260.
99. Fujiyama F, Kuramoto E, Okamoto K, et al. Presynaptic localization of an AMPA-type glutamate receptor in corticostriatal and thalamostriatal axon terminals. *Eur J Neurosci.* 2004; 20(12):3322–3330.
100. Barroso-Chinea P, Castle M, Aymerich MS, Lanciego JL. Expression of vesicular glutamate transporters 1 and 2 in the cells of origin of the rat thalamostriatal pathway. *J Chem Neuroanat.* 2008;35(1):101–107.
101. Barroso-Chinea P, Castle M, Aymerich MS, et al. Expression of the mRNAs encoding for the vesicular glutamate transporters 1 and 2 in the rat thalamus. *J Comp Neurol.* 2007;501(5):703–715.
102. Agnati LF, Zoli M, Stromberg I, Fuxe K. Intercellular communication in the brain: wiring versus volume transmission. *Neuroscience.* 1995;69(3):711–726.
103. Rice ME, Cragg SJ. Dopamine spillover after quantal release: rethinking dopamine transmission in the nigrostriatal pathway. *Brain Res Rev.* 2008;58(2):303–313.
104. Cragg SJ, Rice ME. DAncing past the DAT at a DA synapse. *Trends Neurosci.* 2004;27(5):270–277.
105. Arbuthnott GW, Wickens J. Space, time and dopamine. *Trends Neurosci.* 2007;30(2):62–69.
106. Gonon F. Prolonged and extrasynaptic excitatory action of dopamine mediated by D1 receptors in the rat striatum in vivo. *J Neurosci.* 1997;17(15):5972–5978.
107. Delle Donne KT, Sesack SR, Pickel VM. Ultrastructural immunocytochemical localization of the dopamine D2 receptor within GABAergic neurons of the rat striatum. *Brain Res.* 1997;746(1–2):239–255.
108. Sesack SR, Aoki C, Pickel VM. Ultrastructural localization of D2 receptor-like immunoreactivity in midbrain dopamine neurons and their striatal targets. *J Neurosci.* 1994; 14(1):88–106.
109. Yung KKL, Bolam JP, Smith AD, Hersch SM, Ciliax BJ, Levey AI. Immunocytochemical localization of D1 and D2 dopamine receptors in the basal ganglia of the rat: light and electron microscopy. *Neuroscience.* 1995;65:709–730.

3 | Molecular pharmacology

3.1 | Molecular Pharmacology of the Dopamine Receptors

MICHELE L. RANKIN, LISA A. HAZELWOOD, R. BENJAMIN FREE,
YOON NAMKUNG, ELIZABETH B. REX, REBECCA A. ROOF, AND
DAVID R. SIBLEY

INTRODUCTION

Dopamine receptors are rhodopsin-like seven-transmembrane receptors (also called *G protein–coupled receptors*) that mediate the central and peripheral actions of dopamine. Dopamine receptors are most abundant in pituitary and brain, particularly in the basal forebrain, but are also found in the retina and in peripheral organs such as the kidney. Stimulation of dopamine receptors modulates natriuresis in the kidney, as well as cell division and hormone synthesis and secretion in the pituitary. Brain dopamine receptors regulate movement and locomotion, motivation, and working memory. Five subtypes of mammalian dopamine receptors have been identified that are divided into D1-like (D1, D5) or D2-like (D2, D3, D4) subgroups. The D1-like receptors couple primarily to the G_s family of G proteins (G_s and G_{olf}), whereas the D2-like receptors couple primarily to the $G_{i/o}$ family. This chapter covers the molecular pharmacology of the five dopamine receptor subtypes.

THE D1 DOPAMINE RECEPTOR SUBTYPE

The D1 dopamine receptor (D1DAR; Fig. 3.1.1) subtype (also called the D1A subtype in some literature) belongs to the D1-like family of dopamine receptors, and it is the discovery and characterization of this receptor that gives this family its name.[1–3] The D1DAR (GenBank Accession NP_000785; located on chromosome 5q35.1) exhibits the most conserved sequence of all the DARs, featuring an extended C-terminus and a shortened third intracellular loop (ICL3) when compared to the D2-like receptor family. Of all the individual DAR knockout mice, the D1DAR knockout exhibits the most severe phenotypes, including spatial learning deficits,[4] hyperactivity,[5] and abnormal memory retention,[6] emphasizing the functional importance of this receptor subtype. The D5DAR (also called D1B) is the remaining receptor subtype that comprises the D1-like family and will be discussed in a later section of this chapter.

D1DAR Structure

The D1DAR belongs to the class A or rhodopsin family of G protein–coupled receptors (GPCRs).[7] The past couple of years have yielded some exciting advances in the elucidation of the structural topography of GPCRs. Although GPCRs are notoriously difficult to crystallize, there are now three crystal structures available for this class of receptors: rhodopsin,[8] the β_1-adrenergic receptor,[9] and the β_2-adrenergic receptor.[10,11] The majority of the data gleaned from these structural studies reveals the details of the intramembrane helical interfaces, whereas the more mobile intracellular loops and carboxyl tail region are too disordered to discern. These data provide insight into how ligand binding occurs and alters the protein structure to transduce the activated receptor conformation to activation of G protein and downstream signaling events.[12,13] Given that all clinically used antipsychotic drugs bind to and antagonize dopamine receptors,[14] structural studies will continue to play a major role in research and development for improved therapies that involve the dopaminergic signaling system.

Posttranslational modifications

Many GPCRs have been shown to be palmitoylated at cysteine residues in the carboxyl tail region. This covalent modification results in the formation of a fourth intracellular loop that is folded into an amphiphilic α-helix that lies parallel to the intracellular surface of the plasma membrane.[15,16] Because palmitoylation is a reversible modification and can provide a mechanism for membrane localization, it has been proposed to participate in membrane association and/or trafficking.[16] In addition, palmitoylation

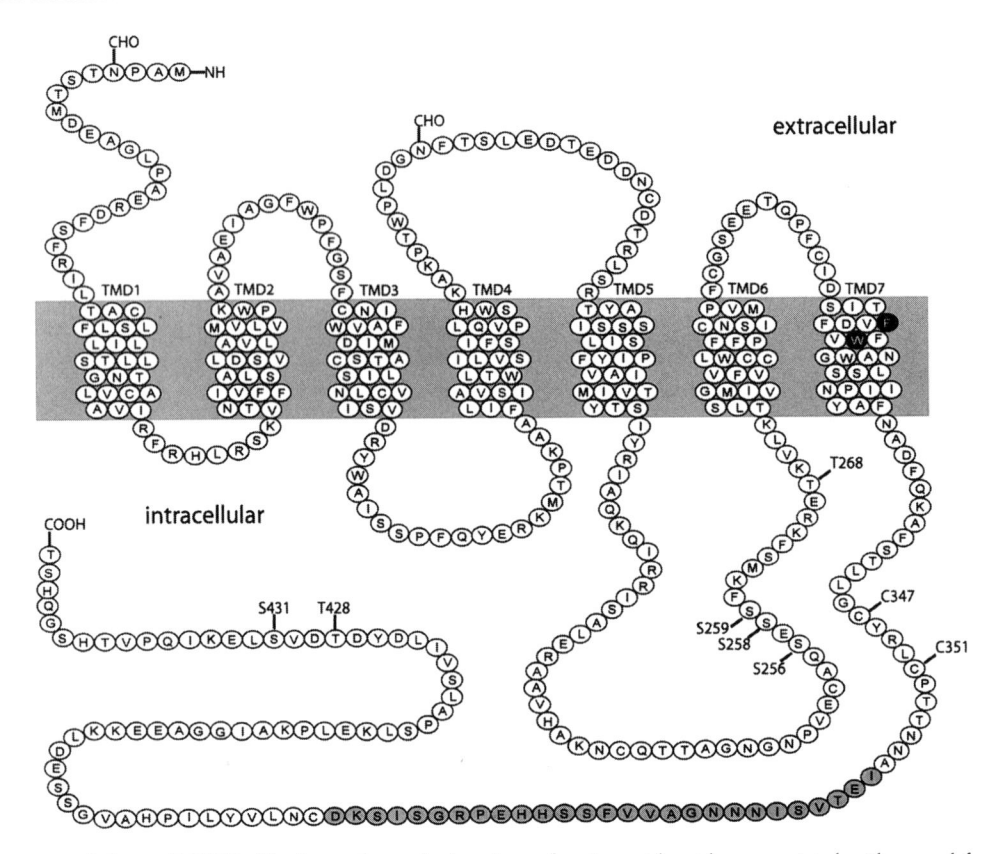

FIGURE 3.1.1. Structure of the rat D1DAR. The figure shows the locations of amino acid residues associated with several functional features. Glycosylation of Asn residues is indicated by "–CHO".[20] Ser256, Ser258, and Ser259 are essential for efficient arrestin association.[40] Thr268 is associated with the rate of desensitization and receptor trafficking.[111,269] Phe313 and Trp318, located in TMD7, are essential for caveolae-mediated endocytosis.[35] Cys347 and Cys351 are palmitoylated.[19] The region within the carboxyl tail represented by gray residues is essential for efficient endocytic recycling.[270] Thr428 and Ser431 are constitutively phosphorylated by GRK4α.[72] See text for details.

of the β2-adrenergic receptor has been shown to govern the accessibility of a protein kinase A (PKA) phosphorylation site in the carboxyl tail that affects desensitization of this receptor.[17,18] The D1DAR is palmitoylated on Cys347 and Cys351 in the proximal portion of the carboxyl tail.[19] The functional significance of this modification remains unknown.

Efficient plasma membrane localization of several GPCRs has been shown to be dependent on N-linked glycosylation. Indeed, the D1DAR is glycosylated at Asn5 and Asn175; however, chemical or mutational inhibition of glycosylation of the D1DAR has no effect on proper targeting of the receptor to the cell surface. In contrast, glycosylation of the D5DAR is essential to proper cell surface trafficking and concomitant ligand binding.[20]

Functional domains of D1DARs

Numerous studies have been performed on the D1-like and D2-like receptors using a combination of mutational strategies and pharmacological approaches to correlate structural domains within receptor proteins to function and subtype specificity. Studies focused on D1-like receptors utilized D1/D5 chimeras to show that the C-terminus of the D1DAR imparts both lower dopamine affinity and reduced constitutive activity to this receptor subtype when compared to that of the D5DAR, while the third extracellular loop (ECL3) mediates reduced D1DAR dopamine potency compared to the D5DAR.[21,22] Furthermore, Sugamori et al.[23] have shown that the C-terminus of the D1DAR is responsible for the actions of benzazepines—antagonists for D1 and partial agonists for D5. In a separate set of experiments, chimeras were designed to study ligand binding and adenylyl cyclase (AC) activation properties associated with D1-like and D2-like receptors. To accomplish this, a chimera composed of the entire proximal portion of the D1DAR extending from the N-terminus to ICL3 was fused to the distal portion of the D2DAR extending from transmembrane domain 6(TMD6) through the C-terminus. Stimulation of cells expressing this chimera with a D2DAR-selective ligand resulted in activation of AC (like D1DARs), indicating that the portion of the receptor imparting D2-selective ligand specificity resides within TMD6 and TMD7.[24]

Early work on the β-adrenergic receptor revealed that the ligand-binding pocket in catecholamine receptors is comprised of TMD3, TMD5, and TMD6, and that Asp102 and a cluster of Ser residues are particularly important.[25] A naturally occurring polymorphism at S199A in TMD5 results in a decreased affinity for the antagonist SCH-23390.[26] More recent site-directed mutagenesis experiments illustrate that Trp99 and Ala195 in D1DAR are important for the interaction with D1-selective, rather then D3-selective, ligands.[27]

Higher order structures

D1DAR interacts with numerous other proteins to form higher order structures including other GPCRs to form both homo-oligomers[28] and hetero-oligomers, such as those observed with D2DAR[29] and A_1 adenosine receptors[30]. Many GPCRs also form complexes with ion channels and transporters including the sodium, potassium adenosine triphosphatase (Na^+, K^+-ATPase)[31] and N-methyl-d-aspartate (NMDA) receptors[32]; adaptors/trafficking proteins such as DRiP78,[33] calnexin,[34] caveolin,[35] N-ethylmaleimide-sensitive factor,[36] and sorting nexin-1[36]; regulatory proteins such as G proteins and arrestins[37]; and kinases.[38] In general, ICL2, ICL3, and the C-terminus of GPCRs are responsible for G protein coupling,[39] while β-arrestin binding to D1DAR is mediated via ICL3 interactions.[40]

D1DAR Pharmacology and Localization in the Brain

Therapeutic potential

The D1DAR is the most highly expressed DAR and is localized within the forebrain in areas such as the caudate putamen, substantia nigra, nucleus accumbens, hypothalamus, frontal cortex, and olfactory bulb. As such, it is implicated in cognitive and motor functions, as well as substance abuse, Parkinson's disease (PD), and schizophrenia.

Although the D2DAR has been the most frequently targeted DAR for the treatment of PD, there are data to suggest that D1DAR agonists may be beneficial. For example, dihydrexidine[41] and other full agonists[42,43] have demonstrated antiparkinson actions in animals, although it should be noted that dihydrexidine is only 10-fold more selective for D1 than for D2; thus, D2 activity could account for some of the antiparkinson activity.[44] While no DAR agonists currently used to treat PD clinically are as effective as levodopa, the agent that comes closest is apomorphine—an agonist that exhibits D1 selectivity.[45]

It is generally agreed that D2 antagonists are effective antipsychotics; however, D1 agonists may also be useful for treatment of the cognitive impairment associated with schizophrenia and may even be neuroprotective.[45] MPTP (1-methy-4-phenyl-1,2,3,6-tetrahydropyridine) is a neurotoxin that causes permanent symptoms of PD by killing certain neurons in the substantia nigra of the brain. One study has shown that the full D1DAR agonist dihydrexidine improved cognitive function in MPTP-treated monkeys,[46] while partial D1 agonists and D2 agonists demonstrated little effect.[47]

Ligand structures

D1DAR ligands can be divided into separate structural classes (Tables 3.1.1 and 3.1.2). The first D1DAR ligands developed were phenyltetrahydrobenzazepines (such as SCH-23390 and SKF-83959), followed by the development of rigid analogs of β-phenyldopamines (such as dihydrexidine). More recently developed D1-selective ligands include constrained phenylbenzazepines and polycyclic analogues of dihydrexidine.[48]

Although few have made it to the market yet, D1DAR ligands have been developed and used in clinical trials for the treatment of drug abuse, sleep disorders, obesity, PD, and schizophrenia.[48] Some of the most common agonists and antagonists for the D1DAR are listed in Tables 3.1.1 and 3.1.2, respectively.

D1DAR Signaling Mechanisms

G protein coupling

D1DAR couples to heterotrimeric (α, β, γ) GTP-binding proteins (G proteins) that activate AC, resulting in an accumulation of the second messenger cyclic adenosine 3′,5′-cyclic monophosphate (cAMP) and a concomitant activation of cAMP-dependent PKA. The domains of the D1DAR involved in G protein interaction have been mapped to ECL2 and ECL3, as well as to the proximal portion of the carboxyl tail.[49] The specific G proteins that couple to the D1DAR are determined by the tissue where the receptor is expressed.

The prototypical Gα protein associated with AC activation is $G\alpha_s$. This G protein subunit is ubiquitously expressed in a variety of tissues and cell culture lines.[50] D1DAR signaling has been studied in a multitude of cell types and displays robust coupling to $G\alpha_s$ upon agonist activation, indicating that D1DAR signaling is effectively mediated by this G protein[50]; however, the

TABLE 3.1.1. *D1DAR Agonists*

D1 Agonist	Structural Class	K_i (nM)	Other Targets	Therapeutic/Experimental Use
Dopamine	Catecholamine	2340	All DAR, $D_5 > D_1$ also αAR	Hemodynamic imbalances
(+)SKF-82526 Fendoldopam	Benzazepine	17	D1- and D2-like	Hypertension
NPA	Nonergoline	1816	D2-like > D1-like	Experimental tool
SKF-38393	Benzazepine	150	D1-like > D2-like	Experimental tool
Dihydrexidine	Benzazepine	33	D1-like ≥ D2-like; also αAR	Clinical trial for cocaine disorders

Note: Some common D1 agonists are listed with K_i values and relative affinities for non-D1 targets from the NIMH Psychoactive Drug Screening Program (PDSP) database.[268] Abreviations: AR, adrenergic receptors; NPA, N-propylnorapomorphine.

TABLE 3.1.2. *D1DAR Antagonists*

D1 Antagonist	Structural Class	K_i (nM)	Other Targets	Therapeutic/Experimental Use
SCH-23390	Benzazepine	0.35	D1-like > 5-HT > D2-like	Experimental tool
SCH-39166 Ecopipam	Benzazepine	3.6	D1-like > D2-like	Experimental tool
SKF-83566	Benzazepine	0.3	D1-like > 5-HT	Experimental tool
Thioridazine	Phenothiazine	100	All DARs, 5-HT, αAR, H1	Antidepressant, antianxiety, antipsychotic
Chlorpromazine	phenothiazine	73	All DARs, 5-HT, MR, AR	Antipsychotic, tranquilizer, antiemetic
Fluphenazine	Phenothiazine	21	All DARs, 5-HT, MR, AR, HT	Antipsychotic

Note: Some common D1 antagonists are listed with K_i values and relative affinities for non-D1 targets from the NIMH Psychoactive Drug Screening Program (PDSP) database. Abbreviations: AR, adrenergic receptors; 5-HT, serotonin receptors; MR, muscarinic receptors; HT, histamine receptors.[268]

neostriatum of the brain is the region where the D1DAR is most abundantly expressed, yet this region lacks robust G$α_s$ expression. G$α_{olf}$, on the other hand, is abundantly expressed in the neostriatum and couples positively to activation of AC.[51] Support for the theory that D1DAR can effectively couple to G$α_{olf}$ was demonstrated by G$α_{olf}$ knockout mice that displayed deficient D1DAR-mediated behaviors.[51]

There is little data identifying the specific β and γ G protein subunits that participate in D1DAR signaling; however, in HEK293 tissue culture cells, depletion of $γ_7$ reduces D1DAR-mediated AC stimulation and decreases the abundance of the $β_1$ subunit.[52] These data suggest that in HEK293 cells the heterotrimeric G proteins that mediate D1DAR signaling are G$α_s β_1 γ_7$.

D1DAR coupling to G$α_q$ remains a controversial topic due to the inability to reconcile contradictory observations reported by several independent research groups. G$α_q$-mediated signal transduction generally involves activation of phospholipase C (PLC) that cleaves phosphatidylinositol-4,5-bisphosphate (PIP$_2$), producing the second messenger diacylglycerol (DAG), that activates several forms of protein kinase C (PKC), and inositol-1,4,5-trisphosphate (IP$_3$) that induces release of calcium from intracellular stores through binding of IP$_3$ receptors (IP$_3$Rs) located on the endoplasmic reticulum. It has been proposed that there exists a new member of the D1-like family of dopamine receptors that couples to G$_q$ signaling pathways but has yet to be identified. Support for this theory stems from the observations that in D1DAR null mice, cAMP signaling and behavior are greatly diminished with the use of D1DAR agonists, while G$_q$ signaling pathways remain intact, as does coimmunoprecipitation of [^{3}H]SCH23390 binding sites with G$α_q$ protein.[53,54] Others disagree with the notion that there exists an unidentified member of the D1-like DARs and propose that G$_q$ signaling represents an alternative signaling pathway for D1DAR.[55] Sahu et al.[56] have looked at IP$_3$ accumulation and DAG production in D5DAR knockout mice and show that accumulation of these second messengers is severely impaired in several brain tissues, diminishing the role of the D1DAR receptor in this pathway. Recent evidence also suggests that G$_q$ coupling may be accomplished by hetero-oligomerization of D1DARs with D2DARs, generating a multi-receptor complex that exhibits altered G protein coupling and requires activation of each receptor in the complex to effect G$_q$ activation.[57] Oligomerization of different receptor proteins within and across G protein–coupled receptor (GPCR) classes to form signaling units with unique pharmacology is an exciting topic in GPCR research; however, it is beyond the scope of this discussion.

D1DAR signaling through cAMP and PKA

Stimulation of D1DAR coupled to G$\alpha_{s/olf}$ pathways results in the activation of AC, production of cAMP, and concomitant activation of PKA. This sequence of events represents the most widely studied signaling paradigm for the D1DAR subtype. The activation of PKA through this signaling paradigm modulates a plethora of downstream targets that contribute to the overall D1DAR response.

One of the most widely studied substrates for activated PKA mediated by dopamine is DARPP-32 (dopamine- and cAMP-regulated phosphoprotein, 32 kDa).[58] DARPP-32 is enriched in the neostriatum and, once activated, amplifies D1DAR signaling by both preventing inactivation of PKA and preventing dephosphorylation, and hence inhibition, of many of the downstream targets of PKA. DARPP-32 is only mentioned briefly here and will be discussed in more detail in later chapters of this volume.

D1DAR-PKA regulation of ion channels

Here, D1DAR regulation of ion channels will only be discussed briefly. For a more in-depth review of DAR regulation of ion channels, the reader is directed to an excellent review by Neve et al.[38] D1DAR regulation of ion channels is achieved by activation of PKA that results in the direct phosphorylation of ion channel subunits and other cellular effectors to alter channel activity. DARPP-32 phosphorylation resulting from D1DAR activation inhibits protein phosphatase 1 (PP1), that in turn enhances phosphorylation of the sodium channel at Ser573 to decrease sodium channel activity.[59,60] D1DAR activation of PKA also decreases inwardly rectifying potassium channels and N- and P/Q-type calcium channels. Conversely, PKA activation increases L-type calcium channels and NMDA and α-amino-3-hydroxy-5-methyl-4-isoxazolepropionic acid (AMPA) currents, as well as modulating GABA currents (reviewed in Neve et al.[38]).

D1DAR-PKA activation of CREB

The signal transduction events that have been discussed so far represent instantaneous signaling events that occur upon initial D1DAR activation. Sustained receptor occupancy that exists during addictive behaviors or therapeutic treatment induces long-term changes in signaling pathways mediated by altered gene expression that result in modifications of neural networks that are translated into behavioral responses.

D1DAR activation of PKA results in phosphorylation of the cAMP response element binding protein (CREB) by PKA on Ser133 of CREB. Phosphorylated CREB dimerizes and acts as a transcription factor by binding to cAMP response element (CRE) DNA sequences located in the upstream promoter regions of a variety of genes, including the immediate early gene family of transcription factors Fos and Jun. Modulation of DAR activity leads to the expression of several neurotransmitter genes such as the those of the neuropeptides enkephalin and neurotensin (discussed in Adams et al.[14]).

D1DAR Regulation

Desensitization

Upon agonist activation, GPCRs undergo desensitization, a process that results in a waning of receptor response under continued agonist stimulation. Desensitization involves phosphorylation of the receptor by G protein receptor kinases (GRKs) and/or second messenger kinases such as PKA or PKC. Phosphorylation of GPCRs is categorized as either homologous or heterologous, with most GPCRs undergoing both types of phosphorylation. Heterologous phosphorylation occurs when a GPCR becomes phosphorylated by kinases activated by a separate signaling pathway. Kinases mediating heterologous desensitization are second-messenger-activated kinases such as PKA and PKC. Homologous desensitization of GPCRs results in the phosphorylation of the receptor by kinases activated as a result of the signaling events initiated upon activation of that receptor. Homologous phosphorylation of GPCRs is primarily mediated by GRKs. A general model of desensitization involves arrestin binding to phosphorylated receptor and concomitant uncoupling of the receptor from G protein, abrogating second messenger production. The arrestin-bound receptor is then internalized via clathrin-coated pits. Once internalized, the receptors are either dephosphorylated by phosphatases within the endocytic vesicle and recycled to the plasma membrane for additional signaling (resensitization) or targeted to lysosomal vesicles for degradation (down regulation). Recent studies have shown that this model for GPCR desensitization contains many layers of complexity, with individual receptors displaying unique modes of regulation, abolishing a universal theme for GPCR desensitization.[61,62]

To date, D1DAR desensitization has been studied using a variety of systems including tissue sections, primary cell cultures, cell cultures expressing endogenous receptor, and heterologous expression of receptors both stably and transiently in various cell culture lines.[63,64] Since the studies were performed in a variety of cellular environments, not all of the data concerning D1DAR regulation are consistent; however, several lines of evidence exist that link D1DAR phosphorylation to desensitization. Gardner et al.[65] showed that the potency for dopamine to induce D1DAR phosphorylation (EC_{50} ~200 nM) was identical to that required to initiate desensitization. Furthermore, D1DAR phosphorylation ($t_{1/2}$ <1 min) preceded receptor desensitization ($t_{1/2}$ ~7 min), while dephosphorylation occurred more quickly ($t_{1/2}$ ~10 min) than the time required for resensitization of the receptor ($t_{1/2}$ ~3 h). Surprisingly, neither concanavalin A nor hypertonic sucrose (inhibitors of receptor internalization) nor okadaic acid or calyculin A (phosphatase inhibitors) prevented D1DAR dephosphorylation. This contradicts the general model described above in which GPCRs are thought to internalize prior to dephosphorylation, since it is within this acidic vesicle that receptors are proposed to be dephosphorylated by phosphatase (G protein receptor phosphatase GRP or PP2A, both sensitive to oakadaic acid and calyculin A).[66] This demonstrates that the D1DAR displays a novel recovery pathway that does not involve internalization or GRP/PP2A phosphatases.

Phosphorylation

Both receptor-specific kinases (GRKs) and second-messenger-activated kinases (PKA) have been shown to mediate desensitization of the D1DAR at specific amino acid residues located within intracellular regions of the receptor protein. Both classes of kinases phosphorylate substrates at Ser and Thr residues. Jiang and Sibley[67] found that mutation of only one (Thr268, located in ICL3) of four possible PKA phosphorylation sites in the D1DAR displayed any deviation from wild-type function: a decreased rate of agonist-induced desensitization in the Thr268 mutant compared to the wild-type receptor. Mason et al.[68] further showed that mutation of Thr268 altered D1DAR trafficking. With respect to GRK phosphorylation of the D1DAR, Tiberi et al.[69] showed that the D1DAR could be phosphorylated in an agonist-dependent manner by GRK2, GRK3, and GRK5 and that this phosphorylation occurred primarily on serine residues within the receptor. In contrast, GRK4α was shown to increase the phosphorylation state of the D1DAR in the absence of agonist stimulation by phosphorylating the receptor on Thr428 and Ser431 in the carboxyl tail, resulting in constitutive desensitization and internalization of the D1DAR.[70–72] The contribution of each GRK to the phosphorylation state of the D1 receptor in the absence and presence of agonist remains unclear; however, phosphorylation-defective D1DAR constructs display reduced receptor expression levels.[40,73,74]

In contrast to the information available for PKA and GRK modification of the D1DAR, there is little available concerning the role for PKC in D1DAR regulation despite the presence of several PKC consensus sequences located in the cytoplasmic region of the receptor.[75,76] Gardner et al.[65] reported that the D1DAR could be phosphorylated upon activation of PKC using the phorbol ester phorbol 12-myristate 13-acetate (PMA), and this phosphorylation was abolished in the presence of the PKC inhibitor bisindolylmaleimide-1 (BIM-1). Interestingly, the basal level of D1DAR phosphorylation was reduced in the presence of BIM-1 alone, implying that PKC can phosphorylate the D1DAR in an agonist-independent fashion. We have since gone on to show that PKC isotypes α, β1, δ, ε, γ, and λ can all phosphorylate the D1DAR; however, the sites of phosphorylation within the receptor and the functional significance of this modification mediated by each PKC isotype are still under investigation.

Jackson et al.[77] examined the effect of phorbol ester treatment (activator of conventional and novel PKCs and PKCμ) on the AC-cAMP accumulation pathway within the D1 family by utilizing HEK293 cells transiently expressing D1DAR and D5DAR individually. Surprisingly, PMA treatment exhibited opposing effects on the ability of each D1-like receptor to stimulate cAMP production within the same cellular environment. PMA treatment potentiated D1DAR dopamine-stimulated cAMP accumulation, whereas in D5DAR-expressing cells, PMA treatment blocked constitutive receptor activity and abrogated dopamine-stimulated cAMP accumulation. It is surprising that two dopamine receptors so similar in sequence, pharmacological profile, and signaling could be individually regulated by phorbol ester treatment within the same cellular environment. This lends credence to the notion that each D1-like receptor does not exist for redundancy, but that each plays a unique role in dopaminergic signaling.

In an attempt to identify the regions within the D1DAR that correspond to particular aspects of receptor regulation, several groups created both truncation and substitution mutants within ICL3 and carboxyl tail regions of the receptor. These mutant receptors were then assayed for binding properties, cAMP production, desensitization, internalization, and incorporation of phosphate. In terms of phosphorylation, one consistent observation arising from these studies is that removal of increasing portions of the C-terminus of the D1DAR results in decreased receptor phosphorylation.[78–80] Furthermore, abolition of phosphorylation in the carboxyl tail precludes phosphorylation of the D1DAR in the ICL3, indicating that phosphorylation of the receptor proceeds in a hierarchical fashion, occurring initially in the carboxyl tail region prior to ICL3. Another striking consistency in the data is that the most severely truncated D1DAR carboxyl tail truncation (Δ351 for Jackson et al.[78] and T347 for Kim et al.[80]) exhibits no detectable phosphorylation yet desensitizes at least as well as wild-type receptor. To explain this observation, we had originally proposed that the truncation of the carboxyl tail allowed access of arrestin protein to binding sites located within ICL3 of the receptor to mediate efficient desensitization. In the wild-type receptor, access to ICL3 would be accomplished by repulsion generated by phosphorylation of the tail and ICL3 regions. This is not the case, as we have extended this observation to include two more D1DAR mutants that represent full-length receptors that are also phosphorylation defective. One mutant has all Ser/Thr replaced in the carboxyl tail region, and one has all Ser/Thr replaced in ICL3 as well as in the carboxyl tail region. Both of these receptors desensitize as efficiently as the wild-type D1DAR, and preliminary evidence suggests that they also efficiently recruit arrestin upon agonist stimulation. These new data demonstrate that phosphorylation is not required for D1DAR desensitization or arrestin binding; thus, the exact role that phosphorylation plays in D1DAR regulation remains unclear.

Receptor endocytosis

D1DAR endocytosis occurs by both a clathrin-mediated pathway[81] and a caveolae-mediated pathway.[35] The specific regions involved in D1DAR endocytosis are still undefined since several groups report contradictory data.[73,74,82,83] The mechanism that regulates which endocytic pathway the D1DAR will take is also unclear; however, the particular cellular environment most likely determines the specific internalization route. Kong et al.[35] determined that in COS7 cells D1DARs internalized preferentially via the caveolae-mediated pathway even though these cells exhibited a robust clatherin-mediated route. Phe313 and Trp318 amino acid residues in TMD7 of the receptor were found to be essential for D1DAR-caveolae endocytosis.

D1DAR Peripheral Localization and Signaling

Nonneuronal D1DAR signaling

D1DAR is most abundantly expressed in the brain, as discussed above; however, dopamine receptors are also localized to the retina, the cardiovascular system, and the kidney.[84,85]

All DAR subtypes are expressed in the kidney, with D1DAR found specifically in the proximal tubule, the medullary, the collecting ducts, the macula densa, and the juxtaglomerular cells (reviewed in Zeng et al.[86]). Dopamine acts in a paracrine fashion to inhibit the transport of sodium at multiple locations along the nephron when sodium intake is moderately high, resulting in sodium excretion (natriuresis). D1DARs specifically block the reuptake of sodium by acting on several transporters including Na^+-K^+-ATPase.[87] Abnormal dopaminergic signaling in the kidney directly participates in hypertension in many animal models including humans. Human essential hypertension was found to be mediated by a single nucleotide polymorphism of GRK4γ that results in increased D1DAR basal phosphorylation and hence desensitization, abrogating dopamine-mediated natriuresis.[71]

THE D2 DOPAMINE RECEPTOR SUBTYPE

D2DAR Structure

The D2DAR (Fig. 3.1.2), located on chromosome 11q22-q23, was first cloned in 1988.[88] It has eight exons[89] and exists as one of two splice variants, $D2_L$ (long) or $D2_S$ (short).[90] The long variant (GenBank Accession NP_000786) has an extra 29 amino acids in the ICL3[90] and is mostly postsynaptically expressed, unlike the $D2_S$DAR (GenBank Accession NP_057658), which exhibits primarily presynaptic expression.[91] The D2DAR is most homologous to the D3DAR, with 75% homology in the TMD segments.[92]

Like the D1DAR, the D2DAR is posttranslationally modified. There are three consensus sites for glycosylation in the N-terminus (Asn5, Asn17, and Asn23)[93] and multiple phosphorylation sites (see below).

Mutational analysis has demonstrated the importance of ECL3,[94] as well as that of selected residues within TMD1, TMD2, TMD3, TMD4, TMD5, and TMD7[95–97] in ligand binding to the D2DAR. Like other class A GPCRs, D2DAR has a DRY motif at the end of TMD3. This motif is known to hold the GPCR in an inactive state as mutations that disrupt salt bridge formation result in constitutive activity.[98] D2DAR basal activity can also be increased with a T343R mutation in the base of TMD6.[95]

FIGURE 3.1.2. Structure of the rat D2$_L$DAR. The 29 amino acids lacking in the D2$_S$DAR isotype are indicated in the figure as gray residues bounded by arrows.

As discussed in the section on the D1DAR receptor subtype, chimeric analysis has been used to evaluate ligand-binding sites in the DARs. Compared to D3, D2 has lower agonist affinity and increased G protein coupling. Lachowicz and Sibley have shown that both D2/D3 chimeras that were split down the center of the ICL3 were able to inhibit AC to near-D2 levels.[99] Interestingly, it has been shown that part or all of the ICL3 of D2 can impart D2-like G protein selectivity to the D3DAR.[100,101]

Like other GPCRs, the D2DAR interacts with several other proteins. It is known to homo-oligomerize[102,103] as well as hetero-oligomerize with D1[104] and D3[105] dopamine receptors and also with the CB1 cannabinoid and adenosine receptors.[106,107] In the case of the D2 homo-oligomer, TMD4 is involved in the interface,[102] while TMD5 and/or ICL3 are responsible for the D2-A2A interaction.[107]

Aside from GPCRs, D2 binds trafficking proteins such as calnexin,[34] scaffolding proteins such as spinophilin,[109] signaling proteins that include regulator of G protein signaling 19 (RGS19),[108] arrestin,[110] and protein kinases,[111] as well as ion exchangers such as Na$^+$,K$^+$-ATPase.[112]

D2DAR Pharmacology

PD can be treated with levodopa, catechol-O-methyltransferase/monoamine oxidase (COMT/MAO) inhibitors, and/or D2-like DAR agonists[113,114] (Table 3.1.3). D2-like agonists can be used as a monotherapy early in disease progression or in combination with levodopa in moderate to advanced PD. Commonly used agonists include pramipexol, ropinirole, cabergoline, bromocriptine, and pergolide.[115] In addition to the short-term therapeutic benefit of increased dopaminergic transmission, many agonists (pramipexole, pergolide, bromocriptine, and others) have demonstrated neuroprotective properties, although the mechanism is unclear.[114]

All known antipsychotic drugs have D2DAR blocking properties,[116] and the clinical potencies of antipsychotic drugs correlate with their D2 dissociation constant.[117] It is of note, however, that antipsychotic drugs not only antagonize DARs, but can also inhibit a series of other GPCRs including adrenergic, serotonergic, histaminergic, cholinergic, and other receptors (Table 3.1.4). Because of the lack of selective ligands, the contribution of other GPCRs to antipsychotic actions is unclear. The extra-pyramidal side effects are thought to be D2-mediated, but only by antagonists with high receptor occupancy.[117] Some common D2DAR agonists and antagonists are listed in Tables 3.1.3 and 3.1.4, respectively.

Aripiprazole is a partial D2 agonist and has been called a prototype of a new third generation of antipsychotics. It is as effective as atypical antipsychotics on the positive symptoms, but also has improved efficacy on negative symptoms and fewer side effects.[118] As a partial agonist, aripiprazole is believed to block excessive dopaminergic input in the mesolimbic areas responsible for the positive symptoms while enhancing D2DAR signaling in brain areas related to memory. This dual mechanism afforded by partial agonism of the D2DAR stabilizes the imbalanced signaling in the patient and represents a dramatic improvement over previous D2DAR antagonist treatments. Other stabilizers of dopaminergic signaling that have shown promise in animal studies include OSU 6162 and ACR 16, although their mechanisms of action have been a matter of debate and warrant further investigation.[119–121]

TABLE 3.1.3. *D2DAR Agonists*

D2 Agonist	K_i (nM) $D2_L$, $D2_S$	Other Targets: Relative Affinity	Therapeutic/Experimental Use
Dopamine	474, 710	All DARs; also AR	Hemodynamic imbalances
Apomorphine	83, 35	D2-like > D1-like > αAR, 5-HT	PD "off-episodes"
Pramipexol	933, 676	D2-like > D1-like > 5-HT	PD, RLS. In clinical trials for depression and OCD
Bromocriptine	15, 5	D2-like = αAR = 5-HT > D1-like	Hyper-prolactinemia
Ropinirole	933, 636	D2-like > D1-like > 5-HT, AR	PD
cabergoline	0.95, 0.62	D2-like > D1-like > 5-HT, AR	Hyper-prolactinemia
Lisuride	0.66, 0.34	α2-AR > D2-like > D1-like. High affinity for other ARs, 5-HT	PD, migraines
Pergolide	26, 32	D2-like > D1-like. Very high affinity for 5-HT and for aα-AR	PD
Aripiprazole	0.74	D2-like > AR = 5-HT > D1-like	Schizophrenia
Quinpirole	1450, 890	D2-like > D1-like	Experimental tool

Note: Some common D2 agonists are listed with K_i values and relative affinities for non-D2 targets from the NIMH Psychoactive Drug Screening Program (PDSP) database.[268] Abbreviations: OCD, obsessive-compulsive disorder; PD, Parkinson's disease; RLS, restless leg syndrome; AR, adrenergic receptors; 5-HT, serotonin receptors.

TABLE 3.1.4. *D2DAR Antagonists*

D2 Antagonist	K_i (nM) $D2_L$	Other Targets	Therapeutic/Experimental Use
Thioridazine	3.3	High affinity 5-HT, AR, MR D2-like > D1-like	Depression, anxiety
Butaclamol	0.8	D2-like > 5-HT and AR	Experimental tool
Clozapine	138	High affinity 5-HT, AR, MR. D1-like = D2-like	Schizophrenia
Fluphenazine	0.5	High affinity 5-HT, AR, MR. D2-like > D1-like	Psychotic disorders
Haloperidol	1	High affinity 5-HT, AR. D2-like > D1-like	Schizophrenia, Tourette's disorder
Spiperone	0.03	D2-like > D1-like. Also, 5-HT, AR, HT	Experimental tool
Sulpiride	31	D2-like > D1-like.	Experimental tool
Risperidone	0.3	High affinity 5-HT, AR. D2-like > D1-like	Schizophrenia

Note: Some common D2 antagonists are listed with K_i values and relative affinities for non-D2 targets from the NIMH Psychoactive Drug Database (PDSP) database.[268] Abbreviations: AR, adrenergic receptors; 5-HT, serotonin receptors; MR, muscarinic receptors; HT, histamine receptors.

Interestingly, GPCR ligands can have different effects depending on the readout used. (S)-3-(3-hydroxyphenyl)-N-propylpiperidine can be an agonist, an antagonist, or an inverse agonist, depending on the G protein being activated.[122] Aripiprazole has been shown to be a partial agonist with regard to inhibition of cAMP accumulation but an antagonist with regard to β-arrestin recruitment, both through the D2$_L$DAR.[123] Dihydrexidine, which has demonstrated antiparkinson effects, has also been suggested to be a functionally selective ligand on the D2DAR.[44] Binding data suggest that it is a D1 and D2 agonist. In the same cell line, it is a full agonist for AC inhibition and has no effect on D2-mediated, G protein–coupled, inwardly rectifying K$^+$ channels.[44]

Although none are selective for D2, there are several allosteric modulators, including Pro-Leu-Gly-NH$_2$, SCH 202676, sodium, zinc, and amiloride.[124]

Recent advances involving more complicated mechanisms such as functional selectivity and allosteric modulation of the D2DAR hold promise for the future of improved therapeutics.

D2DAR Signaling Mechanisms

The D2DAR was first shown to inhibit cAMP accumulation in both pituitary and striatal cells.[125,126] Subsequent cloning[88] and expression[127] of the D2DAR confirmed these early findings. Moreover, mammalian expression studies established that dopamine-induced inhibition of cAMP could be blocked by pertussis toxin,[127] indicating that Gα$_i$ and Gα$_o$ G proteins mediate this response.

Studies employing small synthetic peptides or receptor chimeras show that both ICL2 and ICL3 of the D2DAR are important for G protein coupling. Peptides corresponding to ICL3 could block dopamine-mediated cAMP inhibition in transfected cell lines.[128] Interestingly, studies utilizing D1/D2$_S$ chimeras indicate that both ICL2 and ICL3 of the D2DAR are necessary to convey D2–G$_{i/o}$ coupling.[129] In addition to agonist-mediated signaling, the D2DAR exhibits low constitutive activity or signaling through G$_{i/o}$ proteins in the absence of ligand.[130,131]

Because the sequence variation for the long and short isoforms of the D2DAR lies within the putative ICL3, it has been hypothesized that these receptor variants may differentially couple to G proteins. Indeed, it has been reported that the D2$_S$ receptor inhibits cAMP more effectively than the D2$_L$ receptor.[132–134] These differences in receptor–G protein coupling can likely be attributed to the complement of Gα$_i$ subunits expressed in various cell systems; however, data from these studies suggest that D2$_S$ couples efficiently to either Gα$_{i1}$ or Gα$_{i3}$, while D2$_L$ couples most efficiently to Gα$_{i3}$.[134] When expression of all three Gα$_i$ subtypes was comparable, there was no significant difference in D2$_S$- versus D2$_L$-mediated inhibition of cAMP.

In addition to coupling to G$_{i/o}$, the D2DAR has also been shown to activate the pertussis toxin–insensitive G$_z$ protein. In transfected cells, the D2$_S$ and D2$_L$ receptors both inhibit forskolin-stimulated cAMP levels via G$_z$ coupling.[135] Recently, the generation of a mouse deficient in Gα$_z$ has enabled in vivo testing of the coupling of D2-like receptors to this G protein.[136] Interestingly, treatment of mice lacking Gα$_z$ with the D2 agonist quinpirole did not suppress locomotor activity to the same extent as quinpirole treatment of wild-type mice.[136] The combination of these in vivo data with previous data from heterologous expression systems indicates that the D2DAR can couple to G$_z$ to inhibit cAMP production.

Following receptor activation of G proteins, the Gβγ subunits separate from Gα and can stimulate their own signaling pathways. For the D2DAR, activation of G$_{i/o}$ leads to Gβγ regulation of a variety of effector proteins, including the G protein–coupled inwardly rectifying potassium channels (GIRK or Kir3). Dopamine stimulation of D2DARs robustly increases GIRK activation in mammalian cells.[137] Moreover, D2DARs and GIRKs are known to exist in a stable complex whose formation is dependent upon the Gβγ subunit.[138] D2-activated Gβγ subunits can also inhibit various calcium channels. In neostriatal interneurons, D2 receptor-released Gβγ directly inhibits N-type Ca^{2+} channels,[139] while in medium spiny striatal neurons, the L-type Ca^{2+} channel is inhibited via a Gβγ-mediated PLCβ1 pathway.[140] This D2-regulated PLCβ1 pathway also causes an increase in IP$_3$-induced Ca^{2+} release,[141] thus enabling D2DAR control over a variety of calcium-dependent proteins and pathways (for review see Bergson et al.[142]). PLCβ1 may also contribute to D2 activation of extracellular signal–regulated kinase (ERK),[139] although D2-mediated ERK activation is likely regulated in a cell-dependent manner by a variety of effectors including direct Gβγ activation[143] and transactivation of receptor tyrosine kinases.[144] Finally, Gβγ activation by D2 receptors potentiates arachidonic acid release via phospholipase A$_2$.[145]

In addition to heterotrimeric signaling pathways, the D2 receptor has recently been shown to activate the small G protein, RhoA,[146] leading to phospholipase D stimulation and hydrolysis of phosphatidylcholine.[147] The protein interface mediating D2–RhoA coupling has not yet been elucidated, nor has it been determined if RhoA activation by D2 may lead to other downstream signaling events.

It has recently been demonstrated that D1 and D2 receptors can form functional heterodimers.[29] Upon ligand stimulation, these receptor dimers are capable of activating a Gα$_{q/11}$ pathway leading to calcium release either in transfected cells[29] or in striatum.[57] Both the presence of this heterodimer and its unique signaling profile present a new dopaminergic pharmacology that has yet to be thoroughly investigated.

D2DAR Regulation

Following ligand activation of the D2DAR, the receptor activity and the downstream signal must ultimately be turned off. For the D2DAR, regulator of G protein signaling 9 (RGS9) is responsible for limiting the D2-initiated G$_{i/o}$ signal.[148] Mice lacking RGS9 exhibited increased locomotion and reward responses to cocaine.[148] In addition, overexpression of RGS9 in the rat nucleus accumbens caused a reduction in locomotion after administration of D2 receptor agonists.[148] RGS9 has also been shown to colocalize with the D2 receptor in the striatum.[149] These data implicate RGS9 in the silencing of the D2-stimulated signaling cascade.

In addition to attenuation of the G protein signal, the receptor itself is converted to a nonsignaling state via desensitization. The signaling molecule PKC has been demonstrated to phosphorylate the D2DAR, diminishing its ability to inhibit cAMP levels.[150] GPCR kinases (GRKs) are also involved in D2DAR regulation, most notably GRK2.[151] These regulatory events serve to uncouple the receptor from G proteins and can initiate receptor internalization pathways by initiating binding of accessory proteins, including arrestin or clathrin (for reviews see [152–154]). Once coupled to an internalization pathway, the D2DAR can either be resensitized and recycled back to the plasma membrane or degraded via lysosomal pathways.[152–154]

D2DAR Localization

Dopaminergic neurons project via three main pathways in the brain—the nigrostriatal, mesolimbic, and mesocortical pathways. The D2DAR is found throughout the brain and within each of these pathways. The highest levels of D2DAR mRNA were found to occur within the striatum, the nucleus accumbens, and the rat olfactory tubercule.[155] Lower levels of D2DAR mRNA have been found in the prefrontal cortex, amygdala, ventral tegmental area, hippocampus, hypothalamus, and substantia nigra pars compacta.[155–157] Subsequent studies with specific antibodies have more precisely localized the presence of the D2DAR to medium spiny neurons in the striatum and perikarya and to dendrites within the substantia nigra pars compacta.[158]

The D2$_L$ isoform is more predominant than the D2$_S$ isoform throughout the brain.[159] While there is little functional difference between these subtypes and both act as autoreceptors, the D2$_L$ receptor exists predominantly as a postsynaptic autoreceptor, likely regulating impulse/signal propagation, while the D2$_S$ receptor serves presynaptic autoreceptor functions, likely limiting dopamine release.[91] Interestingly, both D2DAR isoforms are predominantly localized intracellularly in the brain and may be trafficked to the plasma membrane as necessary.[160]

THE D3 DOPAMINE RECEPTOR SUBTYPE

D3DAR Structure

Since its discovery in 1990, there has been considerable interest in the D3 receptor as a potential target for antipsychotic drugs as well as a mediator of reinforcing effects of drugs of abuse and sensitization to psychostimulants. It also has potential importance in PD. The rat D3DAR was first cloned by Sokoloff and colleagues[92] and found to be 52% homologous overall with the rat D2 receptor, exhibiting 75% homology in the transmembrane-spanning domains. Shortly thereafter, the human D3 receptor was cloned (GenBank Accession NP_000787; located on chromosome 3q13.3), exhibiting 79% homology overall to the rat receptor with a 97% homology in the transmembrane domains.[161] The spatial orientation of the conserved amino acids is nearly identical to that of the D2 receptor.

The human D3DAR consists of 400 amino acids with an ICL3 containing 120 residues and a relatively short C-terminus (compared to the long C-terminus seen with D1-like DARs) consisting of only 16 amino acids. The extracellular N-terminus of the receptor has three potential N-linked glycosylation sites likely involved in receptor processing. The D3DAR also contains characteristic cysteine residues in ECL2 and ECL3, as seen in many GPCRs.

D3DAR Pharmacology

Some have postulated that the D3 receptor structure may be more rigid than that of the D2DAR—primarily due to its unique ability to bind agonists with high affinity irrespective of the presence of G proteins.[162] Furthermore, only small differences in ligand affinity are seen between the high- and low-affinity states of the receptor.[163]

Several splice variants of the D3DAR have been identified, including a 113-bp deletion in TMD3 and a down stream frame shift in the coding sequence generating a stop codon producing a 100-amino acid, truncated form of the receptor.[164] There is also another truncated splice variant termed D3nf that has been identified and shown to be present in schizophrenic brains.[165,166] Determination of the pharmacological properties of the D3 receptor has been somewhat hampered by the high degree of sequence similarity between it and the D2 receptor. This has made identification of truly subtype-specific ligands difficult. The problem is further compounded *in vivo* due to low levels of endogenous expression of the D3 receptor and the fact that it is typically coexpressed in brain regions similar to those of the D2 receptor. As a result, the majority of pharmacological data has been obtained using heterologous expression systems, which have allowed for a clean interpretation of D3 pharmacology without interference of the D2 receptor. This approach has led to the development of some relatively selective D3 ligands.

Most D2-like agonist compounds used clinically for the treatment of PD, including some newer D2-like agonists such as ropinirole and pramipexole, also bind to the D3 receptor.[167,168] Fortunately, because of heterologous expression systems coupled with medicinal chemistry efforts, there are now compounds that are relatively selective for the D3DAR. These include compounds with relatively moderate levels of selectivity over D2 including nafadotride, U99194A, and S 14287.[169-171] Newer compounds have recently been developed with higher levels of selectivity, including 7-OH-DPAT, PD 128907, 7-OH-PIPAT, (R)-3-(4-propylmorpholin-2-yl)phenol (an arylmorpholine agonist with >1000-fold functional selectivity over D2DAR),[172] 3-(4-chlorophenyl)-N-(4-(4-(2-fluorophenyl)piperazin-1-yl)butyl)acrylamide (a full agonist with significant binding selectivity over D2DAR),[173] trans-N-{4-[4-(2,3-dichlorophenyl)-1-piperazinyl]cyclohexyl}-3-methoxybenzamide (a full agonist displaying >200-fold binding selectivity over D4DAR, D2DAR, 5-HT1A, and α1-receptors),[174] 7-{[2-(4-phenyl-piperazin-1-yl)ethyl]-propylamino}-5,6,7,8-tetrahydronaphthalen-2-ol,[175] FAUC 346 (a subtype selective partial agonist),[176] and SB-277,011A (a potent and selective D3 receptor antagonist 80- to 100-fold selective for D3 over D2).[177] It is of interest to note that the D3DAR behaves differently from the highly homologous D2DAR in several important aspects. These include the unique binding of dopamine to the D3DAR in a manner that is much less sensitive to GTP or nonhydrolyzable GTP analogs.[178] While the high-affinity state of the D2DAR exhibits about 100-fold higher affinity for agonists compared to the low-affinity state,[179] the D3 receptor shows only about a 5- to 10-fold affinity difference between the two states.[180]

D3DAR Signaling Mechanisms

D3DARs are primarily coupled to G_i/G_o-like proteins, but some studies have also demonstrated their ability to couple to $G\alpha_{q/11}$.[181] Consistent with membership in the D2-like family of DARs, D3DARs inhibit the accumulation of cAMP via the inhibition of AC through G_i/G_o signaling;[182] however, D3DAR-stimulated inhibition of AC is generally weaker, sometimes to the point of being undetectable, compared to that of the D2 signal.[38] Interestingly, it does appear that D3DARs robustly inhibit AC type 5—showing a markedly greater effect on this subtype than others.[183]

Not surprisingly, D3 receptors also modulate other signaling pathways in addition to AC, including ion channels, mitogen-activated protein kinases(MAPKs), and phospholipases. Many of these pathways are regulated by either pertussis toxin-sensitive G_i/G_o signaling and/or G protein βγ subunits released from the D3DAR complex upon receptor activation. Indeed, D3DAR stimulation has marked effects on ion channel activity. D3 activity dose-dependently reduces the outward K$^+$ current—an effect that is blocked by pertussis toxin and is mediated by G_o.[184] Furthermore, inward rectifier potassium channels (GIRK channels) are activated by stimulation of D3, likely through Gβγ signaling.[137,185] The D3DAR is nearly as efficient as the D2$_L$DAR in coupling to homomeric GIRK2—the subtype predominantly expressed by dopamine neurons in the ventral mesencephalon. In addition, inward calcium currents elicited by depolarization of membrane potential are depressed by stimulation of D3 receptors.[186,187] This inhibition of calcium currents is pertussis toxin-sensitive, which suggests the involvement of G_i or G_o. It has also been recently appreciated that D3 receptors are able to modulate GABA$_A$ receptors by increasing the phospho-dependent endocytosis of the receptors—a mechanism likely to be important in reinforcement and reward, particularly for drugs of abuse.[188]

D3 signaling has also been shown to modify the activity of protein kinase cascades in various expression systems. In particular, MAPK has been demonstrated to be stimulated by D3 receptors.[189] This effect involves G_i/G_o signaling and suggests that D3 receptors signal through the MAPK biochemical cascades. In addition, *in vivo* data on c-*fos* indicates that the D3 receptor is also responsible for affecting its expression levels.[190,191] This further suggests that activity of MAPK and CREB (modulators of c-*fos* transcription) may also be modulated by the D3 receptor *in vivo*.

D3DAR Regulation

Investigations into the regulation of the D3 receptor are limited. Receptor desensitization, a common means for rapidly attenuating the agonist signal of GPCRs, typically involves receptor phosphorylation by GRKs and arrestin binding that initiates sequestration. However, in contrast to regulation of the D2 receptor described above, agonist regulation of the D3 receptor is subtle. Furthermore, translocation of arrestin to the membrane and internalization of the receptor are significantly reduced compared to the D2DAR.[192,193] Chimeric proteins made from the D3 receptor with the ICL3 of the D2DAR showed regulatory properties similar to those of the D2 receptor, suggesting that ICL3 is primarily responsible for the receptor's regulatory properties.

D3DAR Localization

The attractiveness of the D3 receptor as a therapeutic target is enhanced by its restricted expression to mesolimbic brain regions, especially the nucleus accumbens.[194] This "limbic" region receives its dopaminergic input from the ventral tegmental area and is known to be associated with cognitive, emotional, and endocrine functions. The D3 receptor is expressed at roughly 10-fold lower abundance than the D2 receptor, and its distribution is more restricted. The highest densities of the D3DAR are found in the islands of calleja and the olfactory bulb, followed by moderate densities in the nucleus accumbens, vestibulocerebellum, and substantia nigra. In contrast, there is relatively little D3 expression in the caudate/putamen.[155,195–198] This is of interest, since low expression of the D3 receptor in striatal regions associated with extrapyramidal effects would suggest that a D3 ligand may display fewer undesirable side effects when used therapeutically.

At least some D3 receptors probably function as presynaptic autoreceptors modulating neuronal firing and dopamine synthesis.[199] Dopamine binds to the D3 receptor with a 20-fold higher affinity than to the D2 receptor, a characteristic expected for autoreceptors. Also, the presence of D3 receptor mRNA in the substantia nigra,[155] the origin of major dopaminergic projection pathways, supports the characterization of the D3 receptor as a presynaptic receptor.

THE D4 DOPAMINE RECEPTOR SUBTYPE

D4DAR Structure

Discovery and molecular cloning of the D4DAR

In 1991, Hubert Van Tol and his colleagues published the first description of the cloning of DNA encoding a novel DAR subtype termed D4.[200] Initially they identified a novel partial complementary DNA (cDNA) clone from a screening of the neuroblastoma SK-N-MC cell cDNA library, under low-stringency conditions, using a full-length D2 receptor cDNA as a probe. Then, using the novel partial cDNA as a probe, the whole genomic clone, encompassing the entire coding region of the novel receptor, was isolated by human genomic library screening. Initial efforts to clone the full-length cDNA were unsuccessful, so a gene/cDNA hybrid was constructed and expressed to study the pharmacological properties of this newly identified DAR subtype. Several approaches have been used to obtain full-length cDNAs.[93] Finally, in 1995, Matsumoto *et al.* succeeded in cloning a "native" human D4DAR cDNA by polymerase chain reaction (PCR), using improved reverse transcriptase PCR (RT-PCR) methods and D4 receptor-enriched retinal polyA RNA as a template.[201]

Molecular structure of the D4DAR protein

The D4 receptor (GenBank Accession NP_000788) was identified as a dopaminergic D2-like receptor based on the predicted amino acid sequence, and displays 41% homology to the D2 receptor and 39% homology to the D3 receptor.[200] Amino acid sequence identity is greatest within the TMD regions, being particularly high in TMD2, TMD3, and TMD7 and lowest in TMD1 and TMD4. Like other $G_{i/o}$ coupled receptors including D2 and D3, the D4 receptor has a relatively long ICL3 and a short carboxyl tail that terminates in Cys. This Cys could serve as a substrate for palmitoylation.[202] There is a potential N-linked glycosylation site at Asn3, and one within the consensus sequence for PKA phosphorylation (R-R-X-S) at Ser234 located toward the amino-terminal end of ICL3.[93]

Genomic structure and polymorphism of the D4DAR gene

The human D4DAR gene possesses four exons and is located on chromosome 11p15.5. The most striking

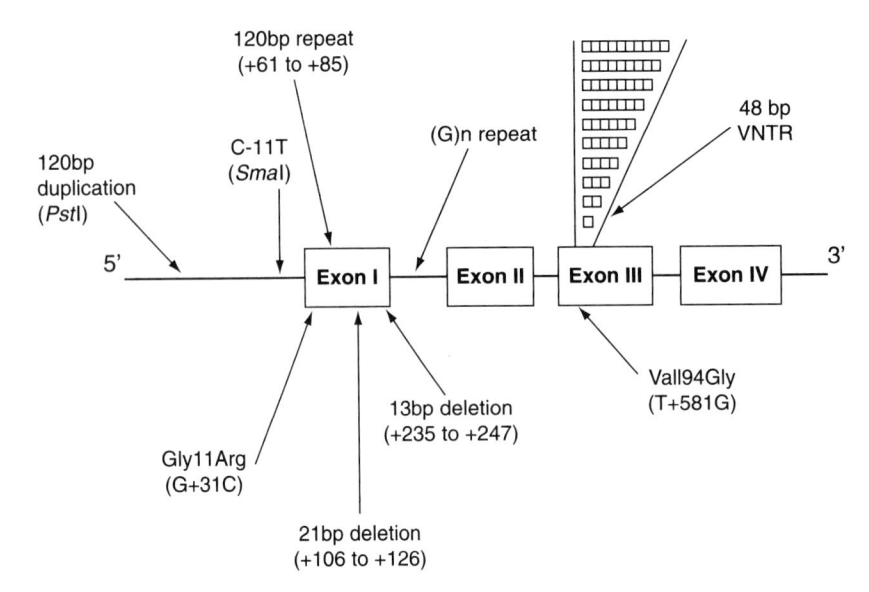

FIGURE 3.1.3. Diagram of the D4DAR gene with polymorphic sites. DRD4 has four exons (I–IV), and nine polymorphisms have been identified; nucleotide positions of polymorphisms are given in parentheses. *Source*: Reference [208].

structural feature of the human D4DAR is the presence of a highly variable number of tandem repeats (VNTR) in exon 3, encoding ICL3 of the receptor. The repeat unit is 48 bp, and at least 35 distinct 48-bp repeats are present in 2 (D4.2) to 11 (D4.11) copies. To date, over 67 different haplotypes have been identified in humans.[203,204] The most common is the 4-repeat (D4.4) allele followed by the D4.7 or D4.2, depending on the population[205] (Fig. 3.1.3). The 48-bp repeat, with intra- and interspecies variations in both the sequences and the number of copies of the repeats, is also found in the D4 receptor of nonhuman primates but not in the rat D4 receptor.

Many studies have attempted to correlate polymorphic human D4DAR variants with psychiatric diseases and personality traits. Evidence does exist to link D4.7 receptor variants to attention deficit hyperactivity disorder (ADHD).[204,206–208]

D4DAR Pharmacology

Pharmacological profile of the D4DAR compared to other D2-like receptors

The pharmacological profile of the D4DAR is generally similar to that of the D2DAR, displaying high affinity for [^{3}H]spiperone,[200] but the D4DAR has several distinct pharmacological features that distinguish it from the D2DAR. First, the classic atypical antipsychotic clozapine binds the D4DAR with an affinity that is 5- to 10-fold higher than its affinity for the D2DAR.[200] This observation suggests that the dopamine D4DAR may mediate atypical features of antipsychotics; however, this still remains to be established.[209,210] Second, the D4DAR displays lower affinity to (+)-butaclamol, fluphenazine, s-sulpiride, and raclopride compared to the D2DARs and D3DARs.[200,211] These features are sometimes used to distinguish D4 function or expression from that of D2DARs or D3DARs. Third, epinephrine and norepinephrine bind with high affinity to the human D4DAR ($K_i = 14$ nM and 33 nM, respectively), although dopamine is more potent ($K_i = 0.9$ nM) and induces D4DAR-mediated signaling.[212–214] These findings suggest that D4DARs may represent a common site of integration of catecholamine signaling in the brain and the periphery.[213] No major pharmacological differences have been reported for the most common variants of this receptor (D4.2, D4.4, and D4.7).

D4DAR-specific ligands

The potential of the D4DAR as an antipsychotic target resulted in the development of various D4DAR-specific ligands.[209] For example, PD168077 has been used as a D4-selective agonist,[215,216] and L-745,870[215,217] and PNU-101,387G[214,218] are used as D4-specific antagonists.

D4DAR Signaling Mechanisms

As with the D2DAR, the signal transduction pathway of the D4DAR is mostly dependent on a pertussis toxin–sensitive G proteins ($G_{i/o}$). Inhibition of AC activity by activation of D4DARs has been shown in heterologous systems and *in vivo*.[209] In the mouse retina, dopamine modulation of the photoreceptor cAMP level has been shown to be pharmacologically consistent with D4DARs,[219] and the genetic ablation of D4DAR resulted in the loss of a quinpirole-elicited decrease in cAMP levels in retina.[220] These findings clearly show physiological coupling of the D4DAR to G protein and AC.

The D4 receptor may also couple to several other second messenger systems. In CHO cells, the D4DAR potentiates ATP- or Ca^{2+} ionophore-stimulated arachidonic acid release and stimulates extracellular acidification through the amiloride-sensitive Na^+/H^+ exchanger. In a mesencephalic cell line, the D4DAR reduces a voltage-dependent outward K^+ current contrary to the D2 and D3DARs that increase this current.[221] Like the D2DAR, the D4DAR inhibits L-type calcium channels in GH4C1 cells[222,223] and AtT-20 cells[222] and activates the MAPK ERK1 (extracellular signal-regulated kinase 1) and ERK2 in CHO-K1 cells.[224] The D4DAR can also activate the inwardly rectifying potassium channel (GIRK) via a Gβγ-dependent pathway, as does the D2DAR.[185,214,215] Recently, it has been shown that the D4DAR up-regulates $Ca^{2+}/$ calmodulin kinase II (CaMKII) activity through the stimulation of phospholipase C (PLC) in prefrontal cortex (PFC) slices, but how the D4 activates the PLC pathway is not clear.[226]

D4DAR Regulation

Regulation of D4DAR expression

In vivo regulation of D4DAR expression has been studied in the context of schizophrenia, antipsychotic drug response, and major depression.[209,227] The mechanism by which dopamine D4DAR expression is regulated is not yet understood. The *in vivo* changes in D4DAR mRNA expression due to antipsychotic and MK-801 treatment suggest that receptor density may be regulated at the level of transcription or RNA stability[209,228]; however, heterologous expression experiments with D4DARs in a variety of cell lines suggest that posttranslational modification can also strongly contribute to expression.[209,229] Little is known about how D4DAR density is regulated by desensitization and/or down-regulation mechanisms *in vivo*, if at all.

Regulation of the D4DAR by protein–protein interactions

The D4DAR has a proline-rich domain in ICL3, particularly the VNTR region and its flanking region. This proline-rich area contains multiple copies of the PXXP motif, which is considered to be the core consensus sequence for SH3 binding domains. It has been shown that the proline-rich domain of the D4DAR can interact with a variety of SH3-domain containing proteins such as Grb2.[224] Recently, it was also found that KLHL12, a BTB-Kelch protein, binds to the polymorphic VNTR region of the D4 receptor and induces D4 receptor ubiquitination by building up an E3 ubiquitination ligase complex.[230] The functional/physiological relevance of these interactions with the D4DAR remains to be elucidated.

D4DAR Localization

Expression of D4DAR mRNA is most abundant in retina,[201,219] followed by prefrontal cortex, amygdala, hippocampus, hypothalamus, and pituitary, and is sparse in the basal ganglia.[231] Immunohistochemical studies in primate brain indicate that the D4DAR is present in both pyramidal and nonpyramidal neurons of the cortex, particularly layer V, and in the hippocampus.[232] Most of the nonpyramidal D4DAR-positive cortical and hippocampal neurons are γ-aminobutyric acid (GABA)-producing neurons. D4DAR-positive neurons in thalamic nuclei, globus pallidus, and substantia nigra pars reticulata are also GABAergic. Since the D4DAR is highly expressed in the prefrontal cortex and to a lesser extent in the basal ganglia, the D4DAR is thought to play a role in the control of cognition, reasoning, perception, and emotion rather than in motor control.[231,232] The expression of D4DARs is not confined to the central nervous system (CNS). D4 receptor expression has been reported in the cardiac atrium,[233,234] in lymphocytes,[235] and in the cortical and medullary collecting ducts of the kidney.[236]

THE D5 DOPAMINE RECEPTOR SUBTYPE

D5DAR Structure and Pharmacology

The human D5DAR was first cloned in 1991 (GenBank Accession NP_000789), followed shortly thereafter by the cloning of the rat homologue (initially referred to as D_{1B}).[237–240] The D5DAR is encoded by an intronless gene located on chromosome 4p16.1; however, two pseudogenes and multiple missense variants of the D5DAR gene have been identified. The pseudogenes encode truncated versions of the receptor that are transcribed but whose function is currently unclear.[241,242]

The D5DAR is structurally and pharmacologically most similar to the D1DAR and is therefore classified as part of the D1-like receptor family. The amino acid sequences encoding the transmembrane regions of the D5DAR are 80% identical to those of the D1DAR, whereas D5DAR-specific sequences are located within the intracellular loops and the C- terminus.[238,243]

The affinities of the D5DAR and D1DAR for antagonists such as (+)-butaclamol and SCH-23390 are similar. This is also true for many of the agonists, although dopamine itself has approximately 10-fold greater affinity for the D5DAR than for the D1DAR.[237,239] The lack of D5DAR-selective ligands has greatly impeded the characterization of the D5 receptor independently of the D1DAR. Recently, a D5DAR-selective antagonist (4-chloro-3-hyrdoxymethyl-5,6,7,8,9,14-hexahydro-dibenz[d,z]azecine) was reported that displays affinity for the D5DAR in the picomolar range versus low nanomolar affinity for the D1DAR.[244] This D5DAR-selective antagonist is a good starting point to distinguish the D5 from the D1 pharmacologically while we await the further development of D5DAR-selective agents.

Several structure-function relationships have been identified, including amino acid residues important for membrane localization, receptor pharmacology, and signaling. N-linked glycosylation within the N-terminus of the D5DAR appears to be important for the correct localization of the receptor at the plasma membrane. For example, tunicamycin treatment (an inhibitor of N-linked glycosylation) of cells expressing the D5DAR prevents membrane localization—a response that is mimicked by mutation of an amino terminal asparagine residue.[20] Interestingly, N-linked glycosylation does not appear to be a prerequisite for D1DAR plasma membrane localization.[20]

As mentioned previously, several human D5DAR missense and nonsense polymorphisms have been reported, some of which significantly impact the pharmacology of the receptor. For example, N351D in the TMD7 decreases the affinity of the receptor for dopamine and SKF-38393 by 10-fold and 3-fold, respectively. However, L88F within TMD2 increases dopamine binding affinity while reducing the affinities of the antagonists SCH-23390 and risperidone.[242] These findings are not surprising as residues within the TMD regions are proposed to form the binding pocket for biogenic amines.

Another important discovery is the role of the C-terminal region of the ICL3 in constitutive D5DAR signaling. Specifically, site-directed mutagenesis of I288F in this region abolishes the constitutive activity of the D5DAR. Moreover, the D5DAR I188F mutant displays pharmacological and signaling properties resembling those of the D1DAR.[245]

D5DAR Signaling Mechanisms

Similar to the D1DAR, the D5DAR couples to $G\alpha_s$ and activates AC, resulting in cAMP accumulation.[238–240] In contrast to the D1DAR, the D5DAR displays agonist-independent coupling to AC, suggesting that this receptor is constitutively active.[246] Interestingly, the D5DAR has been shown to couple to multiple G proteins such as $G\alpha_z$. $G\alpha_z$ coupling decreases AC activity in certain cellular environments[247,248]; however, the signaling cascade elicited by the coupling of the D5DAR to $G\alpha_z$ is unclear.[249]

As with many GPCRs, the D5DAR regulates multiple signaling cascades. For example, cross-talk between the D5DAR and several membrane receptors, such as the γ-aminobutyric acid receptor-A ($GABA_AR$), the N-methyl-D aspartic acid receptor (NMDAR), and the angiotensin II type 1 receptor (AT1), has been reported. The D5DAR interacts with the $GABA_A$ receptor through direct association of the C-terminal domain of the D5DAR with the ICL2 of the $GABA_A$ γ2 receptor subunit.[250] This interaction confers reciprocal diminution of D5DAR and $GABA_AR$ signaling. For example, $GABA_A$ antagonist treatment decreases dopamine-stimulated cAMP accumulation, and dopamine treatment decreases $GABA_AR$-mediated current.[250]

D5DAR signaling has also been implicated in the redistribution of the glutamate-gated NMDAR. The D5DAR-dependent activation of the cAMP/PKA pathway in the ventral tegmental area promotes the incorporation of NMDARs in the membrane and increases NMDA excitatory postsynaptic currents (Fig. 3.1.4).[251] In the brain, D5DAR regulation of both $GABA_AR$ and NMDAR may provide mechanisms that promote synaptic plasticity.

Understanding the role of the D5DAR in disease, and distinguishing its role from that of the D1DAR, has been facilitated by the generation of mice that lack the D5DAR (D5DAR-/-). The D5DAR-/- mice are viable and fertile, with normal neurological reflexes; however, they are hypertensive due to increased sympathetic tone.[252] Moreover, D5DAR mutant mice display an altered startle response, prepulse inhibition, and exploratory locomotion behaviors.[253]

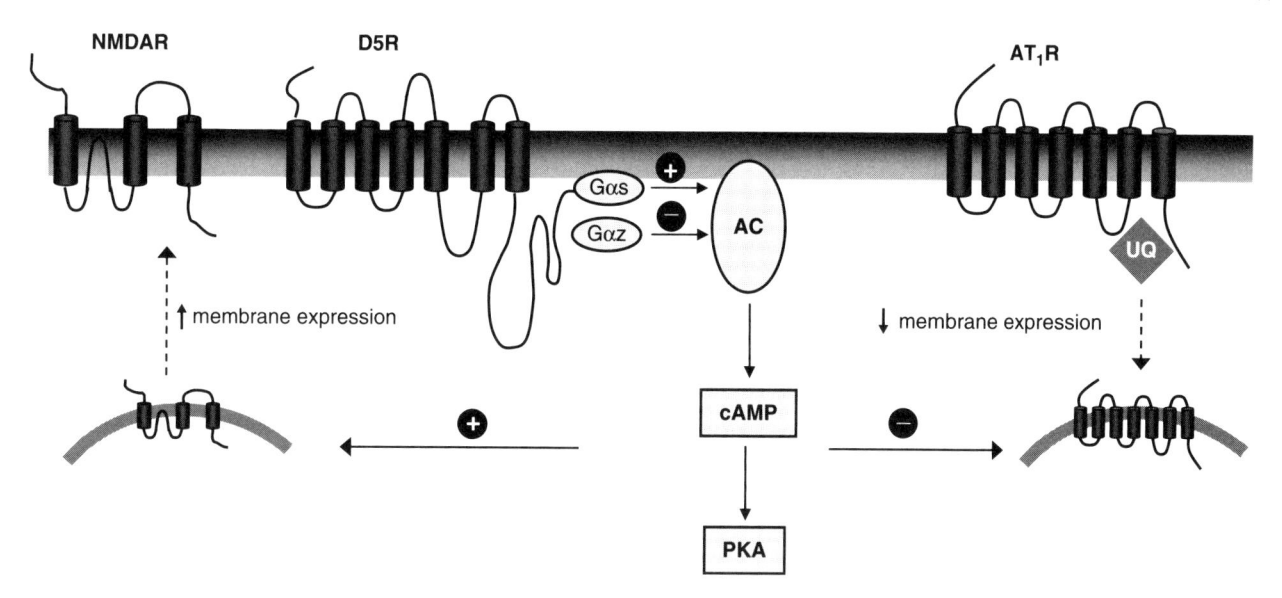

FIGURE 3.1.4. D5DAR coupling to adenylyl cyclase and regulation of NMDA and AT$_1$ receptor signaling. The D5DAR is constitutively active and couples to Gα_s to activate adenylyl cyclase and increase cAMP production. D$_5$DAR also couples to Gα_z. Activation of the D5DAR signaling pathway increases and decreases the membrane expression of the NMDAR and AT$_1$R, respectively. AC, adenylyl cyclase; AT$_1$R, AT$_1$ receptor; PKA, protein kinase A; UQ, ubiquitin.

The D5DAR is an important component for the onset of hypertension—a disease that is associated with the dysregulation of renal sodium balance. Dopamine and angiotensin II have opposing effects on sodium transport in the kidney. Dopamine and angiotensin II increase sodium excretion and transport, respectively.[254–257] In fact, recent studies show that the D5DAR regulates angiotensin$_1$ receptor (AT$_1$R) signaling in the kidney. The D5DAR decreases AT$_1$R expression in the renal proximal tubules; however, the amount of AT$_1$R expressed in D5DAR-/-mice is increased compared to expression in the wild type. Conversely, renal D5DAR expression is increased in AT$_1$R-/- mice.[258] Cross-talk between the two signaling pathways is not mediated by a direct receptor–receptor interaction. Instead, activation of the D5DAR increases the degradation of glycosylated AT$_1$R via a ubiquitin-proteasome pathway[259] (Fig. 3.1.4).

AT$_1$R signaling enhances phospholipase D activity and the generation of reactive oxygen species that contribute to the proliferation of vascular smooth muscle cells and the pathogenesis of hypertension.[260] Interestingly, the D5DAR also regulates phospholipase D (PLD) expression and function. The activity and expression of PLD2 is increased in D5DAR-/-mice. Consistent with these findings, antagonist or agonist treatment of HEK293 cells overexpressing the D5DAR increases or diminishes the expression and activity of PLD2, respectively. The apparent D5DAR regulation of PLCD2 appears to be at the protein level as PLCD2 mRNA levels are not affected.[261] The D5DAR-dependent regulation of PLD2 may be the consequence of the D5DAR-mediated degradation of the AT$_1$R that has recently been reported.[259] Taken together, these findings highlight the importance of the D5DAR-dependent regulation of AT$_1$R and PLD2 signaling pathways and the onset of hypertension.

D5DAR Localization

The D5DAR is expressed in multiple regions of the brain, including the substantia nigra, hypothalamus, striatum, cerebral cortex, nucleus accumbens, and olfactory tubercle.[262] In particular, the D5DAR is expressed within the limbic region, which is rich in dopaminergic innervations and is associated with a number of physiological attributes such as mood, arousal, addiction, and locomotion. Interestingly, postmortem studies of patients with Alzheimer's disease show a significant increase in D5DAR-immunoreactive neurons.[263]

The D5DAR is also expressed in the kidney (renal proximal and distal tubules, cortical collecting ducts, tunica media of arterioles, and ascending limbs of Henle) and in pulmonary and lobar arteries, consistent with the role of the D5DAR in hypertension.[264–266] Additionally, D5DAR mRNA has been identified in peripheral blood lymphocytes and is up-regulated in patients with Tourette's syndrome.[267]

REFERENCES

1. Dearry A, Gingrich JA, Falardeau P, Fremeau RT Jr, Bates MD, Caron MG. Molecular cloning and expression of the gene for a human D1 dopamine receptor. *Nature.* 1990;347(6288):72–76.
2. Monsma FJ Jr, Mahan LC, McVittie LD, Gerfen CR, Sibley DR. Molecular cloning and expression of a D1 dopamine receptor linked to adenylyl cyclase activation. *Proc Natl Acad Sci USA.* 1990;87(17):6723–6727.
3. Sunahara RK, Niznik HB, Weiner DM, et al. Human dopamine D1 receptor encoded by an intronless gene on chromosome 5. *Nature.* 1990;347(6288):80–83.
4. El-Ghundi M, Fletcher PJ, Drago J, Sibley DR, O'Dowd BF, George SR. Spatial learning deficit in dopamine D(1) receptor knockout mice. *Eur J Pharmacol.* 1999;383(2):95–106.
5. Smith DR, Striplin CD, Geller AM, et al. Behavioural assessment of mice lacking D1A dopamine receptors. *Neuroscience.* 1998;86(1):135–146.
6. El-Ghundi M, O'Dowd BF, George SR. Prolonged fear responses in mice lacking dopamine D1 receptor. *Brain Res.* 2001;892(1):86–93.
7. Pierce KL, Premont RT, Lefkowitz RJ. Seven-transmembrane receptors. *Nat Rev Mol Cell Biol.* 2002;3(9):639–650.
8. Palczewski K, Kumasaka T, Hori T, et al. Crystal structure of rhodopsin: a G protein–coupled receptor. *Science.* 2000;289(5480):739–745.
9. Warne T, Serrano-Vega MJ, Baker JG, et al. Structure of a beta1-adrenergic G-protein–coupled receptor. *Nature.* 2008;454(7203):486–491.
10. Cherezov V, Rosenbaum DM, Hanson MA, et al. High-resolution crystal structure of an engineered human beta2-adrenergic G protein–coupled receptor. *Science.* 2007;318(5854):1258–1265.
11. Rasmussen SG, Choi HJ, Rosenbaum DM, et al. Crystal structure of the human beta2 adrenergic G-protein–coupled receptor. *Nature.* 2007;450(7168):383–387.
12. Park JH, Scheerer P, Hofmann KP, Choe HW, Ernst OP. Crystal structure of the ligand-free G-protein–coupled receptor opsin. *Nature.* 2008;454(7201):183–187.
13. Scheerer P, Park JH, Hildebrand PW, et al. Crystal structure of opsin in its G-protein–interacting conformation. *Nature.* 2008;455(7212):497–502.
14. Adams MR, Ward RP, Dorsa DM. Dopamine receptor modulation of gene expression in the brain. In: Neve KA, Neve RL, eds. *The Dopamine Receptors.* Totowa, NJ: Humana Press; 1997:305–342.
15. Jin H, George SR, Bouvier M, O'Dowd BF. Palmitoylation of G protein coupled receptors. In: Benovic JL, Sibley DR, Strader CD, eds. *Regulation of G Protein Coupled Receptor Function and Expression: Receptor Biochemistry and Methodology.* Hoboken, NJ: Wiley; 1999:93–117.
16. Torrecilla I, Tobin AB. Co-ordinated covalent modification of G-protein coupled receptors. *Curr Pharm Des.* 2006;12(14):1797–1808.
17. Moffett S, Mouillac B, Bonin H, Bouvier M. Altered phosphorylation and desensitization patterns of a human beta 2-adrenergic receptor lacking the palmitoylated Cys341. *Embo J.* 1993;12(1):349–356.
18. Moffett S, Adam L, Bonin H, Loisel TP, Bouvier M, Mouillac B. Palmitoylated cysteine 341 modulates phosphorylation of the beta2-adrenergic receptor by the cAMP-dependent protein kinase. *J Biol Chem.* 1996;271(35):21490–21497.
19. Jin H, Xie Z, George SR, O'Dowd BF. Palmitoylation occurs at cysteine 347 and cysteine 351 of the dopamine D(1) receptor. *Eur J Pharmacol.* 1999;386(2–3):305–312.
20. Karpa KD, Lidow MS, Pickering MT, Levenson R, Bergson C. N-linked glycosylation is required for plasma membrane localization of D5, but not D1, dopamine receptors in transfected mammalian cells. *Mol Pharmacol.* 1999;56(5):1071–1078.
21. Demchyshyn LL, McConkey F, Niznik HB. Dopamine D5 receptor agonist high affinity and constitutive activity profile conferred by carboxyl-terminal tail sequence. *J Biol Chem.* 2000;275(31):23446–23455.
22. Tumova K, Iwasiow RM, Tiberi M. Insight into the mechanism of dopamine D1-like receptor activation: evidence for a molecular interplay between the third extracellular loop and the cytoplasmic tail. *J Biol Chem.* 2003;278(10):8146–8153.
23. Sugamori KS, Scheideler MA, Vernier P, Niznik HB. Dopamine D1B receptor chimeras reveal modulation of partial agonist activity by carboxyl-terminal tail sequences. *J Neurochem.* 1998;71(6):2593–2599.
24. MacKenzie RG, Steffey ME, Manelli AM, Pollock NJ, Frail DE. A D1/D2 chimeric dopamine receptor mediates a D1 response to a D2-selective agonist. *FEBS Lett.* 1993;323(1–2):59–62.
25. Strader CD, Sigal IS, Dixon RA. Structural basis of beta-adrenergic receptor function. *FASEB J.* 1989;3(7):1825–1832.
26. Al-Fulaij MA, Ren Y, Beinborn M, Kopin AS. Pharmacological analysis of human D1 AND D2 dopamine receptor missense variants. *J Mol Neurosci.* 2008;34(3):211–223.
27. Lan H, Durand CJ, Teeter MM, Neve KA. Structural determinants of pharmacological specificity between D(1) and D(2) dopamine receptors. *Mol. Pharmacol.* 2006;69(1):185–194.
28. O'Dowd BF, Ji X, Alijaniaram M, et al. Dopamine receptor oligomerization visualized in living cells. *J Biol. Chem.* 2005;280(44):37225–37235.
29. Lee SP, So CH, Rashid AJ, et al. Dopamine D1 and D2 receptor co-activation generates a novel phospholipase C-mediated calcium signal. *J Biol Chem.* 2004;279(34):35671–35678.
30. Gines S, Hillion J, Torvinen M, et al. Dopamine D1 and adenosine A1 receptors form functionally interacting heteromeric complexes. *Proc. Natl Acad Sci USA.* 2000;97(15):8606–8611.
31. Hazelwood LA, Free RB, Cabrera DM, Sibley DR. Dopamine receptor interacting proteins: unraveling the receptor signalplex. *FASEB J.* 2008;22(1):1.
32. Lee FJ, Xue S, Pei L, et al. Dual regulation of NMDA receptor functions by direct protein–protein interactions with the dopamine D1 receptor. *Cell.* 2002;111(2):219–230.
33. Bermak JC, Li M, Bullock C, Zhou QY. Regulation of transport of the dopamine D1 receptor by a new membrane-associated ER protein. *Nat Cell Biol.* 2001;3(5):492–498.
34. Free RB, Hazelwood LA, Cabrera DM, et al. D1 and D2 dopamine receptor expression is regulated by direct interaction with the chaperone protein calnexin. *J Biol Chem.* 2007;282(29):21285–21300.
35. Kong MM, Hasbi A, Mattocks M, Fan T, O'Dowd BF, George SR. Regulation of D1 dopamine receptor trafficking and signaling by caveolin-1. *Mol Pharmacol.* 2007;72(5):1157–1170.
36. Heydorn A, Sondergaard BP, Hadrup N, Holst B, Haft CR, Schwartz TW. Distinct in vitro interaction pattern of

dopamine receptor subtypes with adaptor proteins involved in post-endocytotic receptor targeting. *FEBS Lett.* 2004;556(1–3):276–280.
37. Macey TA, Liu Y, Gurevich VV, Neve KA. Dopamine D1 receptor interaction with arrestin3 in neostriatal neurons. *J Neurochem.* 2005;93(1):128–134.
38. Neve KA, Seamans JK, Trantham-Davidson H. Dopamine receptor signaling. *J Receptor Signal Transduction Res.* 2004;24(3):165–205.
39. Konig B, Arendt A, McDowell JH, Kahlert M, Hargrave PA, Hofmann KP. Three cytoplasmic loops of rhodopsin interact-with transducin. *Proc Natl Acad Sci USA.* 1989;86(18):6878–6882.
40. Kim OJ, Gardner BR, Williams DB, et al. The role of phosphorylation in D1 dopamine receptor desensitization: evidence for a novel mechanism of arrestin association. *J Biol Chem.* 2004;279(9):7999–8010.
41. Taylor JR, Lawrence MS, Redmond DE Jr, et al. Dihydrexidine, a full dopamine D1 agonist, reduces MPTP-induced parkinsonism in monkeys. *Eur J Pharmacol.* 1991;199(3):389–391.
42. Kebabian JW, Britton DR, DeNinno MP, et al. A-77636: a potent and selective dopamine D1 receptor agonist with antiparkinsonian activity in marmosets. *Eur J Pharmacol.* 1992;229(2–3):203–209.
43. Shiosaki K, Jenner P, Asin KE, et al. ABT-431: the diacetyl prodrug of A-86929, a potent and selective dopamine D1 receptor agonist: in vitro characterization and effects in animal models of Parkinson's disease. *J Pharmacol Exp Ther.* 1996;276(1):150–160.
44. Mailman RB. GPCR functional selectivity has therapeutic impact. *Trends Pharmacol Sci.* 2007;28(8):390–396.
45. Lewis MM, Huang X, Nichols DE, Mailman RB. D1 and functionally selective dopamine agonists as neuroprotective agents in Parkinson's disease. *CNS Neurol Disord Drug Targets.* 2006;5(3):345–353.
46. Schneider JS, Sun ZQ, Roeltgen DP. Effects of dihydrexidine, a full dopamine D-1 receptor agonist, on delayed response performance in chronic low dose MPTP-treated monkeys. *Brain Res.* 1994;663(1):140–144.
47. Braun A, Fabbrini G, Mouradian MM, Serrati C, Barone P, Chase TN. Selective D-1 dopamine receptor agonist treatment of Parkinson's disease. *J Neural Transm.* 1987;68(1–2):41–50.
48. Zhang A, Neumeyer JL, Baldessarini RJ. Recent progress in development of dopamine receptor subtype-selective agents: potential therapeutics for neurological and psychiatric disorders. *Chem Rev.* 2007;107(1):274–302.
49. Konig B, Gratzel M. Site of dopamine D1 receptor binding to Gs protein mapped with synthetic peptides. *Biochim Biophys Acta.* 1994;1223(2):261–266.
50. Huff RM. Signaling pathways modulated by dopamine receptors. In: Neve KA, Neve RL, eds. *The Dopamine Receptors.* Totowa, NJ: Humana Press; 1997:167–192.
51. Zhuang X, Belluscio L, Hen R. G(olf)alpha mediates dopamine D1 receptor signaling. *J Neurosci.* 2000;20(16):RC91.
52. Wang Q, Jolly JP, Surmeier JD, et al. Differential dependence of the D1 and D5 dopamine receptors on the G protein gamma 7 subunit for activation of adenyly lcyclase. *J Biol Chem.* 2001;276(42):39386–39393.
53. Clifford JJ, Tighe O, Croke DT, et al. Conservation of behavioural topography to dopamine D1-like receptor agonists in mutant mice lacking the D1A receptor implicates a D1-like receptor not coupled to adenylyl cyclase. *Neuroscience.* 1999;93(4):1483–1489.
54. Friedman E, Jin LQ, Cai GP, et al. D1-like dopaminergic activation of phosphoinositide hydrolysis is independent of D1A dopamine receptors: evidence from D1A knockout mice. *Mol Pharmacol.* 1997;51(1):6–11.
55. Montague DM, Striplin CD, Overcash JS, Drago J, Lawler CP, Mailman RB. Quantification of D1B(D5) receptors in dopamine D1A receptor-deficient mice. *Synapse.* 2001;39(4):319–322.
56. Sahu A, Tyeryar KR, Vongtau HO, Sibley DR, Undieh AS. D5 dopamine receptors are required for dopaminergic activation of phospholipase C. *Mol Pharmacol.* 2009;75(3):447–453.
57. Rashid AJ, So CH, Kong MM, et al. D1-D2 dopamine receptor heterooligomers with unique pharmacology are coupled to rapid activation of Gq/11 in the striatum. *Proc Natl Acad Sci USA.* 2007;104(2):654–659.
58. Le Novere N, Li L, Girault JA. DARPP-32: molecular integration of phosphorylation potential. *Cell Mol Life Sci.* 2008;65(14):2125–2127.
59. Fienberg AA, Hiroi N, Mermelstein PG, et al. DARPP-32: regulator of the efficacy of dopaminergic neurotransmission. *Science.* 1998;281(5378):838–842.
60. Schiffmann SN, Desdouits F, Menu R, et al. Modulation of the voltage-gated sodium current in rat striatal neurons by DARPP-32, an inhibitor of protein phosphatase. *Eur J Neurosci.* 1998;10(4):1312–1320.
61. Kohout TA, Lefkowitz RJ. Regulation of G protein–coupled receptor kinases and arrestins during receptor desensitization. *Mol Pharmacol.* 2003;63(1):9–18.
62. Krupnick JG, Benovic JL. The role of receptor kinases and arrestins in G protein–coupled receptor regulation. *Annu Rev Pharmacol Toxicol.* 1998;38:289–319.
63. Sibley DR, Neve KA. Regulation of dopamine receptor function and expression. In: Neve KA, Neve RL, eds. *The Dopamine Receptors.* Totowa, NJ: Humana Press; 1997:383–424.
64. Demchyshyn LL, O'Dowd BF, George SR. Structure of mammalian D1 and D5 dopamine receptors and their function and regulation in cells. In: Sidhu A, Laruelle M, Vernier P, eds. *Dopamine Receptors and Transporters.* 2nd ed. New York, NY: Marcel Dekker; 2003:45–76.
65. Gardner B, Liu ZF, Jiang D, Sibley DR. The role of phosphorylation/dephosphorylation in agonist-induced desensitization of D1 dopamine receptor function: evidence for a novel pathway for receptor dephosphorylation. *Mol Pharmacol.* 2001;59(2):310–321.
66. Pitcher JA, Payne ES, Csortos C, DePaoli-Roach AA, Lefkowitz RJ. The G-protein–coupled receptor phosphatase: a protein phosphatase type 2A with a distinct subcellular distribution and substrate specificity. *Proc Natl Acad Sci USA.* 1995;92:8343–8347.
67. Jiang D, Sibley DR. Regulation of D1 dopamine receptors with mutations of protein kinase phosphorylation sites: attenuation of the rate of agonist-induced desensitization. *Mol Pharmacol.* 1999;56:675–683.
68. Mason JN, Kozell LB, Neve KA. Regulation of dopamine D1 receptor trafficking by protein kinase A-dependent phosphorylation. *Mol Pharmacol.* 2002;61:806–816.
69. Tiberi M, Nash SR, Bertrand L, Lefkowitz RJ, Caron MG. Differential regulation of dopamine D1A receptor responsiveness by various G protein–coupled receptor kinases. *J Biol Chem.* 1996;271(7):3771–3778.
70. Watanabe H, Xu J, Bengra C, Jose PA, Felder RA. Desensitization of human renal D1 dopamine receptors by G

protein–coupled receptor kinase 4. *Kidney Int.* 2002;62(3):790–798.
71. Felder RA, Sanada H, Xu J, et al. G protein–coupled receptor kinase 4 gene variants in human essential hypertension. *Proc Natl Acad Sci USA.* 2002;99(6):3872–3877.
72. Rankin ML, Marinec PS, Cabrera DM, Wang Z, Jose PA, Sibley DR. The D1 dopamine receptor is constitutively phosphorylated by G protein–coupled receptor kinase 4. *Mol Pharmacol.* 2006;69(3):759–769.
73. Chaar ZY, Jackson A, Tiberi M. The cytoplasmic tail of the D1A receptor subtype: identification of specific domains controlling dopamine cellular responsiveness. *J Neurochem.* 2001;79(5):1047–1058.
74. Jackson A, Iwasiow RM, Chaar ZY, Nantel MF, Tiberi M. Homologous regulation of the heptahelical D1A receptor responsiveness: specific cytoplasmic tail regions mediate dopamine-induced phosphorylation, desensitization and endocytosis. *J Neurochem.* 2002;82(3):683–697.
75. Pinna LA, Ruzzene M. How do protein kinases recognize their substrates? *Biochim Biophys Acta.* 1996;1314(3):191–225.
76. Newton AC. Protein kinase C: structural and spatial regulation by phosphorylation, cofactors, and macromolecular interactions. *Chem Rev.* 2001;101(8):2353–2364.
77. Jackson A, Sedaghat K, Minerds K, James C, Tiberi M. Opposing effects of phorbol-12-myristate-13-acetate, an activator of protein kinase C, on the signaling of structurally related human dopamine D1 and D5 receptors. *J Neurochem.* 2005;95(5):1387–1400.
78. Jackson A, Iwasiow RM, Chaar ZY, Nantel M-F, Tiberi M. Homologous regulation of the heptahelical D1A receptor responsiveness: specific cytoplasmic tail regions mediate dopamine-induced phosphorylation, desensitization, and endocytosis. *J Neurochem.* 2002;82:683–697.
79. Lamey M, Thompson M, Varghese G, et al. Distinct residues in the carboxyl tail mediate agonist-induced desensitization and internalization of the human dopamine D1 receptor. *J Biol Chem.* 2002;277(11):9415–9421.
80. Kim O-J, Gardner B, Williams DB, et al. The role of phosphorylation in D1 dopamine receptor desensitization: evidence for a novel mechanism of arrestin association. *J Biol Chem.* 2003;279:7999–8010.
81. Vickery RG, von Zastrow M. Distinct dynamin-dependent and –independent mechanisms target structurally homologous dopamine receptors to different endocytic membranes. *J Cell Biol.* 1999;144(1):31–43.
82. Lamey M, Thompson M, Varghese G, et al. Distinct residues in the carboxyl tail mediate agonist-induced desensitization and internalization of the human dopamine D1 receptor. *J Biol Chem.* 2002;277(11):9415–9421.
83. Jensen AA, Pedersen UB, Kiemer A, Din N, Andersen PH. Functional importance of the carboxyl tail cysteine residues in the human D1 dopamine receptor. *J Neurochem.* 1995;65(3):1325–1331.
84. Ozono R, O'Connell DP, Wang Z-Q, et al. Localization of the dopamine D1 receptor protein in the human heart and kidney. *Hypertension.* 1997;30(3):725–729.
85. Ariano MA. Distribution of dopamine receptors. In: Neve KA, Sibley DR, eds. *The Dopamine Receptors.* Totowa, NJ: Humana Press; 1997:77–97.
86. Zeng C, Sanada H, Watanabe H, Eisner GM, Felder RA, Jose PA. Functional genomics of the dopaminergic system in hypertension. *Physiol Genomics.* 2004;19(3):233–246.
87. Xu J, Li XX, Albrecht FE, Hopfer U, Carey RM, Jose PA. Dopamine(1) receptor, G(salpha), and Na(+)-H(+) exchanger interactions in the kidney in hypertension. *Hypertension.* 2000;36(3):395–399.
88. Bunzow JR, Van Tol HH, Grandy DK, et al. Cloning and expression of a rat D2 dopamine receptor cDNA. *Nature.* 1988;336(6201):783–787.
89. Gandelman KY, Harmon S, Todd RD, O'Malley KL. Analysis of the structure and expression of the human dopamine D2A receptor gene. *J Neurochem.* 1991;56(3):1024–1029.
90. Monsma FJ Jr, McVittie LD, Gerfen CR, Mahan LC, Sibley DR. Multiple D2 dopamine receptors produced by alternative RNA splicing. *Nature.* 1989;342(6252):926–929.
91. Usiello A, Baik JH, Rouge-Pont F, et al. Distinct functions of the two isoforms of dopamine D2 receptors. *Nature.* 2000;408(6809):199–203.
92. Sokoloff P, Giros B, Martres MP, Bouthenet ML, Schwartz JC. Molecular cloning and characterization of a novel dopamine receptor (D3) as a target for neuroleptics. *Nature.* 1990;347(6289):146–151.
93. Neve KA, Neve RL. Molecular biology of dopamine receptors. In: Neve KA, Neve RL, eds. *The Dopamine Receptor.* Totowa, NJ: Humana Press; 1997:27–76.
94. Shi L, Javitch JA. The second extracellular loop of the dopamine D2 receptor lines the binding-site crevice. *Proc Natl Acad Sci USA.* 2004;101(2):440–445.
95. Wilson J, Lin H, Fu D, Javitch JA, Strange PG. Mechanisms of inverse agonism of antipsychotic drugs at the D(2) dopamine receptor: use of a mutant D(2) dopamine receptor that adopts the activated conformation. *J Neurochem.* 2001;77(2):493–504.
96. Javitch JA, Fu D, Chen J. Residues in the fifth membrane-spanning segment of the dopamine D2 receptor exposed in the binding-site crevice. *Biochemistry (Mosc).* 1995;34(50):16433–16439.
97. Javitch JA, Fu D, Chen J, Karlin A. Mapping the binding-site crevice of the dopamine D2 receptor by the substituted-cysteine accessibility method. *Neuron.* 1995;14(4):825–831.
98. Ballesteros J, Kitanovic S, Guarnieri F, et al. Functional microdomains in G-protein–coupled receptors: the conserved arginine-cage motif in the gonadotropin-releasing hormone receptor. *J Biol Chem.* 1998;273(17):10445–10453.
99. Lachowicz JE, Sibley DR. Chimeric D2/D3 dopamine receptor coupling to adenylyl cyclase. *Biochem Biophys Res Commun.* 1997;237(2):394–399.
100. Lane JR, Powney B, Wise A, Rees S, Milligan G. G protein coupling and ligand selectivity of the D2L and D3 dopamine receptors. *J Pharmacol Exp Ther.* 2008;325(1):319–330.
101. Ilani T, Fishburn CS, Levavi-Sivan B, Carmon S, Raveh L, Fuchs S. Coupling of dopamine receptors to G proteins: studies with chimeric D2/D3 dopamine receptors. *Cell Mol Neurobiol.* 2002;22(1):47–56.
102. Guo W, Shi L, Javitch JA. The fourth transmembrane segment forms the interface of the dopamine D2 receptor homodimer. *J Biol Chem.* 2003;278(7):4385–4388.
103. Guo W, Urizar E, Kralikova M, et al. Dopamine D2 receptors form higher order oligomers at physiological expression levels. *EMBO J.* 2008;27(17):2293–2304.
104. George SR, O'Dowd BF. A novel dopamine receptor signaling unit in brain: heterooligomers of D1 and D2 dopamine receptors. *Sci World J.* 2007;7:58–63.

105. Scarselli M, Novi F, Schallmach E, et al. D2/D3 dopamine receptor heterodimers exhibit unique functional properties. *J Biol Chem.* 2001;276(32):30308–30314.
106. Kearn CS, Blake-Palmer K, Daniel E, Mackie K, Glass M. Concurrent stimulation of cannabinoid CB1 and dopamine D2 receptors enhances heterodimer formation: a mechanism for receptor cross-talk? *Mol Pharmacol.* 2005;67(5):1697–1704.
107. Torvinen M, Kozell LB, Neve KA, Agnati LF, Fuxe K. Biochemical identification of the dopamine D2 receptor domains interacting with the adenosine A2A receptor. *J Mol Neurosci.* 2004;24(2):173–180.
108. Jeanneteau F, Guillin O, Diaz J, Griffon N, Sokoloff P. GIPC recruits GAIP (RGS19) to attenuate dopamine D2 receptor signaling. *Mol Biol Cell.* 2004;15(11):4926–4937.
109. Smith FD, Oxford GS, Milgram SL. Association of the D2 dopamine receptor third cytoplasmic loop with spinophilin, a protein phosphatase-1-interacting protein. *J Biol Chem.* 1999;274(28):19894–19900.
110. Macey TA, Gurevich VV, Neve KA. Preferential Interaction between the dopamine D2 receptor and Arrestin2 in neostriatal neurons. *Mol Pharmacol.* 2004;66(6):1635–1642.
111. Jiang D, Sibley DR. Regulation of D(1) dopamine receptors with mutations of protein kinase phosphorylation sites: attenuation of the rate of agonist-induced desensitization. *Mol Pharmacol.* 1999;56(4):675–683.
112. Hazelwood LA, Free RB, Cabrera DM, Skinbjerg M, Sibley DR. Reciprocal Modulation of Function between the D1 and D2 Dopamine Receptors and the Na$^+$,K$^+$-ATPase. *J Biol Chem.* 2008;283(52):36441–36453.
113. Standaert DG, Young AG. Treatment of central nervous system degenerative disorders. In: Hardman JG, Limbird LE, Gilman AG, eds. *Goodman and Gilman's The Pharmacological Basis of Therapeutics.* New York, NY: McGraw-Hill; 2001:549–569.
114. Chen JJ, Swope DM. Pharmacotherapy for Parkinson's disease. *Pharmacotherapy.* 2007;27(12 pt 2):161S–173S.
115. Constantinescu R. Update on the use of pramipexole in the treatment of Parkinson's disease. *Neuropsychiatr Dis Treat.* 2008;4(2):337–352.
116. Webber MA, Marder SR. Better pharmacotherapy for schizophrenia: what does the future hold? *Curr Psychiatry Rep.* 2008;10(4):352–358.
117. Seeman P. Targeting the dopamine D2 receptor in schizophrenia. *Expert Opinion Ther Targets.* 2006;10(4):515–531.
118. Bhattacharjee J, El-Sayeh HG. Aripiprazole versus typical antipsychotic drugs for schizophrenia. *Cochrane database of systematic reviews (Online).* 2008(3):CD006617.
119. Natesan S, Svensson KA, Reckless GE, et al. The dopamine stabilizers (S)-(-)-(3-methanesulfonyl-phenyl)-1-propyl-piperidine [(-)-OSU6162] and 4-(3-methanesulfonylphenyl)-1-propyl-piperidine (ACR16) show high in vivo D2 receptor occupancy, antipsychotic-like efficacy, and low potential for motor side effects in the rat. *J Pharmacol Exp Ther.* 2006;318(2):810–818.
120. Hadj Tahar A, Ekesbo A, Gregoire L, et al. Effects of acute and repeated treatment with a novel dopamine D2 receptor ligand on L-DOPA-induced dyskinesias in MPTP monkeys. *Eur J Pharmacol.* 2001;412(3):247–254.
121. Lahti RA, Tamminga CA, Carlsson A. Stimulating and inhibitory effects of the dopamine "stabilizer" (-)-OSU6162 on dopamine D(2) receptor function in vitro. *J Neural Transm.* 2007;114(9):1143–1146.
122. Lane JR, Powney B, Wise A, Rees S, Milligan G. Protean agonism at the dopamine D2 receptor: (S)-3-(3-hydroxyphenyl)-N-propylpiperidine is an agonist for activation of Go1 but an antagonist/inverse agonist for Gi1,Gi2, and Gi3. *Mol Pharmacol.* 2007;71(5):1349–1359.
123. Masri B, Salahpour A, Didriksen M, et al. Antagonism of dopamine D2 receptor/beta-arrestin 2 interaction is a common property of clinically effective antipsychotics. *Proc Natl Acad Sci USA.* 2008;105(36):13656–13661.
124. Schetz JA. Allosteric modulation of dopamine receptors. *Mini Rev Med Chem.* 2005;5(6):555–561.
125. De Camilli P, Macconi D, Spada A. Dopamine inhibits adenylate cyclase in human prolactin-secreting pituitary adenomas. *Nature.* 1979;278(5701):252–254.
126. Stoof JC, Kebabian JW. Opposing roles for D-1 and D-2 dopamine receptors in efflux of cyclic AMP from rat neostriatum. *Nature.* 1981;294(5839):366–368.
127. Neve KA, Henningsen RA, Bunzow JR, Civelli O. Functional characterization of a rat dopamine D-2 receptor cDNA expressed in a mammalian cell line. *Mol Pharmacol.* 1989;36(3):446–451.
128. Malek D, Munch G, Palm D. Two sites in the third inner loop of the dopamine D2 receptor are involved in functional G protein-mediated coupling to adenylate cyclase. *FEBS Lett.* 1993;325(3):215–219.
129. Kozell LB, Machida CA, Neve RL, Neve KA. Chimeric D1/D2 dopamine receptors: distinct determinants of selective efficacy, potency, and signal transduction. *J Biol Chem.* 1994;269(48):30299–30306.
130. Heusler P, Newman-Tancredi A, Castro-Fernandez A, Cussac D. Differential agonist and inverse agonist profile of antipsychotics at D2L receptors coupled to GIRK potassium channels. *Neuropharmacology.* 2007;52(4):1106–1113.
131. Strange PG. Agonism and inverse agonism at dopamine D2-like receptors. *Clin Exp Pharmacol Physiol Suppl.* 1999;26:S3–S9.
132. Hayes G, Biden TJ, Selbie LA, Shine J. Structural subtypes of the dopamine D2 receptor are functionally distinct: expression of the cloned D2A and D2B subtypes in a heterologous cell line. *Mol Endocrinol.* 1992;6(6):920–926.
133. Montmayeur JP, Borrelli E. Transcription mediated by a cAMP-responsive promoter element is reduced upon activation of dopamine D2 receptors. *Proc Natl Acad Sci USA.* 1991;88(8):3135–3139.
134. Montmayeur JP, Guiramand J, Borrelli E. Preferential coupling between dopamine D2 receptors and G-proteins. *Mol Endocrinol.* 1993;7(2):161–170.
135. Obadiah J, Avidor-Reiss T, Fishburn CS, et al. Adenylyl cyclase interaction with the D2 dopamine receptor family; differential coupling to Gi, Gz, and Gs. *Cell Mol Neurobiol.* 1999;19(5):653–664.
136. Leck KJ, Blaha CD, Matthaei KI, Forster GL, Holgate J, Hendry IA. Gz proteins are functionally coupled to dopamine D2-like receptors in vivo. *Neuropharmacology.* 2006;51(3):597–605.
137. Kuzhikandathil EV, Yu W, Oxford GS. Human dopamine D3 and D2L receptors couple to inward rectifier potassium channels in mammalian cell lines. *Mol Cell Neurosci.* 1998;12(6):390–402.
138. Lavine N, Ethier N, Oak JN, et al. G protein–coupled receptors form stable complexes with inwardly rectifying potassium channels and adenylyl cyclase. *J Biol Chem.* 2002;277(48):46010–46019.
139. Yan Z, Song WJ, Surmeier J. D2 dopamine receptors reduce N-type Ca^{2+} currents in rat neostriatal cholinergic interneurons

through a membrane-delimited, protein-kinase-C-insensitive pathway. *J Neurophysiol.* 1997;77(2):1003–1015.
140. Hernandez-Lopez S, Tkatch T, Perez-Garci E, et al. D2 dopamine receptors in striatal medium spiny neurons reduce L-type Ca^{2+} currents and excitability via a novel PLC[beta]1-IP3-calcineurin-signaling cascade. *J Neurosci.* 2000;20(24):8987–8995.
141. Kanterman RY, Mahan LC, Briley EM, et al. Transfected D2 dopamine receptors mediate the potentiation of arachidonic acid release in Chinese hamster ovary cells. *Mol Pharmacol.* 1991;39(3):364–369.
142. Bergson C, Levenson R, Goldman-Rakic PS, Lidow MS. Dopamine receptor-interacting proteins: the Ca(2+) connection in dopamine signaling. *Trends Pharmacol Sci.* 2003;24(9):486–492.
143. Ghahremani MH, Forget C, Albert PR. Distinct roles for Galpha(i)2 and Gbetagamma in signaling to DNA synthesis and Galpha(i)3 in cellular transformation by dopamine D2S receptor activation in BALB/c 3T3 cells. *Mol Cell Biol.* 2000;20(5):1497–1506.
144. Oak JN, Lavine N, Van Tol HH. Dopamine D(4) and D(2L) receptor stimulation of the mitogen-activated protein kinase pathway is dependent on trans-activation of the platelet-derived growth factor receptor. *Mol Pharmacol.* 2001;60(1):92–103.
145. Vial D, Piomelli D. Dopamine D2 receptors potentiate arachidonate release via activation of cytosolic, arachidonate-specific phospholipase A2. *J Neurochem.* 1995;64(6):2765–2772.
146. Senogles SE. D2s dopamine receptor mediates phospholipase D and antiproliferation. *Mol Cell Endocrinol.* 2003;209(1–2):61–69.
147. Senogles SE. The D2s dopamine receptor stimulates phospholipase D activity: a novel signaling pathway for dopamine. *Mol Pharmacol.* 2000;58(2):455–462.
148. Rahman Z, Schwarz J, Gold SJ, et al. RGS9 modulates dopamine signaling in the basal ganglia. *Neuron.* 2003;38(6):941–952.
149. Kovoor A, Seyffarth P, Ebert J, et al. D2 dopamine receptors colocalize regulator of G-protein signaling 9–2 (RGS9-2) via the RGS9 DEP domain, and RGS9 knock-out mice develop dyskinesias associated with dopamine pathways. *J Neurosci.* 2005;25(8):2157–2165.
150. Namkung Y, Sibley DR. Protein kinase C mediates phosphorylation, desensitization, and trafficking of the D2 dopamine receptor. *J Biol Chem.* 2004;279(47):49533–49541.
151. Ito K, Haga T, Lameh J, Sadee W. Sequestration of dopamine D2 receptors depends on coexpression of G-protein–coupled receptor kinases 2 or 5. *Eur J Biochem.* 1999;260(1):112–119.
152. Krupnick JG, Benovic JL. The role of receptor kinases and arrestins in G protein–coupled receptor regulation. *Annu Rev Pharmacol Toxicol.* 1998;38:289–319.
153. Tobin S, Newman AH, Quinn T, Shalev U. A role for dopamine D1-like receptors in acute food deprivation-induced reinstatement of heroin seeking in rats. *Int J Neuropsychopharmacol.* 14 2008:1–10.
154. Tobin AB. G-protein–coupled receptor phosphorylation: where, when and by whom. *Br J Pharmacol.* 2008;153(suppl 1):S167–S176.
155. Bouthenet ML, Souil E, Martres MP, Sokoloff P, Giros B, Schwartz JC. Localization of dopamine D3 receptor mRNA in the rat brain using in situ hybridization histochemistry: comparison with dopamine D2 receptor mRNA. *Brain Res.* 1991;564(2):203–219.
156. Meador-Woodruff JH, Mansour A, Bunzow JR, Van Tol HH, Watson SJ Jr, Civelli O. Distribution of D2 dopamine receptor mRNA in rat brain. *Proc Natl Acad Sci USA.* 1989;86(19):7625–7628.
157. Weiner DM, Levey AI, Sunahara RK, et al. D1 and D2 dopamine receptor mRNA in rat brain. *Proc Natl Acad Sci USA.* 1991;88(5):1859–1863.
158. Levey AI, Hersch SM, Rye DB, et al. Localization of D1 and D2 dopamine receptors in brain with subtype-specific antibodies. *Proc Natl Acad Sci USA.* 1993;90(19):8861–8865.
159. Picetti R, Saiardi A, Abdel Samad T, Bozzi Y, Baik JH, Borrelli E. Dopamine D2 receptors in signal transduction and behavior. *Crit Rev Neurobiol.* 1997;11(2–3):121–142.
160. Prou D, Gu WJ, Le Crom S, Vincent JD, Salamero J, Vernier P. Intracellular retention of the two isoforms of the D(2) dopamine receptor promotes endoplasmic reticulum disruption. *J Cell Sci.* 2001;114(pt 19):3517–3527.
161. Livingstone CD, Strange PG, Naylor LH. Molecular modelling of D2-like dopamine receptors. *Biochem J.* 1992;287(pt 1):277–282.
162. Robinson SR, Caron MG. Interactions of Dopamine Receptors with G Proteins. In: Neve KA, Neve RL, eds. *The Dopamine Receptors.* Totawa, NJ: Humana Press; 1997:137–165.
163. Vanhauwe JF, Josson K, Luyten WH, Driessen AJ, Leysen JE. G-protein sensitivity of ligand binding to human dopamine D(2) and D(3) receptors expressed in *Escherichia coli*: clues for a constrained D(3) receptor structure. *J Pharmacol Exp Ther.* 2000;295(1):274–283.
164. Snyder SH. Parkinson's disease. Fresh factors to consider. *Nature.* 1991;350(6315):195.
165. Schmauss C, Haroutunian V, Davis KL, Davidson M. Selective loss of dopamine D3-type receptor mRNA expression in parietal and motor cortices of patients with chronic schizophrenia. *Proc Natl Acad Sci USA.* 1993;90(19):8942–8946.
166. Karpa KD, Lin R, Kabbani N, Levenson R. The dopamine D3 receptor interacts with itself and the truncated D3 splice variant d3nf: D3-D3nf interaction causes mislocalization of D3 receptors. *Mol Pharmacol.* 2000;58(4):677–683.
167. Mierau J, Schneider FJ, Ensinger HA, Chio CL, Lajiness ME, Huff RM. Pramipexole binding and activation of cloned and expressed dopamine D2, D3 and D4 receptors. *Eur J Pharmacol.* 1995;290(1):29–36.
168. Eden RJ, Costall B, Domeney AM, et al. Preclinical pharmacology of ropinirole (SK&F 101468-A), a novel dopamine D2 agonist. *Pharmacol Biochem Behav.* 1991;38(1):147–154.
169. Sautel F, Griffon N, Sokoloff P, et al. Nafadotride, a potent preferential dopamine D3 receptor antagonist, activates locomotion in rodents. *J Pharmacol Exp Ther.* 1995;275(3):1239–1246.
170. Millan MJ, Peglion JL, Vian J, et al. Functional correlates of dopamine D3 receptor activation in the rat in vivo and their modulation by the selective antagonist, (+)-S 14297: 1. Activation of postsynaptic D3 receptors mediates hypothermia, whereas blockade of D2 receptors elicits prolactin secretion and catalepsy. *J Pharmacol Exp Ther.* 1995;275(2):885–898.
171. Haadsma-Svensson SR, Smith MW, Svensson K, Waters N, Carlsson A. The chemical structure of U99194A. *J Neural Transm.* 1995;99(1–3):1.
172. Blagg J, Allerton CM, Batchelor DV, et al. Design and synthesis of a functionally selective D3 agonist and its in vivo delivery via the intranasal route. *Bioorganic Med Chem Lett.* 2007;17(24):6691–6696.
173. Saur O, Hackling AE, Perachon S, Schwartz JC, Sokoloff P, Stark H. N-(4-(4-(2-halogenophenyl)piperazin-1-yl)butyl) substituted cinnamoyl amide derivatives as dopamine

D2 and D3 receptor ligands. *Arch Pharm.* 2007;340(4):178–184.
174. Leopoldo M, Lacivita E, De Giorgio P, et al. Design, synthesis, and binding affinities of potential positron emission tomography (PET) ligands for visualization of brain dopamine D3 receptors. *J Med Chem.* 2006;49(1):358–365.
175. Biswas S, Hazeldine S, Ghosh B, et al. Bioisosteric heterocyclic versions of 7-{[2-(4-phenyl-piperazin-1-yl)ethyl]propylamino}-5,6,7,8-tetrahydronaphthalen-2-ol: identification of highly potent and selective agonists for dopamine D3 receptor with potent in vivo activity. *J Med Chem.* 2008;51(10):3005–3019.
176. Bettinetti L, Schlotter K, Hubner H, Gmeiner P. Interactive SAR studies: rational discovery of super-potent and highly selective dopamine D3 receptor antagonists and partial agonists. *J Med Chem.* 2002;45(21):4594–4597.
177. Reavill C, Taylor SG, Wood MD, et al. Pharmacological actions of a novel, high-affinity, and selective human dopamine D(3) receptor antagonist, SB-277011-A. *J Pharmacol Exp Ther.* 2000;294(3):1154–1165.
178. Ahlgren-Beckendorf JA, Levant B. Signaling mechanisms of the D3 dopamine receptor. *J Receptor Signal Transduction Res.* 2004;24(3):117–130.
179. Grigoriadis D, Seeman P. Complete conversion of brain D2 dopamine receptors from the high- to the low-affinity state for dopamine agonists, using sodium ions and guanine nucleotide. *J Neurochem.* 1985;44(6):1925–1935.
180. Sokoloff P, Andrieux M, Besancon R, et al. Pharmacology of human dopamine D3 receptor expressed in a mammalian cell line: comparison with D2 receptor. *Eur J Pharmacol.* 1992;225(4):331–337.
181. Newman-Tancredi A, Cussac D, Audinot V, Pasteau V, Gavaudan G, Millan MJ. G protein activation by human dopamine D3 receptors in high-expressing Chinese hamster ovary cells: a guanosine-5'-O-(3-[35S]thio)-triphosphate binding and antibody study. *Mol Pharmacol.* 1999;55(3):564–574.
182. Chio CL, Lajiness ME, Huff RM. Activation of heterologously expressed D3 dopamine receptors: comparison with D2 dopamine receptors. *Mol Pharmacol.* 1994;45(1):51–60.
183. Robinson SW, Caron MG. Selective inhibition of adenylyl cyclase type V by the dopamine D3 receptor. *Mol Pharmacol.* 1997;52(3):508–514.
184. Liu LX, Monsma FJ Jr, Sibley DR, Chiodo LA. D2L, D2S, and D3 dopamine receptors stably transfected into NG108-15 cells couple to a voltage-dependent potassium current via distinct G protein mechanisms. *Synapse (New York)* 1996;24(2):156–164.
185. Werner P, Hussy N, Buell G, Jones KA, North RA. D2, D3, and D4 dopamine receptors couple to G protein-regulated potassium channels in *Xenopus* oocytes. *Mol Pharmacol.* 1996;49(4):656–661.
186. Seabrook GR, Kemp JA, Freedman SB, Patel S, Sinclair HA, McAllister G. Functional expression of human D3 dopamine receptors in differentiated neuroblastoma x glioma NG108-15 cells. *Br J Pharmacol.* 1994;111(2):391–393.
187. Kuzhikandathil EV, Oxford GS. Activation of human D3 dopamine receptor inhibits P/Q-type calcium channels and secretory activity in AtT-20 cells. *J Neurosci.* 1999;19(5):1698–1707.
188. Chen G, Kittler JT, Moss SJ, Yan Z. Dopamine D3 receptors regulate GABAA receptor function through a phospho-dependent endocytosis mechanism in nucleus accumbens. *J Neurosci.* 2006;26(9):2513–2521.
189. Cussac D, Newman-Tancredi A, Pasteau V, Millan MJ. Human dopamine D(3) receptors mediate mitogen-activated protein kinase activation via a phosphatidylinositol 3-kinase and an atypical protein kinase C-dependent mechanism. *Mol Pharmacol.* 1999;56(5):1025–1030.
190. Morris BJ, Newman-Tancredi A, Audinot V, Simpson CS, Millan MJ. Activation of dopamine D(3) receptors induces c-fos expression in primary cultures of rat striatal neurons. *J Neurosci Res.* 2000;59(6):740–749.
191. Tremblay M, Rouillard C, Levesque D. The antisense strategy applied to the study of dopamine D3 receptor functions in rat forebrain. *Prog Neuro-Psychopharmacol Biol Psychiatry.* 1998;22(5):857–882.
192. Kim KM, Valenzano KJ, Robinson SR, Yao WD, Barak LS, Caron MG. Differential regulation of the dopamine D2 and D3 receptors by G protein-coupled receptor kinases and beta-arrestins. *J Biol Chem.* 2001;276(40):37409–37414.
193. Oakley RH, Laporte SA, Holt JA, Barak LS, Caron MG. Molecular determinants underlying the formation of stable intracellular G protein-coupled receptor-beta-arrestin complexes after receptor endocytosis. *J Biol Chem.* 2001;276(22):19452–19460.
194. Suzuki M, Hurd YL, Sokoloff P, Schwartz JC, Sedvall G. D3 dopamine receptor mRNA is widely expressed in the human brain. *Brain Res.* 1998;779(1-2):58–74.
195. Levant B, Grigoriadis DE, DeSouza EB. Characterization of [3H]quinpirole binding to D2-like dopamine receptors in rat brain. *J Pharmacol Exp Ther.* 1992;262(3):929–935.
196. Levesque D, Diaz J, Pilon C, et al. Identification, characterization, and localization of the dopamine D3 receptor in rat brain using 7-[3H]hydroxy-N,N-di-n-propyl-2-aminotetralin. *Proc Natl Acad Sci USA.* 1992;89(17):8155–8159.
197. Bancroft GN, Morgan KA, Flietstra RJ, Levant B. Binding of [3H]PD 128907, a putatively selective ligand for the D3 dopamine receptor, in rat brain: a receptor binding and quantitative autoradiographic study. *Neuropsychopharmacology.* 1998;18(4):305–316.
198. Ricci A, Vega JA, Mammola CL, Amenta F. Localisation of dopamine D3 receptor in the rat cerebellar cortex: a light microscope autoradiographic study. *Neurosci Lett.* 1995;190(3):163–166.
199. Shafer RA, Levant B. The D3 dopamine receptor in cellular and organismal function. *Psychopharmacology.* 1998;135(1):1–16.
200. Van Tol HH, Bunzow JR, Guan HC, et al. Cloning of the gene for a human dopamine D4 receptor with high affinity for the antipsychotic clozapine. *Nature.* 1991;350(6319):610–614.
201. Matsumoto M, Hidaka K, Tada S, Tasaki Y, Yamaguchi T. Full-length cDNA cloning and distribution of human dopamine D4 receptor. *Brain Res Mol Brain Res.* 1995;29(1):157–162.
202. O'Dowd BF, Hnatowich M, Caron MG, Lefkowitz RJ, Bouvier M. Palmitoylation of the human beta 2-adrenergic receptor: mutation of Cys341 in the carboxyl tail leads to an uncoupled nonpalmitoylated form of the receptor. *J Biol Chem.* 1989;264(13):7564–7569.
203. Ding YC, Chi HC, Grady DL, et al. Evidence of positive selection acting at the human dopamine receptor D4 gene locus. *Proc Natl Acad Sci USA.* 2002;99(1):309–314.
204. Grady DL, Chi HC, Ding YC, et al. High prevalence of rare dopamine receptor D4 alleles in children diagnosed with attention-deficit hyperactivity disorder. *Mol Psychiatry.* 2003;8(5):536–545.
205. Chang FM, Kidd JR, Livak KJ, Pakstis AJ, Kidd KK. The worldwide distribution of allele frequencies at the human dopamine D4 receptor locus. *Hum Genet.* 1996;98(1):91–101.

206. Ebstein RP, Novick O, Umansky R, et al. Dopamine D4 receptor (D4DR) exon III polymorphism associated with the human personality trait of Novelty Seeking. *Nat Genet.* 1996;12(1):78–80.
207. LaHoste GJ, Swanson JM, Wigal SB, et al. Dopamine D4 receptor gene polymorphism is associated with attention deficit hyperactivity disorder. *Mol Psychiatry.* 1996;1(2):121–124.
208. Paterson AD, Sunohara GA, Kennedy JL. Dopamine D4 receptor gene: novelty or nonsense? *Neuropsychopharmacology.* 1999;21(1):3–16.
209. Oak JN, Oldenhof J, Van Tol HH. The dopamine D(4) receptor: one decade of research. *Eur J Pharmacol.* 2000;405(1-3):303–327.
210. Wilson JM, Sanyal S, Van Tol HH. Dopamine D2 and D4 receptor ligands: relation to antipsychotic action. *Eur J Pharmacol.* 1998;351(3):273–286.
211. Seeman P, Van Tol HH. Dopamine receptor pharmacology. *Trends Pharmacol Sci.* 1994;15(7):264–270.
212. Jovanovic V, Guan HC, Van Tol HH. Comparative pharmacological and functional analysis of the human dopamine D4.2 and D4.10 receptor variants. *Pharmacogenetics.* 1999;9(5):561–568.
213. Lanau F, Zenner MT, Civelli O, Hartman DS. Epinephrine and norepinephrine act as potent agonists at the recombinant human dopamine D4 receptor. *J Neurochem.* 1997;68(2):804–812.
214. Wedemeyer C, Goutman JD, Avale ME, Franchini LF, Rubinstein M, Calvo DJ. Functional activation by central monoamines of human dopamine D(4) receptor polymorphic variants coupled to GIRK channels in *Xenopus* oocytes. *Eur J Pharmacol.* 2007;562(3):165–173.
215. Patel S, Freedman S, Chapman KL, et al. Biological profile of L-745,870, a selective antagonist with high affinity for the dopamine D4 receptor. *J Pharmacol Exp Ther.* 1997;283(2):636–647.
216. Wang X, Zhong P, Yan Z. Dopamine D4 receptors modulate GABAergic signaling in pyramidal neurons of prefrontal cortex. *J Neurosci.* 2002;22(21):9185–9193.
217. Kulagowski JJ, Broughton HB, Curtis NR, et al. 3-((4-(4-Chlorophenyl)piperazin-1-yl)-methyl)-1H-pyrrolo-2,3-b-pyridine: an antagonist with high affinity and selectivity for the human dopamine D4 receptor. *J Med Chem.* 1996;39(10):1941–1942.
218. TenBrink RE, Bergh CL, Duncan JN, et al. (S)-(-)-4-[4-[2-(isochroman-1-yl)ethyl]-piperazin-1-yl] benzenesulfonamide, a selective dopamine D4 antagonist. *J Med Chem.* 1996;39(13):2435–2437.
219. Cohen AI, Todd RD, Harmon S, O'Malley KL. Photoreceptors of mouse retinas possess D4 receptors coupled to adenylate cyclase. *Proc Natl Acad Sci USA.* 1992;89(24):12093–12097.
220. Nir I, Harrison JM, Haque R, et al. Dysfunctional light-evoked regulation of cAMP in photoreceptors and abnormal retinal adaptation in mice lacking dopamine D4 receptors. *J Neurosci.* 2002;22(6):2063–2073.
221. Liu LX, Burgess LH, Gonzalez AM, Sibley DR, Chiodo LA. D2S, D2L, D3, and D4 dopamine receptors couple to a voltage-dependent potassium current in N18TG2 x mesencephalon hybrid cell (MES-23.5) via distinct G proteins. *Synapse.* 1999;31(2):108–118.
222. Kazmi MA, Snyder LA, Cypess AM, Graber SG, Sakmar TP. Selective reconstitution of human D4 dopamine receptor variants with Gi alpha subtypes. *Biochemistry.* 2000;39(13):3734–3744.
223. Seabrook GR, Knowles M, Brown N, et al. Pharmacology of high-threshold calcium currents in GH4C1 pituitary cells and their regulation by activation of human D2 and D4 dopamine receptors. *Br J Pharmacol.* 1994;112(3):728–734.
224. Oldenhof J, Vickery R, Anafi M, et al. SH3 binding domains in the dopamine D4 receptor. *Biochemistry.* 1998;37(45):15726–15736.
225. Pillai G, Brown NA, McAllister G, Milligan G, Seabrook GR. Human D2 and D4 dopamine receptors couple through betagamma G-protein subunits to inwardly rectifying K+ channels (GIRK1) in a *Xenopus* oocyte expression system: selective antagonism by L-741,626 and L-745,870 respectively. *Neuropharmacology.* 1998;37(8):983–987.
226. Gu Z, Yan Z. Bidirectional regulation of Ca^{2+}/calmodulin-dependent protein kinase II activity by dopamine D4 receptors in prefrontal cortex. *Mol Pharmacol.* 2004;66(4):948–955.
227. Xiang L, Szebeni K, Szebeni A, et al. Dopamine receptor gene expression in human amygdaloid nuclei: elevated D4 receptor mRNA in major depression. *Brain Res.* 2008;1207:214–224.
228. Healy DJ, Meador-Woodruff JH. Dopamine receptor gene expression in hippocampus is differentially regulated by the NMDA receptor antagonist MK-801. *Eur J Pharmacol.* 1996;306(1-3):257–264.
229. Knapp M, Wong AH, Schoots O, Guan HC, Van Tol HH. Promoter-independent regulation of cell-specific dopamine receptor expression. *FEBS Lett.* 1998;434(1-2):108–114.
230. Rondou P, Haegeman G, Vanhoenacker P, Van Craenenbroeck K. BTB protein KLHL12 targets the dopamine D4 receptor for ubiquitination by a Cul3-based E3 ligase. *J Biol Chem.* 2008;283(17):11083–11096.
231. Meador-Woodruff JH, Damask SP, Wang J, Haroutunian V, Davis KL, Watson SJ. Dopamine receptor mRNA expression in human striatum and neocortex. *Neuropsychopharmacology.* 1996;15(1):17–29.
232. Mrzljak L, Bergson C, Pappy M, Huff R, Levenson R, Goldman-Rakic PS. Localization of dopamine D4 receptors in GABAergic neurons of the primate brain. *Nature.* 1996;381(6579):245–248.
233. O'Malley KL, Harmon S, Tang L, Todd RD. The rat dopamine D4 receptor: sequence, gene structure, and demonstration of expression in the cardiovascular system. *New Biol.* 1992;4(2):137–146.
234. Ricci A, Bronzetti E, Fedele F, Ferrante F, Zaccheo D, Amenta F. Pharmacological characterization and autoradiographic localization of a putative dopamine D4 receptor in the heart. *J Auton Pharmacol.* 1998;18(2):115–121.
235. Bondy B, de Jonge S, Pander S, Primbs J, Ackenheil M. Identification of dopamine D4 receptor mRNA in circulating human lymphocytes using nested polymerase chain reaction. *J Neuroimmunol.* 1996;71(1-2):139–144.
236. Sun D, Wilborn TW, Schafer JA. Dopamine D4 receptor isoform mRNA and protein are expressed in the rat cortical collecting duct. *Am J Physiol.* 1998;275(5 Pt 2):F742–751.
237. Tiberi M, Jarvie KR, Silvia C, et al. Cloning, molecular characterization, and chromosomal assignment of a gene encoding a second D1 dopamine receptor subtype: differential expression pattern in rat brain compared with the D1A receptor. *Proc Natl Acad Sci USA.* 1991;88(17):7491–7495.
238. Grandy DK, Zhang YA, Bouvier C, et al. Multiple human D5 dopamine receptor genes: a functional receptor and two pseudogenes. *Proc Natl Acad Sci USA.* 1991;88(20):9175–9179.
239. Sunahara RK, Guan HC, O'Dowd BF, et al. Cloning of the gene for a human dopamine D5 receptor with higher affinity for dopamine than D1. *Nature.* 1991;350(6319):614–619.
240. Weinshank RL, Adham N, Macchi M, Olsen MA, Branchek TA, Hartig PR. Molecular cloning and characterization of a high

240. affinity dopamine receptor (D1 beta) and its pseudogene. *J Biol Chem.* 1991;266(33):22427–22435.
241. Sobell JL, Lind TJ, Sigurdson DC, et al. The D5 dopamine receptor gene in schizophrenia: identification of a nonsense change and multiple missense changes but lack of association with disease. *Hum Mol Genet.* 1995;4(4):507–514.
242. Cravchik A, Gejman PV. Functional analysis of the human D5 dopamine receptor missense and nonsense variants: differences in dopamine binding affinities. *Pharmacogenetics.* 1999;9(2):199–206.
243. O'Dowd BF. Structures of dopamine receptors. *J Neurochem.* 1993;60(3):804–816.
244. Mohr P, Decker M, Enzensperger C, Lehmann J. Dopamine/serotonin receptor ligands. 12(1): SAR studies on hexahydrodibenz[d,g]azecines lead to 4-chloro-7-methyl-5,6,7,8,9,14-hexahydrodibenz[d,g]azecin-3-ol, the first picomolar D5-selective dopamine-receptor antagonist. *J Med Chem.* 2006;49(6):2110–2116.
245. Charpentier S, Jarvie KR, Severynse DM, Caron MG, Tiberi M. Silencing of the constitutive activity of the dopamine D1B receptor. Reciprocal mutations between D1 receptor subtypes delineate residues underlying activation properties. *J Biol Chem.* 1996;271(45):28071–28076.
246. Tiberi M, Caron MG. High agonist-independent activity is a distinguishing feature of the dopamine D1B receptor subtype. *J Biol Chem.* 1994;269(45):27925–27931.
247. Wong YH, Conklin BR, Bourne HR. Gz-mediated hormonal inhibition of cyclic AMP accumulation. *Science.* 1992;255(5042):339–342.
248. Kozasa T, Gilman AG. Purification of recombinant G proteins from Sf9 cells by hexahistidine tagging of associated subunits. Characterization of alpha 12 and inhibition of adenylyl cyclase by alpha z. *J Biol Chem.* 1995;270(4):1734–1741.
249. Sidhu A, Kimura K, Uh M, White BH, Patel S. Multiple coupling of human D5 dopamine receptors to guanine nucleotide binding proteins Gs and Gz. *J Neurochem.* 1998;70(6):2459–2467.
250. Liu F, Wan Q, Pristupa ZB, Yu XM, Wang YT, Niznik HB. Direct protein-protein coupling enables cross-talk between dopamine D5 and gamma-aminobutyric acid A receptors. *Nature.* 2000;403(6767):274–280.
251. Schilstrom B, Yaka R, Argilli E, et al. Cocaine enhances NMDA receptor-mediated currents in ventral tegmental area cells via dopamine D5 receptor-dependent redistribution of NMDA receptors. *J Neurosci.* 2006;26(33):8549–8558.
252. Hollon TR, Bek MJ, Lachowicz JE, et al. Mice lacking D5 dopamine receptors have increased sympathetic tone and are hypertensive. *J Neurosci.* 2002;22(24):10801–10810.
253. Holmes A, Hollon TR, Gleason TC, et al. Behavioral characterization of dopamine D5 receptor null mutant mice. *Behav Neurosci.* 2001;115(5):1129–1144.
254. Aperia AC. Intrarenal dopamine: a key signal in the interactive regulation of sodium metabolism. *Annu Rev Physiol.* 2000;62:621–647.
255. Carey RM. Theodore Cooper Lecture: Renal dopamine system: paracrine regulator of sodium homeostasis and blood pressure. *Hypertension.* 2001;38(3):297–302.
256. Chen C, Lokhandwala MF. Potentiation by enalaprilat of fenoldopam-evoked natriuresis is due to blockade of intrarenal production of angiotensin-II in rats. *Naunyn Schmiedebergs Arch Pharmacol.* 1995;352(2):194–200.
257. Hussain T, Abdul-Wahab R, Kotak DK, Lokhandwala MF. Bromocriptine regulates angiotensin II response on sodium pump in proximal tubules. *Hypertension.* 1998;32(6):1054–1059.
258. Zeng C, Yang Z, Wang Z, et al. Interaction of angiotensin II type 1 and D5 dopamine receptors in renal proximal tubule cells. *Hypertension.* 2005;45(4):804–810.
259. Li H, Armando I, Yu P, et al. Dopamine 5 receptor mediates Ang II type 1 receptor degradation via a ubiquitin-proteasome pathway in mice and human cells. *J Clin Invest.* 2008;118(6):2180–2189.
260. Touyz RM, Schiffrin EL. Increased generation of superoxide by angiotensin II in smooth muscle cells from resistance arteries of hypertensive patients: role of phospholipase D-dependent NAD(P)H oxidase-sensitive pathways. *J Hypertens.* 2001;19(7):1245–1254.
261. Yang Z, Asico LD, Yu P, et al. D5 dopamine receptor regulation of phospholipase D. *Am J Physiol Heart Circ Physiol.* 2005;288(1):H55–61.
262. Khan ZU, Gutierrez A, Martin R, Penafiel A, Rivera A, de la Calle A. Dopamine D5 receptors of rat and human brain. *Neuroscience.* 2000;100(4):689–699.
263. Kumar U, Patel SC. Immunohistochemical localization of dopamine receptor subtypes (D1R-D5R) in Alzheimer's disease brain. *Brain Res.* 2007;1131(1):187–196.
264. Ricci A, Mignini F, Tomassoni D, Amenta F. Dopamine receptor subtypes in the human pulmonary arterial tree. *Auton Autacoid Pharmacol.* 2006;26(4):361–369.
265. Amenta F, Ricci A, Tayebati SK, Zaccheo D. The peripheral dopaminergic system: morphological analysis, functional and clinical applications. *Ital J Anat Embryol.* 2002;107(3):145–167.
266. Amenta F. Light microscope autoradiography of peripheral dopamine receptor subtypes. *Clin Exp Hypertens.* 1997;19(1–2):27–41.
267. Ferrari M, Termine C, Franciotta D, et al. Dopaminergic receptor D5 mRNA expression is increased in circulating lymphocytes of Tourette syndrome patients. *J Psychiatr Res.* 2008;43(1):24–29.
268. Roth BL, Kroeze WK, Patel S, Lopez E. The multiplicity of serotonin receptors: useless diverse molecules or an embarresment of riches? *Neuroscientist.* 2000;6:10.
269. Mason JN, Kozell LB, Neve KA. Regulation of dopamine D(1) receptor trafficking by protein kinase A-dependent phosphorylation. *Mol Pharmacol.* 2002;61(4):806–816.
270. Vargas GA, Von Zastrow M. Identification of a novel endocytic recycling signal in the D1 dopamine receptor. *J Biol Chem.* 2004;279(36):37461–37469.

3.2 Role of Dopamine Transporters in Neuronal Homeostasis

MARC G. CARON AND RAUL R. GAINETDINOV

Dopamine (DA) neurotransmission is controlled by several critical processes. A complex homeostatic balance between the amount of DA synthesized, packaged into vesicles, released, reuptaken via plasma membrane transporter, and metabolized determines the overall status of dopaminergic signaling. The plasma membrane dopamine transporter (DAT) provides effective control of both the extracellular and intracellular concentrations of DA by recapturing released neurotransmitter in the presynaptic terminals. The vesicular monoamine transporter 2 (VMAT2) directly controls vesicular storage and release capacity by pumping monoamines from the cytoplasm of neurons into synaptic vesicles. These transporters are primary targets of many psychotropic drugs that potently affect synaptic DA and related physiological processes. In this chapter, we summarize recent advances in the understanding of the molecular and cellular mechanisms involved in the DAT and VMAT2 functions. The role of these transporters in the action of psychostimulant drugs and neurotoxins, as revealed in studies using mutant mice, will also be discussed.

INTRODUCTION

Dopamine neurons innervate major brain regions critically involved in the regulation of movement, emotions, and reward.[1–3] Modulation of the efficiency of DA neurotransmission is believed to be an effective approach to correct abnormalities found in common brain disorders such as Parkinson's disease (PD), schizophrenia, and attention deficit hyperactivity disorder (ADHD).

Synaptic concentrations of DA can be controlled potently through modulation of the mechanisms involved in synaptic vesicle storage and exocytosis.[4,5] Vesicle filling and release mechanisms involve the coordinated activity of numerous synaptic proteins; however, the vesicular monoamine transporters have attracted particular attention as feasible targets for pharmacological interventions. While vesicular monoamine transporter 1 (VMAT1) is primarily responsible for vesicle filling in the periphery, VMAT2 is the major transporter involved in packaging monoamines into synaptic vesicles in neurons that synthesize either dopamine, norepinephrine, serotonin, or histamine. As such, VMATs represent well-established targets for certain drugs that potently affect the functions of monoamines.[5,6] In particular, inhibitors of VMAT2, such as reserpine, are well known as monoamine-suppressing agents that induce profound depletion of DA and other monoamines, both inside the neuron and in the extracellular space.

Extracellular monoamine concentrations are also tightly regulated via a rapid reuptake process mediated by plasma membrane monoamine transporters.[7–9] Selective plasma membrane transporters for each of the monoamines have been identified and characterized.[7–9] The plasma membrane DAT plays an important role in the homeostasis of DA neurons by transporting DA from the extracellular space back into releasing neurons and thus limiting the lifetime and spatial dynamics of extracellular DA.[10] Interaction with this transporter and the resultant elevation of extracellular DA levels is believed to be a primary neurochemical mechanism of action of psychostimulants, such as amphetamines and cocaine, that exert their psychostimulant effects largely via excessive stimulation of dopaminoceptive neurons.[10–12] Plasma membrane monoamine transporters are also well known as molecular gateways for intraneuronal penetration of certain neurotoxins.[13,14] At the same time, it is well established that vesicular sequestration, which is mediated by VMAT2s, provides a critical intraneuronal neuroprotective mechanism.[6,13,14] Thus, alterations in relative activities of plasma membrane DAT and VMAT2 function could have a significant impact on the deleterious potential of dopaminergic neurotoxins.[13]

While the structural organization and biochemistry of these transporters have been characterized extensively over the past decades,[7,9,15] several important issues regarding their roles in neuronal homeostasis, and their contribution to the in vivo mechanisms of action of psychotropic drugs, remain incompletely understood.

Development of mice with targeted mutations in specific genes has provided unique test systems to address these questions. Here we will review recent data on the in vivo functional roles mediated by these transporters, with a particular focus on the recent advances gained by using VMAT2 and DAT mutant mice.

CONTROL OF NEURONAL DA STORAGE AND RELEASE BY VMAT2

The principal storage mechanism of intracellular monoamines available for release involves their transport into small synaptic vesicles by VMAT2. The energy for this transport is derived from the proton gradient generated by adenosine triphosphate (ATP) hydrolysis. It is believed that two protons are released from the storage vesicle in exchange for one monoamine molecule transported into the vesicle.[16] The rat VMAT2 contains 515 amino acids and is predicted to have 12 transmembrane domains.[15] VMAT2 is primarily responsible for packaging DA, serotonin, norepinephrine, and histamine in their respective neurons, while VMAT1 plays a similar role in the periphery and, to some degree, in developing neurons. Both VMAT1 and VMAT2 are members of the toxin-extruding antiporter (TEXAN) gene family, which also includes some bacterial antibiotic resistance genes that extrude potentially toxic substances from bacteria. In eukaryotes, VMATs play a similar protective role. However, this function has been adapted to provide a mechanism to sequester potentially toxic substances including monoamines from the cytoplasm into vesicles, thus preventing interaction of the toxins with the intracellular machinery.

Given the pivotal role of VMAT2 in the transport of monoamines from the cytoplasm into secretory vesicles,[5,6] it is not surprising that mice lacking VMAT2 (VMAT2-KO mice) die within a few days after birth.[17–19] The brains of such mutant mice show a drastic reduction in the storage and vesicular release of DA and other monoamines both in cell cultures and in brain slices.[17,19,20] Heterozygote mice lacking one allele of the VMAT2 gene develop normally into adulthood and display less pronounced neurochemical and behavioral alterations.[18–22] While heart rate and blood pressure are minimally affected in these mice, increased vulnerability to lethal arrhythmias has been observed.[18,23] The brains of VMAT2 heterozygous mice contain significantly lower monoamine tissue levels, and depolarization induces less DA release from mutants both in cell cultures and in vivo in microdialysis studies.[17,19] Furthermore, as might be expected, the VMAT2 inhibitors reserpine and tetrabenazine are less effective in depleting monoamine storage in the brain, while effects of the tyrosine hydroxylase (TH) inhibitor α-methyl-*p*-tyrosine (αMT) are more pronounced.[21]

It is well recognized that amphetamines and related compounds gain access to the vesicular compartment via VMAT2 and eliminate the proton gradient leading to the redistribution of DA from synaptic vesicles into the cytoplasm, a process that is responsible for providing DA for amphetamine-induced reverse efflux.[24,25] Amphetamines caused substantially less elevation of extracellular DA in heterozygous mutants,[19] but in midbrain cultures from VMAT2-KO mice, amphetamines were still able to induce outflow of DA, albeit at a much lower level.[17] Moreover, amphetamines increased movement and prolonged the survival of VMAT2-KO pups, further indicating that VMAT2-mediated vesicular transport is not absolutely necessary for the DA-releasing action of amphetamines.[17] In adult heterozygous mice, a diminished reward to amphetamines in the conditioned place preference (CPP) test was observed.[18] At the same time, the locomotor responses to amphetamines, cocaine, ethanol, and the DA agonist apomorphine were all enhanced in heterozygous VMAT2 mice,[18,19] suggesting that the decreases in VMAT2-dependent DA storage and release can cause pronounced postsynaptic DA receptor supersensitivity.[19] Furthermore, VMAT2 heterozygous mice show alterations in alcohol preference and consumption[26,27] and display depressive-like behaviors.[28]

It has also been observed that methamphetamine (METH)–induced neurotoxicity was increased in VMAT2 heterozygous mice.[21] A neurotoxic regimen of METH administration caused more consistent DA depletion and a greater decrease in DAT immunoreactivity in the striatum of these mice. Importantly, the enhanced neurotoxicity was accompanied by less pronounced increases in both extracellular DA and indices of free radical formation detected by in vivo microdialysis, indicating that intraneuronal DA redistribution from vesicles to the cytoplasm, rather than the excessive extraneuronal DA accumulation, was primarily responsible for this effect.[21] Similar observations were also made in midbrain cell cultures derived from animals lacking VMAT2.[29]

The death of dopaminergic neurons induced by 1-methyl-4-phenyl-1,2,3,6-tetrahydropyridine (MPTP) is a well-established model of PD in rodents and primates.[30–32] Administration of MPTP causes a toxic insult to DA neurons via the intracellular oxidative stress induced by its reactive metabolite 1-methyl-4-phenylpyridium (MPP$^+$).[33] VMAT2 pumps not only monoamine neurotransmitters but also neurotoxins such as MPP$^+$ from the neuronal cytoplasm into the vesicles.[5,15] In fact, both VMAT1 and VMAT2 were cloned based on the ability of these transporters to prevent MPP+ toxicity in chromaffin cells.[6] Accordingly, it has been observed that in VMAT2 heterozygous mice, this neuroprotective function is reduced and administration of MPTP produces a more pronounced DA cell loss.[18] Furthermore, striatal DA content, the levels of DAT protein, and the expression of glial fibrillary acidic protein (GFAP) mRNA, a marker of gliosis, are all more significantly affected in VMAT2 mutant mice.[22]

Mice with a hypomorphic allele of the VMAT2 gene (VMAT2 knockdown) have also been developed.[34] These mice express very low levels of VMAT2 (<5%) and have striking alterations in monoamine homeostasis but survive into adulthood. As might be expected, homozygous mice show drastic reductions in brain tissue monoamines, significant motor impairments, and enhanced locomotor sensitivity to DA agonists. Like other VMAT2 mutant mice, the VMAT2 knockdown mice are also more vulnerable to the neurotoxic effects of MPTP[34] and show exacerbated METH-induced neurodegeneration.[35] Intriguingly, VMAT2 knockdown animals also display age-dependent nigrostriatal degeneration that starts in the terminals and progresses to eventual death of the cell bodies, alpha-synuclein accumulation, and L-DOPA-sensitive behavioral abnormalities, thus recapitulating certain aspects of PD.[36] Overall, these observations stress the importance of VMAT2 in the maintenance of proper control of presynaptic mechanisms of vesicular storage and release. In addition, it is evident that disruption of VMAT2 function may cause not only monoaminergic deficits related to compromised release but also postsynaptic receptor sensitization and enhanced toxin-induced neurodegeneration.

PLASMA MEMBRANE DAT AS A KEY REGULATOR OF DOPAMINERGIC NEUROTRANSMISSION

The DAT belongs to the family of the Na^+/Cl^-–dependent transporters that also includes transporters for such neurotransmitters/neuromodulators as serotonin, norepinephrine, GABA, glycine, proline, creatine, betaine, and taurine.[7–9] The human DAT protein contains 620 residues and consists of 12 transmembrane domains with cytoplasmic localization of both amino and carboxy termini. It is believed that the mechanism of DAT-mediated transport of DA involves sequential binding and cotransport of Na^+ and Cl^- ions. Thus, DAT should transport two Na^+ ions and one Cl^- ion with one molecule of DA.[7,9,37] It was generally believed that DAT functions as a monomeric protein; however, various studies have provided evidence that DAT can also exist in an oligomeric form.[38–41]

The DAT is expressed selectively on dopaminergic neurons.[42,43] In the brain, DAT protein expression is highest in projections to the striatum and nucleus accumbens, followed by the olfactory tubercle, nigrostriatal bundle and lateral habenula, and in medial prefrontal cortex.[42,43] In addition, DAT is found in the peripheral system, including the retina, gastrointestinal tract, lung, pancreas, kidney, and lymphocytes.[44–46] In DA neurons, DAT is mostly associated with intracellular membranes in perikarya and large proximal dendrites, and is also localized on plasma membranes of more distal dendrites and unmyelinated axons.[47] At the ultrastructural level, DAT is mostly localized perisynaptically rather than within the synaptic part of the presynaptic membrane,[47,48] thus providing anatomical evidence for the concept that recapture of DA occurs at a distance from release sites.[49,50] The DAT can be regulated at the level of gene expression and by posttranslational modifications. Several protein kinases and phosphatases may regulate the surface expression and functional properties of DAT, and such regulation may occur during essentially every step of the DAT protein life cycle.[9,51] In addition, DAT is known to interact with several scaffold proteins such as the PSD-95/Dlg-1/ZO-1 (PDZ) domain-containing protein PICK1 (protein interacting with C kinase),[9] the multiple Lin-11, Isl-1, and Mec-3 (LIM) domain-containing adaptor protein Hic-5,[52] and synuclein.[53,54] These interactions may presumably facilitate delivery of the transporter to its sites of action or stabilize the surface expression of DAT.

The DAT is a well-established target of many psychostimulants, such as cocaine and amphetamines, and of certain antidepressants including nomifensine. Numerous pharmacological studies have shown that inhibition of DAT causes significant alterations in extracellular DA dynamics. However, full appreciation of the fundamental role of DAT in the control of DA homeostasis was gained from characterization of the remarkable alterations in DA neurochemistry in genetic strains of mice lacking DAT.[10,55] The DAT knockout (DAT-KO) mice, developed by homologous recombination,[10,56] displayed extreme dopaminergic dysregulation resulting from disruption of the reuptake process. Cyclic voltammetry experiments performed in striatal slices from DAT-KO mice revealed a 300-fold prolonged lifetime of DA in the extracellular space.[10,57] Furthermore, the rate of DA clearance was not altered by inhibition of other plasma membrane monoamine transporters [serotonin transporter (SERT) or norepinephrine transporter (NET)] or major enzymes involved in the metabolism of DA [monoamine oxidase (MAO) and catechol-O-methyl transferase (COMT)].[57] Initial fast-scan cyclic voltammetry (FSCV) measurements revealed that both cocaine and amphetamine were unable to affect the clearance of released DA in striatal slices from DAT-KO mice.[10,57] Taken together, these observations indicated that under experimental conditions involving single-pulse stimulation, diffusion plays the major role in clearance of DA from the extracellular space in the striatal tissue of DAT-KO mice.[57] Similarly, carbon fiber amperometry experiments performed in anaesthetized DAT-KO mice[58] confirmed that the extracellular half-life of DA was about two orders of magnitude longer in the mutant mice. In these experiments, the COMT inhibitor tolcapone had no effect on DA clearance in mutant mice, while the MAO inhibitor pargyline induced a modest prolongation, suggesting that under conditions of multiple stimulations in anesthetized animals, the metabolism of DA by MAO may play some additional, albeit minor, role in the clearance of DA in the striatum.[58] As a result of the disrupted clearance of DA, DAT-KO mice display a fivefold elevation in basal extracellular DA in the striatum, as has been demonstrated by quantitative "no net flux" microdialysis measurements (Figure 3.2.1).[57,59]

Strikingly, a persistent increase in extracellular DA was observed despite the fact that the amplitude of evoked DA release was decreased by 75%–93 %,[10,57,58,60] suggesting that the size of the releasable pool of DA in nerve terminals was limited. Furthermore, total striatal tissue DA levels, which generally represent intraneuronal DA, were also reduced by about 95% in mutant mice (Figure 3.2.1).[57,59] The low striatal tissue levels of DA in DAT-KO mice are extremely sensitive to inhibition of TH, the rate-limiting enzyme in DA synthesis, suggesting that they represent mostly a newly synthesized pool of DA.[57–59] It should be noted that a similar dependence of tissue DA on ongoing synthesis was described previously in frontal cortex neurons of normal animals that have relatively low expression of DAT.[62,48]

In DAT-KO mice, the levels of TH protein in the striatum were also significantly decreased.[57,61] At the same time, the number of TH-positive neurons in the substantia nigra (SN) was not significantly affected, and no alteration in the level of TH mRNA levels per neuron was found. Moreover, striatal levels of another enzyme involved in DA synthesis, L-aromatic acid decarboxylase (L-AADC), and of VMAT2 were little affected in DAT-KO mice, indicating that the number of DA terminals in the striatum is not significantly affected.[61] Thus, depletion of the DA storage pools and the decreased amplitude of DA release in DAT-KO mice cannot be explained by the loss of DA terminals, but rather result from the absence of DAT-mediated inward transport of DA in these mice. Taken together, these observations indicate that large DA storage pools in striatal terminals in normal animals are mostly dependent upon DAT-mediated DA

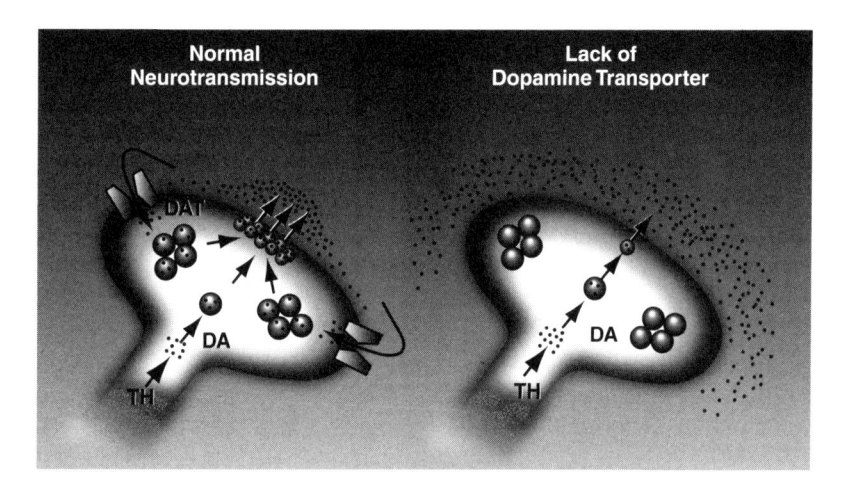

FIGURE 3.2.1. Hypothetical model of striatal DA terminals in normal conditions and in the absence of DAT. Deletion of the DAT results in 5-fold elevated levels of extracellular DA and 20-fold decreased intracellular DA storage. Due to a loss of DAT-mediated inward transport, intraneuronal concentrations of DA in DAT-KO mice are solely dependent on its ongoing synthesis and become extremely sensitive to TH inhibition. Reproduced, with permission, from Gainetdinov and Caron.[55]

recycling rather than ongoing synthesis.[55,57] In fact, in vivo investigations using the *NSD-1015 model*—in which AADC is inhibited—have shown that the rate of DA synthesis in DAT-KO mice is elevated about twofold,[57] despite the low levels of TH protein, suggesting markedly increased activity of fewer TH molecules.[57] Potentially, this could be explained by disinhibition of TH from tonic negative feedback mechanisms in DAT-KO mice.[1,62] Intraneuronal DA levels are greatly reduced in DAT-KO mice, which could indeed result in a disinhibition of TH from end-product inhibition. Alternatively, TH activity might be increased due to a loss of inhibitory autoreceptor function in DAT-KO mice. Accordingly, as a result of persistently increased DA tone, all the major autoreceptor functions, such as regulation of the neuronal firing rate, nerve terminal DA release, and DA synthesis, are significantly reduced in mutant mice.[63]

Dopamine transporter–related alterations of extracellular DA dynamics are known to produce transsynaptic dysregulation of the responsiveness of postsynaptic DA receptors. As might be expected, persistently increased DA tone results in a significant down-regulation of D1 and D2 DA receptors in the striatum.[10] Nevertheless, this down-regulation is not uniform, and certain populations of postsynaptic DA receptors appear to be functionally supersensitive. For example, while decreased mRNA levels of both D1 and D2 receptors were observed in DAT-KO mice, D3 receptor mRNA levels were increased,[64] and in DA-depleted DAT-KO mice, the locomotor effects of the nonselective DA agonist apomorphine were enhanced.[65] It is tempting to speculate that these divergent regulatory pathways might be determined by the specific localization of different receptor populations relative to synaptic and extrasynaptic/perisynaptic membrane compartments where DAT is preferentially localized. Importantly, such transsynaptic regulation can significantly influence postsynaptic cellular organization and plasticity. For example, levels of the scaffold protein postsynaptic density-95 (PSD-95) are reduced in DAT-KO mice, and enhanced long-term potentiation (LTP) of the cortico-accumbal glutamatergic synapses has been observed in mutant mice.[66] Significant alterations in the activity of molecules involved in postsynaptic DA receptor signal transduction, such as dopamine- and 3',5'-cyclic monophosphate-regulated phosphoprotein, 32 kDa (DARPP-32), extracellular signal-regulated protein kinase (ERK), protein kinase B (PKB/Akt), and glycogen synthase kinase 3 (GSK3), have also been observed in DAT-KO mice.[67]

Importantly, many of the neurochemical alterations listed above show gene dose dependence.[57,63] For example, DAT heterozygous mice have a twofold elevation of extracellular DA levels and a proportional reduction in DA clearance rates. At the same time, in DAT knockdown mice expressing only 10% of DAT[68], the alterations in neurochemical parameters are greater than those in DAT heterozygous mice,[57] but the magnitude of the changes is significantly less than that displayed by DAT-KO mice. Conversely, BAC transgenic mice overexpressing DAT (threefold) show doubled rates of DA clearance, about a 40% reduction in striatal extracellular DA levels, and significant up-regulation of postsynaptic DA receptors.[69]

Taken together, these observations highlight the fundamental role of DAT in the regulation of both extracellular and intracellular DA and the maintenance of homeostatic control over both presynaptic and postsynaptic organization and function.[55,57]

Hyperdopaminergic DAT-KO mice are hyperactive, dwarf, display perseverative patterns of locomotion (thigmotaxis) and predominantly perseverative types of errors in cognitive tests. Furthermore, they have disrupted sensorimotor gating and sleep dysregulation, as well as various other behavioral abnormalities mostly related to behavioral inflexibility.[10,55,56,70–76] They also display skeletal abnormalities[77] and alterations in gut motility,[78] indicating an important role of DAT in certain processes in the periphery.

In striking contrast to their effect on wild-type mice, the psychostimulants cocaine, methylphenidate, amphetamine, and METH do not affect significantly clearance or extracellular DA levels in the striatum of DAT-KO mice.[24,57,71,79] Both classical DA reuptake blockers, such as cocaine and methylphenidate, and amphetamines require interaction with DAT for their actions, although their mechanisms of action are different. In contrast to DAT blockers, amphetamine and related compounds enter the DA neuron through the DAT but also by diffusion.[80] Inside the terminal, amphetamine penetrate into storage vesicles primarily via the VMAT2 but also by diffusion, and as "weak bases" disrupt the vesicular pH gradient,[25] which results in a redistribution of DA from vesicles into the cytoplasm. This massive efflux of DA from vesicles to cytoplasm triggers DAT-mediated reverse transport of DA into the extracellular space. While the mechanism of amphetamine-induced reversal of DAT-mediated transport remains unclear, reports have indicated that it might promote phosphorylation of the N terminus of DAT,[81] Ca^{2+}/calmodulin kinase II-alpha (CaMKIIalpha) binding to the DAT C-terminus,[82] and/or interaction with syntaxin 1A.[83] Another important difference between amphetamines and DAT blockers is that amphetamines are potent inhibitors of MAO and can thus directly affect intraneuronal concentrations of monoamines.[80] Accordingly, while amphetamine and DAT blockers do not significantly affect extracellular DA levels in the striatum of DAT-KO mice, amphetamine can still exert its intraneuronal actions in the mutant mice.[24]

However, in the nucleus accumbens of DAT-KO mice, both amphetamine and cocaine further increase extracellular DA and promote rewarding/reinforcing effects.[84,85] Despite lacking the major target of cocaine, DAT-KO mice are still able to self-administer cocaine[85] and to display conditioned place preference (CPP) for both cocaine and amphetamines.[56,86–88] In the search for DAT-independent mechanisms that could explain these unexpected effects of psychostimulants, other known targets of these compounds were explored. In microdialysis studies, an increase in DA extracellular levels in the nucleus accumbens in DAT-KO mice was found after administration of the NET inhibitor reboxetine. This led to the hypothesis that NET contributes to the clearance of DA in this brain region in the absence of DAT, and that the disruption of NET-mediated transport of DA after psychostimulant administration may cause this increase.[89] However, subsequent voltammetry and microdialysis investigations have challenged this hypothesis and suggested that modulation of DA cells in the ventral tegmental area (VTA) via an indirect action on the serotonin (5-HT) system is primarily responsible for the DA-releasing and -rewarding actions of cocaine and amphetamine in mice lacking DAT.[84,87] Furthermore, in double mutant mice lacking both DAT and SERT, no preference for cocaine in CPP tests was observed, suggesting that the interaction of psychostimulants with SERT is sufficient to initiate and maintain a cocaine reward as long as there is a high extracellular DA tone, such as in hyperdopaminergic mice without the DAT.[90]

Furthermore, several psychostimulants, such as amphetamine, methylphenidate, cocaine, 3,4-methylenedioxymethamphetamine (MDMA),[71,91–93] and an "endogenous amphetamine" trace amine β-phenylethylamine,[94] while inducing hyperactivity in normal mice, paradoxically inhibited it in DAT-KO mice. A similar inhibition of hyperactivity by amphetamine was observed in mice with markedly reduced expression of DAT,[68] and the opposite effect, an increase in amphetamine-induced hyperactivity, was observed in mice overexpressing DAT.[69] The paradoxical inhibitory effect of psychostimulants in hyperactive DAT mutant mice suggests that these drugs might be acting on molecular targets other than the DAT. It is well known that amphetamines and other psychostimulants also interact, albeit with different potencies, with other plasma membrane monoamine transporters, such as SERT and NET.

In fact, the hyperactivity of DAT-KO mice could be inhibited potently by drugs affecting serotonergic transmission such as the SERT inhibitor fluoxetine, the 5-HT receptor agonist quipazine, and the 5-HT precursors tryptophan and 5-hydroxytryptophan, but not by the NET inhibitor nisoxetine.[71] Thus, it is likely that indirect modulation of the 5-HT system is responsible, at least in part, for the inhibitory effect of psychostimulants. These observations support the concept of the general inhibitory role of 5-HT in the modulation of DA-dependent hyperactivity proposed more than 30 years ago based on pharmacological investigations.[71,95]

Several lines of evidence suggest that the paradoxical inhibitory action of psychostimulants may be mediated by the frontostriatal glutamatergic pathway,[96] which can modulate activity in a DA-independent manner.[97,98] First, the hyperactivity of DAT mutant mice can be potentiated markedly by the N-methyl-D-aspartate (NMDA) antagonist MK-801 with a potency directly proportional to the differences in basal extracellular DA in DAT-KO and heterozygous mice.[96] Second, compounds that can enhance the efficacy of glutamatergic transmission, AMPAkines,[96] or glycine transporter type 1 (Glyt1) inhibitors suppress hyperactivity in DAT-KO mice. Finally, pretreatment with MK-801 effectively prevented the inhibitory action of psychostimulants and serotonergic compounds on hyperactivity.[96] Thus, striatal DA-mediated responses depend on the intensity of glutamatergic signaling, which, in turn, could be affected by alterations in 5-HT tone in the frontal cortex. In fact, methylphenidate, amphetamine, and cocaine are able to increase c-fos expression or immunoreactivity in the frontal cortex of DAT-KO mice.[85,96]

Over the past decade, several groups have reported an association between a polymorphism in the DAT gene and ADHD.[99] Psychostimulants such as methylphenidate and amphetamines are commonly used to treat the impulsivity, hyperactivity, and inattention of ADHD. As discussed above, DAT-KO mice display hyperactivity, perseverative patterns of locomotion, and cognitive impairments in the eight-arm radial maze and Morris water maze tests.[10,55,71,100] Particularly in the eight-arm radial maze test, DAT-KO mice demonstrate predominantly perseverative types of errors, suggesting that they could have impaired behavioral inhibition.[71] Psychostimulants potently inhibited the hyperactivity of DAT-KO mice to normal levels.[71] We hypothesized that this hyperdopaminergic-related hyperactivity could be controlled by enhancing 5-HT transmission. Thus, psychostimulants, which are known to interact with SERT to increase 5-HT levels, would inhibit the activity of mutant mice by enhancing the inhibitory action of 5-HT on DA-dependent hyperactivity.[71,101,102] This hypothesis has strong support from multiple reports implicating 5-HT in impulsiveness regulation and inhibitory control on external stimuli-induced behavioral activation.[103,104] Several endophenotypes of ADHD, including hyperactivity, cognitive impairments, and paradoxical inhibitory responses to psychostimulants, were observed in DAT-KO mice, suggesting that these mice could be a valuable animal model of this disorder.[71,101,102] Another mutant model, DAT knockdown mice (mice carrying more than a 90% reduction in DAT expression), displayed a similar, although less pronounced, phenotype.[68] In particular, moderate hyperactivity, impaired response habituation, and a paradoxical hypolocomotor effect of amphetamine were observed in these mutant mice. By striking contrast, mice with a 30% increase in DAT expression displayed hypoactivity in a novel environment,[105] while mice with a threefold increase in DAT expression displayed markedly enhanced hyperactivity following amphetamine administration.[69]

Apart from their powerful neurochemical and behavioral effects, some psychostimulants and other compounds interacting with DAT are known to cause various types of neurodegenerative processes. The selective dopaminergic neurotoxicity of MPTP and METH has been used for many years to model processes related to PD, a major human disorder affecting the nigrostriatal system.[13,106,107] Numerous reports have documented that DAT blockade prevents MPTP toxicity, suggesting that MPP$^+$ has to enter DA terminals via DAT in order to exert its deleterious effects.[108,109] Investigations involving DAT-KO mice have provided strong support for this hypothesis by demonstrating a total insensitivity to MPTP-induced neurodegeneration in mice lacking DAT.[110,111] Similarly, DAT plays an important role in the toxic action of amphetamines, particularly in METH toxicity. Like other amphetamines, METH exerts its major action through outward transport of DA from intracellular storage pools to the extracellular space via the DAT.[24] High doses or repeated administration of METH are known to cause toxic damage to dopaminergic and serotonergic neurons in several species.[112–114] It is believed that redistribution of DA within presynaptic terminals from vesicular storage to the cytoplasmic compartment, causing DA auto-oxidation and oxidative stress, is a major cause of METH toxicity.[14,29,113,115] As might be expected, DAT-KO mice were found to be fully resistant to METH-induced dopaminergic neurotoxicity, thus confirming that DAT is necessary for the neurotoxic properties of METH on presynaptic DA terminals.[79] However, an important question remains: can DAT similarly be involved in the neurotoxic actions of other substances, such as pesticides, that have for many years remained potential suspects in the pathogenesis of PD?[13,106]

DAT-KO mice have also been instrumental in demonstrating another aspect of DA-related neurodegeneration: a deleterious outcome of excessive DA receptor signaling.[116,117] It has been noted that, under certain conditions, a number (up to 30%) of DAT-KO mice sporadically develop a progressive locomotor disorder characterized by a loss of spontaneous hyperactivity, development of dyskinetic movements, paralysis, and, eventually, death.[116] This phenotype is similar to that of transgenic mice expressing Huntington's disease (HD)–related variants of huntingtin,[118] suggesting a dysfunction of DA-responsive medium spiny GABA neurons. In fact, the affected mice displayed approximately 30% loss of these neurons, accompanied by the appearance of markers of apoptotic processes and perikaryal accumulations of the hyperphosphorylated microtubule-associated protein tau.[116] Furthermore, significant activation of the major protein kinase, cyclin-dependent kinase 5 (CDK5), which mediates tau hyperphosphorylation in neuropathologies,[119–122] was noted. CDK5 is activated under conditions of excessive DA receptor stimulation.[116] In addition, ΔFosB, which is known to accumulate and enhance CDK5 expression in medium spiny GABA neurons following chronic cocaine administration, was elevated. The CDK5 coactivator p35 was also elevated in the striatum of symptomatic DAT-KO mice.[116] Furthermore, crossing DAT-KO mice to a knock-in mouse model of HD-containing 92 CAG repeats resulted in enhanced motor dysfunctions and neuropathology of mutant huntingtin. In particular, increased stereotypic activity at earlier ages, followed by a progressive decline in locomotion and hyperactivity and enhanced aggregation of mutant huntingtin protein, were observed in these double mutants.[124] Taken together, these observations provide strong support for a role of overactive DA receptor signaling in the development of a postsynaptic neurodegeneration that could amplify the deleterious effects of mutated huntingtin or other pathogenic factors on striatal GABA neurons. Interestingly, DAT-KO mice were also found to be hypersensitive to 3-nitropropionic acid–induced damage to striatal GABA neurons.[125]

CONCLUSIONS

Both VMAT2 and DAT play indispensable roles in supporting DA homeostasis. Elimination of VMAT2-mediated transport drastically disrupts vesicular storage and release, resulting in pronounced postsynaptic supersensitivity.[17–19,34] Elimination of the active DAT-mediated transport results in a fundamental shift in the neurochemistry of DA neurons. Disrupted DA clearance, persistently elevated extracellular levels of DA, drastically depleted storage of DA, and the trans-synaptic changes at the level of postsynaptic DA receptor expression and signaling observed in DAT mutant mice highlight the indispensable role of plasma membrane monoamine transporters in monoamine homeostasis.[55,57] Given this prominent role of transporters in the control of DA dynamics, it is not surprising that pharmacological compounds modulating DAT functions have proven to be very powerful tools for altering behavioral states.

Importantly, both DAT and VMAT2 mutant mice display remarkable depletions of DA tissue stores in the striatum. Intriguingly, the degree of DA depletion was found to be essentially the same in both DAT-KO and VMAT2-KO mice (~95%). While it is not clear whether disruption of these transporters affects the same vesicle-filling machinery at different levels,[126] these observations seem to indicate that proper maintenance of the intraneuronal storage pool requires coordinated activity of both vesicular and plasma membrane monoamine transporters. This conclusion may have several important ramifications. Monoamine depletion by drugs affecting vesicular transport, such as reserpine, has been well established and understood for many years.[4] However, this is not the case with regard to plasma monoamine transporter inhibitors. As a rule, the DA depletion observed after chronic administration of DAT inhibitors or amphetamines was considered evidence of the neurotoxic processes induced by these compounds.[113] However, the observations gained in plasma membrane monoamine transporter mutant mice demonstrate that deficiency of DAT,[57] NET,[127] or SERT[128] function result in marked depletion of cognate monoamines. It is likely that the decreases in monoamine levels following administration of transporter inhibitors do not necessarily reflect damage or loss of neurons but instead may represent depletion of monoamine stores due to prolonged blockade of plasma membrane monoamine uptake. Furthermore, as described above, the remaining DA levels in DAT-KO mice are highly dependent on ongoing DA synthesis.[129,130] Thus, conversely, if and when DA synthesis is compromised, it is highly possible that chronic DAT blockade may induce even more pronounced depletions. This possibility should be taken into account when effects of chronic treatment with monoamine transporter inhibitors are contemplated for therapeutic interventions.

In conclusion, the coordinated action of plasma membrane and vesicular transporters is required for proper maintenance of DA neuronal homeostasis. Disruption of this coordination by amphetamines[115] or other compounds[131] could induce a chain of reactions leading to toxic insults to DA neurons. The indispensable role of DA transporters in the multifaceted regulation of DA functions fuels its continued interest among the most attractive targets for psychopharmacology.

REFERENCES

1. Molinoff PB, Axelrod J. Biochemistry of catecholamines. *Annu Rev Biochem*. 1971;40:465–500.
2. Carlsson A, Waters N, Holm-Waters S, Tedroff J, Nilsson M, Carlsson ML. Interactions between monoamines, glutamate, and GABA in schizophrenia: new evidence. *Annu Rev Pharmacol Toxicol*. 2001;41:237–260.
3. Hornykiewicz O. Biochemical aspects of Parkinson's disease. *Neurology*. 1998;51(2 suppl 2):S2–S9.
4. Carlsson A. Biochemical and pharmacological aspects of Parkinsonism. *Acta Neurol Scand Suppl*. 1972;51:11–42.
5. Schuldiner S, Shirvan A, Linial M. Vesicular neurotransmitter transporters: from bacteria to humans. *Physiol Rev*. 1995;75(2):369–392.
6. Liu Y, Edwards RH. The role of vesicular transport proteins in synaptic transmission and neural degeneration. *Annu Rev Neurosci*. 1997;20:125–156.
7. Amara SG, Sonders MS. Neurotransmitter transporters as molecular targets for addictive drugs. *Drug Alcohol Depend*. 1998;51(1–2):87–96.
8. Giros B, Caron MG. Molecular characterization of the dopamine transporter. *Trends Pharmacol Sci*. 1993;14(2):43–49.
9. Torres GE, Gainetdinov RR, Caron MG. Plasma membrane monoamine transporters: structure, regulation and function. *Nat Rev Neurosci*. 2003;4(1):13–25.
10. Giros B, Jaber M, Jones SR, Wightman RM, Caron MG. Hyperlocomotion and indifference to cocaine and amphetamine in mice lacking the dopamine transporter. *Nature*. 1996;379(6566):606–612.
11. Kuhar MJ, Sanchez-Roa PM, Wong DF, et al. Dopamine transporter: biochemistry, pharmacology and imaging. *Eur Neurol*. 1990;30(suppl 1):15–20.
12. Wise RA. Cocaine reward and cocaine craving: the role of dopamine in perspective. *NIDA Res Monogr*. 1994;145:191–206.
13. Miller GW, Gainetdinov RR, Levey AI, Caron MG. Dopamine transporters and neuronal injury. *Trends Pharmacol Sci*. 1999;20(10):424–429.
14. Uhl GR. Hypothesis: the role of dopaminergic transporters in selective vulnerability of cells in Parkinson's disease. *Ann Neurol*. 1998;43(5):555–560.
15. Chaudhry FA, Edwards RH, Fonnum F. Vesicular neurotransmitter transporters as targets for endogenous and exogenous toxic substances. *Annu Rev Pharmacol Toxicol*. 2008;48:277–301.
16. Kanner BI, Schuldiner S. Mechanism of transport and storage of neurotransmitters. *CRC Crit Rev Biochem*. 1987;22(1):1–38.
17. Fon EA, Pothos EN, Sun BC, Killeen N, Sulzer D, Edwards RH. Vesicular transport regulates monoamine storage and release but

is not essential for amphetamine action. *Neuron.* 1997;19(6):1271–1283.
18. Takahashi N, Miner LL, Sora I, et al. VMAT2 knockout mice: heterozygotes display reduced amphetamine-conditioned reward, enhanced amphetamine locomotion, and enhanced MPTP toxicity. *Proc Natl Acad Sci USA.* 1997;94(18):9938–9943.
19. Wang YM, Gainetdinov RR, Fumagalli F, et al. Knockout of the vesicular monoamine transporter 2 gene results in neonatal death and supersensitivity to cocaine and amphetamine. *Neuron.* 1997;19(6):1285–1296.
20. Travis ER, Wang YM, Michael DJ, Caron MG, Wightman RM. Differential quantal release of histamine and 5–hydroxytryptamine from mast cells of vesicular monoamine transporter 2 knockout mice. *Proc Natl Acad Sci USA.* 2000;97(1):162–167.
21. Fumagalli F, Gainetdinov RR, Wang YM, Valenzano KJ, Miller GW, Caron MG. Increased methamphetamine neurotoxicity in heterozygous vesicular monoamine transporter 2 knock-out mice. *J Neurosci.* 1999;19(7):2424–2431.
22. Gainetdinov RR, Fumagalli F, Wang YM, et al. Increased MPTP neurotoxicity in vesicular monoamine transporter 2 heterozygote knockout mice. *J Neurochem.* 1998;70(5):1973–1978.
23. Itokawa K, Sora I, Schindler CW, Itokawa M, Takahashi N, Uhl GR. Heterozygous VMAT2 knockout mice display prolonged QT intervals: possible contributions to sudden death. *Brain Res Mol Brain Res.* 1999;71(2):354–357.
24. Jones SR, Gainetdinov RR, Wightman RM, Caron MG. Mechanisms of amphetamine action revealed in mice lacking the dopamine transporter. *J Neurosci.* 1998;18(6):1979–1986.
25. Sulzer D, Chen TK, Lau YY, Kristensen H, Rayport S, Ewing A. Amphetamine redistributes dopamine from synaptic vesicles to the cytosol and promotes reverse transport. *J Neurosci.* 1995;15(5 pt 2):4102–4108.
26. Hall FS, Sora I, Uhl GR. Sex-dependent modulation of ethanol consumption in vesicular monoamine transporter 2 (VMAT2) and dopamine transporter (DAT) knockout mice. *Neuropsychopharmacology.* 2003;28(4):620–628.
27. Savelieva KV, Caudle WM, Miller GW. Altered ethanol-associated behaviors in vesicular monoamine transporter heterozygote knockout mice. *Alcohol.* 2006;40(2):87–94.
28. Fukui M, Rodriguiz RM, Zhou J, et al. Vmat2 heterozygous mutant mice display a depressive-like phenotype. *J Neurosci.* 2007;27(39):10520–10529.
29. Larsen KE, Fon EA, Hastings TG, Edwards RH, Sulzer D. Methamphetamine-induced degeneration of dopaminergic neurons involves autophagy and upregulation of dopamine synthesis. *J Neurosci.* 2002;22(20):8951–8960.
30. Kopin IJ. Features of the dopaminergic neurotoxin MPTP. *Ann N Y Acad Sci.* 1992;648:96–104.
31. Snyder SH, D'Amato RJ. MPTP: a neurotoxin relevant to the pathophysiology of Parkinson's disease. The 1985 George C. Cotzias lecture. *Neurology.* 1986;36(2):250–258.
32. Tipton KF, Singer TP. Advances in our understanding of the mechanisms of the neurotoxicity of MPTP and related compounds. *J Neurochem.* 1993;61(4):1191–1206.
33. Heikkila RE, Manzino L, Cabbat FS, Duvoisin RC. Protection against the dopaminergic neurotoxicity of 1–methyl-4-phenyl-1,2,5,6-tetrahydropyridine by monoamine oxidase inhibitors. *Nature.* 1984;311(5985):467–469.
34. Mooslehner KA, Chan PM, Xu W, et al. Mice with very low expression of the vesicular monoamine transporter 2 gene survive into adulthood: potential mouse model for parkinsonism. *Mol Cell Biol.* 2001;21(16):5321–5331.
35. Guillot TS, Shepherd KR, Richardson JR, et al. Reduced vesicular storage of dopamine exacerbates methamphetamine-induced neurodegeneration and astrogliosis. *J Neurochem.* 2008;106(5):2205–2217.
36. Caudle WM, Richardson JR, Wang MZ, et al. Reduced vesicular storage of dopamine causes progressive nigrostriatal neurodegeneration. *J Neurosci.* 2007;27(30):8138–8148.
37. Chen NH, Reith ME, Quick MW. Synaptic uptake and beyond: the sodium- and chloride-dependent neurotransmitter transporter family SLC6. *Pflugers Arch.* 2004;447(5):519–531.
38. Hastrup H, Karlin A, Javitch JA. Symmetrical dimer of the human dopamine transporter revealed by cross-linking Cys-306 at the extracellular end of the sixth transmembrane segment. *Proc Natl Acad Sci USA.* 2001;98(18):10055–10060.
39. Norgaard-Nielsen K, Norregaard L, Hastrup H, Javitch JA, Gether U. Zn(2+) site engineering at the oligomeric interface of the dopamine transporter. *FEBS Lett.* 2002;524(1–3):87–91.
40. Torres GE, Carneiro A, Seamans K, et al. Oligomerization and trafficking of the human dopamine transporter. Mutational analysis identifies critical domains important for the functional expression of the transporter. *J Biol Chem.* 2003;278(4):2731–2739.
41. Sorkina T, Doolen S, Galperin E, Zahniser NR, Sorkin A. Oligomerization of dopamine transporters visualized in living cells by fluorescence resonance energy transfer microscopy. *J Biol Chem.* 2003;278(30):28274–28283.
42. Ciliax BJ, Heilman C, Demchyshyn LL, et al. The dopamine transporter: immunochemical characterization and localization in brain. *J Neurosci.* 1995;15(3 pt 1):1714–1723.
43. Hoffman BJ, Hansson SR, Mezey E, Palkovits M. Localization and dynamic regulation of biogenic amine transporters in the mammalian central nervous system. *Front Neuroendocrinol.* 1998;19(3):187–231.
44. Amenta F, Bronzetti E, Cantalamessa F, et al. Identification of dopamine plasma membrane and vesicular transporters in human peripheral blood lymphocytes. *J Neuroimmunol.* 2001;117(1–2):133–142.
45. Mitsuma T, Rhue H, Hirooka Y, et al. Distribution of dopamine transporter in the rat: an immunohistochemical study. *Endocr Regul.* 1998;32(2):71–75.
46. Gordon J, Barnes NM. Lymphocytes transport serotonin and dopamine: agony or ecstasy? *Trends Immunol.* 2003;24(8):438–443.
47. Nirenberg MJ, Chan J, Vaughan RA, Uhl GR, Kuhar MJ, Pickel VM. Immunogold localization of the dopamine transporter: an ultrastructural study of the rat ventral tegmental area. *J Neurosci.* 1997;17(11):4037–4044.
48. Sesack SR, Hawrylak VA, Guido MA, Levey AI. Cellular and subcellular localization of the dopamine transporter in rat cortex. *Adv Pharmacol.* 1998;42:171–174.
49. Garris PA, Ciolkowski EL, Wightman RM. Heterogeneity of evoked dopamine overflow within the striatal and striatoamygdaloid regions. *Neuroscience.* 1994;59(2):417–427.
50. Garris PA, Wightman RM. Different kinetics govern dopaminergic transmission in the amygdala, prefrontal cortex, and striatum: an in vivo voltammetric study. *J Neurosci.* 1994;14(1):442–450.
51. Mortensen OV, Amara SG. Dynamic regulation of the dopamine transporter. *Eur J Pharmacol.* 2003;479(1–3):159–170.
52. Carneiro AM, Ingram SL, Beaulieu JM, et al. The multiple LIM domain-containing adaptor protein Hic-5 synaptically colocalizes and interacts with the dopamine transporter. *J Neurosci.* 2002;22(16):7045–7054.

53. Lee FJ, Liu F, Pristupa ZB, Niznik HB. Direct binding and functional coupling of alpha-synuclein to the dopamine transporters accelerate dopamine-induced apoptosis. *FASEB J.* 2001; 15(6):916–926.
54. Wersinger C, Sidhu A. Attenuation of dopamine transporter activity by alpha-synuclein. *Neurosci Lett.* 2003;340(3):189–192.
55. Gainetdinov RR, Caron MG. Monoamine transporters: from genes to behavior. *Annu Rev Pharmacol Toxicol.* 2003;43:261–284.
56. Sora I, Wichems C, Takahashi N, et al. Cocaine reward models: conditioned place preference can be established in dopamine- and in serotonin-transporter knockout mice. *Proc Natl Acad Sci USA.* 1998;95(13):7699–7704.
57. Jones SR, Gainetdinov RR, Jaber M, Giros B, Wightman RM, Caron MG. Profound neuronal plasticity in response to inactivation of the dopamine transporter. *Proc Natl Acad Sci USA.* 1998;95(7):4029–4034.
58. Benoit-Marand M, Jaber M, Gonon F. Release and elimination of dopamine in vivo in mice lacking the dopamine transporter: functional consequences. *Eur J Neurosci.* 2000;12(8):2985–2992.
59. Gainetdinov RR, Jones SR, Fumagalli F, Wightman RM, Caron MG. Re-evaluation of the role of the dopamine transporter in dopamine system homeostasis. *Brain Res Brain Res Rev.* 1998;26(2–3):148–153.
60. Budygin EA, John CE, Mateo Y, Jones SR. Lack of cocaine effect on dopamine clearance in the core and shell of the nucleus accumbens of dopamine transporter knock-out mice. *J Neurosci.* 2002;22(10):RC222.
61. Jaber M, Dumartin B, Sagne C, et al. Differential regulation of tyrosine hydroxylase in the basal ganglia of mice lacking the dopamine transporter. *Eur J Neurosci.* 1999;11(10):3499–3511.
62. Bannon MJ, Roth RH. Pharmacology of mesocortical dopamine neurons. *Pharmacol Rev.* 1983;35(1):53–68.
63. Jones SR, Gainetdinov RR, Hu XT, et al. Loss of autoreceptor functions in mice lacking the dopamine transporter. *Nat Neurosci.* 1999;2(7):649–655.
64. Fauchey V, Jaber M, Caron MG, Bloch B, Le Moine C. Differential regulation of the dopamine D1, D2 and D3 receptor gene expression and changes in the phenotype of the striatal neurons in mice lacking the dopamine transporter. *Eur J Neurosci.* 2000;12(1):19–26.
65. Gainetdinov RR, Jones SR, Caron MG. Functional hyperdopaminergia in dopamine transporter knock-out mice. *Biol Psychiatry.* 1999;46(3):303–311.
66. Yao WD, Gainetdinov RR, Arbuckle MI, et al. Identification of PSD-95 as a regulator of dopamine-mediated synaptic and behavioral plasticity. *Neuron.* 2004;41(4):625–638.
67. Beaulieu JM, Sotnikova TD, Gainetdinov RR, Caron MG. Paradoxical striatal cellular signaling responses to psychostimulants in hyperactive mice. *J Biol Chem.* 2006;281(43):32072–32080.
68. Zhuang X, Oosting RS, Jones SR, et al. Hyperactivity and impaired response habituation in hyperdopaminergic mice. *Proc Natl Acad Sci USA.* 2001;98(4):1982–1987.
69. Salahpour A, Ramsey AJ, Medvedev IO, et al. Increased amphetamine-induced hyperactivity and reward in mice overexpressing the dopamine transporter. *Proc Natl Acad Sci USA.* 2008;105(11):4405–4410.
70. Bosse R, Fumagalli F, Jaber M, et al. Anterior pituitary hypoplasia and dwarfism in mice lacking the dopamine transporter. *Neuron.* 1997;19(1):127–138.
71. Gainetdinov RR, Wetsel WC, Jones SR, Levin ED, Jaber M, Caron MG. Role of serotonin in the paradoxical calming effect of psychostimulants on hyperactivity. *Science.* 1999; 283(5400): 397–401.
72. Ralph RJ, Paulus MP, Fumagalli F, Caron MG, Geyer MA. Prepulse inhibition deficits and perseverative motor patterns in dopamine transporter knock-out mice: differential effects of D1 and D2 receptor antagonists. *J Neurosci.* 2001;21(1):305–313.
73. Spielewoy C, Roubert C, Hamon M, Nosten-Bertrand M, Betancur C, Giros B. Behavioural disturbances associated with hyperdopaminergia in dopamine-transporter knockout mice. *Behav Pharmacol.* 2000;11(3–4):279–290.
74. Morice E, Denis C, Macario A, Giros B, Nosten-Bertrand M. Constitutive hyperdopaminergia is functionally associated with reduced behavioral lateralization. *Neuropsychopharmacology.* 2005;30(3):575–581.
75. Rodriguiz RM, Chu R, Caron MG, Wetsel WC. Aberrant responses in social interaction of dopamine transporter knockout mice. *Behav Brain Res.* 2004;148(1–2):185–198.
76. Wisor JP, Nishino S, Sora I, Uhl GH, Mignot E, Edgar DM. Dopaminergic role in stimulant-induced wakefulness. *J Neurosci.* 2001;21(5):1787–1794.
77. Bliziotes M, McLoughlin S, Gunness M, Fumagalli F, Jones SR, Caron MG. Bone histomorphometric and biomechanical abnormalities in mice homozygous for deletion of the dopamine transporter gene. *Bone.* 2000;26(1):15–19.
78. Walker JK, Gainetdinov RR, Mangel AW, Caron MG, Shetzline MA. Mice lacking the dopamine transporter display altered regulation of distal colonic motility. *Am J Physiol Gastrointest Liver Physiol.* 2000;279(2):G311–318.
79. Fumagalli F, Gainetdinov RR, Valenzano KJ, Caron MG. Role of dopamine transporter in methamphetamine-induced neurotoxicity: evidence from mice lacking the transporter. *J Neurosci.* 1998;18(13):4861–4869.
80. Seiden LS, Sabol KE, Ricaurte GA. Amphetamine: effects on catecholamine systems and behavior. *Annu Rev Pharmacol Toxicol.* 1993;33:639–677.
81. Khoshbouei H, Sen N, Guptaroy B, et al. N-terminal phosphorylation of the dopamine transporter is required for amphetamine-induced efflux. *PLoS Biol.* 2004;2(3):E78.
82. Fog JU, Khoshbouei H, Holy M, et al. Calmodulin kinase II interacts with the dopamine transporter C terminus to regulate amphetamine-induced reverse transport. *Neuron.* 2006;51(4):417–429.
83. Binda F, Dipace C, Bowton E, et al. Syntaxin 1A interaction with the dopamine transporter promotes amphetamine-induced dopamine efflux. *Mol Pharmacol.* 2008;74(4):1101–1108.
84. Budygin EA, Brodie MS, Sotnikova TD, et al. Dissociation of rewarding and dopamine transporter-mediated properties of amphetamine. *Proc Natl Acad Sci USA.* 2004;101(20):7781–7786.
85. Rocha BA, Fumagalli F, Gainetdinov RR, et al. Cocaine self-administration in dopamine-transporter knockout mice. *Nat Neurosci.* 1998;1(2):132–137.
86. Hall FS, Li XF, Sora I, et al. Cocaine mechanisms: enhanced cocaine, fluoxetine and nisoxetine place preferences following monoamine transporter deletions. *Neuroscience.* 2002;115(1):153–161.
87. Mateo Y, Budygin EA, John CE, Jones SR. Role of serotonin in cocaine effects in mice with reduced dopamine transporter function. *Proc Natl Acad Sci USA.* 2004;101(1):372–377.
88. Medvedev IO, Gainetdinov RR, Sotnikova TD, Bohn LM, Caron MG, Dykstra LA. Characterization of conditioned place

preference to cocaine in congenic dopamine transporter knockout female mice. *Psychopharmacology (Berl)*. 2005; 180(3):408–413.
89. Carboni E, Spielewoy C, Vacca C, Nosten-Bertrand M, Giros B, Di Chiara G. Cocaine and amphetamine increase extracellular dopamine in the nucleus accumbens of mice lacking the dopamine transporter gene. *J Neurosci*. 2001;21(9): RC141: 141–144.
90. Sora I, Hall FS, Andrews AM, et al. Molecular mechanisms of cocaine reward: combined dopamine and serotonin transporter knockouts eliminate cocaine place preference. *Proc Natl Acad Sci USA*. 2001;98(9):5300–5305.
91. Spielewoy C, Biala G, Roubert C, Hamon M, Betancur C, Giros B. Hypolocomotor effects of acute and daily *d*-amphetamine in mice lacking the dopamine transporter. *Psychopharmacology (Berl)*. 2001;159(1):2–9.
92. Morice E, Denis C, Giros B, Nosten-Bertrand M. Phenotypic expression of the targeted null–mutation in the dopamine transporter gene varies as a function of the genetic background. *Eur J Neurosci*. 2004;20(1):120–126.
93. Powell SB, Lehmann-Masten VD, Paulus MP, Gainetdinov RR, Caron MG, Geyer MA. MDMA "ecstasy" alters hyperactive and perseverative behaviors in dopamine transporter knockout mice. *Psychopharmacology (Berl)*. 2004;173(3–4):310–317.
94. Sotnikova TD, Budygin EA, Jones SR, Dykstra LA, Caron MG, Gainetdinov RR. Dopamine transporter-dependent and -independent actions of trace amine beta-phenylethylamine. *J Neurochem*. 2004;91(2):362–373.
95. Breese GR, Cooper BR, Hollister AS. Involvement of brain monoamines in the stimulant and paradoxical inhibitory effects of methylphenidate. *Psychopharmacologia*. 1975;44(1):5–10.
96. Gainetdinov RR, Mohn AR, Bohn LM, Caron MG. Glutamatergic modulation of hyperactivity in mice lacking the dopamine transporter. *Proc Natl Acad Sci USA*. 2001;98(20):11047–11054.
97. Mohn AR, Gainetdinov RR, Caron MG, Koller BH. Mice with reduced NMDA receptor expression display behaviors related to schizophrenia. *Cell*. 1999;98(4):427–436.
98. Martin P, Waters N, Schmidt CJ, Carlsson A, Carlsson ML. Rodent data and general hypothesis: antipsychotic action exerted through 5-Ht2A receptor antagonism is dependent on increased serotonergic tone. *J Neural Transm*. 1998;105(4–5):365–396.
99. Cook EH Jr, Stein MA, Krasowski MD, et al. Association of attention-deficit disorder and the dopamine transporter gene. *Am J Hum Genet*. 1995;56(4):993–998.
100. Morice E, Billard JM, Denis C, et al. Parallel loss of hippocampal LTD and cognitive flexibility in a genetic model of hyperdopaminergia. *Neuropsychopharmacology*. 2007; 32(10):2108–2116.
101. Gainetdinov RR, Caron MG. An animal model of attention deficit hyperactivity disorder. *Mol Med Today*. 2000;6(1):43–44.
102. Gainetdinov RR, Caron MG. Genetics of childhood disorders: XXIV. ADHD, part 8: hyperdopaminergic mice as an animal model of ADHD. *J Am Acad Child Adolesc Psychiatry*. 2001;40(3):380–382.
103. Winstanley CA, Theobald DE, Dalley JW, Robbins TW. Interactions between serotonin and dopamine in the control of impulsive choice in rats: therapeutic implications for impulse control disorders. *Neuropsychopharmacology*. 2005; 30(4): 669–682.
104. Lucki I. The spectrum of behaviors influenced by serotonin. *Biol Psychiatry*. 1998;44(3):151–162.
105. Donovan DM, Miner LL, Perry MP, et al. Cocaine reward and MPTP toxicity: alteration by regional variant dopamine transporter overexpression. *Brain Res Mol Brain Res*. 1999; 73(1–2):37–49.
106. Betarbet R, Sherer TB, MacKenzie G, Garcia-Osuna M, Panov AV, Greenamyre JT. Chronic systemic pesticide exposure reproduces features of Parkinson's disease. *Nat Neurosci*. 2000;3(12):1301–1306.
107. Le WD, Xu P, Jankovic J, et al. Mutations in NR4A2 associated with familial Parkinson disease. *Nat Genet*. 2003;33(1):85–89.
108. Javitch JA, Snyder SH. Uptake of MPP(+) by dopamine neurons explains selectivity of parkinsonism-inducing neurotoxin, MPTP. *Eur J Pharmacol*. 1984;106(2):455–456.
109. Pifl C, Giros B, Caron MG. Dopamine transporter expression confers cytotoxicity to low doses of the parkinsonism-inducing neurotoxin 1-methyl-4-phenylpyridinium. *J Neurosci*. 1993;13(10):4246–4253.
110. Gainetdinov RR, Fumagalli F, Jones SR, Caron MG. Dopamine transporter is required for in vivo MPTP neurotoxicity: evidence from mice lacking the transporter. *J Neurochem*. 1997;69(3):1322–1325.
111. Bezard E, Gross CE, Fournier MC, Dovero S, Bloch B, Jaber M. Absence of MPTP-induced neuronal death in mice lacking the dopamine transporter. *Exp Neurol*. 1999;155(2):268–273.
112. Gibb JW, Kogan FJ. Influence of dopamine synthesis on methamphetamine-induced changes in striatal and adrenal tyrosine hydroxylase activity. *Naunyn Schmiedebergs Arch Pharmacol*. 1979;310(2):185–187.
113. Seiden LS, Sabol KE. Methamphetamine and methylenedioxymethamphetamine neurotoxicity: possible mechanisms of cell destruction. *NIDA Res Monogr*. 1996;163:251–276.
114. Ricaurte GA, Schuster CR, Seiden LS. Long-term effects of repeated methylamphetamine administration on dopamine and serotonin neurons in the rat brain: a regional study. *Brain Res*. 1980;193(1):153–163.
115. Volz TJ, Hanson GR, Fleckenstein AE. The role of the plasmalemmal dopamine and vesicular monoamine transporters in methamphetamine-induced dopaminergic deficits. *J Neurochem*. 2007;101(4):883–888.
116. Cyr M, Beaulieu JM, Laakso A, et al. Sustained elevation of extracellular dopamine causes motor dysfunction and selective degeneration of striatal GABAergic neurons. *Proc Natl Acad Sci USA*. 2003;100(19):11035–11040.
117. Fernagut PO, Chalon S, Diguet E, Guilloteau D, Tison F, Jaber M. Motor behaviour deficits and their histopathological and functional correlates in the nigrostriatal system of dopamine transporter knockout mice. *Neuroscience*. 2003;116(4):1123–1130.
118. Levine MS, Cepeda C, Hickey MA, Fleming SM, Chesselet MF. Genetic mouse models of Huntington's and Parkinson's diseases: illuminating but imperfect. *Trends Neurosci*. 2004;27(11):691–697.
119. Baumann K, Mandelkow EM, Biernat J, Piwnica-Worms H, Mandelkow E. Abnormal Alzheimer-like phosphorylation of tau-protein by cyclin-dependent kinases cdk2 and cdk5. *FEBS Lett*. 1993;336(3):417–424.
120. Hanger DP, Hughes K, Woodgett JR, Brion JP, Anderton BH. Glycogen synthase kinase-3 induces Alzheimer's disease-like phosphorylation of tau: generation of paired helical filament epitopes and neuronal localisation of the kinase. *Neurosci Lett*. 1992;147(1):58–62.
121. Patrick GN, Zukerberg L, Nikolic M, de la Monte S, Dikkes P, Tsai LH. Conversion of p35 to p25 deregulates Cdk5 activity

and promotes neurodegeneration. *Nature.* 1999;402(6762): 615–622.
122. Beaulieu JM, Julien JP. Peripherin-mediated death of motor neurons rescued by overexpression of neurofilament NF-H proteins. *J Neurochem.* 2003;85(1):248–256.
123. Beaulieu JM, Sotnikova TD, Yao WD, et al. Lithium antagonizes dopamine-dependent behaviors mediated by an AKT/glycogen synthase kinase 3 signaling cascade. *Proc Natl Acad Sci USA.* 2004;101(14):5099–5104.
124. Cyr M, Sotnikova TD, Gainetdinov RR, Caron MG. Dopamine enhances motor and neuropathological consequences of polyglutamine expanded huntingtin. *FASEB J.* 2006;20(14):2541–2543.
125. Fernagut PO, Diguet E, Jaber M, Bioulac B, Tison F. Dopamine transporter knock-out mice are hypersensitive to 3-nitropropionic acid–induced striatal damage. *Eur J Neurosci.* 2002;15(12):2053–2056.
126. Yamamoto H, Kamegaya E, Hagino Y, et al. Genetic deletion of vesicular monoamine transporter-2 (VMAT2) reduces dopamine transporter activity in mesencephalic neurons in primary culture. *Neurochem Int.* 2007;51(2–4):237–244.
127. Xu F, Gainetdinov RR, Wetsel WC, et al. Mice lacking the norepinephrine transporter are supersensitive to psychostimulants. *Nat Neurosci.* 2000;3(5):465–471.
128. Bengel D, Murphy DL, Andrews AM, et al. Altered brain serotonin homeostasis and locomotor insensitivity to 3,4-methylenedioxymethamphetamine ("Ecstasy") in serotonin transporter-deficient mice. *Mol Pharmacol.* 1998;53(4):649–655.
129. Sotnikova TD, Beaulieu JM, Barak LS, Wetsel WC, Caron MG, Gainetdinov RR. Dopamine-independent locomotor actions of amphetamines in a novel acute mouse model of Parkinson disease. *PLoS Biol.* 2005;3(8):e271.
130. Sotnikova TD, Caron MG, Gainetdinov RR. DDD mice, a novel acute mouse model of Parkinson's disease. *Neurology.* 2006;67(7 suppl 2):S12–s17.
131. Hatcher JM, Pennell KD, Miller GW. Parkinson's disease and pesticides: a toxicological perspective. *Trends Pharmacol Sci.* 2008;29(6):322–329.

3.3 Intracellular Dopamine Signaling

GILBERTO FISONE

INTRODUCTION

Starting from the years immediately following its discovery, dopamine has been the subject of intensive studies that have demonstrated its implication in basic physiological processes, as well as in numerous diseases affecting the central nervous system. Several neurodegenerative and neuropsychiatric disorders are currently treated with drugs that affect dopamine transmission. For instance, Parkinson's disease is still largely treated with L-DOPA. Schizophrenia and related disorders are treated with antipsychotic agents, which act, at least in part, as antagonists at specific subtypes of dopamine receptors. The most effective cure for attention deficit hyperactivity disorder (ADHD) is represented by amphetamines and methylphenydate, which act by interfering with the reuptake of dopamine. In addition, virtually all drugs of abuse promote the release of dopamine in specific brain regions, thereby initiating a series of events leading to long-term modification of neuronal function.

The strategies adopted to counteract dysfunctions of dopaminergic transmission are still based on a limited repertoire of approaches, which rely almost exclusively on targeting, directly or indirectly, dopamine receptors or inhibiting the dopamine transporter. One important challenge facing the treatment of dopamine-related disorders is the development of more sophisticated and selective therapies that go beyond the idea of mimicking or repressing the action of dopamine at the membrane level. In this regard, the identification and characterization of intracellular components involved in dopamine signaling will provide essential information for the design of a new generation of dopaminergic drugs.

STRIATAL MEDIUM SPINY NEURONS: A REFERENCE CELL TYPE FOR THE STUDY OF DOPAMINE SIGNALING

Much of our understanding of dopamine transmission in the brain stems from studies of the midbrain dopaminergic system, which consists of two nuclei, the substantia nigra pars compacta and the ventral tegmental area, and their multiple forebrain targets. The striatal formation, which includes the caudate-putamen and the nucleus accumbens, is most densely innervated by midbrain dopaminergic fibers. This structure is the major component of the basal ganglia, a group of subcortical nuclei that control executive, motivational, and cognitive aspects of motor function.

GABAergic medium spiny neurons (MSNs) account for 90%–95% of striatal neurons and express a large number of dopamine receptors, whose main function is to modulate a major glutamatergic input from cortex, thalamus, hippocampus, and amygdala. Although morphologically identical, striatal MSNs can be distinguished based on their ability to express different types of dopamine receptors. This is particularly evident at the level of the dorsal part of the striatum (e.g., the caudate-putamen). In this region, the distribution of dopamine D1 receptors (D1Rs) and dopamine D2 receptors (D2Rs) coincides with the two major efferent pathways formed by MSNs. The MSNs that directly innervate the output structures of the basal ganglia (i.e., internal segment of the globus pallidus and substantia nigra pars reticulata) express D1Rs, whereas the MSNs that project to those structures indirectly (via the external segment of the globus pallidus and the subthalamic nucleus) express D2Rs.[1–3]

The "direct" striatonigral pathway and the "indirect" striatopallidal pathway exert opposite effects on motor function via modulation of thalamocortical neurons. Thus, activation of striatonigral MSNs disinhibits thalamocortical neurons and facilitates motor activity, whereas activation of the striatopallidal MSNs increases inhibition on thalamocortical neurons and suppresses motor activity. The most common model of striatal dopaminergic transmission posits that dopamine promotes motor function by activating the direct pathway, via D1Rs, and inhibiting the indirect pathway, via D2Rs.[4,5] Accumulating evidence indicates that this opposite regulation is exerted by promoting or counteracting glutamatergic transmission via modulation of voltage- and ligand-gated ion channels. An analogous regulation appears to occur at the level of synaptic plasticity. Long-term potentiation (LTP) of corticostriatal synapses requires activation of D1Rs[6–8] and is counteracted by D2Rs,[9] which are instead necessary for the induction of long-term depression (LTD).[9]

The key function played by dopamine in the striatum is dramatically highlighted by the severe motor impairments caused by the degeneration of the substantia nigra pars compacta and the loss of dopamine in the dorsal striatum, which represent the main pathological features of Parkinson's disease. Dopamine is also involved in the motivational system that regulates responses to natural (water, sex, etc.) and artificial (drugs of abuse) reinforcers. Both types of rewarding stimuli activate the neurons of the ventral tegmental area and increase the release of dopamine on the MSNs of the nucleus accumbens. The sequence of events that follow this primary effect has been the subject of intense studies,[10,11] which led to the identification of important mechanisms of signal transduction implicated in dopamine transmission.

The purpose of this chapter is to discuss signaling mechanisms triggered by the activation of dopamine receptors and to describe their impact on the regulation of downstream targets involved in short- and long-term neuronal responses. Because of the prevalent distribution of dopamine receptors in the striatum, particular attention will be given to dopaminergic transmission in MSNs.

GENERAL CLASSIFICATION OF DOPAMINE RECEPTORS INTO D1 AND D2 TYPES

The physiological effects of dopamine on its target cells are exerted through binding to metabotropic, heptahelical receptors coupled to heterotrimeric guanosine triphosphate (GTP) binding proteins called *G proteins*. G protein-coupled receptors (GPCRs) often modulate neuronal activity via complex sequences of biochemical reactions, resulting in a strong amplification of the response. Particularly important in the context of GPCR signaling is that each step in the transduction cascade represents a potential site of control and interaction with other signal transduction pathways.

The elucidation of the molecular mechanisms involved in dopamine signaling began with the observation that low concentrations of this neurotransmitter stimulated adenylyl cyclase (AC), the enzyme responsible for the synthesis of 3′,5′-cyclic monophosphate (cAMP).[12] Subsequent studies revealed that certain effects of dopamine, such as its ability to inhibit the release of prolactin in the pituitary gland, were associated with inhibition rather than activation of AC.[13–15] Dopamine receptors were therefore divided into two different types, D1 and D2, based on their ability to stimulate (D1) or inhibit (D2) the production of cAMP.[16]

This criterion for classification remained valid even after the identification of five different receptors for dopamine, named D1 to D5.[17–24] Thus, the D1 and D5 receptors, which stimulate AC, are referred to as *D1-type receptors*, whereas the D2, D3, and D4 receptors, which inhibit AC, are classified as *D2-type receptors*.[25]

The ability of D1- and D2-type receptors to exert an opposite regulation of cAMP signaling depends on their selective interaction with specific G proteins composed of different combinations of α, β, and γ subunits. The binding of dopamine to the D1-type receptors results in the activation of G proteins containing α subunits (i.e., αs and αolf; cf. below) able to stimulate AC. Conversely, the interaction of dopamine with D2-like receptors leads to activation of αi/o subunits, which are coupled to other types of regulations, including inhibition of AC.

SIGNALING VIA D1-TYPE RECEPTORS: INTERACTION WITH G_S AND G_{OLF} PROTEINS

Cloning of the D1R (also referred to as D1a)[18,19,24] and the D5 receptor (D5R; also named D1b or D1β)[21,26,27] revealed a high sequence homology and primary structures characterized by short third intracellular loops, which are common for receptors interacting with Gαs proteins. Expression in various types of host cells demonstrated that activation of both D1Rs and D5Rs produced a robust stimulation of AC.[18,19,21,24,26,27]

The positive coupling between D1Rs and AC depends on their ability to interact not only with G proteins containing an αs subunit, but also with those containing a highly homologous α subunit, originally described in the olfactory epithelium and named αolf.[28] Signaling via activation of Gαolf is particularly important in striatal MSNs, which express little αs and are enriched in αolf.[29,30] For instance, the ability of dopamine to increase cAMP synthesis is prevented in the striata of Gαolf $^{-/-}$ mice.[31] Moreover, in the same mice, cocaine, an addictive drug that increases dopamine release, fails to induce hyperlocomotion and immediate early gene expression.[32]

The dopaminergic signal transduction machinery of striatal MSNs appears to be particularly sensitive to changes in Gαolf expression. Thus, experiments performed in heterozygous Gαolf knockout mice have shown that a reduction of about 60% in the level of this protein results in a severe impairment of the biochemical and behavioral responses to dopaminergic agonists and psychostimulants.[33] This finding assumes particular relevance in view of the observation that the degeneration of the dopaminergic innervation to the striatum, which is a primary feature of Parkinson's disease, is accompanied by increased Gαolf expression in both humans and rodents.[30,34] This change is most likely responsible for the enhancement in D1R–G protein coupling,[35] the potentiation of D1R-mediated cAMP signaling,[34] and the sensitized response to L-DOPA observed in animal models of Parkinson's disease and related to the development of motor complications or dyskinesias.[36–39]

Whereas the importance of Gαolf for D1R-mediated transmission in the striatum is well established, Gαs is probably the main signal transducer for D1-type receptors in other brain regions. For instance, genetic inactivation of Gαolf does not affect the ability of dopamine to stimulate AC in the frontal cortex.[31] It is therefore likely that, in this region, responses to activation of D1-type receptors are mainly mediated via activation of Gαs.

A receptor with high affinity for SCH23390 (a D1-type receptor antagonist) has been proposed to increase cytoplasmic Ca^{2+} via coupling to a Gαq protein linked to activation of phospholipase C (PLC).[40] The identity of this receptor, which is distinct from that of the D1R,[41] remains to be established.

A few studies have addressed the question of the involvement of other G protein subunits in D1R-mediated transmission. The emerging picture indicates that D1R- but not D5R-mediated activation of AC requires the expression of a specific Gγ7 subunit,[42,43] which is highly expressed in striatal MSNs[44] and upregulated in the striata of parkinsonian patients.[34] The importance of the Gγ7 subunit in striatal dopaminergic signaling remains to be fully evaluated.

REGULATION OF AC AND cAMP

The binding of dopamine to D1Rs catalyzes exchange of guanosine diphosphate (GDP) for GTP on the Gαolf/s protein, which dissociates into Gαolf/s-GTP and Gβγ dimer. The Gαolf/s-GTP complex activates all nine membrane-bound ACs (AC1–AC9).[45] Several isoforms of AC are also regulated by the Gβγ complex.[46] In the striatum, MSNs express high levels of AC5,[47,48] which is activated by Gαolf/s-GTP, inhibited by Gαi/o, and insensitive to Gβγ.[45,46]

In addition to the canonical regulation via Gαolf/s, AC5 is inhibited by low concentrations of Ca^{2+}, which competes with Mg^{2+} for the activation of the catalytic domain of this isoform.[49,50] Furthermore, cAMP-dependent protein kinase (PKA)-catalyzed phosphorylation of recombinant AC5 results in reduced enzyme activity.[51] Studies in transfected cells have shown that this regulation is facilitated by the association of AC5 to the multiassembly scaffold A-kinase anchoring proteins, AKAP79/150[52] (cf. below). AKAP150 is particularly abundant in rat and mouse striatum.[53,54] It is therefore possible that, in striatal MSNs, AKAP150 participates in the regulation of AC by forming a signaling complex incorporating both AC5 and PKA. Such regulation may represent a feedback mechanism that controls D1R-mediated activation of cAMP signaling.

The G protein–mediated stimulation of AC is terminated when GTP is hydrolyzed by the Gαolf/s subunit, leading to reassembly of the heterotrimeric form of the G protein. The inactivation of G proteins is therefore dependent on the intrinsic guanosine triphosphatase (GTPase) activity of their Gα subunits and is accelerated by a family of about 40 GTPase activating proteins, denominated *regulators of G protein signaling* (RGSs) (for a recent review, see[55]). Besides promoting GTPase activity, RGSs interact directly with several effectors of G protein–coupled receptors. For instance, AC5 is inhibited by RGS2,[56–58] an isoform widely expressed in the brain, including the striatum.[59,60]

The physiological relevance of AC5 for D1R-mediated responses has been examined using AC5-deficient mice. In these animals, the ability of D1R agonists to stimulate cAMP accumulation in the striatum is reduced by 90%.[61] Surprisingly, in spite of this major impairment at the biochemical level, AC5 knockout mice show intact or even enhanced behavioral responses to D1R agonists.[61–63] These results contrast with those of studies in which dopamine D1R transmission was examined in mice deficient for phosphodiesterase (PDE) 1B, a Ca^{2+}/calmodulin-dependent PDE isoform principally responsible for the degradation of cAMP in striatal MSNs.[64,65] These animals displayed enhanced D1R-mediated activation of cAMP signaling and enhanced behavioral responses to administration of a D1R agonist.[65] Further studies will be necessary to fully understand the impact of the components that control cAMP production on D1R-mediated responses.

GPCR-KINASE-MEDIATED DESENSITIZATION OF D1Rs AND FRAGILE X MENTAL RETARDATION PROTEIN

Activation of GPCRs is followed by their rapid desensitization. This process begins with the phosphorylation of the receptor by GPCR kinases (GRKs), followed by binding to scaffold proteins named *arrestins* (cf. below) and receptor internalization.[66] In the prefrontal cortex, GRK2-mediated phosphorylation of D1Rs is controlled via interaction of this kinase with the fragile X mental retardation protein (FMRP), an RNA-binding protein that regulates translational efficiency. D1R signaling is impaired in FMRP null mice, which display subcellular redistribution of GRK2 and D1R hyperphosphorylation.[67] Therefore, it appears that FMRP is also able to act as a modulator of D1R desensitization.

PROTEIN KINASE A (PKA) AND A-KINASE ANCHORING PROTEIN (AKAP)

The increase in cAMP synthesis produced by binding of dopamine to D1Rs leads to activation of PKA, a tetrameric protein formed by two regulatory (R) and two catalytic (C) subunits. The binding of four molecules of cAMP to the R subunits promotes a conformational change in PKA that causes the dissociation of the C subunits. The activated C monomers bind ATP and can phosphorylate, in the cytoplasm and the nucleus, seryl and threonyl residues on proteins that contain the appropriate consensus sequence.[68]

The RIIb isoform of PKA has the highest expression in the striatum.[69] The importance of RIIb in dopamine signaling is indicated by the observation that genetic inactivation of this isoform impairs motor learning and reduces the ability of dopaminergic agents to regulate gene expression.[70]

One important characteristic of RIIb is its ability to interact with the three AKAP79/150 horthologues: rat AKAP75,[71,72] mouse AKAP150,[73] and human AKAP79.[74] AKAP79/150 is a constituent of the postsynaptic densitiy (PSD) and is known to interact with several proteins, including protein phosphatase-2B (PP-2B, or calcineurin)[75] and AC5[52] (cf. above). By tethering components of D1R signaling at the level of specific subcellular environments, AKAPs ensure a rapid and selective transduction of the dopaminergic signal. Thus, in hippocampal neurons, disruption of the interaction between PKA RIIb and AKAP15 abolishes the modulation exerted by a D1R agonist on voltage-dependent Na+ channels[76] (cf. below).

DARPP-32

D1R signaling depends not only on PKA-dependent phosphorylation of downstream target proteins, but also on concomitant inhibition of their dephosphorylation. This latter mechanism is mediated by the dopamine- and cAMP-regulated phosphoprotein, 32 kDa (DARPP-32),[77] which is particularly abundant in striatal MSNs.[78,79] DARPP-32 binds to the catalytic subunit of protein phosphatase-1 (PP-1), a ubiquitous threonyl/seryl phosphatase,[80] through a short KKIQF motif consisting of residues 7–11[81] (Fig. 3.3.1). D1R-mediated phosphorylation, catalyzed by PKA on Thr34 allows the interaction of DARPP-32 with the active site of PP-1, thereby preventing access to the phosphorylated substrate and reducing catalytic activity.[82,83] The consequent suppression of the dephosphorylation of downstream targets regulated by PKA intensifies cAMP-mediated responses and plays an important role in D1R-mediated transmission.[84] DARPP-32-mediated inhibition of PP-1 is terminated by dephosphorylation of Thr34, which involves mainly calcineurin[85,86] (Fig. 3.3.1).

DARPP-32 is also regulated via phosphorylation at Thr75, which is catalyzed by cyclin-dependent kinase 5. In this case, DARPP-32 is converted into an inhibitor of PKA.[87] Activation of D1Rs has been shown to reduce the phosphorylation of DARPP-32 on Thr75,[88] most likely via PKA-dependent phosphorylation and activation of protein phosphatase-2A (PP-2A), which is responsible for dephosphorylation of DARPP-32 at Thr75[88–90] (Fig. 3.3.1). This effect has been proposed to further promote D1R-mediated stimulation of the cAMP pathway by removing the inhibition exerted by phosphoThr75-DARPP-32 on PKA.[88]

Recent work has unveiled a further mode of regulation of DARPP-32 based on the control of its nuclear translocation. Phosphorylation catalyzed by casein kinase 2 on Ser97 (Ser102 in the rat), which is located in the vicinity of a nuclear export signal on DARPP-32, is necessary for the translocation of DARPP-32 from the nucleus to the cytoplasm. Interestingly, activation of D1Rs promotes the dephosphorylation of Ser97 via PKA-dependent activation of PP-2A (Fig. 3.3.1). This regulation results in the nuclear accumulation of DARPP-32 and is critical for the control exerted by dopamine on the phosphorylation of histone H3.[91]

In addition to its role in the nuclear export of DARPP-32, phosphoSer97 facilitates PKA-catalyzed phosphorylation of DARPP-32 on Thr34.[92] Thr34 phosphorylation is also promoted via casein kinase 1-dependent phosphorylation of DARPP-32 on Ser130 (Ser137 in the rat), which reduces calcineurin-mediated dephosphorylation on Thr34[93] (Fig. 3.3.1).

The importance of DARPP-32 in D1R-mediated transmission has been demonstrated by extensive studies performed with various types of genetically modified mice. DARPP-32 knockout mice show attenuated short- and long-term responses to several

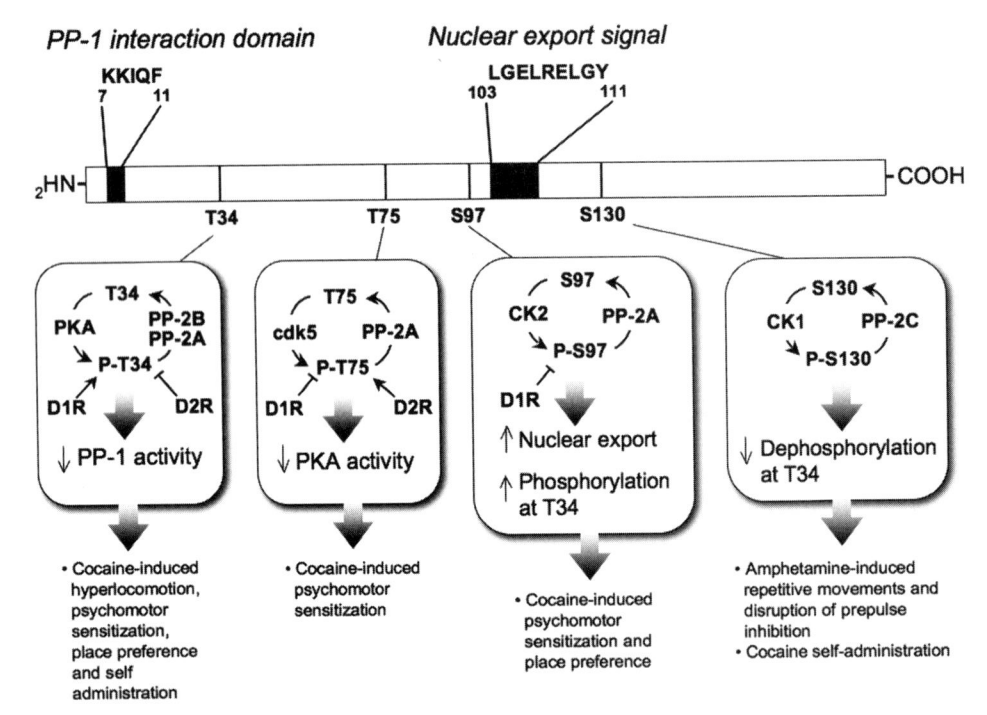

FIGURE 3.3.1. Regulation of DARPP-32 phosphorylation. DARPP-32 is regulated by phosphorylation at Thr34, Thr75, Ser97, and Ser130. Activation of D1Rs leads to PKA-catalyzed phosphorylation at Thr34, which converts DARPP-32 into an inhibitor of PP-1. Phosphorylation at Thr34 is reduced by activation of D2Rs through suppression of PKA signaling and activation of calcineurin (PP-2B). This latter effect is most likely mediated via Gβγ-mediated stimulation of PLC and mobilization of Ca^{2+} from intracellular stores. Phosphorylation at Thr75 is reduced by stimulation of D1Rs via PKA-mediated activation of PP-2A. A similar mechanism has been implicated in the D1R-mediated decrease in Ser97 phosphorylation. Psychostimulants such as cocaine and amphetamines increase phosphorylation at Thr34 and decrease phosphorylation at Thr75, most likely through D1Rs. Amphetamines increase phosphorylation at Ser130 in intact mice; however, the involvement of dopamine receptors in this effect remains to be assessed. Each of the phosphorylation sites has been implicated in various responses to cocaine or amphetamines (cf. the lower part of the diagram). See text for further details. cdk5, cyclin-dependent kinase 5, CK2, casein kinase 2, PKA, cAMP-dependent protein kinase; PP-1, -2A, -2B, protein phosphatase-1, -2A, -2B.

dopaminergic drugs, including cocaine, amphetamines, and L-DOPA.[38,84,94] Disruptions of dopaminergic responses have been also described in mutant mice lacking Thr34, Thr75, Ser97, or Ser130[91,95–98] (cf. Fig. 3.3.1).

REGULATION OF AMPA AND GABA$_A$ TRANSMISSION BY D1Rs

In the striatum, D1R-mediated activation of the cAMP/PKA/DARPP-32 cascade results in increased phosphorylation of the GluR1 subunit of the glutamate α-amino-3-hydroxy-5-methylisoxazole-4-propionic acid receptor (AMPAR).[99] This effect promotes neuronal excitability by increasing AMPA channel conductance[100,101] and cell surface expression.[102] Activation of D1Rs has also been shown to reduce current rundown through the AMPA channel.[103] This type of regulation involves DARPP-32-dependent inhibition of PP-1, which is anchored in proximity to AMPARs by spinophilin, a scaffold protein highly enriched in dendritic spines.[103]

Activation of cAMP/PKA/DARPP-32 signaling is also involved in D1R-mediated inhibition of GABA$_A$ receptors.[104] This effect has been proposed to promote glutamatergic transmission, thereby enhancing the ability of corticostriatal terminals to evoke activity in MSNs.[104]

A different mechanism, based on direct protein–protein interaction, is at the base of the regulation exerted, in the hippocampus, by D5Rs on GABA$_A$ transmission. In this case, the agonist-dependent formation of a complex between the D5R and the GABA$_A$ receptor γ2 subunit results in their reciprocal inhibition.[105]

D1R CONTROL OF NMDA TRANSMISSION VIA G PROTEIN-DEPENDENT AND -INDEPENDENT MECHANISMS

Activation of D1Rs enhances currents through the glutamate N-methyl-D-aspartate receptor (NMDAR) channel.[106] This effect, which is necessary for the induction of striatal LTP,[107] has been proposed to involve PKA- and DARPP-32-dependent phosphorylation of the NMDAR at the NR1 subunit.[84,108,109] The D1R-mediated increase in NR1 phosphorylation at Ser897 results in increased cytosolic Ca^{2+}, which, in association with cAMP/PKA signaling, activates the transcription factor Ca^{2+}/cAMP response element binding protein (CREB) and promotes CRE-dependent gene expression.[110] This synergistic control may be implicated in the actions of drugs of abuse. In fact, amphetamines have been shown to increase NR1 phosphorylation via stimulation of D1Rs in striatal MSNs.[111] Moreover, combined activation of D1Rs and NMDARs has been implicated in the activation by psychostimulants of the two mitogen-activated protein kinases (MAPKs), extracellular signal-regulated kinases 1 and 2 (ERK) (cf. below).[96]

D1Rs may also promote NMDAR transmission indirectly via a PKA/DARPP-32-mediated increase in L-type Ca^{2+} currents.[112] This possibility is suggested by the observation that blockade of L-type Ca^{2+} channels reduces the ability of a D1R agonist to potentiate NMDAR responses.[113,114] The regulation exerted by the cAMP/PKA/DARPP-32 signaling cascade on NMDAR function is likely to affect striatal synaptic plasticity. In fact, corticostriatal LTP and LTD are both prevented in DARPP-32 knockout mice.[6]

In the striatum, activation of D1Rs results in phosphorylation of the NR2B subunit at tyrosine sites.[115,116] This regulation, which occurs together with a rapid translocation of NMDARs to postsynaptic compartments, does not involve the PKA/DARPP-32 cascade but requires Fyn,[117] a member of the Src family of tyrosine kinases, which has been reported to phosphorylate the NR2 subunits[118] and to enhance channel activity.[119]

The positive interaction between D1Rs and NMDARs appears to be reciprocal. Experiments performed in neuronal and organotypic cultures from rat striatum show that activation of NMDARs recruits functional D1Rs to dendritic spines by reducing their mobility.[120,121] In line with these observations, studies performed in heterologous cells show that the D1R binds to NR1 and that, in the presence of NR2B, the formation of this complex promotes D1R trafficking to synaptic sites and D1R-mediated cAMP accumulation.[122,123]

Protein–protein interaction studies have described the existence of a negative modulation exerted on NMDARs by D1Rs, which may represent a mechanism to control excessive glutamatergic activation. Two regions in the carboxy terminal portion of the D1R bind to the NR1 and NR2A subunits of the NMDA receptor, thereby reducing NMDAR-induced excitotoxicity and NMDA-gated currents.[124]

D1Rs can also bind to PSD-95,[125] a protein that anchors the NMDAR to the PSD.[126] Interestingly, PSD-95-deficient mice display an increased locomotor response to cocaine, suggesting that the association of D1Rs with PSD-95 may reduce dopamine efficiency.[127] In line with this observation, it has been shown that the association of D1Rs with PSD-95 increases receptor internalization and reduces D1R signaling.[125] Therefore, it appears that PSD-95 promotes the positive interaction between D1Rs and NMDARs by facilitating their association and, at the same time, controls this phenomenon by curtailing D1R function.

D1R REGULATION OF NA^+ CHANNELS

Studies performed in striatum and hippocampus have shown that stimulation of D1Rs inhibits voltage-dependent Na^+ channels.[128,129] This effect requires activation of PKA[129–131] and, in striatal neurons, DARPP-32-mediated inhibition of PP-1.[132] It has been shown that PKA-catalyzed phosphorylation of Na^+ channels, which occurs at the level of the pore-forming α subunit,[129] promotes a process of slow inactivation that reduces channel availability during sustained depolarization.[131,133] The action of D1Rs on Na^+ channels may therefore represent a mechanism to control and coordinate neuronal excitability in response to strong glutamatergic input, which is thought to shift the membrane potential of MSNs from a "down-state" to a depolarized "up-state," closer to the spike threshold.[134]

D1R CONTROL OF VOLTAGE-DEPENDENT CA^{2+} CHANNELS: cAMP-DEPENDENT AND -INDEPENDENT MECHANISMS

In striatal neurons, D1R-mediated activation of the cAMP/PKA/DARPP-32 cascade enhances opening of L-type Ca^{2+} channels.[112] This effect, together with activation of NMDARs, is thought to promote the transition of MSNs to a higher level of excitability, similar to the up-state.[135] The ability of D1Rs to increase intracellular Ca^{2+} via L-type channels may also be implicated in the activation of ERK signaling and in the regulation of gene expression (cf. below).

D1R-mediated activation of cAMP signaling has been found to inhibit N- and P-type Ca^{2+} channels.[112] It is possible that this effect is involved in the suppression of

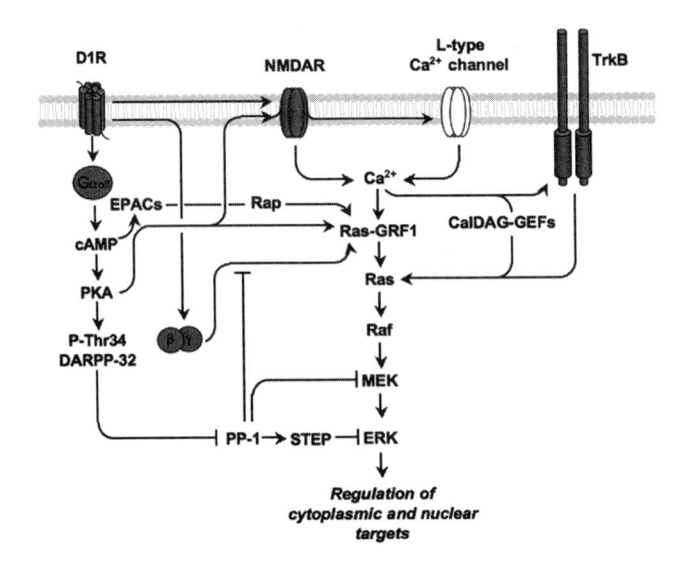

FIGURE 3.3.2. Cross-talk between cAMP/PKA/DARPP-32 and the Ras/ERK signaling cascades. Summary of the various signaling components involved in D1R-mediated activation of ERK. See text for explanation. CalDAG-GEFs, Ras-guanyl nucleotide releasing factors; cAMP, 3′,5′-cyclic adenosine monophosphate; DARPP-32, dopamine- and cAMP-regulated phosphoprotein, 32 kDa; D1R, dopamine 1 receptor; EPAC, exchange protein activated by cAMP; ERK, extracellular signal-regulated kinases 1 and 2; MEK, mitogen-activated protein kinase/extracellular signal-regulated protein kinase kinase; NMDAR, N-methyl-D-aspartate receptor;. PKA, cAMP-dependent protein kinase; PP-1, protein phosphatase-1; Ras-GRF-1, Ras-guanyl nucleotide releasing factor 1; STEP, striatal-enriched protein tyrosine phosphatase; TrkB, tyrosine receptor kinase B.

weak glutamatergic input on down-state, hyperpolarized MSNs[134] (cf. above). The negative control exerted by D1Rs on N-type Ca^{2+} channels has also been described in the pyramidal neurons of the prefrontal cortex. In this case, D1Rs inhibit N-type Ca^{2+} currents by directly interacting with the channel, leading to channel internalization and reduced cell surface expression.[136]

D1Rs AND ERK SIGNALING: A MATTER OF CROSS-TALK

DARPP-32 represents not only a critical feedforward mechanism able to amplify D1R-mediated responses (cf. above), but also a point of interaction between cAMP signaling and other signal transduction pathways. This latter function of DARPP-32 is exemplified by the regulation exerted by D1Rs on ERK, which are involved in multiple functions, including synaptic plasticity.[137] In neuronal cells, ERK signaling is activated by Ca^{2+} influx produced by depolarization and activation of NMDA receptors and L-type voltage-dependent Ca^{2+} channels.[138–140] Ca^{2+} activates the brain-specific exchange factor Ras-guanyl nucleotide releasing factor 1 (Ras-GRF1, or CDC25^{Mm}),[141,142] which promotes the exchange of GDP for GTP on the small G protein Ras.[143] A similar effect is produced by Ca^{2+} (in combination with diacylglycerol) via activation of the Ras-guanyl nucleotide releasing proteins (Ras-GRPs, or CalDAG-GEFs), which are highly expressed in striatal MSNs.[144] Ras-GTP activates the protein kinase Raf, leading to sequential phosphorylation of MAPK/ERK kinase (MEK) and ERK[137] (Fig. 3.3.2).

D1R-mediated activation of ERK is induced by drugs of abuse, such as cocaine and amphetamines, and requires concomitant activation of NMDARs (cf. above).[96,145] In addition, administration of L-DOPA dramatically increases ERK signaling in the dopamine-depleted striatum.[38,39] Studies using the conditioned place preference paradigm have demonstrated the requirement of phosphoERK for both retrieval[145,146] and reconsolidation of cocaine-associated contextual memory.[146] Moreover, inhibition of MEK and blockade of ERK signaling counteract the motor side effects produced by prolonged administration of L-DOPA.[38,147]

In striatal MSNs, activated ERK phosphorylates the transcription factor Elk-1,[148] which targets serum-response element-driven gene expression,[149] and the mitogen- and stress-activated kinase 1 (MSK1).[150] MSK1 is responsible for the activation of CREB[148] and for the phosphorylation of histone H3.[150] Collectively, these various effects lead to modifications of gene expression. Indeed, ERK has been involved in the regulation of several immediate early genes, including

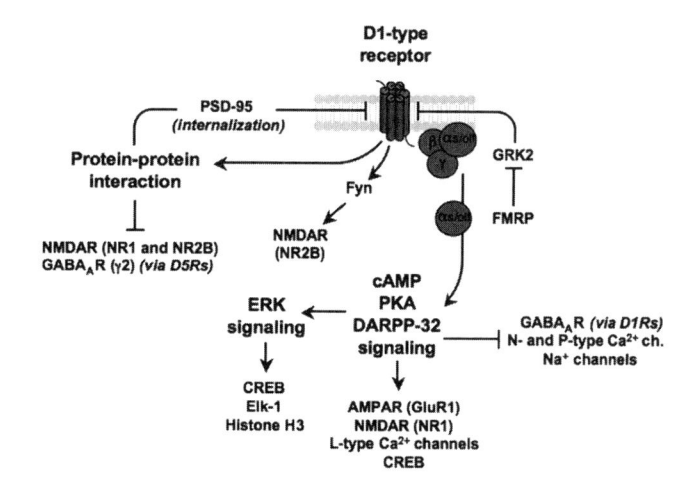

FIGURE 3.3.3. Summary of D1-type receptor-mediated signaling. D1-type dopamine receptors are coupled to Gαs/olf proteins that stimulate adenylyl cyclase, leading to activation of the cAMP/PKA/DARPP-32 signaling pathway. This results in phosphorylation and activation, or inhibition, of downstream targets, including ion channels and the transcription factor CREB. In addition, the cAMP/PKA/DARPP-32 cascade promotes ERK signaling by acting on various components of this intracellular pathway (cf. Fig. 3.3.2). This results in further transcriptional regulation. Activation of D1Rs is attenuated via GRK2-mediated phosphorylation. This regulation is suppressed by GRK2 interaction with FMRP. D1R interaction with PSD-95 leads to receptor internalization, and D1R interaction with NR1 and NR2B results in inhibition of NMDAR channels. This effect may represent a mechanism to control the overall positive regulation exerted by D1Rs on NMDAR-mediated transmission (cf. text). D5 receptors reduce GABAergic transmission by binding to the γ2 subunit of the GABA$_A$ receptor. See text for further details. AMPAR, α-amino-3-hydroxy-5-methylisoxazole-4-propionic acid receptor; cAMP, 3′,5′-cyclic adenosine monophosphate; CREB, Ca^{2+}/cAMP response element binding protein; DARPP-32, dopamine- and cAMP-regulated phosphoprotein, 32 kDa; ERK, extracellular signal-regulated kinases 1 and 2; FMRP, fragile X mental retardation protein; GRK2, G-protein coupled receptor kinase 2; NMDAR, N-methyl-D-aspartate receptor; PKA, cAMP-dependent protein kinase; PSD-95, postsynaptic density-95.

c-fos, fosB, zif-268, and arc.[148,151,152] The potential impact of these changes in the long-term responses to dopamine is indicated by the observation that zif-268 is required for cocaine-induced psychomotor sensitization and conditioned place preference.[153]

Cocaine-induced activation of ERK and Elk-1 depends on D1R and NMDAR transmission and requires PKA-catalyzed phosphorylation of DARPP-32 at Thr34.[96,145] Therefore, one possible mechanism by which D1Rs promote ERK activation is by increasing the intracellular Ca^{2+} concentration through positive modulation of NMDARs and L-type Ca^{2+} channels (cf. above). Furthermore, it has been proposed that phosphoThr34-DARPP-32 promotes ERK phosphorylation via inhibition of PP-1 and reduced dephosphorylation of MEK and of the striatal-enriched protein tyrosine phosphatase (STEP). Increased levels of phosphoMEK result in stimulation of kinase activity and phosphorylation of ERK. Increased levels of phosphoSTEP result in decreased phosphatase activity and suppression of ERK dephosphorylation[150] (Fig. 3.3.2).

A possible additional mechanism by which the cAMP cascade could promote ERK signaling is via PKA-mediated phosphorylation and activation of Ras-GRF1.[154] Interestingly, Ras-GRF1 is also activated by G protein βγ subunits, and this effect is prevented by PP-1.[155] This observation provides a further potential mechanism by which phosphoThr34-DARPP-32, an inhibitor of PP-1 (see above), may promote Ras-GRF1 and ERK signaling. Increased ERK phosphorylation in response to cAMP accumulation may also occur through activation of the exchange protein activated by cAMP 1 and 2 (ECAP1 and 2), which activate the Ras family GTPases, Rap1 and 2.[156]

Recent evidence shows that stimulation of D1Rs increases surface expression of TrkB receptors via a Ca^{2+}-dependent mechanism.[157] Therefore, D1R-mediated phosphorylation of ERK may also occur through transactivation of TrkB receptors, whose stimulation is known to promote MAPK signaling[158] (Fig. 3.3.2).

SIGNALING VIA D2-TYPE RECEPTORS: INTERACTION WITH Gαi/o PROTEINS

Dopamine receptors such as D2R, D3R, and D4R are coupled to a family of G proteins that includes Gαi1, Gαi2, Gαi3, and Gαo and share the ability to inhibit AC. This negative modulation is particularly evident at the level of AC5,[159] which is the isoform preferentially expressed in striatal MSNs.[47,48] In addition, D2Rs regulate ion channels via Gαo and Gβγ. The ability to interact with G proteins involved in the control of multiple effectors suggests a high degree of complexity in D2R signaling, which has been emphasized by several recent studies.

D2R CONTROL OF THE cAMP/PKA/DARPP-32 CASCADE

The inhibition exerted by D2Rs on AC and cAMP synthesis is reflected in a reduction of the phosphorylation of downstream proteins targeted by PKA. For instance, quinpirole, a D2R agonist, reduces the phosphorylation of DARPP-32 at Thr34 and that of GluR1 at Ser845.[160–162] Conversely, in intact animals, administration of dopamine D2R antagonists, such as eticlopride and haloperidol, increases the levels of phosphoThr34-DARPP-32 and phosphoSer845-GluR1.[160,161,163] The negative regulation exerted on GluR1 phosphorylation provides a possible mechanism explaining the ability of D2Rs to decrease the AMPA receptor current.[164]

Administration of haloperidol results in PKA-dependent phosphorylation of NR1, which promotes NMDA transmission.[165] The resulting increase in cytoplasmic Ca^{2+} activates CREB, leading to enhanced *c-fos* and *proenkephalin* gene expression.[166,167] Blockade of D2Rs also increases the phosphorylation of Elk-1[168] and of the acetylated form of histone H3,[169,170] which could result in further changes in gene expression and chromatin remodeling.[171]

Overall, the above observations indicate that, in MSNs, the cAMP/PKA/DARPP-32 cascade and its downstream targets are tonically inhibited by D2Rs. The importance of this regulation is exemplified by the observation that, in mice deficient in the RIIβ of PKA, haloperidol fails to induce the expression of mRNA for c-Fos and neurotensin and to induce catalepsy.[172] A similar reduction in the cataleptic response to a D2R antagonist has been reported in DARPP-32 knockout mice.[84]

D2R REGULATION OF VOLTAGE-DEPENDENT Ca^{2+} AND K^+ CHANNELS: SIGNALING VIA Gβγ

In the striatum, activation of D2Rs inhibits L-type Ca^{2+} channels. This action is opposite to the positive regulation exerted by activation of D1Rs via PKA-catalyzed phosphorylation (see above). Studies performed in isolated MSNs show that the D2R-mediated control of L-type Ca^{2+} currents does not involve suppression of cAMP signaling, but is instead secondary to mobilization of Ca^{2+} from intracellular stores.[173] This effect, in turn, is mediated by Gβγ-dependent activation of PLC and increased production of inositol-1,4,5-trisphosphate, which activates a Ca^{2+}-permeable receptor located on the endoplasmic reticulum. The following transient increase in Ca^{2+} leads to activation of calcineurin and inactivation of L-type channels by dephosphorylation.[173]

The reduction of L-type Ca^{2+} currents produced by D2Rs provides a mechanism explaining the ability of dopamine to inhibit the MSNs of the striatopallidal pathway, which selectively express D2Rs.[4,5] Thus, the increases in gene expression occurring in these neurons, and associated to dopamine deficit or blockade of D2Rs[2,174], could result, at least in part, from lack of D2R-mediated inhibition of L-type Ca^{2+} channels. This possibility is suggested by the observation that, in the striatum, blockade of L-type Ca^{2+} channels interferes with the ability of haloperidol to increase c-Fos expression.[175]

It has been reported that dopamine depletion results in the activation of the Cav1.3 subunit of the L-type channel and in the reduction of glutamatergic synapses on striatopallidal MSNs.[176] This loss of corticostriatal connectivity may have important repercussions in Parkinson's disease and concur to the development of the motor symptoms associated with this condition.

D2-type receptors are known to activate the G protein–regulated inwardly rectifying K^+ channel, leading to decreased cell excitability.[177,178] This effect is most likely produced via increased levels of Gβγ, which directly modulates the channel.[179,180]

D2-TYPE RECEPTOR-MEDIATED REGULATION OF NMDARs

The opposite regulation exerted by D1Rs and D2Rs on the activity of MSNs is also reflected by their opposite control of NMDAR transmission. Thus, whereas activation of D1Rs promotes NMDAR function, activation of D2Rs reduces NMDAR-mediated currents. This effect is mediated by the interaction between the third intracellular loop of the D2R and the carboxy terminal of NR2B, a phenomenon that has been shown to occur following administration of cocaine.[181] The formation of the D2R-NR2B complex in the PSD of corticostriatal synapses prevents the association of the NMDAR with Ca^{2+}/calmodulin-dependent protein kinase II (CaMKII), which normally phosphorylates NR2B at Ser1303. Suppression of CaMKII-dependent phosphorylation results in inhibition of NMDAR currents and cocaine-induced motor stimulation.[181]

A different type of modulation of NMDARs has been described in the hippocampus. In CA1 pyramidal neurons, which express high levels of D2R and D4R,[182] quinpirole reduces excitatory transmission at NMDARs via Gβγ-mediated transactivation of platelet-derived growth factor receptors (PDGFRs) and mobilization of Ca^{2+} from intracellular stores.[183]

CONTROLLING THE EFFICIENCY OF D2R-Gαi/o COUPLING VIA RGS9-2, GRK6, AND CaMKII

As previously discussed, RGS affect GPCR-mediated transmission by acting as GTPase-accelerating proteins and promoting G protein inactivation. RGS9-2, which is particularly enriched in striatal MSNs,[184] has been implicated in D2R-mediated responses.[185] Viral overexpression of RGS9-2 in the nucleus accumbens results in a reduced locomotor response to cocaine. In addition, RGS9-2-overexpressing mice show an impaired motor response to D2R but not to D1R agonists. Conversely, RGS9-2 knockout mice show enhanced locomotor activity, sensitization, and place preference in response to cocaine.[185] The simplest way of interpreting these results is that RGS9-2 counteracts D2R-mediated transmission by accelerating Gαi/o GTPase activity and promoting the formation of the inactive GDP-Gαi/o protein complex.

In the brain, GRK-induced interaction of GPCRs with β-arrestin1 and 2 uncouples the receptor from the G protein and mediates receptor internalization (cf. above).[66] GRKs and arrestins have both been implicated in D2R-mediated signaling. Among the seven GRKs identified in mammalian tissue, GRK6 is the most prominent in the striatum, where it is highly expressed in MSNs.[186] GRK6 knockout mice display an increased response to cocaine and amphetamines, as well as enhanced coupling of D2Rs to Gαi/o protein. In contrast, D1R-mediated transmission is not affected in these animals.[186] These studies indicate that GRK6 is specifically involved in the desensitization of canonical G protein–dependent D2R transmission.

Another mechanism that affects the coupling efficacy of D2Rs to Gαi/o involves the direct interaction of the receptor with calmodulin. D2Rs bind to Ca^{2+}-activated calmodulin through a region located in the third cytoplasmic loop, which is adjacent to that involved in NR2B interaction (cf above). The formation of the D2R-Ca^{2+}/calmodulin complex reduces the ability of the D2R to activate Gαi/o, thereby counteracting the negative regulation on cAMP production.[187] The same motif that allows binding of the D2R to activated calmodulin is also involved in the interaction of the receptor with the prostate apoptosis response-4 protein (Par-4). Association to Par-4 is required for D2R-mediated responses, including inhibition of cAMP synthesis and of CREB phosphorylation. Interestingly, Ca^{2+}/calmodulin competes with Par-4 for binding to the D2R and may therefore reduce D2R-mediated responses by preventing the formation of the D2R-Par-4 complex.[188] Taken together, these observations indicate that an increased Ca^{2+} concentration may reduce D2R-mediated transmission and promote CREB-dependent gene expression by disrupting the interaction of D2Rs with Par-4. The significance of this type of regulation is indicated by the observation that mice with a loss of function mutation in Par-4 display depression-like behavior, which may result from impaired dopamine signaling.[188] Further studies will be necessary to identify the stimuli responsible for the enhancement in Ca^{2+} concentration that leads to dissociation of Par-4 from the D2R.

D2R SIGNALING VIA THE AKT/GSK-3 CASCADE

D2Rs regulate the protein kinase B (PKB, or Akt)/glycogen synthase-3 (GSK-3) signaling cascade via interaction with β-arrestin 2. Akt is activated by phosphorylation mediated through the phosphatidylinositol-3-kinase signaling pathway. Activated Akt inhibits GSK-3α and β via phosphorylation of their N-terminal regulatory domains.[189] Dysregulation of the Akt-GSK-3 cascade has been implicated in many pathological processes, including neurodegenerative diseases, mood disorders, and schizophrenia.[189–191]

Stimulation of D2Rs results in inhibition of Akt and concomitant activation of GSK-3.[192,193] This effect appears to be potentiated by activation of D3Rs.[193] The control exerted by D2Rs on the Akt/GSK-3 cascade is independent of G protein activation and occurs through binding of the D2R to β-arrestin 2, which recruits Akt and PP-2A, leading to dephosphorylation/inactivation of Akt.[194]

The importance of the D2R/Akt/GSK-3 signaling pathway in dopamine transmission is demonstrated by experiments conducted in β-arrestin 2 knockout mice. In these animals, administration of amphetamine or apomorphine (a nonselective dopamine receptor agonist) fails to reduce Akt phosphorylation without affecting cAMP/DARPP-32 signaling. This specific lack of regulation is accompanied by a dramatic decrease in the motor stimulant responses to both drugs.[194]

The ability of D2Rs to control Akt-GSK-3 signaling is particularly interesting in view of the observation that the brains of individuals affected by schizophrenia contain lower levels of the Akt isoform, Akt1, and of phosphorylated GSK3β.[191] These abnormalities may result from enhanced dopamine transmission and may be involved in the disruption of sensorimotor gating associated with schizophrenia.[191] In line with this interpretation, it has been shown that blockade of D2Rs with haloperidol, an antipsychotic drug, increases the phosphorylation of Akt. This effect, which compensates for the decrease in Akt expression by increasing the amount of activated protein, may explain, at least in part, the ability of haloperidol to reduce the symptoms of schizophrenia.

In conclusion, the regulation of the Akt/GSK-3 cascade by D2Rs represents an important G protein–independent signaling mechanism. Interestingly, this intracellular cascade is activated with a slower kinetic, when compared to the canonical G protein–mediated signaling pathway involving cAMP and DARPP-32.[195] It is therefore likely that the formation of the β-arrestin 2/Akt/protein phosphatase-2A complex is preferentially involved in long-lasting D2R-mediated responses. Finally, it has been reported that incubation of striatal neurons with a D1R agonist increases the phosphorylation of Akt,[196] raising the possibility that this pathway is regulated by dopamine via activation of both D1Rs and D2Rs.

PRESYNAPTIC AND POSTSYNAPTIC D2R ISOFORMS

D2Rs are highly expressed in dopaminergic midbrain neurons, where they function as inhibitory autoreceptors. Alternative splicing of the D2R results in the production of a long (D2LR) and a short (D2SR) isoform that differ by 29 amino acids within the third intracellular loop.[197,198] Several lines of evidence indicate that D2LR is the principal D2R at the postsynaptic level (i.e., on striatal MSNs), whereas D2SR is implicated in the presynaptic control of nigrostriatal neurons.[199–201]

One important question regarding D2SRs and D2LRs concerns possible differences in their signaling properties, particularly considering that the third intracellular loop is implicated in receptor–G protein coupling and other types of protein–protein interactions (cf. above). Increased levels of phosphorylated Akt and GSK-3β have been found in the striata of D2LR knockout mice,[193] suggesting that D2LRs may be preferentially coupled to this specific signaling cascade. However, these results may also be explained by a preferential postsynaptic localization of the Akt/ GSK-3β signaling machinery.

Studies performed in pituitary GH4 cells show that D2SRs, but not D2LRs, inhibit the activation of ERK produced by thyrotropin-releasing hormone.[202] Other studies performed in CHO and HEK293 cells indicate that D2LRs and D2SRs promote ERK phosphorylation via distinct mechanisms involving transactivation of PDGFRs and modulation of β-arrestin-dynamin receptor endocytosis, respectively.[203] These various studies are difficult to interpret in view of the contrasting evidence with respect to the type of regulation exerted by D2R on ERK signaling. Experiments performed in brain slices and primary cultures from striatum and midbrain indicate that activation of D2R promotes ERK phosphorylation.[204–206] In contrast, a more recent study indicates that incubation of striatal neurons with a D2R agonist reduces depolarization-induced activation of ERK.[202] In line with this observation, it has been shown that administration of D2R antagonists increases ERK phosphorylation in the striatum.[168,207,208] In conclusion, a clear distinction between D2LRs and D2SRs in terms of G protein coupling and signaling is still lacking. In this regard, it should be noted that the sites of interaction between the D2R and many signaling proteins [e.g., spinophilin,[209] NR2B, CaMKII, Par-4 (see above)], although located on the third cytoplasmic loop, are conserved in both D2R isoforms.

CONCLUSIONS AND FUTURE PERSPECTIVES

During the past four decades, a major effort has been made to elucidate the molecular basis of dopamine signaling. This work has led not only to the identification of important components of the canonical G protein- and cAMP-dependent cascade involved in dopaminergic transmission, but also to the characterization of alternative signaling mechanisms based on cross-talk between the cAMP/PKA/DARPP-32 pathway and other signaling cascades. In addition, increasing information is becoming available with regard to the ability of dopamine receptors to bind to a variety of interacting proteins. These complex mechanisms lead to increased or decreased neuronal excitability, which generally appear to depend on activation of D1Rs and D2Rs, respectively.

One important challenge facing future studies on dopaminergic signal transduction is the transfer of information obtained from the study of heterologous cell systems to more physiologically relevant systems (e.g., primary neuronal cultures, brain slices, or intact animals). This is particularly important, as the expression of specific receptor subtypes and the composition

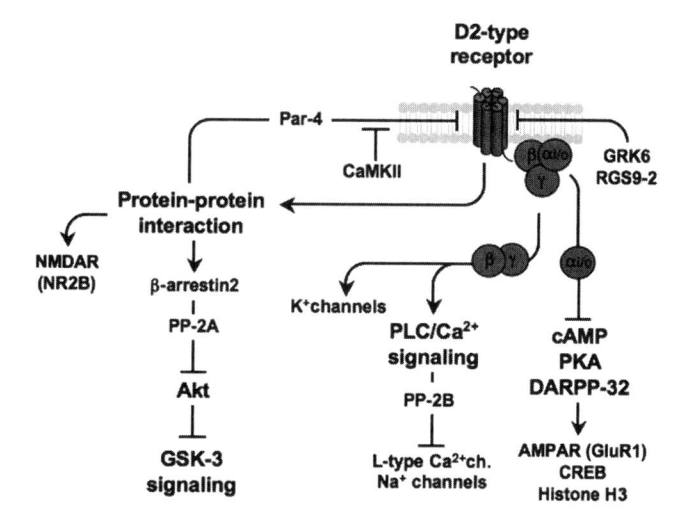

FIGURE 3.3.4. Summary of D2-type receptor-mediated signaling. D2-type dopamine receptors are coupled to a Gαi/o proteins that inhibit adenylyl cyclase and regulate ion channels. The negative control exerted on the cAMP/PKA/DARPP-32 cascade is reflected in decreased phosphorylation of PKA target proteins, such as the GluR1 subunit of the AMPA receptor and CREB. D2Rs also modulate voltage-dependent Ca^{2+}, Na^+, and K^+ channels, most likely via Gβγ and, at least in part, via activation of PLC and stimulation of calcineurin-dependent dephosphorylation. D2R-mediated transmission is reduced by GRK6-mediated desensitization, by RGS9.2-mediated inactivation of Gαi/o and by CaMKII, which competes with the activator Par-4 for binding to the D2R. Activation of D2Rs promotes interaction with β-arrestin2, leading to PP-2A-dependent dephosphorylation of Akt and activation of GSK-3. See text for further details and abbreviations. Akt, protein kinase B; AMPAR, amino-3-hydroxy-5-methylisoxazole-4-propionic acid receptor; CaMKII, Ca^{2+}/calmodulin-dependent protein kinase II; cAMP, 3′,5′-cyclic adenosine monophosphate; CREB, Ca^{2+}/cAMP response element binding protein; DARPP-32, dopamine- and cAMP-regulated phosphoprotein, 32 kDa; GRK6, G-protein coupled receptor kinase 6; GSK-3, glycogen synthase-3; NMDAR, N-methyl-D-aspartate receptor; Par-4, prostate apoptosis response-4 protein; PKA, cAMP-dependent protein kinase; PLC, phospholipase C; PP-2A, -2B protein phosphatase-2A, -2B; RGS9-2, regulator of G protein signaling 9-2.

of the signal transduction machinery in one striatal MSNs, or in any other dopaminoceptive neuron, may not be compatible with a mechanism characterized in transfected cells. For instance, it has been proposed that coactivation of D1Rs and D2Rs shifts the coupling of dopamine receptors to activation of a Gαq protein, which stimulates PLC and increases cytoplasmic Ca^{2+}.[210] This type of regulation is interesting, but its physiological implications remain to be fully assessed, particularly in the striatum, where D1Rs and D2Rs are for the most part expressed in distinct populations of MSNs.[1–3]

Another important question concerns the identification of distinct populations of neurons where changes in signaling produced by manipulation of specific subtypes of dopamine receptors occur. The availability of bacterial artificial chromosome (BAC) vectors as cell targeting tools[211] has greatly simplified this issue, allowing the visualization of distinct subsets of neurons through cell-targeted expression of fluorescent probes.[3] BAC vectors can also be used to drive cell-targeted expression of Cre recombinase[212] and thereby to induce null mutations of selected genes in discrete groups of neurons using, for instance, the Cre/loxP recombination system.[213] This will reduce unwanted general effects produced by systemic gene knockout and allow a more precise characterization of the role played by specific signaling components in dopamine transmission.

REFERENCES

1. Bertran-Gonzalez J, Bosch C, Maroteaux M, et al. Opposing patterns of signaling activation in dopamine D1 and D2 receptor-expressing striatal neurons in response to cocaine and haloperidol. *J Neurosci.* 2008;28:5671–5685.
2. Gerfen CR, Engber TM, Mahan LC, et al. D1 and D2 dopamine receptor-regulated gene expression of striatonigral and striatopallidal neurons. *Science.* 1990;250:1429–1432.
3. Gong S, Zheng C, Doughty ML, et al. A gene expression atlas of the central nervous system based on bacterial artificial chromosomes. *Nature.* 2003;425:917–925.
4. Albin RL, Young AB, Penney JB. The functional anatomy of basal ganglia disorders. *Trends Neurosci.* 1989;12:366–375.
5. Gerfen CR. The neostriatal mosaic: multiple levels of compartmental organization in the basal ganglia. *Annu Rev Neurosc.i* 1992;15:285–320.
6. Calabresi P, Gubellini P, Centonze, D, et al. Dopamine and cAMP–regulated phosphoprotein 32 kDa controls both striatal long-term depression and long-term potentiation, opposing forms of synaptic plasticity. *J Neurosci.* 2000;20:8443–8451.

7. Centonze D, Grande C, Saulle E, et al. Distinct roles of D1 and D5 dopamine receptors in motor activity and striatal synaptic plasticity. *J Neurosc.i* 2003;23:8506–8512.
8. Kerr JN, Wickens JR. Dopamine D-1/D-5 receptor activation is required for long-term potentiation in the rat neostriatum in vitro. *J Neurophysio.l* 2001;85:117–124.
9. Calabresi P, Saiardi A, Pisani A, et al. Abnormal synaptic plasticity in the striatum of mice lacking dopamine D2 receptors. *J Neurosc.i* 1997;17:4536–4544.
10. Nestler EJ. Molecular basis of long-term plasticity underlying addiction. *Nat Rev Neurosci.* 2001;2:119–128.
11. Robbins TW, Everitt BJ. Drug addiction: bad habits add up. *Nature.* 1999;398:567–570.
12. Brown JH, Makman MH. Stimulation by dopamine of adenylate cyclase in retinal homogenates and of adenosine-3′:5′-cyclic monophosphate formation in intact retina. *Proc Natl Acad Sci USA.* 1972;69:539–543.
13. Caron MG, Beaulieu M, Raymond V, et al. Dopaminergic receptors in the anterior pituitary gland. Correlation of [3H]dihydroergocryptine binding with the dopaminergic control of prolactin release. *J Biol Chem.* 1978;253:2244–2253.
14. De Camilli P, Macconi D, Spada A. Dopamine inhibits adenylate cyclase in human prolactin-secreting pituitary adenomas. *Nature.* 1979;278:252–254.
15. Spano PF, Govoni S, Trabucchi M. Studies on the pharmacological properties of dopamine receptors in various areas of the central nervous system. *Adv Biochem Psychopharmacol.* 1978;19:155–165.
16. Kebabian JW, Calne DB. Multiple receptors for dopamine. *Nature.* 1979;277:93–96.
17. Bunzow JR, Van Tol HH, Grandy DK, et al. Cloning and expression of a rat D2 dopamine receptor cDNA. *Nature.* 1988;336:783–787.
18. Dearry A, Gingrich JA, Falardeau P, et al. Molecular cloning and expression of the gene for a human D1 dopamine receptor. *Nature.* 1990;347:72–76.
19. Monsma FJ Jr, Mahan LC, McVittie LD, et al. Molecular cloning and expression of a D1 dopamine receptor linked to adenylyl cyclase activation. *Proc Natl Acad Sci USA.* 1990;87:6723–6727.
20. Sokoloff P, Giros B, Martres MP, et al. Molecular cloning and characterization of a novel dopamine receptor (D3) as a target for neuroleptics. *Nature.* 1990;347:146–151.
21. Sunahara RK, Guan HC, O'Dowd BF, et al. Cloning of the gene for a human dopamine D5 receptor with higher affinity for dopamine than D1. *Nature.* 1991;350:614–619.
22. Sunahara RK, Niznik HB, Weiner DM, et al. Human dopamine D1 receptor encoded by an intronless gene on chromosome 5. *Nature.* 1990;347:80–83.
23. Van Tol HH, Bunzow JR, Guan HC, et al. Cloning of the gene for a human dopamine D4 receptor with high affinity for the antipsychotic clozapine. *Nature.* 1991;350:610–614.
24. Zhou QY, Grandy DK, Thambi L, et al. Cloning and expression of human and rat D1 dopamine receptors. *Nature.* 1990;347:76–80.
25. Missale C, Nash SR, Robinson SW, et al. Dopamine receptors: from structure to function. *Physiol Rev.* 1998;78:189–225.
26. Tiberi M, Jarvie KR, Silvia C, et al. Cloning, molecular characterization, and chromosomal assignment of a gene encoding a second D1 dopamine receptor subtype: differential expression pattern in rat brain compared with the D1A receptor. *Proc Natl Acad Sci USA.* 1991;88:7491–7495.
27. Weinshank RL, Adham N, Macchi M, et al. Molecular cloning and characterization of a high affinity dopamine receptor (D1 beta) and its pseudogene. *J Biol Chem.* 1991;266:22427–22435.
28. Jones DT, Reed RR. Golf: an olfactory neuron specific-G protein involved in odorant signal transduction. *Science.* 1989;244:790–795.
29. Drinnan SL, Hope BT, Snutch TP, et al. G_{olf} in the basal ganglia. *Mol Cell Neurosci.* 1991;2:66–70.
30. Herve D, Levi-Strauss M, Marey-Semper I, et al. G(olf) and Gs in rat basal ganglia: possible involvement of G(olf) in the coupling of dopamine D1 receptor with adenylyl cyclase. *J Neurosci.* 1993;13:2237–2248.
31. Corvol JC, Studler JM, Schonn JS, et al. Galpha(olf) is necessary for coupling D1 and A2a receptors to adenylyl cyclase in the striatum. *J Neurochem.* 2001;76:1585–1588.
32. Zhuang X, Belluscio L, Hen R. G(olf)alpha mediates dopamine D1 receptor signaling. *J Neurosci.* 2000;20:RC91.
33. Corvol JC, Valjent E, Pascoli V, et al. Quantitative changes in Gαolf protein levels, but not D1 receptor, alter specifically acute responses to psychostimulants. *Neuropsychopharmacology.* 2007;32:1109–1121.
34. Corvol JC, Muriel MP, Valjent E, et al. Persistent increase in olfactory type G-protein alpha subunit levels may underlie D1 receptor functional hypersensitivity in Parkinson disease. *J Neurosci.* 2004;24:7007–7014.
35. Aubert I, Guigoni C, Hakansson K, et al. Increased D1 dopamine receptor signaling in levodopa-induced dyskinesia. *Ann Neurol.* 2005;57:17–26.
36. Pavon N, Martin AB, Mendialdua A, et al. ERK phosphorylation and FosB expression are associated with L-DOPA-induced dyskinesia in hemiparkinsonian mice. *Biol Psychiatry.* 2006;59:64–74.
37. Picconi B, Centonze D, Hakansson K, et al. Loss of bidirectional striatal synaptic plasticity in L-DOPA-induced dyskinesia. *Nat Neurosc.i* 2003;6:501–506.
38. Santini E, Valjent E, Usiello A, et al. Critical involvement of cAMP/DARPP-32 and extracellular signal-regulated protein kinase signaling in L-DOPA-induced dyskinesia. *J Neurosci.* 2007;27:6995–7005.
39. Westin JE, Vercammen L, Strome EM, et al. Spatiotemporal pattern of striatal ERK1/2 phosphorylation in a rat model of L-DOPA-induced dyskinesia and the role of dopamine D1 receptors. *Biol Psychiatry.* 2007;62:800–810.
40. Undie AS, Friedman E. Stimulation of a dopamine D1 receptor enhances inositol phosphates formation in rat brain. *J Pharmacol Exp Ther.* 1990;253:987–992.
41. Friedman E, Jin LQ, Cai GP, et al. D1-like dopaminergic activation of phosphoinositide hydrolysis is independent of D1A dopamine receptors: evidence from D1A knockout mice. *Mol Pharmacol.* 1997;51:6–11.
42. Schwindinger WF, Betz KS, Giger KE, et al. Loss of G protein gamma 7 alters behavior and reduces striatal alpha(olf) level and cAMP production. *J Biol Chem.* 2003;278:6575–6579.
43. Wang Q, Jolly JP, Surmeier JD, et al. Differential dependence of the D1 and D5 dopamine receptors on the G protein gamma 7 subunit for activation of adenylylcyclase. *J Biol Chem.* 2001;276:39386–39393.
44. Watson JB, Coulter PM 2nd, Margulies JE, et al. G-protein gamma 7 subunit is selectively expressed in medium-sized neurons and dendrites of the rat neostriatum. *J Neurosci Res.* 1994;39:108–116.
45. Sunahara RK, Dessauer CW, Gilman AG. Complexity and diversity of mammalian adenylyl cyclases. *Annu Rev Pharmacol Toxicol.* 1996;36:461–480.
46. Tang WJ, Gilman AG: Type-specific regulation of adenylyl cyclase by G protein beta gamma subunits. *Science.* 1991;254:1500–1503.

47. Glatt CE, Snyder SH. Cloning and expression of an adenylyl cyclase localized to the corpus striatum. *Nature.* 1993;361:536–538.
48. Mons N, Cooper DM. Selective expression of one Ca(2+)-inhibitable adenylyl cyclase in dopaminergically innervated rat brain regions. *Brain Res Mol Brain Res.* 1994;22:236–244.
49. Guillou JL, Nakata H, Cooper DM. Inhibition by calcium of mammalian adenylyl cyclases. *J Biol Chem.* 1999;274:35539–35545.
50. Hu B, Nakata H, Gu C, et al. A critical interplay between Ca^{2+} inhibition and activation by Mg^{2+} of AC5 revealed by mutants and chimeric constructs. *J Biol Chem.* 2002;277:33139–33147.
51. Iwami G, Kawabe J, Ebina T, et al. Regulation of adenylyl cyclase by protein kinase A. *J Biol Chem.* 1995;270:12481–12484.
52. Bauman AL, Soughayer J, Nguyen BT, et al. Dynamic regulation of cAMP synthesis through anchored PKA-adenylyl cyclase V/VI complexes. *Mol Cell.* 2006;23:925–931.
53. Glantz SB, Amat JA, Rubin CS. cAMP signaling in neurons: patterns of neuronal expression and intracellular localization for a novel protein, AKAP 150, that anchors the regulatory subunit of cAMP-dependent protein kinase II beta. *Mol Biol Cell.* 1992;3:1215–1228.
54. Ostroveanu A, Van der Zee EA, Dolga AM, et al. A-kinase anchoring protein 150 in the mouse brain is concentrated in areas involved in learning and memory. *Brain Res.* 2007;1145:97–107.
55. Abramow-Newerly M, Roy AA, Nunn C, et al. RGS proteins have a signalling complex: interactions between RGS proteins and GPCRs, effectors, and auxiliary proteins. *Cell Signal.* 2006;18:579–591.
56. Roy AA, Baragli A, Bernstein LS, et al. RGS2 interacts with Gs and adenylyl cyclase in living cells. *Cell Signal.* 2006;18:336–348.
57. Salim S, Sinnarajah S, Kehrl JH, et al. Identification of RGS2 and type V adenylyl cyclase interaction sites. *J Biol Chem.* 2003;278:15842–15849.
58. Sinnarajah S, Dessauer CW, Srikumar D, et al. RGS2 regulates signal transduction in olfactory neurons by attenuating activation of adenylyl cyclase III. *Nature.* 2001;409:1051–1055.
59. Grafstein-Dunn E, Young KH, Cockett MI, et al. Regional distribution of regulators of G-protein signaling (RGS) 1, 2, 13, 14, 16, and GAIP messenger ribonucleic acids by in situ hybridization in rat brain. *Brain Res Mol Brain Res.* 2001;88:113–123.
60. Taymans JM, Wintmolders C, Te Riele P, et al. Detailed localization of regulator of G protein signaling 2 messenger ribonucleic acid and protein in the rat brain. *Neuroscience.* 2002;114:39–53.
61. Lee KW, Hong JH, Choi IY, et al. Impaired D2 dopamine receptor function in mice lacking type 5 adenylyl cyclase. *J Neurosci.* 2002;22:7931–7940.
62. Iwamoto T, Okumura S, Iwatsubo K, et al. Motor dysfunction in type 5 adenylyl cyclase-null mice. *J Biol Chem.* 2003;278:16940.
63. Kim KS, Lee KW, Baek IS, et al. Adenylyl cyclase-5 activity in the nucleus accumbens regulates anxiety-related behavior. *J Neurochem.* 2008;107:105–115.
64. Polli JW, Kincaid RL. Expression of a calmodulin-dependent phosphodiesterase isoform (PDE1B1) correlates with brain regions having extensive dopaminergic innervation. *J Neurosci.* 1994;14:1251–1261.
65. Reed TM, Repaske DR, Snyder GL, et al. Phosphodiesterase 1B knock-out mice exhibit exaggerated locomotor hyperactivity and DARPP-32 phosphorylation in response to dopamine agonists and display impaired spatial learning. *J Neurosci.* 2002;22:5188–5197.
66. Shenoy SK, Lefkowitz RJ. Multifaceted roles of beta-arrestins in the regulation of seven-membrane-spanning receptor trafficking and signalling. *Biochem J.* 2003;375:503–515.
67. Wang H, Wu LJ, Kim SS, et al. FMRP acts as a key messenger for dopamine modulation in the forebrain. *Neuron.* 2008;59:634–647.
68. Fimia GM, Sassone-Corsi P. Cyclic AMP signalling. *J Cell Sci.* 2001;114:1971–1972.
69. Cadd G, McKnight GS. Distinct patterns of cAMP-dependent protein kinase gene expression in mouse brain. *Neuron.* 1989;3:71–79.
70. Brandon EP, Logue SF, Adams MR, et al. Defective motor behavior and neural gene expression in RIIbeta-protein kinase A mutant mice. *J Neurosci.* 1998;18:3639–3649.
71. Bregman DB, Hirsch AH, Rubin CS. Molecular characterization of bovine brain P75, a high affinity binding protein for the regulatory subunit of cAMP-dependent protein kinase II beta. *J Biol Chem.* 1991;266:7207–7213.
72. Sarkar D, Erlichman J, Rubin CS. Identification of a calmodulin-binding protein that co-purifies with the regulatory subunit of brain protein kinase II. *J Biol Chem.* 1984;259:9840–9846.
73. Bregman DB, Bhattacharyya N, Rubin CS. High affinity binding protein for the regulatory subunit of cAMP-dependent protein kinase II-B. Cloning, characterization, and expression of cDNAs for rat brain P150. *J Biol Chem.* 1989;264:4648–4656.
74. Carr DW, Stofko-Hahn RE, Fraser ID, et al. Localization of the cAMP-dependent protein kinase to the postsynaptic densities by A-kinase anchoring proteins. Characterization of AKAP 79. *J Biol Chem.* 1992;267:16816–16823.
75. Dell'Acqua ML, Dodge KL, Tavalin SJ, et al. Mapping the protein phosphatase-2B anchoring site on AKAP79. Binding and inhibition of phosphatase activity are mediated by residues 315–360. *J Biol Chem.* 2002;277:48796–48802.
76. Cantrell AR, Tibbs VC, Westenbroek RE, et al. Dopaminergic modulation of voltage-gated Na^+ current in rat hippocampal neurons requires anchoring of cAMP-dependent protein kinase. *J Neurosci.* 1999;19:RC21.
77. Greengard P. The neurobiology of slow synaptic transmission. *Science.* 2001;294:1024–1030.
78. Ouimet CC, Langley-Gullion KC, Greengard P. Quantitative immunocytochemistry of DARPP-32-expressing neurons in the rat caudatoputamen. *Brain Res.* 1998;808:8–12.
79. Ouimet CC, Miller PE, Hemmings HC, et al. DARPP-32, a dopamine- and adenosine 3′:5′-monophosphate-regulated phosphoprotein enriched in dopamine-innervated brain regions. III. Immunocytochemical localization. *J Neurosci.* 1984;4:111–124.
80. Cohen PT. Protein phosphatase 1–targeted in many directions. *J Cell Sci.* 2002;115:241–256.
81. Kwon YG, Huang HB, Desdouits F, et al. Characterization of the interaction between DARPP-32 and protein phosphatase 1 (PP-1): DARPP-32 peptides antagonize the interaction of PP-1 with binding proteins. *Proc Natl Acad Sci USA.* 1997;94:3536–3541.
82. Desdouits F, Cheetham JJ, Huang HB, et al. Mechanism of inhibition of protein phosphatase 1 by DARPP-32: studies with recombinant DARPP-32 and synthetic peptides. *Biochem Biophys Res Commun.* 1995;206:652–658.
83. Hemmings HC Jr, Greengard P, Tung HY, et al. DARPP-32, a dopamine-regulated neuronal phosphoprotein, is a potent inhibitor of protein phosphatase-1. *Nature.* 1984;310:503–505.
84. Fienberg AA, Hiroi N, Mermelstein PG, et al. DARPP-32: regulator of the efficacy of dopaminergic neurotransmission. *Science.* 1998;281:838–842.

85. King MM, Huang CY, Chock PB, et al. Mammalian brain phosphoproteins as substrates for calcineurin. *J Biol Chem*. 1984;259:8080–8083.
86. Nishi A, Snyder GL, Nairn AC, et al. Role of calcineurin and protein phosphatase-2A in the regulation of DARPP-32 dephosphorylation in neostriatal neurons. *J Neurochem*. 1999;72:2015–2021.
87. Bibb JA, Snyder GL, Nishi A, et al. Phosphorylation of DARPP-32 by Cdk5 modulates dopamine signalling in neurons. *Nature*. 1999;402:669–671.
88. Nishi A, Bibb JA, Snyder GL, et al. Amplification of dopaminergic signaling by a positive feedback loop. *Proc Natl Acad Sci USA*. 2000;97:12840–12845.
89. Ahn JH, McAvoy T, Rakhilin SV, et al. Protein kinase A activates protein phosphatase 2A by phosphorylation of the B56delta subunit. *Proc Natl Acad Sci USA*. 2007;104:2979–2984.
90. Usui H, Inoue R, Tanabe O, et al. Activation of protein phosphatase 2A by cAMP-dependent protein kinase-catalyzed phosphorylation of the 74-kDa B″ (delta) regulatory subunit in vitro and identification of the phosphorylation sites. *FEBS Lett*. 1998;430:312–316.
91. Stipanovich A, Valjent E, Matamales M, et al. A phosphatase cascade by which rewarding stimuli control nucleosomal response. *Nature*. 2008;453:879–884.
92. Girault JA, Hemmings HC Jr, Williams KR, et al. Phosphorylation of DARPP-32, a dopamine- and cAMP-regulated phosphoprotein, by casein kinase II. *J Biol Chem*. 1989;264:21748–21759.
93. Desdouits F, Siciliano JC, Greengard P, et al. Dopamine- and cAMP-regulated phosphoprotein DARPP-32: phosphorylation of Ser-137 by casein kinase I inhibits dephosphorylation of Thr-34 by calcineurin. *Proc Natl Acad Sci USA*. 1995;92:2682–2685.
94. Bibb JA, Chen J, Taylor JR, et al. Effects of chronic exposure to cocaine are regulated by the neuronal protein Cdk5. *Nature*. 2001;410:376–380.
95. Svenningsson P, Tzavara ET, Carruthers R, et al. Diverse psychotomimetics act through a common signaling pathway. *Science*. 2003;302:1412–1415.
96. Valjent E, Pascoli V, Svenningsson P, et al. Regulation of a protein phosphatase cascade allows convergent dopamine and glutamate signals to activate ERK in the striatum. *Proc Natl Acad Sci USA*. 2005;102:491–496.
97. Zachariou V, Sgambato-Faure V, Sasaki T, et al. Phosphorylation of DARPP-32 at threonine-34 is required for cocaine action. *Neuropsychopharmacology*. 2006;31:555–562.
98. Zhang Y, Svenningsson P, Picetti R, et al. Cocaine self-administration in mice is inversely related to phosphorylation at Thr34 (protein kinase A site) and Ser130 (kinase CK1 site) of DARPP-32. *J Neurosci*. 2006;26:2645–2651.
99. Snyder GL, Allen PB, Fienberg AA, et al. Regulation of phosphorylation of the GluR1 AMPA receptor in the neostriatum by dopamine and psychostimulants in vivo. *J Neurosci*. 2000;20:4480–4488.
100. Banke TG, Bowie D, Lee H, et al. Control of GluR1 AMPA receptor function by cAMP-dependent protein kinase. *J Neurosci*. 2000;20:89–102.
101. Roche KW, O'Brien RJ, Mammen AL, et al. Characterization of multiple phosphorylation sites on the AMPA receptor GluR1 subunit. *Neuron*. 1996;16:1179–1188.
102. Mangiavacchi S, Wolf ME. D1 dopamine receptor stimulation increases the rate of AMPA receptor insertion onto the surface of cultured nucleus accumbens neurons through a pathway dependent on protein kinase A. *J Neurochem*. 2004;88:1261–1271.
103. Yan Z, Hsieh-Wilson L, Feng J, et al. Protein phosphatase 1 modulation of neostriatal AMPA channels: regulation by DARPP-32 and spinophilin. *Nat Neurosci*. 1999;2:13–17.
104. Flores-Hernandez J, Hernandez S, Snyder GL, et al. D(1) dopamine receptor activation reduces GABA(A) receptor currents in neostriatal neurons through a PKA/DARPP-32/PP1 signaling cascade. *J Neurophysiol*. 2000;83:2996–3004.
105. Liu F, Wan Q, Pristupa ZB, et al. Direct protein-protein coupling enables cross-talk between dopamine D5 and gamma-aminobutyric acid A receptors. *Nature*. 2000;403:274–280.
106. Flores-Hernandez J, Cepeda C, Hernandez-Echeagaray E, et al. Dopamine enhancement of NMDA currents in dissociated medium-sized striatal neurons: role of D1 receptors and DARPP-32. *J Neurophysiol*. 2002;88:3010–3020.
107. Calabresi P, Pisani A, Mercuri NB, et al. Long-term potentiation in the striatum unmasked by removing the voltage-dependent magnesium block of NMDA receptor channels. *Eur J Neurosci*. 1992;4:929–935.
108. Blank T, Nijholt I, Teichert U, et al. The phosphoprotein DARPP-32 mediates cAMP-dependent potentiation of striatal N-methyl-D-aspartate responses. *Proc Natl Acad Sci USA*. 1997;94:14859–14864.
109. Snyder GL, Fienberg AA, Huganir RL, et al. A dopamine/D1 receptor/protein kinase A/dopamine- and cAMP-regulated phosphoprotein (Mr 32 kDa)/protein phosphatase-1 pathway regulates dephosphorylation of the NMDA receptor. *J Neurosci*. 1998;18:10297–10303.
110. Dudman JT, Eaton ME, Rajadhyaksha A, et al. Dopamine D1 receptors mediate CREB phosphorylation via phosphorylation of the NMDA receptor at Ser897-NR1. *J Neurochem*. 2003;87:922–934.
111. Liu Z, Mao L, Parelkar NK, et al. Distinct expression of phosphorylated N-methyl-D-aspartate receptor NR1 subunits by projection neurons and interneurons in the striatum of normal and amphetamine-treated rats. *J Comp Neurol*. 2004;474:393–406.
112. Surmeier DJ, Bargas J, Hemmings HC Jr, et al. Modulation of calcium currents by a D1 dopaminergic protein kinase/phosphatase cascade in rat neostriatal neurons. *Neuron*. 1995;14:385–397.
113. Cepeda C, Colwell CS, Itri JN, et al. Dopaminergic modulation of NMDA-induced whole cell currents in neostriatal neurons in slices: contribution of calcium conductances. *J Neurophysiol*. 1998;79:82–94.
114. Liu JC, DeFazio RA, Espinosa-Jeffrey A, et al. Calcium modulates dopamine potentiation of N-methyl-D-aspartate responses: electrophysiological and imaging evidence. *J Neurosci Res* 76:315–322, 2004.
115. Dunah AW, Standaert DG: Dopamine D1 receptor-dependent trafficking of striatal NMDA glutamate receptors to the postsynaptic membrane. *J Neurosci*. 2001;21:5546–5558.
116. Hallett PJ, Spoelgen R, Hyma BT, et al. Dopamine D1 activation potentiates striatal NMDA receptors by tyrosine phosphorylation-dependent subunit trafficking. *J Neurosc.i* 2006;26:4690–4700.
117. Dunah AW, Sirianni AC, Fienberg AA, et al. Dopamine D1-dependent trafficking of striatal N-methyl-D-aspartate glutamate receptors requires Fyn protein tyrosine kinase but not DARPP-32. *Mol Pharmacol*. 2004;65:121–129.
118. Suzuki T, Okumura-Noji K. NMDA receptor subunits epsilon 1 (NR2A) and epsilon 2 (NR2B) are substrates for Fyn in the postsynaptic density fraction isolated from the rat brain. *Biochem Biophys Res Commun*. 1995;216:582–588.
119. Kohr G, Seeburg PH. Subtype-specific regulation of recombinant NMDA receptor-channels by protein tyrosine kinases of the src family. *J Physiol*. 1996;492(pt 2):445–452.

120. Scott L, Kruse MS, Forssberg H, et al. Selective up-regulation of dopamine D1 receptors in dendritic spines by NMDA receptor activation. *Proc Natl Acad Sci USA*. 2002;99:1661–1664.
121. Scott L, Zelenin S, Malmersjo S, et al. Allosteric changes of the NMDA receptor trap diffusible dopamine 1 receptors in spines. *Proc Natl Acad Sci USA*. 2006;103:762–767.
122. Fiorentini C, Gardoni F, Spano P, et al. Regulation of dopamine D1 receptor trafficking and desensitization by oligomerization with glutamate N-methyl-D-aspartate receptors. *J Biol Chem*. 2003;278:20196–20202.
123. Pei L, Lee FJ, Moszczynska A, et al. Regulation of dopamine D1 receptor function by physical interaction with the NMDA receptors. *J Neurosci*. 2004;24:1149–1158.
124. Lee FJ, Xue S, Pei L, et al. Dual regulation of NMDA receptor functions by direct protein–protein interactions with the dopamine D1 receptor. *Cell*. 2002;111:219–230.
125. Zhang J, Vinuela A, Neely MH, et al. Inhibition of the dopamine D1 receptor signaling by PSD-95. *J Biol Chem*. 2007;282:15778–15789.
126. Kim DS, Palmiter RD, Cummins A, et al. Reversal of supersensitive striatal dopamine D1 receptor signaling and extracellular signal-regulated kinase activity in dopamine-deficient mice. *Neuroscience*. 2006;137:1381–1388.
127. Yao WD, Gainetdinov RR, Arbuckle MI, et al. Identification of PSD-95 as a regulator of dopamine-mediated synaptic and behavioral plasticity. *Neuron*. 2004;41:625–638.
128. Calabresi P, Mercuri N, Stanzione P, et al. Intracellular studies on the dopamine-induced firing inhibition of neostriatal neurons in vitro: evidence for D1 receptor involvement. *Neuroscience*. 1987;20:757–771.
129. Cantrell AR, Smith RD, Goldin AL, et al. Dopaminergic modulation of sodium current in hippocampal neurons via cAMP-dependent phosphorylation of specific sites in the sodium channel alpha subunit. *J Neurosci*. 1997;17:7330–7338.
130. Schiffmann SN, Lledo PM, Vincent JD. Dopamine D1 receptor modulates the voltage-gated sodium current in rat striatal neurones through a protein kinase A. *J Physiol*. 1995;483(pt 1):95–107.
131. Surmeier DJ, Eberwine J, Wilson CJ, et al. Dopamine receptor subtypes colocalize in rat striatonigral neurons. *Proc Natl Acad Sci USA*. 1992;89:10178–10182.
132. Schiffmann SN, Desdouits F, Menu R, et al. Modulation of the voltage-gated sodium current in rat striatal neurons by DARPP-32, an inhibitor of protein phosphatase. *Eur J Neurosci*. 1998;10:1312–1320.
133. Carr DB, Day M, Cantrell AR, et al. Transmitter modulation of slow, activity-dependent alterations in sodium channel availability endows neurons with a novel form of cellular plasticity. *Neuron*. 2003;39:793–806.
134. Wickens JR, Wilson CJ. Regulation of action-potential firing in spiny neurons of the rat neostriatum in vivo. *J Neurophysiol*. 1998;79:2358–2364.
135. Vergara R, Rick C, Hernandez-Lopez S, et al. Spontaneous voltage oscillations in striatal projection neurons in a rat corticostriatal slice. *J Physiol*. 2003;553:169–182.
136. Kisilevsk AE, Mulligan SJ, Altier C, et al. D1 receptors physically interact with N-type calcium channels to regulate channel distribution and dendritic calcium entry. *Neuron*. 2008;58:557–570.
137. Thomas GM, Huganir RL. MAPK cascade signalling and synaptic plasticity. *Nat Rev Neurosci*. 2004;5:173–183.
138. Fiore RS, Murphy TH, Sanghera JS, et al. Activation of p42 mitogen-activated protein kinase by glutamate receptor stimulation in rat primary cortical cultures. *J Neurochem*. 1993;61:1626–1633.
139. Rosen LB, Ginty DD, Weber MJ, et al. Membrane depolarization and calcium influx stimulate MEK and MAP kinase via activation of Ras. *Neuron*. 1994;12:1207–1221.
140. Xia Z, Dudek H, Miranti CK, et al. Calcium influx via the NMDA receptor induces immediate early gene transcription by a MAP kinase/ERK-dependent mechanism. *J Neurosci*. 1996;16:5425–5436.
141. Martegani E, Vanoni M, Zippel R, et al. Cloning by functional complementation of a mouse cDNA encoding a homologue of CDC25, a *Saccharomyces cerevisiae* RAS activator. *Embo J*. 1992;11:2151–2157.
142. Shou C, Farnsworth CL, Neel BG, et al. Molecular cloning of cDNAs encoding a guanine-nucleotide-releasing factor for Ras p21. *Nature*. 1992;358:351–354.
143. Farnsworth CL, Freshney NW, Rosen LB, et al. Calcium activation of Ras mediated by neuronal exchange factor Ras-GRF. *Nature*. 1995;376:524–527.
144. Toki S, Kawasaki H, Tashiro N, et al. Guanine nucleotide exchange factors CalDAG-GEFI and CalDAG-GEFII are colocalized in striatal projection neurons. *J Comp Neurol*. 2001;437:398–407.
145. Valjent E, Corvol JC, Pages C, et al. Involvement of the extracellular signal-regulated kinase cascade for cocaine-rewarding properties. *J Neurosci*. 2000;20:8701–8709.
146. Miller CA, Marshall JF. Molecular substrates for retrieval and reconsolidation of cocaine-associated contextual memory. *Neuron*. 2005;47:873–884.
147. Schuster S, Nadjar A, Guo JT, et al. The 3-hydroxy-3-methyl-glutaryl-CoA reductase inhibitor lovastatin reduces severity of L-DOPA-induced abnormal involuntary movements in experimental Parkinson's disease. *J Neurosci*. 2008;28:4311–4316.
148. Sgambato V, Pages C, Rogard M, et al. Extracellular signal-regulated kinase (ERK) controls immediate early gene induction on corticostriatal stimulation. *J Neurosci*. 1998;18:8814–8825.
149. Treisman R. Regulation of transcription by MAP kinase cascades. *Curr Opin Cell Biol*. 1996;8:205–215.
150. Brami-Cherrier K, Valjent E, Herve D, et al. Parsing molecular and behavioral effects of cocaine in mitogen- and stress-activated protein kinase-1-deficient mice. *J Neurosci*. 2005;25:11444–11454.
151. Waltereit R, Dammermann B, Wulff P, et al. Arg3.1/Arc mRNA induction by Ca^{2+} and cAMP requires protein kinase A and mitogen-activated protein kinase/extracellular regulated kinase activation. *J Neurosci*. 2001;21:5484–5493.
152. Zhang L, Lou D, Jiao H, et al. Cocaine-induced intracellular signaling and gene expression are oppositely regulated by the dopamine D1 and D3 receptors. *J Neurosc.i* 2004;24:3344–3354.
153. Valjent E, Aubier B, Corbille AG, et al. Plasticity-associated gene Krox24/Zif268 is required for long-lasting behavioral effects of cocaine. *J Neurosci*. 2006;26:4956–4960.
154. Mattingly RR. Phosphorylation of serine 916 of Ras-GRF1 contributes to the activation of exchange factor activity by muscarinic receptors. *J Biol Chem*. 1999;274:37379–37384.
155. Mattingly RR, Macara IG. Phosphorylation-dependent activation of the Ras-GRF/CDC25Mm exchange factor by muscarinic receptors and G-protein beta gamma subunits. *Nature*. 1996;382:268–272.
156. Bos JL. Epac proteins: multi-purpose cAMP targets. *Trends Biochem Sci*. 2006;31:680–686.
157. Iwakura Y, Nawa H, Sora I, et al. Dopamine D1 receptor-induced signaling through TrkB receptors in striatal neurons. *J Biol Chem*. 2008;283:15799–15806.

158. Chao MV. Neurotrophins and their receptors: a convergence point for many signalling pathways. *Nat Rev Neurosci.* 2003;4:299–309.
159. Robinson SW, Caron MG. Selective inhibition of adenylyl cyclase type V by the dopamine D3 receptor. *Mol Pharmacol.* 1997;52:508–514.
160. Bateup HS, Svenningsson P, Kuroiwa M, et al. Cell type-specific regulation of DARPP-32 phosphorylation by psychostimulant and antipsychotic drugs. *Nat Neurosci.* 2008;11:932–939.
161. Håkansson K, Galdi S, Hendrick J, et al. Regulation of phosphorylation of the GluR1 AMPA receptor by dopamine D2 receptors. *J Neurochem.* 2006;96:482–488.
162. Nishi A, Snyder GL, Greengard P. Bidirectional regulation of DARPP-32 phosphorylation by dopamine. *J Neurosci.* 1997;17:8147–8155.
163. Svenningsson P, Lindskog M, Ledent C, et al. Regulation of the phosphorylation of the dopamine- and cAMP-regulated phosphoprotein of 32 kDa in vivo by dopamine D1, dopamine D2, and adenosine A2A receptors. *Proc Natl Acad Sci USA.* 2000;97:1856–1860.
164. Cepeda C, Buchwald NA, Levine MS. Neuromodulatory actions of dopamine in the neostriatum are dependent upon the excitatory amino acid receptor subtypes activated. *Proc Natl Acad Sci USA.* 1993;90:9576–9580.
165. Leveque JC, Macias W, Rajadhyaksha A, et al. Intracellular modulation of NMDA receptor function by antipsychotic drugs. *J Neurosci.* 2000;20:4011–4020.
166. Konradi C, Heckers S. Haloperidol-induced Fos expression in striatum is dependent upon transcription factor cyclic AMP response element binding protein. *Neuroscience.* 1995;65:1051–1061.
167. Konradi C, Kobierski LA, Nguyen TV, et al. The cAMP-response-element-binding protein interacts, but Fos protein does not interact, with the proenkephalin enhancer in rat striatum. *Proc Natl Acad Sci USA.* 1993;90:7005–7009.
168. Pozzi L, Håkansson K, Usiello A, et al. Opposite regulation by typical and atypical anti-psychotics of ERK1/2, CREB and Elk-1 phosphorylation in mouse dorsal striatum. *J Neurochem.* 2003;86:451–459.
169. Bertran-Gonzalez J, Håkansson K, Borgkvist A, et al. Histone H3 is under the opposite tonic control of dopamine D2 and adenosine A2A receptors in striatopallidal neurons. *Neuropsychopharmacology.* 2009;34:1710–1720.
170. Li J, Guo Y, Schroeder FA, et al. Dopamine D2-like antagonists induce chromatin remodeling in striatal neurons through cyclic AMP-protein kinase A and NMDA receptor signaling. *J Neurochem.* 2004;90:1117–1131.
171. Nowak SJ, Corces VG. Phosphorylation of histone H3: a balancing act between chromosome condensation and transcriptional activation. *Trends Genet.* 2004;20:214–220.
172. Adams MR, Brandon EP, Chartoff EH, et al. Loss of haloperidol induced gene expression and catalepsy in protein kinase A-deficient mice. *Proc Natl Acad Sci USA.* 1997;94:12157–12161.
173. Hernandez-Lopez S, Tkatch T, Perez-Garci E, et al. D2 dopamine receptors in striatal medium spiny neurons reduce L-type Ca^{2+} currents and excitability via a novel PLC[beta]1-IP3-calcineurin-signaling cascade. *J Neurosci.* 2000;20:8987–8995.
174. Robertson GS, Fibiger HC. Neuroleptics increase c-*fos* expression in the forebrain: contrasting effects of haloperidol and clozapine. *Neuroscience.* 1992;46:315–328.
175. Lee J, Rushlow WJ, Rajakumar N. L-type calcium channel blockade on haloperidol-induced c-Fos expression in the striatum. *Neuroscience.* 2007:149:602–616.
176. Day M, Wang Z, Ding J, et al. Selective elimination of glutamatergic synapses on striatopallidal neurons in Parkinson disease models. *Nat Neurosci.* 2006;9:251–259.
177. Greif GJ, Lin YJ, Liu JC, et al. Dopamine-modulated potassium channels on rat striatal neurons: specific activation and cellular expression. *J Neurosci.* 1995;15:4533–4544.
178. Kuzhikandathil EV, Yu W, Oxford GS. Human dopamine D3 and D2L receptors couple to inward rectifier potassium channels in mammalian cell lines. *Mol Cell Neurosci.* 1998;12:390–402.
179. Hopf FW, Cascini MG, Gordon AS, et al. Cooperative activation of dopamine D1 and D2 receptors increases spike firing of nucleus accumbens neurons via G-protein betagamma subunits. *J Neurosci.* 2003;23:5079–5087.
180. Wickman KD, Iniguez-Lluhl JA, Davenport PA, et al. Recombinant G-protein beta gamma-subunits activate the muscarinic-gated atrial potassium channel. *Nature.* 1994;368:255–257.
181. Liu XY, Chu XP, Mao LM, et al. Modulation of D2R-NR2B interactions in response to cocaine. *Neuron.* 2006;52:897–909.
182. Meador-Woodruff JH, Grandy DK, Van Tol HH, et al. Dopamine receptor gene expression in the human medial temporal lobe. *Neuropsychopharmacology.* 1994;10:239–248.
183. Kotecha SA, Oak JN, Jackson MF, et al. A D2 class dopamine receptor transactivates a receptor tyrosine kinase to inhibit NMDA receptor transmission. *Neuron.* 2002;35:1111–1122.
184. Thomas EA, Danielson PE, Sutcliffe JG. RGS9: a regulator of G-protein signalling with specific expression in rat and mouse striatum. *J Neurosci Res.* 1998;52:118–124.
185. Rahman Z, Schwarz J, Gold SJ, et al. RGS9 modulates dopamine signaling in the basal ganglia. *Neuron.* 2003;38:941–952.
186. Gainetdinov RR, Bohn LM, Sotnikova TD, et al. Dopaminergic supersensitivity in G protein–coupled receptor kinase 6–deficient mice. *Neuron.* 2003;38:291–303.
187. Bofill-Cardona E, Kudlacek O, Yang Q, et al. Binding of calmodulin to the D2-dopamine receptor reduces receptor signaling by arresting the G protein activation switch. *J Biol Chem.* 2000;275:32672–32680.
188. Park SK, Nguyen MD, Fischer A, et al. Par-4 links dopamine signaling and depression. *Cell.* 2005;122:275–287.
189. Cohen P, Frame S. The renaissance of GSK3. *Nat Rev Mol Cell Biol.* 2001;2:769–776.
190. Beaulieu JM, Marion S, Rodriguiz RM, et al. A beta-arrestin 2 signaling complex mediates lithium action on behavior. *Cell.* 2008;132:125–136.
191. Emamian ES, Hall D, Birnbaum MJ, et al. Convergent evidence for impaired AKT1-GSK3beta signaling in schizophrenia. *Nat Genet.* 2004;36:131–137.
192. Beaulieu JM, Sotnikova TD, Yao WD, et al. Lithium antagonizes dopamine-dependent behaviors mediated by an AKT/glycogen synthase kinase 3 signaling cascade. *Proc Natl Acad Sci USA.* 2004;101:5099–5104.
193. Beaulieu JM, Tirotta E, Sotnikova TD, et al. Regulation of Akt signaling by D2 and D3 dopamine receptors in vivo. *J Neurosci.* 2007;27:881–885.
194. Beaulieu JM, Sotnikova TD, Marion S, et al. An Akt/beta-arrestin 2/PP2A signaling complex mediates dopaminergic neurotransmission and behavior. *Cell.* 2005;122:261–273.
195. Beaulieu JM, Gainetdinov RR, Caron MG. The Akt-GSK-3 signaling cascade in the actions of dopamine. *Trends Pharmacol Sci.* 2007;28:166–172.
196. Brami-Cherrier K, Valjent E, Garcia M, et al. Dopamine induces a PI3-kinase-independent activation of Akt in striatal neurons: a new route to cAMP response element-binding protein phosphorylation. *J Neurosci.* 2002;22:8911–8921.

197. Dal Toso R, Sommer B, Ewert M, et al. The dopamine D2 receptor: two molecular forms generated by alternative splicing. *EMBO J*. 1989;8:4025–4034.
198. Monsma FJ Jr, McVittie LD, Gerfen CR, et al. Multiple D2 dopamine receptors produced by alternative RNA splicing. *Nature*. 1989;342:926–929.
199. Centonze D, Usiello A, Gubellini P, et al. Dopamine D2 receptor-mediated inhibition of dopaminergic neurons in mice lacking D2L receptors. *Neuropsychopharmacology*. 2002;27:723–726.
200. Lindgren N, Usiello A, Goiny M, et al. Distinct roles of dopamine D2L and D2S receptor isoforms in the regulation of protein phosphorylation at presynaptic and postsynaptic sites. *Proc Natl Acad Sci USA*. 2003;100:4305–4309.
201. Usiello A, Baik JH, Rouge-Pont F, et al. Distinct functions of the two isoforms of dopamine D2 receptors. *Nature*. 2000;408:199–203.
202. Van-Ham I, Banihashemi B, Wilson AM, et al. Differential signaling of dopamine-D2S and -D2L receptors to inhibit ERK1/2 phosphorylation. *J Neurochem*. 2007;102:1796–1804.
203. Kim SJ, Kim MY, Lee EJ, et al. Distinct regulation of internalization and mitogen-activated protein kinase activation by two isoforms of the dopamine D2 receptor. *Mol Endocrinol*. 2004;18:640–652.
204. Kim SY, Choi KC, Chang MS, et al. The dopamine D2 receptor regulates the development of dopaminergic neurons via extracellular signal-regulated kinase and Nurr1 activation. *J Neurosci*. 2006;26:4567–4576.
205. Wang C, Buck DC, Yang R, et al. Dopamine D2 receptor stimulation of mitogen-activated protein kinases mediated by cell type-dependent transactivation of receptor tyrosine kinases. *J Neurochem*. 2005;93:899–909.
206. Yan Z, Feng J, Fienberg AA, et al. D(2) dopamine receptors induce mitogen-activated protein kinase and cAMP response element-binding protein phosphorylation in neurons. *Proc Natl Acad Sci USA*. 1999;96:11607–11612.
207. Gerfen CR, Miyachi S, Paletzki R, et al. D1 dopamine receptor supersensitivity in the dopamine-depleted striatum results from a switch in the regulation of ERK1/2/MAP kinase. *J Neurosci*. 2002;22:5042–5054.
208. Valjent E, Pages C, Herve D, et al. Addictive and non-addictive drugs induce distinct and specific patterns of ERK activation in mouse brain. *Eur J Neurosci*. 2004;19:1826–1836.
209. Smith FD, Oxford GS, Milgram SL. Association of the D2 dopamine receptor third cytoplasmic loop with spinophilin, a protein phosphatase-1-interacting protein. *J Biol Chem*. 1999;274:19894–19900.
210. Lee SP, So CH, Rashid AJ, et al. Dopamine D1 and D2 receptor co-activation generates a novel phospholipase C-mediated calcium signal. *J Biol Chem*. 2004;279:35671–35678.
211. Heintz N. BAC to the future: the use of bac transgenic mice for neuroscience research. *Nat Rev Neurosci*. 2001;2:861–870.
212. Gong S, Doughty M, Harbaugh CR, et al. Targeting Cre recombinase to specific neuron populations with bacterial artificial chromosome constructs. *J Neurosci*. 2007;27:9817–9823.
213. Kuhn R, Torres RM. Cre/loxP recombination system and gene targeting. *Methods Mol Biol*. 2002;180:175–204.

3.4 Ion Channels and Regulation of Dopamine Neuron Activity

BIRGIT LISS AND JOCHEN ROEPER

INTRODUCTION AND CHAPTER SUMMARY

In this chapter, we will review the central role of ion channels in the generation and regulation of electrical activity of dopamine neurons. We will concentrate on midbrain dopamine neurons located in the nuclei substantia nigra (SN, A9) and the adjacent ventral tegmental area (VTA, A10) (see Chapter 2, this volume). For these dopamine midbrain neurons we possess a detailed picture on the role of different ion channels in the generation and control of electrical activity.

Ion channels are at the heart of generating electrical activity of neurons and coupling it to neurotransmitter release. They comprise a superfamily of transmembrane proteins that form pores through plasma membranes, enabling ions to pass with high efficiency. The open probability of most of these channel pores is controlled either by the cellular membrane potential (voltage-gated) or by ligand binding (ligand-gated).[1] The coupling of electrical activity via calcium ions to neurotransmitter release is one of the basic tenets of cellular neuroscience.[2] The neurotransmitter dopamine mediates its actions via dopamine receptors (see Chapter 3.1, this volume); however, in order to reach its pre- and postsynaptic receptors, dopamine has to be present in the extracellular compartment. The most effective way to enter the extracellular domain is via calcium-mediated exocytosis of dopamine-filled vesicles.[3] The required calcium signal to trigger vesicle release is generated via action potentials, which in turn lead to the opening of voltage-gated ion channels selectively permeable to calcium ions.[2] An action potential (AP) is a rapid voltage change of the cellular membrane potential to positive potentials in the order of about 100 mV for a few milliseconds.[4] A single action potential in a dopamine neuron acts as a short millisecond trigger that activates a several-hundred-fold longer event of elevated dopamine levels in the extracellular domain, thereby controlling the effective spatiotemporal profile of the dopamine concentration—further shaped by diffusion, re-uptake, and catabolic enzymatic processes (see Chapters 2.2, 2.4, and 3.2, this volume)—that mediates its actions via dopamine receptors.[3] The distinct and flexible temporal patterns of action potentials in dopamine-releasing neurons explain why the actions of dopamine can be fully understood only when these patterns of electrical activity are taken into account. This holds true not only for axonal dopamine release but also for action potential–triggered dopamine release from the somatodendritic domain.

As we will discuss in detail below, already low firing rates of about 5 Hz are sufficient to create in vivo steady-state levels of extracellular dopamine concentrations within the low nanomolar range—often called *tonic* dopamine signalling. However, dopamine neurons can also display so-called burst activity, where clusters of several action potentials at high frequency are separated by longer periods of electrical silence. These burst activity patterns are often referred to as *phasic* dopamine signaling, with increased dopamine concentrations in the micromolar range for a few hundred milliseconds.[5] The so-called interspike interval (ISI) not only spans the subthreshold gap between two action potentials but also provides rich information on the synaptic and intrinsic ion channel processes that control the timing and patterning of electrical discharge; a dopamine neuron discharging 3-ms-long action potentials at a mean rate of about 5 Hz will spend more than 98% (985 ms) of the time in this subthreshold range.

To define the functional roles of ion channels in dopamine neurons in a cell-specific manner, we need to consider the different types of electrical activity that these neurons generate in intact brains, as well as the anatomical and functional diversity of the dopamine midbrain system (see Chapters 2.1 and 2.3, this volume). Upstream regulation of ion channels and the convergence of excitatory, inhibitory, and modulatory synaptic inputs on dopamine neurons introduce additional levels of complexity. How electrical activity is controlled in dopamine neurons in the intact brain has yet to be fully understood. In addition, the majority of electrophysiological studies on dopamine midbrain neurons over the last 30 years have been carried out with a focus on a single (often called *classic*) homogeneous electrical phenotype of dopamine midbrain neuron. We will summarize these findings in this chapter but we will also tackle this simplification by describing current advances in understanding the functional diversity of dopamine neuron phenotypes within the midbrain and its implication for health and disease, in particular for Parkinson's disease (PD).

Before we dissect the roles of distinct ion channels in the various aspects of electrophysiological functions of dopamine midbrain neurons, we wish to alert the reader to the respective powers and limitations of the set of distinct in vivo and in vitro electrophysiological preparations and techniques that have been used to study electrical activity and ion channels in dopamine neurons. However, detailed explanations of the molecular biophysics of the discussed ion channels and the respective electrophysiological techniques and concepts are beyond the scope of this chapter. For these we wish to refer the reader to several excellent textbooks.[1,6,7]

METHODOLOGICAL CONSIDERATIONS FOR ANALYZING ACTIVITY OF DOPAMINE NEURONS

An Ideal Electrophysiological Setting for Dopamine Neurons?

Let us first define the ideal electrophysiological setting to study the electrical properties of dopamine neurons in the most relevant and complete context. In this ideal experiment, we would record the changes in membrane potential—at a selected compartment—of single dopamine neurons with the highest temporal resolution without any perturbation, in an awake animal, engaged in a well-defined behavioral task. This ideal experiment would require in vivo intracellular electrophysiological recordings of individual dopamine neurons in an awake, behaving animal, and would provide not only an unbiased temporal sequence of action potentials, but also the characteristics of the membrane potential in the subthreshold range during the ISIs. Unfortunately, this type of ideal experiment has not yet been established for dopamine neurons. However, it is promising to note that intracellular recordings in freely moving rats have indeed been successfully carried out with cortical neurons using in vivo patch-clamp recordings.[8]

Recording Membrane Potentials of Dopamine Neurons

For single dopamine neurons, extracellular in vivo recordings of single-unit activity (i.e., the pattern of action potentials) in awake monkeys have been successfully carried out by Schultz and colleagues during the last 25 years[9] and, more recently, by other groups also in awake rodents.[10,11] In addition, intracellular in vivo recordings of dopamine neurons were made almost 30 years ago by Grace and Bunney in anesthetized rats.[12] These intracellular recordings were necessary to provide a full and continuous stream of in vivo membrane potential changes in the subthreshold range. Throughout the last three decades, this still very challenging approach, using sharp, penetrating microelectrodes, has produced essential information on the subthreshold membrane changes in the intact brain, as we will discuss below. The bulk of intracellular recordings, however—via sharp microelectrodes[13] or, in recent years, predominantly by patch-clamp recordings[14]—has been carried out in acute in vitro preparations—either midbrain slices or acutely dissociated single dopamine neurons.[14] Due to current technical limitations, only the somatodendritic compartment of dopamine neurons is directly accessible for intracellular recordings, both in vivo and in vitro; the much smaller axonal and presynaptic compartments have not yet been studied routinely. Information about the electrical behavior of these latter compartments was inferred by the use of more indirect techniques such as imaging or amperometric methods—often in combination with ion channel pharmacology.[15]

It is obvious that the in vitro preparations, which routinely allow the intracellular recording of dopamine neurons for several hours, have severe shortcomings. In all in vitro preparations, the complex basal ganglia network, which is a major player in controlling the electrical activity of dopamine neurons in the intact brain (see Chapters 2.7, this volume), is almost completely absent. Even if some nuclei and their axonal connections are partially retained (e.g., between GABAergic substantia nigra pars recticulata and dopamine pars compacta neurons), their pattern of activity is still different from those in vivo.[16] Moreover, the dopamine neuron itself is severely truncated in brain slices as well as in acutely dissociated preparations: most of its extensive axon[17] is amputated, and a significant part of its dendritic tree is also lost. Finally, due to technical issues, many studies in brain slices and dissociated single cells have used very young animals (i.e., rodents only 2–3 weeks old), at an age when motor behavior and basal ganglia networks are not yet fully mature. Although, in our ideal imagined experimental setting, these dopamine neurons in vitro are only a sorry functional and morphological remnant in comparison to their extended state in vivo,[17] we owe most direct insights to the role of ion channels in generating and regulating distinct electrical activity patterns to these reduced preparations. However, the large body of in vivo extracellular recordings of action potential discharge of dopamine midbrain neurons during behavior in awake animals constitutes a kind of gold standard on the behaviorally relevant pattern of action potentials—although providing no information about the distinct subthreshold membrane processes that drive these patterns. We utilize this in vivo information to critically assess the types of electrical patterns we observe in reduced, anesthetized in vivo settings as well as in the even more reduced in vitro preparations. As most of this chapter summarizes findings from these in vitro preparations, readers might want to keep in mind the large difference between a contemplated ideal setting and the routinely used preparations for the electrophysiological analysis of dopamine neuron activity.

From Membrane Potential to Ionic Currents

The second major element of electrophysiological analysis of ion channel activity in dopamine midbrain neurons requires active control of membrane potentials—mainly by using whole-cell or cell-attached voltage-clamp[18] or, more recently, dynamic-clamp[19] configurations of the patch-clamp technique.[7] This type of analysis is needed to record the distinct biophysical and pharmacological properties of individual ion channel species in dopamine neurons, either at the microscopic level of individual channel molecules or at the macroscopic level of channel populations that give rise to ionic currents on the whole-cell level. In this context, the truncated in vitro preparations now possess a major advantage: the absence of synaptic network activity and the simplified morphology (at best a single isoelectric compartment) enable us to resolve the kinetics of ionic currents generated by the gating properties of ion channel populations. If these reduced cellular preparations are still not sufficient for kinetic analysis (as in the case of some fast gating potassium or sodium channels), the patch-clamp technique offers more reduced cell-free preparations where the membrane potential can be controlled in a fast and reliable manner in defined membrane patches along the somatodendritic axis.[7]

Finally, the biophysical profiles of sets of ion channels (e.g., the voltage dependence of gating transitions between open, closed, or inactivated states) can be used to create realistic Hodgkin-Huxley-style computer models of electrical behavior of neurons.[7,20] In the case of dopamine neurons, these have been used to create *in silico* single dopamine cell models that are able to reproduce the basic pattern of electrical activity of these neurons.[21,22] In addition, these model conductances can be fed back to real dopamine neurons in the so-called dynamic clamp configuration[23] to study their functional impact on the basic electrical pattern of discharge.[24]

Having reviewed these methodological considerations, we now take a closer look at the distinct types of electrical activity of dopamine midbrain neurons (with a focus on the classic dopamine neurons) in distinct preparations—ranging from intact, behaving animals (the gold standard) to more reduced preparations from in vivo anesthetized animals down to isolated single neurons.

IN VIVO ACTIVITY PATTERNS OF DOPAMINE MIDBRAIN NEURONS IN AWAKE ANIMALS

Action Potential Properties

In vivo, using extracellular single-unit recordings, classic dopamine midbrain neurons are in most cases identified via their long-lasting action potentials (>1 ms) with a multiphasic shape[9] in combination with their inhibition of electrical activity via pharmacological activation of D2-like autoreceptors.[25,26] There are two major temporal patterns of action potential sequences—tonic mode or burst mode—that occur naturally in dopamine midbrain neurons in awake, behaving animals (compare also figure 3.4.1).

The Tonic Activity Mode

When an animal (in most experimental settings, a rodent or a primate) is not actively engaged in a behavioural task (e.g., in a conditioning experiments when the animal is waiting for the next trial), dopamine midbrain neurons show stereotypical electrical activity; they continuously discharge single action potentials in a frequency range of about 0.1–10 Hz.[27,28] In awake rodents (mean rate 4.0 Hz),[11] cats (mean rate 3.6 Hz),[29] or primates (mean rate 3.3/5.7 Hz).[27,28] This low-frequency, continuous baseline discharge is often called the *tonic mode* of dopamine midbrain neurons. The internal temporal order (i.e., spike-to-spike variability) of this tonic mode can vary significantly. While in monkeys[27] the respective spike trains are quite irregular, as indicated by a high coefficient of variation (CV) of the ISI distributions (mean CV = 0.62), respective spike train analysis in awake rodents has demonstrated more regular tonic activity under baseline conditions.[11] In addition, the autocorrelograms of spike activity in rodents have indicated the presence of regular oscillations,[11,30] closely related to the intrinsically generated pacemaker discharge apparent in reduced dopamine cell preparations (see the later section "In Vitro Activity

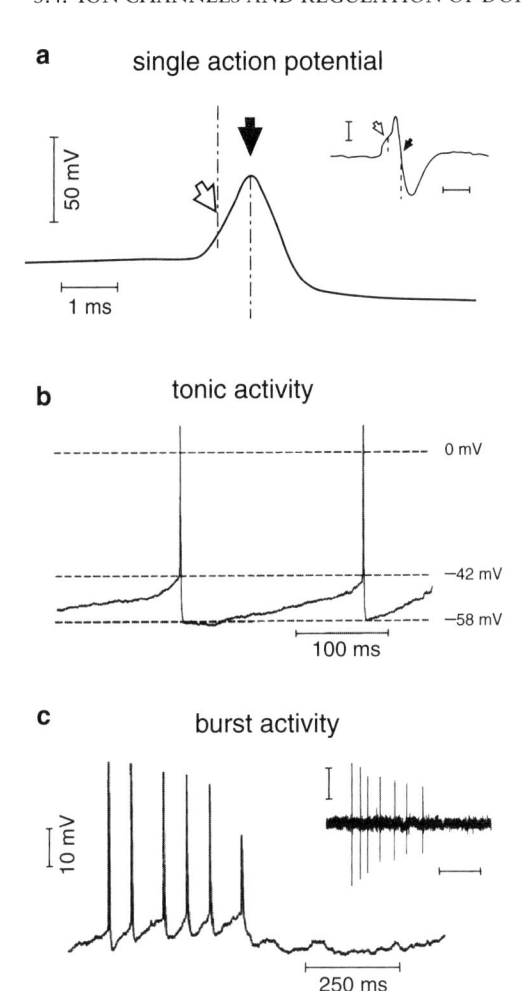

FIGURE 3.4.1. Classic dopamine midbrain neurons recorded in vivo in anesthetized rats. Intracellular recordings of (a) a typical single action potential, (b) tonic activity mode, and (c) burst activity mode. Inserts in (a) and (c) display respective extracellular in vivo recordings, with scale bars representing 0.3 mV (a, c) and 1 ms (a) or 250 ms (c). The inflection in the rising phase of the intracellular action potential in (a) corresponds to the notch in the positive phase of the extracellularly recorded action potential (open arrows). The peak of the intracellular action potential in (a) corresponds to the 0mV crossing of the extracellular action potential (solid arrows). *Source*: Adapted from[31,35,37]

Patterns of Dopamine Midbrain Neurons"). Also, the average tonic continuous single spike discharge frequencies of 3–6 Hz for dopamine midbrain neurons in awake animals, not actively engaged in a task, are only about twofold higher than those of isolated single dopamine cells in the most reduced in vitro preparations (see the section referenced above). This indicates that similar intrinsic ion channel mechanisms play a prominent role in the generation of the tonic discharge mode in vitro as well as in vivo in awake animals. In the functional context, this ongoing tonic electrical activity is likely to be responsible for the time-independent homogeneous basal dopamine levels detected at baseline in vivo—for example, in axonal target areas such as the striatum (see Chapter 2.1, this volume). Thus, tonic discharge of dopamine neurons might serve as the electrical correlate for the so-called volume transmission mode of the dopamine system at rest, describing three-dimensional diffusion of released dopamine across larger distances rather than focal point-to-point communication limited to individual synapses.[3]

The Burst Activity Mode

In awake animals, the continuous low-frequency tonic mode of single action potential discharge is replaced in both rodents and primates by packets of 2–10 action potentials, discharged in a higher frequency range (20–80 Hz) in various behavioural contexts. As first described by Schultz,[28] this so-called burst discharge mode of dopamine midbrain neurons usually occurred time-locked with a fixed latency of about 60 ms (range,

40–100 ms) after the presentation of a novel, salient, rewarding or reward-predicting sensory stimulus. These stimulus-triggered burst discharges contained 2–10 action potentials reaching high instantaneous firing rates of up to 100 Hz.[28] More recently, Bayer et al. carried out a detailed statistical analysis of the properties of the stimulus-triggered burst discharge in awake monkeys.[27] They observed a mean intraburst frequency of about 20 Hz after a sensory stimulus and a frequency of 35 Hz for maximal reward prediction error (see Chapter 6.4, this volume). Also, the burst sizes (i.e., number of intraburst action potentials, ranging from two to seven) and firing rates during the postreward interval (best correlation for a 150-ms interval), but not burst latencies, were correlated with the amplitudes of the positive reward prediction errors. Finally, tonic baseline spike rates and phasic reward interval spike rates showed only a very weak positive correlation.[27,31] In contrast, negative reward prediction errors were associated with increasing length of electrical silence (ranging up to 400 ms). Similar properties of postreward bursting have been observed by Hyland and colleagues in the awake rat,[11] where dopamine neurons fired up to seven intraburst actions potentials with maximums intraburst frequencies of about 30 Hz. Importantly, they found no systematic change in action potential shapes during the postreward burst discharge (but see [31]). This flexible signalling of dopamine neurons via the burst mode in different behavioural contexts has been the most important factor in the development of a computational learning theory of dopamine function coding for a quantitative reward prediction error and has spawned the burst discharge of dopamine neurons to identification of self-generated actions with rewarding outcomes.[32]

The Role of Ion Channels

In summary, dopamine midbrain neurons in awake, behaving animals show two prominent types of electrical discharge—a tonic baseline discharge of about 3–6 Hz and stimulus-locked bursts of about 2–10 action potentials with up to 10-fold higher discharge frequencies or ≤400-ms pauses of electrical activity. The role of distinct ion channels in controlling the frequency and regularity of these two discharge modes of dopamine midbrain neurons has not yet been systematically studied in awake, behaving animals—apart from the described inhibition of electrical activity by stimulation of somatodendritic D2 receptors.[25] Analysis of ion channel subunit gene knockout animals—ideally selective for dopamine midbrain neurons—could facilitate these types of experiments in awake animals.[33,34]

However, for now, we must turn to the anesthetized in vivo rodent preparations for more mechanistic insights into the role of ion channels in generating and controlling the two patterns of discharge in dopamine neurons—the tonic mode and the burst mode.

IN VIVO ACTIVITY PATTERNS OF DOPAMINE MIDBRAIN NEURONS IN ANESTHETIZED ANIMALS

Action Potential Properties

In their landmark studies,[31,35] Grace and Bunney presented a still unsurpassed large body of intracellular recordings of spontaneous bursts and tonic firing of dopamine midbrain neurons in anesthetized rats (see figure 3.4.1). Intracellular in vivo recordings of dopamine midbrain neurons[12,35–38] provided more information on how the action potential shapes itself, as well as the subthreshold trajectory of the membrane potential. Also, for the first time, these recordings allowed the direct neurochemical identification of dopamine neurons via intracellular L-DOPA injection followed by processing for catecholamine fluorescence histochemistry.[12] The passive membrane properties of these identified dopamine neurons revealed an in vivo input resistance of about 30 MΩ with membrane time constants in the range of 12 ms (relatively high values compared to those of other neuronal cell types recorded at the time via sharp intracellular electrodes).[37] Even when we take into account that sharp intracellular electrodes introduce a large somatic shunt by penetrating the plasma membrane compared to the tight gigaseal patch-clamp recordings, and thus might lead to underestimation of the input resistance and time constant, these passive membrane properties indicate that dopamine neurons might respond relatively slowly, and thus would effectively integrate synaptic events and distribute potential changes along the somatodendritic axis.

The action potential of classical dopamine midbrain neurons, intracellularly recorded in vivo, lasted about 2–3 ms and showed a relatively high threshold at around -40 mV, with amplitudes of up to 75 mV after an often biphasic rising phase (figure 3.4.1 a/b). Antidromic activation of action potentials (generated from the axonal region, not the somatodendritic compartment, SD) elicited a shorter (≤1.7 ms) and smaller (≤25 mV) initiation segment (IS) spike compared to the spontaneously occurring longer spike present in the somatodendritic compartment.[38] An in vitro double patch-clamp study confirmed an initial action potential generation in dopamine midbrain neurons at axonal sites (often branching from dendrites rather than the soma) and the robust back propagation of the action potential across the somatodendritic compartment via the recruitment of dendritic ion channels.[39] The complete action potential (IS + SD) was followed by a prominent afterhyperpolarization (AHP), which was sensitive to potassium channel blockers and increased calcium buffering.[35] This strongly suggested a major contribution of Ca^{2+}-activated potassium channels, which were later identified in in vitro studies (see the later section "In Vitro Pacemaker Activity: Role of Ion Channels"[35]). The AHP was followed by a slow pacemaker depolarization back to spike threshold. The properties of this phase, as well as the occurrence of an anomalous rectification in the subthreshold range (so-called sag component) upon injection of hyperpolarizing currents, suggested important contributions of voltage-dependent ion channels in the control of tonic activity that were later studied in detail in vitro. During in vivo burst activity, the speed of the depolarization phase became slower, the peak of the action potential became smaller, and the repolarization became more prolonged in subsequent action potentials. Although the ionic mechanisms were not studied at the time, these changes might be indicative of the cumulative activity-dependent inactivation of voltage-gated sodium or calcium channels.[31]

Tonic and Burst Activity Modes

The single-spike tonic discharge mode (figure 3.4.1 b) appeared to be fairly similar in anesthetized compared to that in awake in vivo preparations. In the burst mode, as spontaneous motor behaviour is absent and sensory processing is severely blunted, stimulus-locked burst discharges are minimal (similar to sleep states[40]) and have not been studied in a meaningful behavioural context. However, visual stimulus-locked burst discharges were reinstated under general anesthesia by local pharmacological disinhibition (by a $GABA_A$ receptor blocker) in the superior colliculus.[41] These burst discharges had basic properties (latency, burst duration, and frequency) similar to those recorded in awake animals. But most studies with anesthetized animals rely on the occurrence of spontaneous bursts, which are not associated with identified sensory events and have to be identified by distinct properties of the spike train itself. The most commonly utilized criteria for burst identification were provided by Grace and Bunney, who defined the start of a burst within a spike train by an ISI of ≤ 80 ms and the end of the burst by an ISI of ≥ 160 ms[31] (compare figure 3.4.1 c; for an alternative approach, see [42]). These spontaneous bursts are studied and utilized in reference to the stimulus-triggered burst phenomenon in vivo in awake animals. Indeed, in awake mobile animals, spontaneous bursts (detected by the same spike train properties) where the relevant sensory source was not identified, are similar to stimulus-locked bursts.[11] However, it is important to note that while the in vivo tonic discharge as well as the burst mode properties (such as number of spikes per burst and intraburst frequencies) are largely unaffected by anesthesia compared to those of awake, behaving animals,[9,11,27,28,43] the frequency of spontaneous bursts is indeed altered by the type and level of anesthesia[43,44]. (With reduction of spontaneous burst firing in animals under chloral hydrate, urethane, and pentobarbital but not ketamine anesthesia compared to restrained or paralyzed, awake rodents).

The Role of Ion Channels

The in vivo work of Grace and Bunney[31,35] allowed for the first time a closer look at the underlying ionic mechanisms of the two main discharge modes of dopamine midbrain neurons: tonic and burst activity. Their intracellular in vivo recordings of spontaneous bursts in anesthetized rats showed that bursts appeared to ride on a depolarizing wave lasting several hundred milliseconds. In addition, these bursts were facilitated by both glutamate application and as Ca^{2+} influx, suggesting an essential contribution of calcium-selective ionotropic glutamate receptors. Indeed, the activation of ionotropic NMDA (N-methyl-D-aspartate) glutamate receptors on dopamine midbrain neurons was shown to be an essential regulator of spontaneous in vivo bursting in anesthetized rats.[45,46] Important synaptic glutamatergic inputs are coming from the prefrontal cortex, subthalamic nucleus, pedunculopontine nucleus, and dorsolateral tegmentum.[47-49] The results of local iontophoresis of ion channel inhibitors with broad specificity (like cobalt, barium, or tetraethylammonium ions) suggested that intrinsic Ca^{2+} and K^+ channels were also involved in the regulation of burst activity.[31] While their predicted Ca^{2+}-inactivated potassium channel has not yet been identified, an important role of Ca^{2+}-activated small-conductance potassium (SK) channels (which are expressed in dopamine midbrain neurons[50]) as negative modulators of burst discharge in vivo has been defined more recently for rats.[51,52] Local iontophoresis of selective SK channel inhibitors (like apamin or N-methyl laudanosine) increased in vivo burst firing, while the systemic application of an SK channel opener (1-ethyl-2-benzimidazolinone) decreased burst discharge.[51,52] Another class of potassium channels expressed in dopamine midbrain neurons,[53,54] the KCNQ channels (also known as Kv7 or M channels), has a similar negative effect on burst activity in vivo, as shown via systemic in vivo application of a selective opener (retigabine) and inhibitor (XE-991) of KCNQ channels.[55] Given the ubiquitous expression of KCNQ channels in the brain and the relative paucity of in vitro studies on these channels in dopamine neurons, more work is necessary to test whether burst regulation via KCNQ channels is dependent on the expression of these channels in dopamine midbrain neurons.

The important discovery by Groves and colleagues[25] that D2 autoreceptor activation induces an inhibition of spontaneous discharge of dopamine midbrain neurons in vivo[25] was further studied by Grace and Bunney using intracellular recordings. They showed that the inhibitory effect of a D2 agonist was mediated by a membrane hyperpolarization of dopamine midbrain neurons, in vivo which could completely silence the respective neurons or (at lower concentrations) reduce their activity and burst discharge mode in response to D2 autoreceptor activation.[37] More recent molecular studies identified G-protein-activated inwardly rectifying potassium channels (GIRK, Kir3; see next section for details) as the downstream target of the somatodendritic D2 receptors in dopamine midbrain neurons.[56] In addition to D2 agonists, Tepper and colleagues showed that GABA$_A$- but not GABA$_B$-receptor antagonists also lead to a prominent increase in burst discharge of mouse dopamine midbrain neurons in vivo.[57,58] Furthermore, changes in dopamine midbrain neuron activity and dopamine release in vivo in response to intravenous glucose application have been described, but the underlying ionic mechanisms were not addressed.[59,60] Similar, the contribution and identification of the variety of additional ion channels in the activity of dopamine midbrain neurons have not yet been carried out in vivo but have been done in vitro and will be discussed in the next section.

IN VITRO ACTIVITY PATTERNS OF DOPAMINE MIDBRAIN NEURONS

The Action Potential

The advent of in vitro brain slice preparations in the mid-1980s in combination with intracellular recording techniques—initially sharp microelectrode recordings[13,61–65] and later patch-clamp approaches[66–69]—dramatically eased the electrophysiological access to dopamine neurons in the midbrain, leading to a large increase in studies and thus in the amount of information on dopamine neuron function. In addition, the electrical properties of acutely isolated[14,70] and cultured dopamine midbrain neurons[71–73] were studied.

Similar to studies in vivo, the broad action potential of classic dopamine midbrain neurons, recorded in vitro from adult mice using patch-clamp techniques, lasted about 2–4 ms, showing relatively depolarized thresholds between −30 and −40 mV and biphasic rising phases to peak amplitudes at around +30 mV followed by a prominent AHP (figure 3.4.2a). Double-patch-clamp studies of classic dopamine midbrain neurons have shown that the broad action potential (carried by the influx of sodium and calcium ions) actively backpropagates along the dendrites.[39] This active action potential invasion of the dendritic tree (SD spike) might also serve as a trigger for the calcium-dependent dopamine release from dendrites.[74,75] The action potential at the initial segment (IS spike) of the axon, which in dopamine midbrain neurons often originates from a dendrite, has distinct properties: a lower threshold and a shorter duration (probably due to a lack of contribution from voltage-gated calcium channels). The coupling between the IS spike and the SD spike is fragile and is itself controlled by membrane potential and dendritic ion channels.[74,76]

Tonic Pacemaker and Burst Activity

In most cases, the activity pattern and underlying ionic mechanisms of dopamine midbrain neurons were studied in isolated cells or in in vitro slice preparations in the presence of inhibitors of fast excitatory and inhibitory synaptic transmission (for review, see[4]). In almost all of these in vitro preparations, one striking attribute of the electrical activity of classical dopamine neurons was now becoming obvious: dopamine midbrain neurons were pacemaker cells, meaning that they generate a tonic, regular, and low-frequency discharge entirely via cell-autonomous ionic mechanisms, even in complete synaptic isolation. Pacemaker cells possess no stable negative resting membrane potential in the absence of synaptic input but create rhythmic oscillations of their membrane potential, sufficiently large to repetitively cross the threshold for action potential generation (figure 3.4.2b). Electrical stimulation experiments demonstrated that dopamine midbrain neurons in vitro are not able to fire at much higher frequencies than those generated spontaneously (in the range of 0.5–10 Hz),[77,78] indicating that intrinsic conductances and morphology determine the narrow frequency bandwidth of their pacemaker.[79] The limited frequency range of tonic pacemaker activity of dopamine neurons in vitro resembled that of the tonic activity recorded in vivo, but quantitative analysis of the ISI histogram distributions revealed that in vitro pacemaker activity was at least fivefold more regular than the single-spike modes observed in vivo (the CV of ISI distributions was about

a single action potential

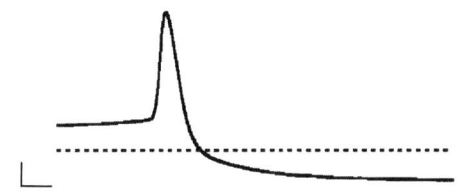

b pacemaker activity

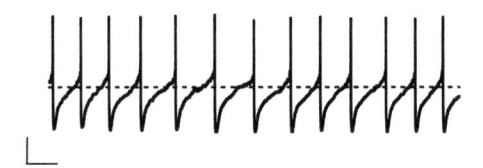

c burst activity

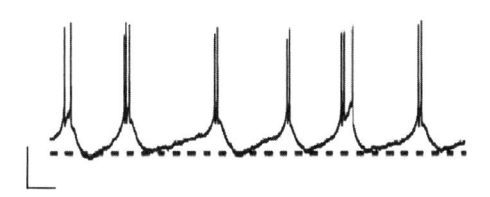

FIGURE 3.4.2 Classic dopamine midbrain neurons recorded in vitro in mouse brain slices. Perforated patch-clamp recordings of (a) a typical single action potential, (b) tonic pacemaker mode, and (c) burst activity mode (induced by pharmacological SK channel inhibition). Dotted lines represent –50 mV. Scale bars represent 10 mV (a–c) and 2 ms (a) or 500 ms (b, c).

0.12 in vitro compared to about 0.6 in vivo).[50,80] Another striking difference from in vivo recordings was the almost complete absence of spontaneously occurring burst activity in dopamine midbrain neurons from in vitro preparations. However, burst-like discharges could be induced in vitro by pharmacological modulation of distinct ion channels (figure 3.4.2c).[13]

The reduction of the more diverse spiking pattern of dopamine midbrain neurons in intact brains to only a single prominent in vitro pacemaker activity mode focused attention on the identification of those distinct ion channels that generate and control the intrinsic pacemaker frequency and regularity and prevent the switch to an in vitro burst mode. A large number of functional electrophysiological studies—in some cases combined with molecular single-cell gene expression approaches[81]—addressed these issues and identified several distinct ion channel subtypes that are involved in pacemaker activity control. The findings of these in vitro studies are summarized in the next sections and in figure 3.4.3.

In Vitro Pacemaker Activity: Role of Ion Channels

Depolarization

A prerequisite for pacemaker activity is a nonstable but spontaneously oscillating "nonresting" membrane potential also called a *spontaneous oscillatory potential* (SOP).[82] Pacemaking neurons possess in general two or more voltage- or Ca^{2+}-gated ion channel types, which build the core of a spontaneous membrane potential oscillator.[6] Each phase of the waxing and waning oscillator is limited by negative feedback mechanisms, either

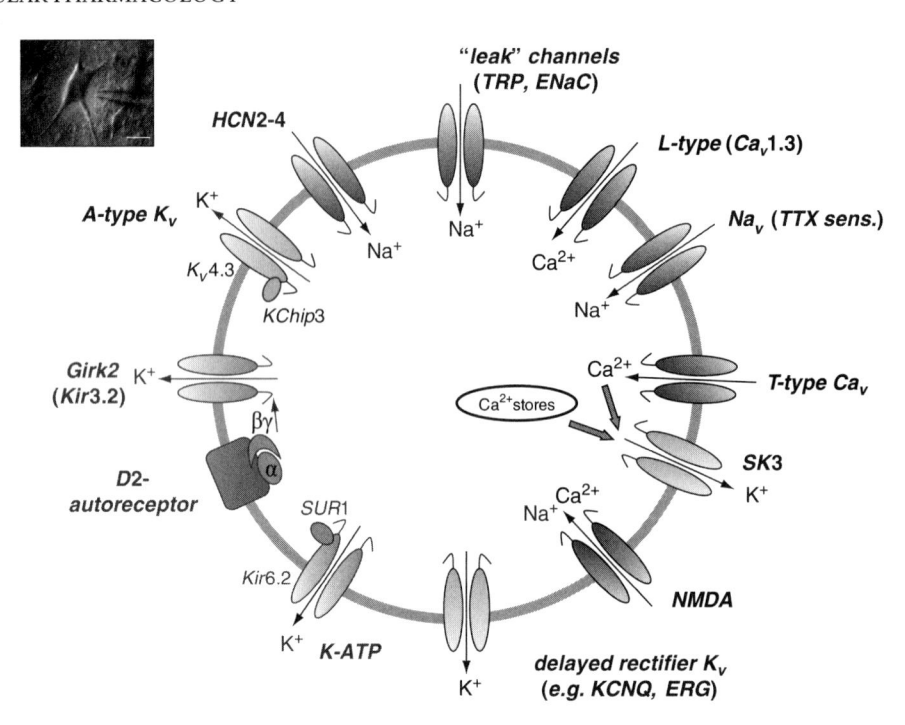

FIGURE 3.4.3. Ion channels generate and modulate spontaneous pacemaker activity and burst activity of classic dopamine midbrain neurons in vitro and in vivo. For details, see text. The insert shows a classic dopamine neuron in an in vitro mouse brain slice preparation before patch clamp recording (visualized via infrared videomicroscopy; note the smaller GABAergic interneuron in the right lower part; scale bar represents 15 μm). Ca$_V$/K$_V$/Na$_V$, voltage gated calcium-, potassium-, sodium-channel; D2, dopamine-receptor subtype 2; ENaC, epithelial sodium channel; ERG, Ether-a-go-go-related gene potassium channel; Girk, G-protein coupled inwardly rectifying potassium channel; HCN, hyperpolarization activated cyclic nucleotide gated cation channel; K-ATP, ATP-sensitive potassium channel; KCNQ, KQT-like potassium channel; Kir, inwardly rectifying potassium channel; NMDA, N-methyl-D-aspartate glutamate receptor; SK, small conductance Ca^{2+} activated potassium channel; SUR, sulfonylurea receptor, TRP, transient receptor potential; TTX, tetrodotoxin. Source: Adapted from.[170] (See Color Plate 3.4.3.)

intrinsic to the channel (e.g., voltage- or Ca^{2+}-mediated inactivation) or caused by subsequent activation of other ion channels with antagonistic effects on the membrane potential. In the latter case, interacting antagonistic ion channel pairs in pacemaker neurons are coupled either with their respective voltage ranges of activation and deactivation or with calcium ions, which flow into the cell through Ca^{2+} channels and in turn open Ca^{2+}-activated potassium channels.[4]

In non-pacemaking neurons (in the absence of synaptic input), so-called leak potassium channels (of the inwardly rectifying[83] or two-pore domain families,[84] in particular Kir2 and TASK channels) are constantly open and dominate the resting conductance, thereby setting a stable resting membrane potential close to the electrochemical equilibrium potential of potassium ions, which at physiological potassium gradients is around −90 mV. Dopamine midbrain neurons show only low densities of constitutively open leak potassium conductances, and gene expression studies show that neither members of the Kir2 channel family[85] nor members of the extended two-pore domain K channel families[86] are expressed in high abundance in dopamine midbrain neurons. Instead, several types of depolarizing, nonselective cation leak or sodium leak channels are expressed (e.g., of the transient receptor potential superfamilies TRP[87] and ENaC,[88] in particular TRPC, TRPV, and ASIC) that are either constitutively active or activated by upstream signalling cascades. Although the detailed contributions of these different leak channel subtypes remain unclear, they are likely to contribute to the relatively positive membrane potential in the range of −55 to −40 mV observed in nonspiking dopamine midbrain neurons.[89–96]

But which types of ion channels build the core of the membrane potential oscillator in pacemaking dopamine midbrain neurons? Several studies have shown that conventional tetrodotoxin (TTX)-sensitive voltage-gated sodium (Na$_V$) channels[97] contribute to the action potentials in dopamine midbrain neurons. The exact molecular composition of their pore-forming α-subunits and auxiliary β-subunits has yet to be determined.[98] However, in a large subpopulation of dopamine midbrain neurons, predominantly located in the SN, the membrane potential continues to oscillate in the presence of TTX.[99,100] Ion substitution experiments and channel pharmacology studies have shown that these TTX-resistant membrane oscillations essentially depend on the activity of voltage-gated L-type calcium channels ($Ca_V1.1$-1.4).[100] The α-subunit $Ca_V1.2$, as well as both major $Ca_V1.3$ splice variants, are expressed in dopamine midbrain neurons[99,101]; however, $Ca_V1.3$ channels that activate at more negative potentials compared to other members of the L-type Ca^{2+} channel family carry the bulk of Ca^{2+} inward currents during the ISI.[99,100] This Ca^{2+} component of the ISI is far more dominant in classic dopamine midbrain neurons compared to other pacemaker neurons in the brain, which mainly rely on interspike Na^+ influx through TTX-sensitive persistent Na^+ channels or hyperpolarization-activated cyclic nucleotide-gated cation (HCN) channels.[4] While most classic dopamine neurons of the SN depend on this Ca^{2+} channel–driven pacemaker, and completely stop firing when Ca^{2+} is replaced by cobalt or when L-type Ca^{2+} channels are blocked,[99,100] the inhibition of Ca^{2+} channels does not prevent firing of other dopamine midbrain neurons, in particular in the neighbouring ventral tegmental areal (VTA); in these dopamine neurons, pacemaker activity is indeed driven by TTX-sensitive Na^+ channels and HCN channels,[99,100] which have been identified in one study as the predominant pacemaker in classical dopamine neurons from young mice.[99]

When the oscillating membrane potential reaches the threshold for generation of action potentials at around −40 mV, TTX-sensitive, voltage-gated Na^+ channels and other high-threshold, voltage-activated Ca^{2+} channels[102] (like L-type $Ca_V1.2$, N-type $Ca_V2.1$, and P/Q-type channels $Ca_V2.2$) are recruited and generate the full somatodendritic action potential (SD spike).[82,100] Again, the exact subunit composition of the relevant species of voltage-gated Ca^{2+} channels has yet to be determined for dopamine neurons. By contrast, the depolarizing phase of the isolated IS spike activated by, for example, antidromic stimulation is likely to be mediated only by TTX-sensitive, voltage-gated Na^+ channels.[13]

Repolarization

The IS and SD action potentials are repolarized by activation of members of the large family of voltage-gated potassium channels (K_V). The pharmacological profile of these so-called delayed rectifier potassium channels (in particular, Kv1-Kv4)[103] has not yet been fully defined, as peptide blockers that selectively inhibit molecularly defined subclasses of these K_V channels (like dendrotoxins) have not yet been systematically analyzed.[14] Voltage-gated Kv7 channels (i.e., the KCNQ family[104]) are also expressed in high abundance in dopamine midbrain neurons,[54] but it is not yet clear whether they contribute to SD or IS excitability.[105] Apart from the Kv4 family (see below),[106] the molecular subunit composition of pore-forming and auxiliary subunits for voltage-gated potassium channels in dopamine neurons have not been resolved.

Due to the exceptionally large Ca^{2+} influx during the SD action potential,[4] which is potentiated by release from intracellular Ca^{2+} pools,[107] it is not surprising that Ca^{2+}-activated potassium channels are also activated and contribute to action potential repolarization in classic dopamine midbrain neurons. While there is currently little evidence for an important functional role of big-conductance Ca^{2+} and voltage-activated potassium (BK) channels,[14] which are known to be responsible for the fast afterhyperpolarization (fAHP) in other central neurons,[108] a prominent role of small-conductance Ca^{2+}-activated potassium (SK) channels is well established.[108] In contrast to BK channels, SK channels are slowly activated, as they are not physically coupled to their respective Ca^{2+} sources.[108] SK channels show a preferential functional coupling to fast-inactivating T-type Ca^{2+} channels[109] that are activated during the action potential of dopamine midbrain neurons.[110,111] But they are also recruited by receptor-mediated Ca^{2+} mobilization through activation of inositol-1,4,5-triphosphate (IP_3)- and cyclic adenosine diphosphate (ADP) ribose-pathways.[107,112,113] SK channels in classic dopamine midbrain neurons are mediated mainly by SK3 subunits[50] and generate a prominent medium afterhyperpolarization (mAHP) with a delayed onset that lasts significantly longer than the SD action potential itself.[80,114] More recently, ether-a-gogo-related voltage-gated potassium channels (also known as ERG or Kv11 channels[115]) have been shown to control a Ca^{2+}-insensitive slow afterhyperpolarization (sAHP) in dopamine midbrain neurons that lasts for several seconds.[116]

The Interspike Interval (ISI)

Both SK and ERG channels are not primarily involved in pacemaker frequency control but are essential for its stability.[117,118] However, glutamatergic mGLUR1- or α1-adrenergic (Gq/IP3/Ca^{2+}) receptor-mediated activation of SK channels does transiently inhibit pacemaker frequency by hyperpolarizing the membrane potential below the action potential threshold.[119,120] The desensitization of the mGLUR1-SK-mediated slow inhibitory postsynaptic potential (IPSP) unravelled a prolonged, Ca^{2+}-independent slow excitation that stimulated pacemaker activity.[120,121] TRPC cation channels are good candidates for mediating the underlying cationic conductance, but given the limited selectivity of TRPC inhibitors, definitive proof is still missing.[92] Treatment with psychostimulants (like amphetamine) reduces receptor-mediated Ca^{2+} release and SK channel activation, which in turn might enhance TRPC-mediated pacemaker stimulation of dopamine neurons.[117,118] These TRPC channels might also be activated by neuropeptides like neurotensin or cholecystokinin, which increase the pacemaker frequency of dopamine neurons by activation of cationic conductances.[89,93,122–124] However, a very recent study suggests that novel channel subunits might be the prime target of neuropeptides in dopamine midbrain neurons.[125]

In general, the ionic currents that are present in the subthreshold range during the ISI and control pacemaker frequency are very small compared to those during the action potential.[4] Most of these subthreshold ion channels are either voltage-independent (like SK and TRP) or already activated in the subthreshold membrane potential range, like fast-inactivating A-type K^+ channels of the Kv4 family[126] (in particular, Kv4.3/KChip3.1 channels), hyperpolarization-activated HCN cation channels (in particular, HCN2-4),[127] and fast-inactivating T-type voltage-gated Ca^+ channels (Ca_V3.1–3.3).[128]

The functional role of A-type K^+ channels (mediating I_A currents) in pacemaker control is opposed to that of HCN channels (mediating I_H currents). In classic SN dopamine neurons, A-type potassium channels are composed of pore-forming Kv4.3L (long splice variant) α-subunits and the auxiliary β-subunit KChip3.1.[129] Their activation during the ISI delays the depolarization toward the action potential threshold and consequently the frequency of the pacemaker.[24,129] Recent studies show that the activation of A-type potassium channels in dopamine midbrain neurons is highly sensitive to the speed of depolarization in the subthreshold range, which further enhances their negative feedback control of pacemaker frequency.[130,131] Quantitative single-cell phenotype-genotype analysis demonstrated strong inverse correlations between the rate of discharge and the density and abundance of Kv4.3/KChip3.1 K^+ channels in individual SN dopamine neurons.[129] This pacemaker frequency tuning by A-type Kv4.3 channels might be predominantly controlled at the transcriptional level on a slower time scale, as shown for the adaptation to chronic challenges of antipsychotic substances.[132,133] Thus, the large variations in A-type K^+ currents and related distinct channel subunit mRNA expression might reflect different regulatory states of a homogeneous type of dopamine neuron.[129] In this context, it is interesting to note that the KChip3 β-subunits are also known as DREAM (DRE agonist modulator), which in the nucleus operates as a Ca^{2+}-regulated transcription repressor.[134]

Slowly gating HCN channels are likely to be composed of HCN2, HCN3, and HCN4 subunits, which are coexpressed in single dopamine neurons.[135] However, their precise subunit composition is unknown. Due to their mixed selectivity for sodium and potassium ions, HCN channel activity increases the speed of subthreshold depolarization during the ISI and thus the pacemaker frequency of classic dopamine neurons from young and adult mice.[136,37,138] As voltage-dependent gating of HCN channels is sensitive to changes in intracellular cyclic adenosine monophosphate (cAMP) or cyclic guanosine monophosphate (cGMP) concentrations,[127] these channels can also translate upstream signalling that controls cyclic nucleotide levels in dopamine midbrain neurons into altered pacemaker activity.[139] Furthermore, similar to L-type Ca^+ channel–driven pacemaking, several studies (in early postnatal and 3-month adult mice) suggest that HCN channels control pacemaker frequency only in a subpopulation of dopamine midbrain neurons (again, mainly in the SN, as discussed in the later section "Distinct Functional Types of Dopamine Midbrain Neurons: Role of Ion Channels").[132,133]

Similarly as shown in vivo, dopamine itself also controls pacemaker activity in vitro in classic dopamine neurons in a negative fashion via D2 autoreceptors. As first described in landmark studies by Lacey, Mercuri, and North,[63,140] the activation of G-protein-coupled inwardly rectifying potassium (Girk, also named Kir3) channels via somatodendritic D2 and $GABA_B$ receptors leads to membrane hyperpolarization and a complete silencing of pacemaker activity. These potassium channels are either homotetramers of Kir3.2 subunits[77,141] or Kir3.2-Kir3.3 heterotetramers.[142,143] The activation of Kir3 channels also limits the action potential propagation along the somatodendritic axis.[74] This form of dynamic electrical compartmentalization might partially dissociate axonal dopamine release (coupled to IS action potentials) from Ca^{2+}-dependent vesicular somatodendritic dopamine release, which depends on the propagation of the action potential into the dendritic domain[74] and, in turn, activates inhibitory D2-autoreceptor-coupled Kir3 channels.[144] D2- and $GABA_B$-mediated Kir3 channel signalling is also important in the context of plasticity and pathophysiological dysfunction of dopamine midbrain neurons: use-dependent long-term depression[145] and drug-induced potentiation[143] of G-protein-coupled receptor-activated Kir3 currents have been reported. Surprisingly, D2 receptors themselves are also voltage-dependent, displaying reduced activity with more depolarized membrane potentials, which further increases the complex role of D2 autoreceptors for pacemaker activity control of dopamine neurons.[146]

In addition to dopamine itself, a multitude of other metabolic signals can modulate the pacemaker activity of dopamine neurons in vivo[59,60] and in vitro.[147,148] For example, elevated extracellular glucose levels and neuropeptide Y inhibit electrical activity of dopamine neurons, while the application of orexin or ghrelin enhance it.[147,148] One key player for integrating and transducing a variety of metabolic signals in dopamine neurons to altered pacemaker activity is the adenosine triphosphate (ATP)–sensitive potassium (K-ATP) channel, which is composed of SUR1 and Kir6.2 subunits in dopamine neurons from adult mice.[149] K-ATP channels are either completely closed or partially activated by endogenous redox signalling in vitro[150] but are stimulated in response to a number of nonphysiological stimuli like removal of extracellular glucose, oxidative stress, or inhibition of energy metabolism.[69,151,152] When activated, K-ATP channels hyperpolarize the membrane potential and can completely suppress pacemaker activity—similar to the action of Kir3 channels.[152] While activation of K-ATP channels by leptin and insulin signalling has been described for hypothalamic neurons,[153] the complex physiological control of K-ATP channel activity in dopamine neurons remains to be defined. Their role in the pathophysiology of PD is discussed below (see the later section "Differential Vulnerability of Dopamine Midbrain Neurons to Degeneration").

In Vitro Burst Activity: Role of Ion Channels

As already mentioned, spontaneous burst discharge is in most cases not observed in dopamine midbrain neurons in in vitro preparations.[48] This indicates that, in contrast to other neuronal cell types,[154] burst firing is not a manifest intrinsic activity pattern of dopamine midbrain neurons. Accordingly, bursting in dopamine neurons is also not elicited by simple release from membrane hyperpolarization or injection of depolarizing current.[20] Apparently, switching to the burst state by simple membrane potential fluctuations is prevented by distinct ion channel activities in dopamine neurons.[22,79,155] However, burst activity of dopamine midbrain neurons in vitro is induced via pulsatile or tonic pharmacological activation of NMDA glutamate receptors[155,156] or selective inhibition of either SK potassium channels[52,80,114,157] or their upstream calcium sources.[111,112,158] (compare figure 3.4.3). The pharmacological inhibition (apamin) of SK channels in vitro prolongs L-type Ca^{2+} channel activity generating long Ca^{2+} plateau potentials, on top of which sodium channel burst activity is generated in a frequency range similar to that during event-locked bursts in vivo.[114] Although NMDA activation and SK channel inhibition could work synergistically to enhance burst discharge,[159] there is also evidence that they control different modes of in vitro busting: while NMDA-induced bursting requires a hyperpolarizing current and is insensitive to L-type Ca^{2+} channel inhibition, SK inhibition-induced bursting is facilitated by depolarization and agonists of L-type Ca^{2+} channels and is prevented by L-type channel inhibitors (e.g., nifedipine).[160] The activation of the electrogenic Na^+/K^+ adenosine triphosphatase (ATPase) has been suggested to provide the hyperpolarizing current necessary for NMDA-induced bursting,[156] while ERG potassium channels might facilitate the repolarization of the long L-type Ca^{2+} channel–mediated plateau potentials in the SK channel block–induced type of in vitro bursting.[116,161] The proteolytic activation of protein kinase M downstream L-type channel activation might serve as a positive feedback mechanism to stabilize this burst mode in dopamine midbrain neurons.[162]

In summary, there is compelling evidence that the burst discharge mode is activated by a combination of synaptic events and an associated change in intrinsic channel gating that modifies the resonant properties of dopamine midbrain neurons.[22,79,163,164] Given the importance of the burst signal in vivo, it is plausible that entering this mode is carefully controlled by (at least) two distinct ion channel types (NMDA and SK3), which have to coincide to activate the burst discharge in dopamine midbrain neurons. However, it remains to be shown which—if any—of these candidate burst mechanisms derived from in vitro studies is operative during the stimulus-induced or reward-related phasic discharge of dopamine neurons in awake, behaving animals.

FUNCTIONAL DIVERSITY OF DOPAMINE MIDBRAIN NEURONS

Distinct Functional Types of Dopamine Midbrain Neurons: A Dual System

As already described, electrophysiological studies of dopamine midbrain neurons have been strongly biased toward a single classic functional phenotype of dopamine neuron, displaying broad action potentials with prominent AHPs, regular low pacemaker frequency, a (HCN-mediated) sag component upon injection of hyperpolarizing current, and the inhibition of pacemaker activity by activation of D2 autoreceptors. These functional criteria were used for identification of a presumed dopaminergic midbrain genotype, without neurochemical verification of the dopaminergic identity—for example, by single-cell reverse transcriptase polymerase chain reaction (RT-PCR) or immunhistochemical analysis of dopaminergic marker gene expression (like that of tyrosine hydroxylase [TH] and the dopamine transporter [DAT]). This heuristic approach does allow reliable identification of most dopamine midbrain neurons within the SN (A9) and the retrorubral area (A8), as about 85% of all neurons in these nuclei indeed conform to the classic functional dopamine phenotype.[50,77,138] However, even in the relatively homogeneous SN, electrophysiological diversity among dopamine neurons has been noted.[50,138,165] Nevertheless, in the VTA (A10), where dopamine neurons comprise only about 50% of the entire neuronal population, the heuristic functional approach is not sufficient for reliable identification of dopamine midbrain neurons for two reasons. First, a large population of dopamine neurons in the VTA displays electrophysiological properties very different from those of classic dopamine neurons.[77] And second, a substantial population of non-dopamine VTA neurons does indeed display electrophysiological properties that are very similar to those used to define the classic dopamine phenotype.[166] In consequence, identification of dopamine midbrain neurons on the basis of their classical functional characteristics can lead to a large number of false-negative or false-positive assignments. In contrast, independent neurochemical identification of dopamine neurons not only allows unbiased functional characterization of all dopamine midbrain neurons, but also enables their somatic localization within the different midbrain nuclei and—in combination with retrograde tracing—the identification of their distinct striatal, limbic, and cortical projections (see Chapter 2.3, this volume).

Using a method involving in vivo retrograde tracing combined with in vitro electrophysiological, molecular, and immunohistochemical analysis, we have recently systematically studied the electrophysiological properties of molecularly defined dopamine midbrain neurons with six distinct axonal projections in the adult mouse (summarized in figure 3.4.4).[77] Using an unsupervised clustering approach of the sampled in vitro electrophysiological profiles, we identified two very distinct basal functional phenotypes of dopamine midbrain neurons.[77] As expected, one of the phenotypes conformed to the well-known classic dopamine midbrain neuron, which has been extensively described in the literature and in the above sections. These classic dopamine midbrain neurons in adult mice were located predominantly in

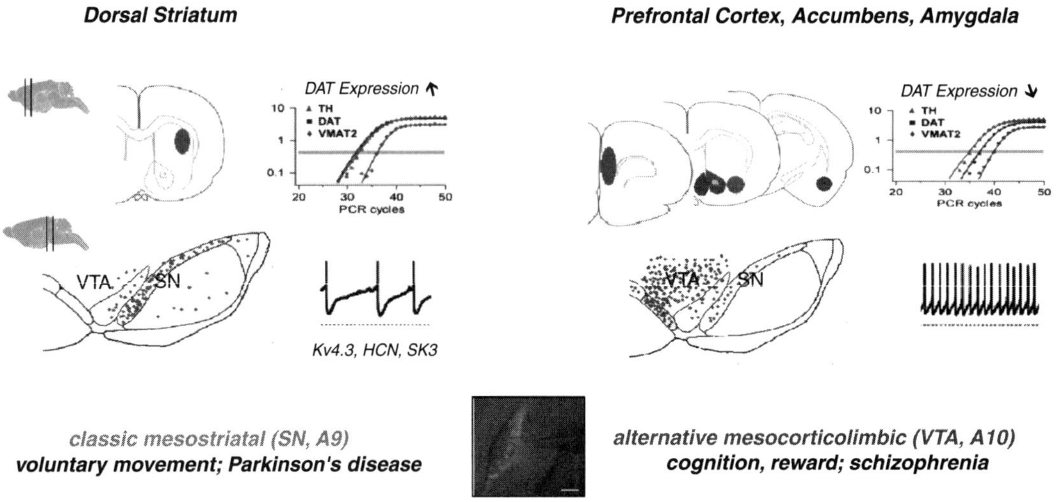

FIGURE 3.4.4. The dual dopaminergic midbrain system of the adult mouse. Classic dopamine midbrain neurons (green dots, coronal midbrain sections) display well-described electrical properties (i.e., low-frequency pacemaker activity (dotted line: –80 mV), controlled by Kv4.3, HCN, and SK3 channels), express high DAT levels, and are located predominantly in the SN and the lateral VTA, projecting to dorsal striatum and to the lateral shell of the nucleus accumbens (green blobs, coronal sections). Alternative dopamine midbrain neurons (red dots) display distinct electrical properties (i.e., higher pacemaker frequency, not controlled by Kv4.3, HCN, or SK3 channels), express lower DAT levels, and are located in the more medial VTA, projecting to the prefrontal cortex, basolateral amygdala, and medial shell and core of the nucleus accumbens (red blobs). Data were obtained by combining in vivo retrograde tracing of six distinct projections of dopamine midbrain neurons with ultraviolet laser microdissection, quantitative single-cell RT-PCR, immunohistochemistry, and patch-clamp recordings of fluorescence-labeled neurons from in vitro brain slices (lower picture, fluorescence beads visualized by infrared videomicroscopy and epifluorescence; scale bar represents 5 μm); DAT, dopamine transporter; SN, substantia nigra; TH, tyrosine hydroxylase; VMAT2, vesicular monoamine transporter2; VTA, ventral tegmental area. Source: Adapted from[77,170]. (See Color Plate 3.4.4.)

the SN and the lateral VTA projecting to the dorsal striatum (SN) and the lateral shell of the nucleus accumbens (VTA), respectively. In young rats, however, classic dopamine midbrain neurons have also been found to project to the cortex.[167–169] Future studies will clarify whether these differences are of a technical nature or reflect genuine developmental or species differences.

The second, non-classic "alternative" electrophysiological phenotype of dopamine midbrain neurons in adult mice was very distinct from the classic one, displaying significantly broader action potentials, with less prominent AHPs, a more irregular pacemaker with higher spontaneous and maximally inducible frequencies, and a less pronounced or completely absent sag component (anomalous rectification) upon injection of hyperpolarizing current.[77] This alternative dopamine phenotype was present predominantly in VTA dopamine neurons, with projections either to the medial prefrontal cortex, to the core or medial shell of the nucleus accumbens, or to the basolateral amygdala. This atypical functional phenotype of dopamine midbrain neurons is not only linked to defined axonal projections and clustered in midbrain subregions, but is also associated with differential gene expression levels of enzymes involved in synthesizing (TH), packaging (vesicular monoamine transporter 2, VMAT2), and uptake (DAT) of dopamine.[77] Based on these functional and molecular findings for adult mice in vitro, we propose that dopamine midbrain neurons are segregated into a dual system that is involved in distinct neuronal networks with different functional roles, and might contribute in vivo to the distinct temporal profiles of behavioural dopamine signals observed in subcortical and cortical target areas in awake, behaving animals (ranging from phasic subsecond peaks of extracellular dopamine to slow, stable increases of extracellular dopamine over many minutes; see Chapters 5 and 6, this volume).

Distinct Functional Types of Dopamine Midbrain Neurons: Role of Ion Channels

Dopamine midbrain neurons that possess the non-classical, alternative phenotype show significant differences in spontaneous and evoked electrical activity in vitro in synaptic isolation. Given the central roles of distinct ion channels just described for defining the intrinsic activity pattern of dopamine neurons[170], these intrinsic functional differences are very likely to reflect intrinsic differences in their ion channel expression. In this section, we summarize how the differential expression and activation of distinct ion channels (again, with a focus on somatodendritic ion channels) contributes to the functional diversity of dopamine midbrain neurons within the just described dual system.

Alternative dopamine midbrain neurons display significantly faster and less regular pacemaker activity than those with the classic phenotype.[77] This more irregular activity is associated with a nearly complete absence of SK channel–mediated AHP,[50,77] which in classic dopamine midbrain neurons is, as discussed, essential for their pacemaker stability. Accordingly, inhibition of SK channels has no effects on the spontaneous pacemaker discharge in alternative dopamine midbrain neurons in vitro.[50] Molecular and immunocytochemical data have confirmed a very low expression of SK3, the dominant SK channel subunit in classical dopamine midbrain neurons.[50] This lack of functional SK channels might also indicate that alternative dopamine midbrain neurons utilize different mechanisms to switch between tonic pacemaker and phasic burst activity. Possibly their distinct resonant properties enable burst firing in response to phasic synaptic excitation without the additional need to modify intrinsic ion channel conductances.[79] Indeed, alternative dopamine midbrain neurons in vitro are able to generate significantly faster sustained maximal firing rates in response to current injections (in the beta range of 15–30 Hz) compared to classic dopamine neurons (with depolarization block above 10 Hz).[77] In vivo depolarization block of dopamine neurons is a candidate mechanism for the action of antipsychotics, but the underlying ionic mechanisms remain unclear.[171]

As pacemaker activity and burst switching are significantly different in the two types of dopamine midbrain neurons, one would speculate that the respective cores of the membrane potential oscillators are built by different sets of ion channels. However, differences in voltage-gated sodium and calcium channels have not yet been systematically compared for the two functional phenotypes of dopamine midbrain neurons. Nevertheless, as pacemaker activity in the homogeneous population of classic dopamine neurons can be generated by two distinct mechanisms (Ca$_V$1.3 or HCN dependent; see above),[99,100] respective differences are also expected for the alternative dopamine neurons.

However, differences between HCN and A-type channels, which antagonistically tune pacemaker frequency in classic dopamine neurons, as well as inhibitory D2-coupled Kir3.2 channels, have been systematically compared between the two types of dopamine midbrain neurons. In accordance with the very small or completely absent (HCN-mediated) sag component upon injection of hyperpolarizing currents in alternative dopamine midbrain neurons, functional HCN channels are either expressed in very low abundance or are absent in these neurons.[77,138] Consequently, HCN channel inhibition does not affect the spontaneous pacemaker frequency of alternative dopamine neurons, as it does in the classic type.[135,138] The gating properties of I_H currents, however, show no significant kinetic differences between the two types of dopamine midbrain neurons.[138] In contrast, functional A-type potassium channels are expressed in classic as well as alternative dopamine midbrain neurons. However, our preliminary data show that these channels in alternative dopamine midbrain neurons have a distinct molecular composition[172] and do not contribute to pacemaker frequency control.[173] Accordingly, A-type potassium channels in alternative dopamine midbrain neurons display biophysical properties distinct from those of classical dopamine neurons; in particular, their inactivation time constant is significantly slower.[173] In summary, none of the three prominent subthreshold channels that control pacemaker frequency and regularity in vitro in classic dopamine midbrain neurons have a similar function in alternative dopamine neurons.

In regard to the D2-Kir3.2-mediated dopamine autoinhibition of pacemaker activity of dopamine midbrain neurons, the alternative subpopulation of VTA dopamine neurons projecting to the prefrontal cortex of adult mice is unique: it is the only subpopulation of classical and alternative dopamine neurons that neither responded to dopamine or D2 agonist nor expressed D2 or Kir3.2 (mRNA/protein) at significant levels.[77] In consequence, pacemaker activity of mesocortcial VTA dopamine neurons in vivo will only be indirectly affected by D2 agonists or D2 antagonists—drugs that are used in the treatment of PD (see Chapter 9, this volume) or schizophrenia (see Chapter 10, this volume). These findings match those of earlier in vivo studies in rats that had identified mesocortical dopamine midbrain neurons, which did not change their impulse rate in response to systemically applied D2 agonists.[174] However, our in vitro findings are in contrast to those of another in vitro study on younger rats, which reported that instead of mesocortical dopamine neurons those dopamine neurons projecting to the amygdala were not responsive to D2 agonists.[167]

In summary, in contrast to classic dopamine neurons, pacemaker activity of alternative dopamine neurons from adult mice is regulated by neither SK nor by HCN or A-type Kv4 channels. The complete absence of D2-Kir3.2 autoinhibition of pacemaker activity, however, is found only in a subpopulation of alternative dopamine neurons projecting to the prefrontal cortex. The borders between these different functional phenotypes of dopamine midbrain neurons might, however, be fluent and flexible and further shaped by synaptic inputs—for example, during ontogenetic development (see Chapter 4, this volume)—as well as due to functional plasticity in health and disease states. Indeed, all key subthreshold ion channels for dopamine midbrain neurons, discussed here in detail (SK, HCN, Kv4, Kir3), have been shown to be subject to plastic changes in other central neurons.[175–178]

Are the respective electrophysiological properties of dopamine subtypes in general stabilized by powerful homeostatic mechanisms like those described recently for classic dopamine neurons from $Ca_V1.3$ channel knockout mice,[99] or are dopamine midbrain neurons able to switch functional phenotypes in a compensatory response to (patho-physiological) challenges? Currently, we have little information on these important questions. However, the first examples of these types of plasticity have already been described for dopamine midbrain neurons in physiological[179] and pathophysiological scenarios.[99,152] In the last section, we will summarize and discuss the contributions of stable or flexible differences between electrophysiological properties of dopamine midbrain subtypes to their different fates in the pathophysiology of PD.

DIFFERENTIAL VULNERABILITY OF DOPAMINE MIDBRAIN NEURONS TO DEGENERATION

Differential Vulnerability of Dopamine Midbrain Neurons in Parkinson's disease

Dopamine midbrain neurons not only display distinct functions and activity patterns, they also have different fates in diseases that target the dopamine system. One prominent example is the different fate of dopamine midbrain neurons in PD and its chronic animal models (see Chapter 9.1–9.4, this volume). Classical dopamine neurons within the SN, projecting to the dorsolateral striatum (mesostriatal pathway), are almost completely lost, while those dopamine midbrain neurons that constitute the mesolimbic or mesocortical dopaminergic pathways are significantly less affected by degeneration throughout the course of the disease.[180] The mesocortical dopamine system might even display hyperactivity at early stages of PD.[181] This differential vulnerability of the dopamine midbrain system reflects an essential problem of current pharmacological strategies that target the dopamine system, for example, in PD or schizophrenia: they do not account for the selective involvement of distinct subpopulations of the dopamine system in disease processes. Consequently, conventional pharmacotherapy often burdens patients with a high rate of side effects.[182] However, although the causes for the selective vulnerability of the dopamine midbrain system are still unknown, electrophysiological differences and related differential activity of ion channels are emerging candidate mechanisms. In particular, findings by Jim Surmeier's group and our own work have identified two ion channels as important players: L-type Ca^{2+} channels and K-ATP channels, both involved in pacemaker control of dopamine midbrain neurons.

Differential Vulnerability of Dopamine Midbrain Neurons to Degeneration: Role of K-ATP and L-Type Calcium Channels

Mitochondrial dysfunction appears to be at the heart of the pathogenesis of idiopathic and toxin-induced cases of PD, as well as that of several monogenic familial PD forms (see Chapter 9.1, this volume). Acute in vitro challenges of dopamine midbrain neurons from adult mice with toxins that perturb mitochondrial function and induce dopaminergic degeneration and parkinsonism in vivo (like rotenone and MPTP/MPP+)[*] revealed distinct acute in vitro responses: while the spontaneous electrical activity of less vulnerable mesolimbic dopamine neurons (displaying the alternative functional phenotype) was not affected by toxin concentrations sufficient to induce neurodegeneration in vivo, the electrical activity of highly vulnerable classic mesostriatal dopamine neurons was dramatically altered.[152] Selective activation of K-ATP channels in vitro in these classic dopamine midbrain neurons tonically hyperpolarized their membrane potential and completely prevented action potential generation in response to PD toxins.[152] Studies in K-ATP channel knockout mice demonstrated that functional K-ATP channels were necessary mediators of this in vitro response.[152] Quantitative single-cell analysis showed that K-ATP channel subunits (mRNA for SUR1 and Kir6.2) were expressed at about twofold higher levels in vulnerable mesostriatal SN dopamine neurons compared to more resistant mesolimbic VTA dopamine neurons.[152] The selective activation of K-ATP channels only in classic SN dopamine midbrain neurons, however, appeared to be controlled by oxidative upstream mechanisms including different degrees of uncoupling of the mitochondrial membrane potential and the differential expression of the uncoupling protein UCP2.[150,152] If maintaining regular electrical activity of dopamine neurons in vivo is important for their survival, the high vulnerability of mesostriatal SN dopamine neurons but not mesolimbic VTA dopamine neurons to cell death in PD should be altered in K-ATP channel knockout mice. This potential role of selective K-ATP channel activation in vivo was validated in a chronic MPTP PD mouse model, where a complete and selective rescue of highly vulnerable SN dopamine neurons in K-ATP knockout mice was demonstrated. By contrast, the minor loss of VTA dopamine neurons in vivo was not affected by the presence or absence of K-ATP channels.[152] A similarly selective but only partial rescue was obtained in a mechanistically independent genetic model of early postnatal neurodegeneration of dopamine neurons, the *weaver* mouse.[141,152]

Chan and colleagues identified a second, channel-based mechanism for differential vulnerability among dopamine midbrain neurons.[99] By analyzing a $Ca_V1.3$ knockout mouse, they showed that classic SN dopamine midbrain neurons from adolescent mice in vitro continued to generate spontaneous pacemaker activity due to a switch from L-type Ca^{2+} to HCN channel–driven pacemaking. They further demonstrated that in vivo treatment with L-type channel inhibitors significantly reduced the loss of these neurons in vivo in a chronic MPTP PD mouse model, presumably via a corresponding drug-induced pacemaker switch ("rejuvenation") in SN dopamine midbrain neurons.[99] These findings match those of earlier work by Kupsch and colleagues, who showed that local L-type Ca^{2+} channel inhibition in vivo in the SN but not in the striatum reduced MPTP toxicity and degeneration of dopamine neurons in MPTP models in mice and monkeys.[183,184]

In summary, selective K-ATP channel activation as well as L-type Ca^{2+} channel activation in highly vulnerable classic SN dopamine midbrain neurons provide functional candidate mechanisms for the differential vulnerability of dopamine midbrain neurons in PD. However, in both studies, general knockout mice were analyzed, and it was found that K-ATP channels as well as L-type Ca^{2+} channels are abundantly expressed in many neurons and other cell types. Thus, it remains to be shown in both cases that L-type Ca^{2+} channels and K-ATP channels present on SN dopamine midbrain neurons are necessary and sufficient to promote the neurodegeneration of these highly vulnerable nerve cells. To address this question, the use of new generations of cell type–selective knockout mice or similar tools is necessary.[33,34] Similarly, the proposed roles of K-ATP and L-type Ca^{2+} channels for the physiological control of firing frequencies and patterns in vivo need to be clarified. Mechanistically, electrical silencing or reduction of pacemaker activity by K-ATP channel activation might be a prerequisite for triggering neurodegeneration of highly vulnerable dopamine midbrain neurons in vivo. Indeed,

---

[*]MPTP: 1-methyl-4-phenyl-1,2,3,6-tetrahydropyridine; MPP+: 1-methyl-4-phenylpyridinium

a previous study has demonstrated that electrical activity and the associated influx of sodium and calcium ions was necessary for the survival of dopamine neurons, at least in vitro.[185] Thus, switching from a Ca^{2+} channel–driven pacemaker to a metabolically less demanding HCN-based pacemaker[99,186] could help to keep K-ATP channels closed and thus to maintain electrical activity of dopamine midbrain neurons in pathophysiologically challenging situations—which might help sustain the survival of vulnerable dopamine neurons throughout the disease process of PD.

ACKNOWLEDGMENTS

This work was supported by grants from the Gemeinnützige Hertie Stiftung (Hertie Foundation), the Nationales Genomforschungs Netzwerk (NGFN-II/-plus), the DFG (SFB 497 and SFB 815), and the Alfried Krupp von Bohlen und Halbach Foundation.

REFERENCES

1. Hille B. *Ion Channels of Excitable Membranes*. 3rd ed. Sunderland, MA: Sinauer Associates; 2001.
2. Sudhof TC. The synaptic vesicle cycle. *Annu Rev Neurosci*. 2004;27:509–547.
3. Rice ME, Cragg SJ. Dopamine spillover after quantal release: rethinking dopamine transmission in the nigrostriatal pathway. *Brain Res Rev*. 2008;58(2):303–313.
4. Bean BP. The action potential in mammalian central neurons. *Nat Rev Neurosci*. 2007;8(6):451–465.
5. Wightman RM. Detection technologies. Probing cellular chemistry in biological systems with microelectrodes. *Science*. 2006;311(5767):1570–1574.
6. Izhikevich EM. *Dynamical Systems in Neuroscience*. Cambridge, MA: MIT Press; 2007.
7. Waltz W. *Patch-Clamp Analysis: Advanced Techniques*. Totowa, NJ: Humana Press; 2007.
8. Lee AK, Manns ID, Sakmann B, Brecht M. Whole-cell recordings in freely moving rats. *Neuron*. 2006;51(4):399–407.
9. Aebischer P, Schultz W. The activity of pars compacta neurons of the monkey substantia nigra is depressed by apomorphine. *Neurosci Lett*. 1984;50(1–3):25–29.
10. Roesch MR, Calu DJ, Schoenbaum G. Dopamine neurons encode the better option in rats deciding between differently delayed or sized rewards. *Nat Neurosci*. 2007;10(12):1615–1624.
11. Hyland BI, Reynolds JN, Hay J, Perk CG, Miller R. Firing modes of midbrain dopamine cells in the freely moving rat. *Neuroscience*. 2002;114(2):475–492.
12. Grace AA, Bunney BS. Nigral dopamine neurons: intracellular recording and identification with L-dopa injection and histofluorescence. *Science*. 1980;210(4470):654–656.
13. Grace AA, Onn SP. Morphology and electrophysiological properties of immunocytochemically identified rat dopamine neurons recorded in vitro. *J Neurosci*. 1989;9(10):3463–3481.
14. Silva NL, Pechura CM, Barker JL. Postnatal rat nigrostriatal dopaminergic neurons exhibit five types of potassium conductances. *J Neurophysiol*. 1990;64(1):262–272.
15. Staal RG, Mosharov EV, Sulzer D. Dopamine neurons release transmitter via a flickering fusion pore. *Nat Neurosci*. 2004;7(4):341–346.
16. Tepper JM, Lee CR. GABAergic control of substantia nigra dopaminergic neurons. *Prog Brain Res*. 2007;160:189–208.
17. Matsuda W, Furuta T, Nakamura KC, et al. Single nigrostriatal dopaminergic neurons form widely spread and highly dense axonal arborizations in the neostriatum. *J Neurosci*. 2009;29(2):444–453.
18. Perkins KL. Cell-attached voltage-clamp and current-clamp recording and stimulation techniques in brain slices. *J Neurosci Methods*. 2006;154(1–2):1–18.
19. Prinz AA, Abbott LF, Marder E. The dynamic clamp comes of age. *Trends Neurosci*. 2004;27(4):218–224.
20. Herz AV, Gollisch T, Machens CK, Jaeger D. Modeling single-neuron dynamics and computations: a balance of detail and abstraction. *Science*. 2006;314(5796):80–85.
21. Amini B, Clark JW Jr, Canavier CC. Calcium dynamics underlying pacemaker-like and burst firing oscillations in midbrain dopaminergic neurons: a computational study. *J Neurophysiol*. 1999;82(5):2249–2261.
22. Wilson CJ, Callaway JC. Coupled oscillator model of the dopaminergic neuron of the substantia nigra. *J Neurophysiol*. 2000;83(5):3084–3100.
23. Milescu LS, Yamanishi T, Ptak K, Mogri MZ, Smith JC. Real-time kinetic modeling of voltage-gated ion channels using dynamic clamp. *Biophys J*. 2008;95(1):66–87.
24. Putzier I, Kullmann PH, Horn JP, Levitan ES. Dopamine neuron responses depend exponentially on pacemaker interval. *J Neurophysiol*. 2009;101(2):926–33.
25. Groves PM, Wilson CJ, Young SJ, Rebec GV. Self-inhibition by dopaminergic neurons. *Science*. 1975;190(4214):522–528.
26. Bunney BS, Walters JR, Roth RH, Aghajanian GK. Dopaminergic neurons: effect of antipsychotic drugs and amphetamine on single cell activity. *J Pharmacol Exp Ther*. 1973;185(3):560–571.
27. Bayer HM, Lau B, Glimcher PW. Statistics of midbrain dopamine neuron spike trains in the awake primate. *J Neurophysiol*. 2007;98(3):1428–1439.
28. Schultz W. Responses of midbrain dopamine neurons to behavioral trigger stimuli in the monkey. *J Neurophysiol*. 1986;56(5):1439–1461.
29. Steinfels GF, Heym J, Jacobs BL. Single unit activity of dopaminergic neurons in freely moving cats. *Life Sci*. 1981;29(14):1435–1442.
30. Wilson CJ, Young SJ, Groves PM. Statistical properties of neuronal spike trains in the substantia nigra: cell types and their interactions. *Brain Res*. 1977;136(2):243–260.
31. Grace AA, Bunney BS. The control of firing pattern in nigral dopamine neurons: burst firing. *J Neurosci*. 1984;4(11):2877–2890.
32. Redgrave P, Gurney K. The short-latency dopamine signal: a role in discovering novel actions? *Nat Rev Neurosci*. 2006;7(12):967–975.
33. Zweifel LS, Argilli E, Bonci A, Palmiter RD. Role of NMDA receptors in dopamine neurons for plasticity and addictive behaviors. *Neuron*. 2008;59(3):486–496.
34. Engblom D, Bilbao A, Sanchis-Segura C, et al. Glutamate receptors on dopamine neurons control the persistence of cocaine seeking. *Neuron*. 2008;59(3):497–508.

35. Grace AA, Bunney BS. The control of firing pattern in nigral dopamine neurons: single spike firing. *J Neurosci.* 1984;4(11):2866–2876.
36. Tepper JM, Sawyer SF, Groves PM. Electrophysiologically identified nigral dopaminergic neurons intracellularly labeled with HRP: light-microscopic analysis. *J Neurosci.* 1987;7(9):2794–2806.
37. Grace AA, Bunney BS. Intracellular and extracellular electrophysiology of nigral dopaminergic neurons–1. Identification and characterization. *Neuroscience.* 1983;10(2):301–315.
38. Grace AA, Bunney BS. Intracellular and extracellular electrophysiology of nigral dopaminergic neurons–2. Action potential generating mechanisms and morphological correlates. *Neuroscience.* 1983;10(2):317–331.
39. Hausser M, Stuart G, Racca C, Sakmann B. Axonal initiation and active dendritic propagation of action potentials in substantia nigra neurons. *Neuron.* 1995;15(3):637–647.
40. Steinfels GF, Heym J, Strecker RE, Jacobs BL. Response of dopaminergic neurons in cat to auditory stimuli presented across the sleep-waking cycle. *Brain Res.* 1983;277(1):150–154.
41. Dommett E, Coizet V, Blaha CD, et al. How visual stimuli activate dopaminergic neurons at short latency. *Science.* 2005;307(5714):1476–1479.
42. Redgrave P, Gurney K, Reynolds J. What is reinforced by phasic dopamine signals? *Brain Res Rev.* 2008;58(2):322–339.
43. Kelland MD, Chiodo LA, Freeman AS. Anesthetic influences on the basal activity and pharmacological responsiveness of nigrostriatal dopamine neurons. *Synapse.* 1990;6(2):207–209.
44. Fa M, Mereu G, Ghiglieri V, Meloni A, Salis P, Gessa GL. Electrophysiological and pharmacological characteristics of nigral dopaminergic neurons in the conscious, head-restrained rat. *Synapse.* 2003;48(1):1–9.
45. Chergui K, Charlety PJ, Akaoka H, et al. Tonic activation of NMDA receptors causes spontaneous burst discharge of rat midbrain dopamine neurons in vivo. *Eur J Neurosci.* 1993;5(2):137–144.
46. Overton P, Clark D. Iontophoretically administered drugs acting at the N-methyl-D-aspartate receptor modulate burst firing in A9 dopamine neurons in the rat. *Synapse.* 1992;10(2):131–140.
47. Chergui K, Akaoka H, Charlety PJ, Saunier CF, Buda M, Chouvet G. Subthalamic nucleus modulates burst firing of nigral dopamine neurones via NMDA receptors. *Neuroreport.* 1994;5(10):1185–1188.
48. Grace AA, Floresco SB, Goto Y, Lodge DJ. Regulation of firing of dopaminergic neurons and control of goal-directed behaviors. *Trends Neurosci.* 2007;30(5):220–227.
49. Lodge DJ, Grace AA. The laterodorsal tegmentum is essential for burst firing of ventral tegmental area dopamine neurons. *Proc Natl Acad Sci USA.* 2006;103(13):5167–5172.
50. Wolfart J, Neuhoff H, Franz O, Roeper J. Differential expression of the small-conductance, calcium-activated potassium channel SK3 is critical for pacemaker control in dopaminergic midbrain neurons. *J Neurosci.* 2001;21(10):3443–3456.
51. Waroux O, Massotte L, Alleva L, et al. SK channels control the firing pattern of midbrain dopaminergic neurons in vivo. *Eur J Neurosci.* 2005;22(12):3111–3121.
52. Ji H, Shepard PD. SK Ca^{2+}-activated K^+ channel ligands alter the firing pattern of dopamine-containing neurons in vivo. *Neuroscience.* 2006;140(2):623–633.
53. Koyama S, Brodie MS, Appel SB. Ethanol inhibition of m-current and ethanol-induced direct excitation of ventral tegmental area dopamine neurons. *J Neurophysiol.* 2007;97(3):1977–1985.
54. Cooper EC, Harrington E, Jan YN, Jan LY. M channel KCNQ2 subunits are localized to key sites for control of neuronal network oscillations and synchronization in mouse brain. *J Neurosci.* 2001;21(24):9529–9540.
55. Sotty F, Damgaard T, Montezinho LP, et al. Antipsychotic-like effect of retigabine, a KCNQ potassium channel opener, via modulation of mesolimbic dopaminergic neurotransmission. *J Pharmacol Exp Ther.* 2009;328(3):951–962.
56. Koyrakh L, Lujan R, Colon J, et al. Molecular and cellular diversity of neuronal G-protein-gated potassium channels. *J Neurosci.* 2005;25(49):11468–11478.
57. Brazhnik E, Shah F, Tepper JM. GABAergic afferents activate both GABAA and GABAB receptors in mouse substantia nigra dopaminergic neurons in vivo. *J Neurosci.* 2008;28(41):10386–10398.
58. Paladini CA, Tepper JM. GABA(A) and GABA(B) antagonists differentially affect the firing pattern of substantia nigra dopaminergic neurons in vivo. *Synapse.* 1999;32(3):165–176.
59. Levin BE. Glucose-regulated dopamine release from substantia nigra neurons. *Brain Res.* 2000;874(2):158–164.
60. Saller CF, Chiodo LA. Glucose suppresses basal firing and haloperidol-induced increases in the firing rate of central dopaminergic neurons. *Science.* 1980;210(4475):1269–1271.
61. Llinas R, Greenfield SA, Jahnsen H. Electrophysiology of pars compacta cells in the in vitro substantia nigra–a possible mechanism for dendritic release. *Brain Res.* 1984;294(1):127–132.
62. Silva NL, Bunney BS. Intracellular studies of dopamine neurons in vitro: pacemakers modulated by dopamine. *Eur J Pharmacol.* 1988;149(3):307–315.
63. Lacey MG, Mercuri NB, North RA. Dopamine acts on D2 receptors to increase potassium conductance in neurones of the rat substantia nigra zona compacta. *J Physiol.* 1987;392:397–416.
64. Kita T, Kita H, Kitai ST. Electrical membrane properties of rat substantia nigra compacta neurons in an in vitro slice preparation. *Brain Res.* 1986;372(1):21–30.
65. Pinnock RD. Hyperpolarizing action of baclofen on neurons in the rat substantia nigra slice. *Brain Res.* 1984;322(2):337–340.
66. Hausser MA, de Weille JR, Lazdunski M. Activation by cromakalim of pre- and post-synaptic ATP-sensitive K^+ channels in substantia nigra. *Biochem Biophys Res Commun.* 1991;174(2):909–914.
67. Mereu G, Costa E, Armstrong DM, Vicini S. Glutamate receptor subtypes mediate excitatory synaptic currents of dopamine neurons in midbrain slices. *J Neurosci.* 1991;11(5):1359–1366.
68. Hicks GA, Henderson G. Lack of evidence for coupling of the dopamine D2 receptor to an adenosine triphosphate–sensitive potassium (ATP-K^+) channel in dopaminergic neurones of the rat substantia nigra. *Neurosci Lett.* 1992;141(2):213–217.
69. Roper J, Ashcroft FM. Metabolic inhibition and low internal ATP activate K-ATP channels in rat dopaminergic substantia nigra neurones. *Pflugers Arch.* 1995;430(1):44–54.
70. Hainsworth AH, Roper J, Kapoor R, Ashcroft FM. Identification and electrophysiology of isolated pars compacta neurons from guinea-pig substantia nigra. *Neuroscience.* 1991;43(1):81–93.
71. Cardozo DL. Midbrain dopaminergic neurons from postnatal rat in long-term primary culture. *Neuroscience.* 1993;56(2):409–421.
72. Liu L, Shen RY, Kapatos G, Chiodo LA. Dopamine neuron membrane physiology: characterization of the transient outward current (IA) and demonstration of a common signal transduction pathway for IA and IK. *Synapse.* 1994;17(4):230–240.
73. Chiodo LA, Kapatos G. Membrane properties of identified mesencephalic dopamine neurons in primary dissociated cell culture. *Synapse.* 1992;11(4):294–309.

74. Gentet LJ, Williams SR. Dopamine gates action potential backpropagation in midbrain dopaminergic neurons. *J Neurosci.* 2007;27(8):1892–1901.
75. Rice ME, Cragg SJ, Greenfield SA. Characteristics of electrically evoked somatodendritic dopamine release in substantia nigra and ventral tegmental area in vitro. *J Neurophysiol.* 1997;77(2):853–862.
76. Nedergaard S. Regulation of action potential size and excitability in substantia nigra compacta neurons: sensitivity to 4-aminopyridine. *J Neurophysiol.* 1999;82(6):2903–2913.
77. Lammel S, Hetzel A, Hackel O, Jones I, Liss B, Roeper J. Unique properties of mesoprefrontal neurons within a dual mesocorticolimbic dopamine system. *Neuron.* 2008;57(5):760–773.
78. Richards CD, Shiroyama T, Kitai ST. Electrophysiological and immunocytochemical characterization of GABA and dopamine neurons in the substantia nigra of the rat. *Neuroscience.* 1997;80(2):545–557.
79. Kuznetsov AS, Kopell NJ, Wilson CJ. Transient high-frequency firing in a coupled-oscillator model of the mesencephalic dopaminergic neuron. *J Neurophysiol.* 2006;95(2):932–947.
80. Shepard PD, Bunney BS. Repetitive firing properties of putative dopamine-containing neurons in vitro: regulation by an apamin-sensitive Ca(2+)-activated K+ conductance. *Exp Brain Res.* 1991;86(1):141–150.
81. Liss B, Roeper J. Correlating function and gene expression of individual basal ganglia neurons. *Trends Neurosci.* 2004;27(8):475–481.
82. Nedergaard S, Flatman JA, Engberg I. Nifedipine- and omega-conotoxin-sensitive Ca^{2+} conductances in guinea-pig substantia nigra pars compacta neurones. *J Physiol.* 1993;466:727–747.
83. Stanfield PR, Nakajima S, Nakajima Y. Constitutively active and G-protein coupled inward rectifier K+ channels: Kir2.0 and Kir3.0. *Rev Physiol Biochem Pharmacol.* 2002;145:47–179.
84. Bayliss DA, Barrett PQ. Emerging roles for two-pore-domain potassium channels and their potential therapeutic impact. *Trends Pharmacol Sci.* 2008;29(11):566–575.
85. Karschin C, Dissmann E, Stuhmer W, Karschin A. IRK(1-3) and GIRK(1-4) inwardly rectifying K+ channel mRNAs are differentially expressed in the adult rat brain. *J Neurosci.* 1996;16(11):3559–3570.
86. Talley EM, Solorzano G, Lei Q, Kim D, Bayliss DA. CNS distribution of members of the two-pore-domain (KCNK) potassium channel family. *J Neurosci.* 2001;21(19):7491–7505.
87. Talavera K, Nilius B, Voets T. Neuronal TRP channels: thermometers, pathfinders and life-savers. *Trends Neurosci.* 2008;31(6):287–295.
88. Rossier BC, Stutts MJ. Activation of the epithelial sodium channel (ENaC) by serine proteases. *Annu Rev Physiol.* 2009;71:361–379.
89. Chien PY, Farkas RH, Nakajima S, Nakajima Y. Single-channel properties of the nonselective cation conductance induced by neurotensin in dopaminergic neurons. *Proc Natl Acad Sci USA.* 1996;93(25):14917–14921.
90. Guatteo E, Chung KK, Bowala TK, Bernardi G, Mercuri NB, Lipski J. Temperature sensitivity of dopaminergic neurons of the substantia nigra pars compacta: involvement of transient receptor potential channels. *J Neurophysiol.* 2005;94(5):3069–3080.
91. Marinelli S, Pascucci T, Bernardi G, Puglisi-Allegra S, Mercuri NB. Activation of TRPV1 in the VTA excites dopaminergic neurons and increases chemical- and noxious-induced dopamine release in the nucleus accumbens. *Neuropsychopharmacology.* 2005;30(5):864–870.
92. Bengtson CP, Tozzi A, Bernardi G, Mercuri NB. Transient receptor potential-like channels mediate metabotropic glutamate receptor EPSCs in rat dopamine neurones. *J Physiol.* 2004;555(Pt 2):323–330.
93. Mercuri NB, Stratta F, Calabresi P, Bernardi G. Neurotensin induces an inward current in rat mesencephalic dopaminergic neurons. *Neurosci Lett.* 1993;153(2):192–196.
94. Kim SH, Choi YM, Jang JY, Chung S, Kang YK, Park MK. Nonselective cation channels are essential for maintaining intracellular Ca^{2+} levels and spontaneous firing activity in the midbrain dopamine neurons. *Pflugers Arch.* 2007;455(2):309–321.
95. Arias RL, Sung ML, Vasylyev D, et al. Amiloride is neuroprotective in an MPTP model of Parkinson's disease. *Neurobiol Dis.* 2008;31(3):334–341.
96. Pidoplichko VI, Dani JA. Acid-sensitive ionic channels in midbrain dopamine neurons are sensitive to ammonium, which may contribute to hyperammonemia damage. *Proc Natl Acad Sci USA.* 2006;103(30):11376–11380.
97. Mechaly I, Scamps F, Chabbert C, Sans A, Valmier J. Molecular diversity of voltage-gated sodium channel alpha subunits expressed in neuronal and non-neuronal excitable cells. *Neuroscience.* 2005;130(2):389–396.
98. Johnson D, Bennett ES. Isoform-specific effects of the beta2 subunit on voltage-gated sodium channel gating. *J Biol Chem.* 2006;281(36):25875–25881.
99. Chan CS, Guzman JN, Ilijic E, et al. "Rejuvenation" protects neurons in mouse models of Parkinson's disease. *Nature.* 2007;447(7148):1081–1086.
100. Puopolo M, Raviola E, Bean BP. Roles of subthreshold calcium current and sodium current in spontaneous firing of mouse midbrain dopamine neurons. *J Neurosci.* 2007;27(3):645–656.
101. Striessnig J, Koschak A, Sinnegger-Brauns MJ, et al. Role of voltage-gated L-type Ca^{2+} channel isoforms for brain function. *Biochem Soc Trans.* 2006;34(pt 5):903–909.
102. Dolphin AC. A short history of voltage-gated calcium channels. *Br J Pharmacol.* 2006;147(suppl 1):S56–62.
103. Yellen G. The voltage-gated potassium channels and their relatives. *Nature.* 2002;419(6902):35–42.
104. Delmas P, Brown DA. Pathways modulating neural KCNQ/M (Kv7) potassium channels. *Nat Rev Neurosci.* 2005;6(11):850–862.
105. Koyama S, Appel SB. Characterization of M-current in ventral tegmental area dopamine neurons. *J Neurophysiol.* 2006;96(2):535–543.
106. Birnbaum SG, Varga AW, Yuan LL, Anderson AE, Sweatt JD, Schrader LA. Structure and function of Kv4-family transient potassium channels. *Physiol Rev.* 2004;84(3):803–833.
107. Morikawa H, Imani F, Khodakhah K, Williams JT. Inositol 1,4,5-triphosphate-evoked responses in midbrain dopamine neurons. *J Neurosci.* 2000;20(20):RC103.
108. Fakler B, Adelman JP. Control of K(Ca) channels by calcium nano/microdomains. *Neuron.* 2008;59(6):873–881.
109. Perez-Reyes E. Molecular characterization of T-type calcium channels. *Cell Calcium.* 2006;40(2):89–96.
110. Cui G, Okamoto T, Morikawa H. Spontaneous opening of T-type Ca^{2+} channels contributes to the irregular firing of dopamine neurons in neonatal rats. *J Neurosci.* 2004;24(49):11079–11087.
111. Wolfart J, Roeper J. Selective coupling of T-type calcium channels to SK potassium channels prevents intrinsic bursting in dopaminergic midbrain neurons. *J Neurosci.* 2002;22(9):3404–3413.
112. Cui G, Bernier BE, Harnett MT, Morikawa H. Differential regulation of action potential- and metabotropic glutamate receptor-induced Ca^{2+} signals by inositol 1,4,5-trisphosphate in dopaminergic neurons. *J Neurosci.* 2007;27(17):4776–4785.
113. Morikawa H, Khodakhah K, Williams JT. Two intracellular pathways mediate metabotropic glutamate receptor-induced

114. Ping HX, Shepard PD. Blockade of SK-type Ca^{2+}-activated K+ channels uncovers a Ca^{2+}-dependent slow afterdepolarization in nigral dopamine neurons. *J Neurophysiol.* 1999;81(3):977–984.
115. Shepard PD, Trudeau MC. Emerging roles for ether-a-go-go-related gene potassium channels in the brain. *J Physiol.* 2008;586(Pt 20):4785–4786.
116. Nedergaard S. A Ca^{2+}-independent slow afterhyperpolarization in substantia nigra compacta neurons. *Neuroscience.* 2004;125(4):841–852.
117. Arencibia-Albite F, Paladini C, Williams JT, Jimenez-Rivera CA. Noradrenergic modulation of the hyperpolarization-activated cation current (Ih) in dopamine neurons of the ventral tegmental area. *Neuroscience.* 2007;149(2):303–314.
118. Paladini CA, Fiorillo CD, Morikawa H, Williams JT. Amphetamine selectively blocks inhibitory glutamate transmission in dopamine neurons. *Nat Neurosci.* 2001;4(3):275–281.
119. Paladini CA, Williams JT. Noradrenergic inhibition of midbrain dopamine neurons. *J Neurosci.* 2004;24(19):4568–4575.
120. Fiorillo CD, Williams JT. Glutamate mediates an inhibitory postsynaptic potential in dopamine neurons. *Nature.* 1998;394(6688):78–82.
121. Mercuri NB, Stratta F, Calabresi P, Bonci A, Bernardi G. Activation of metabotropic glutamate receptors induces an inward current in rat dopamine mesencephalic neurons. *Neuroscience.* 1993;56(2):399–407.
122. Cathala L, Guyon A, Eugene D, Paupardin-Tritsch D. Alpha2-adrenoceptor activation increases a cationic conductance and spontaneous GABAergic synaptic activity in dopaminergic neurones of the rat substantia nigra. *Neuroscience.* 2002;115(4):1059–1065.
123. Wu T, Wang HL. G alpha q/11 mediates cholecystokinin activation of the cationic conductance in rat substantia nigra dopaminergic neurons. *J Neurochem.* 1996;66(3):1060–1066.
124. Pinnock RD. Neurotensin depolarizes substantia nigra dopamine neurones. *Brain Res.* 1985;338(1):151–154.
125. Lu B, Su Y, Das S, et al. Peptide neurotransmitters activate a cation channel complex of NALCN and UNC-80. *Nature.* 2009;457(7230):741–744.
126. Jerng HH, Pfaffinger PJ, Covarrubias M. Molecular physiology and modulation of somatodendritic A-type potassium channels. *Mol Cell Neurosci.* 2004;27(4):343–369.
127. Siu CW, Lieu DK, Li RA. HCN-encoded pacemaker channels: from physiology and biophysics to bioengineering. *J Membr Biol.* 2006;214(3):115–122.
128. Feltz A. Low-threshold-activated Ca channels: from molecules to functions: over 25 years of progress. *Crit Rev Neurobiol.* 2006;18(1–2):169–178.
129. Liss B, Franz O, Sewing S, Bruns R, Neuhoff H, Roeper J. Tuning pacemaker frequency of individual dopaminergic neurons by Kv4.3L and KChip3.1 transcription. *EMBO J.* 2001;20(20):5715–5724.
130. Khaliq ZM, Bean BP. Dynamic, nonlinear feedback regulation of slow pacemaking by A-type potassium current in ventral tegmental area neurons. *J Neurosci.* 2008;28(43):10905–10917.
131. Jackson AC, Bean BP. State-dependent enhancement of subthreshold A-type potassium current by 4-aminopyridine in tuberomammillary nucleus neurons. *J Neurosci.* 2007;27(40):10785–10796.
132. Hahn J, Kullmann PH, Horn JP, Levitan ES. D2 autoreceptors chronically enhance dopamine neuron pacemaker activity. *J Neurosci.* 2006;26(19):5240–5247.
133. Hahn J, Tse TE, Levitan ES. Long-term K+ channel-mediated dampening of dopamine neuron excitability by the antipsychotic drug haloperidol. *J Neurosci.* 2003;23(34):10859–10866.
134. Cheng HY, Pitcher GM, Laviolette SR, et al. DREAM is a critical transcriptional repressor for pain modulation. *Cell.* 2002;108(1):31–43.
135. Franz O, Liss B, Neu A, Roeper J. Single-cell mRNA expression of HCN1 correlates with a fast gating phenotype of hyperpolarization-activated cyclic nucleotide-gated ion channels (Ih) in central neurons. *Eur J Neurosci.* 2000;12(8):2685–2693.
136. Zolles G, Klocker N, Wenzel D, et al. Pacemaking by HCN channels requires interaction with phosphoinositides. *Neuron.* 2006;52(6):1027–1036.
137. Seutin V, Massotte L, Renette MF, Dresse A. Evidence for a modulatory role of Ih on the firing of a subgroup of midbrain dopamine neurons. *Neuroreport.* 2001;12(2):255–258.
138. Neuhoff H, Neu A, Liss B, Roeper J. I(h) channels contribute to the different functional properties of identified dopaminergic subpopulations in the midbrain. *J Neurosci.* 2002;22(4):1290–1302.
139. Nedergaard S, Flatman JA, Engberg I. Excitation of substantia nigra pars compacta neurones by 5-hydroxy-tryptamine in-vitro. *Neuroreport.* 1991;2(6):329–332.
140. Lacey MG, Mercuri NB, North RA. On the potassium conductance increase activated by $GABA_B$ and dopamine D2 receptors in rat substantia nigra neurones. *J Physiol.* 1988;401:437–453.
141. Liss B, Neu A, Roeper J. The weaver mouse gain-of-function phenotype of dopaminergic midbrain neurons is determined by coactivation of wvGirk2 and K-ATP channels. *J Neurosci.* 1999;19(20):8839–8848.
142. Cruz HG, Ivanova T, Lunn ML, Stoffel M, Slesinger PA, Luscher C. Bi-directional effects of GABA(B) receptor agonists on the mesolimbic dopamine system. *Nat Neurosci.* 2004;7(2):153–159.
143. Labouebe G, Lomazzi M, Cruz HG, et al. RGS2 modulates coupling between $GABA_B$ receptors and GIRK channels in dopamine neurons of the ventral tegmental area. *Nat Neurosci.* 2007;10(12):1559–1568.
144. Beckstead MJ, Grandy DK, Wickman K, Williams JT. Vesicular dopamine release elicits an inhibitory postsynaptic current in midbrain dopamine neurons. *Neuron.* 2004;42(6):939–946.
145. Beckstead MJ, Williams JT. Long-term depression of a dopamine IPSC. *J Neurosci.* 2007;27(8):2074–2080.
146. Sahlholm K, Nilsson J, Marcellino D, Fuxe K, Arhem P. Voltage-dependence of the human dopamine D2 receptor. *Synapse.* 2008;62(6):476–480.
147. Abizaid A, Liu ZW, Andrews ZB, et al. Ghrelin modulates the activity and synaptic input organization of midbrain dopamine neurons while promoting appetite. *J Clin Invest.* 2006;116(12):3229–3239.
148. Korotkova TM, Brown RE, Sergeeva OA, Ponomarenko AA, Haas HL. Effects of arousal- and feeding-related neuropeptides on dopaminergic and GABAergic neurons in the ventral tegmental area of the rat. *Eur J Neurosci.* 2006;23(10):2677–2685.
149. Liss B, Bruns R, Roeper J. Alternative sulfonylurea receptor expression defines metabolic sensitivity of K-ATP channels in dopaminergic midbrain neurons. *EMBO J.* 1999;18(4):833–846.
150. Avshalumov MV, Chen BT, Koos T, Tepper JM, Rice ME. Endogenous hydrogen peroxide regulates the excitability of midbrain dopamine neurons via ATP-sensitive potassium channels. *J Neurosci.* 2005;25(17):4222–4231.
151. Liss B, Roeper J. ATP-sensitive potassium channels in dopaminergic neurons: transducers of mitochondrial dysfunction. *News Physiol Sci.* 2001;16:214–217.

152. Liss B, Haeckel O, Wildmann J, Miki T, Seino S, Roeper J. K-ATP channels promote the differential degeneration of dopaminergic midbrain neurons. *Nat Neurosci.* 2005;8(12):1742–1751.
153. Plum L, Ma X, Hampel B, et al. Enhanced PIP3 signaling in POMC neurons causes KATP channel activation and leads to diet-sensitive obesity. *J Clin Invest.* 2006;116(7):1886–1901.
154. Sherman SM. Tonic and burst firing: dual modes of thalamocortical relay. *Trends Neurosci.* 2001;24(2):122–126.
155. Blythe SN, Atherton JF, Bevan MD. Synaptic activation of dendritic AMPA and NMDA receptors generates transient high-frequency firing in substantia nigra dopamine neurons in vitro. *J Neurophysiol.* 2007;97(4):2837–2850.
156. Johnson SW, Seutin V, North RA. Burst firing in dopamine neurons induced by N-methyl-D-aspartate: role of electrogenic sodium pump. *Science.* 1992;258(5082):665–667.
157. Shepard PD, Bunney BS. Effects of apamin on the discharge properties of putative dopamine-containing neurons in vitro. *Brain Res.* 1988;463(2):380–384.
158. Kitai ST, Shepard PD, Callaway JC, Scroggs R. Afferent modulation of dopamine neuron firing patterns. *Curr Opin Neurobiol.* 1999;9(6):690–697.
159. Seutin V, Johnson SW, North RA. Apamin increases NMDA-induced burst-firing of rat mesencephalic dopamine neurons. *Brain Res.* 1993;630(1–2):341–344.
160. Johnson SW, Wu YN. Multiple mechanisms underlie burst firing in rat midbrain dopamine neurons in vitro. *Brain Res.* 2004;1019(1–2):293–296.
161. Canavier CC, Oprisan SA, Callaway JC, Ji H, Shepard PD. Computational model predicts a role for ERG current in repolarizing plateau potentials in dopamine neurons: implications for modulation of neuronal activity. *J Neurophysiol.* 2007;98(5):3006–3022.
162. Liu Y, Dore J, Chen X. Calcium influx through L-type channels generates protein kinase M to induce burst firing of dopamine cells in the rat ventral tegmental area. *J Biol Chem.* 2007;282(12):8594–8603.
163. Canavier CC, Landry RS. An increase in AMPA and a decrease in SK conductance increase burst firing by different mechanisms in a model of a dopamine neuron in vivo. *J Neurophysiol.* 2006;96(5):2549–2563.
164. Komendantov AO, Komendantova OG, Johnson SW, Canavier CC. A modeling study suggests complementary roles for $GABA_A$ and NMDA receptors and the SK channel in regulating the firing pattern in midbrain dopamine neurons. *J Neurophysiol.* 2004;91(1):346–357.
165. Shepard PD, German DC. Electrophysiological and pharmacological evidence for the existence of distinct subpopulations of nigrostriatal dopaminergic neuron in the rat. *Neuroscience.* 1988;27(2):537–546.
166. Margolis EB, Lock H, Hjelmstad GO, Fields HL. The ventral tegmental area revisited: is there an electrophysiological marker for dopaminergic neurons? *J Physiol.* 2006;577(pt 3):907–924.
167. Margolis EB, Mitchell JM, Ishikawa J, Hjelmstad GO, Fields HL. Midbrain dopamine neurons: projection target determines action potential duration and dopamine D(2) receptor inhibition. *J Neurosci.* 2008;28(36):8908–8913.
168. Margolis EB, Lock H, Chefer VI, Shippenberg TS, Hjelmstad GO, Fields HL. Kappa opioids selectively control dopaminergic neurons projecting to the prefrontal cortex. *Proc Natl Acad Sci USA.* 2006;103(8):2938–2942.
169. Ford CP, Mark GP, Williams JT. Properties and opioid inhibition of mesolimbic dopamine neurons vary according to target location. *J Neurosci.* 2006;26(10):2788–2797.
170. Liss B, Roeper J. Individual dopamine midbrain neurons: functional diversity and flexibility in health and disease. *Brain Res Rev.* 2008;58(2):314–321.
171. Grace AA, Bunney BS, Moore H, Todd CL. Dopamine-cell depolarization block as a model for the therapeutic actions of antipsychotic drugs. *Trends Neurosci.* 1997;20(1):31–37.
172. Liss B, An WF, Roeper, J. Differential co-expression of KChip subunits defines inactivation kinetics of A-type potassium channels in dopaminergic VTA neurons. *Soc Neurosci Abstr;* 32th annual meeting, 2002.
173. Krabbe S, Lammel S, Liss B, Roeper, J. A-type potassium channels contribute to the diversity of the dopaminergic midbrain system. *Soc Neurosci Abstr.;* 38th annual meeting, 2008.
174. Chiodo LA, Bannon MJ, Grace AA, Roth RH, Bunney BS. Evidence for the absence of impulse-regulating somatodendritic and synthesis-modulating nerve terminal autoreceptors on subpopulations of mesocortical dopamine neurons. *Neuroscience.* 1984;12(1):1–16.
175. Chung HJ, Ge WP, Qian X, Wiser O, Jan YN, Jan LY. G protein-activated inwardly rectifying potassium channels mediate depotentiation of long-term potentiation. *Proc Natl Acad Sci USA.* 2009;106(2):635–640.
176. Lin MT, Lujan R, Watanabe M, Adelman JP, Maylie J. SK2 channel plasticity contributes to LTP at Schaffer collateral-CA1 synapses. *Nat Neurosci.* 2008;11(2):170–177.
177. Fan Y, Fricker D, Brager DH, et al. Activity-dependent decrease of excitability in rat hippocampal neurons through increases in I(h). *Nat Neurosci.* 2005;8(11):1542–1551.
178. Frick A, Magee J, Johnston D. LTP is accompanied by an enhanced local excitability of pyramidal neuron dendrites. *Nat Neurosci.* 2004;7(2):126–135.
179. Migliore M, Cannia C, Canavier CC. A modeling study suggesting a possible pharmacological target to mitigate the effects of ethanol on reward-related dopaminergic signaling. *J Neurophysiol.* 2008;99(5):2703–2707.
180. Damier P, Hirsch EC, Agid Y, Graybiel AM. The substantia nigra of the human brain. II. Patterns of loss of dopamine-containing neurons in Parkinson's disease. *Brain.* 1999;122 (pt 8):1437–1448.
181. Williams-Gray CH, Hampshire A, Robbins TW, Owen AM, Barker RA. Catechol O-methyltransferase Val158Met genotype influences frontoparietal activity during planning in patients with Parkinson's disease. *J Neurosci.* 2007;27(18): 4832–4838.
182. Weintraub D, Stern MB. Psychiatric complications in Parkinson disease. *Am J Geriatr Psychiatry.* 2005;13(10):844–851.
183. Kupsch A, Gerlach M, Pupeter SC, et al. Pretreatment with nimodipine prevents MPTP-induced neurotoxicity at the nigral but not at the striatal level in mice. *Neuroreport.* 1995;6(4):621–625.
184. Kupsch A, Sautter J, Schwarz J, Riederer P, Gerlach M, Oertel WH. 1-Methyl-4-phenyl-1,2,3,6-tetrahydropyridine-induced neurotoxicity in non-human primates is antagonized by pretreatment with nimodipine at the nigral but not at the striatal level. *Brain Res.* 1996;741(1–2):185–196.
185. Salthun-Lassalle B, Hirsch EC, Wolfart J, Ruberg M, Michel PP. Rescue of mesencephalic dopaminergic neurons in culture by low-level stimulation of voltage-gated sodium channels. *J Neurosci.* 2004;24(26):5922–5930.
186. Bean BP. Neurophysiology: stressful pacemaking. *Nature.* 2007;447(7148):1059–1060.

4 Genes in development

4.1 | Genetic Control of Meso-diencephalic Dopaminergic Neuron Development in Rodents

WOLFGANG WURST AND NILIMA PRAKASH

INTRODUCTION

Meso-diencephalic dopaminergic (mdDA) neurons play a key role in several human brain functions and are thus also involved in the pathophysiology of severe neurological and psychiatric disorders. The prospect of regenerative therapies for some of these disorders has fueled the interest of developmental neurobiologists in deciphering the molecular cues and processes controlling the generation of the mdDA neurons in the vertebrate brain. This section provides a brief summary of the spatiotemporal provenance of the mdDA precursor population giving rise to the dopaminergic (DA) neurons in the adult substantia nigra pars compacta and ventral tegmental area. Next, the secreted and transcription factors known to be involved in mdDA neuron development are described based on their function in the induction, specification, differentiation, or maintenance of the mdDA cell fate. Rodents, in particular the mouse, have served as the classical model organism due to their phylogenetic relationship to humans, their relatively well-characterized mdDA system on both the anatomical and physiological levels, and the propensity of the mouse to undergo genetic manipulation. This review focuses on in vivo data obtained from the analyses of mutant mice, as several reports have indicated that cell culture–based in vitro data do not always recapitulate the in vivo situation. Several of the genes described in this section, however, are also expressed in the developing human ventral midbrain,[1] suggesting that key aspects of mdDA neuron development are conserved among mammalian species.

SPATIAL ORIGIN OF THE mdDA NEURONS DURING DEVELOPMENT

The precursors of the adult retrorubral field (RrF, A8), substantia nigra pars compacta (SNc, A9), and ventral tegmental area (VTA, A10) DA neurons are located in the cephalic flexure of higher vertebrates (reptiles, birds, and mammals), which corresponds to the ventral domain (tegmentum) of the mesencephalon and diencephalon[2,3] (Fig. 4.1.1). Dopaminergic neurons are not found in the midbrain of lower vertebrates (teleost fish and amphibians), but functionally equivalent DA neuron populations are located in diencephalic territories in these species.[3,4] The mesencephalic DA neuronal populations are therefore thought to have their phylogenetic origin in the diencephalon of lower vertebrates, and this aspect of their evolutionary history may be recapitulated during their ontogeny.[5] In fact, the first cells transcribing the *Th* gene encoding the rate-limiting enzyme in DA biosynthesis, tyrosine hydroxylase, are detected in prosomeres (p) 1–3 of the developing mouse embryo,[6] which, according to the prosomeric model of Puelles and Rubenstein,[7] correspond to the diencephalon. Based on this and other evidence, the term *meso-diencephalic dopaminergic neurons* is now being applied to this neuronal population.[8,9]

While recent data have firmly established the ventral midline of the neural tube, the so-called floor plate (FP), as the site along the dorsoventral (D/V) neuraxis giving birth to all mdDA neurons,[10–14] (Fig. 4.1.2) the precise origin of the different mdDA subpopulations along the anteroposterior (A/P) axis of the midbrain/caudal diencephalon is still largely unknown and therefore hotly debated. Some authors have proposed that the SNc DA neurons are generated from a rostrolateral domain in the caudal diencephalon, whereas the VTA DA neurons derive from a caudomedial domain in the mesencephalon.[9,15,16] Postmitotic mdDA precursors and neurons, however, undergo an extensive migration before arriving at their final destinations in the ventral midbrain/caudal diencephalon, further complicating this

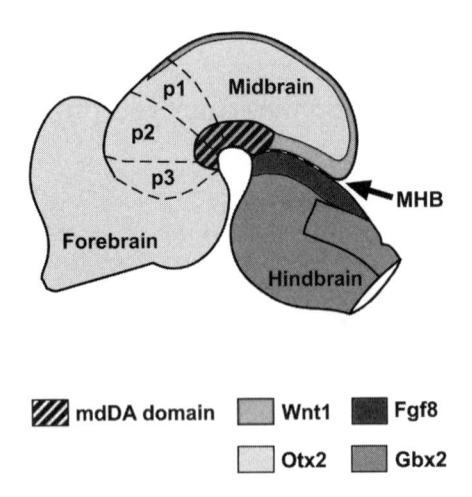

FIGURE 4.1.1. The four dimensions of mdDA neuron development. First dimension: A/P positioning. Sagittal view of the neural tube in a late midgestational (E10.5–E12.5) mouse embryo depicting the domain within the cephalic flexure from where mdDA neurons (black stripes) arise. This area comprises the ventral domain (tegmentum) of the midbrain and caudal diencephalon, corresponding to prosomeres (p) 1–3 according to the prosomeric model of Puelles and Rubenstein.[7] Otx2 (grayish) is expressed in the forebrain and midbrain, whereas Gbx2 (gray) is expressed in the rostral hindbrain. The mid-/hindbrain boundary (MHB) is positioned at the expression interface of these two TFs. At E10.5, the secreted factor Wnt1 (light gray) is expressed in a ring encircling the caudal midbrain rostral to the MHB, in the RP of the midbrain and caudal diencephalon and in two converging stripes within the FP of the midbrain/p1–3. The latter Wnt1 expression domain overlaps with the mdDA progenitor domain. The secreted factor Fgf8 (dark gray) is expressed at E10.5 in a ring encircling the rostral hindbrain caudal to the MHB. Otx2 and Wnt1 are required for the establishment of the mdDA progenitor domain in the ventral midbrain/caudal diencephalon, and the MHB delimits the caudal extent of this progenitor domain. Fgf8, Fibroblast growth factor 8; Gbx2, Gastrulation brain homeobox 2; mdDA, meso-diencephalic dopaminergic; MHB, mid-/hindbrain boundary; Otx2, Orthodenticle homolog 2; p, prosomere; Wnt1, Wingless-related MMTV integration site 1. Source: Modified from Marín et al.[6] (See Color Plate 4.1.1.)

issue (reviewed by [8]). These cells initially migrate out of the ventricular zone (VZ)/subventricular zone (SVZ) containing their progenitors into the mantle zone (MZ) of the neural tube in a radial manner[17] (Fig. 4.1.2). Subsequently, these neurons move tangentially in a mediolateral and probably also a rostrocaudal direction along the pial surface of the mesencephalon/caudal diencephalon to reach their final destinations in the SNc and VTA[17,18] (Fig. 4.1.2).

TIME COURSE OF mdDA NEURON DEVELOPMENT

Although the first appearance of cells expressing Th protein was reported to occur at around day 9.0–9.5 of embryonic development (E9.0–E9.5) in the mouse ventral midbrain using a special fixation procedure,[19] the transcription of Th mRNA is first detected in the diencephalon (p1–3 domain) of the E10.5 mouse embryo.[6] As shown recently, the vast majority of the SNc DA neurons are born (i.e., they undergo their last cycle of cell division before becoming postmitotic) on E12 in the rat,[20] which corresponds approximately to E10.5 in the mouse, thus contradicting earlier reports of their later time of origin at E12.5 in the mouse.[18,21] Nevertheless, this study confirmed the earlier observation of the VTA DA neurons being generated approximately 1 day later than the majority of the SNc DA neurons.[18,20] The different spatiotemporal origins of these two mdDA subpopulations may already determine their distinct molecular and functional makeup in the adult, as it is very likely that the SNc DA neurons are exposed to slightly different factors than the VTA DA neurons during development. However, the question of their precise spatiotemporal origin remains unanswered due to the lack of mdDA subpopulation–specific marker genes at early developmental stages and of cell fate–mapping data for the mdDA subgroups. We therefore refer in general terms to the "mdDA neurons" without distinguishing between the different mdDA subpopulations.

The first mdDA progenitors are detected in the cephalic flexure of the mouse embryo at around E9.5 by the expression of the retinoic acid (RA)–synthesizing enzyme Aldh1a1 (Raldh1, Ahd2).[22] These proliferating progenitors generate immature postmitotic mdDA precursors characterized by the expression of the nuclear receptor Nurr1 (Nr4a2) and several other transcription factors in the E10.5 mouse embryo.[22,23] The Nurr1+ mdDA precursors subsequently differentiate into mature mdDA neurons from E11.5 on; that is, these cells acquire all molecular, morphological, and physiological features characteristic of an adult mdDA neuron.[8,23] Apart from establishing the proper dendritic and axonal connections with their target fields in the midbrain and forebrain and acquiring their adult electrophysiological properties, this includes the expression of all enzymes, transporters, and receptors required for the biosynthesis, synaptic release/reuptake, and autoreceptor control of DA, such as Th, aromatic L-amino acid (or L-DOPA) decarboxylase (Aadc/Ddc), vesicular monoamine transporter 2 (Vmat2/Slc18a2), dopamine transporter (Dat/Slc6a3), and dopamine receptor 2 (D2R/Drd2). The expression of these genes/proteins has therefore been used as a marker for the corresponding stages in mdDA neuron development (Fig. 4.1.2), although recent work has indicated that several other markers, in particular for the mdDA progenitors, may be used in addition but with caution.

E12.5

Legend:
- Nkx2-2
- Nkx6-1
- Otx2, Shh, Foxa1/2, Ngn2, Mash1
- Wnt1, Lmx1a, Msx1, Aldh1a1
- Foxa1/2, Lmx1a/b, Ngn2, Aldh1a1, Nurr1
- Foxa1/2, Lmx1a/b, Aldh1a1, Nurr1, Pitx3, En1/2, Th, Aadc, Vmat2, Dat, D2R

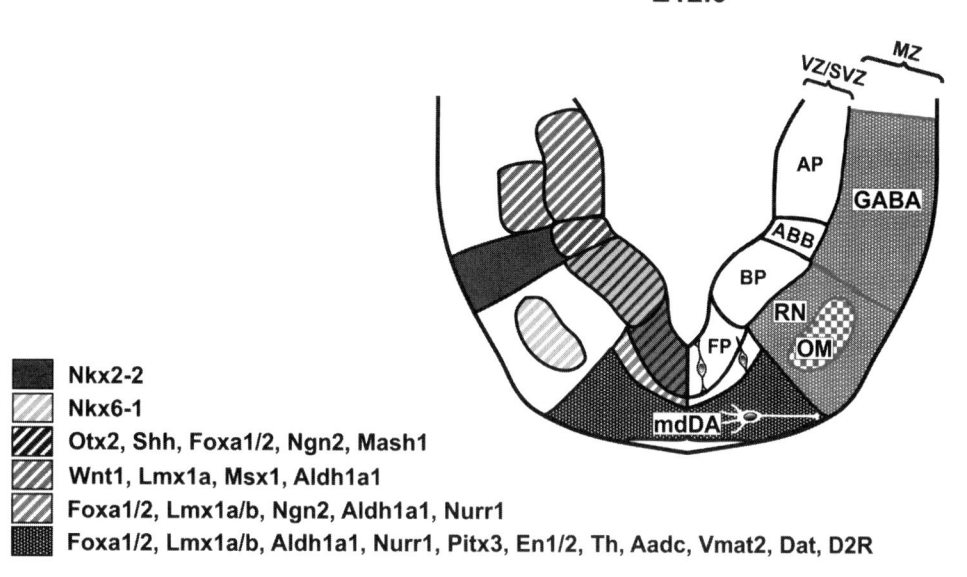

FIGURE 4.1.2. The four dimensions of mdDA neuron development. Second and third dimensions: D/V and mediolateral positioning. Coronal view of the ventral midbrain in a late midgestational (E12.5) mouse embryo depicting the progenitor domain within the FP (black and dark gray stripes) from which mdDA neurons (dotted dark gray) develop. Wnt1, Lmx1a, and Msx1 expression (dark gray stripes) is restricted to the midbrain/p1–3 FP and necessary for proper mdDA neurogenesis from the progenitors located in the VZ/SVZ. Msx1 also represses Nkx6-1 expression (light gray stripes) within the FP. Aldh1a1 expression in these cells serves as a marker for mdDA progenitors. Expression of Otx2, Shh, Foxa1/2, Ngn2, and Mash1 (black stripes) is not restricted to the midbrain FP but is also found in BP progenitors. These secreted factors and TFs, however, play a prominent role in mdDA neurogenesis. Moreover, Otx2 is necessary for the ventral repression of Nkx2-2 (dark gray), a 5-HT neuron-inducing factor. The postmitotic mdDA precursors express Foxa1/2, Lmx1a/b, Ngn2, Nurr1, and Aldh1a1 (gray stripes) and require these TFs for their proper differentiation into mdDA neurons. Differentiating and adult mdDA neurons (dotted dark gray) express Foxa1/2, Lmx1a/b, Aldh1a1, Nurr1, Pitx3, En1/2, the DA biosynthetic enzymes Th and Aadc, the DA transporters Vmat2 and Dat, and the DA autoreceptor D2R. These TFs and the RA-synthesizing enzyme Aldh1a1 are required for the maturation and/or survival of mdDA neurons. The proliferating, radial glia-like mdDA progenitors are located in the VZ/SVZ of the midbrain/p1–3 FP and give birth to the postmitotic mdDA precursors, which migrate radially out of the VZ/SVZ into the MZ and begin differentiation into mdDA neurons. The mature mdDA neurons migrate tangentially in a mediolateral and A/P (not shown) direction to their final destinations in the SNc and VTA. ABB, alar-basal boundary; AP, alar plate; BP, basal plate; FP, floor plate; GABA, γ-aminobutyric acid-synthesizing neurons; mdDA, meso-diencephalic dopaminergic neurons; MZ, mantle zone; OM, oculomotor nucleus; RN, red nucleus; VZ/SVZ, ventricular/subventricular zone. (See Color Plate 4.1.2.)

The generation of a mature mdDA neuron from a pluripotent neuroepithelial stem cell can be subdivided into four distinct steps (Fig. 4.1.3). First, a territory competent to generate mdDA progenitor cells is demarcated in the neural plate/tube during early neural development (i.e., at around E8.0–E9.5) by a process we name *induction*. This initial step is equivalent to the induction of the ventral midbrain/caudal diencephalon in the early mouse embryo.[24] Second, neural precursors within this competent field are committed to the mdDA cell fate at around E9.5–E11.5 by a process we call *specification*. While neural precursors may adopt an alternative cell fate before this step, this is not possible once the mdDA cell fate has been specified, as it simultaneously means the initiation of the differentiation process in these cells. This step is therefore also termed *early differentiation* of immature mDA neurons.[23]

Third, the committed precursors undergo *terminal or late differentiation* into mdDA neurons; that is, these cells acquire all features defining them as neurons in general and as mdDA neurons in particular. This process lasts for several days or even weeks and is thought to start at about E11.5 in the mouse embryo. Fourth, the *maintenance* of the terminally differentiated mdDA neurons in the late gestational (after E14.5) and adult brain is essential for the survival of the whole organism. This includes the prevention of excessive apoptotic cell death in the mdDA subpopulations and their local and retrograde neurotrophic support once their projections have reached their target fields in the forebrain[25,26] (reviewed by [27,28]). The transition between these steps is continuous, and several signaling cascades and regulatory processes active in one step may also be active during the following step. It is therefore not possible to

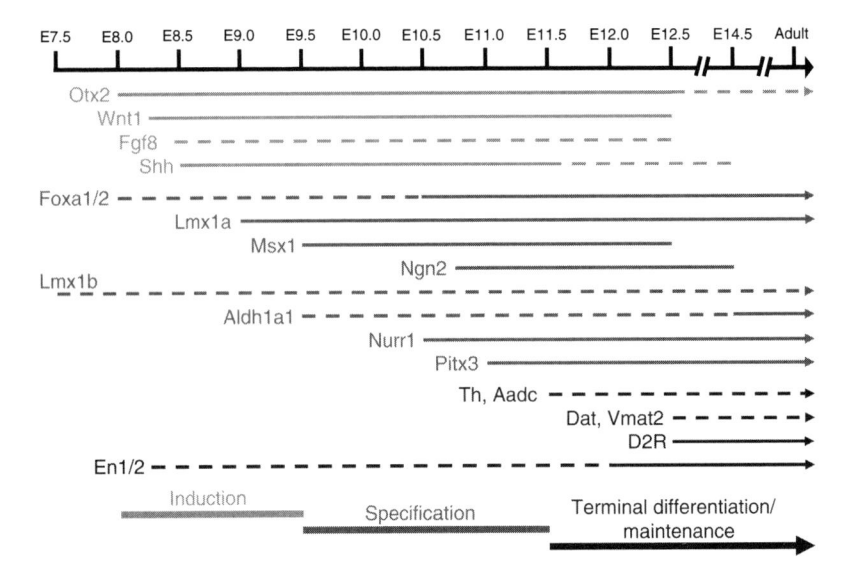

FIGURE 4.1.3. The four dimensions of mdDA neuron development. Fourth dimension: developmental time. A time scale of mouse embryonic development from E7.5 (when neurulation starts) to adulthood with a special focus on early and late midgestational stages is shown at the top. The color-coded (as in Fig. 4.1.2) bars below indicate the onset of expression of the corresponding secreted factor or TF, enzyme, or transporter protein according to the time scale on top and the time interval during which the corresponding gene is transcribed. Solid bars indicate a requirement of the corresponding molecule for proper mdDA neuron development during that time interval. Dotted bars indicate (a) that the corresponding protein is not required for mdDA neuron development (Shh, En1/2, Th, Aadc, Dat, Vmat2) or (b) that a direct requirement has not yet been demonstrated (Otx2, Fgf8, Foxa1/2, Lmx1b, Aldh1a1) during that time interval. Arrows indicate that the corresponding protein is expressed throughout adulthood and required for the maintenance/survival or physiological function of mdDA neurons. The time intervals during which the induction of the mdDA progenitor domain, the specification of the mdDA neuronal fate in postmitotic mdDA precursors, and the terminal differentiation/maintenance of the mdDA neurons take place in the mouse ventral midbrain are depicted at the bottom of the figure. Aldh1a1, aldehyde dehydrogenase family 1, subfamily a1 (Raldh1, Ahd2); Dat, dopamine transporter (Slc6a3); D2R, dopamine receptor 2 (Drd2); En1/2, Engrailed 1 and 2; Fgf8, Fibroblast growth factor 8; Foxa1/2, Forkhead box A1/A2 (Hnf3α/β); Lmx1a/b, LIM homeobox transcription factor 1 alpha/beta; Msx1, Muscle-segment homeobox-like 1; Ngn2, Neurogenin 2 (Neurog2); Nurr1, nuclear receptor subfamily 4, group A, member 2 (Nr4a2); Otx2, Orthodenticle homolog 2; Pitx3, Paired-like homeodomain transcription factor 3; Shh, Sonic hedgehog; Th, tyrosine hydroxylase; Vmat2, vesicular monoamine transporter 2 (Slc18a2); Wnt1, Wingless-related MMTV integration site 1. (See Color Plate 4.1.3.)

distinguish between these steps without taking into consideration the developmental time, the location, and the combinatorial code of transcription factors active in the cell under scrutiny. In the next paragraphs, we will review the signaling cascades and transcription factors characteristic of each individual step and required for the generation of a mature mdDA neuron from an uncommitted progenitor.

Induction of a Competent Field to Generate mdDA Neurons

The induction of the prospective midbrain territory is the initial step necessary for the development of any neuronal population in this region, whose details are not described here (reviewed by [24,29–31]). Three events, however, are crucial in this step. First, the initiation of Otx2 expression in the anterior neuroectoderm defines the prospective forebrain and midbrain territory along the A/P axis. Second, the establishment of the mid-/hindbrain boundary (MHB) as a secondary organizer controls the further development of the midbrain and rostral hindbrain. Third, and although Sonic hedgehog (*Shh*) is required for early D/V patterning of the midbrain, canonical Wnt signaling–mediated repression of *Shh* expression is necessary for the initiation of mdDA neurogenesis from the midbrain FP.

Otx2 and Wnt1 are required for the establishment of the mdDA progenitor domain

The bicoid class homeodomain (HD) transcription factor (TF) Otx2 belongs to the vertebrate orthologues of the *Drosophila* orthodenticle protein whose expression within the central nervous system (CNS) is restricted to the presumptive forebrain and midbrain from presomitic (E8.0) stages on.[32,33] The early loss of all anterior head structures including the entire brain rostral to rhombomere (r) 3 in the $Otx2^{-/-}$ mice impeded the assessment of

Otx2 functions at later embryonic stages,[34–36] and this issue was resolved only when conditional *Otx2* mutants became available. The paralogous gene *Otx1* is expressed in the lateral midbrain but not in the FP.[37] Deletion of *Otx2* in the lateral midbrain of *Otx1*[+/Cre]; *Otx2*[−/flox] mice leads to a dorsal expansion of the *Shh* domain and to the increased proliferation of mdDA progenitors, resulting in an enlarged mdDA neuron population in these mutants at the expense of neighboring neuronal populations.[37] This finding indicated an Otx dose-dependent control of D/V midbrain patterning by antagonizing the ventral Shh signal. Deletion of *Otx2* in the entire midbrain including the FP and BP of *En1*[+/Cre]; *Otx2*[flox/flox] mice, by contrast, results in a strong reduction of the mdDA neuron population and in the ectopic generation of serotonergic (5-HT) neurons in the mutant ventral midbrain.[38] A similar loss of mdDA neurons and ectopic generation of 5-HT neurons is observed in conditional *Nestin-Cre*/+; *Otx2*[flox/flox] mutants, in which *Otx2* is inactivated in neural progenitors at a later developmental stage compared to the *En1*[+/Cre]; *Otx2*[flox/flox] mice.[39] The absence of Otx2 expression in the ventral midbrain of *En1*[+/Cre]; *Otx2*[flox/flox] and *Nestin-Cre*/+; *Otx2*[flox/flox] mice also results in the loss of the ventral HD TF Nkx6-1+ domain and in a ventral expansion of the HD TF Nkx2-2 expression domain, which is normally confined to the alar-basal boundary (ABB) delimiting the ventral from the dorsal midbrain[38,39] (Fig. 4.1.2). No 5-HT neurons are generated in the forebrain and midbrain of wild-type mice, and the most rostral 5-HT neuron populations in the dorsal and medial raphe nuclei derive from an Otx2−, Nkx6-1−, Nkx2-2+ and Shh+ progenitor domain in the rostral hindbrain.[38] The combinatorial TF code of ventral midbrain progenitors in the *En1*[+/Cre]; *Otx2*[flox/flox] and *Nestin-Cre*/+; *Otx2*[flox/flox] mutants is therefore the same as in the rostral hindbrain of wild-type mice, thus explaining the ectopic generation of 5-HT neurons in these conditional mutants. Notably, removal of ectopic Nkx2-2 expression in the mutant ventral midbrain (by crossing the conditional mouse strains with *Nkx2-2*[−/−] mice,[40]) results in a rescue of the mdDA neuron population in the *En1*[+/Cre]; *Otx2*[flox/flox]; *Nkx2-2*[−/−] but not in the *Nestin-Cre*/+; *Otx2*[flox/flox]; *Nkx2-2*[−/−] mutants.[39,41] The reason for this discrepancy has not yet been resolved but may be linked to the different developmental time points of *Otx2* inactivation in these conditional mutants and/or to the later function of Otx2 in mdDA neurogenesis (see below and [23]).

The expression of the secreted lipid-modified glycoprotein Wnt1 encompasses initially (at E8.5) a broad domain corresponding to the presumptive midbrain and is later (at E9.5) restricted to two stripes along the lateral FP of the midbrain merging in the caudal diencephalon and to the MHB, where it is transcribed in a ring encircling the neural tube.[42,43] Wnt1 is also expressed along the dorsal midline (roof plate, RP) of the caudal diencephalon, mesencephalon, caudal rhombencephalon, and spinal cord at E9.5 (Fig. 4.1.1). A striking observation is the loss of the ventral Wnt1 expression domain in the midbrain FP concomitant with the loss of Otx2 expression and of mdDA neurons in the *En1*[+/Cre]; *Otx2*[flox/flox] mice, and the rescue of this Wnt1+ domain concomitant with the rescue of the mdDA neurons in the compound *En1*[+/Cre]; *Otx2*[flox/flox]; *Nkx2-2*[−/−] mutants.[41] This suggested a causal relationship between Wnt1 expression and the generation of mdDA neurons. Indeed, ectopic mdDA neurons are generated in the rostral hindbrain FP of *En1*[+/Wnt1] mice expressing ectopically Wnt1 in the rostral hindbrain and midbrain.[41] Otx2 is also induced ectopically in the rostral hindbrain FP of the *En1*[+/Wnt1] mice, indicating that this hindbrain territory had acquired midbrain identity.[41] More importantly, and using an experimental paradigm similar to that of Ye at al.,[44] ectopic mdDA neurons were not induced in anterior neural tube explant cultures of *Wnt1*[−/−] mouse embryos even in the presence of Shh and Fibroblast growth factor 8 (Fgf8).[41] Several conclusions were drawn from these analyses: (1) Otx2 is required for the repression of *Nkx2-2* in the midbrain FP and BP, thereby maintaining, either directly or indirectly, Wnt1 expression; (2) the generation of mdDA neurons is tightly linked to the expression of Wnt1 in the ventral midbrain; (3) mdDA neurons cannot be induced ectopically in the absence of *Wnt1*. The early function of Otx2 as a homeotic gene conferring forebrain and midbrain identity to E8.0–E9.5 neural progenitors may be similar to the function of *Hox* genes in hindbrain and spinal cord development.[45] Otx2, however, has to interact locally with the Wnt1-signaling pathway to establish the mdDA progenitor domain in the ventral midbrain of the developing mouse embryo at later stages (at around E9.5–E10.5; see below).[41]

The position of the MHB determines the location and size of the mdDA neuron population

The expression of Otx2 not only demarcates the forebrain and midbrain territory, but its posterior (caudal) limit also positions the border between the prospective midbrain and the prospective hindbrain in the embryonic neural tube, the MHB.[46] The MHB is one of the most important secondary signaling centers (also called *organizers*) in the mouse embryo, as it controls the development of the midbrain and anterior hindbrain (reviewed by [31,47,48]). The MHB is established at the expression interface of the two HD TFs Otx2 and Gastrulation brain homeobox 2 (Gbx2) during early neural development, which later becomes visible as the isthmic constriction (Fig. 4.1.1). Subsequently, the expression of different secreted factors and TFs is initiated in a strict spatiotemporal sequence at or across the MHB. Among these are the secreted proteins Wnt1 and Fgf8 and the HD TFs Lmx1b, Engrailed 1 (En1), and Engrailed 2 (En2)[49] (reviewed by [30,31,48]). Wnt1 expression is confined to the anterior border of the MHB in a ring encircling the caudal midbrain and abutting Fgf8 expression, which is confined to the posterior border of the MHB in a ring encircling the rostral hindbrain[42,43,50] (Fig. 4.1.1). Lmx1b and En1/2 are expressed in a broader domain across the MHB encompassing the caudal two-thirds of the mesencephalon and the rostral third of r1.[49,51,52] These factors activate different intracellular signaling cascades that together promote the patterning of the mid-/hindbrain region (MHR), thus having an organizing activity. In functional terms, the MHB is therefore also known as the *mid-/hindbrain organizer* (MHO) or the *isthmic organizer* (IsO).[30,31,47,48]

The posterior shift of the MHB by the ectopic expression of Otx2 in the rostral hindbrain of $En1^{+/Otx2}$ mice results in a caudal expansion of the mdDA neuron population at the expense of the rostral hindbrain 5-HT neurons.[46,53] Conversely, the reduction of Otx dosage in the brain of $Otx1^{-/-}$; $Otx2^{+/-}$ mice results in an anterior repositioning of the MHB at the p2/p3 boundary in the forebrain and in the generation of a reduced number of mdDA neurons rostral to this ectopic position, whereas the rostral hindbrain 5-HT population is expanded rostrally to the new position of the MHB.[53,54] The MHB therefore delimits the caudal extent of the mdDA progenitor domain.[53] Wnt1 is also induced ectopically in the rostral hindbrain FP of the $En1^{+/Otx2}$ mice, confirming a positive feedback of Otx2 on Wnt1 expression.[41]

Most of the patterning activity at the MHB is conferred by only one molecule, the secreted protein Fgf8 (reviewed by [48,55–57]). Using rat embryo explant (in vitro) cultures, Ye et al.[44] showed that the generation of mdDA neurons is inhibited in ventral midbrain explants by blocking the Fgf signal transduction and that ectopic mdDA neurons are induced by the application of an Fgf8-coated bead to ventral forebrain explants. This work established the notion of Fgf8 as one of the key factors in mdDA neuron development, and subsequent studies have tried to confirm this hypothesis in vivo. The early death (at around E9.5) of the Fgf8 null mutants due to severe gastrulation defects impeded the analysis of mdDA neuron development in these mutants.[58] The conditional inactivation of Fgf8 in the $En1^+$ domain across the MHB leads to a progressive loss of mid-/hindbrain tissue including the mdDA neuron population due to a massive cell death in the MHR preceded by the loss of Wnt1, Gbx2, and Fgf17/18 expression at the MHB.[59] The loss of mdDA neurons in these mutants could therefore be an indirect consequence of the loss of Wnt1 expression and subsequent cell death in the MHR. The analysis of conditional mouse mutants for the Fgf receptor (Fgfr) tyrosine kinase family transducing the Fgf signal at the MHB has not proven to be very informative. Only three (Fgfr1, 2, and 3) of the four known Fgfr genes are expressed in the vertebrate CNS.[60,61] Of these, only Fgfr1 and Fgfr2 are expressed in the midbrain FP from early developmental stages on, whereas the expression of Fgfr3 exhibits a gap at the MHB that is closed only at later developmental stages.[60,62,63] Conditional inactivation of Fgfr1 across the MHB of $En1^{+/Cre}$; $FgfR1^{flox/flox}$ mice results in the loss of dorsal but not ventral neural tissue. The mdDA neuron population shows a subtle disorganization as a consequence of expression pattern changes for several MHO genes, including Wnt1 and En1/2, and of the lack of a coherent MHB, but is otherwise unaffected in these conditional mutants.[63,64] Fgfr1 is therefore implicated in the regulation of cell adhesion at the MHB rather than in mdDA neuron development. As was expected from their expression patterns and a possible functional redundancy, the generation of mdDA neurons is unaffected in conditional $En1^{+/Cre}$; $Fgfr2^{lox/lox}$ and in $Fgfr3^{-/-}$ mice.[65] Since none of the Fgfr single mutants recapitulate the full mid-/hindbrain phenotype of the conditional Fgf8 mutants, a redundant function of the Fgfr at the MHB became very likely. The phenotype of $Fgfr1^{cko}$; $Fgfr2^{cko}$; $Fgfr3^{null}$ triple mutant embryos indeed resembles closely the phenotype of the conditional Fgf8 mutants, including the loss of mdDA neurons.[66] A reduction of mdDA precursors and loss of mdDA neurons is also observed in $Fgfr1^{cko}$; $Fgfr2^{cko}$ but neither in $Fgfr1^{cko}$; $Fgfr3^{null}$ nor in $Fgfr2^{cko}$; $Fgfr3^{null}$ double mutants, establishing a hierarchical order of Fgfr1 > Fgfr2 > Fgfr3 requirement for MHR and mdDA neuron development in vivo.[66] Nevertheless, a few mdDA neurons are still generated in the $Fgfr1^{cko}$; $Fgfr2^{cko}$; $Fgfr3^{null}$ triple mutants, indicating that the establishment of the mdDA progenitor domain is not severely disrupted in these mutants. The major phenotype of the triple mutants is in fact the reduced proliferation and premature differentiation of ventral midbrain neural progenitors irrespective of their future neuronal lineage.[66]

The analysis of the conditional *Fgf8* and *Fgfr* mutant mice is hampered by the fact that all mutants display early patterning defects and a loss of mid-/hindbrain tissue in a dose-dependent manner, which prevents the assessment of a cell-autonomous function of Fgf8 signaling in mdDA neuron development in vivo. Nevertheless, all available data indicate a general requirement of Fgf8 signaling for the proper growth and survival of neural tissue in the MHR rather than a cell-specific function in the generation of distinct neuronal populations in this region. Future research therefore awaits the cell-specific ablation of Fgf8 signal transduction in individual neuronal populations within the MHR to clarify this issue.

Canonical Wnt signaling-mediated repression of Shh is necessary for the initiation of mdDA neurogenesis from the midbrain FP

The FP consists of morphologically and functionally specialized cells in the ventral midline of the neural tube extending from the posterior diencephalon to the caudal end of the spinal cord (reviewed by [67]). Although FP cells differ in their molecular and functional properties along the A/P axis of the neural tube, a unifying feature is their expression of the secreted glycoprotein Shh.[67] The majority of the mdDA neurons are generated from the midbrain/caudal diencephalon FP, which acquires neurogenic properties distinct from those of the rest of the neural tube.[10,13,14] Initial in vitro experiments showed that the coculture of FP tissue with ventral midbrain explants induces mdDA neurons in the explants next to the FP tissue.[68] Shh was subsequently identified as the secreted factor mediating the mdDA-inducing effects of the FP and was therefore one of the first factors implicated in mdDA neuron development.[44,69,70]

Shh expression starts at around E8.5 in the midbrain, where it includes not only the FP but also the BP.[71] The Shh protein precursor is cleaved autocatalytically into a bioactive N-terminal fragment that is subsequently lipid-modified by adding a cholesterol moiety to the C terminus and palmitoylation at the N terminus (reviewed by [72,73]). The bioactive Shh fragment binds to the Patched (Ptch) receptor and releases the G-protein-coupled receptor Smoothened (Smo) from its constitutive repression by Ptch. Smo activation then leads to a complex intracellular signaling cascade that ultimately results in the formation of repressor or activator forms of the Gli family of zinc-finger TFs. The initial in vitro findings were confirmed by in vivo gain-of-function (GOF) experiments using transgenic mice ectopically expressing the bioactive N-terminal Shh fragment or the Gli1 TF across the MHB in the dorsal midbrain and in the dorsal hindbrain.[74] In both mouse mutants, mdDA neurons are induced ectopically in the dorsal midbrain but not in the dorsal hindbrain, indicating that additional factors (possibly Otx2 and Wnt1) must confer A/P positional information for the induction of the mdDA progenitor domain in the midbrain. These GOF experiments were complemented by the corresponding loss-of-function (LOF) experiments. mdDA neurons are not generated in $Shh^{-/-}$ mice and are strongly reduced in $En1^{+/Cre}$; $Smo^{flox/-}$ mice, in which transduction of the Shh signal is abolished (by Smo inactivation) in the MHR at E9.0.[75] However, mdDA neurons are induced normally in *Nestin-Cre/+*; $Shh^{flox/flox}$ and *Nestin-Cre/+*; $Smo^{flox/-}$ mice in which Shh signaling is abolished only after E11.5, indicating a requirement of Shh signaling for the early induction but not for the later development of mdDA neurons.[11,75] As Shh is necessary for the induction of all ventral neural progenitor domains in the midbrain and hindbrain/spinal cord,[44,75–78] these findings confirmed the early ventralizing activity of Shh in the neural tube. The reason for the neurogenic capacity of the midbrain FP (in contrast to the hindbrain/spinal cord), however, remained enigmatic until very recently. A study by Joksimovic et al.[79] shows that Shh expression in the neural tube inversely correlates with the neurogenic capacity of the FP at midgestational stages (E9.5–E11.5), and removal of *Shh* in *Shh::cre*; *Shh cKO* mice confers neurogenic capacity to the hindbrain/spinal cord FP, including the generation of mdDA precursors in the hindbrain FP. Remarkably, the same effect is seen after constitutive activation of the canonical Wnt signaling pathway in Shh$^+$ FP cells (*Shh::cre*; $Ctnnb1^{lox(ex3)}$ mice) due to the loss of *Shh* expression in these cells.[79] By contrast, conditional inactivation of *ß-catenin* (*Ctnnb1*; a key component of the canonical Wnt pathway) in *Shh::cre*; *Ctnnb1 cKO* mice maintains *Shh* expression and results in reduced neurogenesis and in the loss of mdDA precursors in the midbrain FP.[79] These findings indicate that *Shh* expression has to be down-regulated by canonical Wnt signaling at midgestational stages to induce mdDA neurogenesis from the midbrain FP.

In summary, the competent field to generate mdDA neurons is delimited by the expression of Otx2 along the A/P axis and the expression of Shh along the D/V axis of the neural tube. These coordinates are sufficient to restrict the mdDA progenitor domain to the anterior (fore-/midbrain) and ventral (FP) neural tube. The precise location of this domain in the diencephalic and mesencephalic tegmentum, however, is determined by the position of the MHB and by the expression of Wnt1 in this region. The mdDA competent field can thus be defined as the Otx2⁺, Shh⁺, and Wnt1⁺ territory in the anterior neural plate of the E8.5–9.5 mouse embryo. Fate mapping of Shh- and Wnt1-expressing cells in the neural tube has indeed shown that these cells give rise to mdDA neurons.[13,80] Although other factors are also expressed within this territory at these stages, their participation in the induction of the mdDA competent field is more controversial, as will be discussed below.

Specification of the mdDA Cell Fate in Neural Precursors (Early Differentiation of mdDA Precursors)

A unifying feature of all mouse mutants and experimental manipulations described in the previous section is the conversion of neural progenitors from one cell fate to another. Apart from uncovering a remarkable plasticity of neural progenitors in the CNS, these experiments also revealed that neither the secreted factors nor TFs described previously are able to impart irreversibly the mdDA neuron fate on these progenitors. We thus distinguish the "inductive" from the "specifying" capacity of a given secreted factor or TF by its ability to convey permanently the mdDA neuron fate on ventral midbrain progenitors/precursors. The LOF of these factors will therefore lead to the lack or death of the corresponding cells but not to a respecification of their cell fate. The expression of the majority of these factors starts somewhat later (at around E9.0–9.5) than the ones discussed in the previous section. Moreover, these factors can be split into two groups, depending on whether they confer generic neuronal (vs. glial) properties or mdDA-specific characteristics to the mdDA progenitors/precursors.

TFs conferring generic neuronal properties to mdDA precursors

Foxa1/2 (Hnf3α/β). One of the Shh target genes in the developing mouse embryo is the forkhead/winged helix TF *Foxa2* (*Hnf3β*).[81] Foxa2 and its paralogue Foxa1 (*Hnf3α*) are expressed in the midbrain FP and BP from E8.0–8.5 on[11,82–84] and therefore satisfy one of the criteria for an inductive factor. The early lethality of the *Foxa2*⁻/⁻ embryos due to gastrulation defects hindered the assessment of Foxa2 function in mdDA neuron development until recently.[85,86] Conditional inactivation of the *Foxa2* gene between E10.5 and E12.5 in *Nestin-Cre/+;Foxa2*$^{flox/flox}$ (*Foxa2cko*) mice on a *Foxa1*⁻/⁻ background revealed the requirement of Foxa1/2 for the generation of different ventral midbrain neuronal populations in a dose-dependent manner,[11] a finding that was reproduced in *Foxa2*⁻/⁻ explant cultures.[13]. Foxa1/2 are expressed in the progenitors not only of the mdDA lineage but also of neighboring motorneurons and interneurons arising from the midbrain BP (Fig. 4.1.2), as well as in their postmitotic offspring except for the motorneurons.[11] The number of postmitotic neurons is strongly reduced in the ventral midbrain of *Foxa1*⁻/⁻; *Foxa2cko* double mutants, correlating with a strong down-regulation of the proneural basic helix-loop-helix (bHLH) TF Neurogenin 2 (Ngn2/Neurog2) in their progenitors.[11] Although the mdDA progenitors still express their lineage-specific markers such as Lmx1a/b, they generate very few postmitotic Nurr1⁺ mdDA precursors and Th⁺ mdDA neurons, indicating that mdDA neurogenesis is disrupted by the loss of *Foxa1/2*.[11] Although a dose-dependent block or delay in the late differentiation of mdDA precursors into mature mdDA neurons is also reported in the *Foxa1/2* single mutants,[11] it is unclear if this phenotype is due to a direct requirement of Foxa1/2 for mdDA neuron differentiation or to the loss of Ngn2 expression in their progenitors. The partial rescue of these defects in the *Foxa1/2* single mutants before birth rather resembles the *Ngn2*⁻/⁻ phenotype (see below) and would argue for the latter possibility. Nevertheless, Foxa2 is also required for the maintenance of mdDA neurons in adulthood, as discussed in the section "Maintenance of Mature mdDA Neurons throughout Late Gestation and Adulthood."

An inductive function of Foxa1/2 in the establishment of the mdDA progenitor domain at earlier embryonic stages still remains elusive due to the rather late inactivation of *Foxa2* in the *Foxa2cko* mutants.[11] Interestingly, Nkx2-2 expression is shifted ventrally into the mdDA progenitor domain in the *Foxa1/2* double but not single mutants, concomitant with a reduction of the adjacent Nkx6-1⁺ domain in the midbrain BP.[11] These patterning defects resemble those of the *En1*$^{+/Cre}$;*Otx2*$^{flox/flox}$ mice[38] and suggest a function of Foxa1/2 downstream of the inductive factors Otx2 and Shh. Although Foxa2 activates the expression of *Shh* in the FP,[87] the inactivation of *Shh* in the E11.5 ventral midbrain of *Nestin-Cre/+; Shh*$^{flox/flox}$ mice or blockade of the Shh signaling pathway in the presence of Foxa2 in embryonic stem (ES) cells does not affect ventral midbrain patterning or the generation of mdDA neurons, indicating that Foxa2 is indeed acting downstream of Shh.[11,13]

Lmx1a and Msx1. Shh was also reported to induce the expression of the LIM HD TF Lmx1a in midbrain explants.[88] The two HD TFs *Lmx1a* and *Msx1* were found independently by two groups in a search for genes that are expressed in the ventral midbrain and involved in mdDA neuron development.[14,88] The expression of both genes is confined to the midbrain/caudal diencephalon FP without extending into neighboring progenitor domains, and translation of *Lmx1a* in this region starts at E9.0, whereas Msx1 expression begins half a day later at E9.5[14,88] (Figs. 4.1.2, 4.1.3). Msx1 is expressed only in the mdDA progenitors located within the VZ/SVZ, while Lmx1a expression extends into the MZ and includes postmitotic mdDA precursors and mature mdDA neurons[14,88] (Fig. 4.1.2). Using different experimental paradigms, Lmx1a was shown to be necessary and sufficient for the generation of mdDA neurons in the chicken embryo.[88] Upon *Lmx1a* overexpression, ectopic mdDA neurons are generated in the ventral but not dorsal aspect of the chicken midbrain, indicating that a ventralizing signal (Shh or another factor) must also be present. Knockdown of *Lmx1a* by RNA interference results in an almost complete loss of mdDA neurons in the chick ventral midbrain.[88] Although Lmx1a appears to mediate its effects through the activation of *Msx1*, overexpression of *Msx1* alone is not sufficient to induce mdDA neurons in the chicken midbrain.[88] Msx1, however, represses the expression of the HD TF Nkx6-1 in the midbrain FP, thereby contributing to the establishment of the mdDA progenitor domain in the chick.[88] The role of these two genes is less clear in the mouse embryo: dreher (*Lmx1a$^{dr/dr}$*) mutant mice carrying a LOF mutation in the *Lmx1a* gene display a milder phenotype (approx. 30% less mdDA neurons), and *Lmx1a* alone is not sufficient for the ectopic induction of mdDA neurons and of *Msx1* in the hindbrain of transgenic mice.[14] The generation of postmitotic neurons is reduced in the midbrain FP of *Lmx1a$^{dr/dr}$* mice, concomitant with a reduced expression of the proneural bHLH TFs Ngn2 and Mash1 (Ascl1) in this domain.[14] The remaining mdDA neurons, however, differentiate properly in the *Lmx1a$^{dr/dr}$* mice, indicating that Lmx1a is not required for the correct specification of the mdDA fate in their progenitors.[14] *Msx1$^{-/-}$* mice display a 40% reduction in Ngn2$^+$ progenitor and mdDA precursor cell numbers[88]. The premature expression of *Msx1* in the midbrain FP of *ShhE-Msx1* transgenic embryos results in the premature repression of Nkx6-1 and induction of Ngn2 expression, leading to the earlier generation of mdDA neurons in the ventral midbrain of these transgenic mice.[88] The reduced generation of mdDA neurons in the *Lmx1a*- and *Msx1*-deficient mice is therefore primarily due to reduced neurogenesis from the mdDA progenitor domain and not to a defective mdDA fate specification in the progeny. The target gene in both cases appears to be the proneural bHLH TF *Ngn2* playing an important role in mdDA neurogenesis (see below). The LOF and GOF experiments, however, also indicate that Lmx1a and Msx1 must interact with additional signals providing A/P and D/V positional information, as either factor alone or together cannot induce mdDA neurons at ectopic dorsal or posterior locations in the chicken and/or mouse neural tube.[14,88] Both studies implicated Shh as the crucial signal providing D/V positional information,[14,88] and Ono et al.[14] established Otx2 as the factor conferring A/P positional information on ventral midbrain progenitors for the generation of mdDA neurons.

Otx2. The ectopic expression of Otx2 in the hindbrain FP is indeed sufficient for the ectopic induction of Lmx1a, Ngn2, and Mash1 and for the ectopic generation of mdDA neurons at this location.[14] Conversely, the deletion of *Otx2* from the ventral midbrain of *Nestin-Cre/+; Otx2$^{flox/flox}$* mice results in a reduced neurogenesis and loss of *Ngn2*, *Mash1*, *Hes5*, and Delta-like 1 (*Dll1*) expression in the medial FP concomitant with the loss of mdDA neurons.[39] These experiments, however, did not establish whether Otx2 has a cell-intrinsic function in the control of mdDA neurogenesis. Very recently, the analysis of mutant mice conditionally overexpressing (*En1$^{+/Cre}$; tOtx2$^{ov/ov}$*) or lacking (*En1$^{+/Cre}$; Otx2$^{flox/flox}$*) Otx2 in the midbrain has shed light on this issue.[89] The proliferation of mdDA progenitors is selectively enhanced and their cell cycle exit is delayed by the overexpression of Otx2 in *En1$^{+/Cre}$; tOtx2$^{ov/ov}$* mutants, whereas the opposite phenotype (strongly reduced mdDA progenitor proliferation and premature cell cycle exit) is seen in the *En1$^{+/Cre}$; Otx2$^{flox/flox}$* mice.[89] Remarkably, proliferation in the adjacent progenitor domains in which Otx2 is also expressed is not affected in these mutant mice, indicating a specific function of Otx2 in mdDA progenitors. Moreover, the response to Otx2 appears to be more pronounced in the caudal midbrain, as the rostral mdDA progenitors and neurons are less affected by the *Otx2* GOF or LOF.[89] In both mutants, the proliferation and cell cycle exit defects correlate with the misexpression of *Lmx1a*, *Msx1*, *Ngn2*, and *Mash1* in the mdDA progenitors, resulting in a reduced generation of mature mdDA neurons after *Otx2* LOF but increased generation after *Otx2* GOF. The overexpression of *Otx2* also causes an expansion of the ventral *Wnt1* domain and an increase of Cyclin D1 (Ccnd1) but a decrease of cyclin-dependent kinase inhibitor (cdki) p27^{Kip1} (Cdkn1b) expression in the midbrain FP of *En1$^{+/Cre}$; tOtx2$^{ov/ov}$* mutants, whereas the loss of *Otx2* in the *En1$^{+/Cre}$; Otx2$^{flox/flox}$* mice results in the opposite phenotype. Since Cyclin D1 and p27^{Kip1} are directly involved in the regulation of cell cycle progression and Cyclin D1 is a known target gene of the canonical Wnt signaling pathway,[90–92] Otx2 may control indirectly the proliferation and cell cycle exit of mdDA progenitors through the regulation of *Wnt1* expression. Thus, Otx2 also acts upstream of the Lmx1a/Msx1/Ngn2-controlled cascade required for the specification of the generic neuronal fate in mdDA progenitors besides its earlier inductive function (see the section "Otx2 and Wnt1 Are Required for the Establishment of the mdDA Progenitor Domain").

Ngn2 (Neurog2). The action of all TFs described previously converges on the induction of Ngn2 in mdDA progenitors/precursors, and Ngn2 is indeed a key regulator of mdDA neurogenesis.[12,93] Ngn2 belongs to the family of proneural bHLH TFs promoting the acquisition of a generic neuronal fate and repressing the alternative glial fate in neural progenitors but also implicated in the specification of neuronal subtypes in the CNS (reviewed by [94]). Ngn2 expression does not begin before E10.75, coinciding with the start of mdDA neurogenesis in the ventral midbrain.[20,88] At E11.5, Ngn2 is coexpressed with Mash1 in the midbrain FP and BP including the mdDA progenitor domain[12] (Fig. 4.1.2). In addition, Ngn2 is expressed in some postmitotic Nurr1+ mdDA precursors but not in mature Th+ mdDA neurons.[12] Nurr1+ mdDA precursor numbers are reduced to 10%–20% in the $Ngn2^{-/-}$ embryos at E11.5 but recover to approximately 40% at E14.5, while Th+/Pitx3+ mdDA neurons are virtually absent at E11.5 and recover to less than 50% of the wild-type numbers in the postnatal $Ngn2^{-/-}$ brain.[12,93] Although Mash1 itself is dispensable for mdDA neuron development, the recovery of mdDA precursor/neuron numbers in the $Ngn2^{-/-}$ embryos is due to a partial compensation by Mash1.[12] Substitution of *Ngn2* expression by *Mash1* rescues about 60% of the mdDA neurons, but less than 10% of the normal mdDA neuron numbers are generated in $Ngn2^{-/-};Mash1^{-/-}$ double mutants.[12] The most notable defect in the $Ngn2^{-/-}$ embryos is an initial accumulation of radial glia-like neural progenitors and absence of neuronal cell bodies in the medial FP correlating with a down-regulation of *Mash1*, *Hes5*, and *Dll1* (two other markers of proneural activity) expression and the reduced generation of postmitotic mdDA precursors, indicating a block of generic neuronal specification in $Ngn2^{-/-}$ mdDA progenitors.[12,93] Since the generation of Th+ mdDA neurons is also delayed in the $Ngn2^{-/-}$ embryos, Ngn2 also appears to be required for the differentiation of Nurr1+ mdDA precursors into Th+ mdDA neurons.[12] Notably, the remaining mdDA neurons in the $Ngn2^{-/-}$ mice differentiate normally into the SNc and VTA subpopulations and establish proper connections with their target fields in the forebrain, indicating that Ngn2 is not required for the acquisition of the mdDA-specific cell fate.[93] In fact, overexpression of Ngn2 in dorsal midbrain progenitors in vitro and in vivo is not sufficient for the ectopic induction of mdDA neurons.[12,93]

Altogether, the analyses of the *Foxa1/2*, *Lmx1a*, *Msx1*, *Otx2*, and *Ngn2* mutant mice indicate that these TFs, rather than conferring specific mdDA neuronal properties to ventral midbrain progenitors, are required for the acquisition of a generic neuronal phenotype by the corresponding progeny. Furthermore, their specific action in mdDA neuron development (in the case of *Lmx1a*, *Msx1*, and *Ngn2*) is best explained by their restricted expression within the mdDA progenitor domain or their functional compensation outside of this domain.

TFs conferring mdDA-specific characteristics to mdDA precursors

Nurr1 (Nr4a2). One of the first TFs implicated in mdDA neuron development is the nuclear receptor family member Nurr1 (reviewed by [95]). *Nurr1* expression starts at E10.5 in postmitotic mdDA precursors and persists in mature Th+ mdDA neurons (Figs. 4.1.2, 4.1.3); however, it is not restricted to the ventral midbrain but is also found in other metencephalic, diencephalic and telencephalic neuronal populations including the forebrain DA neurons.[22,96,97] The mdDA progenitor domain is established correctly in the $Nurr1^{-/-}$ embryos, as judged by the normal expression of Aldh1a1 at E9.5–E10.5[22]. The postmitotic $Nurr1^{-/-}$ mdDA precursors initiate part of their differentiation program including expression of the HD TFs *Pitx3*, En1/2, and *Lmx1b*, but *Th*, *Vmat2*, and *Dat* are not expressed in these precursors, indicating that they do not acquire the mdDA neurotransmitter phenotype.[22,96,98–101] Consequently, the mdDA precursors are lost shortly before birth in the $Nurr1^{-/-}$ embryos due to their apoptotic cell death.[22,99] Several in vitro studies established that Nurr1 binds to specific DNA recognition sites and directly activates the *Th* promoter in a cell line- and context-dependent manner.[102–104] The transcriptional activation of the *Th* gene by Nurr1 must indeed be modulated by other cofactors, as *Th* expression is not abolished in the forebrain DA and hindbrain catecholaminergic neurons of the $Nurr1^{-/-}$ embryos (also expressing *Nurr1* in the wild type).[96,98] Nurr1 lacks a steroid ligand-binding pocket and is therefore unlikely to be activated by these compounds.[105] A Nurr1-interacting protein (NuIP) expressed in adult mdDA neurons was identified recently, potentiating the transcriptional activity of Nurr1 on the *Th* promoter.[106] Another gene target and protein–protein interaction partner of Nurr1 is the cdki p57[Kip2] (Cdkn1c).[107] Expression of *p57[Kip2]* starts at late midgestation in mdDA progenitors and in postmitotic precursors coexpressing Nurr1.[107] The normal differentiation of mdDA precursors into mature Th+/Nurr1+ mdDA neurons is disrupted in the *p57[Kip2]−/−* embryos, leading to their premature cell death.[107] Since p57[Kip2], Nurr1, and related Nurr1 family proteins all induce cell cycle arrest of mdDA precursors and their differentiation into Th+ mdDA neurons in vitro, it appears that these factors coordinate the acquisition of the mature mdDA neurotransmitter phenotype with cell cycle exit of mdDA precursors.[106–108] Nurr1, however, cannot confer the generic neuronal phenotype to in vitro cultured neural precursors, although it enhances Th expression in these precursors, indicating that this aspect is controlled by another transcriptional network (see the section "TFs Conferring Generic Neuronal Properties to mdDA Precursors").[109,110]

Lmx1b. Based on the analysis of *Lmx1b−/−* mice, the LIM HD TF Lmx1b was postulated to control an aspect of mdDA fate specification other than the acquisition of the mdDA neurotransmitter phenotype, in contrast to Nurr1[100]. Expression of *Lmx1b* starts at the head-fold stage (E7.5) in an initially broader domain of the anterior neural plate, but by E9.5 it is confined to the MHB and to the ventral midbrain/caudal diencephalon FP and RP.[49,100] Lmx1b is coexpressed with Lmx1a in mdDA progenitors at this stage, but at E11.5 it is downregulated in these progenitors and becomes restricted to the postmitotic Lmx1a+/Nurr1+ progeny[14,88] (Fig. 4.1.2). At late midgestation and in adulthood, Lmx1b is expressed in Pitx3+/Th+ mdDA neurons.[100,111,112], Nurr1+/Th+ mdDA neurons are found in the *Lmx1b−/−* ventral midbrain at E12.5 but are lost at E15.5, and the HD TF Pitx3 is not expressed in the mutant ventral midbrain at E12.5, suggesting that Pitx3 is not induced in the absence of *Lmx1b*.[49,100] The *Lmx1b−/−* embryos, however, display an early disruption of IsO activity including the early loss of *Wnt1* and *En1* expression and the failure to induce *Fgf8* transcription at the MHB, which results in severe patterning defects in the MHR at later embryonic stages and could lead secondarily to the loss of Pitx3+ and Th+ mdDA neurons.[49] The restitution of *Lmx1b* expression at the MHB is indeed sufficient for the rescue of MHO activity and of mdDA neurons in *Wnt1-Lmx1b; Lmx1b−/−* transgenic mice.[113] Moreover, inactivation of *Lmx1b* in mdDA progenitors at E10.5 or in postmitotic mdDA neurons at E12.5 does not affect the differentiation or survival of these neurons.[113] Nevertheless, a redundant function of Lmx1b and Lmx1a in mdDA fate specification is also possible, as the LIM HD TFs Lhx1/5 (Lim1/2) are misexpressed in some Lmx1b+/Nurr1+ mdDA precursors of the *Lmx1a[dr/dr]* mutants, which will not differentiate into Th+/Pitx3+ mdDA neurons.[14]

Interestingly, Lmx1b maintains *Wnt1* expression at the MHB and induces it at ectopic locations, indicating that Lmx1b is acting upstream of Wnt1 in the development of the MHR.[49,114,115] A few Th+ cells are generated in the *Wnt1−/−* embryos at E11.5–E12.5, but these cells do not express Pitx3 and *Dat/Slc6a3*.[41] Wnt1 may therefore act downstream of Lmx1b in the control of Pitx3 expression in differentiating mdDA precursors.

Pitx3. The paired-like HD TF *Pitx3* was cloned based on its homology to *Pitx2*, another member of the Pitx family, and to bicoid-related homeobox genes[116,117] (reviewed by [118]). Expression of Pitx3 within the CNS starts at E11.0–E11.5, where it is highly specific for the mdDA neuronal population.[117] Pitx3, however, is also expressed in other tissues and organs outside the CNS.[116,119] The question of whether Pitx3 is expressed in all or only in a subpopulation of the mdDA neurons was highly debated at one time,[120,121] but it is now accepted that all SNc and VTA DA neurons of the adult rodent and human brain express Pitx3.[117,119,120] Based on the initiation of Pitx3 expression before or after Th expression, however, two ontogenetically distinct subpopulations can be distinguished in the mdDA lineage: cells located in the ventrolateral part of the rostral mesencephalon/caudal diencephalon initiate Pitx3 expression before Th, whereas cells located in the dorsomedial part express Th before Pitx3.[16] This initial segregation between Pitx3- and Th-expressing cells is not detected in the caudal midbrain and disappears as development proceeds.[16]

The differential expression of Pitx3 in the mdDA system at early developmental stages may provide one explanation for the phenotype of the *Pitx3* mutant mice. The *aphakia* (*ak*) mouse is a naturally occurring mutant with two major deletions close to and within the murine *Pitx3* gene abolishing Pitx3 expression in the homozygote *ak/ak* mice.[120–125] Apart from a disrupted eye lens development (and thus the name),[123] *ak/ak* mice show a progressive loss of the SNc DA neurons, which are nearly absent at birth, but only an approximately 50% reduction of the VTA DA neurons in the adult brain.[120–122,126] The loss of the SNc DA neurons results in a lack of nigrostriatal innervation and a drastically reduced DA content in the striatum of adult *ak/ak* mice, together with a reduced overall and spontaneous locomotor activity of these mice.[120–122,126] Since the expression of genes involved in mdDA neuron development and differentiation (*Nurr1, Lmx1b, En1/2, Aadc, Vmat2, D2R,* and *Dat*) except *Th* is not affected in the remaining mdDA neurons of the *ak/ak* mice,[120,122] Pitx3 appears to regulate the expression of Th in prospective SNc DA neurons. The analysis of a *Pitx3$^{eGFP/eGFP}$* reporter and at the same time a *Pitx3* null mutant mouse revealed that it is indeed the rostral ventrolateral mdDA subpopulation (expressing Pitx3 before Th) that fails to initiate Th expression and is subsequently lost due to the premature apoptotic death of these cells.[16,119] The rostral dorsomedial mdDA subpopulation (expressing Th before Pitx3) is less affected in the *Pitx3$^{eGFP/eGFP}$* mice, in agreement with Th probably not being a direct target of Pitx3 in these cells.[16] Although Pitx3 activates the *Th* promoter in vitro,[127,128] the loss of Th expression in the SNc DA neurons of the *Pitx3$^{eGFP/eGFP}$* mice cannot be the sole reason for their premature death, as a lack of Th expression and consequently of DA production does not lead to the loss of the corresponding cells.[129]

A more recent study identified another target gene of Pitx3, the enzyme Aldehyde dehydrogenase family 1, subfamily a1 (*Aldh1a1/Raldh1/Ahd2*), which catalyzes the oxidation of retinaldehyde into RA.[15] Aldh1a1 is expressed at E9.5–E10.5 in proliferating mdDA progenitors and postmitotic precursors,[22] but from E13.5 on, it is restricted to differentiating mdDA neurons, in particular to those of the SNc[15,130,31] (Fig. 4.1.2). *Aldh1a1* expression is largely abolished in homozygous *ak/ak* and reduced in heterozygous *ak/+* mice.[15] Retinoic acid has a crucial function in many developmental processes including neural patterning, neuronal differentiation, and survival[132]. Therefore, a lack of RA synthesis may underlie the premature loss of mdDA neurons in the *Pitx3*-deficient mice.[15] Maternal RA supplementation of *ak/ak* mice indeed causes a significant but incomplete rescue of mdDA neurons in the rostral midbrain (the prospective SNc) and has no effect on caudal midbrain (the prospective VTA) mdDA neurons, in line with the distinct Aldh1a1 expression pattern in these regions.[15] Although an mdDA neuron phenotype has so far not been reported in *Aldh1a1/Raldh1$^{-/-}$* mice,[133] Aldh1a1 is the only RA-synthesizing enzyme expressed in developing and mature mdDA neurons.[130] Pitx3 therefore also plays an important role in the survival of the SNc DA neurons by directly controlling the production of RA through Aldh1a1 in this region.[15]

Altogether, the analyses of the *Nurr1, Lmx1b,* and *Pitx3* mutant mice indicate that these TFs confer mdDA-specific properties to ventral midbrain precursors by activating the transcription of genes encoding either enzymes/transporters required for DA neurotransmission or other proteins/enzymes required for the maturation and survival of the mdDA but not of other neuronal populations. Their specific action in mdDA precursors/neurons is best explained by their restricted expression in these cells (Lmx1b and Pitx3) or because they have to interact with other mdDA-restricted factors (Nurr1). Mutations in the human *NURR1/NR4A2* and *PITX3* genes are associated with a rare form of familial Parkinson's disease (PD) and with the risk of developing sporadic PD, respectively,[134,135] highlighting the importance of these TFs for the proper maintenance of the human mdDA system.

Terminal (Late) Differentiation of Committed mdDA Precursors into Mature mdDA Neurons

Although a strict distinction between the previous *specification/early differentiation* and the *terminal (late) differentiation* of mdDA neurons is impossible and may be even artificial, we keep this terminology to point out that the postmitotic mdDA precursors expressing the TFs discussed in the previous section still have to acquire their characteristic neuronal as well as mdDA-specific features. This includes the development of axonal and dendritic morphologies and projections to their target areas, the acquisition of their intrinsic electrophysiological properties and firing patterns, the establishment of the proper synaptic contacts in the forebrain and midbrain, and the biosynthesis, storage, release, and reuptake of DA and its negative feedback regulation. Very little is known about how the mdDA neurons establish the proper contacts with their pre- and postsynaptic targets; due to space limitations, the reader is referred to an excellent recent review.[136] Even less, if anything, is known about how these neurons acquire their characteristic electrophysiological properties in the adult. A wealth of data, however, has accumulated on the physiological roles of the different enzymes, transporters, and receptors (such as Th, Aadc, Vmat2, Dat, and D2R) involved in DA biosynthesis and neurotransmission based on the analyses of null mutant mice for the corresponding genes. Notably, the anatomy of the mdDA system is not affected in most of these mutant mice; that is, the mdDA neurons themselves (even though they are not "DA neurons" in a strict sense) arise normally in the SNc and VTA and establish the proper connections with their forebrain target areas (reviewed by [137–139]). The only exception is the *D2R$^{-/-}$* mouse displaying a significant reduction of Th$^+$/Nurr1$^+$ mdDA neurons during development.[140] As D2R activation appears to induce the transcription of

Nurr1, the early loss of mdDA neurons may be due to the reduced expression of Nurr1 in the $D2R^{-/-}$ mice.[140] Although the majority of the DA-synthesizing enzymes, transporters, and receptors do not play a role in mdDA neuron development, they are necessary for postnatal survival, as they control proper DA neurotransmission/homeostasis required for several vital behaviors (reviewed by [137–139]). In addition to these genes, two secreted factors were implicated in the acquisition of the mature mdDA phenotype by controlling the differentiation of Nurr1+ mdDA precursors into Th+ mdDA neurons. However, as summarized here, their precise role remains controversial.

Secreted factors that may promote the differentiation of Nurr1+ mdDA precursors into Th+ mdDA neurons

Wnt5a. Wnt5a, another member of the secreted Wnt family, was reported to promote the differentiation of in vitro cultured Nurr1+ mdDA precursors into Th+ mdDA neurons with very little effect on their proliferation.[141] Wnt5a transcription starts at E8.75 in the cephalic flexure of the mouse embryo, and at midgestation it is expressed predominantly in midbrain FP/BP progenitors.[141,142] At later stages until birth, Wnt5a is expressed in a subpopulation of postmitotic mdDA precursors/neurons.[143] The numbers of Th+ mdDA neurons are not altered in the $Wnt5a^{-/-}$ mice at birth and these neurons also express *Pitx3* and *Dat*, indicating that Wnt5a is not required for the proper differentiation of mdDA neurons in vivo.[143] A transient increase in Th+ mdDA neuron numbers, however, is observed in the $Wnt5a^{-/-}$ embryos at E14.5, correlating with an increased proliferation of FP progenitors and accumulation of Nurr1+ mdDA precursors at E11.5, whose differentiation into Th+ mdDA neurons is delayed in the mutant embryos at E12.5.[143] The major phenotype of the $Wnt5a^{-/-}$ embryos is a distorted morphogenesis of the ventral midbrain from the earliest developmental stages on, in line with a function of Wnt5a as a "noncanonical" Wnt regulating cell orientation, migration, and adhesion rather than cell fate specification and differentiation of proliferating progenitors.[143] The transient deficits in the proliferation and differentiation of mdDA progenitors/precursors therefore appear to be secondary to the morphogenetic defects in the $Wnt5a^{-/-}$ embryos, and the precise role of Wnt5a in mdDA neuron development in vivo remains to be established.

Transforming growth factors (Tgfs). The members of the Tgf superfamily, Tgfα/Tgfa and Tgfβ/Tgfb, are also implicated in the control of mdDA neuron generation[144] (reviewed by [28,145]). The numbers of Th+ SNc neurons are reduced by about 50% on the first postnatal day in $Tgf\alpha^{-/-}$ mice, whereas no differences are detected in the Th+ VTA neurons.[144] The apoptotic death of mdDA precursors/neurons during late midgestation does not appear to be the reason for their reduced numbers at birth, suggesting that Tgfα controls earlier steps in the development of mdDA neurons.[144] However, since the embryonic development of the mdDA system was not studied in the $Tgf\alpha^{-/-}$ mice, the role of this secreted factor in mdDA neurogenesis remains unknown.

The two Tgfβ isoforms, *Tgfβ2/Tgfb2* and *Tgfβ3/Tgfb3*, and one of their receptors, *TbR-II/Tgfbr2*, are expressed in the midgestational ventral midbrain of the rat, albeit predominantly in postmitotic cells of the midbrain BP lying laterally to the mdDA neurogenic FP.[146] Although in ovo and in vitro experiments suggested a requirement of Tgfβ for the induction of the mdDA neuron phenotype,[146] the in vivo analyses of the corresponding mouse mutants are less conclusive.[147] A significant reduction in Th+ mdDA neuron numbers is observed in the $Tgf\beta2^{-/-};Tgf\beta3^{-/-}$ double mutants at E14.5, but the earlier steps in mdDA neuron development were not analyzed in this study.[147] The differentiation of the remaining mdDA precursors into Nurr1+/Th+ neurons, however, is not affected in the $Tgf\beta2^{-/-};Tgf\beta3^{-/-}$ mice, and the loss of Th+ mdDA neurons at a rather late stage (E14.5) may also hint at a requirement of Tgfβ for their proper survival.[146,147]

Maintenance of Mature mdDA Neurons throughout Late Gestation and Adulthood

This step refers to the survival of the newly born and still maturing and of the terminally differentiated mdDA neurons during late gestation and in the adult animal. We summarize here the evidence for some of the TFs described before as being also necessary for the survival of mature mdDA neurons and refer the reader to Chapter 4.2 in this volume for a comprehensive overview of this issue. The most notable feature of these TFs is that their single gene dosage is sufficient for ensuring the proper development of mdDA neurons but not for ensuring their survival in the postnatal and adult brain.

Foxa2 (Hnf3β). Apart from its early mdDA cell fate–specifying function (see the section "TFs Conferring Generic Neuronal Properties to mdDA Precursors"), Foxa2 also appears to control the survival of SNc DA neurons in adulthood. One-third of the heterozygous $Foxa2^{+/-}$ mice display a selective and initially unilateral loss of the SNc but not of the VTA DA neurons after 18 months of age, which is associated with asymmetric locomotor and other behavioral deficits in these mice.[13] Foxa2 haploinsufficiency thus seems to predispose to parkinsonian symptoms in mice, although the reduced penetrance of this phenotype suggests that other genetic modifiers or environmental factors may play an additional role. The downstream targets of Foxa2 providing this neuroprotective effect remain unknown.[13]

Nurr1 (Nr4a2). Heterozygous $Nurr1^{+/-}$ mice have no obvious developmental phenotype, but after 15 months of age these mice display several motor deficits that are associated with a decreased striatal DA content, a reduced number of SNc DA neurons, and reduced expression of *Dat* in these neurons.[148] These findings indicate a requirement of the full (diploid) Nurr1 dosage for the survival of mdDA neurons in adulthood. The tyrosine kinase *Ret* appears to be a gene target of Nurr1 mediating this survival-promoting activity.[96,149] Ret is an essential coreceptor for neurotrophic factors such as glial cell line–derived neurotrophic factor (GDNF), and GDNF is an important trophic factor required for the postnatal survival of mdDA neurons[26] (see Chapter 4.2, this volume). Conditional ablation of *Ret* in adult mice leads to a progressive and late-onset loss of SNc DA neurons.[150] Nurr1 may therefore support the survival of mdDA neurons in the adult midbrain by maintaining the expression of *Ret* in these neurons. Nurr1 also forms heterodimers with the retinoid X receptors (RXRs), and the activation of these heterodimers by cognate RXR ligands promotes the survival of mdDA neurons in vitro.[151] Retinoid signaling is indeed required for the proper maintenance of the mdDA system in the adult brain,[152] suggesting that the interaction between Nurr1 and RXRs may also have a functional relevance in vivo.

En1/2. Apart from its initial broad expression across the MHB, the homeobox gene *En1* is transcribed after birth in all mdDA neurons, whereas its paralogue, *En2*, is restricted to an mdDA subpopulation.[153] En1 expression is initiated progressively in these neurons, so only a few Th⁺ mdDA neurons coexpress En1 at E12 but almost all express En1 at E14.[154] As expected from the functional redundancy of En1/2,[155] Th⁺ mdDA neurons develop normally in $En1^{-/-}$ and $En2^{-/-}$ single mutant mice but are completely lost in newborn $En1^{-/-};En2^{-/-}$ double mutants.[153,154] The mdDA neurons undergo apoptotic cell death between E11 and E14 in the double mutants, in correlation with the onset of En1/2 expression in these neurons.[153,154] En1/2 are thus required cell autonomously for the survival of mdDA neurons after midgestation but not for their initial induction and cell fate specification/differentiation. The En TFs promote the survival of mdDA neurons in a gene dosage-dependent manner: the mdDA neuron population is unaffected in $En1^{+/-};En2^{-/-}$ but severely diminished in $En1^{-/-};En2^{+/-}$ compound heterozygote/homozygote mutants at birth, indicating a stricter requirement of En1 compared to En2.[153] En1 haploinsufficiency also leads to the progressive loss of mdDA neurons in the postnatal and adult brain. The SNc DA neurons degenerate progressively in the $En1^{+/-};En2^{-/-}$ compound mutants during the first 3 months of age, leading to reduced DA release in the striatum and to motor and nonmotor behavioral deficits.[156] Although Sgado et al.[156] reported an intact mdDA system in the postnatal and adult $En1^{+/-}$ mice, the progressive degeneration of SNc and VTA DA neurons starting after the third postnatal week and continuing until 6 months of age, after which no further cell loss is detected, was recently reported in these mice.[157]. As expected, the reduced mdDA neuron numbers in the $En1^{+/-}$ mice correlate with a reduced striatal DA content and with several motor and nonmotor deficits. Notably, the intraparenchymal infusion of En2 protein rescues the progressive mdDA neuron loss in the $En1^{+/-}$ mice, confirming a direct requirement of En proteins for mdDA neuron survival.[157] Although α-*synuclein* was reported as an En target gene whose expression is reduced in $En1^{-/-}$ single mutants and abolished in $En1^{-/-};En2^{-/-}$ double mutants,[153] the precise mechanism of En1/2-controlled mdDA neuron survival remains to be established.

CONCLUDING REMARKS

At the same time that developmental neurobiologists are beginning to unravel the molecular pathways and genetic cascades regulating distinct aspects of mdDA neuron development, it is becoming increasingly clear that the identity of each individual mdDA neuron is specified during development at a unique "node" within a complex four-dimensional (three-dimensional space plus time) network of interacting signaling and TFs. Several of the factors discussed in the previous sections will be reused at a later time point and for a different process during mdDA neuron development, depending on their spatiotemporal expression profile and the availability of different interaction partners or transcriptional targets. The complexity of the regulatory networks acting during mdDA neuron development is also increasing. We cannot assume a linear relationship, but instead have to take into consideration different feedforward and feedback interactions as well as cross-interactions between their individual components. Moreover, recent evidence indicates that microRNAs (miRNAs) add an additional level of control within these genetic networks by regulating the expression of some key factors such as Pitx3 within a negative feedback loop.[158]

The integration of all secreted factors and TFs described previously into one regulatory network, the identification of the yet unknown components of this network, and the resolution of the unclear issues will, it is hoped, advance our understanding of how an uncommitted neuroepithelial stem cell becomes a fully differentiated mdDA neuron with distinct properties, particularly in the light of regenerative therapies for neurological disorders such as PD but also in the attempt to understand the etiology and pathophysiology of severe human psychiatric disorders associated with mdDA neuron dysfunction.

ACKNOWLEDGMENTS

We apologize to colleagues whose work could not be cited due to space limitations, and we refer readers to the quoted reviews for many older and/or primary references. Our work was funded by the Federal Ministry of Education and Research (BMBF) in the framework of the National Genome Research Network (NGFN+ Functional Genomics of Parkinson Disease FKZ 01GS08174), by the Virtual Institute on Neurodegeneration and Ageing (VH-VI-252), by the Initiative and Networking Fund in the framework of the Helmholtz Alliance of Systems Biology and of Mental Health in an Ageing Society (HA-215), Bayerischer Forschungsverbund 'ForNeuroCell' (F2-F2410-10c/20697), European Union (mdDANEURODEV FP7-Health-2007-B-222999, EuTRACC LSHG-CT-2006-037445, EUMODIC LSHG-CT-2005-513769), and Deutsche Forschungsgemeinschaft (DFG) SFB 596, WU 164/3-2 and WU 164/4-1.

REFERENCES

1. Jorgensen JR, Juliusson B, Henriksen KF, et al. Identification of novel genes regulated in the developing human ventral mesencephalon. *Exp Neurol.* 2006;198:427–437.
2. Bjorklund A, Dunnett SB. Dopamine neuron systems in the brain: an update. *Trends Neurosci.* 2007;30:194–202.
3. Marin O, Smeets WJ, Gonzalez A. Evolution of the basal ganglia in tetrapods: a new perspective based on recent studies in amphibians. *Trends Neurosci.* 1998;21:487–494.
4. Smeets WJ, Gonzalez A. Catecholamine systems in the brain of vertebrates: new perspectives through a comparative approach. *Brain Res Brain Res Rev.* 2000;33:308–379.
5. Vernier P, Moret F, Callier S, et al. The degeneration of dopamine neurons in Parkinson's disease: insights from embryology and evolution of the mesostriatocortical system. *Ann NY Acad Sci.* 2004;1035:231–249.
6. Marín F, Herrero M-T, Vyas S, Puelles L. Ontogeny of tyrosine hydroxylase mRNA expression in mid- and forebrain: Neuromeric pattern and novel positive regions. *Dev Dyn.* 2005;234:709–717.
7. Puelles L, Rubenstein JL. Forebrain gene expression domains and the evolving prosomeric model. *Trends Neurosci.* 2003;26:469–476.
8. Smidt MP, Burbach JPH. How to make a mesodiencephalic dopaminergic neuron. *Nat Rev Neurosci.* 2007;8:21–32.
9. Smits SM, Burbach JP, Smidt MP. Developmental origin and fate of meso-diencephalic dopamine neurons. *Prog Neurobiol.* 2006;78:1–16.
10. Bonilla S, Hall AC, Pinto L, et al. Identification of midbrain floor plate radial glia-like cells as dopaminergic progenitors. *Glia* 2008;56:809–820.
11. Ferri ALM, Lin W, Mavromatakis YE, et al. Foxa1 and Foxa2 regulate multiple phases of midbrain dopaminergic neuron development in a dosage-dependent manner. *Development.* 2007;134:2761–2769.
12. Kele J, Simplicio N, Ferri ALM, et al. Neurogenin 2 is required for the development of ventral midbrain dopaminergic neurons. *Development.* 2006;133:495–505.
13. Kittappa R, Chang WW, Awatramani RB, McKay RDG. The foxa2 gene controls the birth and spontaneous degeneration of dopamine neurons in old age. *PLoS Biol.* 2007;5:e325.
14. Ono Y, Nakatani T, Sakamoto Y, et al. Differences in neurogenic potential in floor plate cells along an anteroposterior location: midbrain dopaminergic neurons originate from mesencephalic floor plate cells. *Development.* 2007;134:3213–3225.
15. Jacobs FMJ, Smits SM, Noorlander CW, et al. Retinoic acid counteracts developmental defects in the substantia nigra caused by Pitx3 deficiency. *Development.* 2007;134:2673–2684.
16. Maxwell SL, Ho HY, Kuehner E, et al. Pitx3 regulates tyrosine hydroxylase expression in the substantia nigra and identifies a subgroup of mesencephalic dopaminergic progenitor neurons during mouse development. *Dev Biol.* 2005;282:467–479.
17. Kawano H, Ohyama K, Kawamura K, Nagatsu I. Migration of dopaminergic neurons in the embryonic mesencephalon of mice. *Dev Brain Res.* 1995;86:101–113.
18. Bayer SA, Wills KV, Triarhou LC, Ghetti B. Time of neuron origin and gradients of neurogenesis in midbrain dopaminergic neurons in the mouse. *Exp Brain Res.* 1995;105:191–199.
19. Di Porzio U, Zuddas A, Cosenza-Murphy DB, Barker JL. Early appearance of tyrosine hydroxylase immunoreactive cells in the mesencephalon of mouse embryos. *Int J Dev Neurosci.* 1990;8:523–32.
20. Gates MA, Torres EM, White A, et al. Re-examining the ontogeny of substantia nigra dopamine neurons. *Eur J Neurosci.* 2006;23:1384–1390.
21. Marti J, Wills KV, Ghetti B, Bayer SA. A combined immunohistochemical and autoradiographic method to detect midbrain dopaminergic neurons and determine their time of origin. *Brain Res Protoc.* 2002;9:197–205.
22. Wallen A, Zetterstrom RH, Solomin L, et al. Fate of mesencephalic AHD2-expressing dopamine progenitor cells in *Nurr1* mutant mice. *Exp Cell Res.* 1999;253:737–746.
23. Ang S-L. Transcriptional control of midbrain dopaminergic neuron development. *Development.* 2006;133:3499–3506.
24. Prakash N, Wurst W. Specification of midbrain territory. *Cell Tissue Res.* 2004;318:5–14.
25. Baquet ZC, Bickford PC, Jones KR. Brain-derived neurotrophic factor is required for the establishment of the proper number of

dopaminergic neurons in the substantia nigra pars compacta. *J Neurosci.* 2005;25:6251–6259.
26. Pascual A, Hidalgo-Figueroa M, Piruat JI. et al. Absolute requirement of GDNF for adult catecholaminergic neuron survival. *Nat Neurosci.* 2008;11:755–761.
27. Burke RE. Ontogenic cell death in the nigrostriatal system. *Cell Tissue Res.* 2004;318:63–72.
28. Krieglstein K. Factors promoting survival of mesencephalic dopaminergic neurons. *Cell Tissue Res.* 2004;318:73–80.
29. Liu A, Joyner AL. Early anterior/posterior patterning of the midbrain and cerebellum. *Annu Rev Neurosci.* 2001;24:869–896.
30. Rhinn M, Picker A, Brand M. Global and local mechanisms of forebrain and midbrain patterning. *Curr Opin Neurobiol.* 2006;16:5–12.
31. Wurst W, Bally-Cuif L. Neural plate patterning: upstream and downstream of the isthmic organizer. *Nat Rev Neurosci.* 2001;2:99–108.
32. Boyl PP, Signore M, Annino A, et al. Otx genes in the development and evolution of the vertebrate brain. *Int J Dev Neurosci.* 2001;19:353–363.
33. Simeone A, Puelles E, Acampora D. The Otx family. *Curr Opin Genet Dev.* 2002;12:409–415.
34. Acampora D, Mazan S, Lallemand Y, et al. Forebrain and midbrain regions are deleted in $Otx2^{-/-}$ mutants due to a defective anterior neuroectoderm specification during gastrulation. *Development.* 1995;121:3279–3290.
35. Matsuo I, Kuratani S, Kimura C, et al. Mouse *Otx2* functions in the formation and patterning of rostral head. *Genes Dev.* 1995;9:2646–2658.
36. Ang SL, Jin O, Rhinn M, Daigle N, Stevenson L, Rossant J. A targeted mouse Otx2 mutation leads to severe defects in gastrulation and formation of axial mesoderm and to deletion of rostral brain. *Development.* 1996;122:243–252.
37. Puelles E, Acampora D, Lacroix E, et al. Otx dose-dependent integrated control of antero-posterior and dorso-ventral patterning of midbrain. *Nat Neurosci.* 2003;6:453–460.
38. Puelles E, Annino A, Tuorto F, et al. Otx2 regulates the extent, identity and fate of neuronal progenitor domains in the ventral midbrain. *Development.* 2004;131:2037–2048.
39. Vernay B, Koch M, Vaccarino F, et al. Otx2 regulates subtype specification and neurogenesis in the midbrain. *J Neurosci.* 2005;25:4856–4867.
40. Sussel L, Kalamaras J, Hartigan-O'Connor DJ, et al. Mice lacking the homeodomain transcription factor Nkx2.2 have diabetes due to arrested differentiation of pancreatic beta cells. *Development.* 1998;125:2213–2221.
41. Prakash N, Brodski C, Naserke T, et al. A Wnt1-regulated genetic network controls the identity and fate of midbrain-dopaminergic progenitors in vivo. *Development.* 2006;133:89–98.
42. Parr BA, Shea MJ, Vassileva G, McMahon AP. Mouse *Wnt* genes exhibit discrete domains of expression in the early embryonic CNS and limb buds. *Development.* 1993;119:247–261.
43. Wilkinson DG, Bailes JA, McMahon AP. Expression of the proto-oncogene *int-1* is restricted to specific neural cells in the developing mouse embryo. *Cell.* 1987;50:79–88.
44. Ye W, Shimamura K, Rubenstein JL, et al. FGF and Shh signals control dopaminergic and serotonergic cell fate in the anterior neural plate. *Cell.* 1998;93:755–766.
45. Cordes SP. Molecular genetics of cranial nerve development in mouse. *Nat Rev Neurosci.* 2001;2:611–623.
46. Broccoli V, Boncinelli E, Wurst W. The caudal limit of Otx2 expression positions the isthmic organizer. *Nature.* 1999;401:164–168.
47. Echevarria D, Vieira C, Gimeno L, Martinez S. Neuroepithelial secondary organizers and cell fate specification in the developing brain. *Brain Res Brain Res Rev.* 2003;43:179–191.
48. Rhinn M, Brand M. The midbrain-hindbrain boundary organizer. *Curr Opin Neurobiol.* 2001;11:34–42.
49. Guo C, Qiu H-Y, Huang Y, et al. Lmx1b is essential for Fgf8 and Wnt1 expression in the isthmic organizer during tectum and cerebellum development in mice. *Development.* 2007;134:317–325.
50. Crossley PH, Martin GR. The mouse Fgf8 gene encodes a family of polypeptides and is expressed in regions that direct outgrowth and patterning in the developing embryo. *Development.* 1995;121:439–451.
51. Davis CA, Joyner AL. Expression patterns of the homeobox containing genes *En1* and *En2* and the proto-oncogene *int-1* diverge during mouse development. *Genes Dev.* 1988;2:1736–1744.
52. Davis CA, Noble-Topham SE, Rossant J, Joyner AL. Expression of the homeobox-containing gene *En2* delineates a specific region of the developing mouse brain. *Genes Dev.* 1988;2:361–371.
53. Brodski C, Weisenhorn DM, Signore M, et al. Location and size of dopaminergic and serotonergic cell populations are controlled by the position of the midbrain-hindbrain organizer. *J Neurosci.* 2003;23:4199–4207.
54. Acampora D, Avantaggiato V, Tuorto F, Simeone A. Genetic control of brain morphogenesis through Otx gene dosage requirement. *Development.* 1997;124:3639–3650.
55. Echevarria D, Belo JA, Martinez S. Modulation of Fgf8 activity during vertebrate brain development. *Brain Res Brain Res Rev.* 2005;49:150–157.
56. Sato T, Joyner AL, Nakamura H. How does Fgf signaling from the isthmic organizer induce midbrain and cerebellum development? *Dev Growth Differ.* 2004;46:487–494.
57. Martinez S. The isthmic organizer and brain regionalization. *Int J Dev Biol.* 2001;45:367–371.
58. Sun X, Meyers EN, Lewandoski M, Martin GR. Targeted disruption of *Fgf8* causes failure of cell migration in the gastrulating mouse embryo. *Genes Dev.* 1999;13:1834–1846.
59. Chi CL, Martinez S, Wurst W, Martin GR. The isthmic organizer signal FGF8 is required for cell survival in the prospective midbrain and cerebellum. *Development.* 2003;130:2633–2644.
60. Blak AA, Naserke T, Weisenhorn DM, et al. Expression of Fgf receptors 1, 2, and 3 in the developing mid- and hindbrain of the mouse. *Dev Dyn.* 2005;233:1023–1030.
61. Walshe J, Mason I. Expression of FGFR1, FGFR2 and FGFR3 during early neural development in the chick embryo. *Mech Dev.* 2000;90:103–110.
62. Liu A, Li JY, Bromleigh C, et al. FGF17b and FGF18 have different midbrain regulatory properties from FGF8b or activated FGF receptors. *Development.* 2003;130:6175–6185.
63. Trokovic R, Trokovic N, Hernesniemi S, et al. FGFR1 is independently required in both developing mid- and hindbrain for sustained response to isthmic signals. *EMBO J.* 2003;22:1811–1823.
64. Trokovic R, Jukkola T, Saarimaki J, et al. Fgfr1-dependent boundary cells between developing mid- and hindbrain. *Dev Biol.* 2005;278:428–439.
65. Blak AA, Naserke T, Saarimaki-Vire J, et al. Fgfr2 and Fgfr3 are not required for patterning and maintenance of the midbrain and anterior hindbrain. *Dev Biol.* 2007;303:231–243.

66. Saarimaki-Vire J, Peltopuro P, Lahti L, et al. Fibroblast growth factor receptors cooperate to regulate neural progenitor properties in the developing midbrain and hindbrain. *J Neurosci.* 2007;27:8581–8592.
67. Placzek M, Briscoe J. The floor plate: multiple cells, multiple signals. *Nat Rev Neurosci.* 2005;6:230–240.
68. Hynes M, Poulsen K, Tessier-Lavigne M, Rosenthal A. Control of neuronal diversity by the floor plate: contact-mediated induction of midbrain dopaminergic neurons. *Cell.* 1995;80:95–101.
69. Hynes M, Porter JA, Chiang C, et al. Induction of midbrain dopaminergic neurons by Sonic Hedgehog. *Neuron.* 1995;15:35–44.
70. Wang MZ, Jin P, Bumcrot DA, et al. Induction of dopaminergic neuron phenotype in the midbrain by Sonic hedgehog protein. *Nat Med.* 1995;1:1184–1188.
71. Echelard Y, Epstein DJ, St-Jacques B, et al. Sonic Hedgehog, a member of a family of putative signaling molecules, is implicated in the regulation of CNS polarity. *Cell.* 1993;75:1417–1430.
72. Dessaud E, McMahon AP, Briscoe J. Pattern formation in the vertebrate neural tube: a sonic hedgehog morphogen-regulated transcriptional network. *Development.* 2008;135:2489–2503.
73. Fuccillo M, Joyner AL, Fishell G. Morphogen to mitogen: the multiple roles of hedgehog signalling in vertebrate neural development. *Nat Rev Neurosci.* 2006;7:772–783.
74. Hynes M, Stone DM, Dowd M, et al. Control of cell pattern in the neural tube by the zinc finger transcription factor and oncogene *Gli-1. Neuron.* 1997;19:15–26.
75. Blaess S, Corrales JD, Joyner AL. Sonic hedgehog regulates Gli activator and repressor functions with spatial and temporal precision in the mid/hindbrain region. *Development.* 2006;133:1799–1809.
76. Ericson J, Morton S, Kawakami A, et al. Two critical periods of Sonic Hedgehog signaling required for the specification of motor neuron identity. *Cell.* 1996;87:661–673.
77. Chiang C, Litingtung Y, Lee E, et al. Cyclopia and defective axial patterning in mice lacking Sonic hedgehog gene function. *Nature.* 1996;383:407–413.
78. Fogel JL, Chiang C, Huang X, Agarwala S. Ventral specification and perturbed boundary formation in the mouse midbrain in the absence of Hedgehog signaling. Dev Dyn. 2008;237:1359–1372.
79. Joksimovic M, Yun BA, Kittappa R, et al. Wnt antagonism of Shh facilitates midbrain floor plate neurogenesis. *Nat Neurosc.i* 2009;12:125–131.
80. Zervas M, Millet S, Ahn S, Joyner AL. Cell behaviors and genetic lineages of the mesencephalon and rhombomere 1. *Neuron.* 2004;43:345–357.
81. Sasaki H, Hui C-C, Nakafuku M, Kondoh H. A binding site for Gli proteins is essential for *HNF-3β* floor plate enhancer activity in transgenics and can respond to Shh in vitro. *Development.* 1997;124:1313–1322.
82. Ang SL, Wierda A, Wong D, et al. The formation and maintenance of the definitive endoderm lineage in the mouse: involvement of HNF3/forkhead proteins. *Development.* 1993;119:1301–1315.
83. Monaghan AP, Kaestner KH, Grau E, Schutz G. Postimplantation expression patterns indicate a role for the mouse forkhead/HNF-3 alpha, beta and gamma genes in determination of the definitive endoderm, chordamesoderm and neuroectoderm. *Development.* 1993;119:567–578.
84. Sasaki H, Hogan BL. Differential expression of multiple fork head related genes during gastrulation and axial pattern formation in the mouse embryo. *Development.* 1993;118:47–59.
85. Ang SL, Rossant J. HNF-3 beta is essential for node and notochord formation in mouse development. *Cell.* 1994;78:561–574.
86. Weinstein DC, Ruiz i Altaba A, Chen WS, et al. The winged-helix transcription factor *HNF-3β* is required for notochord development in the mouse embryo. *Cell.* 1994;78:575–588.
87. Jeong Y, Epstein DJ. Distinct regulators of Shh transcription in the floor plate and notochord indicate separate origins for these tissues in the mouse node. *Development.* 2003;130:3891–3902.
88. Andersson E, Tryggvason U, Deng Q, et al. Identification of intrinsic determinants of midbrain dopamine neurons. *Cell.* 2006;124:393–405.
89. Omodei D, Acampora D, Mancuso P, et al. Anterior-posterior graded response to Otx2 controls proliferation and differentiation of dopaminergic progenitors in the ventral mesencephalon. *Development.* 2008;135:3459–3470.
90. Herrup K, Yang Y. Cell cycle regulation in the postmitotic neuron: oxymoron or new biology? Nat Rev Neurosci. 2007;8:368–378.
91. Shtutman M, Zhurinsky J, Simcha I, et al. The cyclin D1 gene is a target of the beta-catenin/LEF-1 pathway. *Proc Natl Acad Sci USA.* 1999;96:5522–5527.
92. Tetsu O, McCormick F. Beta-catenin regulates expression of cyclin D1 in colon carcinoma cells. *Nature.* 1999;398:422–426.
93. Andersson E, Jensen JB, Parmar M, et al. Development of the mesencephalic dopaminergic neuron system is compromised in the absence of neurogenin 2. *Development.* 2006;133:507–516.
94. Bertrand N, Castro DS, Guillemot F. Proneural genes and the specification of neural cell types. *Nat Rev Neurosci.* 2002;3:517–530.
95. Perlmann T, Wallen-Mackenzie A. Nurr1, an orphan nuclear receptor with essential functions in developing dopamine cells. *Cell Tissue Res.* 2004;318:45–52.
96. Zetterstrom RH, Solomin L, Jansson L, et al. Dopamine neuron agenesis in Nurr1-deficient mice. *Science.* 1997;276:248–250.
97. Zetterstrom RH, Solomin L, Mitsiadis T, et al. Retinoid X receptor heterodimerization and developmental expression distinguish the orphan nuclear receptors NGFI-B, Nurr1, and Nor1. *Mol Endocrinol.* 1996;10:1656–1666.
98. Castillo SO, Baffi JS, Palkovits M, et al. Dopamine biosynthesis is selectively abolished in substantia nigra/ventral tegmental area but not in hypothalamic neurons in mice with targeted disruption of the Nurr1 gene. *Mol Cell Neurosci.* 1998;11:36–46.
99. Saucedo-Cardenas O, Quintana-Hau JD, Le WD, et al. Nurr1 is essential for the induction of the dopaminergic phenotype and the survival of ventral mesencephalic late dopaminergic precursor neurons. *Proc Natl Acad Sci USA.* 1998;95:4013–4018.
100. Smidt MP, Asbreuk CH, Cox JJ, et al. A second independent pathway for development of mesencephalic dopaminergic neurons requires Lmx1b. Nat Neurosci. 2000;3:337–341.
101. Smits SM, Ponnio T, Conneely OM. et al. Involvement of Nurr1 in specifying the neurotransmitter identity of ventral midbrain dopaminergic neurons. *Eur J Neurosci.* 2003;18:1731–1738.
102. Iwawaki T, Kohno K, Kobayashi K. Identification of a potential Nurr1 response element that activates the tyrosine hydroxylase gene promoter in cultured cells. *Biochem Biophys Res Commun.* 2000;274:590–595.
103. Kim KS, Kim CH, Hwang DY, et al. Orphan nuclear receptor Nurr1 directly transactivates the promoter activity of the tyrosine hydroxylase gene in a cell-specific manner. *J Neurochem.* 2003;85:622–634.
104. Sakurada K, Ohshima-Sakurada M, Palmer TD, Gage FH. Nurr1, an orphan nuclear receptor, is a transcriptional activator

of endogenous tyrosine hydroxylase in neural progenitor cells derived from the adult brain. *Development*. 1999;126:4017–4026.
105. Wang Z, Benoit G, Liu J, et al. Structure and function of Nurr1 identifies a class of ligand-independent nuclear receptors. *Nature*. 2003;423:555–560.
106. Luo Y, Xing F, Guiliano R, Federoff HJ. Identification of a novel Nurr1-interacting protein. *J Neurosci*. 2008;28:9277–9286.
107. Joseph B, Wallen-Mackenzie A, Benoit G, et al. p57^{Kip2} cooperates with Nurr1 in developing dopamine cells. *Proc Natl Acad Sci USA*. 2003;100:15619–15624.
108. Castro DS, Hermanson E, Joseph B, et al. Induction of cell cycle arrest and morphological differentiation by Nurr1 and retinoids in dopamine MN9D cells. *J Biol Chem*. 2001;276:43277–43284.
109. Kim JY, Koh HC, Lee JY, et al. Dopaminergic neuronal differentiation from rat embryonic neural precursors by Nurr1 overexpression. *J Neurochem*. 2003;85:1443–1454.
110. Park CH, Kang JS, Kim JS, et al. Differential actions of the proneural genes encoding Mash1 and neurogenins in Nurr1-induced dopamine neuron differentiation. *J Cell Sci*. 2006;119:2310–2320.
111. Asbreuk CHJ, Vogelaar CF, Hellemons A, et al. CNS expression pattern of *Lmx1b* and coexpression with *Ptx* genes suggest functional cooperativity in the development of forebrain motor control systems. *Mol Cell Neurosci*. 2002;21:410–420.
112. Dai JX, Hu ZL, Shi M, et al. Postnatal ontogeny of the transcription factor Lmx1b in the mouse central nervous system. *J Comp Neurol*. 2008;509:341–355.
113. Guo C, Qiu HY, Shi M, et al. Lmx1b-controlled isthmic organizer is essential for development of midbrain dopaminergic neurons. *J Neurosci*. 2008;28:14097–14106 (Added in proof).
114. Adams KA, Maida JM, Golden JA, Riddle RD. The transcription factor Lmx1b maintains Wnt1 expression within the isthmic organizer. *Development*. 2000;127:1857–1867.
115. Matsunaga E, Katahira T, Nakamura H. Role of Lmx1b and Wnt1 in mesencephalon and metencephalon development. *Development*. 2002;129:5269–5277.
116. Semina EV, Reiter RS, Murray JC. Isolation of a new homeobox gene belonging to the Pitx/Rieg family: expression during lens development and mapping to the aphakia region on mouse chromosome 19. *Hum Mol Genet*. 1997;6:2109–2116.
117. Smidt MP, van Schaick HS, Lanctot C, et al. A homeodomain gene Pitx3 has highly restricted brain expression in mesencephalic dopaminergic neurons. *Proc Natl Acad Sci USA*. 1997;94:13305–13310.
118. Smidt MP, Smits SM, Burbach JPH. Homeobox gene *Pitx3* and its role in the development of dopamine neurons of the substantia nigra. *Cell Tissue Res*. 2004;318:35–43.
119. Zhao S, Maxwell S, Jimenez-Beristain A, et al. Generation of embryonic stem cells and transgenic mice expressing green fluorescence protein in midbrain dopaminergic neurons. *Eur J Neurosci*. 2004;19:1133–1140.
120. Smidt MP, Smits SM, Bouwmeester H, et al. Early developmental failure of substantia nigra dopamine neurons in mice lacking the homeodomain gene *Pitx3*. *Development*. 2004;131:1145–1155.
121. van den Munckhof P, Luk KC, Ste-Marie L, et al. Pitx3 is required for motor activity and for survival of a subset of midbrain dopaminergic neurons. *Development*. 2003;130:2535–2542.
122. Hwang DY, Ardayfio P, Kang UJ, et al. Selective loss of dopaminergic neurons in the substantia nigra of Pitx3-deficient aphakia mice. *Brain Res Mol Brain Res*. 2003;114:123–131.
123. Rieger DK, Reichenberger E, McLean W, et al. A double-deletion mutation in the *Pitx3* gene causes arrested lens development in aphakia mice. *Genomics*. 2001;72:61–72.
124. Semina EV, Murray JC, Reiter R, et al. Deletion in the promoter region and altered expression of Pitx3 homeobox gene in the aphakia mice. *Hum Mol Genet*. 2000;9:1575–1585.
125. Varnum DS, Stevens LC. Aphakia, a new mutation in the mouse. *J Hered*. 1968;59:147–150.
126. Nunes I, Tovmasian LT, Silva RM, et al. Pitx3 is required for development of substantia nigra dopaminergic neurons. *Proc Natl Acad Sci USA*. 2003;100:4245–4250.
127. Cazorla P, Smidt MP, O'Malley KL, Burbach JP. A response element for the homeodomain transcription factor *Ptx3* in the tyrosine hydroxylase gene promoter. *J Neurochem*. 2000;74:1829–1837.
128. Lebel M, Gauthier Y, Moreau A, Drouin J. Pitx3 activates mouse tyrosine hydroxylase promoter via a high-affinity binding site. *J Neurochem*. 2001;77:558–567.
129. Zhou Q-Y, Palmiter RD. Dopamine-deficient mice are severely hypoactive, adipsic, and aphagic. *Cell*. 1995;83:1197–1209.
130. Niederreither K, Fraulob V, Garnier J-M, et al. Differential expression of retinoic acid-synthesizing (RALDH) enzymes during fetal development and organ differentiation in the mouse. *Mech Dev*. 2002;110:165–171.
131. Westerlund M, Galter D, Carmine A, Olson L. Tissue- and species-specific expression patterns of class I, III, and IV Adh and Aldh1 mRNAs in rodent embryos. *Cell Tissue Res*. 2005;322:227–236.
132. Maden M. Retinoic acid in the development, regeneration and maintenance of the nervous system. *Nat Rev Neurosci*. 2007;8:755–765.
133. Fan X, Molotkov A, Manabe S-I, et al. Targeted Disruption of Aldh1a1 (Raldh1) provides evidence for a complex mechanism of retinoic acid synthesis in the developing retina. *Mol Cell Biol*. 2003;23:4637–4648.
134. Fuchs J, Mueller JC, Lichtner P, et al. The transcription factor PITX3 is associated with sporadic Parkinson's disease. *Neurobiol Aging*. 2007;doi:10.1016/j.neurobiolaging.2007.08.014.
135. Le WD, Xu P, Jankovic J, et al. Mutations in NR4A2 associated with familial Parkinson disease. *Nat Gene.t* 2003;33:85–89.
136. Van den Heuvel DM, Pasterkamp RJ. Getting connected in the dopamine system. *Prog Neurobiol*. 2008;85:75–93.
137. Holmes A, Lachowicz JE, Sibley DR. Phenotypic analysis of dopamine receptor knockout mice; recent insights into the functional specificity of dopamine receptor subtypes. *Neuropharmacology*. 2004;47:1117–1134.
138. Smidt MP, Smits SM, Burbach JP. Molecular mechanisms underlying midbrain dopamine neuron development and function. *Eur J Pharmacol*. 2003;480:75–88.
139. Uhl GR, Li S, Takahashi N, et al. The VMAT2 gene in mice and humans: amphetamine responses, locomotion, cardiac arrhythmias, aging, and vulnerability to dopaminergic toxins. *FASEB J*. 2000;14:2459–2465.
140. Kim SY, Choi KC, Chang MS, et al. The dopamine D2 receptor regulates the development of dopaminergic neurons via extracellular signal-regulated kinase and Nurr1 activation. *J Neurosci*. 2006;26:4567–4576.
141. Castelo-Branco G, Wagner J, Rodriguez FJ, et al. Differential regulation of midbrain dopaminergic neuron development by Wnt-1, Wnt-3a, and Wnt-5a. *Proc Natl Acad Sci USA*. 2003;100:12747–12752.

142. Yamaguchi TP, Bradley A, McMahon AP, Jones S. A Wnt5a pathway underlies outgrowth of multiple structures in the vertebrate embryo. *Development*. 1999;126:1211–1223.
143. Andersson ER, Prakash N, Cajanek L, et al. Wnt5a regulates ventral midbrain morphogenesis and the development of A9-A10 dopaminergic cells in vivo. *PLoS ONE* 2008;3:e3517.
144. Blum M. A null mutation in TGF-alpha leads to a reduction in midbrain dopaminergic neurons in the substantia nigra. *Nat Neurosci*. 1998;1:374–377.
145. Roussa E, Krieglstein K. Induction and specification of midbrain dopaminergic cells: focus on SHH, FGF8, and TGF-ß. *Cell Tissue Res*. 2004;318:23–33.
146. Farkas LM, Dunker N, Roussa E, et al. Transforming growth factor-βs are essential for the development of midbrain dopaminergic neurons *in vitro* and *in vivo*. J Neurosci. 2003;23:5178–5186.
147. Roussa E, Wiehle M, Dunker N, et al. Transforming growth factor ß is required for differentiation of mouse mesencephalic progenitors into dopaminergic neurons in vitro and in vivo: ectopic induction in dorsal mesencephalon. *Stem Cells*. 2006;24:2120–2129.
148. Jiang C, Wan X, He Y, et al. Age-dependent dopaminergic dysfunction in Nurr1 knockout mice. *Exp Neurol*. 2005;191:154–162.
149. Wallen AA, Castro DS, Zetterstrom RH, et al. Orphan nuclear receptor Nurr1 is essential for Ret expression in midbrain dopamine neurons and in the brain stem. *Mol Cell Neurosci*. 2001;18:649–663.
150. Kramer ER, Aron L, Ramakers GM, et al. Absence of Ret signaling in mice causes progressive and late degeneration of the nigrostriatal system. *PLoS Biol*. 2007;5:e39.
151. Wallen-Mackenzie A, Mata de Urquiza A, Petersson S, et al. Nurr1-RXR heterodimers mediate RXR ligand-induced signaling in neuronal cells. *Genes Dev*. 2003;17:3036–3047.
152. Krezel W, Ghyselinck N, Samad TA, et al. Impaired locomotion and dopamine signaling in retinoid receptor mutant mice. *Science*. 1998;279:863–867.
153. Simon HH, Saueressig H, Wurst W, et al. Fate of midbrain dopaminergic neurons controlled by the engrailed genes. *J Neurosci*. 2001;21:3126–3134.
154. Alberi L, Sgado P, Simon HH. Engrailed genes are cell-autonomously required to prevent apoptosis in mesencephalic dopaminergic neurons. *Development*. 2004;131:3229–3236.
155. Hanks M, Wurst W, Anson-Cartwright L, et al. Rescue of the En-1 mutant phenotype by replacement of En-1 with En-2. *Science*. 1995;269:679–682.
156. Sgado P, Alberi L, Gherbassi D, et al. Slow progressive degeneration of nigral dopaminergic neurons in postnatal Engrailed mutant mice. *Proc Natl Acad Sci USA*. 2006;103:15242–15247.
157. Sonnier L, Le Pen G, Hartmann A, et al. Progressive loss of dopaminergic neurons in the ventral midbrain of adult mice heterozygote for Engrailed1. *J Neurosci*. 2007;27:1063–1071.
158. Kim J, Inoue K, Ishii J, et al. A microRNA feedback circuit in midbrain dopamine neurons. *Science*. 2007;317:1220–1224.

4.2 | Factors Shaping Later Stages of Dopamine Neuron Development

ROBERT E. BURKE

Following their birth in the prenatal period, dopamine neurons of the mesencephalon undergo a complex series of cellular events, in response to external cues, that ultimately result in the establishment of their phenotype, as reviewed in Chapter 4.1. In addition to their birth[1,2] and specification[3,4] during the prenatal period, these neurons undergo migration along radial glia from the subventricular zone, the site of their birth, to their positions in the mature nervous system in the ventral mesencephalon.[5,6] After these prenatal events (Fig. 4.2.1) that establish the individual identity of dopamine neurons and their group identify as the A9 and A10 nuclei of the ventral mesencephalon, these neurons confront very different challenges in the postnatal period. It is during this time that they must establish relationships with the rest of the brain. Their final adult number must be determined, so as to be appropriate for the size and number of neurons within anatomically related structures. They must send out axons to the appropriate targets and make synaptic contacts that are correct in their location and numbers. They must also receive afferent inputs from projecting nuclei that are correct to permit precise functional regulation. Surely the factors that regulate and determine these characteristics of mesencephalic dopamine neurons may have important relevance to the many neurological and psychiatric conditions in which these neurons may play a role, including Parkinson's disease, schizophrenia, and addictive and satiety behaviors, to name but a few. The many cellular responses that mediate important events in the postnatal maturation of dopamine neurons are numerous and include programmed cell death, axon guidance and pathfinding, axon sprouting and pruning, and dendrite formation and maturation. An attempt to adequately describe all of these important cellular responses and their regulation is beyond the scope of this brief chapter. We will focus on a single important event in the postnatal development of mesencephalic dopamine neurons: the determination of their final adult number.

NATURALLY OCCURRING CELL DEATH IN MESENCEPHALIC DOPAMINE NEURONS

Like most other neuronal populations, the dopamine neurons of the mesencephalon form a larger population during development than ultimately exists in adulthood.[7] During the postnatal period, these neurons undergo a naturally occurring cell death (NCD) event (also known as *developmental cell death*). Natural cell death has been identified in the substantia nigra (SN) in both rats[8,9] and mice.[10] The morphology of this death event has been identified as apoptotic by electron microscopy and light microscopy, by both terminal deoxyribonucleotide transferase mediated dUTP-digoxigenin nick end labeling (TUNEL) labeling and immunostaining for the activated form of caspase-3, in conjunction with nuclear chromatin counterstaining.[10] These studies have validated the use of the light microscope to identify and quantify these apoptotic profiles, following either thionin staining[11] or suppressed silver staining,[12] in order to clearly and distinctively label the intranuclear chromatin clumps characteristic of apoptosis (Fig. 4.2.2). Whereas in other developmental settings other nonapoptotic morphologies of cell death have been identified, including cytoplasmic and autophagic forms,[13] these forms have not been identified in the SN by either electron microscopy or the suppressed silver stain. The latter is a sensitive technique for screening neuron populations at the light microscopic level for alternate morphologies of cell death.[14]

In order to determine the time course of NCD specifically for dopamine neurons of the SN, we have used immunohistochemistry for tyrosine hydroxylase (TH) to define the dopaminergic phenotype in combination with a thionin counterstain to identify apoptosis[9] (Fig. 4.2.2). The NCD event in the SN begins on embryonic day 20 (E20) in rats and reaches a peak on postnatal day 2 (PND2), defined as the day after birth. The event reaches a nadir by PND8 to PND12 but then resurges on PND14 before ceasing on PND28 (Fig. 4.2.3). Thus, in the rat, the event is largely postnatal and is biphasic, with the major phase being the

4.2: FACTORS SHAPING LATER STAGES OF DOPAMINE NEURON DEVELOPMENT

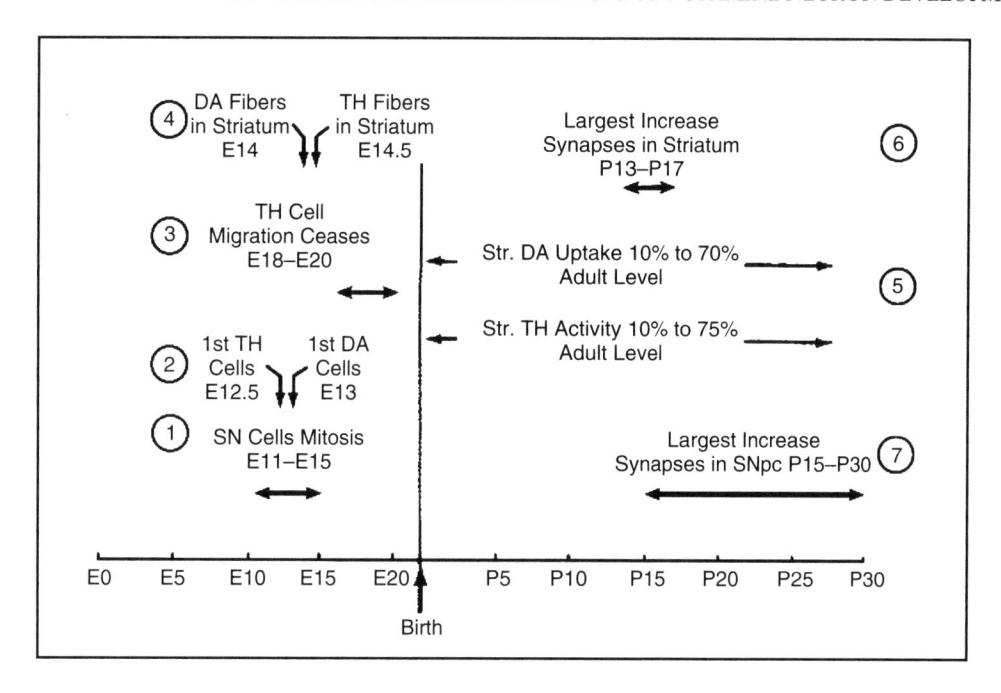

FIGURE 4.2.1. Important milestones in the development of the nigrostriatal dopaminergic system in rats. (1) Substantia nigra dopamine neurons are born between E11 and E15, with a peak on E13.[1,2] (2) Immunoreactivity for TH is first observed on E12.5[3] and for dopamine on E13.[4] (3) Prior to E18, dopaminergic neurons can be observed from the aqueduct to the ventral pial surface of the mesencephalon in association with radial glia; they are likely to be migrating from their locus of origin to their final positions in the mesencephalon.[6] By E20, dopamine neurons assume a topography similar to that of the adult brain, so it is likely that extranigral migration has ceased. (4) Dopaminergic fibers are first observed in the striatum by TH immunohistochemistry at E14.5[3] and by dopamine immunohistochemistry at E14.[4] (5) Differentiation of dopamine terminals takes place postnatally, indicated by large increases in TH activity and dopamine uptake between birth and PND30.[20] (6) Synapses form in the striatum postnatally, with the most rapid increase occurring between PND 13 and 17.[19] (7) The largest increase in synapses in SNpc occurs between P15 and 30.[15]

first, just after birth. The time course of this event is similar in the mouse.[10] It should not be assumed, based on these data, that apoptosis occurs in these species in the developing dopaminergic population exclusively within the perinatal and postnatal periods. Our analysis began at E19, so it remains possible that there is an earlier independent NCD event. The postnatal event does, however, occur after mitosis has ceased among these neurons,[1,15] and so it is believed to determine their final adult number.

The mechanistic basis for this biphasic time course is not known, but distinct developmental events are probably involved. The first phase of NCD occurs at a time when the nigral dopaminergic innervation of the striatum is being completed; it is partial and localized to the ventrolateral striatum at E18 and is essentially complete by PND4.[16] Therefore, the magnitude of this death event may be regulated by early target contact and support, and by competition among projecting dopamine neurons for this support, as envisioned by classic neurotrophic theory.[7,17,18] Several important developmental events occur within the nigrostriatal system during the second phase of cell death. There is a maximal level of production of synapses within the striatum[19] and within the substantia nigra pars compacta (SNpc),[15] the latter indicating the maturation of afferent projections to the SNpc (Fig. 4.2.1). While the mechanisms underlying the biphasic time course are unknown, it has an important role in determining the final adult number of these neurons and therefore in the interpretation of developmental studies of this system. As will be detailed and illustrated below, we have observed instances in which the second phase appears to permit a "fine tuning" of the final adult number; that is, while a particular experimental manipulation may alter the number of surviving neurons after the first phase of NCD, the number is "retuned" to normal control values after the second phase.

The question often arises regarding the magnitude of the NCD event in dopamine neurons: how many are lost? Unfortunately, this is not precisely known due to methodological limitations. First, It is not possible to use information about the number of apoptotic profiles in sections during NCD to derive the number of neurons that are lost, because the duration of persistence of a given apoptotic profile in living mammalian brain is not

162   GENES IN DEVELOPMENT

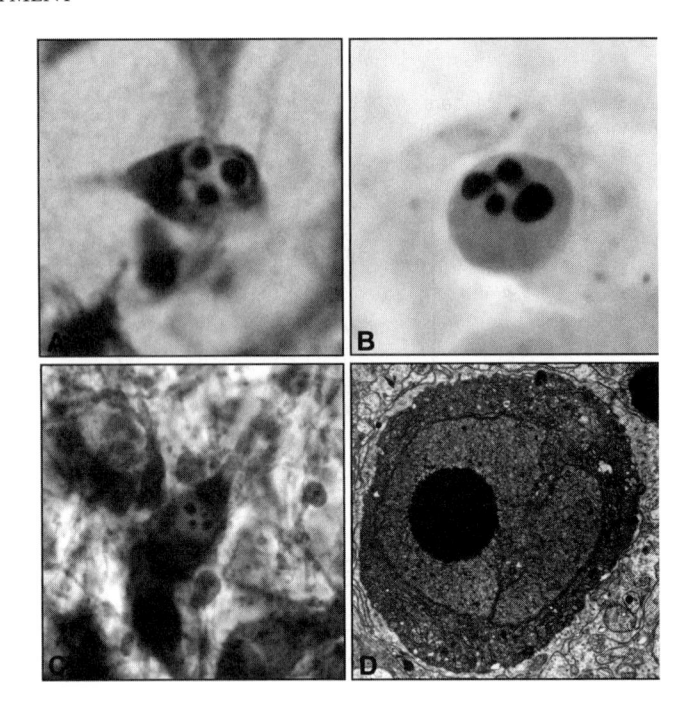

FIGURE 4.2.2 Apoptosis in SNpc during postnatal development. (A) Thionin stain of the SNpc of a normal rat at PND8. Within the nucleus of this neuron are three intensely and homogeneously stained round chromatin clumps with sharp, clearly defined edges. These chromatin clumps are highly characteristic of apoptosis at the light microscopic level. Note that this profile, in spite of the presence of apoptotic chromatin in its nucleus, has some preservation of neuronal morphology, including a polygonal shape and a dendrite. (B) Suppressed silver stain of an apoptotic profile at PND2 in a normal rat. Four intensely argyrophilic chromatin clumps are observed. *Source*: Adapted from[8]. (C) Immunoperoxidase stain for TH, with a thionin counterstain, in the SNpc 24 hours following axon-sparing excitotoxic striatal target lesioning at PND7. The brown reaction product identifies this neuron as dopaminergic. The four intranuclear chromatin clumps, stained by the thionin counterstain, are characteristic of apoptosis. *Source*: Adapted from[44]. (D) An electron micrograph of an apoptotic profile in the SNpc 24 hours following excitotoxic striatal target lesioning at PND7. The single intensely and homogeneously electron-dense clump of chromatin within the nucleus is a defining feature of apoptosis. The intact nuclear and cellular membranes in this degenerating profile are also characteristic of apoptotic cell death.[44] (See Color Plate 4.2.2.)

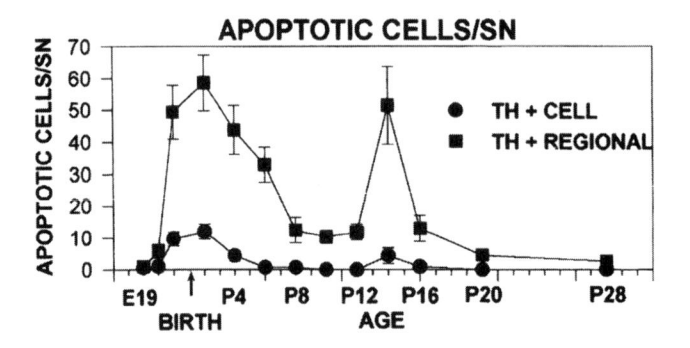

FIGURE 4.2.3. The time course of NCD in dopaminergic neurons of the SNpc (E embryonic, P postnatal). Natural cell death among dopamine neurons in rats is largely postnatal and biphasic, with an initial major peak just after birth and a second minor peak at PND14. Some of the apoptotic profiles express TH within their cytoplasm, as shown in Figure 4.2.2C, and therefore are identifiable as dopaminergic at the cellular level. However, the majority of apoptotic profiles in the SNpc have lost surrounding cytoplasm and are found in close proximity to TH-positive dopaminergic neurons. These apoptotic profiles are identified as within the SNpc at the regional level. The counts of apoptotic profiles by either cellular or regional criteria demonstrate the same postnatal time course.[9]

precisely known. Second, if one attempts to determine the number of neurons lost by simply counting Nissl-stained neuronal profiles, the problem is that not all dopamine neurons of the SN are confined to a single well-delineated somatotopic location, unlike motor neuron nuclei of the brainstem and spinal cord. Even where they are most concentrated, in the SNpc, they are not the only population present; some GABAergic neurons are also found here.[10] Third, If one tries to count the number of immunostained, TH-positive neurons, there is also a methodological concern, because the level of phenotypic markers within each cell increases during this developmental period,[20] making the number of profiles detected by immunohistochemistry steadily increase even as these neurons undergo NCD. When such counts of TH-positive profiles have been performed,[21] they show decrements in TH-positive neuron number, confirming NCD, but these are almost certainly underestimates of the number of neurons lost. In addition, as discussed below, when the antiapoptotic protein Bcl-2 is overexpressed specifically within catecholamine neurons, there is suppression of NCD in SNpc,[10] and this results in a 30% increase in the adult number of SN dopamine neurons. If we assume that other antiapoptotic proteins may compensate for this early genetic deletion of Bcl-2, then this again may be an underestimate.

The occurrence of apoptotic NCD in mesencephalic dopamine neurons has been observed in a primate species, the African green (vervet) monkey.[22] Apoptosis was identified by the formation of distinct nuclear chromatin clumps in TH-positive neurons, by TUNEL labeling, and by immunostaining for the activated form of caspase-3. Interestingly, these investigators showed that the single peak of NCD in this species, at E80, corresponded to the period of maximal development of striatal contact, estimated by striatal dopamine levels, and TH-positive fiber staining. These investigators point out that although the timing of the NCD in the vervet, in terms of gestational age, is quite different from that in rodents, its timing in terms of the maturational state of the brain is quite similar. In the rhesus monkey, a species with the same gestational period as the vervet, E78 is equivalent to PND2 in the rat. Thus, NCD in mesencephalic dopamine neurons occurs during the period of maximal development of striatal contact in both species, as would be anticipated based on concepts of classic neurotrophic theory.

Limited information exists about the molecular pathways mediating programmed cell death during NCD in dopamine neurons. At the risk of oversimplification, programmed cell death mechanisms can be thought of as being mediated by three major interacting pathways: the intrinsic pathway, through which caspase-9 is activated by cytochrome c release from mitochondria; the extrinsic pathway, in which the interaction of ligands with cell surface receptors leads to the activation of caspase-8; and the endoplasmic reticulum (ER) stress pathway, which is postulated to result in activation of caspase-12.[23] Endoplasmic reticulum stress seems unlikely to be involved in NCD, as there is no expression of CCAAT/enhancer-binding protein-homologous protein (CHOP), an important mediator of apoptosis in that context.[24,25] To date, there is no information about the possible role of the extrinsic pathway.

However, several lines of evidence indicate that components of the intrinsic pathway play a role. Members of the Bcl-2 family take an important part in controlling the release of cytochrome c and other cell death mediators from mitochondria, with the ensuing activation of the caspase cascade leading to cell death.[26,27]. When the antiapoptotic protein Bcl-2 is overexpressed specifically within catecholamine neurons under the control of the TH promoter in transgenic mice, there is suppression of NCD in SNpc,[10] resulting in a 30% increase in the adult number of SN dopamine neurons. The related antiapoptotic protein Bcl-x also appears to play a developmental role in the determination of the final adult number of SN dopamine neurons. When Bcl-x is selectively knocked out within dopaminergic neurons, about 30% fewer neurons survive at 1 month of age, and this deficit persists into adulthood.[28] Homozygous Bax null mice show diminished levels of apoptotic NCD; however, the null mutation does not result in an increased adult number of SN dopaminergic neurons.[29] This result suggests that other pro-apoptotic members of the Bcl-2 family may be able to mediate death in the absence of Bax. In normal rats, an increase occurs in the ratio of Bax to Bcl-2 in the nigra during the NCD period, supporting the possibility of a role for these proteins in regulating this death event.[30]

Caspases of the intrinsic pathway are involved in NCD of dopamine neurons. During development, the activated form of caspase-9 can be identified within apoptotic profiles in the SNpc.[31] The activated form of the downstream effector, caspase-3, can also be identified,[32] as can protein cleavage products of caspase-3.[33]

SYSTEMS REGULATION OF NCD IN MESENCEPHALIC DOPAMINE NEURONS

Classic neurotrophic theory postulates that neuronal populations are created in excess numbers during embryogenesis and undergo an NCD event that determines their final adult number. It proposes that the magnitude of this event is regulated by competition among members of the neuronal population for support by their target, and that a component of this competition is for limiting protein neurotrophic factors provided by the target.[7,14,34] Thus, a neuron within a developing population that fails to contact its target, or to succeed in competing for the relevant trophic factor, or to successfully transport the neurotrophic survival signal retrogradely to the neuron cell body, will undergo NCD. Classically, it has been proposed that this competitive strategy serves two principal purposes: to correctly match the number of neurons in a projecting neuronal population with its target and to eliminate projecting neurons with incorrect target connections.[17] It is important to recognize that these classic concepts rest largely on experiments performed on neural systems with peripheral targets.[17,35] Nevertheless, evidence exists that central neurons also are likely to match their numbers to the size of their targets. A well-characterized example is the matching of cerebellar granule cell numbers with the numbers of their target Purkinje cells.[36,37]

Many early in vitro studies suggested that developing dopaminergic neurons of the mesencephalon were supported, both in their viability and differentiation, by preparations derived from their target, the striatum. Prochiantz et al.[38] first demonstrated that primary dissociated striatal cells grown in coculture with embryonic mesencephalic dopamine neurons enhanced their differentiation. Hemmendinger et al.[39] subsequently showed that embryonic dopamine neurons formed the appropriate number and type of axons in coaggregate culture only in the presence of the appropriate target tissue. This group also went on to show that the striatum in coculture with dopamine neurons increased their viability.[40] In keeping with classic neurotrophic theory, Tomozawa and Appel[41] demonstrated that a soluble factor purified from rat striatum was able to support the viability and differentiation of embryonic mesencephalic dopamine neurons.

Studies performed in vivo have also provided evidence that the striatal target is likely to influence the development of SN dopamine neurons. We have observed that an axon-sparing lesion of the striatum, made during development, results in a smaller number of SNpc dopamine neurons in adulthood[42] (Fig. 4.2.4). This decrease occurs in the absence of significant injury to striatal dopaminergic terminals[43] or any direct injury to the nigra itself. We subsequently showed that, as neurotrophic theory would predict, this loss of the striatal target during development results in a striking augmentation of the nigral NCD event.[44] Since the lesioned striatal neurons may provide not only retrograde support to dopamine neurons but also afferent projections to them, it was important to selectively assess the role of retrograde support by ablating dopaminergic axon terminals within the striatum. This was done by intrastriatal injection of the selective catecholaminergic neurotoxin 6-hydroxydopamine (6-OHDA). This selective lesion abrogates striatal retrograde support via the nigrostriatal projection but spares striatal afferents to nigra. This lesion does indeed result in an augmentation of the nigral NCD event,[45] supporting the conclusion that retrograde influences are likely to regulate the death event, at least in part. In further keeping with neurotrophic theory, axotomy lesioning of the nigrostriatal axons within the medial forebrain bundle also induces apoptosis among dopamine neurons of the SNpc.[46]

Both the excitotoxic striatal target lesion model and the 6-OHDA terminal destruction model show a developmental dependence in their ability to augment NCD among dopamine neurons of the SNpc. In the target injury model, the effect is entirely limited to the first 2 postnatal weeks.[47] In the 6-OHDA model, the effect is largely also limited to this time, but unlike the target injury model, some apoptosis can also be induced in adulthood due to a direct toxic effect, as discussed further below.[45,48] Thus, the developmental period of principal death induction by these two lesions, both of which interfere with target support of developing dopaminergic neurons, corresponds to the period of NCD. Such a correspondence between the period of target dependence and NCD has been observed for other systems.[17]

In all of these studies of the effect of selective lesions on the development of dopamine neurons of the SNpc, the light microscopic morphology of cell death was apoptotic and was no different from that observed during NCD. In this respect, dopamine neuron developmental death differs from some other systems in which the morphology of induced death may differ from the natural form.[35] It is important to note, however, that although apoptotic profiles visualized by thionin and silver staining do not differ among these lesion models of induced death and NCD, differences in morphology can be observed between the 6-OHDA model and the others when profiles are visualized by immunostaining for the activated form of caspase-3 and its protein cleavage products.[32,33] In the 6-OHDA model, immunostaining is localized to the cytoplasm of some apoptotic neurons as well as in the nucleus, whereas in NCD, the striatal target lesion and axotomy models, it is localized strictly to the nucleus. Therefore, in the 6-OHDA model, induced apoptosis is quite likely to be due not only to loss of target support during the first 2 postnatal weeks, but also to a direct toxic effect. This possibility is supported by the observation that intrastriatal 6-OHDA is capable of inducing apoptotic death in SNpc dopamine neurons even in adult rodents, long after loss of target is capable of doing so.[47,48]

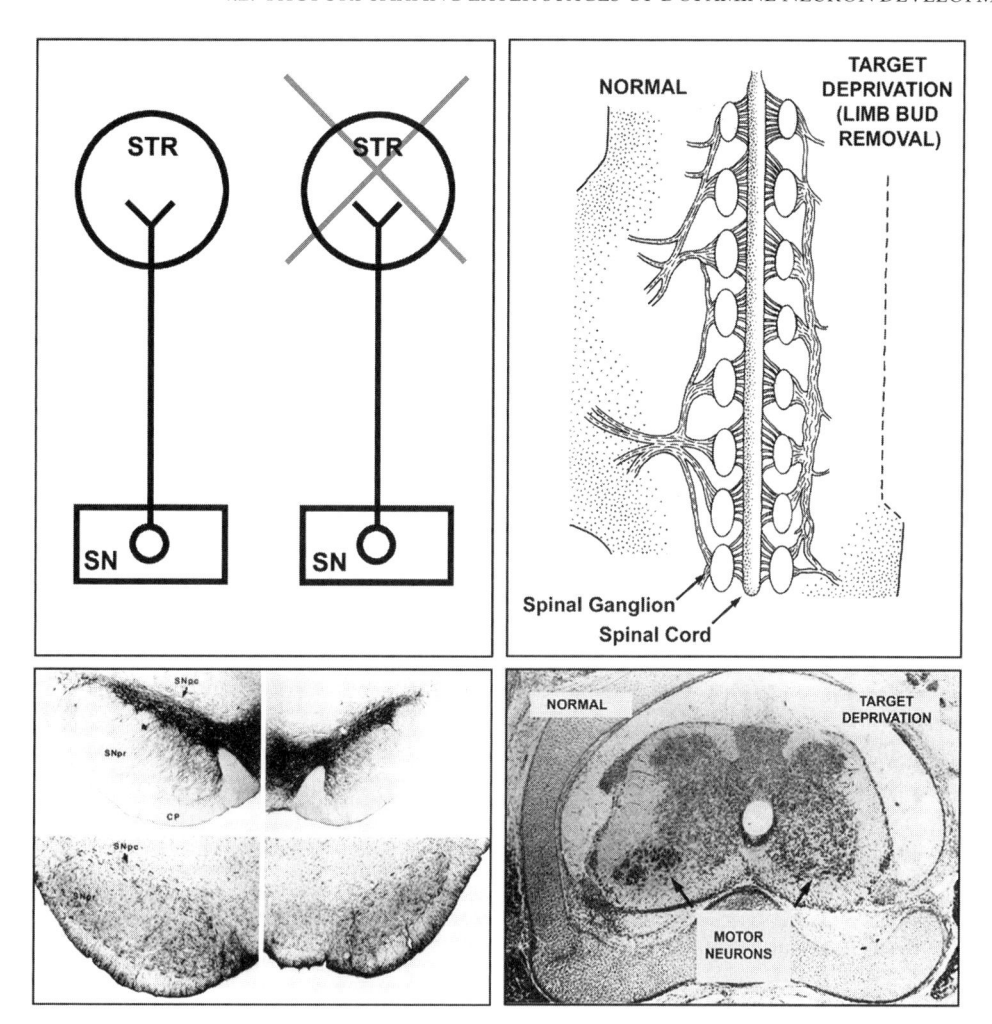

FIGURE 4.2.4. Early postnatal axon-sparing injury to the striatum, the target of the mesencephalic dopaminergic projection, results in a diminished number of dopaminergic neurons and a decrease in the size of the SNpc in adulthood. The top left panel is a schematic representation of a unilateral striatal lesion, made with the excitotoxin quinolinic acid, at PND7. This lesion destroys neurons intrinsic to the striatum but spares dopaminergic terminals. This lesion results in a 10-fold induction of postnatal NCD[44] and a reduction in the adult number of dopaminergic neurons. The bottom left panel shows a reduced number of SNpc dopamine neurons, demonstrated by TH immunostaining (upper) in an adult rat following striatal lesion at PND7. The bottom left lower panel shows a reduced size of the entire SN, demonstrated by GFAP immunostain and thionin counterstain.[42] The right-hand panels are adapted from Hamburger.[142] The top right panel is a schematic representation of a unilateral limb bud extirpation in a chick. This procedure results in a diminished adult number of motor neurons in the adult on the operated side, as shown in the bottom right panel.

The developmental NCD event and the determination of the mature number of neurons within a nucleus are not determined exclusively by interactions with the target. There is abundant experimental evidence to suggest that these events are also regulated by afferent anterograde influences on developing neurons (see [49]). These afferent influences may operate by a variety of mechanisms, including release of neurotrophic factors or release of specific neurotransmitters, resulting in changes in intracellular calcium stores. These afferent effects may be provided by anterograde neural projections, by local glia, or in cell autonomous fashion by the developing neurons themselves.[49] Relatively little is known about the possible role of afferent systems in the regulation of dopamine neurons. The possibility of regulation of NCD in mesencephalic dopamine neurons by an afferent projection from the locus ceruleus (LC) was proposed by Alonso-Vanegas and coworkers based on studies in transgenic mice with augmented release of brain-derived neurotrophic factor (BDNF) by this LC projection to SNpc.[50] These studies will be presented in greater detail below in the section on BDNF.

MOLECULAR REGULATION OF NCD IN MESENCEPHALIC DOPAMINE NEURONS

Regulation by Extrinsic Molecules

GDNF (glial cell line–derived neurotrophic factor)

Since its discovery, GDNF has been considered a candidate neurotrophic factor for SN dopamine neurons.[51] In support of the possibility that it may serve as a striatal target-derived factor, its mRNA is expressed in striatum, most abundantly during the early postnatal period.[52–58] At a cellular level, developmental expression of GDNF mRNA in the striatum occurs exclusively within medium-sized neurons. In spite of its name, there is no detectable expression of GDNF in striatal glia during normal postnatal development.[59] GDNF protein is expressed within the postnatal striatum.[60] At a cellular level, its protein expression can be identified rarely in medium-sized striatal neurons, but most expression is identified within the neuropil, some of which is positive for TH. In this location, it is likely that GDNF is undergoing retrograde transport to the cell bodies of dopamine neurons of the SNpc. Specific retrograde transport of GDNF by the dopaminergic nigrostriatal system has been demonstrated.[61] Given that GDNF mRNA is more abundant in striatum than in SN[59] and that mRNA for the GDNF receptor GFRα1 is more abundant in SN than in striatum,[62] it would be predicted that GDNF protein undergoes retrograde transport from striatum to SNpc, as envisioned by neurotrophic theory.

Within the SNpc, the abundant expression of mRNA for the GDNF receptor GFRα1 and its signaling tyrosine kinase Ret[63–65] offer additional support for the concept that developing SN dopamine neurons are receptive to GDNF. In further keeping with such a possibility, GFRα1 protein is identified within TH-positive fibers of the striatum, where it is likely to be undergoing anterograde transport from, and retrograde transport to, the cell bodies of dopamine neurons of the SNpc.[62] The latter has been demonstrated directly in sympathetic neurons in vitro[66] and is likely to occur in nigrostriatal dopaminergic axons in vivo as well.

The principal evidence that has been marshaled against a possible neurotrophic role for GDNF is that mice homozygous null for GDNF or GFRα1 show no decrease in the number of SN dopamine neurons on the day of birth.[67–71] However, these mice die shortly after birth because of developmental abnormalities of the kidney and the enteric nervous system. Therefore, they die before much of the NCD event has occurred (see Fig. 4.2.3), making it impossible to observe a relevant postnatal phenotype. Furthermore, these conventional null mutations were not temporally regulated, so it is quite possible that compensatory changes may have taken place during prenatal development. These considerations offer ample grounds for regarding the negative observations in the homozygous null mice with skepticism as far as a phenotype affecting the SN dopaminergic system is concerned.

To further evaluate the role of GDNF, we assessed its ability to support the viability of mesencephalic dopamine neurons in a unique *postnatal* primary culture model.[72] The critical feature of this approach is that the culture model is established when these neurons would normally undergo NCD, and therefore effects on viability would be expected to have more relevance to the endogenous regulation of this event. We found that among factors that had been reported up to that time to support mesencephalic dopamine neurons in *embryonic* culture, including BDNF, transforming growth factors (TGF) β1, 2, and 3, neurotrophin 3, β-fibroblast growth factor, TGFα, and epidermal growth factor, only GDNF augmented survival, and it did so by suppressing apoptosis.[73]

These observations generalized to the in vivo context. Direct injection of GDNF protein into the striatum at PND2 suppresses the level of NCD in SN dopamine neurons by 60%.[74] To explore whether endogenous GDNF may play a role, we investigated the effect of passive immunization by direct injection of neutralizing antibodies into the striatum. In these experiments, a threefold induction of NCD in dopamine neurons was observed.[74] This ability of anti-GDNF antibodies to induce NCD was limited to the first postnatal week. Therefore, although our earlier lesion experiments had suggested that SN dopamine neurons were dependent on striatal target until PND14, that is, throughout the first and second phases of NCD, the dependence on GDNF was observed only during the first phase. The dependence of postnatal dopamine neurons on GDNF for their survival is supported by the observations of Granholm and colleagues based on monitoring viability following transplant to adult wild-type mice. Implants from GDNF null mice show improved survival if they are pre-treated with GDNF prior to implantation.[75] Thus, in its ability to regulate acutely the NCD event of SN dopamine neurons both in vitro and in vivo, GDNF fulfills important criteria for an endogenous neurotrophic factor for these neurons.

Classic neurotrophic theory would also predict that a sustained increase in the supply of a limiting target-derived factor during the NCD period should augment the number of neurons that survive and result in an increased number of neurons in adulthood. It is important to emphasize that an adequate test of this prediction requires a *sustained* increase in expression. Single intrastriatal injections of GDNF at PND2 have been shown not to have a lasting effect on the number of surviving dopamine neurons.[76] However, single, early developmental injections are exceedingly unlikely to have a lasting effect, given that the NCD event takes place over a 2-week period. Therefore, In order to achieve a sustained overexpression of GDNF specifically in the target regions of the mesencephalic dopaminergic projections, we utilized a double transgenic approach.[77] Mice transgenic for calcium/calmodulin-dependent protein kinase II-tetracycline-dependent transcription activator (CaMKII-tTA) permit regionally specific expression of the transactivator tTA within the cortex, striatum, and hippocampus based on the regionally selective expression of CaMKII. When these mice are crossed with mice transgenic for BiTetO-LacZ-ratGDNF (rGDNF), a regionally selective (and regulatable) increase in GDNF expression can be achieved. These double transgenic mice (CaMKII-tTA-BiTetO-LacZ- rGDNF double transgenic, or CBLG-DT) demonstrate staining for LacZ specifically in the striatum (where it was most abundant), hippocampus, and cortex, as previously described for other CaMKII-tTA double transgenic mice.[77,78] The selective overexpression of GDNF in these regions not only permits evaluation of GDNF specifically as a target-derived factor, but also avoids a detrimental effect of GDNF on SN dopamine neuron development when it is expressed *within* these neurons under the regulation of the TH promoter.[79] The precise mechanism of the detrimental effect of this cell autonomous expression of GDNF within SNpc dopamine neurons, which results in a decrease in their number and size, is unknown. Nevertheless, the occurrence of this effect clearly necessitates a regionally specific overexpression of GDNF in target structures alone in order to meaningfully assess target effects. The CBLG-DT mice overexpress GDNF in forebrain structures throughout the period of NCD,[80] and within the striatum, at the cellular level, β-galactosidase reporter expression is observed strictly within medium striatal neurons, as it is for endogenous GDNF.[80] Increased expression of GDNF within striatal medium- sized neurons throughout development leads to a 46% increase in the number of SN dopaminergic neurons surviving the first phase of NCD. This increase does not, however, persist into adulthood. We therefore conclude, based on these studies in the double transgenic mice and the aforementioned studies with neutralizing antibodies, that although striatal GDNF is both necessary and sufficient for the regulation of SN dopamine neuron survival during the first phase of NCD, it alone is not sufficient to lead to a lasting increase in their adult number. We postulate that at some time between PND7 and adulthood, the number of these neurons is regulated back to their normal wild-type number by mechanisms that we do not yet understand. This regulation does not seem to be due to a "rebound" phenomenon in which there is an augmented level of NCD during the second phase of death on PND14. On the contrary, levels of apoptosis are still reduced in the double transgenic mice on that postnatal day. Therefore, the time course and mechanism of normalization of the adult number of SN dopamine neurons in the CBLG-DT mice at a later time in postnatal development are unknown.

Just as there is no increase in the adult number of SN dopamine neurons, there is also no increase in dopaminergic innervation of the striatum in adult CBLG-DT mice. We assessed the morphological features of TH-positive and dopamine transporter (DAT)-positive fibers, TH and vesicular monoamine transporter (VMAT2) protein expression, biochemical measures of dopamine and its metabolites, and physiological measures of dopamine release and reuptake, and no changes in the CBLG-DT mice were found. However, the response of the ventral tegmental area (VTA) dopaminergic system (A10) to sustained overexpression of GDNF in its targets in the CBLG-DT mice was quite different from that of SN dopamine neurons (A9). In the VTA, there was a 55% increase in the number of adult dopamine neurons compared with wild-type controls.[80] In addition, adult CBLG-DT mice demonstrated increased dopaminergic innervation of cortical regions that are targets of A10 mesencephalic dopamine neurons, assessed by both TH and DAT-positive fiber analysis. This morphological phenotype was accompanied by a behavioral phenotype as well: the CBLG-DT mice demonstrated an augmented motor activity response to amphetamine. Thus, there are fundamental differences between the SN (A9) and VTA (A10) dopaminergic systems in their developmental response to GDNF expression in target structures.

Thus, for the first phase of NCD in SN dopamine neurons, GDNF fulfils many of the criteria required by classic neurotrophic theory for a target-derived neurotrophic factor. GDNF mRNA and protein are both expressed in the striatal target, maximally during the early postnatal period, and both the GDNF receptor GFRα1 and its signaling tyrosine kinase Ret are abundantly expressed in the SNpc. The acute experiments that we have described suggest an ability of GDNF activity in striatal target to regulate the postnatal NCD event, and chronic studies in the CBLG-DT mice also suggest that striatal GDNF is a limiting factor for the survival of dopamine neurons of the SNpc during the first phase of NCD. The concept that limited availability of GDNF-GFRα1-Ret signaling during postnatal development regulates the final adult number of mesencephalic dopamine neurons is also supported by the observations of Mijatovic and colleagues, who showed that expression of a constitutively active mutant of Ret (Met918Thr) results in a 26% increase in their number.[81]

In spite of this evidence, recent investigations of mice with a regionally selective knockout of the Ret tyrosine kinase in mesencephalic dopamine neurons raise questions about the precise role of GDNF-GFRα1-Ret signaling in the development of these neurons. Jain and colleagues achieved a conditional deletion of Ret in mesencephalic dopamine neurons by crossbreeding mice with a floxed wild-type human Ret cDNA, targeted to the mouse Ret locus, with mice that have Cre targeted to the DAT locus. This strategy results in an excision of Ret in all dopaminergic neurons by the time of birth.[82] In spite of the absence of Ret throughout the NCD period, these mice have normal numbers of dopamine neurons in both the SNpc and VTA and normal patterns of striatal dopaminergic innervation in adulthood. Using a similar approach, Kramer and colleagues likewise observed no effect of selective ablation of Ret in mesencephalic dopamine neurons on their number or their innervation of the striatum at 3 months of age.[83] Although they observed a decline in these measures at 12 months of age and later, these changes were attributed to a late degeneration of the nigrostriatal system, not a developmental effect.

There are several considerations that may reconcile these negative observations with the aforementioned studies that support a role for GDNF in the development of mesencephalic dopamine neurons. First, it is possible that in the absence of Ret, alternate signaling pathways by GDNF play a role.[84] GDNF signaling independent of Ret has been reported to activate a GFRα1-associated Src-like kinase.[85,86] In Ret-deficient kidney epithelial cells, GDNF is capable of inducing Met receptor tyrosine kinase by acting through Src-kinase activity and mediating a biological response.[87] GFRα1 has been reported to bind to neural cell adhesion molecule (NCAM), thereby promoting its ability to bind GDNF, leading to activation of protein tyrosine kinases Fyn and FAK.[88] Chao and colleagues have shown that another cell adhesion molecule, integrin αv, is coexpressed with NCAM in mesencephalic dopamine neurons, and both are up-regulated upon treatment with GDNF.[89] Blocking integrin αv activity with neutralizing antibodies abrogated the effects of GDNF in dopamine neurons to promote survival and differentiation.

Alternatively, even if GDNF signaling in mesencephalic neurons is absolutely dependent on Ret, it is possible that elimination of GDNF-Ret signaling during embryogenesis results in a compensatory activation of other neurotrophic influences on the early viability of dopamine neurons. For example, in the Jain study, following elimination of Ret during embryogenesis, there was no alteration in the number of dopamine neurons in the SNpc in adult mice at 6–12 months of age.[82] However, Pascual and colleagues have demonstrated that after the acute ablation of GDNF by the use of a tamoxifen-inducible Esr1-Cre transgene, there is a 60% loss of SNpc dopamine neurons within 7 months.[90] These results indicate that if GDNF is indeed entirely dependent on Ret signaling within mesencephalic dopamine neurons, then relevant phenotypes may be obscured by the abrogation of critical gene function during embryogenesis.[90]

In conclusion, there is much evidence that GDNF is a candidate target–derived neurotrophic factor for dopamine neurons of the SNpc during early postnatal development. However, confirmation of such a role at the gene level is required, and such confirmation will require regional specificity and temporally regulated approaches that are rapidly inducible to minimize compensatory changes in this complex system.

Neurturin

Among the other members of the GDNF family of ligands, attention has focused on neurturin as a possible neurotrophic factor for dopamine neurons. It was originally cloned based on its ability to support the survival of sympathetic neurons in culture.[91] It was subsequently shown in many studies both to protect and to restore SN dopamine neurons in vitro and in vivo.[92–94] Its highest levels of mRNA expression in the striatum are at PND15,[92] suggesting that perhaps it plays a role during the second phase of NCD. We have found, however, that patterns of neurturin mRNA expression are not highly suggestive of a role as a target-derived factor for SN dopamine neurons. Unlike GDNF, neurturin mRNA is much more abundant in SN than it is in striatum.[95] Therefore, rather than serving as a target-derived factor, neurturin might be considered to serve in a local nigral autocrine or paracrine role. However, no developmental regulation of neurturin expression occurs within the SN.[95] In addition, although the neurturin receptor, GFRα2, is highly expressed in the SN and is developmentally regulated,[95] it does not appear to colocalize with SN dopamine neurons.[93] Thus, the precise physiological role of neurturin, if any, in regulating the normal development of dopamine neurons remains to be defined.

BDNF

A novel neurotrophic activity, distinct from nerve growth factor (NGF), was first identified in the conditioned medium of a glioma cell line on the basis of effects on the viability of chick dorsal root ganglion neurons.[96] Using this bioassay, BDNF was purified[97] and subsequently cloned by Liebrock, Barde and colleagues[98] (reviewed by Lindsay[99]). BDNF was the first purified molecule demonstrated to directly support the viability of embryonic dopamine neurons.[100] Subsequently, it was shown to provide neuroprotection for embryonic mesencephalic dopamine neurons against neurotoxins.[101] Nevertheless, the physiological role of endogenous BDNF in regulating the development of mesencephalic dopamine neurons has been uncertain, in part because conventional null mutations of either BDNF or its receptor TrkB are incompatible with long-term survival.[102–104] The possibility of a physiologically relevant role for BDNF in the development of these neurons is supported by the expression of mRNA[105,106] and protein[107] for the BDNF receptor, TrkB, in SNpc neurons, indicating that these neurons are likely to be receptive to BDNF.

It is unlikely that BDNF plays a role as a target-derived neurotrophic factor, as envisioned by neurotrophic theory,[7] because it is not expressed in the striatum during development.[108] Alternatively, it has been proposed that BDNF may serve as an afferent projection–derived factor[109,110] for these neurons. This possibility has been supported by observations in dopamine β-hydroxylase-BDNF transgenic mice, in which there is increased BDNF protein expression within the LC afferent projection to the SNpc. In these mice there is a 50% increase in the adult number of dopamine neurons in the SNpc.[50] These investigators postulated that this effect is due to a suppression of postnatal NCD in dopamine neurons by BDNF, resulting in an increase in their adult number. The potential physiological relevance of these observations is supported by the expression of BDNF mRNA in the LC[111] and BDNF protein in the SNpc [112].

An alternative, or additional, possible role for BDNF in regulating the development of dopamine neurons of the SN is based on the observation that they express BDNF,[113,114] and it may therefore serve in an autocrine fashion to support their development.[115] Such a possibility is supported by the observations of Baquet et al.,[114] who examined the effects of local mesencephalic-hindbrain deletion of BDNF by means of a Wnt1-Cre transgene expressed in BDNF$^{fl/fl}$ mice. These investigators observed a diminished number of TH-positive neurons at birth. However, this effect appeared to be due principally to loss of phenotype, because there was no alteration in the number of SNpc neurons based on NeuN staining.

In order to examine the possible role of endogenous BDNF in dopamine neuron development, and to circumvent the problem of perinatal lethality in conventional nulls, we have used approaches based on acute blockade of BDNF activity in the postnatal SNpc, by local injection of either neutralizing antibodies or a competitive antagonist ligand for the TrkB receptor.[116] Intranigral injection of a neutralizing antibody to BDNF induced apoptosis in SNpc, in keeping with a role for endogenous local BDNF in regulating this postnatal NCD. These observations were confirmed by the use of a conformationally constrained synthetic peptide competitive antagonist of BDNF.[117] This antagonist, L2-8, is based on the incorporation of cysteines into the Lys41 and Lys50 positions of the amino acid sequence of the second β-hairpin loop (Loop 2) of BDNF and the formation of a disulfide bond between them. This bond constrains the Loop 2 peptide sequence and mimics its native structure. Since Loop 2 is essential for interaction with the TrkB receptor, this peptide has competitive antagonist properties.[117] Like neutralizing antibodies, L2-8, when injected locally into the postnatal SNpc, also induces apoptosis in dopamine neurons of the SNpc. It therefore appears likely that local endogenous BDNF does indeed regulate NCD in postnatal dopamine neurons. A developmental time course analysis reveals that BDNF appears to play a role only during the first phase of NCD in these neurons, not during the second phase.[116]

To confirm that the induction of apoptosis by neutralizing antibodies was affecting dopamine neurons in these studies, we determined their number 6–7 days after injection (at PND12), before the occurrence of the second phase of developmental cell death at PND14. As expected, induction of apoptosis does result in a decreased number of postmitotic dopamine neurons. However, this decrease in the number of neurons at PND12 does not persist; following postnatal injection of neutralizing antibodies, the number of neurons in adulthood is normal. These observations provide yet another example, like that described above for GDNF, of the ability of the nigrostriatal system to normalize its adult numbers in spite of an alteration after the first phase of NCD. As discussed for GDNF, the mechanism of this normalization is not known, but these observations with acute, early developmental BDNF neutralization provide a second example of the ability of the nigrostriatal system to "fine tune" the final number of dopamine neurons between the termination of the first phase of NCD and adulthood.

In spite of this evidence from acute studies that local BDNF regulates the first phase of developmental cell death in dopamine neurons, we found that elimination of BDNF in brain during the postnatal period by a knockout approach in BDNF$^{fl/fl}$:Nestin-Cre mice did not affect the number of dopamine neurons surviving after the first phase of NCD or in adulthood. These results therefore provide yet another example of a discordance between studies utilizing acute ablation approaches and those using genetic ablation of neurotrophic molecules during development. As discussed above for GDNF, one possible explanation for these disparate results is that compensatory changes occur following gene knockout during embryonic development. Such compensation may take the form of an enhanced role of other neurotrophins, rather than regulation at the level of the TrkB receptor, because mice hypomorphic for TrkB (with only 25% of wild-type protein levels) show a 40% reduction in the adult number of SN dopamine neurons.[118] However, it is uncertain whether this reported alteration is due to a developmental deficiency or an adult degenerative phenomenon. In addition, this alteration may not be due to loss of BDNF-TrkB signaling directly within SNpc dopamine neurons, because other investigators have demonstrated that regionally selective deletion of the TrkB receptor within these neurons does not result in a change in their adult number.[83]

Although we did not observe alterations in the adult number of dopamine neurons in BDNF$^{fl/f}$:Nestin-Cre mice, we did observe disruption of their anatomical organization. In adulthood, the BDNF$^{fl/f}$:Nestin-Cre mice showed a loss of definition of the SNpc-SNpr boundary and the appearance of ectopic dopaminergic neurons in the SNpr. Similar abnormalities had been depicted in the studies by Baquet and colleagues in BDNF$^{fl/fl}$:Wnt1-Cre mice[114] and by Baker et al. in BDNF$^{-/-}$ mice at PND14.[119]

Although these investigations suggest a role for local BDNF in SN in regulating the first phase of NCD, they do not identify its source. It may be provided by an afferent projection from the LC, as postulated by Alonso-Vanegas and colleagues,[50] or on an autocrine basis from SNpc neurons themselves.[114] These studies do suggest that an acute genetic ablation or knock-down approach will be required to fully illuminate the role of BDNF in regulating the development of dopamine neurons of the SNpc, as discussed above in relation to GDNF signaling.

Other neurotrophic factors

Although many neurotrophic factors have been reported to have effects on the development of SN dopamine neurons in embryonic primary culture, for the purposes of this review, we will consider only factors that have been reported to have effects on development in vivo.

TGFβ. GDNF is a member of the transforming growth factor-β superfamily.[84] Within this family, TGFβ2 and -3 have also been demonstrated to support the viability of embryonic (E14) mesencephalic dopamine neurons.[53] It has been proposed that, rather than acting directly and independently as neurotrophic molecules, the TGFβs act to increase the neurotrophic potency of other neurotrophic factors, including GDNF,[120] sonic hedgehog and fibroblast growth factor (FGF-8).[121] However, these observations were made in embryonic mesencephalic cultures. In postnatal primary mesencephalic cultures, established during the rodent NCD period, none of the TGFβs supported the viability of dopamine neurons, and the ability of GDNF to do so was demonstrable in the absence of both serum and glia. Whether any of the in vivo effects of GDNF to suppress NCD in dopamine neurons, as presented above, requires any of the TGFβs is unknown.

The possibility that TGFβs may play a role in vivo in the development of mesencephalic dopamine neurons is suggested by studies of mice null for homeodomain interacting protein 2 (HIPK2).[122] HIPK2 is a transcriptional cofactor that directly interacts with receptor-regulated Smads (R-Smads), which, in turn, regulate TGFβ signaling.[123] Mice homozygous null for HIPK2 have normal numbers of mesencephalic dopamine neurons at E12.5, but by the time of birth, throughout the postnatal period, and in adulthood, their numbers are reduced by about 40%.[122] This reduction is not due to an impairment in the neurogenesis of these neurons, but rather to an augmentation of apoptosis during NCD. The possibility that these effects are due to abrogation of TGFβ3 signaling is supported by the observations that HIPK2 is essential for TGFβ3-mediated survival in tissue culture and that mice homozygous null for TGFβ3 have a similar reduction in the number of dopamine neurons associated with an increase in NCD.[122]

TGFα. TGFα mRNA is highly expressed in the striatum[124,125] and reaches its maximal level of expression on PND1.[124] It has therefore been evaluated for a possible role as a target-derived neurotrophic factor for SN dopamine neurons. Alexi and Hefti[126] have demonstrated that it is capable of supporting the differentiation and survival of embryonic mesencephalic dopamine neurons. However, it is not capable of supporting the viability of SN dopamine neurons in postnatal primary culture.[73] Nevertheless, TGFα remains of interest in relation to the development of SN dopamine neurons, because homozygous null mice have only about 50% as many of these neurons as wild-type controls.[127] This difference is attributable neither to diminished phenotype expression, because it was also observed for counts of Nissl-stained profiles, nor to an accentuation of natural cell death, because the difference is present at PND1. The latter observation would suggest that TGFα influences the prenatal ontogeny of dopamine neurons during either their proliferation or successful migration. Such possible effects remain to be defined.

Regulation by Intrinsic Molecules

Relatively little is known about the cell signaling pathways within mesencephalic dopamine neurons that mediate the trophic responses that suppress developmental apoptosis or induce axon sprouting, particularly in the in vivo context. One candidate pathway for a role in these responses is phosphatidylinositol-3 kinase (PI3K) and Akt/protein kinase B (PKB) activation.[128,129] In studies of neural cells in tissue culture, PI3K/Akt signaling has been implicated in the survival effects of nerve growth factor, platelet-derived growth factor, and insulin-like growth factor.[130,131] Activation of PI3K/Akt signaling has been demonstrated following GDNF-GFRα1 binding and Ret activation.[132] A number of trophic effects of GDNF have been attributed to PI3K/Akt activation, including cell survival,[133,134] neurite differentiation,[135,136] and neuroprotection.[137]

In support of the possibility that PI3K/Akt signaling may serve in vivo to mediate developmental trophic effects of either GDNF or BDNF in dopamine neurons, mRNA for all three isoforms of Akt is expressed during development in the SNpc, and phospho-Akt(Ser473) protein can be identified within dopamine neurons.[138] Using an adeno-associated viral (AAV) vector approach to transduce SN dopamine neurons with either constitutively active or dominant negative forms of Akt, we have explored the cell autonomous role that it plays to regulate three aspects of the development of these neurons: the magnitude of the NCD event, neuron size, and axon growth.

In keeping with a role for endogenous Akt in the regulation of developmental apoptosis in SNpc neurons, we have found that transduction on PND5-6 with a dominant negative form (AAV DN-Akt(PH)) results in a 100% augmentation in the number of apoptotic profiles during the second phase of developmental cell death on PND14.[139] This increase in the magnitude of the NCD event by DN-Akt(PH) results in a decrease in the final adult number of these neurons. Conversely, transducing these neurons with a constitutively active form of Akt (AAV Myr-Akt) results in an increase in their number (Fig. 4.2.5). We therefore conclude that Akt

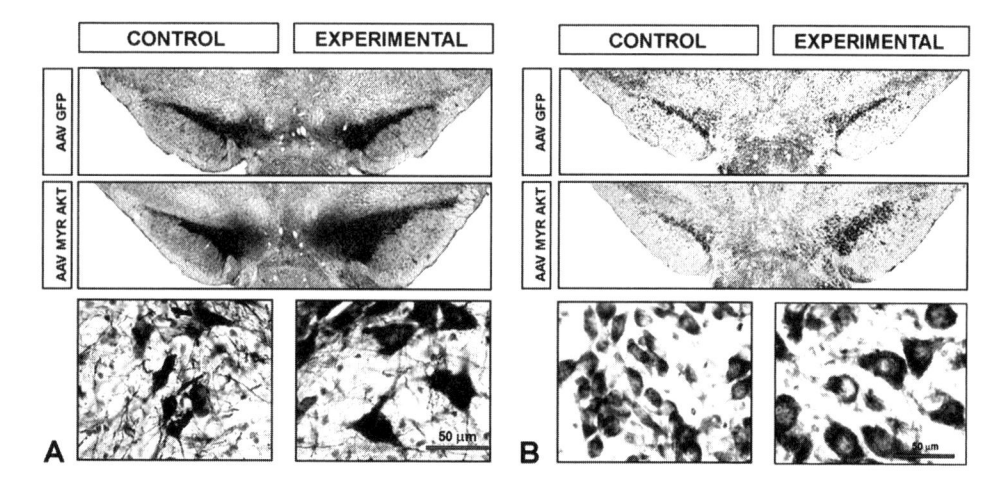

FIGURE 4.2.5. Transduction of SNpc neurons during postnatal development with Myr-Akt results in an increase in their size and number. (A) Immunoperoxidase labeling of TH within the SNpc at 28 days postinjection (PND33) on one side of the brain of either AAV GFP or AAV Myr-Akt on PND5. The low-power photomicrographs in the top panel show, at a regional level, that there is no difference between the injected side (Experimental) and the uninjected side (Control) in the AAV GFP-treated animals, whereas the Experimental side of the AAV Myr-Akt-injected animals demonstrates a markedly increased extent of TH immunostaining in the SNpc. The higher-power micrographs shown in the lower panels demonstrate at a cellular level that this increased extent of TH staining is due primarily to a marked increase in the size of SNpc TH-positive neurons. (B) Nissl stain of the ventral mesencephalon at 28 days postinjection (PND33) of either AAV GFP or AAV Myr-Akt on PND5. The low-power photomicrographs in the top panel show that the apparent increase in the size and number of dopamine neurons in the SNpc is observed independently of the expression of TH. An increase in the size of Nissl-stained neurons is observed at the cellular level in the lower panels. (See Color Plate 4.2.5.)

signaling endogenous to dopamine neurons of the SNpc regulates postnatal NCD and thereby regulates the final adult number of these neurons.

The most pronounced developmental effect that we observed following transduction of neurons of the SNpc with Myr-Akt was a striking increase in their individual cell size (Fig. 4.2.5). Endogenous Akt is likely to play a role in regulating cell size, because transduction with the dominant negative form, DN-Akt(PH), induces a decrease. These observations that Akt plays a role in the regulation of SNpc neuron cell size are in keeping with prior observations made in mice with deletion of the tumor suppressor phosphatase and tensin homologue (PTEN) gene in select neurons postnatally.[140,141] PTEN negatively regulates Akt activation by dephosphorylating phosphatidylinositol-3,4,5-triphosphate, thereby inhibiting interaction between the pleckstrin homology domain of Akt and the inner plasma membrane. In mice with postnatal PTEN deletion in cerebellar and dentate gyrus neurons, there is a striking increase in their size.

In addition to regulating the developmental apoptosis and cell size of mesencephalic dopamine neurons, Akt regulates their axon growth. Following transduction with Myr-Akt, there is an increase in the density of dopaminergic axon and terminal TH immunostaining in the striatum. Conversely, there is a decrease following DN-Akt(PH).[139]

In conclusion, these studies provide evidence that Akt plays a role in the regulation of apoptosis, cell size, and axon growth during postnatal development in dopamine neurons of the SNpc. Akt may therefore mediate the effects of GDNF, BDNF, or other neurotrophic factors on the development of these neurons. We have previously shown that in the SN, unlike the striatum and cortex, Akt mRNA remains highly expressed after development,[138] raising the possibility that it plays a role in the adult maintenance of these neurons. In this regard, it is of interest to recall the findings of Pascual and colleagues[90] demonstrating that GDNF is essential for the maintenance of the viability of SNpc dopamine neurons in adulthood.

CONCLUSIONS

The postnatal development of mesencephalic dopamine neurons follows the fundamental principles of classic neurotrophic theory. There is an apoptotic NCD event that is maximal in both rodents and primates during the period of maximal development of target contact. As proposed by classic theory, this NCD event is regulated by target contact and retrograde neurotrophic support. In addition, there is evidence that it may also be regulated by afferent anterograde influences and autocrine control.

It is also clear, however, that developmental trophic support of mesencephalic dopamine neurons is much more complex than that of the simple peripheral neuron systems in which classic theory was first established. Although there is much evidence, for example, that GDNF may be a target-derived neurotrophic factor for these neurons, it remains difficult to reconcile that possibility with the lack of a phenotype in mice with selective deletion of Ret, a principal mediator of GDNF signaling, in these neurons. It is likely that this disparity is due to redundancies and compensatory mechanisms that have evolved to ensure the proper development of this critically important neural system. Similarly, although acute blockade and chronic overexpression experiments suggest an important local role for BDNF in the development of these neurons, it is again difficult to reconcile these observations with the minimal phenotype observed in mice with regionally selective embryonic deletion of BDNF or its receptor TrkB. Again, these disparate results suggest that functional redundancy and compensatory mechanisms exist.

In conclusion, although we have made strides in our understanding of the postnatal development of mesencephalic dopamine neurons, a more complete understanding will depend on the use of more temporally and regionally selective tools. In addition, there are almost certainly other important neurotrophic factors and signaling mechanisms that remain to be identified. Given the critical importance of mesencephalic dopamine neurons to so many neurological and psychiatric conditions, as presented in this volume, future advances in our understanding of the development and maintenance of these neurons is certain to bring rewards in the treatment of human disease.

ACKNOWLEDGMENTS

The author is supported by NIH NINDS awards NS26836 and NS38370 and by the RJG Foundation. The author expresses his gratitude to the Parkinson's Disease Foundation for their enduring loyalty and support.

REFERENCES

1. Marchand R, Poirer LJ. Isthmic origin of neurons of the rat substantia nigra. *Neuroscience*. 1983;9:373–381.
2. Lauder JM, Bloom FE. Ontogeny of monoamine neurons in the locus coeruleus, raphe nuclei and substantia nigra of the rat. *J Comp Neurol*. 1974;155:469–482.
3. Specht LA, Pickel VM, Joh TH, Reis DJ. Light-microscopic immunocytochemical localization of tyrosine hydroxylase in

prenatal brain. I. Early ontogeny. *J Comp Neurol.* 1981;199:233–253.
4. Voorn P, Kalsbeek A, Jorritsma-Byham B, Groenewegen HJ. The pre- and postnatal development of the dopaminergic cell groups in the ventral mesencephalon and the dopaminergic innervation of the striatum of the rat. *Neuroscience.* 1988;25(3):857–887.
5. Kawano H, Ohyama K, Kawamura K, Nagatsu I. Migration of dopaminergic neurons in the embryonic mesencephalon of mice. *Brain Res Dev Brain Res.* 1995;86(1-2):101–113.
6. Shults CW, Hasimoto R, Brady RM, Gage FH. Dopaminergic cells align along radial glia in the developing mesencephalon of the rat. *Neuroscience.* 1990;38:427–436.
7. Oppenheim RW. Cell death during development of the nervous system. *Annu Rev Neurosci.* 1991;14:453–501.
8. Janec E, Burke RE. Naturally occurring cell death during postnatal development of the substantia nigra of the rat. *Mol Cell Neurosci.* 1993;4:30–35.
9. Oo TF, Burke RE. The time course of developmental cell death in phenotypically defined dopaminergic neurons of the substantia nigra. *Dev Brain Res.* 1997;98:191–196.
10. Jackson-Lewis V, Vila M, Djaldetti R, Guegan C, Liberatore G, Liu J, et al. Developmental cell death in dopaminergic neurons of the substantia nigra of mice. *J Comp Neurol.* 2000;424: 476–488.
11. Clarke PGH, Oppenheim RW. Neuron death in vertebrate development: *In vivo* methods. In: Schwartz LM, Osborne BA, eds. *Methods in Cell Biology: Cell Death.* New York: Academic Press; 1995:277–321.
12. Gallyas F, Wolff JR, Bottcher H, Zaborsky L. A reliable and sensitive method to localize terminal degeneration and lysosomes in the central nervous system. *Stain Tech.* 1980;55:299–306.
13. Clarke PGH. Developmental cell death: morphological diversity and multiple mechanisms. *Anat Embryol.* 1990;181: 195–213.
14. Oo TF, Blazeski R, Harrison SMW, Henchcliffe C, Mason CA, Roffler-Tarlov S, et al. Neuron death in the substantia nigra of weaver mouse occurs late in development and is not apoptotic. *J Neurosci.* 1996; 6:6134–6145.
15. Lauder JM, Bloom FE. Ontogeny of monoamine neurons in the locus coeruleus, raphe nuclei and substantia nigra of the rat. *J Comp Neurol.* 1975;163:251–264.
16. Kalsbeek A, Voorn P, Buijs RM. Development of dopamine-containing systems in the CNS. In: Bjorklund A, Hokfelt T, Tohyama M, eds. *Handbook of Chemical Neuroanatomy, Vol. 10: Ontogeny of Transmitters and Peptides in the CNS.* Amsterdam, The Netherlands. Elsevier; 1992:63–112.
17. Clarke PGH. Neuronal death in the development of the vertebrate nervous system. *Trends Neurosci.* 1985;8:345–349.
18. Cowan WM, Fawcett JW, O'Leary DDM, Stanfield BB. Regressive events in neurogenesis. *Science.* 1984;225:1258–1265.
19. Hattori T, McGeer PL. Synaptogenesis in the corpus striatum of infant rat. *Exp Neurol.* 1973;38:70–79.
20. Coyle JT. Biochemical aspects of neurotransmission in the developing brain. *Int Rev Neurobiol.* 1977;20:65–102.
21. Tepper JM, Damlama M, Trent F. Postnatal changes in the distribution and morphology of rat substantia nigra dopaminergic neurons. *Neuroscience.* 1994;60:469–477.
22. Morrow BA, Roth RH, Redmond DE Jr, Sladek JR Jr, Elsworth JD. Apoptotic natural cell death in developing primate dopamine midbrain neurons occurs during a restricted period in the second trimester of gestation. *Exp Neurol.* 2007;204(2): 802–807.
23. Burke RE. Programmed cell death and new discoveries in the genetics of parkinsonism. *J Neurochem.* 2008;104(4): 875–890.
24. Zinszner H, Kuroda M, Wang X, Batchvarova N, Lightfoot RT, Remotti H, et al. CHOP is implicated in programmed cell death in response to impaired function of the endoplasmic reticulum. *Genes Dev.* 1998;12(7):982–995.
25. Silva RM, Ries V, Oo TF, Yarygina O, Jackson-Lewis V, Ryu EJ, et al. CHOP/GADD153 is a mediator of apoptotic death in substantia nigra dopamine neurons in an in vivo neurotoxin model of parkinsonism. *J Neurochem.* 2005;95:974–986.
26. Kluck RM, Bossy-Wetzel E, Green DR, Newmeyer DD. The release of cytochrome c from mitochondria: a primary site for Bcl-2 regulation of apoptosis. *Science.* 1997;275: 1132–1136.
27. Scorrano L, Korsmeyer SJ. Mechanisms of cytochrome c release by proapoptotic BCL-2 family members. *Biochem Biophys Res Commun.* 2003;304(3):437–444.
28. Savitt JM, Jang SS, Mu W, Dawson VL, Dawson TM. Bcl-x is required for proper development of the mouse substantia nigra. *J Neurosci.* 2005;25(29):6721–6728.
29. Vila M, Jackson-Lewis V, Vukosavic S, Djaldetti R, Liberatore G, Offen D, et al. Bax ablation prevents dopaminergic neurodegeneration in the 1-methyl- 4-phenyl-1,2,3,6-tetrahydropyridine mouse model of Parkinson's disease. *Proc Natl Acad Sci USA.* 2001;98(5):2837–2842.
30. Groc L, Bezin L, Jiang H, Jackson TS, Levine RA. Bax, Bcl-2, and cyclin expression and apoptosis in rat substantia nigra during development. *Neurosci Lett.* 2001;306(3):198–202.
31. Ganguly A, Oo TF, Rzhetskaya M, Pratt R, Yarygina O, Momoi T, et al. CEP11004, a novel inhibitor of the mixed lineage kinases, suppresses apoptotic death in dopamine neurons of the substantia nigra induced by 6-hydroxydopamine. *J Neurochem.* 2004;88(2):469–480.
32. Jeon BS, Kholodilov NG, Oo TF, Kim S, Tomaselli KJ, Srinivasan A, et al. Activation of caspase-3 in developmental models of programmed cell death in neurons of the substantia nigra. *J Neurochem.* 1999;73:322–333.
33. Oo TF, Siman R, Burke RE. Distinct nuclear and cytoplasmic localization of caspase cleavage products in two models of induced apoptotic death in dopamine neurons of the substantia nigra. *Exp Neurol.* 2002;175:1–9.
34. Barde YA. Trophic factors and neuronal survival. *Neuron.* 1989;2:1525–1534.
35. Purves D, Lichtman JW. *Principles of Neural Development.* Sunderland, MA: Sinauer; 1985.
36. Wetts R, Herrup K. Direct correlation between Purkinje and granule cell number in the cerebella of lurcher chimeras and wild-type mice. *Dev Brain Res.* 1983;10:41–47.
37. Herrup K, Sunter K. Numerical matching during cerebellar development: quantitative analysis of granule cell death in staggerer mouse chimeras. *J Neurosci.* 1987;7(3):829–836.
38. Prochiantz A, di Porzio U, Kato A, Berger B, Glowinski J. In vitro maturation of mesencephalic dopaminergic neurons from mouse embryos is enhanced in presence of their striatal target cells. *Proc Natl Acad Sci USA.* 1979;76:5387–5391.
39. Hemmendinger LM, Garber BB, Hoffmann PC, Heller A. Target neuron-specific process formation by embryonic mesencephalic dopamine neurons in vitro. *Proc Natl Acad Sci USA.* 1981;78:1264–1268.
40. Hoffmann PC, Hemmendinger LM, Kotake C, Heller A. Enhanced dopamine cell survival in reaggregates containing target cells. *Brain Res.* 1983;274:275–281.

41. Tomozawa Y, Appel SH. Soluble striatal extracts enhance development of mesencephalic dopaminergic neurons in vitro. *Brain Res.* 1986;399:111–124.
42. Burke RE, Macaya A, DeVivo D, Kenyon N, Janec EM. Neonatal hypoxic-ischemic or excitotoxic striatal injury results in a decreased adult number of substantia nigra neurons. *Neuroscience.* 1992;50:559–569.
43. Coyle JT, Schwarcz R. Lesion of striatal neurones with kainic acid provides a model for Huntington's chorea. *Nature.* 1976;263:244–246.
44. Macaya A, Munell F, Gubits RM, Burke RE. Apoptosis in substantia nigra following developmental striatal excitotoxic injury. *Proc Natl Acad Sci USA.* 1994;91:8117–8121.
45. Marti MJ, James CJ, Oo TF, Kelly WJ, Burke RE. Early developmental destruction of terminals in the striatal target induces apoptosis in dopamine neurons of the substantia nigra. *J Neurosci.* 1997;17:2030–2039.
46. El-Khodor BF, Burke RE. Medial forebrain bundle axotomy during development induces apoptosis in dopamine neurons of the substantia nigra and activation of caspases in their degenerating axons. *J Comp Neurol.* 2002;452:65–79.
47. Kelly WJ, Burke RE. Apoptotic neuron death in rat substantia nigra induced by striatal excitotoxic injury is developmentally dependent. *Neurosci Lett.* 1996;220:85–88.
48. Marti MJ, Saura J, Burke RE, Jackson-Lewis V, Jimenez A, Bonastre M, et al. Striatal 6-hydroxydopamine induces apoptosis of nigral neurons in the adult rat. *Brain Res.* 2002;958:185–191.
49. Linden R. The survival of developing neurons: a review of afferent control. *Neuroscience.* 1994;58:671–682.
50. Alonso-Vanegas MA, Fawcett JP, Causing CG, Miller FD, Sadikot AF. Characterization of dopaminergic midbrain neurons in a DBH:BDNF transgenic mouse. *J Comp Neurol.* 1999;413(3):449–462.
51. Lin L-FH, Doherty DH, Lile JD, Bektesh S, Collins F. GDNF: a glial cell line–derived neurotrophic factor for midbrain dopaminergic neurons. *Science.* 1993;260:1130–1132.
52. Schaar DG, Sieber BA, Dreyfus CF, Black IB. Regional and cell specific expression of GDNF in rat brain. *Exp Neurol.* 1993;124:368–371.
53. Poulsen KT, Armanini MP, Klein RD, Hynes MA, Phillips HS, Rosenthal A. TGFb2 and TGFb3 are potent survival factors for midbrain dopaminergic neurons. *Neuron.* 1994;13:1245–1252.
54. Stromberg I, Bjorklund L, Johansson M, Tomac A, Collins F, Olson L, et al. Glial cell line derived neurotrophic factor Is expressed in the developing but not adult striatum and stimulates developing dopamine neurons in vivo. *Exp Neurol.* 1993;124:401–412.
55. Blum M, Weickert CS. GDNF mRNA expression in normal postnatal development, aging, and in weaver mutant mice. *Neurobiol Aging.* 1995;16:925–929.
56. Choi-Lundberg DL, Bohn MC. Ontogeny and distribution of glial cell line–derived neurotrophic factor (GDNF) mRNA in rat. *Dev Brain Res.* 1995;85:80–88.
57. Golden JP, DeMaro JA, Osborne PA, Milbrandt J, Johnson EMJ. Expression of neurturin, GDNF, and GDNF family-receptor mRNA in the developing and mature mouse. *Exp Neurol.* 1999;158:504–528.
58. Cho J, Kholodilov NG, Burke RE. The developmental time course of glial cell line-derived neurotrophic factor (GDNF) and GDNF receptor alpha-1 mRNA expression in the striatum and substantia nigra. *Ann NY Acad Sci.* 2003;991:284–287.
59. Oo TF, Ries V, Cho J, Kholodilov N, Burke RE. Anatomical basis of glial cell line-derived neurotrophic factor expression in the striatum and related basal ganglia during postnatal development of the rat. *J Comp Neurol.* 2005;484(1):57–67.
60. Lopez-Martin E, Caruncho HJ, Rodriguez-Pallares J, Guerra MJ, Labandeira-Garcia JL. Striatal dopaminergic afferents concentrate in GDNF-positive patches during development and in developing intrastriatal striatal grafts. *J Comp Neurol.* 1999;406:199–206.
61. Tomac A, Widenfalk J, Lin LH, Kohno T, Ebendal T, Hoffer BJ, et al. Retrograde axonal transport of glial cell line-derived neurotrophic factor in the adult nigrostriatal system suggests a trophic role in the adult. *Proc Natl Acad Sci USA.* 1995;92:8274–8278.
62. Cho JW, Yarygina O, Kholodilov N, Burke RE. Glial cell line-derived neurotrophic factor receptor GFRá-1 is expressed in the rat striatum during postnatal development. *Mol Br Res.* 2004;127:96–104.
63. Widenfalk J, Nosrat C, Tomac A, Westphal H, Hoffer B, Olson L. Neurturin and glial cell line-derived neurotrophic factor receptor-beta (GDNFR-beta), novel proteins related to GDNF and GDNFR-alpha with specific cellular patterns of expression suggesting roles in the developing and adult nervous system and in peripheral organs. *J Neurosci.* 1997;17:8506–8519.
64. Yu T, Scully S, Yu Y, Fox GM, Jing S, Zhou R. Expression of GDNF family receptor components during development: implications in the mechanisms of interaction. *J Neurosci.* 1998;18:4684–4696.
65. Sarabi A, Hoffer BJ, Olson L, Morales M. GFRalpha-1 mRNA in dopaminergic and nondopaminergic neurons in the substantia nigra and ventral tegmental area. *J Comp Neurol.* 2001;441(2):106–117.
66. Coulpier M, Ibanez CF. Retrograde propagation of GDNF-mediated signals in sympathetic neurons. *Mol Cell Neurosci.* 2004;27(2):132–139.
67. Treanor JJS, Goodman L, deSauvage F, Stone DM, Poulsen KT, Beck CD, et al. Characterization of a multicomponent receptor for GDNF. *Nature.* 1996;382:80–83.
68. Pichel JG, Shen L, Sheng HZ, Granholm A-C, Drago J, Grinberg A, et al. Defects in enteric innervation and kidney development in mice lacking GDNF. *Nature.* 1996;382:73–76.
69. Sanchez MP, Silos-Santiago I, Frisen J, He B, Lira SA, Barbacid M. Renal agenesis and the absence of enteric neurons in mice lacking GDNF. *Nature.* 1996;382:70–73.
70. Cacalano G, Farinas I, Wang LC, Hagler K, Forgie A, Moore M, et al. GFRalpha1 is an essential receptor component for GDNF in the developing nervous system and kidney. *Neuron.* 1998;21:53–62.
71. Enomoto H, Araki T, Jackman A, Heuckeroth RO, Snider WD, Johnson EMJ, et al. GFR alpha1-deficient mice have deficits in the enteric nervous system and kidneys. *Neuron.* 1998;21:317–324.
72. Rayport S, Sulzer D, Shi WX, Sawasdikosol S, Monaco J, Batson D, et al. Identified postnatal mesolimbic dopamine neurons in culture morphology and electrophysiology. *J Neurosci.* 1992;12:4264–4280.
73. Burke RE, Antonelli M, Sulzer D. Glial cell line-derived neurotrophic growth factor inhibits apoptotic death of postnatal substantia nigra dopamine neurons in primary culture. *J Neurochem.* 1998;71:517–525.
74. Oo TF, Kholodilov N, Burke RE. Regulation of natural cell death in dopaminergic neurons of the substantia nigra by striatal GDNF in vivo. *J Neurosci.* 2003;23:5141–5148.

75. Granholm AC, Reyland M, Albeck D, Sanders L, Gerhardt G, Hoernig G, et al. Glial cell line-derived neurotrophic factor is essential for postnatal survival of midbrain dopamine neurons. *J Neurosci.* 2000;20:3182–3190.
76. Beck KD, Irwin I, Valverde J, Brennan T, Langston JW, Hefti F. GDNF induces a dystonia-like state in neonatal rats and stimulates dopamine and serotonin synthesis. *Neuron.* 1996;16:665–673.
77. Mayford M, Bach ME, Huang YY, Wang L, Hawkins RD, Kandel ER. Control of memory formation through regulated expression of a CaMKII transgene. *Science.* 1996;274:1678–1683.
78. Yamamoto A, Lucas JJ, Hen R. Reversal of neuropathology and motor dysfunction in a conditional model of Huntington's disease. *Cell.* 2000;101:57–66.
79. Chun HS, Yoo MS, DeGiorgio LA, Volpe BT, Peng D, Baker H, et al. Marked dopaminergic cell loss subsequent to developmental, intranigral expression of glial cell line-derived neurotrophic factor. *Exp Neurol.* 2002;173(2):235–244.
80. Kholodilov N, Yarygina O, Oo TF, Zhang H, Sulzer D, Dauer WT, et al. Regulation of the development of mesencephalic dopaminergic systems by the selective expression of glial cell line-derived neurotrophic factor in their targets. *J Neurosci.* 2004;24(13):3136–3146.
81. Mijatovic J, Airavaara M, Planken A, Auvinen P, Raasmaja A, Piepponen TP, et al. Constitutive Ret activity in knock-in multiple endocrine neoplasia type B mice induces profound elevation of brain dopamine concentration via enhanced synthesis and increases the number of TH-positive cells in the substantia nigra. *J Neurosci.* 2007;27(18):4799–4809.
82. Jain S, Golden JP, Wozniak D, Pehek E, Johnson EM Jr, Milbrandt J. RET is dispensable for maintenance of midbrain dopaminergic neurons in adult mice. *J Neurosci.* 2006;26(43):11230–11238.
83. Kramer ER, Aron L, Ramakers GM, Seitz S, Zhuang X, Beyer K, et al. Absence of Ret signaling in mice causes progressive and late degeneration of the nigrostriatal system. *PLoS Biol.* 2007;5(3):e39.
84. Airaksinen MS, Saarma M. The GDNF family: signalling, biological functions and therapeutic value. *Nat Rev Neurosci.* 2002;3(5):383–394.
85. Trupp M, Scott R, Whittemore SR, Ibanez CF. Ret-dependent and -independent mechanisms of glial cell line-derived neurotrophic factor signaling in neuronal cells. *J Biol Chem.* 1999;274(30):20885–20894.
86. Poteryaev D, Titievsky A, Sun YF, Thomas-Crusells J, Lindahl M, Billaud M, et al. GDNF triggers a novel Ret-independent Src kinase family-coupled signaling via a GPI-linked GDNF receptor alpha1. *FEBS Lett.* 1999;463(1-2):63–66.
87. Popsueva A, Poteryaev D, Arighi E, Meng X, Angers-Loustau A, Kaplan D, et al. GDNF promotes tubulogenesis of GFRalpha1-expressing MDCK cells by Src-mediated phosphorylation of Met receptor tyrosine kinase. *J Cell Biol.* 2003;161(1):119–129.
88. Paratcha G, Ledda F, Ibanez CF. The neural cell adhesion molecule NCAM is an alternative signaling receptor for GDNF family ligands. *Cell.* 2003;113(7):867–879.
89. Chao CC, Ma YL, Chu KY, Lee EH. Integrin alphav and NCAM mediate the effects of GDNF on DA neuron survival, outgrowth, DA turnover and motor activity in rats. *Neurobiol Aging.* 2003;24(1):105–116.
90. Pascual A, Hidalgo-Figueroa M, Piruat JI, Pintado CO, Gomez-Diaz R, Lopez-Barneo J. Absolute requirement of GDNF for adult catecholaminergic neuron survival. *Nat Neurosci.* 2008;11(7):755–761.
91. Kotzbauer PT, Lampe PA, Heuckeroth RO, Golden JP, Creedon DJ, Johnson EMJ, et al. Neurturin, a relative of glial-cell-line-derived neurotrophic factor. *Nature.* 1996;384:467–470.
92. Akerud P, Alberch J, Eketjall S, Wagner J, Arenas E. Differential effects of glial cell line-derived neurotrophic factor and neurturin on developing and adult substantia nigra dopaminergic neurons. *J Neurochem.* 1999;73:70–78.
93. Horger BA, Nishimura MC, Armanini MP, Wang LC, Poulsen KT, Rosenblad C, et al. Neurturin exerts potent actions on survival and function of midbrain dopaminergic neurons. *J Neurosci.* 1998;18:4929–4937.
94. Oiwa Y, Yoshimura R, Nakai K, Itakura T. Dopaminergic neuroprotection and regeneration by neurturin assessed by using behavioral, biochemical and histochemical measurements in a model of progressive Parkinson's disease. *Brain Res.* 2002;947(2):271–283.
95. Cho J, Kholodilov NG, Burke RE. Patterns of developmental mRNA expression of neurturin and GFRalpha2 in the rat striatum and substantia nigra do not suggest a role in the regulation of natural cell death in dopamine neurons. *Brain Res Dev Brain Res.* 2004;148(1):143–149.
96. Barde YA, Edgar D, Thoenen H. Sensory neurons in culture: changing requirements for survival factors during embryonic development. *Proc Natl Acad Sci USA.* 1980;77(2):1199–1203.
97. Barde YA, Edgar D, Thoenen H. Purification of a new neurotrophic factor from mammalian brain. *EMBO J.* 1982;1(5):549–553.
98. Leibrock J, Lottspeich F, Hohn A, Hofer M, Hengerer B, Masiakowski P, et al. Molecular cloning and expression of brain-derived neurotrophic factor. *Nature.* 1989;341(6238):149–152.
99. Lindsay RM. Brain-derived neurotrophic factor: an NGF-related neurotrophin. In: Loughlin SE, Fallon JH, eds. *Neurotrophic Factors.* San Diego, CA: Academic Press; 1993:257–284.
100. Hyman C, Hofer M, Barde YA, Juhasz M, Yancopoulos GD, Squinto SP, et al. BDNF is a neurotrophic factor for dopaminergic neurons of the substantia nigra. *Nature.* 1991;350:230–232.
101. Spina MB, Squinto SP, Miller J, Lindsay RM, Hyman C. Brain-derived neurotrophic factor protects dopamine neurons against 6-hydroxydopamine and N-methyl-4-phenylpyridinium ion toxicity: involvement of the glutathione system. *J Neurochem.* 1992;59(1):99–106.
102. Ernfors P, Kuo-Fen L, Jaenisch R. Mice lacking brain-derived neurotrophic factor develop with sensory deficits. *Nature.* 1994;368:147–150.
103. Jones KR, Farinas I, Backus C, Reichardt LF. Targeted disruption of the BDNF gene perturbs brain and sensory neuron development but not motor neuron development. *Cell.* 1994;76:989–999.
104. Klein R, Smeyne RJ, Wurst W, Long LK, Auerbach BA, Joyner AL, et al. Targeted disruption of the trkB neurotrophin receptor gene results in nervous system lesions and neonatal death. *Cell.* 1993;75(1):113–122.
105. Numan S, Seroogy KB. Expression of trkB and trkC mRNAs by adult midbrain dopamine neurons: a double-label in situ hybridization study. *J Comp Neurol.* 1999;403(3):295–308.
106. Numan S, Gall CM, Seroogy KB. Developmental expression of neurotrophins and their receptors in postnatal rat ventral midbrain. *J Mol Neurosci.* 2005;27(2):245–260.
107. Yan Q, Radeke MJ, Matheson CR, Talvenheimo J, Welcher AA, Feinstein SC. Immunocytochemical localization of TrkB in the central nervous system of the adult rat. *J Comp Neurol.* 1997;378(1):135–157.

108. Maisonpierre PC, Belluscio L, Friedman B, Alderson RF, Wiegand SJ, Furth ME, et al. NT-3, BDNF, and NGF in the developing rat nervous system: parallel as well as reciprocal patterns of expression. *Neuron.* 1990;5:501–509.
109. Altar CA, Cai N, Bliven T, Juhasz M, Conner JM, Acheson AL, et al. Anterograde transport of brain-derived neurotrophic factor and its role in the brain. *Nature.* 1997;389(6653):856–860.
110. Altar CA, DiStefano PS. Neurotrophin trafficking by anterograde transport. *Trends Neurosci.* 1998;21(10):433–437.
111. Castren E, Thoenen H, Lindholm D. Brain-derived neurotrophic factor messenger RNA is expressed in the septum, hypothalamus and in adrenergic brain stem nuclei of adult rat brain and is increased by osmotic stimulation in the paraventricular nucleus. *Neuroscience.* 1995;64(1):71–80.
112. Conner JM, Lauterborn JC, Yan Q, Gall CM, Varon S. Distribution of brain-derived neurotrophic factor (BDNF) protein and mRNA in the normal adult rat CNS: evidence for anterograde axonal transport. *J Neurosci.* 1997;17(7):2295–2313.
113. Seroogy KB, Lundgren KH, Tran TM, Guthrie KM, Isackson PJ, Gall CM. Dopaminergic neurons in rat ventral midbrain express brain-derived neurotrophic factor and neurotrophin-3 mRNAs. *J Comp Neurol.* 1994;342(3):321–334.
114. Baquet ZC, Bickford PC, Jones KR. Brain-derived neurotrophic factor is required for the establishment of the proper number of dopaminergic neurons in the substantia nigra pars compacta. *J Neurosci.* 2005;25(26):6251–6259.
115. Acheson A, Conover JC, Fandl JP, DeChiara TM, Russell M, Thadani A, et al. A BDNF autocrine loop in adult sensory neurons prevents cell death. *Nature.* 1995;374(6521):450–453.
116. Oo TF, Marchionini DM, Yarygina O, O'Leary PD, Hughes RA, Kholodilov N, et al. Brain-derived neurotrophic factor regulates early postnatal developmental cell death of dopamine neurons of the substantia nigra in vivo. *Mol Cell Neurosci.* 2009; 41(4):440–447.
117. O'Leary PD, Hughes RA. Structure–activity relationships of conformationally constrained peptide analogues of loop 2 of brain-derived neurotrophic factor. *J Neurochem.* 1998;70(4):1712–1721.
118. Zaman V, Nelson ME, Gerhardt GA, Rohrer B. Neurodegenerative alterations in the nigrostriatal system of trkB hypomorphic mice. *Exp Neurol.* 2004;190(2):337–346.
119. Baker SA, Stanford LE, Brown RE, Hagg T. Maturation but not survival of dopaminergic nigrostriatal neurons is affected in developing and aging BDNF-deficient mice. *Brain Res.* 2005;1039(1-2):177–188.
120. Krieglstein K, Henheik P, Farkas L, Jaszai J, Galter D, Krohn K, et al. Glial cell line-derived neurotrophic factor requires transforming growth factor-beta for exerting its full neurotrophic potential on peripheral and CNS neurons. *J Neurosci.* 1998; 18: 9822–9834.
121. Roussa E, Farkas LM, Krieglstein K. TGF-beta promotes survival on mesencephalic dopaminergic neurons in cooperation with Shh and FGF-8. *Neurobiol Dis.* 2004;16(2):300–310.
122. Zhang J, Pho V, Bonasera SJ, Holtzman J, Tang AT, Hellmuth J, et al. Essential function of HIPK2 in TGFbeta-dependent survival of midbrain dopamine neurons. *Nat Neurosci.* 2007;10(1):77–86.
123. Massague J, Seoane J, Wotton D. Smad transcription factors. *Genes Dev.* 2005;19(23):2783–2810.
124. Lazar LM, Blum M. Regional distribution and developmental expression of epidermal growth factor and transforming growth factor-alpha mRNA in mouse brain by a quantitative nuclease protection assay. *J Neurosci.* 1992;12(5):1688–1697.
125. Wilcox JN, Derynck R. Localization of cells synthesizing transforming growth factor-alpha mRNA in the mouse brain. *J Neurosci.* 1988;8(6):1901–1904.
126. Alexi T, Hefti F. Trophic actions of transforming growth factor a on mesencephalic dopaminergic neurons developing in culture. *Neuroscience.* 1993;55:903–918.
127. Blum M. A null mutation in TGF-alpha leads to a reduction in midbrain dopaminergic neurons in the substantia nigra. *Nat Neurosci.* 1998;1(5):374–377.
128. Downward J. PI 3-kinase, Akt and cell survival. *Semin Cell Dev Biol.* 2004;15(2):177–182.
129. Manning BD, Cantley LC. AKT/PKB signaling: navigating downstream. *Cell.* 2007;129(7):1261–1274.
130. Yao R, Cooper GM. Requirement for phosphatidylinositol-3 kinase in the prevention of apoptosis by nerve growth factor. *Science.* 1995;267(5206):2003–2006.
131. Dudek H, Datta SR, Franke TF, Birnbaum MJ, Yao R, Cooper GM, et al. Regulation of neuronal survival by the serine-threonine protein kinase Akt. *Science.* 1997;275(5300):661–665.
132. Besset V, Scott RP, Ibanez CF. Signaling complexes and protein–protein interactions involved in the activation of the Ras and phosphatidylinositol 3-kinase pathways by the c-Ret receptor tyrosine kinase. *J Biol Chem.* 2000;275(50):39159–39166.
133. Soler RM, Dolcet X, Encinas M, Egea J, Bayascas JR, Comella JX. Receptors of the glial cell line-derived neurotrophic factor family of neurotrophic factors signal cell survival through the phosphatidylinositol 3-kinase pathway in spinal cord motoneurons. *J Neurosci.* 1999;19:9160–9169.
134. Mograbi B, Bocciardi R, Bourget I, Busca R, Rochet N, Farahi-Far D, et al. Glial cell line-derived neurotrophic factor-stimulated phosphatidylinositol 3-kinase and Akt activities exert opposing effects on the ERK pathway: importance for the rescue of neuroectodermal cells. *J Biol Chem.* 2001;276(48):45307–45319.
135. Encinas M, Tansey MG, Tsui-Pierchala BA, Comella JX, Milbrandt J, Johnson EM Jr. c-Src is required for glial cell line-derived neurotrophic factor (GDNF) family ligand-mediated neuronal survival via a phosphatidylinositol-3 kinase (PI-3K)-dependent pathway. *J Neurosci.* 2001;21(5):1464–1472.
136. Pong K, Xu RY, Baron WF, Louis JC, Beck KD. Inhibition of phosphatidylinositol 3-kinase activity blocks cellular differentiation mediated by glial cell line-derived neurotrophic factor in dopaminergic neurons. *J Neurochem.* 1998;71(5):1912–1919.
137. Ugarte SD, Lin E, Klann E, Zigmond MJ, Perez RG. Effects of GDNF on 6-OHDA-induced death in a dopaminergic cell line: modulation by inhibitors of PI3 kinase and MEK. *J Neurosci Res.* 2003;73(1):105–112.
138. Ries V, Henchcliffe C, Kareva T, Rzhetskaya M, Bland RJ, During MJ, et al. Oncoprotein Akt/PKB: trophic effects in murine models of Parkinson's Disease. *Proc Natl Acad Sci USA.* 2006;103(49):18757–18762.
139. Ries V, Cheng H, Baohan A, Kareva T, Oo TF, Rzhetskaya M, et al. Akt/protein kinase B regulates the postnatal development of dopamine neurons of the substantia nigra in vivo. *J Neurochem.* 2009;110(1):23–33.
140. Kwon CH, Zhu X, Zhang J, Knoop LL, Tharp R, Smeyne RJ, et al. Pten regulates neuronal soma size: a mouse model of Lhermitte-Duclos disease. *Nat Genet.* 2001;29(4):404–411.
141. Backman SA, Stambolic V, Suzuki A, Haight J, Elia A, Pretorius J, et al. Deletion of Pten in mouse brain causes seizures, ataxia and defects in soma size resembling Lhermitte-Duclos disease. *Nat Genet.* 2001;29(4):396–403.
142. Hamburger V. Regression versus peripheral control of differentiation in motor hypoplasia. *Am J Anat.* 1958;102(3):365–409.

4.3 | Postnatal Maturation of Dopamine Actions in the Prefrontal Cortex

PATRICIO O'DONNELL AND KUEI Y. TSENG

INTRODUCTION

To understand the modulation of prefrontal cortical activity by dopamine (DA), it is critical to consider not only different receptor subtypes and the cell type DA acts upon, but also complex changes that occur postnatally, sometimes as late as during adolescence. A large body of literature deals with DA actions on physiological properties of the prefrontal cortex (PFC), ranging from recordings in cultured neurons and brain slices to anesthetized animals and awake, freely moving animals. All these levels of analysis offer unique perspectives on the complex pattern of DA actions; combined, they have produced a reasonable understanding of how this modulator affects function in this critical brain region. However, many divergent views persist, and many of them arise from the use of different techniques on animals at different postnatal developmental stages. For example, cellular physiology studies using the whole-cell technique typically rely on slices from very young animals, in many cases obtained before weaning, while behavioral and anatomical studies are conducted mainly in adult animals. In this chapter, we will summarize recent work bridging those age groups, highlighting the maturation of DA electrophysiological actions in the PFC during adolescence.

There is ample evidence that the anatomical and molecular organization of cortical microcircuits change during adolescence. Human imaging studies using diffusion tensor imaging reveal that cortical connectivity changes during puberty and adolescence in a manner that correlates with cognitive maturation.[1] In nonhuman primates, the density of tyrosine hydroxylase (TH)-positive axon terminals reaches a peak during puberty and then declines,[2] and DA receptor mRNA levels peak during adolescence in the human PFC.[3] Markers of GABA transmission within the primate PFC also change during adolescence, with parvalbumin (PV)-containing terminals showing a rapid rise before being pruned to adult levels.[4] In rodents, the density of DA receptors increases postnatally, with D_1 receptors reaching adult levels by postnatal day (PD) 60.[5] Dramatic processes of cell overproduction and elimination take place during adolescence in cortical regions (see [6] for review). Furthermore, cognitive functions that depend on prefrontal DA, such as decision making and working memory,[7] change with the transition to adulthood.[8] Neurophysiological measures such as error-related negativity and event-related potentials mature well into late adolescence,[9] suggesting that the neural substrate of cognitive functions is being refined at that time. Despite emerging information highlighting critical changes in DA, GABA, and glutamate neurotransmission in the PFC during adolescence, little is known about how the modulation of cortical physiology matures during this time.

ELECTROPHYSIOLOGICAL ACTIONS OF DA IN PFC CIRCUITS

The actions of DA on PFC physiology are complex and controversial, with effects that can be described as both excitatory and inhibitory (see chapter 5.2 by Floresco and chapter 5.3 by Arnsten et al. in this volume). Many factors can account for such diversity of responses, including which synaptic processes and cell types are being modulated by DA.[10] In vivo intracellular recordings from adult rats reveal that ventral tegmental area (VTA) stimulation with trains of pulses mimicking DA cell burst firing depolarizes PFC pyramidal neurons while suppressing firing in the vast majority of neurons.[11] A similar result had been obtained earlier with local administration of DA with iontophoresis[12] and can be seen with intra-VTA injection of N-methyl-D-aspartate (NMDA),[11] a procedure that causes sustained bursting in DA neurons[13,14] and evokes phasic DA release.[15] The reduction of firing in target neurons has been interpreted as phasic DA reducing overall activity in the PFC and nucleus accumbens, allowing only strongly activated neurons to overcome such inhibition and fire action potentials; in short, DA can increase the signal-to-noise ratio in the system,[16,17] thereby highlighting reward-related or salient stimuli.

Dopamine exerts its effects primarily by modulating fast synaptic responses (i.e., those of glutamate and GABA). For the most part, PFC D1 receptors have been reported to enhance NMDA currents[18,19] and potentiate NMDA effects on cell excitability[20,21] in slice recordings. This action of D1 receptors in the PFC can be blocked by protein kinase A (PKA) antagonists and by interfering with Ca^{2+} signaling,[20,21] suggesting a dependence on G_s activation and Ca^{2+}. D2 receptors, on the other hand, reduce pyramidal cell excitability and attenuate α-amino-3-hydroxyl-5-methyl-4-isoxazole-propionate (AMPA) and NMDA responses in pyramidal neurons.[20,22] Several mechanisms could mediate the D2 inhibition of AMPA/kainate synaptic transmission in the PFC (Fig. 4.3.1), including direct postsynaptic activation of phospholipase C-inositol 1,4,5-triphosphate (IP_3) and inhibition of PKA signaling pathways.[20] The D2 inhibition of NMDA responses in pyramidal neurons could also occur indirectly, via an enhancement of local GABAergic tone. In fact, increased levels of GABA have been observed with the D2 agonist quinpirole,[23] a treatment that in slices from adult rats increases GABA interneuron excitability.[20,24] A potentiation of NMDA responses by D1 receptors would allow an excitatory action of DA only on already depolarized PFC neurons, thereby reinforcing ongoing behaviorally relevant cortical activity. In vivo, D1 agonists enhance PFC long-term potentiation (LTP),[25] and suppression of VTA activity impairs hippocampal-PFC LTP.[26] Furthermore, PFC D1 receptors improve memory retrieval and working memory performance,[27,28] and D1-NMDA coactivation in the PFC is required for appetitive instrumental learning in adult rats.[29] These results suggest that a D1 potentiation of NMDA responses is critical for several PFC-dependent functions. On the other hand, if phasic DA encounters pyramidal neurons at their resting membrane potential, a state in which NMDA receptors are not effectively activated, the dominant effect may be a D2-mediated reduction of glutamate responses. Thus, DA actions in the PFC seem to be a combination of excitatory and inhibitory effects, with the net result being a D1 reinforcement of strongly activated neurons and a D2-dependent attenuation of weakly driven neurons.

An additional layer of complexity is provided by DA effects on local inhibitory interneurons. Juxtacellular recordings show that the reduction in pyramidal cell firing in vivo is accompanied by an increase in firing by fast-spiking interneurons with a similar time course,[30] suggesting that the strong inhibitory effect of VTA stimulation on pyramidal neurons may involve activation of local interneurons. In slices, DA modulates GABA inputs to pyramidal neurons,[31] with a strong D1 excitation of interneurons in slices from young animals.[32] In slices from adult rats, D2 receptors also increase interneuron excitability, and the D2-mediated attenuation of NMDA responses involves GABA-A receptors.[20,24] Thus, the combination of DA actions on pyramidal neurons and interneurons may contribute to the increase in signal-to-noise ratio that DA causes on PFC information processing.

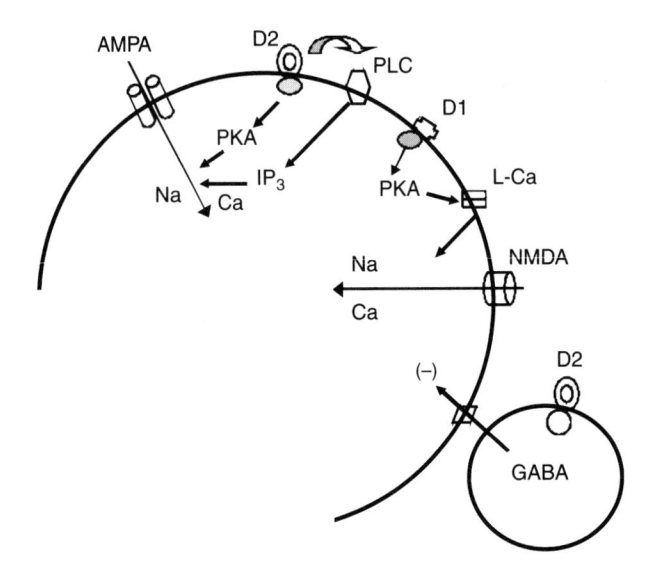

FIGURE 4.3.1. Schematic representation of pathways governing DA control of pyramidal neuron excitability in the PFC. AMPA receptors are Na or Na/Ca channels (depending on the receptor subtype involved), and their effect on excitability is down-regulated by D2 agonists by either suppression of PKA activity or activation of phospholipase C (PLC) and inositol-3-phosphate (IP_3). No regulation of AMPA responses by D1 agonists was observed. N-methyl-D-aspartate (NMDA) receptors are Na/Ca channels and can be up-regulated by D1 receptors via activation of PKA, an effect that requires L-type calcium channels (L-Ca). D2 agonists attenuate the effects of NMDA receptors on pyramidal cell excitability by a mechanism that can be blocked by GABA-A receptor blockade. This suggests that D2 receptors may activate interneurons causing GABA release in the vicinity of pyramidal neurons.

CHANGES IN DA MODULATION OF PYRAMIDAL NEURONS DURING ADOLESCENCE

Dopamine's effects on pyramidal neurons become refined during adolescence. Recordings from our lab show age differences in the modulation of pyramidal cell excitability by AMPA, NMDA, and D1 agonists, as well as in D1–NMDA interactions. In slices obtained from prepubertal rats (PD < 35), AMPA, NMDA, and the D1 agonist SKF38393 enhanced pyramidal cell excitability in response to intrasomatic current injection.[21] Similar recordings in slices from late adolescent or adult rats (i.e., older than 55 days) revealed similar effects, but with dose-response curves shifted to the left,[20] an indication of higher potency of these agents in the adult brain. Furthermore, dendritic Na^+ and Ca^{2+} regenerative potentials in pyramidal neurons, which are important players in synaptic plasticity, become effective in coupling distal apical dendrites with somata at PD 42,[33] a time in which NMDA receptor subunit expression changes.[34] These observations indicate that DA and glutamate become more efficient in driving pyramidal cell firing as the animals mature through adolescence. The interactions between DA and glutamate also change during this critical period. In slices from juvenile rats, a D1 agonist potentiated NMDA responses in a synergistic manner.[21] In slices from young adult rats, such synergism is capable of yielding persistent depolarizations similar to the up states that are observed in vivo.[11] Coadministering SKF38393 and NMDA caused a series of plateau depolarizations lasting hundreds of milliseconds, but only in slices from adult rats[35] (Fig.

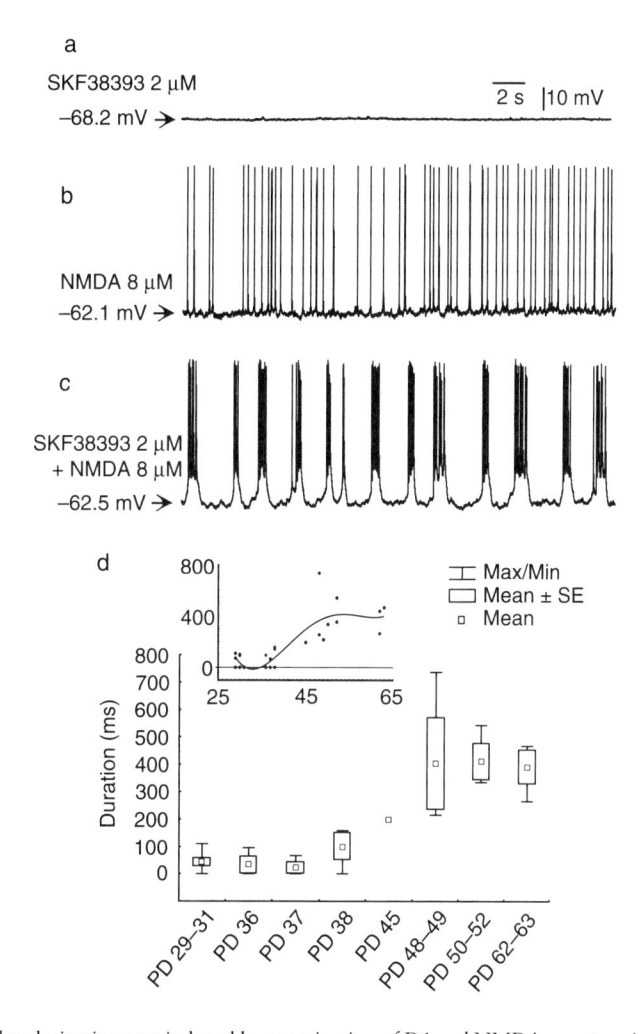

FIGURE 4.3.2. Spontaneous plateau depolarizations are induced by coactivation of D1 and NMDA receptors in the medial PFC from adult but not prepubertal rats. (a) Tracing of spontaneous activity in the presence of 2 μM SKF38393. (b) Trace showing a steady depolarization without plateaus in the presence of NMDA. (c) Combining the D1 agonist with NMDA yielded frequent plateau depolarizations. (d) Box plots summarizing the duration of D1-NMDA-induced depolarizing plateaus recorded in PFC pyramidal neurons from prepubertal (PD 29–38) and adult (PD > 45) rat brain slices. The inset shows a second-order polynomial regression for all data points. *Source:* From [35] with permission.

4.3.2). Before that age, all D1+NMDA-induced depolarizations are in the range of tens of milliseconds, suggesting simple synaptic responses and not persistent activity.[35] The D1+NMDA plateaus observed in adult slices are likely driven by enhanced glutamatergic activity in the local network, as they disappear with administration of tetrodotoxin or the AMPA antagonist CNQX.[35] A D1 facilitation of plateau depolarizations in the PFC would provide a temporal window during which context-relevant inputs can drive pyramidal neuron firing and NMDA-dependent synaptic plasticity would be enabled. Because activation of mesocortical DA is context-dependent and related to attention and salient stimuli,[36–38] the relevant ongoing activity in the PFC, that is, that mediated by AMPA and NMDA receptors, would therefore become enhanced. Thus, the maturation of cortical networks during adolescence results in a state in which persistent activity can be more readily driven and reinforced by DA.

CHANGES IN DA MODULATION OF GABA INTERNEURONS DURING ADOLESCENCE

The DA modulation of local interneurons also changes dramatically during adolescence. In slices from juvenile rats, the D1 agonist SKF38393 increases interneuron excitability,[32] while the D2 agonist quinpirole does not have a major effect.[20,32] These actions are balanced by a DA-dependent attenuation of GABA synaptic responses

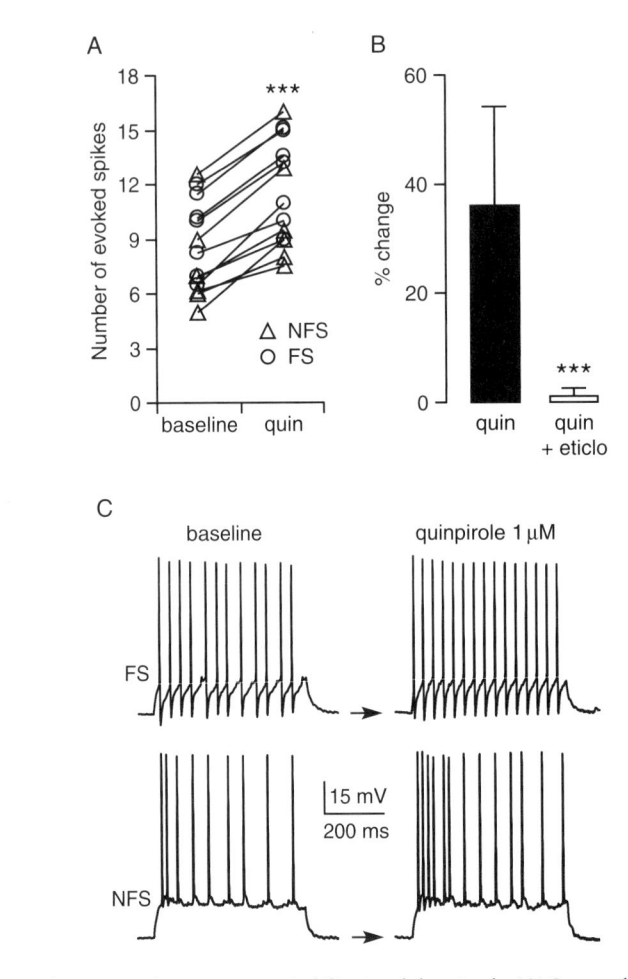

FIGURE 4.3.3. The D_2 agonist quinpirole increases interneuron excitability in adult animals. (A) Scatterplot showing the effect of 1 μM quinpirole on fast-spiking interneuron (FS; open circles) and non-FS interneuron (NFS; triangles) excitability. The data shown are the number of evoked spikes to intracellular current pulse injection and were obtained in the PFC of adult (PD > 50) rats. Quinpirole increased the number of evoked spikes in all PFC interneurons (***$P < 0.0001$, paired Student's t-test). (B) Bar graph summarizing the effect of quinpirole and its blockade by 20 μM of the D2 antagonist eticlopride (indicated as percent change to baseline; ***$P < 0.0001$, repeated measures ANOVA). (C) Two representative traces of FS and NFS interneurons illustrating the number of evoked spikes before (baseline) and its increase after 5 min of bath application of quinpirole (1 μM) in both interneuron types. Source: From [24] with permission.

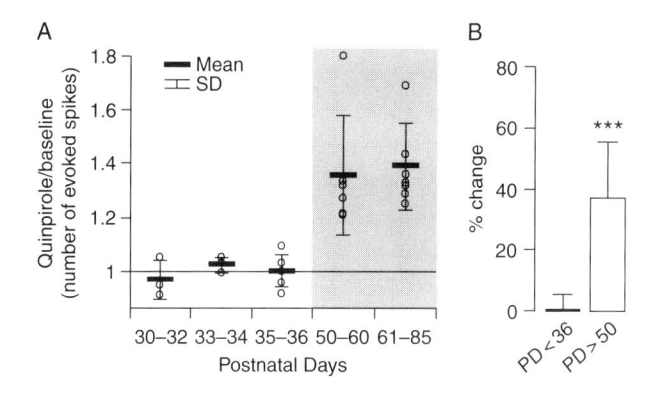

FIGURE 4.3.4. The excitatory effect of the D2 agonist is observed only in slices from adult animals. (A) Scatterplot showing the effect of quinpirole on PFC interneuron excitability from prepubertal to adult ages, expressed as the ratio between evoked firing after and before quinpirole administration. Open circles are individual neurons recorded from each group, and the data were grouped into age ranges. The excitatory effect of quinpirole (1 μM) was observed only in developmentally mature rats. (B) Bar graph (mean ± SD) summarizing the effect of quinpirole on prepubertal PFC interneurons (PD < 36) and those observed in the PFC of adult (PD > 50) animals (***$P < 0.0001$, repeated measures ANOVA). Source: From [24] with permission.

in pyramidal neurons.[31,39] On the other hand, whole-cell recordings in slices from adult rats reveal an excitatory effect of quinpirole on fast-spiking interneurons.[24] This effect is only observed in slices from rats older than PD 45[24] (Figs. 4.3.3, 4.3.4). In adult slices, quinpirole also increases spontaneous firing of fast-spiking interneurons.[20] Thus, during adolescence, DA becomes strongly excitatory on interneurons by virtue of both D1 and D2 receptors increasing their excitability. The cellular or synaptic changes responsible for this late maturation remain to be determined. It can be speculated that they could depend on changes in the receptor subtypes expressed by interneurons (D2 vs. D4), the G protein they are coupled to (Gi vs. Gq), or the dimerization with other receptors.[40]

The changes in D2 modulation of interneurons during adolescence affect the DA modulation of cortico-cortical information. The emergence of a D2 up-regulation of interneuron excitability during adolescence contributes to the D2 attenuation of NMDA effects on pyramidal cell excitability that we observed in slices from adult rats, as blocking GABA-A receptors prevented the D_2 modulation of NMDA responses.[20] Furthermore, D_2 recruitment of interneurons has an impact on intracortical synaptic activity. Electrical stimulation of cortico-cortical fibers by placing an electrode in superficial layers (I or II) about 1 mm lateral to the apical dendrite of the deep layer pyramidal neuron being recorded evokes AMPA-dependent excitatory postsynaptic potentials (EPSPs). Adding quinpirole attenuates the EPSPs by a dual mechanism in slices from adult rats: (1) an early component that is not blocked by GABA-A antagonists and therefore may be due to a direct effect on D2 receptors on the pyramidal neuron being recorded and (2) a slow component that lasts several minutes and is blocked by GABA-A antagonists.[41] In juvenile rats, only the early, direct component is observed,[41] consistent with the notion that D2 receptors activate interneurons only in late adolescent or adult tissue. The maturation of DA actions on interneurons is therefore important for appropriate information processing in the PFC and may balance the increase in responsivity to D1 and NMDA activation. Thus, the excitation–inhibition balance responsible for proper PFC processing of salient information becomes refined during adolescence, and such refinement could contribute to establishing a more efficient PFC in the transition to adulthood.

ABNORMAL PERIADOLESCENT MATURATION OF DA ACTIONS IN DEVELOPMENTAL ANIMAL MODELS OF SCHIZOPHRENIA

Adolescence is a critical period for several psychiatric disorders. In schizophrenia, for example, although there are some early cognitive traits,[42] the full onset of hallucinations and delusions does not occur until late adolescence or early adulthood.[43] On the other hand, there is a clear genetic predisposition toward this disorder,[44] suggesting that early developmental anomalies may be present. How can early developmental deficits cause such delayed symptom onset? Several animal models were developed to directly assess this issue. Perhaps one of the most extensively studied is the neonatal ventral hippocampal lesion (NVHL), developed by Barbara Lipska

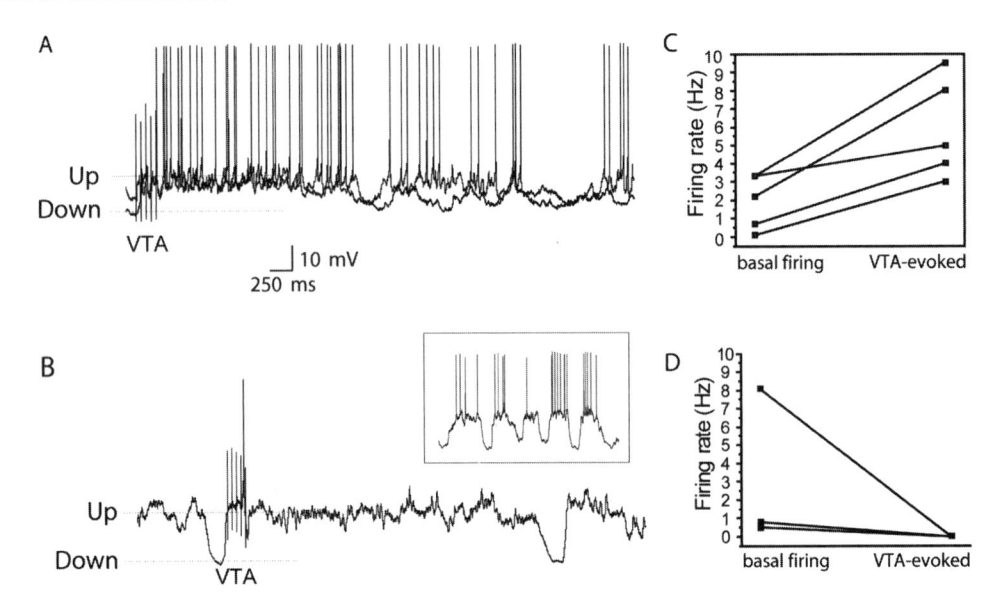

FIGURE 4.3.5. Burst stimulation of the VTA evoked a prolonged up state, along with an increase in cell firing in neonatally lesioned animals tested as adults. (A) Overlay of two traces illustrating the depolarization in response to a 20-Hz train of five stimuli delivered to the VTA and the increased firing following the stimulation. These traces were recorded from a medial PFC neuron in a rat lesioned at PD6 and tested as an adult. (B) Tracing from a sham-operated rat illustrating the prolonged up state concomitant with a reduced firing that was characteristic of untreated rats. Inset: representative tracing from the same neuron, illustrating its baseline firing before the VTA was stimulated. (C) Plot of spontaneous firing rate and VTA-evoked firing in all five neurons tested with VTA stimulation in this group, revealing an increase in firing by VTA stimulation. (D) Similar plot from all neurons in the neonatal sham group showing the typical decrease in firing by VTA stimulation. Source: From [61] with permission.

and Danny Weinberger[45] to determine whether early hippocampal deficits would have an impact on adult behaviors. Indeed, rats with a NVHL present behavioral, molecular, and electrophysiological anomalies, most of which emerge during adolescence. Specifically, adult NVHL rats become hyperactive,[46] show enhanced reactivity to stress, psychostimulants, and NMDA antagonists,[45,47,48] and exhibit sensorimotor gating deficits in the form of reduced prepulse inhibition (PPI) of the acoustic startle response.[49] Social interactions are also altered,[50] there is an increased liability to addictive behaviors,[51,52] and nucleus accumbens (NA) neurons respond to VTA stimulation with an abnormal increase in firing[53] instead of the typical decrease.[54] Thus, the NVHL is a useful tool to study periadolescent changes secondary to earlier developmental manipulations.

Several findings point to the PFC, and more specifically PFC interneurons, as being affected in this model. A PFC lesion in adult rats with a NVHL blocks the hyperlocomotion[55] and the abnormal responses of NA neurons to VTA stimulation.[56] Furthermore, several PFC-dependent behaviors, such as working memory, are affected in NVHL rats[57,58] and primates,[59] and there is a reduction in GAD67 in the PFC of NVHL rats.[60] Many anomalies in the NVHL model cannot be reproduced if the lesion is produced when the animals are already adults,[61] suggesting that the altered responses may reflect abnormal postnatal developmental changes within the mesocortical-PFC pathway. Thus, an abnormal PFC is likely to underscore alterations in the NVHL model.

The DA modulation of glutamate and GABA responses is altered in the PFC of NVHL rats. In vivo intracellular recordings revealed that VTA stimulation with bursts of pulses caused the transition to an up state in pyramidal neurons from adult rats, but instead of the normal decrease in firing,[11] pyramidal neurons increased their firing,[61] suggesting the possibility that interneuron activation by DA was impaired in this model (Fig. 4.3.5). Whole-cell recordings reveal that adult PFC pyramidal neurons are hyperexcitable in response to NMDA and D1 activation,[62] and fast-spiking interneurons are less active in slices from adult rats with a NVHL.[63] Furthermore, the periodolescent maturation of DA effects on PFC interneurons fails to occur in NVHL rats (Fig. 4.3.6). Quinpirole increases interneuron excitability in slices from adult sham-treated rats, as it does in naive rats, but does not yield an increase in excitability in slices from adult NVHL rats.[63] In many NVHL recordings, quinpirole actually reduces excitability. The NVHL procedure therefore causes an alteration in interneuron

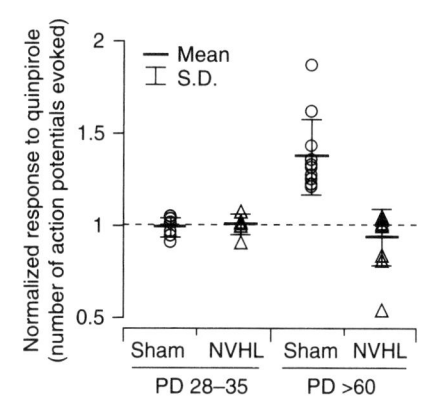

FIGURE 4.3.6. Plot of normalized responses in cell excitability changes by quinpirole in NVHL and sham rats before and after adolescence. An excitatory action of quinpirole was observed only in the PFC of adult sham rats. In contrast, the majority of adult NVHL interneurons recorded remained unchanged after quinpirole administration, resembling the response observed in the PFC of prepubertal rats, while others showed a decrease in excitability.

development such that the maturation of responses to DA during adolescence either does not occur or takes a wrong direction. Thus, even though the neonatal lesion may have caused abnormal development of PFC circuits, the functional impact of such an anomaly is minimal in the immature brain; it is only when the normal periadolescent maturation fails to occur that symptoms become florid.

Other models also point to a deficit in cortical interneurons. Raising rats in social isolation also yields abnormal responses in adult PFC pyramidal neurons to VTA stimulation.[64] Treating pregnant rats with the antimitotic agent methylazoxymethanol (MAM) causes a decrease in PV-expressing neurons in the hippocampus[65] and PFC,[66] which is associated with loss of gamma oscillations in the electroencephalogram (EEG). Some of the emerging genetic models also display interneuron deficits. For example, a dominant-negative form of disrupted-in-schizophrenia-1 (DISC1) shows a reduction in PV interneurons.[67] An immune challenge in pregnant rats has been proposed to mimic the impact of maternal infection. In our hands, injecting the bacterial endotoxin lipopolysaccharide (LPS) in the ventral hippocampus at the same age that the lesions are typically made also causes an abnormal maturation of PFC interneurons during adolescence. In slices from LPS-treated rats, quinpirole failed to increase interneuron excitability.[68] This indicates that the deleterious effects of the NVHL on interneuron development are not related to the lesion, but to abnormal activity or inactivation in the ventral hippocampus impairing development of PFC circuits. The convergence in interneuron deficits across several different models is remarkable, and it highlights the possibility that several different mechanisms may share a common interference with postnatal maturation of local inhibition in cortical circuits.

IMPLICATIONS FOR SCHIZOPHRENIA PATHOPHYSIOLOGY AND NOVEL TREATMENTS

The periadolescent maturation of PFC circuits is likely to have a strong impact on schizophrenia pathophysiology. Although several candidate genes that may confer a predisposition for the disease have been identified,[44] symptoms do not emerge until late adolescence or early adulthood. Our work with NVHL rats suggests that PFC circuits rendered abnormal by early manipulations may become evident when the late periadolescent maturation fails to occur. It is possible that a combination of predisposing genes and epigenetic factors contributes to establishing abnormally wired cortical circuits, perhaps characterized by altered interneuron function. It is with the critical maturation during adolescence that the impact of such miswiring on behavior becomes evident (Fig. 4.3.7). Thus, the protracted maturation of inhibitory circuits in the cortex may serve as a bridge between the early developmental nature of predisposing factors and the late onset of symptoms.

The involvement of such delayed maturation of inhibitory circuits in symptom onset offers opportunities for new approaches to drug treatment for schizophrenia. The traditional approaches to treating this disease have been DA antagonists, mostly targeting D2 receptors. There is indeed evidence that their clinical efficacy correlates with their ability to block D2 receptors.[69] Both classical and atypical antipsychotic drugs reverse some of the abnormal behaviors and

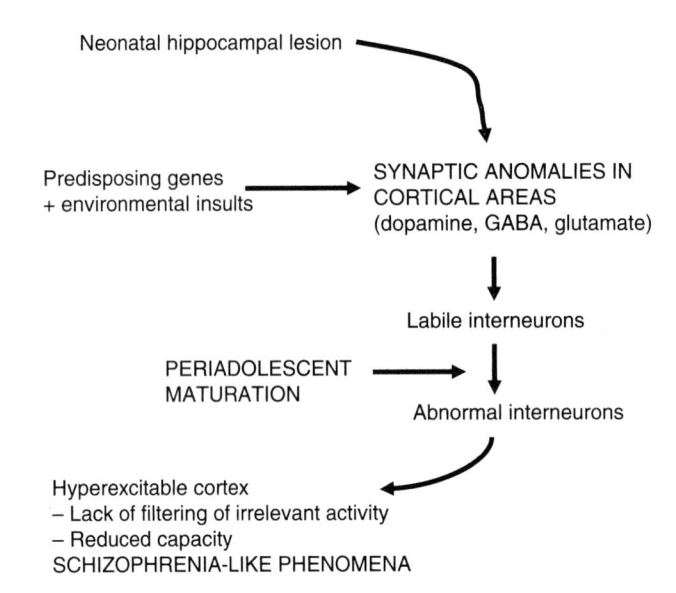

FIGURE 4.3.7. Diagram illustrating the hypothetical sequence of events that may be common to the NVHL and other developmental animal models of schizophrenia and the impact that predisposing genes and epigenetic factors may have. Altered early development of cortical circuits may be affected, but the full set of endophenotypes may emerge only after the periadolescent maturation of interneuron function fails to occur or takes the wrong turn. A possible deleterious effect of stress in such critical periods can account for that emergence.

electrophysiological deficits associated with the NVHL,[53,70,71] indicating that the lesion model may be reproducing pathophysiological changes that are targeted by antipsychotic drugs. Because neuroleptics have disabling side effects and compliance is poor, there has been an intense search for novel therapeutic approaches. For many years, this search has focused on different compounds that retained the D2 antagonism. More recently, however, the conceptualization that excitation–inhibition balance in cortical regions may be a critical factor and that dopamine dysregulation may occur downstream to altered cortical function led the field to consider approaches targeting glutamate and GABA receptors. A recent clinical trial revealed that restoring such balance with a metabotropic glutamate 2/3 agonist (which may reduce the levels of glutamate release) has efficacy similar to that of olanzapine.[72] This was the first non-dopaminergic compound with proven efficacy. The consideration that excitation–inhibition balance matures during adolescence should guide drug discovery, and in particular calls for consideration of external factors to which adolescents seem vulnerable and may contribute to triggering symptom onset, such as stress. In short, the maturation of dopamine effects in the PFC and other cortical areas is likely to determine whether a particular component in those circuits is vulnerable and may settle into an abnormal configuration that may lead to symptoms.

ACKNOWLEDGMENTS

This work was supported by NIH grant MH57683.

REFERENCES

1. Casey BJ, Giedd JN, Thomas KM. Structural and functional brain development and its relation to cognitive development. *Biol Psychol.* 2000;54(1-3):241–257.
2. Rosenberg DR, Lewis DA. Postnatal maturation of the dopaminergic innervation of monkey prefrontal and motor cortices: a tyrosine hydroxylase immunohistochemical analysis. *J Comp Neurol.* 1995;358(3):383–400.
3. Weickert CS, Webster MJ, Gondipalli P, et al. Postnatal alterations in dopaminergic markers in the human prefrontal cortex. *Neuroscience.* 2007;144(3):1109–1119.
4. Lewis DA, Hashimoto T, Volk DW. Cortical inhibitory neurons and schizophrenia. *Nat Rev Neurosci.* 2005;6(4):312–324.
5. Tarazi FI, Baldessarini RJ. Comparative postnatal development of dopamine D(1), D(2) and D(4) receptors in rat forebrain. *Int J Dev Neurosci.* 2000;18(1):29–37.
6. Andersen SL. Trajectories of brain development: point of vulnerability or window of opportunity? *Neurosci Biobehav Rev.* 2003;27(1-2):3–18.
7. Funahashi S. Neuronal mechanisms of executive control by the prefrontal cortex. *Neurosci Res.* 2001;39(2):147–165.
8. Bunge SA, Wright SB. Neurodevelopmental changes in working memory and cognitive control. *Curr Opin Neurobiol.* 2007;17(2):243–250.
9. Segalowitz SJ, Davies PL. Charting the maturation of the frontal lobe: an electrophysiological strategy. *Brain Cogn.* 2004;55(1):116–133.

10. Seamans JK, Yang CR. The principal features and mechanisms of dopamine modulation in the prefrontal cortex. *Prog Neurobiol.* 2004;74(1):1–58.
11. Lewis BL, O'Donnell P. Ventral tegmental area afferents to the prefrontal cortex maintain membrane potential "up" states in pyramidal neurons via D1 dopamine receptors. *Cereb Cortex.* 2000;10:1168–1175.
12. Bernardi G, Cherubini E, Marciani MG, Mercuri N, Stanzione P. Responses of intracellularly recorded cortical neurons to the iontophoretic application of dopamine. *Brain Res.* 1982;245:268–274.
13. Overton P, Clark D. Iontophoretically administered drugs acting at the N-methyl-d-aspartate receptor modulate burst firing in A9 dopamine neurons in the rat. *Synapse.* 1992;10:431–440.
14. Chergui K, Charléty PJ, Akaoka H, et al. Tonic activation of NMDA receptors causes spontaneous burst discharge of rat midbrain dopamine neurons in vivo. *Eur J Neurosci.* 1993;5:137–144.
15. Suaud-Chagny MF, Chergui K, Chouvet G, Gonon F. Relationship between dopamine release in the rat nucleus accumbens and the discharge activity of dopaminergic neurons during local in vivo application of amino acids in the ventral tegmental area. *Neuroscience.* 1992;49:63–72.
16. DeFrance JF, Sikes RW, Chronister RB. Dopamine action in the nucleus accumbens. *J. Neurophysiol.* 1985;54:1568–1577.
17. O'Donnell P. Dopamine gating of forebrain neural ensembles. *Eur J Neurosci.* 2003;17(3):429–435.
18. Seamans JK, Durstewitz D, Christie BR, Stevens CF, Sejnowski TJ. Dopamine D1/D5 receptor modulation of excitatory synaptic inputs to layer V prefrontal cortex neurons. *Proc Natl Acad Sci USA.* 2001;98(1):301–306.
19. Chen G, Greengard P, Yan Z. Potentiation of NMDA receptor currents by dopamine D1 receptors in prefrontal cortex. *Proc Natl Acad Sci USA.* 2004;101(8):2596–2600.
20. Tseng KY, O'Donnell P. Dopamine–glutamate interactions controlling prefrontal cortical pyramidal cell excitability involve multiple signaling mechanisms. *J Neurosci.* 2004;24(22):5131–5139.
21. Wang J, O'Donnell P. D_1 dopamine receptors potentiate NMDA-mediated excitability increase in rat prefrontal cortical pyramidal neurons. *Cereb Cortex.* 2001;11:452–462.
22. Gulledge AT, Jaffe DB. Dopamine decreases the excitability of layer V pyramidal cells in the rat prefrontal cortex. *J Neurosci.* 1998;18:9139–9151.
23. Grobin AC, Deutch AY. Dopaminergic regulation of extracellular gamma-aminobutyric acid levels in the prefrontal cortex of the rat. *J Pharmacol Exp Ther.* 1998;285(1):350–357.
24. Tseng KY, O'Donnell P. Dopamine modulation of prefrontal cortical interneurons changes during adolescence. *Cereb Cortex.* 2007;17:1235–1240.
25. Gurden H, Takita M, Jay TM. Essential role of D1 but not D2 receptors in the NMDA receptor-dependent long-term potentiation at hippocampal-prefrontal cortex synapses In vivo. *J Neurosci.* 2000;20(22):RC106.
26. Gurden H, Tassin J-P, Jay T. Integrity of the mesocortical dopaminergic system is necessary for complete expression of in vivo hippocampal-prefrontal cortex long-term potentiation. *Neuroscience.* 1999;94:1019–1027.
27. Floresco SB, Phillips AG. Delay-dependent modulation of memory retrieval by infusion of a dopamine D1 agonist into the rat medial prefrontal cortex. *Behav Neurosci.* 2001;115(4):934–939.
28. Seamans JK, Floresco SB, Phillips AG. D1 receptor modulation of hippocampal-prefrontal cortical circuits integrating spatial memory with executive functions. *J. Neurosci.* 1998;18:1613–1621.
29. Baldwin AE, Sadeghian K, Kelley AE. Appetitive instrumental learning requires coincident activation of NMDA and dopamine D1 receptors within the medial prefrontal cortex. *J Neurosci.* 2002;22(3):1063–1071.
30. Tseng KY, Mallet N, Toreson KL, Le Moine C, Gonon F, O'Donnell P. Excitatory response of prefrontal cortical fast-spiking interneurons to ventral tegmental area stimulation in vivo. *Synapse.* 2006;59(7):412–417.
31. Seamans JK, Gorelova N, Durstewitz D, Yang CR. Bidirectional dopamine modulation of GABAergic inhibition in prefrontal cortical pyramidal neurons. *J Neurosci.* 2001;21(10):3628–3638.
32. Gorelova N, Seamans JK, Yang CR. Mechanisms of dopamine activation of fast-spiking interneurons that exert inhibition in rat prefrontal cortex. *J Neurophysiol.* 2002;88(6):3150–3166.
33. Zhu JJ. Maturation of layer 5 neocortical pyramidal neurons: amplifying salient layer 1 and layer 4 inputs by Ca^{2+} action potentials in adult rat tuft dendrites. *J Physiol.* 2000;526 Pt 3:571–587.
34. Monyer H, Burnashev N, Laurie DJ, Sakmann B, Seeburg PH. Developmental and regional expression in the rat brain and functional properties of four NMDA receptors. *Neuron.* 1994;12(3):529–540.
35. Tseng KY, O'Donnell P. Post-pubertal emergence of prefrontal cortical up states induced by D_1-NMDA co-activation. *Cereb Cortex.* 2005;15(1):49–57.
36. Cohen JD, Braver TS, Brown JW. Computational perspectives on dopamine function in prefrontal cortex. *Curr Opin Neurobiol.* 2002;12(2):223–229.
37. Horvitz JC. Mesolimbocortical and nigrostriatal dopamine responses to salient non-reward events. *Neuroscience.* 2000;96(4):651–656.
38. Tobler PN, Fiorillo CD, Schultz W. Adaptive coding of reward value by dopamine neurons. *Science.* 2005;307(5715):1642–1645.
39. Trantham-Davidson H, Neely LC, Lavin A, Seamans JK. Mechanisms underlying differential D1 versus D2 dopamine receptor regulation of inhibition in prefrontal cortex. *J Neurosci.* 2004;24(47):10652–10659.
40. Franco R, Casado V, Mallol J, et al. The two-state dimer receptor model: a general model for receptor dimers. *Mol Pharmacol.* 2006;69(6):1905–1912.
41. Tseng KY, O'Donnell P. D_2 dopamine receptors recruit a GABA component for their attenuation of excitatory synaptic transmission in the adult rat prefrontal cortex. *Synapse.* 2007;61:843–850.
42. Nuechterlein KH, Dawson ME, Green MF. Information-processing abnormalities as neuropsychological vulnerability indicators for schizophrenia. *Acta Psychiatr Scand.* 1994;90 (suppl. 384):71–79.
43. Thompson JL, Pogue-Geile MF, Grace AA. Developmental pathology, dopamine, and stress: a model for the age of onset of schizophrenia symptoms. *Schizophr Bull.* 2004;30(4):875–900.
44. Harrison PJ, Weinberger DR. Schizophrenia genes, gene expression, and neuropathology: on the matter of their convergence. *Mol Psychiatry.* 2005;10(1):40–68.
45. Lipska BK, Jaskiw GE, Weinberger DR. Postpuberal emergence of hyperresponsiveness to stress and to amphetamine after neonatal excitotoxic hippocampal damage: a potential animal model of schizophrenia. *Neuropsychopharmacology.* 1993;90:67–75.
46. Lipska BK, Weinberger DR. Genetic variation in vulnerability to the behavioral effects of neonatal hippocampal damage in rats. *Proc Natl. Acad Sci USA.* 1995;92:8906–8910.

47. Swerdlow NR, Lipska BK, Weinberger DR, Braff DL, Jaskiw GE, Geyer MA. Increased sensitivity to the sensorimotor gating-disruptive effects of apomorphine after lesions of medial prefrontal cortex or ventral hippocampus in adult rats. *Psychopharmacology (Berl)*. 1995;122(1):27–34.

48. Al-Amin HA, Weinberger DR, Lipska BK. Exaggerated MK-801-induced motor hyperactivity in rats with the neonatal lesion of the ventral hippocampus. *Behav Pharmacol*. 2000;11(3-4):269–278.

49. Lipska BK, Swerdlow NR, Geyer MA, Jaskiw GE, Braff DL, Weinberger DR. Neonatal excitotoxic hippocampal damage in rats cause post-pubertal changes in prepulse inhibition of startle and its disruption by apomorphine. *Psychopharmacology*. 1995;132:303–310.

50. Sams-Dodd F, Lipska BK, Weinberger DR. Neonatal lesions of the rat ventral hippocampus result in hyperlocomotion and deficits in social behavior in adulthood. *Psychopharmacology*. 1997;132:303–310.

51. Chambers RA, Self DW. Motivational responses to natural and drug rewards in rats with neonatal ventral hippocampal lesions: an animal model of dual diagnosis schizophrenia. *Neuropsychopharmacology*. 2002;27(6):889–905.

52. Brady AM, McCallum SE, Glick SD, O'Donnell P. Enhanced methamphetamine self-administration in a neurodevelopmental rat model of schizophrenia. *Psychopharmacology (Berl)*. 2008;200:205–215.

53. Goto Y, O'Donnell P. Delayed mesolimbic system alteration in a developmental animal model of schizophrenia. *J Neurosci*. 2002;22(20):9070–9077.

54. Goto Y, O'Donnell P. Network synchrony in the nucleus accumbens in vivo. *J Neurosci*. 2001;21(12):4498–4504.

55. Lipska B, al-Amin H, Weinberger D. Excitotoxic lesions of the rat medial prefrontal cortex. Effects on abnormal behaviors associated with neonatal hippocampal damage. *Neuropsychopharmacology*. 1998;19:451–464.

56. Goto Y, O'Donnell P. Prefrontal lesion reverses abnormal mesoaccumbens response in an animal model of schizophrenia. *Biol Psychiatry*. 2004;55(2):172–176.

57. Chambers RA, Moore J, McEvoy JP, Levin ED. Cognitive effects of neonatal hippocampal lesions in a rat model of schizophrenia. *Neuropsychopharmacology*. 1996;15:587–594.

58. Lipska BK, Aultman JM, Verma A, Weinberger DR, Moghaddam B. Neonatal damage of the ventral hippocampus impairs working memory in the rat. *Neuropsychopharmacology*. 2002;27(1):47–54.

59. Heinz A, Saunders RC, Kolachana BS, et al. Striatal dopamine receptors and transporters in monkeys with neonatal temporal limbic damage. *Synapse*. 1999;32(2):71–79.

60. Lipska BK, Lerman DN, Khaing ZZ, Weickert CS, Weinberger DR. Gene expression in dopamine and GABA systems in an animal model of schizophrenia: effects of antipsychotic drugs. *Eur J Neurosci*. 2003;18(2):391–402.

61. O'Donnell P, Lewis BL, Weinberger DR, Lipska BK. Neonatal hippocampal damage alters electrophysiological properties of prefrontal cortical neurons in adult rats. *Cereb Cortex*. 2002;12(9):975–982.

62. Tseng KY, Lewis BL, Lipska BK, O'Donnell P. Post-pubertal disruption of medial prefrontal cortical dopamine–glutamate interactions in a developmental animal model of schizophrenia. *Biol Psychiatry*. 2007;62:730–738.

63. Tseng KY, Lewis BL, Hashimoto T, et al. A neonatal ventral hippocampal lesion causes functional deficits in adult prefrontal cortical interneurons. *J Neurosci*. 2008;28:12691–12699.

64. Peters YM, O'Donnell P. Social isolation rearing affects prefrontal cortical response to ventral tegmental area stimulation. *Biol Psychiatry*. 2005;57(10):1205–1208.

65. Penschuck S, Flagstad P, Didriksen M, Leist M, Michael-Titus AT. Decrease in parvalbumin-expressing neurons in the hippocampus and increased phencyclidine-induced locomotor activity in the rat methylazoxymethanol (MAM) model of schizophrenia. *Eur J Neurosci*. 2006;23(1):279–284.

66. Lodge D, Behrens M, Grace AA. A loss of parvalbumin containing interneurons is associated with diminished gamma oscillatory activity in an animal model of schizophrenia. *Schizophr Res*. 2008;102(suppl 2):112.

67. Hikida T, Jaaro-Peled H, Seshadri S, et al. Dominant-negative DISC1 transgenic mice display schizophrenia-associated phenotypes detected by measures translatable to humans. *Proc Natl Acad Sci USA*. 2007;104(36):14501–14506.

68. O'Donnell P, Tseng KY, Feleder C. Periadolescent emergence of impaired dopamine modulation of prefrontal GABA circuits in developmental animal models of schizophrenia. *Int J Neuropsychopharmacol*. 2008;11(S1):37.

69. Seeman P. Dopamine receptors and the dopamine hypothesis of schizophrenia. *Synapse*. 1987;1:133–152.

70. Le Pen G, Moreau JL. Disruption of prepulse inhibition of startle reflex in a neurodevelopmental model of schizophrenia: reversal by clozapine, olanzapine and risperidone but not by haloperidol. *Neuropsychopharmacology*. 2002;27(1):1–11.

71. Lipska BK, Weinberger DR. Subchronic treatment with haloperidol and clozapine in rats with neonatal excitotoxic hippocampal damage. *Neuropsychopharmacology*. 1994;10:199–205.

72. Patil ST, Zhang L, Martenyi F, et al. Activation of mGlu2/3 receptors as a new approach to treat schizophrenia: a randomized Phase 2 clinical trial. *Nat Med*. 2007;13(9):1102–1107.

4.4 Genetic Dissection of Dopamine-Mediated Prefrontal-Striatal Mechanisms and Its Relationship to Schizophrenia

HAO-YANG TAN AND DANIEL R. WEINBERGER

INTRODUCTION

Abnormalities in dopamine (DA) function have long been implicated in etiological hypotheses of schizophrenia and associated cognitive deficits. These cognitive changes, the precise mechanisms of which remain unclear, account for high morbidity and costs.[1] Recent real-world clinical trials have highlighted the need to improve the efficacy of antipsychotic medications in symptom management and patient acceptability,[2,3] as well as their effect on the cognitive deficits central to the illness.[4] The clinical data, nevertheless, support the canonical finding that the brain DA system, the common target of all antipsychotic drugs (of limited efficacy though they may be), is a system that must be related to the treatment and pathophysiology of psychosis. However, it also follows that older conceptualizations of schizophrenia as being simply about too much brain DA should be and indeed have been superseded by more sophisticated understandings of the role of DA in distinct cortical microcircuit functions related to cognition and psychosis. For example, cerebrospinal fluid homovanillic acid, a measure of cortical DA turnover, predicted cortical activation measured during an executive cognitive task in patients with schizophrenia, suggesting that cortical deficits associated with this illness might be at least in part related to reduced cortical DA function.[5] This led to a hypothesis[6] that cortical DA hypofunction could underlie cortical deficits associated with schizophrenia and drive subcortical DA hyperactivity, long believed to correlate with psychosis. These and other hypotheses of DA mechanisms have undergone modifications and elaborations as data emerged on further finer-grained parcellation of these DA effects on cortical-striatal microcircuit processing, driven by work on genes influencing DA function, basic experiments on animals, and other human studies (for reviews, see [7,11]).

This knowledge, together with emerging insights into the genetics of human brain processing in schizophrenia, appear to be critical in enhancing future new treatments based on a precise mechanistic perspective of DA function in the human brain.

The genetic diathesis of schizophrenia has been well documented by classical twin, adoption, and family studies.[12,13] What has remained challenging is the elucidation of specific genes and their relationship with human mechanisms associated with this complex neuropsychiatric condition or the attendant changes in cognition and functional outcome. Some promising results have emerged from recent whole genome association studies involving multiple centers and thousands of patients, including studies combining patients with schizophrenia and bipolar disorders to examine common disease genetics.[14] Replicable signals may emerge. To date, these whole-genome association studies have not (yet) implicated genes directly related to DA function, but they could identify novel targets and stimulate paradigm shifts in the research. However, recent meta-analyses of genetic association studies have also pointed, for instance, to a number of DA receptor genes (e.g., DRD1, DRD2), among others, that could play significant roles in psychosis[15] and presumably in brain function related to psychosis. While the literature can partly be driven by bias,[16] this would still be a difficult argument against the weight of evidence-based psychotropic treatments for psychosis, which point largely to the DA system, and the voluminous animal and human data that have accumulated about DA function in brain. Moving forward, genes related to the DA system should continue to serve as logical starting points in the study of the genetics and dissection of cognitive brain mechanisms relevant to psychosis. A goal of this chapter is thus to examine recent developments integrating DA processing in human neuroimaging, cognition, and genetics, and the elucidation of putative genetic mechanisms of human prefrontal cognitive functions relevant to disorders such as schizophrenia.

'INTERMEDIATE PHENOTYPE' APPROACH IN HUMAN GENETICS TO DISSECT COMPONENT BRAIN MECHANISMS

At the fundamental level, genes encode proteins; they do not directly encode any psychopathological manifestations, be they hallucinations, delusions, sadness, or anxiety. To the extent that genes are implicated—for example, with the constellation of symptoms we call schizophrenia or bipolar disorder—they do so by affecting the development and function of brain cells and neural systems that mediate the expression of such diverse behavioral and perceptual phenomena. Patients with neuropsychiatric disease have changes in cognition, brain function, and structure. These brain abnormalities are also found more frequently in their unaffected siblings, including unaffected monozygotic twins, than in control subjects without such a family history. These various deviations therefore represent biological expressions of increased genetic risk rather than a disease entity per se. Since these biological changes are found in individuals who are carrying a greater genetic risk but who do not manifest a DSM-IV diagnosis,[17-22] it follows that these brain changes are susceptibility-related phenotypes, intermediate between the cellular effects of susceptibility genes and manifest psychopathology.[7,8,23]

Such genetic links, particularly with quantifiable, reliable, and heritable measures of brain function, should facilitate the elucidation of the underlying neural mechanisms of these genetic brain processes. This is a principal value of studying specific, well-designed intermediate brain phenotypes and their genetic associations. These studies characterize, in the live human brain, the neural system mechanisms of the clinical genetic associations.[7,8,23] An illustrative example that we will develop in this chapter is prefrontal cortically mediated working memory, one of several dopaminergic brain processes implicated in the cognitive abnormalities of psychosis and its putative genetic mechanisms.

PREFRONTAL CORTEX, WORKING MEMORY, AND SCHIZOPHRENIA

Working memory is a limited-capacity system that enables us to temporarily hold, update, and work with relevant information; it underlies almost all higher-order thinking, language, and goal-directed behavior.[24]

Critically engaging prefrontal cortical brain systems, working memory has been shown to be an important component underlying many cognitive deficits observed in schizophrenia.[25-27] An extensive body of functional imaging experiments is also consistent with prefrontal cortical physiological dysfunction in schizophrenia (e.g., see reviews by [28-30]). The recent literature has also clarified that the directionality of the imaging findings (i.e., too much or too little prefrontal activity) depends on the specific task paradigm utilized, disease-related differences in behavioral performance, capacity constraints of the task load,[30-34] and the engagement of a system of dysfunctional and compensatory brain circuits.[35,36] In particular, absent task performance as a confounder, the increased prefrontal activation and physiological inefficiency phenotype has often been observed with schizophrenia and with an increased genetic risk for schizophrenia.[8,18,37]

Several cognitive abnormalities associated with schizophrenia have consistently been found with increased prevalence in healthy siblings of patients with schizophrenia, including healthy monozygotic cotwins. The evidence from twin studies suggests that the cognitive deficits related to IQ, attention, and working memory are particularly heritable and risk-associated traits.[19,22] With functional neuroimaging, prefrontal cortical changes during the use of working memory and cognitive control also have been observed to be familial and heritable.[18,38,39] Thus, the genetic mechanisms of these intermediate working memory brain processes are potentially tractable using these neuroimaging paradigms, a particular focus in this discussion on DA mechanisms in prefrontal brain systems.

NEUROBIOLOGICAL AND COMPUTATIONAL MODELS OF DA, WORKING MEMORY, AND PREFRONTAL SIGNAL-TO-NOISE PROCESSING

Working memory has been studied extensively in animal and computational models (see also reviews by [10,40,41]). Seminal work on single-unit recordings in nonhuman primates has demonstrated that neurons located around the principal sulcus in the dorsolateral prefrontal cortex exhibit activity corresponding to maintaining information in an active state during the delay period of the working memory paradigm.[42,43] Numerous functional neuroimaging studies have since elaborated the key role of the prefrontal cortex in human working memory. It is also clear that the neural system supporting working memory engages a distributed network of cortical areas. These include the posterior parietal cortex, inferotemporal cortex, cingulate gyrus, and hippocampus—all of which have anatomical connections with the prefrontal cortex.[44,50]

Neural mechanisms of working memory are critically dependent on dopaminergic modulation of glutamatergic and GABAergic brain systems during the processing of delay-period maintenance and manipulation of information. Experiments with DA D1 receptor agonists and antagonists modulate memory-dependent prefrontal neural firing, with optimal tuning of activity following an inverted-U curve across incremental D1 stimulation.[51–53] These data support the hypothesis that locally sustained firing of prefrontal cortical neurons crucial in the maintenance of relevant information during the delay period of working memory is stabilized against distracters through DA D1 receptors.[53] These D1 receptors allow a focused augmentation of the task-relevant signal[54,55] by enhancing N-methyl-D-aspartate (NMDA)-receptor-mediated postsynaptic currents in prefrontal pyramidal neurons, which are also active during the delay period.[56–58] Concurrently, D1 receptors also trigger a tonic increase in the firing of GABAergic inhibitory interneurons acting farther afield, reducing irrelevant firing activity while allowing the focused increase in task-relevant activity, thus optimizing the neural signal-to-noise.[54]

On the other hand, DA D2 receptor signaling appears to play a crucial complementary role in working memory. D2 signals are hypothesized to enable new information to be rapidly updated and/or manipulated online by transiently reducing the barriers for new information signals to be established in the cortical networks. This D2 process is key in ensuring that salient new information from cognitive computations of task goal-directed or reward information is rapidly processed and updated online; this information processing network involves the prefrontal cortex, the posterior cortex, and, importantly, the striatum.[40,59–61] Thus, dopaminergic, glutamatergic, and GABAergic systems are finely tuned to act in concert to maintain neural signal-to-noise critical for effective information processing via these cortical-striatal microcircuits.

IMAGING AND GENETIC DISSECTION OF DOPAMINERGIC MECHANISMS IN HUMAN PREFRONTAL CORTEX

Dopaminergic and glutamatergic systems, other neurotransmitter systems, and related genes play principal roles in prefrontal function and have been implicated in the prefrontal neuropathology of schizophrenia.[62] Given that the human prefrontal cortex is onto- and phylogenetically late developing, is biologically complex, and is implicated in neuropsychiatric disease, understanding the mechanisms by which genetic susceptibility impacts active human prefrontal function is a clinically important challenge. Combining neuroimaging and genetics experiments has been shown to be a promising strategy for elucidating such brain mechanisms.[7,8,23,63] In what follows, we will elaborate on recent imaging investigations on the effects of genetic variation on component brain systems engaged in working memory. We will also examine epistatic interactions of genes reflecting the nonadditive activity-dependent cross-talk across various neurotransmitter systems and intracellular signaling mechanisms that have begun to translate basic neurobiology and genetic risk into human systems neuroscience.

Initial experiments on the impact of dopaminergic gene variation on cortical function examined catechol-O-methyltransferase (COMT). Decades of early research on COMT focused on DA metabolism in the striatum, where little role for COMT had been found. In contrast, recent studies have demonstrated that COMT is a major enzyme in prefrontal synaptic DA catabolism with a critical role in prefrontal cortical DA signaling in synapses because of the relative lack of DA transporters within synapses in this region.[64–66] A common polymorphism in the COMT gene resulting from a valine-to-methionine Val(108/158)Met substitution gives rise to a significant reduction in its enzymatic activity.[64,67,68] This was found to correspond to reduced prefrontal DA in proportion to the Val-allele load. Located on chromosome 22q11, COMT is also deleted in velocardiofacial syndrome, a condition that has a 20-fold increased risk for psychosis.[69] However, while this susceptibility locus has been implicated in some meta-analyses of linkage to schizophrenia,[70,71] the effect on the risk for schizophrenia of the specific COMT Val(108/158)Met polymorphism is small and inconsistent.[15,72,73] This is not surprising given the manifold factors associated with schizophrenia pathogenesis, such as the involvement of combinations of single nucleotide polymorphisms or haplotypes within the gene,[74,75] interactions across different susceptibility genes,[76] and interactions between genes and the environment.[77]

In contrast to the weak effect of genetic variation in COMT on behavioral syndromes, the effect of the COMT Val(108/158)Met polymorphism on intermediate measures of human brain function has been more reflective of predictions from the basic cellular models of prefrontal DA described earlier. Reduced prefrontal DA in COMT Val carriers should lead to decreased tonic D1-receptor activation. This might result, firstly, in a reduced cortical signal-to-noise ratio; secondly, in a relatively inefficient prefrontal cortical activation pattern if performance accuracy is still maintained; and ultimately, in reduced performance outcomes in working memory and executive function tasks. Each of these predicted effects has been observed in replicated experiments. Catechol-O-methyltransferase genotypes account for a small but significant (about 3%–4%) variance in performance on frontal lobe tests, with poorer performance by subjects carrying the Val rather than the Met allele, even among subjects without neuropsychiatric disease.[78–82] Analogous results were also obtained in healthy children.[83] Correspondingly, using functional magnetic resonance imaging (fMRI) to study cortical activity, healthy COMT Val allele carriers were found to engage a relatively greater extent of prefrontal cortical activation to perform the working memory task with the same speed and accuracy as those with the Met allele; this finding is consistent with the interpretation that Val carriers are relatively less efficient without advantages in performance accuracy or reaction time.[79,84,85] As might be predicted by the greater dependence on dorsolateral prefrontal cortical processing for more complex executive tasks,[86–89] the COMT effects in dorsolateral prefrontal cortex were more prominent at higher working memory loads.[84] The study by Mattay et al.[84] also demonstrated that the inefficient prefrontal cortical activation in Val homozygotes could be improved by administering dopamimetic amphetamine. In these less efficient Val-homozygous individuals, amphetamine resulted in a more focused reduction of dorsolateral prefrontal activation, presumably shifting these individuals closer to the peak on the inverted-U DA tuning curve that characterizes the relationship between cortical DA levels and cortical function. Conversely, Met homozygotes, already closer to peak dopaminergic tuning at baseline, became relatively inefficient after amphetamine administration and thus appeared to have been shifted off the peak of the curve by amphetamine (Fig. 4.4.1).

Cortical information processing, especially in prefrontal cortex, impacts on the regulation of DA activity in the mesencephalon, as demonstrated in many experiments in rodents, nonhuman primates, and humans.[9] This is important in learning, as reward signals emanating from brainstem DA neuronal firing should correspond to prefrontal cortical executive action for learning to take place. Consistent with these basic studies, COMT, presumably via its actions at the cortical level, appears to impact on DA activity in the brainstem. In a study of normal postmortem human brainstem, individuals with a COMT Val/Val genotype had twice the normal expression of the mRNA for tyrosine hydroxylase, the rate-limiting biosynthetic enzyme for DA.[90] Remarkably, this relationship has been confirmed in a positron emission tomography (PET) imaging study of normal living subjects. Catechol-O-methyltransferase Val carriers were found to have relatively increased midbrain DA synthesis (measured with f-18 flurodopa uptake) that correlated negatively with N-back dorsolateral prefrontal cortical activation (measured with O-15 H_2O regional cerebral blood flow), while prefrontal activation in Met homozygotes correlated positively with midbrain DA synthesis.[91] These reciprocal relationships fit tightly to the inverted-U DA tuning curve, whereby increased midbrain DA synthesis in COMT-Val carriers with lower prefrontal DA tended to restore efficient cortical activation, while COMT-Met homozygotes became less efficient with increased DA synthesis. Furthermore, the anatomical and receptor specificities of these findings were recently elaborated in another PET study on D1-receptor availability in relation to COMT Val/Met.[92] Here, it was demonstrated that putatively decreased levels of cortical DA were associated with up-regulation of D1 receptor availability, as measured with the PET radiotracer [11C]NNC112. Subjects with Val/Val alleles and presumably reduced cortical DA had significantly higher cortical [11C]NNC 112 binding compared with Met carriers, but the genotype groups did not differ in striatal binding. These results confirmed the prominent role of COMT in regulating D1 transmission in cortex but not striatum. These multimodality PET imaging findings support conceptualizations of DA's role in the fine tuning of neural circuits,[55,90,93] as do the fMRI data on amphetamine and COMT.[84]

Clearly, COMT did not evolve to do neuropsychological tests, but rather to modulate DA-related tuning of intrinsic cortical circuitry engaged in these cognitive functions. In animal models, DA signaling has been shown to be critical for several executive processes, including maintenance and interference control,[41] but this has not been explored in detail in humans. Less is known, for example, about how these functional dopaminergic effects correspond to the differing prefrontal-parietal-striatal networks during specific human working memory component processes in the hierarchically organized lateral prefrontal cortex. It has been proposed that the dorsal and anterior prefrontal regions (e.g., dorsolateral prefrontal cortex [DLPFC]: Brodmann areas BA 9, 10, and 46) are engaged in higher-order processing, such as in manipulating information or applying it in context, while the ventrolateral prefrontal cortex (VLPFC: Brodmann areas 44, 45, and 47) is engaged during simpler operations.[47,88,89,94,95] If, indeed, DA tuning of cortical neural assemblies is critical for their effective function in working memory,[53,93,96] it might be expected that these DA effects would also be observed in these specific, hierarchically organized prefrontal cortical

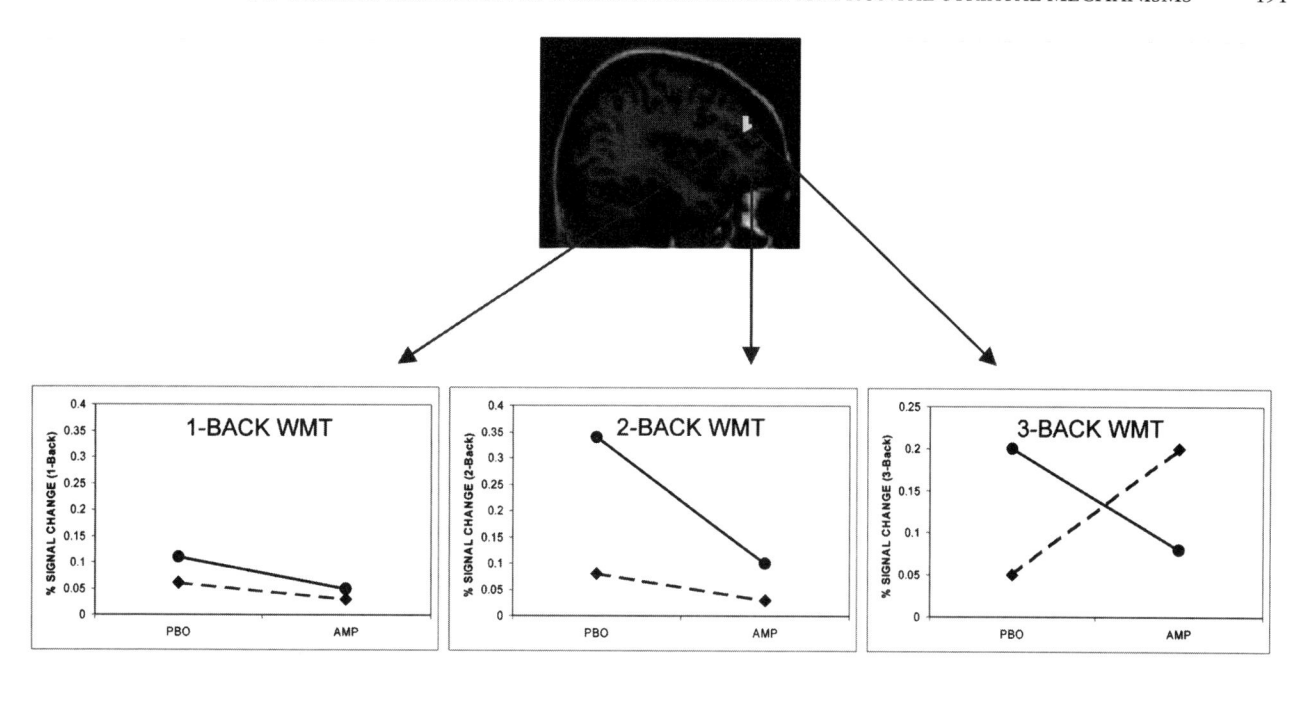

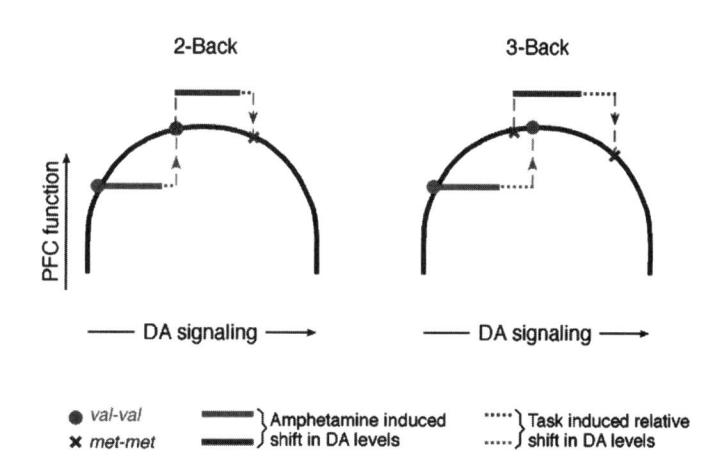

FIGURE 4.4.1. The fMRI activation signal was extracted from the dorsal prefrontal cortex (PFC) (top panel) in the presence of amphetamine (AMP) or placebo (PBO) administration at differing working memory task (WMT) loads as a function of the COMT genotype (middle panel). In COMT-Val homozygote individuals (who have relatively less cortical DA; solid lines, middle panel), AMP improved PFC efficiency (lower activation). In contrast, in individuals homozygous for the *met* allele (who have relatively greater cortical DA; dashed lines, middle panel), AMP had deleterious effects on PFC efficiency (greater activation) at a three-back WMT load (rightmost graph in the middle panel). These results suggest that individuals homozygous for the COMT-*val* allele have PFC functional efficiency on the up slope of the normal range, whereby AMP could increase DA signaling to more optimal levels closer to the peak (bottom panel). On the other hand, individuals homozygous for the COMT-*met* allele appear to already be near peak PFC functional efficiency, so increased DA signaling from AMP shifts PFC function onto the down slope of the inverted-U efficiency curve (bottom panel). *Source*: Adapted from Mattay et al. [84]; courtesy of Venkata S Mattay. (See Color Plate 4.4.1.)

functions. Moreover, if DA function is especially implicated in the updating and stabilization of representations,[40,61,93] then executive working memory tasks emphasizing encoding, manipulating, and temporally integrating information should be disproportionately more dependent on changes in cortical DA signaling than tasks emphasizing simple retrieval of already stabilized information. Some of the former processes are also likely to involve the DA-rich striatum, which has been postulated to have intimate connections to cortex in implementing the selective gating of information during rapid updating and manipulation in working memory.[60,97-99] For example, manipulating numerical representations should engage DA-dependent processes in the striatum, prefrontal cortex, and number-sensitive regions in the parietal cortex.[100,101] On the other hand, anterior regions in the DLPFC might be more specifically engaged during DA-dependent processing of higher-order temporal or episodic aspects of working memory.[89,94]

Indeed, using genetic imaging of COMT, it has been observed that dopaminergic modulation integral to differing levels of working memory processing occurs with a degree of spatial and process specificity within the human prefrontal-parietal-striatal network.[102] In an event-related fMRI task dissociating component numerical working memory processes, baseline numerical size comparison engaged VLPFC activation that correlated with the COMT Val-allele load (COMT Val>Met), while further performance of arithmetic transformations engaged this genotype effect in DLPFC, as well as in parietal and striatal regions (Fig. 4.4.2). Critically, additional temporal integration of information in working memory disproportionately engaged greater COMT Val>Met effects only at the DLPFC. Catechol-O-methyltransferase Val>Met effects were also observed in DLPFC during encoding of new information into working memory but not at its subsequent retrieval. Thus, temporal updating operations, but less so the retrieval of already encoded representations, engaged relatively specific dopaminergic tuning at the DLPFC. Manipulating and rapidly updating representations were sensitive to dopaminergic modulation of neural signaling in a larger prefrontal-parietal-striatal network. The relatively specific engagement of prefrontal-parietal-striatal dopaminergic modulation during these computational tasks supports their role in the effective control of rapid switching and stabilization processes intrinsic in such tasks that engage the manipulation of information. This is also consistent with models predicting basal ganglia coupling of prefrontal cortex and modality-specific (e.g., numerical) regions in the posterior cortex in order to effect this highly selective information transformation and updating; these models also propose that DA is critical in the implementation of these targeted gating processes in the human brain.[97,98]

On the other hand, processes involved in the manipulation of information might be distinguished from those engaged in the temporal integration of information in working memory. The latter are associated with more prominent dopaminergic modulation within the anterior DLPFC rather than in the striatum or the posterior parietal cortex. This observation suggests that dopaminergic processes in these DLPFC regions might more critically mediate higher-order temporal processes, such as when contextual information is encoded for future operations or when new probe information needs to be integrated with that encoded previously. It has been proposed that these higher-order processes engage more overall inhibitory[95] or biasing[103] cognitive control mechanisms putatively engaging greater D1 than D2 activity,[55,93] the former postulated to predominate in the prefrontal cortex.[104] To the extent that D1 functions are differentially regulated by COMT Val/Met,[92] one might speculate that our systems-level findings at these DLPFC regions during live human cognition could reflect greater D1 dopaminergic modulation during higher-order temporal integration of information. Conversely, rapid updating and information manipulation involving the DLPFC, striatum and posterior cortex might reflect the engagement of predominantly D2 mechanisms.[59,60,93,97,98] In an analogous cognitive model, three DA system genes—DARPP-32, DRD2, and COMT—have been shown to impact differentially on specific processes in prefrontal striatally mediated reinforcement learning from positive and negative outcomes.[105]

EPISTATIC GENE MECHANISMS OF HUMAN PREFRONTAL CORTICAL-STRIATAL FUNCTION

Glutamatergic abnormalities, in addition to DA, are important in schizophrenia and working memory function. The NMDA receptor system is a critical partner in working memory processes,[57,93,106] and disease-related changes in glutamate signaling could well impair working memory. For example, the gene *GRM3* on chromosome 7q21-22, which encodes the metabotropic glutamate receptor mGluR3, modulates NMDA receptor transmission.[62,107,108] mGluR3 regulates synaptic glutamate via a presynaptic mechanism and by regulating the expression of the glial glutamate transporter, which inactivates synaptic glutamate. A polymorphism in intron 2 and related haplotypes were significantly associated with schizophrenia in several samples,[107,109-111] though negative studies also have been reported.[112] Risk variants in *GRM3* may also influence alternative splicing of *GRM3* mRNA and its product.[113] In postmortem brain, the risk allele is associated with a reduction in the prefrontal glial glutamate transporter EAAT2, a protein modulating synaptic glutamate.[107] Consistent with the role of the glutamatergic system in schizophrenia and working memory, the risk allele was associated with inefficient prefrontal cortical fMRI activation and reduced working memory performance even in normal subjects.[107]

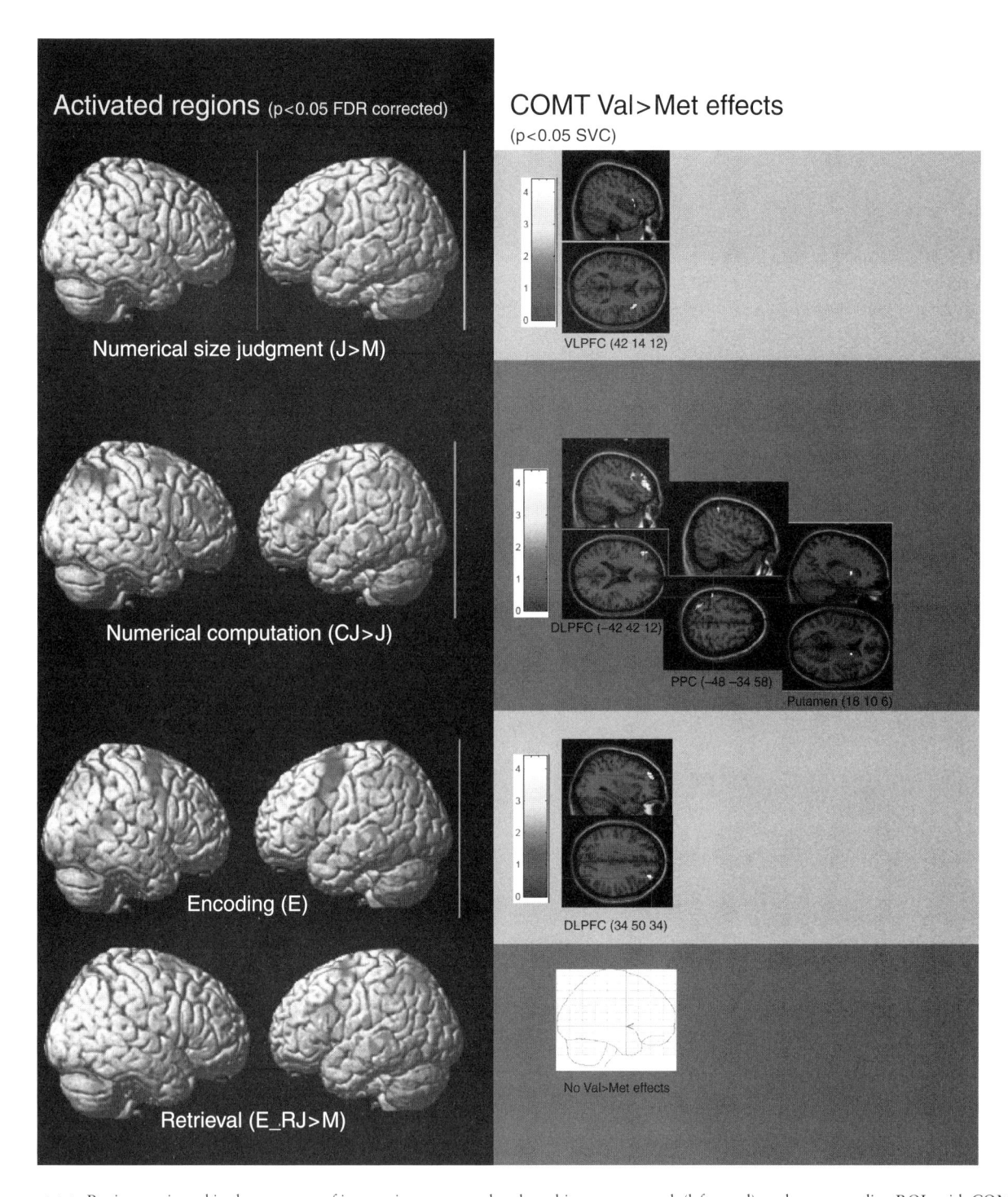

FIGURE 4.4.2. Regions activated in the contrasts of interest in an event-related working memory task (left panel), and corresponding ROIs with COMT Val>Met effects (right panel). During baseline numerical size judgment, subjects engaged COMT effects at the ventrolateral prefrontal cortex (VLPFC). During encoding into working memory, COMT effects were observed in the dorsolateral prefrontal cortex (DLPFC) but not in the striatum. During numerical computations engaging rapid updating of new information, COMT effects were observed in the prefronto-parietal-striatal network. During simple retrieval in working memory, no suprathreshold COMT effects were observed. SVC: small volume correction for multiple comparisons. PFC, prefrontal cortex; PPC, posterior parietal cortex; ROI, region of interest. *Source*: Adapted from Tan et. al.[102] (See Color Plate 4.4.2.)

Importantly, given the tight relationships governing dopaminergic and glutamatergic (and GABAergic) dynamics in the biology of working memory,[51,53,58,93,114,115] and their putatively greater involvement in executive aspects of working memory at the DLPFC,[84,86–89,102] we would expect that higher-order working memory processes taxing the DLPFC might be more vulnerable to the combined effect of suboptimal dopaminergic and glutamatergic influence. Consistent with the interplay of cortical macrocircuits suggested by these possibilities, a recent fMRI study revealed that the integrity of higher executive areas in the DLPFC could be disproportionately compromised and inefficient in the presence of combined relatively deleterious COMT and *GRM3* genotypes in normal subjects[116] (Fig. 4.4.3).

Elegant extensions of interacting receptor systems on cortico-striatal working memory brain systems have also examined specific genetic variations involving the D2 receptor and the dopamine transporter (DAT).[117,118] The D2 and DAT variants studied were previously shown to influence the expression of these proteins in vivo.[118,119] The results in brain imaging confirmed the nonlinear relationship of DA effects via D2 and DAT on striatal-frontal function during executive working memory, where differential DAT expressed indexed by the 3'-Variable Number Tandem Repeat variant affected striatal-frontal brain activity predominantly in the context of the variant associated with reduced D2 expression.[117] When these D2 and DAT genotype groups were ordered from putatively less DA reuptake to greater reuptake and release, a nonlinear inverted-U relationship between the compound genotype and the blood-oxygen-level-dependent (BOLD) response was obtained,[117] mirroring earlier findings on DA effects via COMT reviewed above.[84,91] Thus, genetic variation impacting important nodes in the DA and glutamatergic systems at a molecular level, when combined, had disproportionate or nonadditive influence on executive cognitive brain function at the human systems level. If so, we might also expect key nodes at the intracellular signaling cascades of these receptor systems to play similarly important roles as we extend our search for molecules impacting variation in the dissection of human prefrontal function.

HUMAN PREFRONTAL FUNCTIONAL GENETIC LINKS WITH DA-ASSOCIATED INTRACELLULAR SIGNALING MOLECULES

Classically, D1 receptors, implicated in the maintenance of relevant information during the working memory delay period,[53] couple through $G\alpha_s$ to stimulate the

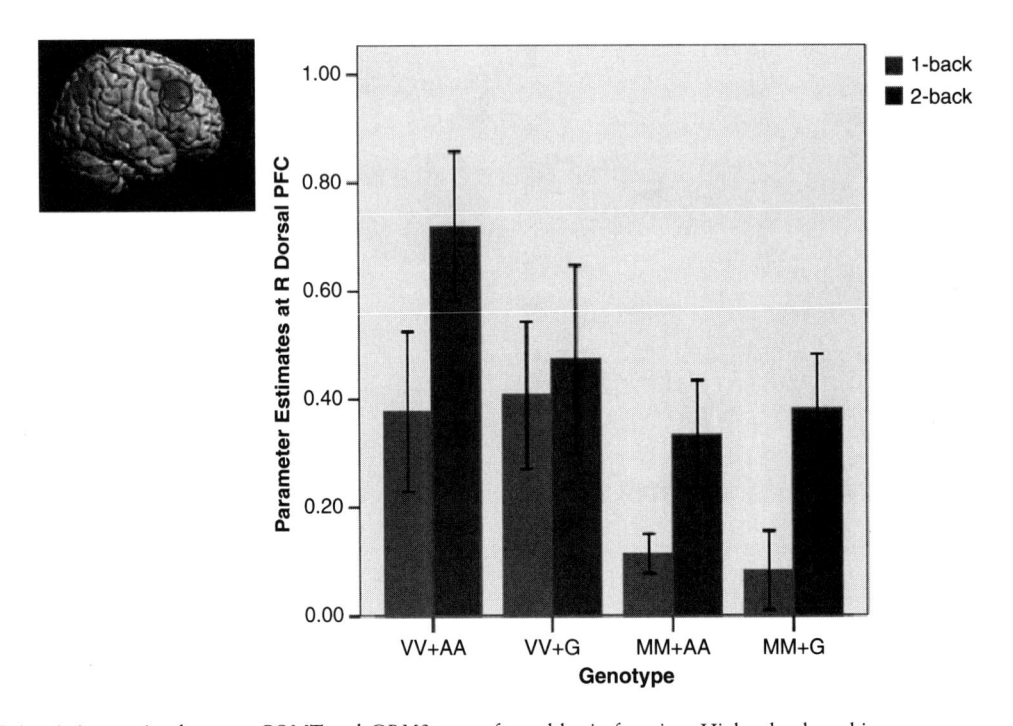

FIGURE 4.4.3. Epistatic interaction between COMT and *GRM3* on prefrontal brain function. Higher-load working memory processes engaging the dorsolateral prefrontal cortex (PFC) was disproportionately inefficient in the context of combined suboptimal COMT and *GRM3* risk alleles ($F_{1,25} = 4.47$, $p = 0.045$). *Source*: Adapted from Tan et al.[116] (See Color Plate 4.4.3.)

production of cyclic adenosine monophosphate (cAMP) and the activity of protein kinase A (PKA).[120] Conversely, D2 receptors, which in neural models play critical roles marking salience, prediction errors, and updating and manipulating new information,[40] couple through $G\alpha_{i/o}$ to reduce cAMP production and PKA activity.[120] Downstream from PKA, DA- and cAMP-regulated phosphoprotein of molecular weight 32 (DARPP-32) is a key signaling integrator that regulates an array of subsequent neurophysiological processes.[120] Indeed, human genetic variation in *DARPP-32* has been found to impact normal human variation in frontostriatal cognitive performance, in neostriatal volume, and in physiological activation and functional connectivity between the striatum and prefrontal cortex, as well as the risk for schizophrenia.[121]

In addition to the cAMP-PKA pathway, D2 receptors may also signal through an AKT1 (protein kinase B)-GSK-3 signaling cascade via β-arrestin 2.[122] Of note, this AKT-GSK-3 pathway influences the expression of DA-associated psychomotor behaviors that, in transgenic mice models, have been predictably modulated by dopaminergic drugs; this pathway also appears to be independent of the cAMP-associated one, and represents a novel means by which D2 receptor signaling and associated cognitive and neuropsychiatric effects could be mediated.[122–124] *AKT1* knockout mice, in particular, evidenced abnormal prepulse inhibition of startle[125] and poorer working memory performance under dopaminergic agonist challenge, as well as concurrent changes in prefrontal pyramidal dendritic ultrastructure, possibly mediated by downstream alterations in the expression of genes controlling neuronal development in prefrontal cortex.[124]

We developed a strategy to examine the genetic association of *AKT1* with human brain phenotypes related to DA function.[126] In examining the effect of a genetic variant in *AKT1* that consistently affected the expression of AKT1 protein levels,[126,127] we found that this same single nucleotide polymorphism (SNP) influenced, even in healthy individuals, frontostriatal cognitive tasks taxing processing speed, IQ and executive cognitive control, and cardinal DA-mediated cognitive functions,[99,128–130] as well as "tuning" of the prefrontal physiological phenotype previously linked to cortical DA function during working memory.[79,84,102,131] The same genotype also predicted a reduced gray matter volume in parts of the frontostriatal network.[126] In these studies, the allele associated with reduced AKT1 expression predicted relatively reduced measures in all of these DA-related phenotypes. As a further test that the AKT1 effects were linked to dopaminergic function, we found that the same prefrontal regions showing *AKT1* main effects in fMRI and MRI volumetry evidenced epistasis with the *COMT* Val/Met.[126] Individuals with a COMT-Val homozygous genetic background had exaggerated (i.e., nonadditive) *AKT1* effects with "inefficient" activation in replicated datasets at the prefrontal cortex (Fig. 4.4.4). In terms of brain structure, gray matter volume from the prefrontal cortex also showed the *AKT1*-by-*COMT* interaction in that individuals with combined deleterious *AKT1* minor and *COMT* Val alleles had disproportionately reduced gray matter volume. Thus, these results provided multiple lines of converging evidence implicating *AKT1* gene effects that influence protein expression as well as system-level human prefrontal structure and function. The results were consistent with pre-clinical evidence that coupled AKT1 to dopaminergic signaling and downstream effects on prefrontal cellular structure and cognition,[122,124,125,132] and suggest that these brain mechanisms impacted the biology of active human cognitive function. The data also suggest that the mechanisms of prior associations of *AKT1* with psychosis, a condition associated with DA abnormalities in brain[9,133] and treated with antidopaminergic drugs, could involve these biological processes.

CONCLUSION

In this chapter, we have examined findings through which heritable human neuroimaging intermediate phenotypes could provide a window to examine genetic mechanisms of active prefrontal cognitive processing related to DA. Genetic variation influencing task-related prefrontal cortical function was consistent with fundamental predictions based on the biology of DA tuning in cortical microcircuits. These findings also extended the basic biological data to implicate molecules impacting variation in active human brain function, potentially mirroring component disease-related brain processes in schizophrenia. The findings of interacting genetic elements consistent with the cross-talk within and across DA and glutamatergic systems, and their intracellular signaling pathways, arguably contribute further empirical validation to the strategy to identify molecules whose genetic variation could be of substantial combined influence on human brain function at the network or systems level. Indeed, an increasing number of recent pharmacological and gene-based human studies,[134–139] including those in patients with Parkinson's disease,[131,140] have also been consistent with the main findings of DA-dependent gene effects on prefrontal function highlighted in this review.

Ultimately, it is suggested that the complexity of human brain function in health and disease could be systematically dissected with combinations of multiple

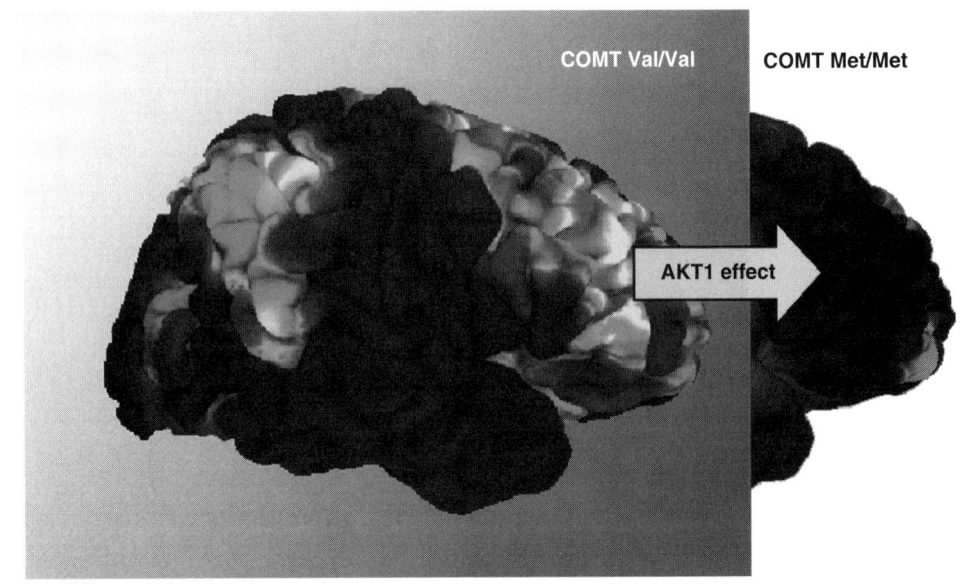

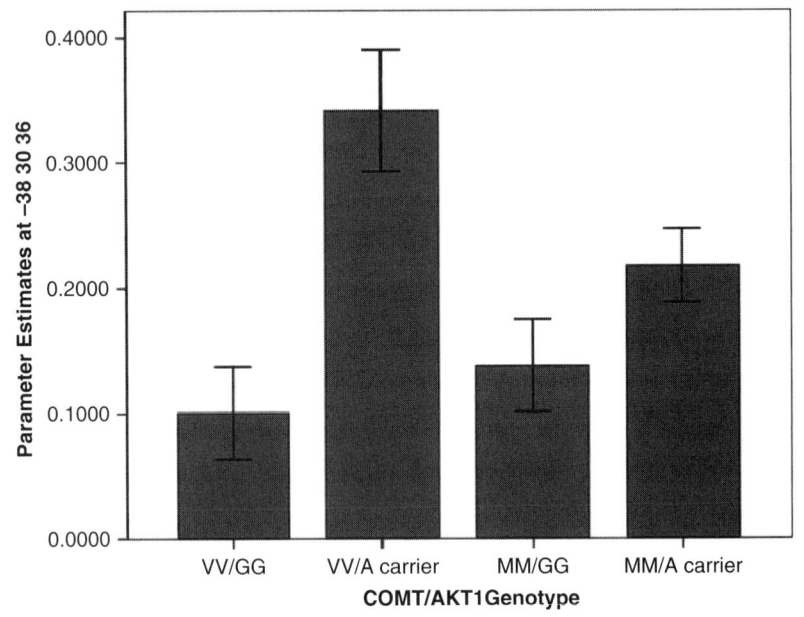

FIGURE 4.4.4. Epistatic interaction between AKT1 and COMT. Here, individuals with the AKT1 allele associated with reduced gene expression showed disproportionately inefficient DPFC activity in the background of a relatively deleterious COMT Val allele ($F_{1,42} = 4.466$, $p = 0.041$). *Source*: Adapted from Tan et al.[126] (See Color Plate 4.4.4.)

neuroimaging paradigms and genetic markers. The discovery of new treatments that could improve cognitive function is yet to be, although it might be speculated that extensions of human neuroimaging and genetic paradigms could be attractive strategies in the discovery of sets of key molecules influencing active human brain function relevant to disease pathophysiology and potential treatment. Preliminary evidence in normal subjects suggests that central nervous system penetrant COMT inhibitors may enhance working memory without stimulant effects, particularly in individuals with COMT val/val genotypes.[134] Encouraging parallels might also be drawn from recent independent data suggesting that novel treatments targeting metabotropic glutamate receptors are potentially beneficial in treating symptoms of schizophrenia.[141] It remains an open question whether treatments targeting, for example, combinations of *GRM3* or AKT1 could impact cognitive brain processes such as working memory, perhaps in concert with dopaminergic modulation and as a function of individual genotype status. Nevertheless, systematically elucidating these functional genetic networks could lead to the identification of critical sets of nodes linked to disease mechanisms that will bring us closer to rational treatment development to improve the lives of patients and their families.

REFERENCES

1. Green MF. What are the functional consequences of neurocognitive deficits in schizophrenia? *Am J Psychiatry*. 1996; 153(3):321–330.
2. Lieberman JA, Stroup TS, McEvoy JP, et al. Effectiveness of antipsychotic drugs in patients with chronic schizophrenia. *N Engl J Med*. 2005;353(12):1209–1223.
3. Jones PB, Barnes TRE, Davies L, et al. Randomized controlled trial of the effect on quality of life of second- vs first-generation antipsychotic drugs in schizophrenia: Cost Utility of the Latest Antipsychotic Drugs in Schizophrenia Study (CUtLASS 1). *Arch Gen Psychiatry*. 2006;63(10):1079–1087.
4. Goldberg TE, Weinberger DR. Effects of neuroleptic medications on the cognition of patients with schizophrenia: a review of recent studies. *J Clin Psychiatry*. 1996;57(suppl 9):62–65.
5. Weinberger DR, Berman KF, Illowsky BP. Physiological dysfunction of dorsolateral prefrontal cortex in schizophrenia.3. A new cohort and evidence for a monoaminergic mechanism. *Arch Gen Psychiatry*. 1988;45(7):609–615.
6. Weinberger DR. Implications of normal brain development for the pathogenesis of schizophrenia. *Arch Gen Psychiatry*. 1987;44(7):660–669.
7. Meyer-Lindenberg AS, Weinberger DR. Intermediate phenotypes and genetic mechanisms of psychiatric disorders. *Nat Rev Neurosci*. 2006;7(10):818–827.
8. Weinberger DR, Egan MF, Bertolino A, et al. Prefrontal neurons and the genetics of schizophrenia. *Biol Psychiatry*. 2001;50:825–844.
9. Winterer G, Weinberger DR. Genes, dopamine and cortical signal-to-noise ratio in schizophrenia. *Trends Neurosci*. 2004;27(11):683–690.
10. Rolls ET, Loh M, Deco G, Winterer G. Computational models of schizophrenia and dopamine modulation in the prefrontal cortex. *Nat Rev Neurosci*. 2008;9(9):696–709.
11. Tan HY, Callicott JH, Weinberger DR. Prefrontal cognitive systems in schizophrenia – towards human genetic brain mechanisms. *Cogn Neuropsychiatry*. 2009;in press).
12. Kety SS, Rosenthal D, Wender PH, Schulsinger F. Studies based on a total sample of adopted individuals and their relatives: why they were necessary, what they demonstrated and failed to demonstrate. *Schizophr Bull*. 1976;2:413–428.
13. Gottesman II, Shields J. *Schizophrenia and Genetics. A Twin Study*. Vantage Point, NY: Academic Press; 1972.
14. O'Donovan MC, Craddock N, Norton N, et al. Identification of loci associated with schizophrenia by genome-wide association and follow-up. *Nat Genet*. 2008;40(9):1053–1055.
15. Allen NC, Bagade S, McQueen MB, et al. Systematic meta-analyses and field synopsis of genetic association studies in schizophrenia: the SzGene database. *Nat Genet*. 2008;40(7):827–834.
16. Ioannidis JPA. Why most published research findings are false. *PLoS Med*. 2005;2(8):e124.
17. Blackwood DH, St Clair DM, Muir WJ, Duffy JC. Auditory P300 and eye tracking dysfunction in schizophrenic pedigrees. *Arch Gen Psychiatry*. 1991;48(10):899–909.
18. Callicott JH, Egan MF, Mattay VS, et al. Abnormal fMRI response of the dorsolateral prefrontal cortex in cognitively intact siblings of patients with schizophrenia. *Am J Psychiatry*. 2003;160(4):709–719.
19. Cannon TD, Huttunen MO, Lonnqvist J, et al. The inheritance of neuropsychological dysfunction in twins discordant for schizophrenia. *Am J Hum Genet*. 2000;67:369–382.
20. Goldberg TE, Torrey EF, Gold JM, et al. Genetic risk of neuropsychological impairment in schizophrenia: a study of monozygotic twins discordant and concordant for the disorder. *Schizophr Res*. 1995;17(1):77–84.
21. Goldman AL, Pezawas L, Mattay VS, et al. Heritability of brain morphology related to schizophrenia: a large-scale automated magnetic resonance imaging segmentation study. *Biol Psychiatry*. 2008;63(5): 475–483.
22. Toulopoulou T, Picchioni M, Rijsdijk F, et al. Substantial genetic overlap between neurocognition and schizophrenia: genetic modeling in twin samples. *Arch Gen Psychiatry*. 2007;64(12):1348–1355.
23. Tan HY, Callicott JH, Weinberger DR. Intermediate phenotypes in schizophrenia genetics redux: Is it a no brainer? *Mol Psychiatry*. 2008;13(3):233–238.
24. Baddeley AD. Working memory: looking back and looking forward. *Nat Rev Neurosci*. 2003;4(10):829–839.
25. Goldberg TE, Weinberger DR, Berman KF, Pliskin NH, Podd MH. Further evidence for dementia of the prefrontal type in schizophrenia? A controlled study of teaching the Wisconsin Card Sorting Test. *Arch Gen Psychiatry*. 1987;44(11):1008–1014.
26. Goldman-Rakic PS. Working memory dysfunction in schizophrenia. *J Neuropsychiatry Clin Neurosci*. 1994;6:348–357.
27. Silver H, Feldman P, Bilker WB, Gur RC. Working memory deficit as a core neuropsychological dysfunction in schizophrenia. *Am J Psychiatry*. 2003;160(10):1809–1816.
28. Callicott JH, Weinberger DR. Neuropsychiatric dynamics: the study of mental illness using functional magnetic resonance imaging. *Eur J Radiol*. 1999;30(2):95–104.
29. Tan HY, Callicott JH, Weinberger DR. Dysfunctional and compensatory prefrontal cortical systems, genes and the pathogenesis of schizophrenia. *Cereb Cortex*. 2007;17:i171-i181.
30. Manoach DS. Prefrontal cortex dysfunction during working memory performance in schizophrenia: reconciling discrepant findings. *Schizophr Res*. 2003;60(2-3):285–298.
31. Ramsey NF, Koning HA, Welles P, Cahn W, Van Der Linden JA, Kahn RS. Excessive recruitment of neural systems subserving logical reasoning in schizophrenia. *Brain*. 2002;125(pt 8): 1793–1807.
32. Callicott JH, Mattay VS, Bertolino A, et al. Physiological characteristics of capacity constraints in working memory as revealed by functional MRI. *Cereb Cortex*. 1999;9(1):20–26.
33. Callicott JH, Bertolino A, Mattay VS, et al. Physiological dysfunction of the dorsolateral prefrontal cortex in schizophrenia revisited. *Cereb Cortex*. 2000;10:1078–1092.
34. Callicott JH, Mattay VS, Verchinski BA, Marenco S, Egan MF, Weinberger DR. Complexity of prefrontal cortical dysfunction in schizophrenia: more than up or down. *Am J Psychiatry*. 2003;160(12):2209–2215.
35. Tan HY, Choo WC, Fones CSL, Chee MWL. fMRI study of maintenance and manipulation processes within working memory in first-episode schizophrenia. *Am J Psychiatry*. 2005;162(10):1849–1858.

36. Tan HY, Sust S, Buckholtz JW, et al. Dysfunctional prefrontal regional specialization and compensation in schizophrenia. *Am J Psychiatry*. 2006;163:1969–1977.
37. Glahn DC, Ragland JD, Abramoff A, et al. Beyond hypofrontality: a quantitative meta-analysis of functional neuroimaging studies of working memory in schizophrenia. *Hum Brain Mapp*. 2005;25:60–69.
38. MacDonald AW, Pogue-Geile MF, Johnson MK, Carter CS. A specific deficit in context processing in the unaffected siblings of patients with schizophrenia. *Arch Gen Psychiatry*. 2003;60(1):57–65.
39. Fusar-Poli P, Perez J, Broome M, et al. Neurofunctional correlates of vulnerability to psychosis: a systematic review and meta-analysis. *Neurosci Biobehav Rev*. 2007;31(4):465–484.
40. O'Reilly RC. Biologically based computational models of high-level cognition. *Science*. 2006;314:91–94.
41. Robbins TW, Roberts AC. Differential regulation of fronto-executive function by the monoamines and acetylcholine. *Cereb Cortex*. 2007;17(suppl_1):i151-i160.
42. Goldman-Rakic PS. Circuitry of primate prefrontal cortex and regulation of behavior by representational memory. In: Plum F, Mountcastle V, eds. *Handbook of Physiology: The Nervous System, Vol 5*. Bethesda, MD: American Physiological Society; 1987:373–417.
43. Goldman-Rakic PS. Cellular and circuit basis of working memory in prefrontal cortex of nonhuman primates. *Prog Brain Res*. 1990;85:325–336.
44. Cavada C, Goldman-Rakic PS. Posterior parietal cortex in rhesus monkey: II. Evidence for segregated corticocortical networks linking sensory and limbic areas with the frontal lobe. *J Comp Neurol*. 1989;287(4):422–445.
45. Selemon LD, Goldman-Rakic PS. Longitudinal topography and interdigitation of corticostriatal projections in the rhesus monkey. *J Neurosci*. 1985;5(3):776–794.
46. Goldman-Rakic PS. The physiological approach: functional architecture of working memory and disordered cognition in schizophrenia. *Biol Psychiatry*. 1999;46(5):650–661.
47. Fuster JM. *The Prefrontal Cortex—Anatomy, Physiology, and Neuropsychology of the Frontal Lobe*: Lippincott-Raven; Philadelphia, PA 1997.
48. Goldman-Rakic PS. Topography of cognition: parallel distributed networks in primate association cortex. *Annu Rev Neurosci*. 1988;11:137–156.
49. Jonides J, Schumacher EH, Smith EE, et al. The role of parietal cortex in verbal working memory. *J Neurosci*. 1998;18:5026–5034.
50. Petrides M, Pandya DN. Projections to the frontal cortex from the posterior parietal region in the rhesus monkey. *J Comp Neurol*. 1984;228(1):105–116.
51. Sawaguchi T, Goldman-Rakic PS. D1 dopamine receptors in prefrontal cortex: involvement in working memory. *Science*. 1991;251:947–950.
52. Sawaguchi T, Goldman-Rakic PS. The role of D1-dopamine receptor in working memory: local injections of dopamine antagonists into the prefrontal cortex of rhesus monkeys performing an oculomotor delayed-response task. *J Neurophysiol*. 1994;71:515–528.
53. Williams GV, Goldman-Rakic PS. Modulation of memory fields by dopamine D1 receptors in prefrontal cortex. *Nature*. 1995;376:572–575.
54. Seamans JK, Gorelova N, Durstewitz D, Yang CR. Bidirectional dopamine modulation of GABAergic inhibition in prefrontal cortical pyramidal neurons. *J Neurosci*. 2001;21(10):3628–3638.
55. Durstewitz D, Seamans JK, Sejnowski TJ. Neurocomputational models of working memory. *Nat Neurosci*. 2000;3:1184–1191.
56. Seamans JK, Durstewitz D, Christie BR, Stevens CF, Sejnowski TJ. Dopamine D1/D5 receptor modulation of excitatory synaptic inputs to layer V prefrontal cortex neurons. *PNAS*. 2001;98(1):301–306.
57. Wang J, O'Donnell P. D1 dopamine receptors potentiate NMDA-mediated excitability increase in layer V prefrontal cortical pyramidal neurons. *Cereb Cortex*. 2001;11(5):452–462.
58. Wang X-J. Synaptic basis of cortical persistent activity: the importance of NMDA receptors to working memory. *J Neurosci*. 1999;19(21):9587–9603.
59. Mink JW. The basal ganglia: focused selection and inhibition of competing motor programs. *Prog Neurobiol*. 1996;50(4):381–425.
60. Goldman-Rakic PS. Cellular basis of working memory. *Neuron*. 1995;14(3):477–485.
61. Tanaka SC, Doya K, Okada G, Ueda K, Okamoto Y, Yamawaki S. Prediction of immediate and future rewards differentially recruits cortico-basal ganglia loops. *Nat Neurosci*. 2004;7(8):887–893.
62. Harrison PJ, Weinberger DR. Schizophrenia genes, gene expression, and neuropathology: on the matter of their convergence. *Mol Psychiatry*. 2005;10:40–68.
63. Egan MF, Goldberg TE, Gscheidle T, et al. Relative risk for cognitive impairments in siblings of patients with schizophrenia. *Biol Psychiatry*. 2001;50(2):98–107.
64. Chen J, Lipska BK, Halim N, et al. Functional analysis of genetic variation in catechol-O-methyltransferase (COMT): effects on mRNA, protein, and enzyme activity in postmortem human brain. *Am J Hum Genet*. 2004;75(5):807–821.
65. Seasack SR, Hawrylak VA, Matus C, Guido MA, Levey AI. Dopamine axon varicosities in the prelimbic division of the rat prefrontal cortex exhibit sparse immunoreactivity for the dopamine transporter. *J Neurosci*. 1998;18:2697–2708.
66. Lewis DA, Melchitzky DS, Seasack SR, Whitehead RE, Sungyoung AUH, Sampson A. Dopamine transporter immunoreactivity in monkey cerebral cortex. *J Comp Neurol*. 2001;432:119–136.
67. Lachman HM, Papolos DF, Saito T, Yu YM, Szumlanski CL, Weinshilboum RM. Human catechol-O-methyltransferase pharmacogenetics: description of a functional polymorphism and its potential application to neuropsychiatric disorders. *Pharmacogenetics*. 1996;6(3):243–250.
68. Lotta T, Vidgren J, Tilgmann C, et al. Kinetics of human soluble and membrane-bound catechol-O-methyltransferase: a revised mechanism and description of the thermolabile variant of the enzyme. *Biochemistry (Mosc)*. 1995;34(13):4202–4210.
69. Murphy KC. Schizophrenia and velo-cardio-facial syndrome. *Lancet*. 2002;359(9304):426–430.
70. Badner JA, Gershon ES. Meta-analysis of whole-genome linkage scans of bipolar disorder and schizophrenia. *Mol Psychiatry*. 2002;7(4):405–411.
71. Lewis CM, Levinson DF, Wise LH, et al. Genome scan meta-analysis of schizophrenia and bipolar disorder, part II: schizophrenia. *Am J Hum Genet*. 2003;73(1):34–48.
72. Glatt SJ, Faraone SV, Tsuang MT. Association between a functional catechol-O-methyltransferase gene polymorphism and schizophrenia: meta-analysis of case-control and family-based studies. *Am J Psychiatry*. 2003;160(3):469–476.

73. Fan JB, Zhang CS, Gu NF, et al. Catechol-O-methyltransferase gene Val/Met functional polymorphism and risk of schizophrenia: a large-scale association study plus meta-analysis. *Biol Psychiatry.* 2005;57:139–144.

74. Shifman S, Bronstein M, Sternfeld M, et al. A highly significant association between a COMT haplotype and schizophrenia. *Am J Hum Genet.* 2002;71(6):1296–1302.

75. Palmatier MA, Pakstis AJ, Speed W, et al. COMT haplotypes suggest P2 promoter region relevance for schizophrenia. *Mol Psychiatry.* 2004;9(9):859–870.

76. Nicodemus K, Kolachana B, Vakkalanka R, et al. Evidence for statistical epistasis between catechol-O-methyltransferase (COMT) and polymorphisms in RGS4, G72 (DAOA), GRM3, and DISC1: influence on risk of schizophrenia. *Hum Genet.* 2007;120(6):889–906.

77. Caspi A, Moffitt TE, Cannon M, et al. Moderation of the effect of adolescent-onset cannabis use on adult psychosis by a functional polymorphism in the catechol-O-methyltransferase gene: longitudinal evidence of a gene X environment interaction. *Biol Psychiatry.* 2005;57:1117–1127.

78. Goldberg T, Egan M, Gscheidle T, et al. Executive subprocesses in working memory: relationship to catechol-O-methyltransferase Val158Met genotype and schizophrenia. *Arch Gen Psychiatry.* 2003;60(9):889–896.

79. Egan MF, Goldberg TE, Kolachana BS, et al. Effect of COMT Val108/158 Met genotype on frontal lobe function and risk for schizophrenia. *Proc Natl Acad Sci USA.* 2001;98(12):6917–6922.

80. Malhotra AK, Kestler LJ, Mazzanti C, Bates JA, Goldberg T, Goldman D. A functional polymorphism in the COMT gene and performance on a test of prefrontal cognition. *Am J Psychiatry.* 2002;159(4):652–654.

81. de Frias CM, Annerbrink K, Westberg L, Eriksson E, Adolfsson R, Nilsson LG. Catechol-O-methyltransferase Val158Met polymorphism is associated with cognitive performance in nondemented adults. *J Cogn Neurosci.* 2006;17(7):1018–1025.

82. Nolan KA, Bilder RM, Lachman HM, Volavka J. Catechol-O-methyltransferase Val158Met polymorphism in schizophrenia: differential effects of Val and Met alleles on cognitive stability and flexibility. *Am J Psychiatry.* 2004;161(2):359–361.

83. Diamond A, Briand L, Fossella J, Gehlbach L. Genetic and neurochemical modulation of prefrontal cognitive functions in children. *Am J Psychiatry.* 2004;161(1):125–132.

84. Mattay VS, Goldberg TE, Fera F, et al. Catechol-O-methyltransferase val158-met genotype and individual variation in the brain response to amphetamine. *Proc Natl Acad Sci USA.* 2003;100(10):6186–6191.

85. Meyer-Lindenberg AS, Nichols T, Callicott JH, et al. Impact of complex genetic variation in COMT on human brain function. *Mol Psychiatry.* 2006;11(9):867–877.

86. Deco G, Rolls ET, Horwitz B. "What" and "where" in visual working memory: a computational neurodynamical perspective for integrating fMRI and single-neuron data. *J Cogn Neurosci.* 2004;16(4):683–701.

87. Deco G, Rolls ET. Sequential memory: a putative neural and synaptic dynamical mechanism. *J Cogn Neurosci.* 2005;17(2):294–307.

88. D'Esposito M, Postle BR, Ballard D, Lease J. Maintenance versus manipulation of information held in working memory: an event-related fMRI study. *Brain Cogn.* 1999;41:66–86.

89. Koechlin E, Ody C, Kouneiher FT. The architecture of cognitive control in the human prefrontal cortex. *Science.* 2003;302(5648):1181–1185.

90. Akil M, Kolachana BS, Rothmond DA, Hyde TM, Weinberger DR, Kleinman JE. Catechol-O-methyltransferase genotype and dopamine regulation in the human brain. *J Neurosci.* 2003;23(6):2008–2013.

91. Meyer-Lindenberg A, Kohn PD, Kolachana B, et al. Midbrain dopamine and prefrontal function in humans: interaction and modulation by COMT genotype. *Nat Neurosci.* 2005;8(5):594–596.

92. Slifstein M, Kolachana B, Simpson EH, et al. COMT genotype predicts cortical-limbic D1 receptor availability measured with 11C NNC112 and PET. *Mol Psychiatry.* 2008;13(8):821–827.

93. Seamans JK, Yang CR. The principal features and mechanisms of dopamine modulation in the prefrontal cortex. *Prog Neurobiol.* 2004;74:1–57.

94. Sakai K, Passingham RE. Prefrontal interactions reflect future task operations. *Nat Neurosci.* 2002;6(1):75–81.

95. Deco G, Rolls ET. Attention, short-term memory, and action selection: a unifying theroy. *Prog Neurobiol.* 2005;76:236–256.

96. Vijayraghavan S, Wang M, Birnbaum SG, Williams GV, Arnsten AFT. Inverted-U dopamine D1 receptor actions on prefrontal neurons engaged in working memory. *Nat Neurosci.* 2007;10(3):376–384.

97. O'Reilly RC, Frank MJ. Making working memory work: a computational model of learning in the prefrontal cortex and basal ganglia. *Neural Comput.* 2006;18:283–328.

98. Gruber AJ, Dayan P, Gutkin BS, Solla SA. Dopamine modulation in the basal ganglia locks the gate to working memory. *J Comput Neurosci.* 2006;20:153–166.

99. Alexander GE, DeLong MR, Strick PL. Parallel organization of functionally segregated circuits linking basal ganglia and cortex. *Annu Rev Neurosci.* 1986;9(1):357–381.

100. Hubbard EM, Piazza M, Pinel P, Dehaene S. Interactions between number and space in parietal cortex. *Nat Rev Neurosci.* 2005;6:435–448.

101. Dehaene S, Piazza M, Pinel P, Cohen L. Three parietal circuits for number processing. *Cogn Neuropsychol.* 2003;20:487–506.

102. Tan HY, Chen Q, Goldberg TE, et al. Catechol-O-methyltransferase Val158Met modulation of prefrontal-parietal-striatal brian systems during arithmetic and temporal transformations in working memory. *J Neurosci.* 2007;27(49):13393–13401.

103. Miller EK, Cohen JD. An integrative theory of prefrontal cortex function. *Annu Rev Neurosci.* 2001;24:167–202.

104. Goldman-Rakic PS, Lidow MS, Gallager DW. Overlap of dopaminergic, adrenergic, and serotoninergic receptors and complementarity of their subtypes in primate prefrontal cortex. *J Neurosci.* 1990;10(7):2125–2138.

105. Frank MJ, Moustafa AA, Haughey HM, Curran T, Hutchison KE. Genetic triple dissociation reveals multiple roles for dopamine in reinforcement learning. *Proc Natl Acad Sci USA.* 2007;104(41):16311–16316.

106. Jackson ME, Homayoun H, Moghaddam B. NMDA receptor hypofunction produces concomitant firing rate potentiation and burst activity reduction in the prefrontal cortex. *Proc Natl Acad Sci USA.* 2004;101(22):8467–8472.

107. Egan MF, Straub RE, Goldberg TE, et al. Variation in GRM3 affects cognition, prefrontal glutamate, and risk for schizophrenia. *Proc Natl Acad Sci USA.* 2004;101(34):12604–12609.

108. Moghaddam B, Adams BW. Reversal of phencyclidine effects by a group II metabotropic glutamate receptor agonist in rats. *Science.* 1998;281(5381):1349–1352.

109. Fallin MD, Lasseter VK, Avramopoulos D, et al. Bipolar I disorder and schizophrenia: a 440-single-nucleotide polymorphism

screen of 64 candidate genes among Ashkenazi Jewish case-parent trios. *Am J Hum Genet.* 2005;77(6):918–936.
110. Martí SB, Cichon S, Propping P, Nöthen M. Metabotropic glutamate receptor 3 (GRM3) gene variation is not associated with schizophrenia or bipolar affective disorder in the German population. *Am J Med Genet.* 2002;114(1):46–50.
111. Chen Q, He G, Chen Q, et al. A case-control study of the relationship between the metabotropic glutamate receptor 3 gene and schizophrenia in the Chinese population. *Schizophr Res.* 2005;73(1):21–26.
112. Norton N, Williams H, Dwyer S, et al. No evidence for association between polymorphisms in GRM3 and schizophrenia. *BMC Psychiatry.* 2005;5(1):23.
113. Sartorius LJ, Nagappan G, Lipska BK, et al. Alternative splicing of human metabotropic glutamate receptor 3. *J Neurochem.* 2006;96(4):1139–1148.
114. Lisman JE, Fellous JM, Wang XJ. A role for NMDA-receptor channels in working memory. *Nat Neurosci.* 1998;1(4):273–275.
115. Durstewitz D, Seamans JK. The computational role of dopamine D1 receptors in working memory. *Neural Networks.* 2002;15(4-6):561–572.
116. Tan HY, Chen Q, Sust S, et al. Epistasis between catechol-O-methyltransferase and type II metabotropic glutamate receptor 3 genes in working memory brain function. *Proc Natl Acad Sci USA.* 2007;104(30):12536–12541.
117. Bertolino A, Fazio L, Di Giorgio A, et al. Genetically determined interaction between the dopamine transporter and the D2 receptor on prefronto-striatal activity and volume in humans. *J Neurosci.* 2009;29(4):1224–1234.
118. Zhang Y, Bertolino A, Fazio L, et al. Polymorphisms in human dopamine D2 receptor gene affect gene expression, splicing, and neuronal activity during working memory. *Proc Natl Acad Sci USA.* 2007;104(51):20552–20557.
119. Heinz A, Goldman D, Jones DW, et al. Genotype influences in vivo dopamine transporter availability in human striatum. *Neuropsychopharmacology.* 2000;22(2):133–139.
120. Greengard P. The neurobiology of slow synaptic transmission. *Science.* 2001;294(5544):1024–1030.
121. Meyer-Lindenberg A, Straub RE, Lipska BK, et al. Genetic evidence implicating DARPP-32 in human frontostriatal structure, function, and cognition. *J Clin Invest.* 2007;117(3):672–682.
122. Beaulieu JM, Sotnikova TD, Marion S, Lefkowitz RJ, Gainetdinov RR, Caron MG. An Akt/β-arrestin 2/PP2A signaling complex mediates dopaminergic neurotransmission and behavior. *Cell.* 2005;122(2):261–273.
123. Beaulieu JM, Sotnikova TD, Yao WD, et al. Lithium antagonizes dopamine-dependent behaviors mediated by an AKT/glycogen synthase kinase 3 signaling cascade. *Proc Natl Acad Sci USA.* 2004;101(14):5099–5104.
124. Lai WS, Xu B, Westphal KGC, et al. Akt1 deficiency affects neuronal morphology and predisposes to abnormalities in prefrontal cortex functioning. *Proc Natl Acad SciUSA.* 2006;103(45):16906–16911.
125. Emamian ES, Hall D, Birnbarum MJ, Karayiogou M, Gogos JA. Convergent evidence for impaired AKT1-GSK3b signaling in schizophrenia. *Nat Genet.* 2004;36(2):131–137.
126. Tan HY, Nicodemus KK, Chen Q, et al. Genetic variation in AKT1 is linked to dopamine-associated prefrontal cortical structure and function in humans. *J Clin Invest.* 2008;118:2200–2208.
127. Harris SL, Gil G, Robins H, et al. Detection of functional single-nucleotide polymorphisms that affect apoptosis. *Proc Natl Acad Sci U S A.* 2005;102(45):16297–16302.
128. Salthouse TA. The processing-speed theory of adult age differences in cognition. *Psychol Rev.* 1996;103(3):403–428.
129. Pantelis C, Barnes TR, Nelson HE, et al. Frontal-striatal cognitive deficits in patients with chronic schizophrenia. *Brain.* 1997;120(10):1823–1843.
130. Kane MJ, Engle RW. The role of prefrontal cortex in working-memory capacity, executive attention, and general fluid intelligence: an individual-differences perspective. *Psychonom Bull Rev.* 2002;9:637–671.
131. Mattay VS, Tessitore A, Callicott JH, et al. Dopaminergic modulation of cortical function in patients with Parkinson's disease. *Ann Neurol.* 2002;51(2):156–164.
132. Wei Y, Williams JM, Dipace C, et al. Dopamine transporter activity mediates amphetamine-induced inhibition of Akt through a Ca^{2+}/calmodulin-dependent kinase II-dependent mechanism. *Mol Pharmacol.* 2007;71(3):835–842.
133. Laruelle M, Kegeles LS, Abi-Dargham A. Glutamate, dopamine, and schizophrenia: from pathophysiology to treatment. *Ann NY Acad Sci.* 2003;1003(1):138–158.
134. Apud JA, Mattay V, Chen J, et al. Tolcapone improves cognition and cortical information processing in normal human subjects. *Neuropsychopharmacology.* 2007;32(5):1011–1020.
135. Bertolino A, Blasi G, Latorre V, et al. Additive effects of genetic variation in dopamine regulating genes on working memory cortical activity in human brain. *J Neurosci.* 2006;26(15):3918–3922.
136. Gothelf D, Eliez S, Thompson T, et al. COMT genotype predicts longitudinal cognitive decline and psychosis in 22q11.2 deletion syndrome. *Nat Neurosci.* 2005/11//print 2005;8(11):1500–1502.
137. McIntosh AM, Baig BJ, Hall J, et al. Relationship of catechol-O-methyltransferase variants to brain structure and function in a population at high risk of psychosis. *Biol Psychiatry.* 2007;61(10):1127–1134.
138. Bertolino A, Caforio G, Blasi G, et al. Interaction of COMT Val108/158 Met genotype and olanzapine treatment on prefrontal cortical cunction in patients with schizophrenia. *Am J Psychiatry.* 2004;161:1798–1805.
139. Bertolino A, Fazio L, Caforio G, et al. Functional variants of the dopamine receptor D2 gene modulate prefronto-striatal phenotypes in schizophrenia. *Brain.* 2009 (In press).
140. Williams-Gray CH, Hampshire A, Robbins TW, Owen AM, Barker RA. Catechol-O-methyltransferase val158met genotype influences frontoparietal activity during planning in patients with Parkinson's disease. *J Neurosci.* 2007;27(18):4832–4838.
141. Patil ST, Zhang L, Martenyi F, et al. Activation of mGlu2/3 receptors as a new approach to treat schizophrenia: a randomized Phase 2 clinical trial. *Nat Med.* 2007;13(9):1102–1107.

5 | Dopamine in prefrontal cortex and cognition

5.1 | From Behavior to Cognition: Functions of Mesostriatal, Mesolimbic, and Mesocortical Dopamine Systems

TREVOR W. ROBBINS

INTRODUCTION

The seminal mapping of the mesencephalic dopamine (DA) pathways into ramifying mesostriatal, mesolimbic, and mesocortical projections, as well as the identification of several DA receptors and their signaling pathways, have raised important questions about the functions of this important neuromodulatory neurotransmitter. The possibly misleading triadic division of these projections has suggested discrete and even parallel functions in movement (e.g., Parkinson's disease, dorsal striatum), reward (e.g., drugs of abuse, nucleus accumbens), and cognition (e.g., schizophrenia and attention deficit hyperactivity disorder [ADHD], prefrontal cortex [PFC]). However, although this parcellation is attractively parsimonious, there is considerable evidence for overlapping roles—for example, of cognitive function in the caudate-putamen and of aspects of reinforcement in the orbitofrontal cortex. Similarly, the mediation of positive reinforcement by DA-dependent functions of the nucleus accumbens also entails an implication in learning and decision-making processes. A key issue is, under what states or conditions are the central DA systems active and how does this activity affect cognition, behavior, and movement? As there are considerable neurochemical data indicating that central DA is affected by such factors as stress, this question may equate to understanding the relationship between such states as stress or mood and behavior. A particularly useful principle, applying especially to the understanding of the relationship between DA and behavioral or cognitive output, is the Yerkes-Dodson principle,[1] which generally takes the form of an inverted-U-shaped function linking level of arousal with behavioral performance (Fig. 5.1.1). Thus, whereas performance at low or high values of arousal is relatively poor, at intermediate values it is optimal.

When discussing the functions of the dopamine system, we have employed the term *activation* to describe an "energetic" construct similar to that of arousal, which is, however, meant to capture how dopamine affects the rate and vigor of behavioral (and cognitive, e.g., thinking) output. Unlike *arousal, activation* does not connote a simple wakefulness construct associated with neocortical changes—for example, in encephalography (EEG). As posited in our 1992 review[2] of the considerable empirical data already available, activation is induced by many related states or stimuli, including food deprivation, stress, psychomotor stimulant drugs, aversive stimuli such as tail pinch and foot shock, novelty, and conditioned stimuli, including predictors of appetitive events such as food provision and also aversive events. The function of activation is to enhance behavior in preparation for the presentation of a goal or reinforcer (whether appetitive or aversive). Activation affects processing in target structures innervated by the mesolimbic, mesocortical, and mesostriatal pathways, essentially in *gain-amplificatory* mode. In the mesolimbic projections—for example, to the ventral striatum, including the nucleus accumbens—the role of enhanced DA activity is to increase responsivity to cues paired with reinforcement and thus also to enhance the appetitive approach to the goal. This is very similar to Berridge's concept of *incentive salience*[3] and is related to other earlier writings on the role of DA in motivation.[4] Another major empirical advance has been the recognition that the fast phasic firing of cells in the ventral tegmental area and substantia nigra appears to model an error prediction signal relevant to Pavlovian or temporal difference learning models.[5] There is an evident need to understand the relative functional contribution of such phasic responses that are implicated in plasticity and in new, mainly appetitive learning of Pavlovian associations, with the tonic mode of action of the same DA systems that we assume underlie the activational effects of DA.[6,7]

The Yerkes-Dodson principle has been often criticized in experimental psychology for its apparent capacity to account too readily for diverse data sets.

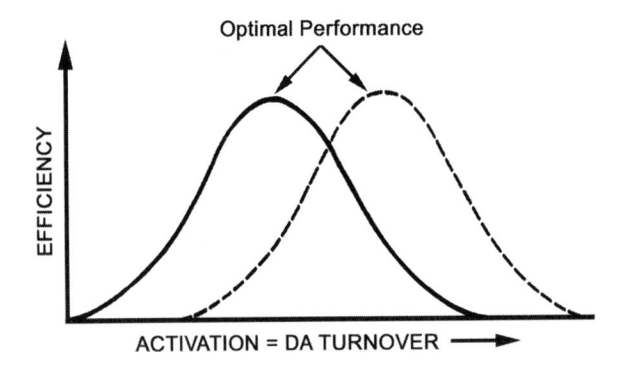

FIGURE 5.1.1. The generalized Yerkes-Dodson relationship, showing how (1) performance efficiency may vary as a function of activation (e.g., DA turnover) and (2) that optimal levels of activation for different tasks (hypothetically, which vary in difficulty) may differ.

However, it does conform to many dose–response relationships observed in drug effects on behavior, which often have the characteristic inverted-U-shaped functions. The principle was applied initially to important data suggesting that the level of DA D1 receptor activity produced Yerkes-Dodson-like effects on working memory in both rats and monkeys.[8] A more recent manifestation of the principle was shown in work on the catechol-O-methyl transferase polymorphism that hypothetically modulates PFC DA function and produces a predictable pattern of effects on working memory performance.[9,10] However, these data raise several exciting issues: (1) Is the function relating DA to performance the same for all forms of behavior? The finding of Yerkes and Dodson[1] that easy tasks were optimally performed at higher levels of arousal than difficult tasks suggests that it might not be. (2) Are there inverted-U-shaped functions for the subcortical systems, as well as for prefrontal DA D1 receptors? (3) How do these systems interact? And (4) in comparable brain regions, are there distinct inverted-U-shaped functions for other neuromodulators, such as the other monoamines and acetylcholine?

The mechanism underlying the effect of DA in its terminal regions is probably via an increase in signal-to-noise processing. However, the molecular syntax by which these effects are produced is quite complex and depends upon a number of discrete actions. For example, in the prefrontal cortex, the signal-to-noise-enhancing effect of DA at postsynaptic D1 receptors depends inter alia on boosting of N-methyl-D-aspartate (NMDA) receptor and GABA-A receptor currents, the former serving to preserve neuronal activity, the latter inhibiting interrupts from incoming glutamatergic traffic.[11] The net effect is to optimize the output of the pyramidal output cells of the PFC. The additional contribution of DA D2 receptors in the PFC and their interaction within the PFC is not entirely clear at present, but there are promising attempts to model this interaction in terms of the overall level of DA activity within the PFC. It is likely that analogous actions occur within the striatum, with a coupling of DA D1 and NMDA receptor activity in the so-called up-states, particularly with respect to hippocampal input and DA D2 receptors "gating" an inhibitory top-down influence of the PFC.[12] Having discussed the presumed generalities of the functioning of the mesencephalic DA system, we will now examine how such a neuromodulatory influence is expressed functionally in the context of the mesostriatal, mesolimbic, and mesocortical domains—and also relate it to the burgeoning evidence on the functioning of the human DA systems.

MESOSTRIATAL DA SYSTEM

The activational effects of the mesostriatal DA system are captured most vividly by the effects of nigrostriatal DA in Parkinson's disease, leading to akinesia and motor rigidity. A prominent model for this function has been the unilateral lesioning technique pioneered by Ungerstedt[13] using the selective neurotoxin 6-hydroxydopamine (6-OHDA) to produce a profound unilateral depletion of striatal DA, resulting in circling behavior. Follow-up studies suggested that such unilateral DA loss, when limited to the dorsal striatum, impaired the capacity of rats to initiate a lateralized head movement into space contralateral to the side of the lesion.[14] Moreover, the lateralized motor readiness to respond was also impaired: if the rat was required to make this response after unpredictable delays, responding was "primed" or speeded—an effect probably resulting from enhanced motor readiness. Thus, the animal had prepared the response optimally in terms of adjusting its posture—for example, orienting toward the target and producing the lateralized head

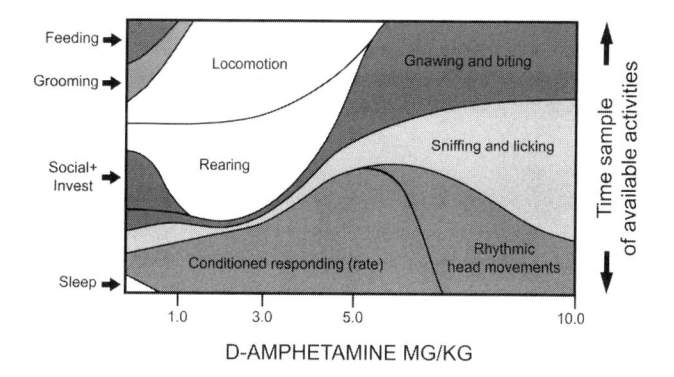

FIGURE 5.1.2. A schematic showing the distribution of different activities in the rat as a function of amphetamine dose, taken from Lyon and Robbins.[16] The gradual trend toward behavioral stereotypy ("an increasing rate of responses within a reduced number of response categories") is evident with increasing dose. This pattern of effects can hypothetically be related to the Yerkes-Dodson model shown in Figure 5.1.1, as d-amphetamine is an indirect DA receptor agonist. *Source*: Lyon and Robbins.[16]

movement that serves to demonstrate its detection. Brown and Robbins[15] found that lateralized dorsal striatal DA loss abolished the delay-dependent speeding effect, suggesting that it normally subserves lateralized activation in terms of motor readiness.

The opposite effect of *overactivation* within the DA systems is illustrated most obviously by the effects of d-amphetamine, a DA transporter blocker and DA release enhancer. An early synthesis[16] suggested that amphetamine-like stimulants produce an increase in responding in a reduced number of response categories (Fig. 5.1.2). Typically in the rat, in an unstructured environment, this is shown by increased signs of psychomotor stimulation such as locomotor hyperactivity breaking up long sequences of behavior such as grooming before, at higher doses, the behavior becomes increasingly repetitive and focused into a restricted space and form of response (generally repetitive sniffing and head movements) – a profile termed stereotypy.[16] This evolution of behavior as a function of the dose is reminiscent of a succession of inverted-U-shaped functions that each describe the effect of the drug on individual response sequences: the parallel with the Yerkes-Dodson principle is clear. The descending limb for each response arises from competition from an alternative response or responses. Following administration of amphetamine (and other DA agonists such as apomorphine), it appears that the simpler (i.e., shorter, requiring less sensory feedback) responses are the ones dominating the behavioral output profile at higher doses of the drug. Importantly, if environmental contingencies are more structured, for example in an operant chamber, elements of conditioned behavior can become stereotyped at high doses of the drug—either the instrumental lever press, or the approach to the magazine normally made to collect delivered food pellets (Fig. 5.1.3). In both cases, the behavior resembles compulsive response patterns in which responding no longer has any apparent consequence. This may be compatible with observations that the dorsal striatum is especially implicated in habit (stimulus-response) learning, which similarly is less dependent on the occurrence of primary reinforcers such as food[17]—and also with the hypothesis that drug-seeking behavior develops compulsive or habitual properties that powerfully contribute to the addictive process.[18]

There is considerable evidence that the initial phase of locomotor activation after treatment with amphetamine is mediated by DA release in the nucleus accumbens, probably under hippocampal modulation,[12,19–21] whereas the more stereotyped phase is mediated primarily by DA release in the dorsal striatum.[20] The competitive nature of the outputs of the dorsal and ventral striatum can be seen from the fact that if the stereotyped behavior is reduced by dorsal striatal DA, the hyperactivity produced by amphetamine is greatly potentiated.[22] This competitive interaction between different responses following global DA release produced by amphetamine may also underlie the so-called rate-dependent effects of amphetamine whereby low rates of operant responding are enhanced by the drug and high levels are reduced.[23] Again, the rate-decreasing effects of the drug are assumed to result from the stimulation of other, competing responses, whereas the stimulation of low rates arises from the behavioral activating effects of the drug.

MESOLIMBIC DA SYSTEM

The rate-increasing effects of amphetamine on instrumental responding almost certainly depend primarily on

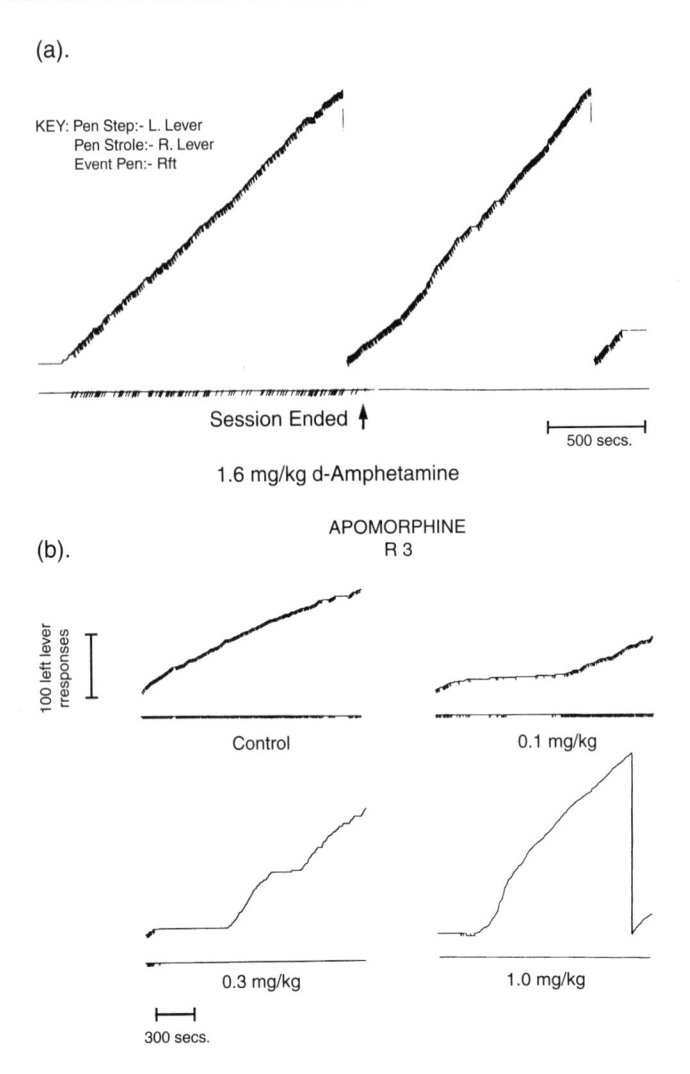

FIGURE 5.1.3. Examples of stereotyped operant behavior following administration of dopaminergic agents. Top: Cumulative records show how rats normally switch responses between two levers (pen ticks and increments) to obtain food in a magazine that they visit to collect the earned food (bottom event record). This rat, with a moderate dose of d-amphetamine, continued to switch between the two levers after the end of the session when food was no longer delivered. Thus, this behavior could be defined as having compulsive and stereotyped qualities despite its complexity, as the responding became divorced from its original goal. Bottom: Dose-related effects of the DA receptor agonist apomorphine in rats on the same schedule of food reinforcement. Note how apomorphine actually restricts persistent responding to one lever, resulting in omission of food reinforcement.

DA release in the ventral striatum.[24–26] Thus, intra-accumbens infusions of d-amphetamine have long been known to increase the control over behavior exerted by stimuli previously paired in a Pavlovian fashion with appetitive reinforcement (conditioned reinforcers).[25] This potentiation of conditioned reinforcement is evidently related to enhanced *incentive salience*, as posited by Berridge.[3] Moreover, approach to the magazine where the primary reinforcer (water or food) is formerly presented, signaled by the conditioned stimulus (CS) may actually diminish following treatment with the stimulant drug. Hence, the appetitive reinforcing effects of the CS are enhanced, while its discriminative effects on behavior are reduced. On the other hand, there is much about this behavior that it is pathological since the drug greatly enhances responding in extinction under the control of the conditioned reinforcer, though in the absence of the goal or primary reinforcer (i.e., food or water). This can thus be considered as an obvious example of maladaptive perseverative behavior.

The capacity of a cue to function as a conditioned reinforcer depends upon input from the basolateral amygdala and the integrity of the core region of the nucleus accumbens. In contrast, the response-enhancing effects of d-amphetamine depend on a circuitry including the shell region of the accumbens, the ventral

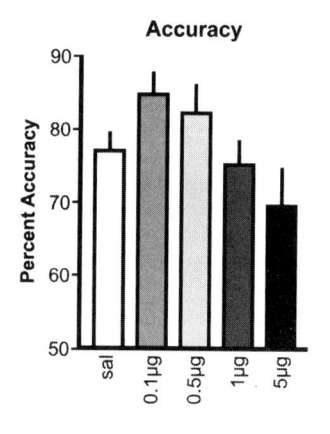

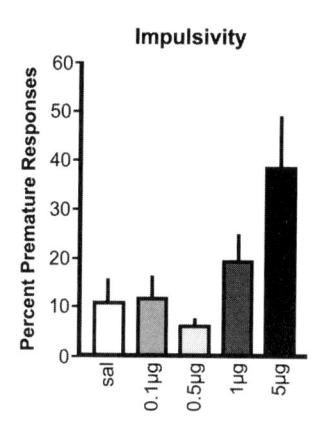

FIGURE 5.1.4. The DA D1 partial agonist SKF-38393 administered intra-accumbens produces dose-related improvement at a low dose and then deficits at higher doses in visual target detection on the 5CSRTT, that is, an inverted-U-shaped curve. This is paralleled by an increase in premature responses that show dose-related increases over the range tested. Source: Data redrawn from Pezze et al.[31]

subiculum, and the central nucleus of the amygdale, as well as the ascending mesolimbic projection itself.[27]. Wyvell and Berridge[28] have shown that intra-accumbens d-amphetamine, targeted primarily at the shell region, also affects Pavlovian-instrumental transfer in the sense that the effects of a noncontingently presented appetitive CS to increase responding on an operant baseline are enhanced. The effects of intra-accumbens amphetamine on responding with conditioned reinforcement are evidently more specific behaviorally since responding is selectively enhanced when it is contingent upon the presentation of the food- or water-paired conditioned reinforcer. However, it is plausible that in both cases, the rate-increasing effects of amphetamine arise from an exaggeration of a Pavlovian arousal process to which we have applied the more general term *activation*, to include the entire functional spectrum of the mesencephalic dopamine system. A corollary hypothesis is that this activational state has affective properties, such that certain levels are found to be rewarding or reinforcing for the animal if allowed to exert self-regulation—for example, by the self-administration of drugs (see Chapter 8.2 in this volume for an account of the role of DA in addiction).

Intriguingly, it does appear that a behavioral index of activation in the mesolimbic DA system may predict the propensity of rats to self-administer cocaine[29] (see also Chapter 8.2 in this volume). This behavioral index is the tendency of rats to respond prematurely in a five-choice serial reaction time task (5CSRTT) in which mildly food-deprived rats are trained to detect brief visual targets. Dalley et al.[29] found that there were large, though stable, individual differences in this inappropriate, premature responding, which can be thought of as a measure of impulsivity (the tendency to respond prematurely without foresight, often with adverse consequences). Moreover, the premature responding is significantly related to reduced D2/D3 binding in the ventral (but not dorsal) striatum. However, equally, it appears to depend on DA release in the nucleus accumbens, as (1) d-amphetamine infused there increases premature responding, which is (2) reduced by DA depletion in the accumbens produced by 6-OHDA.[30] This behavioral impulsivity, like the enhanced responding with conditioned reinforcement also produced by the drug, also has maladaptive aspects, as it results in the omission of food reward.

In a recent study, we showed explicitly how the Yerkes-Dodson principle applies to performance affected by DA-ergic agents administered to the ventral striatum. Pezze et al.[31] infused either D1 or D2 agents directly into the nucleus accumbens and found that low doses of the D1 agonist SKF-38393 produced significant improvements in the accuracy of detecting food-related visual targets in the 5CSRTT described above, perhaps as a consequence of enhancing their incentive salience. However, high doses, which also significantly increased premature responding, impaired accuracy (see Fig. 5.1.4). It is not, of course, clear whether the increased impulsivity actually caused the impaired accuracy (e.g., they may both be linked to a third factor). However, it is clear that an inverted-U-shaped function relating DA D1 receptor activity to performance may apply in the nucleus accumbens, as well as in the PFC (c.f.[32]).

MESOCORTICAL DA SYSTEM

There is already considerable evidence favoring an inverted-U-shaped function relating D1 receptor function to the efficiency of working memory in the rhesus macaque[8] and the rat.[33] However, these observations focused on the descending limb of the function—that is, the deficits associated with large doses. Granon et al.[34] provided some of the first evidence that performance could be enhanced in normal animals by a DA D1 receptor agonist. Thus, infusion of the partial D1 agonist SKF 38393 into the medial PFC (mPFC) in rats improved the accuracy of detecting visual targets on the 5CSRTT task, but only in rats whose performance was at a relatively low level. The hypothesis was that the high-performing rats had already "recruited" the D1 system to attain optimal performance and so were not susceptible to further boosting of accuracy. This was supported by the observation that these rats, unlike those with lower baseline accuracy, were impaired by infusions of a D1 receptor antagonist (SCH23390). However, it should be noted that this study did not report any decremental effects of the D1 receptor antagonist on attentional performance. These findings were consistent with those of a later study by Chudasama and Robbins[35] showing that a full D1 receptor agonist (SKF-81297) dose-dependently improved attentional accuracy when infused into the mPFC. This study also included a working memory component of the task, the inverted-U-shaped relationship with performance being more obvious in this component; low doses tended to improve performance, whereas higher doses tended to impair it, particularly at short delays (Fig. 5.1.5). These findings are compatible with conclusions reached earlier by Floresco and Phillips[36] using a rather different working memory paradigm: that the effects of DA on working memory are baseline-dependent in the sense that performance at longer delays generally is improved by D1 agents, whereas performance at shorter delays (which is generally superior) is made worse, even at the same dose. Phillips et al.[37] provide a convincing explanation for this pattern of findings, depending on the fluctuation of DA levels within the mPFC during the memory delay, as measured using in vivo microdialysis.

The findings of Chudasama and Robbins[35] show no obvious relationship between the effects of the D1 agent on attention and working memory, raising the possibility that there are different optimal levels of D1 receptor activation for distinct cognitive tasks, in accordance with a Yerkes-Dodson formulation. This hypothesis is also supported by other findings in the rat by Floresco et al.,[38] who found that tests requiring set shifting or cognitive flexibility do not demonstrate the same benefits on performance as tests of working memory. Specifically, performance was improved more by D2 receptor agonists or D4 receptor antagonists infused into the mPFC than by D1 agents, which only had inconsistent effects. What remains unclear is how these receptor agents interact with the overall level of DA-ergic activity in the PFC, although some possibilities are discussed by Seamans and Robbins.[39]

A similar pattern of findings has been shown in studies of nonhuman primates in terms of manipulations of DA differentially affecting performance according to the nature of the task. Thus, for example, DA depletion from the PFC produces different effects in the marmoset on spatial delayed response, self-ordered working memory, and set formation and shifting performance. Whereas spatial delayed response with distracting stimuli was impaired, self-ordered working memory performance, surprisingly, was not, despite being

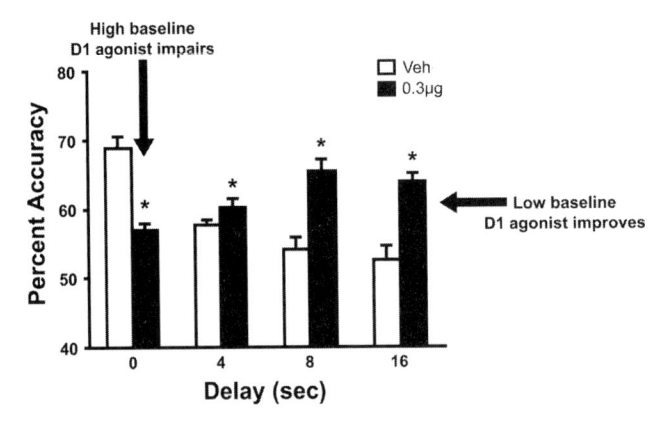

FIGURE 5.1.5. Data redrawn from Chudasama and Robbins[35] to show both deficits in performance and improvements at the same dose of a DA agonist in a test of spatial working memory at different delays. The most obvious account of the data is that they are baseline-dependent; several such effects can be seen following treatment with DA-ergic drugs in experimental animals and human volunteers. The same dose of the drug significantly improved performance in the attentional phase of the task (data not shown).

sensitive to PFC damage.[40] Parallel to the studies of PFC DA function in rats, there were also differential effects on set formation and set-shifting performance as a consequence of PFC DA depletion in marmosets. Thus, shifting between perceptual dimensions (extradimensional shifting) was actually improved, and reversal learning (when the contingencies are reversed for two-choice discrimination) was unaffected.[41] Subsequent studies[42] elucidated the likely basis of these apparently surprising findings. The marmosets with PFC DA loss exhibited problems with maintaining set and resisting distraction, suggesting that there were problems with the stabilization of representations (e.g., of task rules), which would also account for the working memory deficits. The difficulty in maintaining set may well account for the relative ease of disengaging set during the extradimensional shift.

Different sectors of the primate PFC appear to subserve different functions, a notable example being the dissociation between extradimensional shifting and reversal learning following lesioning of the lateral and orbital PFC.[43] This finding is matched by a neurochemical dissociation in that, while PFC DA depletion impairs extradimensional shifting but not reversal learning, PFC serotonin (5-HT) depletion has the opposite pattern of effects.[44] This result clearly shows how the different ascending neurochemical systems contribute to different types of processing within the PFC. The finding that selective depletion of DA in the orbitofrontal cortex has no effect on reversal learning does not mean that DA is without function in this region, as single-response extinction learning is greatly retarded by similar DA depletion within the orbitofrontal cortex.[45] The latter finding may indicate that DA signaling normally conveys a prediction error to this sector of the PFC. At any rate, these findings indicate that the mesocortical DA system may impact on several aspects of cognition and behavior.

THE ROLE OF DA IN HUMAN COGNITION AND BEHAVIOR

Many of the effects of DA agents on human cognition are consistent with the Yerkes-Dodson model and relate quite clearly to the animal studies reviewed above. Unfortunately, there is a paucity of information on the effects of manipulating the D1 receptor because of the relative unavailability of D1 agents, whether agonists or antagonist, for human studies, and so several of the hypotheses emanating from animal studies remain untested. Considerable attention has focused on stimulant drugs such as methylphenidate, doubtless in view of their success in the treatment of ADHD. These drugs enhance presynaptic DA function but also affect other monoamine neurotransmitter systems. The relatively selective DA D2 receptor agonist bromocriptine has also been a major focus of study.

For example, Kimberg et al.[46] found that the effects of bromocriptine on working memory in humans were analogous to the "rate-dependent" effect in experimental animals; low levels of performance were enhanced by the drug, but high levels were impaired in different individuals. Mehta et al.[47] found that bromocriptine also improved some aspects of spatial working memory (Cambridge Neuropsychological Test Automated Battery [CANTAB] spatial span) while impairing reversal learning—illustrating once more that optimal levels of DA activity for some tasks will not be optimal for others. Both of these findings are, of course, consistent with the Yerkes-Dodson inverted-U-shaped formulation. A third recent study also focused on the effects of bromocriptine on another aspect of executive function, task-set switching, as well as on distractibility in a group of human volunteers who varied considerably in their propensity to exhibit impulsivity. The drug enhanced task-switching performance in those individuals scoring high in impulsivity on the Barratt Scale but, if anything, impaired performance in low-impulsive subjects.[48] The effects in high impulsives were correlated with enhanced striatal activity during functional magnetic resonance imaging but were not present in low impulsives. The drug also tended to reduce distractibility and its concomitant frontal activation in high-impulsive volunteers.

In a study on self-ordered spatial working memory from the CANTAB battery, Mehta et al.[49] found that methylphenidate improved performance in healthy volunteers while reducing regional cerebral blood flow within frontoparietal circuitry, consistent with the hypothesis that this drug can enhance the efficiency of PFC processing, perhaps by enhancing the signal-to-noise ratio. The cognitive enhancing effect of methylphenidate depended on basal working memory performance (digit span)—with lower basal scores increasing to a greater extent after methylphenidate administration. These findings too are consistent with the Yerkes-Dodson model and also hint at the possibility of genetically endowed differences in working memory capacity determining the efficacy of the stimulant in its memory-enhancing capabilities.

Abnormalities in PFC processing, including cognitive functioning, have recently been associated with functional polymorphisms of the catechol-O-methyl transferase (COMT) gene. By modifying the enzyme's activity, these polymorphisms appear to have a special impact on prefrontal DA and can affect performance on fronto-executive-type tasks.[9] The evidence for specific effects on prefrontal DA comes in part from experiments in animals, such as COMT knockout mice, which have increased prefrontal DA (but not increased noradrenaline).[50] Moreover, pharmacological experiments with both rats and monkeys have implicated COMT in the regulation of extracellular DA in the PFC.[51,52] In addition, COMT inhibitors such as talcapone have been reported to improve working memory[53] and extradimensional shift performance[54] in rats. The theory is that COMT assumes a much greater role in regulating DA in the PFC because of the relative paucity of DA synaptic transporters there.

The COMT gene may influence prefrontal DA function because it contains a single nucleotide polymorphism at position 472 (guanine-to-adenine substitution), which is a valine-to-methionine alteration, resulting in reduction of COMT activity. Humans with the val/val genotype have hypothetically more rapid inactivation of released PFC DA than those with the met/met genotype, with those with the val/met heterozygote intermediate between these two. These changes in PFC DA function should be associated with relatively impaired performance on tests of cognition sensitive to frontal lobe dysfunction. This prediction has been confirmed for the Wisconsin Card Sort Test (WCST) and working memory (n-back tasks) performance, with the COMT genotype predicting 4% of WCST performance.[9,55] Furthermore, the val/val individuals benefited most from the enhancing effects of amphetamine on performance, whereas the met-met individuals tended to perform worse under the drug, as might have been predicted by the Yerkes-Dodson inverted-U-shaped function.

As an extension of the predictions arising from the COMT polymorphism data, it might be predicted that tolcapone, a COMT inhibitor, would improve cognitive performance in humans with val-val alleles. This prediction has recently been tested[54] using a range of tests of executive function. The most interesting finding was that the drug improved performance on the *intradimensional* shift test (Fig. 5.1.6), which is precisely that part of the CANTAB (intradimensional/extradimensional shift test [ID/ED]) test that is susceptible to PFC DA loss in marmosets (see above) and which provides a test of the capability to maintain sets or rules. This converging evidence provides strong corroborative support for the hypothesis that PFC DA is especially implicated in the stabilization of representations.[11]

The COMT phenotype modulates the L-DOPA response in Parkinson's disease (PD). Somewhat surprisingly, it is the met-met individuals who exhibit the greatest degree of cognitive deficit in PD, as measured by tests of planning and recognition memory from the CANTAB battery, especially in response to dopaminergic medications.[56,57] Recent studies have further shown that it is only PD patients with COMT met-met alleles relatively early in the course of the disease who show such deficits.[57,58] More severely impaired patients show the normal val-val deficit, and so it seems that the COMT polymorphism modulates a transient dysregulation of DA function with repercussions for cognition.

Other evidence helps to explain the variable effects of L-DOPA on cognition in terms of Yerkes-Dodson considerations. Gotham et al.[59] proposed a hypothesis that related the effects of L-DOPA to the pattern and course of DA loss within the striatum in PD. Regions with extensive DA depletion, such as the putamen, would have their functions optimally titrated by DA medication. By contrast, regions relatively spared in the early stages, such as the caudate and ventral striatum, would potentially be disrupted by medication, as the level of DA function would presumably be influenced supraoptimally by the drug. This hypothesis thus invokes the same Yerkes-Dodson principle invoked above to explain the disruptive effects of excessive PFC DA activity. Further evidence to support the Gotham et al. hypothesis comes from a study by Swainson et al.[60] that showed that mildly impaired, medicated PD patients performed poorly on tests of probability reversal learning associated with ventral striatal and orbitofrontal function,[61] while these same PD patients were relatively improved on tests of CANTAB spatial working memory function. These findings have recently been confirmed in a detailed study of effects of L-DOPA withdrawal using parallel, matched groups of PD patients.[62] This study compared effects of L-DOPA withdrawal in three tests of cognitive flexibility: task-set switching, attentional set shifting (the CANTAB ID/ED test, analogous to the discrimination tests used in marmoset monkeys; see above), and probability reversal. The drug selectively improved task-set switching, although it had no effect on extradimensional performance on the ID/ED task. These findings are consistent with the effects of caudate DA depletion in monkeys shown by Collins et al.[63] However, L-DOPA withdrawal actually resulted in *improved* probability reversal performance in PD patients, a test associated with ventral striatal-orbitofrontal circuitry on the basis of both monkey lesion[43] and human neuroimaging[61] findings. Cools et al.[62] interpreted these findings in terms of the pattern of DA depletion in frontostriatal circuits. Specifically, DA loss is greater in the more dorsal, caudate PFC than in the more ventral striatal "loops." Consequently, "overdosing" of these ventral loops via systemically administered L-DOPA is

control condition
Compound discrimination

shape to line
line to shape
Extra-dimensional shift

Intra-dimensional shift

Reversal

shape to shape
line to line

stimuli stay the same;
reward shifts

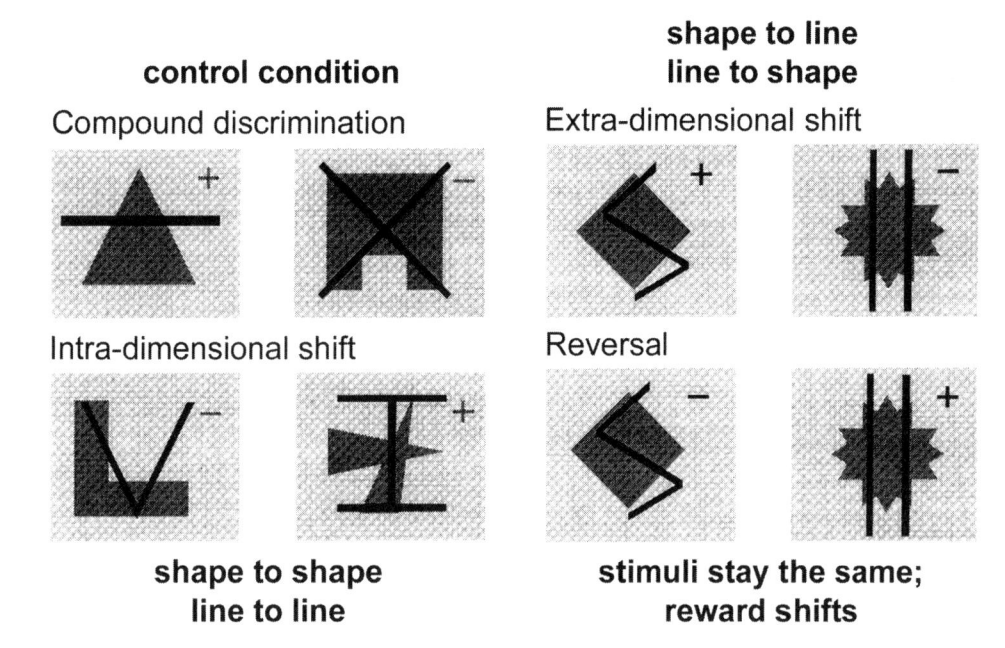

FIGURE 5.1.6. Typical stimuli from the ID/ED shifting test. Subjects are trained to respond to one dimension only of the compound stimulus (top left), consisting of line and shape dimensions. The intradimensional shift (ids) occurs when the stimulus exemplars are changed but the previously reinforced dimension (e.g., line) continues to be reinforced, whereas in the extradimensional shift, the relevant dimension is changed, and during reversal learning, the stimuli stay the same but which subject is rewarded is switched. Reproduced by permission of the publishers from Dias et al [43].

more likely, according to the Yerkes-Dodson inverted-U-shaped function (Fig. 5.1.7).

This hypothesis has recently been tested in a functional imaging study[64] in which PD patients were tested on the probabilistic reversal task, either on or off L-DOPA medication. The findings were consistent with the overdosing hypothesis. A blood oxygenation level-dependent (BOLD) signal, corresponding to the time point at which the reversal response is made, was present in the region of the nucleus accumbens in PD patients withdrawn from L-DOPA medication but not in the patients on medication. Therefore, the strong implication is that L-DOPA is producing the deficit in probabilistic reversal by obliterating the signal to switch responses. This effect of L-DOPA also extends to measures of impulsive gambling, with an increase in the amount bet in a gambling task[65]—and is reminiscent of reports of compulsive gambling following medication with DA D2 receptor antagonists (surveyed in [66]).

What exactly is meant by the *overdosing effect*? One obvious implication is that the drug occludes the signal provided by phasic burst firing in the mesolimbic DA

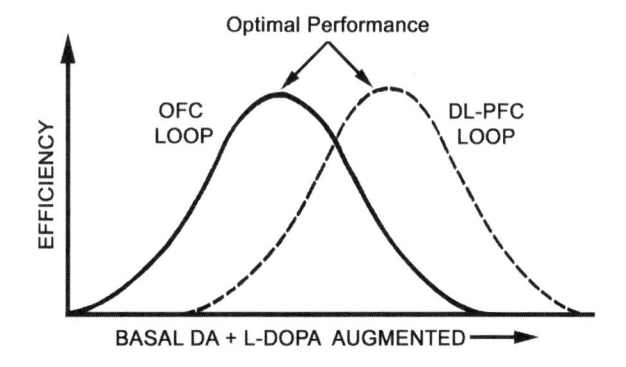

FIGURE 5.1.7. The Yerkes-Dodson relationship applied to the functioning of different frontostriatal loops, a dorsolateral prefrontal cortex (DL-PFC) loop and an orbitofrontal (OFC) loop. Various studies[49,60,62] suggest that the level of DA that optimizes functioning in the DL-PFC loop, which is often implicated in working memory and task-set switching, may impair performance in tasks such as reversal learning and gambling that recruit the orbitofrontal loop (see text for further explanation).

system. The effect in humans may be paralleled by the result described earlier in which drugs such as *d*-amphetamine enhance control by conditioned reinforcers, producing perseverative responding in extinction (c.f.[26]).

CONCLUSIONS

Strong themes run through this brief review of the role of DA in mesostriatal, mesolimbic, and mesocortical systems in experimental animals and humans. It appears that each of these systems is "tuned" according to an inverted-U-shaped function, such that either too low or too high levels of DA activity will produce impaired performance, whether in the motor, behavioral, or cognitive domains. This tuning probably varies among the major terminal domains, each of which may function optimally at a different level of DA activity. Related to this observation is the evidence that different cognitive tasks also appear to be performed optimally at different levels of DA function. In addition to this complexity, it appears that individuals vary in their degree of dopaminergic tuning, at least partly because of factors such as genetic polymorphisms (such as COMT). Challenges for the future include testing the Yerkes-Dodson hypothesis for the central DA systems with a range of techniques and conditions and also determining the relative roles of the different DA receptors in the same region, especially the D1-like and D2-like receptors, which probably function optimally at different levels of tonic activity of the DA systems. This approach has already yielded some relevant clinical observations, and this relevance is expected to become even more evident in future studies.

ACKNOWLEDGMENTS

Thanks to all of my collaborators for their contributions to these reviewed studies. The Behavioural and Clinical Neuroscience Institute is supported by a joint award from the Medical Research Council and the Wellcome Trust.

REFERENCES

1. Yerkes RM, Dodson JD. The relation of strength of stimulus to rapidity of habit-formation. *J Comp Neurol Psychol.* 1908;18:459–482.
2. Robbins TW, Everitt BJ. Functions of DA in the dorsal and ventral striatum. In: Robbins TW, ed. *Seminars in the Neurosciences*, Vol 4. London: Saunders; 1992;119–127.
3. Berridge KC. What is the role of dopamine in reward today? *Psychopharmacology.* 2006; 191:391–432.
4. Crow TJ. Specific monoamine systems as reward pathways. In: Wauquier A, Rolls ET, eds. *Brain-Stimulation Reward.* Amsterdam: North-Holland; 1976;211–238.
5. Schultz W. Getting formal with dopamine and reward. *Neuron.* 2002;36:241–253.
6. Grace A. Phasic versus tonic dopamine release and the modulation of dopamine system responsivity: a hypothesis for the etiology of schizophrenia. *Neuroscience.* 1991;41:1–24.
7. Niv Y, Daw ND, Joel D, Dayan P. Tonic dopamine: opportunity costs and the control of response vigor. *Psychopharmacology.* 2006;191:507–520.
8. Williams GV, Goldman-Rakic PS. Modulation of memory fields by DA D1 receptors in prefrontal cortex. *Nature.* 1995;376: 572–575.
9. Egan MF, Goldberg TE, Kolachana BS, et al. Effect of COMT Val108/158 Met genotype on frontal lobe function and risk for schizophrenia. *Proc Natl Acad Sci USA.* 2001;98: 6917–6922.
10. Mattay VS, Callicott JH, Fera F, et al. Catechol-O-methyltransferase val (158)-met genotype and individual variation in the response to amphetamine. *Proc Natl Acad Sci USA.* 2003;100: 6186–6191.
11. Durstewitz D, Seamans JK, Sejnowski T. Dopamine-mediated stabilization of delay-period activity in a network model of prefrontal cortex. *J Neurophysiol.* 2000;19:2807–2822.
12. Goto Y, Grace AA. Dopaminergic modulation of limbic and cortical drive of nucleus accumbens in goal-directed behavior. *Nat Neurosci.* 2005;8:805–812.
13. Ungerstedt U. Striatal DA release after amphetamine or nerve regeneration revealed by rotational behaviour. *Acta Physiol Scand.* 1971;367:49–68.
14. Carli M, Evenden JL, Robbins TW. Depletion of unilateral striatal dopamine impairs initiation of contralateral actions and not sensory attention. *Nature.* 1985;313:679–682.
15. Brown VJ, Robbins TW. Simple and choice reaction time performance following unilateral striatal DA depletion in the rat. *Brain.* 1991;114:513–525.
16. Lyon M, Robbins TW. The action of central nervous system stimulant drugs: a general theory concerning amphetamine effects. In: Essman W, Valzelli L, eds. *Current Developments in Psychopharmacology*, Vol. 2. New York: Spectrum; 1975:79–163.
17. Yin HH, Knowlton BJ, Balleine BW. Lesions of dorsolateral striatum preserve outcome expectancy but disrupt habit formation in instrumental learning. *Eur J Neurosci.* 2004;19:181–189.
18. Everitt BJ, Robbins TW. Neural systems of reinforcement for drug addiction: from actions to habits to compulsion. *Nat Neurosci.* 2005;8:1481–1489.
19. Burns LH, Robbins TW, Everitt BJ. Differential effects of excitotoxic lesions of the basolateral amygdala, ventral subiculum and medial prefrontal cortex on responding with conditioned reinforcement and locomotor activity potentiated by intra-accumbens infusions of *d*-amphetamine. *Behav Brain Res.* 1993;55:167–184.
20. Kelly PH, Seviour PW, Iversen SD. Amphetamine and apomorphine responses in the rat following 6-OHDA lesions of the nucleus accumbens septi and corpus striatum. *Brain Res.* 1975;94;507–522.
21. Wilkinson LS, Mittleman G, Torres E, et al. Enhancement of amphetamine-induced locomotor activity and dopamine release in nucleus accumbens following excitotoxic lesions of the hippocampus. *Behav Brain Res.* 1993;55:143–150.
22. Joyce EM, Iversen SD. Dissociable effects of 6-OHDA-induced lesions of neostriatum on anorexia, locomotor activity and

stereotypy—the role of behavioral competition. *Psychopharmacology*. 1984;83:363–366.
23. Robbins TW. Behavioural determinants of drug action; rate-dependency revisited. In: Cooper SJ, ed. *Theory in Psychopharmacology*, Vol. 1. London: Academic Press. 1981;1–63.
24. Robbins TW, Roberts DCS, Koob GF. Effects of *d*-amphetamine and apomorphine upon operant behavior and schedule-induced licking in rats with 6-hydroxydopamine-induced lesions of the nucleus accumbens. *J Pharm Exp Ther*. 1983;224:662–673.
25. Robbins TW, Taylor JR, Cador M, Everitt BJ. Limbic–striatal interactions and reward-related processes. *Neurosci Biobehav Rev*. 1989;13:155–162.
26. Parkinson JA, Olmstead MC, Burns LH, et al. Dissociation in effects of lesions of the nucleus accumbens core and shell on appetitive Pavlovian approach behavior and the potentiation of conditioned reinforcement and locomotor activity by D-amphetamine. *J Neurosci*. 1999;19:2401–2411.
27. Everitt BJ, Parkinson JA, Olmstead MC, et al. Associative processes in addiction and reward: the role of amygdala-ventral striatal subsystems. *Ann NY Acad Sci*. 1999;877:412–438.
28. Wyvell CL, Berridge KC. Incentive-sensitization by previous amphetamine exposure: increased cue-triggered "wanting" for sucrose reward. *J Neurosci*. 2001;21:7831–784.
29. Dalley JW, Fryer TD, Brichard L, et al. Nucleus accumbens D2/3 receptors predict trait impulsivity and cocaine reinforcement. *Science*. 2007;315:1267–1270.
30. Cole BJ, Robbins TW. Effects of 6-hydroxydopamine lesions of the nucleus accumbens septi on peformance of a 5-choice serial reaction-time task in rats: implications for theories of selective attention and arousal. *Behav Brain Res*. 1989;33:165–179.
31. Pezze MA, Dalley JW, Robbins TW. Differential roles of dopamine D1 and D2 receptors in the nucleus accumbens in attentional performance on the five-choice serial reaction time task. *Neuropsychopharmacology*. 2007; 32:273–283.
32. Arnsten AFT. Catecholaminergic regulation of the prefrontal cortex. *J Psychopharmacol*. 1997;11:151–162.
33. Zahrt J, Taylor JR, Mathew RG, et al Supra-normal stimulation of D1 DA receptors in the rodent prefrontal cortex impairs working memory performance. *J Neurosci*. 1997;17:8528–8535.
34. Granon S, Passetti F, Thomas KL, et al. Enhanced and impaired attentional performance following infusion of DA receptor agents into rat prefrontal cortex. *J Neurosci*. 2000;20:1208–1215.
35. Chudasama Y, Robbins TW. Dopaminergic modulation, visual attention and working memory in the rodent prefrontal cortex. *Neuropsychopharmacology*. 2004;29:1628–1636.
36. Floresco SB, Phillips AG. Delay-dependent of memory retrieval by infusion of a dopamine D1 agonist into the rat medial prefrontal cortex. *Behav Neurosci*. 2001;115:934–939.
37. Phillips A, Ahn S, Floresco S. Magnitude of dopamine release in medial prefrontal cortex predicts accuracy of memory on a delayed response task. *J Neurosci*. 2004;14:547–553.
38. Floresco SB, Magyar O, Ghods-Sharifi S, et al. Multiple dopamine receptor subtypes in the medial prefrontal cortex of the rat regulate set-shifting. *Neuropsychopharmacology*.2006;31:297–309.
39. Seamans JK, Robbins TW. Dopamine receptors and cognitive function. In: Neve K, ed. *Dopamine Receptors*. 2nd ed. NY: Humana Press; 2009.
40. Collins P, Roberts AC, Dias R, et al. Perseveration and strategy in a novel spatial self-ordered sequencing task for nonhuman primates: effects of excitotoxic lesions and DA depletions of the prefrontal cortex. *J Cogn Neurosci*. 1998;10:332–354.
41. Roberts AC, De Salvia MA, Wilkinson LS, et al. 6-Hydroxydopamine lesions of the prefrontal cortex in monkeys enhance performance on an analogue of the Wisconsin Card Sorting Test: possible interactions with subcortical DA. *J Neurosci*. 1994;14:2531–2544.
42. Crofts HS, Dalley JW, Collins P, et al. Differential effects of 6-OHDA lesions of the prefrontal cortex and caudate nucleus on the ability to acquire an attentional set. *Cereb Cortex*. 2001;11:1015–1026.
43. Dias R, Robbins TW, Roberts AC. Dissociation in prefrontal cortex of affective and attentional shifts. *Nature*. 1996;380: 69–72.
44. Clarke HF, Walker SC, Dalley JW, et al Cognitive inflexibility after prefrontal serotonin depletion is behaviourally and neurochemically specific. *Cereb Cortex*. 2006;17:18–27.
45. Walker S, Clarke HF, Robbins TW, et al. Differential contributions of dopamine and serotonin to orbitofrontal cortex function in the marmoset. *Cereb Cortex*. 2009;19:889–898.doi:10.1093/cercor/bhn36.
46. Kimberg DY, D'Esposito M, Farah MJ. Effects of bromocriptine on human subjects depend on working memory capacity. *Neuroreport*. 1997;8:3581–3585.
47. Mehta MA, Swainson R, Ogilivie AD, Sahakian BJ, Robbins TW. Improved short-term memory but impaired reversal learning following the D2 agonist bromocriptine in human volunteers. *Psychopharmacology*. 2001;159:12–20.
48. Cools R, Sheridan M, Jacobs E, et al. Impulsive personality predicts dopamine-dependent changes in fronto-striatal activity during component processes of working memory. *J Neurosci*. 2007;27:5506–5514.
49. Mehta MA, Owen AM, Sahakian BJ, et al Methylphenidate enhances working memory by modulating discrete frontal and parietal lobe regions in the human brain. *J Neurosci*. 2000;20: RC651–6.
50. Gogos JA, Morgan M, Luine V, et al. Catechol-O-methyltransferase-deficient mice exhibit sexually dimorphic changes in catecholamine levels and behavior. *Proc Natl Acad Sci USA*. 1998;95:9991–9996.
51. Elsworth JD, Leahy DJ, Roth RH, et al. Homovanillic acid concentrations in brain, CSF and plasma as indicators of central dopamine function in primates. *J Neural Transm*. 1987;68:51–62
52. Karoun F, Chrpusta SJ, Egan MF. 3-Methoxytyramine is the major metabolite of released dopamine in the rat frontal cortex: reassessment of the effects of antipsychotics on the dynamics of dopamine release and metabolism in the frontal cortex, nucleus accumbens, and striatum by a simple two pool model. *J Neurochem*. 1994;63:972–979.
53. Apud JA, Mattay V, Chen JS, et al. Tolcapone improves cognition and cortical information processing in normal human subjects. *Neuropsychopharmamcology*. 2007;32:1011–1020.
54. Tunbridge EM, Bannerman DM, Sharp T, et al. Catechol-O-methyltransferase inhibition improves set-shifting performance and elevates stimulated dopamine release in the rat prefrontal cortex. *J Neurosci*. 2004;24:5331–5335.
55. Barnett JH, Jones PB, Robbins TW, et al. Effects of the catechol-O-methyltransferase Val (158)Met polymorphism on executive function: a meta-analysis of the Wisconsin Card Sort Test in schizophrenia and healthy controls. *Mol Psychiatry*. 2007;5:502–509.
56. Foltynie T, Goldberg TE, Lewis SGJ, et al. Planning ability in Parkinson's disease is influenced by the COMT val158met polymorphism. *Mov Disord*. 2004;19:885–891.
57. Williams-Gray CH, Hampshire A, Robbins TW, et al. Catechol-O-methyltransferase Val158Met genotype influences frontoparietal activity during planning in patients with Parkinson's disease. *J Neurosci*. 2007;18:4832–4838.

58. Williams-Gray CH, Goris AN, Foltynie T. The two cognitive syndromes in Parkinson's disease: 5 year follow-up of a population-based incident cohort (the CamPaiGN study). Submitted for publication, 2009.
59. Gotham AM, Brown RG, Marsden CD. "Frontal" cognitive function in patients with Parkinson's disease "on" and "off" levodopa. *Brain.* 1988;111:299–321.
60. Swainson R, Rogers RD, Sahakian BJ, et al. Probabilistic learning and reversal deficits in patients with Parkinson's disease or frontal or temporal lobe lesions: possible adverse effects of DA-ergic medication. *Neuropsychologia.* 1999;38:596–612.
61. Cools R, Clark L, Owen AM, et al. Defining the neural mechanisms of probabilistic reversal learning using event-related functional magnetic resonance imaging. *J Neurosci.* 2002;22:4563–4567.
62. Cools R, Barker R, Sahakian BJ, et al. Enhanced or impaired cognitive function in Parkinson's disease as a function of dopaminergic medication and task demands. *Cereb Cortex.* 2001;11:1136–1143.
63. Collins P, Wilkinson LS, Everitt BJ, et al. The effect of dopamine depletion from the caudate nucleus of the common marmoset (*Callithrix jacchus*) on tests of prefrontal cognitive function. *Behav Neurosci.* 2000;114:3–17.
64. Cools R, Lewis SJ, Clark L, et al. L-DOPA disrupts activity in the nucleus accumbens during reversal learning in Parkinson's disease. *Neuropsychopharmacology.* 2007;32:180–189.
65. Cools R, Barker R, Sahakian BJ, et al. L-Dopa medication remediates cognitive inflexibility but increases impulsivity in patients with Parkinson's disease. *Neuropsychologia.* 2003;41:1431–1441.
66. Dagher A, Robbins TW. Personality, addiction, dopamine: insights from Parkinson's disease. *Neuron.* 2009;61:502–510.

5.2 | Contributions of Mesocorticolimbic Dopamine to Cognition and Executive Function

STAN B. FLORESCO

The seminal findings of Brozoski and colleagues[1] revealed that depletion of dopamine (DA) in the primate prefrontal cortex (PFC) produces impairments in delayed response tasks of a magnitude similar to those observed following complete removal of the frontal lobes. Since this initial report, a considerable amount of psychopharmacological research has been devoted to elucidating the functional role that mesocortical DA plays in complex forms of cognition and the specific DA receptor subtypes through which these actions are mediated.

Dopamine exerts its effects on PFC neural activity via multiple receptor subtypes. Both D1-like and D2-like (D2, D4) receptors are localized within the PFC, although the subcellular localization of these receptors differs. Expression of D1-like receptors on principal pyramidal neurons in the PFC appears to be substantially greater than that of D2-like (D2, D4) receptors,[2] whereas both types of DA receptors have been localized on GABAergic interneurons and may also reside on presynaptic excitatory glutamate terminals.[3-6] Numerous studies have shown that activation of DA receptors exerts dissociable electrophysiological actions on the activity of different classes of PFC neurons (reviewed in[7]). Yet, until recently, the majority of studies focusing on the role of DA in executive functioning have focused on the contribution of D1-like receptors in mediating working memory functions. However, it is becoming increasingly apparent that mesocortical DA transmission contributes to other forms of executive function regulated by the frontal lobes distinct from working memory processes. These studies have revealed that the specific DA receptor mechanisms that facilitate these processes appear to vary substantially across different functions. Thus, a primary purpose of this review is to present a summary of studies that have investigated the contribution of PFC DA transmission to higher-order cognition, and to compare and contrast the specific DA receptor mechanisms that regulate different types of executive function.

The dorsal and ventral regions of the striatum are major outputs of the PFC[8,9] and also receive dense dopaminergic innervation from the substantial nigra and ventral tegmental area in the midbrain. Whereas the dorsal striatum has traditionally be linked to motor learning, the nucleus accumbens (NAc) region of the ventral striatum has long been implicated in reward-related processes.[10] In addition, studies in rodents have revealed that the NAc also appears to make a critical contribution to behaviors requiring executive processing mediated by the PFC, including working memory[11,12] and behavioral flexibility.[13] Yet, despite the fact that lesions of either the PFC or its striatal outputs can exert similar disruptions in behavior, manipulations of striatal DA transmission can in some instances have different effects on executive functioning than the effects caused by similar manipulations of PFC DA. Thus, a secondary purpose of this review will be to highlight some of the similarities and differences in the contributions that mesocortical and striatal (primarily mesoaccumbens) DA make to different forms of cognition mediated by these circuits.

WORKING MEMORY

Primate Studies

The initial finding that lesions of DA terminals in the primate PFC impaired delayed responding tasks led to a number of pharmacological studies utilizing local administration of DA receptor agents. An elegant series of studies conducted by Goldman-Rakic and colleagues demonstrated that local administration of D1 receptor antagonists into the dorsolateral PFC of monkeys induced pronounced deficits on an oculomotor delayed response task[14,15] (see also Chapter 5.3, this volume). However, blockade of D2-like receptors with either sulpiride or raclopride did not impair performance on this task,[16] indicating that, in primates, the modulation by mesocortical DA of working memory processes is mediated primarily via D1 receptors. Subsequent studies combining local administration of DA receptor agents with neurophysiological recordings from awake, behaving monkeys revealed that a primary function of D1 receptor activation is to enhance and stabilize task-related activity in PFC neurons. Specifically, iontophoretic application of high doses of D1 antagonists attenuated the sustained firing displayed by PFC neurons during the delay component of a delayed response task, whereas administration of DA enhanced this activity.[15,17] These effects of D1 receptor stimulation are mediated through a number of different cellular pathways in both pyramidal and GABAergic interneurons within the PFC. Some of these include D1-mediated activation of persistent Na^+ and L-type Ca^{2+} currents, suppression of certain types of K^+ currents,[18,19] and alteration in both glutamate- and GABA-mediated synaptic currents.[20–25] Thus, DA, acting on D1 receptors, augments the effect that stronger excitatory inputs have on pyramidal neuron firing, thereby facilitating recurrent excitation within networks of PFC neurons that mediate working memory.[19,26]

More recent research has demonstrated that D2 receptors appear to mediate saccade (response)-related neural firing, as local application of D2 agonists or antagonists augments or attenuates the normal increase in PFC neural activity associated with the "response" component an of oculomotor delayed response task.[27] However, the functional significance of this action of D2 receptors in the mediation of working memory remains unclear, given that blockade of these receptors in the PFC does not impair performance.[16] In this regard, there have been a number of reports that systemic administration of D2 antagonists can impair working memory in both humans and primates.(e.g.,[28,29]) These findings seem to suggest that D2 receptors in the PFC may also play a role in mediating working memory. However, it is equally likely that these effects may be due to blockade of D2 receptors in other regions, such as the striatum, given that DA depletion in the caudate nucleus also impairs spatial delayed response.[30] Moreover, administration of selective D2 antagonists would be expected to block DA autoreceptors and increase DA extracellular levels in the PFC. This would, in turn, lead to an excessive stimulation of D1 receptors, which can also exert detrimental effects on working memory functions (see below). Thus, although systemic blockade of D2 receptors can impair delayed responding, the fact remains that local blockade of these receptors in the PFC does not perturb working memory.

Rodent Studies

Studies in rodents investigating the role that DA neurotransmission in the PFC plays in working memory have yielded findings that complement some of those obtained from primates. From an anatomical perspective, the medial regions of the PFC in rodents (e.g., the prelimbic, infralimbic, and anterior cingulate) share similar patterns of efferent and afferent connectivity to the medial PFC in primates.[31] Recent studies have indicated that the medial PFC in rodents may mediate processes such as conflict resolution and decision making, functions served by the cingulate cortex in primates and humans.[32,33] However, it is important to emphasize that lesions to the medial PFC in rats also produce impairments in working memory and behavioral flexibility that resemble those observed following damage to the dorsolateral PFC in primates and humans.[34,35] Thus, it has been suggested that the medial PFC in rodents may also share a *functional* homology to the dorsolateral PFC in primates. In keeping with this notion, depletion of DA in the medial PFC of rats impairs the learning of a delayed alternation task on a T-maze but does not affect short-term memory when no delay is inserted between choices,[36] indicating that intact PFC DA transmission is essential for the initial acquisition of working memory tasks. A subsequent study employing an operant version of a delayed matching-to-position paradigm confirmed that DA receptors in the medial PFC also mediate working memory performance in well-trained rats. Microinfusion of either the nonselective DA antagonist flupenthixol or the D1 antagonist SCH 23390 produced what was termed *delay-independent* impairments in performance when delays between sample and response phases ranged from 1.5 to 30 s.[37] However, these effects were also accompanied by an increase in response latencies, interpreted to mean that the effects of DA receptor blockade were due to a general disruption in performance that could not "be attributed to a specific impairment in short-term memory." Interestingly, another study using a delayed alternation procedure on a T-maze found no impairments in working memory following local blockade of either D1 or D2 receptors in the medial PFC of well-trained rats.[38] In that study, blockade of glutamate receptors did impair performance, indicating that working memory assessed using this type of delayed responding is dependent on the integrity of excitatory transmission in the medial PFC. Yet, infusions of SCH 23390 or the D2 antagonist sulpiride affected neither accuracy nor response latencies on this task when a delay of 10 s was used. Thus, the fact that blockade of DA receptors in the rat medial PFC does not appear to impair delayed alternation performance suggests that this form of delayed responding may not be the most sensitive paradigm to use in assessing the role of PFC DA receptors in the mediation of working memory functions in rodents.

It is important to note that working memory is not a unitary phenomenon, but a collection of different cognitive operations that work in concert to guide behavior. One component is the "online" retention of information over a brief delay period, which may be subserved by firing in local PFC circuits and modulated by D1 receptor activity.[17] However, other components of working memory are executive functions that mediate the manipulation and retrieval of trial-unique information to guide behavior across contexts or delays, independent of how long this information has been stored.[39–41] Our approach to investigating the role of the medial PFC in working memory processes has utilized a delayed response variant of the radial-arm maze task. The delayed spatial win-shift task consists of a training phase and a test phase separated by a delay (Fig. 5.2.1A). During the training phase, four out of eight arms of the maze are randomly selected each day and baited, while the remaining arms are blocked. The rat is required to retrieve the four pieces of food from the open arms, after which it is removed from the maze for the delay, which in this version is typically much longer than the length of delays used in the standard delayed response tasks (30 min). Following the delay, the rat is placed back in the maze with all eight arms open, but now, only the arms that were previously blocked contain food. Thus, during the test phase, the animal must recall previously acquired trial-unique information about which arms were blocked during the training phase in order to obtain food efficiently. Given that this task uses substantially longer delays, it is unlikely that this type of delayed responding measures the online maintenance of information that is required for accurate performance of the classic delayed response task. Indeed, it is likely that the storage of relevant information during the delay component of this task is mediated by the dorsal

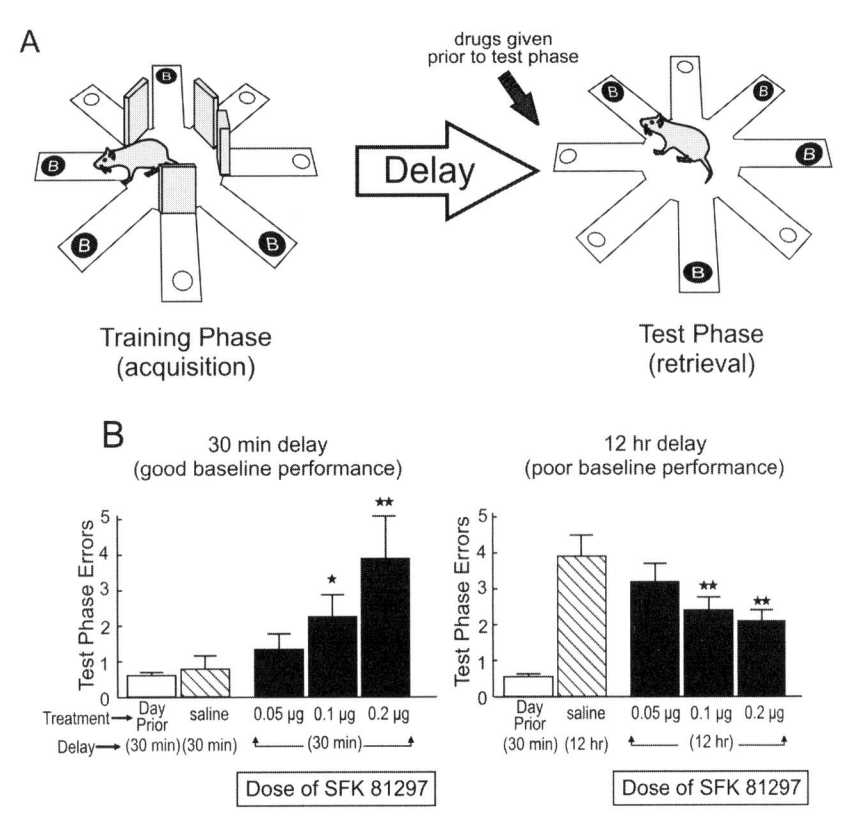

FIGURE 5.2.1. (A) The delayed response variant of the radial-arm maze task consists of a training (acquisition) phase and a test (retrieval) phase. During the training phase, the rat must retrieve four pieces of food from four randomly selected arms, with the four remaining arms blocked. The rat is then removed from the maze for a delay and then placed back in the maze for the test phase. The arms that were blocked previously are now open and baited. The studies summarized in this review administered DA agents into the PFC prior to the test phase. (B) In well-trained rats performing this delayed response task, infusions of D1 agonist SKF 81297 into the PFC significantly impaired performance when infusions were made after a relatively short delay (30 min; left panel), when performance of the rats was good. However, similar infusions improved performance when the D1 receptor agonist was administered after an extended 12-hr delay, when performance of the rats was poor (right panel; *$p < .05$, **$p < .01$. vs saline). *Source*: Adapted from Floresco and Phillips.[41]

hippocampus.[42] We have shown previously that infusion of the local anesthetic lidocaine into the medial PFC severely disrupts performance on this task when inactivations are administered prior to the test phase but not the training phase, indicating that this region of the PFC is selectively involved in the manipulation and retrieval of trial-unique information.[43] Furthermore, efficient retrieval of information during the test phase is dependent on the serial transfer of information from both the ventral hippocampus and mediodorsal thalamus converging in the medial PFC.[12,44] In addition, bilateral inactivation of the NAc, or disconnection between the PFC and the NAc, disrupts working memory performance assessed in this manner.[11,12] This finding indicates that this corticostriatal circuit plays an essential role in the transformation of working memory processes mediated by the temporal and frontal lobes into an efficient search strategy.

A separate series of experiments assessed the importance of D1 and D2 receptor signaling in the PFC on performance of this type of working memory task. In keeping with the findings from primate studies, blockade of D1 receptors in the medial PFC with SCH 23390 prior to the test phase impaired this form of delayed responding, whereas similar infusions of the D2 antagonist sulpiride had no effect.[45] These same manipulations did not impair performance on a single-phase version of the task where rats had no prior knowledge about the location of food in the maze, indicating that the effects of D1 receptor blockade could not be attributed to impairments in motivational, motor, or spatial navigation processes. A detailed analysis of the types of errors committed during the delayed task revealed a delay-independent deficit; rats were just as likely to reenter arms visited initially during the training phase 30 min earlier (across-phase errors) as they were to reenter arms recently entered during the test phase a few seconds earlier (within-phase errors), a finding that has been replicated subsequently by other groups.[46] We have interpreted this pattern of deficits to indicate that blockade of PFC D1 receptors induces a complete disruption in the search strategy that rats use normally to guide foraging behavior when they have received prior information about the probable location of food. The specific mechanisms by which D1 receptor activity may facilitate these processes appear to include modulation of hippocampal inputs to the PFC, as has been demonstrated electrophysiologically.[47-49] In a separate experiment incorporating an asymmetrical infusion design, unilateral inactivation of ventral subicular outputs combined with contralateral infusion of SCH 23390 into the PFC also disrupted performance on the delayed task in a manner similar to that observed following bilateral infusions of the D1 receptor blocker.[45] Thus, as has been observed in primate studies, it appears that D1 receptor activation is required for efficient working memory performance, whereas D2 receptors do not appear to play a role in these processes. Moreover, given that D1 receptor blockade impairs memory retrieval at either short (seconds[14,15]) or long (30 min[45]) delays, it is apparent that D1 receptor activity in the PFC mediates different components of working memory, which include both short-term online maintenance of information and the manipulation and retrieval of trial-unique information over longer delays.

It is interesting to note that although inactivation of the NAc via infusion of local anesthetics impairs working memory in a manner similar to that observed following similar manipulations of the PFC,[11,12,44] disruption of DA transmission in this nucleus does not appear to have the same effect. Infusions of the DA antagonist haloperidol did not affect retrieval of information during the test phase of the delayed win-shift task, although these manipulations did impair search behavior on a simpler one-phase random foraging task.[50] This finding suggests that mesoaccumbens DA activity may play a more prominent role in facilitating simpler forms of exploratory search behavior guided by short-term memory, mediated by hippocampal-NAc circuitry.[43] Furthermore, it highlights the fact that, under some conditions, the contribution of the NAc to different types of executive functions mediated by the PFC is not dependent on DA transmission in this nucleus. However, DA input to the dorsal striatum does appear to contribute to some forms of working memory, as depletion of DA in the caudate nucleus of marmosets markedly impairs performance on a spatial delayed response task.[30]

Whereas the above-mentioned studies have shown that blockade of D1 receptors in the PFC can perturb working memory, excessive stimulation of PFC D1 receptors also impairs performance on delayed response tasks. Earlier studies reported that systemic administration of the indirect DA agonist amphetamine impaired delayed responding in both primates[51] and rats.[52] Subsequent investigations revealed that these effects are mediated by excessive stimulation of D1 receptors, because impairments in working memory induced by increased PFC DA transmission can be alleviated by coadministration of a D1 receptor antagonist.[53] Zahrt and colleagues[54] later confirmed that infusions of the full D1 receptor agonist SKF 81297 into the medial PFC of rats impaired delayed alternation. Interestingly, although supranormal stimulation of D1 receptors in the PFC impairs working memory assessed in this fashion, blockade of these receptors in the PFC does not impair delayed alternation, as mentioned previously.[38]. Nevertheless, the fact that performance of other types of delayed response tasks can be impaired by local blockade of D1 receptors in the PFC[45] has led to the hypothesis that there is an optimum range of D1 receptor activation in the rat medial PFC for efficient working memory performance. Deviations from this range alter PFC neural activity, which in turn impairs working memory functions, in agreement with studies in primates (see Chapter 5.3, this volume).

Electrophysiological studies have provided further support for this notion. Iontophoretic application of very low concentrations of D1 receptor antagonists enhanced task-related activity in PFC pyramidal cells of monkeys performing an oculomotor delayed response task, whereas at higher concentrations, these antagonists disrupted task-related firing.[17] Further elucidation of the cellular mechanisms by which D1 receptor activation mediates these biphasic effects comes from in vitro electrophysiological studies. A reduction in D1 receptor activity would be expected to attenuate the normal signal sharpening or "gain-amplifying" effect over task-relevant inputs to PFC neurons, whereas suprathreshold activation of D1 receptors can exert a number of actions on pre- and postsynaptic neurons that lead to a reduction in PFC neural excitability.[7,18,19,26] Thus, it is apparent that PFC D1 receptor modulation of working memory takes the form of an inverted-U-shaped curve,[26,54,55] where too little or too much D1 receptor stimulation can hamper patterns of activity in PFC neural networks that normally mediate efficient working memory processes. Although the effects of D2 receptor stimulation on working memory have yet to be explored fully, one notable study has shown that infusions of a D2 agonist into the rat medial PFC also disrupt performance on a delayed response task. conducted in a U-maze, whereas infusions of the D2 antagonist sulpiride improved performance.[56] This finding indicates that under some conditions, D2 receptor activity in the PFC may also modulate working memory functions, but its contribution to these processes is distinct from its contribution to the processes regulated by D1 receptors.

Impairments in working memory induced by supranormal PFC D1 receptor stimulation are typically observed in animals that have been very well trained, where their performance on the task is optimal under baseline conditions. However, more recent work in rodents has demonstrated that exogenous stimulation of medial PFC D1 receptors can exert differential effects on working memory, and in some situations may exert a beneficial effect on PFC function. Specifically, under conditions where baseline levels of performance are poor, administration of D1 receptor agonists may enhance PFC function, whereas when performance is good, similar treatments may impede working memory processes. One manner in which baseline levels of performance may be altered is by adjusting the delay between the acquisition and response phases. Another study addressed this possibility by utilizing the delayed response variant of the radial-arm maze task that is sensitive to blockade of D1 receptors in the PFC.[41] In this experiment, baseline levels of performance were altered by adjusting the delay between the acquisition and response phases. We observed that intra-PFC infusions of SKF 81297 at doses known to impair delayed alternation (0.1–0.2 µg) also impaired memory retrieval in well-trained rats when the delay between acquisition and retrieval of information was relatively short (30 min; Fig. 5.2.1B, left panel). However, if performance was degraded by extending the delay from 30 min to 12 hr, the same doses of SKF 81297 improved performance relative to that of saline-treated animals tested at a 12-hr delay (Fig. 5.2.1B, right panel). Similar results have been reported using a combined attention/memory task that used shorter delays (0–16 s[57]). Using a within-subjects design, the authors demonstrated that infusions of the same D1 receptor agonist at doses of 0.06 or 0.3 µg improved attentional performance, whereas the 0.3-µg dose impaired working memory over short delays but improved it over longer delays. Thus, the heuristic of the inverted-U-shaped function underlying D1 receptor modulation of working memory appears to be valid only in situations where the cognitive and neural processes mediating performance are functioning optimally. In contrast, when mnemonic information used by the PFC to guide behavior has been degraded, performance may be improved by exogenous stimulation of D1 receptors in the PFC.

The ability of exogenous stimulation of PFC D1 receptors to improve performance in situations where memory has been degraded appears to be related to changes in the profile of mesocortical DA release. In a subsequent in vivo microdialysis study, we observed that increasing the duration between the acquisition and retrieval phases attenuates DA release in the PFC that is normally required for effective recall of information to guide search behavior following a delay.[58] We observed that when rats were tested at a relatively short 30-min delay, they displayed good levels of performance (approximately one error). Under these conditions, extracellular levels of PFC DA increased during both the acquisition of information in the training phase and the retrieval phase of this task, whereas DA levels returned to baseline during the delay period (Fig. 5.2.2A). This finding indicates that increased mesocortical DA

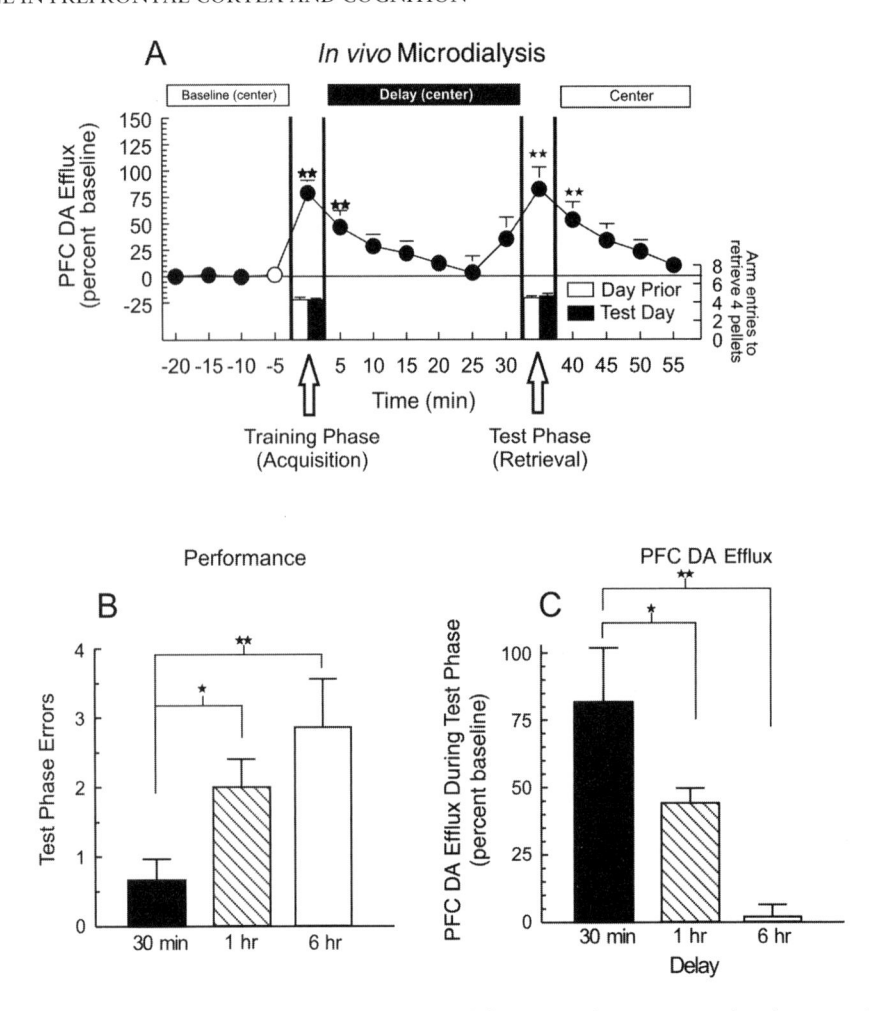

FIGURE 5.2.2. (A) Acquisition and retrieval of information during a delayed foraging task are associated with increased DA efflux in the PFC measured with in vivo microdialysis. Circles represent percent change in basal DA extracellular levels in the PFC over 5-min samples during (1) baseline, (2) the training phase, (3) a 30-min delay period, and (4) in the retrieval test phase, throughout a delayed response trial. Bars represent the total number of choices required to retrieve the four food reward pellets during each phase on the day before (white bars) and the microdialysis test day (black bar). *$p < .05$, **$p < .01$ vs baseline (white circle). *Source*: Adapted from Phillips et al.[58] (B) In the same study, extending the delay between training and test phases to 1 hr or 6 hr was accompanied by an impairment in performance, indexed by an increase in errors and (C) reduced levels of DA efflux in the PFC during the recall phase (*$p < .05$, **$p < 01$).

neurotransmission does not appear to be necessary for the active maintenance and storage of information during a delay period, but instead may be particularly important during the retrieval phase of this task. However, when the delay between acquisition and retrieval was extended to either 1 hr or 6 hr, we observed[1] an increase in the number of errors committed during the test phase (Fig. 5.2.2B) and[2] a marked attenuation of DA efflux during this phase of the task (Fig. 5.2.2C). Indeed, in the 6-hr delay condition, there was no change in extracellular DA levels in the PFC, despite the fact that animals actively explored the maze and readily consumed food placed in the maze arms. Taken together, the results of this study demonstrate that the magnitude of DA efflux in the PFC during the retrieval phase of a delayed response task is predictive of the accuracy of recall, with lower levels of DA efflux associated with poorer performance. Moreover, the fact that stimulation of D1 receptors in the PFC can restore working memory that is disrupted at a time when mesocortical DA release would be perturbed (i.e., by an extended delay[41,57]) further supports the contention that the magnitude of DA release and the accuracy of working memory are causally linked. Thus, perturbations in working memory that occur following particularly long delays between acquisition and recall may be due in part to attenuated mesocortical DA release that is normally required for efficient memory retrieval. This, in turn, would lead to suboptimal levels of D1 receptor activation in the PFC, disrupting neurophysiological patterns of activity in PFC circuits associated with memory retrieval.[14,44] Under these conditions, exogenous stimulation of D1 receptors via local infusion of SKF 81297 into the PFC would be expected to normalize levels of D1 receptor activity and improve performance.

To summarize, psychopharmacological studies using both primates and rats have shown that mesocortical DA exerts its effects on working memory functions primarily by acting on D1 receptors in the PFC. In contrast, although neural activity in the NAc is also an essential component of efficient working memory retrieval, disruption of DA transmission in this nucleus does not appear to interfere with this form of behavior. Moreover, exogenous stimulation of D1 receptors can exert differential effects on working memory, depending on the baseline levels of performance. Interestingly, blockade of D2-like receptors in the PFC does not appear to disrupt working memory functions, despite the fact that these receptors exert a number of electrophysiological effects on PFC neural activity.[59–61] However, recent findings point to a role for these receptors in other types of functions governed by the frontal lobes that are independent of working memory.

BEHAVIORAL FLEXIBILITY

The ability to use information flexibly and to execute appropriate adaptive behaviors in response to changes in one's environment is an essential survival skill. Different regions of the mammalian PFC have been strongly implicated in enabling an organism to alter its behavior strategy in response to changing task demands.[34,62] It is important to note that behavioral flexibility, like working memory, is not a unitary phenomenon, but rather may be viewed as a hierarchical process, and recent studies indicate that different forms of flexibility are dependent on anatomically distinct subregions of the PFC. For example, extinction entails the suppression of a conditioned response elicited by a stimulus that no longer predicts reinforcement and appears to be dependent in part on the ventral infralimbic region of the PFC in rats.[63,64] Reversal learning requires switching between stimulus–reinforcement associations when an organism must discriminate between different stimuli. This form of flexibility is severely impaired following lesioning of the orbitofrontal region of the PFC.[62,65,66] On the other hand, set shifting is a more complex process that entails shifts between strategies, rules, or attentional sets, requiring that attention be paid to multiple aspects of complex environmental stimuli. In humans, an inability to shift strategies is epitomized by impairments on the Wisconsin Card Sorting task. Patients with frontal lobe damage are initially able to sort cards by one dimension (e.g., color) but have great difficulty in altering their strategy when required to organize cards by another dimension (e.g., number or shape), perseverating to the now incorrect strategy. Similarly, manipulations of the dorsolateral PFC in primates or the prelimbic region of the medial PFC in rats do not affect initial discrimination learning, but they profoundly impair the ability to inhibit an old strategy and utilize a new one.[34,62,65,67]

Contributions by Mesocortical DA to Aspects of Behavioral Flexibility

Primate studies

Studies with experimental animals have provided important insight into the role that PFC DA plays in the mediation of more complex forms of behavioral flexibility. An initial report by Roberts and colleagues[68] showed that depletion of DA in the PFC of marmosets actually improved extradimensional set shifting while disrupting performance on a spatial delayed response task. This surprising finding was later attributed to an impairment in the formation of an attentional set, because in a subsequent study, 6-hydroxydopamine (6-OHDA) lesions of the PFC resulted in an impaired ability to perform repeated shifts within the same stimulus dimension (intradimensional shifts[69]). This finding was interpreted to suggest that impairments in the formation of an initial attentional set induced by DA lesions of the PFC led to improved performance when animals were required to shift attention to another stimulus dimension (extradimensional shift, EDS). However, this effect was observed for only one type of EDS, notably when animals were required to shift responding from a more difficult "lines" dimension to a "shapes" dimension. Nevertheless, these data indicate that mesocortical DA serves to stabilize representations, facilitating the ability to attend to relevant stimuli.[70] Interestingly, in these above-mentioned studies, reversal learning was unimpaired in these animals, even though these manipulations did result in depletion of DA in the orbital regions of the PFC. This lack of effect implies that DA transmission in the PFC does not appear to play a role in this simpler form of behavioral flexibility. Rather, recent evidence indicates that serotonin inputs to the orbital PFC may be the monoamine neurotransmitter that is of primary importance in modulating reversal learning.[71,72]

The finding that permanent lesions of DA terminals in the PFC impair the formation of an attentional set makes it difficult to ascertain whether mesocortical DA may also contribute to processes that mediate shifting from one discrimination strategy to another. Furthermore, studies of this kind preclude an assessment of the specific DA receptors that may be involved in these functions. However, recent studies using rodents in combination with local infusions of selective DA agents into the medial PFC have provided important information on the specific role that DA plays in facilitating complex forms of flexibility (i.e., set shifting) and the specific DA receptor subtypes that mediate these effects.

Rodent studies

One manner in which to assess set-shifting ability in rodents uses a strategy-shifting paradigm conducted in a cross-maze. In this task, rats initially learn to use either an egocentric response (e.g., always turn left) or a visual-cue discrimination strategy (e.g., always approach the arm with a visual cue, located in the left or right arm with equal frequency) to locate food in a cross-maze (Fig. 5.2.3A). During the set shift, rats are now required to shift from the previously acquired response or visual-cue-based strategy and learn the alternate discrimination. We and others have shown previously that this form of strategy set shift engages attentional set-shifting functions mediated by the medial PFC[67,73] and does not appear to require cognitive operations entailing a reversal of stimulus–reward associations mediated by the orbital PFC.[74] In addition, a key advantage of this task is that it permits a detailed analysis of the different types of errors that rats may commit during the set shift, providing further insight into whether impairments in behavior are due to enhanced perseverative responding or a deficit in acquiring and maintaining a new strategy. Studies using this protocol have shown that reversible inactivation of the medial PFC does not impair the initial acquisition of either a response or visual discrimination, but causes robust perseverative-type deficits when rats must shift from one strategy to another.[67,75]

Akin to its importance in working memory, D1 receptor activity in the PFC also plays an essential role in mediating this form of strategy set shifting. Ragozzino[76] reported that infusions of SCH 23390 into the medial PFC severely disrupted the ability to shift between response- and visual-cue-based strategies without affecting acquisition of either discrimination. However, a subsequent series of experiments utilizing a similar behavioral protocol performed in our laboratory revealed that the DA receptor mechanisms that mediate this form of behavioral flexibility and those that underlie working memory differ in a number of respects.[77] We assessed the functional role of D2 and D4 receptors in the PFC using selective antagonists for these receptors, as well as the effects of local stimulation of D1, D2, and D4 receptors using selective agonists for each of these targets. In contrast to what has been observed in studies of working memory, blockade of D2 receptors in the PFC via infusions of eticlopride impaired the ability to shift from one discrimination strategy to another (Fig. 5.2.3B). Furthermore, the nature of this impairment was similar to that induced by SCH 23390, in that it caused a selective increase in perseverative errors without affecting the acquisition or maintenance of a new strategy. From these data, it is apparent that set-shifting functions mediated by the medial PFC are dependent on a cooperative interaction between D1 and D2 receptors. Although the specific roles that each of these receptors play in facilitating the suppression of a previously acquired strategy remains unclear, recent electrophysiological data provide important information that may clarify the nature of the interactions between these receptors. D1 receptor activation is thought to maintain persistent levels of activity in PFC neural networks that may mediate the stabilization of particular representations.[7,70] By contrast, D2 receptor activation is thought to decrease inhibition of PFC pyramidal neurons.[7,22,25] This would be expected to place networks of PFC neurons in a more labile state, allowing them to process multiple stimuli and representations. Thus, activation of D2 receptors may facilitate the ability of PFC networks to disengage from the previous strategy and compare the viability of alternative response options, whereas D1 receptor activation may facilitate the stabilization of a novel strategy.[7,77]

Further differences in the DA receptor pharmacology that underlies working memory and set shifting were observed following administration of DA agonists. Infusions of the D1 agonist SKF 81297 at doses that have been shown to differentially alter performance on delayed response tasks neither impaired nor improved strategy set shifting (Fig. 5.2.3C). Likewise, infusions of the D2 agonist quinpirole did not alter performance during the set shift. This lack of effects was surprising, considering that stimulation of DA receptors in the medial PFC can alter working memory performance.[41,54,56] However, another study utilizing a perceptual set-shifting task designed by Birrell and Brown[78] showed that infusion of the partial D1 agonist SKF 38393 does not alter set shifting, although these manipulations do alleviate impairments induced by repeated

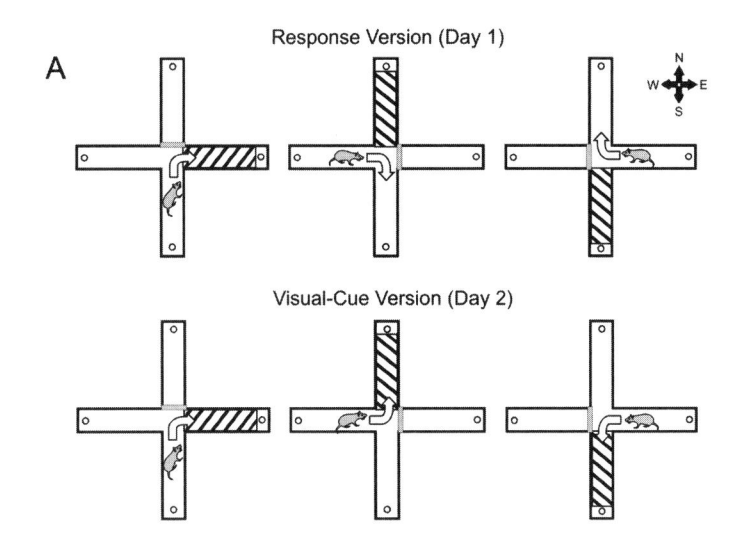

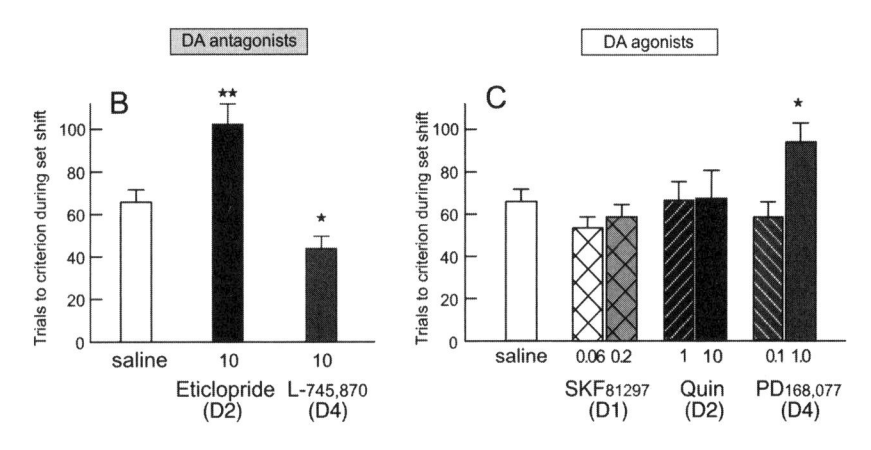

FIGURE 5.2.3. (A) On the set-shifting task conducted in a cross-maze, rats are initially trained to make a 90° right turn to receive food reinforcement (top panel). A black and white striped visual cue is randomly placed in one of the choice arms on each trial but does not reliably predict the location of food. During the set shift (bottom panel), the rat is now required to use a visual cue discrimination strategy, entering the arm with the visual cue, requiring either a right or left turn. Thus, the rat must shift from the old strategy and approach the previously irrelevant cue in order to obtain reinforcement. (B) Blockade of D2 receptors in the PFC with eticlopride significantly impairs strategy set shifting, as shown by an increase in the number of trials required to achieve criterion performance of 10 correct choices in a row relative to animals receiving saline infusions. Blockade of D4 receptors with L-745,870 significantly improves performance. (C) Infusions of either the D1 agonist SKF 81297 or the D2 agonist quinpirole (Quin) did not affect this type of set shifting, but stimulation of D4 receptors in the PFC with PD 168,077 impaired performance during the set shift ($*p < .05$, $**p < .01$). Numbers underneath each bar represent the drug dose in micrograms. *Source:* Adapted from Floresco et al. (2006a).[77]

amphetamine treatments.[79] Collectively, these findings indicate that the construct of an inverted-U-shaped function underlying D1 receptor modulation of working memory does not appear to hold true for set-shifting functions mediated by the PFC. These observations further highlight the differences between the DA receptor mechanisms that mediate these distinct PFC functions.

In contrast to the above-mentioned findings, infusion of D4 receptor agents revealed symmetrical effects on set shifting. Intra-PFC administration of the D4 agonist PD-168,077 impaired performance (Fig. 5.2.3C), whereas blockade of D4 receptors with L-745,870 improved shifting from one strategy to another (Fig. 5.2.3B). These findings indicate that D4 receptor activity may act to antagonize the effects that D1 and D2 receptor activity exerts over behavioral flexibility. The specific actions of D4 receptors may be mediated in part via neurophysiological modulation of N-methyl-D-aspartate (NMDA) receptor activity, given that stimulation of D4 receptors reduces NMDA receptor–mediated transmission in PFC pyramidal neurons[27] and blockade of NMDA receptors also impairs set shifting.[75] The improvements in set shifting induced by D4 receptor antagonism are in accordance with the results of other studies showing that systemic blockade of these receptors improves performance on tasks mediated by the PFC, either in intact animals or in those in which behavior has been disrupted by other pharmacological treatments.[80–82]

Further insight into the contributions of prefrontal DA to set shifting comes from in vivo microdialysis studies conducted in freely behaving rats. In an important series of experiments, Stefani and Moghaddam[83] measured DA release from rats performing a strategy set-shifting task in a cross maze. These experiments also included two key control groups. A yoked reward group consisted of rats that obtained reward on an intermittent schedule matched to rats that performed the set-shifting task but were not required to discriminate between arms or switch strategies. A second, reward retrieval control condition included rats that obtained food on every trial, regardless of their arm choice. Extracellular levels of DA in the PFC increased during both initial discrimination learning and the set shift, and were negatively correlated with the number of trials required to achieve criterion performance. Interestingly, rats in the yoked reward group (but not those in the reward retrieval group) displayed a similar profile of release, although the correlations between the magnitude of this change and behavioral performance were not as robust. This suggests that mesocortical DA levels are particularly sensitive to unpredictable situations, where the availability of reward is uncertain. Yet, the fact that blockade of DA receptors in the PFC induces a selective deficit in shifting between strategies, but not in the initial acquisition of a rule, suggest that DA release in this region plays a selective role in facilitating this form of behavioral flexibility.

Contributions by Mesostriatal DA to Aspects of Behavioral Flexibility

Different forms of behavioral flexibility mediated by anatomically dissociable regions of the PFC are critically dependent on interactions between these regions and their striatal targets. The electrophysiological actions of DA on striatal neural activity are complex; DA can act to either inhibit or augment neural excitability of neurons in the dorsal striatum and NAc, depending on a number of experimental conditions.[84] These opposing actions suggest that DA activity in the striatum may mediate the integration and gating of different afferent sources of information, amplifying one subset of inputs while concurrently inhibiting activation of NAc neurons evoked by other afferent projections.[84–86] It follows that mesoaccumbens DA transmission may play a particularly important role in mediating behaviors in situations where there is ambiguity about the environmental stimuli that may have motivational relevance (i.e., requiring behavioral flexibility).

There is evidence to suggest that simpler forms of behavioral flexibility, such as reversal learning, are sensitive to global disruptions in DA activity and appear to be critically dependent on D2, rather than D1 receptor activity. Systemic blockade of D2 receptors with either haloperidol or raclopride in primates induces perseverative deficits in reversal learning of visual discriminations,[87,88;] but see[89] although these effects are apparent only if animals are provided with a retention session immediately prior to the reversal. Blockade of D1 receptors with SCH 23390 at doses that do not cause gross motoric impairments does not affect this form of flexibility.[88] D2 receptor–deficient mice also display impairments in two-odor reversal learning.[90] Despite these findings, identification of the specific neural region where DA may be acting to facilitate this form of flexibility has remained elusive. As mentioned above, neurotoxic lesions of DA terminals in the PFC that include the orbital regions do not affect reversal learning.[68,69] This would suggest that DA activity in certain regions of the striatum may play a more prominent role in facilitating shifting between stimulus–reward associations within a particular stimulus dimension, given that lesions of the dorsal striatum impair different forms of reversal learning in both rodents and primates.[91–93] Yet, studies of the role of striatal DA have also yielded discrepancies. Dopamine lesions of the dorsomedial striatum in rats have been reported to impair reversal learning.[94] However, earlier studies in marmosets[30,69] did not observe impairments in reversal learning with DA lesions of the caudate nucleus, although these animals were slower to reengage a previously relevant perceptual dimension. In a similar vein, excitotoxic or dopaminergic lesions of the core and shell subregions of the NAc induced prior to behavioral training impaired reversal learning of a spatial discrimination in a T-maze; however, interpretation of these data was complicated by the fact that these lesions also impaired learning of the initial discrimination.[95,96] Subsequent studies revealed that lesions of the NAc that did not affect initial discrimination learning also did not impair learning of motor, odor, or visual reversals.[97,98] Moreover, some studies reported no impairment[99] or a delayed impairment on subsequent, but not the first, spatial reversal.[97] Thus, it is apparent that although the NAc and mesoaccumbens DA may be important for some forms of spatial discrimination learning, it does not appear to make a critical contribution to simpler forms of behavioral flexibility. Viewed in a broader context, these discrepancies clearly indicate that more research is required to elucidate the specific role that DA transmission plays in reversal learning and the specific terminal regions where it may be acting to facilitate these types of shifts.

In contrast to the somewhat inconclusive findings mentioned above, the dorsal and ventral aspects of the striatum appear to play a prominent role in shifting between discrimination strategies. Nucleus accumbens lesions cause robust, nonperseverative impairments when rats must reverse from a matching to a nonmatching strategy in an operant chamber, indicating an impairment of "higher-order response organization."[100] A subsequent series of experiments conducted in our laboratory investigated the role of different subregions of the NAc on strategy set shifting.[13] Reversible inactivations of the NAc core did not affect the initial learning of either a response- or a visual cue–based strategy, nor did it affect performance on the set shift conducted on the following day. In contrast, inactivations administered prior to the set shift impaired the ability to shift from one discrimination strategy to another. A detailed analysis of the types of errors rats committed during the set shift revealed a pattern of deficits distinct from those observed following similar manipulations of the PFC. Specifically, inactivation of the NAc core did not enhance perseveration, but instead increased "regressive" errors in a manner similar to that observed following inactivations of the dorsomedial striatum.[101] This pattern of errors suggests that these striatal outputs of the PFC facilitate the maintenance of novel behavioral responses that conflict with previously acquired strategies. In addition, inactivation of the NAc core increased "never-reinforced" errors, which may be viewed as an index of how readily rats are able to ascertain a novel strategy upon changes in reinforcement contingencies. This indicates that the NAc core also mediates the elimination of inappropriate response options, enabling an organism to reorganize its behavior in order to obtain reward in an optimal manner.

Although a number of studies have investigated the effects of lesions or inactivations of the NAc in more complex forms of behavioral flexibility, surprisingly few studies have looked at the role of mesoaccumbens DA transmission in processes such as set shifting. A recent report by Goto and Grace[102] noted that unilateral blockade of D1 receptors in the NAc combined with contralateral inactivation of the ventral hippocampus impaired shifting from a visual cue—to a response-based strategy, although these manipulations also impaired initial discrimination learning. Interestingly, these authors also observed that unilateral inactivation of the PFC combined with a contralateral infusion of the D2 agonist quinpirole induced a perseverative impairment in strategy set shifting. Preliminary studies in our laboratory have obtained similar results using bilateral infusions of DA agonists and antagonists in the NAc core using a strategy set-shifting task conducted in an operant chamber.[103] Infusions of a D1 antagonist into the NAc core induced nonperseverative deficits in shifting between strategies in a manner similar to inactivation of this nucleus.[13] In addition, we observed that blockade of D2 receptors in the NAc with eticlopride did not impair strategy set shifting, although this manipulation did decrease locomotion and increased response latencies. Moreover, whereas DA receptor blockade did not affect reversal learning, stimulation of D2 receptors again impaired performance. These findings demonstrate that D1 receptor modulation of NAc neuronal activity is an essential component of the neural circuitry that underlies behavioral flexibility mediated by PFC-NAc circuitry. Furthermore, they show that increased D2 receptor activity induces a deficit distinct from that induced by either inactivation or DA receptor blockade in the NAc core. This latter effect may be attributable to a suppression of PFC inputs to the NAc, as activation of D2 receptors attenuates PFC-evoked activity in this region of the ventral striatum.[102]

The above-mentioned findings are complemented by recent microdialysis studies measuring changes in DA efflux in the ventral and dorsal striatum in rats performing a set-shifting task. In the study by Stefani and Moghaddam[83] described earlier, these authors reported that DA efflux in the NAc increased slightly when rats were learning an initial discrimination, but showed a much more pronounced increase during a set shift. This indicates that tonic mesoaccumbens DA levels play a more prominent role in shifting between different rules as opposed to simple rule acquisition. In contrast, changes in DA levels in the dorsal striatum were more related to motoric aspects of behavior than to cognitive- or reward-related aspects of the task.

It is interesting to compare the differential effects of dopaminergic manipulations in the NAc on set shifting to those obtained following similar manipulations in the medial PFC. Blockade of D1 receptors in either region impairs the ability to shift between strategies, but in distinctly different manners. Mesocortical D1 receptor antagonism induces perseveration, whereas blockade of D1 signaling in the NAc core impairs the maintenance of a novel strategy after perseveration has ceased. By contrast, D2 receptors in the PFC play a critical role in suppressing the use of a previously relevant strategy, whereas in the NAc, D2 receptor activity appear to be more important for motivational aspects of performance, but not specifically in facilitating behavioral flexibility. Furthermore, excessive stimulation of DA receptors in the PFC and NAc produces differential effects, in that DA agonists neither improve nor impair flexibility when administered in the PFC, yet D2 receptor stimulation in the NAc induces a pronounced perseverative deficit. Profiles of DA release in these two regions also differ; mesocortical DA increases uniformly under conditions of reward uncertainty, whereas mesoaccumbens DA levels are more sensitive to changes in reward contingencies. Collectively, these findings highlight the fact that both mesocortical and mesoaccumbens DA plays a critical role in facilitating complex forms of behavioral flexibility. However, the specific functions of DA and receptor mechanisms through which it facilitates these processes vary considerably between these two regions.

SUMMARY

Viewed collectively, the findings reviewed here make it apparent that dopaminergic input to the forebrain, including the frontal lobes and the dorsal and ventral striatum, forms an essential component of the neural circuits that mediate a variety of cognitive and executive functions, including working memory and different forms of behavioral flexibility. Both of these executive functions engage distinct types of cognitive operations and functional neural circuits. Therefore, it is not surprising that the receptor mechanisms by which DA exerts its effects are not unitary across these functions; instead, each type of process relies on different patterns of activation of DA receptors in the PFC and the striatum. A primary purpose of this review has been to highlight this fact, and to put forth the argument that the principles of operation underlying mesocortical and mesoaccumbens DA modulation of one function do not necessarily apply to other types of executive functions mediated by the PFC. To summarize:

1. D1 receptor activity in the PFC is of primary importance in the mediation of working memory, whereas D1 and D2 receptors act cooperatively to mediate behavioral flexibility.
2. Although there is clear evidence that dopaminergic modulation of working memory takes the form of an inverted-U-shaped function, these biphasic effects of DA are not necessarily shared with other PFC functions related to behavioral flexibility. Furthermore, the effects of D1 receptor activation on working memory are critically dependent on baseline levels of performance.
3. Striatal DA activity, particularly in the NAc, may make a more prominent contribution to different forms of behavioral flexibility compared to working memory, with D2 receptors regulating simpler processes such as reversal learning, while D1 receptor activity in the NAc facilitates more complex processes related to set shifting.
4. Excessive activation of D2 receptors in the NAc, but not in the PFC, induces a pronounced deficit in different forms of behavioral flexibility.

The finding that different DA receptors mediate different types of executive functions is particularly relevant when devising novel pharmacotherapeutic treatment strategies for the cognitive dysfunction present in a number of neuropsychiatric diseases where dysfunction in DA signaling has been implicated as an underlying cause. Given that different disease states present distinct clusters of cognitive deficits (including deficits in executive functions mediated by the PFC), pharmacological treatment strategies that include dopaminergic agents must take into account the specific types of executive functions that are impaired in these patients, as well as the beneficial or deleterious effects that these drugs may have on different cognitive functions. Further elucidation of the specific DA receptor mechanisms that contribute to different types of executive functions will facilitate the development of more selective and effective treatments for specific domains of cognitive dysfunction.

ACKNOWLEDGMENTS

Some of the research reviewed in this chapter was supported by a Natural Sciences and Engineering Research Council of Canada Discovery Grant, a Canadian Institutes of Health Research Operating Grant, and a National Alliance for Research on Schizophrenia and Depression Young Investigator Award to SBF.

REFERENCES

1. Brozoski TJ, Brown RM, Rosvold HE, et al. Cognitive deficits caused by regional depletion of dopamine in prefrontal cortex of rhesus monkey. *Science*. 1979;205:929–932.
2. Gaspar P, Bloch B, LeMoine C. D1 and D2 receptor gene expression in rat frontal cortex: cellular localization in different classes of efferent neurons. *Eur J Neurosci*. 1995;7:1050–1063.
3. Mrzijak L, Bergson C, Pappy M, et al. Localization of dopamine D4 receptors in GABAergic neurons of the primate brain. *Nature*. 1996;381:245–248.

4. Muly EC, Szigeti K, Goldman-Rakic PS. D1 receptor in interneurons of macaque prefrontal cortex: distribution and subcellular localization. *J Neurosci*. 1998;18:10553–10565.
5. Sesack SR, King SW, Bressler CN. Electron microscopic visualization of dopamine D2 receptors in the forebrain: cellular, regional, and species comparisons. *Soc Neurosci Abstr*. 21:365, 1995.
6. Wedzony K, Chocyk A, Mackowiak M, et al. Cortical localization of dopamine D4 receptors in the rat brain-immunocytochemical study. *J Physiol Pharmacol*. 2001;51:205–221.
7. Seamans JK, Yang CR. The principal features and mechanisms of dopamine modulation in the prefrontal cortex. *Prog Neurobiol*. 2004;74:1–58.
8. Brog JS, Salyapongse A, Deutch A, et al. The pattern of afferent innervation of the core and shell in the "accumbens" part of the ventral striatum: immunohistochemical detection of retrogradely transported fluoro-gold. *J Comp Neurol*. 1993;338:255–278.
9. Sesack SR, Deutch AY, Roth RH, et al. Topographical organization of the efferent projections of the medial prefrontal cortex in the rat: an anterograde tract-tracing study with *Phaseolus vulgaris* leucoagglutinin. *J Comp Neurol*. 1989;290:213–242.
10. Robbins TW, Everitt BJ. Neurobehavioural mechanisms of reward and motivation. *Curr Opin Neurobiol*. 1996;6:228–236.
11. Seamans JK, Phillips AG. Selective memory impairments produced by transient lidocaine-induced lesions of the nucleus accumbens in rats. *Behav Neurosci*. 1994;108:456–468.
12. Floresco SB, Braaksma DN, Phillips AG. Thalamic-cortical-striatal circuitry subserves working memory during delayed responding on a radial arm maze. *J Neurosci*. 1999;19:11061–11071.
13. Floresco, SB, Ghods-Sharifi S, Vexelman C, et al. Dissociable roles for the nucleus accumbens core and shell in regulating set shifting. *J Neurosci*. 2006;26:2449–2457.
14. Sawaguchi T, Goldman-Rakic PS. D1 dopamine receptors in prefrontal cortex: involvement in working memory. *Science*. 1991;251:947–950.
15. Sawaguchi T, Matsumura M, Kubota K. Dopamine enhances the neuronal activity of spatial short-term memory performance in the primate prefrontal cortex. *Neurosci Res*. 1988;5:465–473.
16. Sawaguchi T, Goldman-Rakic PS. The role of D1-dopamine receptor in working memory: local injections of dopamine antagonists into the prefrontal cortex of rhesus monkeys performing an oculomotor delayed-response task. *J Neurophysiol*. 1994;71:515–528.
17. Williams GV, Goldman-Rakic PS. Modulation of memory fields by dopamine D1 receptors in prefrontal cortex. *Nature*. 1995;376:572–575.
18. Yang CR, Seamans JK, Gorelova N. Developing a neuronal model for the pathophysiology of schizophrenia based on the nature of electrophysiological actions of dopamine in the prefrontal cortex. *Neuropsychopharmacology*. 1999;21:161–194.
19. Yang CR, Seamans JK. Dopamine D1 receptor actions in layers V–V1 rat prefrontal cortex neurons in vitro: modulation of dendritic-somatic signal integration. *J Neurosci*. 1996;16:1922–1935.
20. Yang CR, Mogenson GJ. Dopaminergic modulation of cholinergic responses in rat medial prefrontal cortex: and electrophysiological study. *Brain Res*. 1990;524:271–281.
21. Pirot S, Godbout R, Mantz J, et al. Inhibitory effects of ventral tegmental area stimulation on the activity of prefrontal cortical neurons: evidence for involvement of both dopaminergic and GABAergic components. *Neuroscience*. 1992;49:857–865.
22. Seamans JK, Gorelova N, Durstewitz D, et al. Bidirectional dopamine modulation of GABAergic inhibition in prefrontal cortical pyramidal neurons. *J Neurosci*. 2001;21:3628–3638.
23. Seamans JK, Durstewitz D, Christie B, et al. Dopamine D1/D5 receptor modulation of excitatory synaptic inputs to layer V prefrontal cortex neurons. *Proc Natl Acad Sci USA*. 2001;98:301–306.
24. Wang J, O'Donnell P. D1 dopamine receptors potentiate NMDA-mediated excitability increase in layer V prefrontal cortical pyramidal neurons. *Cereb Cortex*. 2001;11:452–462.
25. Trantham-Davidson H, Neely LC, Lavin A, et al. Mechanisms underlying differential D1 versus D2 dopamine receptor regulation of inhibition in prefrontal cortex. *J Neurosci*. 2004;24:10652–10659.
26. Williams GV, Castner SA. Under the curve: critical issues for elucidating D_1 receptor function in working memory. *Neuroscience*. 2006;139:263–276.
27. Wang X, Zhong P, Gu Z, et al. Regulation of NMDA receptors by dopamine D4 signaling in prefrontal cortex. *J Neurosci*. 2003;23:9852–9861.
28. Mehta, MA, Manes FF, Magnolfi G, et al. Impaired set-shifting and dissociable effects on tests of spatial working memory following the dopamine D2 receptor antagonist sulpiride in human volunteers. *Psychopharmacology*. 2004;176:331–342.
29. Von Huben SN, Davis SA, Lay CC, et al. Differential contributions of dopaminergic D(1)- and D(2)-like receptors to cognitive function in rhesus monkeys. *Psychopharmacology*. 2006;188:586–596.
30. Collins P, Wilkinson LS, Everitt BJ, et al. The effect of dopamine depletion from the caudate nucleus of the common marmoset (*Callithrix jacchus*) on tests of prefrontal cognitive function. *Behav Neurosci*. 2000;114:3–17.
31. Öngür D, Price JL. The organization of networks within the orbital and medial prefrontal cortex of rats, monkeys and humans. *Cereb Cortex*. 2000;10:206–219.
32. de Wit S, Kosaki Y, Balleine BW, Dickinson A. Dorsomedial prefrontal cortex resolves response conflict in rats. *J Neurosci*. 2006;26:5224–5229.
33. Rushworth MF, Buckley MJ, Behrens TE, Walton ME, Bannerman DM. Functional organization of the medial frontal cortex. *Curr Opin Neurobiol*. 2007;17:220–227.
34. Brown VJ, Bowman EM. Rodent models of prefrontal cortical function. *Trends Neurosci*. 2002;25:340–343.
35. Uylings HB, Groenewegen HJ, Kolb B. Do rats have a prefrontal cortex? *Behav Brain Res*. 2003;146:3–17.
36. Bubser M, Schmidt WJ. 6-Hydroxydopamine lesion of the rat prefrontal cortex increases locomotor activity, impairs acquisition of delayed alternation tasks, but does not affect uninterrupted tasks in the radial maze. *Behav Brain Res*. 1990;37:157–168.
37. Broersen LM, Heinsbroek RP, deBruin JPC, et al. Effects of local application of dopaminergic drugs into the dorsal part of the medial prefrontal cortex of rats in a delayed matching to position task: comparison with cholinergic blockade. *Brain Res*. 1995;645:113–122.
38. Romanides AJ, Duffy P, Kalivas PW. Glutamatergic and dopaminergic afferents to the prefrontal cortex regulate spatial working memory in rats. *Neuroscience*. 1999;92:97–106.
39. Baddeley AD. *Working Memory*. Oxford: Clarendon Press; 1987.
40. Mizumori SJ, Channon V, Rosenzweig MR, et al. Short- and long-term components of working memory in the rat. *Behav Neurosci*. 1987;101:782–789.
41. Floresco SB, Phillips AG. Delay-dependent modulation of memory retrieval by infusion of a dopamine D1 agonist into the rat medial prefrontal cortex. *Behav Neurosci*. 2001;115:934–939.

42. Packard MG. Dissociations of multiple memory systems by post-training intracerebral injections of glutamate. *Psychobiology.* 1999;127:40–50.
43. Seamans JK, Floresco SB, Phillips AG. Functional differences between the prelimbic and anterior cingulate regions of the rat prefrontal cortex. *Behav Neurosci.* 1995;109:1063–1073.
44. Floresco SB, Seamans JK, Phillips AG. Selective roles for hippocampal prefrontal cortical, and ventral striatal circuits in radial-arm maze tasks with or without a delay. *J Neurosci.* 1997;17:1880–1890.
45. Seamans JK, Floresco SB, Phillips AG. D1 receptor modulation of hippocampal-prefrontal cortical circuits integrating spatial memory with executive functions in the rat. *J Neurosci.* 1998;18:1613–1621.
46. Aujla H, Beninger RJ. Hippocampal-prefrontocortical circuits: PKA inhibition in the prefrontal cortex impairs delayed nonmatching in the radial maze in rats. *Behav Neurosci.* 2001;115:1204–1211.
47. Jay TM, Glowinski J, Thierry AM. Inhibition of hippocampo-prefrontal cortex excitatory responses by the mesocortical DA system. *Neuroreport.* 1995;6:1845–1848.
48. Gurden H, Tassin J-P, Jay TM. Integrity of the mesocortical dopaminergic system is necessary for complete expression of in vivo hippocampal-prefrontal cortex long-term potentiation. *Neuroscience.* 1999;94:1019–1027.
49. Floresco SB, Grace AA. Gating of hippocampal-evoked activity in prefrontal cortical neurons by inputs from the mediodorsal thalamus and ventral tegmental area. *J Neurosci.* 2003;23:3930–3943.
50. Floresco SB, Seamans JK, Phillips AG. A selective role for dopamine in the nucleus accumbens of the rat in random foraging but not delayed spatial win-shift-based foraging. *Behav Brain Res.* 1996;80:161–168.
51. Bauer RH, Fuster JM. Effects of d-amphetamine and prefrontal cortical cooling on delayed matching-to-sample behavior. *Pharmacol Biochem Behav.* 1978;8:243–249.
52. Kesner RP, Bierley RA, Pebbles P. Short-term memory: the role of d-amphetamine. *Pharmacol Biochem Behav.* 1981;15:673–676.
53. Murphy BL, Arnsten AF, Goldman-Rakic PS, et al. Increased dopamine turnover in the prefrontal cortex imapirs spatial working memory performance in rats and monkeys. *Proc Natl Acad Sci USA.* 1996;93:1325–1329.
54. Zahrt J, Taylor JR, Mathew RG, et al. Supranormal stimulation of D1 dopamine receptors in the rodent prefrontal cortex impairs working memory performance. *J Neurosci.* 1997;17:8528–8535.
55. Arnsten, AF. Catecholamine regulation of the prefrontal cortex. *J Psychopharmacol.* 1997;11:151–162.
56. Druzin MY, Kurzina NP, Malinina EP, et al. The effects of local application of D2 selective dopaminergic drugs into the medial prefrontal cortex of rats in a delayed spatial choice task. *Behav Brain Res.* 2000;109:99–111.
57. Chudasama Y, Robbins TW. Dopaminergic modulation of visual attention and working memory in the rodent prefrontal cortex. *Neuropsychopharmacology.* 2004;29:1628–1636.
58. Phillips AG, Ahn S, Floresco SB. Magnitude of dopamine release in medial prefrontal cortex predicts accuracy of memory on a delayed response task. *J Neurosci.* 2004;24:547–553.
59. Gulledge AT, Jaffe DB. Dopamine decreases the excitability of layer V pyramidal cells in the rat prefrontal cortex. *J Neurosci.* 1998;18:9139–9151.
60. Tseng KY, O'Donnell P. Dopamine–glutamate interactions controlling prefrontal cortical pyramidal cell excitability involve multiple signaling mechanisms. *J Neurosci.* 2004;24:5131–5139.
61. Wang M, Vijayraghavan S, Goldman-Rakic PS. Selective D2 receptor actions on the functional circuitry of working memory. *Science.* 2004;303:853–856.
62. Dias R, Robbins TW, Roberts AC. Primate analogue of the Wisconsin Card Sorting Test: effects of excitotoxic lesions of the prefrontal cortex in the marmoset. *Behav Neurosci.* 1996;110:872–886.
63. Quirk GJ, Russo GK, Barron JL, et al. The role of ventromedial prefrontal cortex in the recovery of extinguished fear. *J Neurosci.* 2000;20:6225–6231.
64. Rhodes SE, Killcross S. Lesions of rat infralimbic cortex enhance recovery and reinstatement of an appetitive Pavlovian response. *Learn Mem.* 2004;11:611–616.
65. Dias R, Robbins TW, Roberts AC. Dissociable forms of inhibitory control within prefrontal cortex with an analog of the Wisconsin Card Sort Test: restriction to novel situations and independence from "on-line" processing. *J Neurosci.* 1997;17:9285–9297.
66. McAlonan K, Brown VJ. Orbital prefrontal cortex mediates reversal learning and not attentional set shifting in the rat. *Behav Brain Res.* 2003;146:97–103.
67. Ragozzino ME, Detrich S, Kesner RP. Involvement of the prelimbic-infralimbic areas of the rodent prefrontal cortex in behavioral flexibility for place and response learning. *J Neurosci.* 1999;19:4585–4594.
68. Roberts AC, De Salvia MA, Wilkinson LS, et al. 6-Hydroxydopamine lesions of the prefrontal cortex in monkeys enhance performance on an analog of the Wisconsin Card Sort Test: possible interactions with subcortical dopamine. *J Neurosci.* 1994;14:2531–2544.
69. Crofts HS, Dalley JW, Collins P, et al. Differential effects of 6-OHDA lesions of the frontal cortex and caudate nucleus on the ability to acquire attentional set. *Cereb Cortex.* 2001;11:1015–1026.
70. Robbins TW. Chemistry of the mind: neurochemical modulation of prefrontal cortical function. *J Comp Neurol.* 2005;493:140–146.
71. Clarke HF, Walker SC, Crofts HS, et al. Prefrontal serotonin depletion affects reversal learning but not attentional set shifting. *J Neurosci.* 2005;25:532–538.
72. Clarke HF, Dalley JW, Crofts HS, et al. Cognitive inflexibility after prefrontal serotonin depletion. *Science.* 2004;304:878–880.
73. Floresco SB, Block AE, Tse MT. Inactivation of the medial prefrontal cortex of the rat impairs strategy set-shifting, but not reversal learning, using a novel, automated procedure. *Behav Brain Res.* 2008;190:85–96.
74. Ghods-Sharifi S, Haluk DM, Floresco SB. Differential effects of inactivation of the orbitofrontal cortex on strategy set shifting and reversal learning. *Neurobiol Learn Mem.* 2008;89:567–573.
75. Stefani MR, Groth K, Moghaddam B. Glutamate receptors in the rat medial prefrontal cortex regulate set-shifting ability. *Behav Neurosci.* 2003;117:728–737.
76. Ragozzino ME. The effects of dopamine D1 receptor blockade on the prelimbic-infralimbic areas on behavioral flexibility. *Learn Mem.* 2002;9:18–28.
77. Floresco SB, Magyar O, Ghods-Sharifi S, et al. Multiple dopamine receptor subtypes in the medial prefrontal cortex of the rat regulate set-shifting. Neuropsychopharmacology. 2006;31:297–309.
78. Birrell JM, Brown VJ. Medial frontal cortex mediates perceptual attentional set shifting in the rat. *J Neurosci.* 2000;20:4320–4324.
79. Fletcher PJ, Tenn CC, Rizos Z, et al. Sensitization to amphetamine, but not PCP, impairs attentional set shifting: reversal by a

D1 receptor agonist injected into the medial prefrontal cortex. *Psychopharmacology*. 2005;183:190–200.

80. Jentsch JD, Taylor JR, Redmond DE, et al. Dopamine D4 receptor antagonist reversal of subchronic phencyclidine-induced object retrieval/detour deficits in monkeys. *Psychopharmacology*. 1999;142:78–84.

81. Arnsten AF, Murphy B, Merchant K. The selective dopamine D4 receptor antagonist, PNU-101387G, prevents stress-induced cognitive deficits in monkeys. *Neuropsychopharmacology*. 2000;23:405–410.

82. Zhang K, Grady CJ, Tsapakis EM, et al. Regulation of working memory by dopamine D4 receptor in rats. *Neuropsychopharmacology*. 2004;29:1648–1655.

83. Stefani MR, Moghaddam B. Rule learning and reward contingency are associated with dissociable patterns of dopamine activation in the rat prefrontal cortex, nucleus accumbens, and dorsal striatum. *J Neurosci*. 2006;26:8810–8818.

84. Floresco SB. Dopaminergic regulation of limbic-striatal interplay. *J Psychiatry Neurosci*. 2007;32:400–411.

85. Oades RD. The role of noradrenaline in tuning and dopamine in switching between signals in the CNS. *Neurosci Biobehav Rev*. 1985;9:261–282.

86. Pennartz CMA, Groenewegen HJ, Lopes Da Silva FH. The nucleus accumbens as a complex of functionally distinct neuronal ensembles: an integration of behavioural, electrophysiological and anatomical data. *Prog Neurobiol*. 1994;42:719–761.

87. Ridley RM, Haystead TA, Baker HF. An analysis of visual object reversal learning in the marmoset after amphetamine and haloperidol. *Pharmacol Biochem Behav*. 1981;75:345–351.

88. Lee B, Groman S, London ED, et al. Dopamine D2/D3 receptors play a specific role in the reversal of a learned visual discrimination in monkeys. *Neuropsychopharmacology*. 2007;32:2125–2134.

89. Smith AG, Neill JC, Costall B. The dopamine D3/D2 receptor agonist 7-OH-DPAT induces cognitive impairment in the marmoset. *Pharmacol Biochem Behav*. 1999;63:201–211.

90. Kruzich PJ, Grandy DK. Dopamine D$_2$ mediate two-odor discrimination and reversal learning in C57/BL6 mice. *BMC Neurosci*. 2004;5:12.

91. Divac I, Rosvold HE, Szwarcbart MK. Behavioral effects of selective ablation of the caudate nucleus. *J Comp Physiol Psychol*. 1967;63:184–190.

92. Ragozzino ME. The contribution of the medial prefrontal cortex, orbitofrontal cortex, and dorsomedial striatum to behavioral flexibility. *Ann NY Acad Sci*. 1121:355–375, 2007.

93. Man MS, Clarke HF, Roberts AC. The role of the orbitofrontal cortex and medial striatum in the regulation of prepotent responses to food rewards. *Cereb Cortex*. 2009; 19:889–906.

94. O'Neill M, Brown VJ. The effect of striatal dopamine depletion and the adenosine A2A antagonist KW-6002 on reversal learning in rats. *Neurobiol Learn Mem*. 2007;88:75–81.

95. Taghzouti K, Louilot A, Herman, JP, et al. Alternation behavior, spatial discrimination, and reversal disturbances following 6-hydroxydopamine lesions in the nucleus accumbens of the rat. *Behav Neural Biol*. 1985;44:354–363.

96. Annett LE, McGregor A, Robbins TW. The effects of ibotenic acid lesions of the nucleus accumbens on spatial learning and extinction in the rat. *Behav Brain Res*. 1989;31:231–242.

97. Stern CE, Passingham RE. The nucleus accumbens in monkeys (*Macaca fascicularis*). III. Reversal learning. *Exp Brain Res*. 1995;106:239–247.

98. Schoenbaum G, Setlow B. Lesions of nucleus accumbens disrupt learning about aversive outcomes. *J Neurosci*. 2003;23:9833–9841.

99. Burk JA, Mair RG. Effects of dorsal and ventral striatal lesions on delayed matching trained with retractable levers. *Behav Brain Res*. 2001;122:67–78.

100. Reading PJ, Dunnett SB. The effects of excitotoxic lesions of the nucleus accumbens on a matching to position task. *Behav Brain Res*. 1991;46:17–29.

101. Ragozzino ME, Ragozzino KE, Mizumori SJ, et al. Role of the dorsomedial striatum in behavioral flexibility for response and visual cue discrimination learning. *Behav Neurosci*. 2002;116:105–115.

102. Goto Y, Grace AA. Dopaminergic modulation of limbic and cortical drive of nucleus accumbens in goal-directed behavior. *Nat Neurosci*. 2005;8:805–812.

103. Haluk DM, Floresco SB. Ventral striatal dopamine modulation of different forms of behavioral flexibility. *Neuropsychopharmacology*. 2009; 34:2041–2052..

5.3 | Dopamine's Influence on Prefrontal Cortical Cognition: Actions and Circuits in Behaving Primates

AMY F. T. ARNSTEN, SUSHEEL VIJAYRAGHAVAN, MIN WANG, NAO J. GAMO, AND CONSTANTINOS D. PASPALAS

In 1979, Brozoski et al.[1] published their landmark paper showing that depletion of dopamine (DA) from the prefrontal cortex (PFC) in monkeys was as detrimental to cognition as ablation of the cortex itself. This was the first evidence that a neuromodulator could have such a powerful role in cortical function. Since then, we have learned that DA has detrimental as well as beneficial actions, and that either too little or too much stimulation of the D1 family of DA receptors can be harmful to PFC function—the so-called inverted-U dose-response curve. The inverted U has been highly relevant to our understanding of human cognition, clarifying disparities in cognitive abilities arising from genetic and environmental alterations, including changes in mental illnesses such as schizophrenia and attention deficit hyperactivity disorder (ADHD). More recent research has revealed that PFC DA also has remarkable actions through the D2 family of its receptors, which may be relevant to the etiology and treatment of symptoms such as hallucinations. The following review summarizes our current knowledge of DA and DA receptor localization in primate PFC, and the powerful influences of DA on PFC physiology and cognitive function.

OVERVIEW OF PFC FUNCTION, PHYSIOLOGY, AND CIRCUITRY

The PFC is key for the regulation of thought, action, and emotion, providing the "mental sketchpad" and "central executive" essential to human cognition (reviewed in[2–4]). Networks of PFC neurons maintain information in mind, or bring to mind information from long-term stores, in order to guide a thoughtful response in the absence of external stimuli. This process is often referred to as *working memory* (WM). The ability of the PFC to represent goals is key for the regulation of attention (focusing, dividing, shifting, or maintaining attention), for inhibiting inappropriate thoughts, actions, and feelings, for insight, judgment, and decision making, and for planning for the near or distant future. The PFC has extensive connections with other cortical and subcortical regions and thus is ideally positioned to orchestrate behavior, thoughts, and feelings.[5] These projections can engage both inhibitory and excitatory neurons,[6] such that the PFC is able to attenuate or promote responses based on its representational knowledge.

The PFC consists of specialized yet interconnected subregions. In primates, the dorsolateral PFC guides attention and action, while the ventromedial PFC guides affect.[7] The ventral surface is often called the *orbital* PFC, as it sits on top of the eye orbits. In rodents, the PFC is much smaller; there is a ventromedial portion (prelimbic and infralimbic medial PFC) that is most similar in its connections to the medial PFC of primates and subserves cognitive functions (e.g., spatial WM and attentional set shifting) and ventrolateral regions that share affective functions similar to those performed by the orbital PFC in primates.[8]

The PFC is able to represent information through networks of pyramidal neurons engaged in recurrent excitation.[9] Such microcircuits have been the focus of the most intensive study in area 46 of the rhesus monkey dorsolateral PFC, which receives visuospatial inputs from the parietal association cortex and is specialized for spatial WM.[5] In particular, a model for the microcircuitry underlying spatial WM was proposed by Patricia Goldman-Rakic[9] and is presented in Figure 5.3.1. Pyramidal neurons in PFC with shared inputs from parietal cortex interconnect to represent a spatial position (e.g., 90°), designated the *preferred direction* of the neurons. These networks are "tuned" by inhibitory interneurons,[10] which reduce firing to other spatial locations—that is, they inhibit responses to nonpreferred directions—in order to provide spatial specificity (e.g., the basket B cell in Fig. 5.3.1D). Single-unit recordings in

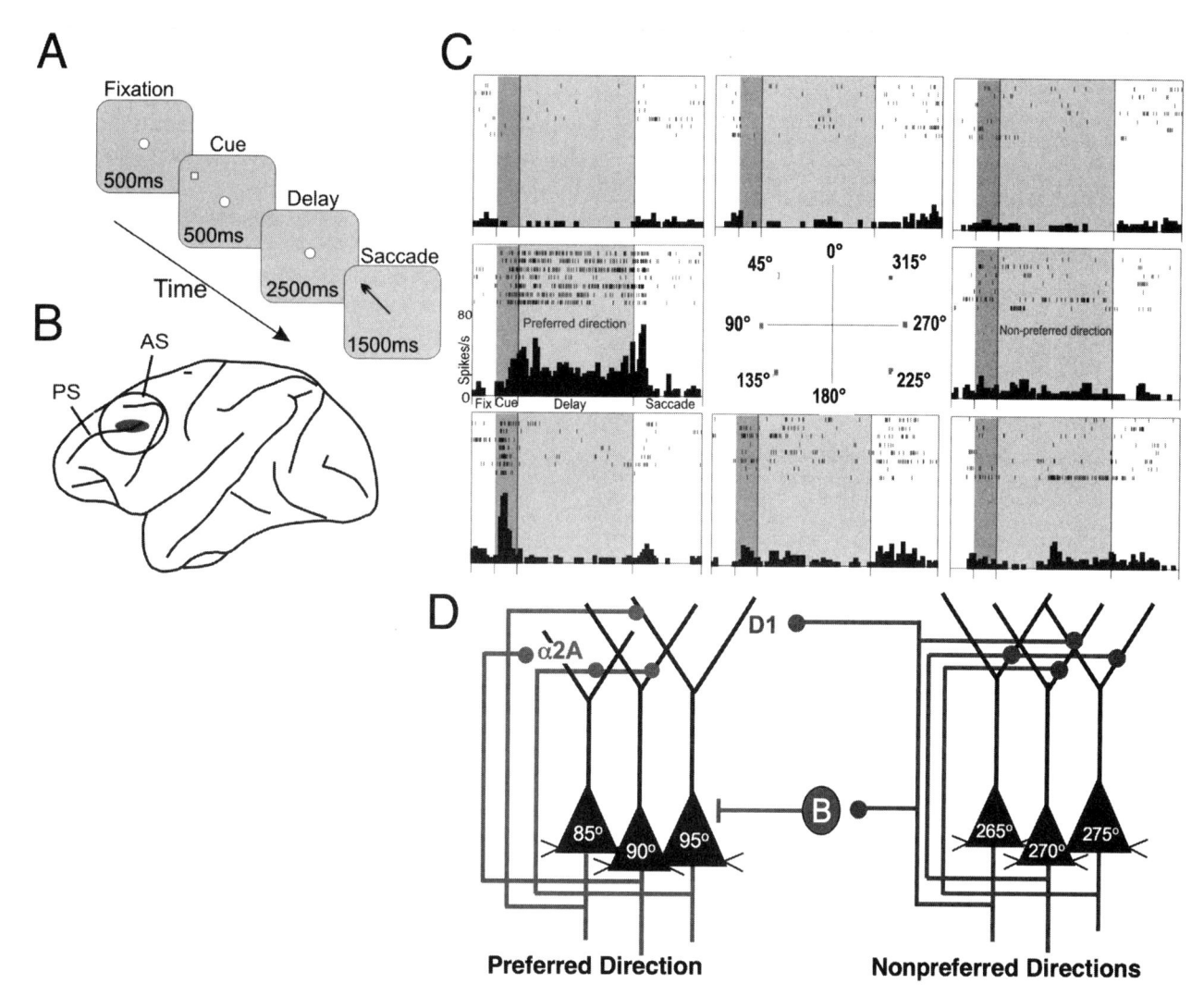

FIGURE 5.3.1. Spatial working memory in monkeys is often tested using the ODR task, illustrated in panel A (see text for description). Single-unit recordings from the principal sulcal PFC in the region illustrated in panel B show spatial mnemonic tuning as the monkey performs the task. Rasters from a typical neuron are illustrated in panel C. This neuron shows increased firing during the delay period if the cue had been at 90° (the preferred direction for the neuron), but does not show increased firing if the cue had been at other spatial locations (nonpreferred directions). Panel D illustrates the microcircuits in PFC thought to underlie the spatially tuned mnemonic firing (based on Goldman-Rakic, 1995[9]). The persistent firing during the delay period is thought to arise from recurrent excitation among similarly tuned pyramidal cells, while the spatial tuning arises from GABAergic inhibition, (e.g., the basket cell B illustrated here). Prefrontal cortex neuronal firing is also powerfully modulated by catecholamines. Under optimal conditions, α2A-AR stimulation strengthens delay-related firing for the preferred direction, while DA D1R stimulation suppresses responses to non-preferred directions. (See Color Plate 5.3.1.)

monkeys have shown that PFC neurons are able to maintain firing for the preferred modality-specific information "online" during a delay and use this represented information to guide behavior in the absence of environmental cues. A unique feature of PFC neurons is their ability to maintain information during a delay in the presence of distracting stimuli.[11]

Prefrontal cortical neurons can also represent other types of information. For example, they can fire in relationship to an abstract rule that is used to govern action.[12] Delay-related firing also can serve as a basis for behavioral inhibition, such as having to look away from a remembered visual stimulus, or for the reversal of reward contingencies.[13] Furthermore, PFC network activity can convey complex decisions and prediction errors (e.g., see[14,15]). However, these network properties have not been examined in regard to neurochemical modulation. In contrast, the neurochemical influences on spatial WM networks in dorsolateral PFC have been studied extensively and thus will be the focus of this review.

THE ANATOMY OF DA AND ITS RECEPTORS IN THE PRIMATE PFC

DA Projections and Synapses in the Primate PFC

The frontal lobes of humans and nonhuman primates receive a wealth of DAergic input from the mesencephalon, with the highest density in motor areas and a more delicate distribution in the PFC.[16,17] Dopamine projections follow a bilaminar distribution pattern in the upper and deeper cortical layers and rarely ramify, such that DA varicosities are organized *en passant* along the unmyelinated axon.[18] As with other monoaminergic afferents, the synaptic nature of the DA mesocortical system has been the subject of debate. Smiley and Goldman-Rakic[19] estimated that almost 40% of DA axon varicosities in PFC neuropil form identifiable, predominantly symmetric synapses, with the remaining likely to contribute to extrasynaptic DA release. Although the inconspicuous junctional specializations of DA synapses could be responsible for the purported low synaptic incidence, it is now well acknowledged that DA in PFC has a synaptic as well as a nonsynaptic signaling component.

Synaptic DA is released predominantly onto spine membranes and hence selectively targets the distal dendritic field of pyramidal neurons.[19] Dendritic spines are the exact site of pyramid-to-pyramid communication in prefrontal microcircuits. Therefore, synaptic DA is positioned to modulate recurrent excitation between pyramidal neurons mediated via axospinous synapses, which is commonly perceived as the cellular basis of WM processes (see above). Along these lines, a synaptic triad similar to that found in striatum was described in PFC to engage paired DA and glutamate synapses onto single dendritic spines.[20] Dopamine synapses may also target pyramidal and nonpyramidal dendritic stems. Sesack et al.[21] demonstrated the specificity of DA axodendritic synapses for parvalbumin (PV)- as opposed to calretinin-expressing neurons in PFC. This again translates into spatially selective modulation of the pyramidal principal cells, since PV neurons target the pyramidal perisomatic region. Thus, synaptically released DA is able to modulate the pyramidal neuron distally, via axospinous and to a lesser extent axodendritic synapses, as well as proximally, via a proxy interneuron providing key perisomatic inhibition.[22,23]

Extrasynaptically released DA or DA escaping from synapses could also have key roles in modulating the PFC. Two lines of indirect anatomical evidence support a function of DA as a volume neurotransmitter. Dopamine transporters (DATs) are typically found at a distance from DA varicosities and DA synapses, permitting neurotransmitter diffusion in the intercellular space.[17,24] If DA were allowed to diffuse, then it would be capable of reaching receptors on remote membranes. Indeed, electron microscopy has repeatedly demonstrated that the majority of identified DA receptors (reviewed in the next section) are distributed along nonsynaptic portions of plasma membranes.

DA Receptors in the Primate PFC: D1 and D2 Receptor Families

Based on their affinity for specific ligands and their signal transduction mechanisms, the G protein–coupled DA receptors are categorized into D1 and D2 receptor families. The original view is that D1-like receptors (D1R and D5R subtypes) coupled to Gαs proteins activate adenylyl cyclase and elevate cyclic adenosine monophosphate (cAMP), whereas D2-like receptors (D2R, D3R, and D4R subtypes) increase phosphodiesterase activity and suppress cAMP production via coupling to Gαi proteins.[25]

Almost two decades of research on DA receptor anatomy have produced maps of their cellular and subcellular distribution in human and particularly in monkey PFC. Early studies placed emphasis on D1R and D2R, the prototypic subtypes of the D1 and D2 receptor families. Particular interest focuses on the need to dissociate D1R and D5R signaling components, but the lack of specific ligands has hitherto precluded consideration of their potentially distinct roles within the PFC circuitry. Very little is known about the localization of D3Rs and D4Rs in primate PFC. All DARs present a functional (i.e., dopaminoceptive) component on plasma membranes as well as intracellular pools, consistent with synthesis and/or trafficking dynamics.

Localization of D1Rs and D5Rs in primate PFC

Smiley and colleagues first used immunoelectron microscopy to localize D1Rs in primate cerebral cortex.[26] They described a prominent expression on pyramidal dendritic spines and noted a perisynaptic distribution on membranes flanking asymmetric axospinous synapses. This pattern is now recognized as a salient feature of D1R anatomical organization in PFC (Fig. 5.3.2A) and also includes dendritic stems of pyramidal and nonpyramidal neurons.[27]

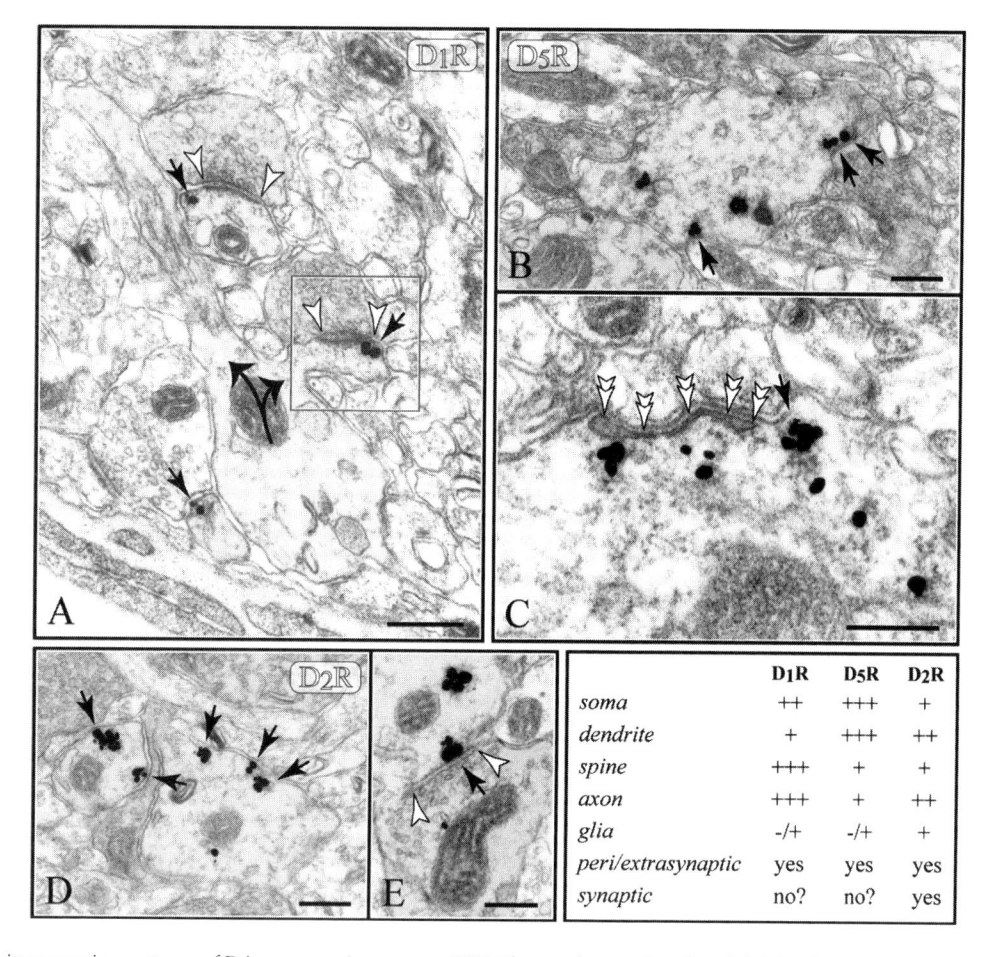

FIGURE 5.3.2. Main expression patterns of DA receptors in macaque PFC. Please refer to color plate 5.3.2 for clarity. Arrows point to immunogold labeling; synaptic specializations are between the arrowheads. Dendrites, axons, and somata are pseudocolored in blue, red, and yellow, respectively. (A) Perisynaptic expression on spine membranes (frame) is a salient feature of the D1R; curved arrows point to emerging spines. (B) Dendritic stems are the prevalent D5R-immunoreactive profiles. (C) In the pyramidal soma, nonsynaptic D5Rs are affiliated with subsurface cisterns (double arrowheads) that hold Ca^{2+} stores. Medium and fine dendrites exhibit nonsynaptic D2Rs (D) and D2Rs embedded in the postsynaptic density of symmetric axodendritic synapses (E). The table summarizes the expression patterns of individual receptor subtypes in the PFC neuropil. Scale bars: 200 nm. From[29,30,33,36]. (See Color Plate 5.3.2.)

D5 receptors have a rather complementary expression in PFC. They predominate in dendritic stems (Fig. 5.3.2B) and only infrequently populate the spines, suggesting that the D1-like receptor subtypes are for the most part segregated in the neuropil[28,29]; (see the table in Fig. 5.3.2). Thus, compartmentalization of D1Rs versus D5Rs could subserve subtype-specific signaling in PFC. The first such paradigm was revealed in the perisomatic region of pyramidal neurons (Fig. 5.3.2C), where D5Rs—but not D1Rs—associate with subsurface cisterns (SSCs) that hold inositol trisphosphate receptor (IP_3R)–gated Ca^{2+} stores.[30] Thus SSC-lined plasma membranes are equipped to function as junctional and signaling microdomains linking extrasynaptic D5R activation to internal Ca^{2+} stores via the phosphoinositide signal transduction system.[31]

Both D1Rs and D5Rs are present in axons of pyramidal and nonpyramidal neurons.[28,32] Moreover, presynaptic D1Rs in infragranular PFC are selectively aimed at pyramid-to-pyramid synapses,[33] suggesting target specificity and a selective involvement of D1R in modulating recurrent excitation in PFC.[34] It is obvious that D5Rs may partially account for the D1-like presynaptic effects in PFC, yet it is not clear whether D1Rs and D5Rs are on the same axons. Recently, Muly's group used a combined labeling approach to reason by deduction that the two receptor subtypes are in fact extensively colocalized on spines and axons,[35] which apparently contradicts the complementary patterns reported earlier. This intriguing finding remains to be verified.

There are no reports to date of either D1R or D5R being localized within the active zone of synapses. In dendritic spines and stems, and in axons, the receptors are distributed on perisynaptic membranes and extrasynaptically. However, the dense protein matrix of the junctional specializations might have hindered immunodetection if the receptor proteins were embedded within the synapse per se; hence, we should not exclude the possibility of a true synaptic D1-like receptor functioning in PFC.

Localization of D2Rs in primate PFC

Compared to D1Rs, the anatomy of D2Rs in PFC is poorly understood, and often data are difficult to replicate and interpret. A central feature of D2Rs is their predominant expression on dendritic stems of both pyramidal and nonpyramidal neurons (Fig. 5.3.2D). There is a substantial intracellular pool on dendrites that associates with the clathrin endocytosis pathway, consistent with D2R internalization and postendocytotic shorting in the cytoplasm.[36] It was reported that interaction with Neuronal Calcium Sensor-1 (NCS-1) modulates agonist-induced endocytosis of D2R, and hence DA neurotransmission, by altering net receptor availability. NCS-1 is up-regulated in patients with schizophrenia and bipolar disorder,[37] which would further suggest that alterations in internalization properties of D2Rs may contribute to DAergic pathologies.

D2 receptors are unique in that they are the only DA receptor subtype demonstrated in synapses with electron microscopy (Fig. 5.3.2E). Despite a predominant nonsynaptic distribution in the neuropil, D2Rs are also found within the postsynaptic density (PSD) of symmetric axo-dendritic synapses in monkey PFC[29] and with the PSD fraction from rodent neocortex (N. Kabbani and C. Paspalas, unpublished data, 2009). Future work will determine whether D2Rs or D2L and D2S receptor isoforms are associated with specific types of synapses and whether those are DA synapses.

Besides dendrites, axons are the major D2R-expressing component in PFC. Presynaptic D2Rs are commonly implicated in autoreceptor functions to regulate DA release, a notion that is for the most part inferred from findings concerning other brain areas rather than based on direct anatomical evidence in PFC. Relatively few monoaminergic-like axons express D2Rs in primate PFC. In fact, most studies report D2Rs in glutamatergic varicosities, where they apparently function as heteroreceptors, and in myelinated axonal segments, which is consistent with the presence of the receptor in PFC efferents.[29,36,38,39] Moreover, it was shown that unlike the nigrostriatal DA system, the field of origin of mesocorticolimbic projections does not encode for D2Rs in primates.[40] Yet, it could be that D2 autoreceptor levels in certain DAergic axons are below the threshold of immunodetection or that a single D2R isoform (D2S;[41]) functions as an autoreceptor; this, however, cannot explain why antibodies against both isoforms would not label a dense plexus of axons in PFC.

There have been sporadic reports of glial DA receptors in primate PFC, including D1-like receptors.[35] A D2R astrocytic component was consistently demonstrated with immunoelectron microscopy, and was corroborated by biochemical and physiological analyses.[36,39,42]

Localization of D4Rs in primate PFC

D4 receptors are of particular interest because of their higher affinity for atypical antipsychotics, but relatively little is known about their subcellular expression patterns in primate PFC. In monkey, D4Rs are expressed predominantly by GABAergic interneurons and are found in a subset of pyramidal neurons.[43] Plasmalemmal receptors appear both at the soma and at dendritic processes.

THE EFFECTS OF DA ON PFC PHYSIOLOGY AND FUNCTION

The Oculomotor Delayed Response Task

Much of our understanding of DA's physiological effects in PFC has arisen from research employing the oculomotor delayed response (ODR) task, a test of spatial WM performance that requires an awake, behaving monkey to remember the most recently cued spatial location despite massive proactive interference from numerous trials. In this task, monkeys are required to make a memory-guided saccade to a visuospatial target (Fig. 5.3.1A). A trial begins with the monkey fixating on a spot at the center of the screen. A cue briefly illuminates in one of eight possible locations, followed by a delay period of several seconds. At the end of the delay period the fixation spot disappears, and the monkey makes a saccade to the location of the cue and receives a juice reward. The monkey must update its memory for the cue location for each trial, thus using spatial WM during the delays. Goldman-Rakic and colleagues discovered that PFC neurons in the caudal principal sulcus (Fig. 5.3.1B) exhibit spatially tuned firing during the delay period; that is, they maintain firing if the cue had been at the preferred direction for the neuron, but do not fire during the delay period if the cue had been at other, nonpreferred spatial locations (Fig. 5.3.1C[44]). Thus, for optimal spatial WM function, PFC neurons must be able to both (1) maintain persistent activity over a delay and (2) be spatially tuned. As described above, Goldman-Rakic proposed that persistent activity arises from recurrent excitation between pyramidal cells with shared spatial characteristics, while spatial tuning arises from GABAergic inhibition (Fig. 5.3.1D). In addition to firing during the delay period, some PFC neurons fire in relationship to the memory-guided eye movement, so-called response-related firing.[45] Physiological data indicate that DA can alter many aspects of these firing patterns and thus can powerfully modulate WM performance.

DA Has Powerful Inverted-U Actions through the D1R

D1R effects on spatial WM

Following the seminal study that demonstrated the effects of DA depletion on primate PFC,[1] physiological investigations by Sawaguchi and colleagues revealed that injection of D1-like receptor antagonists SCH23390 and SKF39166 into monkey PFC caused disruptions in memory-guided saccade performance in the contralateral hemifield during the ODR task; D2-like receptor antagonists had negligible effects on mnemonic saccadic accuracy.[46] These findings focused the field on the D1 family of receptors, as it appeared that the powerful beneficial effects of DA on PFC WM function arose from D1R actions (please note that *D1R* will signify *D1-like receptor* in the remainder of this chapter, as there are currently no pharmacological manipulations that can distinguish D1Rs from D5Rs). It was initially assumed that DA had only beneficial actions through the D1R. Indeed, when biochemists discovered that exposure to mild uncontrollable stress selectively increased DA release in PFC, it was presumed that these higher DA levels would engage D1Rs and improve spatial WM.[47] However, a parallel set of studies by Arnsten and Goldman-Rakic showed that increasing D1R stimulation did not have a linear enhancing effect, as expected, but rather produced an inverted-U dose response. Higher levels of D1R stimulation induced either by a D1R agonist or through stress exposure impaired spatial WM, and these impairments could be reversed by D1R blockade.[48–52] This inverted-U dose response was subsequently appreciated in genetic and pharmacological studies of PFC function in humans as well (see below), and in physiological studies of PFC neurons (Fig. 5.3.3).

D1R effects on PFC neuronal physiology

What is the physiological substrate for DA's powerful effects on PFC spatial WM functions? The actions of D1Rs on cortical circuits have been studied extensively ex vivo in rodent slice preparations, and to a moderate extent in vivo in anesthetized rats, and in monkeys performing the ODR task. Studies in rodents have been particularly helpful in identifying DAergic influences on intracellular signaling events, while studies of awake, behaving monkeys have been most useful for linking DA actions to higher cognitive operations. Awake animals performing cognitive tasks are absolutely essential to observing the modulatory effects of DA through interactions with a cognitively engaged neural network. The following section provides a brief synthesis of this field.

D1R inverted-U physiological actions in monkeys performing the ODR task

The first insights into DA and D1R effects on PFC neurons in awake, behaving animals came from the micro-iontophoretic study of Sawaguchi et al.,[53] where DA was shown to enhance the activity of PFC cells engaged in spatial WM. Further, this enhancement appeared impervious to application of the D2R antagonist, sulpiride. A shortcoming of the study, however, was that D1R-selective antagonists of adequate specificity were unavailable at the time, and whether this effect was truly D1R-dependent was not clear. In a subsequent study, Williams and Goldman-Rakic[54] micro-iontophoresed the highly selective D1R antagonist, SCH39166, on PFC cells engaged in WM. They found that low doses increased neuronal firing in the preferred direction, while higher doses suppressed neuronal activation completely (Fig. 5.3.3, far left side of curve). These findings were the first physiological evidence of the inverted-U response previously shown in behavioral studies.

While the effects of antagonists in vivo shed some light on D1R actions, the effects of D1R stimulation on PFC neurons engaged in a cognitive task had not been systematically analyzed until recently. A physiological appraisal of the effects of overstimulation of D1Rs could explain more fully the inverted-U effects of D1R stimulation on mnemonic performance. Iontophoretic studies were undertaken in the Arnsten laboratory to address the effects of D1R agonists on the activity of PFC neurons.[55] We found that, while the overwhelming physiological effect of D1Rs on PFC cells with memory fields was a dramatic suppression of delay-related firing, suppression at lower doses was mostly in nonpreferred directions in the spatial WM task, thus sculpting and refining the spatial tuning and information capacity of PFC memory fields. This is shown in Figure 5.3.3. We have proposed this spatially asymmetric suppression to be the cellular and

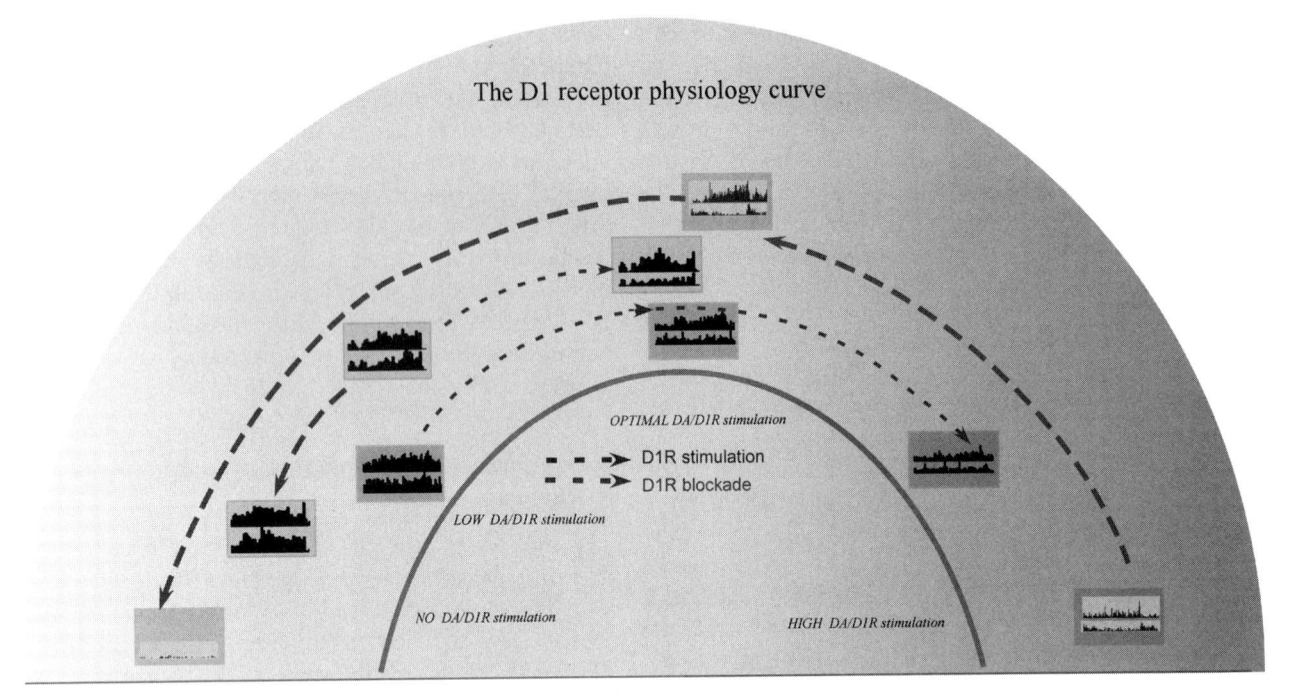

FIGURE 5.3.3. D1 receptor stimulation shows an inverted-U dose/response on the physiological profiles of neurons in the principal sulcal PFC in monkeys performing a spatial WM task. Please refer to 5.3.3 color plate for clarity. The neuron's firing patterns to its preferred direction is shown above the response to its nonpreferred direction. Pink shading from[54]; blue or beige shading from[55]. See text for explanation. (See Color Plate 5.3.3.)

physiological basis of the behavioral inverted-U effect. Higher doses of D1R agonists lead to a collapse of activity in the preferred directions as well, resulting in a detuning of the neuronal memory field (Fig. 5.3.3, far right side of curve). We propose that this effect may be the cellular substrate of the detrimental effects of high levels of D1R stimulation, induced either pharmacologically or in psychiatric contexts. Both partial and full D1R agonists caused suppression, and the effects were both D1R-specific and dependent on the activation of the cAMP pathway.[55] Thus, inhibition of cAMP signaling prevented and restored PFC memory fields previously disrupted by high levels of D1R engagement, while agents that mimic or increase cAMP suppress network firing.[56]

We are currently examining the ionic basis for cAMP's suppressive effects on PFC network firing in monkeys. At least some of these effects appear to arise from opening of HCN (hyperpolarization-activated cyclic nucleotide gated) channels. These channels pass both Na$^+$ and K$^+$ when they are opened (the h-current), reducing membrane resistance. We have localized HCN channels on the spines of PFC pyramidal cells, where they are ideally positioned to gate network inputs. We have further shown that α2A adrenergic receptor (α2A-AR) stimulation enhances PFC network firing in monkeys by inhibiting cAMP-HCN channel signaling, increasing network firing for the preferred direction of the neuron.[56] Preliminary data indicate that at least a component of D1R-mediated suppressive actions arise from cAMP opening of HCN channels, as the D1R response is prevented by blocking the channels (M. Wang, N.J. Gamo, and A. Arnsten, unpublished data, 2009). These findings suggest that D1R may gate network inputs by regulating the open state of HCN channels on spines (discussed in detail below). D1 receptor activation of cAMP signaling may also suppress firing through presynaptic inhibition of network inputs[57] and through general effects on excitability—for example, cAMP inhibition of the Ca(2+)-activated, non-selective (CAN) current, as suggested by Wang.[58]

It is interesting to note that D1R-mediated suppression effects in monkey PFC were not restricted to regular-spiking, putative pyramidal neurons. Fast-spiking, putative PV-positive interneurons were also suppressed upon agonist application (Fig. 5.3.4). This is particularly intriguing, given that in vitro slice studies of monkey[59] and rat[60] PFC have shown that fast-spiking cells are activated rather than inhibited by DA and D1R agonists. The findings suggest that network dynamics in the cognitively engaged PFC are very powerful and may override basic influences observed in quieter slice preparations.

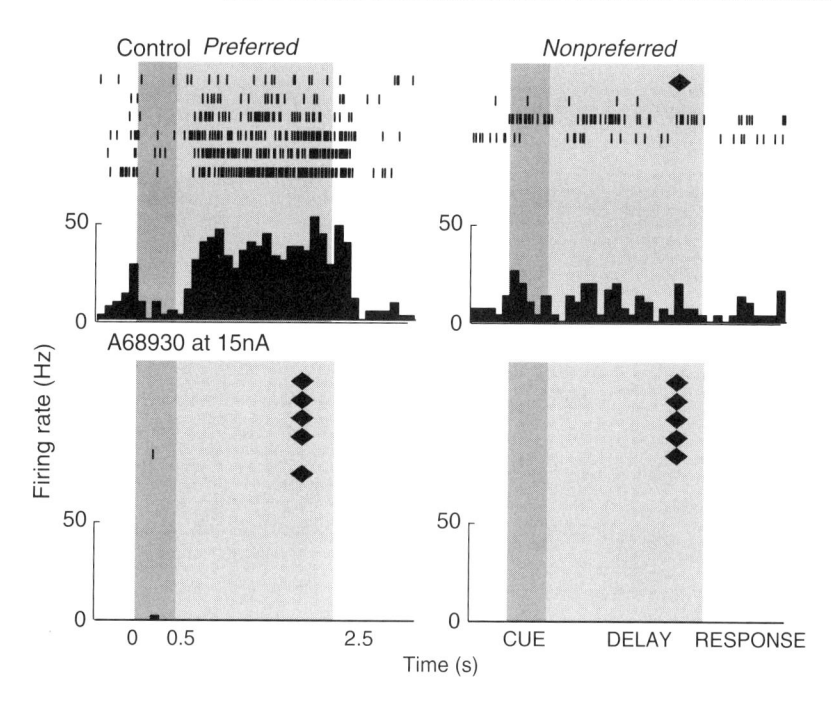

FIGURE 5.3.4. The D1R agonist, A68930, applied at 20 nA, completely suppressed a putative GABAergic fast-spiking unit engaged in the ODR task in both preferred (left panels) and nonpreferred (right panels) directions. Rasters and histograms are aligned at cue onset; black diamonds indicate trials where no spikes occurred. Similar suppression was observed in all other fast-spiking units (n = 9). Similar effects were observed in regular-spiking neurons.[55]

Mechanistic studies in rat PFC

Research in rodents has also observed inhibitory effects of DA on PFC physiology. Studies in anesthetized rats have examined the effects of stimulating DA cell bodies in the ventral tegmental area (VTA) on the prelimbic PFC cell firing patterns. Ventral tegmental area stimulation caused inhibition of PFC neurons, an effect that could be blocked by 6-hydroxydopamine (6-OHDA) lesions of monoaminergic neurons.[61,62] However, since the VTA also contributes a GABAergic input to the cortex, the cause of the inhibition could also be due to GABA mechanisms.[63] Depletion of DA stores reduced the number of inhibited PFC units to a third of the control level,[61,62] arguing that DAergic stimulation of PFC had inhibitory actions. VTA stimulation and DA application can gate long-latency responses in the mediodorsal thalamus–PFC bidirectional circuit while sparing short-latency responses.[64] The studies described above were not specific to any DA receptor subtype. However, neurochemical results suggest that D1R stimulation in PFC can suppress PFC network activity, as direct D1R agonist application reduced glutamate and GABA release measured by microdialysis in the medial PFC.[62]

Recordings from rat PFC slices have also provided evidence for D1R-mediated inhibitory actions. These studies show that fast-spiking interneurons are activated rather than inhibited by DA and D1R agonists.[59] D1 receptor stimulation depolarized interneurons by suppressing a voltage-independent "leak" K$^+$ current.[60] How do we explain the differences between the inhibitory effects of D1R on fast-spiking cells in behaving monkeys and the excitatory effects in rodent PFC slices? We hypothesize that the suppression observed following D1R stimulation in the behaving monkey PFC may arise from loss of recurrent excitation due to the collapse of network excitability. Thus, although the evoked excitability of these interneurons may be higher, this is not manifest in the overall effects of D1R agonists on the cortical column. The comparable effects of the D1R agonist on interneurons and pyramids appear to diminish the possibility that the suppressive effects we observed in the memory fields of regular-spiking pyramids were due to increased GABAergic inhibition, although a contribution of this component from non-fast-spiking interneurons cannot be ruled out. Thus, mechanisms intrinsic to pyramidal cells appear to be responsible for the dose-dependent suppression in response to D1R stimulation in cognitively engaged animals.

Although brain studies *in situ* generally have observed inhibitory effects of DA D1R stimulation, *in vitro* recordings from rat PFC slices have also revealed important excitatory effects on pyramidal cell physiology that may be obscured in the intact animal by endogenous arousal mechanisms. These studies have also observed complex effects of D1R stimulation on dendritic integration. Early investigations showed D1R-mediated excitatory or facilitatory effects on prefrontal pyramids.[65] A more recent study showed that D1R stimulation appears to increase the evoked excitability of pyramidal cells while reducing spontaneous excitability by acting in concert on the persistent Na^+ current, high-threshold dendritic Ca^{2+} spikes and a slowly inactivating K^+ current.[66] Another potential mechanism by which D1R activation could increase signal-to-noise characteristics in PFC neurons was examined,[67] wherein D1R agonists increased N-methyl-D-aspartate (NMDA) receptor–mediated synaptic currents while slightly reducing AMPA receptor currents, leading to enhancement of sustained excitatory postsynaptic potential (EPSP) trains. D1 receptor agonists also appear to shift the activation curve of the persistent Na^+ current to more hyperpolarized potentials, thus contributing further to signal-to-noise improvements.[68,69] D1 receptor actions also suppress the slowly inactivating K^+ current[70] through cAMP intracellular actions. D1 receptor stimulation increases L-type Ca^{2+} currents while attenuating N/P/Q type channels, thus altering integration along the dendritic arbor.[71] Finally, intracellular recordings from layer II/III PFC pyramidal neurons have shown that D1R stimulation may promote excitation by increasing the amplitude of excitatory postsynaptic currents[72] and reducing the amplitude of inhibitory postsynaptic currents in PFC cells[73] through a protein kinase A (PKA) mechanism. Thus, D1R stimulation may enhance excitability of PFC pyramidal cells but increase the threshold for response to neural inputs.

Synthesis of in vitro and in vivo findings

Figure 5.3.3 summarizes our current speculations on how varying levels of DA D1R stimulation alter PFC network firing in a monkey performing a spatial WM task. On the far left, delay-related firing for the preferred and nonpreferred directions is shown for a neuron under conditions of no D1R stimulation, that is, following high-dose SCH39166 administration. This neuron exhibits little firing, as it has lost the fundamental excitatory influences of D1R stimulation that have been especially evident in slice preparations *in vitro*. As we continue rightward along the inverted U, we observe the effects of a lower dose of D1R antagonist. We hypothesize that under these conditions there is adequate D1R stimulation for cell excitability but insufficient stimulation for proper sculpting of inputs. Thus, we have a "noisy" neuron. Under optimal conditions for spatial WM (the crest of the inverted U), moderate levels of D1R stimulation suppress firing to nonpreferred inputs, but firing to preferred inputs remains intact. Current data suggest that these sculpting actions likely involve D1R/cAMP opening of HCN channels on spines receiving dissimilar network inputs (see below), and may also involve other processes of dendritic integration (e.g., the closing of n or p Ca^{2+} channels, as described above) and D1R facilitation of GABAergic tuning.[74] Finally, at the far right of the inverted U, a neuron with excessive D1R stimulation has suppressed firing for all directions. Current data suggest that this global suppression arises from excessive cAMP opening of HCN channels on all spines, disconnecting the cell from all network inputs. As discussed above, these global suppressive effects may also arise from inhibition of presynaptic signaling[57] and possibly from inhibition of depolarizing currents (e.g., the CAN current). The relationship of the inverted U to arousal state and mental illness are discussed below.

Finally, it should be noted that D1R stimulation has additional and quite different effects on longer-term memory operations in PFC. Although this review has focused on spatial WM D1R influences on PFC network representation of spatial information on the order of seconds, D1R stimulation in PFC is also important for memory consolidation over much longer delays. Thus, when there are delays of 30 min or longer, D1R stimulation in PFC is needed to facilitate hippocampus–PFC interactions.[75] These long-term changes require increases in cAMP signaling[76] and appear to be similar to classic plasticity mechanisms, such as those studied in hippocampus.[77] Similarly, gradually learned changes in habits require cAMP stimulation in PFC just as they do in striatum (J.R. Taylor, personal communication). Indeed, inhibition of cAMP actions in PFC impairs long-term memory consolidation[76] but improves WM.[78] These findings may be related to D1R-cAMP-PKA enhancement of EPSCs in PFC.[72] Thus, the D1R mechanisms modulating WM operations are likely distinct from those necessary for long-term plastic changes in cerebral cortex.

D2R Actions

The D2 family of DA receptors has been implicated in schizophrenia, especially in the positive symptoms of the disorder, since the therapeutic potency of antipsychotic drugs directly correlates with their affinity for D2Rs.[79] The D2R agonist, bromocryptine, has been reported to facilitate spatial WM performance in normal human subjects with weak WM.[80] However, in contrast to the role of D1R in the mnemonic processes of the PFC, iontophoretic application of D2R antagonists has only a minor inhibitory effect or no effect at all on the memory fields of PFC neurons.[45] It is of particular interest that D2R binding and transcripts are most prominent in layer V, which contains the output pyramidal neurons of the cortex, indicating that D2R actions in cortical circuits might be associated with particular motor control functions.[81] Consistent with this hypothesis, we have observed that the saccade-related activity of PFC neurons can be modulated by D2R stimulation in monkeys performing the ODR task.[45] A subpopulation of PFC cells was active at around the time of saccade execution and was often highly selective for saccade direction. These neurons are said to have *saccadic activity* (also called *response-related firing*). Iontophoretic application of the D2R antagonist, raclopride, eliminated the pre- and perisaccadic activity for the preferred direction without affecting overall activity levels or responses to the nonpreferred direction.[45] Conversely, the D2R agonist, quinpirole, increased perisaccadic activity[45].

Although response-related firing of PFC neurons often precedes the saccade, and likely contributes to the signal for eye movement per se, there can also be response-related firing that occurs after the eye movement has started. These postsaccadic responses may serve as a corollary discharge. Sommer and Wurtz[82] have shown that when a saccade is produced, movement neurons in the superior colliculus send an efference copy (corollary discharge) of the motor command signal back to the frontal eye field through the mediodorsal thalamus. This efference copy lets the brain know that it has made a response; thus, timing of this firing is critical. We have observed that D2R stimulation alters the timing as well as the amplitude of saccadic activation, including the postsaccadic firing that may represent corollary discharge in the PFC. Figure 5.3.5 illustrates a PFC neuron with significant postsaccadic firing under control conditions; firing was initiated 110 ms after the saccade was initiated. Following application of the D2R agonist, quinpirole, the postsaccadic activity began at only 70 ms after saccade generation. This speeding of response-related firing could disrupt the timing of the message conveying that a motor command was self-generated. A deficit in corollary discharge has been proposed as a mechanism for explaining hallucinations in schizophrenia, whereby inadequate efference copies do not allow inner voices to be tagged as self-generated.[83] Thus, these D2R influences in PFC may be relevant to the generation of hallucinations.

D4R Actions

The D4R has high affinity for both DA and norepinephrine (NE), and its role in PFC function remains intriguing but poorly understood. Receptor blockade appears to be beneficial for the treatment of schizophrenia, as D4R antagonism is a common feature of atypical antipsychotic medications such as clozapine.[84] As described above, D4Rs are concentrated on GABAergic interneurons in primate PFC, and studies in rodent PFC slices indicate that stimulation of the receptor inhibits these cells.[85] Consistent with these data, our unpublished findings have indicated that iontophoresis of D4R antagonists generally decreases the activity of regular-spiking neurons (presumed pyramidal cells). Recordings from certain pairs of regular-spiking and fast-spiking neurons (presumed interneurons) indicated complementary responses to D4R blockade: D4R antagonists increased the activity of fast-spiking cells and decreased the activity of regular-spiking cells (M. Wang and A. Arnsten, unpublished, 2009). These results are consistent with endogenous D4R stimulation inhibiting the interneurons that normally suppress pyramidal cell firing. However, there were also exceptions to this rule—for example, fast-spiking neurons inhibited by D4R antagonists and some regular-spiking neurons excited by these compounds. Nonetheless, the predominant response to D4R antagonists was suppression of the PFC neuronal response. As polymorphisms of the D4R gene result in weaker D4R actions, subjects with this polymorphism may similarly have insufficient pyramidal cell activity in PFC, contributing to weaker executive functioning (e.g., in ADHD).[86] In contrast, D4R blockade may be helpful in schizophrenia if there is deficient GABAergic transmission in PFC.

DA DYNAMICALLY REGULATES PFC TUNING BASED ON THE STATE OF AROUSAL

Dopamine D1R stimulation appears to play a key role in gating network inputs to PFC pyramidal neurons. As described above, single-unit recordings from monkeys performing a spatial WM task indicate that D1R stimulation triggers cAMP signaling to suppress neuronal responses to nonpreferred directions, thus sharpening the spatial tuning of the cell.[55] Conversely, α2A-AR stimulation strengthens delay-related firing for the preferred direction via inhibition of cAMP–HCN signaling.[56] In the presence of cAMP, HCN channels pass both Na^+ and K^+ (the so-called h-current or I_h). Electron microscopy has localized HCN channels next

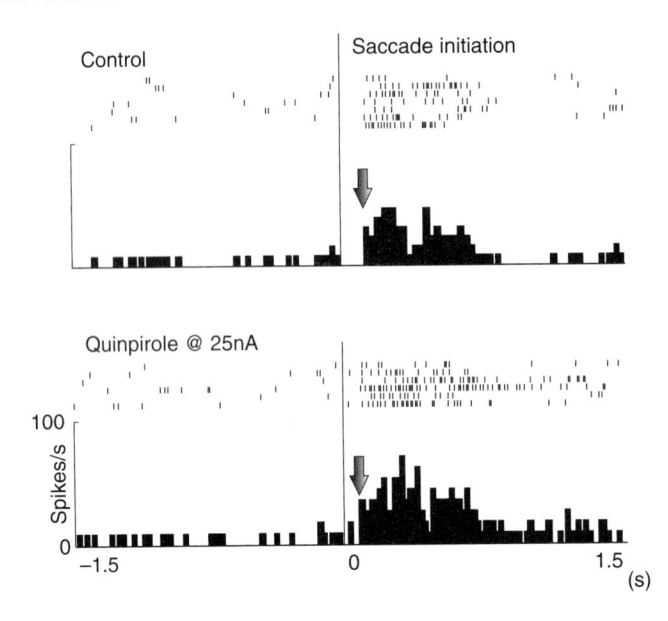

FIGURE 5.3.5. D2 receptor stimulation in PFC increases the amplitude of response-related firing[45] but also speeds response-related firing, as shown here. Under control conditions, this neuron fires *after* the animal has initiated the eye movement, consistent with corollary discharge. Iontophoresis of the D2R agonist, quinpirole, speeds the response. As the corollary discharge must be precisely timed, speeding of the response (e.g., due to loss of RGS4 regulation of D2R signaling in PFC) may contribute to disruptions of corollary discharge in patients with schizophrenia.

to α2A-ARs on dendritic spines of pyramidal neurons in layers II/III of the monkey PFC, the layers that participate in cortico-cortical networks.[56] As channels typically flank excitatory synapses on spine heads and appear on the spine neck region, they are in a key position for gating inputs to the spine. Physiological and behavioral data are consistent with cAMP–HCN signaling in spines gating synaptic inputs to the neuron, allowing dynamic regulation of the strength of those synapses.[56] Thus, when HCN channels are open in the presence of cAMP, inputs to the spines are weakened and delay-related firing and WM performance are impaired, whereas blockade of the HCN channels restores firing and improves cognitive performance.[56]

We propose that DA D1 and NE α2A receptor stimulation play complementary roles in regulating PFC network connectivity, with α2A-ARs increasing signals and D1Rs decreasing noise. This model is shown in Figure 5.3.1D and in Figure 5.3.6. The physiological results point to a structural organization whereby α2A-ARs are localized on spines receiving inputs from neurons with shared characteristics (e.g., other neurons that are also tuned to 90°). Thus, α2A-AR stimulation would inhibit cAMP–HCN signaling in these spines and strengthen the primary network representation of 90°. Conversely, D1Rs are localized on spines receiving inputs from neurons with dissimilar characteristics (e.g., nearby PFC neurons in principal sulcus tuned to 45°) and possibly from more distant neurons as well (e.g., neurons in ventrolateral PFC that respond primarily to faces; Fig. 5.3.6). D1 receptors would thus be positioned to regulate dynamically the breadth of network inputs to the neuron based on cognitive demands, reward evaluation, and the general state of arousal.

Optimal D1R Actions in PFC

Under optimal arousal conditions, when the subject is alert but not stressed, DA neurons fire to stimuli associated with reward, such as the visuospatial cues in the ODR task.[87,88] These firing patterns indicate that there would be a brief release of DA in PFC during and after the cue, which would provide necessary D1R actions during the delay period when the spatial position is maintained in WM. When an optimal number of D1Rs are stimulated, cAMP would be generated in those spines receiving nonpreferred inputs, weakening their effect on the cell and thus sharpening the spatial tuning of the neuron. This would improve WM for spatial location.

Under conditions where DA levels are depleted or inadequate, insufficient D1R stimulation would permit inappropriate inputs to influence the neuron, and the cell's response would be "noisy" with poor tuning. We have observed this firing pattern in response to a D1R

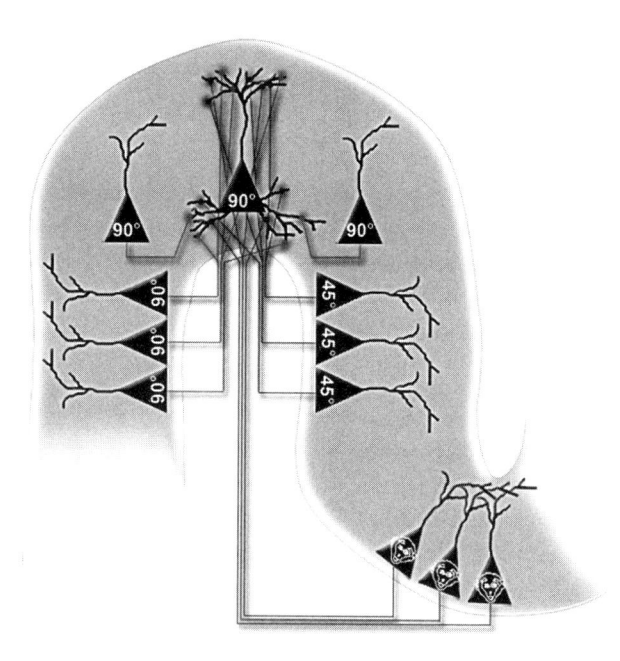

FIGURE 5.3.6. Hypothetical model illustrating how D1R and α2A-AR may dynamically regulate network connections to a pyramidal neuron in the principal sulcal PFC, in which α2A-ARs gate isodirectional inputs, and D1Rs gate contradirectional and other inputs. Please refer to color plate 5.3.6 for clarity. Red axons indicate network inputs from pyramidal neurons with shared spatial tuning properties (i.e., the best response to 90°). These inputs appear to be modulated by α2A-ARs, as receptor stimulation increases delay-related firing for the preferred direction. Spatial inputs from nearby PFC neurons with tuning for other spatial directions (e.g., 45°) are illustrated in blue. These network inputs appear to be gated by D1R stimulation. We hypothesize that other dissimilar inputs (e.g., ventral PFC neurons that respond to the memory of faces, shown in green) would similarly be gated by D1R stimulation. Thus, with greater D1R stimulation, the neuron would become more narrowly tuned. This would be helpful during some cognitive demands (e.g., spatial WM for a precise location) but would be detrimental under conditions where flexibility and breadth are required (e.g., set shifting, creative insights). (See Color Plate 5.3.6.)

antagonist, but it likely occurs under endogenous conditions in response to fatigue or boredom and in pathological conditions such as Parkinson's disease or ADHD. This may explain why a low dose of stimulant medication can normalize PFC cognitive abilities in patients with ADHD and enhance the efficiency of PFC blood-oxygen-level dependent (BOLD) activity in functional magnetic resonance imaging (fMRI) studies.[89–91]

It is intriguing to speculate that endogenous DA levels may vary in PFC according to cognitive demands. If a cognitive challenge requires broad tuning (e.g., representation of a large face extending from 45° to 90°), there would be less DA release, while cognitive challenges that require sharp tuning (e.g., only 90°, as in the ODR task) would evoke greater DA release, perhaps because the more difficult cognitive challenge is more arousing. Although the PFC projects back to the DA cell bodies in the midbrain,[92] it is not clear whether such a circuit would support this fine-grained regulation. It is more likely that DA firing and release would be regulated by arousal and reward, whereby tonic-firing levels would be influenced by the arousal state, and phasic responses to stimuli would be based on the association of those stimuli with reward. Increasing levels of DA release with increasing levels of arousal/reward may explain some of the interactions between arousal state and cognitive performance. With increasing levels of arousal, there would be increasing DA release in the PFC and increasingly fewer inputs influencing the cell. This model fits with classic human behavior studies showing that attentional focus becomes increasingly narrow (and increasingly labile) with increasing arousal.[93] This narrowing of inputs would be helpful if cognitive demands at that moment required very focal memory (e.g., the precise spatial memory demands in the ODR task) but would not be helpful if one needed broad or flexible inputs (e.g., as occurs in set switching). Indeed, Crofts et al.[94] have found that DA is needed to establish an attentional set but that it is actually helpful to have low DA levels in PFC when switching attentional set. Under optimal conditions, endogenous DA levels may be dynamically regulated to meet changing cognitive demands. This flexibility may be lost if DA is depleted or an individual is stressed, causing constant high levels of release.[95] This model may also explain why creative insights happen when one is relaxed but not stressed,[96] as D1R stimulation during even mild stress may shunt the broad inputs onto PFC neurons that are likely necessary for novel, insightful solutions. This may also explain why stimulant medications can limit creative thought and mental flexibility if the dose is too high.[97] Similarly, D1R agonists may not be ideal cognitive enhancers, as these drugs would provide static levels of D1R stimulation irrespective of cognitive demands.

It is also interesting to speculate that the high DA levels in the motor and premotor cortices of certain species perform a similar function for complex motor abilities. Dopamine does not innervate the motor cortices in simple species such as rodents that do not have intricate use of fingers. In contrast, in rhesus monkeys and in humans, the densest DA projections are to the motor cortices: the supplementary motor area, the premotor cortex, and the primary motor cortex.[98–100] Both rhesus monkeys and humans have control over individual digits and are able to perform intricate finger movements. Might DA D1R stimulation be needed in motor cortices for dynamic regulation of inputs to coordinate changing representations of fingers, similar to what is seen in prefrontal areas with dynamic representations of spatial positions? There have been very few studies of DA actions on neuronal responses in motor cortices (e.g.,[101]), and this will be an important area for future research.

Excessive D1R Stimulation During Stress

In contrast to the essential beneficial effects of DA under nonstress conditions, WM performance is markedly impaired by exposure to stress.[49–51] Even mild, uncontrollable stress induces high levels of DA release in PFC.[95] Recent data indicate that the PFC determines whether a subject feels in control over a stressor and inhibits brainstem stress responses if there is even the illusion of control.[102] With mild, uncontrollable stress, DA is released in the PFC, but less so in striatum.[95] High DA levels in PFC would engage large numbers of D1Rs and may generate high levels of cAMP that would diffuse throughout the dendrite, weakening all network inputs to the neuron. Physiological evidence is consistent with this model; high levels of D1R stimulation or cAMP induce a collapse in network firing that is restored by inhibition of cAMP or blockade of HCN channels[55,56]; N. J. Gamo, A. Arnsten, and M. Wang, unpublished data, 2009). Behavioral findings are also consistent with this model, as stress-induced WM impairment is rescued by agents that inhibit cAMP[103] or block HCN channels (B. Ramos, N. J. Gamo and A. Arnsten, unpublished data, 2009).

In normal subjects, the stress response would be inhibited by enzymes that catabolize cAMP, that is, phosphodiesterases such as PDE4B. This important negative feedback regulation is controlled by DISC1 (Disrupted In SChizophrenia), which "senses" high cAMP levels and unleashes PDE4B activity.[104] DISC1 has recently been linked to families with a high incidence of mental illness, and is important both to the development of PFC and to phosphodiesterase regulation of cAMP signaling.[105–107] In these families, disc1 has a loss-of-function translocation, which would make them especially vulnerable to cAMP buildup during stress. It is highly relevant in this regard that patients with severe mental illness are particularly susceptible to stress exposure.[108]

RELEVANCE TO MENTAL ILLNESS

Even quite subtle changes in DA transmission in PFC play a large role in several neuropsychiatric disorders in relation to both etiology and treatment.

Parkinson's Disease

The hallmark of Parkinson's disease is degeneration of the midbrain DA system. Although most clinicians focus on the motor deficits arising from loss of DA in striatum, the DA (and NE) cells that project to the PFC degenerate as well. Patients with Parkinson's disease have cognitive deficits that likely arise from loss of catecholamines in both caudate and PFC.[109] Unfortunately, the doses of DAergic medications needed to normalize motor function (i.e., to compensate for the degenerating nigrostriatal DA system) are too high for restoration of the cognitive functioning of the PFC. Thus, medications such as L-DOPA and apomorphine can actually make cognitive deficits worse.[110]

Attention Deficit Hyperactivity Disorder

ADHD is another disorder associated with inadequate DA in PFC. Although imaging studies are unable to visualize the delicate catecholamine innervation of PFC, one study suggests that there is a reduction of catecholamine terminals in the PFC of adults with ADHD.[111] Imaging techniques can be used to visualize reliably the more extensive DA innervation of striatum. Positron emission tomography (PET) studies of adult patients with ADHD have shown evidence of reduced DA release in the striatum.[112] As DA loss in striatum leads to reduced locomotor activity,[113] while DA loss in PFC leads to hyperactivity,[114] it is likely that the latter predominates in ADHD. In some patients, the reduction in catecholamine actions may have a genetic basis, for the disorder has been associated with a number of catecholamine-related molecules, such as D1, D5, and D4 receptors and the DAT.[115] However, as reduced

DAT function would lead to increased, not decreased, DA levels, and as there are relatively few DATs in PFC, this relationship is not clear. Treatments for ADHD also suggest a role for DA, as the stimulant medications normalize PFC functions in patients with ADHD. However, it should be noted that (1) stimulants also improve PFC cognitive function in normal individuals[116] and (2) stimulants increase both NE and DA in PFC, and actually have more effect on NE than on DA when studied at therapeutic doses.[117] There are also associations with NE genes; for example, changes in the gene encoding for DA β-hydroxylase, the rate-limiting enzyme in NE synthesis, are associated with impairments in PFC executive functions.[118] Finally, ADHD can be treated with agents that mimic NE or selectively block the NE transporter (guanfacine and atomoxetine, respectively), indicating that NE is as important as DA in the etiology and treatment of ADHD.

Drug Abuse

Much of the research on drug abuse has focused on DAergic mechanisms in the nucleus accumbens. Drugs of abuse induce high levels of DA release in accumbens and markedly alter affective habits, increasing drug craving and compulsive drug seeking.[119] However, these drugs also cause high levels of DA release in PFC, where they impair its function and cause loss of inhibitory control.[120] Thus, subjects are unable to inhibit their compulsions and to guide their behavior effectively. An excellent review of this topic can be found in Jentsch and Taylor.[120] Alpha-2 adrenergic receptor agonists used to treat drug abuse and withdrawal (e.g., clonidine, lofexidine) may contribute to improvement by strengthening PFC function, as well as decreasing catecholamine overflow during withdrawal.

Schizophrenia

Dopamine changes in the PFC of patients with schizophrenia are complex. Postmortem evaluations show subtle reductions in tyrosine hydroxylase immunostaining in layer VI only, which likely reflect reduced DA input to the PFC (but could result from DA overflow decreasing the expression of this synthetic enzyme).[121] Imaging studies are unable to visualize the delicate DA input to PFC, but PET studies of D1R occupancy indicate increased numbers of functional D1Rs, which correlate with impaired PFC cognitive function.[122] Receptor up-regulation may be a compensatory change in response to reduced DA levels. Support for this hypothesis arises from the finding that amphetamine can improve WM performance in patients with schizophrenia who are taking neuroleptic medications,[123] although this could also result from increased NE release in PFC, which is most sensitive to stimulant actions.[117] Genetic findings are also consistent with inadequate DA actions in the PFC of patients with schizophrenia. Individuals with a polymorphism that substitutes methionine for valine (Val158Met) in the catabolic enzyme, catechol-O-methyltransferase (COMT), have weaker degradation of DA. Both normal subjects and patients with schizophrenia with the methionine substitution have more efficient PFC function than those with the native COMT genotype.[124] Taken together, all of these findings are consistent with reduced DA actions in the PFC of patients with schizophrenia.

However, patients with schizophrenia also show signs of exaggerated DA-like actions. For example, they show heightened sensitivity to stress, and stress exposure can greatly exacerbate symptoms.[125,126] Similarly, stimulant use can worsen or elicit symptoms, and high doses of stimulants can mimic symptoms of the illness.[79,127] It is possible that schizophrenia is associated with inadequate DA release/actions under basal conditions, but once DA is released—for example, under conditions of stress—the response is exaggerated by the increased numbers of D1Rs in PFC. Thus, the shape of the inverted-U D1R response would be altered in these patients, as illustrated in Figure 5.3.7. The detrimental effects of excessive D1R stimulation would also be exacerbated in patients with loss-of-function mutations in *disc*1. As DISC1 normally provides negative feedback on cAMP signaling,[104] loss of DISC1 function would dysregulate D1R–cAMP signaling, leading to network disconnection and loss of PFC network firing. Thus, patients may quickly go from too little to too much, without optimal regulation of PFC network activity.

Finally, as described above, D2Rs regulate response-related firing of PFC neurons, which may include modulation of efference copy and/or corollary discharge (i.e., the messages relaying that the brain has elicited an action). Studies of patients with schizophrenia have indicated that a weakened efference copy emanating from the PFC to Wernicke's cortical area may contribute to auditory hallucinations (i.e., the inability to tag an inner voice as self-generated).[83] As D2R stimulation alters the timing and amplitude of the PFC response signals, altered D2R signaling in schizophrenia may disrupt these key signals and contribute to hallucinations. For example, RGS4, a key regulator of G protein signaling that likely regulates D2R signal transduction, is greatly reduced in the PFC of patients with schizophrenia,[128] which could disrupt the timing of the efference copy emanating from the PFC. Although this idea is currently speculative, it deserves further research.

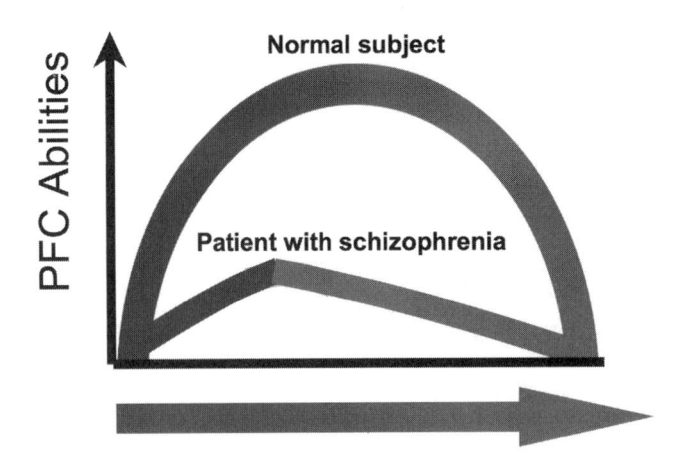

FIGURE 5.3.7. Hypothetical illustration of the D1R inverted U in normal individuals versus patients with schizophrenia. Patients may have reduced levels of DA in PFC, but may be more responsive to the detrimental effects of D1R due to up-regulation of D1Rs and loss of enzymes that hold intracellular stress pathways in check (e.g., loss of DISC1). Thus, they may have reduced D1R beneficial actions, as well as potentiation of detrimental actions. (See Color Plate 5.3.7.)

FUTURE DIRECTIONS

Finally, very recent data suggest that DA cells in midbrain may have very heterogeneous responses to rewarding vs. aversive stimuli, with a subset of cells increasing their response to aversive stimuli[129]. As distinct pools of DA neurons project to selective regions of frontal lobe[99], it is possible that DA may be released under differing conditions in each subregion, for example, DA neurons that fire to reward projecting to orbital PFC and/or the supplementary motor area, while a separate group of neurons that fire to aversive stimuli could project to anterior cingulate to enhance error detection. Alternatively, DA may be released in a PFC subregion under either rewarding or punishing conditions, but the DA innervation may terminate on distinct layers or subcellular compartments, for example, reward-responsive DA neurons terminating near PFC spines, and aversive-responsive DA neurons terminating near the dendritic stem. Future research will need to align specific midbrain dopamine cell response profiles with their corresponding actions in terminal cortical regions.

SUMMARY

Dopamine's powerful influence on PFC networks remains an intriguing topic of research. Although much has been learned in the last three decades, the complexity of DA actions demands much further research.

Studies in monkeys performing higher cognitive tasks provide a unique opportunity to observe the modulatory effects of DA on PFC networks engaged in cognitive operations. The data from these studies have been invaluable in illuminating DA influences on WM, including powerful effects on human cognitive performance.

ACKNOWLEDGMENTS

Supported by P50MH068789, NIA MERIT AG06036, the Kavli Neuroscience Institute at Yale, and a NARSAD Distinguished Investigator Award to AFTA.

REFERENCES

1. Brozoski T, Brown RM, Rosvold HE, et al. Cognitive deficit caused by regional depletion of dopamine in prefrontal cortex of rhesus monkey. *Science*. 1979;205:929–931.
2. Fuster JM. The prefrontal cortex, mediator of cross-temporal contingencies. *Hum Neurobiol*. 1985;4:169–
3. Goldman-Rakic PS. The prefrontal landscape: implications of functional architecture for understanding human mentation and the central executive. *Philos Trans R Soc Lond*. 1996;351:1445–
4. Robbins TW. Dissociating executive functions of the prefrontal cortex. *Philos Trans R Soc Lond*. 1996;351:1463.
5. Goldman-Rakic PS. Circuitry of the primate prefrontal cortex and the regulation of behavior by representational memory In: Plum F, ed. *Handbook of Physiology: The Nervous System,*

Higher Functions of the Brain, Vol. V. Bethesda, MD: American Physiological Society; 1987:373–417.

6. Barbas H, Medalla M, Alade O, et al., Relationship of prefrontal connections to inhibitory systems in superior temporal areas in the rhesus monkey. *Cereb Cortex.* 2005;15:1356–

7. Dias R, Robbins TW, Roberts AC. Dissociable forms of inhibitory control within prefrontal cortex with an analog of the Wisconsin Card Sort Test: restriction to novel situations and independence from "on-line" processing. *J Neurosci.* 1997;17:9285–

8. Kolb B. Prefrontal cortex. In: Kolb B, Tees RC, eds. *The Cerebral Cortex of the Rat.* Cambridge, MA: MIT Press; 1990:437–458.

9. Goldman-Rakic PS. Cellular basis of working memory. *Neuron.* 1995;14:477–

10. Rao SG, Williams GV, Goldman-Rakic PS. Destruction and creation of spatial tuning by disinhibition: GABA(A) blockade of prefrontal cortical neurons engaged by working memory. *J Neurosci.* 2000;20:485–

11. Miller EK, Erickson CA, Desimone R, et al. Neural mechanisms of visual working memory in prefrontal cortex of the macaque. *J Neurosci.* 1996;16:5154–

12. Wallis JD, Anderson KC, Miller EK. Single neurons in prefrontal cortex encode abstract rules. *Nature.* 2001;411:953–

13. Funahashi S, Chafee MV, Goldman-Rakic PS. Prefrontal neuronal activity in rhesus monkeys performing a delayed anti-saccade task. *Nature.* 1993;365:753–

14. Pasupathy A, Miller EK. Different time courses of learning-related activity in the prefrontal cortex and striatum. *Nature.* 2005; 433:873–

15. Lee D, Rushworth MF, Walton ME, et al. Functional specialization of the primate frontal cortex during decision making. *J Neurosci.* 2007;27:8170–

16. Goldman-Rakic PS, Bergson C, Krimer LS, et al. The primate mesocortical dopamine system In: Bloom FE, Björklund A, Hökfelt T, eds. *Handbook of Chemical Neuroanatomy: The Primate Nervous System.* Amsterdam: Elsevier; 1999:403–428.

17. Lewis DA, Melchitzky DA, Sesack SR, et al., Dopamine transporter immunoreactivity in monkey cerebral cortex: regional, laminar, and ultrastructural localization. *J Comp Neurol.* 2001;432:119–

18. Goldman-Rakic PS, Lidow MS, Smiley JF, et al. The anatomy of dopamine in monkey and human prefrontal cortex. *J Neural Transm Suppl.* 1992;36:163–

19. Smiley JF, Goldman-Rakic PS. Heterogeneous targets of dopamine synapses in monkey prefrontal cortex demonstrated by serial section electron microscopy: a laminar analysis using the silver-enhanced diaminobenzidine sulfide (SIDS) immunolabeling technique. *Cereb Cortex.* 1993;3:233–

20. Goldman-Rakic PS, Leranth C, Williams SM, et al. Dopamine synaptic complex with pyramidal neurons in primate cerebral cortex. *Proc Natl Acad Sci USA.* 1989;86:9015–

21. Sesack SR, Hawrylak VA, Melchitzky DS, et al. Dopamine innervation of a subclass of local circuit neurons in monkey prefrontal cortex: ultrastructural analysis of tyrosine hydroxylase and parvalbumin immunoreactive structures. *Cereb Cortex.* 1998;8:614–

22. Sesack SR, Carr DB, Omelchenko N, et al. Anatomical substrates for glutamate–dopamine interaction: evidence for specificity of connections and extrasynaptic actions. *Ann NY Acad Sci.* 2003;1003:36–

23. Lewis DA, Hashimoto T, Volk DW. Cortical inhibitory neurons and schizophrenia. *Nat Rev Neurosci.* 2005;6:312–

24. Pickel VM, Garzón M, Mengual E. Electron microscopic immunolabeling of transporters and receptors identifies transmitter-specific functional sites envisioned in Cajal's neuron. *Prog Brain Res.* 2002;136:145–

25. Missale C, Nash SR, Robinson SW, et al. Dopamine receptors: from structure to function. *Physiol Rev.* 1998;78:189–

26. Smiley JF, Levey AI, Ciliax BJ, et al. D1 dopamine receptor immunoreactivity in human and monkey cerebral cortex: predominant and extrasynaptic localization in dendritic spines. *Proc Natl Acad Sci USA.* 1994;91:5720–

27. Goldman-Rakic PS, Muly EC, Williams GV. D_1 receptors in prefrontal cells and circuits. *Brain Res Rev.* 2000;31:295–

28. Bergson C, Mrzljak L, Smiley JF, et al., Regional, cellular and subcellular variations in the distribution of D1 and D5 dopamine receptors in primate brain. *J Neurosci.* 1995;15:7821–

29. Paspalas CD, Goldman-Rakic PS. Receptor compartmentalization for defining input-specificity of dopamine volumetric signaling: a study of D1, D2 and D5 receptor subtypes in primate prefrontal cortex. *Soc Neurosci Abstr.* 2004;30:2779–

30. Paspalas CD, Goldman-Rakic PS. Microdomains for dopamine volume neurotransmission in primate prefrontal cortex. *J Neurosci.* 2004;24:5292–

31. Delmas P, Brown DA. Junctional signaling microdomains: bridging the gap between the neuronal cell surface and Ca^{2+} stores. *Neuron.* 2002;36:787–

32. Muly EC, Szigeti K, Goldman-Rakic PS. D_1 receptor in interneurons of macaque prefrontal cortex: distribution and subcellular localization. *J Neurosci.* 1998;18:10553–

33. Paspalas CD, Goldman-Rakic PS. Presynaptic D1 dopamine receptors in primate prefrontal cortex: target-specific expression in the glutamatergic synapse. *J Neurosci.* 2005;25:1260–

34. Gao WJ, Goldman-Rakic PS. Selective modulation of excitatory and inhibitory microcircuits by dopamine. *Proc Natl Acad Sci USA.* 2003;100:2836–

35. Bordelon-Glausier JR, Khan ZU, Muly EC. Quantification of D_1 and D_5 dopamine receptor localization in layers I, III and V of *Macaca mulatta* prefrontal cortical area 9: coexpression in dendritic spines and axon terminals. *J Comp Neurol.* 2008;508:893–

36. Paspalas CD, Rakic P, Goldman-Rakic PS. Internalization of D2 dopamine receptors is clathrin-dependent and select to dendro–axonic appositions in primate prefrontal cortex. *Eur J Neurosci.* 2006;24:1395–

37. Koh PO, Undie AS, Kabbani N, et al. Up-regulation of neuronal calcium sensor-1 (NCS-1) in the prefrontal cortex of schizophrenic and bipolar patients. *Proc Natl Acad Sci USA.* 2003;100:313–

38. Wang H, Pickel VM. Dopamine D2 receptors are present in prefrontal cortical afferents and their targets in patches of the rat caudate-putamen nucleus. *J Comp Neurol.* 2002; 442:392–

39. Negyessy L, Goldman-Rakic PS. Subcellular localization of the dopamine D_2 receptor and coexistence with the calcium-binding protein neuronal calcium sensor-1 in the primate prefrontal cortex. *J Comp Neurol.* 2005;488:464–

40. Meador-Woodruff JH, Damask SP, Watson SJJ. Differential expression of autoreceptors in the ascending dopamine systems of the human brain. *Proc Natl Acad Sci USA.* 1994;91:8297–

41. Khan ZU, Mrzljak L, Gutierrez A, et al. Prominence of the dopamine D2 short isoforms in dopaminergic pathways. *Proc Natl Acad Sci USA.* 1998;95:7731–

42. Khan ZU, Koulen P, Rubinstein M, et al. An astroglia-linked dopamine D2-receptor action in prefrontal cortex. *Proc Natl Acad Sci USA.* 2001;98:1964–

43. Mrzljak L, Bergson C, Pappy M, et al. Localization of dopamine D4 receptors in GABAergic neurons of the primate brain. *Nature.* 1996;381:245–
44. Funahashi S, Bruce CJ, Goldman-Rakic PS. Mnemonic coding of visual space in the monkey's dorsolateral prefrontal cortex. *J Neurophysiol.* 1989;61:331–
45. Wang M, Vijayraghavan S, Goldman-Rakic PS. Selective D2 receptor actions on the functional circuitry of working memory. *Science.* 2004;303:853–
46. Sawaguchi T, Goldman-Rakic PS. The role of D1-dopamine receptors in working memory: local injections of dopamine antagonists into the prefrontal cortex of rhesus monkeys performing an oculomotor delayed response task. *J Neurophysiol.* 1994;71:515–
47. Roth RH, Tam S-Y, Ida Y, et al. Stress and the mesocorticolimbic dopamine systems. *Ann NY Acad Sci.* 1988;537:138–
48. Arnsten AFT, Goldman-Rakic PS. Stress impairs prefrontal cortex cognitive function in monkeys: role of dopamine. *Soc Neurosci Abstr.* 1990;16:164–
49. Murphy BL, Arnsten AFT, Goldman-Rakic PS, et al. Increased dopamine turnover in the prefrontal cortex impairs spatial working memory performance in rats and monkeys. *Proc Natl Acad Sci USA.* 1996;93:1325–
50. Arnsten AFT, Goldman-Rakic PS. Noise stress impairs prefrontal cortical cognitive function in monkeys: evidence for a hyperdopaminergic mechanism. *Arch Gen Psychiatry.* 1998;55:362–
51. Arnsten AFT. The biology of feeling frazzled. *Science.* 1998;280:1711–
52. Zahrt J, Taylor JR, Mathew RG, et al., Supranormal stimulation of dopamine D1 receptors in the rodent prefrontal cortex impairs spatial working memory performance. *J Neurosci.* 1997;17:8528–
53. Sawaguchi T, Matsumura M, Kubota K. Dopamine enhances the neuronal activity of spatial short-term memory task in the primate prefrontal cortex. *Neurosci Res.* 1988;5:465–
54. Williams GV, Goldman-Rakic PS. Blockade of dopamine D1 receptors enhances memory fields of prefrontal neurons in primate cerebral cortex. *Nature.* 1995;376:572–
55. Vijayraghavan S, Wang M, Birnbaum SG, et al. Inverted-U dopamine D1 receptor actions on prefrontal neurons engaged in working memory. *Nat Neurosci.* 2007;10:376–
56. Wang M, Ramos BP, Paspalas CD, et al. Alpha2A-adrenoceptor stimulation strengthens working memory networks by inhibiting cAMP-HCN channel signaling in prefrontal cortex. *Cell.* 2007;129:397–
57. Gao WJ, Krimer LS, Goldman-Rakic PS. Presynaptic regulation of recurrent excitation by D1 receptors in prefrontal circuits. *Proc Natl Acad Sci USA.* 2001;98:295–
58. Tegner J, Compte A, Wang XJ. The dynamical stability of reverberatory neural circuits. *Biol Cybern.* 2002;87:471–
59. Kroner S, Krimer LS, Lewis DA, et al., Dopamine increases inhibition in the monkey dorsolateral prefrontal cortex through cell type-specific modulation of interneurons. *Cereb Cortex.* 2007;17:1020–
60. Gorelova NA, Seamans JK, Yang CR. Mechanisms of dopamine activation of fast-spiking interneurons that exert inhibition in rat prefrontal cortex. *J Neurophysiol.* 2002;88:150–
61. Ferron A, Thierry AM, Le Douarin C, et al. Inhibitory influence of the mesocortical dopaminergic system on spontaneous activity or excitatory response induced from the thalamic mediodorsal nucleus in the rat medial prefrontal cortex. *Brain Res.* 1984;302:257–
62. Abekawa T, Ohmori T, Ito K, et al. D1 dopamine receptor activation reduces extracellular glutamate and GABA concentrations in the medial prefrontal cortex. *Brain Res.* 2000;867:250–
63. Pirot S, Godbout R, Mantz J, et al., Inhibitory effects of ventral tegmental area stimulation on the activity of prefrontal cortical neurons: evidence for the involvement of both dopaminergic and GABAergic components. *Neuroscience.* 1992;49:857–
64. Pirot S, Glowinski J, Thierry AM. Mediodorsal thalamic evoked responses in the rat prefrontal cortex: influence of the mesocortical DA system. *Neuroreport.* 1996;7:1437–
65. Penit-Soria J, Audinat E, Crepel F. Excitation of rat prefrontal cortical neurons by dopamine: an in vitro electrophysiological study. *Brain Res.* 1987;425:263–
66. Yang CR, Seamans JK. Dopamine D1 receptor actions in layers V–VI rat prefrontal cortex neurons in vitro: modulation if dendritic-somatic signal integration. *J Neurosci.* 1996;16:1922–
67. Seamans JK, Durstewitz D, Christie BR, et al. Dopamine D1/D5 receptor modulation of excitatory synaptic inputs to layer V prefrontal cortex neurons. *Proc Natl Acad Sci USA.* 2001;98:301–
68. Gorelova NA, Yang CR. Dopamine D1/D5 receptor activation modulates a persistent sodium current in rat prefrontal cortical neurons in vitro. *J Neurophysiol.* 2000;84:75–
69. Franceschetti S, Taverna S, Sancini G, et al. Protein kinase C-dependent modulation of Na$^+$ currents increases the excitability of rat neocortical pyramidal neurones. *J Physiol.* 2000;528:291–
70. Dong Y, White FJ. Dopamine D1-class receptors selectively modulate a slowly inactivating potassium current in rat medial prefrontal cortex pyramidal neurons. *J Neurosci.* 2003;23:2686–
71. Young CE, Yang CR. Dopamine D1/D5 receptor modulates state-dependent switching of soma-dendritic Ca^{2+} potentials via differential protein kinase A and C activation in rat prefrontal cortical neurons. *J Neurosci.* 2004;24:8–
72. Gonzalez-Islas C, Hablitz JJ. Dopamine enhances EPSCs in layer II–III pyramidal neurons in rat prefrontal cortex. *J Neurosci.* 2003;23:867–
73. Gonzalez-Islas C, Hablitz JJ. Dopamine inhibition of evoked IPSCs in rat prefrontal cortex. *J Neurophysiol.* 2001;86:2911–
74. Seamans JK, Gorelova N, Daniel D, et al. Bidirectional dopamine modulation of GABAergic inhibition in prefrontal cortical pyramidal neurons. *J Neurosci.* 2001;21:3628–
75. Seamans JK, Floresco SB, Phillips AG. D1 receptor modulation of hippocampal-prefrontal cortical circuits integrating spatial memory with executive functions in the rat. *J Neurosci.* 1998;18:1613–
76. Runyan JD, Dash PK. Distinct prefrontal molecular mechanisms for information storage lasting seconds versus minutes. *Learn Mem.* 2005;12:232–
77. Kandel ER. The molecular biology of memory storage: a dialogue between genes and synapses. *Science.* 2001;294:1030–
78. Ramos B, Birnbaum SB, Lindenmayer I, et al. Dysregulation of protein kinase A signaling in the aged prefrontal cortex: new strategy for treating age-related cognitive decline. *Neuron.* 2003;40:835–
79. Seeman P. Dopamine receptors and the dopamine hypothesis of schizophrenia. *Synapse.* 1987;1:133–
80. Kimberg DY, D'Esposito M, Farah MJ. Effects of bromocriptine on human subjects depend on working memory capacity. *Neuroreport.* 1997;8:3581–
81. Lidow M, Goldman-Rakic P, Rakic P, et al. Dopamine D2 receptors in the cerebral cortex: distribution and

pharmacological characterization with 3H-raclopride. *Proc Natl Acad Sci USA.* 1989;86:6412–
82. Sommer MA, Wurtz RH. Brain circuits for the internal monitoring of movements. *Annu Rev Neurosci.* 2008;31:317–
83. Ford JM, Mathalon DH, Whitfield S, et al. Reduced communication between frontal and temporal lobes during talking in schizophrenia. *Biol Psychiatry.* 2002;51:485–
84. Van Tol HHM, Bunzow JR, Guan H-C, et al. Cloning of the gene for a human dopamine D4 receptor with high affinity for the antipsychotic clozapine. *Nature.* 1991;350:610–
85. Wang X, Zhong P, Yan Z. Dopamine D4 receptors modulate GABAergic signaling in pyramidal neurons of prefrontal cortex. *J Neurosci.* 2002;22:9185–
86. LaHoste GJ, Swanson JM, Wigal SB, et al. Dopamine D4 receptor gene polymorphism is associated with attention deficit hyperactivity disorder. *Mol Psychiatry.* 1996;1:121–
87. Schultz W, Apicella P, Ljungberg T. Responses of monkey dopamine neurons to reward and conditioned stimuli during successive steps of learning a delayed response task. *J Neurosci.* 1993;13–
88. Schultz W. The phasic reward signal of primate dopamine neurons. *Adv Pharmacol.* 1998;42:686–
89. Mehta MA, Owen AM, Sahakian BJ, et al. Methylphenidate enhances working memory by modulating discrete frontal and parietal lobe regions in the human brain. *J Neurosci.* 2000;20:RC651–
90. Mehta MA, Goodyer IM, Sahakian B. Methylphenidate improves working memory and set-shifting in AD/HD: relationships to baseline memory capacity. *J Child Psychol Psychiatry.* 2004;45:293–
91. Turner DC, Blackwell AD, Dowson JH, et al. Neurocognitive effects of methylphenidate in adult attention-deficit/hyperactivity disorder. *Psychopharmacology.* 2005;178:286–
92. Carr DB, Sesack SR. Projections from the rat prefrontal cortex to the ventral tegmental area: target specificity in the synaptic associations with mesoaccumbens and mesocortical neurons. *J Neurosci.* 2000;20:3864–
93. Hockey GRJ. Effect of loud noise on attentional selectivity. *Q J Exp Psychol.* 1970;22:28–
94. Crofts HS, Dalley JW, Collins P, et al. Differential effects of 6-OHDA lesions of the prefrontal cortex and caudate nucleus on the ability to acquire an attentional set. *Cereb Cortex.* 2001;11:1015–
95. Deutch AY, Roth RH. The determinants of stress-induced activation of the prefrontal cortical dopamine system. *Prog Brain Res.* 1990;85:367–
96. Subramaniam K, Kounios J, Parrish TB, et al. A brain mechanism for facilitation of insight by positive affect. *J Cogn Neurosci.*:2009; 21;415.
97. Tannock R, Schachar R. Methylphenidate and cognitive perseveration in hyperactive children. *J Child Psychol Psychiatry.* 1992;33:1217–
98. Berger B, Gaspar P, Verney C. Dopaminergic innervation of the cerebral cortex: unexpected differences between rodents and primates. *Trends Neurosci.* 1991;14:21–
99. Williams SM, Goldman-Rakic PS. Widespread origin of the primate mesofrontal dopamine system. *Cereb Cortex.* 1998;8:321–
100. Lewis DA. The catecholamine innervation of primate cerebral cortex In: Solanto MV, Arnsten AFT, Castellanos FX, eds. *Stimulant Drugs and ADHD: Basic and Clinical Neuroscience.* New York, NY: Oxford University Press; 2001:77–103.
101. Matsumura M, Sawaguchi T, Kubota K. Modulation of neuronal activities by iontophoretically applied catecholamines and acetylcholine in the primate motor cortex during a visual reaction-time task. *Neurosci Res.* 1990;8:138–
102. Amat J, Paul E, Zarza C, et al. Previous experience with behavioral control over stress blocks the behavioral and dorsal raphe nucleus activating effects of later uncontrollable stress: role of the ventral medial prefrontal cortex. *J Neurosci.* 2006;26:13264–
103. Birnbaum SG, Podell DM, Arnsten AFT. Noradrenergic alpha-2 receptor agonists reverse working memory deficits induced by the anxiogenic drug, FG7142, in rats. *Pharmacol Biochem Behav.* 2000;67:397–
104. Millar JK, Mackie S, Clapcote SJ, et al. Disrupted in schizophrenia 1 and phosphodiesterase 4B: towards an understanding of psychiatric illness. *J Physiol.* 2007;584:401–
105. Millar JK, Pickard BS, Mackie S, et al. DISC1 and PDE4B are interacting genetic factors in schizophrenia that regulate cAMP signaling. *Science.* 2005;310:1187–
106. Millar JK, Wilson-Annan JC, Anderson SL, et al. Disruption of two novel genes by a translocation co-segregating with schizophrenia. *Hum Mol Genet.* 2000;9:1415–
107. Ishizuka K, Paek M, Kamiya A, et al. A review of Disrupted-In-Schizophrenia-1 (DISC1): neurodevelopment, cognition, and mental conditions. *Biol Psychiatry.* 2006;59:1189–
108. Mazure CM, ed. *Does Stress Cause Psychiatric Illness?* Vol. 46 Washington, DC: American Psychiatric Press, 1995:270.
109. Owen AM, James M, Leigh PN, et al. Fronto-striatal cognitive deficits at different stages of Parkinson's disease. *Brain.* 1992;115:1727.
110. Gotham AM, Brown RG, Marsden CD. "Frontal" cognitive function in patients with Parkinson's disease "on" and "off" levodopa. *Brain.* 1988;111:299–
111. Ernst M, Zametkin AJ, Matochik JA, et al. DOPA decarboxylase activity in attention deficit disorder adults. A [fluorine-18]fluorodopa positron emission tomographic study. *J Neurosci.* 1998;18:5901–
112. Volkow ND, Wang GJ, Newcorn J, et al. Brain dopamine transporter levels in treatment and drug naïve adults with ADHD. *Neuroimage.* 2007;34:1182–
113. Makanjuola RO, Ashcroft GW. Behavioural effects of electrolytic and 6-hydroxydopamine lesions of the accumbens and caudate-putamen nuclei. *Psychopharmacology.* 1982;76:33–
114. Simon, H. Dopaminergic A10 neurons and the frontal system. *J Physiol (Paris).* 1981;77:81–
115. Faraone SV, Perlis RH, Doyle AE, et al. Molecular genetics of attention-deficit/hyperactivity disorder. *Biol Psychiatry.* 2005;57:1313–
116. Elliott R, Sahakian BJ, Matthews K, et al. Effects of methylphenidate on spatial working memory and planning in healthy young adults. *Psychopharmacology.* 1997;131:196–
117. Berridge CW, Devilbiss DM, Andrzejewski ME, et al. Methylphenidate preferentially increases catecholamine neurotransmission within the prefrontal cortex at low doses that enhance cognitive function. *Biol Psychiatry.* 2006;60:1111–
118. Bellgrove MA, Hawi Z, Gill M, et al. The cognitive genetics of attention deficit hyperactivity disorder (ADHD): Sustained attention as a candidate phenotype. *Cortex.* 2006;42:838–
119. Nestler EJ. Molecular basis of long-term plasticity underlying addiction. *Nat Rev Neurosci.* 2001;2:119–
120. Jentsch JD, Taylor JR. Impulsivity resulting from frontostriatal dysfunction in drug abuse: implications for the control of

behavior by reward related stimuli. *Psychopharmacology.* 1999;146:373–

121. Akil M, Pierri JN, Whitehead RE, et al., Lamina-specific alterations in the dopamine innervation of the prefrontal cortex in schizophrenic subjects. *Am J Psychiatry.* 1999; 156:1580–

122. Abi-Dargham A, Mawlawi O, Lombardo I, et al. Prefrontal dopamine D1 receptors and working memory in schizophrenia. *J Neurosci.* 2002;22:3708–

123. Daniel DG, Weinberger DR, Jones DW, et al. The effect of amphetamine on regional cerebral blood flow during cognitive activation in schizophrenia. *J Neurosci.* 1991;11:1907–

124. Egan MF, Goldberg TE, Kolachana BS, et al. Effect of COMT Val108/158 Met genotype on frontal lobe function and risk for schizophrenia. *Proc Natl Acad Sci USA.* 2001;98:6917–

125. Breier A, Wolkowitz O, Pickar D. Stress and schizophrenia: Advances in neuropsychiatry and psychopharmacology In: Tamminga C, Schult S, eds. *Schizophrenia Research*, Vol 1. New York, NY: Raven Press, 1991:141–152.

126. Dohrenwend BP, Shrout PE, Link BG, et al. Life events and other possible risk factors for episodes of schizophrenia and major depression: a case-control study. In: Mazure CM, ed. *Does Stress Cause Psychiatric Illness?* Washington, DC: American Psychiatric Press; 1995:43–65.

127. Castner SA, Goldman-Rakic PS. Long-lasting psychotomimetic consequences of repeated low-dose amphetamine exposure in rhesus monkeys. *Neuropsychopharmacology.* 1999;20:10–

128. Mirnics K, Middleton FA, Stanwood GD, et al. Disease-specific changes in regulator of G-protein signaling 4 (RGS4) expression in schizophrenia. *Mol Psychiatry.* 2001;6:293–

129. Matsumoto M, Hikosaka O. Two types of dopamine neuron distinctly convey positive and negative motivational signals. *Nature* 2009;459:837–

5.4 | Dopaminergic Modulation of Flexible Cognitive Control in Humans

ROSHAN COOLS AND MARK D'ESPOSITO

INTRODUCTION

One of the most fascinating aspects of our environment is that it is changing constantly. The ability to adapt flexibly to these constant changes is a capacity that humans are uniquely good at. Its importance is illustrated by the current turmoil in the financial markets. For instance, while a certain government might have previously voted against a particular (rescue) plan, it might later completely reverse its behavior and vote in favor of this same plan when changes in the environment become sufficiently salient. Such flexible minds are essential for preventing disastrous outcomes like collapsing banking systems. So, how do our minds adapt to the changes around us? This is not a straightforward issue, because only some of the changes around us are relevant and require cognitive flexibility. Most other changes are irrelevant and should be ignored. In the latter case, adaptive behavior depends on cognitive stability rather than cognitive flexibility. What we need is an ability to regulate dynamically the balance between these two processes, depending on current task demands.

The higher order cognitive control processes necessary for this ability have been associated most commonly with the anterior pole of the brain, the prefrontal cortex (PFC).[1–4] The PFC is highly sensitive to its neurochemical environment, which is not surprising given the diffuse ascending inputs from the major neurochemical systems of dopamine (DA), noradrenaline (NA), serotonin, and acetylcholine.[5] These neurotransmitters are of fundamental importance to the etiology of neuropsychiatric abnormalities such as Parkinson's disease (PD), attention deficit hyperactivity disorder (ADHD), and drug addiction. Indeed, the importance of studying adaptive behavior is further illustrated by these disorders, in which flexible cognitive control goes awry, often leading to inflexibility, impulsivity, and/or compulsivity. But failures of the flexible mind occur not only in neurochemical disorders or after extended drug abuse. Prolonged or severe periods of stress and fatigue also lead the mind to be inflexible or unfocused. Thus, a better understanding of the flexible mind will provide insights not only into the abnormal mind but also into the usually adaptive but at time maladaptive healthy mind. Accumulating evidence from research with monkeys has revealed that one particular family of neurotransmitters, the catecholamines (DA and NA), plays an important role in these failures of cognitive control.[6] While appreciating that flexible cognitive control implicates other neurotransmitters, we have focused here on the role of DA, partly because PFC (and striatal) function appears to be particularly sensitive to modulation by DA.

THE PARADOX OF THE FLEXIBLE MIND

Demands for cognitive flexibility and stability appear to be reciprocal. If we are too flexible, we are likely to become distracted and our behavior unstable. Conversely, if we are too stable, our behavior is likely to become inflexible and unresponsive to new information. A pure form of reciprocity would imply that we need only a single mechanism that can be adjusted dynamically, depending on task demands. However, we propose that a single mechanism does not suffice. Indeed, we often need to be both flexible and stable at the same time, at least at the global level. That is, while we should be flexible in response to relevant changes, we should be simultaneously stable as long as the changes are irrelevant. To resolve this apparent paradox, it is more plausible to postulate two separate mechanisms that nevertheless work together. The need for two separate mechanisms is also illustrated by the observation that various disorders, such as ADHD, are accompanied by a combination of inflexible as well as unstable behavior and distraction.

So, what might these separate mechanisms be and how is the balance between them regulated? The brain structure that has been associated most commonly with such complex cognitive requirements is the PFC. However, we also know that this region does not act in isolation to bias cognitive control, but rather interacts with a set of deep brain subcortical structures, in particular the striatum. One purpose of the present review is to highlight the importance for cognitive control not only of the PFC, but also of the striatum, which has been traditionally associated primarily with movement control. Here we elaborate on a previously proposed working hypothesis,[7–9] which states that the balance between cognitive flexibility and stability depends on an adjustment of processing in circuits connecting the PFC with the striatum by the neurotransmitter DA.

The hypothesis that DA in particular is implicated in the regulation of flexibility and stability concurs with findings that cognitive inflexibility and instability in disorders like PD and ADHD can be remedied with DA-enhancing drugs.[10,11] This example also further highlights the above-mentioned paradox: How can drugs that enhance DA improve cognitive flexibility in some individuals while improving cognitive stability in others? Partly inspired by such paradoxical effects of dopaminergic drugs, we (and others) have put forward the working hypothesis that the effects of DA on cognitive control depend on two interactive factors.[9,10] First, its effects depend on the brain region that it innervates, so that it will enhance some cognitive functions (e.g., cognitive flexibility) by modulating the striatum while enhancing other cognitive functions (e.g., cognitive stability) by acting at the level of the PFC. Second, its effects will depend on the baseline levels of DA in the brain region at which it acts, so that it will remedy function in brain regions with low baseline levels of DA while detrimentally overdosing function in brain regions with already optimized baseline levels of DA. According to this hypothesis, the effects of DA are both regionally specific and baseline-dependent. For instance, DA-enhancing drugs in PD might shift the balance toward flexibility at the expense of stability by remediating severely depleted DA levels in the striatum but simultaneously detrimentally overdosing relatively intact DA levels in the PFC. However, in subjects with the opposite profile, the same drugs may have rather different consequences, improving stability at the expense of flexibility.

BASELINE DEPENDENCY OF DOPAMINERGIC DRUG EFFECTS

The insight that drug effects are baseline-dependent stems from as early as the 1950s, when Wilder[12] first observed that (the intensity and direction of) drug effects on blood pressure and pulse rate depend on the preexperimental level of the function tested (*Law of Initial Value*). Discoveries that methamphetamine in pigeons *reduced high* rates of responding but increased low rates of responding led to the notion that drug effects on motor activity can also be predicted partly from the initial state of the system.[13,14]

More recent evidence indicates that similar baseline dependency exists for the effects of dopaminergic drugs on cognitive function[15–18] (Fig. 5.4.1). This evidence comes primarily from studies on working memory, defined as the ability to maintain and update currently relevant information "in mind" during a short delay.[19] In 1979, a landmark study by Brozoski et al. revealed that DA and NA depletion in the PFC caused severe working memory impairment in monkeys,[20] and subsequent work in both animals and humans has substantiated the necessity of DA for working memory.[21–26] Nevertheless, consistent with the notion of baseline dependency, the relationship between DA and working memory is complex: There is large variability in the direction and extent of dopaminergic drug effects on working memory. Thus psychopharmacological studies in humans have shown that the effects of the administration of DA receptor agents on cognition (as well as serum prolactin levels) depend on baseline levels of working memory capacity as measured with the listening span test,[27,28] with diametrically opposite effects in subjects with high and low listening spans.[15,18,29–33] Specifically, administration of DA receptor agonists improves cognitive performance (e.g., set shifting,[15,30,33] working memory updating,[33,34] and working memory retrieval[31]) in subjects with a low span but impairs performance in subjects with a high span (Fig. 5.4.1c).

Research with experimental animals has indicated that these contrasting effects of DA agents might reflect differential baseline levels of DA.[6,35–38;see also 39] For instance, Phillips et al.[35] have shown that poor performance on a difficult (working) memory task (with a long delay) was accompanied by low DA levels in the PFC, while good performance on an easy task (with a shorter delay) was accompanied by high DA levels in the PFC. Interestingly, performance on the difficult task was improved by administration of a DA D1 receptor agonist, whereas good performance on the easy task was impaired.[40] This study provided the first direct evidence in animals for the hypothesis that the dependency of drug effects on basal performance levels reflects differences in basal DA levels. Evidence for a similar mechanism underlying contrasting effects of DA receptor agents in humans came from a recent neurochemical positron emission tomography (PET) study[41] (see Fig. 5.4.1a). In this study, a subgroup of high- and a subgroup of low-span subjects underwent a PET scan with the radiotracer 6-[^{18}F]fluoro-L-m-tyrosine (FMT). This substance is a substrate of DA synthesis capacity, and uptake of the tracer reflects the degree to which DA is synthesized in the striatum. Subjects with a low listening span had significantly lower DA synthesis capacity in the left caudate nucleus than did subjects with a

(a) Basal DA synthesis capacity in the striatum

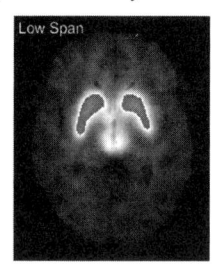

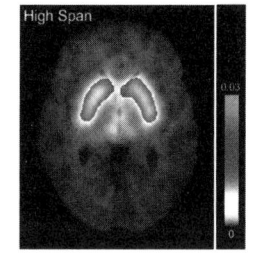

(b) Effect of bromocriptine on switch-related BOLD signal in the striatum

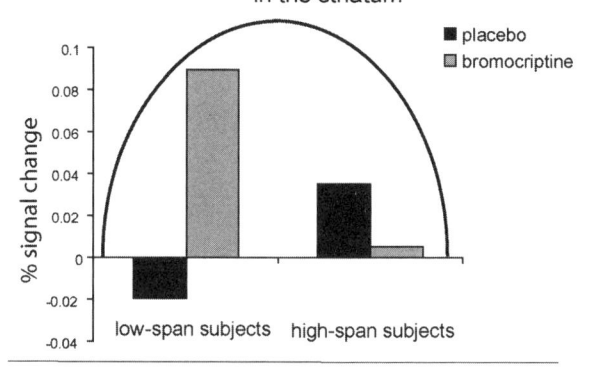

(c) Switch-related error rates

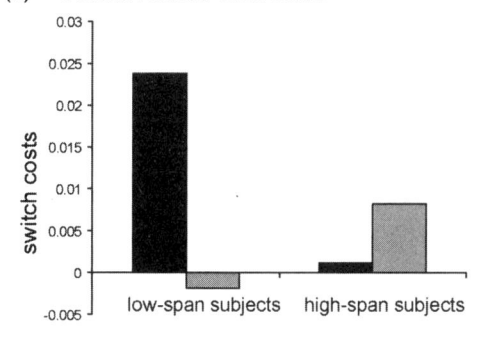

FIGURE 5.4.1. The effects of DA receptor stimulation depend on baseline working memory capacity as measured with the listening span test,[30] which correlates with baseline DA synthesis capacity in the striatum.[41] (a) Subjects with a high listening span had lower DA synthesis capacity, as measured with neurochemical PET imaging, than did subjects with a low listening span; (b) bromocriptine had opposite effects on neural activity measured with fMRI and (c) performance in high- and low-span subjects, consistent with an inverted-U-shaped relationship between DA receptor stimulation and cognitive performance. (See Color Plate 5.4.1.)

high listening span. Dopamine synthesis capacity was also lower for low-span subjects in the left putamen, the right caudate nucleus, and the right putamen, but these effects did not reach significance, with the left lateralization of the effect possibly reflecting the verbal nature of the task. These data provide empirical evidence for the pervasive but hitherto untested hypothesis that the dependency of dopaminergic drug effects on baseline working memory capacity reflects differential baseline levels of DA function.

In sum, there is an optimal level of DA for cognitive function, with both excessive as well as insufficient DA levels impairing working memory performance.

FUNCTIONAL SPECIFICITY OF DOPAMINERGIC DRUG EFFECTS

The above-reviewed studies have suggested that the large variability in drug effects can be explained partly by variation in basal levels of DA between different individuals. However, dopaminergic drug effects vary not only between individuals, but also within individuals between different tasks. Indeed, the particular cognitive demand might be the critical determinant of where the optimal DA level is set for the task under study. For example, in marmosets, DA depletion in the PFC impaired performance on a delayed response task with high demands for maintenance of information.[42] Conversely, PFC DA depletion improved performance on an attentional set-shifting task, requiring the ability to alter behavior according to changes in dimensional relevance of multidimensional stimuli.[43] The improved set shifting was subsequently accounted for by enhanced distractibility during the earlier set-formation and set-maintenance stages of the task.[44] Enhanced distractibility might well underlie the impairment on the delayed-response task. Thus, there might be different optimum levels of DA for different forms of cognitive processing, even if this processing reflects the output of the same brain region (here the PFC). Whereas certain (high) levels of DA in the PFC optimize the maintenance of task-relevant representations, other (low) levels of DA in the PFC optimize the flexible updating of (i.e., shifting between) information. In humans, administration of the DA D2 receptor agonist bromocriptine to healthy volunteers improved performance on a spatial memory task but impaired performance on a task of reversal shifting according to changes in reward values.[45] Furthermore Mehta et al.[46] showed that the DA D2 receptor antagonist sulpiride improved performance on a delayed response task that required the maintenance of information in the face of task-irrelevant distraction but, by contrast, impaired performance on task switching.

Candidate gene studies have provided further evidence for distinct optimal DA levels for different functions. For example, Nolan and others[47] have assessed the effects on reversal learning of the Val^{158}Met polymorphism of the catechol-O-methyltransferase (COMT) gene. This polymorphism regulates the expression of COMT, an enzyme that breaks down DA released into the synapse and is thought to have regionally selective effects on DA in the PFC. The Met allele of the Val[158]Met polymorphism has been associated with reduced activity of the COMT enzyme and thus *higher DA in the PFC* than the Val allele.[48,49] Relative to Met/Met homozygotes, Val/Val homozygotes exhibited better performance at the reversal stage but poorer performance at the acquisition stage of a reversal learning task. This performance pattern was interpreted to reflect enhanced cognitive flexibility but reduced cognitive stability in subjects with low levels of DA in the PFC.[8]

In both monkeys and rats, the impairments following injection of DA-enhancing drugs into the PFC have been characterized as perseverative or persistent, such that the animal repeats the previous response inappropriately.[37,50] It is perhaps not difficult to appreciate that this tendency toward persistence can be beneficial if task demands emphasize robust maintenance in the face of distracting new input. It is detrimental only if the task also requires the updating of current working memory representations.

In vitro studies have further highlighted the cellular basis of the effects of DA D1 receptor stimulation in the PFC.[51,52] Specifically, DA D1 receptor stimulation increases the impact of the NMDA (N-methyl-D-aspartate) component of excitatory synaptic input on PFC neurons, thought to be essential for the maintenance of current PFC activity; reduces calcium currents, which convey information from dendrites to cell bodies of pyramidal PFC neurons; and increases the excitability of inhibitory GABAergic inter-neurons, thereby attenuating the strength of further excitatory input. These cellular mechanisms restrict activity to the most strongly active cells, resulting in a strengthening of working memory representations and increased resistance to subsequent (distracting) inputs. These same mechanisms, however, also lead to greater difficulty with switching between various high-activity (active memory) states. Thus, while the D1-induced increase in NMDA currents promotes the currently active memory state by boosting recurrent excitation within cell assemblies, the increase in GABA currents leads to fiercer competition among different active ensembles of neurons, thus limiting the set of items encoded in working memory.

It may be noted that the consequences of D1 receptor stimulation are quite different from those of D2 receptor stimulation in the PFC. Partly based on these distinct cellular effects, Durstewitz and Seamans[51] have put forward the dual state theory of PFC DA function, according to which working memory maintenance is D1-dependent but flexible updating is D2-dependent (see below). Contrasting effects of DA receptor agents might be accounted for by differential sensitivity to DA of these distinct receptor types. However, as mentioned below, it is currently difficult to test this hypothesis in humans due to a lack of receptor-selective drugs available for research with healthy human volunteers. Therefore, we have focused here on other factors, such as regional specificity and baseline dependency, to account for contrasting drug effects.

The general principle of distinct optimum DA levels for different functions is also illustrated by research in a different cognitive domain: reinforcement learning. Thus, we have recently combined FMT PET with psychopharmacology in healthy volunteers. Specifically, we have assessed the effects of the DA D2 receptor agonist bromocriptine in humans on reward- and punishment-based reversal learning in relation to baseline levels of DA synthesis capacity in the striatum. As predicted, the results indicate that the effects of bromocriptine depended on baseline synthesis capacity: Bromocriptine improved reward-based reversal learning in subjects with low baseline synthesis capacity but impaired it in subjects with high baseline synthesis capacity. Remarkably, the opposite relationship was observed for punishment-based reversal learning: Bromocriptine impaired punishment-based reversal learning in subjects with low baseline synthesis capacity but improved it in subjects with high baseline synthesis capacity. The finding that the effects on punishment-based reversal learning contrasted with those on reward-based reversal learning suggests that, unlike reward-based learning, punishment-based learning benefits from *low* rather than *high* synaptic DA levels in the striatum. This concurs with prior neuropsychological evidence showing that PD patients (characterized by severe striatal DA depletion) exhibited a bias away from selecting reward-associated stimuli toward avoiding punishment-associated stimuli, whereas DA-enhancing medication in these patients induced a bias away from punishment toward reward.[53,54]

REGIONAL SPECIFICITY OF DOPAMINERGIC DRUG EFFECTS

The above-reviewed studies suggest that different optimum levels of DA exist for different forms of cognitive processing, even if this processing reflects the output of a single brain region. However, as mentioned above, a single mechanism that can be adjusted depending on task demands might not suffice to account for the observation that certain states are characterized by a

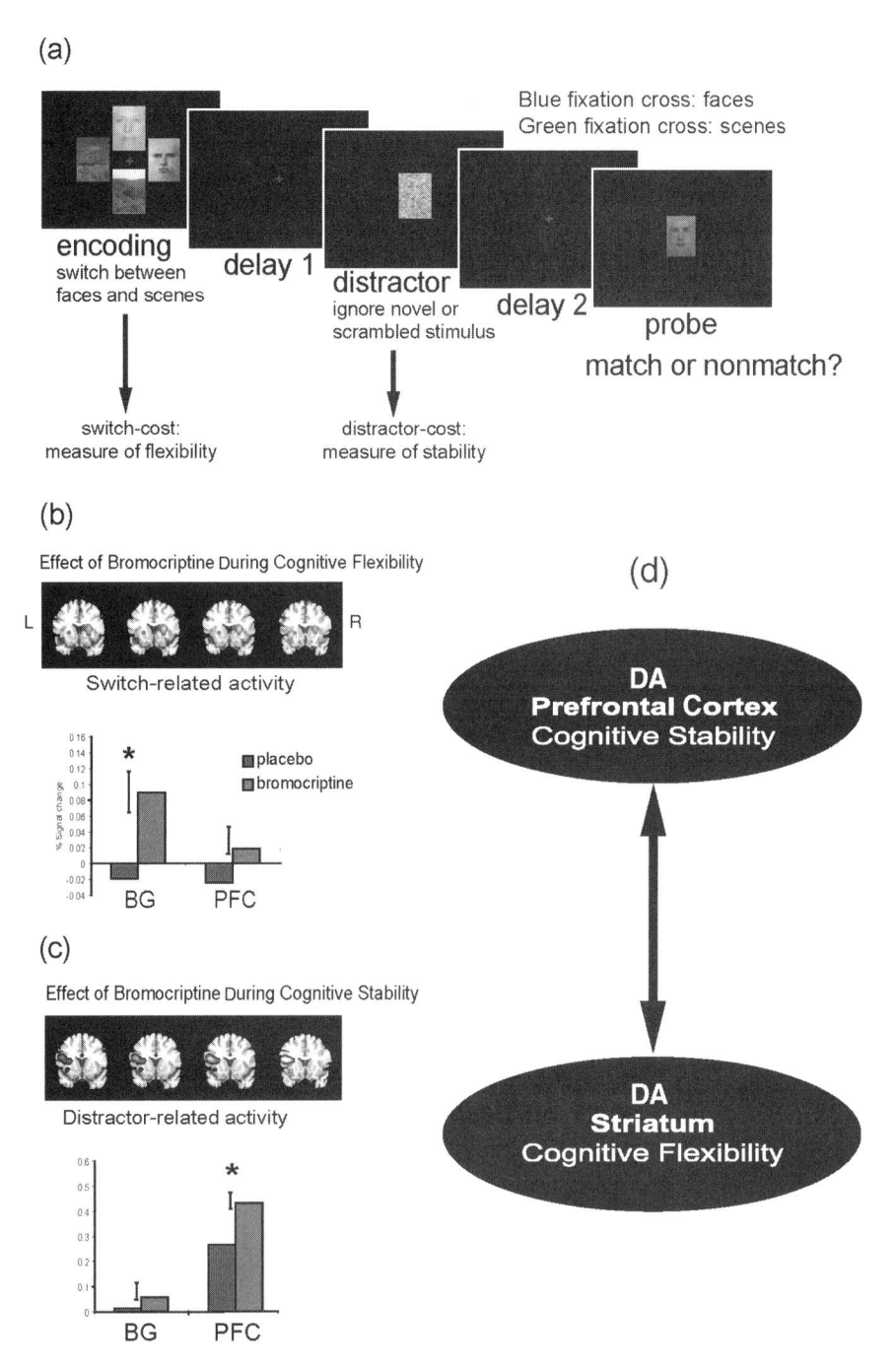

FIGURE 5.4.2. The effects of DA receptor stimulation depend on task demands and the neural site of modulation. (a) A delayed match-to-sample (DMS) task was used that provided a measure of cognitive flexibility (cognitive switching during encoding) as well as a measure of cognitive stability (distractor resistence during the delay). Subjects memorized faces or scenes, depending on the color of the fixation cross. Subjects occasionally switched between encoding faces and scenes. A distractor was presented during a delay. Subjects were instructed to ignore this distractor. (b) Top panel: Effects of bromocriptine on striatal activity during switching as a function of group (the group x drug interaction effect, whole-brain contrast values (>25) are overlaid on four coronal slices [slice numbers displayed on top] from the Montreal Neurological Institute high-resolution single-subject magnetic resonance image; L, left; R, right. Bottom panel: Effects of bromocriptine on switch-related activity in the striatum and left PFC in low-span subjects only. (c) Top panel: Effects of bromocriptine on frontal activity during distraction as a function of the group (the group x drug interaction effect; all contrast values >25 shown); Bottom panel: Effects of bromocriptine on distractor-related activity in the striatum and left PFC in low-span subjects only. (d) Schematic representation of the hypothesis that DA modulates cognitive flexibility by acting at the level of the striatum while modulating cognitive stability by acting at the level of the frontal cortex.

combination of inflexible and unstable behavior. Therefore, we have hypothesized that cognitive flexibility and stability are mediated by two separate mechanisms that nevertheless work together: Dopamine would promote stability or flexibility, depending on the neural site of modulation (Fig. 5.4.2). Specifically, DA receptor stimulation in the PFC is hypothesized to promote stability by increasing distractor resistance.[55] Conversely, DA in the striatum is hypothesized to promote flexibility by allowing the updating of newly relevant representations.[56] The proposal that the striatum plays an important role in flexible cognitive control concurs with recent results and theorizing[7,8,57–61] as well as with the classic view of behavioral neuroscientists that striatal DA is essential for behavioral flexibility and switching.[62,63]

Thus, high levels of striatal DA might be good for flexibility but bad for stability, whereas high levels of PFC DA might be good for stability but bad for flexibility.

The functional opponency between stability and flexibility maps well onto the neurochemical reciprocity between DA in the PFC and the striatum: Increases and decreases in PFC DA lead to decreases and increases in striatal DA, respectively.[64–66] One implication of this model is that stability and flexibility trade off in the healthy brain, where DA levels interact dynamically. However, in the diseased brain, abnormal DA regulation may independently disrupt flexibility and stability, sometimes causing the apparently paradoxical combination of distractibility and inflexibility (e.g., in ADHD).

Empirical evidence for regional specificity of the effects of dopaminergic drugs comes from a range of studies with experimental animals and human volunteers as well as patients.

For example, in contrast to the increased distractibility and the maintenance impairment observed following DA lesioning of the PFC (see above), DA lesions from the striatum in marmosets induced greater focusing on the relevant perceptual dimension during the maintenance of an attentional set within the same paradigm.[44] Animals with striatal DA lesions were significantly less distractible than control monkeys. Functional neuroimaging studies on the effects of candidate genes on working memory have also provided results that are consistent with the hypothesis. Complementary changes in neural activity were seen as a function of genetic variation in DA metabolism in the PFC, mediated by COMT, and in the striatum, mediated by the DA transporter (DAT).[67,68] The 10-repeat allele of the DAT gene has been associated with *lower DA in the striatum* relative to the 9-repeat allele. On the other hand, as described above, the Met allele of the Val[158]Met polymorphism in the COMT gene has been associated with *higher DA in the PFC* relative to the Val allele. Remarkably, Bertolino and others[67] have observed similar effects on neuronal activity of the 10-repeat allele of the DAT1 gene and the Met allele of the COMT gene, so that the activity pattern of subjects with putatively low striatal DA levels resembled that seen in subjects with putatively high DA levels in the PFC: Both alleles induced more focused activity in the PFC during the n-back task.

The hypothesis that the PFC and the striatum mediate different effects of DA was recently strengthened by an event-related functional magnetic resonance imaging (fMRI) study with healthy volunteers. In this study, subjects were scanned on two occasions, once after intake of an oral dose of the catecholamine enhancer methylphenidate (40 mg) and once after intake of placebo. During scanning they performed a probabilistic reversal learning task,[70] previously shown to be sensitive to manipulation of DA levels by dopaminergic medication withdrawal in PD patients.[71,72] The task enabled the separate measurement of neural activity during negative feedback that signaled the need to switch responding flexibly to the previously unrewarded stimulus and of neural activity during misleading negative feedback that required the maintenance of current response strategies. Dodds et al. found that, like dopaminergic medication in PD patients,[72] methylphenidate abolished neural activity in the (ventral) striatum during negative feedback events that led to behavioral switching. Conversely, the same drug modulated activity in the PFC during negative feedback events that required the maintenance of response strategies. These opposing effects of DA on striatal and frontal activity underline the possible competition and coordination between the PFC and the striatum during maintenance and shifting. However, that study was not specifically designed to contrast cognitive flexibility and cognitive stability; accordingly, the results speak more to the domain of reward-based learning and punishment processing than to the need to resist distraction. Furthermore, methylphenidate is not specific for DA and also affects NA.

In a recent study, we tested more directly the hypothesis that dissociable brain regions mediate the dopaminergic modulation of cognitive stability and cognitive flexibility[30] (Fig. 5.4.2). To this end, a group of 23 young, healthy volunteers were scanned with fMRI on two occasions, once after intake of an oral dose (1.25 mg) of the DA D2 receptor agonist bromocriptine and once after intake of placebo (in a double-blind, cross-over design). During scanning, subjects performed a novel delayed match-to-sample (DMS) paradigm that allowed the separate investigation of cognitive flexibility and cognitive stability. In this task, subjects had to encode, maintain, and retrieve visual stimuli. Four such stimuli (two faces and two scenes) were presented during the encoding period, followed by a delay period during which subjects had to maintain the relevant stimuli in memory. Following this initial delay period, another stimulus was presented, which subjects were instructed to ignore. This distractor was either a scrambled image (the nondistractor) or a novel face or scene (the congruent distractor). It was followed by a second delay, after which subjects were probed to respond with the right or left finger, depending on whether the probe stimulus matched one of the two task-relevant encoding stimuli (Fig. 5.4.2a). Critically, subjects were instructed on each trial to attend to either the faces or the scenes. If the fixation cross was blue, they had to memorize the faces; if it was green, they had to memorize the scenes. The blue face trials and the green scene trials were randomized within blocks, enabling measurement of the flexible switching of attention between faces and scenes. The critical measure of cognitive flexibility was the switch cost, which was calculated by subtracting performance (error rates and reaction times measured at probe) on nonswitch trials from that on switch trials. The critical measure of cognitive stability was the distractor cost, which was calculated by subtracting performance (measured at probe) after scrambled nondistractors from that after congruent distractors.

The first aim of this study was to test the hypothesis that bromocriptine would modulate activity in the PFC during cognitive stability (as a function of distractor type) but in the striatum during cognitive flexibility (as a function of switching). Second, we also predicted that the effects of bromocriptine would depend on the baseline levels of DA. Bromocriptine would remedy the function of brain regions with low baseline levels of DA while detrimentally overdosing the function of brain regions with already optimized baseline levels of DA.

To test for individual differences in baseline levels of DA, we assessed drug effects separately in two groups of subjects. These groups differed in terms of their baseline working memory capacity, as measured with the listening span test (as well as trait impulsivity, not discussed further here). Variation in the listening span had previously been shown to reflect variation in basal levels of DA (see[73]). The results were consistent with our hypotheses: Bromocriptine modulated distinct brain regions, the striatum and the lateral PFC, during switching and distractor resistance, respectively (Fig. 5.4.2b,c). Critically, these effects depended on individual differences in working memory capacity (Fig. 5.4.1b,c). Specifically, when tested on placebo, the low-span subjects exhibited a numerically larger switch cost than did the high-span subjects. Interestingly, the effects of bromocriptine also depended on the listening span. Bromocriptine attenuated the switch cost, that is, it improved switching in the low-span subjects. By contrast, the same drug enhanced the switch cost in the high-span subjects, albeit nonsignificantly. The next question was whether these behavioral effects were accompanied by changes in neural activity, in particular in the striatum. Consistent with this prediction, there was a significant group-by-drug interaction for switch-related activity, thus paralleling the behavioral switch-costs, and this interaction was found only in the striatum (Fig. 5.4.1b,c). Again, the drug had contrasting effects in the low- and high-span subjects. In the low-span subjects, bromocriptine significantly potentiated striatal activity during switching. By contrast, the same drug nonsignificantly attenuated striatal activity during switching in the high-span subjects. Therefore, a drug-induced improvement in switching was accompanied by a drug-induced potentiation of striatal activity in the low-span subjects. Conversely, a drug-induced (nonsignificant) impairment in switching was accompanied by a drug-induced (nonsignificant) attenuation of striatal activity in the high-span subjects.

Our next question was whether these effects were regionally specific. We had hypothesized that the dopaminergic effects on switching would be mediated by the striatum but not by the lateral PFC (Fig. 5.4.2d). Therefore, we also assessed switch-related activity in the lateral PFC. As predicted, the lateral PFC was not modulated by bromocriptine during switching (Fig. 5.4.2b). This lack of effect was not due to insufficient power to detect changes in lateral frontal activity. Interestingly, lateral PFC activity was modulated by bromocriptine during a different task period, namely, during distraction. Specifically, activity in the lateral PFC was potentiated by bromocriptine in the low-span subjects (Fig. 5.4.2c) while remaining unaltered in the high-span subjects. Similar effects were not observed in the striatum. Together, these data concur with the hypothesis that cognitive flexibility and cognitive stability are mediated by dopaminergic modulation of the striatum and the PFC, respectively (Fig. 5.4.2d). Furthermore, the effects of bromocriptine were not only regionally specific as a function of task demands, but also baseline-dependent, as illustrated by the opposite effects in high- and low-span subjects.

A different approach to studying the role of DA in human cognition involves investigating patients with PD. Parkinson's disease is a progressive neurodegenerative disorder characterized by severe DA depletion in the striatum. Dopamine levels in the PFC are relatively intact, at least in the early stages of the disease.[74,75] Interestingly, there are some reports that mild PD might be accompanied by compensatory up-regulation of frontal DA levels.[76,77] Accordingly, mild PD provides a particularly good model for assessing the regionally specific and baseline-dependent effects of DA.

According to our working hypothesis, the disease should be accompanied by a shift in the balance between flexibility and stability, leading, on the one hand, to an inflexible state, due to low striatal DA levels, that is, however, also abnormally stable due to high frontal DA levels. Thus, on our delayed match-to-sample paradigm, we expected patients with mild PD to exhibit *enhanced* switch costs while showing abnormally *reduced* distractor costs. On the other hand, treatment with dopaminergic medication was predicted to shift this balance away from cognitive stability back to cognitive flexibility. Evidence for the first part of this hypothesis, namely, impairments in the domain of cognitive flexibility, is overwhelming. Set-shifting difficulties have been shown on a variety of tasks ranging from Wisconsin Card Sorting Test (WCST)–like discrimination learning tasks to more rapid task-switching paradigms.[78-81] For example, several studies have revealed a selective deficit at the extradimensional set-shifting stage of the intradimensional/extradimensional (ID/ED) shifting task in mild PD patients.[82-84] In addition, a number of researchers have employed the task-switching paradigm, in which the acquisition of task sets is rapidly learned beforehand and switches are externally cued, thereby minimizing demands for working memory and trial-and-error learning.[85-88] The paradigm requires subjects to switch continuously between two tasks, A and B, and the sequence of trials (e.g., AABBAA and so on) enables the measurement of switching against a baseline of nonswitching. The critical measure, the switch cost, is calculated by subtracting the performance on nonswitch trials from that on switch trials. Using such a paradigm, we showed that mild PD patients exhibited significantly enhanced switch costs compared with matched control subjects.[88] Moreover, the deficit in switching between attentional and/or task sets was alleviated by administration of dopaminergic medication.[71,90,91] Notably, several studies have revealed that these beneficial effects occur in the context of detrimental effects of the same medication in the same patients on other cognitive tasks[71,91] and, therefore, cannot be accounted for by global effects on motor symptoms, arousal, and/or motivation. Thus, the hypothesis that even mild PD, characterized primarily by striatal DA depletion (and perhaps frontal up-regulation), is accompanied by cognitive inflexibility is uncontroversial. The question we posed is whether this deficit was accompanied by a benefit, that is, enhanced distractor resistance.[92]

To test this second part of our hypothesis, we adapted our DMS paradigm so that it was suitable for use in patients.[92] To this end, we made two changes. First, we reduced the delays from 8 s to 3 s. Second, we increased the duration of the encoding period from 1 s to 3 s. Fifteen patients were tested on two occasions, once on and once off their normal dopaminergic medication, and their performance was compared with that of 14 age- and education-matched control subjects. On each test session we administered not only our experimental paradigm, but also the Unified Parkinson's Disease Rating Scale (UPDRS). As expected, medication withdrawal significantly worsened their movement ratings on the UPDRS, but on the DMS task all patients performed well above chance levels (chance accuracy 50%): There was no significant difference between patients and controls in terms of error rates or mean reaction times across the task as a whole. However, intriguingly, there was a significant difference between patients off their medication and controls in terms of the distractor cost. Specifically, patients off medication exhibited *reduced* distractor costs compared with controls, who responded more slowly after a congruent distractor than after a scrambled nondistractor. Thus, when they were off their medication, patients were less distracted by the congruent distractor in the delay than were controls. This pattern of performance of the patients in their off state was particularly striking given their significantly increased motor symptoms. Further, the reduced distractor cost was normalized when the same patients were tested on their normal dopaminergic medication, so that the distractor cost of the patients no longer differed from that of controls when they were on medication.

In terms of cognitive flexibility, patients off their medication were again shown to exhibit greater switch costs than did controls, although this did not reach significance in this experiment. This lack of effect likely reflects a failure of sensitivity to switch costs of the current version of the paradigm. Indeed, the lengthening of the encoding period from 1 s (in the prior fMRI study) to 3 s (in the current patient study) prevented the surfacing of a switch cost even in healthy volunteers.

Together with abundant prior evidence for task-switching deficits in mild PD, these data confirm that the DA-depleted state of PD is accompanied by changes in cognitive control. However, PD seems to confer either deficits *or benefits*, depending on the precise task demands under study. Whereas they suffer enhanced switch costs, they also show reduced distractibility. Based on their anatomical pattern of DA depletion as well as the fMRI data reviewed above, we hypothesize that the combination of poor flexibility and good stability in PD patients OFF medication reflects depletion of striatal DA and up-regulation of DA in the PFC, respectively. An intriguing possibility is that the restoration of switch and distractor costs by dopaminergic medication reflects a normalization of the balance between frontal and striatal DA.

RECEPTOR SPECIFICITY OF DOPAMINERGIC DRUG EFFECTS

The reviewed data demonstrate that the effects of dopaminergic drugs are complex, so that contrasting effects are seen, depending not only on task demands and associated brain regions, but also on basal levels of DA in those underlying brain regions. A third factor that might contribute to the variability in the effects of dopaminergic drugs is their receptor specificity. For example, the dual-state theory of prefrontal cortex DA function, put forward by Durstewitz, Seamans, and Yang,[51,52] states that a D1-dominated state favors robust online maintenance of information, while a D2-dominated state is beneficial for flexible and fast switching among representational states. This theory is based on studies at the cellular and synaptic levels as well as on biophysically realistic computational models, which have shown that the effects of DA in the PFC via D1- and D2-class receptors are often apparently opposing. Although this receptor-specificity model and the regional selectivity model described above are not necessarily mutually exclusive (e.g., due to predominance of D2 receptors in the striatum), it is currently difficult to test its predictions in humans due to a lack of receptor-selective drugs available for human research. Specifically, there are no D1-selective drugs that are safe for use with healthy human volunteers, whereas so-called D2 receptor agonists, like bromocriptine, are not selective and also stimulate D1 family receptors.[93,94] The study of polymorphisms in the D1 or D2 receptor genes will hold promise for the future.

CONCLUSION

What brain mechanisms underlie flexible cognitive control? The present review illustrates the importance of approaching this question by investigating cognitive control in terms of its subcomponent processes. It is a multifactorial phenomenon that requires a dynamic balance between cognitive flexibility and cognitive stability. We propose that these distinct components implicate the striatum and the PFC, respectively, the functional outputs of which are adjusted by DA in order to direct behavioral output to current goals. This chapter highlights the complexity not only of cognitive control but also of the dopaminergic system. Manipulation of DA has contrasting effects on the expression of function, depending on, among other factors, the brain region that is implicated by the type of function under study and the baseline levels of DA in that brain region.

REFERENCES

1. Fuster J. *The Prefrontal Cortex*. New York: Raven Press; 1989.
2. Miller E, Cohen J. An integrative theory of prefrontal cortex function. *Annu Rev Neurosci.* 2001;24:167–202.
3. Chao L, Knight R. Human prefrontal lesions increase distractibility to irrelevant sensory inputs. *Neuroreport.* 1995;6(12):1605–1610.
4. Miller B, D'Esposito M. Searching for "the top" in top-down control. *Neuron.* 2005;48(4):535–538.
5. Robbins TW. Chemical neuromodulation of frontal-executive functions in humans and other animals. *Exp Brain Res.* 2000;133(1):130–138.
6. Arnsten AFT. Catecholamine modulation of prefrontal cortical cognitive function. *Trends Cogn Sci.* 1998;2(11):436–446.
7. Frank M, Loughry B, O'Reilly R. Interactions between frontal cortex and basal ganglia in working memory: a computational model. *Cogn Affect Behav Neurosci.* 2001;1(2):137–160.
8. Bilder R, Volavka K, Lachman H, Grace A. The catechol-O-methyltransferase polymorphism: relations to the tonic-phasic dopamine hypothesis and neuropsychiatric phenotypes. *Neuropsychopharmacology.* 2004;29(11):1943–1961.
9. Cools R, Robbins TW. Chemistry of the adaptive mind. *Philos Transact A Math Phys Eng Sci.* 2004;362(1825):2871–2888.
10. Cools R. Dopaminergic modulation of cognitive function-implications for L-DOPA treatment in Parkinson's disease. *Neurosci Biobehav Rev.* 2006;30(1):1–23.
11. Kramer A, Cepeda N, Cepeda M. Methylphenidate effects on task-switching performance in attention-deficit/hyperactivity disorder. *J Am Acad Child Adolesc Psychiatry.* 2001;40(11):1277–1284.
12. Wilder J. Paradoxic reactions to treatment. *NY State J Med.* 1957;57:3348–3352.
13. Dews P. Rate-dependency hypothesis. *Science.* 1977;198:1182–1183.
14. Dews PB. Studies on behavior. IV. Stimulant actions of methamphetamine. *J Pharmacol Exp Ther.* 1958;122(137–147).
15. Kimberg DY, D'Esposito M, Farah MJ. Effects of bromocriptine on human subjects depend on working memory capacity. *Neuroreport.* 1997;8:3581–3585.

16. Granon S, Passetti F, Thomas KL, Dalley JW, Everitt BJ, Robbins T. Enhanced and impaired attentional performance after infusion of D1 dopaminergic receptor agents into rat prefrontal cortex. *J Neurosci.* 2000;20(3):1208–1215.
17. Mehta M, Owen AM, Sahakian BJ, Mavaddat N, Pickard JD, Robbins TW. Methylphenidate enhances working memory by modulating discrete frontal and parietal lobe regions in the human brain. *J Neurosci.* 2000(RC 65):1–6.
18. Mattay VS, Callicot JH, Bertolino A, et al. Effects of dextroamphetamine on cognitive performance and cortical activation. *Neuroimage.* 2000;12(3):268–275.
19. Baddeley AD. *Working Memory.* Oxford: Clarendon Press; 1986.
20. Brozoski TJ, Brown R, Rosvold HE, Goldman PS. Cognitive deficit caused by regional depletion of dopamine in the prefrontal cortex of rhesus monkeys. *Science.* 1979;205:929–931.
21. Sawaguchi T, Goldman-Rakic PS. D1 dopamine receptors in prefrontal cortex: involvement in working memory. *Science.* 1991;251:947–950.
22. Sawaguchi T, Matsumura M, Kubota K. Effects of dopamine antagonists on neuronal activity related to a delated response task in monkey prefrontal cortex. *J Neurophysiol.* 1990;63(6): 1401–1412.
23. Sawaguchi T, Matsumura M, Kubota K. Catecholaminergic effects on neuronal activity related to a delayed response task in monkey prefrontal cortex. *J Neurophysiol.* 1990;63(6): 1385–1399.
24. Luciana L, Collins PF, Depue RA. Opposing roles for dopamine and serotonin in the modulation of human spatial working memory functions. *Cereb Cortex.* 1998;8:218–226.
25. Luciana M, Collins P. Dopaminergic modulation of working memory for spatial but not object cues in normal volunteers. *J Cogn Neurosci.* 1997;9:330–347.
26. Luciana M, Depue RA, Arbisi P, Leon A. Facilitation of working memory in humans by a D2 dopamine receptor agonist. *J Cogn Neurosci.* 1992;4(1):58–68.
27. Daneman M, Carpenter P. Individual differences in working memory and reading. *J Verbal Learn Verbal Behav.* 1980;19: 450–466.
28. Salthouse T, Babcock R. Decomposing adult age differences in working memory. *Dev Psychol.* 1991;27(762–766).
29. Kimberg D, D'Esposito M. Cognitive effects of the dopamine receptor agonist pergolide. *Neuropsychologia.* 2003;41(8): 1020–1007.
30. Cools R, Sheridan M, Jacobs E, D'Esposito M. Impulsive personality predicts dopamine-dependent changes in frontostriatal activity during component processes of working memory. *J Neurosci.* 2007;27(20):5506–5514.
31. Gibbs SE, D'Esposito M. Individual capacity differences predict working memory performance and prefrontal activity following dopamine receptor stimulation. *Cogn Affect Behav Neurosci.* 2005;5(2):212–221.
32. Gibbs SE, D'Esposito M. A functional magnetic resonance imaging study of the effects of pergolide, a dopamine receptor agonist, on component processes of working memory. *Neuroscience.* 2006;139(1):359–371.
33. Frank MJ, O'Reilly RC. A mechanistic account of striatal dopamine function in human cognition: psychopharmacological studies with cabergoline and haloperidol. *Behav Neurosci.* 2006;120(3):497–517.
34. Mehta M, Calloway P, Sahakian B. Amelioration of specific working memory deficits by methylphenidate in a case of adult attention deficit/hyperactivity disorder. *J Psychopharmacol.* 2000;14(3):299–302.
35. Phillips A, Ahn S, Floresco S. Magnitude of dopamine release in medial prefrontal cortex predicts accuracy of memory on a delayed response task. *J Neurosci.* 2004;14(2):547–553.
36. Williams GV, Goldman-Rakic PS. Modulation of memory fields by dopamine D1 receptors in prefrontal cortex. *Nature.* 1995;376:572–575.
37. Zahrt J, Taylor JR, Mathew RG, Arnsten AFT. Supranormal stimulation of D1 dopamine receptors in the rodent prefrontal cortex impairs spatial working memory performance. *J Neurosci.* 1997;17(21):8528–8535.
38. Vijayraghavan S, Wang M, Birnbaum S, Williams G, Arnsten A. Inverted-U dopamine D1 receptor actions on prefrontal neurons engaged in working memory. *Nat Neurosci.* 2007;10(3): 176–184.
39. Mattay V, Goldberg T, Fera F, et al. Catechol-O-methyltransferase Val^{158}-met genotype and individual variation in the brain response to amphetamine. *Proc Natl Acad Sci USA.* 2003;100(10):6186–6191.
40. Floresco S, Phillips A. Delay-dependent modulation of memory retrieval by infusion of a dopamine D1 agonist into the rat medial prefrontal cortex. *Behav Neurosci.* 2001;115(4): 934–939.
41. Cools R, Frank M, Gibbs S, Miyakawa A, Jagust W, D'Esposito M. Striatal dopamine synthesis capacity predicts dopaminergic drug effects on flexible outcome learning. *J Neurosci.* 2009;29:1538–1543.
42. Collins P, Roberts AC, Dias R, Everitt BJ, Robbins TW. Perseveration and strategy in a novel spatial self-ordered sequencing task for nonhuman primates: effects of excitotoxic lesions and dopamine depletions of the prefrontal cortex. *J Cogn Neurosci.* 1998;10(3):332–354.
43. Roberts AC, De Salvia MA, Wilkinson LS, et al. 6-Hydroxydopamine lesions of the prefrontal cortex in monkeys enhance performance on an analog of the Wisconsin Card Sort Test: possible interactions with subcortical dopamine. *J Neurosci.* 1994;14(5):2531–2544.
44. Crofts HS, Dalley JW, Van Denderen JCM, Everitt BJ, Robbins TW, Roberts AC. Differential effects of 6-OHDA lesions of the frontal cortex and caudate nucleus on the ability to acquire an attentional set. *Cereb Cortex.* 2001;11(11):1015–1026.
45. Mehta MA, Swainson R, Ogilvie AD, Sahakian BJ, Robbins TW. Improved short-term spatial memory but impaired reversal learning following the dopamine D2 agonist bromocriptine in human volunteers. *Psychopharmacology.* 2001;159:10–20.
46. Mehta M, Manes F, Magnolfi G, Sahakian B, Robbins T. Impaired set-shifting and dissociable effects on tests of spatial working memory following the dopamine D2 receptor antagonist sulpiride in human volunteers. *Psychopharmacology.* 2004; 176(3–4):331–42.
47. Nolan K, Bilder R, Lachman H, Volavka K. Catechol-O-methyltransferase Val158Met polymorphism in schizophrenia: differential effects of Val and Met alleles on cognitive stability and flexibility. *Am J Psychiaty..* 2004;161:359–361.
48. Lotta T, Vidgren J, Tilgmann C, et al. Kinetics of human soluble and membrane-bound catechol-O-methyltransferase: a revised mechanism and description of the thermolabile variant of the enzyme. *Biochemistry.* 1995;34:4202–4210.
49. Chen J, Lipska B, Halim N, et al. Functional analysis of genetic variation in catechol-O-methyltransferase (COMT): effects on mRNA, protein, and enzyme activity in postmortem human brain. *Am J Genet.* 2004;75:807–821.
50. Druzin M, Kurzina N, Malinina E, Kozlov A. The effects of local application of D2 selective dopaminergic drugs into the medial

prefrontal cortex of rats in a delayed spatial choice task. *Behav Brain Res.* 2000;109(1):99–111.
51. Durstewitz D, Seamans J. The dual-state theory of prefrontal cortex dopamine function with relevance to catechol-O-methyl-transferase genotypes and schizophrenia. *Biol Psychiatry.* 2008;64(9):739–749.
52. Seamans JK, Yang CR. The principal features and mechanisms of dopamine modulation in the prefrontal cortex. *Prog Neurobiol.* 2004;74(1):1–58.
53. Frank MJ, Seeberger LC, O'Reilly RC. By carrot or by stick: cognitive reinforcement learning in parkinsonism. *Science.* 2004;306(5703):1940–1943.
54. Cools R, Altamirano L, D'Esposito M. Reversal learning in Parkinson's disease depends on medication status and outcome valence. *Neuropsychologia.* 2006;44(10):1663–1673.
55. Durstewitz D, Seamans J, Sejnowski T. Dopamine-mediated stabilization of delay-period activity in a network model of prefrontal cortex. *J Neurophysiol.* 2000;83:1733–1750.
56. Frank MJ. Dynamic dopamine modulation in the basal ganglia: a neurocomputational account of cognitive deficits in medicated and nonmedicated Parkinsonism. *J Cogn Neurosci.* 2005;17(1):51–72.
57. Zhang Y, Bertolino A, Fazio L, et al. Polymorphisms in human dopamine D2 receptor gene affect gene expression, splicing, and neuronal activity during working memory. *Proc Natl Acad Sci USA.* 2007;104(51):20552–20557.
58. Meyer-Lindenberg A, Straub RE, Lipska BK, et al. Genetic evidence implicating DARPP-32 in human frontostriatal structure, function, and cognition. *J Clin Invest.* 2007;117(3):672–682.
59. Gruber AJ, Dayan P, Gutkin BS, Solla SA. Dopamine modulation in the basal ganglia locks the gate to working memory. *J Comput Neurosci.* 2006;20(2):153–166.
60. McNab F, Klingberg T. Prefrontal cortex and basal ganglia control access to working memory. *Nat Neurosci.* 2008;11(1):103–107.
61. Leber A, Turk-Browne N, Chun M. Neural predictors of moment-to-moment fluctuations in cognitive flexibility. *Proc Natl Acad Sci USA.* 2008;105(36):13592–13597.
62. Cools AR. Role of the neostriatal dopaminergic activity in sequencing and selecting behavioural strategies: facilitation of processes involved in selecting the best strategy in a stressful situation. *Behav Brain Res.* 1980;1:361–378.
63. Lyon M, Robbins TW. The action of central nervous system stimulant drugs: a general theory concerning amphetamine effects. In: Essman W, ed. *Current Developments in Psychopharmacology,* Vol 2. New York: Spectrum; 1975:79–163.
64. Pycock CJ, Kerwin RW, Carter CJ. Effect of lesion of cortical dopamine terminals on subcortical dopamine receptors in rats. *Nature.* 1980;286:74–77.
65. Meyer-Lindenberg A, Kohn PD, Kolachana B, et al. Midbrain dopamine and prefrontal function in humans: interaction and modulation by COMT genotype. *Nat Neurosci.* 2005;8(5):594–596.
66. Akil M, Kolachana BS, Rothmond DA, Hyde TM, Weinberger DR, Kleinman JE. Catechol-O-methyltransferase genotype and dopamine regulation in the human brain. *J Neurosci.* 15 2003;23(6):2008–2013.
67. Bertolino A, Blasi G, Latorre V, et al. Additive effects of genetic variation in dopamine regulating genes on working memory cortical activity in human brain. *J Neurocsci.* 2006;26(15):3918–3922.
68. Caldu X, Vendrell P, Bartres-Faz D, et al. Impact of the COMT Val108/158 Met and DAT genotypes on prefrontal function in healthy subjects. *Neuroimage.* 2007;37(4):1437–1444.
69. Dodds CM, Muller U, Clark L, van Loon A, Cools R, Robbins TW. Methylphenidate has differential effects on blood oxygenation level-dependent signal related to cognitive subprocesses of reversal learning. *J Neurosci.* 2008;28(23):5976–5982.
70. Cools R, Clark L, Owen AM, Robbins TW. Defining the neural mechanisms of probabilistic reversal learning using event-related functional magnetic resonance imaging. *J Neurosci.* 2002;22(11):4563–4567.
71. Cools R, Barker RA, Sahakian BJ, Robbins TW. Enhanced or impaired cognitive function in Parkinson's disease as a function of dopaminergic medication and task demands. *Cereb Cortex.* 2001;11:1136–1143.
72. Cools R, Lewis S, Clark L, Barker R, Robbins TW. L-DOPA disrupts activity in the nucleus accumbens during reversal learning in Parkinson's disease. *Neuropsychopharmacology.* 2007;32:180–189.
73. Cools R, Gibbs S, Miyakawa A, Jagust W, D'Esposito M. Working memory capacity predicts dopamine synthesis capacity in the human striatum. *J Neurosci.* 2008;28:1208–1212.
74. Sawamoto N, Piccini P, Hotton G, Pavese N, Thielemans K, Brooks D. Cognitive deficits and striato-frontal dopamine release in Parkinson's disease. *Brain* 2008;131(5):1294–1302.
75. Agid Y, Ruberg M, Javoy-Agid F, et al. Are dopaminergic neurons selectively vulnerable to Parkinson's disease? *Adv Neurol.* 1993;60:148–164.
76. Rakshi J, Uema T, Ito K, et al. Frontal, midbrain and striatal dopamergic function in early and advanced Parkinson's disease. A 3D [(18)F]dopa-PET study. *Brain.* 1999;122(1637–1650).
77. Kaasinen V, Nurmi E, Bruck A, et al. Increased frontal [(18)F]fluorodopa uptake in early Parkinson's disease: sex differences in the prefrontal cortex. *Brain.* 2001;124:1125–1130.
78. Bowen FP, Kamienny RS, Burns MM, Yahr MD. Parkinsonism: effects of levodopa treatment on concept formation. *Neurology.* 1975;25:701–704.
79. Lees AJ, Smith E. Cognitive deficits in the early stages of Parkinson's disease. *Brain.* 1983;106:257–270.
80. Taylor A, Saint-Cyr J. The neuropsychology of Parkinson's disease. *Brain Cogn.* 1995;28:281–296.
81. Cools AR, Van Den Bercken JHL, Horstink MWI, Van Spaendonck KPM, Berger HJC. Cognitive and motor shifting aptitude disorder in Parkinson's disease. *J Neurol Neurosurg Psychiatry.* 1984;47:443–453.
82. Downes JJ, Roberts AC, Sahakian BJ, Evenden JL, Morris RG, Robbins TW. Impaired extra-dimensional shift performance in medicated and unmedicated Parkinson's disease: evidence for a specific attentional dysfunction. *Neuropsychologia.* 1989;27:1329–1343.
83. Owen AM, James M, Leigh JM, et al. Fronto-striatal cognitive deficits at different stages of Parkinson's disease. *Brain.* 1992;115:1727–1751.
84. Owen AM, Sahakian BJ, Hodges JR, Summers BA, Polkey CE, Robbins TW. Dopamine-dependent frontostriatal planning deficits in early Parkinson's disease. *Neuropsychology.* 1995;9:126–140.
85. Pollux P, Robertson C. Reduced task-set inertia in Parkinson's disease. *J Clin Exp Neuropsychol.* 2002;24(8):1046–1056.
86. Pollux P. Advance preparation of set-switches in Parkinson's disease. *Neuropsychologia.* 2004;42(7):912–919.
87. Rogers RD, Sahakian BJ, Hodges JR, Polkey CE, Kennard C, Robbins TW. Dissociating executive mechanisms of task control following frontal lobe damage and Parkinson's disease. *Brain.* 1998;121:815–842.

88. Cools R, Barker RA, Sahakian BJ, Robbins TW. Mechanisms of cognitive set flexibility in Parkinson's disease. *Brain.* 2001;124:2503–2512.
89. Rogers RD, Monsell S. Costs of a predictable switch between simple cognitive tasks. *J Exp Psychol.* 1995;124:207–231.
90. Hayes AE, Davidson MC, Keele SW. Towards a functional analysis of the basal ganglia. *J Cogn Neurosci.* 1998;10(2):178–198.
91. Cools R, Barker RA, Sahakian BJ, Robbins TW. L-Dopa medication remediates cognitive inflexibility, but increases impulsivity in patients with Parkinson's disease. *Neuropsychologia.* 2003;41:1431–1441.
92. Cools R, Miyakawa A, D'Esposito M. Dopaminergic modulation of distractor-resistant working memory in healthy volunteers and Parkinson's disease. Comparison with frontal lobe lesions. Paper presented at The Society for Neuroscience Annual Meeting, 2007; San Diego, CA.
93. Gerlach M, Double K, Arzberger T, Leblhuber F, Tatschner T, Riederer P. Dopamine receptor agonists in current clinical use: comparative dopamine receptor binding profiles defined in the human striatum. *J Neural Transm.* 2003;110(10):1119–1127.
94. De Keyser J, De Backer J-P, Wilczack N, Herroelen L. Dopamine agonists used in the treatment of Parkinson's disease and their selectivity for the D1, D2, and D3 dopamine receptors in human striatum. *Prog Neuropsychopharmacol Biol Psychiatry.* 1995;19:1147–1154.

5.5 | Neurocomputational Analysis of Dopamine Function

DANIEL DURSTEWITZ

The brain is a computational system in the sense that it processes information derived from environmental inputs to compute adaptive behavioral outputs. One major goal of computational neuroscience is to explore the nature of these computational operations that link inputs to outputs, and how they are implemented at the physiological and anatomical/ morphological levels. The latter task is particularly daunting, as the brain is a highly complex system consisting of thousands of interacting feedback loops at many different levels of organization (molecules, neurons, areas, etc.) from which computational processes emerge. The link between any two variables within this system, say N-methyl-D-aspartate (NMDA) receptor binding and activation of Ca^{2+}-dependent channels, is subject to modulation by the many often highly nonlinear feedback loops within which these variables are embedded in vivo, for example the level of spiking coherence among neurons that will regulate NMDA activation but will in turn be regulated by Ca^{2+}-dependent channels. As a result, it is often impossible to predict intuitively the functional implications that changes in particular variables will have, and it is very difficult to understand how experimentally dissected feedback loops will operate as parts of an integrated whole. Thus, to date, we have a largely correlational understanding of brain function. We know, for instance, that a particular transmitter is involved in a particular cognitive function, but we often lack a truly mechanistic understanding of how this function is conveyed by the transmitter's specific role in neural network dynamics. Computational neuroscience provides some of the tools that allow us to tackle these problems, exploiting the fact that in a computer simulation, unlike experiments, one can monitor and manipulate every single variable of the system simultaneously and independently. Obviously, however, such an approach only works in close collaboration with experiments that provide the data input to simulated models or test their predictions.

Current computational models of dopamine (DA) modulation have worked either from a more abstract neuroalgorithmic level, starting with specific assumptions about DA's computational role and then working out its implications at a higher cognitive level, or have used a more biophysical/physiological implementation to unravel the dynamic and functional consequences of DA's effects on voltage-gated and synaptic ion channels. This chapter will focus on the latter, and in addition will specifically review models of DA-innervated target regions rather than models of ventral tegmental area/substantia nigra (VTA/SN) DA neurons themselves (see, e.g.,[1,2]). The chapter will start with a brief discussion of how DA may change the input/output functions of single striatal and cortical neurons in the first section, move on to the network level and the potential computational role of DA in higher cognitive functions in the second section, review DA-based models of reinforcement learning in the third section, and close with some conclusions in the fourth section.

DA AND THE SINGLE CELL INPUT/OUTPUT (I/O) FUNCTION

Neurons receive spatiotemporal patterns of synaptic inputs and convert them into temporal series of action potentials. This transformation is a complex and active process, achieved through multiple voltage- and molecularly gated ion channels operating in different voltage regimes and on different time scales, and depends on the current state of the postsynaptic neuron and its own ongoing dynamics (e.g., bursting). We still do not know the true nature of these transformations, and often simple approximations are made that cover various aspects of them. For instance, one may characterize I/O behavior in terms of the average output spiking rate as a function of the average rate of excitatory and inhibitory synaptic inputs. This is basically the level of description used in abstract connectionist-like neural networks where a unit's output is a monotonically increasing and saturating (e.g., sigmoid-like) function of the summed input (Fig. 5.5.1A).

One of the earliest and still very influential computational proposals regarding DA function is that it

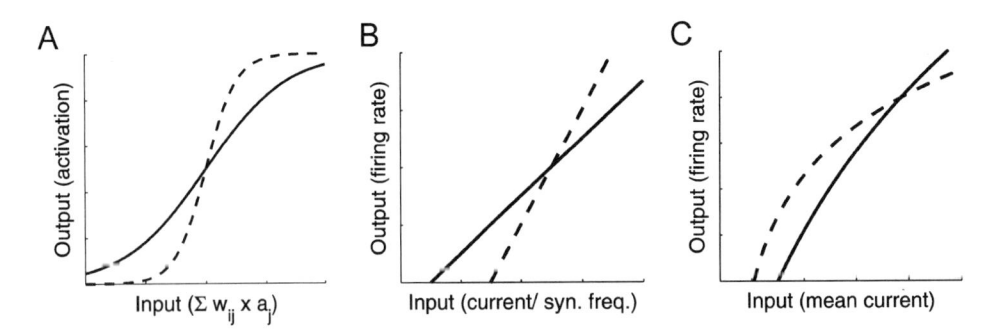

FIGURE 5.5.1. Dopamine/D1R modulation of the single-cell I/O function. (A) In connectionist-like abstract model networks, the output (activation) is usually assumed to be a sigmoid function of the weighted sum of presynaptic activities (a_j). One of the earliest computational proposals was that DA increases the slope (gain) of this sigmoid I/O function,[3] thereby also enhancing the S/N ratio of the unit. (B) Simulations of D1R effects in a biophysically and morphologically highly realistic 189-compartment representation of a striatal MSN provided evidence for this idea,[4] although I/O functions for the simulated MSN were almost linear (schema). (C) Input/output functions under control and DA/D1R activation in PFC pyramidal cells in vitro stimulated with a fluctuating somatic current input (schema).[5] Note that despite the initial increase in gain, the D1R effects on low and high inputs are the opposite of those in (A) and (B). Solid lines, control; dashed lines, D1R activation in all graphs.

increases the gain of this firing rate I/O function[3] (Fig. 5.5.1A), thereby increasing a neuron's signal-to-noise (S/N) ratio by depressing weaker and enhancing stronger inputs. A similar result was recently obtained in a biophysically highly realistic 189-compartment representation (Fig. 5.5.2) of a striatal medium spiny neuron (MSN) with 15 active ion channels distributed across its soma and dendrites.[4] Based on a number of published physiological results on DA's modulation of active ionic conductances, simulated D1-class receptor (D1R) activation led to an increased slope and a right shift of the single-cell I/O function (Fig. 5.5.1B), where the input was either a somatic current injection or was provided by hundreds of synapses distributed across the dendritic tree. Hence, as in the abstract Servan-Schreiber et al. model,[3] D1 activation in this biophysically highly realistic model resulted in depression of small and enhancement of larger inputs. This behavior was not observed with D2-class receptor (D2R) activation, however, which caused a more uniform enhancement at all input levels. Recent experimental observations in prefrontal cortex (PFC) pyramidal neurons recorded in brain slices in vitro also support a D1-induced increase of the single-neuron I/O gain,[5] but its manifestation is different from what has been proposed in the models (Fig. 5.5.1C). Mimicking synaptic input from a network of presynaptic neurons by a fluctuating (noisy) somatic current injection, DA induced more curvature (stronger nonlinearity) in the I/O function (Fig. 5.5.1C), and this effect was blocked by the D1R antagonist SCH23390. But in this case, lower input currents are enhanced while very strong inputs tend to produce diminished output, contrary to the model's assumptions/results, yet consistent with a number of previous electrophysiological reports that D1 stimulation generally enhances excitability of pyramidal cells and interneurons.[6–9] Hence, while an increased neural gain induced by DA via D1R stimulation is a consistent outcome of different approaches, it is less clear whether this also goes hand in hand with an enhanced S/N ratio at the single-neuron level.

Another proposal based on both experimental[7,10] and modeling[11] results has been that DA induces bistability in striatal cells and prefrontal cortex (PFC) neurons. The term *bistability* refers to the fact that a dynamical system may exhibit two stable (steady) states that exist simultaneously, and among which external stimuli might switch the system back and forth (Fig. 5.5.3A; see also Fig. 5.5.4). In particular, the idea is that single neurons may exhibit two stable membrane potentials instead of the usual single resting potential (Fig. 5.5.3A). Two or even multiple stable membrane potential and firing rate levels have indeed been demonstrated experimentally in single synaptically isolated entorhinal and cerebellar neurons, for instance.[12,13] Experimentally, the effects of D1R stimulation on isolated striatal MSN and on PFC pyramidal cells are state-dependent, reducing or leaving unaffected excitability at more hyperpolarized membrane potential levels while increasing it at more depolarized levels.[7,10] These differential state-dependent effects are rooted in the different membrane voltage operating regimes of DA-affected intrinsic currents in striatal MSN, especially an inwardly rectifying potassium current (I_{KIR}) and a high-voltage-activated calcium current (I_{CaL}) which are both enhanced by D1 stimulation (Fig. 5.5.3B): While at relatively hyperpolarized levels I_{KIR} dominates and hence the D1 effect is mainly suppressive, I_{CaL}

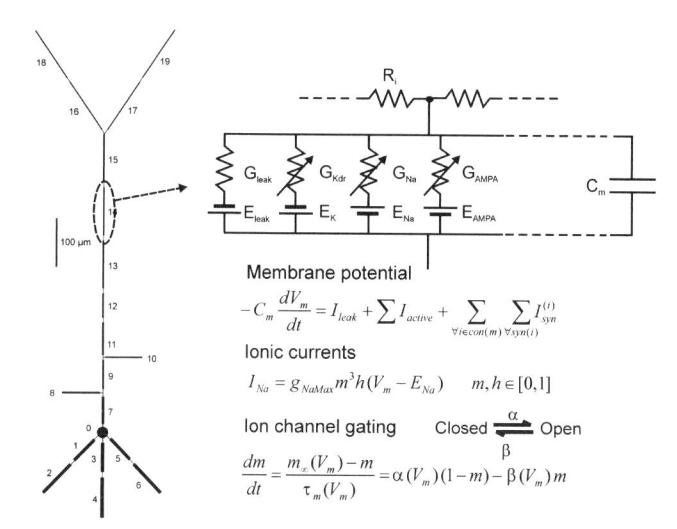

FIGURE 5.5.2. Biophysical modeling in a nutshell. In this computational approach to single-neuron and network function, the morphologies of real neurons are first translated into a structure of connected compartments (20 in this example), each of them in turn being represented by an equivalent electrical circuit that captures all the passive and active (ligand-, voltage-, or ion-gated) currents flowing across the cell membrane. Each of these membrane currents is generated by a static (passive) or adjustable (active) conductance (the zigzag lines) in series with an ionic battery driving that current. All currents are in parallel to the membrane capacitance (C_m), and patches of membrane are connected through the intracellular (cytoplasmatic) resistance (R_i). The operation of these circuits is described by a set of nonlinear differential equations for the membrane voltages in all somatodendritic compartments, and the gating variables mimicking the voltage-dependent transitions between different states of the underlying ion channel. The voltage is regulated by the sum of all passive, active, and synaptic currents, which in turn are given by Ohm's law and the product of different gating (activation and inactivation) variables. The whole system of differential equations describing a network of such neurons is implemented on a computer and then integrated numerically, yielding solutions of all system variables as functions of time (see Figs. 5.5.4 and 5.5.5). *Source*: Reprinted from[64] with permission from Elsevier (copyright 2008).

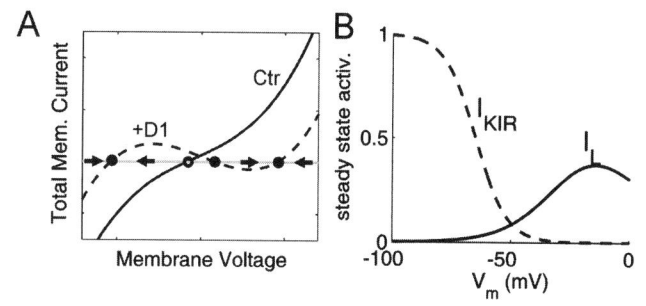

FIGURE 5.5.3. D1 receptor modulation of cellular bistability. (A) Schema of the steady-state relationship (i.e., given all ionic gating variables being in their voltage-dependent steady states) between total membrane current (I_m) and membrane voltage (V_m). Note that the change in membrane voltage (dV_m/dt) is proportional to the membrane current; hence, whenever $I_m = 0$ (the intersections of the black solid and dashed lines with the gray line), the cell is in a steady state (fixed point) where $dV_m/dt = 0$. Under control conditions (Ctr), this cell exhibits only one fixed point (open circle), while according to Gruber et al.,[11] under D1R stimulation it may exhibit three (solid circles). Only the two outermost of these three fixed points are stable, however, with regard to small voltage deflections. As indicated by the arrows, in these two cases V_m is driven back to the fixed points after a small perturbation since the membrane current is depolarizing below and hyperpolarizing above these points. (B) Schema of the steady-state voltage dependence of K⁺ inwardly rectifying (I_{KIR}) and L-type Ca²⁺ (I_L) conductances (see[11]). *Source*: Part (B) slightly modified from[125] with permission from Georg Thieme Verlag (copyright 2006).

increasingly activates with membrane depolarization and finally overrides the I_{KIR}-mediated D1 effects. Thereby D1 stimulation causes increased nonlinearity in the current–voltage relationship that ultimately might give rise to bistability[11] (Fig. 5.5.3A). However, Moyer et al.[4] could not reproduce D1-induced bistability in their highly realistic 189-compartment representation of a striatal MSN. D1 activation still moved the cellular dynamic in this direction, yet the cell fell short of attaining true bistability. Various other hypotheses regarding DA modulation of dendritic signal integration were tested by these authors, but none of them could be conclusively supported by their simulation results.

In conclusion, both experimental and modeling results support the idea that D1R activation increases the I/O (or frequency over current, f/I) gain of cortical and striatal neurons, yet at least in prefrontal neurons it may not be accompanied by an enhanced S/N ratio, and it may fall short of inducing true bistability, although there might be a tendency to move in this direction. Bistability may manifest under some physiological conditions, however—for instance, if a background of synaptic inputs in combination with D1R activation lifts cells that are not intrinsically bistable up to the level of true bistability (see, e.g.,[14,15]). As described next, both D1-induced bistability and an increased neuronal f/I gain may help to stabilize working memory.

DA AND NEURAL NETWORK DYNAMICS: FROM PHYSIOLOGICAL MODELS TO COGNITIVE IMPLICATIONS

Early connectionist-type models suggested that a putative DA-mediated increase in the single-neuron gain may explain some of the cognitive and attentional deficits in schizophrenic patients,[16,17] and later on proposed a role for DA D2R activation in gating of information into the PFC that could account for behavioral findings in different cognitive settings like the Stroop task.[18–20] Although connectionist models can provide interesting insights into the cognitive dynamics resulting from a specific set of neural givens, it is important to note that they already start with quite explicit assumptions about DA's computational function. During the 1990s, more and more data became available on the detailed D1R and D2R modulation of voltage-gated and synaptic ion channels using patch-clamp techniques that went beyond earlier studies on cell excitability[6,8,9,21–29] and paved the way for detailed biophysical models of DA modulation.[30–34] Biophysical models often represent the dendritic structure of real neurons by a set of connected compartments, each modeled by a set of differential equations that describe the evolution of the membrane potential according to various voltage-gated, Ca^{2+}-gated, and synaptically gated ionic currents (see, e.g.,[35]). These, in turn, are often modeled by Hodgkin-Huxley-like gating kinetics as first formulated by Hodgkin and Huxley[36] in their Nobel Prize–winning work on action potential generation. Hence, this approach translates neuronal structures into equivalent electrical circuits that mimic current flow across active and passive membrane channels and between neuronal compartments, as illustrated in Figure 5.5.2. The appeal of these models is their close relation to biophysical quantities measured electrophysiologically. They allow effects of DA measured in vitro to be implemented rather directly, with no or only few additional assumptions. For instance, the ~40% change in NMDA conductances revealed in vitro may translate into a ~40% change in the parameter regulating the maximum NMDA conductance in the model.[31,34]

D1R Modulation of Working Memory

Some of the first simulation studies investigating DA function in biophysical network models focused on working memory and the online maintenance of information within the PFC.[30–32] A dysregulated DA system within the PFC has been proposed to play a key role in schizophrenia, and working memory and other deficits of executive control are among the most prominent cognitive symptoms of schizophrenic patients.[37–40] Studies dating back to the 1970s demonstrated the fundamental importance of DA in working memory functions,[41,42] while in the 1990s, the specific regulation through D1R and D2R moved into the spotlight,[43–46] alongside the detailed cellular studies of D1R/D2R modulation of prefrontal neurons and synapses.[6,8,9,22,24–29] At the cellular level, DA via D1Rs and D2Rs has multiple and diverse effects on a variety of voltage- and ligand-gated ion channels in striatal[10,21,23] and cortical neurons. In PFC cells, D1 stimulation enhances persistent Na^+ channels,[6,25; but see 26] depresses slow K^+ and presumably N-type Ca^{2+} channels,[6,9] has mixed effects on L-type Ca^{2+} channels,[29] enhances I_h,[47] and enhances both $GABA_A$- and NMDA-mediated synaptic inputs through various mechanisms,[24,27,28] yet reduces presynaptic release probability.[27,48] The functional outcome of this combination of different and partly apparently opposing effects is difficult to predict intuitively, even at the single-neuron level (see the previous section), but obviously the problem is much more complicated if network interactions are added. Biophysical models can be an invaluable tool for shedding light on the functional implications of this diverse complexity of DA-modulated cellular and synaptic processes, and thus help to establish specific links between the biophysical and cognitive levels.

Single-unit recordings from primate PFC during working memory tasks had suggested that prefrontal neurons keep an active (online) memory of goal-related items by elevating their firing rates in a stimulus-specific manner during the delay periods of this task.[49–54] These are the periods intervening between the presentation of a cue stimulus and a choice situation during which mental maintenance of task-related information is required in the absence of external cues. Accordingly, the computational models for probing DA involvement in working memory were set up to reproduce the low- and high-activity states associated with spontaneous (baseline) activity and stimulus-specific delay activity in the in vivo recordings[49,51] (Fig. 5.5.4B). This is achieved by embedding *cell assemblies*—that is, groups of functionally related neurons that share a strong excitatory connectivity with each other—in the network (Fig. 5.5.4A). Once the firing activity within a cell assembly is driven across a certain threshold by an external stimulus, the assembly can maintain activity autonomously due to this strong recurrent excitation that is mainly supported by slowly decaying NMDA currents (Fig. 5.5.4B), an idea first made explicit by Wang[55] and experimentally supported, for instance, by Seamans et al.[56] Formally, as illustrated in Figure 5.5.4C and explained in the corresponding legend, the spontaneous rest state and the stimulus-specific high-activity states correspond to different attractor states of the system and represent a form of network bistability or multistability (as opposed to the single-cell bistability discussed in the previous section).

Starting from such a configuration that mimics important functional characteristics of a working memory network, it can now be investigated in detail how the D1R- or D2R-mediated changes reported in vitro affect the network dynamics. The results of such simulations revealed that the combined effects of D1-induced conductance modulations led to a change in network dynamics that made it more difficult to

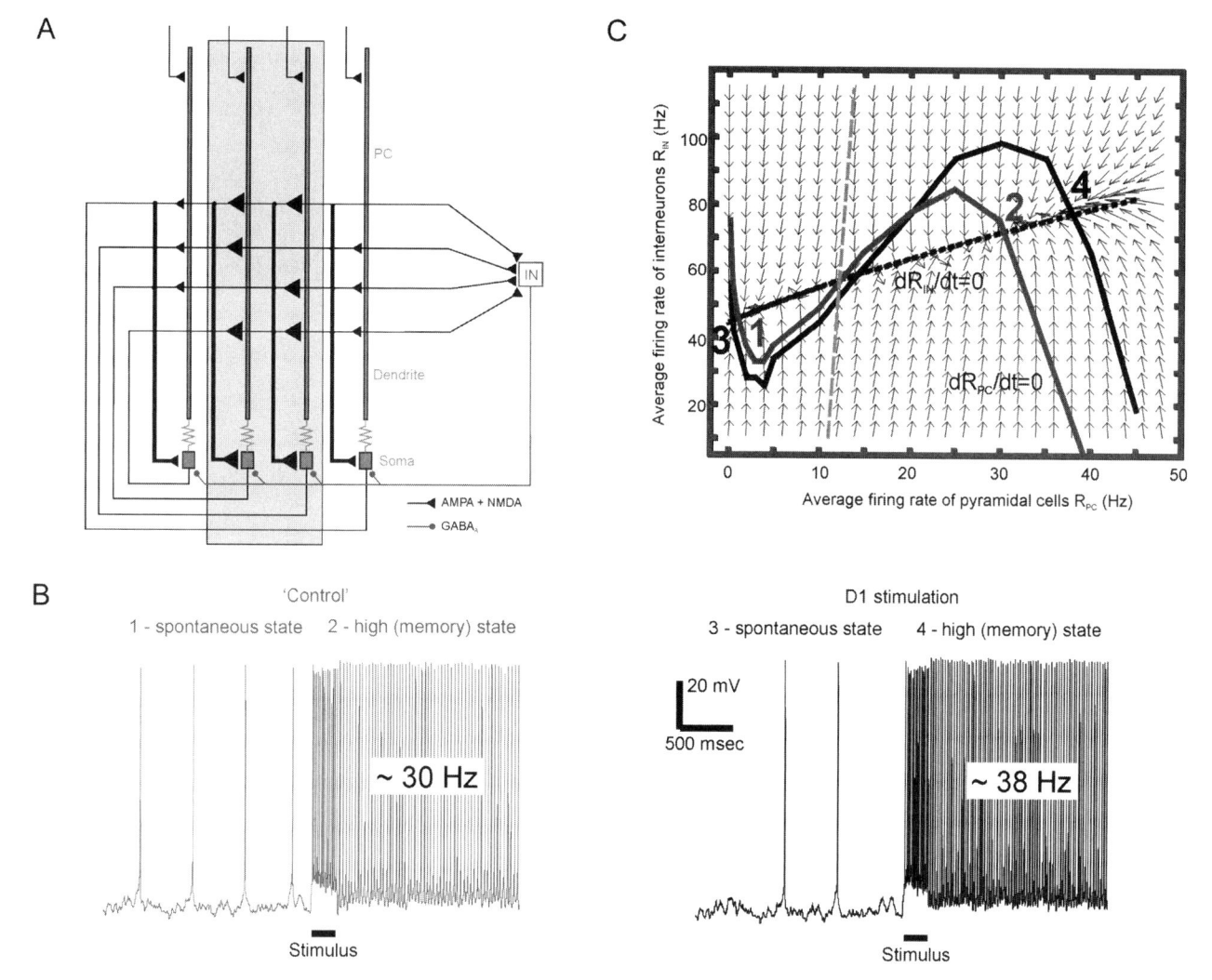

FIGURE 5.5.4. Stimulus-selective persistent activity, presumably underlying working memory (see text), and D1 modulation in a PFC network model. (A) Structure of a network model underlying simulations such as those shown in (B) and (C) and in Figure 5.5.5. Pyramidal cells consist of a somatic and a dendritic compartment and are recurrently coupled via both AMPA and NMDA excitatory synapses. They also excite a population of interneurons that feeds back inhibition mediated by $GABA_A$ synaptic conductances into all pyramidal cells. Stimulus-specific persistent (delay) activity could be produced by cell assemblies (highlighted by the gray square) consisting of subpopulations of pyramidal neurons with stronger than average mutual synaptic connections. Once a threshold of activation is crossed, these strong, recurrent excitatory inputs could keep high firing levels within a stimulated cell assembly going, and thus maintain stimulus-specific enhanced firing rates, as observed experimentally. *Source*: Slightly modified from[34] and reprinted from[64] with permission from Elsevier (copyright 2002, 2008). (B) An external stimulus switches a cell assembly from a low, spontaneous (1) into a stimulus-specific high, persistent (2) activity state that is maintained even after withdrawal of the initiating stimulus (until terminated by some other event), thus encoding an online memory of the stimulus, in agreement with experimental observations.[49,51] D1 stimulation differentially suppresses low (spontaneous) and enhances high (stimulus-selective memory) activity in these simulations. *Source*: Reprinted from[34] with permission from Elsevier (copyright 2002). (C) The dynamical basis of these phenomena revealed by a numerically derived state space representation of the model dynamics: The graph shows a two-dimensional plane spanned by the average firing rate of the pyramidal cells (x-axis) and the average firing rate of the interneurons (y-axis) within a cell assembly. Arrows indicate the flow, that is, give the direction of change of average firing rates as a function of the current state of the network (the length of all arrows was normalized to 1). Following these arrows, one sees that firing rates converge to either one of two points (labeled "1" and "2"), corresponding to the low, spontaneous and high, persistent firing rates illustrated in (B). These points of convergence are called *attractor states* of the system dynamics, and they are more formally given by the intersection of two lines (called *nullclines*), one giving the steady-state firing rates of the pyramidal cells as a function of a fixed average rate of the interneurons (dark gray solid curve), and the other showing the steady-state firing rate of the interneurons as a function of a fixed pyramidal cell rate (black dotted curve). Hence, where these lines intersect, both pyramidal neurons and interneurons are in their steady states, yielding a "fixed point." The two regions of convergence for the low and high firing rate attractors are separated by the light gray dashed line and are called their *basins of attraction*. This line separating the basins of attraction can be seen as a threshold: To elicit memory activity, a stimulus has to drive the network from its low, spontaneous state across this border between the basins such that it converges toward one of the high-activity states. Within this representation, D1 stimulation leads to a stretching of the pyramidal cell nullcline (black solid curve) along the x- and y-axes that underlies the increase in energy barriers among activity states illustrated more explicitly in (B). *Source*: Reprinted from[34] with permission from Elsevier (copyright 2002).

switch between various high-activity (active memory) states—that is, to an increase in the "energy barrier" between different discrete states of network activity[5,30–33] (Fig. 5.5.5A). These effects are partly rooted in the differential contribution of various D1-modulated currents to different activity regimes (see also Fig. 5.5.3B): While the D1-induced increase in NMDA and other voltage-dependent currents[6,24,27] fosters the currently active memory state by boosting recurrent excitation within cell assemblies (see also[57]), the concomitant increase in GABA$_A$ currents[28] leads to fiercer competition among different active ensembles of neurons, thereby limiting the set of items encoded in working memory. At the same time, this D1-mediated enhancement of GABA$_A$ currents, as well as the reduction in glutamate release probability,[27,48] make it harder to evoke activity in cell assemblies in the first place.[31] This increased differentiation among attractor states under D1R action has important functional implications: It strongly boosts the robustness of items in working memory and protects them from distracting stimuli and noise as it becomes harder to switch the system among different activity states.[5,30–33] In this manner, by unraveling how changes in D1R-modulated ionic conductances shape the network dynamics, these computational studies helped to link the cellular and synaptic effects of D1R stimulation reported in vitro[6,9,24,25,27,28] to their functional implications for working memory as demonstrated in psychopharmacological studies.[43–46] The—in neurodynamical terms—increased energy barrier among different PFC activity states may be seen as a form of

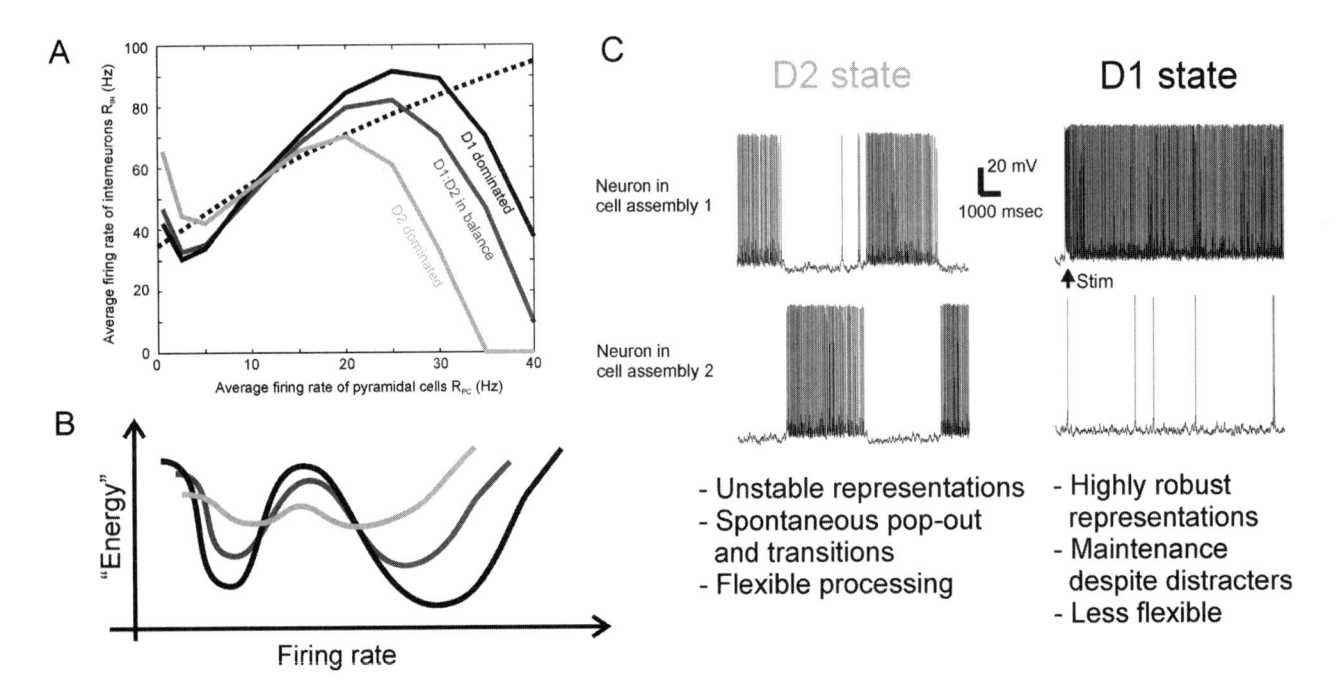

FIGURE 5.5.5. D1-state and D2-state dynamics in the biophysical model network. (A) Top: State space representation of the network dynamics (see Fig. 5.5.4C for explanation) in D1-dominated, balanced, and D2-dominated regimes (only the nullclines are shown). While D1 stimulation leads to a stretching of the pyramidal cell nullcline along the *x*- and *y*-axes, D2 stimulation leads to a contraction along both dimensions. Bottom: Representation of the same information (with corresponding line colors) in terms of an "energy landscape" (note that this graph is just a schema). Minima of the energy correspond to the fixed point attractors in the top graph, and the state of the system may be envisioned as a ball rolling down into the nearest minimum. The local slopes in this graph depend on the sign and magnitude of the derivatives of the underlying system, as indicated by the flow field in Figure 5.5.4C. The graph makes it clear that it becomes much harder to switch between different attractor states in the D1-dominated regime as the troughs move apart and the "valleys" become much steeper. Conversely, in the D2-dominated regime, the valleys become so flat and nearby that noise may easily push the system from one state into the other. Also note that the ease of attractor hopping could in principle be regulated purely by the steepness of the valley slopes, without any change in the position of the minima, and hence without any change in average firing rates. For simplicity, the energy landscape is shown just as a two-dimensional graph. For a two-dimensional state space, as in the top graph and Figure 5.5.4C, the full energy landscape would be a surface in a three-dimensional space (the axes of the state space plus the energy axis) that may be obtained, for instance, by integrating along the flow field. (B) Network simulation illustrating the fact that the system spontaneously switches or cycles among different attractors (neural representations) in the D2-dominated regime while robustly maintaining a once elicited attractor in the D1-dominated regime. *Source*: Slightly modified from[80] and reprinted from[64] with permission from Springer Science+Business Media (copyright 2007) and Elsevier (copyright 2008), respectively.

enhanced S/N ratio and in this sense fits well with many other reports and proposals that DA enhances S/N.[3,38,58,59] It is important to note, however, that the term *S/N ratio* often refers to quite different phenomena in different contexts. An enhanced differentiation among network attractor states is not the same as the previously proposed S/N amplification conveyed by an increased single-unit gain,[3] although these effects might support each other.

Empirically, a crucial role for D1Rs in PFC-dependent working memory has been well documented,[43–46] although its functional involvement is more complex, with both sub- and supranormal levels of D1R stimulation being detrimental.[44,45,60,61] In principle, the biophysical models reviewed above could account for these behavioral observations by noting that while subnormal D1 stimulation would easily lead to loss of information from working memory due to interference, supranormal stimulation might lead to overly stable representations that resist task-related updating processes.[31] Behaviorally, this may result in perseveration with a once activated representation maintaining control over the behavior, and this has in fact been described as the major source of errors in rats with supranormal D1 stimulation.[44,62] On the other hand, D1R stimulation has been reported to have inverted-U-shaped dose/response curves even with regard to electrophysiological observables.[63] This could have many possible explanations rooted either in important differences in the D1 agonist dose dependency of various ion channels or in dynamic network interactions still unaccounted for by the models described above (see [64] for a more detailed discussion).

There is also in vivo electrophysiological support for the biophysical models. One indication of the D1-induced changes in network dynamics is an increased differentiation of firing rates associated with target- or memory-related activity as compared to nontarget, background, and spontaneous activity. This D1-mediated differentiation reflects the underlying dynamical changes that cause the increased robustness of working memory representations. Indeed, these effects in the model replicate both early electrophysiological observations suggesting a (relative to baseline) stronger amplification of delay- and response-related single-unit activity by DA during working memory,[65,66] and very recent findings suggesting that D1 agonists *diminish* nontarget related activity to a much larger degree than target-related activity.[67,68] In both cases, the outcome is an increased differentiation among target and nontarget activity, as predicted by the computational models (cf. Fig. 5.5.4). Similarly, stimulation of the origin of the DA pathway in the VTA increased current pulse-evoked high-rate firing while decreasing spontaneous low firing of PFC neurons recorded intracellularly in vivo.[69] Therefore, in accord with simulations, DA in vivo appears to enhance the differentiation among low and high firing rate states either through diminishing the former, amplifying the latter, or both, or—in terms of the dynamical models—through an increase in the energy barrier between different activity states.

How does the D1R regulation of single-neuron dynamics reviewed in the previous section fit into this picture? A shift of the single-neuron dynamic toward cellular bistability[11] would be expected to complement and boost the D1R effects on network bistability reviewed above. In fact, in some of the network simulation studies, D1R effects on intracellular currents like I_{NaP} were taken into account and contributed to the overall change in attractor dynamics,[30,31] although a specific role for cellular bistability was not assessed explicitly. In a recent study, Thurley et al.[5] reported that the experimentally established cellular gain increase mediated by D1R also helped to separate and stabilize low and high rate attractor states at the network level. However, the crucial factor here was more the increased nonlinearity (curvature) of the single-cell f/I function (Fig. 5.5.1C) rather than the gain increase per se or a change in the cellular S/N ratio (see also[70]).

D2 Modulation of PFC Attractor Dynamics and Function

Most models of DA function in PFC have focused on D1R for a number of reasons. First, D1Rs are 5- to 10-fold higher in density in PFC than D2Rs,[71,72] which was taken to indicate their much higher physiological relevance. Based on the comparable physiological effectiveness of D1 and D2 agonists in PFC slices, however, this inference may be contested.[24,28] Second, behavioral-pharmacological experiments demonstrated a clear role for D1R in working memory but failed to do so for D2R, which in many studies seemed not to influence performance significantly[43–46,60] or neural activity during delay periods.[63,65,66,73] Probably for these reasons, D1R effects on prefrontal neurons received more attention from in vitro electrophysiologists during the 1990s than those mediated by D2R. While D2R effects on working memory may be modest, a number of recent studies have supported their role in certain aspects of executive function, notably in tasks that require a high degree of flexibility as assessed, for instance, by the ability to switch easily among rule sets (i.e., set-shifting tasks).[74–76] And although D2Rs, unlike D1Rs, do not seem to have much impact on delay activity itself, they do modulate response-related activity in working memory tasks.[73]

In PFC neurons in vitro, D2 agonists act oppositely from D1R (even within the same cells) by reducing NMDA and GABA$_A$ currents[24,27,77] as well as pyramidal cell excitability[6,22] rather than enhancing these characteristics. Such opposing effects of D1 versus D2 stimulation have also been observed for various molecular markers of intracellular cascades, like cyclic adenosine monophosphate (cAMP) production and phosphorylation of DA- and cAMP-regulated phosphoprotein, 32 kDa (DARPP-32).[78,79] Consequently, in PFC attractor models, the dynamic implications of simulated D2R activation are just the opposite of those obtained for simulated D1R effects: D2 activation *reduces* the barrier among activity states in the model networks (Fig. 5.5.5); that is, the valleys of the energy landscape become so flat and move towards each other such that noise may easily push the system from one state into the other.[31,64,80,81] This would cause spontaneous (i.e., stimulus-unrelated) activation of memory states due to noisy fluctuations, highly unstable representations, and fast and spontaneous transitions between many different activity states, as illustrated in Figure 5.5.4C.[28,31,64] Functionally, it is clear that this would imply less stable working memory (persistent activity). On the other hand, it might favor functions that require a high degree of cognitive flexibility, as it should facilitate transitions among representations as required in a set-shifting task. Hence, the opposing effects of D1R and D2R on PFC attractor landscapes as revealed by biophysical models might explain the dualism of D1R and D2R involvement in PFC cognition on a biophysical basis. Again, a related conceptual role for D2R has been anticipated in connectionist-type models where the function of gating information into the PFC was assigned to these receptors.[19] Although the computational purpose is similar, namely, allowing a switch among PFC representations, the dynamical mechanisms are different in the two cases, however.

In summary, implementation of D1R- and D2R-mediated effects, as measured in vitro in biophysically realistic network models, revealed some of the dynamical mechanisms through which they may impact higher cognitive functions. These studies suggested that D1R activation may increase the energy barrier among different attractor states of the PFC dynamics, preventing noise- or distracter-induced transitions among representational states and thus boosting working memory. D2 receptor activation, on the other hand, may lower the energy barrier among discrete states of network activity, potentially favoring cognitive functions that require fast and flexible switching among representational states at the expense of stable maintenance of information. Although, as reviewed above, a number of behavioral and pharmacological studies are well in line with these predictions, they still await more direct electrophysiological and behavioral testing in which D1R and D2R agents are applied locally to the PFC while behavioral measures in both sets of cognitive tasks are combined with electrophysiological recordings.

DA-BASED MODELS OF REINFORCEMENT LEARNING

A long-standing behavioral literature relates DA to reinforcement learning and reward processing in a number of different Pavlovian, operant, and working memory (delayed response and reversal tasks) learning tasks, using a variety of psychopharmacological, electrophysiological, and DA measurements via microdialysis or voltammetry approaches.[82–90] In some tasks, interfering with the DA system has a greater effect on acquisition than on behavioral expression.[82,91] In vitro findings furthermore support the conclusion that DA regulates long-term potentiation (LTP) and depression (LTD) at various cortical and striatal synapses through D1Rs and D2Rs.[92–95] Starting in the early 1990s, a series of in vivo electrophysiological studies by Schultz and colleagues[96–99] led to a completely new perspective on DA's role in reinforcement learning: These authors observed in operant conditioning and delayed response tasks that putative dopaminergic neurons recorded in the VTA and SN transiently (<200 ms) increased their firing rates during the presentation of unpredicted rewards, but these responses vanished during the course of training as the animal learned to predict the forthcoming rewards. Instead, these phasic responses transferred to the preceding conditioned stimulus (CS). Impressively, in those cases where a predicted reward was omitted, VTA/SN neurons responded with a brief cessation of activity at around the time at which the reward would have appeared on previous trials.[100] These findings were interpreted to imply that phasic firing of VTA/SN neurons did not signal rewarding events per se but rather the *deviation* of the actually perceived reward value from the one expected, that is, a reward prediction error (see Chapter 6.4 by Tobler in this volume for an in-depth discussion of these findings).

Since the dawn of neural modeling,[101,102] learning through modification of connection strengths (*synaptic weights*) has always been one of the topics, if not the central topic, of at least the more abstract (connectionist-type) neural network theories. Theories of animal learning also played an important role in an area of artificial intelligence called *machine learning*.[103] Hence, the literature of DA's involvement in reinforcement learning and plasticity received a lot of attention from computational approaches and vice versa.[84,100,104–108] Before the empirical studies of Schultz and his colleagues, DA was often conceived as a direct reward signal that may act in synaptic plasticity rules like $\Delta w_{ij} = \gamma\, r\, a_i\, a_j$,[109] $r \in \{-1,0,1\}$, where γ is a learning rate and w_{ij} is the synaptic weight between neurons i and j that is modified according to a Hebb-like correlation rule where $a_{i,j}$ represents the activities of the connected units. In the presence of the reward signal ($r = +1$), the weights will be increased if the activities of the units are positively correlated but will be decreased if punishment is signaled ($r = -1$). Hence, importantly, learning only takes place in the presence of a reinforcement signal, and it will be such as to strengthen current representations and cue–response associations if the unconditioned stimulus (US) was rewarding and to weaken them if the US was punishing.

However, according to the findings of Schultz and colleagues,[84,96–100] DA signals a prediction error, a mismatch between what was expected and what really happened, not reward or punishment per se. These findings tie in extremely nicely with a concept in machine learning and artificial neural networks called *temporal-difference-error learning* (TDE-L).[103,110] The idea behind TDE-L is that animals (or machines) should strive to maximize a weighted sum of all future rewards $\sum_{k=0}^{\infty} \alpha^k r(t+k)$ with $0 \leq \alpha < t$. The weights α^k are exponentially decaying in time—that is, rewards are temporally discounted the further ahead in the future they lie, which makes sense biologically, as future rewards are more uncertain and the lifetime of an animal is limited. Behavioral studies have demonstrated temporal discounting in humans and animals.[111,112] To maximize this quantity (the temporally discounted sum of future rewards), the animal has to be able to predict the total reward associated with different behavioral options and then choose accordingly (such a mapping is also called a *policy* in machine learning). Hence, the learning problem is one of adjusting predictions according to empirical observations rather than reinforcing links between stimuli and responses directly. A learning rule for this problem can be derived by noting a basic consistency requirement: Predictions $P(t+1)$ for time $t+1$ have to be consistent with what was predicted for the previous time t, $P(t)$; otherwise, the prediction for time t must be wrong. Hence, within this learning scheme, synaptic weights are adjusted to reduce the temporal difference (prediction) error $TDE(t) = r(t) + \alpha P(t+1) - P(t)$. The basic insight of Schultz, Dayan, Montague, and others[100,105] was that the response properties of VTA/SN neurons adhere quite closely to what is required according to the TDE-L theory, that is, the phasic response of DA neurons encodes $TDE(t)$ (note that in the simplest case, one may take $\alpha = 0$, so that $TDE(t)$ would just reflect the difference between the actual and the predicted reward for time t). This basic insight has triggered a substantial amount of computational work. Some studies focused on explaining, within the TDE-L framework, the development and temporal transfer of phasic VTA/SN responses themselves during the experimental paradigms employed by Schultz and others.[105] Other studies exploited the putative DA TDE signal for learning action sequences or delayed response tasks, for instance,[106,107] or for explaining the emergence of anticipatory responses in striatal or cortical neurons.[113]

Most of the work using the TDE-L framework was performed at a more abstract or conceptual level where single units were characterized simply by their average firing rate or represented as even more abstract conceptual entities. Processes of synaptic plasticity or transmission were not explicitly modeled; hence, how precisely the DA signal led to adaptive modification of synaptic connections was not addressed. More recently, more physiologically oriented models have focused on the experimental phenomenon of spike-timing dependent plasticity (STDP), where the change in synaptic strength is a function of the difference between the timing of the pre- and postsynaptic spikes (Fig. 5.5.6; reviewed in[114]). In particular, if the postsynaptic neuron emits a spike before receiving presynaptic input, LTD is the consequence, while the postsynaptic neuron firing shortly after the arrival of a presynaptic input results in LTP of the excitatory postsynaptic potential (EPSP) elicited by that input. These effects can be traced partly to the supralinear amount of Ca^{2+} influx through NMDA- and high-voltage-activated Ca^{2+} channels triggered if a dendritic backpropagating spike arrives on top of an EPSP.[115–117] Recently, STDP at corticostriatal synapses was demonstrated to be modulated by D1R and D2R activation, with somewhat different results from different groups.[93,118] Pawlak and Kerr[118] showed that both LTP and LTD produced by an STDP protocol depend on D1R activation, while D2R antagonists have no effects on LTP/LTD amplitude. Shen et al.[93] similarly found that D1R antagonists abolish LTP in an STDP protocol in D1R-expressing MSN. However, they did not observe the LTD part of the STDP curve under control conditions (see Fig. 5.5.6), yet LTD was revealed when D1Rs were blocked. Hence, rather than enhancing both the LTP and LTD parts of the STDP curve, as reported by Pawlak and Kerr,[118] in the Shen et al.[93] study D1R stimulation seemed to shift the whole STDP curve upward, enhancing LTP and diminishing LTD. Furthermore, Shen et al.[93] found that D2R

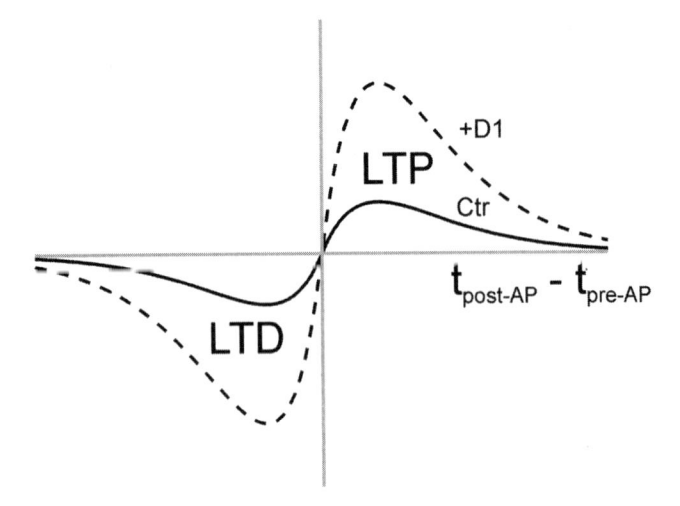

FIGURE 5.5.6. D1 receptor modulation of STDP (schema). In STDP protocols, the long-term change in synaptic efficacy (i.e., LTP vs. LTD; solid curve) is determined by the order of spiking in the pre- and the postsynaptic neuron, that is, the temporal difference between the spike times of the pre- (t_{pre-AP}) and postsynaptic ($t_{post-AP}$) action potentials (Ctr = control). In the Izhikevich[119] model, DA is assumed to increase both the LTP and the LTD part of the STDP curve (dashed line), in agreement with some experimental findings.[92,118]

antagonists blocked LTP in D2R-expressing MSN, while D2 agonists, on the other hand, converted LTP into LTD when the presynaptic spike led the postsynaptic one. The fact that Pawlak and Kerr[118] used mainly single-spike-like stimulation while Shen et al.[93] used burst-like stimulation of afferent fibers and, in addition, were able to reliably differentiate D1R- and D2R-expressing MSN populations may account for these partial differences. A recent physiology-oriented computational study of a network of spiking neurons[119] showed that a combination of an STDP-rule with a DA-mediated reinforcement signal, implemented as an enhancement of both the LTP and LTD parts of the STDP curve (Fig. 5.5.6; as observed by Pawlak and Kerr[118]), could solve the "distal reward problem"[110]: The connection between a presynaptic stimulus and a postsynaptic response could be strengthened by a reward-indicating DA signal even seconds later. The reason is that the precise timing between the pre- and postsynaptic spikes exploited by STDP is so unlikely to occur by chance in a spontaneously spiking network that in combination with an *eligibility trace* (synaptic tagging[120]), it might leave a unique signature that survives for many seconds. The study by Izhikevich[119] therefore provides the first specific link between the phasic DA response as a reward or TDE signal and DA-modulated synaptic plasticity in a physiology-based network model.

CONCLUSIONS

The computational role of DA has been investigated at many different levels, from detailed cellular studies of D1/D2 modulation of dendritic signal integration to abstract connectionist-type large-scale networks implementing cognitive functions and learning. While the biophysically oriented neural models are commonly more concerned with working out the computational function of DA itself,[4,5,11,30,31,33] the connectionist models usually already start with specific assumptions about DA's computational role (e.g., gating or signaling TD errors) and then evaluate how these assumptions will affect cognitive processing and learning at a higher level.[19,20,106,107,121]

Although there are many empirical and computational questions and apparent contradictions that still need to be resolved, some of which are discussed below, it is interesting to note that there seems to be some functional coherence among the computational roles assigned to DA, at least those linked to D1R: At the cellular level, D1R stimulation may cause an increased nonlinearity of the single cell f/I curve and may shift the single-neuron dynamic toward bistability. Both of these factors could favor the online maintenance of active representations in the face of noise and distractors.[5,122] Such functionality has been explicitly demonstrated at the network level: Experimentally revealed D1R effects on intrinsic voltage-gated and synaptic ion channels acted jointly in biophysical network models to render active memory representations more robust to noise and distraction.[30–33] More generally, underlying this effect is an increased differentiation (barrier) among neural attractor states that makes hopping between them harder. Favoring the short-term maintenance of goal states or recent relevant inputs this way might help to establish links across time between different behavioral and environmental events and consequences—for example, the relationship between temporally preceding predictive stimuli, specific behavioral responses, and rewards or punishments occurring later. The D1R effects on synaptic plasticity, boosting both LTP and LTD,[118] as well as recently demonstrated LTP of intrinsic excitability by D1R,[123] might further help to engrave the detected temporal links from short-term memory in long-term memory. In this way, all D1R-mediated effects on voltage-gated and synaptic ion channels and plasticity might come together within the superordinate function of detecting and memorizing predictive relations within the stream of environmental inputs, directing long-term behavior toward the most rewarding events and drawing it away from punishing events. This overall picture also complies well with the response properties of DA midbrain neurons, as reported by Schultz and colleagues.[84,100] Although this consistency among different D1R-mediated functions as well as the task-related responses of DA neurons is very compelling, however, one issue that still remains to be worked out is the different time courses of DA effects.[86,124] At least in vitro, many D1R effects on voltage-gated and synaptic channels take minutes to set in and then persist throughout the recording session (e.g.,[6,27,28]). Part of the D1R effects on synaptic plasticity are likely to be mediated as well by the D1R modulation of NMDA and voltage-gated Ca^{2+} channels.[6,24,29] Thus, under in vivo conditions, the time course and control of D1R activation induced by phasic DA midbrain responses are still important issues that need to be resolved.

For D2R, the picture is much less clear than for D1R, and at least with regard to PFC function at each level of description (in vitro physiology, functional in vivo physiology, computational models, and cognition), the data on D2R involvement are less extensive. Findings that D2R stimulation inverts D1R effects on NMDA and $GABA_A$ currents,[24,28] on cell excitability,[22] and on various parameters of the DA-triggered intracellular signaling cascade[79] suggest that its functional/computational role might be just the opposite of what D1Rs implement. In fact, in biophysical network simulations, D2R stimulation is predicted to decrease differentiation (barriers) among attractor states, thus favoring hopping among different representational states.[31,80,81] Thus, PFC networks were suggested to operate in two fundamentally different dynamic regimes, a "D1-dominated" and a "D2-dominated" one, each associated with particular computational advantages and disadvantages.[64] Whereas a D1-dominated regime favors the maintenance of active representations, a D2-dominated regime may aid cognitive flexibility by promoting fast switches among representational states.[74,75] The dynamic regime would hence be adjusted via the D1:D2 receptor activation ratio, which in turn might be an inverted-U-type function of DA concentration (see[64] for details). Behavioral findings dissociating D1 and D2 effects in PFC support such a distinction.[43–46,74,75] In terms of synaptic plasticity, at least at corticostriatal synapses, D2R stimulation may actually convert LTP into LTD,[93] which again may favor cognitive flexibility by breaking down existing long-term-memory representations rather than imprinting them, as with D1R stimulation. Hence, although the empirical evidence for the D2 side of this "dual-state" theory of DA function is certainly much weaker, a number of findings at the in vitro electrophysiological and behavioral levels, as well as with regard to synaptic plasticity, consistently point in the direction of D2R involvement in computational functions just the opposite of those mediated by D1R, namely, flexibility and change rather than persistence and long-term representation.

I conclude with a list of some of the most important open issues from a computational point of view. First, the problem of the time course of D1 effects was pointed out above. In contrast to the delayed and long-lasting effects of D1R stimulation, D2R-mediated effects seem to set in earlier and are more transient in PFC slices.[22,28,77] To what degree this is the case in vivo still needs to be worked out (see[124]), but these findings suggest that there might be temporally unfolding effects of DA stimulation whose functional meaning is not yet understood (see[125] for some suggestions). Second, both empirically and computationally, most effects have been characterized in terms of "D1R-class" and "D2R-class" functions, but there are quite a number of experimental hints that within each class, DA receptor effects can be further dissociated at the behavioral level.[75] Third, interactions with neuromodulatory systems other than the fast glutamate and GABA signaling systems have not yet been studied computationally. The joint action of several modulatory substances (e.g., dopamine and acetylcholine) might yield nonlinear interactions on computational dynamics and cognitive function that could play an important yet poorly studied role in vivo. Fourth, most computational models of DA function so far rest on a quite simple view of cortical and striatal neurodynamics. They focus on bistability or multistability of firing rate attractors as supposedly underlying working memory, but they do not consider more complex dynamical phenomena like oscillations in various frequency bands or quasi-attracting states embedded in a chaotic ground state. The latter are probably a more realistic scenario for cortical dynamics,[126] and oscillations in various bands have been implied in working memory, for instance.[127–130] Fifth, very few computational models, at least those specified at a more physiological level, have investigated the full feedback dynamics between DA-innervated brain areas and the DA midbrain neurons themselves, or more generally among different DA-modulated and DA-level-affecting brain areas. For instance, PFC activation of VTA/SN neurons[131–133] may have important implications for DA levels in other areas, such as the striatum, as well. Finally, it would be interesting to see whether there are important commonalities in DA function within various cortical and subcortical/basal ganglia regions, between species, or even between phyla. For instance, it might turn out that an important general computational function of D1R activation is to stabilize active goal states, whether these are basal motoric states (like a specific arm position), specific motivational states represented in striatal structures, or high-level behavioral goal states in PFC. Indeed, one overarching theme seems to be that DA is primarily involved in functions of behavioral organization, like prediction learning, reward evaluation, working memory, or motor planning, rather than in early sensory processes. Indeed even in invertebrates, DA is primarily associated with motor functions.[134]

ACKNOWLEDGMENTS

This work was funded by grants from the Deutsche Forschungsgemeinschaft to DD (Du 354/5-1 & Du 354/6-1).

REFERENCES

1. Komendantov AO, Komendantova OG, Johnson SW, Canavier CC. A modeling study suggests complementary roles for GABAA and NMDA receptors and the SK channel in regulating the firing pattern in midbrain dopamine neurons. *J Neurophysiol.* 2004;91(1):346–357.
2. Canavier CC, Oprisan SA, Callaway JC, Ji H, Shepard PD. Computational model predicts a role for ERG current in repolarizing plateau potentials in dopamine neurons: implications for modulation of neuronal activity. *J Neurophysiol.* 2007;98(5): 3006–3022.
3. Servan-Schreiber D, Printz H, Cohen JD. A network model of catecholamine effects: gain, signal-to-noise ratio, and behavior. *Science.* 1990;249:892–895.
4. Moyer JT, Wolf JA, Finkel LH. Effects of dopaminergic modulation on the integrative properties of the ventral striatal medium spiny neuron. *J Neurophysiol.* 2007;98(6):3731–3748.
5. Thurley K, Senn W, Lüscher HR. Dopamine increases the gain of the input-output response of rat prefrontal pyramidal neurons. *J Neurophysiol.* 2008;99(6):2985–2997.
6. Yang CR, Seamans JK. Dopamine D1 receptor actions in layer v-vi rat prefrontal cortex neurons in vitro: modulation of dendritic-somatic signal integration. *J Neurosci.* 1996;16:1922–1935.
7. Lavin A, Grace AA. Stimulation of D1-type dopamine receptors enhances excitability in prefrontal cortical pyramidal neurons in a state-dependent manner. *Neuroscience.* 2001;104(2):335–346.
8. Gorelova N, Seamans JK, Yang CR. Mechanisms of dopamine activation of fast-spiking interneurons that exert inhibition in rat prefrontal cortex. *J Neurophysiol.* 2002;88(6):3150–3166.
9. Dong Y, White FJ. Dopamine D1-class receptors selectively modulate a slowly inactivating potassium current in rat medial prefrontal cortex pyramidal neurons. *J Neurosci.* 2003;23(7): 2686–2695.
10. Hernández-López S, Bargas J, Surmeier DJ, Reyes A, Galarraga E. D1 receptor activation enhances evoked discharge in neostriatal medium spiny neurons by modulating an L-type Ca^{2+} conductance. *J Neurosci.* 1997;17(9):3334–3342.
11. Gruber AJ, Solla SA, Surmeier DJ, Houk JC. Modulation of striatal single units by expected reward: a spiny neuron model displaying dopamine-induced bistability. *J Neurophysiol.* 2003;90(2):1095–1114.
12. Egorov AV, Hamam BN, Fransen E, Hasselmo ME, Alonso AA. Graded persistent activity in entorhinal cortex neurons. *Nature.* 2002;420:173–178.
13. Loewenstein Y, Mahon S, Chadderton P, Kitamura K, Sompolinsky H, Yarom Y, Häusser M. Bistability of cerebellar Purkinje cells modulated by sensory stimulation. *Nat Neurosci.* 2005;8(2):202–211.
14. Lisman JE, Fellous JM, Wang XJ. A role for NMDA-receptor channels in working memory. *Nat Neurosci.* 1998;1:273–275.
15. Durstewitz D, Seamans JK, Sejnowski TJ. Neurocomputational models of working memory. *Nat Neurosci.* 2000;3(suppl): 1184–1191.
16. Cohen JD, Servan-Schreiber D. Context, cortex, and dopamine: a connectionist approach to behavior and biology in schizophrenia. *Psychol Rev.* 1992;99(1):45–77.
17. Cohen JD, Servan-Schreiber D. A theory of dopamine function and its role in cognitive deficits in schizophrenia. *Schizophr Bull.* 1993;19(1):85–104.
18. Braver TS, Barch DM. A theory of cognitive control, aging cognition, and neuromodulation. *Neurosci Biobehav Rev.* 2002;26(7):809–817.
19. Cohen JD, Braver TS, Brown JW. Computational perspectives on dopamine function in prefrontal cortex. *Curr Opin Neurobiol.* 2002;12(2):223–229.
20. O'Reilly RC, Noelle DC, Braver TS, Cohen JD. Prefrontal cortex and dynamic categorization tasks: representational organization and neuromodulatory control. *Cereb Cortex.* 2002;12(3):246–257.
21. Surmeier DJ, Kitai ST. D1 and D2 dopamine receptor modulation of sodium and potassium currents in rat neostriatal neurons. *Prog Brain Res.* 1993;99:309–324.
22. Gulledge AT, Jaffe DB. Dopamine decreases the excitability of layer V pyramidal cells in the rat prefrontal cortex. *J Neurosci.* 1998;18:9139–9151.
23. Nisenbaum ES, Mermelstein PG, Wilson CJ, Surmeier DJ. Selective blockade of a slowly inactivating potassium current in striatal neurons by (+/-) 6-chloro-APB hydrobromide (SKF82958). *Synapse.* 1998;29(3):213–224.
24. Zheng P, Zhang XX, Bunney BS, Shi WX. Opposite modulation by cortical N-Methyl-D-aspartate receptor-mediated responses by low and high concentrations of dopamine. *Neuroscience.* 1999;91:527–535.
25. Gorelova NA, Yang CR. Dopamine D1/D5 receptor activation modulates a persistent sodium current in rat prefrontal cortical neurons in vitro. *J Neurophysiol.* 2000;84:75–87.
26. Maurice N, Tkatch T, Meisler M, Sprunger LK, Surmeier DJ. D1/D5 dopamine receptor activation differentially modulates rapidly inactivating and persistent sodium currents in prefrontal cortex neurons. *J Neurosci.* 2001;21(7): 2268–2277.
27. Seamans JK, Durstewitz D, Christie BR, Stevens CF, Sejnowski TJ. Dopamine D1/D5 receptor modulation of excitatory synaptic inputs to layer V prefrontal cortex neurons. *Proc Natl Acad Sci USA.* 2001;98:301–306.
28. Seamans JK, Gorelova N, Durstewitz D, Yang CR. Bidirectional dopamine modulation of GABAergic inhibition in prefrontal cortical pyramidal neurons. *J Neurosci.* 2001;21:3628–3638.
29. Young CE, Yang CR. Dopamine D1/D5 receptor modulates state-dependent switching of soma-dendritic Ca^{2+} potentials via differential protein kinase A and C activation in rat prefrontal cortical neurons. *J Neurosci.* 2004;24:8–23.
30. Durstewitz D, Kelc M, Güntürkün O. A neurocomputational theory of the dopaminergic modulation of working memory functions. *J Neurosci.* 1999;19:2807–2822.

31. Durstewitz D, Seamans JK, Sejnowski TJ. Dopamine-mediated stabilization of delay-period activity in a network model of prefrontal cortex. *J Neurophysiol*. 2000;83:1733–1750.
32. Compte A, Brunel N, Goldman-Rakic PS, Wang XJ. Synaptic mechanisms and network dynamics underlying spatial working memory in a cortical network model. *Cereb Cortex*. 2000;10(9):910–923.
33. Brunel N, Wang XJ. Effects of neuromodulation in a cortical network model of object working memory dominated by recurrent inhibition. *J Comput Neurosci*. 2001;11(1):63–85.
34. Durstewitz D, Seamans JK. The computational role of dopamine D1 receptors in working memory. *Neural Networks*. 2002;15:561–572.
35. Koch C, Segev I, eds. *Methods in Neuronal Modeling*. 2nd ed. Cambridge, MA: MIT Press; 1997.
36. Hodgkin AL, Huxley AF. A quantitative description of membrane current and its application to conduction and excitation in nerve. *J Physiol (Lond)*. 1952;117:500–544.
37. Goldman-Rakic PS, Muly EC 3rd, Williams GV. D(1) receptors in prefrontal cells and circuits. *Brain Res Brain Res Rev*. 2000;31:295–301.
38. Winterer G, Weinberger DR. Genes, dopamine and cortical signal-to-noise ratio in schizophrenia. *Trends Neurosci*. 2004;27(11):683–690.
39. Barch DM. The cognitive neuroscience of schizophrenia. *Annu Rev Clin Psychol*. 2005;1:321–353.
40. Beck AT, Rector NA. Cognitive approaches to schizophrenia: theory and therapy. *Annu Rev Clin Psychol*. 2005;1:577–606.
41. Brozoski TJ, Brown RM, Rosvold HE, Goldman PS. Cognitive deficit caused by regional depletion of dopamine in prefrontal cortex of rhesus monkey. *Science*. 1979;205:929–932.
42. Simon H, Scatton B, Moal ML. Dopaminergic A10 neurones are involved in cognitive functions. *Nature*. 1980;286(5769):150–151.
43. Sawaguchi T, Goldman-Rakic PS. The role of D1-dopamine receptor in working memory: local injections of dopamine antagonists into the prefrontal cortex of rhesus monkeys performing an oculomotor delayed-response task. *J Neurophysiol*. 1994;71:515–528.
44. Zahrt J, Taylor JR, Mathew RG, Arnsten AFT. Supranormal stimulation of D1 dopamine receptors in the rodent prefrontal cortex impairs spatial working memory performance. *J Neurosci*. 1997;17:8528–8535.
45. Seamans JK, Floresco SB, Phillips AG. D1 receptor modulation of hippocampal-prefrontal cortical circuits integrating spatial memory with executive functions in the rat. *J Neurosci*. 1998;18:1613–1621.
46. Müller U, von Cramon DY, Pollmann S. D1- versus D2-receptor modulation of visuospatial working memory in humans. *J Neurosci*. 1998;18:2720–2728.
47. Rosenkranz JA, Johnston D. Dopaminergic regulation of neuronal excitability through modulation of Ih in layer V entorhinal cortex. *J Neurosci*. 2006;26:3229–3244.
48. Gao WJ, Krimer LS, Goldman-Rakic PS. Presynaptic regulation of recurrent excitation by D1 receptors in prefrontal circuits. *Proc Natl Acad Sci USA*. 2001;98:295–300.
49. Fuster JM. Unit activity in prefrontal cortex during delayed-response performance: neuronal correlates of transient memory. *J Neurophysiol*. 1973;36:61–78.
50. Quintana J, Yajeya J, Fuster JM. Prefrontal representation of stimulus attributes during delay tasks. I. Unit activity in cross-temporal integration of sensory and sensory-motor information. *Brain Res*. 1988;474:211–221.
51. Funahashi S, Bruce CJ, Goldman-Rakic PS. (1989) Mnemonic coding of visual space in the monkey's dorsolateral prefrontal cortex. *J Neurophysiol*. 1989;61:331–349.
52. Miller EK, Erickson CA, Desimone R. Neural mechanisms of visual working memory in prefrontal cortex of the macaque. *J Neurosci*. 1996;16:5154–5167.
53. Rainer G, Asaad WF, Miller EK. Selective representation of relevant information by neurons in the primate prefrontal cortex. *Nature*. 1998;393:577–579.
54. Fuster JM, Bodner M, Kroger JK. Cross-modal and cross-temporal association in neurons of frontal cortex. *Nature*. 2000;405:347–351.
55. Wang XJ. Synaptic basis of cortical persistent activity: the importance of NMDA receptors to working memory. *J Neurosci*. 1999;19:9587–9603.
56. Seamans JK, Nogueira L, Lavin A. Synaptic basis of persistent activity in prefrontal cortex in vivo and in organotypic cultures. *Cereb Cortex*. 2003;13:1242–1250.
57. Fellous JM, Sejnowski TJ. (2003) Regulation of persistent activity by background inhibition in an in vitro model of a cortical microcircuit. *Cereb Cortex*. 2003;13(11):1232–1241.
58. Rolls ET, Thorpe SJ, Boytim M, Szabo I, Perrett DI. Responses of striatal neurons in the behaving monkey. 3. Effects of iontophoretically applied dopamine on normal responsiveness. *Neuroscience*. 1984;12:1201–1212.
59. Winterer G, Ziller M, Dorn H, Frick K, Mulert C, Dahhan N, Herrmann WM, Coppola R. (1999) Cortical activation, signal-to-noise ratio and stochastic resonance during information processing in man. *Clin Neurophysiol*. 1999;110:1193–1203.
60. Arnsten AF, Cai JX, Murphy BL, Goldman-Rakic PS. Dopamine D1 receptor mechanisms in the cognitive performance of young adult and aged monkeys. *Psychopharmacology (Berl)*. 1994;116(2):143–151.
61. Cai JX, Arnsten AF. Dose-dependent effects of the dopamine D1 receptor agonists A77636 or SKF81297 on spatial working memory in aged monkeys. *J Pharmacol Exp Ther*. 1997;283(1):183–189.
62. Floresco SB, Phillips AG. Delay-dependent modulation of memory retrieval by infusion of a dopamine D1 agonist into the rat medial prefrontal cortex. *Behav Neurosci*. 2001;115:934–939.
63. Williams GV, Goldman-Rakic PS. Modulation of memory fields by dopamine D1 receptors in prefrontal cortex. *Nature*. 1995;376:572–575.
64. Durstewitz D, Seamans JK. The dual-state theory of prefrontal cortex dopamine function with relevance to COMT genotypes and schizophrenia. *Biol Psychiatry*. 2008;64(9):739–749.
65. Sawaguchi T, Matsumara M, Kubota K. Dopamine enhances the neuronal activity of spatial short-term memory task in the primate prefrontal cortex. *Neurosci Res*. 1988;5:465–473.
66. Sawaguchi T, Matsumara M, Kubota K. Catecholaminergic effects on neuronal activity related to a delayed response task in monkey prefrontal cortex. *J Neurophysiol*. 1990;63:1385–1400.
67. Arnsten AF. Catecholamine and second messenger influences on prefrontal cortical networks of "representational knowledge": a rational bridge between genetics and the symptoms of mental illness. *Cereb Cortex*. 2007;suppl 1:i6–i15.
68. Vijayraghavan S, Wang M, Birnbaum SG, Williams GV, Arnsten AF. Inverted-U dopamine D1 receptor actions on prefrontal neurons engaged in working memory. *Nat Neurosci*. 2007;10(3):376–384.

69. Lavin A, Nogueira L, Lapish CC, Wightman RM, Phillips PE, Seamans JK. Mesocortical dopamine neurons operate in distinct temporal domains using multimodal signaling. *J Neurosci.* 2005;25(20):5013–5023.
70. Brunel N. Persistent activity and the single-cell frequency-current curve in a cortical network model. *Network.* 2000;11(4):261–280.
71. Lidow MS, Goldman-Rakic PS, Gallager DW, Rakic P. Distribution of dopaminergic receptors in the primate cerebral cortex: quantitative autoradiographic analysis using [3H]raclopride, [3H]spiperone and [3H]SCH23390. *Neuroscience.* 1991;40(3):657–671.
72. Joyce JN, Goldsmith S, Murray A. Neuroanatomical localization of D1 versus D2 receptors: similar organization in the basal ganglia of the rat, cat and human and disparate organization in the cortex and limbic system. In: Waddington Jl, ed. *D1:D2 Dopamine Receptor Interactions.* London: Academic Press; 1993:23–49.
73. Wang M, Vijayraghavan S, Goldman-Rakic PS. Selective D2 receptor actions on the functional circuitry of working memory. *Science.* 2004;303:853–856.
74. Floresco SB, Magyar O, Ghods-Sharifi S, Vexelman C, Tse MT. Multiple dopamine receptor subtypes in the medial prefrontal cortex of the rat regulate set-shifting. *Neuropsychopharmacology.* 2006;31:297–309.
75. Floresco SB, Magyar O. Mesocortical dopamine modulation of executive functions: beyond working memory. *Psychopharmacology (Berl).* 2006;188:567–585.
76. Lee B, Groman S, London ED, Jentsch JD. Dopamine D2/D3 receptors play a specific role in the reversal of a learned visual discrimination in monkeys. *Neuropsychopharmacology.* 2007;32:2125–2134.
77. Trantham-Davidson H, Neely LC, Lavin A, Seamans JK. Mechanisms underlying differential D1 versus D2 dopamine receptor regulation of inhibition in prefrontal cortex. *J Neurosci.* 2004;24(47):10652–10659.
78. Nishi A, Snyder GL, Greengard P. Bidirectional regulation of DARPP-32 phosphorylation by dopamine. *J Neurosci.* 1997;17:8147–8155.
79. Greengard P. (2001) The neurobiology of dopamine signaling. *Biosci Rep.* 2001;21(3):247–269.
80. Durstewitz D. Dopaminergic modulation of prefrontal cortex network dynamics. In Tseng K-Y, Atzori M, eds. *Monoaminergic Modulation of Cortical Excitability.* New York, NY: Springer; 2007:217–234.
81. Loh M, Rolls ET, Deco G. A dynamical systems hypothesis of schizophrenia. *PLoS Comput Biol.* 2007;3(11):e228.
82. Beninger RJ. Role of D1 and D2 receptors in learning. In: Waddington Jl, eds. *D1:D2 Receptor Interactions.* London: Academic Press; 1993:115–157.
83. Beninger RJ. Dopamine and incentive learning: a framework for considering antipsychotic medication effects. *Neurotox Res.* 2006;10(3–4):199–209.
84. Schultz W. Predictive reward signal of dopamine neurons. *J Neurophysiol.* 1998;80(1):1–27.
85. Schultz W. Behavioral theories and the neurophysiology of reward. *Annu Rev Psychol.* 2006;57:87–115.
86. Schultz W. Multiple dopamine functions at different time courses. *Annu Rev Neurosci.* 2007;30:259–288.
87. Wickens JR, Reynolds JN, Hyland BI. Neural mechanisms of reward-related motor learning. *Curr Opin Neurobiol.* 2003;13(6):685–690.
88. Wise RA. Dopamine, learning and motivation. *Nat Rev Neurosci.* 2004;5(6):483–494.
89. Mingote S, de Bruin JP, Feenstra MG. Noradrenaline and dopamine efflux in the prefrontal cortex in relation to appetitive classical conditioning. *J Neurosci.* 2004;24(10):2475–2480.
90. van der Meulen JA, Joosten RN, de Bruin JP, Feenstra MG. (2007) Dopamine and noradrenaline efflux in the medial prefrontal cortex during serial reversals and extinction of instrumental goal-directed behavior. *Cereb Cortex.* 2007;17(6):1444–1453.
91. Ploeger GE, Spruijt BM, Cools AR. Spatial localization in the Morris water maze in rats: acquisition is affected by intra-accumbens injections of the dopaminergic antagonist haloperidol. *Behav Neurosci.* 1994;108(5):927–934.
92. Otani S, Daniel H, Roisin MP, Crepel F. Dopaminergic modulation of long-term synaptic plasticity in rat prefrontal neurons. *Cereb Cortex.* 2003;13(11):1251–1256.
93. Shen W, Flajolet M, Greengard P, Surmeier DJ. Dichotomous dopaminergic control of striatal synaptic plasticity. *Science.* 2008;321(5890):848–851.
94. Stuber GD, Klanker M, de Ridder B, Bowers MS, Joosten RN, Feenstra MG, Bonci A. Reward-predictive cues enhance excitatory synaptic strength onto midbrain dopamine neurons. *Science.* 2008;321(5896):1690–1692.
95. Wickens JR. Synaptic plasticity in the basal ganglia. *Behav Brain Res.* 2008;199(1):119–128.
96. Schultz W, Romo R. Dopamine neurons of the monkey midbrain: contingencies of responses to stimuli eliciting immediate behavioral reactions. *J Neurophysiol.* 1990;63(3):607–624.
97. Ljungberg T, Apicella P, Schultz W. Responses of monkey dopamine neurons during learning of behavioral reactions. *J Neurophysiol.* 1992;67(1):145–163.
98. Schultz W, Apicella P, Ljungberg T. Responses of monkey dopamine neurons to reward and conditioned stimuli during successive steps of learning a delayed response task. *J Neurosci.* 1993;13(3):900–913.
99. Schultz W, Romo R, Ljungberg T, Mirenowicz J, Hollermann JR, Dickinson A. (1995) Reward-related signals carried by dopamine neurons. In: Houk JC, David JL, Beiser DG, eds. *Models of Information Processing in the Basal Ganglia.* Cambridge, MA: MIT Press; 1995:233–248.
100. Schultz W, Dayan P, Montague PR. A neural substrate of prediction and reward. *Science.* 1997;275(5306):1593–1599.
101. Rosenblatt F. *Principles of Neurodynamics.* New York, NY: Spartan Books; 1962.
102. Rumelhart DE, McClelland JL. *Parallel Distributed Processing: Explorations in the Microstructure of Cognition.* Cambridge, MA: MIT Press; 1986.
103. Sutton RS, Barto AG. *Reinforcement Learning.* Cambridge, MA: MIT Press; 1998.
104. Wickens J. Striatal dopamine in motor activation and reward-mediated learning: steps towards a unifying model. *J Neural Transm Gen Sect.* 1990;80(1):9–31.
105. Montague PR, Dayan P, Sejnowski TJ. A framework for mesencephalic dopamine systems based on predictive Hebbian learning. *J Neurosci.* 1996;16(5):1936–1947.
106. Suri RE, Schultz W. Learning of sequential movements by neural network model with dopamine-like reinforcement signal. *Exp Brain Res.* 1998;121(3):350–354.
107. Suri RE, Schultz W. A neural network model with dopamine-like reinforcement signal that learns a spatial delayed response task. *Neuroscience.* 1999;91(3):871–890.
108. Daw ND, Doya K. The computational neurobiology of learning and reward. *Curr Opin Neurobiol.* 2006;16(2):199–204.
109. Klopf AH. *The Hedonistic Neuron: A Theory of Memory, Learning, and Intelligence.* Washington, DC: Hemisphere; 1982.

110. Barto AG. (1995) Reinforcement learning. In: Arbib MA, ed. *The Handbook of Brain Theory and Neural Networks.* Cambridge, MA: MIT Press; 1995:804–809.
111. Rachlin H, Green L. Commitment, choice and self-control. *J Exp Anal Behav.* 1972;17(1):15–22.
112. Madden GJ, Petry NM, Badger GJ, Bickel WK. Impulsive and self-control choices in opioid-dependent patients and non-drug-using control participants: drug and monetary rewards. *Exp Clin Psychopharmacol.* 1997;5(3):256–262.
113. Suri RE, Schultz W. Temporal difference model reproduces anticipatory neural activity. *Neural Comput.* 2001;13(4):841–862.
114. Bi G, Poo M. Synaptic modification by correlated activity: Hebb's postulate revisited. *Annu Rev Neurosci.* 2001;24:139–166.
115. Magee JC, Johnston D. A synaptically controlled, associative signal for Hebbian plasticity in hippocampal neurons. *Science.* 1997;275:209–212.
116. Koester HJ, Sakmann B. Calcium dynamics in single spines during coincident pre- and postsynaptic activity depend on relative timing of back-propagating action potentials and subthreshold excitatory postsynaptic potentials. *Proc Natl Acad Sci USA.* 1998;95:9596–9601.
117. Schiller J, Schiller Y, Clapham DE. NMDA receptors amplify calcium influx into dendritic spines during associative pre- and postsynaptic activation. *Nat Neurosci.* 1998;1:114–118.
118. Pawlak V, Kerr JN. Dopamine receptor activation is required for corticostriatal spike-timing-dependent plasticity. *J Neurosci.* 2008;28(10):2435–2446.
119. Izhikevich EM. Solving the distal reward problem through linkage of STDP and dopamine signaling. *Cereb Cortex.* 2007;17(10):2443–2452.
120. Frey U, Morris RG. Synaptic tagging and long-term potentiation. *Nature.* 1997;385(6616):533–6.
121. Vitay J, Hamker FH. Sustained activities and retrieval in a computational model of the perirhinal cortex. *J Cogn Neurosci.* 2008;20(11):1993–2005.
122. Fransen E, Alonso AA, Hasselmo ME. Simulations of the role of the muscarinic-activated calcium-sensitive nonspecific cation current INCM in entorhinal neuronal activity during delayed matching tasks. *J Neurosci.* 2002;22(3):1081–1097.
123. Chen L, Bohanick JD, Nishihara M, Seamans JK, Yang CR. Dopamine D1/5 receptor-mediated long-term potentiation of intrinsic excitability in rat prefrontal cortical neurons: Ca^{2+}-dependent intracellular signaling. *J Neurophysiol.* 2007;97(3):2448–2464.
124. Lapish CC, Kroener S, Durstewitz D, Lavin A, Seamans JK. The ability of the mesocortical dopamine system to operate in distinct temporal modes. *Psychopharmacology (Berl).* 2007;191:609–625.
125. Durstewitz D. A few important points about dopamine's role in neural network dynamics. *Pharmacopsychiatry.* 2006;39(S1):S72–S75.
126. Durstewitz D, Deco G. Computational significance of transient dynamics in cortical networks. *Eur J Neurosci.* 2008;27:217–227.
127. Raghavachari S, Kahana MJ, Rizzuto DS, Caplan JB, Kirschen MP, Bourgeois B, Madsen JR, Lisman JE. Gating of human theta oscillations by a working memory task. *J Neurosci.* 2001;21:3175–3183.
128. Lutzenberger W, Ripper B, Busse L, Birbaumer N, Kaiser J. Dynamics of gamma-band activity during an audiospatial working memory task in humans. *J Neurosci.* 2002;22:5630–5638.
129. Pesaran B, Pezaris JS, Sahani M, Mitra PP, Andersen RA. Temporal structure in neuronal activity during working memory in macaque parietal cortex. *Nat Neurosci.* 2002;5:805–811.
130. Lee H, Simpson GV, Logothetis NK, Rainer G. Phase locking of single neuron activity to theta oscillations during working memory in monkey extrastriate visual cortex. *Neuron.* 2005;45:147–156.
131. Tong Z-Y, Overton PG, Clark D. Stimulation of the prefrontal cortex in the rat induces patterns of activity in midbrain dopaminergic neurons which resemble natural burst events. *Synapse.* 1996;22:195–208.
132. Tong Z-Y, Overton PG, Clark D. Antagonism of NMDA but not AMPA/kainate receptors blocks bursting in dopaminergic neurons induced by electrical stimulation of the prefrontal cortex. *J Neural Transm.* 1996;103:889–905.
133. Gao M, Liu CL, Yang S, Jin GZ, Bunney BS, Shi WX. Functional coupling between the prefrontal cortex and dopamine neurons in the ventral tegmental area. *J Neurosci.* 2007;27:5414–5421.
134. Puhl JG, Mesce KA. Dopamine activates the motor pattern for crawling in the medicinal leech. *J Neurosci.* 2008;28(16):4192–4200.

6 Striatum and midbrain—motor and motivational functions

6.1 | Dopamine and Motor Function in Rat and Mouse Models of Parkinson's Disease

TIMOTHY SCHALLERT AND SHEILA M. FLEMING

INTRODUCTION

Ascending dopaminergic systems are involved in a wide array of essential behavioral functions including movement, reward, and learning. In particular, the relationship between dopamine (DA) and movement has been studied in depth since the discovery of DA as a neurotransmitter in the brain in the late 1950s and its association with the disorder Parkinson's disease (PD) soon thereafter.[1] Substantia nigra DA neurons that terminate in the striatum modulate cortical, basal ganglia, and connected regions involved in voluntary motor initiation and execution.[2,3]

DA Required for Some Motor Functions in Rats

The effects of drugs that severely interfere substantially and bilaterally with DA synaptic transmission in rats and mice are dramatic. There is little or no spontaneous movement or response to tactile, ingestive, or most other (but not all) types of sensory stimuli. During moments of postural stability, total akinesia occurs. For example, when treated with a high dose of the DA receptor antagonist haloperidol, reserpine, or severe bilateral neurotoxin-induced DA deficiency, a rat will stand in the middle of an open field without moving until the drug begins to wear off. The rat will cling in awkward postures for hours (catalepsy) as long as it can maintain static stable equilibrium. However, DA is not required for certain types of movement to take place. Thus, if the rat is held upside down and then dropped, it will quickly right itself before being caught by the experimenter.[4] If the animal is picked up and the experimenter tries to move his/her hands in multiple ways to prevent stability, the rat will efficiently adjust its grasping limbs and posture to prevent falling. Any imposed perturbation of postural stability results in immediate stability-regaining reactions that appear quite normal. When the experimenter-imposed challenge to postural stability is discontinued, however, there is no doubt even to a casual observer that the animal has lost the capacity for initiating self-activated movement. But importantly, the animal retains normal or near-normal non-DA motor subsystems organized to achieve and maintain static, stable equilibrium. Only with highly sensitive movement analyses can one detect that the haloperidol-resistant motor responses to maintain stable equilibrium are very slightly slower than normal and that the threshold level of stability perturbation needed to trigger a response is slightly higher than that of control animals.[4–6]

Although DA plays an essential role in voluntary movements, it is not likely to directly mediate the movements. Bilateral inactivation of a small part of the pontine tegmental area of the brain (transiently by infusing a GABA agonist or chronically by infusing a cell body neurotoxin to induce cell loss) completely prevents akinesia,[7] suggesting that DA transmission may not be required even for spontaneous movement initiation. Rather than mediating motor dysfunction directly, blocking DA transmission appears to cause excessive excitation of one or more non-DAergic brain regions, at least one being remote from the basal ganglia. Furthermore, salient sensory input such as aversive sensations, the odor of the opposite sex, or a return to the home cage can often cause otherwise akinetic animals to move immediately and dramatically (paradoxical kinesia).[8–10] Thus, although DA replacement is a highly effective treatment for akinesia and related motor impairments following a loss of DA terminals, the role of DA in motor function may be reasonably understood as indirect rather than direct, the major motor effects of severe DA deficiency (without non-DA neuron loss) as highly selective rather than global, and the deficits as subject to override.

In PD, DA neurons in the substantia nigra pars compacta (SNc) progressively degenerate. This disruption in nigrostriatal DA transmission results in many motor abnormalities. A number of neurotoxins have been used to induce DA cell loss to model PD, with the most extensively studied models being 6-hydroxydopamine (6-OHDA) in the rat and 1-methyl-4-phenyl-1,2,3,6-tetrahydropyridine (MPTP) in the mouse. More recently, the discovery of genetic forms of PD has led to the development of genetic mouse models of parkinsonism. In all models, sensorimotor tests that are sensitive to dysfunction and loss of nigrostriatal DA neurons have been developed to provide important endpoint measures for preclinical testing of potential therapeutic treatments for PD. This chapter reviews many of the tests used in the unilateral 6-OHDA rat and in mice with mutations associated with PD and/or the development of DA neurons.

THE UNILATERAL 6-OHDA RAT

The toxicity of 6-OHDA is relatively specific to catecholaminergic cells and, by preventing transport of the neurotoxin into noradrenergic neurons, it has been used for over 30 years to study the functional consequences of nigrostriatal dopaminergic cell loss.[11–14] 6-Hydroxydopamine kills cells by forming cytotoxic products like hydrogen peroxide, superoxide, and hydroxy radicals.[15–17] 6-OHDA is found in the urine of PD patients and may be linked to the pathology of the disease, although this has yet to be firmly established.[18] Animals with bilateral DA depletion must be fed by gavage in order to survive, so they are not optimal for research. Rats with unilateral 6-OHDA-induced nigrostriatal DA depletion can survive without tube feeding or watering, and they show sensorimotor impairments of the limbs that are contralateral to the side of the lesion.[19–25]

Several reliable non-drug-induced behavioral measures have been developed for the unilateral 6-OHDA rat model of PD. Among the most useful are tests for limb-use asymmetry for weight shifting and support, forelimb movement initiation, and somatosensory dysfunction that are not affected by differences in the amount of experience associated with repeated testing.[20,21,23,26,27] They are sensitive to varying degrees of nigrostriatal DA neuron loss and have been used extensively to evaluate the efficacy of various types of therapies, including drugs, motor enrichment, cell transplants and viral vectors that deliver beneficial agents.[10,28–32] Rats with unilateral DA depletion also show deficits in food pellet reaching and fine digit use during handling of dry strands of pasta,[33,34] the capacity to disengage from ongoing ingestive behavior in response to distractive sensory stimulation,[35,36] forelimb reaction time,[37] forelimb placing,[38,39] head-orienting deficits to tactile von Frey hair stimulation,[19,40,41] turning asymmetries,[20,26] and frequency modulated (50-kHz range) ultrasonic vocalization.[42,43] Three common tests that most likely model key characteristics of parkinsonian akinesia are highlighted in the next sections. These motor deficits are highly related but represent distinct aspects of akinesia, one of the most consistent hallmarks of PD.

Limb-Use Asymmetry During Wall Exploration

The limb-use asymmetry test is a measure of limb use for weight shifting and maintaining stability during vertical exploration of the walls of an enclosure.[23,27,44–48] Forelimb use during wall explorative activity can be assessed by videotaping rats in a transparent cylinder (20-cm diameter and 30-cm height) until 20 limb usages occur. A mirror placed behind the cylinder at an angle enables the rater to record forelimb movements when the animal is turned away from the camera. The cylindrical shape encourages vertical exploration of the walls with the forelimbs and prevents the variability associated with enclosures that have corners. Several behaviors are scored to determine the extent of the asymmetry in forelimb use displayed by the animal. These behaviors include independent and simultaneous or rapidly alternating use of the left and/or right forelimb for contacting the wall during a full rear (see www.schallertlab.org for a movie). This test has been shown to be highly sensitive to varying degrees of DA cell loss[23,47] and is widely used to assess the efficacy of potential treatment interventions.[28–31,49] With severe unilateral DA deficiency, the animal relies primarily on the ipsilateral forelimb for landing on the wall and for lateral wall-based movements. In contrast, the contralateral forelimb primarily is used simultaneously or alternating with the ipsilateral forelimb, and almost never is used independently in successive steps. Methods for scoring and quantifying limb-use asymmetry can be found in detail in previous publications.[27,50] The degree of limb-use asymmetry is correlated with the level of DA terminal loss in the striatum and may be influenced by deficits in the capacity both to initiate weight shifting and to make adjusting steps to deviations in the center of gravity (described in the next two sections).

Movement initiation (single forelimb akinesia)

Voluntary movement initiation can also be easily measured in rats with nigrostriatal DA depletion. Rats are largely front-wheel drive in that they will walk readily on the forelimbs when the hindlimbs are lifted above the ground but not on the hindlimbs when the forelimbs are lifted and vibrissae contact with a horizontal surface is not provided.[51] In this test, the rat is held by its torso with its hindlimbs and one forelimb lifted above the surface of a table so that the weight of its body is supported by one forelimb alone. The number of self-initiated steps occurring in, for example, a 10-s trial is recorded for each forelimb for two trials and then averaged. The time required to initiate the first step with each forelimb can also be measured.[21,24,25,45,52,53] This test is sensitive to moderately severe or severe degrees of DA cell loss.[47]

For decades researchers assumed that, by comparison with bilateral DA depletion, a severe unilateral loss of DA terminals had no detectable effect on initiation of limb stepping. Indeed, when placed in an open field or observed in the home cage, the animals appeared to walk normally, with no hesitation in stepping even with the contralateral limbs. However, careful analysis of behavior indicates that stepping in the unilateral rat model is initiated by the ipsilateral forelimb, and the contralateral limb responds to the weight shift by making a catch-up (adjusting) step to maintain the center of gravity, which leads to the false appearance that both limbs are stepping normally. Moreover, each step of the ipsilateral limb activates brain control mechanisms that promote a more responsive catch-up step in the contralateral limb so that it does not brace or drag behind. This can be demonstrated most simply by examining single limb catchup/adjusting steps, a method described in the following section.

Catch-up (adjusting) steps

Rats with bilateral DA depletion show short steps while walking, and bracing reactions in the limbs rather than stepping when pushed by the experimenter forward or laterally on a smooth surface.[54,55] This action is levodopa reversible, although dyskinesias may occur.[55] Rats with unilateral DA depletion show dragging only of the contralateral limb or a delayed adjusting step, both of which contribute to fewer steps being made when the animal is pushed over a set distance.[6,21,23,56,57] To assess this, each forelimb is examined separately by holding the other three limbs off the smooth surface so that the tested forelimb bears all the weight of the animal while the animal is slowly moved laterally. This test may be comparable to the push-pull test used to observe postural instability in PD patients, who fail to step adequately or quickly enough to maintain their center of gravity. If the surface is rough (e.g., sandpaper) and the animal's center of mass is slowly shifted by the experimenter, the contralateral limb does not brace but instead reacts to the imposed shift of weight by a stepping reaction that is delayed relative to that of the ipsilateral forelimb,[6] resulting in the distance of the weight shift in the contralateral limb being longer. The size of the steps taken by the ipsilateral forelimb is shorter than in control animals, suggesting that the sensitivity to deviation from the center of mass is enhanced by some mechanism that compensates for the contralateral limb deficit.

If both forelimbs in the unilateral DA-deficient rat are placed on a rough surface and the animal is moved slowly forward, the ipsilateral limb steps first, followed immediately by the same size step, rather than a delayed longer step, in the contralateral forelimb. Thus, the movement of the ipsilateral limb appears to facilitate normal stepping in the contralateral limb, and this cross-midline normalization effect occurs only during a 1- to 2-s time window. This may contribute to the normal appearance of spontaneous ground walking or forelimb stepping that had led researchers to assume that locomotor deficits were not present in the unilateral PD model. If, instead, the ipsilateral limb steps and the imposed weight shift is paused for more than about 2 s, the contralateral forelimb fails to match the ipsilateral limb. As a result, the ipsilateral limb makes two steps in a row before the imposed weight shift is large enough to trigger a contralateral limb step, and this contralateral adjusting step size is abnormally large due to a delay in reactivity to the weight shift.

GENETIC MOUSE MODELS OF PARKINSONISM

Within the past decade, several genetic mutations causing rare familial cases of PD have been identified. Mutations in the presynaptic protein alpha-synuclein were some of the first to be described.[58–61] Soon afterward, it was shown that alpha-synuclein is also a major component of Lewy bodies,[62] a pathological hallmark of PD, indicating that alpha-synuclein plays a significant role in both familial and sporadic forms of PD. Several lines of mice overexpressing human alpha-synuclein have been generated, including mice that overexpress human wild-type alpha-synuclein under the Thy1 promoter.[63] In addition to mice with mutations associated with familial forms of PD, mice have also been generated that have mutations that interfere with the development of nigrostriatal DA neurons.[64] The nigrostriatal system is altered to different degrees in these mice; however, it has been difficult to detect motor deficits using traditional automated tests for mice.

Sensorimotor Tests in Genetic Mouse Models of PD

With the development of new genetic mouse models of PD, sensitive and reliable sensorimotor tests like those described for the unilateral 6-OHDA rat are needed for detection of bilateral deficits in mice, which may be subtle. Therefore, based on the tests established for rats, several novel tests have been developed for mice that are sensitive to varying levels of dopaminergic dysfunction in mice. These include tests of motor performance and coordination, response to sensory stimuli, spontaneous activity, and nest building.

Challenging beam traversal

Following brain injury or degeneration, both humans and animals use compensatory strategies to perform tasks accurately, making it difficult to detect impairments especially in the early stages of the damage.[41,65] Therefore, it is important to challenge the animals to the limit of their abilities to uncover early effects of the mutations. The challenging beam traversal test measures motor performance and coordination and is sensitive to even subtle alterations within the nigrostriatal DA system.[64,66–68] Briefly, the beam consists of four sections (25 cm each, 1 m total length), each section having a different width. The beam starts at a width of 3.5 cm and gradually narrows to 0.5 cm by 1-cm increments. Animals are trained to traverse the length of the beam, starting at the widest section and ending at the narrowest, most difficult, section. Animals receive 2 days of training prior to testing. On the day of the test, a mesh grid (1-cm squares) of corresponding width is placed above the beam surface, leaving approximately a 1-cm space between the grid and the beam surface, which serves as a crutch to prevent compensatory motor learning that can mask extant deficits.[69] Animals are then videotaped while traversing the grid-surfaced beam. Videotapes are rated for errors, number of steps made by each animal, and time required to traverse. This test has been shown to be highly sensitive in mice with mutations associated with familial PD[66,67] and in mice with a developmental loss of nigrostriatal DA neurons.[64] In addition, impairments in mice with DA cell loss can be reversed with levodopa.[64]

Pole test and inverted grid

The pole and inverted grid tests are also good measures of motor performance and coordination. For the inverted grid test, animals are placed upside down on a grid above the ground and the number of steps made along the grid and slips through the grid are counted for each mouse, while for pole test, animals are placed head up on the top of a pole, and the time required to orient the body downward as well as the time needed to descend are measured. Both tests are sensitive measures of deficits in alpha-synuclein–overexpressing mice[67] and in mice with a loss of DA neurons.[64,70]

Response to sensory stimuli

Because genetic mouse models of PD have bilateral deficits, we adapted the "dot test" of somatosensory neglect to assess mice with bilateral deficits. In this test, small adhesive stimuli (Avery adhesive-backed labels, 1/4 in. round, or "Tough Spots") are placed on the snout of the mouse, and the time required to make contact and remove the adhesive stimulus is recorded. If the animal does not remove the stimulus within 60 s, then the experimenter removes it, and the trial for the next mouse begins. Stimulus contact and removal times are calculated for each animal. This test has been shown to be sensitive in many genetic mouse models of PD including mice that overexpress alpha-synuclein,[67] DJ-1 knockout mice,[71] parkin knockout mice,[66] and, most recently, mice with a parkin Q113X mutation.[72]

Spontaneous activity

Spontaneous movement is always an important behavior to assess when characterizing novel models of movement disorders. Here spontaneous activity is measured by placing animals in a small, transparent cylinder (height, 15.5 cm, diameter, 12.7 cm).[64,67,68] The cylinder is placed on a piece of glass with a mirror positioned at an angle beneath the cylinder to allow a clear view of motor movements along the ground and walls of the cylinder. The number of rears, forelimb and hindlimb steps, and time spent grooming are measured. Videotapes are viewed and rated in slow motion by an experimenter blind to the mouse genotype. In this test, alpha synuclein–overexpressing mice showed significant reductions in spontaneous activity that persisted over time; in particular, they showed a robust reduction in hindlimb stepping.[67] A similar pattern of reduced hindlimb stepping has also been observed in mice with a developmental loss of DA neurons and is reversed by levodopa.[64]

Nest building

Nest building is a natural mouse behavior related to thermoregulation and pup survival.[73–75] Analysis of nest-building behavior has been used to assess nigrostriatal sensorimotor function in rodents.[76–79] These movements are DA-dependent and can be reduced with low doses of DA antagonists that do not disrupt other motor behaviors[76] or by injection of viral vectors containing DA in DA-deficient mice.[79] In this test, cotton material for nest building is weighed and then placed in the feeder bin of the animal's home cage. Because the feeder bin is positioned high off the floor, animals must rear up and pull the material from the feeder. The amount of cotton used is measured by re-weighing after a 24-hr period, although rodents build nests primarily during their dark cycle. A control test with nesting material in the cage should always be conducted to rule out a decrease in nest-building motivation. Impairments in nest building can first be detected in mice overexpressing alpha-synuclein between 4 and 8 months of age.[67]

CONCLUSION

Many motor functions are greatly affected directly or indirectly by pathological, genetic, or pharmacological manipulations of dopaminergic neuronal activity. The tests reviewed in this chapter reflect measures of motor function that are sensitive to varying levels of dysfunction or cell loss within the nigrostriatal DA system and provide useful endpoint targets for testing potential treatments for PD.[80,81] Although this review has focused on the influence of DA on control of movement, DA is also an important regulator of cognitive flexibility, habit formation, and goal-directed behaviors. Furthermore, the nigrostriatal, mesocortical, and mesolimbic DA projections interact dynamically with non-DA pathways to regulate behavioral control. Finally, movement initiation deficits associated with severe DA cell loss may be, at least in part, a consequence of disinhibition of non-DA pathways and increased thresholds for sensory activation of movement. Unlike motor impairments caused by stroke, deficits caused by nigrostriatal DA degeneration can be overridden by salient sensory stimuli or manipulation of non-DA brain regions. Thus, research exploring therapeutic targets for akinesia in the late stages of PD should continue to include avenues other than DA replacement or neuroprotection.

References

1. Carlsson A. The occurrence, distribution and physiological role of catecholamines in the nervous system. *Pharmacol Rev.* 1959;11(2, pt 2):490–493.
2. Albin RL, Young AB, Penney JB. The functional anatomy of basal ganglia disorders. *Trends Neurosci.* 1989;2(10):366–375.
3. DeLong MR. Primate models of movement disorders of basal ganglia origin. *Trends Neurosci.* 1990;13(7):281–285.
4. Schallert T, Teitelbaum P. Haloperidol, catalepsy, and equilibrating functions in the rat: antagonistic interaction of clinging and labyrinthine righting reactions. *Physiol Behav.* 1981;27:1077–1083.
5. De Ryck R, Schallert T, Teitelbaum P. Morphine versus haloperidol catalepsy in the rat: a behavioral analysis of postural support mechanisms. *Brain Res.* 1980;201:143–172.
6. Woodlee MT, Kane JR, Chang J, Cormack LK, Schallert T. Enhanced gait function in the good forelimb of hemi-parkinson rats: compensatory adaptation for contralateral postural instability. *Exp Neurol.* 2008;211:511–517.
7. Cheng JT, Schallert T, De Ryck M, Teitelbaum P. Galloping induced by pontile tegmental damage or GABA in rats: a form of Parkinsonian "festination" not blocked by haloperidol. *PNAS.* 1981;78:3279–3283.
8. Robinson TE, Whishaw IQ. The effects of posterior hypothalamic lesions on voluntary behavior and hippocampal electroencephalograms in the rat. *J. Comp. Physiol. Psychol.* 1974;86:768–786.
9. Keefe KA, Salamone JD, Zigmond MJ, Stricker EM. Paradoxical kinesia in parkinsonism is not caused by dopamine release. Studies in an animal model. *Arch Neurol.* 1989;46(10):1070–1075.
10. Cenci MA, Whishaw IQ, Schallert T. Animal models of neurological deficits: how relevant is the rat? *Nat Rev Neurosci.* 2002;3:574–579.
11. Ungerstedt U. 6-Hydroxy-dopamine induced degeneration of central monoamine neurons. *Eur J Pharmacol.* 1968;5(1):107–110.
12. Ungerstedt U. Adipsia and aphagia after 6-hydroxydopamine induced degeneration of the nigro-striatal dopamine system. *Act Physiol Scand Suppl.* 1971;367:95–122.
13. Kostrzewa RM, Jacobwitz DM. Pharmacological actions of 6-hydroxydopamine. *Pharmacol Rev.* 1974;26(3):199–288.
14. Schallert T, Wilcox RE. Neurotransmitter-selective brain lesions. In: Boulton AA, Baker GB, eds. *Neuromethods (Series 1: Neurochemistry), General Neurochemical Techniques*. Totowa, NJ: Humana Press; 1985:343–387.
15. Heikkila R, Cohen G. Inhibition of biogenic amine uptake by hydrogen peroxide: a mechanism for toxic effects of 6-hydroxydopamine. *Science.* 1971;172(989):1257–1258.
16. Heikkila R, Cohen G. 6-Hydroxydopamine: evidence for superoxide radical as an oxidative intermediate. *Science.* 1973;181(98):456–457.
17. Cohen G, Heikkila RE. The generation of hydrogen peroxide, superoxide radical, and hydroxyl radical by 6-hydroxydopamine, dialuric acid, and related cytotoxic agents. *J Biol Chem.* 1974;249(8):2447–2452.
18. Andrew R, Watson DG, Best SA, Midgley JM, Wenlong H, Petty RK. The determination of hydroxydopamines and other trace amines in the urine of parkinsonian patients and normal controls. *Neurochem Res.* 1993;18(11):1175–1177.
19. Ljungberg T, Ungerstedt U. Sensory inattention produced by 6-hydroxydopamine-induced degeneration of ascending dopamine neurons in the brain. *Exp Neurol.* 1976;53(3):585–600.
20. Schallert T, Upchurch M, Lobaugh N, et al. Tactile extinction: distinguishing between sensorimotor and motor asymmetries in rats with unilateral nigrostriatal damage. *Pharmacol Biochem Behav.* 1982;16(3):455–462.
21. Schallert T, Norton D, Jones TA. A clinically relevant unilateral rat model of parkinsonian akinesia. *J Neural Trans Plasticity.* 1992;3:332–333.
22. Schwarting RK, Huston JP. Unilateral 6-hydroxydopamine lesions of meso-striatal dopamine neurons and their physiological sequelae. *Prog Neurobiol.* 1996;49(3):215–266.
23. Schallert T, Tillerson JL. Intervention strategies for degeneration of dopamine neurons in parkinsonism: optimizing behavioral assessment of outcome. In: Emerich DF, Dean RL III, Sandberg PR, eds. *Central Nervous System Diseases*. Totowa, NJ: Humana Press; 2000:131–151.
24. Tillerson JL, Cohen AD, Philhower J, Miller GW, Zigmond MJ, Schallert T. Forced limb-use effects on the behavioral and

24. neurochemical effects of 6-hydroxydopamine. *J Neurosci.* 2001;21(12):4427–4435.
25. Tillerson JL, Cohen AD, Caudle WM, Zigmond MJ, Schallert T, Miller GW. Forced nonuse in unilateral parkinsonian rats exacerbates injury. *J Neurosci.* 2002;22(15):6790–6799.
26. Schallert T, Upchurch M, Wilcox RE., Vaughn DM. Posture-independent sensorimotor analysis of inter-hemispheric receptor asymmetries in neostriatum. *Pharmacol Biochem Behav.* 1983;18(5):753–759.
27. Schallert T, Woodlee MT. Orienting and placing. In: Whishaw IQ, Kolb B, eds. *The Behavior of the Laboratory Rat.* New York: Oxford University Press; 2005:129–140.
28. Connor B, Kozlowski DA, Schallert T, Tillerson JL, Davidson BL, Bohn MC. Differential effects of glial cell line-derived neurotrophic factor (GDNF) in the striatum and substantia nigra of the aged Parkinsonian rat. *Gene Ther.* 1999;6(12):1936–1951.
29. Kozlowski DA, Connor B, Tillerson JL, Schallert T, Bohn MC. Delivery of a GDNF gene into the substantia nigra after a progressive 6-OHDA lesion maintains functional nigrostriatal connections. *Exp Neurol.* 2000;166(1):1–15.
30. Yang M, Stull ND, Berk MA, Snyder EY, Iacovitti L. Neural stem cells spontaneously express dopaminergic traits after transplantation into the intact or 6-hydroxydopamine-lesioned rat. *Exp Neurol.* 2002;177(1):50–60.
31. Luo J, Kaplitt MG, Fitzsimons HL, et al. Subthalamic GAD gene therapy in a Parkinson's disease rat model. *Science.* 2002;298(5592):425–429.
32. Yasuhara T, Matsukawa N, Hara K, et al. Transplantation of human neural stem cells exerts neuroprotection in a rat model of Parkinson's disease. *J Neurosci.* 2006;26:12497–12511.
33. Whishaw IQ. Loss of the innate cortical engram for action patterns used in skilled reaching and the development of behavioral compensation following motor cortex lesions in the rat. *Neuropharmacology.* 2000;39(5):788–805.
34. Allred R, Adkins D, Woodlee MT, Kane JR, Schallert T, Jones TA. Vermicelli handling: a test for fine dexterous forelimb function in rat models of stroke and parkinsonism. *J Neurosci Meth.* 2008;170:229–244.
35. Hall S, Schallert T. Striatal dopamine and the interface between orienting and ingestive functions. *Physiol Behav.* 1988;44(4-5):469–471.
36. Schallert T, Hall S. "Disengage" sensorimotor deficit following apparent recovery from unilateral dopamine depletion. *Behav Brain Res.* 1988;30:15–24.
37. Spirduso WW, Gilliam P, Schallert T, Upchurch M, Wilcox RE. Reactive capacity: a sensitive behavioral marker of movement initiation and nigrostriatal dopamine function. *Brain Res.* 1985;335:45–54.
38. Woodlee MT, Asseo-Garcia AM, Zhao X, Liu S-J, Jones TA, Schallert T. Testing forelimb placing "across the midline" reveals distinct, lesion-dependent patterns of recovery in rats. *Exp Neurol.* 2005;191:310–317.
39. Anstrom KA, Schallert T, Woodlee MT, Shattuck A, Roberts DCS. Repetitive vibrissae-elicited forelimb placing before and immediately after unilateral 6-hydroxydopamine improves outcome in a model of Parkinson's disease. *Behav Brain Res.* 2007;179:183–191.
40. Marshall JF, Gotthelf T. Sensory inattention in rats with 6-hydroxydopamine-induced degeneration of ascending dopaminergic neurons: apomorphine-induced reversal of deficits. *Exp Neurol.* 1979;65(2):398–411.
41. Schallert T. Aging-dependent emergence of sensorimotor dysfunction in rats recovered from dopamine depletion sustained early in life. *Ann NY Acad Sci.* 1988;515:108–120.
42. Ciucci M, Ma TS, Fox C, Kane JR, Ramig L, Schallert T. Qualitative changes in ultrasonic vocalization in rats after unilateral dopamine depletion or haloperidol. *Behav Brain Res.* 2007;182:284–289.
43. Ciucci MR, Ahrens A, Ma ST. Reduction of dopamine synaptic activity: degradation of 50-kHz ultrasonic vocalization in rats. *Behav Neurosci.* 2009;123(2):328–336.
44. Schallert T, Kozlowski DA, Humm JL, Cocke RR. Use-dependent structural events in recovery of function. *Adv Neurol.* 1997;73:229–238.
45. Schallert T, Fleming SM, Leasure JL, Tillerson JL, Bland ST. CNS plasticity and assessment of forelimb sensorimotor outcome in unilateral rat models of stroke, cortical ablation, parkinsonism and spinal cord injury. *Neuropharmacology.* 2000;39(5):777–787.
46. Johnson RE, Schallert T, Becker JB. Akinesia and postural abnormality after unilateral dopamine depletion. *Behav Brain Res.* 1999;104(1-2):189–196.
47. Fleming SM, Delville Y, Schallert T. An intermittent, controlled-rate, slow progressive degeneration model of Parkinson's disease: antiparkinson effects of sinemet and protective effects of methylphenidate. *Behav Brain Res.* 2005;156(2):201–213.
48. Ariano MA, Grissell AE, Littlejohn FC, et al. Partial dopamine loss enhances activated caspase-3 activity: differential outcomes in striatal projection systems. *J Neurosci Res.* 2005;82:387–396.
49. Shi LH, Woodward DJ, Luo F, Anstrom K, Schallert T, Chang JY. High-frequency stimulation of the subthalamic nucleus reverses limb-use asymmetry in rats with unilateral 6-hydroxydopamine lesions. *Brain Res.* 2004;1013(1):98–106.
50. Schallert T. Behavioral tests for preclinical intervention assessment. *NeuroRx.* 2006;3(4):497–504.
51. Schallert T, Woodlee MT. Brain-dependent movements and cerebral-spinal connections: key targets of cellular and behavioral enrichment in CNS injury models. *J Rehab Res Dev.* 2003;40(4):9–18.
52. Olsson M, Nikkhah G, Bentlage C, Bjorklund A. Forelimb akinesia in the rat Parkinson model: differential effects of dopamine agonists and nigral transplants as assessed by a new stepping test. *J Neurosci.* 1995;15(5 pt 2):3863–3875.
53. Lindner MD, Winn SR, Baetge EE, et al. Implantation of encapsulated catecholamine and GDNF-producing cells in rats with unilateral dopamine depletions and parkinsonian symptoms. *Exp Neurol.* 1996;132(1):62–76.
54. Schallert T, Whishaw IQ, Ramirez VD, Teitelbaum P. Compulsive, abnormal walking caused by anticholinergics in akinetic, 6-hydroxydopamine-treated rats. *Science.* 1978;199:1461–1463.
55. Schallert T, De Ryck M, Whishaw IQ, Ramirez VD, Teitelbaum P. Excessive bracing reactions and their control by atropine and L-DOPA in an animal analog of Parkinsonism. *Exp Neurol.* 1979;64(1):33–43.
56. Chang JW, Wachtel SR, Young D, Kang UJ. Biochemical and anatomical characterization of forepaw adjusting steps in rat models of Parkinson's disease: studies on medial forebrain bundle and striatal lesions. *Neuroscience.* 1999;88(2):617–628.
57. Fleming SM, Zhu C, Fernagut P-O, et al. Behavioral and immunohistochemical effects of chronic intravenous and subcutaneous infusions of varying doses of rotenone. *Exp Neurol.* 2004;187:418–429.

58. Polymeropoulos MH, Lavedan C, Leroy E, et al. Mutation in the alpha-synuclein gene identified in families with Parkinson's disease. *Science*. 1997;276:2045–2047.
59. Kruger R, Kuhn W, Muller T, et al. Ala30Pro mutation in the gene encoding alpha-synuclein in Parkinson's disease. *Nat Genet*. 1998;18:106–108.
60. Singleton AB, Farrer M, Johnson J, et al. Alpha-synuclein locus triplication causes Parkinson's disease. *Science*. 2003;302:841.
61. Chartier-Harlin MC, Kachergus J, Roumier C, et al. Alpha-synuclein locus duplication as a cause of familial Parkinson's disease. *Lancet*. 2004;364:1167–1169.
62. Spillantini MG, Schmidt ML, Lee VM, Trojanowski JQ, Jakes R, Goedert M. Alpha-synuclein in Lewy bodies. *Nature*. 1997;388:839–840.
63. Rockenstein E, Mallory M, Hashimoto M, et al. Differential neuropathological alterations in transgenic mice expressing alpha-synuclein from the platelet-derived growth factor and Thy-1 promoters. *J Neurosci Res*. 2002;68:568–578.
64. Hwang DY, Fleming SM, Ardayfio P, et al. 3,4-Dihydroxyphenylalanine reverses the motor deficits in Pitx3-deficient aphakia mice: behavioral characterization of a novel genetic model of Parkinson's disease. *J Neurosci*. 2005;25(8):2132–2137.
65. LeVere TE. Neural system imbalances and the consequences of large brain injuries. In: Finger S, LeVere TE, Almi CR, Stein DG, eds. *Brain Injury and Recovery, Theoretical and Controversial Issues*. New York: Plennum Press; 1988.
66. Goldberg MS, Fleming SM, Palacino JJ, et al. Parkin-deficient mice exhibit nigrostriatal deficits but not loss of dopaminergic neurons. *J Biol Chem*. 2003;278:43628–43635.
67. Fleming SM, Salcedo J, Fernagut PO, et al. Early and progressive sensorimotor anomalies in mice overexpressing wild-type human alpha-synuclein. *J Neurosci*. 2004;24:9434–9440.
68. Fleming SM, Salcedo J, Hutson CB, et al. Behavioral effects of dopaminergic agonists in transgenic mice overexpressing human wild-type alpha-synuclein. *Neuroscience*. 2006;142(4):1245–1253.
69. Schallert T, Woodlee MT, Fleming SM. Disentangling multiple types of recovery from brain injury. In: Krieglstein J, Klumpp S, eds. *Pharmacology of Cerebral Ischemia*. Stuttgart: Medpharm Scientific Publishers; 2002:201–216.
70. Tillerson JL, Caudle WM, Reveron ME, Miller GW. Detection of behavioral impairments correlated to neurochemical deficits in mice treated with moderate doses of 1-methyl-4-phenyl 1,2,3,6-tetrahydropyridine. *Exp Neurol*. 2002;178:80–90.
71. Chen L, Cagniard B, Matthews T, et al. Age-dependent motor deficits and dopaminergic dysfunction in DJ-1 null mice. *J Biol Chem*. 2005;280(22):21418–21426.
72. Lu XH, Fleming SM, Meurers B, et al. BAC mice expressing a truncated mutant parkin exhibit progressive motor deficits and late-onset dopaminergic neuron degeneration. *J Neurosci*. 2009; 29(7): 1962–1976.
73. Lynch CB. Response to divergent selection for nesting behavior in *Mus musculus*. *Genetics*. 1980;96(3):757–765.
74. Broida J, Svare B. Strain-typical patterns of pregnancy-induced nestbuilding in mice: maternal and experiential influences. *Physiol Behav*. 1982;29(1):53–57.
75. Crawley JN. *What's Wrong with My Mouse? Behavioral Phenotyping of Transgenic and Knockout Mice*. New York, NY: Wiley-Liss; 2000.
76. Upchurch M, Schallert T. A behavior analysis of the offspring of "haloperidol-sensitive" and "haloperidol-resistant" gerbils. *Behav Neural Biol*. 1983;39(2):221–228.
77. Sedelis M, Schwarting RK, Huston JP. Behavioral phenotyping of the MPTP mouse model of Parkinson's disease. *Behav Brain Res*. 2001;125:109–125.
78. Hofele K, Sedelis M, Auburger GW, Morgan S, Huston JP, Schwarting RK. Evidence for a dissociation between MPTP toxicity and tyrosinase activity based on congenic mouse strain susceptibility. *Exp Neurol*. 2001;168(1):116–122.
79. Szczypka MS, Kwok K, Brot MD, et al. Dopamine production in the caudate putamen restores feeding in dopamine-deficient mice. *Neuron*. 2001;30(3):819–828.
80. Pienaar IS, Schallert T, Hattingh, Daniels WMU. Behavioral and quantitative mitochondrial proteome analyses of the effects of simvastatin: Implications for models of neural degeneration. *J Neural Transm* 2009. doi: 10.1007/s007-009-0247-4.
81. Byler SL, Boehm GW, Karp JD, et al. Systemic lipopolysaccharide plus MPTP as a model of dopamine loss and gait instability in C57Bl/6J mice. *Behav. Brain. Res*. 2009;198:434–439.

6.2 Involvement of Nucleus Accumbens Dopamine in Behavioral Activation and Effort-Related Functions

JOHN D. SALAMONE

INTRODUCTION

Scientific progress is marked by more than just the accumulation of novel pieces of data. As well as providing basic discoveries, scientists are constantly organizing their findings, testing hypotheses, offering theories, and articulating general conceptual frameworks. Over the last several decades, the behavioral functions of nucleus accumbens dopamine (DA) have been conceptualized according to various organizing principles. These diverse perspectives have included an emphasis on aspects of incentive motivation, reinforcement, "reward," and motor function. One of the most influential and persistent general frameworks for summarizing the functions of nucleus accumbens was offered by Mogenson and others several decades ago (e.g.,[1,2]). In an important review paper, Mogenson and colleagues[1] suggested that nucleus accumbens served as a "limbic-motor" interface, which was thought to be critical for translating motivational and cognitive information into action. This way of thinking about the functions of nucleus accumbens has been very useful for organizing the emerging body of anatomical and physiological evidence related to nucleus accumbens and the role of DA in this structure, and also has helped to set the stage for recent conceptual developments in the field. According to current anatomical and physiological models, nucleus accumbens is a nodal point for filtering and integrating the flow of information from limbic and prefrontal regions regulating motivational, emotional, and cognitive processes to those brain systems that are more directly involved in the control of behavioral output.[3–12] Distinct subregions and cell groups in nucleus accumbens appear to act as "gates" that allow multiple channels of information to be processed, and the neurotransmitter DA, along with GABA, glutamate, acetylcholine, adenosine, and other substances, regulate the physiological responses of accumbens neurons, which in turn influence the eventual impact of this information on structures that generate and control behavior.[5–12] Because nucleus accumbens is a component of the larger striatal complex, these functions of the accumbens appear to be a specific case of the general principle that striatal DA modulates the ability of various telencephalic inputs to regulate behavioral output[12]; this includes the classic sensorimotor functions of the lateral neostriatum (i.e., putamen in primates), sensorimotor gating processes,[13] and the variety of other motivational and cognitive functions regulated by different striatal subregions.

EMPIRICAL AND CONCEPTUAL PROBLEMS WITH THE TRADITIONAL FORMULATION OF THE DA HYPOTHESIS OF REWARD

One of the factors that has contributed to the recent conceptual restructuring in the field has been the gradual demise of the traditional form of the DA hypothesis of reward.[12,14,15] A full review of this hypothesis, and the overwhelming body of evidence demonstrating its shortcomings, is beyond the scope of the present chapter. Nevertheless, it is worthwhile to provide a brief overview. Several recent review articles have detailed many of the problems with the hypothesis that mesolimbic DA directly mediates the primary motivational properties of natural stimuli such as food.[12,15] In summary, there is substantial evidence that nucleus accumbens DA does not mediate primary food motivation (i.e., appetite for food) and that interference with accumbens DA transmission does not produce effects that closely resemble extinction or withdrawal of reinforcement ([15]; see additional discussion below). Nucleus accumbens DA release is not uniquely related to pleasure,[12,15,16] and in fact, there is little evidence that interference with accumbens DA transmission results in a loss of markers of hedonia such as appetitive taste reactivity.[16] Accumbens DA is not exclusively involved in appetitive forms of instrumental learning, to the exclusion of aversive forms.[12,15] Another set of problems is related to the persistent use of the term *reward* to describe, in a simple or direct way, the major function being modulated by mesolimbic DA. The term *reward* often is used to refer to the aspects of reinforcement that are most closely associated with emotional processes, such as subjective pleasure, or with primary appetitive motivation.[15] In this regard, the empirical and conceptual basis of the DA hypothesis of reward is highly problematic. It is particularly ironic that those aspects of incentive motivation that are generally conveyed by the use of the term *reward* (i.e., pleasure and primary motivation for natural stimuli) are the very aspects for which there appears to be little direct evidence of mediation by accumbens DA.[15] Within the last few years, the traditional emphasis on hedonia and primary reward that has been so prevalent in the literature has gradually yielded to diverse lines of research that focus on aspects of instrumental learning (both appetitive and aversive), Pavlovian/instrumental interactions, reinforcer prediction, incentive salience, and behavioral activation. Because the DA hypothesis of reward has been so predominant for so many years, penetrating widely into media as varied as science textbooks, the popular press, the Internet, and even film, these recent changes in the field have been characterized as representing a kind of Kuhnian paradigm shift ([14,15]; see[17] for a more complete discussion of paradigm shifts in the history and philosophy of science). Yet in many ways, they also represent a return to concepts that should be familiar to anyone who read those Mogenson articles several years ago, although the current iteration of these ideas is supported by a richer body of empirical findings, and the scope of functions being emphasized is much broader.[12,15,18–22]

This chapter will focus on the behavioral activation functions of nucleus accumbens DA, and in particular will emphasize how these functions appear to be engaged in such a way as to promote the exertion of effort in motivated behavior. In addition, the chapter will discuss the role of accumbens DA in enabling animals to overcome work-related constraints that separate them from significant stimuli, and the involvement of DA in effort-related choice behavior that is based upon the allocation of responses to various alternatives. Finally, the role of accumbens DA will be placed in an overall anatomical and neurochemical context by discussing other brain areas and neurotransmitters as well. However, before this discussion progresses, it should be emphasized that DA in nucleus accumbens, as well as in other structures, obviously participates in multiple behavioral processes. Thus, as noted previously, a discussion of the role of accumbens DA in behavioral activation or effort-related processes is not inconsistent with the hypothesized involvement of DA in other functions[15]; this point will be discussed further in the final section of this chapter.

BEHAVIORAL ACTIVATION FUNCTIONS OF NUCLEUS ACCUMBENS DA

The term *motivation* refers to the behaviorally relevant processes that enable organisms to regulate their internal and external environment.[22] Organisms engage in actions that regulate the availability, proximity, or probability of delivery of a diverse array of stimuli with potential biological significance. As with any complex set of processes, psychologists have found it useful to break down motivational functions into various components; one such division is the distinction between *directional* and *activational* aspects of motivated behavior.[15,18–23] Directional aspects of motivation refer to the observation that behavior is directed toward or away from specific motivational stimuli, and also is directed in relation to the activities that involve interacting with those stimuli. Activational aspects of motivation reflect the observation that motivated behavior is characterized by vigor, persistence, the instigation and maintenance of substantial activity, and high levels of work output. Organisms are usually separated from significant stimuli such as food or water by work-related response costs or constraints. In the natural environment, foraging animals typically must use a great deal of energy and spend a considerable amount of time to obtain access to these stimuli. The discussion of activational aspects of motivation has been a recurring feature of the literature in several areas, including animal behavior, psychology, and even psychiatry. Because the amount of effort or time expended to obtain motivational stimuli is seen as an important determinant of choice behavior, optimal foraging theory has been a useful conceptual tool for several decades.[24] In the psychology literature, behavioral activation processes also have been seen as important for supporting the energy requirements that are necessary for engaging in vigorous instrumental behaviors. The concepts of *drive* and *incentive* were used by neobehaviorists such as Hull and Spence[25,26] to emphasize that motivational conditions such as deprivation, or the presentation of conditioned stimuli, can produce energizing effects on behavior. An *anticipation-invigoration mechanism,* which was thought to be triggered by the presentation of conditioned stimuli, and which then served to invigorate instrumental behavior, was proposed by Cofer and Apley several years ago.[18] The work requirements of an instrumental task have been shown to be important determinants of behavioral output,[27] and *behavioral economic* perspectives[28–31] have focused upon how factors such as work requirements, time constraints, reinforcement availability, and preference jointly influence choice behavior. Furthermore, activational aspects of motivated behavior are thought to have enormous clinical significance. Psychomotor slowing, anergia and fatigue are considered by psychiatrists and clinical psychologists to be important features of depression and other psychiatric or neurological conditions.[15,21]

Activational aspects of motivation are marked by various indices of behavioral performance, including the vigorous output of instrumental behavior and the heightened behavioral responses to various stimuli. Locomotor activity is commonly employed as an index of a specific aspect of motor function but also is used to provide a measure of behavioral activation that can be related to motivation. The induction of high levels of locomotor activity is one of the defining characteristics of the behavioral response to psychomotor stimulant drugs, and considerable research has demonstrated that DA in nucleus accumbens is a critical mediator of the locomotor effects of major stimulants such as amphetamine or cocaine.[32–36] Furthermore, nucleus accumbens DA is seen as participating in the behavioral activation induced by repeated presentation of natural motivational stimuli. Periodic noncontingent presentation of reinforcers such as food can induce various "schedule-induced" activities, including wheel-running, locomotion, drinking, and licking.[37–41] Depletions of DA in nucleus accumbens impaired schedule-induced drinking and wheel running.[42–45] Schedule-induced locomotor activity (i.e., the increase in locomotion induced by periodic food presentation) was blocked by systemic administration of the DA antagonist haloperidol[19] and also by nucleus accumbens DA depletions.[46] Increases in accumbens DA release, as measured by microdialysis[46] and voltammetry,[47] have been shown to accompany the production of schedule-induced behavior. In summary, several decades of research have demonstrated that nucleus accumbens DA is critically involved in the induction of motor activity induced by novelty or stimulant drugs, as well as in various forms of schedule-induced behavior. These observations, as well as several other lines of research, have provided critical support for the idea that accumbens DA is an important part of the brain circuitry involved in activational aspects of motivated behavior.[17,22,33,36,43,48,49]

NUCLEUS ACCUMBENS DA IS INVOLVED IN THE EXERTION OF EFFORT IN FOOD-MOTIVATED INSTRUMENTAL BEHAVIOR

Interference with DA transmission in nucleus accumbens can have selective effects on particular components of motivated behavior, impairing some processes while sparing others and effectively dissociating these components from each other.[12,15,22,50] For example, it has been suggested that low doses of DA antagonists, or depletions of DA in nucleus accumbens, impair activational aspects of food motivation but leave directional aspects fundamentally intact.[15,19–23] This point is critical for interpreting the literature on the effects of dopaminergic manipulations of food-reinforced instrumental behaviors. Across many conditions, food-motivated tasks that have low response requirements tend to be relatively insensitive to the effects of DA antagonism or accumbens DA depletions, while tasks that have more substantial response costs, such as operant conditioning schedules with high ratio requirements (i.e., large numbers of lever presses are required for each reinforcer), have generally been demonstrated to be more sensitive to dopaminergic manipulations.

For several decades, it has been evident that the effects of DA antagonism depend greatly upon the task being performed, and interact powerfully with the response requirements that allow access to reinforcers. Doses of haloperidol that produce massive reductions in food-reinforced fixed ratio lever pressing (0.4 mg/kg) were shown to have little effect upon the instrumental response of simply being in proximity to the food delivery dish on an interval schedule.[19,51] Despite the persistence of the reinforced instrumental response in the face of this drug-induced challenge, this dose of haloperidol dramatically reduced schedule-induced locomotion in the same chamber.[51] In contrast, extinction did produce a substantial reduction in the reinforced response, an effect that differed dramatically from that produced by haloperidol.[51] The effects of haloperidol on food-reinforced lever pressing differed markedly across different operant schedules (i.e., fixed ratio 1 vs. progressive ratio).[52] The effects of accumbens DA depletions also vary greatly, depending upon which ratio requirement is in use. Depletions of accumbens DA that substantially reduced fixed ratio 5 (FR5) lever pressing had no significant effects on FR1 performance.[53] Aberman and Salamone[54] employed a wide range of ratio schedules from FR1 to FR64 to assess the effects of accumbens DA depletions. Responding on the FR64 schedule was severely impaired and FR16 lever pressing was moderately impaired, while FR4 responding was only affected transiently and FR1 responding was basically intact. In a subsequent study,[55] rats were tested across a range of ratio schedules as high as FR300 (i.e., FR5 to FR300), under conditions in which the macroscopic density of food delivered per lever press was kept constant (i.e., kept at an FR50 density). FR20 and FR50 responding was slowed by DA depletion, and at very high ratio levels such as FR200 and FR300, DA-depleted rats showed *ratio strain*, essentially ceasing to respond altogether.[55] Taken together, these studies demonstrated that the magnitude of the ratio requirement appears to be a critical determinant of sensitivity to the effects of accumbens DA depletions, with larger ratios making rats more sensitive to the disruption in DA transmission.

Of course, a number of factors other than work requirements could potentially be contributing to the different pattern of effects seen in animals with DA depletions, and these factors need to be investigated. Baseline response rates generated by the schedule being used appear to contribute to the response slowing shown by DA-depleted rats, and across a large group of schedules there is an overall relation between baseline response rate and the degree of suppression produced by accumbens DA depletions, with the higher rate schedules being more sensitive.[21,56,57] Nevertheless, this particular factor does not appear to be the primary determinant of the "crashing" or ratio strain shown by rats with accumbens DA depletions when ratio requirements are very high.[55] Another important factor that has been studied is time.[55,58] It generally takes more time to complete a schedule with a high ratio requirement than it does to complete a schedule with a lower requirement. Therefore, it is possible that the degree of intermittence of a schedule (i.e., the lengths of time without primary reinforcement) could be a factor that contributes to the schedule dependency shown by animals with accumbens DA depletions or DA antagonism. To assess this issue, some studies compared the effects of accumbens DA depletions on the performance of standard variable interval (VI) schedules versus VI schedules that have an additional ratio requirement attached (i.e., tandem VI/FR schedules). These procedures allow one to assess the effects of DA depletions on schedules that have different ratio requirements but the same degree of intermittence. Depletions of accumbens DA significantly impaired responding on a VI 30-s schedule that had a FR5 component attached (i.e., a tandem VI 30-s/FR5 schedule) but had no effect on the conventional VI 30-s schedule with the lower response requirement.[59] In a subsequent investigation, it was shown that accumbens DA depletions did not significantly affect VI 60- or 120-s performance when a minimal (i.e., FR1) requirement was attached, but they did suppress the lever pressing rate on the two tandem schedules that had FR10 requirements added.[60] Dopamine depletions also produced signs of response slowing (i.e., reductions in the number of short interresponse times) and response fragmentation, as indicated by increases in the number of pauses. These studies demonstrate that ratio requirements make rats sensitive to the effects of accumbens DA depletions, independently of any contribution that interval requirements may have. This conclusion is supported by additional studies demonstrating that responding on a progressive interval schedule was not impaired by intra-accumbens DA antagonism[61] and that delay discounting was not significantly altered by accumbens DA depletions.[62] Recent experiments using operant discounting procedures demonstrated that DA antagonism affected effort discounting in a manner that was independent of any effects on delay discounting.[63]

In summary, ratio requirements present a significant challenge to animals with impaired DA transmission in nucleus accumbens. This represents at least one dimension of work output and effort expenditure that is highly dependent upon the integrity of nucleus accumbens DA transmission. Other aspects of work, such as force or weight requirements, appear to be less dependent upon accumbens DA.[53,64] Additional factors such as time requirements (i.e., intermittence), despite being important determinants of instrumental behavior, cannot explain on their own why animals with accumbens DA depletions are so sensitive to schedules with high ratio requirements. Furthermore, the effects of accumbens DA depletions on ratio schedules do not closely resemble the effects of extinction and do not appear to be dependent upon changes in appetite or primary food motivation. Although the FR1 schedule is sensitive to extinction, appetite suppressant drugs, and reinforcer devaluations such as prefeeding to reduce food motivation, this schedule is relatively insensitive to the effects of accumbens DA depletions.[53,54] The effects of prefeeding on operant responding are easily distinguishable from the effects of DA depletions when ratio performance is studied across a broad range of ratio values.[54] In consideration of these findings and those of additional related studies, it is reasonable to conclude that a major function of accumbens DA is to enable organisms to overcome work-related response costs that separate them from significant stimuli.[19–22,49,54,65–69]

NUCLEUS ACCUMBENS DA IS INVOLVED IN EFFORT-RELATED CHOICE BEHAVIOR

Significant stimuli that are necessary for survival often are not easily accessible. Therefore, organisms must frequently exert effort in order to overcome response constraints, including physical distance and other obstacles, that separate them from these stimuli. Moreover, there often are a number of potential paths that can lead to reinforcers; for this reason, organisms must constantly make effort-related decisions involving cost/benefit assessments across a wide variety of stimuli and response requirements.[22,57,66–70] As well as being involved in the exertion of effort, as described above, evidence indicates that accumbens DA is part of the forebrain circuitry involved in effort-related choice behavior. A large number of studies have shown that interference with accumbens DA transmission alters the outcome of cost/benefit analyses that involve trade-offs between work-related response costs and the value of the reinforcers that can be obtained.

Several behavioral tasks have been used to evaluate the effects of dopaminergic manipulations on effort-related choice behavior. The underlying hypothesis for these studies has been that dopaminergic manipulations should alter effort-related choice, biasing animals toward lower-cost alternatives. One task that has been employed is a T-maze procedure that was developed to assess the effects of DA antagonists and accumbens DA depletions on effort-related choice behavior.[71] The two arms of the maze can have different reinforcement densities that offer choices to the animal (e.g., four vs. two food pellets, or four vs. zero), and under some conditions, a vertical 44-cm barrier is positioned in the arm with the higher reward density to provide a work-related challenge to the rat. During conditions in which no barrier was present in the arm with the high reinforcement density, untreated rats strongly preferred that arm. Treatment with the DA antagonist haloperidol and depletion of accumbens DA failed to alter arm preference when no barrier was present.[71] In addition, when the arm with the barrier was loaded with four reinforcement pellets but the other arm contained no food (i.e., the only way to get food was to climb the barrier), rats with accumbens DA depletions ran more slowly than control animals, but still chose the high-density arm, climbed the barrier, and ate the food pellets.[72] However, systemic injections of haloperidol and accumbens DA depletions dramatically altered choice behavior when the high-density arm (four pellets) had the barrier in place and the arm without the barrier contained an alternative food source (two pellets) but no barrier. Under these conditions, DA depletions or DA antagonism decreased the choice of the high-density arm and increased the choice of the low-density arm.[71,72] In other words, interference with DA transmission altered choice behavior, shifting rats from the high-cost alternative to the low-cost alternative. These results support the hypothesis that interference with DA transmission can cause animals to alter their instrumental response selection based upon the work requirements of the task.[12,20,22,57,66]

Another procedure that has been used to assess effort-related choice behavior is an operant choice task[49] that offers rats the option of selecting between lever pressing to obtain a relatively preferred food (e.g. Bioserve or other high-carbohydrate operant pellets) versus approaching and consuming a less preferred food (standard lab chow) that is concurrently available in the chamber. When the lever pressing ratio requirement is either FR1 or FR5, rats responding on baseline days, or under control conditions, typically get most of their food by lever pressing, and they generally consume only small amounts of the chow.[49] Several drugs have been tested using the concurrent FR5/chow intake version of this task; DA antagonists with different patterns of receptor subtype selectivity, including cis-flupenthixol, haloperidol, raclopride, eticlopride, SCH 23390, ecopipam, and SKF83566, all have been shown to decrease lever pressing for the preferred food but substantially increase intake of the concurrently available chow.[49,73–77] The drug-induced shifts from lever pressing to chow intake in these studies are generally characterized by a high inverse correlation between these two variables (i.e., high lever pressing is associated with low chow intake, and vice versa[76,77]).

In order to understand the various factors that contribute to performance on this procedure, and thereby interpret the significance of the findings involving DA antagonists, the task has undergone an extensive amount of behavioral and pharmacological validation. For example, rats are sensitive to the ratio requirement of the operant behavior component of the task, as increases in the ratio requirement up to FR10 and FR20 lead to shifts in choice behavior such that lever pressing decreased and chow intake increase.[22,66] Another study has indicated that lever pressing in untreated rats performing this task remains under the control of appetite-related factors. Switching the alternative food from chow to the preferred pellets leads to decreases in lever pressing for pellets and an increase in consumption of the freely available pellets.[22] These behavioral manipulations demonstrate that performance on this task involves assessments of both response costs and reinforcement value. However, dopaminergic manipulations that affect choice behavior do not appear to be acting to alter the reinforcing value of food or the appetite for food. The low dose of haloperidol that reliably produces the shift in behavior from lever pressing to chow intake (0.1 mg/kg) did not alter food intake or preference in free-feeding choice tests.[49,75] Moreover, microinjections of the D1 family antagonist SCH 23390 or the D2 family antagonist sulpiride directly into the nucleus accumbens also failed to alter preference between the two food types in free-feeding preference tests.[75] Although DA antagonists that are nonselective, or selective for D1 or D2 family receptors, consistently have been shown to reduce FR 5 lever pressing and increase chow intake, the serotonergic appetite suppressant fenfluramine decreased both lever pressing and chow intake,[77] an effect similar to that produced by prefeeding to reduce food motivation.[49] More recently, the cannabinoid CB1 antagonist AM 4113 and the CB1 inverse agonist AM 251, which are thought reduce food intake either by suppressing appetite or by producing food aversions, failed to increase chow intake at doses that suppressed lever pressing.[76] Together with the studies cited above, these results are consistent with the hypothesis that low doses of DA antagonists do not suppress lever pressing simply because they reduce appetite or primary food motivation.[15,22]

Considerable research has focused upon identifying the specific terminal fields at which dopaminergic manipulations could shift effort-related choice behavior. Dopamine depletions in anterior/medial neostriatum dorsal to the nucleus accumbens had no effect on the performance of the concurrent FR5/chow intake task.[78] Ventrolateral neostriatal DA depletions reduced food intake and impaired various aspects of food handling, but did not shift behavior from lever pressing to chow intake and instead decreased both behaviors.[78] Based upon several published studies, nucleus accumbens is the striatal region in which pharmacological or neurotoxic disruption of DA transmission mimics the effects of low doses of systemic DA antagonists in rats performing the concurrent choice task. Injections of D1 or D2 family antagonists directly into the accumbens, as well as accumbens DA depletions, have been shown repeatedly to decrease lever pressing and increase chow intake.[49,67,74,75,78–80] Nucleus accumbens has been divided by anatomists into distinct subregions,[4] and the shift from lever pressing to chow intake has been demonstrated to occur after injections of either D1 or D2 family antagonists into the medial core, lateral core, or dorsomedial shell subregions of the accumbens.[49,80] A summary of the studies in this area that employed the concurrent lever pressing/chow feeding procedure is shown in Table 6.2.1. When tested on the same task, DA transporter knockdown mice that have enhanced DA transmission displayed increased selection of lever pressing relative to chow intake.[81]

In summary, a large body of evidence gathered from studies using different behavioral tasks has demonstrated that rats with impaired DA transmission remain directed toward the acquisition and consumption of

TABLE 6.2.1. *Summary of Results with Concurrent Lever Pressing/Chow Feeding Choice Procedure: Increase in Chow Consumption/Decrease in Lever Pressing*

Systemic DA antagonism	
Cis-flupentixol (no selective)	Cousins et al.[74]
SCH 23390 (D1)	Cousins et al.[74]
SKF 83566 (D1)	Salamone et al.[77]
Ecopipam (SCH 39166; D1)	Sink et al.[76]; Worden et al.[100]
Haloperidol (D2)	Cousins et al.[74]; Salamone et al.[49,73]
Raclopride (D2)	Salamone et al.[77]
Eticlopride (D2)	Sink et al.[76]; Worden et al.[100]
Intra-accumbens DA antagonism: medial core/adjacent shell	
Haloperidol (D2)	Salamone et al.[49]
SCH 23390 (D1)	Koch et al.[75]
Sulpiride (D2)	Koch et al.[75]
Intra-accumbens DA antagonism: lateral core vs. dorsomedial shell	
SCH 23390 (D1)	Nowend et al.[80]
Raclopride (D2)	Nowend et al.[80]
Accumbens DA depletions: medial core/adjacent shell	
6-OHDA	Cousins et al.[78]; Cousins and Salamone[79]; Salamone et al.[49]
Accumbens DA depletions: lateral core vs. dorsomedial shell	
6-OHDA	Sokolowski and Salamone[67]

food, but nevertheless display a markedly reduced tendency to emit responses with a high rate or speed. When faced with the challenge presented by high response costs in effort-related choice tasks, rats with compromised DA function in the accumbens show a compensatory reallocation of behavior, selecting a relatively lower-cost alternative path to a different food source (i.e., the available chow or the arm with less food). Taken together, these studies have led to the suggestion that mesolimbic DA is a critical component of the forebrain circuitry regulating effort-related processes.[15,22]

DA AND ADENOSINE INTERACT IN THE CONTROL OF BEHAVIORAL ACTIVATION, EXERTION OF EFFORT AND EFFORT-RELATED CHOICE

As described above, nucleus accumbens DA appears to be an important component of the brain circuitry regulating effort-related processes. The empirical findings described above have stimulated considerable additional research in this area, and also have been modeled by researchers using various computational approaches.[82,83] Furthermore, it has become evident that other transmitters and brain areas in addition to nucleus accumbens DA also are involved. One of these components, which appears to interact strongly with dopaminergic mechanisms, is the purine neuromodulator adenosine. Minor stimulants such as caffeine and theophylline are nonselective adenosine antagonists. Moreover, there is a well-characterized interaction between DA and adenosine A_{2A} receptors in neostriatum and nucleus accumbens.[84–87] These striatal areas are rich in adenosine A_{2A} receptors, with DA D2 family receptors and adenosine A_{2A} receptors showing a high degree of colocalization on the same medium spiny neurons.[84] The DA–adenosine interaction in striatal regions has been most commonly studied using animal models of neostriatal motor functions related to parkinsonism.[84,88–95] For example, the adenosine A_{2A} antagonists KF17837, KW6002, and MSX-3 all were shown to suppress the oral tremor induced by DA antagonism and DA depletion, an effect that appears to involve actions on A_{2A} receptors in ventrolateral neostriatum.[93,94] Consistent with this line of research, adenosine A_{2A} receptor antagonists are being assessed for their potential antiparkinsonian effects in human clinical trials.[90,91] In addition to these studies related to the role of adenosine A_{2A} receptors in modulating neostriatal functions, research has focused on the role of nucleus accumbens adenosine A_{2A} receptors. Local injections of the adenosine A_{2A} receptor agonist CGS 21680 into nucleus accumbens decreased locomotor activity.[96,97] Haloperidol-induced suppression of locomotion was reversed by injections of the adenosine A_{2A} antagonist MSX-3 into the nucleus accumbens core, although injections into the accumbens shell or the ventrolateral neostriatum were ineffective.[98]

Based upon the possibility that DA and adenosine receptors interact to regulate effort-related processes, recent studies were conducted to study the ability of the adenosine A_{2A} receptor antagonist MSX-3 to reverse the behavioral effects of DA antagonism. MSX-3 increased lever pressing in rats coadministered with haloperidol, and also reversed the haloperidol-induced shift from lever pressing to chow intake in rats performing the concurrent FR5/chow intake procedure.[99] Doses of MSX-3 that produced a significant reversal of the effects of haloperidol had no effect when administered alone. Thus, there appears to be a functional interaction between DA and adenosine A_{2A} receptors that is involved in the regulation of instrumental response output and effort-related choice behavior. Further investigations have indicated that MSX-3 may interact differently with D1 and D2 family antagonists. Although MSX-3 completely reversed the behavioral effects of the D2 family antagonist eticlopride in rats tested on the concurrent FR5/chow intake task, in the same dose range MSX-3 produced only modest effects on the suppression of lever pressing induced by the D1-selective drug ecopipam.[100] Additional studies have used the T-maze choice task described above to investigate this interaction between DA D2 receptors and adenosine receptors. Haloperidol-induced reductions in selecting the high-cost arm (i.e., the arm with the barrier) in the T-maze were reversed by MSX-3 but not by the adenosine A_1 antagonist DPCPX.[101] Taken together with the lever pressing data, these results highlight the importance of specific interactions between drugs that act on DA D2 receptors and those acting upon adenosine A_{2A} receptors, which may in part be related to the colocalization of these two subtypes of receptors on the same population of medium spiny cells.

In view of the recent findings indicating that adenosine A_{2A} receptor antagonists can reverse the effects of DA D2 antagonists, studies with adenosine A_{2A} receptor agonists also have been conducted to determine if these drugs could produce effects similar to those produced by DA antagonism or DA depletion. Intra-accumbens injections of the adenosine A_{2A} receptor agonist CGS 21680 substantially impaired performance on a VI 60-s operant schedule with a FR10 requirement attached, but not when the interval schedule had a minimal (i.e., FR1) requirement attached,[102] a pattern of effects similar to that previously shown to occur after accumbens DA depletions.[60] Local injections of CGS21680 into the accumbens also decreased lever pressing and increased chow intake in rats performing the concurrent choice operant task.[103] For both of these studies, injections of CGS 21680 into a control site dorsal to the accumbens had no significant effects.[102,103] Together with those findings related to the effects of adenosine A_{2A} receptor antagonists, these studies with intra-accumbens injections of CGS 21680 indicate that adenosine and DA in nucleus accumbens jointly regulate operant response output and effort-related choice. Blockade of adenosine A_{2A} receptors is able to reverse the effects of DA antagonism. Conversely, stimulation of adenosine A_{2A} receptors in the accumbens produces effects that closely resemble those resulting from accumbens DA depletions or antagonism.

ACCUMBENS DA IS A COMPONENT OF THE BROADER FOREBRAIN CIRCUITRY INVOLVED IN EFFORT-RELATED PROCESSES

As evidence implicating the nucleus accumbens in effort-related processes has continued to accumulate, recent studies also have examined the role of other related brain areas, including prefrontal cortex and amygdala, using the T-maze task that was originally developed in our laboratory to assess the effects of accumbens DA depletions. The effects of large lesions of medial frontal cortex that included the prelimbic, infralimbic, and anterior cingulate cortex were assessed by Walton et al.[104] Large medial frontal cortex lesions shifted the behavior of the rats away from the arm with the barrier that contained the high density of reinforcement to the arm with no barrier. Multiple areas of frontal cortex were investigated further in a subsequent study.[105] Anterior cingulate cortex lesions produced the same changes in effort-related choice that had been shown previously with the larger lesions, while lesions of prelimbic and infralimbic cortex had no effect on choice behavior.[105] The effects of anterior cingulate cortex lesions appear to be task-dependent, in the sense that they altered effort-related choice in the T-maze task but not in operant choice tasks.[106] Large depletions of DA in anterior cingulate cortex also were shown to impair effort-related decision making in the T-maze.[107] Floresco and Ghods-Sharifi[108] reported that bilateral inactivation of the basolateral amygdala by injections of the local anesthetic bupivacaine reduced the preference for the high-barrier arm with the higher reinforcement density. Furthermore, these authors used *disconnection* methodology to demonstrate that unilateral inactivation of the basolateral amygdala combined with contralateral inactivation of anterior cingulate cortex also disrupted effort-based decision making. These results suggest that serial transfer of information between basolateral amygdala and anterior cingulate cortex is involved in work-related choice. A recent disconnection experiment also has demonstrated that combined contralateral lesions of anterior cingulate corex and nucleus accumbens core, as well as bilateral core lesions, altered effort-related choice behavior.[109]

The ventral pallidum, which receives a profuse GABAergic innervation from the accumbens, is another potentially important part of the circuitry involved in effort-related processes.[2,3,110] GABAergic neurons from ventral pallidum project to several brainstem motor areas and also to the mediodorsal thalamic nucleus.[2,3,110] Ventral pallidum is thought to act as a relay station that conveys output from nucleus accumbens and to integrate information related to diverse striatal and limbic inputs.[111] Stimulation of ventral pallidal GABA receptors, either with local injections of GABA itself or with injections of the $GABA_A$ agonist muscimol, has been shown to suppress spontaneous or novelty-induced locomotor activity.[112–114] More recently, the effort-related functions of ventral pallidal GABA have become the subject of several experiments. Consistent with the studies summarized above, it was hypothesized that stimulation of $GABA_A$ receptors in ventral pallidum should produce many of the same behavioral effects as DA depletion in accumbens. With rats responding on the concurrent FR5 lever pressing/chow intake procedure described above, infusions of the $GABA_A$ agonist muscimol into the lateral ventral pallidum decreased lever pressing for the preferred food and produced a corresponding increase in the consumption of the less preferred chow.[115] Muscimol injected into the ventral pallidum did not alter food preference, and injections of muscimol into a control site dorsal to the ventral pallidum had no significant behavioral effects. These results indicate that the ventral pallidum, like the nucleus accumbens, is a component of the brain circuitry regulating effort-related processes, and that it may be a critical link in the transfer of effort-related information from the accumbens to other brain areas.

Recent studies combining anatomical, neurochemical, and behavioral methods have investigated the functional relation between adenosine A_{2A} receptors in nucleus accumbens and GABA transmission in the ventral pallidum.[102] Double-labeling methods that involved both immunohistochemistry for adenosine receptors and track-tracing methods demonstrated that ventral striatopallidal neurons expressed A_{2A} receptor immunoreactivity. In a microdialysis study, local intra-accumbens injections of the adenosine A_{2A} receptor agonist CGS 21680 elevated extracellular levels of

GABA in the ventral pallidum. An additional experiment involved disconnection methods that assessed the combined and separate effects of unilateral injections of CGS 21680 into the accumbens and contralateral injections of muscimol into the ventral pallidum. Unilateral injections of CGS 21680 into the nucleus accumbens, combined with contralateral ventral pallidal injections of the $GABA_A$ agonist muscimol, produced a synergistic effect that dramatically suppressed responding on an interval lever pressing schedule that also had a high ratio requirement (VI60/FR10). Thus, combined stimulation of adenosine A_{2A} receptors in the accumbens on one side of the brain and $GABA_A$ receptors in ventral pallidum on the other side altered the exertion of effort in a manner that is similar to the effects of interference with DA transmission.[102] These results indicate that nucleus accumbens and ventral pallidum appear to be components of the forebrain circuitry regulating behavioral activation and effort-related functions.[22]

In summary, it is clear that nucleus accumbens DA is one component of a broader system that involves several interconnected brain areas (ventral pallidum, anterior cingulate cortex, basolateral amygdala) and multiple transmitters and neuromodulators. Nucleus accumbens receives inputs from frontal cortex and limbic areas that are interconnected with each other, and also receives DA inputs from the ventral tegmental area that form part of the mesolimbic DA system (Fig. 6.2.1). GABAergic medium spiny neurons that contain both DA and adenosine A_{2A} receptors project from the nucleus accumbens to the ventral pallidum. In turn, the ventral pallidum sends projections to thalamic nuclei that relay information to neocortex. Based upon the research summarized above, the nucleus accumbens, ventral pallidum, frontal cortex, and basolateral amygdala appear to be critical components of the circuitry regulating effort-related processes.[15,102,108,109,115–117] Further studies are needed to investigate the role played by other neurotransmitters and additional brain structures (e.g., dorsomedial thalamic nucleus).

SUMMARY AND CONCLUSIONS

Over the last few years, ideas about the behavioral functions of nucleus accumbens DA have continued to evolve. Several lines of evidence have pointed to both empirical and conceptual weaknesses in the traditional DA hypothesis of reward.[12,15,66] For example, low doses of DA antagonists and depletions of nucleus accumbens DA have been shown to produce effects that do not closely resemble extinction,[51,66,118,119] prefeeding,[49,54,56] or appetite suppression.[74,76,77] Although some researchers have identified a role for DA systems in instrumental learning,[120–123] other studies have posed problems for this view.[16,124,125] Furthermore, the potential role of DA systems in instrumental behavior is not limited to situations in which appetitive stimuli are used, because striatal mechanisms in general, and mesolimbic DA in particular, also participate in aspects of aversive learning and aversive motivation.[15,126–131] The idea that nucleus accumbens DA mediates the pleasure associated with positive reinforcers has been strongly challenged,[15,16,124] while physiological and neurochemical studies indicate that DA neuron activity is not simply tied to the delivery of primary reinforcers. In fact, DA neuron activity and DA release can be activated by a diverse array of conditions, with varying time scales, including tonic, slow phasic, and fast phasic signaling.[12,19,21,83,129–137] As suggested in recent computational models,[82,83] even if fast phasic DA neuron activity is often activated by stimuli predicting reinforcers, or by better than expected outcomes, it is possible that the major functions of this heightened DA release could include support for high rates of responding, optimization of action selection, and alterations in the threshold of cost expenditure. Moreover, important issues remain in terms of the functional significance of slow metabotropic signal transduction changes induced by DA activity and how they modulate long-term postsynaptic responses.[135,136]

In parallel with this ongoing evolution of thinking about DA, there has been an enormous expansion of our understanding of the brain circuitry involved

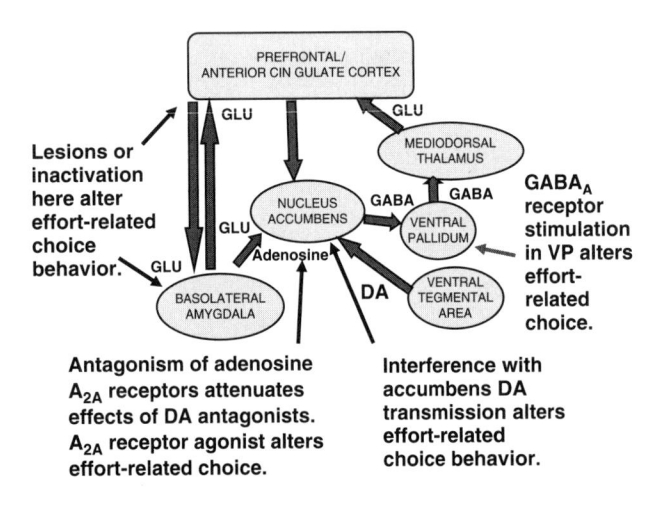

FIGURE 6.2.1. Schematic circuit diagram showing some of the anatomical connections linking cortical/limbic/striatal structures that are involved in effort-related processes. The projection patterns of distinct accumbens core and shell subregions are not shown.

in behavioral activation and effort-related functions.[12,15,22]. Low doses of DA antagonists and accumbens DA depletions blunt the tendency to respond to the challenge presented by high ratio schedules and bias animals toward alternative paths to reinforcement that require less effort.[12,15,66] Accumbens DA is a critical participant in this circuitry, but it is only one part; several neurotransmitters present in multiple brain areas also are involved. Some of these brain areas may be more directly involved in the exertion of effort (i.e., response output in the face of high response costs), while others may be more selectively regulating effort-related decision making or the perception of effort. Further research is necessary to distinguish between those specific functions and to identify the relevant brain structures involved. Nevertheless, it already is evident that research in this area has helped to clarify our understanding of important aspects of natural motivation. Of course, in addition to further subdividing effort-related processes into various components, it is important to consider behavioral activation and effort in relation to other features of motivation that involve mesolimbic DA. As emphasized in recent papers,[12,15] it is clear that accumbens DA does not merely perform one behavioral function. For that reason, evidence in favor of the hypothesis that DA is involved in the exertion of effort or effort-related choice behavior does not argue against the involvement of this system in processes related to instrumental learning,[120–123] incentive salience,[16,124,138,139] aversive motivation,[12,15,66,126–131] action selection and engagement,[140–142], or Pavlovian-instrumental transfer.[48,125,143–146] Moreover, the observation that mesolimbic DA participates in several behavioral processes related to motivation is consistent with the idea that nucleus accumbens appears to be organized into assemblies of task-specific neurons that are modulated by DA.[8–10,12,146] Generally speaking, the decline of the traditional form of the DA hypothesis of reward has led to a period of rich conceptual restructuring in the field. Studies of the role of nucleus accumbens in behavioral activation and effort-related processes, together with studies of other functions, are leading to a greater understanding of the brain mechanisms regulating distinct aspects of motivation. They also serve to emphasize the fundamental relation between motivational processes and the regulation of action.

In addition to being important for understanding basic scientific principles related to aspects of motivation, identification of the brain systems involved in regulating behavioral activation and effort-based choice in animals may have substantial clinical significance. This research has provided important clues regarding the brain systems that are involved in clinical psychopathologies related to psychomotor retardation, fatigue, or anergia in depression, parkinsonism, and other disorders.[15,21,147] Indeed, there is a striking similarity between the brain systems known to participate in effort-related functions in animals and those involved in psychomotor dysfunction in humans.[15,21] Moreover, the activational functions of nucleus accumbens DA are not only critical for aspects of motivation for natural stimuli; they also appear to be important for drug-seeking behavior. Drug use and abuse involve numerous psychological functions, including reinforcement, learning, motivation, emotion, habit formation, and compulsiveness, but exertion of effort also is a critical feature of the self-administration process and the persistence of drug seeking. Over the last few years, there has been a growing emphasis upon the effort-related processes involved in drug seeking behavior.[137,148–152] Clinical studies have demonstrated that withdrawal-related deficits in DA function following exessive drug taking in addicts appear to be related to motivational impairments such as psychomotor slowing.[153] Thus, research on the activational functions of mesolimbic DA and related brain systems may yield important information about the neural basis of various forms of normal and pathological motivation.

ACKNOWLEDGMENTS

Much of the work cited in this review was supported by grants to JDS from the U.S. NSF and NIH/NIMH, NIDA and NINDS. Many thanks to Dra. Merce Correa for her many helpful comments.

REFERENCES

1. Mogenson GJ, Jones DL, Yim CY. From motivation to action: functional interface between the limbic system and the motor system. *Prog Neurobio.l* 1980;14:69–97.
2. Yang CR, Mogenson GJ. An electrophysiological study of the neural projections from the hippocampus to the ventral pallidum and the subpallidal areas by way of the nucleus accumbens. *Neuroscience.* 1985;15:1015–1024.
3. Nauta WJ, Smith JP, Faull RL, Domesick VB. Efferent connections and nigral afferents of the nucleus accumbens septi in the rat. *Neuroscience.* 1978;3:385–401.
4. Brog JS, Salyapongse A, Deutch AY, Zahm DS. The patterns of afferent innervation of the core and shell in the accumbens part of the rat ventral striatum—immunohistochemical detection of retrogradely transported fluoro-gold. *J Comp Neurol.* 1993;338:255–278.
5. Groenewegen HJ, Wright CI, Beijer AV. The nucleus accumbens: gateway for limbic structures to reach the motor system? *Prog Brain Res.* 1996;107:485–511.

6. Steiniger-Brach B, Kretschmer BD. Different function of pedunculopontine GABA and glutamate receptors in nucleus accumbens dopamine, pedunculopontine glutamate and operant discriminative behavior. *Eur J Neurosci.* 2005;22:1720–1730.
7. Zahm DS. An integrative neuroanatomical perspective on some subcortical substrates of adaptative responding with emphasis on the nucleus accumbens. *Neurosci Biobeh Rev.* 2000;24:85–105.
8. Pennartz CM, Groenewgen HJ, Lopez de Silva FH. The nucleus accumbens as a complex of functionally distinct neuronal ensembles: an integration of behavioral, electrophysiological and anatomical data. *Prog Neurobiol.* 1994;42:719–761.
9. O'Donnell P. Dopamine gating of forebrain neural ensembles. *Eur J Neurosci.* 2003;17:429–435.
10. Carelli RM, Wondolowski J. Selective encoding of cocaine versus natural rewards by nucleus accumbens neurons is not related to chronic drug exposure. *J Neurosci.* 2003;23:11214–11223.
11. Nicola SM, Yun IA, Wakabayashi KT, Fields HL. Cue-evoked firing of nucleus accumbens neurons encodes motivational significance during a discriminative stimulus task. *J Neurophysiol.* 2004;91:1840–1865.
12. Salamone JD, Correa M, Mingote SM, Weber SM. Beyond the reward hypothesis: alternative functions of nucleus accumbens dopamine. *Curr Opin Pharmacol.* 2005;5:34–41.
13. Swerdlow NR, Mansbach RS, Geyer MA, Pulvirenti L, Koob GF, Braff DL. Amphetamine disruption of prepulse inhibition of acoustic startle is reversed by depletion of mesolimbic dopamine. *Psychopharmacology.* 1990;100:413–416.
14. Salamone JD. Functions of mesolimbic dopamine: changing concepts and shifting paradigms. *Psychopharmacology.* 2007;191:389.
15. Salamone JD, Correa M, Farrar A, Mingote SM. Effort-related functions of nucleus accumbens dopamine and associated forebrain circuits. *Psychopharmacology.* 2007;191:461–482.
16. Berridge KC, Kringlebach ML. Affective neuroscience of pleasure: reward in humans and animals. *Psychopharmacology.* 2008;199:457–480.
17. Kuhn TS. *The Structure of Scientific Revolutions.* Chicago, IL: University of Chicago Press; 1962.
18. Cofer CN, Appley MH. *Motivation: Theory and Research.* New York, NY: Wiley; 1964.
19. Salamone JD. Dopaminergic involvement in activational aspects of motivation: effects of haloperidol on schedule induced activity, feeding and foraging in rats. *Psychobiology.* 1988;16:196–206.
20. Salamone JD. Behavioral pharmacology of dopamine systems: A new synthesis. In: Willner P, Scheel-Kruger J, eds. *The Mesolimbic Dopamine System: From Motivation to Action.* Cambridge, England: Cambridge University Press; 1991:599–613.
21. Salamone JD, Correa M, Mingote SM, Weber SM, Farrar AM. Nucleus accumbens dopamine and the forebrain circuitry involved in behavioral activation and effort-related decision making: implications of understanding anergia and psychomotor slowing and depression. *Curr Psychiatry Rev.* 2006;2:267–280.
22. Salamone J, Correa M, Font L, Pennarola A, Farrar AM, Mingote S. Nucleus accumbens and the neurochemical interactions regulating effort-related processes. In: David HN, ed. *The Nucleus Accumbens: Neurotransmitters and Related Behaviours.* Research Signpost, Keralia, India: Research Signpost; 2008:195–217.
23. Barbano MF, Cador M. Opioids for hedonic experience and dopamine to get ready for it. *Psychopharmacology.* 2007;191:497–506.
24. Krebs JR. Optimal foraging: theory and experiment. *Nature.* 1977;268:583–584.
25. Hull CL. *Principles of Behaviour.* New York, NY: Appleton-Century-Crofts; 1943.
26. Spence KW. *Behavior Theory and Conditioning.* New Haven, CT: Yale University Press; 1956.
27. Collier GH, Jennings W. Work as a determinant of instrumental performance. *J Comp Physiol Psychol.* 1969;68:659–662.
28. Epstein LH, Roemmich JN, Paluch RA, Raynor HA. Physical activity as a substitute for sedentary behavior in youth. *Ann Behav Med.* 2005;29:200–209.
29. Lea SEG. The psychology and economics of demand. *Psychol Bull.* 1978;85:441–466.
30. Hursh SR, Raslear TG, Shurtleff D, Bauman R, Simmons L. A cost-benefit analysis of demand for food. *J Exp Anal Behav.* 1988;50:419–440.
31. Bickel WK, Marsch LA, Carroll ME. Deconstructing relative reinforcing efficacy and situating the measures of pharmacological reinforcement with behavioral economics: a theoretical proposal. *Psychopharmacology.* 2000;153:44–56.
32. Pijnenburg AJ, Honig WM, Van Rossum JM. Inhibition of d-amphetamine-induced locomotor activity by injection of haloperidol into the nucleus accumbens of the rat. *Psychopharmacologia.* 1975;41:87–95.
33. Kelly PH, Seviour PW, Iversen SD. Amphetamine and apomorphine responses in the rat following 6-OHDA lesions of the nucleus accumbens septi and corpus striatum. *Brain Res.* 1975;94:507–522.
34. Koob GF, Riley SJ, Smith SC, Robbins TW. Effects of 6-hydroxydopamine lesions of the nucleus accumbens septi and olfactory tubercle on feeding, locomotor activity, and amphetamine anorexia in the rat. *J Comp Physiol Psychol.* 1978;92:917–927.
35. Delfs JM, Schreiber L, Kelley AE. Microinjection of cocaine into the nucleus accumbens elicits locomotor activation in the rat. *J Neurosci.* 1990;10:303–310.
36. Koob GF, Swerdlow NR. The functional output of the mesolimbic dopamine system. *Ann NY Acad Sci.* 1988;537:216–227.
37. Falk JL. The nature and determinants of adjunctive behavior. *Physiol Behav.* 1971;6:577–588.
38. Killeen PR. Incentive theory. In: Bernstein D, ed. *Response Structure and Organization.* Lincoln: University of Nebraska Press; 1982;38:169–216.
39. Killeen PR. On the temporal control of behavior. *Psychol Rev.* 1975;82:89–115.
40. Killeen PR, Hanson SJ, Osborne SR. Arousal: its genesis and manifestation as response rate. *Psychol Rev.* 1978;85:571–581.
41. Lopez-Crespo G, Rodriguez M, Pellon R, Flores P. Acquisition of schedule-induced polydipsia by rats in proximity to upcoming food delivery. *Learn Behav.* 2004;32:491–499.
42. Robbins TW, Koob GF. Selective disruption of displacement behaviour by lesions of the mesolimbic dopamine system. *Nature.* 1980;285:409–412.
43. Wallace M, Singer G, Finlay J, Gibson S. The effect of 6-OHDA lesions of the nucleus accumbens septum on schedule-induced drinking, wheelrunning and corticosterone levels in the rat. *Pharmacol Biochem Behav.* 1983;18:129–136.
44. Robbins TW, Roberts DC, Koob GF. Effects of d-amphetamine and apomorphine upon operant behavior and schedule-induced licking in rats with 6-hydroxydopamine-induced lesions of the nucleus accumbens. *J Pharmacol Exp Ther.* 1983;224:662–673.

45. Mittleman G, Whishaw IQ, Jones GH, Koch M, Robbins TW. Cortical, hippocampal, and striatal mediation of schedule-induced behaviors. *Behav Neurosci*. 1990;104:399–409.

46. McCullough LD, Salamone JD. Involvement of nucleus accumbens dopamine in the motor activity induced by periodic food presentation: a microdialysis and behavioral study. *Brain Res*. 1992;592:29–36.

47. Weissenborn R, Blaha CD, Winn P, Phillips AG. Schedule-induced polydipsia and the nucleus accumbens: electrochemical measurements of dopamine efflux and effects of excitotoxic lesions in the core. *Behav Brain Res*. 1996;75:147–158.

48. Robbins TW, Everitt B. A role for mesencephalic dopamine in activation: a commentary on Berridge (2007). *Psychopharmacology*. 2007;191:433–437.

49. Salamone JD, Steinpreis RE, McCullough LD, Smith P, Grebel D, Mahan K. Haloperidol and nucleus accumbens dopamine depletion suppress lever pressing for food but increase free food consumption in a novel food choice procedure. *Psychopharmacology*. 1991;104:515–521.

50. Berridge KC, Robinson TE. Parsing reward. *Trends Neurosc*. 2003;26:507–513.

51. Salamone JD. Different effects of haloperidol and extinction on instrumental behaviours. *Psychopharmacology*. 1986;88:18–23.

52. Caul WF, Brindle NA. Schedule-dependent effects of haloperidol and amphetamine: multiple-schedule task shows within-subject effects. *Pharmacol Biochem Behav*. 2001;68:53–63.

53. Ishiwari K, Weber SM, Mingote S, Correa M, Salamone JD. Accumbens dopamine and the regulation of effort in food-seeking behavior: modulation of work output by different ratio or force requirements. *Behav Brain Res*. 2004;151:83–91.

54. Aberman JE, Salamone JD. Nucleus accumbens dopamine depletions make rats more sensitive to high ratio requirements but do not impair primary food reinforcement. *Neuroscience*. 1999;92:545–552.

55. Salamone JD, Wisniecki A, Carlson BB, Correa M. Nucleus accumbens dopamine depletions make animals highly sensitive to high fixed ratio requirements but do not impair primary food reinforcement. *Neuroscience*. 2001;105:863–870.

56. Salamone JD, Aberman JE, Sokolowski JD, Cousins MS. Nucleus accumbens dopamine and rate of responding: Neurochemical and behavioral studies. *Psychobiology*. 1999;27:236–247.

57. Salamone JD, Correa M, Mingote S, Weber SM. Nucleus accumbens dopamine and the regulation of effort in food-seeking behavior: implications for studies of natural motivation, psychiatry, and drug abuse. *J Pharmacol Exp Ther*. 2003;305:1–8.

58. Cardinal RN, Robbins TW, Everitt BJ. The effects of d-amphetamine, chlordiazepoxide, alpha-flupenthixol and behavioural manipulations on choice of signalled and unsignalled delayed reinforcement in rats. *Psychopharmacology*. 2000;152:362–375.

59. Correa M, Carlson BB, Wisniecki A, Salamone JD. Nucleus accumbens dopamine and work requirements on interval schedules. *Behav Brain Res*. 2002;137:179–187.

60. Mingote S, Weber SM, Ishiwari K, Correa M, Salamone JD. Ratio and time requirements on operant schedules: effort-related effects of nucleus accumbens dopamine depletions. *Eur J Neurosci*. 2005;21:1749–1757.

61. Wakabayashi KT, Fields HL, Nicola SM. Dissociation of the role of nucleus accumbens dopamine in responding to reward-predictive cues and waiting for reward. *Behav Brain Res*. 2004;154:19–30.

62. Winstanley CA, Theobald DE, Dalley JW, Robbins TW. Interactions between serotonin and dopamine in the control of impulsive choice in rats: therapeutic implications for impulse control disorders. *Neuropsychopharmacology*. 2005;30:669–682.

63. Floresco SB, Tse MT, Ghods-Sharifi S. Dopaminergic and glutamatergic regulation of effort- and delay-based decision making. *Neuropsychopharmacology*. 2008;33:1966–1979.

64. Fowler SC, LaCerra MM, Ettenberg A. Behavioral functions of nucleus accumbens dopamine: empirical and conceptual problems with the anhedonia hypothesis. *Pharmacol Biochem Behav*. 1986;25:791–796.

65. Salamone JD, Kurth PA, McCullough LD, Sokolowski JD, Cousins MS. The role of brain dopamine in response initiation: effects of haloperidol and regionally specific dopamine depletions on the local rate of instrumental responding. *Brain Res*. 1993;628:218–226.

66. Salamone JD, Cousins MS, Snyder BJ. Behavioral functions of nucleus accumbens dopamine: empirical and conceptual problems with the anhedonia hypothesis. *Neurosci Biobehav Rev*. 1997;21:341–359.

67. Sokolowski JD, Salamone JD. The role of accumbens dopamine in lever pressing and response allocation: effects of 6-OHDA injected into core and dorsomedial shell. *Pharmacol Biochem Behav*. 1998;59:557–566.

68. Aberman JE, Ward SJ, Salamone JD. Effects of dopamine antagonists and accumbens dopamine depletions on time-constrained progressive-ratio performance. *Pharmacol Biochem Behav*. 1998;61:341–348.

69. Salamone JD, Correa M. Motivational views of reinforcement: implications for understanding the behavioral functions of nucleus accumbens dopamine. *Behav Brain Res*. 2002;137:3–25.

70. Van den Bos R, Van der Harst J, Jonkman S, Schilders M, Spruijt B. Rats assess costs and benefits according to an internal standard. *Behav Brain Res*. 2006;171:350–354.

71. Salamone JD, Cousins MS, Bucher S. Anhedonia or anergia? Effects of haloperidol and nucleus accumbens dopamine depletion on instrumental response selection in a T-maze cost/benefit procedure. *Behav Brain Res*. 1994;65:221–229.

72. Cousins MS, Atherton A, Turner L, Salamone JD. Nucleus accumbens dopamine depletions alter relative response allocation in a T-maze cost/benefit task. *Behav Brain Res*. 1996;74:189–197.

73. Salamone JD, Cousins MS, Maio C, Champion M, Turski T, Kovach J. Different behavioral effects of haloperidol, clozapine and thioridazine in a concurrent lever pressing and feeding procedure. *Psychopharmacology*. 1996;125:105–112.

74. Cousins MS, Wei W, Salamone JD. Pharmacological characterization of performance on a concurrent lever pressing/feeding choice procedure: effects of dopamine antagonist, cholinomimetic, sedative and stimulant drugs. *Psychopharmacology*. 1994;116:529–537.

75. Koch M, Schmid A, Schnitzler HU. Role of nucleus accumbens dopamine D1 and D2 receptors in instrumental and Pavlovian paradigms of conditioned reward. *Psychopharmacology*. 2000;152:67–73.

76. Sink KS, Vemuri VK, Olszewska T, Makriyannis A, Salamone JD. Cannabinoid CB1 antagonists and dopamine antagonists produce different effects on a task involving response allocation and effort-related choice in food-seeking behavior. *Psychopharmacology*. 2008;196:565–574.

77. Salamone JD, Arizzi MN, Sandoval MD, Cervone KM, Aberman JE. Dopamine antagonsts alter response allocation

78. Cousins MS, Sokolowski JD, Salamone JD. Different effects of nucleus accumbens and ventrolateral striatal dopamine depletions on instrumental response selection in the rat. *Pharmacol Biochem Behav*. 1993;46:943–951.
79. Cousins MS, Salamone JD. Nucleus accumbens dopamine depletions in rats affect relative response allocation in a novel cost/benefit procedure. *Pharmacol Biochem Behav*. 1994;49:85–91.
80. Nowend KL, Arizzi M, Carlson BB, Salamone JD. D1 or D2 antagonism in nucleus accumbens core or dorsomedial shell suppresses lever pressing for food but leads to compensatory increases in chow consumption. *Pharmacol Biochem Behav*. 2001;69:373–382.
81. Cagniard B, Balsam PD, Brunner D, Zhuang X. Mice with chronically elevated dopamine exhibit enhanced motivation, but not learning, for a food reward. *Neuropsychopharmacology*. 2006;31:1362–1370.
82. Niv Y, Daw ND, Joel D, Dayan P. Tonic dopamine: opportunity costs and the control of response vigor. *Psychopharmacology*. 2007;191:507–520.
83. Phillips PE, Walton ME, Jhou TC. Calculating utility: preclinical evidence for cost-benefit analysis by mesolimbic dopamine. *Psychopharmacology*. 2007;191:483–495.
84. Svenningsson P, Le Moine C, Fisone, G, Fredholm BB. Distribution, biochemistry and function of striatal adenosine A_{2A} receptors. *Prog Neurobiol*. 1999;59:355–396.
85. Wang WF, Ishiwata K, Nonaka H, Ishii S, Kiyosawa M, Shimada J, Suzuki F, Senda M. Carbon-11-labeled KF21213: a highly selective ligand for mapping CNS adenosine A(2A) receptors with positron emission tomography. *Nucl Med. Biol*. 2000;27:541–546.
86. Hettinger BD, Lee A, Linden J, Rosin DL. Ultrastructural localization of adenosine A2A receptors suggests multiple cellular sites for modulation of GABAergic neurons in rat striatum. *J Comp Neurol*. 2001;431:331–346.
87. Chen JF, Moratalla R, Impagnatiello F, Grandy DK, Cuellar B, Rubinstein M, Beilstein MA, Hackett E, Fink JS, Low MJ, Ongini E, Schwarzschild MA. The role of the D2 dopamine receptor (D2R) in A2a adenenosine-receptor (A2aR) mediated behavioral and cellular responses as revealed by A2a and D2 receptor knockout mice. *Proc Natl Acad Sci USA*. 2001;98:1970–1975.
88. Ferre S, Fredholm BB, Morelli M, Popoli P, Fuxe K. Adenosine-dopamine receptor-receptor interactions as an integrative mechanism in the basal ganglia. *Trends Neurosci*. 1997;20:482–487.
89. Ferré S, Popoli P, Giménez-Llort L, Rimondini R, Müller CE, Strömberg I, Ögren SO, Fuxe K. Adenosine/dopamine interaction: implications for the treatment of Parkinson's disease. *Parkinsonism Relat Disord*. 2001;7:235–241.
90. Jenner P. A_{2A} antagonists as novel non-dopaminergic therapy for motor dysfunction in PD. *Neurology*. 2003;61:S32-S38.
91. Jenner P. Istradefylline, a novel adenosine A2A receptor antagonist, for the treatment of Parkinson's disease. *Expert Opin Invest Drugs*. 2005;14:729–738.
92. Hauber W, Neuscheler P, Nagel J, Müller CE. Catalepsy induced by a blockade of dopamine D1 or D2 receptors was reversed by a concomitant blockade of adenosine A2a receptors in the caudate putamen of rats. *Eur J Neurosci*. 2001;14:1287–1293.
93. Correa M, Wisniecki A, Betz A, Dobson DR, O'Neill MF, O'Neill MJ, Salamone JD. The adenosine A2A antagonist KF17837 reverses the locomotor suppression and tremulous jaw movements induced by haloperidol in rats: possible relevance to parkinsonism. *Behav Brain Res*. 2004;148:47–54.
94. Salamone JD, Betz AJ, Ishiwari K. Tremorolytic effects of adenosine A2A antagonists: implications for parkinsonism. *Front Biosci*. 2008;13:3594–3605.
95. Pinna A, Wardas J, Simola N, Morelli M. New therapies for the treatment of Parkinson's disease: adenosine A_{2A} receptor antagonists. *Life Sci*. 2005;77:325932–325967.
96. Barraco RA, Martens KA, Parizon M, Normile HJ. Adenosine A2a receptors in the nucleus accumbens mediate locomotor depression. *Brain Res Bull*. 1993;31:397–404.
97. Barraco RA, Martens KA, Parizon M, Normile HJ. Role of adenosine A2a receptors in the nucleus accumbens. *Prog Neuropsychopharmacol Biol Psychiatry*. 1994;18:545–553.
98. Ishiwari K, Madson LJ, Farrar AM, Mingote SM, Valenta JP, DiGianvittorio MD, Frank LE, Correa M, Hockemeyer J, Müller C, Salamone JD. Injections of the selective adenosine A_{2A} antagonist MSX-3 into the nucleus accumbens core attenuate the locomotor suppression induced by haloperidol in rats. *Behav Brain Res*. 2007;178:190–199.
99. Farrar AM, Pereira M, Velasco F, Hockemeyer J, Müller C, Salamone JD. Adenosine A(2A) receptor antagonism reverses the effects of dopamine receptor antagonism on instrumental output and effort-related choice in the rat: implications for studies of psychomotor slowing. *Psychopharmacology*. 2007;191:579–586.
100. Worden LT, Shahriari M, Farrar AM, Sink KS, Hockemeyer J, Müller C, Salamone JD. The adenosine A_{2A} antagonist MSX-3 reverses the effort-related effects of dopamine blockade: differential interaction with D1 and D2 family antagonists. *Psychopharmacology*. 2009;203:489–499.
101. Mott AM, Nunes EJ, Collins LE, Port RG, Sink KS, Hockemeyer J, Müller CE, Salamone JD. The adenosine A_{2A} antagonist MSX-3 reverses the effects of the dopamine antagonist haloperidol on effort-related decision making in a T-maze cost/benefit procedure. *Psychopharmacology*. 2009;204:103–112.
102. Mingote SM, Font L, Farrar AM, Vontell R, Worden LT, Stopper CM, Port RG, Sink KS, Bunce JG, Chrobak JJ, Salamone JD. Nucleus accumbens adenosine A2A receptors regulate exertion of effort by acting on the ventral striatopallidal pathway. *J Neurosci*. 2008;28:9037–9046.
103. Font L, Mingote S, Farrar AM, Pereira M, Worden L, Stopper C, Port RG, Salamone JD. Intra-accumbens injections of the adenosine A(2A) agonist CGS 21680 affect effort-related choice behavior in rats. *Psychopharmacology*. 2008;199:515–526.
104. Walton ME, Bannerman DM, Rushworth MF. The role of rat medial frontal cortex in effort-based decision making. *J Neurosci*. 2002;22:10996–11003.
105. Walton ME, Bannerman DM, Alterescu K, Rushworth MF. Functional specialization within medial frontal cortex of the anterior cingulate for evaluating effort-related decisions. *J Neurosci*. 2003;23:6475–6479.
106. Schweimer J, Hauber W. Involvement of the rat anterior cingulate cortex in control of instrumental responses guided by reward expectancy. *Learn Mem*. 2005;12: 334–342.
107. Schweimer J, Saft S, Hauber W. Involvement of catecholamine neurotransmission in the rat anterior cingulate in effort-related decision making. *Behav Neurosci*. 2005;119:1687–1692.
108. Floresco SB, Ghods-Sharifi S. Amygdala-prefrontal cortical circuitry regulates effort-based decision making. *Cereb Cortex*, 2007;17:251–260.

109. Hauber W, Sommer S. Prefrontostriatal circuitry regulates effort-related decision making. *Cereb Cortex*, in press.
110. Groenewegen HJ, Berendse HW, Haber SN. Organization of the output of the ventral striatopallidal system in the rat: ventral pallidal efferents. *Neuroscience*. 1993;57:113–142.
111. Kretschmer BD. Functional aspects of the ventral pallidum. *Amino Acids*. 2000;19:201–210.
112. Jones DL, Mogenson GJ. Nucleus accumbens to globus pallidus GABA projection subserving ambulatory activity. *Am J Physiol*. 1980;238:R65-R69.
113. Austin MC, Kalivas PW. Enkephalinergic and GABAergic modulation of motor activity in the ventral pallidum. *J Pharmacol Exp Ther*. 1990;252:1370–1377.
114. Hooks MS, Kalivas PW. The role of mesoaccumbens-pallidal circuitry in novelty-induced behavioral activation. *Neuroscience*. 1995;64:587–597.
115. Farrar AM, Font L, Pereira M, Mingote SM, Bunce JG, Chrobak JJ, Salamone JD. Forebrain circuitry involved in effort-related choice: injections of the GABA_A agonist muscimol into ventral pallidum alters response allocation in food-seeking behavior. *Neuroscience*. 2008;152:321–330.
116. Rushworth MF, Walton ME, Kennerley SW, Bannerman DM. Action sets and decisions in the medial frontal cortex. *Trends Cogn Sci*. 2004;8:410–417.
117. Walton ME, Kennerley SW, Bannerman DM, Phillips PE, Rushworth MF. Weighing up the benefits of work: behavioral and neural analyses of effort-related decision making. *Neural Netw*. 2006;19:1302–1314.
118. Rick JH, Horvitz JC, Balsam PD. Dopamine receptor blockade and extinction differentially affect behavioral variability. *Behav Neurosci*. 2006;120:488–492.
119. Salamone JD, Kurth P, McCullough LD, Sokolowski JD. The effects of nucleus accumbens dopamine depletions on continuously reinforced operant responding: contrasts with the effects of extinction. *Pharmacol Biochem Behav*. 1995;50:437–443.
120. Wise RA. Dopamine, learning and motivation. *Nat Rev Neurosci*. 2004;5:483–494.
121. Beninger RJ, Gerdjikov T. The role of signaling molecules in reward-related incentive learning. *Neurotox Res*. 2004;6:91–104.
122. Kelley AE, Baldo BA, Pratt WE, Will MJ. Corticostriatal-hypothalamic circuitry and food motivation: integration of energy, action and reward. *Physiol Behav*. 2005;86:773–795.
123. Baldo BA, Kelley AE. Distinct neurochemical coding of discrete motivational processes: insights from nucleus accumbens control of feeding. *Psychopharmacology*. 2007;191:439–450.
124. Berridge KC. The debate over dopamine's role in reward: the case for incentive salience. *Psychopharmacology*. 2007;191:391–431.
125. Yin HH, Ostlund SB, Balleine BW. Reward-guided learning beyond dopamine in the nucleus accumbens: the integrative functions of cortico-basal ganglia networks. *Eur J Neurosci*. 2009;9:65–73.
126. Salamone JD. The involvement of nucleus accumbens dopamine in appetitive and aversive motivation. *Behav Brain Res*. 1994;61:117–133.
127. Delgado MR, Li J, Schiller D, Phelps EA. The role of the striatum in aversive learning and aversive prediction errors. *Philos Trans R Soc*. 2008;363:3787–3800.
128. Levita L, Hare TA, Voss HU, Glover G, Ballon DJ, Casey BJ. The bivalent side of the nucleus accumbens. *Neuroimage*. 2009;44:1178–1187.
129. Marinelli S, Pascucci T, Bernardi G, Puglisi-Allegra S, Mercuri NB. Activation of TRPV1 in the VTA excites dopaminergic neurons and increases chemical- and noxious-induced dopamine release in the nucleus accumbens. *Neuropsychopharmacology*. 2005;30:864–875.
130. Anstrom KK, Woodward DJ. Restraint increases dopaminergic burst firing in awake rats. *Neuropsychopharmacology*. 2005;30:1832–1840.
131. Faure A, Reynolds SM, Richard JM, Berridge KC. Mesolimbic dopamine in desire and dread: enabling motivation to be generated by localized glutamate disruptions in nucleus accumbens. *J Neurosci*. 2008;28:7184–7192.
132. Schultz W. Multiple dopamine functions at different time courses. *Annu Rev Neurosci*. 2007;30:259–288.
133. Schultz W. Behavioral dopamine signals. *Trends Neurosc.i* 2007;30:203–210.
134. Roitman MF, Stuber GD, Phillips PE, Wightman RM, Carelli RM. Dopamine operates as a subsecond modulator of food seeking. *J Neurosci*. 2004;24:1265–1271.
135. Lavin A, Nogueira L, Lapish CC, Wightman RM, Phillips PE, Seamans JK. Mesocortical dopamine neurons operate in distinct temporal domains using multimodal signaling. *J Neurosci*. 2005;25:5013–5023.
136. Lapish CC, Kroener S, Durstewitz D, Lavin A, Seamans JK. The ability of the mesocortical dopamine system to operate in distinct temporal modes. *Psychopharmacology*. 2007;191:609–625.
137. Marinelli M, Barrot M, Simon H, Oberlander C, Dekeyne A, Le Moal M, Piazza PV. Pharmacological stimuli decreasing nucleus accumbens dopamine can act as positive reinforcers but have a low addictive potential. *Eur J Neurosci*. 1998;10:3269–3275.
138. Wyvell CL, Berridge KC. Incentive sensitization by previous amphetamine exposure: increased cue-triggered "wanting" for sucrose reward. *J Neurosci*. 2007;21:7831–7840.
139. Berridge KC, Robinson TE, Aldridge JW. Dissecting components of reward: 'liking', 'wanting', and learning. *Curr Opin Pharmacol*. 139:65–73.
140. Nicola SM. The nucleus accumbens as part of a basal ganglia action selection circuit. *Psychopharmacology*. 2007;191:521–550.
141. Prescott TJ, Montes González FM, Gurney K, Humphries MD, Redgrave P. A robot model of the basal ganglia: behavior and intrinsic processing. *Neural Netw*. 2006;19:31–61.
142. Redgrave P, Gurney K, Reynolds J. What is reinforced by phasic dopamine signals? *Brain Res Rev*. 2008;58:322–339.
143. Everitt BJ, Robbins TW. Neural systems of reinforcement for drug addiction: from actions to habits to compulsion. *Nat Neurosci*. 2005;8:1481–1489.
144. Di Ciano P, Cardinal RN, Cowell RA, Little SJ, Everitt BJ. Differential involvement of NMDA, AMPA/kainate, and dopamine receptors in the nucleus accumbens core in the acquisition and performance of pavlovian approach behavior. *J Neurosci*. 2001;21:9471–9477.
145. Parkinson JA, Dalley JW, Cardinal RN, Bamford A, Fehnert B, Lachenal G, Rudarakanchana N, Halkerston KM, Robbins TW, Everitt BJ. Nucleus accumbens dopamine depletion impairs both acquisition and performance of appetitive Pavlovian approach behaviour: implications for mesoaccumbens dopamine function. *Behav Brain Res*. 2002;137:149–163.
146. Day JJ, Wheeler RA, Roitman MF, Carelli RM. Nucleus accumbens neurons encode Pavlovian approach behaviors: evidence from an autoshaping paradigm. *Eur J Neurosci*. 2006;23:1341–1351.

147. Stahl SM. The psychopharmacology of energy and fatigue. *J Clin Psychiatry.* 2002;63:7–8.
148. Nadal R, Armario A, Janak PH. Positive relationship between activity in a novel environment and operant ethanol self-administration in rats. *Psychopharmacology.* 2002;62:333–338.
149. Vezina P, Lorrain DS, Arnold GM, Austin JD, Suto N. Sensitization of midbrain dopamine neuron reactivity promotes the pursuit of amphetamine. *J Neurosci.* 2002;22:654–4662.
150. Czachowski CL, Santini LA, Legg BH, Samson HH. Separate measures of ethanol seeking and drinking in the rat: effects of remoxipride. *Alcohol.* 2002;28:39–46.
151. Colby CR, Whisler K, Steffen C, Nestler EJ, Self DW. Striatal cell type-specific overexpression of DeltaFosB enhances incentive for cocaine. *J Neurosci.* 2003;23:2488–2493.
152. Correa M, Salamone JD. Implicación del componente hedónico en el uso y abuso de drogas. In: Juarez J, ed. *Neurobiología del Hedonismo en la Conducta.* Mexico City, Mexico: Manual Moderno; 2007:187–206.
153. Volkow ND, Chang L, Wang G, Fowler JS, Leonido-Yee M, Franceschi D, Sedler MJ, Gatley SJ, Hitzemann R, Ding YS, Logan J, Wong C, Miller EN. Association of dopamine transporter reduction with psychomotor impairment in methamphetamine abusers. *Am J Psychiatry.* 2001;158:377–382.

6.3 | Functional Heterogeneity in Striatal Subregions and Neurotransmitter Systems: Implications for Understanding the Neural Substrates Underlying Appetitive Motivation and Learning

BRIAN A. BALDO AND MATTHEW E. ANDRZEJEWSKI

AUTHORS' NOTE

This chapter is meant to present a general discussion of the literature on the regional heterogeneity of striatal function with regard to behavioral processes. We have provided, however, a particular emphasis on the work of the late Prof. Ann E. Kelley, who is widely regarded to have made one of the greatest contributions to this area of knowledge. Ann was scheduled to attend the June 2007 meeting in Goteborg, Sweden, upon which this volume is based, but had to cancel due to complications arising from her battle with cancer. Although disappointed that she could not attend, she was greatly moved by the outpouring of love from her dear colleagues and friends, many of whom communicated with her from the meeting itself. She died later that summer.

Our goal here was to emphasize the research themes most important to Ann and highlight some of her major contributions within the context of a scholarly review, based upon our own fallible understanding of her thought. We have, hopefully, done some measure of justice to the elegance of her hypotheses and the remarkable thematic cohesiveness that characterized her research program. As students of Ann, we can attest to both her scientific brilliance and her deeply caring and selfless mentorship, rare traits both, and especially so when found within the same person. The authors, standing with generations of trainees spanning four decades, miss her profoundly and hope that this chapter can stand as one small testament to her remarkable and lasting contributions to our understanding of the neural basis of motivation.

OVERVIEW

The Nobel Prize in Physiology or Medicine for the year 2000 was awarded to three prominent neuroscientists: Arvid Carlson, Paul Greengard, and Eric Kandel. Dr. Carlson was recognized for the identification of the ascending dopaminergic pathways in the brain and for the appreciation of dopamine's role in extrapyramidal motor movement disorders such as parkinsonism. Paul Greengard was a corecipient of the Prize for his contributions to understanding the role of second messenger signaling cascades and phosphoproteins in neural function, including work on dopamine- and cyclic adenosine monophosphate (cAMP)–regulated phosphoprotein, 32 kDa (DAARP-32), a phosphoprotein expressed in striatal medium spiny neurons (a major target of the ascending dopaminergic projections) and regulated by dopamine. The juxtaposition of these two awards highlights the following point, now so widely accepted as to be almost taken for granted: that our understanding of dopamine's function in behavioral processes has developed in lock step with our understanding of the primary forebrain target of the ascending dopamine projections, the striatum.

This chapter will focus on how advances in the study of striatal anatomy and physiology have informed our appreciation of dopamine's role in appetitive motivation, with an emphasis on studies of feeding behavior, food-reinforced operant behavior, and striatal gene expression under different motivational conditions. In particular, we will outline the position that striatal dopamine plays a dual role in *augmenting* the various types of motor output associated with appetitively motivated behavior by modulating information flow through functionally differentiable corticostriatal circuits, and in *selecting/strengthening* reinforced behavior by regulating intracellular plasticity within a corticostriatal network. Evidence indicates that while these functions are expressed throughout the striatum, the behavioral domains that are affected depend upon the unique information-processing roles of anatomically distinct striatal territories. Finally, we will discuss the additional layer of complexity conferred by the heterogeneous functions of discrete neurochemical systems within a given striatal territory.

DISTINGUISHABLE DOPAMINE-MEDIATED BEHAVIORAL PROCESSES MAP ONTO ANATOMICALLY DISTINCT STRIATAL SUBDIVISIONS

Overview of Striatal Circuitry

Although the microcircuitry of the striatum is discussed in detail elsewhere in this volume, it will be useful to briefly review several salient points. First, the main intrinsic cell type in the striatum is the GABAergic medium spiny neuron. This cell type makes up 95% of all striatal neurons and represents the source of all projections leaving the striatum. In addition to the biochemical machinery to synthesize and utilize GABA as a transmitter, medium spiny neurons also synthesize several neuropeptides, including preproenkephalin and substance P, which are expressed heterogeneously among populations of striatal neurons. The remaining cells consist of GABAergic and cholinergic interneurons. Analysis of the distribution of peptide markers within medium spiny neurons has contributed to the identification of several important (and interrelated) organizational principles of striatal anatomy, including the patch/matrix organization of striatal medium spiny neurons,[1,2] the segregation to different cell populations of pallidal versus nigral efferent projections,[3,4] and the division of the nucleus accumbens into distinguishable core and shell compartments.[2,5,6] This last issue will be discussed in detail later in this chapter.

Another fundamental principle of striatal organization is based upon the basic wiring diagram of striatal inputs and outputs, in which the stereotypic circuit is formed by cortico- and thalamostriatal glutamatergic inputs impinging upon medium spiny neurons that in turn project to the pallidal and nigral complexes.[7-9] Both the corticostriatal glutamatergic inputs and the GABAergic pallidal outputs are topographically organized, resulting in some degree of segregation among circuits originating in distinct areas of cortex.[7,10-13] An enormous advance in the understanding of striatal anatomy came when it was discovered that these principles of caudate/putamen organization could also be applied to ventral striatum [nucleus accumbens (Acb)] and olfactory tubercles; key observations in this regard were that the ventral striatum receives cortical input from "limbic" allocortical and prefrontal areas and that and that the ventral pallidum represents the pallidal output field of ventral striatal territories.[14-19]

Within the striatum, individual medium spiny neurons act as points of convergence between glutamate-coded corticostriatal and thalamostriatal inputs and a wide variety of neuromodulators; prominent examples of these neuromodulators include ascending monoamine projections, GABA terminals associated with local axonal collaterals, acetylcholine arising from local interneurons, and opioid peptides, which are found within local medium spiny neuron axonal collaterals. Of these, by far the most extensively studied has been dopamine. In accord with early theoretical models of striatal information processing, it is now well established that dopamine acts to modulate glutamate-mediated electrophysiological effects and signal transduction mechanisms in medium spiny neurons.[20-24] A comprehensive review of this literature is beyond the scope of this chapter, but for the present discussion, it is useful to note that considerable electrophysiological evidence indicates that medium spiny neurons are ideally suited to act as "coincidence detectors" for convergent glutamate and dopamine signals onto the same unit[25]; the degree of prevailing dopamine tone exerts an important influence upon the ability of impinging phasic glutamate signals to activate the normally hyperpolarized medium spiny neurons.[26]

In a broad sense, this idea that dopamine exerts a modulatory effect upon functionally segregated, glutamate-coded corticotriatal circuits provides a strong heuristic with which to understand the striatum's role in behavioral regulation. If, as this model implies, the information processing roles of striatal territories differ regionally, depending upon the type of cortical input, then it would be predicted that local stimulation of dopamine transmission would produce heterogeneous behavioral effects across striatal subregions based upon its modulation of these cortical inputs.

Mapping Dopamine's Effects in Striatal Subregions

In the 1970s and 1980s, a revolution in the use of microinfusion techniques to introduce dopamine-selective neurotoxins, dopamine receptor antagonists, or dopamine-releasing agents (such as d-amphetamine) directly into the brain established the obligatory role of forebrain dopamine transmission in normative motor function. For example, many of the symptoms of parkinsonism were reproduced by chemical depletion of dopamine from wide areas of the forebrain, as achieved by 6-hydroxydopamine (6-OHDA)-induced lesions of the mesencephalic dopamine cell bodies in the ventral tegmental area (VTA) and substantia nigra pars compacta (SNc).[27-30]

Along with motor deficits, however, it was noted that dopamine-compromising manipulations, such as receptor blockade with neuroleptic drugs, also produced marked suppression of responding for food reward and rewarding electrical brain stimulation.[31-34] These observations led to the important hypothesis that in addition to their important role in motor control, central dopamine systems played a role in modulating reward per se, not just the performance aspects of reward-related behaviors.[35] This *reward/motor* dichotomy complemented the prevailing idea that the ascending dopamine pathways were organized into functionally distinct nigrostriatal and mesolimbic projections, and was further supported by the observation that the ventral striatum receives innervation from allocortical constituents of the "limbic lobe," including the amygdala and hippocampus (see the previous discussion). In this vein, Mogenson et al. proposed a highly influential hypothesis stating that the Acb represented a "limbic-motor" interface subserving the role of connecting affective states (including that associated with reward) with adaptive voluntary motor output.[36]

Several findings bolstered the idea of distinguishable behavioral effects of dopamine transmission in the caudate versus Acb,[37–39] and a *Acb/dorsal striatum* distinction began to emerge to complement the reward/motor dichotomy. It was noted, however, that many early studies did not attempt to differentiate among striatal subregions outside of the Acb; in many cases, "...drug injections were made into the 'middle' of the striatum, without regard to this structure's size or heterogeneous input."[40] The "heterogeneous input" in question referred to the segregation of corticostriatal inputs to different striatal territories. For example, careful anatomical tracing studies showed that limbic corticostriatal projections arising in the amygdala extended beyond the Acb into areas of the ventrolateral and ventromedial striatum.[18] These observations called into question the validity of a pure Acb/dorsal striatum dichotomy, suggesting instead a gradient of overlapping neocortical and limbic corticostriatal projection territories extending out from the Acb toward the posterior dorsolateral striatum, throughout which a wide variety of specialized functions representing combinations of affect-related, spatial, and purely sensory information processing could putatively be found. Hence, the question arose: could more refined microinfusion mapping approaches reveal hitherto unappreciated behavioral roles for discrete striatal territories, corresponding to their unique complements of cortical inputs?

This question was addressed in an elegant series of studies, which aimed to map the effects of local intrastriatal injections of d-amphetamine and other dopamine-active drugs on feeding behavior, motor stereotypies, and operant responding for conditioned reward.[41–44] A careful microinfusion mapping study using d-amphetamine revealed that the striatal site subserving psychostimulant-induced orofacial stereotypies resides within a restricted ventral lateral territory near the fundus of the striatum.[40] Microinfusions of d-amphetamine into this site, termed the *ventrolateral striatum* (VLS), produced intense, compulsive gnawing and biting behaviors not seen with infusions just 1–2 mm anterior or dorsal (right side of Figure 6.3.1). Because this region receives input from insular cortical sites in the vicinity of gustatory cortex and from the amygdala,[18,45] it was hypothesized that the VLS subserves fine motor behaviors in the context of feeding, especially oromotor control in the context of chewing and tongue movements. This conclusion has been upheld by many studies.[46–52] Importantly, because the motor stereotypies produced by large doses of systemically administered psychostimulants had not, to that point, been recapitulated with central microinfusions into the middle of the striatum, this study helped to validate the utility of careful mapping studies for the appreciation of regional variations in striatal function. Moreover, these results highlighted the often startlingly sharp demarcations that characterize striatal "hot spots" for the modulation of a particular behavioral process.

Guided by these findings, it was proposed that the mapping approach, by virtue of its ability to uncover discretely localized and potentially incompatible behavioral processes, had the potential to resolve conflicting effects from studies of systemically administered dopamine agonists and antagonists on feeding behavior. Several studies had suggested that dopamine transmission was essential for food reward, as indicated by purportedly extinction-like effects of dopamine antagonist administration on food-reinforced instrumental responding and food intake under certain types of scheduled feeding.[34,53,54] In contrast, other studies showed that dopamine antagonism or mesolimbic dopamine depletion did not affect or actually augmented free feeding.[55–58] It was proposed that these conflicting results could be based (in part) on the unique contributions of various striatal subregions to putatively dissociable components of feeding behavior, such as approach, reward, food handling, and oromotor control.[42,43]

To test this hypothesis, three sites with contrasting patterns of cortical innervation were chosen for study: the Acb and VLS (both described above) and the dorsolateral striatum (DLS), a control site, receiving innervation by sensorimotor cortex but not limbic structures such as the amygdala or hippocampus. As predicted, infusions of the dopamine receptor antagonist, haloperidol, or the dopamine releaser, d-amphetamine, into these structures produced clearly differentiable behavioral effects on food deprivation–driven chow intake, feeding microstructure, and associated locomotor activity. In the VLS, both drugs reduced feeding while leaving general locomotor activity intact; moreover, careful analyses revealed that intra-VLS amphetamine-induced effects on food intake were associated with high rates of spillage and competing oral stereotypies. In contrast, locomotor activity was strongly suppressed by intra-Acb haloperidol and enhanced by intra-Acb d-amphetamine infusions. Strikingly, food intake and feeding duration were *increased* by dopamine receptor antagonism in the Acb, while d-amphetamine significantly depressed food intake. This profile, clearly different from that seen with intra-VLS drug infusion, was interpreted as supporting a role for dopamine transmission in the Acb in eliciting approach/foraging responses in the presence of proximal food goals and regulating the process of switching between competing behavioral tendencies (e.g., locomotion and feeding). None of these parameters were affected by drug infusions into the region receiving somatosensory but not limbic input, the DLS. It was suggested that many of the paradoxical findings from the literature on feeding-related effects of systemically administered dopamine-active drugs could be explained partly by the different degrees to which varying behavioral testing procedures stressed preparatory/approach behaviors versus consummatory behaviors; the former would preferentially tax information processing in the Acb and the latter the control of oral motor behaviors in the VLS.[43]

The results of these locomotor activity and feeding studies also suggested that the behavioral effects of psychostimulants are closely aligned to the striatal areas receiving limbic hippocampal and/or amygdalar inputs. Would a similar anatomical segregation be seen in behavioral paradigms designed to probe more complex reward-related psychostimulant effects? To answer this question, an extensive microinfusion mapping study was carried out to explore the potentially dissociable contributions of distinct striatal subregions to conditioned reinforcement (CR).[41] This experimental paradigm is designed to test the control over instrumental behavior of cues that have acquired motivational significance through classical conditioning, and is considered a sensitive probe for the role of Acb dopamine transmission in the behavioral expression of incentive learning. In a study of seven striatal sites, it was found that a sharp gradient for d-amphetamine-induced potentiation of CR exists. The most consistent and sensitive effects (with regard to dose) were found in the Acb and nearby ventromedial caudate (Fig. 6.3.1, left side). In the VLS, responding rates were very high; however, consistent with the stereotypy-producing effects of d-amphetamine at this site, the responding was not specific to the active lever and rats were observed biting the levers. Infusions placed in posterior striatal sites were completely ineffective.

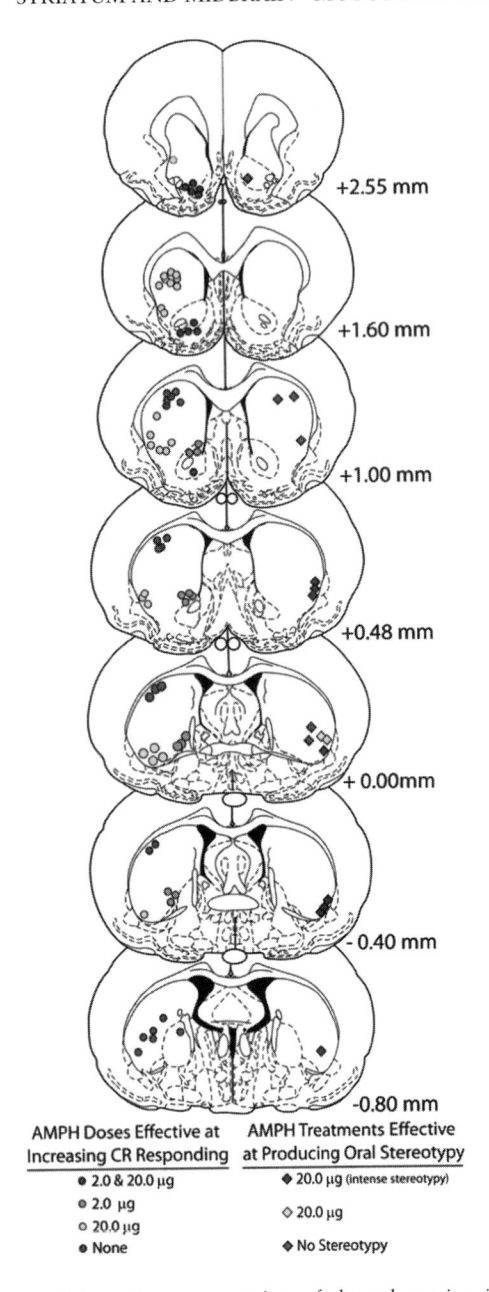

FIGURE 6.3.1. Schematic representation of d-amphetamine infusion sites from Kelley et al.[40] (right side) and Delfs and Kelley[48] (left side). On the right, oral stereotypies (in yellow) and intense stereotypies (in red) were produced by 20-μg infusions in the VLS but not in other regions of the striatum, notably the Acb. Conversely, both 2.0-μg and 20-μg infusions (in red) robustly enhanced responding for a conditioned reinforcer (CR) in the Acb. The lower dose used, 2.0 μg, was also effective in increasing CR responding in the posterior medial striatum (in orange), while the higher dose (20 μg) was effective in a gradient that went from the anterior dorsal portion to more posterior ventral and lateral portions (in yellow). Neither dose was effective in the most posterior and dorsal sites (in green). Note the differential effectiveness within the same site of the same drug on oral stereotypies and CR responding. Line drawings of brain sections were adapted from the atlas of Paxinos and Watson, with permission from Elsevier. (See Color Plate 6.3.1.)

These results are consistent with the interpretation that specific reward-related psychostimulant effects (not just spontaneous activity or food approach) are supported by striatal sites with uniquely overlapping distributions of hippocampal, amygdalar, and prefrontal cortical inputs, such as the Acb, while the lateral sectors of ventral striatum, though innervated by the basolateral amygdala, appear to mediate dopamine-dependent functions related to fine motor control, particularly in the context of oral behaviors. The striatal areas receiving sparse limbic input, however, such as posterior and dorsolateral sectors of striatum, do not appear to play a major role in the dopamine-mediated effects of psychostimulants.

Differentiating the Functional Roles of the Acb Core and Shell

The same critique that was initially applied to microinfusion studies of the dorsal striatum can also be applied to early studies of the Acb. As with the dorsal striatum, the Acb contains distinct zones with distinguishable afferent and efferent connectivity; of these, the most extensively studied are areas in the mediolateral plane corresponding to the histologically identified *core* and *shell* subregions. Immunohistochemical staining for substance P, calbindin, and several other markers reveals a centrally located Acb core surrounding the anterior commissure, with characteristics similar to those of the overlying striatum; the shell surrounds the core medially and laterally.[59] Both the core and shell receive projections from the prefrontal cortex, amygdala, and hippocampus; however, these are topographically organized such that the medial shell receives preferential input from the most ventral aspects of prefrontal cortex (such as the infralimbic region, which modulates autonomic function) and ventral subiculum, and from caudal aspects of the basolateral amygdala, while the core receives input from the prelimbic and anterior cingulate regions of frontal cortex, dorsal subiculum, and more rostral parts of the basolateral amygdaloid complex.[59–62] The lateral shell, like the VLS, receives input from insular cortex, although from more anterior regions.[14] Even more striking are inputs to the shell that are unique among striatal regions and more similar to the "extended amygdala." Examples include the shell-specific noradrenergic innervation from the A1 and A2 cell groups in the brainstem[63,64] and lateral hypothalamic projections containing the arousal-related peptide, hypocretin.[65,66] The shell also possesses a dense concentration of receptors for the pancreatic peptide, amylin[67,68]; this peptide is coreleased with insulin and is thought to cross into the brain to modulate feeding behavior at several levels of the neuraxis including the Acb. Receptors for this peptide are far less abundant in the Acb core. On the output side, the shell sends a unique (among striatal sites) projection to regions of the lateral hypothalamus involved in feeding, arousal, and autonomic activation.[69,70]

These observations support the hypothesis that the shell represents a functionally unique *viscero-endocrine* part of the striatum and the prediction that the functions of this area would be closely aligned to unconditioned behaviors in the context of altered arousal states and/or homeostatic drives.[71,72] This idea has been upheld by numerous studies. Among the first demonstrations of core/shell differences using drug microinfusions was a study showing that infusions of the N-methyl-D-aspartate (NMDA) receptor antagonist, AP-5, produced stronger deficits in locomotor activity and novel object exploration when injected into the core versus the shell.[73] In contrast, blockade of AMPA receptors or stimulation of GABA receptors in the Acb shell, but not the core, released a dramatic hyperphagia that was found to be dependent upon a disinhibition of the lateral hypothalamus.[74,75] This feeding response appeared to reflect a release of fixed-action feeding patterns, in that similar pharmacological manipulations of the shell do not produce enhanced operant responding for food reward[76] or enable the acquisition of a food-reinforced lever-press response in *ad libitum*–fed animals.[77] At more posterior levels of the Acb shell, GABA receptor stimulation elicits fixed-action patterns resembling defensive treading; such effects are absent in the core.[78] The effects of dopamine manipulations also differ between the shell and core; for example, the shell exhibits far greater sensitivity to the locomotor-stimulatory effects of D1 receptor agonists,[79] and the unconditioned dopamine release induced by amphetamine and morphine is greater in the shell than in the core.[80] However, microdialysis studies have shown that dopamine release in association with palatable feeding occurs in the shell only upon the first presentation of the food; when the food is resampled 24 hr later, elevations in dopamine efflux are seen only in the core.[81] These findings are consistent with the idea that the shell has a preferential role in mediating unconditioned behaviors in association with nonspecific changes in arousal and/or autonomic activation.

In contrast, the Acb core appears to play a more prominent role than the shell in controlling motor output associated with complex, changing contexts or with incentive control as established by prior associative learning. These functions would seem consistent with the strongly convergent dorsal hippocampal, prelimbic cortical, and basolateral amygdalar inputs to the core. In several studies, it has been shown that the ability of

Pavlovian associations to influence behavior depends upon an intact core. For example, lesions of the Acb core, but not the shell, impair conditioned approach responses to a stimulus predicting food[82] and diminish the ability of drug-associated Pavlovian cues to support operant lever pressing.[83] In the CR paradigm, lesions of the core but not of the shell disrupt CR-maintained lever pressing, while the shell is important for the augmentation of lever pressing by d-amphetamine.[82]

These studies of functional distinctions between the Acb core and shell are in excellent accord with the striatal microinfusion mapping studies outlined in the preceding section, which uncovered a gradient for the striatal mediation of psychostimulant effects centered on the Acb and tapering off in the posterior, dorsal, and lateral directions. A recent influential model of striatal organization combines these behavioral findings with the results of extensive anatomical mapping studies to propose that regional specialization of striatal function is conferred by areas of overlapping cortical and thalamic inputs arranged in columns angled in a dorsomedial-to-ventrolateral direction.[13] It is proposed that functional distinctions between the core and shell can also be understood within this context. When considered thus, it is apparent that the sites most sensitive to psychostimulant effects correspond to overlapping areas of prefrontal, hippocampal, and amygdalar input, including extra-Acb dorsomedial striatal sites as targeted in the Kelley and Delfs studies on d-amphetamine-potentiated CR responding. Because the shell receives projections more closely aligned to basic arousal and autonomic functions, such as those from the infralimbic cortex, and connects reciprocally with behavioral control modules in the lateral hypothalamus and sites even further downstream, it would appear well positioned to regulate the expression of unconditioned behavioral responses.

In conclusion, mapping studies reviewed in these last two sections provide a clear picture of the remarkable heterogeneity of function in the mammalian striatum. They also strongly support the idea that dopamine's behavioral effects can be understood within the context of its modulation of regionally specific information-processing modules deriving from unique complements of cortical innervation.

STRIATAL ROLE IN APPETITIVELY MOTIVATED LEARNING

It has recently become apparent that in addition to regulating motor output, striatal dopamine release and consequent modulation of glutamate transmission may also play a role in mechanisms of intracellular plasticity that may contribute to the development of adaptive motor responses through the process of reinforcement. Hence, the question arises: is dopamine's participation in appetitively motivated learning and cellular plasticity also expressed differentially across striatal subregions?

It has been noted that the rudiments of motivated behavior are phylogenetically ancient, as are the neurochemical systems that promote motor behavior; for example, it has been shown that dopamine is involved in place conditioning shown by crayfish, and the related molecule, octopamine, modulates adaptive navigational responses in honeybees. Interestingly, the intracellular signaling cascades involved in synaptic plasticity are also evolutionarily well conserved. A comprehensive review of the molecular basis of neuroplasticity is beyond the scope of this review, but germane to this discussion is the observation that numerous molecular components of plasticity-related second messenger systems, synaptic docking proteins, and transcriptional activators are significantly engaged by dopamine via the ascending dopaminergic projections, most likely as modulators of glutamate receptor activation. The evidence for glutamate–dopamine interactions is quite extensive, beginning with the observation that long-term enhancement of synaptic strength occurs when corticostriatal glutatmate excitation and dopaminergic activation are temporally coordinated.[84] Indeed, substantial data suggest that coincident NMDA receptor (NMDAR) and dopamine D1 receptor (D1R) activation plays a critical role in shaping synaptic configurations, and likely predominant neural ensembles, that underlie reinforcement-based learning.[85] Evidence for the role of dopamine in NMDA-dependent long-term potentiation (LTP), a putative in vitro model of neural plasticity, comes from data showing that D1 but not D2 antagonists block LTP in striatal slices.[86] In in vivo models, LTP in hippocampal-prefrontal cortex synapses depends on coactivation of NMDA and D1Rs, as well as intracellular cascades involving protein kinase A (PKA).[87–89] Moreover, in both striatum and prefrontal cortex, D1 activation potentiates NMDAR-mediated responses.[20,90,91] Finally, hippocampal-evoked spiking activity of Acb neurons requires cooperative action of both D1Rs and NMDARs, while a similar synergism is observed for the amygdalo-accumbens pathway.[92,93] Molecular studies complement these findings, showing NMDAR dependence of D1-mediated phosphorylation of cAMP response element binding protein (CREB),[94,95] a transcription factor thought to be an evolutionarily conserved modulator of memory processes and a key protein in cellular pathways affected by addictive drugs.[96,97] More recent data suggest that glutamate and dopamine signals, via NMDA and D1 activation, converge to induce extracellular signal-regulated protein kinase (ERK) activation in the hippocampus and striatum, thereby reconfiguring networks involved in learning and drug use.[98,99] It is important to note that these cascades are the ones proposed to be modified in the addictive process.[100]

A summary of the intracellular convergence of NMDAR and dopamine D1R signals and the effects on plasticity-related mechanisms is displayed in Figure 6.3.2. The schematic illustrates the prevailing hypothesis that glutamate-coded sensory/information processing signals activate NMDAR, leading to Ca^{2+} influx. D1 receptor activation engages adenyl cyclase (AC), and, in turn, cAMP. The two signaling pathways interact in several places, for example, as calmodulin (CaM), induced by NMDAR activation, influences AC (although this is a somewhat oversimplified representation). Protein kinase A activates Ras proximate-1 protein (Rap-1) but also inhibits Ras, suggesting that not only do the pathways converge, but also may compete for signal dominance. Several points of possible convergence are demonstrated, most notably the activation of mitogen-activated protein (MAP)/ERK kinase (MEK), ERK and CREB. Moreover, critical plasticity-related effects are also demonstrated, like the CREB-dependent transcription of immediate early genes (IEGs) *Arc*, *Homer1a*, and *zif268*. Recently, emerging data have suggested an important role for ERK in AMPAR-subunit insertion and regulation of L-type voltage-gated calcium channels (VGCC). (Fig. 6.3.2) The neuromolecular convergence of information from cortico-striatal-limbic NMDAR and dopamine D1R activation provides a likely substrate for plasticity in reward-based learning.

The functional heterogeneity of striatal regions has also been demonstrated repeatedly for instrumental learning. Results from these *behavioral* studies corroborate the *anatomical* distinctions previously shown. Instrumental learning has been described as one of the most elementary forms of behavioral adaptation.[101] Through interchange with its environment, an animal is able to learn about the *consequences of its actions*, and thereby modify the current environment through new behaviors to produce more favorable conditions.[102] The behavior of an organism is altered by response-reinforcer contingencies, and these effects are long-lasting.

Critical and functionally heterogeneous roles for striatal NMDAR activation in instrumental learning have been demonstrated. Hernandez et al. (2005) found that AP-5, an NMDAR antagonist, infused into the core region of Acb, impaired the acquisition of lever pressing for sucrose pellets in rats, while infusions of the same drug into the

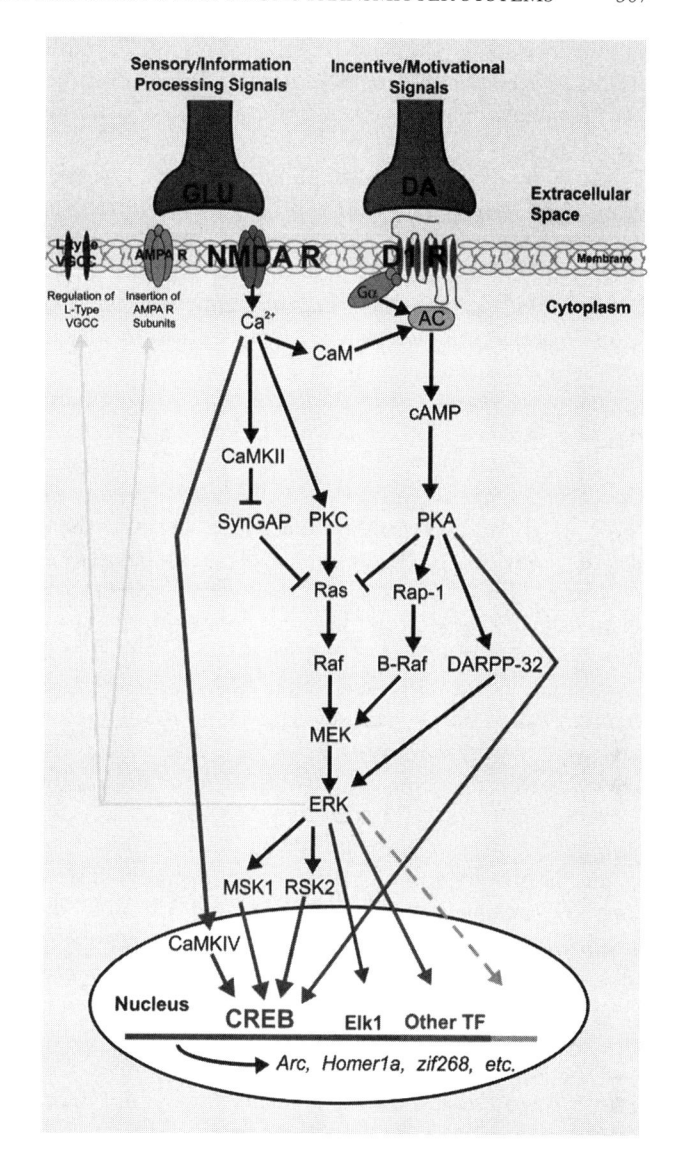

FIGURE 6.3.2. One prominent and influential hypothesis concerning the functional and structural changes involved in neural plasticity implicates coordinated NMDAR and dopamine (DA) D1R activation throughout cortical-striatal-limbic networks. This figure summarizes the prevailing models of convergence and divergence of intracellular signals, following NMDAR and DA D1R activation, leading to activation and/or phosphorylation of key enzymes, inhibition of particular signals, and transcription of critical immediate early genes. CamK, calcium/calmodulin-dependent protein kinase; SynGAP, Synaptic Ras guanosine-5′-triphosphatase activating protein; TF, transcription factor. See text for other abbreviations. *Source:* based on [99,147–149]. (See Color Plate 6.3.2.)

shell region produced a much smaller effect. Blockade of dopamine D1R with SCH-23390 in the core also impaired the acquisition of instrumental lever pressing but produced profound motor deficits as well.[103] Interestingly, coinfusions of AP-5 and SCH-23390, in doses that individually produced no effects on learning or motor behavior, were shown to impair initial instrumental learning,[104] implicating coordinated actions of glutamate- and dopamine-coded signals in the ventral striatum as critical for learning.

The role of coordinated glutamate-dopamine signaling has been investigated in other striatal subregions where a heterogeneous functional organization that parallels the anatomical organization has been demonstrated. For example, detailed mapping studies have found critical roles for NMDAR activation in instrumental learning in the VLS, DMS, and posterior lateral striatum (PLS) but not in the DLS.[105,106] Once again, the VLS, DMS, and PLS receive overlapping projections from prefrontal cortex, amygdala, and VTA, whereas the DLS does not. Moreover, while AP-5 infusions in three of these sites produced learning deficits, those deficits were likely the result of different behavioral processes. In other words, careful experimentation demonstrated that food-directed behavior was profoundly disrupted by infusions in the PLS. Additionally, it seems likely that fine motor control, reminiscent of the key role in orofacial motor behavior, was impaired by infusions in the VLS. Infusions in the DMS, however, produced only impairments during the learning phase, thereby suggesting this site as a key site of plasticity in instrumental learning.

The importance of plasticity mechanisms in instrumental learning has been investigated through the expression of certain IEGs, most notably *Homer1a* and *Zif268*, that were up-regulated in discrete striatal regions following four sessions of instrumental training. In accordance with data from both anatomical studies and functional pharmacological behavioral studies, IEG expression increased in the DMS and VLS but not in the DLS or posterior regions of the striatum. Most interestingly, even after extensive training, IEG expression was elevated in the VLS, suggesting a continuing dynamic role for this structure in the mediation of instrumental learning–dependent plasticity.[107]

FUNCTIONAL SPECIFICITY CONFERRED BY DISCRETE NEUROCHEMICAL SYSTEMS WITHIN A STRIATAL REGION

In addition to the regional segregation of behavioral function conferred by distinct corticostiatal projection fields, it is the case that even within a given striatal subregion, different neurochemical systems may code for behavioral processes that are subtly yet importantly dissociable. This section will focus on one such example: the modulation of incentive-motivational versus hedonic properties of food, as differentially mediated by dopamine and opioid systems in the Acb core.

The influence of transmission through Acb dopamine and mu opioid receptors has been studied extensively in regard to the modulation of reward-related processes. In several behavioral assays, the effects of stimulating Acb dopamine and opioid receptors are quite similar, and indeed, there is considerable evidence for interactions between these two systems within the Acb. For example, infusions of either dopamine or opioid agonists into the Acb produce a similar profile of increased spontaneous motor activity.[108–111] Both drug self-administration and electrical brain stimulation-reward (BSR) are modulated by dopamine and opioid receptors in the Acb. For example, dopamine and opioid agonists are both self-administered directly into the Acb,[112–115] and intra-Acb infusion of d-amphetamine or morphine lowers the threshold for BSR (i.e., heightens sensitivity to the rewarding stimulation).[116,117] Indicative of reciprocal interactions between these two systems, it has been shown that the BSR threshold-lowering effect of systemic morphine is attenuated by 6-OHDA lesions of the VTA, while the threshold-lowering effect of systemic d-amphetamine or cocaine is significantly attenuated by the opioid receptor antagonist, naloxone.[118,119] Opioid–dopamine interactions have also been shown at the level of receptor regulation and intracellular signaling pathways. Thus, dopamine-depleting lesions of the VTA or Acb, or chronic dopamine receptor blockade (both of which up-regulate postsynaptic dopamine receptors), markedly potentiate the motor-activating effects of intra-Acb opioid peptide infusions.[120,121] With regard to intracellular mechanisms, chronic exposure to either cocaine or morphine produces a similar up-regulation of the postsynaptic cAMP-dependent signaling cascade within the Acb,[122] and of the function of the PKA-regulated transcriptional regulator CREB and the target gene delta-Fos B.[123]

These findings suggest that Acb dopamine and opioid systems impinge upon a common function or set of functions relevant to the regulation of reward. Recently, however, evidence has accumulated to suggest that incentive control of arousal and goal-seeking behavior can be pharmacologically "pulled apart" from the affective/interoceptive or "hedonic" aspects of interaction with an appetitive goal object. Much of this evidence comes from studies of Acb control of feeding behavior and instrumental responding for food and food-associated stimuli. For example, as discussed previously, it has been shown that pharmacologically augmenting or blocking dopamine transmission in the Acb has inconsistent effects on food intake. As shown in Figure 6.3.3, some studies have indicated that

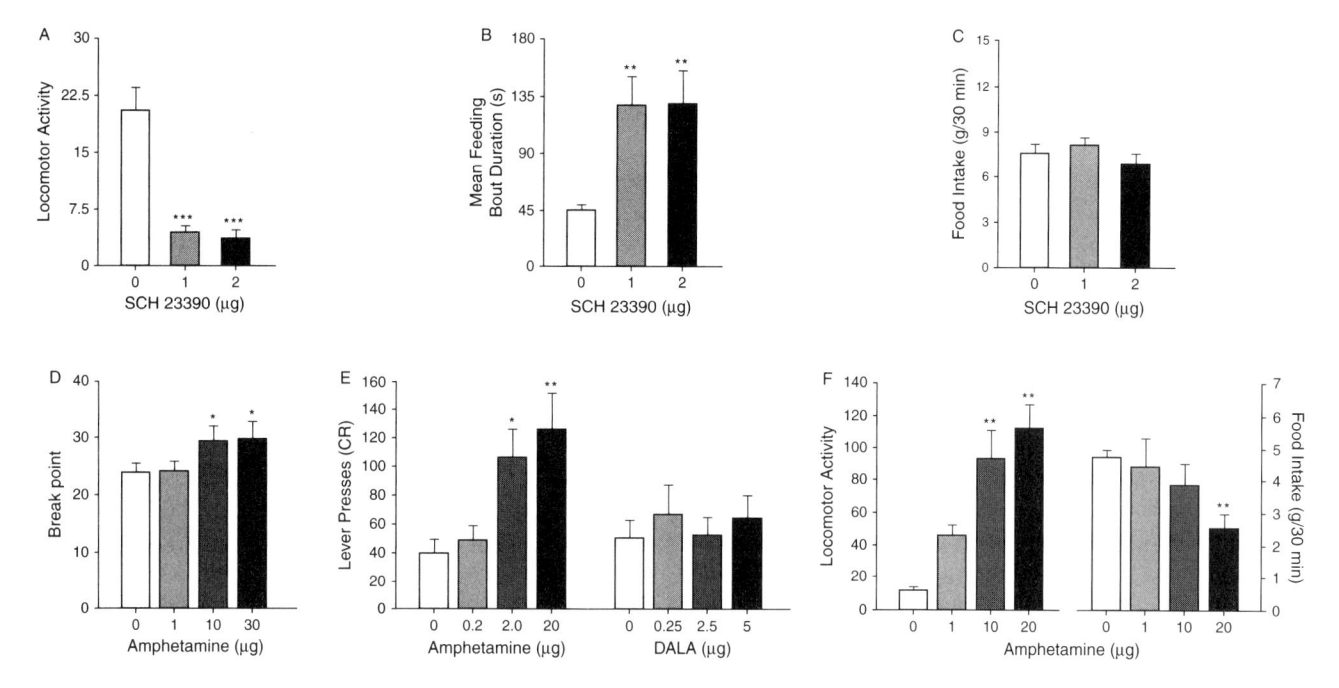

FIGURE 6.3.3. The effects of dopaminergic manipulations following drug infusions into the Acb. Dopamine antagonism with SCH-23390 (A) decreases locomotor activity and (B) increases feeding bout duration but (C) does not alter overall food intake. Dopamine agonism via infusions of amphetamine (D) increases the breakpoint in the progressive ratio, (E) increases lever presses for a conditioned reinforcer (CR), and (F) increases locomotor activity but also decreases food intake. Note the specificity of the CR effect in that the enkephalin analog, [D-Alanine] methionine-enkephalin (DALA), does not enhance responding for a CR (E).

intra-Acb dopamine receptor blockade enhances certain parameters of feeding microstructure, and, conversely, that d-amphetamine infusions reduce food intake. In contrast, dopamine receptor blockade has been found to blunt conditioned hyperactivity associated with food expectation, while intra-Acb infusion of d-amphetamine enhances operant responding for food reinforcement[76] and food-associated conditioned stimuli.[41]

This behavioral profile differs in important ways from the effects observed with intra-Acb opioid receptor stimulation. Most obviously, intra-Acb infusion of mu-opioid agonists markedly and reliably increases food intake.[124–126] This effect is seen with all types of food, but particularly strong augmentation is observed with palatable sweet/fat foods, and when several foods are concurrently available, intra-Acb opioid receptor stimulation selectively enhances the intake of fat-enriched foods.[127–130] Conversely, intra-Acb infusions of opioid receptor antagonists reduce the intake of sugar and saccharine solutions at doses that do not alter the intake of standard chow.[131] With regard to operant responding, intra-Acb mu-opioid receptor stimulation augments the breakpoint for sucrose pellet reinforcement in a progressive ratio task[76] but does not produce consistent effects on CR. In two studies, intra-Acb infusions of the mu-opioid-selective peptide, [D-Ala2-N-MePhe4, Gly-ol]-enkephalin (DAMGO), at doses that produced considerable hyperactivity, failed to augment responding for a sucrose-paired conditioned stimulus.[109] Nevertheless, a third study found augmented operant responding for a conditioned stimulus.[132] At the very least, one can conclude that the effects of intra-Acb opioid receptor stimulation on CR are far less consistent than the effects of intra-Acb d-amphetamine.

Taken together, these results indicate that the effects of intra-Acb mu-opioid manipulations are most reliably observed in behavioral tests in which food is actually encountered and eaten, suggesting that opioid transmission in the Acb is closely linked to the internal affective state arising from palatable sensory inputs. In contrast, dopamine transmission is posited to be more crucial when behavior is controlled by distal environmental cues, stimuli that have acquired conditioned reinforcing properties, or the expectation of imminent reward. In an influential theory of dopamine function, the incentive-salience theory, it is suggested that these two processes map onto discrete motivational functions, 'liking' and 'wanting', respectively. Accordingly, hyperphagia elicited by intra-Acb mu-opioid stimulation is not blocked by coadministration of D1 or D2 dopamine receptor antagonists, providing strong evidence for the dissociation of opioid- and dopamine-mediated control of feeding.[134] This dissociation finds further support in an important series of studies on Acb control of *taste reactivity*, a term that refers to stereotyped orofacial motor reactions to administration of sapid or aversive tastants to the oral cavity. It has been argued that these taste reactions provide a window into the affective state engendered by the tastants as determined by hedonic evaluations. Opioid peptide infusion into the Acb, in particular the medial shell, reliably enhances appetitive taste reactions elicited by sucrose administration.[135] Intra-Acb infusion of d-amphetamine does not produce this effect, even at doses that augment responding for a sucrose-associated stimulus.[136] A conceptually related result was observed in an important series of experiments on intake of palatable food and standard chow, food-seeking behavior in a runway paradigm, and conditioned hyperactivity associated with food expectation. It was found that systemic administration of dopamine and opioid antagonists produced distinct outcomes in these experiments: dopamine antagonist treatment reduced anticipatory hyperactivity at doses that did not affect runway performance or food intake, while opioid antagonist administration selectively diminished palatable food intake and runway performance but did not alter conditioned hyperactivity.[137] These findings support the idea that food-associated motivational arousal and gustatory reward are governed by distinguishable neurochemical processes.

Several questions arise concerning this apparently dissociable regulation of the incentive-motivational versus hedonic properties of food. First and foremost, one may wonder about the mechanistic basis for the behavioral distinctions between dopamine- and opioid-mediated effects in the Acb, particularly given that these two systems interact so closely at the postsynaptic level. An obvious possibility would be a putative differential timing of transmitter dynamics in the Acb, such that dopamine release is more tightly linked to exposure to affectively salient stimuli (as suggested by the incentive-salience theory; see[138]) and vigorous goal-seeking activities (see discussions in[139]). In support of this hypothesis, it has been shown in voltammetry studies that elevations in the dopamine signal in rats lever pressing for food reinforcement are highest during performance of the instrumental response and decline to baseline during consumption of the food.[140,141] The dynamics of opioid release in the striatum, however, have yet to be characterized fully due to the technical difficulties associated with assaying peptide release in vivo. It should be emphasized that although dopamine and mu-opioid-mediated effects can be dissociated pharmacologically, it is widely thought that under physiological conditions these two systems act in a closely cooperative fashion; for example, it has been suggested that the intra-Acb DAMGO-induced increase in progressive-ratio responding for sucrose reward depends upon the opioid-induced amplification of taste reward "feeding into" a dopaminergic substrate for energizing lever-pressing behavior.[76,142] Along these lines, in studies that failed to show intra-Acb opioid peptide enhancement of lever pressing for a sucrose-paired conditioned reinforcer, a history of repeated infusions of these agonists dramatically sensitized the d-amphetamine-induced potentiation of CR responding.[143]

CONCLUSIONS AND FUTURE DIRECTIONS

We have reviewed evidence supporting three broad conclusions about striatal function. First, dissociations in dopamine-mediated behavioral processes among striatal subregions are conferred by the unique complements of glutamate-coded afferents from functionally distinct cortical areas. Second, the same dopamine–glutamate interactions that underlie the elaboration of goal-directed actions also promote intracellular plasticity. It has been proposed that this convergent organization of the systems that generate motor behaviors and the molecules that promote neural plasticity provides an ideal substrate for the selection of successful motor acts based on reinforcing outcomes. More specifically, it has been proposed that medium spiny striatal output neurons represent the individual "processing units" for cellular plasticity, embedded within a corticostriatal network that is responsible for elaborating goal-directed behaviors. We have reviewed evidence that, in the context of appetitive instrumental learning, this network for plasticity appears to overlap the network for generating operant behaviors such that the two processes conform to the same anatomical boundaries. Third, it is apparent that information processing functions within a given striatal subregion are also heterogeneous with regard to neurotransmitter control.

These broad themes open up important avenues of future inquiry. One important area for future study is a more complete mapping of the striatal network that subserves instrumental learning. In particular, it seems important to determine whether plasticity occurs in a regionally specific way for other types of motor processes. For example, would learning in the context of skilled fine-motor control be more reliant upon the VLS than the Acb? In a similar vein, the heterogeneity conferred by different neurotransmitter systems also raises the question of how these systems contribute to cellular plasticity and learning, and in what contexts. For example, do the unique neuromodulators, peptides, and receptors for humoral factors in the Acb shell participate in plasticity that is linked to unconditioned, homeostatically driven behaviors?

Finally, this review has emphasized the regional heterogeneity of striatal function; however, it may be the case that under certain conditions the striatum acts as a unified whole. For example, it has been suggested that certain homeostatic drive states produce a functional coordination of opioid transmission throughout the striatum that serves to bias the organism toward certain types of behavioral responses. For example, a state-dependent enhancement of opioid function throughout the striatum would have the effect of promoting homeostatically driven feeding, but also enabling feeding to exceed acute homeostatic needs to build up an energy reserve for future times of possible food scarcity.[144] In general support of this idea, it has been shown that preproenkephalin expression throughout the striatum tracks the motivational state of the animal with regard to whether or not its acute energy needs have recently been met—in other words, whether or not it has just eaten.[145]

An anatomical circuit for coordinating processing throughout the striatum has been proposed[144]; this circuit is based on the observation that the thalamic regions innervating striatum receive a considerable input from the orexin/hypocretin peptide system in the lateral hypothalamus. This system has been shown to regulate cortical arousal in the context of feeding drives, and its activity is regulated, in part, by energy-balance signaling systems in the mediobasal hypothalamus. Hence, putative orexin/hypocretin modulation of corticostriatal projections could represent a means for energy balance–related information to reach broad regions of the striatum in a coordinated way. As mentioned previously, thalamostriatal projections reach medium spiny neurons, but they also impinge upon cholinergic striatal interneurons. A further component of the model proposes that these interneurons, by virtue of their extensive, interacting axonal processes and influence (through muscarinic receptors) upon enkephalin-containing medium spiny neurons, can act as a reticular coordinating system for enkephalin synthesis, release, and consequent activation of mu-opioid receptors. Accordingly, it has been shown that discrete microinfusions of the muscarinic receptor antagonist, scopolamine, alter expression of preproenkephalin mRNA throughout the striatum.[146] Although at present this model is speculative, it yields testable hypotheses pertaining to the routes of control through which striatal function can be coordinated across large areas.

In conclusion, it is clear that additional knowledge of the information processing parameters of the striatum will provide a more sophisticated context within which to understand the role of dopamine in the control of behavior.

REFERENCES

1. Chesselet MF, Graybiel AM. Striatal neurons expressing somatostatin-like immunoreactivity: evidence for a peptidergic interneuronal system in the cat. *Neuroscience*. 1986;17(3):547–571.
2. Voorn P, Gerfen CR, Groenewegen HJ. Compartmental organization of the ventral striatum of the rat: immunohistochemical distribution of enkephalin, substance P, dopamine, and calcium-binding protein. *J Comp Neurol*. 1989;289(2):189–201.
3. Gerfen CR. The neostriatal mosaic: multiple levels of compartmental organization. *Trends Neurosci*. 1992;14:133–138.
4. Reiner A, Anderson KD. The patterns of neurotransmitter and neuropeptide co-occurrence among striatal projection neurons: conclusions based on recent findings. *Brain Res Brain Res Rev*. 1990;15(3):251–265.
5. Jongen-Relo AL, Voorn P, Groenewegen HJ. Immunohistochemical characterization of the shell and core territories of the nucleus accumbens in the rat. *Eur J Neurosci*. 1994;6(8):1255–1264.
6. Zaborszky L, Alheid GF, Beinfeld MC, Eiden LE, Heimer L, Palkovits M. Cholecystokinin innervation of the ventral striatum: a morphological and radioimmunological study. *Neuroscience*. 1985;14(2):427–453.
7. Alexander GE, DeLong MR, Strick PL. Parallel organization of functionally segregated circuits linking basal ganglia and cortex. *Annu Rev Neurosci*. 1986;9:357–381.
8. Gerfen CR. Synaptic organization of the striatum. *J Electron Microsc Tech*. 1988;10(3):265–281.
9. Nauta WJH. Reciprocal links of the corpus striatum with the cerbral cortex and limbic system: a common substrate for movement and thought? In: Mueller J, ed. *Neurology and Psychiatry: A Meeting of the Minds*. Basel: Karger; 1989:43–63.
10. Haber SN. The primate basal ganglia: parallel and integrative networks. *J Chem Neuroanat*. 2003;26(4):317–330.
11. McGeorge AJ, Faull RLM. The organization of the projection from the cerebral cortex to the striatum in the rat. *Neuroscience*. 1989;29:503–537.
12. Nauta WJ, Domesick VB. Afferent and efferent relationships of the basal ganglia. *Ciba Found Symp*. 1984;107:3–29.
13. Voorn P, Vanderschuren LJ, Groenewegen HJ, Robbins TW, Pennartz CM. Putting a spin on the dorsal-ventral divide of the striatum. *Trends Neurosci*. 2004;27(8):468–474.
14. Berendse HW, Galis-de Graaf Y, Groenewegen HJ. Topographical organization and relationship with ventral striatal compartments of prefrontal corticostriatal projections in the rat. *J Comp Neurol*. 1992;316(3):314–347.
15. Goldman PS, Nauta WJ. An intricately patterned prefronto-caudate projection in the rhesus monkey. *J Comp Neurol*. 1977;72(3):369–386.
16. Heimer L, Wilson RD. The subcortical projections of allocortex: similarities in the neuronal associations of the hippocampus, the piriform cortex, and the neocortex. In: Santini M, ed. *Golgi Centennial Symposium Proceedings*. New York: Raven Press; 1975:173–193.

17. Kelley AE, Domesick VB. The distribution of the projection from the hippocampal formation to the nucleus accumbens in the rat: an anterograde- and retrograde-horseradish peroxidase study. *Neuroscience.* 1982;7:2321–2335.
18. Kelley AE, Domesick VB, Nauta WJH. The amygdalostriatal projection in the rat—an anatomical study by anterograde and retrograde tracing methods. *Neuroscience.* 1982;7:615–630.
19. Russchen FT, Bakst I, Amaral DG, Price JL. The amygdalostriatal projections in the monkey. An anterograde tracing study. *Brain Res.* 1985;329(1-2):241–257.
20. Cepeda C, Buchwald NA, Levine MS. Neuromodulatory actions of dopamine in the neostriatum are dependent upon the excitatory amino acid receptor subtypes activat' *Proc Natl Acad Sci USA.* 1993;90(20):9576–9580.
21. Floresco SB, Blaha CD, Yang CR, Phillips AG. Dopamine D1 and NMDA receptors mediate potentiation of basolateral amygdala-evoked firing of nucleus accumbens neurons. *J Neurosci.* 2001;21(16):6370–6376.
22. Kiyatkin EA, Rebec GV. Dopaminergic modulation of glutamate-induced excitations of neurons in the neostriatum and nucleus accumbens of awake, unrestrained rats. *J Neurophysiol.* 1996;75(1):142–153.
23. Nicola SM, Surmeier J, Malenka RC. Dopaminergic modulation of neuronal excitability in the striatum and nucleus accumbens. *Annu Rev Neurosci.* 2000;23:185–215.
24. West AR, Floresco SB, Charara A, Rosenkranz JA, Grace AA. Electrophysiological interactions between striatal glutamatergic and dopaminergic systems. *Ann NY Acad Sci.* 2003;1003:53–74.
25. Kotter R. Postsynaptic integration of glutamatergic and dopaminergic signals in the striatum. *Prog Neurobiol.* 1994;44(2):163–196.
26. O'Donnell P. Dopamine gating of forebrain neural ensembles. *Eur J Neurosci.* 2003;17(3):429–435.
27. Carey RJ. Differential effects of limbic versus striatal dopamine loss on motoric function. *Behav Brain Res.* 1983;7(3):283–296.
28. Dunnett SB, Bjorklund A, Stenevi U, Iversen SD. Grafts of embryonic substantia nigra reinnervating the ventrolateral striatum ameliorate sensorimotor impairments and akinesia in rats with 6-OHDA lesions of the nigrostriatal pathway. *Brain Res.* 1981;229(1):209–217.
29. Koob GF, Stinus L, Le Moal M. Hyperactivity and hypoactivity produced by lesions to the mesolimbic dopamine system. *Behav Brain Res.* 1981;3(3):341–359.
30. Mendez JS, Finn BW. Use of 6-hydroxydopamine to create lesions in catecholamine neurons in rats. *J Neurosurg.* 1975;42(2):166–173.
31. Fouriezos G, Wise RA. Pimozide-induced extinction of intracranial self-stimulation: response patterns rule out motor or performance deficits. *Brain Res.* 1976;103(2):377–380.
32. Liebman JM, Butcher LL. Effects on self-stimulation behavior of drugs influencing dopaminergic neurotransmission mechanisms. *Naunyn Schmiedebergs Arch Pharmacol.* 1973;277(3):305–318.
33. Lippa AS, Antelman SM, Fisher AE, Canfield DR. Neurochemical mediation of reward: a significant role for dopamine? *Pharmacol Biochem Behav.* 1973;1(1):23–28.
34. Wise RA, Spindler J, DeWit H, Gerber GJ. Neuroleptic-induced "anhedonia" in rats: pimozide blocks reward quality of food. *Science.* 1978;201:262–264.
35. Wise RA. Neuroleptics and operant behavior: the anhedonia hypothesis. *Behav Brain Sci.* 1982;5:39–87.
36. Mogenson GJ, Jones DL, Yim CY. From motivation to action: functional interface between the limbic system and the motor system. *Prog Neurobiol.* 1980;14:69–97.
37. Kelly PH, Seviour PW, Iversen SD. Amphetamine and apomorphine responses in the rat following 6-OHDA lesions of the nucleus accumbens septi and corpus striatum. *Brain Res.* 1975;94:507–522.
38. Amalric M, Koob GF. Depletion of dopamine in the caudate nucleus but not in nucleus accumbens impairs reaction-time performance in rats. *J Neurosci.* 1987;7(7):2129–2134.
39. Carr GD, White NM. Conditioned place preference from intra-accumbens but not intra-caudate amphetamine injections. *Life Sci.* 1983;33:2551–2557.
40. Kelley AE, Lang CG, Gauthier AM. Induction of oral stereotypy following amphetamine microinjection into a discrete subregion of the striatum. *Psychopharmacology.* 1988;95:556–559.
41. Kelley AE, Delfs JM. Dopamine and conditioned reinforcement. I. Differential effects of amphetamine microinjections into striatal subregions. *Psychopharmacology.* 1991;103:187–196.
42. Bakshi VP, Kelley AE. Dopaminergic regulation of feeding behavior: I. Differential effects of haloperidol microinfusion into three striatal subregions. *Psychobiology.* 1991;19:223–232.
43. Bakshi VP, Kelley AE. Dopaminergic regulation of feeding behavior: II. Differential effects of amphetamine microinfusion into three striatal subregions. *Psychobiology.* 1991;19:233–242.
44. Kelley AE, Gauthier AM, Lang CG. Amphetamine microinjections into distinct striatal subregions cause dissociable effects on motor and ingestive behavior. *Behav Brain Res.* 1989;35:27–39.
45. Beckstead RM. An autoradiographic examination of corticocortical and subcortical projections of the mediodorsal-projection (prefrontal) cortex in rats. *J Comp Neurol.* 1979;84:43–62.
46. Pisa M. Motor functions of the striatum in the rat: critical role of the lateral region in tongue and forelimb reaching. *Neuroscience.* 1988;24:453–463.
47. Kelley AE, Gauthier AM, Lang CG. Induction of oral stereotypy following amphetamine microinjection into a discrete subregion of the striatum. *Psychopharmacology.* 1989;95:556–559.
48. Delfs JM, Kelley AE. The role of D-1 and D-2 dopamine receptors in oral stereotypy induced by dopaminergic stimulation of the ventrolateral striatum. *Neuroscience.* 1990;39:59–67.
49. Salamone JD, Johnson CJ, McCullough LD, Steinpreis RE. Lateral striatal cholinergic mechanisms involved in oral motor activities in the rat. *Psychopharmacology (Berl).* 1990;102(4):529–534.
50. Salamone JD, Mahan K, Rogers S. Ventrolateral striatal dopamine depletions impair feeding and food handling in rats. *Pharmacol Biochem Behav.* 1993;44:605–610.
51. Cousins MS, Salamone JD. Skilled motor deficits in rats induced by ventrolateral striatal dopamine depletions: behavioral and pharmacological characterization. *Brain Res.* 1996;732(1-2):186–194.
52. Salamone JD, Ishiwari K, Betz AJ, et al. Dopamine/adenosine interactions related to locomotion and tremor in animal models: possible relevance to parkinsonism. *Parkinsonism Relat Disord.* 2008;14(suppl 2):S130-S134.
53. Wise RA, Spindler J, Legault L. Major attenuation of food reward with performance-sparing doses of pimozide in the rat. *Can J Psychol.* 1978;32(2):77–85.
54. Wise RA, Colle LM. Pimozide attenuates free feeding: best scores analysis reveals a motivational deficit. *Psychopharmacology.* 1984;84:446–451.
55. Tombaugh TN, Tombaugh J, Anisman H. Effects of dopamine receptor blockade on alimentary behaviors: home cage food consumption, magazine training, operant acquisition, and performance. *Psychopharmacology.* 1979;66:219–225.

56. Blackburn JR, Phillips AG, Fibiger HC. Dopamine and preparatory behavior: I. Effects of pimozide. *Behav Neurosci.* 1987;101:352–360.
57. Kelley AE, Stinus L. Disappearance of hoarding behavior after 6-hydroxy lesions of the mesolimbic dopamine neurons and its reinstatement with L-Dopa. *Behav Neurosci.* 1985;99:531–545.
58. Koob GF, Riley SJ, Smith SC, Robbins TW. Effects of 6-hydroxydopamine lesions of the nucleus accumbens septi and olfactory tubercle on feeding, locomotor activity, and amphetamine anorexia in the rat. *J Comp Physiol Psychol.* 1978; 92(5):917–927.
59. Heimer L, Alheid GF, de Olmos JS, et al. The accumbens: beyond the core–shell dichotomy. *J Neuropsychiatry Clin Neurosci.* 1997;9(3):354–381.
60. Vertes RP. Differential projections of the infralimbic and prelimbic cortex in the rat. *Synapse.* 2004;51(1):32–58.
61. Wright CI, Groenewegen HJ. Patterns of convergence and segregation in the medial nucleus accumbens of the rat: relationships of prefrontal cortical, midline thalamic, and basal amygdaloid afferents. *J Comp Neurol.* 1995;361(3):383–403.
62. Brog JS, Salyapongse A, Deutch AY, Zahm DS. The patterns of afferent innervation of the core and shell in the "accumbens" part of the rat ventral striatum: immunohistochemical detection of retrogradely transported fluoro-gold. *J Comp Neurol.* 1993;338(2):255–278.
63. Delfs JM, Zhu Y, Druhan JP, Aston-Jones GS. Origin of noradrenergic afferents to the shell subregion of the nucleus accumbens: anterograde and retrograde tract-tracing studies in the rat. *Brain Res.* 1998;806(2):127–140.
64. Berridge CW, Stratford TL, Foote SL, Kelley AE. Distribution of dopamine-ß-hydroxylase(DBH)-like immunoreactivie fibers within the shell of the nucleus accumbens. *Synapse.* 1997;27:230–241.
65. Baldo BA, Daniel RA, Berridge CW, Kelley AE. Overlapping distributions of orexin/hypocretin- and dopamine-beta-hydroxylase immunoreactive fibers in rat brain regions mediating arousal, motivation, and stress. *J Comp Neurol.* 2003;464(2):220–237.
66. Peyron C, Tighe DK, van den Pol AN, et al. Neurons containing hypocretin (orexin) project to multiple neuronal systems. *J Neurosci.* 1998;18(23):9996–10015.
67. van Rossum D, Menard DP, Fournier A, St-Pierre S, Quirion R. Autoradiographic distribution and receptor binding profile of [125I]Bolton Hunter-rat amylin binding sites in the rat brain. *J Pharmacol Exp Ther.* 1994;270(2):779–787.
68. Sexton PM, Paxinos G, Kenney MA, Wookey PJ, Beaumont K. In vitro autoradiographic localization of amylin binding sites in rat brain. *Neuroscience.* 1994;62(2):553–567.
69. Heimer L, Zahm DS, Churchill L, Kalivas PW, Wohltmann C. Specificity in the projection patterns of accumbal core and shell in the rat. *Neuroscience.* 1991;41:89–125.
70. Zahm DS, Jensen SL, Williams ES, Martin JR 3rd. Direct comparison of projections from the central amygdaloid region and nucleus accumbens shell. *Eur J Neurosci.* 1999;11(4):1119–1126.
71. Kelley AE. Neural integrative activities of nucleus accumbens subregions in relation to motivation and learning. *Psychobiology.* 1999;27:198–213.
72. Kelley AE. Functional specificity of ventral striatal compartments in appetitive behaviors. *Ann NY Acad Sci.* 1999; 877:71–90.
73. Maldonado-Irizarry CS, Kelley AE. Differential behavioral effects following microinjection of an NMDA antagonist into nucleus accumbens subregions. *Psychopharmacology.* 1994; 166:65–72.
74. Maldonado-Irizarry CS, Swanson CJ, Kelley AE. Glutamate receptors in the nucleus accumbens shell control feeding behavior via the lateral hypothalamus. *J Neurosci.* 1995;15:6779–6788.
75. Stratford TR, Kelley AE. Feeding elicited by inhibition of neurons in the nucleus accumbens shell depends on activation of neurons in the lateral hypothalamus. *Soc Neurosci Abstr.* 1997;23:577.
76. Zhang M, Balmadrid C, Kelley AE. Nucleus accumbens opioid, GABAergic, and dopaminergic modulation of palatable food motivation: contrasting effects revealed by a progressive ratio study in the rat. *Behav Neurosci.* 2003;117(2):202–211.
77. Hanlon EC, Baldo BA, Sadeghian K, Kelley AE. Increases in food intake or food-seeking behavior induced by GABAergic, opioid, or dopaminergic stimulation of the nucleus accumbens: is it hunger? *Psychopharmacology (Berl).* 2004;172(3):241–247.
78. Reynolds SM, Berridge KC. Fear and feeding in the nucleus accumbens shell: rostrocaudal segregation of GABA-elicited defensive behavior versus eating behavior. *J Neurosci.* 2001;21(9):3261–3270.
79. Swanson CJ, Heath S, Stratford TR, Kelley AE. Differential behavioral responses to dopaminergic stimulation of nucleus accumbens subregions in the rat. *Pharmacol Biochem Behav.* 1997;58:933–945.
80. Pontieri FE, Tanda G, Di Chiara G. Intravenous cocaine, morphine, and amphetamine preferentially increase extracellular dopamine in the "shell" as compared with the "core" of the rat nucleus accumbens. *Proc Natl Acad Sci USA.* 1995;92:12304–12308.
81. Bassareo V, De Luca MA, Di Chiara G. Differential expression of motivational stimulus properties by dopamine in nucleus accumbens shell versus core and prefrontal cortex. *J Neurosci.* 2002;22(11):4709–4719.
82. Parkinson JA, Olmstead MC, Burns LH, Robbins TW, Everitt BJ. Dissociation in effects of lesions of the nucleus accumbens core and shell in appetitive Pavlovian approach behavior and the potentiation of conditioned reinforcement and locomotor activity by d-amphetamine. *J Neurosci.* 1999;19:2401–2411.
83. Ito R, Robbins TW, Everitt BJ. Differential control over cocaine-seeking behavior by nucleus accumbens core and shell. *Nat Neurosci.* 2004;7(4):389–397.
84. Wickens JR, Begg AJ, Arbuthnott GW. Dopamine reverses the depression of rat corticostriatal synapses which normally follows high-frequency stimulation of cortex in vitro. *Neuroscience.* 1996;70(1):1–5.
85. Jay TM. Dopamine: a potential substrate for synaptic plasticity and memory mechanisms. *Prog Neurobiol.* 2003;69(6):375–390.
86. Kerr JN, Wickens JR. Dopamine D-1/D-5 receptor activation is required for long-term potentiation in the rat neostriatum in vitro. *J Neurophysiol.* 2001;85(1):117–124.
87. Jay TM, Rocher C, Hotte M, Naudon L, Gurden H, Spedding M. Plasticity at hippocampal to prefrontal cortex synapses is impaired by loss of dopamine and stress: importance for psychiatric diseases. *Neurotox Res.* 2004;6(3):233–244.
88. Gurden H, Tassin JP, Jay TM. Integrity of the mesocortical dopaminergic system is necessary for complete expression of in

vivo hippocampal-prefrontal cortex long-term potentiation. *Neuroscience.* 1999;94(4):1019–1027.
89. Gurden H, Takita M, Jay TM. Essential role of D1 but not D2 receptors in the NMDA receptor-dependent long-term potentiation at hippocampal-prefrontal cortex synapses in vivo. *J Neurosci.* 2000;20(22):RC106.
90. Seamans JK, Durstewitz D, Christie BR, Stevens CF, Sejnowski TJ. Dopamine D1/D5 receptor modulation of excitatory synaptic inputs to layer V prefrontal cortex neurons. *Proc Natl Acad Sci USA.* 2001;98(1):301–306.
91. Wang J, O'Donnell P. D(1) dopamine receptors potentiate NMDA-mediated excitability increase in layer V prefrontal cortical pyramidal neurons. *Cereb Cortex.* 2001;11(5):452–462.
92. Floresco SB, Blaha CD, Yang CR, Phillips AG. Modulation of hippocampal and amygdalar-evoked activity of nucleus accumbens neurons by dopamine: cellular mechanisms of input selection. *J Neurosci.* 2001;21(8):2851–2860.
93. Floresco SB, Blaha CD, Yang CR, Phillips AG. Dopamine D1 and NMDA receptors mediate potentiation of basolateral amygdala-evoked firing of nucleus accumbens neurons. *J Neurosci.* 2001;21(16):6370–6376.
94. Konradi C, Leveque JC, Hyman SE. Amphetamine- and dopamine-induced immediate early gene expression in striatal neurons depends on postsynaptic NMDA receptors and calcium. *J Neurosci.* 1996;16(13):4231–4239.
95. Das S, Grunert M, Williams L, Vincent SR. NMDA and D1 receptors regulate the phosphorylation of CREB and the induction of c-*fos* in striatal neurons in primary culture. *Synapse.* 1997;25(3):227–233.
96. Silva AJ, Kogan JH, Frankland PW, Kida S. CREB and memory. *Annu Rev Neurosci.* 1998;21:127–148.
97. Nestler EJ. Molecular basis of long-term plasticity underlying addiction. *Nat Rev Neurosci.* 2001;2(2):119–128.
98. Kaphzan H, O'Riordan KJ, Mangan KP, Levenson JM, Rosenblum K. NMDA and dopamine converge on the NMDA-receptor to induce ERK activation and synaptic depression in mature hippocampus. *PLoS One.* 2006;1:e138.
99. Valjent E, Pascoli V, Svenningsson P, et al. Regulation of a protein phosphatase cascade allows convergent dopamine and glutamate signals to activate ERK in the striatum. *Proc Natl Acad Sci USA.* 2005;102(2):491–496.
100. Hyman SE, Malenka RC. Addiction and the brain: the neurobiology of compulsion and its persistence. *Nat Rev Neurosci.* 2001;2(10):695–703.
101. Rescorla RA. A note on depression of instrumental responding after one trial of outcome devaluation. *Q J Exp Psychol B.* 1994;47(1):27–37.
102. Skinner BF. *Science and Human Behavior.* Vol New York, NY: Macmillan; 1953.
103. Hernandez PJ, Andrzejewski ME, Sadeghian K, Panksepp JB, Kelley AE. AMPA/kainate, NMDA, and dopamine D1 receptor function in the nucleus accumbens core: a context-limited role in the encoding and consolidation of instrumental memory. *Learn Mem.* 2005;12(3):285–295.
104. Smith-Roe SL, Kelley AE. Coincident activation of NMDA and dopamine D1 receptors within the nucleus accumbens core is required for appetitive instrumental learning. *J Neurosci.* 2000;20(20):7737–7742.
105. Andrzejewski ME, Sadeghian K, Kelley A. Central amygdalar and dorsal striatal NMDA-receptor involvement in instrumental learning and spontaneous behavior. *Behav Neurosci.* 2004;118(4):715–729.
106. McKee BL, Harris RL, Feit EC, Andrzejewski ME. The role of glutamate receptors in the orbital frontal cortex and dorsomedial striatum in the acquisition of instrumental learning. Poster presented at the meeting of the Society for Neuroscience, 2008; Washington, DC.
107. Hernandez PJ, Schiltz CA, Kelley AE. Dynamic shifts in corticostriatal expression patterns of the immediate early genes *Homer 1a* and *Zif268* during early and late phases of instrumental training. *Learn Mem.* 2006;13(5):599–608.
108. Cunningham ST, Finn MA, Kelley AE. Sensitization of the locomotor response to psychostimulants after repeated opiate exposure: role of the nucleus accumbens. *Neuropsychopharmacology.* 1997;16:147–155.
109. Cunningham ST, Kelley AE. Opiate infusion into nucleus accumbens: contrasting effects on motor activity and responding for conditioned reward. *Brain Res.* 1992;588:104–114.
110. Di Chiara G, Imperato A. Drugs abused by humans preferentially increase synaptic dopamine concentrations in the mesolimbic system of freely moving rats. *Proc Natl Acad Sci USA.* 1988;85:5274–5278.
111. Vezina P, Kalivas PW, Stewart J. Sensitization occurs to the locomotor effects of morphine and the specific μ opioid receptor agonist, DAGO, administered repeatedly to the ventral tegmental area but not to the nucleus accumbens. *Brain Res.* 1987;417:51–58.
112. Phillips GD, Robbins TW, Everitt BJ. Bilateral intra-accumbens self-administration of d-amphetamine: antagonism with intra-accumbens SCH-23390 and sulpiride. *Psychopharmacology.* 1994;114:477–485.
113. Wise RA, Hoffman DC. Localization of drug reward mechanisms by intracranial injections. *Synapse.* 1992;10(3):247–263.
114. Goeders NE, Lane JD, Smith JE. Self-administration of methionine enkephalin into the nucleus accumbens. *Pharmacol Biochem Behav.* 1984;20(3):451–455.
115. Hoebel BG, Monaco AP, Hernandez L, Aulissi E, Stanley BG, Lenard L. Self-injection of amphetamine directly into the brain. *Psychopharmacology.* 1983;81:158–163.
116. Ranaldi R, Beninger RJ. The effects of systemic and intracerebral injections of D1 and D2 agonists on brain stimulation reward. *Brain Res.* 1994;651(1-2):283–292.
117. Johnson PI, Goodman JB, Condon R, Stellar JR. Reward shifts and motor responses following microinjections of opiate-specific agonists into either the core or shell of the nucleus accumbens. *Psychopharmacology (Berl).* 1995;120(2):195–202.
118. Esposito RU, Perry W, Kornetsky C. Effects of d-amphetamine and naloxone on brain stimulation reward. *Psychopharmacology (Berl).* 1980;69(2):187–191.
119. Bain GT, Kornetsky C. Naloxone attenuation of the effect of cocaine on rewarding brain stimulation. *Life Sci.* 1987;40(11):1119–1125.
120. Stinus L, Winnock M, Kelley AE. Chronic neuroleptic treatment and mesolimbic dopamine denervation induce behavioural supersensitivity to opiates. *Psychopharmacology.* 1985;85:323–328.
121. Stinus L, Nadaud D, Jauregui J, Kelley AE. Chronic treatment with five different neuroleptics elicits behavioral supersensitivity to opiate infusion into the nucleus accumbens. *Biol Psychiatry.* 1986;21(1):34–48.
122. Nestler EJ. Cellular responses to chronic treatment with drugs of abuse. *Crit Rev Neurobiol.* 1993;7(1):23–39.
123. Nestler EJ, Barrot M, Self DW. DeltaFosB: a sustained molecular switch for addiction. *Proc Natl Acad Sci USA.* 2001;98(20):11042–11046.

124. Bakshi VP, Kelley AE. Feeding induced by opioid stimulation of the ventral striatum: role of opiate receptor subtypes. *J Pharmacol Exp Ther*. 1993;265:1253–1260.
125. Mucha RF, Iversen SD. Increased food intake after opioid microinjections into nucleus accumbens and ventral tegmental area of rat. *Brain Res*. 1986;397:214–224.
126. Majeed NH, Przewlocka B, Wedzony K, Przewlocki R. Stimulation of food intake following opioid microinjection into the nucleus accumbens septi in rats. *Peptides*. 1986;7:711–716.
127. Zhang M, Gosnell BA, Kelley AE. Intake of high-fat food is selectively enhanced by mu opioid receptor stimulation within the nucleus accumbens. *J Pharmacol Exp Ther*. 1998;285(2):908–914.
128. Zhang M, Kelley AE. Opiate agonists microinjected into the nucleus accumbens enhance sucrose drinking in rats. *Psychopharmacology*. 1997;132:350–360.
129. Zhang M, Kelley AE. Intake of saccharin, salt, and ethanol solutions is increased by infusion of a mu opioid agonist into the nucleus accumbens. *Psychopharmacology (Berl)*. 2002;159(4):415–423.
130. Evans KR, Vaccarino FJ. Amphetamine- and morphine-induced feeding: evidence for involvement of reward mechanisms. *Neurosci Biobeh Rev*. 1990;14:9–22.
131. Kelley AE, Bless EP, Swanson CJ. An investigation of the effects of opiate antagonists infused into the nucleus accumbens on feeding and sucrose drinking in rats. *J Pharmacol Exp Ther*. 1996;278:1499–1507.
132. Phillips GD, Robbins TW, Everitt BJ. Mesoaccumbens dopamine–opiate interactions in the control over behaviour by a conditioned reinforcer. *Psychopharmacology*. 1994;114:345–349.
133. Berridge KC. Food reward: brain substrates of wanting and liking. *Neurosci Biobehav Rev*. 1996;20(1):1–25.
134. Will MJ, Pratt WE, Kelley AE. Pharmacological characterization of high-fat feeding induced by opioid stimulation of the ventral striatum. *Physiol Behav*. 2006;89(2):226–234.
135. Doyle TG, Berridge KC, Gosnell BA. Morphine enhances hedonic taste palatability in rats. *Pharmacol Biochem Behav*. 1993;46(3):745–749.
136. Wyvell CL, Berridge KC. Intra-accumbens amphetamine increases the conditioned incentive salience of sucrose reward: enhancement of reward "wanting" without enhanced "liking" or response reinforcement. *J Neurosci*. 2000;20(21):8122–8130.
137. Barbano MF, Cador M. Differential regulation of the consummatory, motivational and anticipatory aspects of feeding behavior by dopaminergic and opioidergic drugs. *Neuropsychopharmacology*. 2006;31(7):1371–1381.
138. Berridge KC. The debate over dopamine's role in reward: the case for incentive salience. *Psychopharmacology (Berl)*. 2007;191(3):391–431.
139. Salamone JD, Correa M. Motivational views of reinforcement: implications for understanding the behavioral functions of nucleus accumbens dopamine. *Behav Brain Res*. 2002;137(1-2):3–25.
140. Richardson NR, Gratton A. Behavior-relevant changes in nucleus accumbens dopamine transmission elicited by food reinforcement: an electrochemical study in rat. *J Neurosci*. 1996;16(24):8160–8169.
141. Roitman MF, Stuber GD, Phillips PE, Wightman RM, Carelli RM. Dopamine operates as a subsecond modulator of food seeking. *J Neurosci*. 2004;24(6):1265–1271.
142. Baldo BA, Kelley AE. Discrete neurochemical coding of distinguishable motivational processes: insights from nucleus accumbens control of feeding. *Psychopharmacology (Berl)*. 2007;191(3):439–459.
143. Cunningham ST, Kelley AE. Evidence for opiate-dopamine cross-sensitization in nucleus accumbens: studies of conditioned reward. *Brain Res Bull*. 1992;29:675–680.
144. Kelley AE, Baldo BA, Pratt WE. A proposed hypothalamic-thalamic-striatal axis for the integration of energy balance, arousal, and food reward. *J Comp Neurol*. 2005;493:72–85.
145. Will MJ, VanderHeyden W, Kelley AE. Striatal opioid peptide gene expression differentially tracks short-term satiety but does not vary with negative energy balance, in a manner opposite to hypothalamic NPY. *Am J Physiol. Regul. Integr. Comp. Physiol*. 2007;292:R217–226.
146. Pratt WE, Kelley AE. Striatal muscarinic receptor antagonism reduces 24-h food intake in association with decreased preproenkephalin gene expression. *Eur J Neurosci*. 2005;22(12):3229–3240.
147. Haberny SL, Carr KD. Food restriction increases NMDA receptor-mediated calcium-calmodulin kinase II and NMDA receptor/extracellular signal-regulated kinase 1/2-mediated cyclic amp response element-binding protein phosphorylation in nucleus accumbens upon D-1 dopamine receptor stimulation in rats. *Neuroscience*. 2005;132(4):1035–1043.
148. Kelley AE, Berridge KC. The neuroscience of natural rewards: relevance to addictive drugs. *J Neurosci*. 2002;22(9):3306–3311.
149. Sweatt JD. The neuronal MAP kinase cascade: a biochemical signal integration system subserving synaptic plasticity and memory. *J Neurochem*. 2001;76(1):1–10.

6.4 | Behavioral Functions of Dopamine Neurons

PHILIPPE N. TOBLER

INTRODUCTION

The extracellular study of single dopamine neurons of the behaving nonhuman primate started in the early 1980s. It produced novel and unexpected insights suggesting that subsecond responses of dopamine neurons mediate well-defined behavioral functions. These functions clearly go beyond movement processing, contrary to what could be expected based on the obvious motor impairments of patients with Parkinson's disease. The behavioral contributions of dopamine neurons include a role in reward processing, learning, economic decision making, and attention. The neuronal study of reward and attention is intrinsically behavioral because, contrary to sensory information, there are no dedicated receptors for reward and attention. Instead, we infer what is rewarding or attention-grabbing from behavior, in the context of behavioral theories, such as learning theory and microeconomic decision theory. These theories describe how organisms form associations between reward and stimuli or actions and how they determine the subjective value of, and choose between, choice options. The present chapter reviews the extracellular studies of dopamine neurons in behaving animals. The behavioral theories are introduced as far as necessary for the interpretation of the findings. For a fuller introduction into learning and microeconomic decision theory, the reader is referred to [1-5].

MOTOR FUNCTIONS OF DOPAMINE NEURONS

Among the most obvious symptoms of Parkinson's disease are movement-related impairments, particularly problems with the voluntary initiation of movements (see Chapter 9.1 in this volume). Early extracellular recording studies of single dopamine neurons in nonhuman primates were therefore expected to find movement-related correlates of dopamine firing. In one example of these studies,[6] animals sit in a primate chair and hold a key with one hand. After a variable interval, the experimenter opens a box, located at eye level of the animals. The animals then release the key and reach into the box for a small piece of apple placed there. Once behavior is stable, single dopamine neurons of the substantia nigra are recorded with glass-insulated and platinum-plated tungsten electrodes. Outside of the task, dopamine neurons discharge at relatively low frequencies (0.5–8 impulses/s), with polyphasic waveforms of relatively long durations (1.5–5.0 ms) and with wide action potentials (see Chapter 10.6 in this volume). Dopamine firing increases in a sustained but moderate fashion in about 30% of the cells during the triggered arm movement into the box[6] (Fig. 6.4.1). Only about 10% of dopamine neurons show increased activity (median of 91%) before self-initiated reaching movements into the food box.[7] These data suggest a moderate contribution of sustained changes in dopamine firing to motor functions.

REWARD FUNCTIONS OF DOPAMINE NEURONS

In the behavioral situation described above, the inside of the box can be hidden from the animals' view by putting a little cover in front of the opening. From time to time, the animals then reach into the box to check whether a morsel of food has been hidden. Surprisingly, a strong (median increase of about 200%) phasic activation occurs in up to 90% of the cells at around the time when the animals touch the food[7] (Fig. 6.4.2, top, left and right). This activation is not time-locked to the movement onset but rather to the touch of the food, occurs with a latency and duration of about 100 ms, shows no discrimination between different types of similarly rewarding food, and is not present (or even replaced by depression; see below and Fig. 6.4.2, bottom, left and right) when the animals touch the bare wire that normally holds the food morsels in place or when they touch nonfood items hidden within the box. Dopamine neurons respond not only to food but also to liquid reward delivered through a spout in front of the animal's mouth (e.g.,[8]), and the responses increase with the size of the drops delivered[9] (reward

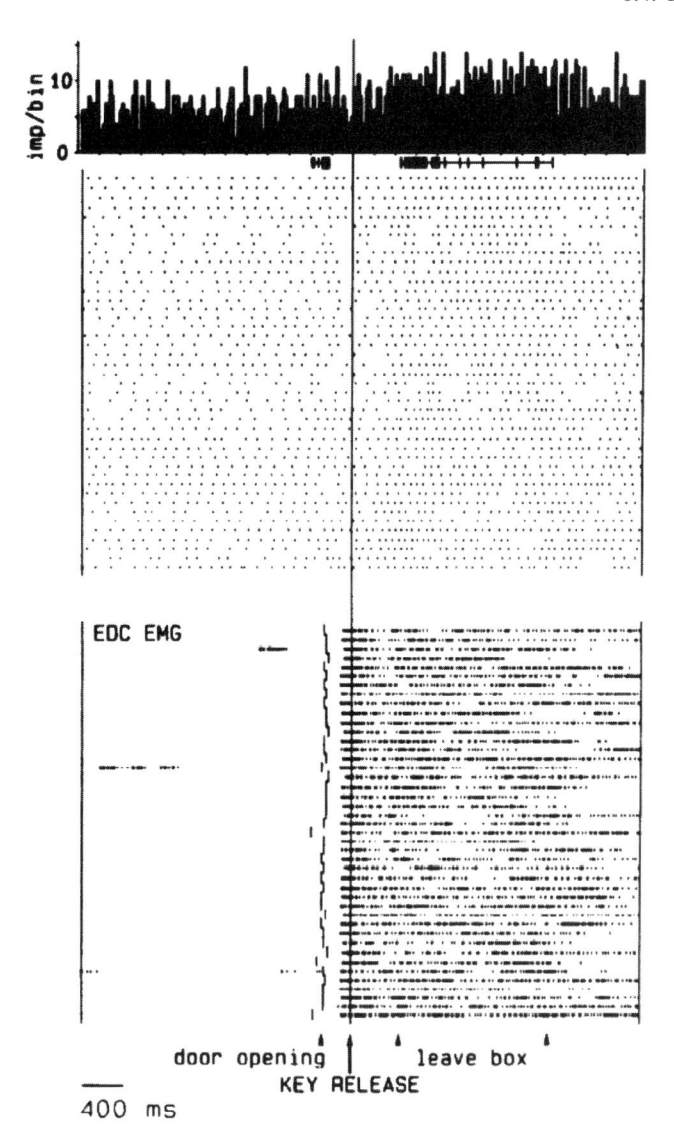

FIGURE 6.4.1 Moderate increase in dopamine activity during execution of arm movement. Top: the animal responded with an arm movement to the opening of a door of a box hiding a morsel of food. Activation of a single neuron is shown as perievent histogram and as dot display. In this and the following figures of this kind, each line corresponds to one trial and each dot to the time of a neuronal impulse; the original trial sequence is from top to bottom. Bottom: electromyographic activity of the extensor digitorum communis muscle recorded in the same trials. *Source*: Adapted with permission from the American Physiological Society.[6]

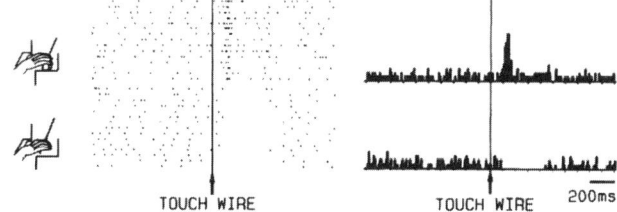

FIGURE 6.4.2 Reward coding of dopamine neurons during self-initiated movement trials. Activation (top, left and right) occurs only when animals touch a morsel of food in the box but not when they touch wire (bottom, left and right) or other nonfood items. When no food was present, some neurons showed depression (bottom, left and right). The contents of the box were hidden from view; movements were self-initiated. The two situations alternated randomly. *Source*: Adapted with permission from the American Physiological Society.[7]

magnitude; Fig. 6.4.3). Responses to the touch of food with the hand and delivery of liquid to the mouth are similarly effective in activating dopamine neurons, but liquid can be quantified more easily and is therefore often the reward of choice in more recent studies. The reward responses do not differ significantly between the dopamine neurons of the substantia nigra pars compacta (A9), ventral tegmental area (A10), and retrorubral area (A8), although occasionally a response gradient has been observed with more medial regions showing stronger reward responses than more lateral regions (e.g.,[6,10]). Due to the relative homogeneity of dopamine responding, figures of single-neuron firing are usually representative of the population of dopamine neurons (e.g., Fig. 6.4.3, bottom). Taken together, these data suggest a role of dopamine firing in general reward processing and in processing of the mircoeconomic parameter of reward magnitude. Indeed, the value of choice options increases monotonically with reward magnitude in all major economic decision theories.[11–13]

REWARD LEARNING FUNCTIONS OF DOPAMINE NEURONS

When the availability of the reward in the food box task is repeatedly and consistently signaled by the sound of the box door opening, the phasic activation no longer occurs at the time when the monkeys touch the food but instead transfers to the time of the sound (Fig. 6.4.4A). The latency, duration, and magnitude of sound-induced activations do not differ significantly between ipsi- and contralateral locations of the food box.[14] Compared to the reward activation, the sound activation is slightly reduced and occurs in about 70% of all the neurons tested (although this proportion may vary with reward value). Instead of auditory stimuli, most recent experiments use visual stimuli as conditioned stimuli, presented on a screen in front of the animals (exemplified in Figs. 6.4.9 and 6.4.11). For stimuli of both modalities, phasic dopamine responding transfers from reward to reward-predicting stimuli during learning (e.g.,[15–17]; Fig. 6.4.4B). Thus, reward activates dopamine neurons only when it is unpredicted.[18–20] If a reward-predicting stimulus is itself preceded by another,

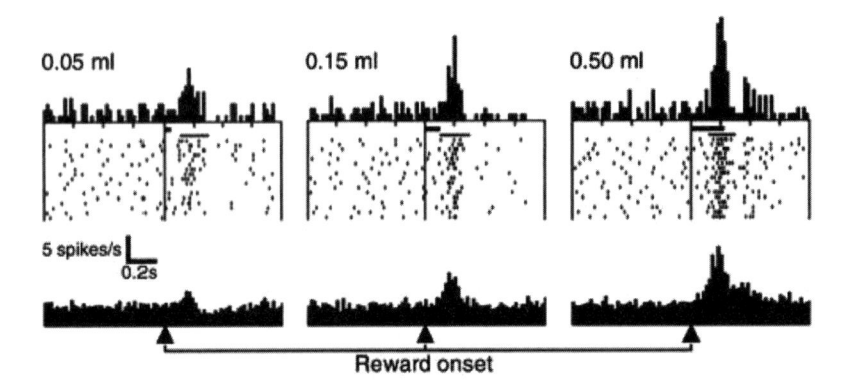

FIGURE 6.4.3 Reward magnitude coding of dopamine neurons. Activation increases in a single neuron (top) and in population of dopamine neurons (bottom; n = 55 neurons) reflecting increases in liquid volume (left to right). Three liquid volumes were delivered outside of any task in pseudorandom alternation (separated for display purposes). For the population histogram, all neurons fulfilling the electrophysiological firing criteria of dopamine neurons were recorded and are shown, irrespective of response properties. In this and following figures, population histograms show the mean firing rate in each condition. *Source*: Adapted with permission from AAAS.[9]

earlier, stimulus, then the phasic activation of dopamine neurons transfers back to this earlier stimulus[15] (Fig. 6.4.5). Thus, dopamine neurons respond to the earliest reward-predicting stimulus. Taken together, these findings suggest that the role of phasic dopamine signals is not restricted to reward processing but fundamentally concerns reward prediction and predictability.

The transfer of the activation from reward to a reward-predicting stimulus is almost complete with stimulus–reward intervals of about 2 s.[8] Further increases in the interval between stimulus and reward lead to decreases in the stimulus-induced response[21,22] (Fig. 6.4.6) and concomitant reemergence of the response to reward. Even though dopamine neurons respond primarily to the reward-predicting stimulus rather than the predicted reward at stimulus–reward intervals of about 2 s, if reward is delivered in an unpredicted manner outside a well-established task, dopamine neurons continue responding to reward. Moreover, they show depression at the usual time of reward when a predicted reward fails to occur (Fig. 6.4.7), either because the animal performs erroneously or the experimenter deliberately withholds the reward or reduces reward magnitude or reward probability. Thus, dopamine neurons appear to compute an error in the prediction of reward with activation induced by positive prediction errors (an unpredicted reward or a larger reward than predicted), no change in firing rate induced by absent prediction errors (reward occurs as predicted), and depression induced by negative prediction errors (the predicted reward fails to occur or is smaller than predicted).[9,10,15–19,23–32] Dopamine neurons show activation at the new time of reward when it occurs earlier or later than usual and depression at the usual time only when it occurs later than usual (Fig. 6.4.8). These data suggest that dopamine neurons internally code the predicted time of reward and that occurrence of an earlier reward can suppress the prediction at the usual time of reward.

The notion that dopamine neurons code errors in the prediction of reward suggested that they may contribute to learning associations between conditioned stimuli and reward[23,33–36] because prediction errors play a central role in formal learning theories[37,38] (for subsequent developments in the application of formal learning theories and other models on dopamine activity, see, e.g., [17,27,39–45] and Chapter 5.5 in this volume). These theories propose that reward learning occurs whenever there is a difference between predicted and actual rewards. Depending on the exact formula, the prediction error term is weighed by learning rate parameters or discount factors of future rewards.[37,38] Sutton and Barto,[38] for example, capture learning in their temporal difference (TD) model, where learning occurs whenever there is reward prediction error across successive time steps,

$$\delta(t) = r(t) + \gamma V(t+1) - V(t), \quad (1)$$

where $r(t)$ corresponds to reward at time (t), γ is a discount factor that weighs future rewards less than sooner rewards, and $V(t)$ is the prediction of reward at time t. $\delta(t)$ is the TD error and serves as a real-time prediction error. The response of dopamine neurons reflects TD errors at each moment in time.[23] More generally, and irrespective of the exact learning formula used, reward learning consists of gradually adjusting our predictions of future reward occurrence until they match actual reward occurrence. In line with this intuition, reward-induced phasic dopamine activations gradually decrease

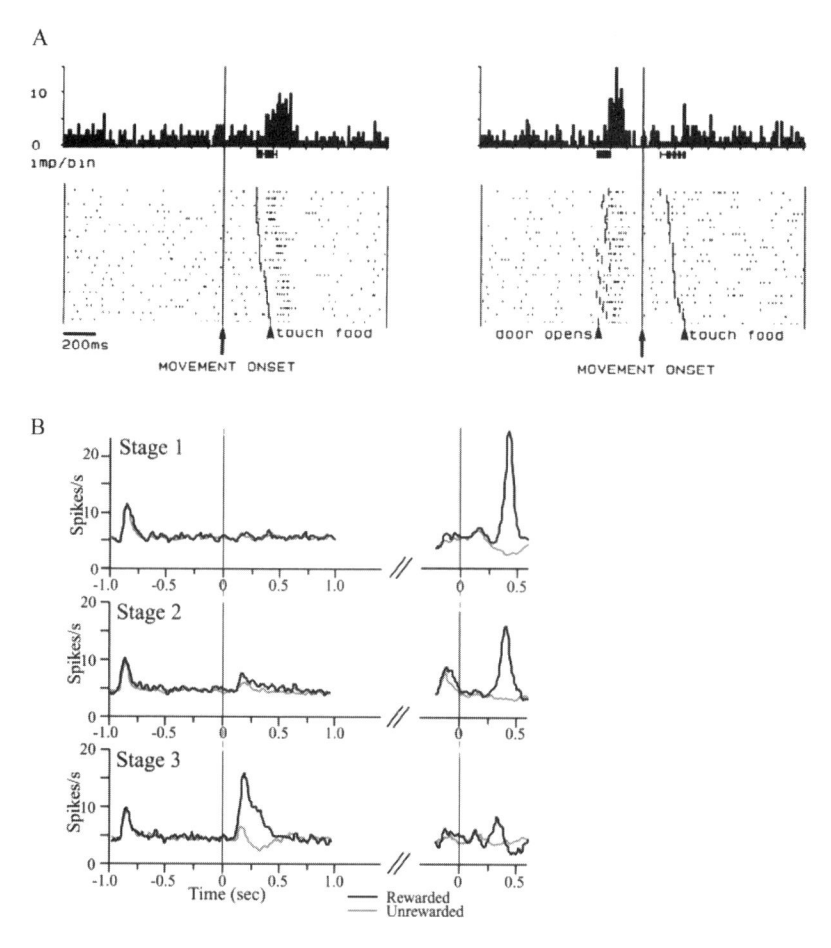

FIGURE 6.4.4 Response transfer of dopamine activation from reward to reward-predicting stimuli. (A) Response of dopamine neurons transfers from reward to reward-predicting visual and auditory stimuli. Left: self-initiated movements into a box in which a morsel of food is hidden (same task as in Fig. 6.4.1). The neuron responds to the touch of food. Right: response of the same neuron to visible and audible opening of the food box door, but absence to touch of food. *Source*: Adapted from [7]; see also [6]. (B) Population response of dopamine neurons transfers from reward to a reward-predicting stimulus in an asymmetrically rewarded saccade task. Throughout blocks of 60 trials, one out of four possible directions was rewarded. A fixation spot was presented 1 s before target onset. After a delay of 1–1.5 s, animals performed a memory-guided saccade to the remembered location of the target. Behavioral discrimination between rewarded and unrewarded saccades, as measured by saccade latency, gradually improved over the course of the experiment (for this monkey, stages 1, 2, and 3 comprised 24, 19, and 10 60-trial blocks, respectively). Simultaneously, neuronal discriminations became more pronounced, whereas reward responses decreased. Activity is aligned to the onset of the fixation spot, which was a 25% predictor of reward, thus explaining the moderate activation (see, e.g., Fig. 6.4.10). After learning (stage 3), unrewarded stimuli (gray lines) primarily depressed dopamine neurons, whereas rewarded stimuli (black lines) activated dopamine neurons, in line with a notion of dopamine neurons coding the prediction error between the 25% reward prediction of the fixation spot and the stimulus (rewarded stimuli predict reward at 100% and elicit a 75% positive prediction error; unrewarded stimuli predict reward at 0% and elicit a 25% negative prediction error. *Source*: Adapted with permission from the American Physiological Society.[16]

together with increasing behavioral learning of the stimulus–reward association[15,19] (see also Fig. 6.4.4B for population responses before, during, and after learning).

A more formal test of the role of prediction errors in learning arises in situations where previous learning prevents prediction errors and thus new learning. In such situations, a new stimulus is added to a previously learned one but the reward occurs just as predicted by the previously learned stimulus. Thus, the reward elicits no prediction error and, according to the theory, we should learn nothing about the new stimulus. In other words, predicted reward "blocks" learning about the new stimulus and the learning situation is therefore called the *blocking* paradigm.[46] In agreement with the crucial role of prediction errors in learning, monkeys show considerably less conditioned licking to stimuli that are blocked from learning than to control stimuli that are learned and paired with a reward prediction error.[24] Similarly, dopamine neurons respond less to

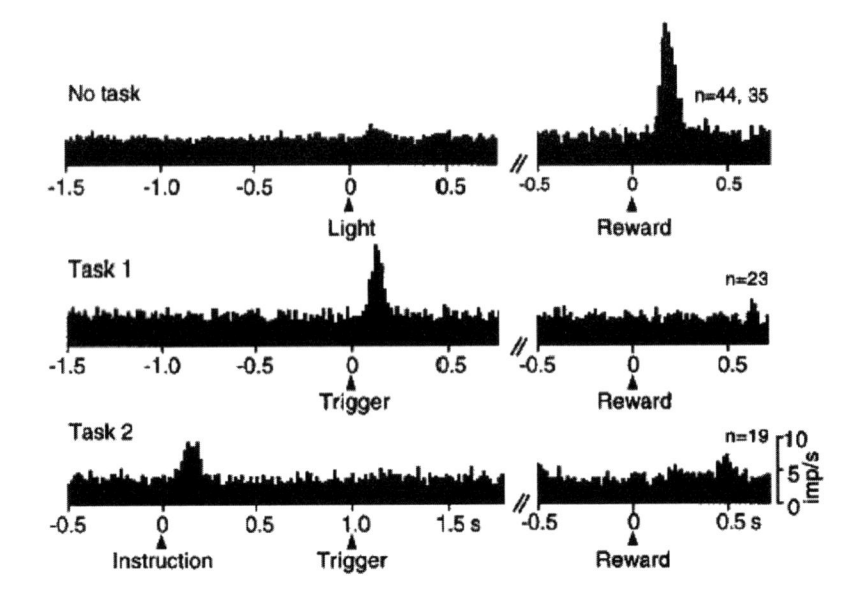

FIGURE 6.4.5 Transfer of dopamine activation to the earliest reward-predicting stimulus. Top: no response to a light but considerable response to an unpredicted reward delivered outside the task. Middle: response to a reward-predicting trigger stimulus but no response to a predicted reward during established task performance. Bottom: response to a reward-predicting instruction cue preceding the trigger stimulus by 1 s but not to the trigger stimulus during established task performance. All displays show averaged population histograms of $n = 19$ to $n = 44$ neurons. The time base is split because of varying intervals between reward-predicting stimuli and reward. *Source*: Used with permission from MIT Press.[35]

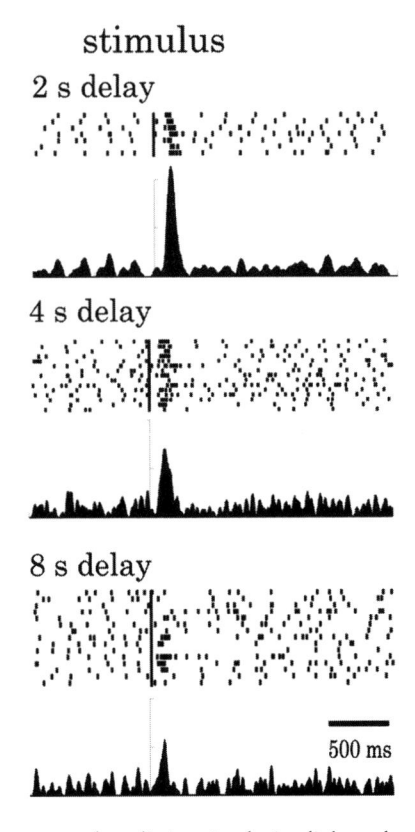

FIGURE 6.4.6 The response of dopamine neurons to reward-predicting visual stimuli depends on the stimulus–reward interval (delay). Neuronal activity is aligned to the onset of stimuli, each of which predicted reward after different delays (2, 4, and 8 s). Stimuli were presented in the middle of the screen, and reward occurred irrespective of the animal's behavioral responses. *Source*: Adapted with permission from the Society for Neuroscience.[22]

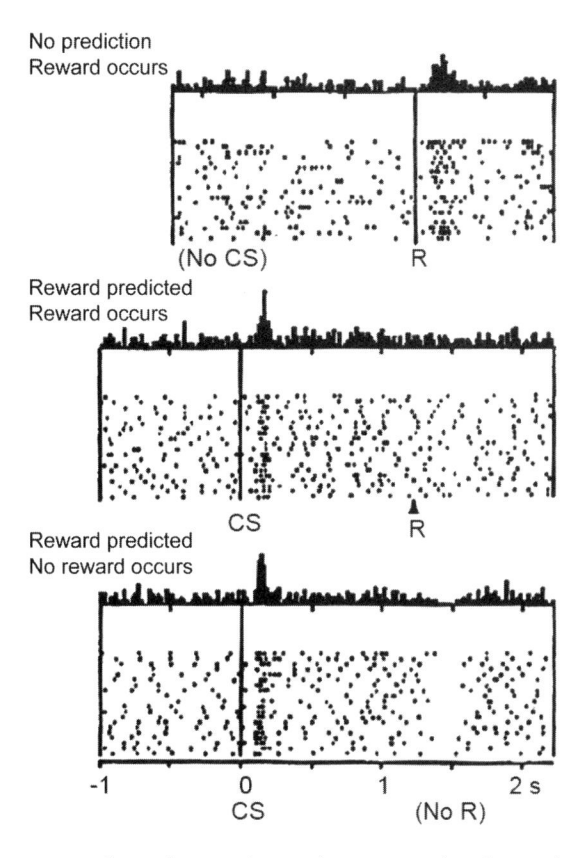

FIGURE 6.4.7 Dopamine neurons report errors in the prediction of reward. Top: unpredicted reward, constituting a positive prediction error, activates dopamine neurons. Middle: unpredicted, reward-predicting conditioned stimulus (left), constituting a positive prediction error, but not reward occurring as predicted (right), constituting no prediction error, activates dopamine neurons. Bottom: unpredicted conditioned stimulus (left), constituting a positive prediction error, again activates dopamine neurons but reward withheld due to erroneous performance (right), constituting a negative prediction error, depresses dopamine neurons. Thus, the response of dopamine neurons corresponds to a subtraction of occurrence (of reward or conditioned stimulus) − prediction (of reward or conditioned stimulus). CS, conditioned stimulus; R, reward. *Source*: Used with permission from AAAS.[23]

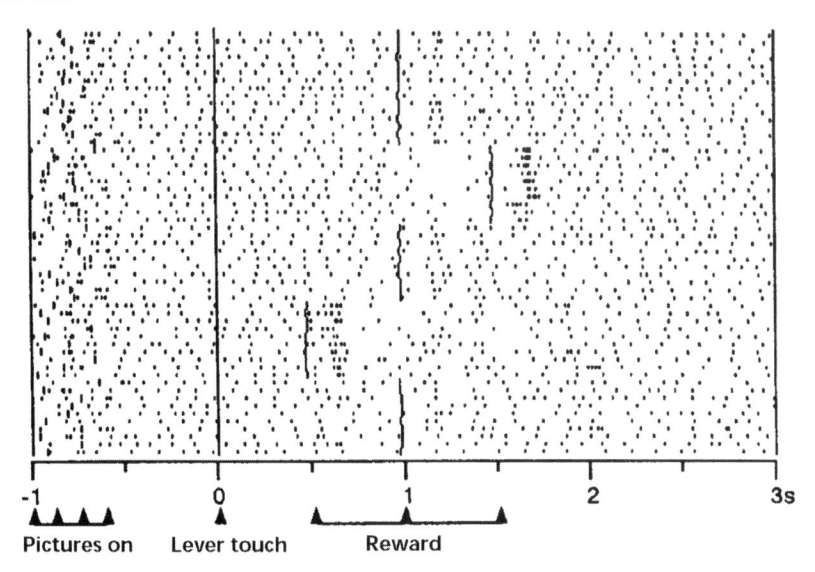

FIGURE 6.4.8 Reward prediction of dopamine neurons carries temporal information on reward occurrence. In established task performance, when reward occurs at the usual time (1.0 s after lever touch), dopamine neurons show no activation but do so when reward occurs earlier (0.5 s after lever touch) or later (1.5 s after lever touch) than usual. A depression occurs at the usual time of reward when it is delivered later but not when it is delivered earlier than usual. Onset of visual pictures is indicated by a small vertical line in each trial (pictures served as reward-predicting stimuli; accordingly, they activated this dopamine neuron in many trials). *Source*: Used with permission from the Nature Publishing Group.[19]

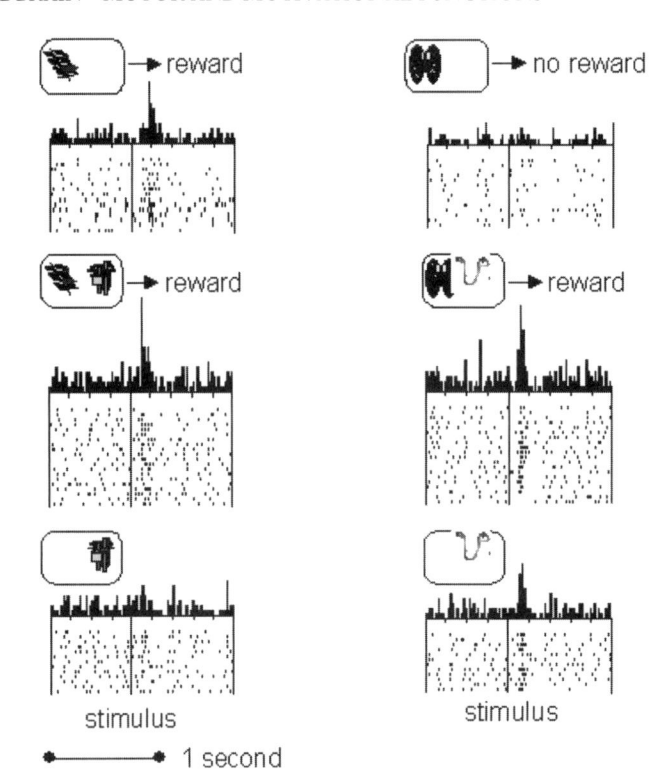

FIGURE 6.4.9 Dopamine responses develop only to conditioned stimuli eliciting prediction errors. In the blocking paradigm (left insets), a pretrained rewarded stimulus (top) blocks learning of an added stimulus (middle), as evidenced when tested alone (bottom). Blocking of reward learning arises because the preexisting reward prediction prevents occurrence of a prediction error when the added stimulus is introduced. In a control situation (right insets), a previously unrewarded stimulus (top) is added to another stimulus in a rewarded compound (middle). Conditioned responding occurs to the reward-predicting stimulus when tested alone (bottom) as a consequence of a prediction error in the initial compound stage (middle) when reward delivery was unpredicted. The example dopamine neuron shows activation (top) to the reward-predicting stimulus (left) but not to the unrewarded stimulus (right). Middle: acquired (right) and maintained (left) activation to reward-predicting compound stimuli. Bottom: no activation to the blocked stimulus (left) but an acquired response to the reward-predicting control stimulus (right). Some neurons showed small activation followed by depression to the blocked stimulus. *Source*: Adapted with permission from the Nature Publishing Group.[24]

blocked than to control stimuli (Fig. 6.4.9). Thus, the phasic activity of dopamine neurons follows the notion of a formal prediction error signal in the central test for the role of such prediction errors in reward learning. By extension, phasic dopamine signals appear to be ideally suited to learn stimulus–reward associations according to the mechanisms suggested by formal learning theories.

Reward prediction error coding by dopamine neurons has been well confirmed, quantified, and extended by subsequent studies. The dopamine neurons of monkeys that have not learned to predict reward show continued positive and negative prediction errors at the time of reward or reward omission. By contrast, the dopamine neurons of monkeys that have learned to predict reward well show conditioned stimulus responses indicative of learning in an asymmetrically rewarded saccade task.[28] In behavioral situations with contingencies changing about every 100 trials, dopamine neurons code the difference between current reward and reward history weighted by the last six to seven trials.[29] Reward occurrence (positive prediction error) or omission (negative prediction error) activates or depresses dopamine neurons in a quantitative fashion, depending on the probability with which conditioned stimuli predict reward[25–27] (Fig. 6.4.10). Prediction error coding does not require reward delivery or omission. When a 25% predictor of reward is followed by either a stimulus predicting reward at 100% or another stimulus predicting 0%, the second stimulus activates or depresses dopamine neurons, respectively[16] (Fig. 6.4.4B, bottom). This finding suggests that all stimulus-induced activation of dopamine neurons could reflect prediction errors. In most experiments, the probability of reward at each moment in time is low due to relatively long and variable intertrial intervals. Reward-predicting stimuli induce positive prediction errors relative to the low background probability. Thus, dopamine neurons

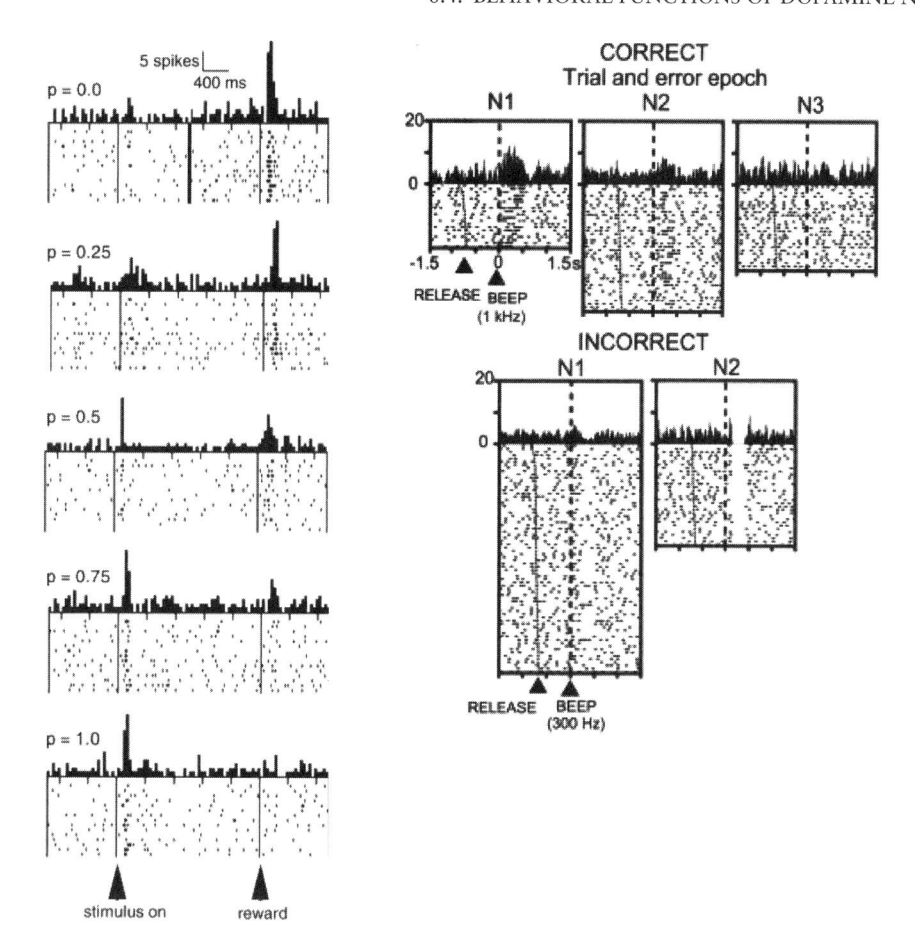

FIGURE 6.4.10 Responses of dopamine neurons reflect reward probability. Left: with increasing reward probability (top to bottom), stimulus-induced responses increase and reward-induced responses decrease, in agreement with probability-dependent prediction error processing. The thick vertical line in the middle of the top panel ($p = 0$) indicates that the stimulus-induced activity on the left and the reward-induced response on the right were from separate trial types. For intermediate probabilities, only rewarded trials are shown. Five distinct visual stimuli predicted reward at indicated probabilities after a 2-s delay; monkeys were not required to perform any action for reward to be delivered (Pavlovian task). *Source*: Adapted with permission from AAAS[25]. Right: responses to a sound reinforcer depend on the probability of a correct response. In each trial, monkeys chose one out of three buttons. Once identified by trial and error, the correct button yielded a further liquid reward in two subsequent trials. Thus, reward delivery depended on monkeys performing an action (instrumental task). Within a block of trials, the probability of choosing the correct button was set to $p = 0.2$ (instead of the chance $p = 0.33$) for the first trial-and-error trial (N1) and was close to $p = 0.5$ and $p = 0.9$ for the second and third trial-and-error trials (N2, N3). Different sounds differentially reinforced correct behavior and errors. Positive reinforcement activated dopamine neurons more at lower than at higher probabilities (correct trials, left to right). Conversely, negative reinforcement depressed dopamine neurons less at lower than at higher probabilities (incorrect trials, left to right). *Source*: Adapted with permission from the Society for Neuroscience.[26]

appear to code TD-like errors in the prediction of reward at each moment in time (equation 1).

Depressions of the firing of dopamine neurons reflect negative prediction errors as quantified, for example, by the difference between predicted and obtained milliliters of liquid reward.[23] Negative prediction error coding occurs not only with omission of predicted rewards[25] but also when rewards are smaller than predicted.[9] Indeed, the duration of depressions correlates with the size of the negative prediction errors.[31] Stimuli predicting reward omission induce negative prediction errors, particularly when they occur together with stimuli predicting reward occurrence. Reward omission–predicting stimuli are called *conditioned inhibitors* because they allow animals to inhibit reward-predictive responses to reward-predicting stimuli.[47,48] In contrast to reward-predicting stimuli, conditioned inhibitors elicit primarily depression in dopamine neurons, reflecting negative prediction errors and prediction of reward omission[10] (Fig. 6.4.11). Taken together, although the dynamic range for depressions is smaller than that for activations due to the low base rate of dopamine neurons, the data suggest that reductions in dopamine neuronal

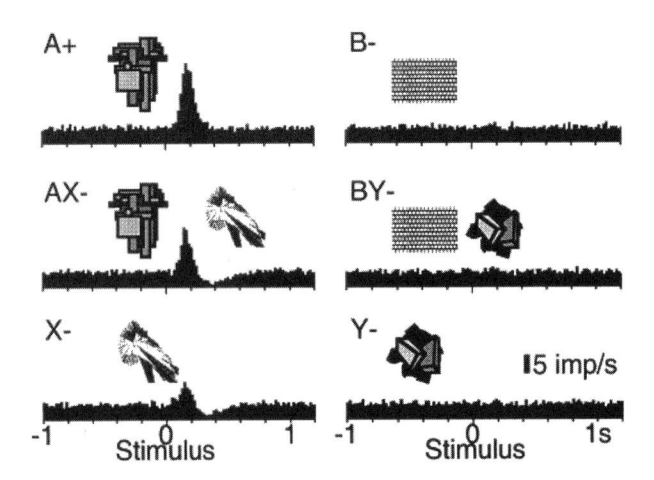

FIGURE 6.4.11 Responses of dopamine neurons in the conditioned inhibition paradigm. Letters denote picture stimuli presented on a screen. Stimulus A+ predicted reward and elicited behavioral and neuronal activations. Stimulus X- was the conditioned inhibitor; it predicted reward omission, inhibited behavior, and primarily depressed neuronal activity (despite its attention-eliciting properties). Stimulus X acquired its properties in unrewarded compound trials of A and X (AX-). Stimuli B-, Y- and their combination (BY-) served as controls (not associated with reward or attention). Dopamine activity is shown as population histograms averaged over 69 neurons. All six trial types alternated semirandomly and were separated for display. *Source*: Adapted with permission from the Society of Neuroscience.[10]

firing show considerable quantitative relations with negative prediction errors.

Adaptive mechanisms may compensate for limitations in the dynamic range of prediction error-coding dopamine neurons, both for negative and positive prediction errors. When different visual stimuli predict different binary combinations of large or small reward with equal probability, the larger (smaller) magnitude in each combination always elicits a similar positive (negative) response, even though absolute reward magnitudes differ substantially and are fully discriminated when delivered in an unpredicted fashion.[9] Thus, prediction error responses adapt to information conveyed by conditioned stimuli, such as predicted mean or standard deviation of reward. As a result of this adaptation, the neural responses of dopamine neurons discriminate similarly between the two equiprobable outcomes, indicating an adjustment of the dynamic range to the most sensitive part of the input (prediction error)–output (spikes) function.

ECONOMIC VALUE FUNCTIONS OF DOPAMINE NEURONS

More probable, larger, and more immediate rewards are economically more valuable than less probable, smaller, and later rewards. We have seen above that dopamine neurons code reward magnitude and delay[9,22] (Figs. 6.4.3 and 6.4.6). Phasic dopamine responses to reward-predicting stimuli increase also with the probability with which such stimuli predict reward[25] (Fig. 6.4.10). If reward probability increases with the number of previously unrewarded trials, conditioned stimulus–induced phasic dopamine activations increase in parallel with this conditional probability.[27] Dopamine neurons also show stronger phasic activations in response to stimuli predicting larger reward magnitudes.[9] Stimulus-related coding of reward probability, magnitude, and delay occurs not only in Pavlovian or instrumental conditioning paradigms (for probability: Fig. 6.4.10, left and right, respectively) but also in choice situations.[30,32] The stimulus-induced phasic activations combine, and are similarly sensitive to, reward probability and magnitude[9] and reward delay and magnitude.[22] Thus, the phasic stimulus-related prediction error responses of dopamine neurons reflect a wealth of economic decision parameters entailed by reward predictive stimuli.

Microeconomic theories suggest that the value of choice options increases monotonically with reward magnitude and probability and decreases monotonically with increasing interval or delay to reward.[11–13,49,50] To determine the value of choice options, these parameters should be integrated. For example:

$$EV = \Sigma(p_i * m_i), \qquad (2)$$

$$EU = \Sigma(p_i * u(m_i)), \text{ or} \qquad (3)$$

$$PT = \Sigma(w(p_i) * u(m_i)). \qquad (4)$$

In these three different formulations of value, magnitude (m), or a utility function (u) of m, is weighted by

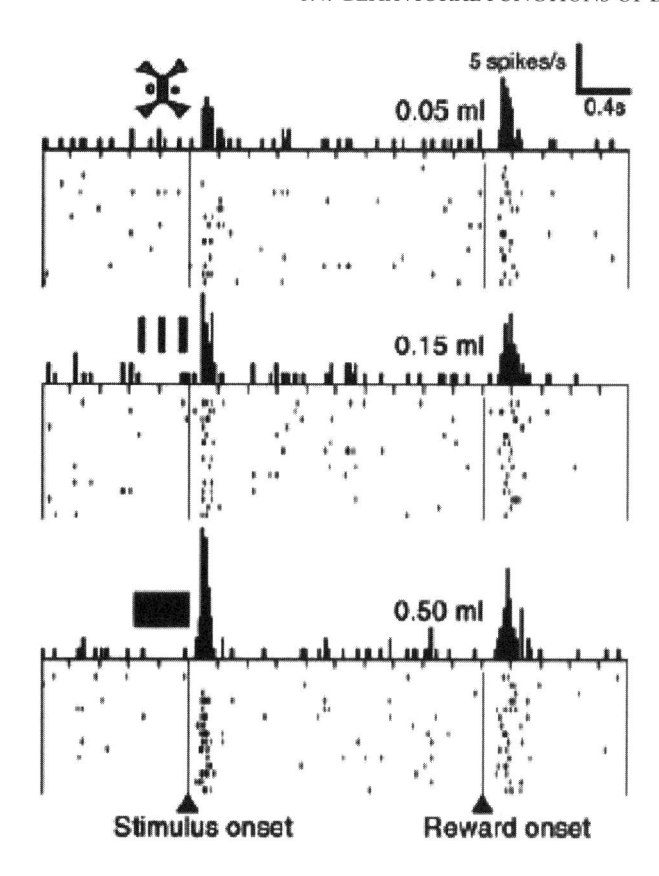

FIGURE 6.4.12 Neural sensitivity to liquid reward volume adapts to predictions entailed by conditioned stimuli. The responses to three liquid reward volumes spanning a 10-fold range are nearly identical. When such rewards occur without conditioned stimuli, responses increase with magnitude (Fig. 6.4.3). Each of the three visual stimuli (shown on the left) was followed by one of two liquid reward volumes at $p = 0.5$ (top, 0.0 or 0.05 ml; middle, 0.0 or 0.15 ml; bottom, 0.0 or 0.5 ml). Responses after onset of visual stimuli increased with their associated expected reward values. Only rewarded trials are shown. *Source*: Adapted with permission from AAAS.[9]

probability (p) or a distortion function (w) of p. *EV* corresponds to expected value,[50] *EU* to expected utility,[11] and *PT* to prospect.[13] The functions u and w take into account that individuals often show nonlinearities in the processing of magnitude and probability. Although it is currently unclear which exact formula the phasic responses of dopamine neurons follow, the core intuition of combining magnitude and probability seems to be fulfilled.[9]

Reward risk is another important economic reward parameter. It follows an inverted-U function of reward probability, is highest at $p = 0.5$, gradually decreases with decreasing and increasing probability, and is zero at $p=0$ and $p=1$. Formal measures for risk include variance and standard deviation. Contrary to risk, mean or expected value increases monotonically with probability (Fig. 6.4.13A; equation 2). Mean corresponds to the first moment of a distribution of reward outcomes, variance to the second. In this view, the value of a choice option is approximated by its moments.[12]

For risk-averse individuals, the value of a choice option decreases with increasing risk; for risk-seeking individuals, it increases. As discussed above (Fig. 6.4.10), when different visual stimuli indicate a fixed amount of reward at different probabilities, the phasic responses of dopamine neurons immediately following stimuli and outcomes scale with reward probability. However, about one-third of dopamine neurons show a more sustained activation that gradually increases toward the time of risky rewards[25] (Fig. 6.4.13B). This activation is strongest for stimuli predicting reward at $p = 0.5$ and decreases with increasing and decreasing probabilities. It occurs with risky rewards and when stimuli remain present until reward delivery but not with risky small visual stimuli and when there is a temporal gap between stimulus and reward delivery. The risk-related sustained responses reach more moderate amplitudes (median increase of about 70%) than the phasic probability-, magnitude-, and delay-related responses (median increase of up to 200%). The lower amplitudes

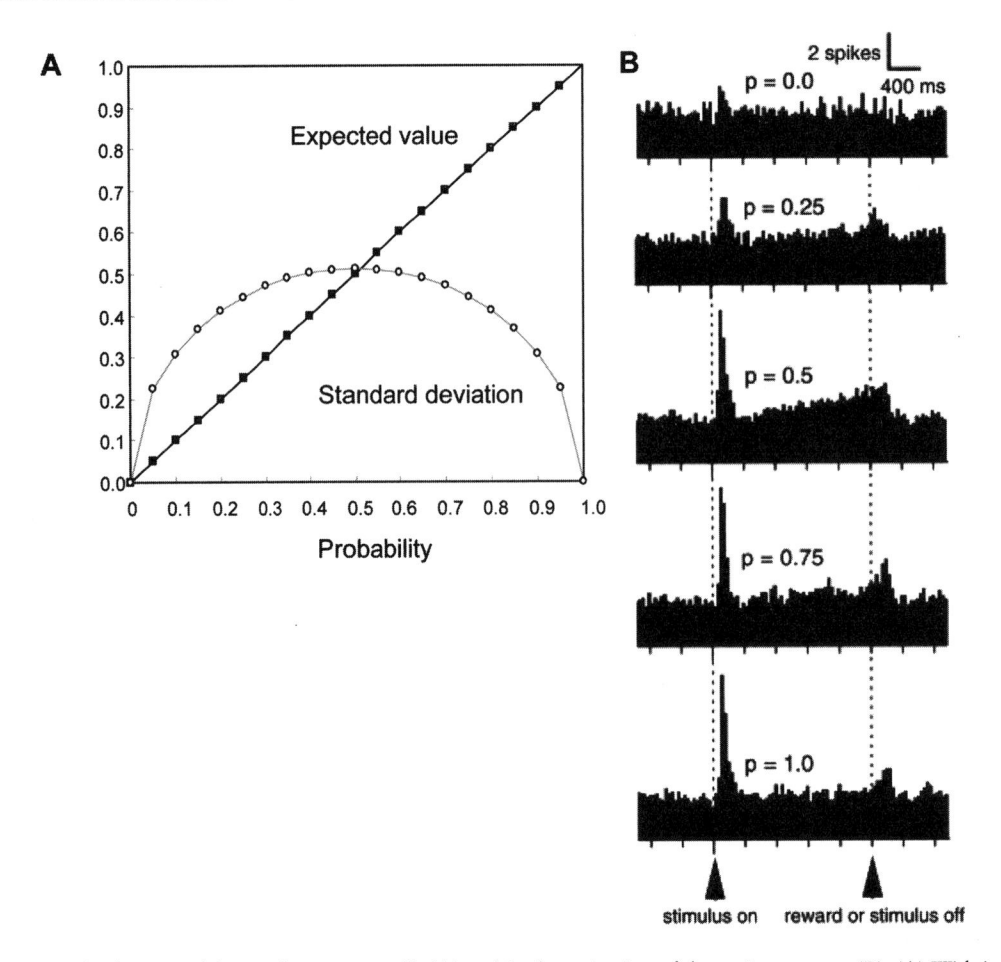

FIGURE 6.4.13 Risk can be dissociated from value conceptually (A) and in the activation of dopamine neurons (B). (A) With increasing reward probability, the value of an option increases (black line), whereas the risk, measured here as the standard deviation, increases from 0 at $p = 0.0$ to maximal values at $p = 0.5$ and then decreases again to reach 0 at $p = 1.0$ (gray line). (B) Dopamine neurons coded reward risk and probability with distinct sustained and phasic activations. Population histograms of 35 to 44 neurons tested with reward probabilities of $p = 0$ (top) to $p = 1$ (bottom). Phasic responses after stimulus onset follow the black line in (A); sustained responses before the time of reward follow the gray line in (A). This is the same experiment as for a single neuron shown in Figure 6.4.10 (left), but here not only rewarded but also unrewarded trials are included at intermediate probabilities. *Source*: Adapted with permission from AAAS.[25]

of risk-related responses may primarily stimulate high-affinity dopamine D2 receptors, whereas the higher amplitudes of phasic responses may be appropriate to stimulate the low-affinity D1 receptors.[51] Thus, it is conceivable that postsynaptically, the sustained, risk-related responses of dopamine neurons could be separated from the phasic, value-related responses.

ATTENTION AND NOVELTY FUNCTIONS OF DOPAMINE NEURONS

The evidence reviewed so far suggests a prime involvement of dopamine neurons in the processing of reward value and risk. However, rewards also induce attention. The question therefore arises whether dopamine neurons respond to stimuli that are not associated with primary reinforcement, and whether they distinguish between rewarding and attention-inducing stimuli. The issue is also raised by early findings that dopamine neurons respond to novel and intense stimuli. For example, the novel sound of a door opening, which had never been paired with reward, elicits activation usually followed by depression in dopamine neurons during the first 20–40 trials[8] (Fig. 6.4.14). The response subsides together with the animals' orienting response to the source of the sound (Fig. 6.4.14). Intense unrewarded stimuli (loud clicks or large pictures) elicit strong activations that decay more slowly without disappearing entirely after more than 1000 trials and are usually followed by depressions.[52,53] Novelty and intensity responses could suggest a primary involvement of phasic dopamine activity in attention functions,[54] but could also reflect the rewarding properties of novel and

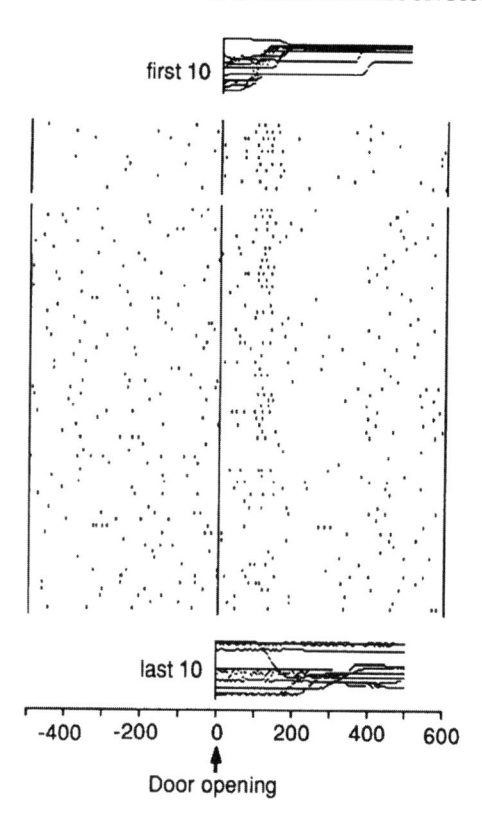

FIGURE 6.4.14 Responses to novel opening of a door in two dopamine neurons. With repeated presentation of a novel stimulus, novelty responses decrease if the stimulus remains unrewarded. Traces above and below rasters show the horizontal components of electrooculograms in the first and last 10 trials, respectively. Upward deflections correspond to rightward saccades. Thus, animals reacted to door opening initially. Note that novelty responses are followed by depressions in both neurons, possibly qualifying the stimulus as (hitherto) unrewarded. The two neurons are separated vertically; the box was empty in all trials; the actual trial sequence is from top to bottom. *Source*: Adapted with permission from the American Physiological Society.[8]

intense stimulation[20,55–58] or a combination of attention and reward functions.[42,59] In any case, the prominent depressions following novelty and intensity responses hint at the possibility that dopamine neurons distinguish between rewarding and attention-inducing stimuli, perhaps particularly at longer latencies.

Attention and reward can be disentangled by testing punishment or stimuli predicting punishment. Punishment produces avoidance behavior, reduces the behavior leading to its occurrence, and increases the behavior leading to its avoidance (negative reinforcement). Thus, punishment has motivationally opposite effects to reward but also induces attention. Dopamine neurons primarily show depressions to punishments, such as air puffs, noxious pinch, hypertonic saline, and electric shock, and to stimuli predicting such punishment[60–62] (Fig. 6.4.15). Depressions occur both in the behaving animal and under anesthesia. They can be long-lasting (tonic), as with long-lasting noxious pinch (Fig. 6.4.16; 51% of depressed dopamine neurons, as in[60], or more, as in[63]; see also[64]). Conversely, only a small proportion of neurons show punishment-induced activation (17% of neurons in[60]; 11–14% in[61]). Some neurons activated by noxious pinch stimulation may not be dopaminergic.[65] These data indicate that the phasic activity of dopamine neurons codes primarily the motivational rather than the attention-inducing properties of reward and punishment, with activations reflecting the positive value of rewards and depressions the negative value of punishments. The few dopamine neurons showing phasic activations to punishment such as air puff and foot shock[62,66] (Fig. 6.4.16C) are located

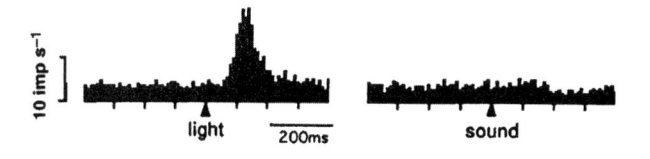

FIGURE 6.4.15 Dopamine neurons show activation to appetitive but not aversive conditioned stimuli. Activation occurred only after conditioned light eliciting a movement for juice reward (left) but not after a conditioned sound eliciting a movement for air puff avoidance (right). Averaged population histograms of 31 neurons. *Source*: Adapted with permission from the Nature Publishing Group.[16]

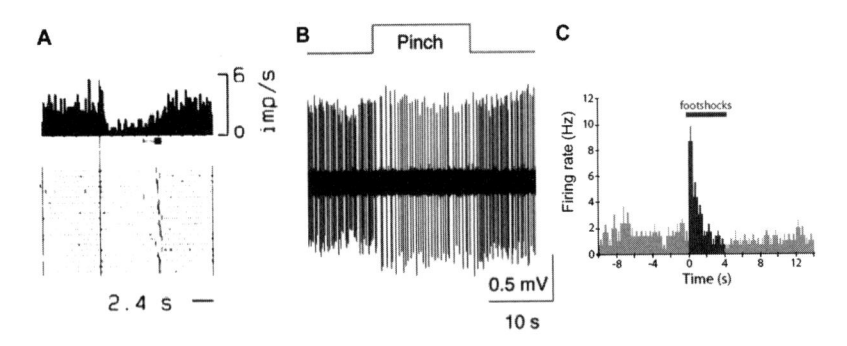

FIGURE 6.4.16 Under anesthesia, dopamine neurons show primarily depression to intense pinch stimulation. (A) Example activity of a single monkey substantia nigra dopamine neuron in response to foot pinch. The vertical line denotes the time of pinch onset; short markers below the histogram and in dot displays denote pinch offset. *Source*: Adapted with permission from the American Physiological Society[60]. (B) Example firing of a single ventral tegmental area dopamine neuron in response to foot pinch (rat). *Source*: Adapted with permission from AAAS[65]. (C) Phasic activation of a single ventral tegmental area dopamine neuron showing phasic activation to foot shock (rat; ventral part of the ventral tegmental area). *Source*: Adapted with permission from the National Academy of Sciences, U.S.A.[66]. Note that in (A) and (B), time scales are much longer than in other figures and in (C), thereby illustrating the sustained nature of the response.

primarily in the dorsolateral part of the substantia nigra in the awake monkey and the ventral part of the ventral tegmental area in the anesthetized rat. Thus, it remains possible that distinct subgroups of dopamine neurons code primarily attention and reward.

Behavioral reward and attention functions can also be disentangled with the conditioned inhibition task introduced above. In order to successfully inhibit responding upon presentation of a conditioned inhibitor, the animal has to attend to the conditioned inhibitor even though it is associated with reward absence. It is worth noting that conditioned inhibitors share motivational properties with punishments and punishment-predicting conditioned stimuli. Just like punishments, conditioned inhibitors are negatively reinforcing, and animals work to avoid them. As with punishments, dopamine neurons are primarily depressed rather than activated by conditioned inhibitors, particularly about 200–500 ms after stimulus onset[10] (Fig. 6.4.11). Thus, similar to the depressions following novelty responses, particularly the late part of phasic changes in dopamine firing reflects the reward omission-predicting properties rather than the attention-inducing properties of conditioned inhibitors. Further research is necessary to determine whether the early, moderate activations to attention-inducing conditioned inhibitors and novel and intense stimuli reflect higher-order associations with reward-predicting stimuli,[10] generalization, or general attention.

In contrast to (the late part of the) phasic responses, the slower risk-related changes in dopamine firing (Fig. 6.4.13) can be described as fulfilling the quite specific functions put forward by an attention-based theory of learning. This theory proposes that individuals pay most attention to stimuli that are associated with risky rewards because such stimuli provide the biggest potential for additional learning.[67] The absolute value of prediction errors is used to determine attention, and reflects how easily conditioned stimuli form an association with rewards (associability):

$$\alpha_i = |\lambda - \Sigma V_i|, \quad (5)$$

where α_i corresponds to the associability of stimulus i, λ to the maximal processing that the presented reward can sustain, and ΣV_i to what has been learned so far (sum of associative strengths of the stimuli present on the previous trial in which stimulus i occurred). Similar to risk, the associability term follows an inverted-U function of reward probability (highest at $p = 0.5$ due to constant intermediate prediction errors arising in rewarded and unrewarded trials, zero at $p = 0.0$ and $p = 1.0$). Thus, the sustained response of dopamine neurons may code the associability term of attentional learning functions.

CONCLUSIONS

Electrophysiological studies of dopamine neurons have come a long way from the early findings of moderate relations with movement. Subsequent studies have shown that dopamine neurons contribute to reward learning by coding errors in the prediction of reward; process and combine economic reward parameters, such as reward magnitude, delay, probability, and risk; adapt their prediction error responses to the predictions entailed by conditioned stimuli; process stimulus novelty and intensity; and discriminate between differently rewarding but similarly attention-inducing stimuli, such as conditioned inhibitors or punishments and reward. Thereby dopamine neurons provide postsynaptic neurons in the striatum and cortex with detailed information about the predicted distribution of future rewards, information that these regions may use to plan and execute profitable behaviors and decisions well in advance of actual reward occurrence and to learn about even earlier reliable predictors of reward. The sustained reward risk-related responses are compatible with the associability term of attention-based learning theories. Taken together, the extracellular perspective suggests that dopamine neurons contribute to a variety of adaptive behavioral functions.

ACKNOWLEDGMENTS

I thank Wolfram Schultz, the Janggen-Poehn Foundation, the Swiss National Science Foundation, the Roche Research Foundation, and the Wellcome Trust for support.

REFERENCES

1. Dickinson A. *Contemporary Animal Learning Theory.* Cambridge, England: Cambridge University Press; 1980.
2. Pearce JM. *Animal Learning and Cognition.* Hove, UK: Psychology Press; 1997.
3. Dickinson A, Balleine B. Motivational control of goal-directed action. *Anim Learn Behav.* 1994;22:1–18.
4. Huang C-F, Litzenberger RH. *Foundations for Financial Economics.* Upper Saddle River, NJ: Prentice-Hall; 1988.
5. Kreps DM. *A Course in Microeconomic Theory.* Princeton, NJ: Princeton University Press; 1990.
6. Schultz W. Responses of midbrain dopamine neurons to behavioral trigger stimuli in the monkey. *J Neurophysiol.* 1986;56:1439–1461.
7. Romo R, Schultz W. Dopamine neurons of the monkey midbrain: contingencies of responses to active touch during self-initiated arm movements. *J Neurophysiol.* 1990;63:592–606.
8. Ljungberg T, Apicella P, Schultz W. Responses of monkey dopamine neurons during learning of behavioral reactions. *J Neurophysiol.* 1992;67:145–163.
9. Tobler PN, Fiorillo CD, Schultz W. Adaptive coding of reward value by dopamine neurons. *Science.* 2005;307:1642–1645.
10. Tobler PN, Dickinson A, Schultz W. Coding of predicted reward omission by dopamine neurons in a conditioned inhibition paradigm. *J Neurosci.* 2003;23:10402–10410.
11. Von Neumann JV, Morgenstern O. *Theory of Games and Economic Behavior.* Princeton, NJ: Princeton University Press; 1944.
12. Markowitz HM. *Portfolio Selection: Efficient Diversification of Investments.* New York, NY: Wiley; 1959.
13. Kahneman D, Tversky A. Prospect theory: An analysis of decision under risk. *Econometrica.* 1979;47:263–291.
14. Schultz W, Romo R. Dopamine neurons of the monkey midbrain: contingencies of responses to stimuli eliciting immediate behavioral reactions. *J Neurophysiol.* 1990;63:607–624.
15. Schultz W, Apicella P, Ljungberg T. Responses of monkey dopamine neurons to reward and conditioned stimuli during successive steps of learning a delayed response task. *J Neurosci.* 1993;13:900–913.
16. Takikawa Y, Kawagoe R, Hikosaka O. A possible role of midbrain dopamine neurons in short- and long-term adaptation of saccades to position-reward mapping. *J Neurophysiol.* 2004;92:2520–2529.
17. Pan WX, Schmidt R, Wickens JR, Hyland BI. Dopamine cells respond to predicted events during classical conditioning: evidence for eligibility traces in the reward-learning network. *J Neurosci.* 2005;25:6235–6242.
18. Mirenowicz J, Schultz W. Importance of unpredictability for reward responses in primate dopamine neurons. *J Neurophysiol.* 1994;72:1024–1027.
19. Hollerman JR, Schultz W. Dopamine neurons report an error in the temporal prediction of reward during learning. *Nat Neurosci.* 1998;1:304–309.
20. Schultz W. Predictive reward signal of dopamine neurons. *J Neurophysiol.* 1998;80:1–27.
21. Fiorillo CD, Newsome WT, Schultz W. The temporal precision of reward prediction in dopamine neurons. *Nat Neurosci.* 2008;11:966–973.
22. Kobayashi S, Schultz W. Influence of reward delays on responses of dopamine neurons. *J Neurosci.* 2008;28:7837–7846.
23. Schultz W, Dayan P, Montague RR. A neural substrate of prediction and reward. *Science.* 1997;275:1593–1599.
24. Waelti P, Dickinson A, Schultz W. Dopamine responses comply with basic assumptions of formal learning theory. *Nature.* 2001;412:43–48.
25. Fiorillo CD, Tobler PN, Schultz W. Discrete coding of reward probability and uncertainty by dopamine neurons. *Science.* 2003;299:1898–1902.
26. Satoh T, Nakai S, Sato T, Kimura M. Correlated coding of motivation and outcome of decision by dopamine neurons. *J Neurosci.* 2003;23:9913–9923.
27. Nakahara H, Itoh H, Kawagoe R, Takikawa Y, Hikosaka O. Dopamine neurons can represent context-dependent prediction error. *Neuron.* 2004;41:269–280.
28. Kawagoe R, Takikawa Y, Hikosaka O. Reward-predicting activity of dopamine and caudate neurons–a possible mechanism of motivational control of saccadic eye movement. *J Neurophysiol.* 2004;91:1013–1024.
29. Bayer HM, Glimcher PW. Midbrain dopamine neurons encode a quantitative reward prediction error signal. *Neuron.* 2005;47:129–141.
30. Morris G, Nevet A, Arkadir D, Vaadia E, Bergman H. Midbrain dopamine neurons encode decisions for future action. *Nat Neurosci.* 2006;9:1057–1063.
31. Bayer HM, Lau B, Glimcher PW. Statistics of midbrain dopamine neuron spike trains in the awake primate. *J Neurophysiol.* 2007;98:1428–1439.
32. Roesch MR, Calu DJ, Schoenbaum G. Dopamine neurons encode the better option in rats deciding between differently delayed or sized rewards. *Nat Neurosci.* 2007;10:1615–1624.
33. Barto AG. Adaptive critics and the basal ganglia. In: Houk JC, Davis JL, Beiser DG, eds. *Models of Information Processing in the Basal Ganglia.* Boston, MA: MIT Press; 1995:215–232.
34. Houk JC, Adams JL, Barto AG. A model of how the basal ganglia generate and use neural signals that predict reinforcement. In:

Houk JC, Davis JL, Beiser DG, eds. *Models of Information Processing in the Basal Ganglia*. Boston, MA: MIT Press; 1995:249–270.
35. Schultz W, Romo R, Ljungberg T, Mirenowicz J, Hollerman JR, Dickinson A. Reward-related signals carried by dopamine neurons. In: Houk JC, Davis JL, Beiser DG, eds. *Models of Information Processing in the Basal Ganglia*. Boston, MA: MIT Press; 1995:233–248.
36. Montague PR, Dayan P, Sejnowski TJ. A framework for mesencephalic dopamine systems based on predictive Hebbian learning. *J Neurosci*. 1996;16:1936–1947.
37. Rescorla RA, Wagner AR. A theory of Pavlovian conditioning: variations in the effectiveness of reinforcement and nonreinforcement. In: Black A, Prokasy WF, eds. *Classical Conditioning II: Current Research and Theory*, New York, NY: Appleton-Century-Crofts; 1972:64–99.
38. Sutton RS, Barto AG. Time-derivative models of Pavlovian reinforcement. In: Gabriel M, Moore J, eds. *Learning and Computational Neuroscience: Foundations of Adaptive Networks*, Boston, MA: MIT Press; 1990:497–537.
39. Suri RE, Schultz W. Learning of sequential movements by neural network model with dopamine-like reinforcement signal. *Exp Brain Res*. 1998;121:350–354.
40. Suri RE, Schultz W. A neural network model with dopamine-like reinforcement signal that learns a spatial delayed response task. *Neuroscience*. 1999;91:871–890.
41. Doya K. Metalearning and neuromodulation. *Neural Netw*. 2002;15:495–506.
42. Kakade S, Dayan P. Dopamine: generalization and bonuses. *Neural Netw*. 2002;15:549–559.
43. Daw ND, Courville AC, Touretzky DS. Representation and timing in theories of the dopamine system. *Neural Comput*. 2006;18:1637–1677.
44. Bertin M, Schweighofer N, Doya K. Multiple model-based reinforcement learning explains dopamine neuronal activity. *Neural Netw*. 2007;20:668–675.
45. Tan CO, Bullock D. A local circuit model of learned striatal and dopamine cell responses under probabilistic schedules of reward. *J Neurosci*. 2008;28:10062–10074.
46. Kamin LJ. Predictability, surprise, attention and conditioning. In: Campbell BA, Church RM, eds. *Punishment and Aversive Behavior*, New York, NY: Appleton-Century-Crofts; 1969:279–296.
47. Pavlov IP. *Conditional Reflexes*. New York, NY: Dover Publications; 1927/1960 (the 1960 edition is an unaltered republication of the 1927 translation by Oxford University Press).
48. Rescorla RA. Pavlovian conditioned inhibition. *Psychol Bull*. 1969;72:77–94.
49. Samuelson PA. Some aspects of the pure theory of capital. *Q J Econ*. 1937;51:469–496.
50. Pascal B. *Great Shorter Works* (translated from French by E. Cailliet and J. C. Blankenagel). Philadelphia, PA: Westminster; 1623-1662/1948.

51. Gonon F. Prolonged and extrasynaptic excitatory action of dopamine mediated by D1 receptors in the rat striatum in vivo. *J Neurosci*. 1997;17:5972–5978.
52. Steinfels GF, Heym J, Strecker RE, Jacobs BL. Response of dopaminergic neurons in cat to auditory stimuli presented across the sleep-waking cycle. *Brain Res*. 1983;277:150–154.
53. Horvitz JC, Stewart T, Jacobs BL. Burst activity of ventral tegmental dopamine neurons is elicited by sensory stimuli in the awake cat. *Brain Res*. 1997;759:251–258.
54. Redgrave P, Prescott TJ, Gurney K. Is the short-latency dopamine response too short to signal reward error? *Trends Neurosci*. 1999;22:146–151.
55. Eisenberger R. Explanation of rewards that do not reduce tissue needs. *Psychol Bull*. 1972;77:319–339.
56. Humphrey NK. "Interest" and "pleasure": two determinants of a monkey's visual preferences. *Perception*. 1972;1:395–416.
57. Mishkin M, Delacour J. An analysis of short-term visual memory in the monkey. *J Exp Psychol: Anim Behav Process*. 1975;1:326–334.
58. Washburn DA, Hopkins WD, Rumbaugh DM. Perceived control in rhesus monkeys (*Macaca mulatta*): enhanced video-task performance. *J Exp Psychol: Anim Behav Process*. 1991;17:123–129.
59. Horvitz JC. Mesolimbocortical and nigrostriatal dopamine responses to salient non-reward events. *Neuroscience*. 2000;96:651–656.
60. Schultz W, Romo R. Responses of nigrostriatal dopamine neurons to high-intensity somatosensory stimulation in the anesthetized monkey. *J Neurophysiol*. 1987;57:201–217.
61. Mirenowicz J, Schultz W. Preferential activation of midbrain dopamine neurons by appetitive rather than aversive stimuli. *Nature*. 1996;379:449–451.
62. Matsumoto M, Hikosaka O. Excitatory and inhibitory responses of midbrain dopamine neurons to cues predicting aversive stimuli. *Soc. Neurosci. Abstr*. 2008;691.24.
63. Gao DM, Hoffman D, Benabid AL. Simultaneous recording of spontaneous activities and nociceptive responses from neurons in the pars compacta of substantia nigra and in the lateral habenula. *Eur J Neurosci*. 1996;8:1474–1478.
64. Tsai CT, Nakamura S, Iwama K. Inhibition of neuronal activity of the substantia nigra by noxious stimuli and its modification by the caudate nucleus. *Brain Res*. 1980;195:299–311.
65. Ungless MA, Magill PJ, Bolam JP. Uniform inhibition of dopamine neurons in the ventral tegmental area by aversive stimuli. *Science*. 2004;303:2040–2042.
66. Brischoux F, Chakraborty, S, Brierley DI, Ungless MA. Phasic excitation of dopamine neurons in ventral tegmental area by noxious stimuli. *Proc Natl Acad Sci USA*. 2009;106:4894–4899.
67. Pearce JM, Hall G. A model of Pavlovian learning: variations in the effectiveness of conditioned but not of unconditioned stimuli. *Psychol. Rev*. 1980;87:532–552.

7 | Plasticity of forebrain dopamine systems

7.1 Dynamic Templates for Neuroplasticity in the Striatum

ANN M. GRAYBIEL

Fifty years on, we can ask again: What does dopamine do? From the broad scope of the contributions to this celebratory volume, reflecting current research on the functions of dopamine-containing systems of the brain, we can surely conclude that this single molecule has profound effects on functions ranging from the modulation of motor action to the modulation of cognition. How can this be so? One clue comes from evidence that dopamine-containing neural pathways reach not only the basal ganglia, which affect movement, but also many parts of the subcortical and cortical limbic systems, which affect emotion and motivation, and regions of the neocortex that influence the tone and information content of mental life. A second clue is that dopamine can act through a diverse set of receptor molecules that, in turn, influence multiple second messenger and higher-order signaling pathways in neurons within these regions. A third clue is that dopamine, by way of these receptor-mediated signaling systems, has crucial effects on the efficacy of the synapses that control the step-by-step operation of information flow through these multiple neural systems.

DOPAMINE AND SYNAPTIC PLASTICITY

Few guessed, at the time that dopamine was identified as a neuromodulator, that dopamine would so potently and broadly controls synaptic function. And for many years, researchers in this field struggled with the question of whether dopamine is "excitatory" or "inhibitory". It now seems that this question may have been ill-posed, as it was rooted in models of synaptic connectivity that pictured synapses as bimodal (on or off) or as passive and/or gates that allow or disallow unidirectional information flow, depending on the excitation or lack of excitation of that connection.

Studies of dopamine helped to change these models of synaptic function. We now envision synapses as dynamic gates with hundreds of molecules on the postsynaptic side being influenced by large numbers of molecules on the presynaptic side, all adjusting the efficacy of each synapse.

We know that there are retrograde signals working from post- to presynaptic effects in addition to conventional pre- to postsynaptic effects and, crucially, we are beginning to appreciate that all of these events are dependent on the relative timing of molecular events on the two sides of the synapse. Extrasynaptic receptor functions mediated by dopamine receptors are also now recognized as important for the control of information flow. Thus, the effects of dopamine can be viewed from the perspective of a dynamic molecular modulator of functional connectivity across the linkages that make up the brain's trafficking systems. Dopamine no longer is thought to have a single function but, at the molecular level, as having many functions. As most of the signaling systems triggered by dopamine lead to changes in gene expression, the field of dopamine research now has gained a new focus on how dopamine affects the molecular biology of the cell.

DOPAMINE AND SYSTEMS-LEVEL NEUROPLASTICITY

The recognition of dopamine's importance for synaptic plasticity has coincided with an equally remarkable evolution in our ideas about the behavioral effects of dopamine signaling in the brain. Dopamine-mediated signaling has effects not only on our ability to move, but also on memory, on cognitive competence, on emotional states, and on motivational tone. Could these apparently disparate effects be related to dopamine's role as a plasticity molecule?

In the 1990s, the answer to this question became, in outline, clear, at least for functions related to the nigrostriatal system. The dopamine-containing neurons of the macaque midbrain were shown in conditioning experiments to carry signals related to reward and reward expectation,[1] and dopamine-recipient neurons in the macaque striatum were shown to undergo learning-related changes in their responses that depended on striatal dopamine.[2,3] Many studies have now shown that dopamine-containing neurons in the midbrain carry signals related to saliency, reward expectancy, and the uncertainty of this expectation.[4] The striatal neurons analyzed in these conditioning experiments have also been shown in further experiments to be linked to saliency and to aspects of both positive and negative reinforcement.[5,6] They are called "tonically active neurons" (TANs) because of their tendency to fire at low spontaneous rates, and they develop a pause in their firing at the time that the dopamine-containing neurons show a phasic burst of activity.[7,8] The TANs have now been shown to correspond to the cholinergic interneurons of the striatum. Thus, the early studies demonstrating the crucial role of dopamine in behavioral plasticity were paralleled by studies demonstrating that the cholinergic interneurons of the striatum are directly influenced by this dynamically changing, dopamine-dependent input and also undergo learning-related changes in activity.

THE DOPAMINERGIC–CHOLINERGIC BALANCE IN THE STRIATUM RECONSIDERED

It is highly likely that these dynamic dopaminergic–cholinergic interactions underlie at least in part the "dopaminergic–cholinergic balance" long-recognized by clinicians. Our laboratory first began to study the relationship between striatal cholinergic systems and the nigrostriatal dopamine system when we found that cholinergic markers in the striatum were not uniformly distributed. Instead, the generally intense anatomical staining for these markers was interrupted by pockets of low staining distributed at fairly regular intervals, roughly 1 mm in the human brain.[9] We called these regions "striosomes" and referred to the large cholinergic-rich striatal regions around them as the "extrastriosomal matrix". The link between this striosomal organization and the dopamine-containing innervation became clear when we began to look at the development of striosomes. We found that early in striatal development, striosomes corresponded to the "dopamine islands" that had been described in the first series of pioneering papers demonstrating the distribution of dopamine-containing fiber systems.[10,11] This demonstration required that we use a permanent marker for the striosomal system, which we did by using ^{3}H thymidine to mark the striatal neurons in striosomes.[12] We had found that the neurons in striosomes share common birth dates, so that pulse labeling with ^{3}H thymidine during embryonic development clearly marked the striosomes throughout life[13] (Fig. 7.1.1).

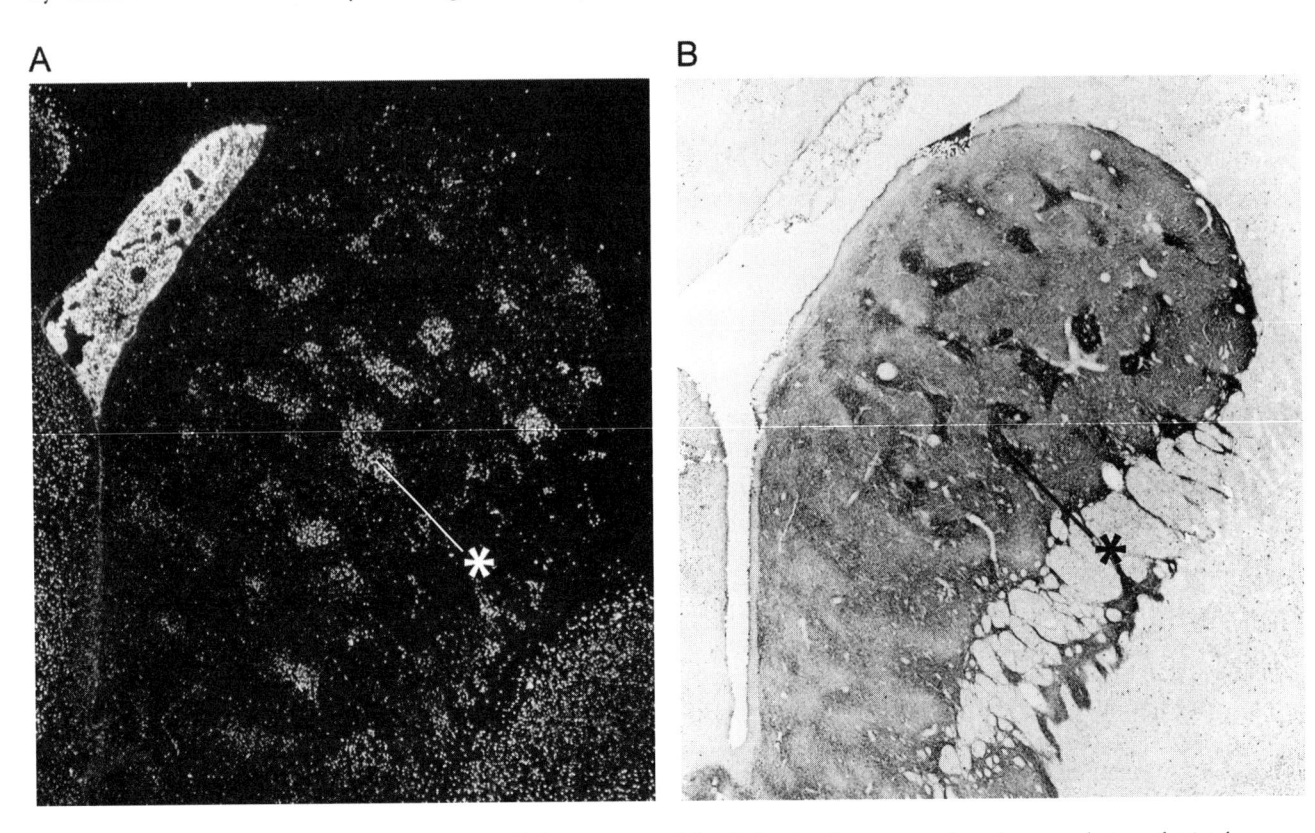

FIGURE 7.1.1. Transverse sections through the striatum of a kitten (postnatal day 8) showing the correspondence between clusters of striatal neurons pulse labeled with ^{3}H thymidine to mark striosomes (A) and developing dopamine islands marked by tyrosine hydroxylase immunohistochemistry (B). The asterisk indicates an example of a striosome. *Source*: Reprinted from the Journal of Neuroscience.[12]

These findings set up a series of further observations that may prove to be key to understanding neural plasticity in cortico-basal ganglia circuits. First, the striatal cholinergic receptors are differentially distributed between the striosome and matrix compartments. This compartmental biasing of receptors is vividly apparent in preparations showing the distribution of M1 muscarinic cholinergic receptors, which are highly enriched in striosomes, especially during development[14,15] (Fig. 7.1.2). Second, dopamine receptors are differentially distributed across the two compartments. These findings encourage the viewpoint that interactions between dopamine and acetylcholine are by no means uniform in the striatum, but instead are compartment-selective from early on in development through adulthood. They are also cell-type specific. Within the large matrix compartment, D1-class and D2-class receptors are sharply divided between the neurons that give rise to the direct and indirect pathways of the basal ganglia (see Chapter 2.1, this volume), and cholinergic receptors are nonuniform. Gradients in expression levels are also evident for molecules related to dopaminergic and cholinergic transmission.

COMPARTMENTAL INTERACTIONS BETWEEN DOPAMINE, ACETYLCHOLINE, AND MU OPIOID RECEPTOR FUNCTIONS

The cholinergic interneurons of the striatum—the TANs identified electrophysiologically—account for most of the cholinergic neuropil of the striatum. These neurons not only lie mainly in the cholinergic-rich matrix compartment, but also occur disproportionately at striosome–matrix borders. This anatomical distribution was shown in experiments in which we identified TANs in the striatum of squirrel monkeys and afterward determined the anatomical location of these TANs.[16] Because we had demonstrated with our colleagues Aosaki and Kimura that the acquired responses of TANs depend on the presence of dopaminergic input,[2] this positioning of TANs at striosome–matrix borders raised the possibility that the TANs might link activity across the borders in a dynamic way, depending on inputs from the dopaminergic midbrain.

This line of experiments has now been related to another striking characteristic of striosomes: they are highly enriched in mu opioid receptors, as illustrated by Pert and her colleagues[17] in their first anatomical demonstration of these receptors. In slice experiments, Miura and his colleagues[18] have now shown that opioid receptor blockade can differentially lift inhibition in striosomes, thus differentially activating striosomes. Moreover, they suggest that enkephalin (coexpressed in indirect pathway neurons) acts on mu opioid receptors to depolarize cholinergic interneurons.[19] Mu opioid receptors are even expressed in the cholinergic neurons themselves, so that bidirectional cross-talk between striosomes and matrix could in part depend on activation of opioid receptors, concentrated in striosomes. These experiments support the idea, originally proposed because of the similarly timed pause responses of widely distributed TANs, that the acquired pause responses of TANs might allow discretely timed windows of striatal plasticity.[3]

CIRCUIT-LEVEL PLASTICITY IN STRIATAL GENE ACTIVATION IN STRIOSOMES AND MATRIX

If neuroplasticity in the striatum is strongly influenced by the striosome–matrix compartmentalization, then it might follow that the effects of dopaminergic drugs are different in the two compartments. This is proving to be

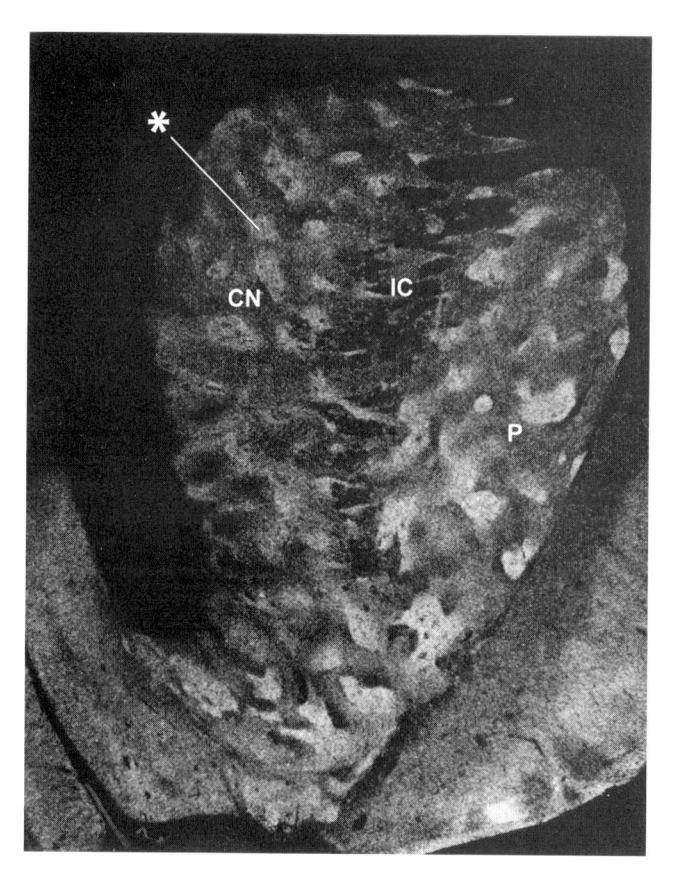

FIGURE 7.1.2. Transverse section through a human fetus at 22 weeks of gestation showing (white) the distribution of autoradiographic labeling for M1 muscarinic acetylcholine receptors. The asterisk indicates an example of a striosome. CN, caudate nucleus; IC, internal capsule; P, putamen. Source: Reprinted from Journal of Comparative Neurology.[14]

the case. When animals are given single doses of psychomotor stimulants such as amphetamine, early response genes are activated noticeably more strongly in striosomes than in the surrounding matrix, especially in the rostral striatum. Moreover, when such indirect dopamine receptor agonists are given repeatedly at doses inducing increasing stereotypic behaviors in the animals, the striosome predominance of early gene activation is even more obvious in some parts of the striatum.[20–23] These differential activation patterns provided the first demonstration of functional differences between striosomes and matrix, and suggested that the functions might be important for circuit-level changes in gene expression.

These experiments also provided a way to test for a correlation between striosome predominance and behavior. In both rodents and monkeys, we found that the degree of striosome predominance of the early gene activation by psychomotor stimulants is highly correlated with the levels of stereotypic, repetitive behaviors exhibited by the animals.[21,22] The stereotypies were increased with repeated exposures, as was the expression of the early response genes. Remarkably, however, this relationship could be broken by neurotoxin-induced ablation of the cholinergic (and nitric oxide synthase [NOS]-containing) interneurons of the striatum.[24] After the neurotoxin treatments, we found no difference in the stereotypy scores of the animals, despite the loss of striosome-predominant expression of the early gene response. The reasons for this breakdown are not clear, but the result raises the possibility that after the instatement of the plastic changes leading to increased stereotypic behaviors, the changed patterns can become independent of further differential gene activation in the two compartments, perhaps because the downstream consequences of the early-gene activation are set. Other conditions can also break this link, such as deletion of D2 dopamine receptors.[25] A strong interaction between the cholinergic and dopaminergic striatal systems controlling repetitive behavior is suggested by these results. A further link between stereotypic behaviors has come from experiments on mice with double knockout of retinoic acid receptors.[26] These animals have lost the rostral part of the striosomal system and exhibit specific changes in stereotypic behaviors.

Many genes are highly expressed in the striatum, and most of these are differentially expressed in either striosomes or matrix. We identified two novel striatum-enriched genes, *CalDAG-GEF1* and *CalDAG-GEF2*, named for their having binding sites for calcium and diacylglycerol input motifs and guanine nucleotide exchange factor (GEF) effector motifs targeting Ras superfamily molecules.[27] CalDAG-GEF1 is enriched in the matrix compartment of the striatum, and CalDAG-GEF2 is enriched in striosomes.[28] Genetic deletion of the matrix-enriched *CalDAG-GEF1*, we reasoned, might disadvantage the matrix relative to striosomes and lead to increased psychomotor stimulant-induced stereotypic behavior in the knockouts relative to the wild types. Our evidence to date suggests that this does happen.[29;in prep.] This evidence adds strength to the possibility that the effects of dopamine on striatum-dependent behaviors are modulated in compartment-dependent ways. We have also found that the striosome-enriched *CalDAG-GEF2* and the matrix-enriched *CalDAG-GEF1* are oppositely regulated at the RNA and protein level in proportion to the dyskinesias induced by repeated L-DOPA treatment in a rodent model of parkinsonism.[28] Evidence suggests that *CalDAG-GEF1* is important for M1 muscarinic cholinergic modulation of signaling in cell cultures[30] and in the striatum, suggesting that interactions between dopaminergic and cholinergic function in striosomes and matrix may be important for the neuroplasticity evidenced in L-DOPA-induced dyskinesias, as they seem to be for stereotypic behaviors and for hyperactivity induced by these drugs.[31;in prep.]

CLINICAL DISORDERS AFFECTING THE STRIATUM AND DIFFERENTIAL VULNERABILITY OF STRIOSOMES AND MATRIX

If the differentiation of dopaminergic functions in the striatum is influenced by striosome–matrix compartmentalization, it might also be predicted that these compartments would have different vulnerability in clinical disorders associated with basal ganglia dysfunction. Evidence is increasingly suggesting that this may be so. Differential loss of either striosomes or extrastriosomal matrix has been reported for Huntington's disease and for X-linked dystonia parkinsonism (e.g.,[32–34]). Differential loss of dopamine markers in matrix or in striosomes has also been found in animal models of Parkinson's disease and dopa-responsive dystonia (e.g.,[35,36,and refs. therein]). Much more work needs to be done to study the clinical significance of striosome–matrix compartmentalization, but these findings raise the possibility that differential processing in and between these compartments contributes to behavioral dysfunction when the balance between the compartments is disturbed.

A STRIATAL GRIDWORK FOR VALUE

Evidence to date suggests that striosomes are differentially connected with regions of the limbic system, and with neocortical regions implicated in the control of mood, motivation, and emotion, including the pregenual anterior cingulate cortex and the caudal orbitofrontal cortex. Further, the dopamine-containing input to striosomes appears to arise in a particular subregion of the pars compacta, and striosomes have at most modest outputs to the main direct and indirect output pathways of the striatum that lead into thalamocortical circuits. Evidence also suggests that striosomes project either directly or indirectly to the substantia nigra pars compacta. These findings (Fig. 7.1.3) suggest that the striosomal system may in part serve to process information related to value or estimated value and to influence, in turn, dopaminergic subsystems according to such calculated value. This connectivity stands in contrast to that of the large matrix compartment, which in general receives input predominantly from sensorimotor and associative regions of the neocortex and projects predominantly to the pallidum and substantia nigra through the direct and indirect pathways leading out of the basal ganglia.

These anatomical considerations have led to the idea that the striosomal system might represent the critic in actor-critic models of the basal ganglia.[37,38] Indirect experimental evidence has suggested that striosomes might be particularly related to reward or saliency processing or to other aspects of state valuation.[7,39,40] Evidence that striosomes project to the lateral habenula, which in turn sends inverse reward signals to the dopaminergic neurons of the pars compacta,[41] raises the possibility that striosomes may engage in negative as well as positive value processing. This viewpoint is compatible with the hypothesis that the striosomes might be responsible for inhibition of dopaminergic neurons in the pars compacta when expected reward does not come,[4,40] and would be compatible with the apparent association of striosomes and cholinergic interneurons, which are sensitive to aversive as well as to rewarding condition cues.[5,6] It is only now becoming possible technically to test these and other ideas about the functions of the striosomal system. It is of great interest to think that the striosomes, distributed at fairly regular spacing within the surrounding matrix, could form a gridwork for transferring evaluative signals to the sensorimotor processing networks of the striatal matrix.

ACKNOWLEDGMENTS

The author expresses gratitude for the support of the recent and ongoing experiments described here provided by the National Institutes of Health (NICHD 28341) and the Stanley H. and Sheila G. Sydney Fund.

REFERENCES

1. Romo R, Schultz W. Dopamine neurons of the monkey midbrain: contingencies of response to active touch during self-initiated arm movements. *J Neurophysiol*. 1990;63:592–606.
2. Aosaki T, Graybiel AM, Kimura M. Effects of the nigrostriatal dopamine system on acquired neural responses in the striatum of behaving monkeys. *Science*. 1994;265:412–415.
3. Graybiel AM, Aosaki T, Flaherty AW, Kimura M. The basal ganglia and adaptive motor control. *Science*. 1994;265:1826–1831.
4. Schultz W. Multiple dopamine functions at different time courses. *Annu Rev Neurosci*. 2007;30:259–288.
5. Blazquez P, Fujii N, Kojima J, Graybiel AM. A network representation of response probability in the striatum. *Neuron*. 2002;33:973–982.
6. Apicella P. Leading tonically active neurons of the striatum from reward detection to context recognition. *Trends Neurosci*. 2007;30(6):299–306.
7. Aosaki T, Tsubokawa H, Ishida A, Watanabe K, Graybiel AM, Kimura M. Responses of tonically active neurons in the primate's striatum undergo systematic changes during behavioral sensorimotor conditioning. *J Neurosci*. 1994;14(6):3969–3984.
8. Morris G, Arkadir D, Nevet A, Vaadia E, Bergman H. Coincident but distinct messages of midbrain dopamine and striatal tonically active neurons. *Neuron*. 2004;43(1):133–143.
9. Graybiel AM, Ragsdale CW Jr. Histochemically distinct compartments in the striatum of human, monkey, and cat demonstrated by acetylthiocholinesterase staining. *Proc Natl Acad Sci USA*. 1978;75(11):5723–5726.
10. Olson L, Seiger A, Fuxe K. Heterogeneity of striatal and limbic dopamine innervation: highly fluorescent islands in developing and adult rats. *Brain Res*. 1972;44:283–288.

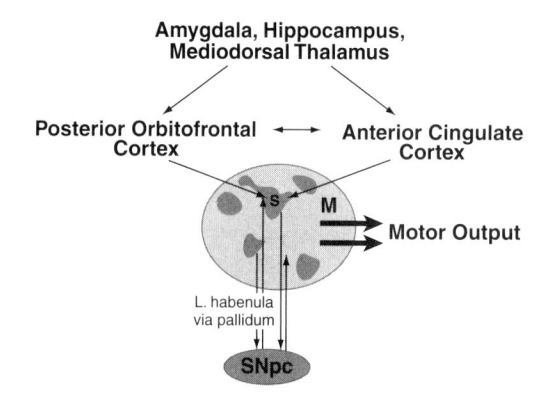

FIGURE 7.1.3. Highly schematic diagram illustrating the striatum (gray oval) with striosomes (dark gray, S) and extrastriosomal matrix (M). The diagram shows the preferential inputs to striosomes from the posterior orbitofrontal cortex and anterior cingulate cortex, and the output leading from the striosomal system toward the substantia nigra pars compacta (SNpc) directly or via the pallidum and lateral habenula. Bold arrows at the right schematically indicate outputs from the matrix leading into the direct and indirect pathways of the basal ganglia. Modified from Graybiel.[40]

11. Tennyson VM, Barrett RE, Cohen G, Cote L, Heikkila R, Mytilneou C. The developing neostriatum of the rabbit: correlation of fluorescence histochemistry, electron microscopy, endogenous dopamine levels, and [3H] dopamine uptake. *Brain Res.* 1972;46:251–285.
12. Graybiel AM. Correspondence between the dopamine islands and striosomes of the mammalian striatum. *Neuroscience.* 1984;13(4):1157–1187.
13. Graybiel AM, Hickey TL. Chemospecificity of ontogenetic units in the striatum: demonstration by combining [³H] thymidine neuronography and histochemical staining. *Proc Natl Acad Sci USA.* 1982;79:198–202.
14. Nastuk MA, Graybiel AM. Autoradiographic localization and biochemical characteristics of M1 and M2 muscarinic binding sites in the striatum of the cat, monkey, and human. *J Neurosci.* 1988;8(3):1052–1062.
15. Nastuk MA, Graybiel, AM. Patterns of muscarinic cholinergic binding in the striatum and their relation to dopamine islands and striosomes. *J Comp Neurology.* 1985;237:176–194.
16. Aosaki T, Kimura M, Graybiel AM. Temporal and spatial characteristics of tonically active neurons of the primate's striatum. *J Neurophysiol.* 1995;73(3):1234–1252.
17. Pert CB, Kuhar MJ, Snyder SH. Opiate receptors: autoradiographic localization in rat brain. *Proc Natl Acad Sci USA.* 1976;73:3729–3733.
18. Miura M, Saino-Saito S, Masuda M, Kobayashi K, Aosaki T. Modulation by mu-opioid receptors on the excitability of cholinergic interneurons in the striosomes/matrix compartments of the striatum. Program No. 514.16. 2007 Neuroscience Meeting Planner. San Diego, CA: Society for Neuroscience, 2007, Online.
19. Miura M, Saino-Saito S, Masuda M, Kobayashi K, Aosaki T. Compartment-specific modulation of GABAergic synaptic transmission by mu-opioid receptor in the mouse striatum with green fluorescent protein-expressing dopamine islands. *J Neurosci.* 2007;27(36):9721–9728.
20. Moratalla R, Elibol B, Vallejo M, Graybiel AM. Network-level changes in expression of inducible Fos-Jun proteins in the striatum during chronic cocaine treatment and withdrawal. *Neuron.* 1996;17:147–156.
21. Canales JJ, Graybiel AM. A measure of striatal function predicts motor stereotypy. *Nat Neurosci.* 2000;3:377–383.
22. Saka E, Goodrich C, Harlan P, Madras BK, Graybiel AM. Repetitive behaviors in monkeys are linked to specific striatal activation patterns. *J Neurosci.* 2004;24:7557–7565.
23. Canales JJ. Stimulant-induced adaptations in neostriatal matrix and striosome systems: transiting from instrumental responding to habitual behavior in drug addiction. *Neurobiol Learn Mem.* 2005;83(2):93–103.
24. Saka E, Iadarola M, Fitzgerald DJ, Graybiel AM. Local circuit neurons in the striatum regulate neural and behavioral responses to dopaminergic stimulation. *Proc Natl Acad Sci USA.* 2002;99:9004–9009.
25. Glickstein SB, Schmauss C. Effect of methamphetamine on cognition and repetitive motor behavior of mice deficient for dopamine D2 and D3 receptors. *Ann NY Acad Sci.* 2004;1025:110–118.
26. Liao WL, Tsai HC, Wang HF, et al. Modular patterning of structure and function of the striatum by retinoid receptor signaling. *Proc Natl Acad Sci USA.* 2008;105(18):6765–6770.
27. Kawasaki H, Springett GM, Toki S, et al. A Rap guanine nucleotide exchange factor enriched highly in the basal ganglia. *Proc Natl Acad Sci USA.* 1998;95:13278–13283.
28. Crittenden JR, Cantuti-Castelvetri I, Saka E, et al. Dysregulation of CalDAG-GEFI and CalDAG-GEFII predicts the severity of motor side-effects induced by anti-parkinsonian therapy. *PNAS.* 2009;106:2892–2896.
29. Crittenden JR, Picconi B, Ghiglieri V, et al. CalDAG-GEFI is required for sensitization to amphetamine-induced stereotypy and cortico-striatal LTP, but not for locomotor sensitization and LTD. Program No. 56.5. 2006 Abstract Viewer/Itinerary Planner. Washington, DC: Society for Neuroscience, 2006, Online.
30. Guo F, Kumahara E, Saffen D. A CalDAG-GEFI/Rap1/B-Raf cassette couples M(1) muscarinic acetylcholine receptors to the activation of ERK1/2. *J Biol Chem.* 2001;276:25568–25581.
31. Gerber DJ, Sotnikova TD, Gainetdinov RR, Huang SY, Caron MG, Tonegawa S. Hyperactivity, elevated dopaminergic transmission, and response to amphetamine in M1 muscarinic acetylcholine receptor-deficient mice. *Proc Natl Acad Sci USA.* 2001;98(26):15312–15317.
32. Hedreen JC, Folstein SE. Early loss of neostriatal striosome neurons in Huntington's disease. *J Neuropathol Exp Neurol.* 1995;54:105–120.
33. Tippett LJ, Waldvogel HJ, Thomas SJ, et al. Striosomes and mood dysfunction in Huntington's disease. *Brain.* 2007;130:206–221.
34. Goto S, Lee LV, Munoz EL, et al. Functional anatomy of the basal ganglia in X-linked recessive dystonia-parkinsonism. *Ann Neurol.* 2005;58(1):7–17.
35. Moratalla R, Quinn B, DeLanney LE, Irwin I, Langston JW, Graybiel AM. Differential vulnerability of primate caudate-putamen and striosome-matrix dopamine systems to the neurotoxic effects of 1-methyl-4-phenyl-1,2,3,6-tetrahydropyridine. *Proc Natl Acad Sci USA.* 1992;89:3859–3863.
36. Sato K, Sumi-Ichinose C, Kaji R, et al. Differential involvement of striosome-matrix dopamine systems in transgenic model for dopa-responsive dystonia: a predominant loss of striosomal dopaminergic inputs. *PNAS.* 2008;105:12551–12556.
37. Doya K. Metalearning and neuromodulation. *Neural Netw.* 2002;15(4–6):495–506.
38. Houk JC, Adams JL, Barto AG. A model of how the basal ganglia generate and use neural signals that predict reinforcement. In: Houk J, Davis J, Beiser D, eds. *Models of Information Processing in the Basal Ganglia.* Cambridge, MA: MIT Press; 1995:249–270.
39. White NM, Hiroi N. Preferential localization of self-stimulation sites in striosomes/patches in the rat striatum. *Proc Natl Acad Sci USA.* 1998;95(11):6486–6491.
40. Graybiel AM. Habits, rituals and the evaluative brain. *Annu Rev Neurosci.* 2008;31:359–387.
41. Matsumoto M, Hikosaka O. Lateral habenula as a source of negative reward signals in dopamine neurons. *Nature.* 2007;447:1111–1115.

7.2 | Dopamine and Synaptic Plasticity in Mesolimbic Circuits

F. WOODWARD HOPF, ANTONELLO BONCI, AND ROBERT C. MALENKA

INTRODUCTION

The mesolimbic system consists of the ventral tegmental area (VTA), a major source of dopamine (DA) for limbic structures, and the nucleus accumbens (NAcb, also termed the *ventral striatum*), which is a major target of VTA DA. This system is generally considered a limbic-motor interface in which motivationally relevant stimuli are able to influence initiation of behavior.[1–6] The NAcb is composed of two major subregions, the core and shell, with the NAcb core implicated in appetitive learning and cued control of behavior, and the NAcb shell implicated in processing of primary rewards as well as novelty. In addition, the NAcb is likely formed by multiple cell populations analogous to the direct and indirect pathways of the dorsal striatum, which control activation and inhibition of movement, respectively.[7]

The VTA and NAcb receive extensive glutamatergic inputs from the prefrontal cortex (PFC) and other brain areas, and these excitatory inputs have been considered critical for establishing and expressing addictive and other motivated behaviors.[1–6] Thus, many studies using glutamate receptor antagonists or GABA receptor agonists suggest that NAcb inactivation prevents the expression of a variety of motivated and goal-directed behaviors.[3,8–11] In addition, DA receptor signaling through D1-type (D1R, D1 or D5) and/or D2-type (D2R, D2, D3 or D4) receptors is required for a wide range of functions of the NAcb.[3,9,12–15]

This chapter will review our current understanding of how DA might modulate glutamatergic synaptic plasticity in mesolimbic brain regions. This topic will be examined in the context of in vitro brain slice experiments and plasticity induction in the anesthetized animal. We will also discuss the possibility that DA modulation of glutamatergic signaling could occur in the awake animal and contribute to the expression of motivated behavior.

SYNAPTIC PLASTICITY IN THE MESOLIMBIC SYSTEM: GENERAL CONCEPTS

Several forms of synaptic plasticity have been identified in the dorsal and ventral striatum and VTA using the in vitro brain slice model. As detailed below, many studies have found that high-frequency stimulation (HFS, e.g., 100 Hz for 1 s) leads to long-term depression (LTD) of evoked AMPA receptor (AMPAR) currents[16,17] (for dorsal striatum, see Chapter 7.3 in this volume). However, there is some diversity within the underlying mechanisms reported to contribute to LTD induction. This may depend in part on the frequency of stimulation and other details of the LTD induction protocol. In addition, some studies have observed a HFS- and N-methyl-D-aspartate (NMDA) receptor (NMDAR)–dependent long-term potentiation (LTP) in the NAcb and VTA.[18–23] Activation of NMDARs and the subsequent increase in postsynaptic calcium are required for LTP in many nonmesolimbic brain areas, such as the hippocampus,[24] suggesting that there could be mechanistic similarities between LTP induction mechanisms across brain regions.

Another important theme related to synaptic plasticity is that changes in AMPAR signaling can be associated with differential trafficking and cell surface expression of different AMPAR subunits.[24,25] Studies of excitatory synaptic transmission in many brain regions support the idea that most synaptic AMPARs under control conditions contain the specific AMPAR subunit GluR2 that forms heteromeric receptors with either GluR1 or GluR3 (i.e., GluR1/2 or GluR2/3 receptors). In contrast, there are few GluR2-lacking AMPARs (i.e., GluR1/1 or GluR1/3 AMPARs, which we term *GluR1 type*.[26] One exception may be in VTA DA neurons[17] (but see[20]). GluR1-type AMPARs have greater single-channel conductance than AMPARs containing GluR2 and are permeable to calcium, perhaps facilitating future calcium-dependent signaling events.[24] Interestingly, some forms of LTP in the VTA and other brain regions are associated with increased surface expression of GluR1-type AMPARs lacking GluR2.[17,20,24,27,28] Such studies have been greatly aided by biochemical cross-linking methods that only affect receptors on the cell surface, allowing precise determination of surface expression of particular GluR subunits. Also useful have been AMPAR subunit–selective peptide antagonists that allow delineation of the relative contribution of GluR1 and GluR2 in vitro and in vivo.

It is also important to note that repeated electrical stimulation in a brain slice can release factors other than glutamate, such as acetylcholine or DA.[29,30] Thus, repeated stimulation in the brain slice is not likely to be identical to strong phasic glutamatergic excitation in the intact, behaving animal. Nonetheless, the brain slice preparation represents an immensely valuable approach for investigating the detailed molecular mechanisms that contribute to synaptic plasticity at excitatory synapses.

DA AND SYNAPTIC PLASTICITY IN THE VTA

Excitatory synapses on VTA DA neurons exhibit both LTP[31,32] and LTD.[26,33] Ventral tegmental area LTP requires NMDARs and postsynaptic calcium,[21,23,31,32,34–36] similar to LTP in other brain areas.[24] Several groups have reported that LTD can be generated in VTA DA neurons but, as with LTD in other brain regions,[24] the mechanisms underlying LTD may differ, depending on the induction protocol. Long-term depression can be triggered by activation of voltage-dependent calcium channels and does not require NMDAR activation,[26,33] but it can also be triggered by activation of metabotropic glutamate receptors (mGluRs).[17,27,28] Both of these forms of LTD appear to involve a decrease in cell surface GluR1-containing AMPARs.[17,27,28,37] Finally, DA receptor inhibition does not block induction of VTA LTD,[26] although increased DA signaling through D2Rs suppresses LTD.[24,26,37] Thus, DA is not necessary for VTA LTD or LTP, but it can modulate LTD induction.

DA AND SYNAPTIC PLASTICITY IN THE NAcb

Many studies have examined plasticity at excitatory synapses on NAcb medium spiny neurons in vitro after repeated stimulation of glutamatergic afferents and multiple forms of LTD and LTP have been identified. Several groups have found that NAcb LTD is not modulated by DA receptor activation.[18,26,38] These findings are in contrast with those in the dorsal striatum, where LTD induction requires DA receptors (Chapter 7.3 in this volume). However, LTD induction in both dorsal striatum and NAcb involves mGluRs, postsynaptic calcium increases, and endocannabinoids, although there are likely additional mechanisms through which mGluRs can induce LTD.[39–41]

In some studies, HFS has been shown to generate LTP in the NAcb, and this LTP requires NMDARs and postsynaptic calcium,[18,42–44] like LTP in many other brain areas.[24] However, mixed results have been observed for regulation of NAcb LTP by DA receptors, with reports of no regulation by DA receptors,[18] a requirement for DA receptors,[22,45] or inhibition of LTP induction by DA receptors.[19] In the anesthetized, intact animal, one study in the NAcb[45] reports a complex pattern of modulation of hippocampal and cortical inputs by HFS, D1Rs, and D2Rs, highlighting the importance of these receptors in fine tuning the contribution of limbic and cortical inputs to goal-directed behaviors. It is also interesting that induction of striatal LTP in the intact animal can be achieved with several different induction procedures[46,47] that are perhaps more likely to produce LTD in the brain slice.

Thus, there are some consistent findings in studies of synaptic plasticity in the NAcb in vitro, but some mixed results as well. There can be many reasons for such discrepancies, such as differences in the recorded cell population, differences in the stimulation procedure used to induce LTD or LTP, and discrepancies in other methodological details such as the age and species of animal utilized. The recent development of sophisticated molecular approaches to visually identify specific subgroups of medium spiny neurons in the NAcb[48] may help resolve such discrepancies.

SYNAPTIC PLASTICITY AFTER DRUG EXPOSURE: GENERAL CONCEPTS

In addition to direct, acute effects of DA on synaptic plasticity in mesolimbic circuits, DA might also influence excitatory synaptic transmission indirectly by supporting behaviors related to drug exposure or associative learning.[49] In this context, a number of recent studies have examined the impact of repeated in vivo exposure to drugs of abuse on excitatory synaptic signaling in the NAcb and VTA. There is great interest in understanding how neuroadaptations at glutamatergic synapses could develop during repeated drug exposure and persist across long periods of abstinence, since these long-lasting changes may facilitate the expression and persistence of addictive behaviors.

Many studies in rodents have focused on one of two models of drug exposure: (1) behavioral sensitization or (2) self-administration and reinstatement. Behavioral sensitization is traditionally defined as an increase in the behavioral response to a drug after the first exposure to the drug. Behavioral sensitization can be very long-lasting and can increase subsequent drug self-administration; thus, it is considered a behavioral indicator of enhanced drug seeking during abstinence.[2,50] Drug-related sensitization has been observed in humans as well, and can contribute to enhancement of psychoses with repeated psychostimulant exposure.[51,52] However, human drug intake is typically active and voluntary, and associative learning between drug taking and reinforcing or negative consequences may be a critical component in the development of addiction.[3,4,50,53–57] Thus, self-administration and reinstatement protocols are thought to mimic more closely many aspects of human drug addiction, and they represent an extremely valuable model for examining the ability of different forms of stimuli to drive relapse to drug seeking.

It is also important to note that neuroadaptations during abstinence from drug exposure can occur across different time frames.[6] Synaptic plasticity can occur early during abstinence but disappear within a few days.[58] Other changes are apparent 1 day after self-administration and last for months.[57] In addition, neuroadaptations that are not present shortly after self-administration can develop across the first weeks of abstinence, a so-called incubation effect.[6,59–61]

Finally, repeated drug exposure likely activates a number of signaling systems, leading to multiple secondary homeostatic changes.[24,41,61] This underscores the importance of determining which of the observed neuroadaptations might represent the critical mediators of increased drug seeking after drug exposure.

SYNAPTIC PLASTICITY IN THE VTA INDUCED BY EXPOSURE TO DRUGS OF ABUSE

A number of studies have examined how in vivo administration of drugs of abuse might alter excitatory synaptic signaling in the VTA. Remarkably, a single exposure to a wide variety of abused drugs (e.g., cocaine, morphine, nicotine) but not nonabused drugs (e.g., fluoxetine) leads to an increase in AMPAR-mediated synaptic responses in VTA DA cells, a modification that appears to share mechanisms with LTP at these same synapses.[17,27,28,34,35,62,63] This LTP-like increase in AMPAR signaling in the VTA lasts 5 but not 10 days after single cocaine administration[34] and requires NMDARs, D1Rs, and orexin receptor activity, since blocking any of these receptors inhibits both the expression of behavioral sensitization and the cocaine-triggered potentiation of AMPARs.[20,34,35,64,65]

Additional studies of the mechanisms underlying the increase in AMPAR-mediated synaptic responses induced by cocaine reported that they occlude spike-timing-dependent LTP[20,36] and are associated with an increase in the proportion of GluR1-containing, GluR2-lacking AMPARs.[17,20,27,28] Interestingly, cocaine-induced LTP can be reversed by mGluR-dependent LTD, which appears to lead to a replacement of GluR2-lacking AMPARs with GluR2-containing ones.[17,27,28]

How does cocaine administration lead to LTP? Recently, it was found that application of cocaine directly to the VTA brain slice leads to potentiation of AMPAR-mediated synaptic responses through activation of D1Rs,[20] which enhances NMDAR-mediated synaptic currents in VTA DA cells within minutes of acute cocaine exposure.[20,65] Such potentiation of NMDARs by cocaine is short-lasting, disappearing within 3 hr.[20] Consistent with these findings, an increase in NMDAR subunit expression in the VTA has been observed 1 hr after cocaine exposure.[66] Thus, a single dose of cocaine produces an early, short-lasting potentiation of NMDARs, which is replaced by a long-lasting LTP-like enhancement of AMPAR-mediated synaptic currents that emerges as early as 3 hr after the acute cocaine exposure.[20]

Additional studies have examined the potentiation of AMPAR-mediated synaptic responses in VTA DA neurons after repeated rather than single administration of psychostimulants. Surprisingly, repeated passive cocaine injection leads to increased AMPAR-mediated synaptic responses in VTA DA cells lasting 5 but not 10 days,[6,64] similar to the time course after a single cocaine exposure.[34] Biochemical analysis of AMPAR subunits also generally supports a shorter-term increase in AMPAR signaling in VTA DA cells after passive or active exposure to psychostimulants or other drugs,[58,67–72] with changes also being reported in NMDARs but not in mGluRs.[72] These results are in agreement with those of early studies showing that glutamate-induced firing of VTA DA cells in intact, anesthetized animals is greater after sensitization.[73,74]

Synaptic plasticity in VTA DA neurons has also been studied in the context of operant responding for cocaine. In stark contrast to the consequences of passive cocaine administration, cocaine self-administration enhanced AMPAR-mediated responses in VTA DA neurons for at least 3 months of abstinence.[57] Interestingly, self-administration of natural rewards increased AMPAR signaling for only 1 week,[23,57] suggesting that learning in relation to natural rewards has much shorter-lasting effects on VTA function than does drug self-administration. Chen and colleagues[57] also observed that several patterns of passive cocaine exposure through an i.v. catheter did not alter AMPAR signaling in VTA DA cells. This suggests not only that repeated cocaine exposure per se does not affect VTA AMPARs, but also that the increased AMPAR signaling in VTA DA cells after repeated cocaine injection[64] must involve some aspect of the animal's experience of being handled and injected, perhaps stress or handling-related cues.[75]

Despite all these studies on drug-induced synaptic modifications in VTA DA neurons, the exact behavioral relevance of the drug-induced LTP in these cells remains unclear. For example, the cocaine-induced synaptic modification in VTA DA neurons is absent in knockout mice lacking GluR1, yet behavioral sensitization appears normal.[76] Furthermore, the level of sensitization exhibited behaviorally by an animal does not correlate with the increase in AMPAR signaling in that animal.[64] On the other hand, Kim et al.[77] showed that a single cocaine injection promotes subsequent conditioned place preference (CPP) to morphine, and that this facilitatory effect of cocaine is present only during the first 5 days, at a time when the cocaine-dependent LTP is expressed.[34] Also consistent with a role of the cocaine-triggered potentiation of AMPARs in VTA DA cells in promoting CPP are the findings that blockade of NMDARs in the VTA during cocaine exposure prevents facilitation of CPP[77] and that CPP in response to cocaine is diminished or absent in GluR1 knockout mice.[76] Thus, cocaine-induced increases in AMPAR signaling in VTA DA cells may promote certain forms of learning associated with the drug experience. It is also appropriate to note that increasing GluR1 in the rostral VTA via viral overexpression has been reported to enhance morphine reward, while increasing GluR1 in the caudal VTA leads to aversion to morphine.[78,79] Thus, future experiments should determine whether increased AMPAR function in vitro after drug exposure may vary across different regions of the VTA. In addition, there is evidence that synaptic plasticity in VTA DA neurons may play a role early in the learning of reward-related behaviors,[23] providing further evidence that increases in VTA AMPAR signaling can modulate a variety of behaviors related to reward and motivation.

We should also note that there is recent evidence that individual VTA DA neurons project to different single target regions such as the PFC or NAcb.[80,81] Most studies of VTA neurons do not distinguish between mesolimbic and nonmesolimbic VTA DA neurons; therefore, our understanding of the relationship between experience-dependent synaptic plasticity in these cells and their specific projection targets is incomplete. However, even with this caveat, many studies have shown that a majority of VTA DA neurons exhibit a given plastic change (e.g., an increase in AMPAR signaling in vitro or in vivo[20,23,34,36,57]), raising the possibility that both mesolimbic and mesocortical VTA DA neurons undergo experience-dependent plasticity after exposure to drugs of abuse.

Finally, drug exposure can result in other important forms of synaptic plasticity in the VTA, for example by affecting GABAergic signaling.[82–84] While a comprehensive discussion of these other forms of synaptic plasticity goes beyond the scope of this chapter, it is crucial to incorporate all these forms of plasticity into a unitary model in order to gain a proper understanding of the role of VTA neurons in modulating addictive behaviors after exposure to drugs of abuse.

SYNAPTIC PLASTICITY IN THE NAcb INDUCED BY EXPOSURE TO DRUGS OF ABUSE

Unlike the VTA, a single in vivo cocaine exposure does not alter AMPAR-mediated synaptic transmission in NAcb medium spiny neurons.[6,62,85] However, the ability to induce endocannabinoid-mediated LTD in the NAcb is abolished after a single exposure to tetrahydrocannabinol (THC) or cocaine, likely due to decreased surface expression of mGluR5.[39–41,86] Interestingly, after chronic THC administration, LTD can now be induced in the NAcb, but through an mGluR2/3-dependent mechanism different from that normally recruited in the NAcb.[41]

In contrast to the modest effects of a single exposure to drugs of abuse on synaptic function in the NAcb, repeated passive or active drug exposure can potently modulate excitatory synaptic transmission in this brain region. This has been examined primarily after repeated exposure to psychostimulants such as cocaine and amphetamine. A number of lines of evidence suggest that AMPAR-mediated synaptic signaling is reduced during early withdrawal from repeated drug exposure.[6] Although biochemical studies of NAcb GluR levels during early withdrawal have produced mixed results,[58,60,70,87,88] electrophysiological studies in vitro show reduced AMPAR-mediated synaptic currents during early withdrawal.[44,89] Furthermore, the reduced AMPAR levels associated with LTD can allow a greater magnitude of LTP,[24,90] and NAcb LTP induction is enhanced during early withdrawal.[43] There are also several changes in NAcb ion channels after repeated drug exposure, leading to greatly decreased intrinsic excitability.[91] This reduction in AMPAR signaling and intrinsic excitability may explain the decreased AMPA-mediated NAcb excitation in anesthetized animals after sensitization.[74]

Repeated drug exposure can also result in decreases in NAcb glutamatergic signaling that are long-lasting during abstinence. After cocaine self-administration but not yoked cocaine exposure, there is a long-lasting disruption of LTD induction in NAcb medium spiny neurons,[55] while sensitization is associated with disrupted induction of synaptic plasticity in hippocampus inputs to the NAcb.[45] Also, sensitization leads to increased D1R inhibition of glutamate release,[92] although DA receptor inhibition of NAcb LTP, which could be mediated by presynaptic effects of DA on glutamate release, is lost after sensitization.[19] Together, these studies suggest that there can be short- and long-lasting inhibitory neuroadaptations in the NAcb during abstinence following repeated drug exposure.

In contrast to these results, other studies report an increase in NAcb AMPAR-mediated synaptic signaling after longer withdrawal periods following repeated administration of drugs of abuse. Biochemical studies generally find increased GluR cell surface expression and/or total GluR levels in the NAcb after sensitization or self-administration.[58,60,70,87,88,93,94] Furthermore, analyses of cell surface AMPARs show increases in both GluR1 and GluR2 at the cell surface following drug administration protocols that elicit sensitization.[60] This finding agrees with those of electrophysiological studies in vitro showing increased AMPAR-mediated synaptic currents in the NAcb after sensitization but no change in the relative levels of GluR1-type and GluR2-containing surface AMPARs.[89] In contrast, after drug self-administration and withdrawal, there are increased cell surface levels of GluR1-type AMPAR subunits,[61,95] a finding confirmed by in vitro electrophysiological studies.[61] Thus, there is some consensus that weeks of abstinence after either passive or active drug exposure lead to increased AMPAR signaling in NAcb medium spiny neurons, although perhaps through different cellular mechanisms.

In addition to postsynaptic increases in AMPAR-mediated synaptic responses, repeated drug administration has been reported to alter the regulation of glutamate release in the NAcb. These findings include decreased D2R and mGluR2/3 inhibition of glutamate release and increased PFC excitability, leading to greatly enhanced NAcb glutamate concentrations during drug exposure and reinstatement.[72,91,96] In addition, reduced glial uptake of glutamate after drug exposure decreases basal NAcb glutamate levels, and normalizing resting glutamate levels reduces reinstatement.[97] Although this may seem paradoxical given other evidence for increased NAcb glutamatergic activity after repeated drug exposure, it has been suggested that reduced basal glutamate signaling and reduced intrinsic excitability in the NAcb may be responsible for homeostatic secondary increases in AMPAR signaling and glutamate release.[59,91]

Of potentially great behavioral and clinical relevance are recent reports that excitatory synaptic transmission in the NAcb can be acutely and dynamically regulated by single cocaine exposure during abstinence weeks following repeated cocaine administration. Cocaine reexposure during abstinence switches the increase in AMPAR-mediated synaptic responses seen after sensitization protocols to a decrease that may share mechanisms with LTD.[85,89] These in vitro observations of enhanced and reduced AMPAR signaling before and after acute cocaine reexposure during abstinence are strongly validated by studies of locomotor activity elicited by intra-NAcb AMPA infusions[15] and by biochemical studies of AMPAR expression.[98] Thus, the exact details of an animal's experience could strongly and rapidly impact the type of synaptic plasticity observed at excitatory synapses in the NAcb. In this context, evidence for reduced AMPAR signaling during early withdrawal from repeated drug exposure may simply reflect the consequence of recent drug exposure.

Along with altered AMPAR signaling, recent work has shown the importance of reductions in Homer proteins, scaffolding proteins that bind mGluRs and NMDARs, as critical neuroadaptations that can drive cocaine seeking.[72] Repeated cocaine administration and abstinence are associated with reduced NAcb protein levels of Homer isoforms and group I mGluRs (mGluR1 and mGluR5). Furthermore, activation of group I mGluRs can increase NAcb glutamate levels and produce locomotor activation, and these effects are blunted during abstinence from cocaine.

Changes in NAcb AMPAR signaling have also been observed in relation to drugs other than psychostimulants.[72] For example, repeated morphine administration decreases surface AMPARs[99] and prevents LTD induction in the NAcb.[16] These findings can be viewed as consistent with an observed decrease in NAcb dendritic spine density following morphine administration,[100] a finding that contrasts with the increase in spine density observed following repeated administration of psychostimulants.[101–103] However, a decrease in presynaptic markers after repeated psychostimulant administration has also been observed.[104]

Given the often bewildering array of neuroadaptations that can occur in the NAcb during and following exposure to drugs of abuse, it is of primary importance to understand the behavioral relevance of the observed changes in NAcb AMPAR signaling. One approach has used AMPAR-subunit-selective peptides to examine the contribution of synaptic plasticity in the NAcb to drug-related motivation. Infusion of agents selective for GluR1-type AMPARs into the NAcb has been reported to significantly reduce reinstatement driven by cocaine-related cues[61] or a priming dose of cocaine.[95] Since cocaine self-administration and abstinence are associated with increased GluR1-type AMPAR signaling, these studies strongly suggest that increased NAcb GluR1-type AMPARs are causally related to the motivation to seek drugs. Consistent with this hypothesis, in vivo electrophysiology studies in behaving animals find that longer periods of abstinence are associated with an increased number of NAcb neurons firing in response to cocaine-predictive cues.[105]

Other studies also suggest that altered glutamatergic signaling in the NAcb after repeated administration of drugs of abuse and abstinence can significantly influence the expression of behaviors associated with the drug experience. Increased levels of surface AMPARs could explain the increased ability of AMPA infusion into the NAcb to enhance locomotion[8] or induce reinstatement[106] after sensitization. In contrast, a peptide that prevents the expression of one form of LTD in the NAcb in vitro disrupts the expression of behavioral sensitization when injected into the NAcb, suggesting that this form of NAcb LTD is necessary for this drug-induced behavioral adaptation.[107] Furthermore, strong increases in GluR1 in the NAcb by viral overexpression can actually reduce reward processing,[108,109] while increased NAcb GluR2 can increase reward processing.[108,110]

Taken together, these results suggest that the regulation of basic reward processing and sensitization by neuroadaptations in the NAcb are complex. However, it is clear that increased AMPAR signaling in the NAcb after self-administration facilitates reinstatement,[61,95] while reversal of cocaine-dependent neuroadaptations that affect glutamate release in the NAcb can also prevent cocaine reinstatement.[91]

IS EXCITATION OR INHIBITION OF NAcb CELL ACTIVITY REQUIRED FOR BEHAVIORAL RESPONDING OR REWARD PROCESSING?

The studies reviewed above show that repeated drug exposure and abstinence are associated with both excitatory and inhibitory neuroadaptations in the NAcb, any of which could contribute to pathological drug seeking during abstinence. Indeed, there has been some controversy about whether activation or inhibition of NAcb neurons reflects encoding of rewards. For example, it has been proposed that inhibition of NAcb cell activity, in particular a D2R inhibition of indirect pathway NAcb neurons, encodes basic reward processing.[109,111] In agreement with this proposal, NAcb firing in vivo is primarily inhibited during reward consumption,[13,112] NAcb inhibition with opioids enhances consumption of highly palatable foods,[113] and rewards are primarily encoded by decreases in NAcb firing, while aversive stimuli result in increased NAcb firing.[114] Furthermore, reduced firing in a subset of NAcb neurons may enable cue-driven behavioral responding for a reward.[115] These findings can be viewed as consistent with the observation that NAcb LTD is required for the expression of behavioral sensitization.[107]

In contrast, there is also a literature supporting the hypothesis that increased NAcb GluR levels and increased excitation of NAcb cells by synaptic inputs mediate a pathological motivation for drug rewards. A possible resolution for the discrepancies in the literature is that some NAcb neurons encode a motor-activating signal, while other neurons encode a motor-inhibiting signal.[1,5,116,117] This idea is similar to the proposed role of direct and indirect pathways in the dorsal striatum in motor activation and inhibition, respectively.[7]

The presence of multiple information channels in the NAcb could explain some apparent paradoxes in the behavioral literature.[117] For example, inhibition of DA receptors in the NAcb reduces responding for a reward-predicting cue, while strong NAcb inactivation leads to hyperactivity[11,118] but does not prevent responding to reward-related cues.[13,117] A similar pattern is observed for cocaine-induced reinstatement, where block of DA receptors in the NAcb shell inhibits reinstatement[14] but strong inactivation of the NAcb shell does not.[3] Thus, the NAcb may contain a DA-dependent behavioral excitatory signal, in concert with a DA-independent behavioral inhibitory signal (e.g., decreased NAcb firing during reward consumption, which may suppress alternate exploratory behavior). In this case, a DA receptor antagonist would only interfere with the excitatory signal, while strong NAcb inactivation would block both the excitatory and inhibitory outputs. Without the strong inhibitory signal, the rest of the brain may have access to the information necessary to perform cue-directed behaviors, but the animal makes more errors and exhibits general hyperactivity.

There has also been controversy about whether DA excites or inhibits NAcb firing,[119,120] although there is a consensus that DA is primarily modulatory, requiring ongoing activity to influence neuronal excitability. Inhibition of NAcb DA signaling greatly reduces both cue-directed behavioral responding and NAcb firing in response to reward-related cues.[13] Furthermore, NAcb neurons can show either phasic increases or decreases in firing during cue presentation, and both firing patterns disappear during VTA inactivation. Finally, other NAcb neurons fire during aspects of the behavioral task not related to the reward-predicting cues (e.g., the decreased firing during reward consumption), and these firing patterns are not affected by VTA inactivation. Thus, this study suggests that NAcb DA may influence only certain aspects of responding for a reward, and that DA can both excite and inhibit different NAcb neurons during facilitation of cue-induced responding.

Taken together, these results suggest that both excitation and inhibition in the NAcb could play an important role in encoding of rewards, reward-related cues, and behavioral activation in relation to these rewards and cues. However, the exact contribution of excitation and inhibition may depend on the specific behavior and on the time of the observation (e.g., acquisition versus expression).

CONCLUSION

The studies briefly reviewed in this chapter show that DA can acutely modulate several forms of synaptic plasticity in the mesolimbic system. This may be particularly important for the modulation of excitatory synaptic transmission during and following exposure to drugs of abuse. In addition to providing information about the critical neural adaptations underlying reward-associated learning and maladaptive drug seeking, the types of studies reviewed here will likely produce a variety of novel targets for therapeutic drugs aimed at improving the treatment of substance abuse and addiction.

REFERENCES

1. Mogenson GJ, Jones DL, Yim CY. From motivation to action: functional interface between the limbic system and the motor system. *Prog Neurobiol.* 1980;14:69–97.
2. Robinson TE, Berridge KC. The neural basis of drug craving: an incentive-sensitization theory of addiction. *Brain Res Brain Res Rev.* 1993;18:247–291.
3. Kalivas PW, McFarland K. Brain circuitry and the reinstatement of cocaine-seeking behavior. *Psychopharmacology.* 2003;168:44–56.
4. Everitt BJ, Robbins TW. Neural systems of reinforcement for drug addiction: from actions to habits to compulsion. *Nat Neurosci.* 2005;8:1481–1489.
5. Meredith GE, Baldo BA, Andrzejewski ME, Kelley AE. The structural basis for mapping behavior onto the ventral striatum and its subdivisions. *Brain Struct Funct.* 2008;213:17–27.
6. Thomas MJ, Kalivas PW, Shaham Y. Neuroplasticity in the mesolimbic dopamine system and cocaine addiction. *Br J Pharmacol.* 2008;154:327–342.
7. Gerfen CR. Basal ganglia. In: Paxinos G, ed. *The Rat Nervous System.* 3rd ed. San Diego, CA: Elsevier Academic Press; 2004:455–508.
8. Pierce RC, Bell K, Duffy P, Kalivas PW. Repeated cocaine augments excitatory amino acid transmission in the nucleus accumbens only in rats having developed behavioral sensitization. *J Neurosci.* 1996;16:1550–1560.
9. Di Ciano P, Cardinal RN, Cowell RA, Little SJ, Everitt BJ. Differential involvement of NMDA, AMPA/kainate, and dopamine receptors in the nucleus accumbens core in the acquisition and performance of pavlovian approach behavior. *J Neurosci.* 2001;21:9471–9477.
10. Parkinson JA, Olmstead MC, Burns LH, Robbins TW, Everitt BJ. Dissociation in effects of lesions of the nucleus accumbens core and shell on appetitive pavlovian approach behavior and the potentiation of conditioned reinforcement and locomotor activity by D-amphetamine. *J Neurosci.* 1999;19:2401–2411.
11. See RE, Elliott JC, Feltenstein MW. The role of dorsal vs. ventral striatal pathways in cocaine-seeking behavior after prolonged abstinence in rats. *Psychopharmacology.* 2007;194:321–331.
12. Parkinson JA, Dalley JW, Cardinal RN, Bamford A, Fehnert B, Lachenal G, Rudarakanchana N, Halkerston KM, Robbins TW, Everitt BJ. Nucleus accumbens dopamine depletion impairs both acquisition and performance of appetitive Pavlovian approach behaviour: implications for mesoaccumbens dopamine function. *Behav Brain Res.* 2002;137:149–163.
13. Yun IA, Wakabayashi KT, Fields HL, Nicola SM. The ventral tegmental area is required for the behavioral and nucleus accumbens neuronal firing responses to incentive cues. *J Neurosci.* 2004;24:2923–2933.
14. Schmidt HD, Pierce RC. Cooperative activation of D1-like and D2-like dopamine receptors in the nucleus accumbens shell is required for the reinstatement of cocaine-seeking behavior in the rat. *Neuroscience.* 2006;142:451–461.
15. Bachtell RK, Self DW. Renewed cocaine exposure produces transient alterations in nucleus accumbens AMPA receptor-mediated behavior. *J Neurosci.* 2008;28:12808–12814.
16. Robbe D, Bockaert J, Manzoni OJ. Metabotropic glutamate receptor 2/3-dependent long-term depression in the nucleus accumbens is blocked in morphine withdrawn mice. *Eur J Neurosci.* 2002;16:2231–2235.
17. Bellone C, Lüscher C. mGluRs induce a long-term depression in the ventral tegmental area that involves a switch of the subunit composition of AMPA receptors. *Eur J Neurosci.* 2005;21:1280–1288.
18. Pennartz CM, Ameerun RF, Groenewegen HJ, Lopes da Silva FH. Synaptic plasticity in an in vitro slice preparation of the rat nucleus accumbens. *Eur J Neurosci.* 1993;5:107–117.
19. Li Y, Kauer JA. Repeated exposure to amphetamine disrupts dopaminergic modulation of excitatory synaptic plasticity and neurotransmission in nucleus accumbens. *Synapse.* 2004;51:1–10.
20. Argilli E, Sibley DR, Malenka RC, England PM, Bonci A. Mechanism and time course of cocaine-induced long-term potentiation in the ventral tegmental area. *J Neurosci.* 2008;28:9092–9100.
21. Nugent FS, Hwong AR, Udaka Y, Kauer JA. High-frequency afferent stimulation induces long-term potentiation of field potentials in the ventral tegmental area. *Neuropsychopharmacology.* 2008;33:1704–1712.

22. Schotanus SM, Chergui K. Dopamine D1 receptors and group I metabotropic glutamate receptors contribute to the induction of long-term potentiation in the nucleus accumbens. *Neuropharmacology*. 2008;54:837–844.
23. Stuber GD, Klanker M, de Ridder B, Bowers MS, Joosten RN, Feenstra MG, Bonci A. Reward-predictive cues enhance excitatory synaptic strength onto midbrain dopamine neurons. *Science*. 2008;321:1690–1692.
24. Kauer JA, Malenka RC. Synaptic plasticity and addiction. *Nat Rev Neurosci*. 2007;8:844–858.
25. Lüscher C, Xia H, Beattie EC, Carroll RC, von Zastrow M, Malenka RC, Nicoll RA. Role of AMPA receptor cycling in synaptic transmission and plasticity. *Neuron*. 1999;24:649–658.
26. Thomas MJ, Malenka RC, Bonci A. Modulation of long-term depression by dopamine in the mesolimbic system. *J Neurosci*. 2000;20:5581–5586.
27. Bellone C, Lüscher C. Cocaine triggered AMPA receptor redistribution is reversed in vivo by mGluR-dependent long-term depression. *Nat Neurosci*. 2006;9:636–641.
28. Mameli M, Balland B, Lujan R, Lüscher C. Rapid synthesis and synaptic insertion of GluR2 for mGluR-LTD in the ventral tegmental area. *Science*. 2007;317:530–533.
29. Partridge JG, Apparsundaram S, Gerhardt GA, Ronesi J, Lovinger DM. Nicotinic acetylcholine receptors interact with dopamine in induction of striatal long-term depression. *J Neurosci*. 2002;22:2541–2549.
30. Benoit-Marand M, O'Donnell P. Cortico-accumbens fiber stimulation does not induce dopamine release in the nucleus accumbens in vitro. *Brain Struct Funct*. 2008;213:177–182.
31. Bonci A, Malenka RC. Properties and plasticity of excitatory synapses on dopaminergic and GABAergic cells in the ventral tegmental area. *J Neurosci*. 1999;19:3723–3730.
32. Overton PG, Richards CD, Berry MS, Clark D. Long-term potentiation at excitatory amino acid synapses on midbrain dopamine neurons. *Neuroreport*. 1999;10:221–226.
33. Jones S, Kornblum JL, Kauer JA. Amphetamine blocks long-term synaptic depression in the ventral tegmental area. *J Neurosci*. 2000;20:5575–5580.
34. Ungless MA, Whistler JL, Malenka RC, Bonci A. Single cocaine exposure in vivo induces long-term potentiation in dopamine neurons. *Nature*. 2001;411:583–587.
35. Saal D, Dong Y, Bonci A, Malenka RC. Drugs of abuse and stress trigger a common synaptic adaptation in dopamine neurons. *Neuron*. 2003;37:577–582.
36. Luu P, Malenka RC. Spike timing-dependent long-term potentiation in ventral tegmental area dopamine cells requires PKC. *J Neurophysiol*. 2008;100:533–538.
37. Gutlerner JL, Penick EC, Snyder EM, Kauer JA. Novel protein kinase A-dependent long-term depression of excitatory synapses. *Neuron*. 2002;36:921–931.
38. Manzoni O, Michel JM, Bockaert J. Metabotropic glutamate receptors in the rat nucleus accumbens. *Eur J Neurosci*. 1997;9:1514–1523.
39. Fourgeaud L, Mato S, Bouchet D, Hemar A, Worley PF, Manzoni OJ. A single in vivo exposure to cocaine abolishes endocannabinoid-mediated long-term depression in the nucleus accumbens. *J Neurosci*. 2004;24:6939–6945.
40. Mato S, Chevaleyre V, Robbe D, Pazos A, Castillo PE, Manzoni OJ. A single in-vivo exposure to delta 9THC blocks endocannabinoid-mediated synaptic plasticity. *Nat Neurosci*. 2004;7:585–586.
41. Mato S, Robbe D, Puente N, Grandes P, Manzoni OJ. Presynaptic homeostatic plasticity rescues long-term depression after chronic Delta 9-tetrahydrocannabinol exposure. *J Neurosci*. 2005;25:11619–11627.
42. Kombian SB, Malenka RC. Simultaneous LTP of non-NMDA- and LTD of NMDA-receptor-mediated responses in the nucleus accumbens. *Nature*. 1994;368:242–246.
43. Yao WD, Gainetdinov RR, Arbuckle MI, Sotnikova TD, Cyr M, Beaulieu JM, Torres GE, Grant SG, Caron MG. Identification of PSD-95 as a regulator of dopamine-mediated synaptic and behavioral plasticity. *Neuron*. 2004;41:625–638.
44. Schramm-Sapyta NL, Olsen CM, Winder DG. Cocaine self-administration reduces excitatory responses in the mouse nucleus accumbens shell. *Neuropsychopharmacology*. 2006;31:1444–1451.
45. Goto Y, Grace AA. Dopamine-dependent interactions between limbic and prefrontal cortical plasticity in the nucleus accumbens: disruption by cocaine sensitization. *Neuron*. 2005;47:255–266.
46. Charpier S, Deniau JM. In vivo activity-dependent plasticity at cortico–striatal connections: evidence for physiological long-term potentiation. *Proc Natl Acad Sci USA*. 1997;94:7036–7040.
47. Charpier S, Mahon S, Deniau JM. In vivo induction of striatal long-term potentiation by low-frequency stimulation of the cerebral cortex. *Neuroscience*. 1999;91:1209–1222.
48. Gong S, Zheng C, Doughty ML, Losos K, Didkovsky N, Schambra UB, Nowak NJ, Joyner A, Leblanc G, Hatten ME, Heintz N. A gene expression atlas of the central nervous system based on bacterial artificial chromosomes. *Nature*. 2003;425:917–25.
49. Di Chiara G. Nucleus accumbens shell and core dopamine: differential role in behavior and addiction. *Behav Brain Res*. 2002;137:75–114.
50. Zernig G, Ahmed SH, Cardinal RN, Morgan D, Acquas E, Foltin RW, Vezina P, Negus SS, Crespo JA, Stockl P, Grubinger P, Madlung E, Haring C, Kurz M, Saria A. Explaining the escalation of drug use in substance dependence: models and appropriate animal laboratory tests. *Pharmacology*. 2007;80:65–119.
51. Pierce RC, Kalivas PW. A circuitry model of the expression of behavioral sensitization to amphetamine-like psychostimulants. *Brain Res Brain Res Rev*. 1997;25:192–216.
52. Bonci A, Singh V. Dopamine dysregulation syndrome in Parkinson's disease patients: from reward to penalty. *Ann Neurol*. 2006;59:852–858.
53. Larimer ME, Palmer RS, Marlatt GA. Relapse prevention. An overview of Marlatt's cognitive-behavioral model. *Alcohol Res Health*. 1999;23:151–160.
54. Epstein DH, Preston KL, Stewart J, Shaham Y. Toward a model of drug relapse: an assessment of the validity of the reinstatement procedure. *Psychopharmacology*. 2006;189:1–16.
55. Martin M, Chen BT, Hopf FW, Bowers MS, Bonci A. Cocaine self-administration selectively abolishes LTD in the core of the nucleus accumbens. *Nat Neurosci*. 2006;9:868–869.
56. Sanchis-Segura C, Spanagel R. Behavioural assessment of drug reinforcement and addictive features in rodents: an overview. *Addict Biol*. 2006;11:2–38.
57. Chen BT, Bowers MS, Martin M, Hopf FW, Guillory AM, Carelli RM, Chou JK, Bonci A. Cocaine but not natural reward self-administration nor passive cocaine infusion produces persistent LTP in the VTA. *Neuron*. 2008;59:288–297.
58. Churchill L, Swanson CJ, Urbina M, Kalivas PW. Repeated cocaine alters glutamate receptor subunit levels in the nucleus accumbens and ventral tegmental area of rats that develop behavioral sensitization. *J Neurochem*. 1999;72:2397–2403.

59. Bossert JM, Ghitza UE, Lu L, Epstein DH, Shaham Y. Neurobiology of relapse to heroin and cocaine seeking: an update and clinical implications. *Eur J Pharmacol.* 2005;526: 36–50.
60. Boudreau AC, Wolf ME. Behavioral sensitization to cocaine is associated with increased AMPA receptor surface expression in the nucleus accumbens. *J Neurosci.* 2005;25:9144–9151.
61. Conrad KL, Tseng KY, Uejima JL, Reimers JM, Heng LJ, Shaham Y, Marinelli M, Wolf ME. Formation of accumbens GluR2-lacking AMPA receptors mediates incubation of cocaine craving. *Nature.* 2008;454:118–121.
62. Grignaschi G, Burbassi S, Zennaro E, Bendotti C, Cervo L. A single high dose of cocaine induces behavioural sensitization and modifies mRNA encoding GluR1 and GAP-43 in rats. *Eur J Neurosci.* 2004;20:2833–2837.
63. Wanat MJ, Sparta DR, Hopf FW, Bowers MS, Melis M, Bonci A. Strain specific synaptic modifications on VTA dopamine neurons after ethanol exposure. *Biol Psychiatry.* 2008;65:646–653.
64. Borgland SL, Taha SA, Sarti F, Fields HL, Bonci A. Orexin A in the VTA is critical for the induction of synaptic plasticity and behavioral sensitization to cocaine. *Neuron.* 2006;49:589–601.
65. Schilstrom B, Yaka R, Argilli E, Suvarna N, Schumann J, Chen BT, Carman M, Singh V, Mailliard WS, Ron D, Bonci A. Cocaine enhances NMDA receptor-mediated currents in ventral tegmental area cells via dopamine D5 receptor-dependent redistribution of NMDA receptors. *J Neurosci.* 2006;26:8549–8558.
66. Schumann J, Michaeli A, Yaka R. Src-protein tyrosine kinases are required for cocaine-induced increase in the expression and function of the NMDA receptor in the VTA. *J Neurochem.* 2008;461:159–162.
67. Fitzgerald LW, Ortiz J, Hamedani AG, Nestler EJ. Drugs of abuse and stress increase the expression of GluR1 and NMDAR1 glutamate receptor subunits in the rat ventral tegmental area: common adaptations among cross-sensitizing agents. *J Neurosci.* 1996;16:274–282.
68. Ferrari R, Le Novère N, Picciotto MR, Changeux JP, Zoli M. Acute and long-term changes in the mesolimbic dopamine pathway after systemic or local single nicotine injections. *Eur J Neurosci.* 2002; 15:1810–1818.
69. >Lu W, Monteggia LM, Wolf ME. Repeated administration of amphetamine or cocaine does not alter AMPA receptor subunit expression in the rat midbrain. *Neuropsychopharmacology.* 2002;26:1–13.
70. Lu L, Grimm JW, Shaham Y, Hope BT. Molecular neuroadaptations in the accumbens and ventral tegmental area during the first 90 days of forced abstinence from cocaine self-administration in rats. *J Neurochem.* 2003;85:1604–1613.
71. Tang W, Wesley M, Freeman WM, Liang B, Hemby SE. Alterations in ionotropic glutamate receptor subunits during binge cocaine self-administration and withdrawal in rats. *J Neurochem.* 2004;89:1021–1033.
72. Szumlinski KK, Ary AW, Lominac KD. Homers regulate drug-induced neuroplasticity: implications for addiction. *Biochem Pharmacol.* 2008;75:112–133.
73. Henry DJ, Greene MA, White FJ. Electrophysiological effects of cocaine in the mesoaccumbens dopamine system: repeated administration. *J Pharmacol Exp Ther.* 1989;251:833–839.
74. White FJ, Hu XT, Zhang XF, Wolf ME. Repeated administration of cocaine or amphetamine alters neuronal responses to glutamate in the mesoaccumbens dopamine system. *J Pharmacol Exp Ther.* 1995;273:445–454.
75. Vezina P, Leyton M. Conditioned cues and the expression of stimulant sensitization in animals and humans. *Neuropharmacology.* 2008;56:160–168.
76. Dong Y, Saal D, Thomas M, Faust R, Bonci A, Robinson T, Malenka RC. Cocaine-induced potentiation of synaptic strength in dopamine neurons: behavioral correlates in GluRA(-/-) mice. *Proc Natl Acad Sci USA.* 2004;101:14282–14287.
77. Kim JA, Pollak KA, Hjelmstad GO, Fields HL. A single cocaine exposure enhances both opioid reward and aversion through a ventral tegmental area-dependent mechanism. *Proc Natl Acad Sci USA.* 2004;101:5664–5669.
78. Carlezon WA Jr, Boundy VA, Haile CN, Lane SB, Kalb RG, Neve RL, Nestler EJ. Sensitization to morphine induced by viral-mediated gene transfer. *Science.* 1997;277:812–814.
79. Carlezon WA Jr, Haile CN, Coppersmith R, Hayashi Y, Malinow R, Neve RL, Nestler EJ. Distinct sites of opiate reward and aversion within the midbrain identified using a herpes simplex virus vector expressing GluR1. *J Neurosci.* 2000;20:RC62.
80. Margolis EB, Lock H, Hjelmstad GO, Fields HL. The ventral tegmental area revisited: is there an electrophysiological marker for dopaminergic neurons? *J Physiol.* 2006;577:907–924.
81. Lammel S, Hetzel A, Hackel O, Jones I, Liss B, Roeper J. Unique properties of mesoprefrontal neurons within a dual mesocorticolimbic dopamine system. *Neuron.* 2008;57:760–773.
82. Liu QS, Pu L, Poo MM. Repeated cocaine exposure in vivo facilitates LTP induction in midbrain dopamine neurons. *Nature.* 2005;437:1027–1031.
83. Nugent FS, Kauer JA. LTP of GABAergic synapses in the ventral tegmental area and beyond. *J Physiol.* 2008;586:1487–1493.
84. Pan B, Hillard CJ, Liu QS. Endocannabinoid signaling mediates cocaine-induced inhibitory synaptic plasticity in midbrain dopamine neurons. *J Neurosci.* 2008;28:1385–1397.
85. Thomas MJ, Beurrier C, Bonci A, Malenka RC. Long-term depression in the nucleus accumbens: a neural correlate of behavioral sensitization to cocaine. *Nat Neurosci.* 2001;4:1217–1223.
86. Hoffman AF, Oz M, Caulder T, Lupica CR. Functional tolerance and blockade of long-term depression at synapses in the nucleus accumbens after chronic cannabinoid exposure. *J Neurosci.* 2003;23:4815–4820.
87. Lu W, Wolf ME. Repeated amphetamine administration alters AMPA receptor subunit expression in rat nucleus accumbens and medial prefrontal cortex. *Synapse.* 1999;32:119–131.
88. Hemby SE, Tang W, Muly EC, Kuhar MJ, Howell L, Mash DC. Cocaine-induced alterations in nucleus accumbens ionotropic glutamate receptor subunits in human and non-human primates. *J Neurochem.* 2005;95:1785–1793.
89. Kourrich S, Rothwell PE, Klug JR, Thomas MJ. Cocaine experience controls bidirectional synaptic plasticity in the nucleus accumbens. *J Neurosci.* 2007;27:7921–7928.
90. Fino E, Glowinski J, Venance L. Bidirectional activity-dependent plasticity at corticostriatal synapses. *J Neurosci.* 2005;25: 11279–11287.
91. Kalivas PW, Hu XT. Exciting inhibition in psychostimulant addiction. *Trends Neurosci.* 2006;29:610–616.
92. Beurrier C, Malenka RC. Enhanced inhibition of synaptic transmission by dopamine in the nucleus accumbens during behavioral sensitization to cocaine. *J Neurosci.* 2002;22:5817–5822.
93. Zhang X, Lee TH, Davidson C, Lazarus C, Wetsel WC, Ellinwood EH. Reversal of cocaine-induced behavioral sensitization and associated phosphorylation of the NR2B and GluR1 subunits of the NMDA and AMPA receptors. *Neuropsychopharmacology.* 2007;32:377–387.
94. Sutton MA, Schmidt EF, Choi KH, Schad CA, Whisler K, Simmons D, Karanian DA, Monteggia LM, Neve RL, Self DW.

Extinction-induced upregulation in AMPA receptors reduces cocaine-seeking behaviour. *Nature*. 2003;421:70–75.
95. Anderson SM, Famous KR, Sadri-Vakili G, Kumaresan V, Schmidt HD, Bass CE, Terwilliger EF, Cha JH, Pierce RC. CaMKII: a biochemical bridge linking accumbens dopamine and glutamate systems in cocaine seeking. *Nat Neurosci*. 2008;11:344–353.
96. Brady AM, O'Donnell P. Dopaminergic modulation of prefrontal cortical input to nucleus accumbens neurons in vivo. *J Neurosci*. 2004;24:1040–1049.
97. Baker DA, McFarland K, Lake RW, Shen H, Tang XC, Toda S, Kalivas PW. Neuroadaptations in cystine-glutamate exchange underlie cocaine relapse. *Nat Neurosci*. 2003;6:743–749.
98. Boudreau AC, Reimers JM, Milovanovic M, Wolf ME. Cell surface AMPA receptors in the rat nucleus accumbens increase during cocaine withdrawal but internalize after cocaine challenge in association with altered activation of mitogen-activated protein kinases. *J Neurosci*. 2007;27:10621–10635.
99. Glass MJ, Lane DA, Colago EE, Chan J, Schlussman SD, Zhou Y, Kreek MJ, Pickel VM. Chronic administration of morphine is associated with a decrease in surface AMPA GluR1 receptor subunit in dopamine D1 receptor expressing neurons in the shell and non-D1 receptor expressing neurons in the core of the rat nucleus accumbens. *Exp Neurol*. 2008;210:750–761.
100. Robinson TE, Gorny G, Savage VR, Kolb B. Widespread but regionally specific effects of experimenter- versus self-administered morphine on dendritic spines in the nucleus accumbens, hippocampus, and neocortex of adult rats. *Synapse*. 2002;46:271–279.
101. Li Y, Acerbo MJ, Robinson TE. The induction of behavioural sensitization is associated with cocaine-induced structural plasticity in the core (but not shell) of the nucleus accumbens. *Eur J Neurosci*. 2004;20:1647–1654.
102. Lee KW, Kim Y, Kim AM, Helmin K, Nairn AC, Greengard P. Cocaine-induced dendritic spine formation in D1 and D2 dopamine receptor-containing medium spiny neurons in nucleus accumbens. *Proc Natl Acad Sci USA*. 2006;103:3399–3404.
103. Toda S, Shen HW, Peters J, Cagle S, Kalivas PW. Cocaine increases actin cycling: effects in the reinstatement model of drug seeking. *J Neurosci*. 2006;26:1579–1587.
104. Subramaniam S, Marcotte ER, Srivastava LK. Differential changes in synaptic terminal protein expression between nucleus accumbens core and shell in the amphetamine-sensitized rat. *Brain Res*. 2001;901:175–183.
105. Hollander JA, Carelli RM. Abstinence from cocaine self-administration heightens neural encoding of goal-directed behaviors in the accumbens. *Neuropsychopharmacology*. 2005;30:1464–1474.
106. Suto N, Tanabe LM, Austin JD, Creekmore E, Pham CT, Vezina P. Previous exposure to psychostimulants enhances the reinstatement of cocaine seeking by nucleus accumbens AMPA. *Neuropsychopharmacology*. 2004;29:2149–2159.
107. Brebner K, Wong TP, Liu L, Liu Y, Campsall P, Gray S, Phelps L, Phillips AG, Wang YT. Nucleus accumbens long-term depression and the expression of behavioral sensitization. *Science*. 2005;310:1340–1343.
108. Todtenkopf MS, Parsegian A, Naydenov A, Neve RL, Konradi C, Carlezon WA Jr. Brain reward regulated by AMPA receptor subunits in nucleus accumbens shell. *J Neurosci*. 2006;26:11665–11669.
109. Bachtell RK, Choi KH, Simmons DL, Falcon E, Monteggia LM, Neve RL, Self DW. Role of GluR1 expression in nucleus accumbens neurons in cocaine sensitization and cocaine-seeking behavior. *Eur J Neurosci*. 2008;27:2229–2240.
110. Kelz MB, Chen J, Carlezon WA Jr, Whisler K, Gilden L, Beckmann AM, Steffen C, Zhang YJ, Marotti L, Self DW, Tkatch T, Baranauskas G, Surmeier DJ, Neve, RL, Duman RS, Picciotto MR, Nestler EJ. Expression of the transcription factor deltaFosB in the brain controls sensitivity to cocaine. *Nature*. 1999;401:272–276.
111. Carlezon WA Jr, Thomas MJ. Biological substrates of reward and aversion: a nucleus accumbens activity hypothesis. *Neuropharmacology*. 2008;56:122–132.
112. Taha SA, Fields HL. Encoding of palatability and appetitive behaviors by distinct neuronal populations in the nucleus accumbens. *J Neurosci*. 2005;25:1193–1202.
113. Will MJ, Pratt WE, Kelley AE. Pharmacological characterization of high-fat feeding induced by opioid stimulation of the ventral striatum. *Physiol Behav*. 2006;89:226–234.
114. Roitman MF, Wheeler RA, Carelli RM. Nucleus accumbens neurons are innately tuned for rewarding and aversive taste stimuli, encode their predictors, and are linked to motor output. *Neuron*. 2005;45:587–597.
115. Taha SA, Fields HL. Inhibitions of nucleus accumbens neurons encode a gating signal for reward-directed behavior. *J Neurosci*. 2006;26:217–222.
116. Maurice N, Deniau JM, Menetrey A, Glowinski J, Thierry AM. Prefrontal cortex–basal ganglia circuits in the rat: involvement of ventral pallidum and subthalamic nucleus. *Synapse*. 1998;29:363–370.
117. Yun IA, Nicola SM, Fields HL. Contrasting effects of dopamine and glutamate receptor antagonist injection in the nucleus accumbens suggest a neural mechanism underlying cue-evoked goal-directed behavior. *Eur J Neurosci*. 2004;20:249–263.
118. Zahm DS. Functional-anatomical implications of the nucleus accumbens core and shell subterritories. *Ann NY Acad Sci*. 1999;877:113–128.
119. Gonon F, Sundstrom L. Excitatory effects of dopamine released by impulse flow in the rat nucleus accumbens in vivo. *Neuroscience*. 1996;75:13–18.
120. Nicola SM, Surmeier J Malenka RC. Dopaminergic modulation of neuronal excitability in the striatum and nucleus accumbens. *Annu Rev Neurosci*. 2000;23:185–215.

7.3 Dopaminergic Modulation of Striatal Glutamatergic Signaling in Health and Parkinson's Disease

D. JAMES SURMEIER, MICHELLE DAY, TRACY S. GERTLER, C. SAVIO CHAN, AND WEIXING SHEN

INTRODUCTION

Dopamine (DA) has long been known to be a critical modulator of striatal processing of cortical and thalamic signals carried by glutamatergic synapses on the principal neurons of the striatum—medium spiny neurons (MSNs). Dopamine regulation of these neurons is important for an array of psychomotor functions ascribed to the basal ganglia, including associative learning and action selection.[1–3] In spite of its significance, an understanding of the physiological principles underlying MSN regulation has developed slowly. One of the major obstacles to unraveling the DA puzzle in the striatum has been the lack of homogeneity in the MSN class; there are at least two major subsets of MSN that differ in their expression of DA receptors.[4,5] Furthermore, both cell types are embedded in a rich neuronal network involving both MSNs and interneurons that is modulated by DA. This has made it extremely difficult to sort out the direct and indirect effects of DA on network properties. The recent development of mouse lines in which neurons "report" their expression of D1 or D2 receptors by coexpressing enhanced green fluorescent protein (EGFP) has made it largely possible to overcome this obstacle.

Another impediment is that DA receptors are primarily found in dendrites that are inaccessible with electrodes (the principal tool of electrophysiologists), making direct study of their actions on glutamatergic signaling and dendritic excitability difficult. Optical techniques, like two photon laser scanning microscopy (2PLSM), are making these regions more accessible. While these approaches are still in their infancy, they are providing fundamental new insights into the dendritic physiology of MSNs and how DA modulates these regions.

This review focuses on four topics: (1) the intrinsic differences between MSNs expressing D1 and D2 dopamine receptors, (2) how DA modulates postsynaptic properties that influence glutamatergic synaptic events and their integration by MSNs in the dorsal striatum, (3) how DA influences the induction of long-term synaptic plasticity, and (4) how DA depletion in Parkinson's disease (PD) models remodels glutamatergic signaling. Only MSNs in the dorsal striatum will be considered. Even with this rather narrow focus, it is impossible to faithfully summarize what has become an enormous literature in the last decade. The reader is referred to several other recent reviews.[6–8] Moreover, there is a rich literature characterizing the impact of glutamate on dopaminergic neurons and DA release that won't be touched.[9,10]

THE DICHOTOMY BETWEEN D1 AND D2 MSNs

Medium spiny neurons have long been thought to be homogeneous in their somatodendritic morphology and physiology. However, recent studies using D1 and D2 BAC transgenic mice have revealed that D1 MSNs were less excitable than D2 MSNs over a broad range of developmental time points.[11] A straightforward explanation for the dichotomy in excitability of D1 and D2 MSNs is that they differ in surface area. To test this hypothesis, D1 and D2 MSNs were identified by epifluorescence in slices from BAC mice and then patched with electrodes containing biocytin.[12] After filling, slices were processed and recorded MSNs were reconstructed, preserving as much three-dimensional architecture as possible (Fig. 7.3.1a,b). A GABAergic interneuron is included for comparison (Fig. 7.3.1c). Dendritic length and branching pattern were measured in a population of D1 and D2 MSNs. A three-dimensional Sholl analysis was performed to determine the number of dendritic processes in concentric shells centered on the soma (Fig. 7.3.1d,e). D1 MSNs had more intersections than D2 MSNs located 10–135µm from the soma. From the Sholl analysis, the cumulative dendritic length within spheres of increasing diameter was measured and averaged to determine where branching diverged.

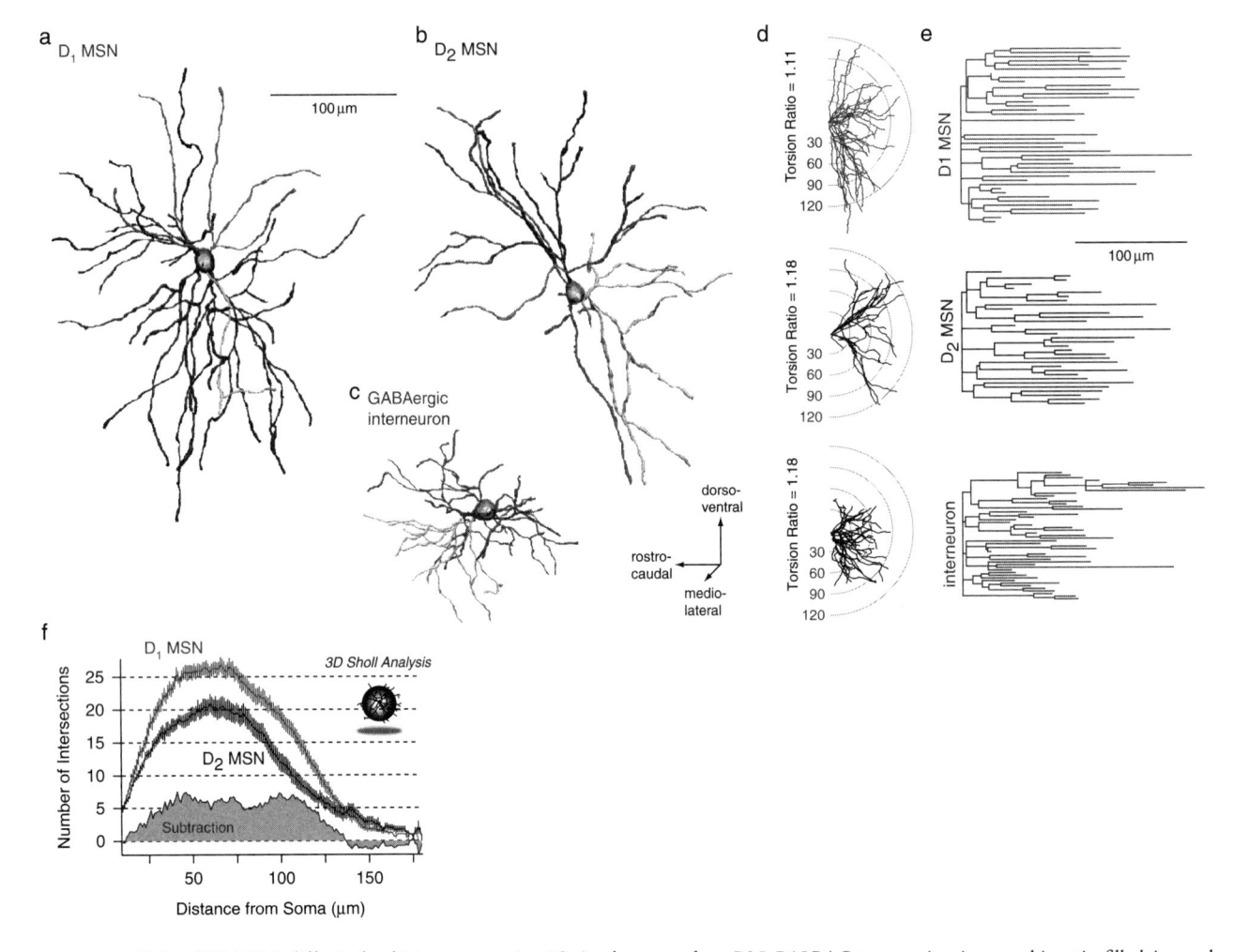

FIGURE 7.3.1. D1 and D2 MSNs differ in dendritic anatomy. (a–c) Striatal neurons from P35–P45 BAC transgenic mice were biocytin-filled, imaged, and reconstructed in three dimensions. A GABAergic interneuron is included for comparison. (d) Fan-in diagrams displayed no apparent preferred orientation in either the D1 or D2 MSN populations. (e) Dendrograms displaying in two dimensions the length, number, and connectivity of dendritic segments in sample neurons. (f) Three-dimensional Sholl analysis of biocytin-filled and reconstructed neurons from P35–P45 BAC transgenic mice. Data are shown as the mean (+/− SEM) number of intersections at 1-μm eccentricities from the soma for 15 D1 and 16 D2 MSNs. D1 MSNs have a more highly branched dendritic tree, as indicated by the increased number of intersections and positive subtracted area (gray shading). (See Color Plate 7.3.1.)

Approximately 25 μm from the soma, the difference in cumulative dendritic length reached ~20% and remained constant (Fig. 7.3.1f). Total dendritic length was positively correlated with whole-cell capacitance, confirming the expected relationship between the electrical and anatomical measurements.

The difference in total dendritic length was attributable to a difference in the number of primary dendrites, as the mean tree length (i.e., total dendritic length/ number of primary dendrites) was similar in the two types of MSNs. D1 MSNs had significantly more branch points and tips, but this was due to their having more primary dendrites. The mean number and length of dendritic segments as a function of branch order was not significantly different between D1 and D2 MSNs. A convex hull analysis was used to estimate the three-dimensional space occupied by dendritic trees (this algorithm takes into account the three-dimensional space occupied by a set of dendritic processes, allowing for a more complex polygonal surface rendering than assuming a cubic or spherical distribution). D1 MSNs occupied significantly more space, though there was no significant difference in the span of the dendritic trees from D1 and D2 MSNs. Taken together, the anatomical analyses showed that, on average, D1 MSNs have more primary dendrites than D2 MSNs.

A basic question is whether this difference in dendritic anatomy depends upon intrinsic (cell autonomous) or extrinsic (environmental) factors. A simple way to begin to examine this question is to see if the differences can be recapitulated in a simple system, such as a two-dimensional, dissociated corticostriatal culture where the normal striatal environment and the topography of cortical connections with MSNs have been disrupted. Medium spiny neurons in these cultures develop a relatively normal dendritic morphology, including spines.[13] Medium spiny neurons cultured from P0 D2 BAC mouse striata and wild-type cerebral cortices were maintained for 3 weeks in vitro. Cultures were then fixed; D2 MSNs were identified by eGFP expression, and D1 MSNs were identified by immunoreactivity for D1 receptors. Although the average branching pattern of D1 and D2 MSNs differed from that seen in vivo, the total dendritic length was significantly greater in D1 MSNs, as found in vivo.

Given the differences in the dendritic anatomy of D1 and D2 MSNs, it's natural to ask whether there are physiological parallels. To answer this question, D1 and D2 MSNs were identified visually in tissue slices from BAC transgenic mice and 2PLSM was used to monitor changes in dendritic Ca^{2+} concentration evoked by somatically generated, backpropagating action potentials (bAPs).[14] In agreement with previous studies,[15,16] bAPs evoked reliable Ca^{2+} signals in both shafts and spines in the proximal dendritic tree. The dichotomy between D1 and D2 MSNs became apparent only when the more distal (i.e., more than 60 μm from the soma) dendrites were examined. We found that in D1 MSNs, single bAPs frequently failed to evoke a detectable Ca^{2+} transient at distal dendritic sites (Fig. 7.3.2a), whereas in D2 MSNs, dendritic Ca^{2+} transients were readily detected at this distance and beyond (Fig. 7.3.2b).

To examine more closely the disparity in the bAP-evoked Ca^{2+} transients between the two populations of MSNs, bAP-evoked Ca^{2+} transients from each cell type were scanned at varying distances from the soma (Fig. 7.3.2c). Here, the amplitudes of bAP-evoked Ca^{2+} transients from each scan point in each cell type were normalized to the most proximal location scanned and then plotted as a function of distance from the soma. These findings show differences in somatodendritic excitability between MSNs, with the D2 MSNs showing less attenuation of bAP-evoked Ca^{2+} transients in distal spines and dendrites than D1 MSNs. To test the possibility that the loss in bAP response was attributable to declining dendritic Ca^{2+} channel density, D1 MSNs were loaded with Cs^+ (to improve voltage control of distal dendrites) and the somatic membrane was briefly stepped to a depolarized potential. In this situation, there was no detectable attenuation of the Ca^{2+} transient with distance from the soma (Fig. 7.3.2d), arguing that the loss of the bAP-evoked Ca^{2+} transient was not due to diminished Ca^{2+} channel density. Further evidence that this phenomenon does not simply reflect the loss of Ca^{2+} channels in distal dendrites is that strong depolarization (1 s) and trains of action potentials (10X 10 Hz) consistently evoked Ca^{2+} transients in the distal process of all MSNs tested.

Although single bAPs were not propagated efficiently into the distal dendrites of D1 MSNs, bursts of somatic action potentials were able to evoke Ca^{2+} transients in more distant dendritic regions. Three spike bursts (50 Hz) delivered at a theta frequency reliably evoked shaft and spine Ca^{2+} transients in both D1 and D2 MSN dendrites 100–120 μm from the soma (Fig. 7.3.2e). The Ca^{2+} signals evoked by successive bursts summed in a sublinear fashion (Fig. 7.3.2e,f). This sublinearity was more pronounced in D1 MSNs than in D2 MSNs (Fig. 7.3.2f). Moreover, consistent with the response to single bAPs, the relative elevation in Ca^{2+} evoked by somatically generated theta bursts was smaller in amplitude and area in D1 MSNs (Fig. 7.3.2f).

The differences between striatonigral and striatopallidal MSNs in their anatomy and excitability are not coincidental in our view. As described below, the functional linkages of the DA receptors these MSNs express counterbalance the intrinsic properties of MSNs.

DA MODULATION OF INTRINSIC EXCITABILITY

The now classical model of how DA shapes striatal activity was advanced almost two decades ago by Albin, Young, and Penny.[3] In this model, D1 receptors excite MSNs of the *direct* striatonigral pathway, whereas D2 receptors inhibit MSNs of the *indirect* striatopallidal pathway. These effects were envisioned as acute and readily reversible. The evidence for this model stemmed almost entirely from indirect measures of neuronal activity (e.g., alterations in gene expression, glucose utilization, or receptor binding). Subsequent work has proven to be largely consistent with the general principles of this model, revealing that DA activation of G protein–coupled receptors (GPCRs) *excites* or *inhibits* MSNs by modulating the gating and trafficking of voltage-dependent and ligand-gated (ionotropic) ion channels, essentially altering cellular excitability.

D1 receptors expressed by striatonigral MSNs are positively coupled to adenylyl cyclase through G_{olf}.[17] Elevation of cytosolic cyclic adenosine monophosphate

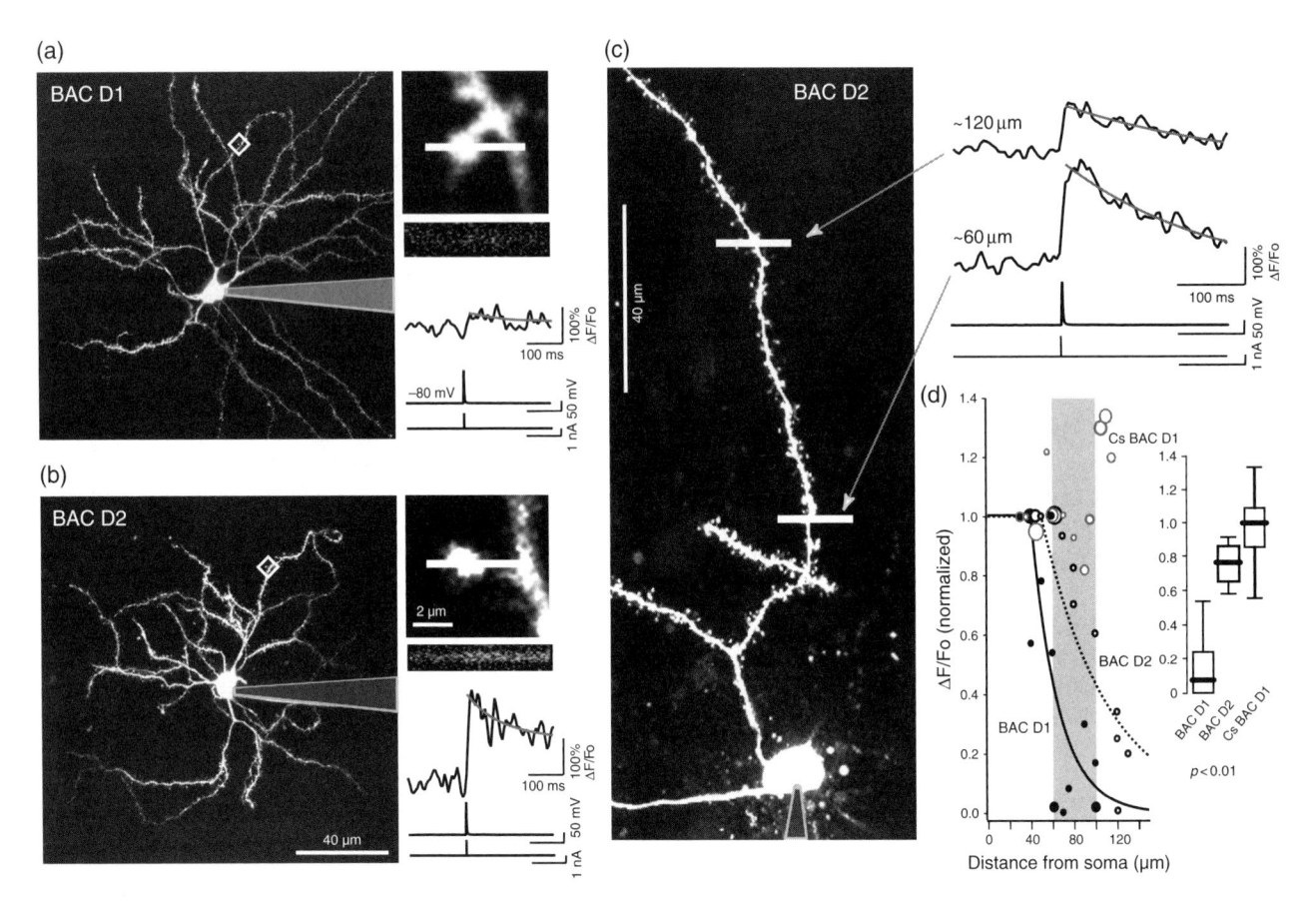

FIGURE 7.3.2. BAP-evoked Ca^{2+} transients are readily detected in the distal dendrites and spines of the D2 population of MSNs. (a, b) 2PLSM images of MSNs in 275-μm-thick corticostriatal slices from (a) a BAC D1 and (b) a BAC D2 mouse. Neurons were visualized with Alexa Fluor 568 (50 μM) by filling through the patch pipette (patch pipettes are grayed out for presentation). Maximum projection images of the somas and dendritic fields (left panels a and b) and high-magnification projections of dendrite segments from the regions outlined by the boxes are shown (top right panels, a and b). BAP-evoked Ca^{2+} transients were detected by line scanning through the spine in the region indicated by the line. Fluorescence traces were generated from the pseudocolor image (lower panels, a and b) by calculating $\Delta F/F_o$ (top black trace). The fluorescence image, $\Delta F/F_o$ trace, action potential (middle trace), and current pulse (bottom trace) are shown in temporal registration. (c) Maximum projection image of a soma and a dendritic branch from a D2 MSN. Line scans were acquired at two eccentricities, 120 and 60 μm, as indicated by the grey arrows. (d) Graph of the change in amplitude with distance from the soma calculated by normalizing scans taken at distal points to the most proximal scan point in each MSN. The magnitude of the Ca^{2+} transients decrements more in the D1 MSNs (D1 MSNs = filled black circles; D2 MSNs = open black circles). This decrementation is not seen in MSNs loaded with Cs^+-based internals (open grey circles). The points were scaled to represent the number of cells scanned at each point (smallest points = one cell; largest points = four cells). The data, fit from the median distance of the most proximal point, show that the magnitude of the Ca^{2+} transients decrements more in the D1 MSNs (n = 11, black line) versus the D2 MSNs (n = 6, dashed line) [Kruskal-Wallis ANOVA, $p < 0.01$]. (e) Maximum projection image of the soma and dendritic field of a D2 MSN. A high-magnification image of the dendritic segment outlined in the box is shown in the inset. Scale bars in (b) apply to both images. The pseudocolor image, $\Delta F/F_o$ trace, action potential (middle trace), and current pulse (bottom trace) are shown in temporal registration. Arrows indicate the timing of current pulses delivered to initiate action potentials. (f) Average peak $\Delta F/F_o$ values after each of the five pulses constituting the theta burst bAP protocol. Values are from distal dendritic spines (100–120 μm from the soma) and normalized to the maximum peak $\Delta F/F_o$ value measured in a proximal spine (60–80 μm) of the same dendrite in response to the first burst of the same theta burst protocol. The area under the $\Delta F/F_o$ plot was calculated for each cell type in response to the entire theta burst protocol; in line with larger peak Ca^{2+} transients, the box plots to the right demonstrate significantly larger Ca^{2+} transient areas in the D2 versus the D1 MSNs [Kruskal-Wallis ANOVA, $p < 0.05$]. *Source:* Reprinted from [14]. (See Color Plate 7.3.2.)

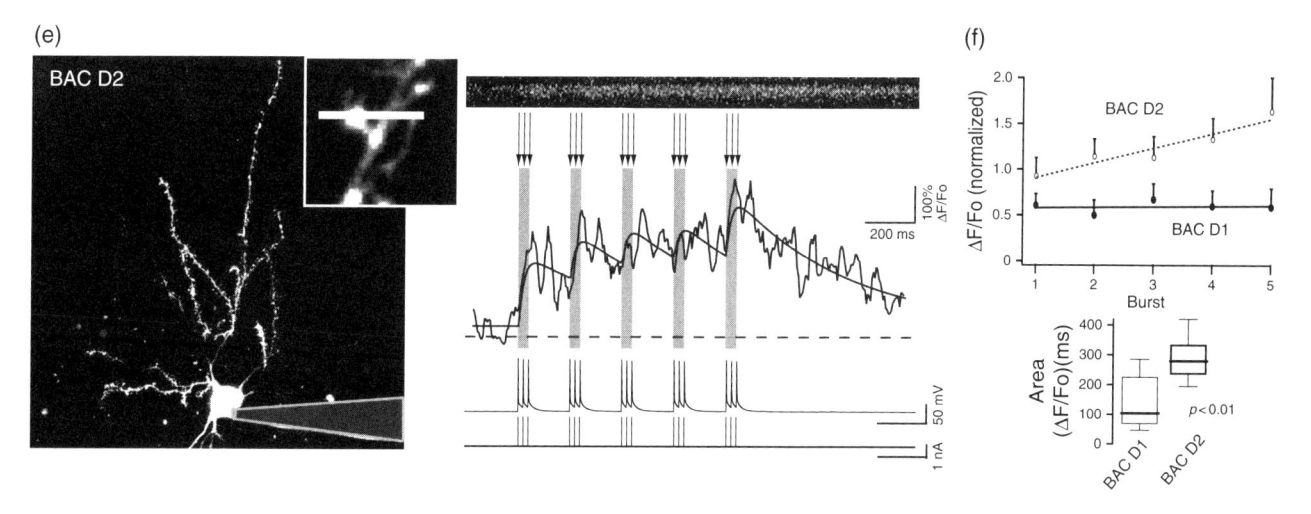

FIGURE 7.3.2. (Continued)

(cAMP) levels leads to the activation of protein kinase A (PKA) and phosphorylation of a variety of intracellular targets, like the dual-function phosphoprotein DA- and cAMP-regulated phosphoprotein, 32 kDa (DARPP-32),[18] altering cellular function. A growing number of studies suggest that the D1/PKA cascade has direct effects on AMPA and N-methyl-D-aspartate (NMDA) receptor function and trafficking. For example, D1 receptor activation of PKA enhances surface expression of both AMPA and NMDA receptors.[19–21] The precise mechanisms underlying the trafficking are still being pursued, but the tyrosine kinase Fyn and the protein phosphatase STEP (striatal-enriched phosphatase) appear to be important regulators of surface expression of glutamate receptors.[22] Trafficking and localization might also be affected by a direct interaction between D1 and NMDA receptors.[23,24]

What is less clear is whether D1 receptor stimulation has rapid effects on glutamate receptor gating. Although PKA phosphorylation of the NR1 subunit is capable of enhancing NMDA receptor currents,[25] the presence of this modulation in MSNs is controversial. In neurons where the engagement of dendritic voltage-dependent ion channels has been minimized by dialyzing the cytoplasm with cesium ions, D1 receptor agonists have little or no discernible effect on AMPA or NMDA receptor–mediated currents in dorsal striatum.[26] However, in MSNs where this has not been done, D1 receptor stimulation rapidly enhances currents evoked by NMDA receptor stimulation.[27] The difference between these results suggests that the effect of D1 receptors on NMDA receptor currents is indirect and mediated by voltage-dependent dendritic conductances that are taken out of play by blocking K$^+$ channels and clamping dendritic voltage. Indeed, blocking L-type Ca^{2+} channels, which open in the same voltage range as NMDA receptors (Mg^{2+} unblock), attenuate the D1 receptor–mediated enhancement of NMDA receptor currents.[28] The mechanisms underlying this effect are not clear, however, as direct application of D1 agonists or DA to the dendrites of D1 MSNs failed to alter bAP-evoked Ca^{2+} transients attributable at least in part to L-type Ca^{2+} channels.[14,15] It is possible that the prolonged whole cell dialysis required for these measurements disrupted the signaling machinery necessary for the modulation.

Voltage-dependent Na$^+$ channels were the first well-characterized targets of the D1 receptor signaling pathway in MSNs. Confirming inferences drawn from earlier work in tissue slices,[29] voltage clamp work showed that D1 receptor signaling led to a reduction in Na$^+$ channel availability without altering the voltage dependence of fast activation or inactivation.[30] Subsequent work has shown that PKA phosphorylation of the pore-forming subunit of the Na$^+$ channel promotes activity-dependent entry into a nonconducting, slow-inactivated state that can be reversed only by membrane hyperpolarization.[31] It is likely that the D1 receptor modulation is mediated by phosphorylation of somatic Nav1.1 channels, as Nav1.6 channels are not efficiently phosphorylated by PKA.[32] This conclusion is consistent with the apparent absence of a significant Na$^+$ channel investment of distal dendrites of MSNs.[14]

When the somatic membrane potential is held for several hundred milliseconds near the up-state (~−60 mV),[33] D1 receptor stimulation has a quite different effect than when it is held at nominal down-state potentials (~−80 mV). At this up-state membrane potential, the personality of the MSN is transformed, as the constellation of ion channels governing activity is reconfigured. Perhaps the most dramatic change is the closure or inactivation of Kir2, Kv1, and Kv4 K+ channels that oppose the depolarizing influences of glutamate receptors. In this state, D1 receptor stimulation elevates (rather than lowers) the response to intrasomatic current injection.[34] The augmented response is attributable in part to enhanced opening of L-type Ca^{2+} channels following PKA phosphorylation.[35,36] L-type channels with a pore-forming Cav1.3 subunit are likely to be major targets of this modulation; these channels have a voltage threshold near −60 mV and are anchored near glutamatergic synapses in spines through a scaffolding interaction with Shank.[37] Enhanced opening of these channels and NMDA receptors[27,38-40] accounts for the ability of D1 receptor stimulation to promote synaptically driven plateau potentials of MSNs (resembling up-states in vivo) in corticostriatal slices,[41] as in cortical pyramidal neurons.[42] D1 receptor stimulation also reduces the opening of Cav2 Ca^{2+} channels that couple to SK K+ channels,[43] potentially further augmenting dendritic electrogenesis.

Taken together, these results suggest that D1 receptor signaling through PKA elevates the responsiveness of striatonigral neurons to sustained synaptic release of glutamate generating up-states but reduces the response to transient or uncoordinated glutamate release that fails to significantly depolarize the dendritic membrane for more than a few tens of milliseconds from the down-state.

MODULATION OF INTRINSIC EXCITABILITY AND GLUTAMATERGIC SIGNALING BY D2 RECEPTORS

D2 receptors couple to G$_{i/o}$ proteins, leading to inhibition of adenylyl cyclase through Gα$_i$ subunits.[44] In parallel, released Gβγ subunits are capable of reducing Cav2 Ca^{2+} channel opening and of stimulating phospholipase Cβ isoforms, generating diacylglycerol (DAG) and protein kinase C (PKC) activation as well as inositol trisphosphate (IP$_3$) liberation and the mobilization of intracellular Ca^{2+} stores.[45,46] D2 receptors also are capable of transactivating tyrosine kinases.[47]

As with D1 receptor signaling, there are a number of studies showing that D2 receptor signaling alters glutamate receptor function in dorsal striatal MSNs. Activation of D2 receptors has been reported to decrease AMPA receptor currents of MSNs recorded in tissue slices.[27] Subsequent work using acutely isolated neurons and voltage clamp techniques supports direct action on dendritic AMPA receptors.[48] D2 receptor signaling leads to dephosphorylation of S845 of the GluR1 subunit, which should promote trafficking of AMPA receptors out of the synaptic membrane.[49] D2 receptor stimulation also diminishes presynaptic release of glutamate[50]; however, it is not clear whether this is mediated by presynaptically or postsynaptically positioned D2 receptors.[51]

Studies of voltage-dependent channels are largely consistent with the proposition that D2 receptors act to reduce the excitability of striatopallidal neurons and their response to glutamatergic synaptic input. D2 receptor–mediated mobilization of intracellular Ca^{2+} leads to negative modulation of Cav1.3 Ca^{2+} channels through a calcineurin-dependent mechanism.[37,45] D2 receptor activation also reduces the opening of voltage-dependent Na+ channels, presumably by a PKC-mediated enhancement of slow inactivation.[30] In addition, D2 receptors promote the opening of K+ channels, diminishing dendritic excitability.[14,52] This coordinated modulation of ion channels provides a mechanistic foundation for the ability of D2 receptor agonists to reduce the responsiveness of MSNs in slices at up-state membrane potentials.[45]

DOPAMINERGIC MODULATION OF LONG-TERM SYNAPTIC PLASTICITY

As mentioned above, DA is thought to exert its principal effects in dendrites where glutamatergic synapses are formed. Although it modulates short-term cellular excitability, DA's role in associative learning and action selection is commonly thought to be in the regulation of corticostriatal synaptic plasticity. The best-studied form of synaptic plasticity in the striatum is long-term depression (LTD). When postsynaptic depolarization is paired with high-frequency stimulation (HFS) of glutamatergic fibers, a long-lasting reduction in the synaptic strength of glutamatergic synapses is seen in most MSNs. Unlike LTD induced by low-frequency stimulation in the ventral striatum,[53] LTD induction in the dorsal striatum is not NMDA dependent. This form of LTD (HFS-LTD) is initiated postsynaptically but expressed through a presynaptic reduction in glutamate release. There is general agreement that striatal LTD requires activation of Cav1.3 L-type Ca^{2+} channels, G$_q$-linked mGluR1/5 receptors, and the generation of endocannabinoids (ECs). Endocannabinoids exert their effect presynaptically by acting at CB1 receptors.[54-56] There is less agreement that activation of D2 receptors is necessary for LTD induction. Activation of D2 receptors is a very potent stimulus for EC production,[57] and the ability of D2 receptors to activate PLC[45] certainly is consistent with a direct involvement in EC production.

However, attempts to test for the necessity of D2 receptor expression using D2 BAC mice have met with mixed results.[58,59] Kreitzer and Malenka[58] reported that LTD was inducible only in striatopallidal MSNs using a minimal local stimulation. However, our group and Lovinger's found that HFS-LTD was inducible in both striatonigral and striatopallidal MSNs using macroelectrode stimulation of the cortex,[59] consistent with the high probability of induction seen in previous work.[60] We have reproduced the Kreitzer-Malenka finding using minimal local stimulation, suggesting that the method of induction is important. This result underscores the difficulties inherent in stimulation paradigms that activate not just glutamatergic fibers, but also a heterogeneous population of dopaminergic, cholinergic, and interneuronal fibers that might influence the induction of plasticity. An example of how we've attempted to sort this out is given below.

One strategy for gaining better control over which fibers are activated in studies of plasticity is to develop in vitro preparations that preserve connectivity between nuclei. Consider the glutamatergic synapses formed on MSNs. Most reviews have focused almost entirely on the cortical innervation of MSNs, leaving the thalamic input as a virtual footnote. Studies using white matter or cortical stimulation of coronal brain slices typically assume that the glutamatergic fibers being stimulated are of cortical origin, but very few of these fibers are left uncut in this preparation.[61] The thalamic innervation of MSNs is similar in magnitude to that of the cerebral cortex, perhaps constituting as much as 40% of the total glutamatergic input to MSNs, terminating on both shafts and spines.[62] Anatomical studies suggest that the intralaminar nuclei target primarily striatonigral neurons in primate striatum; however, this might not be the case in rodents,[63] whereas "motor" nuclei [(ventroanterial (VA) and ventrolateral (VL) nuclei] project primarily to striatopallidal neurons.[64,65] This apparent dichotomy between motor and "associative" inputs is consistent with recent studies suggesting that the input to striatopallidal neurons comes largely from pyramidal neurons contributing to descending motor control circuits, whereas the input to striatonigral neurons comes from neurons whose axons are largely intratelencephalic.[66] Recently, several studies have shown that parahorizontal slices can preserve both cortical and thalamic connectivity, allowing each to be selectively stimulated.[67,68] However, to date, these preparations have not been used to study the rules governing the induction of plasticity at these two types of synapse.

Much less is known about the mechanisms controlling induction of long-term potentiation (LTP) than LTD. Studies in tissue slices have argued that LTP induced by HFS of corticostriatal glutamatergic inputs (HFS-LTP) depends upon coactivation of D1 and NMDA receptors.[69,70] As noted above, D1 receptor stimulation enhances NMDA receptor currents both directly and indirectly by enhancing L-type Ca^{2+} channels located nearby,[28,36] although "boosting" by L-type channels appears not to be necessary for LTP induction.[71] There was some question about the physiological relevance of LTP in MSNs, but this issue has been resolved by the demonstration that it is readily inducible in vivo.[72] The discrepancy presumably stemmed from the difficulty of depolarizing MSN dendrites enough to overcome the Mg^{2+} block of NMDA receptors with focal stimulation in a brain slice. How HFS-LTP is expressed has not been carefully examined. As with HFS-LTD, the dependence of a nominally widespread form of synaptic plasticity upon a receptor with restricted distribution is puzzling. BAC transgenic mice in which D1 and D2 receptor–expressing MSNs are labeled should be helpful in sorting this issue out.

As is apparent from the presentation thus far, there are several obstacles that have slowed progress toward a sound understanding of the dopaminergic modulation of synaptic plasticity in the striatum. Cellular heterogeneity has been the biggest of these in our view. The development of D1 and D2 receptor BAC transgenic mice has made this problem tractable. Another issue is the induction protocol. Until very recently, plasticity studies have not attempted to engage the postsynaptic membrane and dendrites in a physiological way during the induction of synaptic plasticity (e.g., Cs^+ loading cells and voltage clamping).

Why is this important? Most learning theories postulate that changes in synaptic strength reflect the precise temporal relationship between presynaptic and postsynaptic activity. Hebb's classic postulate asserts that excitatory glutamatergic synaptic activity that consistently leads to postsynaptic spiking induces a strengthening or potentiation of the active synapses. An unstated corollary is that presynaptic activity that follows postsynaptic activity (and hence cannot be causally linked to spiking) should be weakened or depressed. Dendrites are an integral part of this learning equation, forming the conduit between the axon initial segment where spikes are initiated and synaptic sites where plasticity is induced. Dopamine receptors richly invest dendrites of MSNs,[73] putting them in a position to modulate this linkage. The extended Hebbian postulate has been tested in several types of neuron by examining how the temporal relationship between presynaptic and postsynaptic spiking influences lasting changes in synaptic strength.[74–76] Spike-timing-dependent plasticity

(STDP) of this sort depends upon bAPs that serve to depolarize synaptic regions before, during, or after glutamate release. At most synapses, Hebb's postulate appears to be correct. That is, when presynaptic activity precedes postsynaptic spiking, LTP is induced, whereas reversing the order induces LTD.[77–80]

Using perforated-patch recordings (to preserve intracellular signaling mechanisms) and minimal local electrical stimulation of glutamatergic afferent fibers in tissue slices from BAC transgenic mice, we have used STDP protocols to examine the rules governing the induction of plasticity at striatonigral and striatopallidal MSN synapses.[81] These studies have revealed a set of rules that are largely consistent with those inferred from studies using conventional induction protocols (see above), but they pushed us beyond our current conceptual model by showing that DA controls the induction of Hebbian synaptic plasticity in a receptor- and cell-type-specific manner.

Specifically, D1 receptor signaling in striatonigral MSNs was necessary for the induction of Hebbian LTP, whereas D2 receptor signaling in striatopallidal MSNs was necessary for the induction of Hebbian LTD. More importantly, our studies demonstrate that DA, in concert with adenosine and glutamate, makes STDP at MSN glutamatergic synapses bidirectional and Hebbian.[81] In striatopallidal MSNs (Fig. 7.3.3a), repeated pairing of a synaptic stimulation with a postsynaptic spike later (positive timing) resulted in LTP of the synaptic response (Fig. 7.3.3b). In contrast, preceding synaptic stimulation with a short burst of postsynaptic spikes (negative timing) induced LTD (Fig. 7.3.3c). The timing-dependent LTP relies upon activation of NMDA and A2a receptors, as blocking them disrupts the potentiation of the synaptic response in striatopallidal MSNs (Fig. 7.3.3d). Like conventional LTD, timing-dependent LTD is disrupted by antagonizing mGluR5, CB1, or D2 receptors (Fig. 7.3.3d). The bidirectionality of STDP appeared to be controlled by a balanced interaction between "opponent" GPCR signaling cascades controlling the induction of LTP and LTD.[79,82,83] D2 and A2a receptor signaling cascades have long been known to oppose one another at several levels.[84,85] In the STDP paradigm, elevating D2 receptor stimulation by bath application of quinpirole resulted in a robust LTD even when postsynaptic activity followed presynaptic activity, a protocol that would normally induce LTP. In contrast, elevating A2a receptor signaling by bath application of CGS21680 restored LTP, even when presynaptic activity followed postsynaptic activity (Fig. 7.3.3d).

In striatonigral MSNs (Fig. 7.3.4a), pairing presynaptic activity with a trailing postsynaptic spike induced

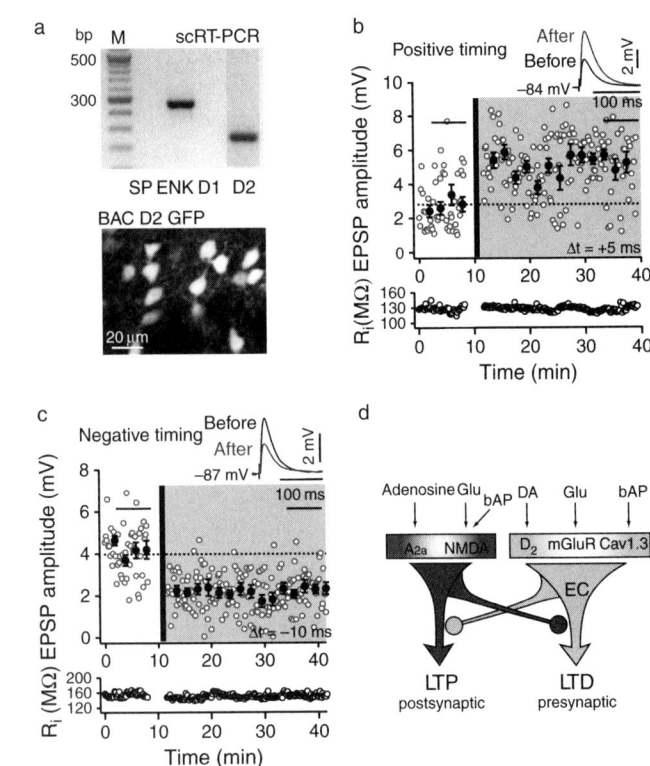

FIGURE 7.3.3. Striatopallidal MSNs displayed bidirectional STDP dependent upon D2 and A2a receptors. (a) Top: single-cell reverse transcriptase-polymerase chain reaction (scRT-PCR) amplicons from an individual BAC D2 eGFP-labeled neuron confirmed coexpression of enkephalin and D2 receptor mRNA. M, marker; SP, substance P; ENK, enkephalin. Bottom: a 2PLSM image of eGFP-labeled MSNs in a slice from a BAC D2 mouse. (b) Long-term potentiation induced in eGFP-labeled striatopallidal MSN by a positive timing pairing. Plots show EPSP amplitude and input resistance as a function of time in a single cell. The dashed line shows the average EPSP amplitude before induction. The induction was performed at the vertical bar. The filled symbol shows the averages of 12 trials (± SEM). The averaged EPSP traces before and after induction are shown at the top. (c) Long-term depression induced by a negative timing pairing. Plots and EPSP traces as in (b). (d) Schematic illustration shows that activation of A2a and NMDA receptors leads to LTP, and activation of D2 and mGluR5 receptors and Cav1.3 channels leads to LTD. Moreover, A2a and D2 receptor activation oppose each other in inducing plasticity. Glu, glutamate; EC, endocannabinoid. Source: Reprinted from [81].

robust LTP (Fig. 7.3.3b). As in striatopallidal MSNs, STDP LTP was dependent upon NMDA receptors (Fig. 7.3.4d). However, when presynaptic activity followed postsynaptic spiking, EPSP amplitude did not change. In light of the opponent signaling hypothesis, we reasoned that this failure could be due to the activation of the GPCR responsible for LTP induction. To test this hypothesis, D1 receptors were blocked by SCH23390.

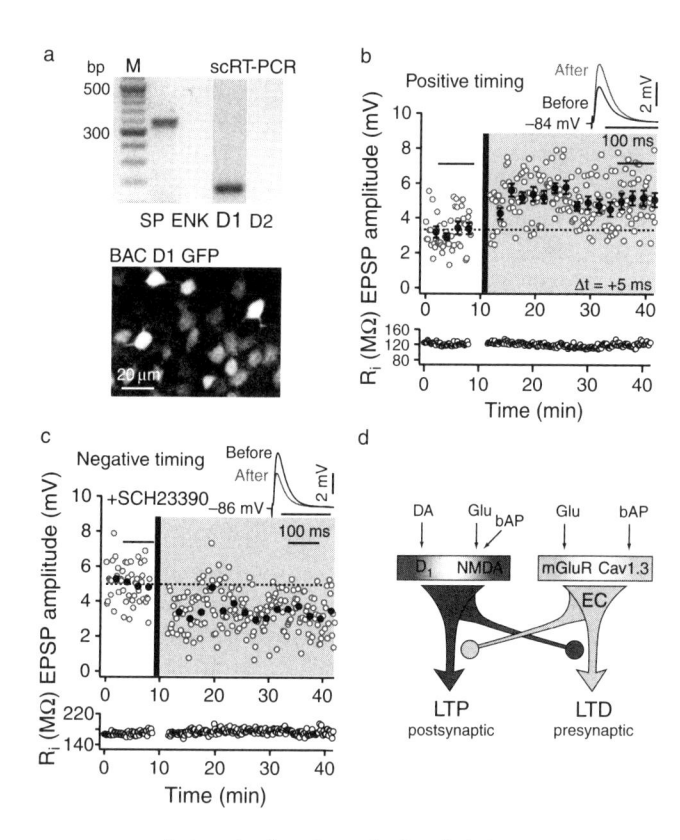

FIGURE 7.3.4. Striatonigral MSNs displayed bidirectional STDP dependent upon D1 receptors. (a) Top: single-cell RT-PCR amplicons from an individual eGFP-labeled neuron from a BAC D1 mouse confirmed coexpression of substance P and D1 receptor mRNA. M, marker; SP, substance P; ENK, enkephalin. Bottom: two-photon image of eGFP-labeled MSNs in a slice from a BAC D1 mouse. (b) Long-term potentiation induction in a labeled striatonigral neuron by a positive timing pairing protocol (+5 ms) coupled with postsynaptic depolarization to −70 mV. The EPSP amplitude and input resistance of the recorded cell were plotted as a function of time. The dashed line shows the average of EPSP amplitude before induction. The induction was performed at the vertical bar. The filled symbol shows the averages of 12 trials (± SEM). The averaged EPSP traces before and after induction are shown at the top. (c) In the presence of SCH23390, a negative timing pairing revealed robust LTD. Plots and EPSP traces are from a single cell, as in (b). (d) Schematic drawing shows that activation of D1 and NMDA receptors evokes LTP, and activation of the mGluR5 receptor and Cav1.3 channels evokes LTD. Moreover, D1 and mGluR5 receptor activation oppose each other in inducing plasticity. Glu, glutamate; EC, endocannabinoid. Source: Reprinted from [81].

In the absence of D1 receptor activity, pairing postsynaptic spiking with a presynaptic volley led to a robust LTD (Fig. 7.3.4c). Moreover, the CB1 receptor antagonist AM251 blocked the LTD, establishing a mechanistic parallel to LTD in striatopallidal MSNs. To determine whether attenuating D1 receptor signaling altered the timing dependence of plasticity, the effects of the positive timing protocol (presynaptic activity followed by postsynaptic activity) were reexamined. In control conditions, this protocol induced a robust LTP (Fig. 7.3.4b). Blocking D1 receptors not only prevented LTP induction, it led to the induction of LTD (Fig. 7.3.4d).

These studies suggest that while DA makes STDP in striatal MSNs bidirectional and Hebbian, it is not necessary for the induction of synaptic plasticity. This has fundamental implications for striatal models of incentive-based action selection.[1,86,87] Pairing reward with action increases the probability of that action; in contrast, pairing action with punishment or the omission of an expected reward diminishes the probability of that action. The physiological principles outlined above are consistent with a computational model of the basal ganglia that simulates this behavior.[86] In the model, transient elevations in DA following reward enhance the activity of "go" MSNs and promote the strengthening of corticostriatal synapses driven by the cortical networks responsible for the action. In contrast, transient drops in DA following punishment, or no reward, release "no-go" MSNs from tonic inhibition, strengthening corticostriatal synapses associated with the unrewarded action. Based upon their network connectivity and receptor expression, the go circuit is built around striatonigral MSNs, whereas the no-go circuit is built around striatopallidal MSNs. Traditional models of plasticity have been difficult to reconcile with this model, in part because there was no way to induce both LTP and LTD in neurons that do not colocalize D1 and D2 receptors. Our work provides a simple way out of this dilemma. Burst firing of DA neurons following reward presentation should briefly elevate activation of low-affinity D1 receptors on striatonigral MSNs, promoting the long-term strengthening of cortical synapses responsible for postsynaptic spiking in the go circuit. At the same time, elevated D2 receptor activation should weaken cortical connections to striatopallidal MSNs in the no-go circuit. Conversely, negative stimuli that diminish the activity of DA neurons should promote the strengthening of cortical synapses on striatopallidal MSNs in the no-go circuit while enabling the induction of synaptic depression in striatonigral MSNs in the go circuit. This bidirectional regulation of go and no-go networks in principle provides much more precise control of action selection.[86]

CHOLINERGIC INTERNEURONS AND DA

In thinking about how DA influences MSN activity, it is impossible to ignore the contribution of interneurons. Most, if not all, of the three types of striatal interneuron express DA receptors.[88–91] A review of this literature is beyond the scope of this chapter but a few comments are called for, particularly in the context of D2 receptor signaling. The best-characterized interneuron is the giant aspiny cholinergic interneuron. In primates, cholinergic interneurons are important determinants of associative and motor learning,[92] which are presumably mediated by alterations in the strength of MSN glutamatergic synapses. D2 receptor signaling diminishes acetylcholine (Ach) release both by reducing autonomous interneuron spiking and by inhibiting the Ca^{2+} entry necessary for exocytosis.[93,94]

Acetylcholine has a plethora of intrastriatal targets, including DA terminals, glutamatergic terminals, and MSNs.[95–97] Five muscarinic receptors have been identified. M1-like receptors (M1, M3, and M5) are coupled to Gq/11, mobilization of intracellular Ca^{2+}, stores and activation of phospholipase C (PLC) and protein kinase C (PKC) signaling. M2-like receptors (M2 and M4) are coupled to Gi/o proteins that inhibit adenylyl cyclase isoforms and reduce the opening of voltage-dependent Cav2 Ca^{2+} channels. Within the striatum, M1 and M4 receptors are the major muscarinic receptors expressed in MSNs [98,99]. Nicotinic ACh receptors are expressed on glutamatergic and dopaminergic terminals but are absent in MSNs.[97]

M1 receptors are highly expressed in both direct and indirect pathway MSNs.[99] Unlike D1 and D2 dopamine receptors, there is little evidence that M1 receptor activation modulates postsynaptic glutamatergic synapses. In contrast, M1 receptor activation elevates postsynaptic excitability by modulating voltage-dependent ion channels. M1 receptor activation reduces the opening of Kv4 channels (A-type potassium channels) throughout the somatodendritic membrane.[14,100] The reduction of Kv4 channel current might be mediated by PKC.[101] In addition, M1 receptor activation coupled to PLCβ and PKC leads to membrane depletion of PIP2, decreasing the opening of KCNQ (M-channel) and Kir2 (inward-rectifying potassium channel) channels.[102,103] M1 receptor activation also modulates MSNs by modulating Cav channels.[104] M1 receptor activation decreases the opening of somatic Cav1.3 and Cav2 Ca^{2+} channels.[37,96,104] These effects on somatic Ca^{2+} channels appear to work in concert with the suppression of KCNQ and Kir2 K^+ channels by diminishing the opening of Ca^{2+}-dependent K^+ (SK) channels that regulate repetitive spiking. The coordinated modulation of K^+ and Ca^{2+} channels leads to increased excitability in both the dendritic and somatic regions, enhancing synaptic integration and the translation of that input to spiking.

M2-like receptors are located both presynaptically and postsynaptically. M2/3 receptors are expressed on presynaptic glutamatergic terminals,[105] whereas M4 receptors are expressed postsynaptically in MSNs and have higher expression levels in striatonigral neurons than those in striatopallidal neurons.[99] M4 receptor activation inhibits Cav2 Ca^{2+} channels (as D2 receptors do) and therefore shapes the spiking and up-state transitions in MSNs.[96,104] Presynaptically, M2/3 receptors reduce the release probability at glutamatergic synapses, tuning them to repetitive cortical activity rather than a single isolated spike.[105–108]

How cholinergic interneurons and dopaminergic regulation of them factors into long-term synaptic plasticity has yet to be worked out. Our work is consistent with the proposition that M1 muscarinic receptors have a role in opposing the induction of LTD; conversely, work by Calabresi et al.[109] and unpublished work from our group is consistent with the contention that M1 receptor stimulation is necessary for LTP induction. Testing this proposition and sorting out how this signaling interacts with DA and adenosine in the control of plasticity is a challenge that awaits.

DOPAMINERGIC MODULATION OF GLUTAMATERGIC SIGNALING IN PD

The relationship between DA and glutamate in PD has long been the subject of speculation. Early work in animal models of PD found alterations in short-term synaptic integration and dendritic morphology, at least in some MSNs.[110–116] BAC transgenic mice in which these MSN populations are labeled have changed the experimental landscape. The first study of DA depletion using these animals revealed a stark asymmetry between striatopallidal and striatonigral MSNs in their response to the loss of DA.[117] Dopamine depletion led to the loss of glutamatergic synapses and spines of striatopallidal MSNs (Fig. 7.3.5). In contrast, DA depletion had no discernible morphological or physiological effect on synaptic function in neighboring striatonigral MSNs. In parallel with the elimination of glutamatergic synaptic contacts, the dendritic trees of striatopallidal neurons shrank, suggesting that the overall loss of glutamatergic synaptic input was even more profound. Unlike other adaptations in PD models,[118] the extent of the loss did not appear to be significantly different 1 month following DA depletion, suggesting that the regulatory processes controlling synapse elimination are complete within days and are dependent upon the loss of DA, not the death of dopaminergic neurons. Although spine and glutamatergic synapse loss

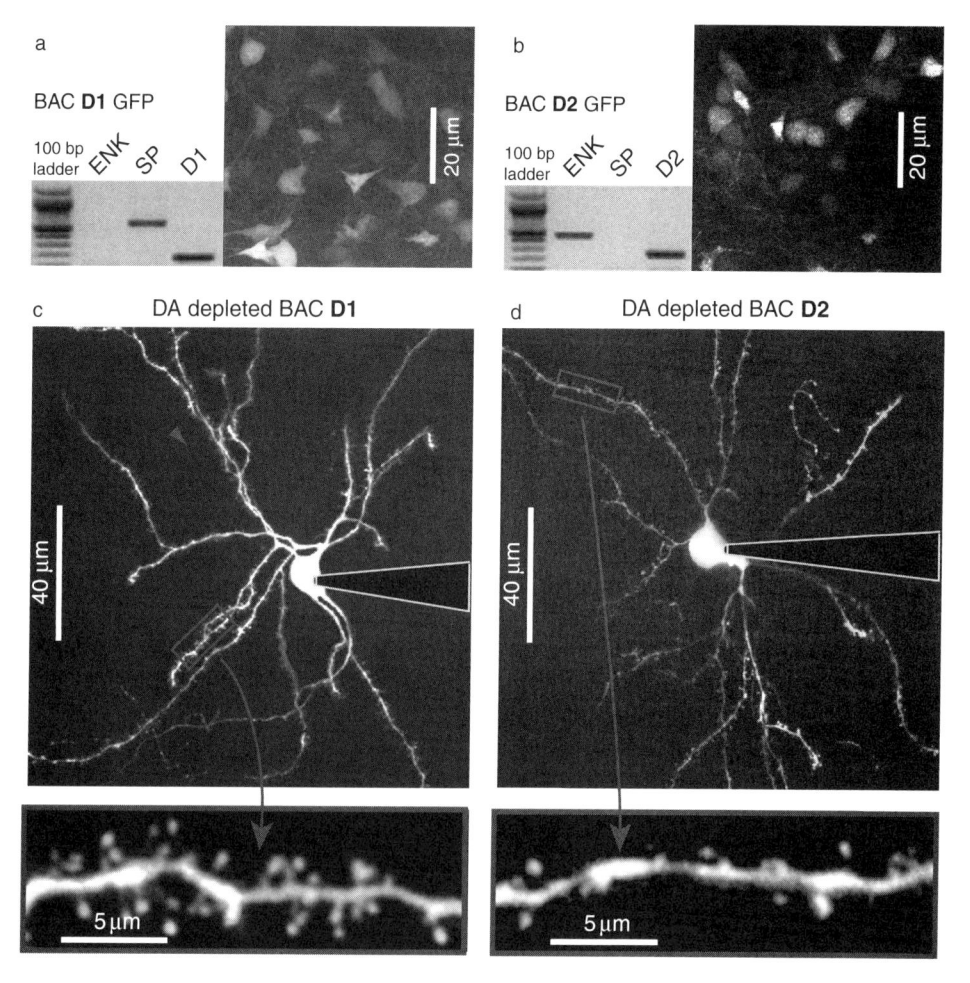

FIGURE 7.3.5. Dopamine depletion causes a reduction in spine density in the D2 receptor expressing–but not the D1 receptor expressing-MSNs. (a) A 2PLSM projection shows EGFP-labeled MSNs in a slice from a BAC D1 EGFP mouse. Green signals (500–550 nm) were acquired from EGFP-labeled D1 BAC neurons (a, right panel and c) using 810-nm excitation, while EGFP-labeled D2 BAC neurons (b, right panel and d) required 900-nm excitation. Amplicons from an individual EGFP-labeled neuron (scRT-PCR, a and b, left panels) show coexpression of SP (616 bp) and D1 receptor (234 bp) mRNAs. (b) A 2PLSM projection shows EGFP-labeled MSNs in a slice from a BAC D2 GFP mouse. Single-cell RT-PCR studies from these EGFP-labeled neurons shows coexpression of ENK (477 bp) and D2 receptor (264 bp) mRNAs. (c) Following DA depletion (reserpine, 5 days), EGFP-labeled MSNs from BAC D1 mice appear normal (projections acquired as per Fig. 7.3.2). (d) EGFP-labeled MSNs from BAC D2 mice show a reduction in the number of spines. (e) Traces taken from a control BAC D1 (top) and a DA-depleted BAC D1 (bottom) show that mEPSCs are similar in frequency and amplitude. (f) Cumulative probability plots illustrate the invariance in the interevent interval of mEPSCs between the control BAC D1 and the DA-depleted BAC D1. (g) Recordings taken from a control (top) and a DA-depleted BAC D2 (bottom) show a reduction in mEPSC frequency. (h) Cumulative probability plots of the DA-depleted D2 BAC shows an increase in interevent interval compared to BAC D2 controls. (i, left panel) Box plots showing that reserpine DA depletion produces a decrease in spine density in the D2 MSN population measured with 2PLSM (wild-type median = 9, n = 11; BAC D2 control median = 8.7, n = 6; DA-depleted D1 median = 8, n = 7; DA-depleted D2 median = 4.5, n = 5; Kruskal-Wallis ANOVA/Mann-Whitney test $P < 0.01$). (i, right panel) Box plots showing that DA depletion produces a decrease in mEPSC frequency in the D2 MSN population (BAC D1 control median = 1.9, n = 11; BAC D1 DA-depleted median = 1.8, n = 12; BAC D2 control median = 2.0, n = 11; BAC D2 DA-depleted median = 0.8, n = 7; Kruskal-Wallis ANOVA/Mann-Whitney test $P < 0.001$). *Source*: Reprinted from [117].

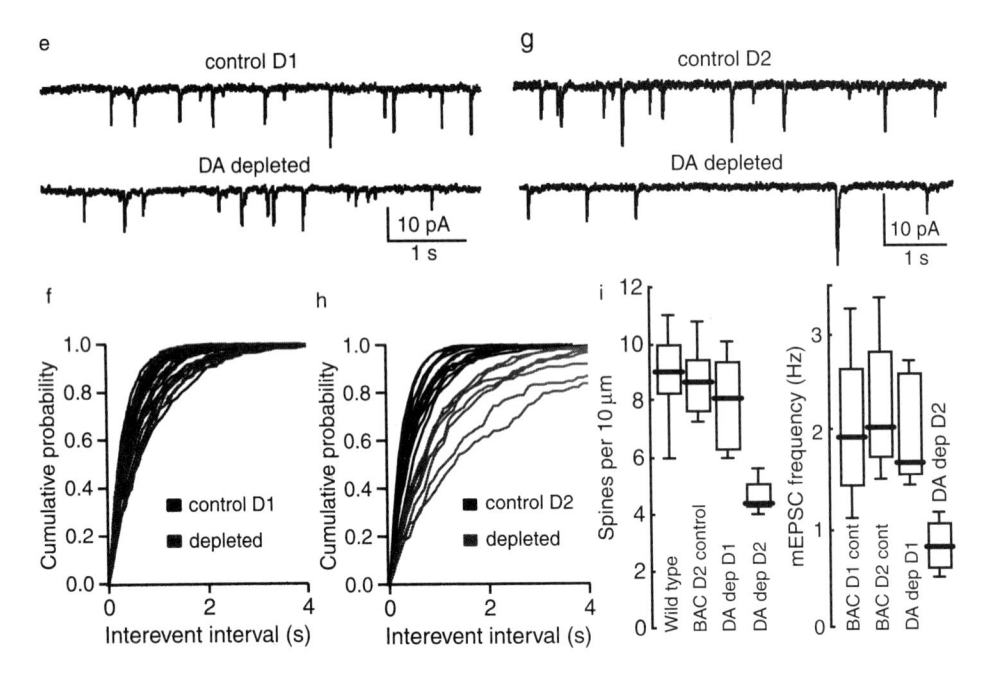

FIGURE 7.3.5. (Continued)

following DA depletion had been seen in animal models of PD and in PD patients,[110,112,119] the speed, selectivity, and magnitude of the loss were not expected.

Some of the determinants of synaptic pruning have been identified. Genetic deletion or pharmacological blockade of L-type Cav1.3 Ca^{2+} channels prevents the loss of spines and synapses following DA depletion. As noted above, these channels are strategically positioned at spiny glutamatergic synapses.[37] L-type channels contribute to the rise of intraspine Ca^{2+} concentration particularly in response to bAPs.[15] Dopamine depletion, by eliminating the D2 receptor "brake" on somatodendritic excitability,[120] could enhance intraspine Ca^{2+} entry. Falling DA levels also increase interneuron ACh release and M1 muscarinic receptor activity in striatopallidal MSNs, further elevating dendritic responsiveness to glutamatergic input.[102,103] Thus, by increasing the dendritic excitability and Ca^{2+} entry associated with excitatory glutamatergic input, DA depletion appears to trigger a homeostatic mechanism aimed at normalizing activity (measured by Ca^{2+} entry).

To pursue this question directly, BAC D2 mice were DA depleted for 5 days using reserpine, and the bAP-evoked Ca^{2+} transient was mapped in the dendrites of D2 MSNs.[14] As described above, the amplitude of the fluorescence change ($\Delta F/F_0$) at distal dendritic sites was normalized by the proximal fluorescence signal. In D2 MSNs from DA-depleted mice, the relative amplitude of bAP-evoked Ca^{2+} transient in dendritic shafts and spines fell less steeply with distance from the soma than in untreated neurons (Fig. 7.3.6a). At distal dendritic locations (100 and 150 μm from the soma), DA depletion significantly increased the relative amplitude of the Ca^{2+} transient evoked by a single bAP (Fig. 7.3.6b). In fact, in all of the neurons examined following DA depletion, bAP-associated Ca^{2+} transients were detectable as far out on the dendrites as we were capable of imaging (~150 μm from the soma). The simplest interpretation of these results is that the loss of spines and dendritic surface area following DA depletion diminished the capacitative load of the dendrites, improving bAP invasion into distal regions. Although consistent with theoretical and experimental examination of other neurons,[121] this hypothesis was tested in an anatomically accurate model of an MSN; to this end, neuron simulations were conducted in which the surface area of spiny dendrites was decreased and the effects on the bAP were examined. These simulations corroborated the inference that spine loss enhances dendritic bAP invasion, showing enhanced bAP propagation, enhanced opening of voltage-dependent Ca^{2+} channels, and an elevation of bAP-evoked change in the intracellular Ca^{2+} concentration at distal dendritic locations. A second explanation is that DA depletion facilitates an increase in ACh tone, as has long been hypothesized to underlie some of the disorders seen in patients with PD. We tested this prospect and found that in BAC D2 MSNs, bath application of the muscarinic antagonist

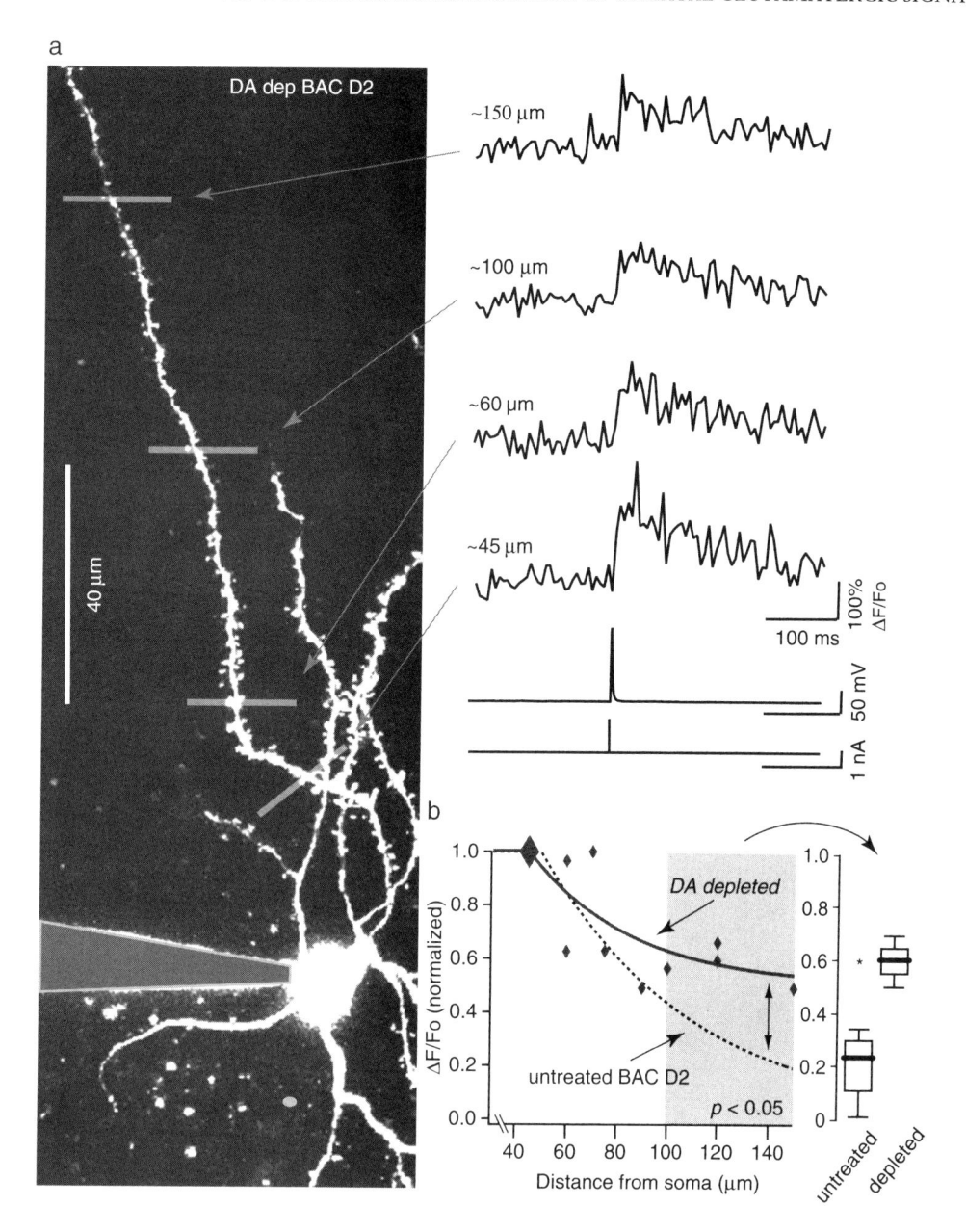

FIGURE 7.3.6. Dopamine depletion enhances excitability in distal dendrites in D2 MSNs. (a) Maximum projection image of a D2 MSN soma and dendrite from a DA-depleted BAC D2 mouse (left). The traces show the bAP-evoked Ca^{2+} transient recorded at four different eccentricities along this dendrite (45, 60, 100, 150 μm, right). (b) Plot of the amplitude of the bAP-evoked Ca^{2+} transient normalized to the most proximal recording in each cell (diamonds, line). For comparison, the fit line from the D2 untreated MSNs (Fig.7.3.2d, dashed line) is added to the plot. The box plot demonstrates the increase in the amplitude of the normalized bAP-evoked Ca^{2+} in the distal regions of the DA-depleted D2 MSN dendrites compared to controls (untreated D2 = 0.24, n = 4; DA-depleted D2 = 0.6, n = 4; Kruskal-Wallis ANOVA, $p < 0.05$). Source: Reprinted from [14].

scopolamine (20 μM) significantly suppressed the bAP-evoked Ca^{2+} transient in the DA-depleted mice compared to untreated controls. This finding suggests that cholinergic tone is elevated in the DA-depleted mice, leading to a down-regulation of Kv4 channels. The down-regulation of Kv4 channels could be sufficient to enhance dendritic excitability in the DA-depleted D2 MSNs or it could synergize with the decrease in spine density to render the dendrites even more excitable. We also considered the possibility that the decrease in the decrement of the bAP-evoked Ca^{2+} transients seen following reserpine treatment reflected a

D2-mediated disinhibition Ca^{2+} channels. To test this hypothesis, we compared the amplitude of bAP-evoked Ca^{2+} transients in untreated BAC D2 MSNs recorded before and after bath application of the D2 antagonist sulpiride (10 µM). We did not detect any significant differences in the distal dendrites before and after D2 blockade, indicating that tonic DA tone is an unlikely contributor to differences in dendritic excitability seen following DA depletion. However, it is too early to exclude other mechanisms. Preliminary quantitative reverse transcriptase-polymerase chain reaction (RT-PCR) studies have shown that while Kv4 mRNA is down-regulated in D2 MSNs by DA depletion, Cav1.3 and Cav3.2-3 mRNA are up-regulated—suggesting that the adaptations are multidimensional.

The loss of D2 receptor stimulation will also handicap the induction of LTD that might serve to normalize global activity without eliminating synapses.[58–60] Recent work by our group[81] has revealed that the loss of D2 receptor stimulation in PD models not only prevents the induction of LTD in D2 MSNs, but also promotes LTP induction through adenosine A2a receptor signaling mechanisms. This interaction is mediated by an antagonism between the signaling mechanisms promoting LTD (D2 receptor dependent) and those promoting LTP (A2a receptor dependent). The loss of D2 receptor signaling disrupts the balance between these two processes, leading to strengthening of synaptic connections in what appear to be inappropriate situations. This maladaptive response to DA depletion, together with the elevation in dendritic excitability attributable to functional down-regulation of Kir2 and Kv4 K$^+$ channels, might provide a partial explanation for the anomalous increase in glutamatergic mini EPSC (mEPSC) frequency seen in several studies of MSNs in PD models.[111,122,123] That said, other mechanisms cannot be excluded at this point; preliminary studies in our lab have shown that prolonged DA depletion induces a reorganization of D2 MSN dendrites (resulting in more dendritic surface close to the somatic compartment) and a clear elevation in the probability of glutamate release—both of which could contribute to a change in mEPSC frequency. In contrast, the loss of DA in PD models prevented the induction of LTP in D1 receptor–expressing striatonigral MSNs but, importantly, promoted the induction of LTD. Again, the loss of a balance between DA and non-DA signaling processes was critical to the change. This disruption should lead to inappropriate weakening of synaptic connections between this part of the striatal network and cortical command structures.

Although the majority of the glutamatergic synapses formed on dendritic spines are of cortical origin, many are not.[124] The thalamic innervation of MSNs is similar in magnitude to that of the cerebral cortex, perhaps constituting as much as 40% of the total glutamatergic input to MSNs, terminating on both shafts and spines. Anatomical studies suggest that the intralaminar nuclei target primarily striatonigral neurons in primate striatum, though this might not be the case in rodents.[63] Motor nuclei (VA, VL) project primarily to striatopallidal neurons.[64,65] This apparent dichotomy between motor and associative inputs is consistent with recent studies suggesting that input to striatopallidal neurons comes largely from pyramidal neurons contributing to descending motor control circuits, whereas input to striatonigral neurons comes from neurons whose axons are largely intratelencephalic.[66] Thus, it would seem that the regions of the brain most directly linked to motor functions become disconnected from the so-called indirect pathway. While the most parsimonious hypothesis is that it is primarily the cortical projection that is cut, this has yet to be rigorously tested.

FUNCTIONAL IMPLICATIONS FOR THE PATHOPHYSIOLOGY IN PD

There are two interrelated lines of inferences that can be drawn from these studies. The control of motor behavior is obviously profoundly disrupted by DA depletion, but why? Our results provide some grist for the classical model, showing that the striatal network will be strongly biased toward action suppression, regardless of the consequences of that action. There are two components of this shift. In striatopallidal MSNs that anchor the no-go pathway,[86] the loss of D2 receptor signaling will lead to elevated intrinsic excitability and inappropriate LTP of corticostriatal inputs; this should lead to widespread and inappropriate suppression of actions. In striatonigral MSNs that anchor the go pathway, precisely the opposite should occur, resulting in a diminished capacity of cortex to activate the striatal circuits necessary for action selection. In addition, to the extent that the pattern and strength of cortical connectivity with striatopallidal MSNs reflect motor memory, the loss of striatal DA and the pruning of connections should induce a form of memory loss.

The other line of inferences has to do with downstream targets of the striatum. Striatopallidal MSNs are clearly important in the expression of PD motor symptoms.[125,126] Perhaps the most compelling piece of evidence on this point is the finding that the activity of neurons they control is dramatically altered in people suffering from PD, as well as in animal models of the disease. In particular, neurons in the globus pallidus and in the reciprocally connected subthalamic nucleus begin to discharge in anomalous rhythmic bursts that are often synchronized. Silencing this abnormal patterning with lesions or deep brain stimulation provides dramatic relief from motor symptoms.[127,128] Computer simulations grounded in experimental observation suggest that this rhythmic bursting is an intrinsic property of the pallido-subthalamic circuitry that is normally suppressed by striatopallidal GABAergic inhibition.[129] Ineffectively timed or patterned striatopallidal activity could release this circuitry, allowing it to display activity patterns like those seen in PD. Because striatopallidal MSNs depend upon highly convergent glutamatergic synaptic inputs from cortical and thalamic motor command centers,[130] the loss of a substantial portion of this input should profoundly disrupt movement-related patterned activity and, in so doing, limit their ability to control the emergence of synchronous bursting in the pallido-subthalamic circuit. The failure to control the pallido-subthalamic circuit should lead to unwanted movements and the cardinal symptom of PD—the inability to translate thought into efficient movement.

CONCLUDING REMARKS

Although we are still some way from a clear understanding of how DA shapes the activity of striatal circuits, some tentative conclusions can be drawn. Acting principally through D2 receptors, DA reduces glutamate release as well as the postsynaptic responsiveness of striatopallidal MSNs to released glutamate. This short-term modulation is complemented by D2 receptor–dependent promotion of LTD of glutamatergic synaptic transmission. Our understanding of how DA modulates striatonigral MSNs is less secure. Acting principally at postsynaptic D1 receptors in striatonigral MSNs, DA appears to depress weak, asynchronous synaptic signals but to augment the response to strong, coordinated glutamatergic input, promoting NMDA receptor opening and up-state transitions. In addition, D1 receptor signaling facilitates LTP of glutamatergic signaling, enhancing network connections, which are consistently active during important environmental events that trigger phasic DA release. This pattern of cellular physiological effects is consistent with higher-order models of cortically driven striatal action selection[1,3,86,87] built upon the conjecture that activity in striatopallidal MSNs serves to suppress action, whereas activity in striatonigral MSNs serves to promote action.[3,86,131]

ACKNOWLEDGMENTS

This work was supported by the Picower Foundation, the Hartman Foundation, and the NIH.

REFERENCES

1. Schultz W. Behavioral theories and the neurophysiology of reward. *Annu Rev Psychol.* 2006;57:87–115.
2. Wickens JR, Reynolds JN, Hyland BI. Neural mechanisms of reward-related motor learning. *Curr Opin Neurobiol.* 2003;13(6):685–690.
3. Albin RL, Young AB, Penney JB. The functional anatomy of basal ganglia disorders. *Trends Neurosci.* 1989;12(10):366–375.
4. Gerfen CR. The neostriatal mosaic: multiple levels of compartmental organization in the basal ganglia. *Annu Rev Neurosci.* 1992;15:285–320.
5. Surmeier DJ, Song WJ, Yan Z. Coordinated expression of dopamine receptors in neostriatal medium spiny neurons. *J Neurosci.* 1996;16(20):6579–6591.
6. Nicola SM, Surmeier J, Malenka RC. Dopaminergic modulation of neuronal excitability in the striatum and nucleus accumbens. *Annu Rev Neurosci.* 2000;23:185–215.
7. Surmeier DJ. Microcircuits in the striatum: cell types, intrinsic membrane properties and neuromodulation. In: Grillner S, Graybiel AM, eds. *Microcircuits: The Interface between Neurons and Global Brain Function.* Berlin, Germany: MIT Press; 2004:105–126.
8. Arbuthnott GW, Wickens J. Space, time and dopamine. *Trends Neurosci.* 2006;30:62–69.
9. David HN, Ansseau M, Abraini JH. Dopamine-glutamate reciprocal modulation of release and motor responses in the rat caudate-putamen and nucleus accumbens of "intact" animals. *Brain Res Brain Res Rev.* 2005;50(2): 336–360.
10. Morari M, Marti M, Sbrenna S, Fuxe K, Bianchi C, Beani L. Reciprocal dopamine-glutamate modulation of release in the basal ganglia. *Neurochem Int.* 1998;33(5):383–397.
11. Gertler TS, Chan CS, Surmeier DJ. Dichotomous anatomical properties of adult striatal medium spiny neurons. *J Neurosci.* 2008;28(43):10814–10824.
12. Horikawa K, Armstrong WE. A versatile means of intracellular labeling: injection of biocytin and its detection with avidin conjugates. *J Neurosci Methods.* 1988;25(1):1–11.
13. Segal M, Greenberger V, Korkotian E. Formation of dendritic spines in cultured striatal neurons depends on excitatory afferent activity. *Eur J Neurosci.* 2003;17(12):2573–2585.
14. Day M, Wokosin D, Plotkin JL, Tian X, Surmeier DJ. Differential excitability and modulation of striatal medium spiny neuron dendrites. *J Neurosci.* 2008;28(45):11603–11614.
15. Carter AG, Sabatini BL. State-dependent calcium signaling in dendritic spines of striatal medium spiny neurons. *Neuron.* 2004;44(3):483–493.
16. Kerr JN, Plenz D. Action potential timing determines dendritic calcium during striatal up-states. *J Neurosci.* 2004;24(4):877–885.
17. Herve D, Rogard M, Levi-Strauss M. Molecular analysis of the multiple Golf alpha subunit mRNAs in the rat brain. *Brain Res Mol Brain Res.* 1995;32(1):125–134.

18. Svenningsson P, Nishi A, Fisone G, Girault JA, Nairn AC, Greengard P. DARPP-32: an integrator of neurotransmission. *Annu Rev Pharmacol Toxicol.* 2004;44:269–296.
19. Snyder GL, Allen PB, Fienberg AA, et al. Regulation of phosphorylation of the GluR1 AMPA receptor in the neostriatum by dopamine and psychostimulants in vivo. *J Neurosci.* 2000;20(12):4480–4488.
20. Hallett PJ, Spoelgen R, Hyman BT, Standaert DG, Dunah AW. Dopamine D1 activation potentiates striatal NMDA receptors by tyrosine phosphorylation-dependent subunit trafficking. *J Neurosci.* 2006;26(17):4690–4700.
21. Mangiavacchi S, Wolf ME. D1 dopamine receptor stimulation increases the rate of AMPA receptor insertion onto the surface of cultured nucleus accumbens neurons through a pathway dependent on protein kinase A. *J Neurochem.* 2004;88(5):1261–1271.
22. Braithwaite SP, Paul S, Nairn AC, Lombroso PJ. Synaptic plasticity: one STEP at a time. *Trends Neurosci.* 2006;29(8):452–458.
23. Lee FJ, Xue S, Pei L, et al. Dual regulation of NMDA receptor functions by direct protein–protein interactions with the dopamine D1 receptor. *Cell.* 2002;111(2):219–230.
24. Scott L, Zelenin S, Malmersjo S, et al. Allosteric changes of the NMDA receptor trap diffusible dopamine 1 receptors in spines. *Proc Natl Acad Sci USA.* 2006;103(3):762–767.
25. Blank T, Nijholt I, Teichert U, et al. The phosphoprotein DARPP-32 mediates cAMP-dependent potentiation of striatal N-methyl-D-aspartate responses. *Proc Natl Acad Sci USA.* 1997;94(26):14859–14864.
26. >Nicola SM, Malenka RC. Modulation of synaptic transmission by dopamine and norepinephrine in ventral but not dorsal striatum. *J Neurophysiol.* 1998;79(4):1768–1776.
27. Cepeda C, Buchwald NA, Levine MS. Neuromodulatory actions of dopamine in the neostriatum are dependent upon the excitatory amino acid receptor subtypes activated. *Proc Natl Acad Sci USA.* 1993;90(20):9576–9580.
28. Liu JC, DeFazio RA, Espinosa-Jeffrey A, Cepeda C, de Vellis J, Levine MS. Calcium modulates dopamine potentiation of N-methyl-D-aspartate responses: electrophysiological and imaging evidence. *J Neurosci Res.* 2004;76(3):315–322.
29. Calabresi P, Mercuri N, Stanzione P, Stefani A, Bernardi G. Intracellular studies on the dopamine-induced firing inhibition of neostriatal neurons in vitro: evidence for D1 receptor involvement. *Neuroscience.* 1987;20(3):757–771.
30. Surmeier DJ, Eberwine J, Wilson CJ, Cao Y, Stefani A, Kitai ST. Dopamine receptor subtypes colocalize in rat striatonigral neurons. *Proc Natl Acad Sci USA.* 1992;89(21):10178–10182.
31. Carr DB, Day M, Cantrell AR, et al. Transmitter modulation of slow, activity-dependent alterations in sodium channel availability endows neurons with a novel form of cellular plasticity. *Neuron.* 2003;39(5):793–806.
32. Scheuer T, Catterall WA. Control of neuronal excitability by phosphorylation and dephosphorylation of sodium channels. *Biochem Soc Trans.* 2006;34(Pt 6):1299–1302.
33. Wickens JR, Wilson CJ. Regulation of action-potential firing in spiny neurons of the rat neostriatum in vivo. *J Neurophysiol.* 1998;79(5):2358–2364.
34. Hernandez-Lopez S, Bargas J, Surmeier DJ, Reyes A, Galarraga E. D1 receptor activation enhances evoked discharge in neostriatal medium spiny neurons by modulating an L-type Ca^{2+} conductance. *J Neurosci.* 1997;17(9):3334–3342.
35. Gao T, Yatani A, Dell'Acqua ML, et al. cAMP-dependent regulation of cardiac L-type Ca^{2+} channels requires membrane targeting of PKA and phosphorylation of channel subunits. *Neuron.* 1997;19(1):185–196.
36. Surmeier DJ, Bargas J, Hemmings HC Jr, Nairn AC, Greengard P. Modulation of calcium currents by a D1 dopaminergic protein kinase/phosphatase cascade in rat neostriatal neurons. *Neuron.* 1995;14(2):385–397.
37. Olson PA, Tkatch T, Hernandez-Lopez S, et al. G-protein-coupled receptor modulation of striatal CaV1.3 L-type Ca^{2+} channels is dependent on a Shank-binding domain. *J Neurosci.* 2005;25(5):1050–1062.
38. Levine MS, Altemus KL, Cepeda C, et al. Modulatory actions of dopamine on NMDA receptor-mediated responses are reduced in D1A-deficient mutant mice. *J Neurosci.* 1996;16(18):5870–5882.
39. Snyder GL, Fienberg AA, Huganir RL, Greengard P. A dopamine/D1 receptor/protein kinase A/dopamine- and cAMP-regulated phosphoprotein (Mr 32 kDa)/protein phosphatase-1 pathway regulates dephosphorylation of the NMDA receptor. *J Neurosci.* 1998;18(24):10297–10303.
40. Flores-Hernandez J, Cepeda C, Hernandez-Echeagaray E, et al. Dopamine enhancement of NMDA currents in dissociated medium-sized striatal neurons: role of D1 receptors and DARPP-32. *J Neurophysiol.* 2002;88(6):3010–3020.
41. Vergara R, Rick C, Hernandez-Lopez S, et al. Spontaneous voltage oscillations in striatal projection neurons in a rat corticostriatal slice. *J Physiol.* 2003;553(Pt 1):169–182.
42. Tseng KY, O'Donnell P. Dopamine–glutamate interactions controlling prefrontal cortical pyramidal cell excitability involve multiple signaling mechanisms. *J Neurosci.* 2004;24(22):5131–5139.
43. Vilchis C, Bargas J, Ayala GX, Galvan E, Galarraga E. Ca^{2+} channels that activate Ca^{2+}-dependent K^+ currents in neostriatal neurons. *Neuroscience.* 2000;95(3):745–752.
44. Stoof JC, Kebabian JW. Two dopamine receptors: biochemistry, physiology and pharmacology. *Life Sci.* 1984;35(23):2281–2296.
45. Hernandez-Lopez S, Tkatch T, Perez-Garci E, et al. D2 dopamine receptors in striatal medium spiny neurons reduce L-type Ca^{2+} currents and excitability via a novel PLC[beta]1-IP3-calcineurin-signaling cascade. *J Neurosci.* 2000;20(24):8987–8995.
46. Nishi A, Snyder GL, Greengard P. Bidirectional regulation of DARPP-32 phosphorylation by dopamine. *J Neurosci.* 1997;17(21):8147–8155.
47. Kotecha SA, Oak JN, Jackson MF, et al. A D2 class dopamine receptor transactivates a receptor tyrosine kinase to inhibit NMDA receptor transmission. *Neuron.* 2002;35(6):1111–1122.
48. Hernandez-Echeagaray E, Starling AJ, Cepeda C, Levine MS. Modulation of AMPA currents by D2 dopamine receptors in striatal medium-sized spiny neurons: are dendrites necessary? *Eur J Neurosci.* 2004;19(9):2455–2463.
49. Hakansson K, Galdi S, Hendrick J, Snyder G, Greengard P, Fisone G. Regulation of phosphorylation of the GluR1 AMPA receptor by dopamine D2 receptors. *J Neurochem.* 2006;96(2):482–488.
50. Bamford NS, Zhang H, Schmitz Y, et al. Heterosynaptic dopamine neurotransmission selects sets of corticostriatal terminals. *Neuron.* 2004;42(4):653–663.
51. Yin HH, Lovinger DM. Frequency-specific and D2 receptor-mediated inhibition of glutamate release by retrograde

51. endocannabinoid signaling. *Proc Natl Acad Sci USA.* 2006;103(21): 8251–8256.
52. Greif GJ, Lin YJ, Liu JC, Freedman JE. Dopamine-modulated potassium channels on rat striatal neurons: specific activation and cellular expression. *J Neurosci.* 1995;15(6): 4533–4544.
53. Brebner K, Wong TP, Liu L, et al. Nucleus accumbens long-term depression and the expression of behavioral sensitization. *Science.* 2005;310(5752):1340–1343.
54. Centonze D, Picconi B, Gubellini P, Bernardi G, Calabresi P. Dopaminergic control of synaptic plasticity in the dorsal striatum. *Eur J Neurosci.* 2001;13(6):1071–1077.
55. Kreitzer AC, Malenka RC. Dopamine modulation of state-dependent endocannabinoid release and long-term depression in the striatum. *J Neurosci.* 2005;25(45):10537–10545.
56. Lovinger DM, Tyler EC, Merritt A. Short- and long-term synaptic depression in rat neostriatum. *J Neurophysiol.* 1993;70(5):1937–1949.
57. Giuffrida A, Parsons LH, Kerr TM, Rodriguez de Fonseca F, Navarro M, Piomelli D. Dopamine activation of endogenous cannabinoid signaling in dorsal striatum. *Nat Neurosci.* 1999;2(4):358–363.
58. Kreitzer AC, Malenka RC. Endocannabinoid-mediated rescue of striatal LTD and motor deficits in Parkinson's disease models. *Nature.* 2007;445(7128):643–647.
59. Wang Z, Kai L, Day M, et al. Dopaminergic control of corticostriatal long-term synaptic depression in medium spiny neurons is mediated by cholinergic interneurons. *Neuron.* 2006;50(3): 443–452.
60. Calabresi P, Picconi B, Tozzi A, Di Filippo M. Dopamine-mediated regulation of corticostriatal synaptic plasticity. *Trends Neurosci.* 2007;30(5):211–219.
61. Kawaguchi Y, Wilson CJ, Emson PC. Intracellular recording of identified neostriatal patch and matrix spiny cells in a slice preparation preserving cortical inputs. *J Neurophysiol.* 1989;62(5): 1052–1068.
62. Wilson CJ. Basal ganglia. In: Shepherd GM, ed. *The Synaptic Organization of the Brain.* Vol 5. Oxford, England: Oxford University Press; 2004:361–414.
63. Bacci JJ, Kachidian P, Kerkerian-Le Goff L, Salin P. Intralaminar thalamic nuclei lesions: widespread impact on dopamine denervation-mediated cellular defects in the rat basal ganglia. *J Neuropathol Exp Neurol.* 2004;63(1):20–31.
64. Smith Y, Raju DV, Pare JF, Sidibe M. The thalamostriatal system: a highly specific network of the basal ganglia circuitry. *Trends Neurosci.* 2004;27(9):520–527.
65. Hoshi E, Tremblay L, Feger J, Carras PL, Strick PL. The cerebellum communicates with the basal ganglia. *Nat Neurosci.* 2005;8(11):1491–1493.
66. Lei W, Jiao Y, Del Mar N, Reiner A. Evidence for differential cortical input to direct pathway versus indirect pathway striatal projection neurons in rats. *J Neurosci.* 2004;24(38): 8289–8299.
67. Ding J, Peterson JD, Surmeier DJ. Corticostriatal and thalamostriatal synapses have distinctive properties. *J Neurosci.* 2008;28(25):6483–6492.
68. Smeal RM, Gaspar RC, Keefe KA, Wilcox KS. A rat brain slice preparation for characterizing both thalamostriatal and corticostriatal afferents. *J Neurosci Methods.* 2007;159(2): 224–235.
69. Centonze D, Grande C, Saulle E, et al. Distinct roles of D1 and D5 dopamine receptors in motor activity and striatal synaptic plasticity. *J Neurosci.* 2003;23(24):8506–8512.
70. Kerr JN, Wickens JR. Dopamine D-1/D-5 receptor activation is required for long-term potentiation in the rat neostriatum in vitro. *J Neurophysiol.* 2001;85(1):117–124.
71. Calabresi P, Centonze D, Gubellini P, et al. Synaptic transmission in the striatum: from plasticity to neurodegeneration. *Prog Neurobiol.* 2000;61(3):231–265.
72. Mahon S, Deniau JM, Charpier S. Corticostriatal plasticity: life after the depression. *Trends Neurosci.* 2004;27(8):460–467.
73. Hersch SM, Ciliax BJ, Gutekunst CA, et al. Electron microscopic analysis of D1 and D2 dopamine receptor proteins in the dorsal striatum and their synaptic relationships with motor corticostriatal afferents. *J Neurosci.* 1995;15(7 Pt 2):5222–5237.
74. Dan Y, Poo MM. Spike timing-dependent plasticity of neural circuits. *Neuron.* 2004;44(1):23–30.
75. Kampa BM, Letzkus JJ, Stuart GJ. Dendritic mechanisms controlling spike-timing-dependent synaptic plasticity. *Trends Neurosci.* 2007;30(9):456–463.
76. Sjostrom PJ, Rancz EA, Roth A, Hausser M. Dendritic excitability and synaptic plasticity. *Physiol. Rev.* 2008;88(2):769–840.
77. Sjostrom PJ, Nelson SB. Spike timing, calcium signals and synaptic plasticity. *Curr Opin Neurobiol.* 2002;12(3):305–314.
78. Letzkus JJ, Kampa BM, Stuart GJ. Learning rules for spike timing-dependent plasticity depend on dendritic synapse location. *J. Neurosci.* 2006;26(41):10420–10429.
79. Nevian T, Sakmann B. Spine Ca^{2+} signaling in spike-timing-dependent plasticity. *J. Neurosci.* 2006;26(43):11001–11013.
80. Pawlak V, Kerr JN. Dopamine receptor activation is required for corticostriatal spike-timing-dependent plasticity. *J. Neurosci.* 2008;28(10):2435–2446.
81. Shen W, Flajolet M, Greengard P, Surmeier DJ. Dichotomous dopaminergic control of striatal synaptic plasticity. *Science.* 2008;321(5890):848–851.
82. Seol GH, Ziburkus J, Huang S, et al. Neuromodulators control the polarity of spike-timing-dependent synaptic plasticity. *Neuron.* 2007;55(6):919–929.
83. Tzounopoulos T, Rubio ME, Keen JE, Trussell LO. Coactivation of pre- and postsynaptic signaling mechanisms determines cell-specific spike-timing-dependent plasticity. *Neuron.* 2007;54(2): 291–301.
84. Schwarzschild MA, Agnati L, Fuxe K, Chen JF, Morelli M. Targeting adenosine A_{2a} receptors in Parkinson's disease. *Trends Neurosci.* 2006;29(11):647–654.
85. Fuxe K, Marcellino D, Genedani S, Agnati L. Adenosine A_{2A} receptors, dopamine D_2 receptors and their interactions in Parkinson's disease. *Mov. Disord.* 2007;22(14):1990–2017.
86. Frank MJ. Dynamic dopamine modulation in the basal ganglia: a neurocomputational account of cognitive deficits in medicated and nonmedicated Parkinsonism. *J Cogn Neurosci.* 2005;17(1): 51–72.
87. Robbins TW, Everitt BJ. Neurobehavioural mechanisms of reward and motivation. *Curr Opin Neurobiol.* 1996;6(2): 228–236.
88. Tepper JM, Koos T, Wilson CJ. GABAergic microcircuits in the neostriatum. *Trends Neurosci.* 2004;27(11):662–669.
89. Yan Z, Surmeier DJ. D5 dopamine receptors enhance Zn^{2+}-sensitive GABA(A) currents in striatal cholinergic interneurons through a PKA/PP1 cascade. *Neuron.* 1997;19(5): 1115–1126.
90. Centonze D, Grande C, Usiello A, et al. Receptor subtypes involved in the presynaptic and postsynaptic actions of dopamine on striatal interneurons. *J Neurosci.* 2003;23(15): 6245–6254.

91. Bracci E, Centonze D, Bernardi G, Calabresi P. Dopamine excites fast-spiking interneurons in the striatum. *J Neurophysiol.* 2002;87(4):2190-2194.
92. Graybiel AM, Aosaki T, Flaherty AW, Kimura M. The basal ganglia and adaptive motor control. *Science.* 1994;265(5180):1826-1831.
93. Salgado H, Tecuapetla F, Perez-Rosello T, et al. A reconfiguration of CaV2 Ca^{2+} channel current and its dopaminergic D2 modulation in developing neostriatal neurons. *J Neurophysiol.* 2005;94(6):3771-3787.
94. Maurice N, Mercer J, Chan CS, et al. D2 dopamine receptor-mediated modulation of voltage-dependent Na^+ channels reduces autonomous activity in striatal cholinergic interneurons. *J Neurosci.* 2004;24(46):10289-10301.
95. Dodt HU, Misgeld U. Muscarinic slow excitation and muscarinic inhibition of synaptic transmission in the rat neostriatum. *J Physiol.* 1986;380:593-608.
96. Perez-Rosello T, Figueroa A, Salgado H, et al. Cholinergic control of firing pattern and neurotransmission in rat neostriatal projection neurons: role of CaV2.1 and CaV2.2 Ca^{2+} channels. *J Neurophysiol.* 2005;93(5):2507-2519.
97. Zhou FM, Wilson CJ, Dani JA. Cholinergic interneuron characteristics and nicotinic properties in the striatum. *J Neurobiol.* 2002;53(4):590-605.
98. Bernard V, Normand E, Bloch B. Phenotypical characterization of the rat striatal neurons expressing muscarinic receptor genes. *J Neurosci.* 1992;12(9):3591-3600.
99. Yan Z, Flores-Hernandez J, Surmeier DJ. Coordinated expression of muscarinic receptor messenger RNAs in striatal medium spiny neurons. *Neuroscience.* 2001;103(4):1017-1024.
100. Akins PT, Surmeier DJ, Kitai ST. Muscarinic modulation of a transient K^+ conductance in rat neostriatal neurons. *Nature.* 1990;344(6263):240-242.
101. Nakamura TY, Coetzee WA, Vega-Saenz De Miera E, Artman M, Rudy B. Modulation of Kv4 channels, key components of rat ventricular transient outward K^+ current, by PKC. *Am J Physiol.* 1997;273(4 pt 2):H1775-1786.
102. Shen W, Hamilton SE, Nathanson NM, Surmeier DJ. Cholinergic suppression of KCNQ channel currents enhances excitability of striatal medium spiny neurons. *J Neurosci.* 2005;25(32):7449-7458.
103. Shen W, Tian X, Day M, et al. Cholinergic modulation of Kir2 channels selectively elevates dendritic excitability in striatopallidal neurons. *Nat Neurosci.* 2007;10(11):1458-1466.
104. Howe AR, Surmeier DJ. Muscarinic receptors modulate N-, P-, and L-type Ca^{2+} currents in rat striatal neurons through parallel pathways. *J Neurosci.* 1995;15(1 pt 1):458-469.
105. Alcantara AA, Mrzljak L, Jakab RL, Levey AI, Hersch SM, Goldman-Rakic PS. Muscarinic m1 and m2 receptor proteins in local circuit and projection neurons of the primate striatum: anatomical evidence for cholinergic modulation of glutamatergic prefronto-striatal pathways. *J Comp Neurol.* 2001;434(4):445-460.
106. Barral J, Galarraga E, Bargas J. Muscarinic presynaptic inhibition of neostriatal glutamatergic afferents is mediated by Q-type Ca^{2+} channels. *Brain Res Bull.* 1999;49(4):285-289.
107. Calabresi P, Centonze D, Gubellini P, Pisani A, Bernardi G. Blockade of M2-like muscarinic receptors enhances long-term potentiation at corticostriatal synapses. *Eur J Neurosci.* 1998;10(9):3020-3023.
108. Pakhotin P, Bracci E. Cholinergic interneurons control the excitatory input to the striatum. *J Neurosci.* 2007;27(2):391-400.
109. Centonze D, Gubellini P, Pisani A, Bernardi G, Calabresi P. Dopamine, acetylcholine and nitric oxide systems interact to induce corticostriatal synaptic plasticity. *Rev Neurosci.* 2003;14(3):207-216.
110. Dunah AW, Wang Y, Yasuda RP, et al. Alterations in subunit expression, composition, and phosphorylation of striatal N-methyl-D-aspartate glutamate receptors in a rat 6-hydroxydopamine model of Parkinson's disease. *Mol Pharmacol.* 2000;57(2):342-352.
111. Gubellini P, Picconi B, Bari M, et al. Experimental parkinsonism alters endocannabinoid degradation: implications for striatal glutamatergic transmission. *J Neurosci.* 2002;22(16):6900-6907.
112. Ingham CA, Hood SH, Taggart P, Arbuthnott GW. Plasticity of synapses in the rat neostriatum after unilateral lesion of the nigrostriatal dopaminergic pathway. *J Neurosci.* 1998;18(12):4732-4743.
113. Nisenbaum ES, Stricker EM, Zigmond MJ, Berger TW. Long-term effects of dopamine-depleting brain lesions on spontaneous activity of type II striatal neurons: relation to behavioral recovery. *Brain Res.* 1986;398(2):221-230.
114. Pang Z, Ling GY, Gajendiran M, Xu ZC. Enhanced excitatory synaptic transmission in spiny neurons of rat striatum after unilateral dopamine denervation. *Neurosci Lett.* 2001;308(3):201-205.
115. Picconi B, Centonze D, Hakansson K, et al. Loss of bidirectional striatal synaptic plasticity in L-DOPA-induced dyskinesia. *Nat Neurosci.* 2003;6(5):501-506.
116. Tseng KY, Kasanetz F, Kargieman L, Riquelme LA, Murer MG. Cortical slow oscillatory activity is reflected in the membrane potential and spike trains of striatal neurons in rats with chronic nigrostriatal lesions. *J Neurosci.* 2001;21(16):6430-6439.
117. Day M, Wang Z, Ding J, et al. Selective elimination of glutamatergic synapses on striatopallidal neurons in Parkinson disease models. *Nat Neurosci.* 2006;9(2):251-259.
118. Zigmond MJ, Hastings TG. Neurochemical responses to lesions of dopaminergic neurons: implications for compensation and neuropathology. *Adv Pharmacol.* 1998;42:788-792.
119. McNeill TH, Brown SA, Rafols JA, Shoulson I. Atrophy of medium spiny I striatal dendrites in advanced Parkinson's disease. *Brain Res.* 1988;455(1):148-152.
120. Mallet N, Ballion B, Le Moine C, Gonon F. Cortical inputs and GABA interneurons imbalance projection neurons in the striatum of parkinsonian rats. *J Neurosci.* 2006;26(14):3875-3884.
121. Wilson CJ. *Single Neuron Computation.* San Diego, CA: Academic Press; 1992.
122. Galarraga E, Bargas J, Martinez-Fong D, Aceves J. Spontaneous synaptic potentials in dopamine-denervated neostriatal neurons. *Neurosci Lett.* 1987;81(3):351-355.
123. Picconi B, Gardoni F, Centonze D, et al. Abnormal Ca^{2+}-calmodulin-dependent protein kinase II function mediates synaptic and motor deficits in experimental parkinsonism. *J Neurosci.* 2004;24(23):5283-5291.
124. Wilson CJ. Basal ganglia. In: Shepherd GM, ed. *The Synaptic Organization of the Brain.* 5th ed. New York, NY: Oxford University Press; 2004.
125. Baik JH, Picetti R, Saiardi A, et al. Parkinsonian-like locomotor impairment in mice lacking dopamine D2 receptors. *Nature.* 1995;377(6548):424-428.
126. Wichmann T, DeLong MR. Functional neuroanatomy of the basal ganglia in Parkinson's disease. *Adv Neurol.* 2003;91:9-18.

127. Gross CE, Boraud T, Guehl D, Bioulac B, Bezard E. From experimentation to the surgical treatment of Parkinson's disease: prelude or suite in basal ganglia research? *Prog Neurobiol.* 1999;59(5):509–532.
128. Hutchison WD, Dostrovsky JO, Walters JR, et al. Neuronal oscillations in the basal ganglia and movement disorders: evidence from whole animal and human recordings. *J Neurosci.* 2004;24(42):9240–9243.
129. Terman D, Rubin JE, Yew AC, Wilson CJ. Activity patterns in a model for the subthalamopallidal network of the basal ganglia. *J Neurosci.* 2002;22(7):2963–2976.
130. Wilson CJ, Kawaguchi Y. The origins of two-state spontaneous membrane potential fluctuations of neostriatal spiny neurons. *J Neurosci.* 1996;16(7):2397–2410.
131. Mink JW. The basal ganglia and involuntary movements: impaired inhibition of competing motor patterns. *Arch Neurol.* 2003;60(10):1365–1368.

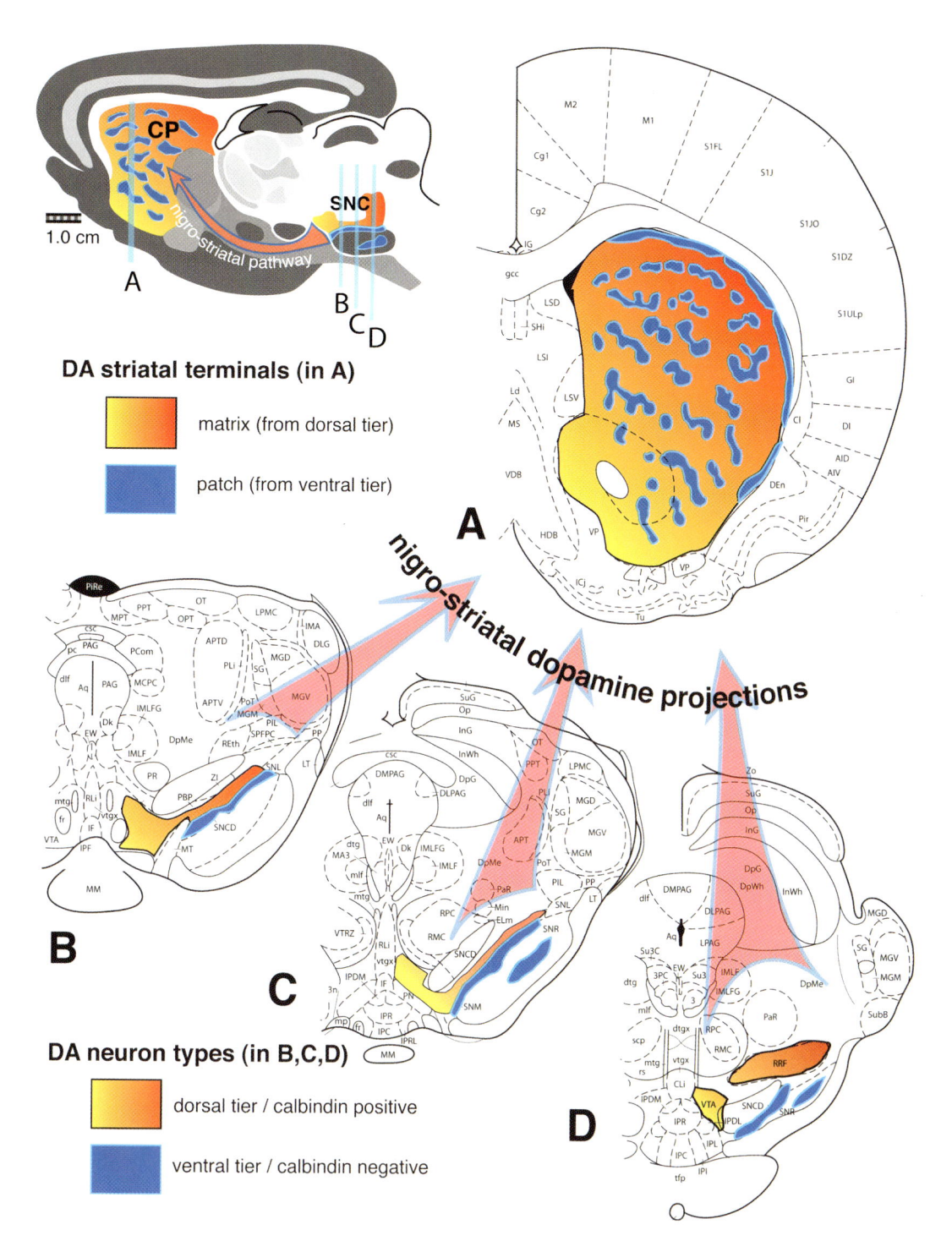

PLATE 2.1.1. The organization of the nigrostriatal dopamine (DA) pathway from the midbrain to the striatum (sagittal diagram at upper right) is diagrammed to show the organization of this system to the striatal patch and matrix compartments. Coronal sections at three levels through the striatum (A) are depicted to show the innervation of the patch and matrix compartments from different subsets of midbrain DA neurons from three levels (B, C, D). Neurons providing inputs to the striatal matrix compartment (white in B, C, D) are located in the ventral tegmental area (VTA, A10 DA cell group), in the dorsal tier of the substantia nigra pars compacta (in B,C: SNCD, A9), and in the retrorubral area (in D: RRF, A8 DA cell group). Neurons providing input to the striatal patch compartment are located in the ventral tier of the substantia nigra pars compacta (in B, C, D: DA neurons) and project from A9 DA cells located in the substantia nigra pars reticulata (in C and D). There is a general topography in that medially located cells project to the ventral striatum and laterally located cells project to the dorsal striatum. Neurons at each rostral-caudal level in the midbrain project rather extensively throughout the rostral-caudal extent of the striatum. (See Figure 2.1.1.)

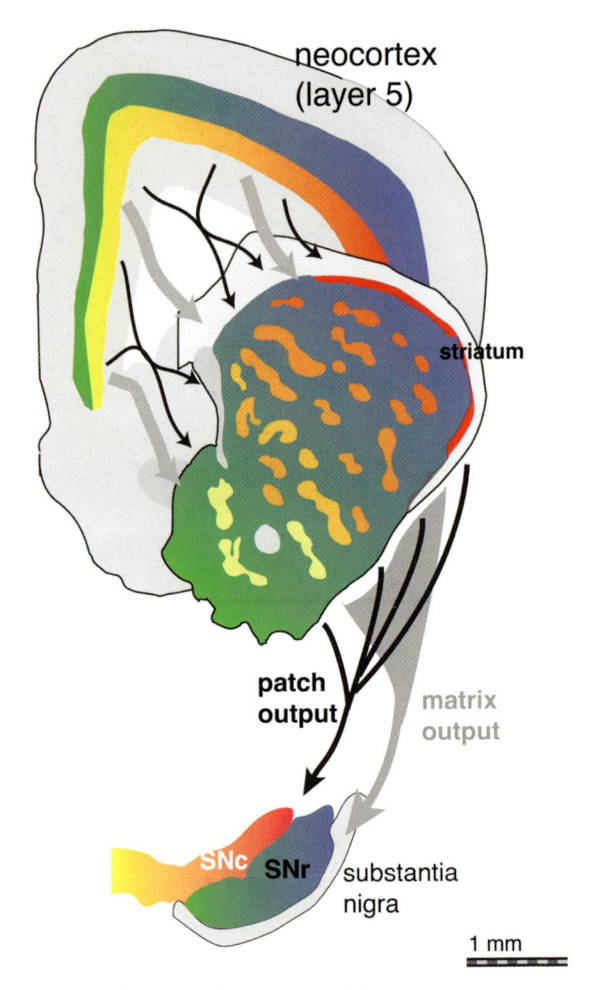

Striatal patch-matrix compartment connections

PLATE 2.1.2. Organization of the striatal patch-matrix compartments provides parallel pathways from the cerebral cortex through the striatum that provide differential input to the dopamine and GABA neurons in the substantia nigra. Deep layer 5 corticostriatal neurons provide selective inputs to the striatal patch compartment, whose neurons provide inputs targeting dopamine neurons in the substantia nigra pars compacta. Superficial layer 5 corticostriatal neurons provide inputs to the striatal matrix compartment, whose neurons project to the substantia nigra pars reticulata, which contains the GABAergic output neurons of the basal ganglia. This organization arises from most neocortical areas, although there is a gradient such that those areas closer to the allocortex provide greater input to the patch compartment, whereas primary sensorimotor areas provide greater input to the matrix compartment. (See Figure 2.1.2.)

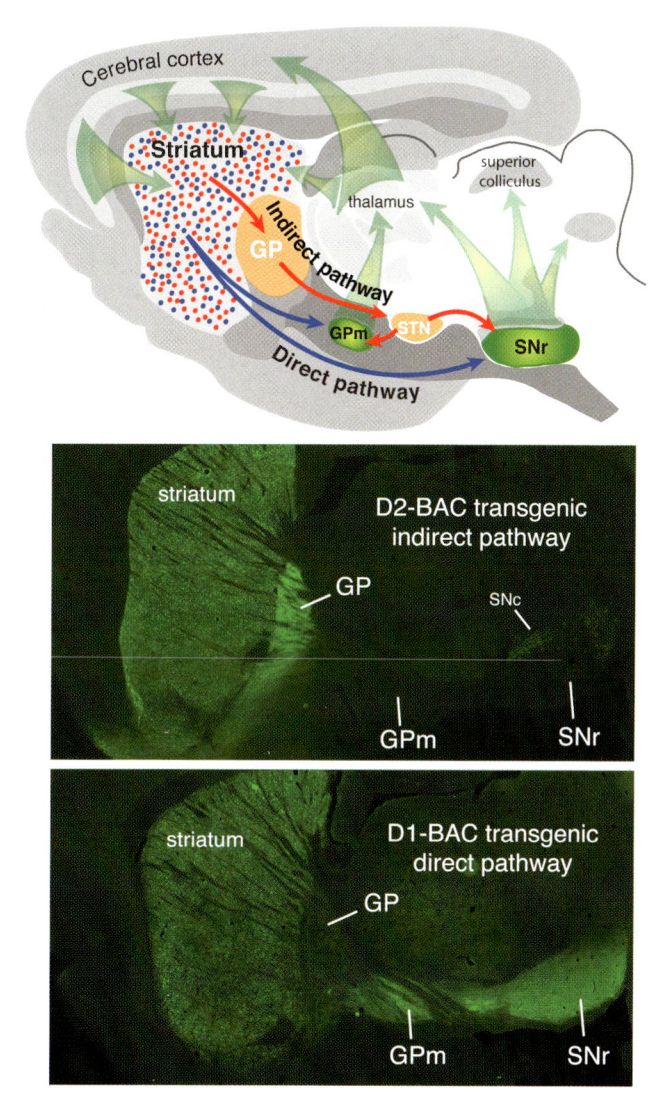

PLATE 2.1.3. Circuitry involved in Parkinson's disease. Upper diagram: Imbalances in the function of direct and indirect pathways of the basal ganglia in Parkinson's disease, shown in a sagittal brain section of the mouse. The cerebral cortex and thalamus provide excitatory inputs (green arrows) to the striatum, the main input nucleus of the basal ganglia. The output of the basal ganglia originates from the medial globus pallidus (GPm) and the substantia nigra pars reticulata (SNr) and is directed primarily to thalamic nuclei, which project to frontal areas of the cerebral cortex. The direct pathway originates from striatal projection neurons (blue) whose axons (dark blue arrows) extend directly to the GPm and SNr output nuclei. The indirect pathway originates from striatopallidal neurons (orange) whose axons terminate within the globus pallidus (GP). Neurons in the GP, in turn, project to the subthalamic nucleus (STN), which projects to the GPm and SNr. Thus, striatopallidal neurons are connected indirectly, through the GP and STN, with the output of the basal ganglia. Lower images: D1 and D2 dopamine neurons are segregated to direct- and indirect-pathway neurons, respectively. Sagittal sections from BAC transgenic mice in which these receptors are labeled with enhanced green fluorescence protein (EGFP) show labeling of the neuron cell bodies in the striatum as well as their axonal projections. D2-BAC transgenic mice show labeling of the indirect-pathway neurons (these axon projections terminate in the GP), whereas D1-BAC mice show labeling of the direct pathway, as seen by labeling of axon terminals in the GPm and SNr.[30,31] (See Figure 2.1.3.)

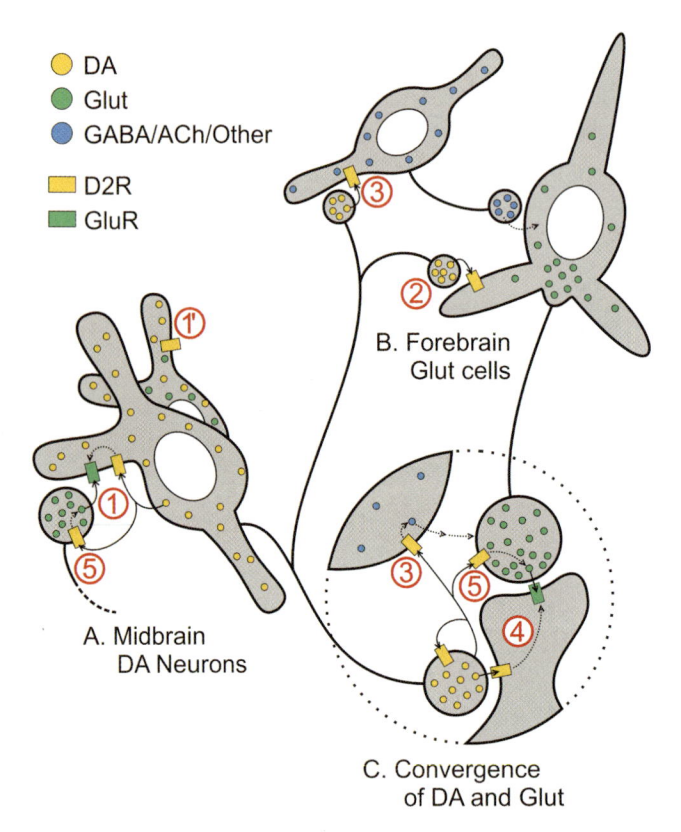

PLATE 2.2.1. Dopamine (DA) acting on D2 receptors (D2Rs) modulates glutamate (Glut) transmission at multiple levels. Solid lines indicate binding of transmitters to their respective receptors; dashed lines indicate secondary actions subsequent to transmitter binding. (A1) Midbrain DA neurons express D2 autoreceptors and receive synaptic Glut afferents. Glut-mediated depolarization would release DA from dendrites to act on D2Rs, which would counteract further excitatory influence from Glut. D2 receptors might also directly modulate Glut transmission at membrane sites where both D2Rs and Glut receptors (GluRs) are distributed in close proximity. (A1′) A subpopulation of DA neurons coexpresses Glut; these neurons may also contain D2Rs that could modulate Glut release from these neurons. (B2) Dopamine axons directly innervate Glut cells in several target areas, most notably the cortex, amygdala, hippocampus, and thalamus. Modulatory actions on D2Rs at these sites can directly alter the excitability of Glut neurons. (B3, C3) D2 receptors can also influence Glut transmission indirectly via actions on local circuit neurons (e.g., GABA cells in the cortex and cholinergic cells in the striatum). (C4) Dopamine and Glut axons often form closely convergent synapses on target neurons throughout the forebrain, especially the cortex and basal ganglia. Postsynaptic cells expressing D2Rs bind DA following synaptic release or via volume transmission. Activation of D2Rs subsequently modulates Glut transmission, primarily via regulatory actions on AMPA receptors. (A5, C5) Many Glut axon terminals express D2Rs, allowing DA to modulate Glut release presynaptically. (See Figure 2.2.1.)

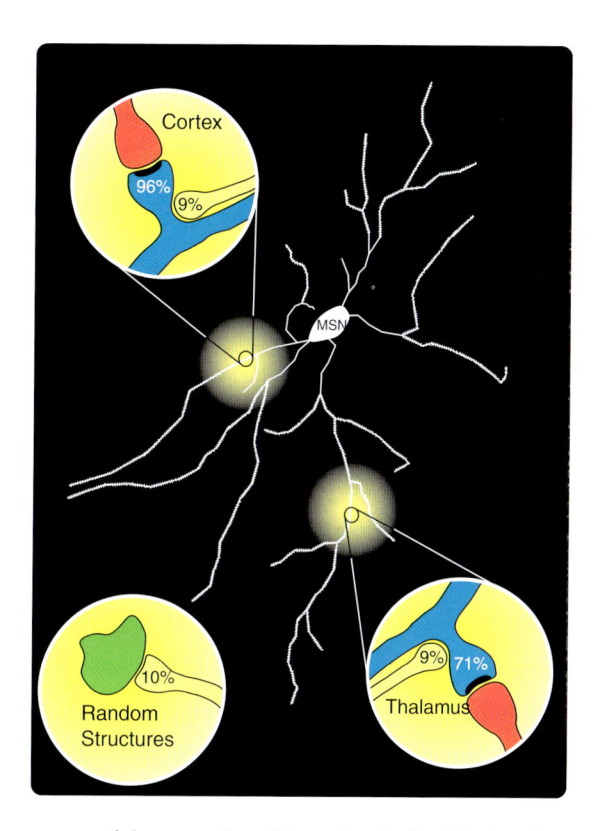

PLATE 2.4.5. Summary diagram of the convergence of glutamatergic and dopaminergic signals in the striatum and its nonselective nature. Cortical and thalamic afferents to the striatum (red) make asymmetrical synaptic contact with dendritic structures (blue) of a medium-sized spiny projection neuron (MSN, white). The majority of these contacts are with dendritic spines (cortical, 96%; thalamic, 71%), of which 9% receive a second input from a dopaminergic axon from the substantia nigra pars compacta (yellow). This, however, is no different from the proportions of random striatal structures (green) contacted by dopaminergic axons (10%), which demonstrates the nonselective nature of the relationship. In addition, dopamine (yellow clouds) spill over from the synapse and diffuse in concentrations capable of activating dopamine receptors for up to 8 µm. (See Figure 2.4.5.)

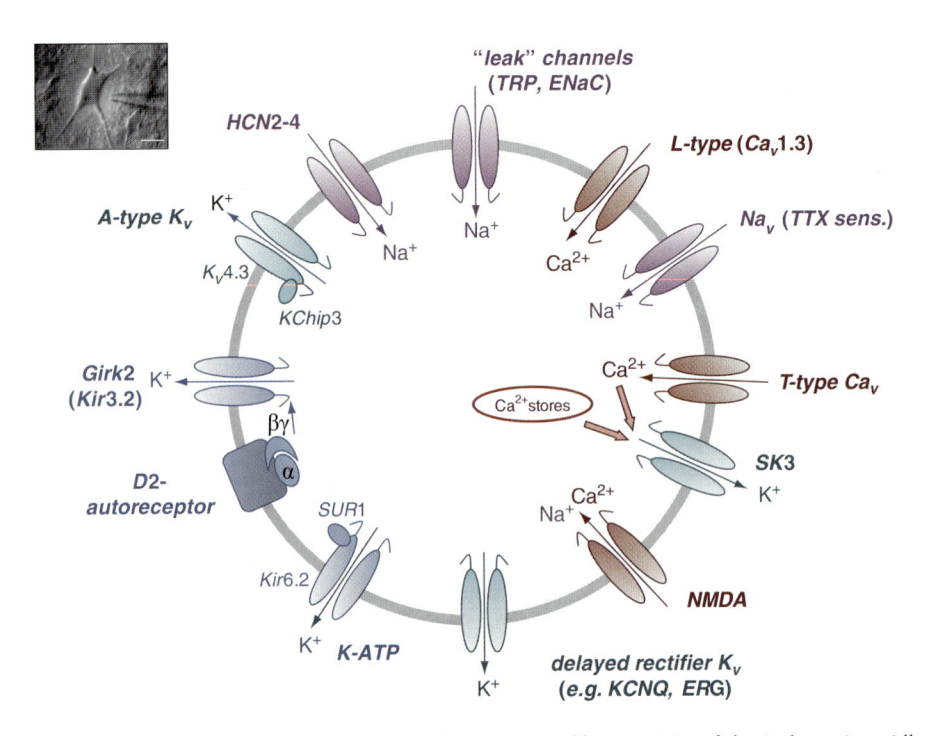

PLATE 3.4.3. Ion channels generate and modulate spontaneous pacemaker activity and burst activity of classic dopamine midbrain neurons in vitro and in vivo. For details, see text. The insert shows a classic dopamine neuron in an in vitro mouse brain slice preparation before patch clamp recording (visualized via infrared videomicroscopy; note the smaller GABAergic interneuron in the right lower part; scale bar represents 15 μm). $Ca_V/K_V/Na_V$, voltage gated calcium-, potassium-, sodium-channel; D2, dopamine-receptor subtype 2; ENaC, epithelial sodium channel; ERG, Ether-a-go-go-related gene potassium channel; Girk, G-protein coupled inwardly rectifying potassium channel; HCN, hyperpolarization activated cyclic nucleotide gated cation channel; K-ATP, ATP-sensitive potassium channel; KCNQ, KQT-like potassium channel; Kir, inwardly rectifying potassium channel; NMDA, N-methyl-D-aspartate glutamate receptor; SK, small conductance Ca^{2+} activated potassium channel; SUR, sulfonylurea receptor, TRP, transient receptor potential; TTX, tetrodotoxin. Source: Adapted from.[170] (See Figure 3.4.3.)

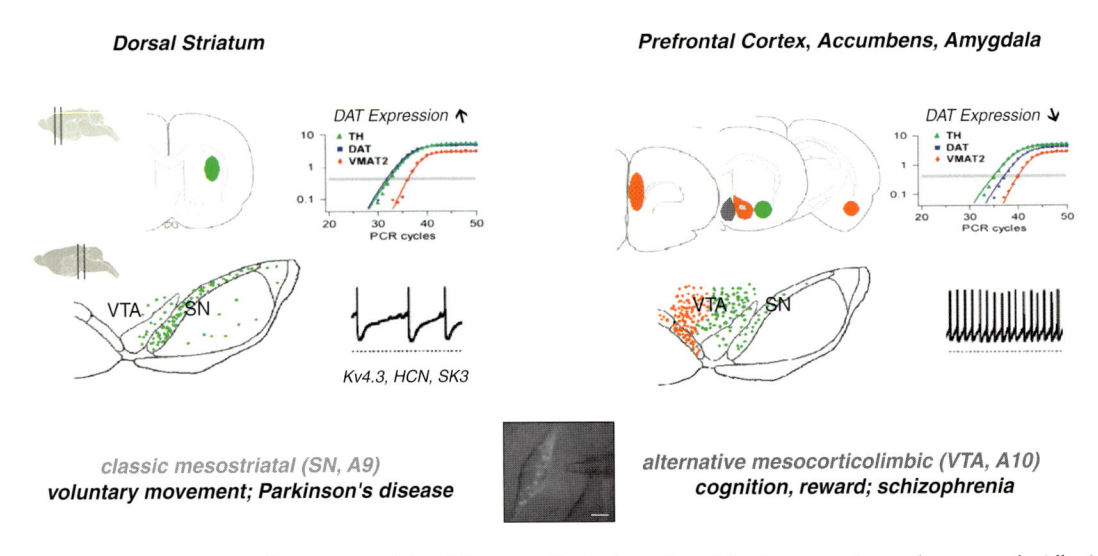

PLATE 3.4.4. The dual dopaminergic midbrain system of the adult mouse. Classic dopamine midbrain neurons (green dots, coronal midbrain sections) display well-described electrical properties (i.e., low-frequency pacemaker activity (dotted line: –80 mV), controlled by Kv4.3, HCN, and SK3 channels), express high DAT levels, and are located predominantly in the SN and the lateral VTA, projecting to dorsal striatum and to the lateral shell of the nucleus accumbens (green blobs, coronal sections). Alternative dopamine midbrain neurons (red dots) display distinct electrical properties (i.e., higher pacemaker frequency, not controlled by Kv4.3, HCN, or SK3 channels), express lower DAT levels, and are located in the more medial VTA, projecting to the prefrontal cortex, basolateral amygdala, and medial shell and core of the nucleus accumbens (red blobs). Data were obtained by combining in vivo retrograde tracing of six distinct projections of dopamine midbrain neurons with ultraviolet laser microdissection, quantitative single-cell RT-PCR, immunohistochemistry, and patch-clamp recordings of fluorescence-labeled neurons from in vitro brain slices (lower picture, fluorescence beads visualized by infrared videomicroscopy and epifluorescence; scale bar represents 5 μm); DAT, dopamine transporter 2; SN, substantia nigra; TH, tyrosine hydroxylase; VMAT2, vesicular monoamine transporter; VTA, ventral tegmental area. Source: Adapted from[77,170]. (See Figure 3.4.4.)

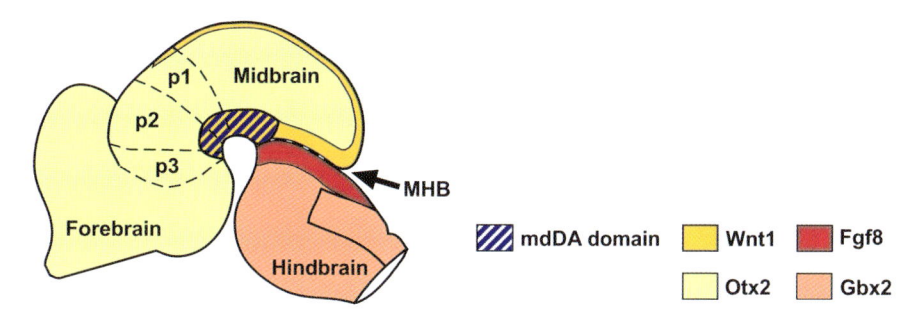

PLATE 4.1.1. The four dimensions of mdDA neuron development. First dimension: A/P positioning. Sagittal view of the neural tube in a late midgestational (E10.5–E12.5) mouse embryo depicting the domain within the cephalic flexure from where mdDA neurons (dark blue) arise. This area comprises the ventral domain (tegmentum) of the midbrain and caudal diencephalon, corresponding to prosomeres (p) 1–3 according to the prosomeric model of Puelles and Rubenstein.[7] Otx2 (light yellow) is expressed in the forebrain and midbrain, whereas Gbx2 (light red) is expressed in the rostral hindbrain. The mid-/hindbrain boundary (MHB) is positioned at the expression interface of these two TFs. At E10.5, the secreted factor Wnt1 (dark yellow) is expressed in a ring encircling the caudal midbrain rostral to the MHB, in the RP of the midbrain and caudal diencephalon and in two converging stripes within the FP of the midbrain/p1–3. The latter Wnt1 expression domain overlaps with the mdDA progenitor domain. The secreted factor Fgf8 (dark red) is expressed at E10.5 in a ring encircling the rostral hindbrain caudal to the MHB. Otx2 and Wnt1 are required for the establishment of the mdDA progenitor domain in the ventral midbrain/caudal diencephalon, and the MHB delimits the caudal extent of this progenitor domain. Fgf8, Fibroblast growth factor 8; Gbx2, Gastrulation brain homeobox 2; mdDA, meso-diencephalic dopaminergic; MHB, mid-/hindbrain boundary; Otx2, Orthodenticle homolog 2; p, prosomere; Wnt1, Wingless-related MMTV integration site 1. *Source*: Modified from Marín et al.[6] (See Figure 4.1.1.)

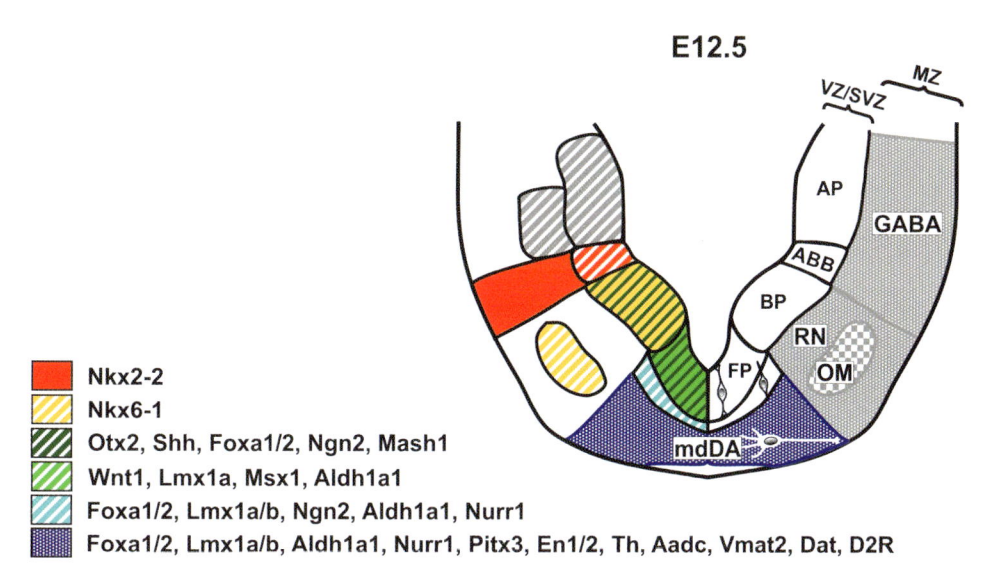

PLATE 4.1.2. The four dimensions of mdDA neuron development. Second and third dimensions: D/V and mediolateral positioning. Coronal view of the ventral midbrain in a late midgestational (E12.5) mouse embryo depicting the progenitor domain within the FP (green) from which mdDA neurons (blue) develop. Wnt1, Lmx1a, and Msx1 expression (light green) is restricted to the midbrain/p1–3 FP and necessary for proper mdDA neurogenesis from the progenitors located in the VZ/SVZ. Msx1 also represses Nkx6-1 expression (yellow) within the FP. Aldh1a1 expression in these cells serves as a marker for mdDA progenitors. Expression of Otx2, Shh, Foxa1/2, Ngn2, and Mash1 (dark green) is not restricted to the midbrain FP but is also found in BP progenitors. These secreted factors and TFs, however, play a prominent role in mdDA neurogenesis. Moreover, Otx2 is necessary for the ventral repression of Nkx2-2 (red), a 5-HT neuron-inducing factor. The postmitotic mdDA precursors express Foxa1/2, Lmx1a/b, Ngn2, Nurr1, and Aldh1a1 (light blue) and require the TFs for their proper differentiation into mdDA neurons. Differentiating and adult mdDA neurons (dark blue) express Foxa1/2, Lmx1a/b, Aldh1a1, Nurr1, Pitx3, En1/2, and the DA biosynthetic enzymes Th and Aadc, the DA transporters Vmat2 and Dat, and the DA autoreceptor D2R. These TFs and the RA-synthesizing enzyme Aldh1a1 are required for the maturation and/or survival of mdDA neurons. The proliferating, radial glia-like mdDA progenitors are located in the VZ/SVZ of the midbrain/p1–3 FP and give birth to the postmitotic mdDA precursors, which migrate radially out of the VZ/SVZ into the MZ and begin differentiation into mdDA neurons. The mature mdDA neurons migrate tangentially in a mediolateral and A/P (not shown) direction to their final destinations in the SNc and VTA. ABB, alar-basal boundary; AP, alar plate; BP, basal plate; FP, floor plate; GABA, γ-aminobutyric acid-synthesizing neurons; mdDA, meso-diencephalic dopaminergic neurons; MZ, mantle zone; OM, oculomotor nucleus; RN, red nucleus; VZ/SVZ, ventricular/subventricular zone. (See Figure 4.1.2.)

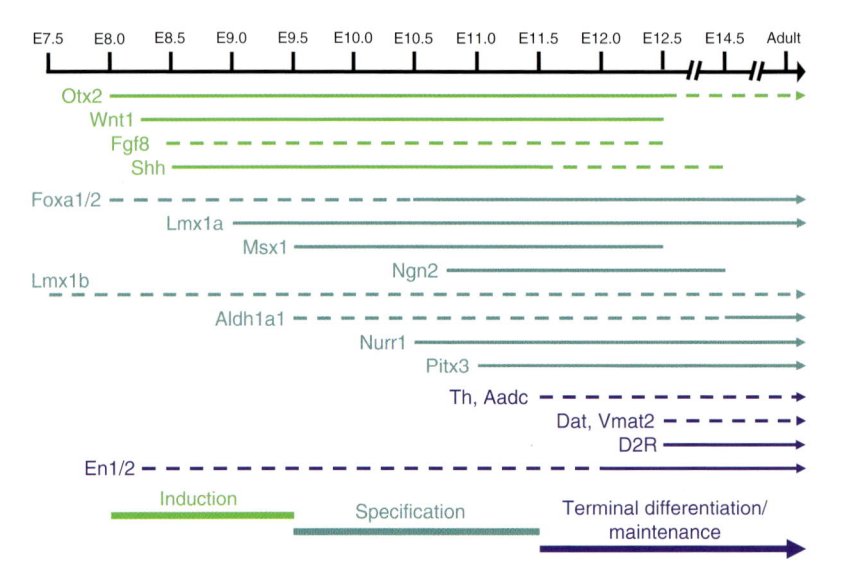

PLATE 4.1.3. The four dimensions of mdDA neuron development. Fourth dimension: developmental time. A time scale of mouse embryonic development from E7.5 (when neurulation starts) to adulthood with a special focus on early and late midgestational stages is shown at the top. The color-coded (as in Fig. 4.1.2) bars below indicate the onset of expression of the corresponding secreted factor or TF, enzyme, or transporter protein according to the time scale on top and the time interval during which the corresponding gene is transcribed. Solid bars indicate a requirement of the corresponding molecule for proper mdDA neuron development during that time interval. Dotted bars indicate (a) that the corresponding protein is not required for mdDA neuron development (Shh, En1/2, Th, Aadc, Dat, Vmat2) or (b) that a direct requirement has not yet been demonstrated (Otx2, Fgf8, Foxa1/2, Lmx1b, Aldh1a1) during that time interval. Arrows indicate that the corresponding protein is expressed throughout adulthood and required for the maintenance/survival or physiological function of mdDA neurons. The time intervals during which the induction of the mdDA progenitor domain, the specification of the mdDA neuronal fate in postmitotic mdDA precursors, and the terminal differentiation/maintenance of the mdDA neurons take place in the mouse ventral midbrain are depicted at the bottom of the figure. Aldh1a1, aldehyde dehydrogenase family 1, subfamily a1 (Raldh1, Ahd2); Dat, dopamine transporter (Slc6a3); D2R, dopamine receptor 2 (Drd2); En1/2, Engrailed 1 and 2; Fgf8, Fibroblast growth factor 8; Foxa1/2, Forkhead box A1/A2 (Hnf3α/β); Lmx1a/b, LIM homeobox transcription factor 1 alpha/beta; Msx1, Muscle-segment homeobox-like 1; Ngn2, Neurogenin 2 (Neurog2); Nurr1, nuclear receptor subfamily 4, group A, member 2 (Nr4a2); Otx2, Orthodenticle homolog 2; Pitx3, Paired-like homeodomain transcription factor 3; Shh, Sonic hedgehog; Th, tyrosine hydroxylase; Vmat2, vesicular monoamine transporter 2 (Slc18a2); Wnt1, Wingless-related MMTV integration site 1. (See Figure 4.1.3.)

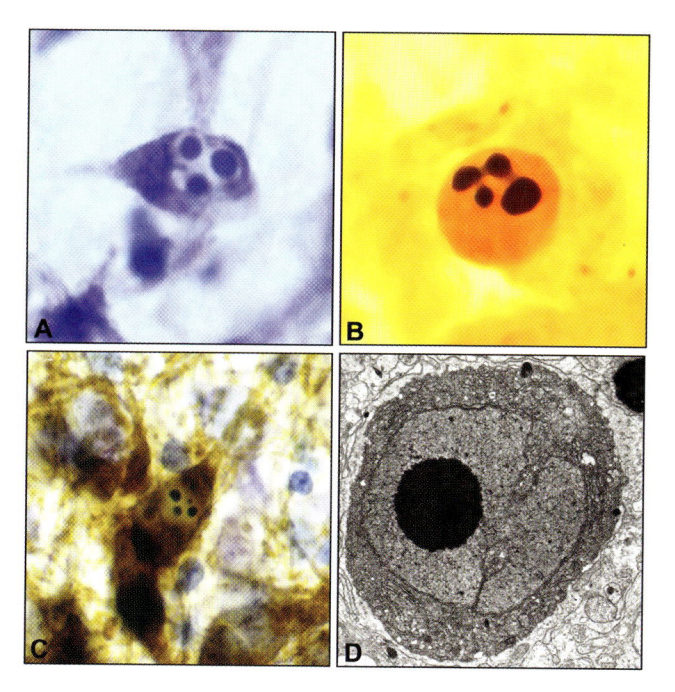

PLATE 4.2.2. Apoptosis in SNpc during postnatal development. (A) Thionin stain of the SNpc of a normal rat at PND8. Within the nucleus of this neuron are three intensely and homogeneously stained round chromatin clumps with sharp, clearly defined edges. These chromatin clumps are highly characteristic of apoptosis at the light microscopic level. Note that this profile, in spite of the presence of apoptotic chromatin in its nucleus, has some preservation of neuronal morphology, including a polygonal shape and a dendrite. (B) Suppressed silver stain of an apoptotic profile at PND2 in a normal rat. Four intensely argyrophilic chromatin clumps are observed. *Source*: Adapted from [8]. (C) Immunoperoxidase stain for TH, with a thionin counterstain, in the SNpc 24 hours following axon-sparing excitotoxic striatal target lesioning at PND7. The brown reaction product identifies this neuron as dopaminergic. The four intranuclear chromatin clumps, stained by the thionin counterstain, are characteristic of apoptosis. *Source*: Adapted from [44]. (D) An electron micrograph of an apoptotic profile in the SNpc 24 hours following excitotoxic striatal target lesioning at PND7. The single intensely and homogeneously electron-dense clump of chromatin within the nucleus is a defining feature of apoptosis. The intact nuclear and cellular membranes in this degenerating profile are also characteristic of apoptotic cell death.[44] (See Figure 4.2.2.)

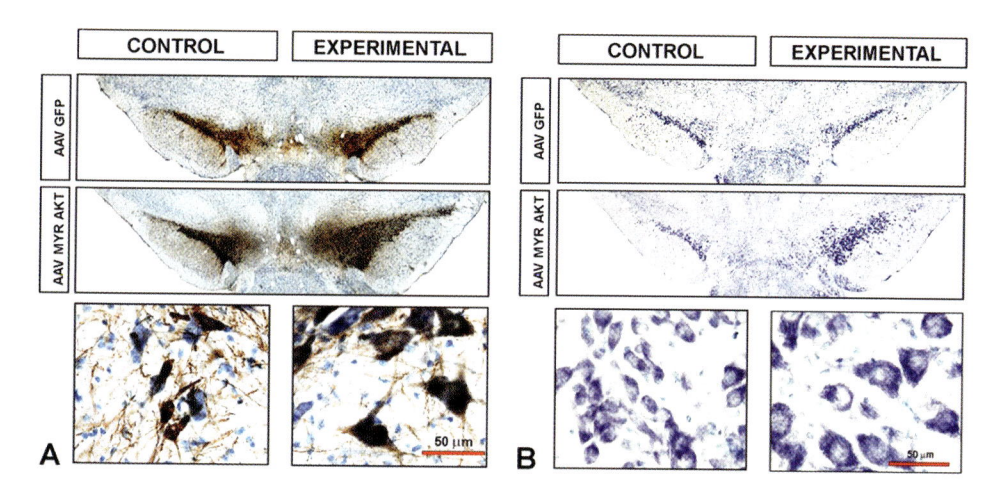

PLATE 4.2.5. Transduction of SNpc neurons during postnatal development with Myr-Akt results in an increase in their size and number. (A) Immunoperoxidase labeling of TH within the SNpc at 28 days postinjection (PND33) on one side of the brain of either AAV GFP or AAV Myr-Akt on PND5. The low-power photomicrographs in the top panel show, at a regional level, that there is no difference between the injected side (Experimental) and the uninjected side (Control) in the AAV GFP-treated animals, whereas the Experimental side of the AAV Myr-Akt-injected animals demonstrates a markedly increased extent of TH immunostaining in the SNpc. The higher-power micrographs shown in the lower panels demonstrate at a cellular level that this increased extent of TH staining is due primarily to a marked increase in the size of SNpc TH-positive neurons. (B) Nissl stain of the ventral mesencephalon at 28 days postinjection (PND33) of either AAV GFP or AAV Myr-Akt on PND5. The low-power photomicrographs in the top panel show that the apparent increase in the size and number of dopamine neurons in the SNpc is observed independently of the expression of TH. An increase in the size of Nissl-stained neurons is observed at the cellular level in the lower panels. (See Figure 4.2.5.)

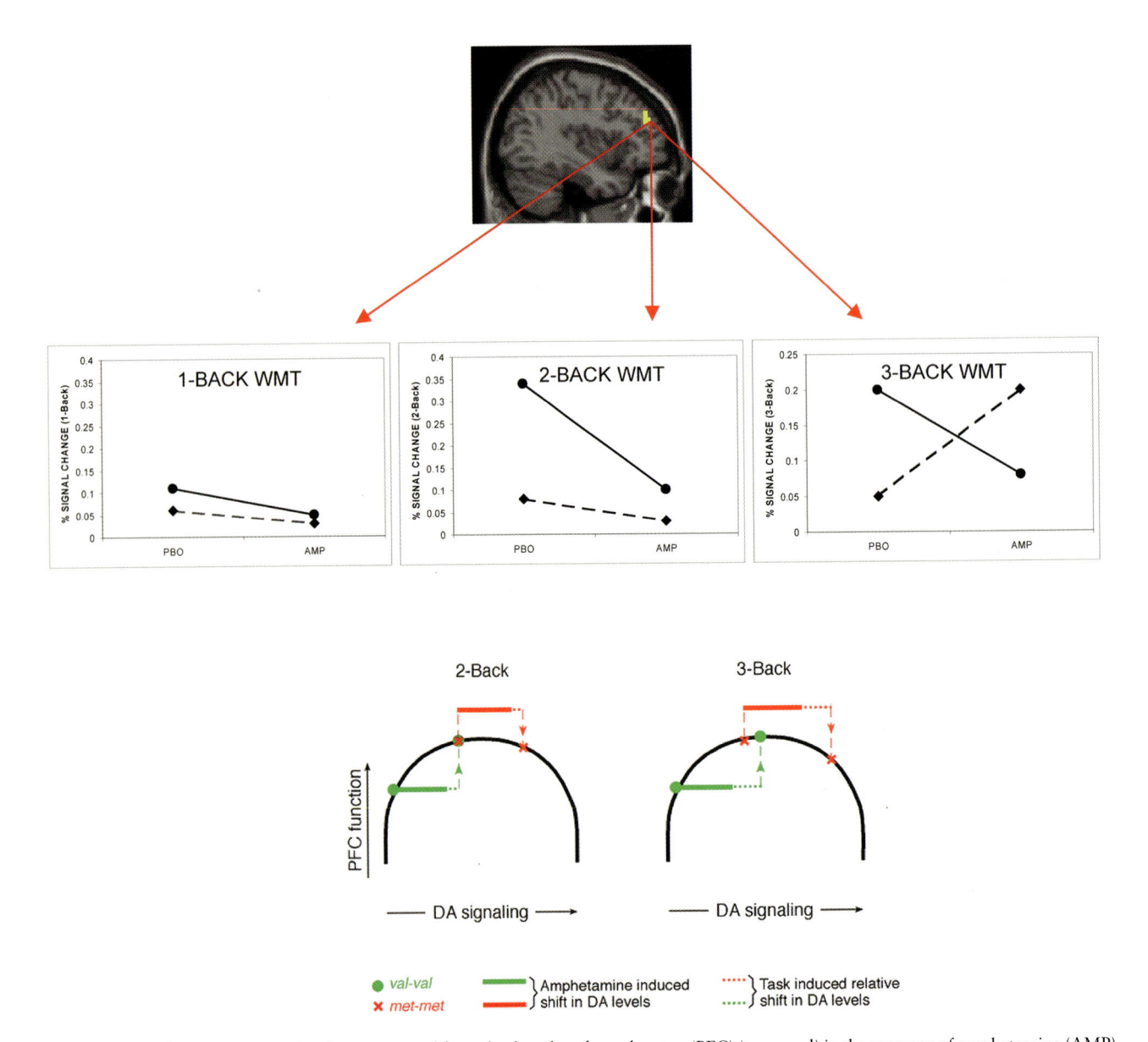

PLATE 4.4.1. The fMRI activation signal was extracted from the dorsal prefrontal cortex (PFC) (top panel) in the presence of amphetamine (AMP) or placebo (PBO) administration at differing working memory task (WMT) loads as a function of the COMT genotype (middle panel). In COMT-Val homozygote individuals (who have relatively less cortical DA; solid lines, middle panel), AMP improved PFC efficiency (lower activation). In contrast, in individuals homozygous for the *met* allele (who have relatively greater cortical DA; dashed lines, middle panel), AMP had deleterious effects on PFC efficiency (greater activation) at a three-back WMT load (rightmost graph in the middle panel). These results suggest that individuals homozygous for the COMT-*val* allele have PFC functional efficiency on the up slope of the normal range, whereby AMP could increase DA signaling to more optimal levels closer to the peak (bottom panel). On the other hand, individuals homozygous for the COMT-*met* allele appear to already be near peak PFC functional efficiency, so increased DA signaling from AMP shifts PFC function onto the down slope of the inverted-U efficiency curve (bottom panel). *Source*: Adapted from Mattay et al.[84]; courtesy of Venkata S Mattay. (See Figure 4.4.1.)

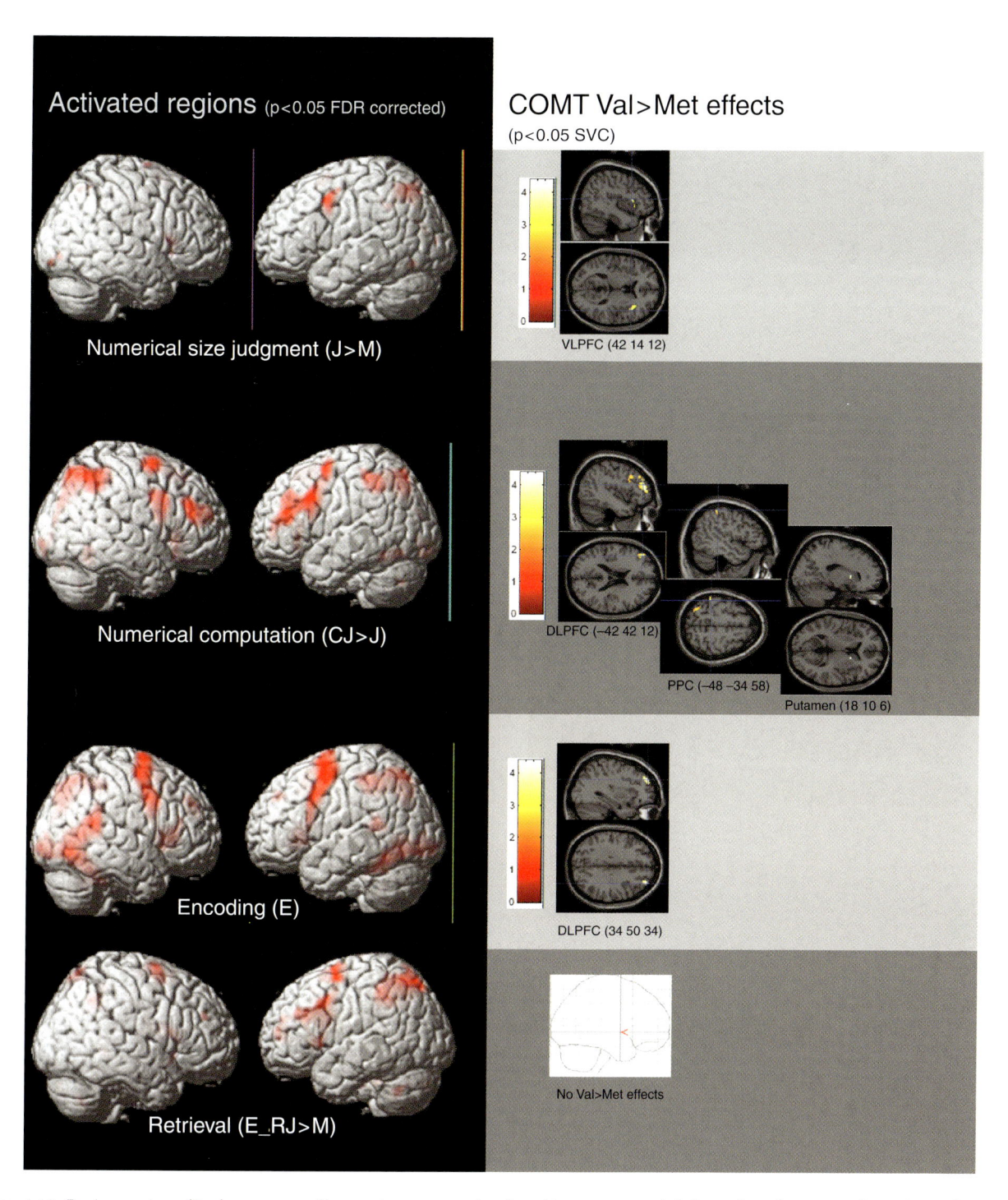

PLATE 4.4.2. Regions activated in the contrasts of interest in an event-related working memory task (left panel), and corresponding ROIs with COMT Val>Met effects (right panel). During baseline numerical size judgment, subjects engaged COMT effects at the ventrolateral prefrontal cortex (VLPFC). During encoding into working memory, COMT effects were observed in the dorsolateral prefrontal cortex (DLPFC) but not in the striatum. During numerical computations engaging rapid updating of new information, COMT effects were observed in the prefronto-parietal-striatal network. During simple retrieval in working memory, no suprathreshold COMT effects were observed. SVC: small volume correction for multiple comparisons. PFC, prefrontal cortex; PPC, posterior parietal cortex; ROI, region of interest. *Source*: Adapted from Tan et al.[102] (See Figure 4.4.2.)

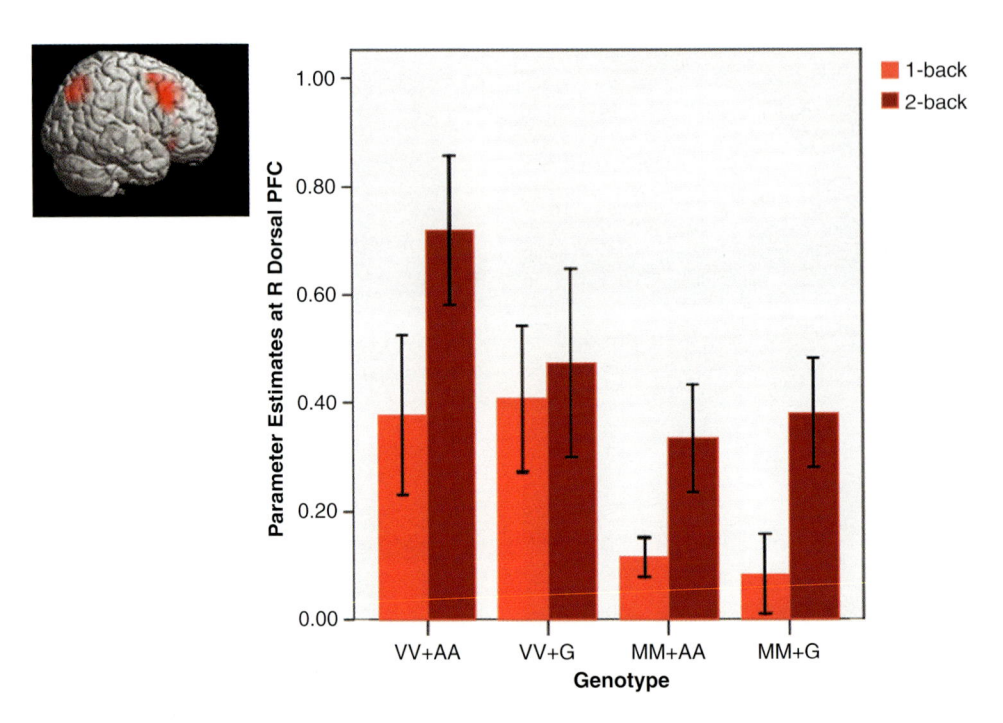

PLATE 4.4.3. Epistatic interaction between COMT and *GRM3* on prefrontal brain function. Higher-load working memory processes engaging the dorsolateral prefrontal cortex (PFC) was disproportionately inefficient in the context of combined suboptimal COMT and *GRM3* risk alleles ($F_{1,25} = 4.47$, $p = 0.045$). *Source*: Adapted from Tan et al.[116] (See Figure 4.4.3.)

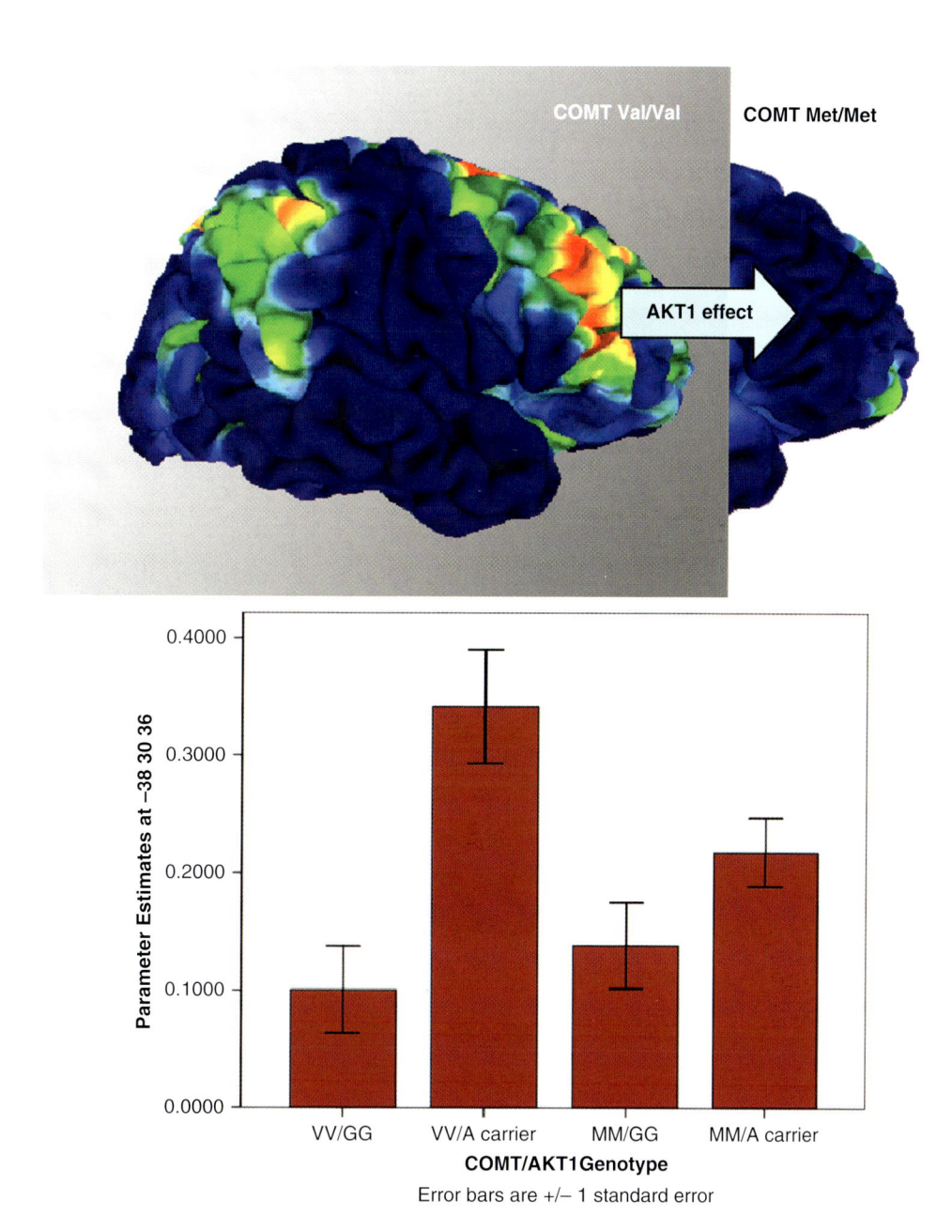

PLATE 4.4.4. Epistatic interaction between AKT1 and COMT. Here, individuals with the AKT1 allele associated with reduced gene expression showed disproportionately inefficient DPFC activity in the background of a relatively deleterious COMT Val allele ($F_{1,42} = 4.466$, $p = 0.041$). *Source*: Adapted from Tan et al.[126] (See Figure 4.4.4.)

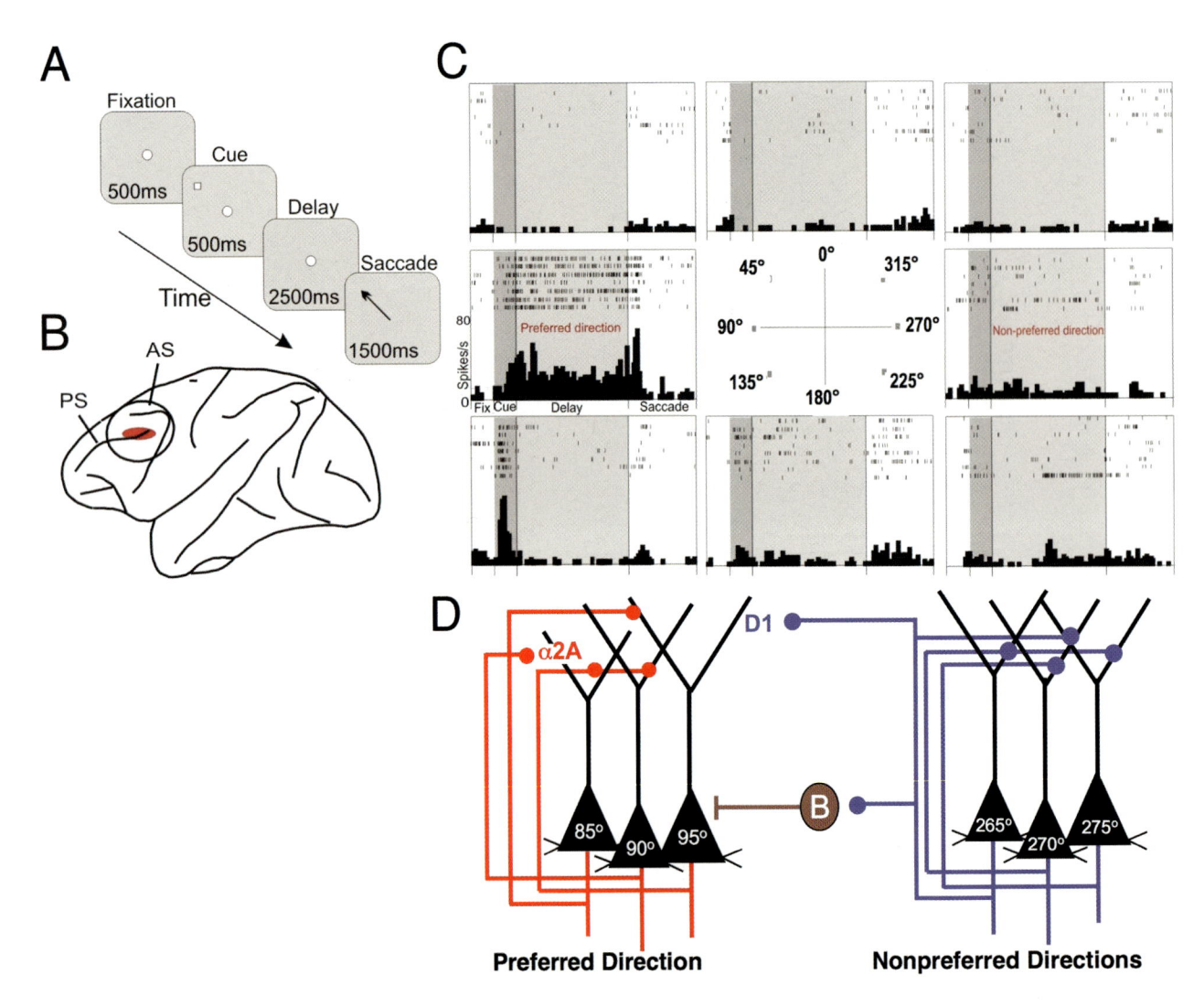

PLATE 5.3.1. Spatial working memory in monkeys is often tested using the ODR task, illustrated in panel A (see text for description). Single-unit recordings from the principal sulcal PFC in the region illustrated in panel B show spatial mnemonic tuning as the monkey performs the task. Rasters from a typical neuron are illustrated in panel C. This neuron shows increased firing during the delay period if the cue had been at 90° (the preferred direction for the neuron), but does not show increased firing if the cue had been at other spatial locations (nonpreferred directions). Panel D illustrates the microcircuits in PFC thought to underlie the spatially tuned mnemonic firing (based on Goldman-Rakic, 1995[9]). The persistent firing during the delay period is thought to arise from recurrent excitation among similarly tuned pyramidal cells, while the spatial tuning arises from GABAergic inhibition, (e.g., the basket cell B illustrated here). Prefrontal cortex neuronal firing is also powerfully modulated by catecholamines. Under optimal conditions, α2A-AR stimulation strengthens delay-related firing for the preferred direction, while DA D1R stimulation suppresses responses to non-preferred directions. (See Figure 5.3.1.)

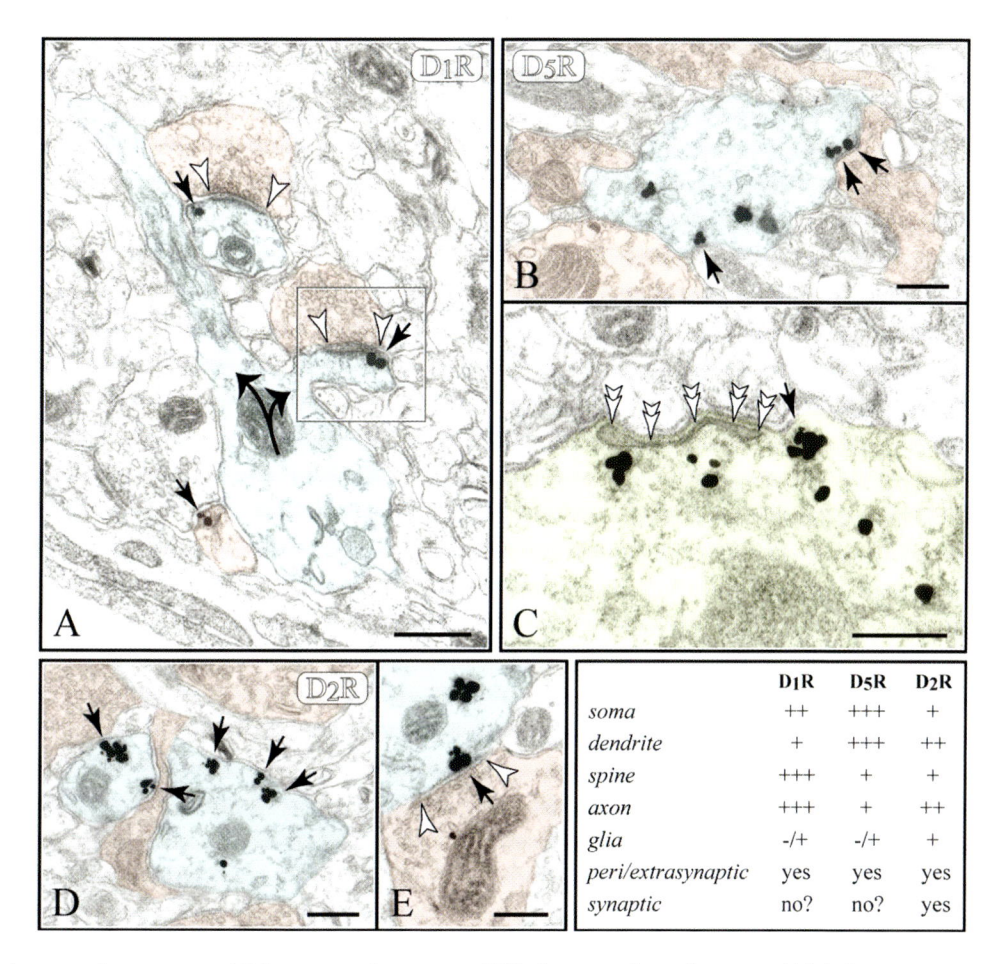

PLATE 5.3.2. Main expression patterns of DA receptors in macaque PFC. Arrows point to immunogold labeling; synaptic specializations are between the arrowheads. Dendrites, axons, and somata are pseudocolored in blue, red, and yellow, respectively. (A) Perisynaptic expression on spine membranes (frame) is a salient feature of the D1R; curved arrows point to emerging spines. (B) Dendritic stems are the prevalent D5R-immunoreactive profiles. (C) In the pyramidal soma, nonsynaptic D5Rs are affiliated with subsurface cisterns (double arrowheads) that hold Ca^{2+} stores. Medium and fine dendrites exhibit nonsynaptic D2Rs (D) and D2Rs embedded in the postsynaptic density of symmetric axodendritic synapses (E). The table summarizes the expression patterns of individual receptor subtypes in the PFC neuropil. Scale bars: 200 nm. From[29,30,33,36]. (See Figure 5.3.2.)

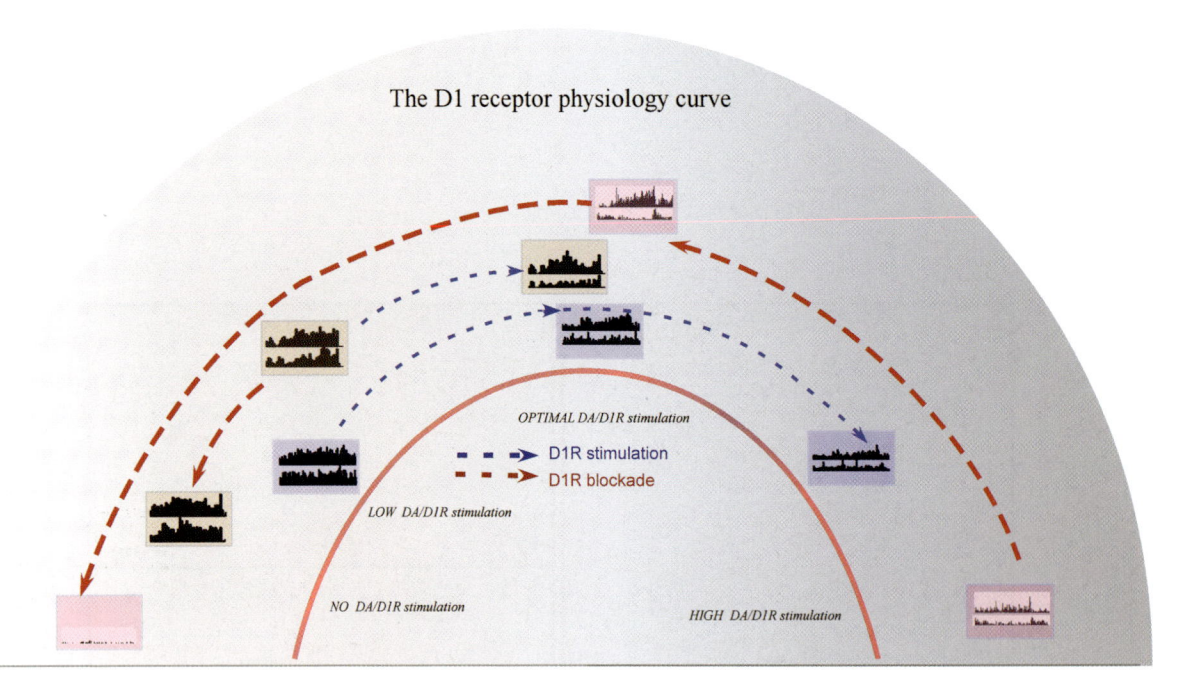

PLATE 5.3.3. D1 receptor stimulation shows an inverted-U dose/response on the physiological profiles of neurons in the principal sulcal PFC in monkeys performing a spatial WM task. The neuron's firing patterns to its preferred direction is shown above the response to its nonpreferred direction. Pink shading from[54]; blue or beige shading from[55]. See text for explanation. (See Figure 5.3.3.)

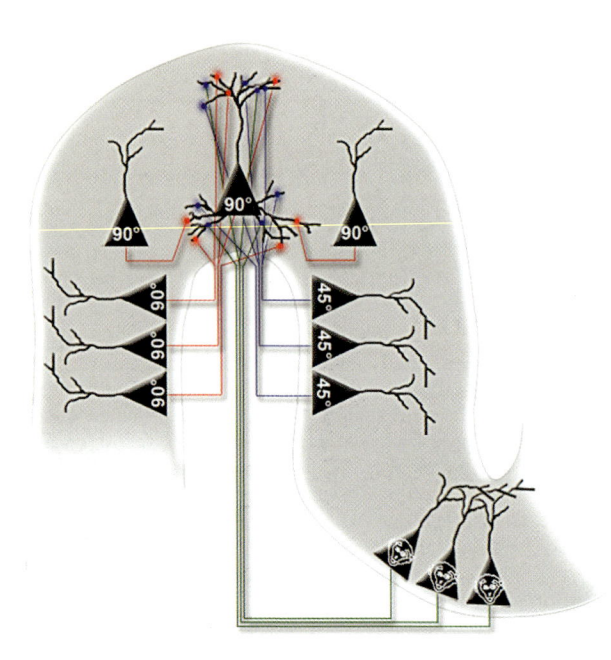

PLATE 5.3.6. Hypothetical model illustrating how D1R and α2A-AR may dynamically regulate network connections to a pyramidal neuron in the principal sulcal PFC, in which α2A-ARS gate isodirectional inputs, and D1Rs gate contradirectional and other inputs. Red axons indicate network inputs from pyramidal neurons with shared spatial tuning properties (i.e., the best response to 90°). These inputs appear to be modulated by α2A-ARs, as receptor stimulation increases delay-related firing for the preferred direction. Spatial inputs from nearby PFC neurons with tuning for other spatial directions (e.g., 45°) are illustrated in blue. These network inputs appear to be gated by D1R stimulation. We hypothesize that other dissimilar inputs (e.g., ventral PFC neurons that respond to the memory of faces, shown in green) would similarly be gated by D1R stimulation. Thus, with greater D1R stimulation, the neuron would become more narrowly tuned. This would be helpful during some cognitive demands (e.g., spatial WM for a precise location) but would be detrimental under conditions where flexibility and breadth are required (e.g., set shifting, creative insights). (See Figure 5.3.6.)

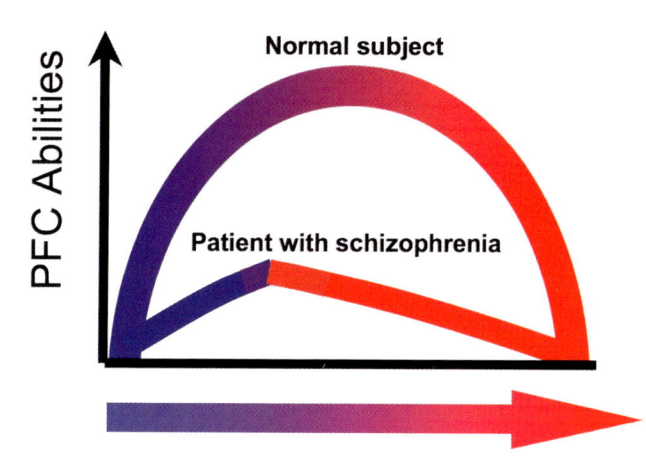

PLATE 5.3.7. Hypothetical illustration of the D1R inverted U in normal individuals versus patients with schizophrenia. Patients may have reduced levels of DA in PFC, but may be more responsive to the detrimental effects of D1R due to up-regulation of D1Rs and loss of enzymes that hold intracellular stress pathways in check (e.g., loss of DISC1). Thus, they may have reduced D1R beneficial actions, as well as potentiation of detrimental actions. (See Figure 5.3.7.)

(a) Basal DA synthesis capacity in the striatum

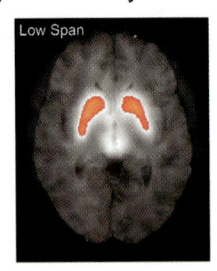

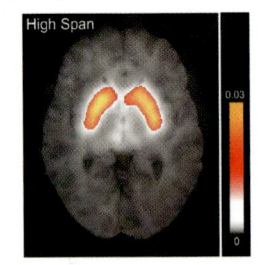

(b) Effect of bromocriptine on switch-related BOLD signal in the striatum

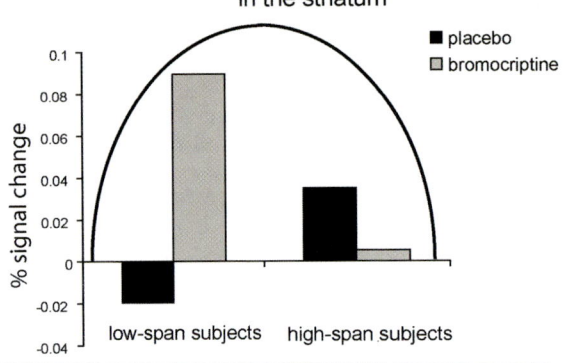

(c) Switch-related error rates

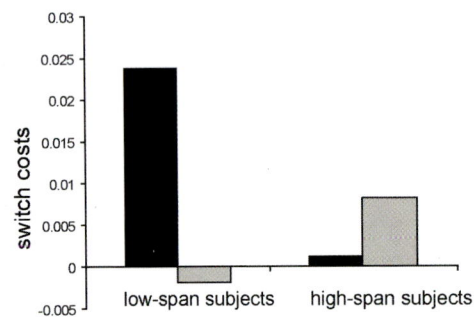

PLATE 5.4.1. The effects of DA receptor stimulation depend on baseline working memory capacity as measured with the listening span test,[30] which correlates with baseline DA synthesis capacity in the striatum.[41] (a) Subjects with a high listening span had lower DA synthesis capacity, as measured with neurochemical PET imaging, than did subjects with a low listening span; (b) bromocriptine had opposite effects on neural activity measured with fMRI and (c) performance in high- and low-span subjects, consistent with an inverted-U-shaped relationship between DA receptor stimulation and cognitive performance. (See Figure 5.4.1.)

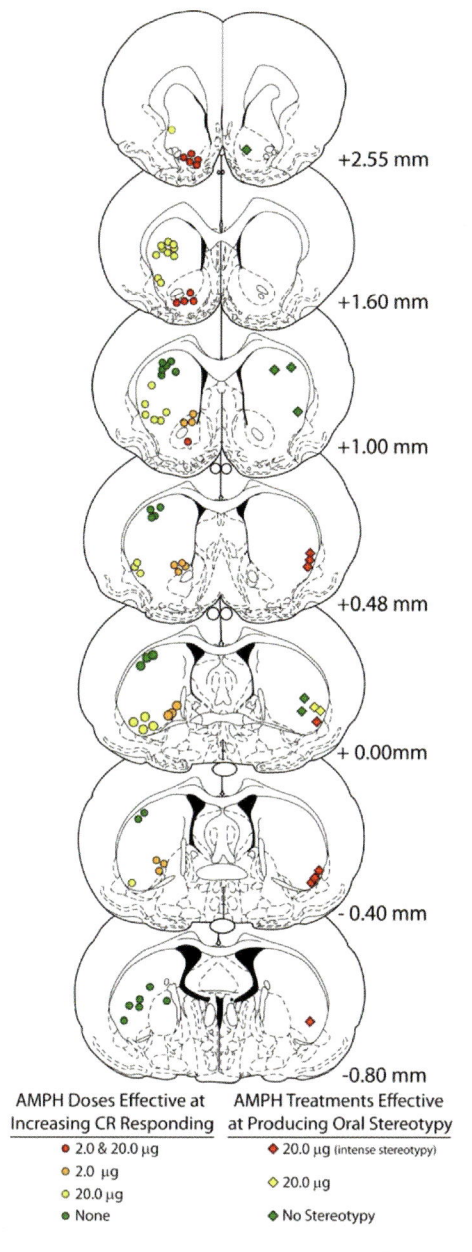

PLATE 6.3.1. Schematic representation of d-amphetamine infusion sites from Kelley et al.[40] (right side) and Delfs and Kelley[48] (left side). On the right, oral stereotypies (in yellow) and intense stereotypies (in red) were produced by 20-μg infusions in the VLS but not in other regions of the striatum, notably the Acb. Conversely, both 2.0-μg and 20-μg infusions (in red) robustly enhanced responding for a conditioned reinforcer (CR) in the Acb. The lower dose used, 2.0 μg, was also effective in increasing CR responding in the posterior medial striatum (in orange), while the higher dose (20 μg) was effective in a gradient that went from the anterior dorsal portion to more posterior ventral and lateral portions (in yellow). Neither dose was effective in the most posterior and dorsal sites (in green). Note the differential effectiveness within the same site of the same drug on oral stereotypies and CR responding. Line drawings of brain sections were adapted from the atlas of Paxinos and Watson, with permission from Elsevier. (See Figure 6.3.1.)

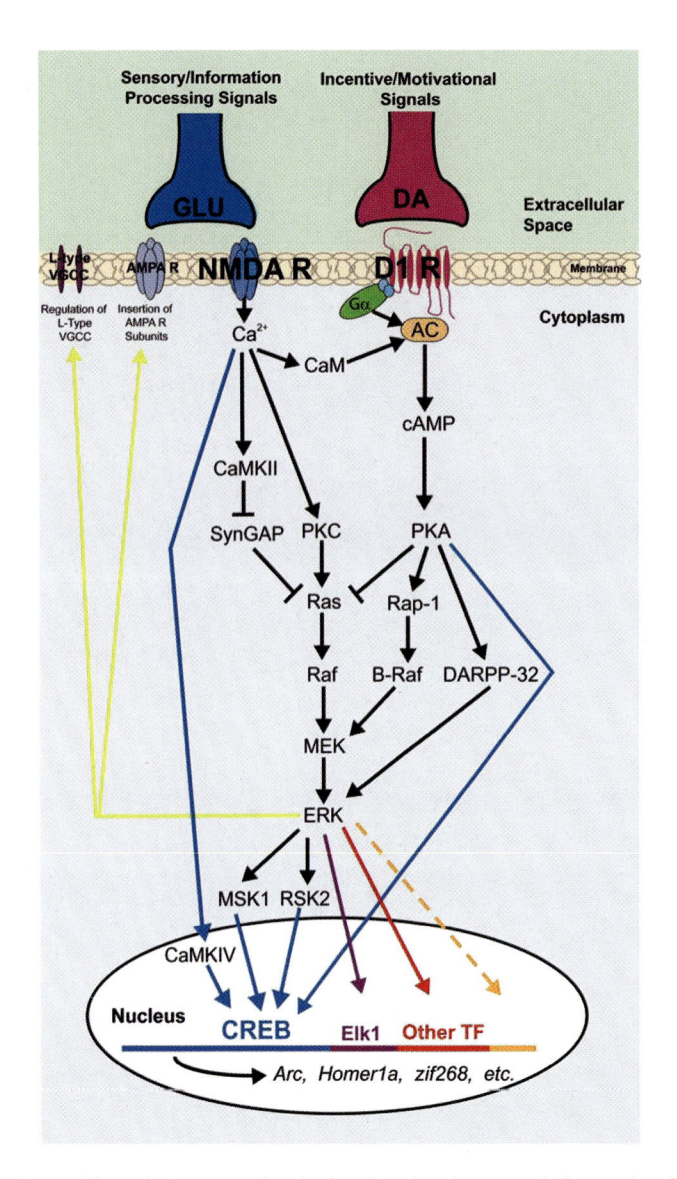

PLATE 6.3.2. One prominent and influential hypothesis concerning the functional and structural changes involved in neural plasticity implicates coordinated NMDAR and dopamine (DA) D1R activation throughout cortical-striatal-limbic networks. This figure summarizes the prevailing models of convergence and divergence of intracellular signals, following NMDAR and DA D1R activation, leading to activation and/or phosphorylation of key enzymes, inhibition of particular signals, and transcription of critical immediate early genes. CamK, calcium/calmodulin-dependent protein kinase; SynGAP, Synaptic Ras guanosine-5′-triphosphatase activating protein; TF, transcription factor. See text for other abbreviations. *Source:* based on[99,147–149]. (See Figure 6.3.2.)

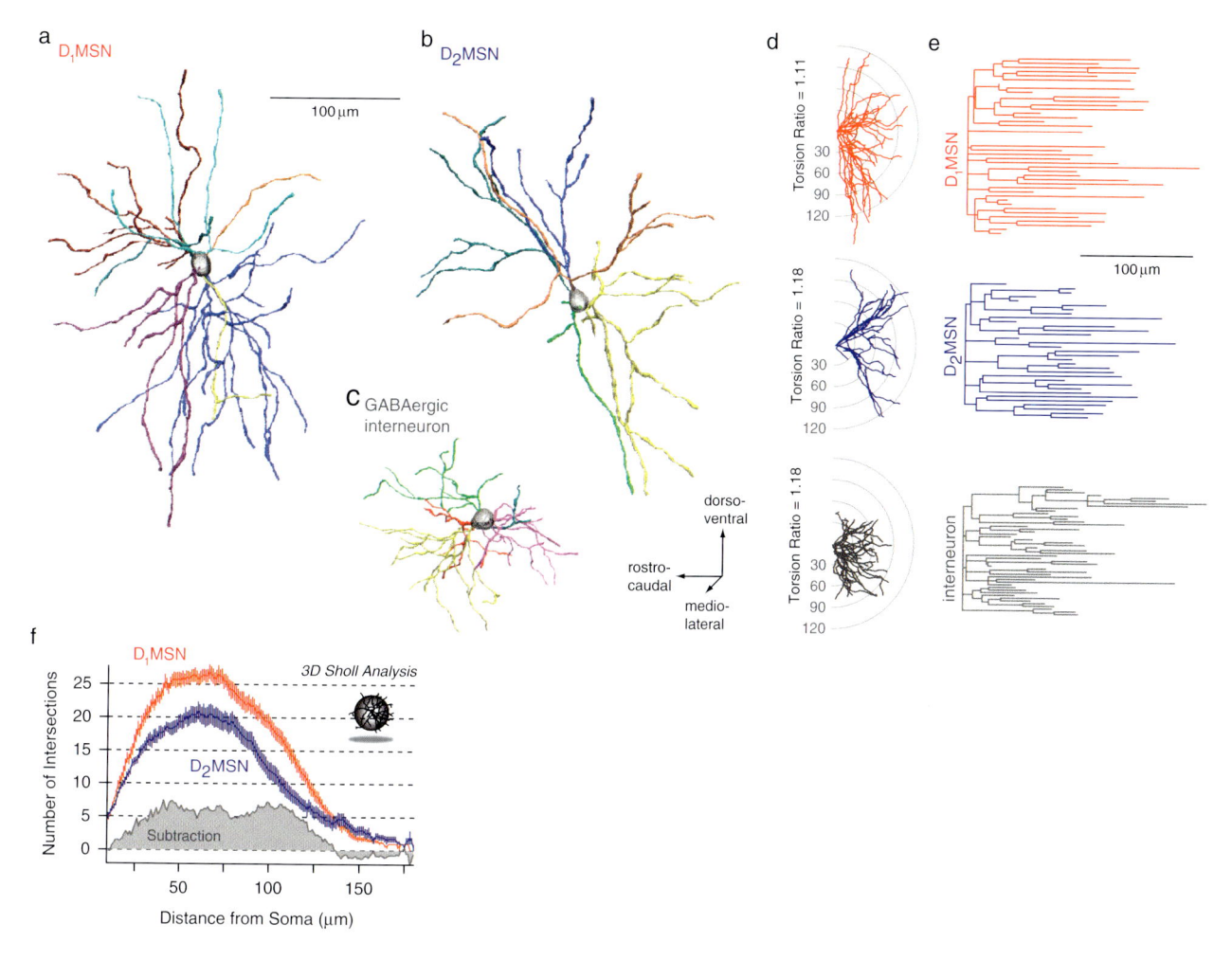

PLATE 7.3.1. D1 and D2 MSNs differ in dendritic anatomy. (a–c) Striatal neurons from P35–P45 BAC transgenic mice were biocytin-filled, imaged, and reconstructed in three dimensions. A GABAergic interneuron is included for comparison. (d) Fan-in diagrams displayed no apparent preferred orientation in either the D1 or D2 MSN populations. (e) Dendrograms displaying in two dimensions the length, number, and connectivity of dendritic segments in sample neurons. (f) Three-dimensional Sholl analysis of biocytin-filled and reconstructed neurons from P35–P45 BAC transgenic mice. Data are shown as the mean (+/- SEM) number of intersections at 1-μm eccentricities from the soma for 15 D1 and 16 D2 MSNs. D1 MSNs have a more highly branched dendritic tree, as indicated by the increased number of intersections and positive subtracted area (gray shading). (See Figure 7.3.1.)

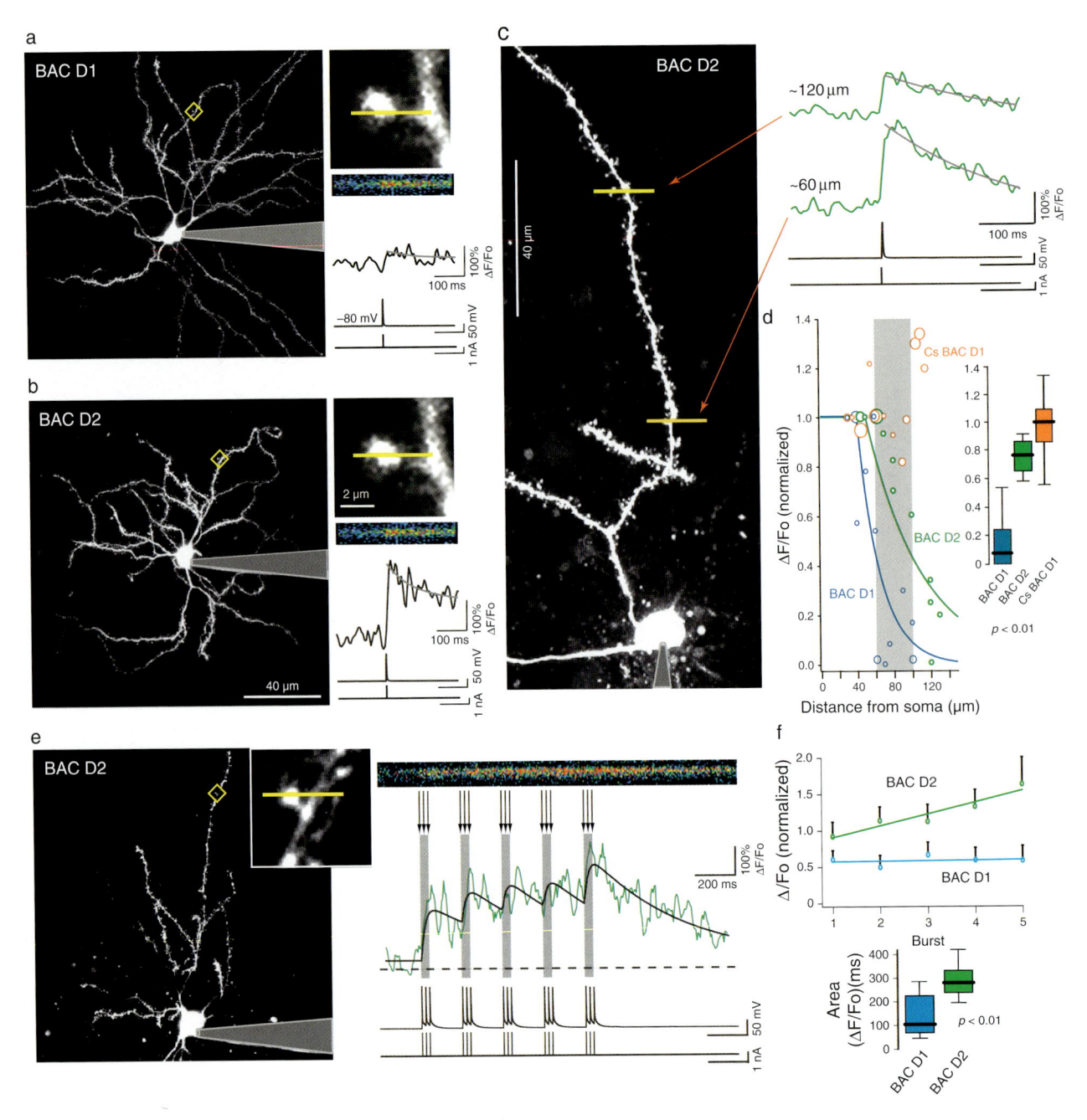

PLATE 7.3.2. BAP-evoked Ca^{2+} transients are readily detected in the distal dendrites and spines of the D2 population of MSNs. (a, b) 2PLSM images of MSNs in 275-μm-thick corticostriatal slices from (a) a BAC D1 and (b) a BAC D2 mouse. Neurons were visualized with Alexa Fluor 568 (50 μM) by filling through the patch pipette (patch pipettes are grayed out for presentation). Maximum projection images of the somas and dendritic fields (left panels a and b) and high-magnification projections of dendrite segments from the regions outlined by the yellow boxes are shown (top right panels, a and b). BAP-evoked Ca^{2+} transients were detected by line scanning through the spine in the region indicated by the yellow line. Fluorescence traces were generated from the pseudocolor image (lower panels, a and b) by calculating $\Delta F/F_o$ (top black trace). The fluorescence image, $\Delta F/F_o$ trace, action potential (middle trace), and current pulse (bottom trace) are shown in temporal registration. (c) Maximum projection image of a soma and a dendritic branch from a D2 MSN. Line scans were acquired at two eccentricities, 120 and 60 μm, as indicated by the red arrows. (d) Graph of the change in amplitude with distance from the soma calculated by normalizing scans taken at distal points to the most proximal scan point in each MSN. The magnitude of the Ca^{2+} transients decrements more in the D1 MSNs (D1 MSNs = open blue circles; D2 MSNs = open green circles). This decrementation is not seen in MSNs loaded with Cs^+-based internals (open orange circles). The points were scaled to represent the number of cells scanned at each point (smallest points = one cell; largest points = four cells). The data, fit from the median distance of the most proximal point, show that the magnitude of the Ca^{2+} transients decrements more in the D1 MSNs ($n = 11$, blue line) versus the D2 MSNs ($n = 6$, green line) [Kruskal-Wallis ANOVA, $p < 0.01$]. (e) Maximum projection image of the soma and dendritic field of a D2 MSN. A high-magnification image of the dendritic segment outlined in the yellow box is shown in the inset. Scale bars in (b) apply to both images. The pseudocolor image, $\Delta F/F_o$ trace, action potential (middle trace), and current pulse (bottom trace) are shown in temporal registration. Arrows indicate the timing of current pulses delivered to initiate action potentials. (f) Average peak $\Delta F/F_o$ values after each of the five pulses constituting the theta burst bAP protocol. Values are from distal dendritic spines (100–120 μm from the soma) and normalized to the maximum peak $\Delta F/F_o$ value measured in a proximal spine (60–80 μm) of the same dendrite in response to the first burst of the same theta burst protocol. The area under the $\Delta F/F_o$ plot was calculated for each cell type in response to the entire theta burst protocol; in line with larger peak Ca^{2+} transients, the box plots to the right demonstrate significantly larger Ca^{2+} transient areas in the D2 versus the D1 MSNs [Kruskal-Wallis ANOVA, $p < 0.05$]. *Source:* Reprinted from[14]. (See Figure 7.3.2.)

Neurochemical Neurocircuits in Drug Reward

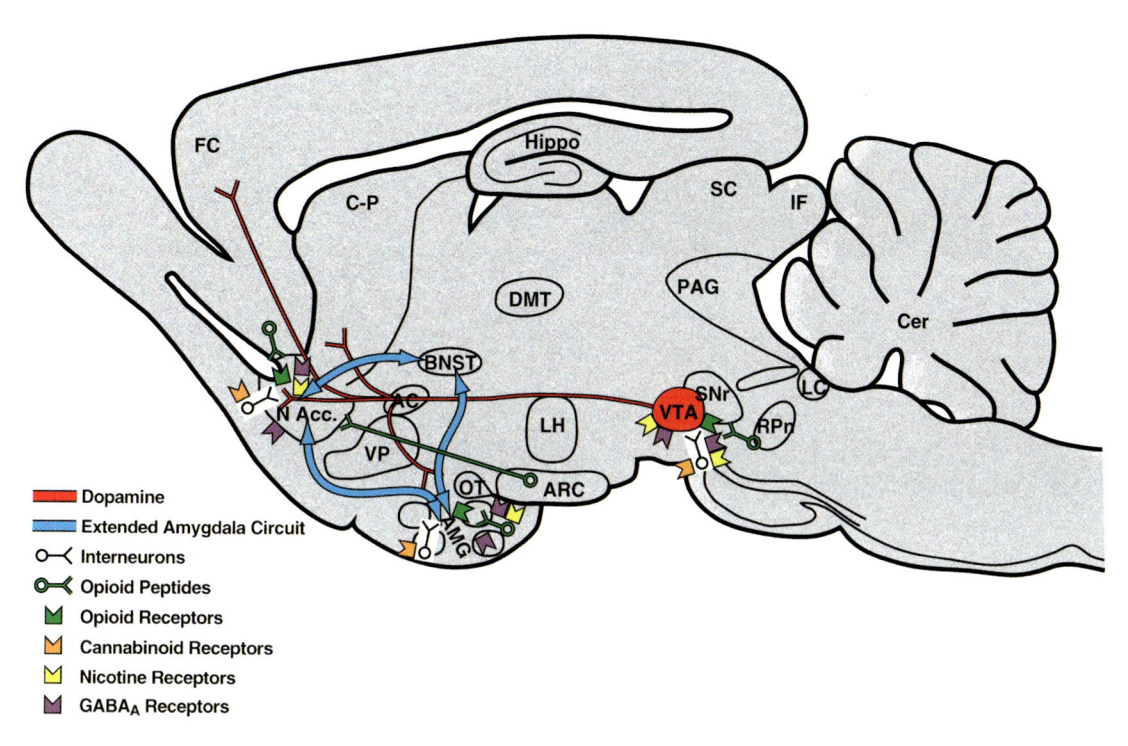

PLATE 8.1.1. Sagittal section through a representative rodent brain illustrating the pathways and receptor systems implicated in the acute reinforcing actions of drugs of abuse. Cocaine and amphetamines activate the release of dopamine in the nucleus accumbens and amygdala via direct actions on dopamine terminals. Opioids activate opioid receptors in the ventral tegmental area, nucleus accumbens, and amygdala via direct actions on interneurons. Opioids facilitate the release of dopamine in the nucleus accumbens via an action either in the ventral tegmental area or the nucleus accumbens but also are hypothesized to activate elements independent of the dopamine system. Alcohol activates γ-aminobutyric acid-A (GABA$_A$) receptors in the ventral tegmental area, nucleus accumbens, and amygdala either by direct actions at the GABA$_A$ receptor or through indirect release of GABA. Alcohol is hypothesized to facilitate the release of opioid peptides in the ventral tegmental area, nucleus accumbens, and central nucleus of the amygdala. Alcohol facilitates the release of dopamine in the nucleus accumbens via an action either in the ventral tegmental area or the nucleus accumbens. Nicotine activates nicotinic acetylcholine receptors in the ventral tegmental area, nucleus accumbens, and amygdala, either directly or indirectly, via actions on interneurons. Nicotine also may activate opioid peptide release in the nucleus accumbens or amygdala independent of the dopamine system. Cannabinoids activate cannabinoid CB$_1$ receptors in the ventral tegmental area, nucleus accumbens, and amygdala via direct actions on interneurons. Cannabinoids facilitate the release of dopamine in the nucleus accumbens via an action either in the ventral tegmental area or the nucleus accumbens, but also are hypothesized to activate elements independent of the dopamine system. Endogenous cannabinoids may interact with postsynaptic elements in the nucleus accumbens involving dopamine and/or opioid peptide systems. The blue arrows represent the interactions within the extended amygdala hypothesized to have a key role in drug reinforcement. AC, anterior commissure; AMG, amygdala; ARC, arcuate nucleus; BNST, bed nucleus of the stria terminalis; Cer, cerebellum; C-P, caudate-putamen; DMT, dorsomedial thalamus; FC, frontal cortex; Hippo, hippocampus; IF, inferior colliculus; LC, locus coeruleus; LH, lateral hypothalamus; N Acc., nucleus accumbens; OT, olfactory tract; PAG, periaqueductal gray; RPn, reticular pontine nucleus; SC, superior colliculus; SNr, substantia nigra pars reticulata; VP, ventral pallidum; VTA, ventral tegmental area. *Source*: Reprinted with permission from[187]. (See Figure 8.1.1.)

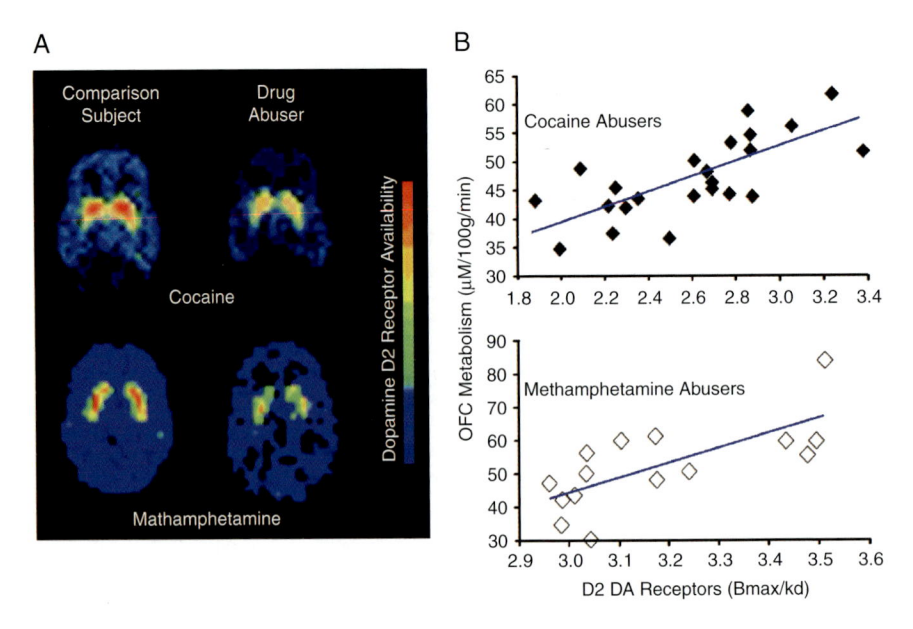

PLATE 8.3.1. (A) Normalized volume distribution of [^{11}C] raclopride binding in the striatum of cocaine and methamphetamine abusers and non-drug-abusing control subjects. (B) Correlation of DA receptor availability (B_{max}/K_d) in the striatum with a measure of metabolic activity in the orbitofrontal cortex (OFC) in cocaine (closed diamonds) and methamphetamine (open diamonds) abusers. *Source*: Modified from[65,66] with permission. (See Figure 8.3.1.)

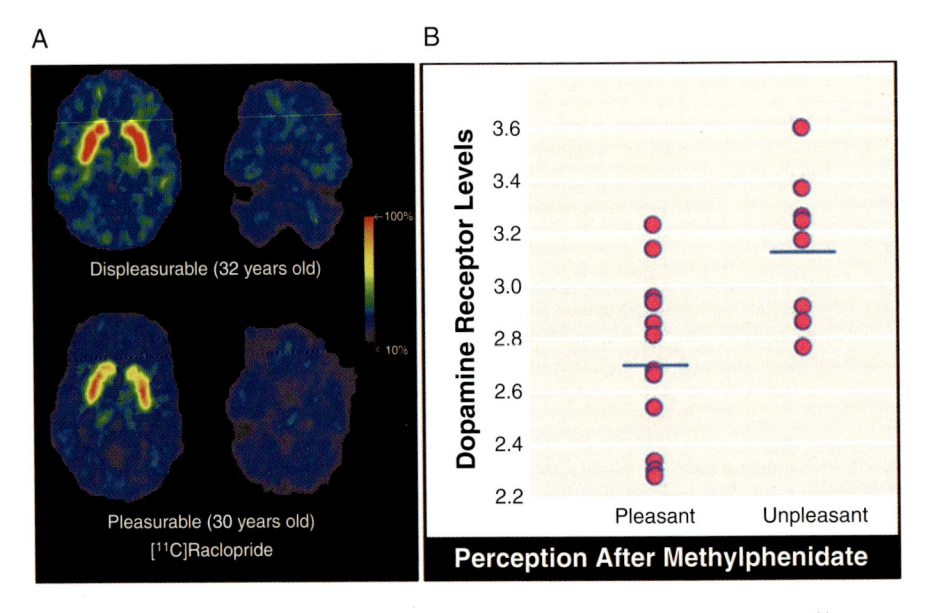

PLATE 8.3.2. Striatal D2 DA is predictive of methylphenidate liking in humans (A) Distribution volume images of [^{11}C]raclopride at the levels of the striatum (left) and cerebellum (right) in a healthy male subject who reported the effects of methylphenidate as pleasant and in a healthy male subject who reported them as unpleasant. (B) D2 DA receptor levels (bmax/kd) in 21 healthy male subjects who reported the effects of methylphenidate as pleasant or unpleasant. *Source*: Modified from[91] with permission. (See Figure 8.3.2.)

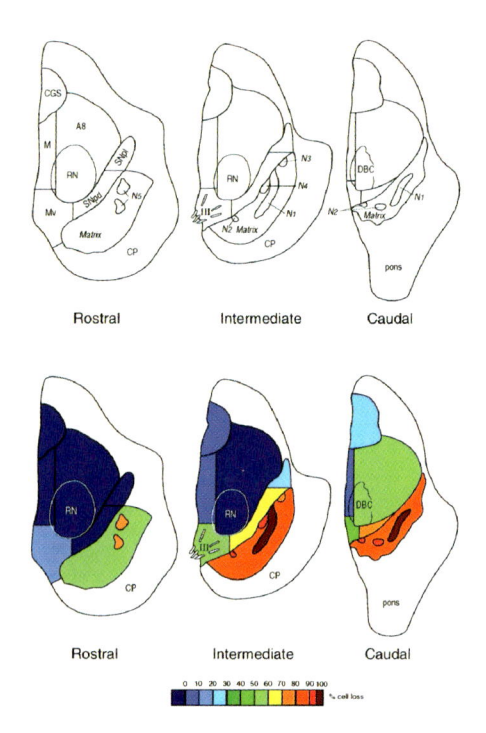

PLATE 9.1.2. Regional and intranigral loss of dopamine-containing neurons in PD. A8, dopaminergic cell group A8; CGS, central gray substance; CP, cerebral peduncle; DBC, decussation of brachium conjunctivum; M, medial group; Mv, medioventral group; N, nigrosome; RN, red nucleus; SNpd, substantia nigra pars dorsalis; SNpl, substantia nigra pars lateralis; III, exiting fibers of the third cranial nerve. The colorimetric scale indicates the estimated amount of cell loss (least = blue; most = red). Across the mesencephalon, dopaminergic cell loss was weak in the central gray substance and the red nucleus and intermediate in the dopaminergic cell group A8 and the medioventral group of the ventral tegmental area. Within the substantia nigra pars compacta, dopamine-containing neurons in the calbindin-rich regions (*matrix*) and in five calbindin-poor pockets (*nigrosomes*) were identified. The spatiotemporal progression of neuronal loss in the substantia nigra pars compacta is as follows: depletion begins in the main pocket (nigrosome 1) and then spreads to other nigrosomes and the matrix along rostral, medial, and dorsal axes of progression. This pattern is reflected clinically by corresponding somatotopic progression of symptoms. *Source*: Adapted from[30]. (See Figure 9.1.2.)

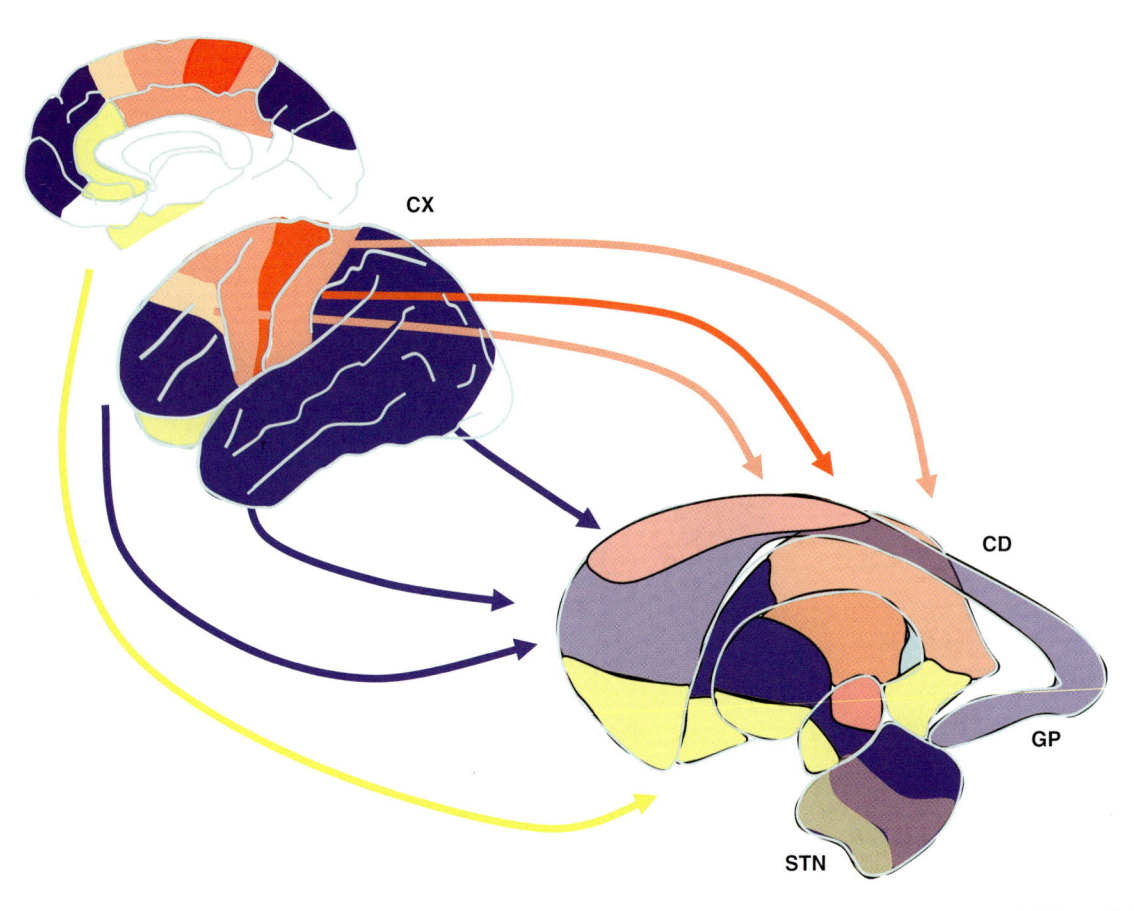

PLATE 9.1.3. Schematic illustration of the convergence of projections from the cerebral cortex (CX), caudate nucleus (CD), and globus pallidus (GP) on the subthalamic nucleus (STN). These projections can be divided into three functional territories—sensorimotor (green), associative (mauve), and limbic (yellow)—and converge in the STN. *Source*: Adapted from[2]. (See Figure 9.1.3.)

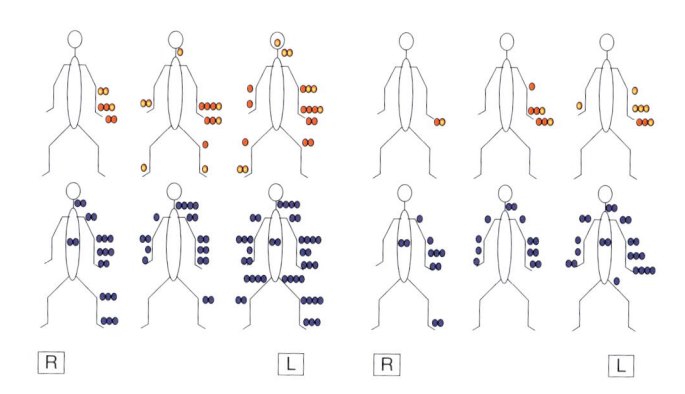

PLATE 9.1.5. Illustration of the topographical progression of rest tremor (upper row) and rigidity (lower row) in two de novo PD patients, from M0 (left) to M12 (right) (Schüpbach et al., in press). Symptom severity is indicated by the number of dots (red/orange for tremor; blue for rigidity) according to the 6-point rating scale. Red dots stand for rest tremor under relaxed conditions; orange dots indicate the worsening of rest tremor during a mental task. R, right; L, left. (See Figure 9.1.5.)

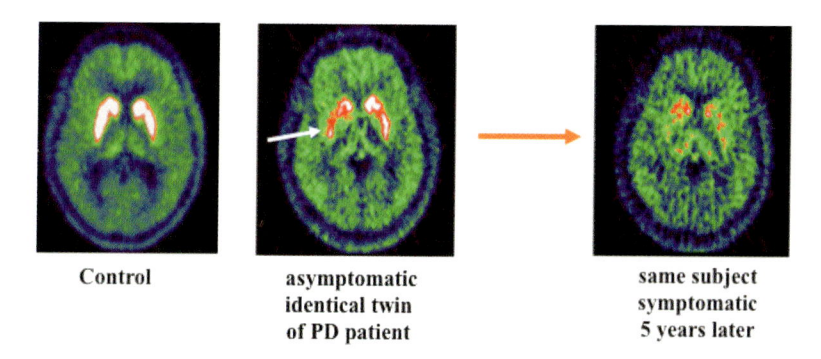

PLATE 9.3.2. ^{18}F-DOPA PET images of a healthy control and a monozygotic twin of a PD patient when asymptomatic at baseline and 5 years later when parkinsonian. The twin shows subclinical reduction of left putamen ^{18}F-DOPA uptake when asymptomatic and further bilateral reductions 5 years later. *Source*: Picture courtesy of P. Piccini. (See Figure 9.3.2.)

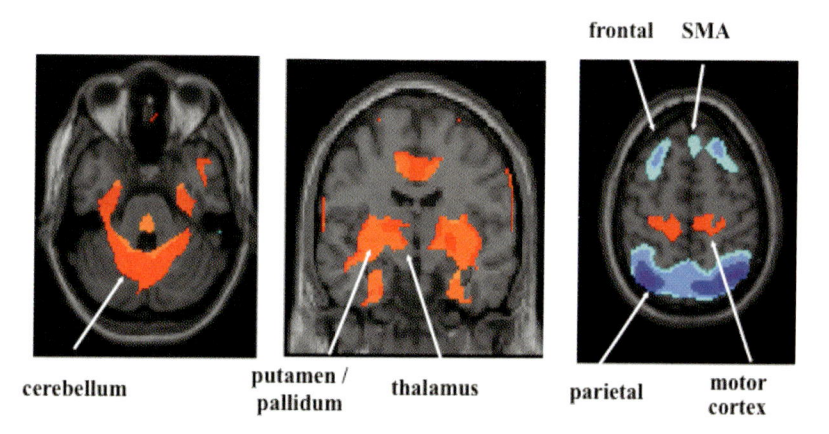

PLATE 9.3.3. The PDRP of abnormal FDG uptake with relatively raised basal ganglia and reduced frontal and parietal glucose metabolism. *Source*: Picture courtesy of D. Eidelberg. (See Figure 9.3.3.)

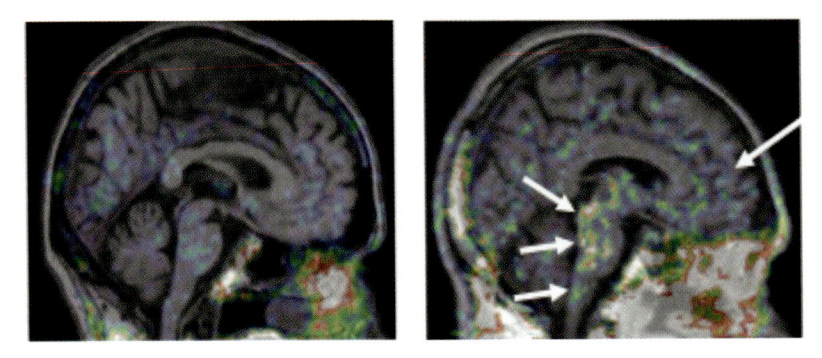

PLATE 9.3.4. ^{11}C-PK11195 PET images of an elderly normal subject (left) and a PD patient (right). The PD patient shows extensive microglial activation of the entire brainstem, striatum, and frontal cortex in line with the distribution of Lewy body pathology described by Braak et al.[2] *Source*: Picture courtesy of A. Gerhard. (See Figure 9.3.4.)

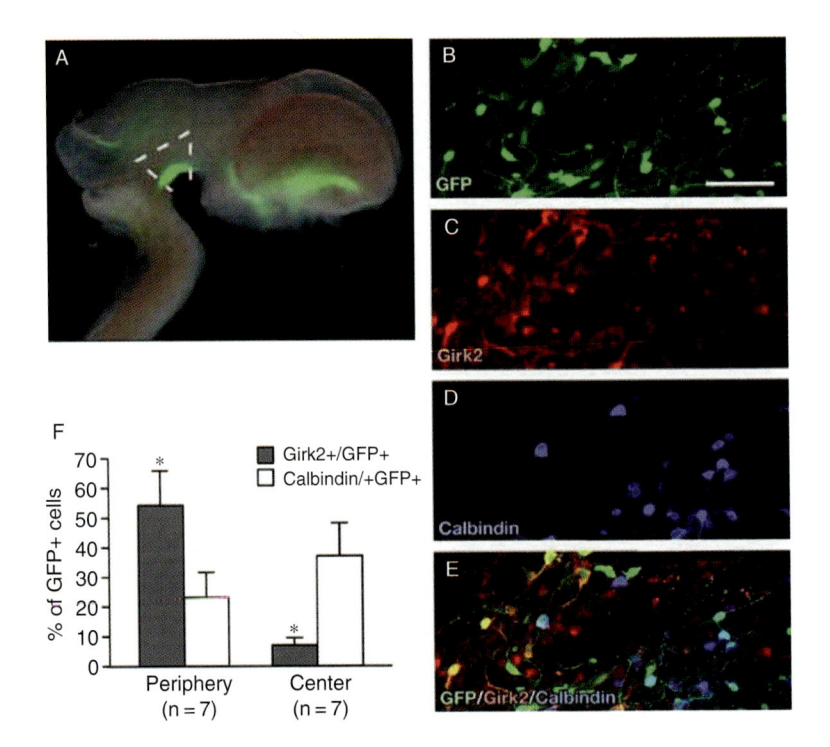

PLATE 9.4.3. Distribution of nigral and VTA cell types in fetal VM transplants. Nigral and VTA neuron subtypes can be distinguished using antibodies to Girk2 (expressed primarily in the neurons of the SNc) and calbindin (expressed mostly in the neurons in the VTA). The intrastriatal fetal VM grafts contain both subtypes, in about equal numbers, but they are differentially localized: the Girk2+, SN-type neurons mostly in the periphery of the transplants (C, F) and the calbindin+, VTA-type neurons in the core (D, F). The VM tissue were dissected from a TH-GFP-expressing transgenic mouse, as illustrated in (A), which allowed the TH-expressing cells to be detected by staining for the GFP reporter (B, E). Data from Thompson et al.[33] (See Figure 9.4.3.)

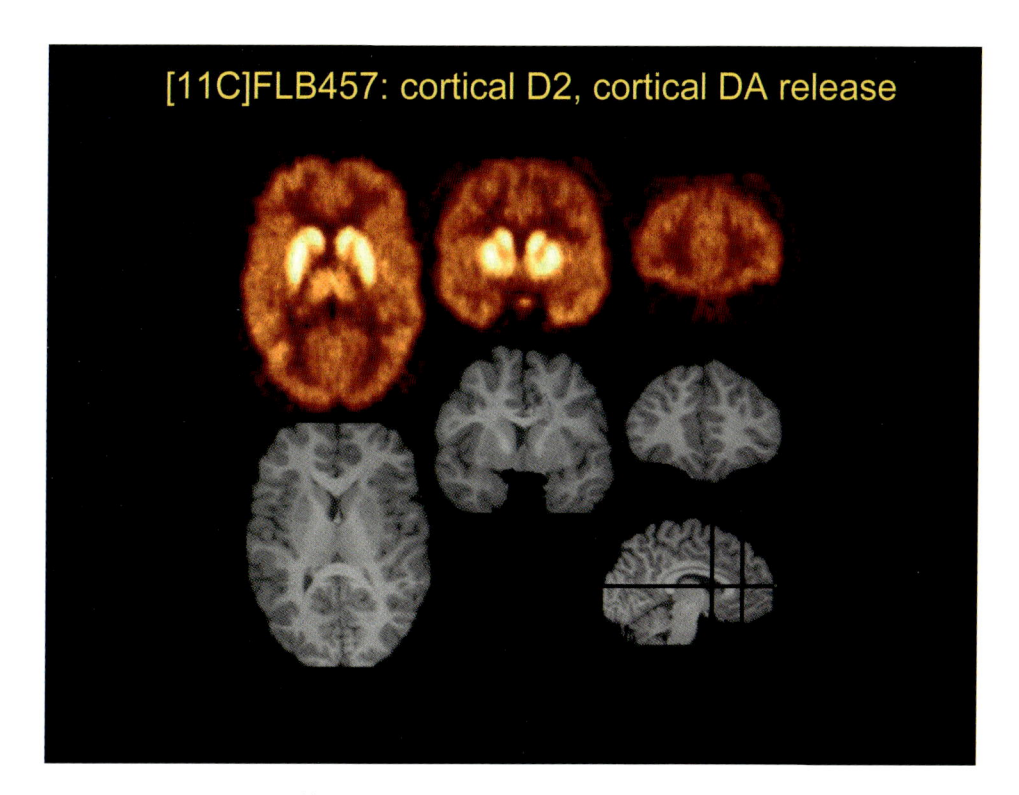

PLATE 10.1.1. Imaging cortical D2 receptors with [^{11}C] FLB 457 scan: here, a scan under baseline conditions with coregistered (SPGR, Spoiled gradient recalled; MRI, Magnetic resonance imaging). Shown are a transverse slice (left) at the level of the striatum and thalamus and coronal slices at the level of the striatum (center) and frontal cortex (right). The sagittal slice (bottom right, not to scale) has lines through the three slice levels. Note the pituitary gland in the center coronal slice. (See Figure 10.1.1.)

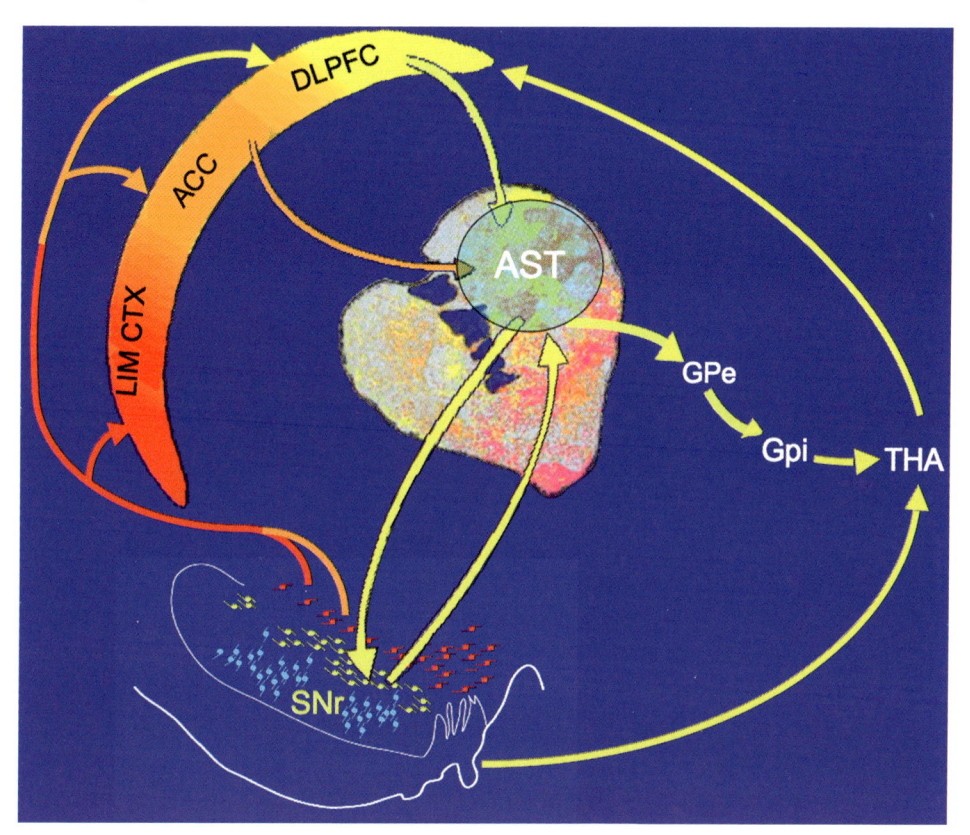

PLATE 10.1.2. Schematic representation of the circuitry involved in the DA dysregulation in schizophrenia. Cortical DA dysfunction and DA hyperactivity at D2 receptors in striatum (left), more specifically preDCA in the associative striatum (AST) (circle), may be linked. The anatomical substrates and circuits that may mediate this relationship are illustrated in this figure. ACC, anterior cingulate prefrontal cortex; DLPFC, dorsolateral prefrontal cortex; GPe, globus pallidum external; GPi, globus pallidum internal; LIM, LIMBIC; CTX; SNR, substantia nigra pars reticulta; THA thalamus. Source: Adapted from[59]. (See Figure 10.1.2.)

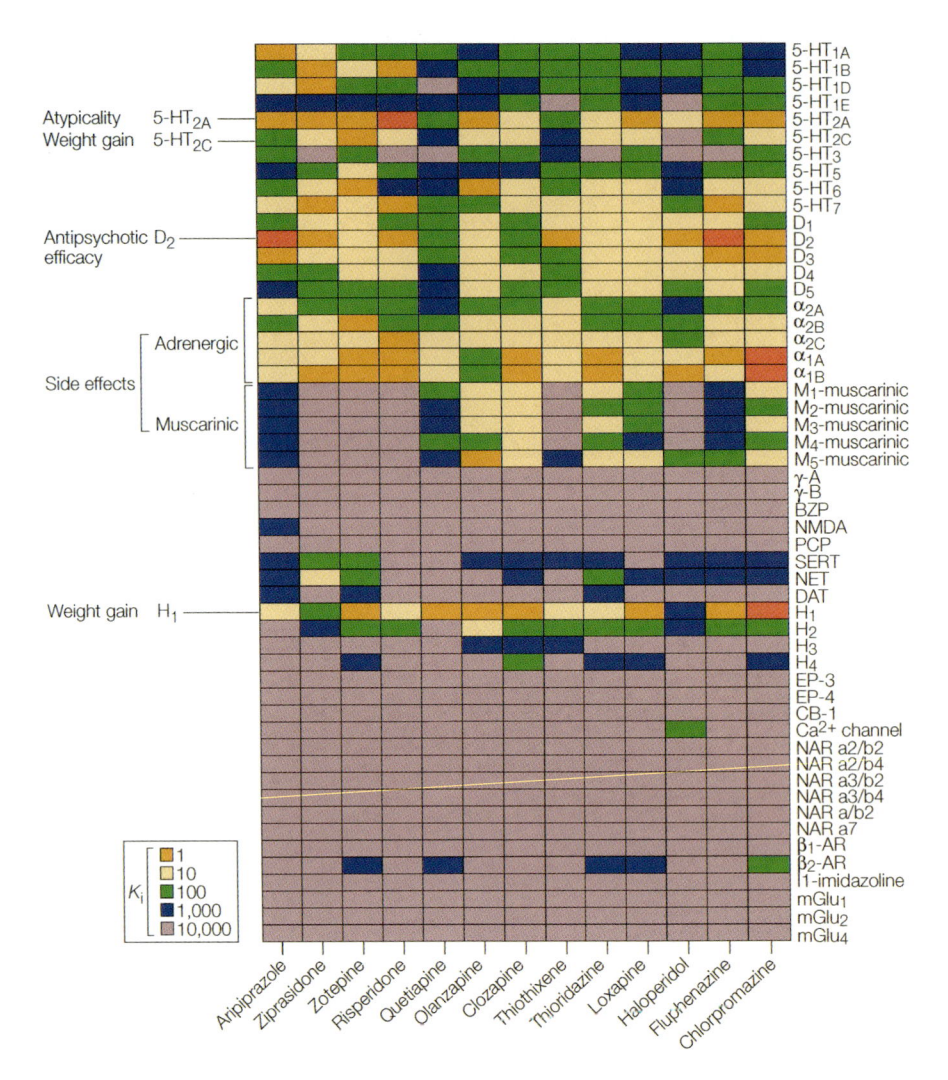

PLATE 10.2.2. Screening the receptorome reveals multiple molecular targets implicated in antipsychotic drug actions. The affinity (K_i) values for clozapine and a large number of other biologically active compounds at various receptors can be found at the PDSP K_i Database (pdsp.med.unc.edu); the database is part of the National Institute of Mental Health Psychoactive Drug Screening Program and represents the largest database of its kind in the public domain. At present, the PDSP K_i database has >47,000 K_i values for more than 700 targets. *Source:* Reprinted from[3] with permission. (See Figure 10.2.2.)

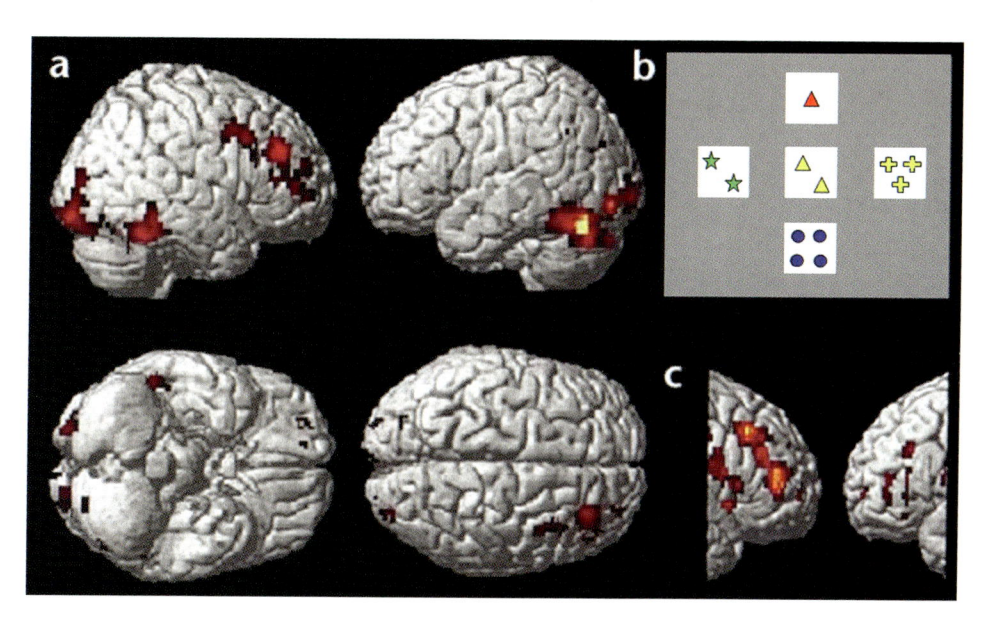

PLATE 10.4.1. Neural mechanisms of cognitive dysfunction. Statistical maps of regional cerebral blood flow during the performance of the WCST in schizophrenia patients and healthy controls. (a) Conjunction analysis showing voxels with significantly ($p < 0.01$, voxel level) higher regional cerebral blood flow (rCBF) during WCST performance than the control task. (b) Computer screen showing the WCST stimuli. (c) Voxels showing significantly ($p < 0.05$) higher rCBF in the task-minus-control contrast in the frontal lobes of controls compared to patients. *Source*: Reprinted with permission from [12]. (See Figure 10.4.1.)

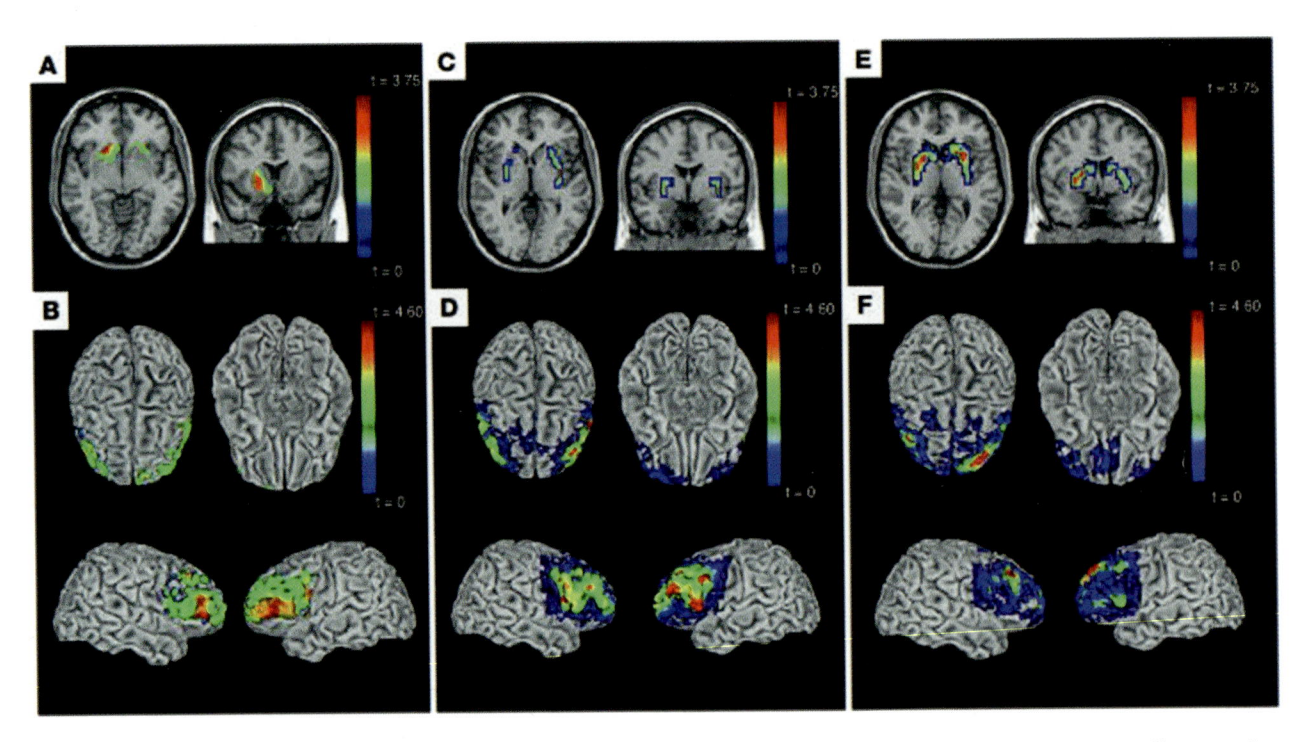

PLATE 10.4.3. Effects of genetic variation in *PPP1R1B* on human brain morphology and function. The top row shows haplotype effects on volume (A) or activation (C and E) in the striatum; the bottom row shows haplotype effects on structural (B) and functional (D and F) connectivity of the striatum to the PFC. Voxel-based morphometry: (A) significantly reduced volume in the striatum ($P < 0.05$) for carriers of the frequent (CGCACTC) haplotype; (B) greater structural connectivity between PFC and striatum for homozygotes for the frequent (CGCACTC) haplotype. Functional MRI, n-back task: (C) significantly reduced reactivity in putamen ($P < 0.05$) for carriers of the frequent (CGCACTC) haplotype; (D) greater functional connectivity between PFC and striatum for homozygotes for the frequent (CGCACTC) haplotype. Functional MRI, face-matching task; (E) significantly reduced reactivity in striatum ($P < 0.05$) for carriers of the frequent (CGCACTC) haplotype; (F) greater functional connectivity between PFC and striatum for homozygotes for the frequent (CGCACTC) haplotype. *Source*: Reprinted with permission from[98]. (See Figure 10.4.3.)

8 | Dopamine mechanisms in addiction

8.1 | The Role of Dopamine in the Motivational Vulnerability to Addiction

GEORGE F. KOOB AND MICHEL LE MOAL

INTRODUCTION

Dopamine has long been hypothesized to have a key role in addiction because of its hypothesized role in mediating incentive salience, motivated responding, and the psychostimulant properties of psychostimulant drugs. Prominent discoveries over the past 50 years of dopamine research have revealed that the mesocorticolimbic dopamine system has an essential role in the acute reinforcing effects of psychostimulant drugs, motivational dependence on psychostimulant drugs, and relapse to psychostimulant drug use. Such actions on key elements of addiction extend to other nonpsychostimulant drugs of abuse in a contributory role. Dopamine also has a prominent role in individual differences for the acquisition of and vulnerability to addiction.

CONCEPTUAL FRAMEWORK: MOTIVATIONAL VIEW OF ADDICTION

Drug addiction, also known as *Substance Dependence*, as currently defined by the *Diagnostic and Statistical Manual of Mental Disorders* (DSM-IV), 4th edition,[1] is a chronically relapsing disorder characterized by (1) compulsion to seek and take the drug, (2) loss of control in limiting intake, and (3) emergence of a negative emotional state (e.g., dysphoria, anxiety, irritability) reflecting a motivational withdrawal syndrome when access to the drug is prevented (defined here as dependence).[2] Addiction is assumed to be identical to the syndrome of *Substance Dependence*. Clinically, the occasional but limited use of a drug with the *potential* for abuse or dependence is distinct from escalated drug intake and the emergence of a chronic drug-dependent state.

Drug addiction has been conceptualized as a disorder that involves elements of both positive reinforcement, which drives the construct of impulsivity, and negative reinforcement, which drives the construct of compulsivity. *Positive reinforcement* is defined as the process by which an event (e.g., drug delivery) increases the probability of a response (i.e., positive reinforcement). *Negative reinforcement* is defined as the process by which drug taking alleviates a negative emotional state. *Impulsivity* can be defined by an increasing sense of tension or arousal before committing an impulsive act and pleasure, gratification, or relief at the time of committing the act, thus involving motivation to seek drugs through largely positive reinforcement. *Compulsivity* can be defined by anxiety and stress before committing a compulsive, repetitive behavior and relief from the stress by performing the compulsive behavior,[1] thus involving motivation to seek drugs through largely negative reinforcement.

The development of the aversive emotional state that drives the negative reinforcement of addiction has been defined as the "dark side" of addiction[3,4] and is hypothesized to be the *b-process* of the hedonic dynamic known as the *opponent process* when the *a-process* is euphoria.[5] Two processes are hypothesized to form the neurobiological basis for the *b-process*: loss of function in reward systems (within-system neuroadaptation) and recruitment of a negative emotional state via the brain stress or antireward systems (between-system neuroadaptation).[2,6]

Binge, Withdrawal, Preoccupation/Anticipation

Collapsing the cycles of impulsivity and compulsivity yields a composite addiction cycle comprising three stages: *preoccupation/anticipation*, *binge/intoxication*, and *withdrawal/negative affect*, in which impulsivity often dominates at the early stages and compulsivity dominates at the terminal stages. As an individual moves from impulsivity to compulsivity, a shift occurs from positive reinforcement driving the motivated behavior to negative reinforcement driving the motivated behavior.[7] These three stages are conceptualized as interacting with each other, becoming more intense, and ultimately leading to the pathological state known as addiction.[2]

A progressive increase in the frequency and intensity of drug use is one of the major behavioral phenomena characterizing the development of addiction and has face validity with the DSM-IV criteria for addiction.[1] A framework with which to model the transition from drug use to drug addiction can be found in recent animal models of prolonged access to intravenous cocaine self-administration. The effects of differential access to intravenous cocaine self-administration on cocaine-seeking in rats were explored by allowing rats to intravenously self-administer cocaine for 1 or 6 hr per day.[8] One-hour access (short access) to intravenous cocaine per session produced low and stable intake similar to that observed previously. In contrast, 6-hr access (long access) to cocaine produced drug intake that gradually escalated over days. When animals were allowed access to different doses of cocaine, both the long- and short-access animals titrated their cocaine intake, but the long-access rats consistently self-administered almost twice as much cocaine at any dose tested, further suggesting an upward shift in the set point for cocaine reward in the escalated animals.[9-11] Animals implanted with intravenous catheters and allowed differential access to intravenous self-administration of cocaine or heroin showed increases in reward thresholds that progressively increased in long-access rats but not in short-access or control rats across successive self-administration sessions.[12,13] Such increased self-administration in dependent animals has now been observed with cocaine, methamphetamine, nicotine, heroin, and alcohol.[8,14-17]

A reflection of the change in motivation associated with a transition to dependence is a measure of reinforcement efficacy measured by changes in progressive-ratio responding.[18] Extended access to drugs resulting in escalation also is associated with an increase in the breakpoint for cocaine in a progressive-ratio schedule of reinforcement, suggesting an enhanced motivation to seek cocaine or an enhanced efficacy of cocaine reward.[19,20] Similar results have been observed with withdrawal-induced drinking in rats made dependent with ethanol vapor.[21] Additionally, animals with extended access to cocaine or with a vulnerability to excessive cocaine intake showed a persistence of drug intake in the face of punishment, further supporting the increased motivation to seek the drug.[10,22] Thus, chronic extended drug access or individual differences in vulnerability can produce compulsive drug intake that has face validity for the human condition.

Different drugs produce different patterns of addiction that emphasize different components of the addiction cycle. A pattern of intravenous or smoked drug taking evolves, including intense intoxication, the development of tolerance, escalation in intake, and profound dysphoria, physical discomfort, and somatic withdrawal signs during abstinence. Intense preoccupation with obtaining drugs (craving) develops that is linked not only to stimuli associated with obtaining the drug but also to stimuli linked to internal and external states of stress. Different drugs of abuse follow variations of this pattern and may involve more the *binge/intoxication* stage (e.g., psychostimulants and alcohol) or less *binge/intoxication* and more *withdrawal/negative affect* and *preoccupation/anticipation* stages (e.g., nicotine and cannabinoids) or all three of these stages (e.g., opioids).

The present review focuses on the role of the dopamine system in (1) the rewarding effects of drugs of abuse (*binge/intoxication* stage), (2) the loss of function in the reward system with dependence (*withdrawal/negative affect* stage), and (3) vulnerability to initiate drug seeking.

ROLE OF DOPAMINE IN THE REWARDING EFFECTS OF DRUGS OF ABUSE

The hypothesis of the existence of a brain reward system has a long history and was given great impetus by the discovery of electrical brain stimulation reward or intracranial self-stimulation by Olds and Milner.[23] Brain stimulation reward involves widespread neurocircuitry in the brain, but the most sensitive sites defined by the lowest thresholds involve the trajectory of the medial forebrain bundle that connects the ventral tegmental area with the basal forebrain.[23-26] All drugs of abuse, when administered acutely, decrease brain stimulation reward thresholds[27] and when administered chronically increase reward thresholds during withdrawal (see below). Although much emphasis was focused initially on the role of the ascending monoamine systems in the medial forebrain bundle in brain stimulation reward, other nondopaminergic systems in the medial forebrain bundle clearly have a key role.[28-31] Indeed, much work suggests that activation of the mesolimbic dopamine system gives incentive salience to stimuli in the environment[32] to drive performance of goal-directed behavior[33-35] or activation in general,[36,37] and work with the acute reinforcing effects of drugs of abuse supports that hypothesis.

Lesion Studies

Activation of the mesolimbic dopamine system has long been known to be critical for the acute rewarding properties of psychostimulant drugs, but it is not necessarily critical for the acute reinforcing effects of other drugs of abuse.[38-42] Neurotoxin-selective lesions of the mesocorticolimbic dopamine system block the reinforcing effects of cocaine.[43–45] Rats trained to self-administer cocaine intravenously and subjected to a 6-hydroxydopamine lesion of the nucleus accumbens exhibit an extinction-like response pattern (i.e., high levels of responding at the beginning of each session and a gradual decline in responding over sessions) and a long-lasting decrease in responding. Neurotoxin-selective lesions of the mesocorticolimbic dopamine system in the nucleus accumbens also block the reinforcing effects of d-amphetamine.[46]

In a series of studies using intravenous self-administration of cocaine with progressive-ratio schedules, lesions of terminal areas[47] with neurotoxin-specific lesions of the central nucleus of the amygdala and medial prefrontal cortex facilitated responding on the progressive-ratio schedule (i.e., increased the reinforcing action of cocaine). Increased sensitivity to stimulants and facilitation of acquisition of self-administration have also been reported after the same selective dopamine terminal lesion within the amygdala, effects that were hypothesized to be attributable to selective interactions between this region and the nucleus accumbens.[48]

Although the dopamine system is activated by opioids, ethanol, nicotine, and Δ^9-tetrahydrocannabinol (Δ^9-THC), much evidence shows that dopamine-independent reinforcement occurs at the level of the nucleus accumbens,[40] suggesting multiple inputs to the activation of critical reinforcement circuitry in the nucleus accumbens/ventral striatum.[36,49,50] Neurochemically specific lesions of dopamine in the nucleus accumbens with 6-hydroxydopamine fail to block heroin or ethanol self-administration, supporting this hypothesis.[42,51–54]

Thus, multiple neurochemical systems have been hypothesized to be involved in the initial reinforcing or rewarding actions of drugs of abuse: dopamine, opioid, and γ-aminobutyric acid (GABA). For indirect sympathomimetics, such as cocaine and amphetamines, the mesolimbic dopamine system is critical. For opioids, the μ opioid receptor was hypothesized to be a critical first step in the reinforcing actions of opioid drugs and for sites both pre- and postsynaptic to the mesolimbic dopamine system in the nucleus accumbens and ventral tegmental area.[55] For alcohol, the $GABA_A$ receptor was hypothesized to be an initial site of action in the reinforcing actions of alcohol, with a prominent role for $GABA_A$ receptors in the ventral tegmental area, nucleus accumbens and amygdala.[56] Data from knockout mice provide key insights into the role of dopamine in the rewarding effects of drugs of abuse. Psychostimulants, such as cocaine, bind directly to the dopamine transporter to inhibit dopamine reuptake and elevate extracellular dopamine, presumably to produce cocaine's reinforcing effects. Genetically altered mice homozygous for a lack of the dopamine transporter protein, with increased extracellular dopamine, decreased dopamine stores, and decreased dopamine receptors, continued to self-administer cocaine,[57] but transgenic animals that expressed DAT but did not bind cocaine did not show cocaine reward.[58] Drugs of abuse have also been suggested to sensitize serotonergic and noradrenergic neurons via a nondopaminergic mechanism.[59,60] Moreover, genetically engineered dopamine-deficient mice continue to exhibit morphine-induced reward measured by conditioned place preference.[41]

Based on this synthesis, an early neurobiological circuit for drug reward was proposed. The starting point for the reward circuit was the medial forebrain bundle, composed of myelinated fibers connecting the olfactory tubercle and nucleus accumbens with the hypothalamus and ventral tegmental area,[61] and with ascending monoamine pathways such as the mesocorticolimbic dopamine system[38] (Fig. 8.1.1). Interestingly, for these classic anatomists, the ventral tegmental area was part of a larger group of regions, including posteriorly in the brainstem the gudden nuclei, raphe nuclei, and some parts of the central gray matter, for which Nauta coined the term *limbic midbrain area*, which was linked to classic forebrain limbic regions. These brainstem regions were considered to be parts of the reticular formation (arousal system).[62,pp308–311] These two brain areas, rhombencephalic and prosencephalic, were linked bidirectionally by the medial forebrain bundle and ascending aminergic fibers.[63]

Drug reward was hypothesized to depend on dopamine release in the nucleus accumbens for cocaine and amphetamine, opioid peptide receptor activation in the ventral tegmental area (via dopamine activation) and nucleus accumbens (independent of dopamine activation) for opiates, and $GABA_A$ receptors in the amygdala for alcohol. The nucleus accumbens was situated strategically to receive important limbic information from the amygdala, frontal cortex, and hippocampus that could be converted to motivational action via its connections with the extrapyramidal motor system. Thus, an early critical role for dopamine was established for the acute reinforcing effects of psychostimulant drugs, with a less critical role for opioids and sedative hypnotics.

Microdialysis Studies

The combination of intravenous and oral drug self-administration and in vivo microdialysis has provided

Neurochemical Neurocircuits in Drug Reward

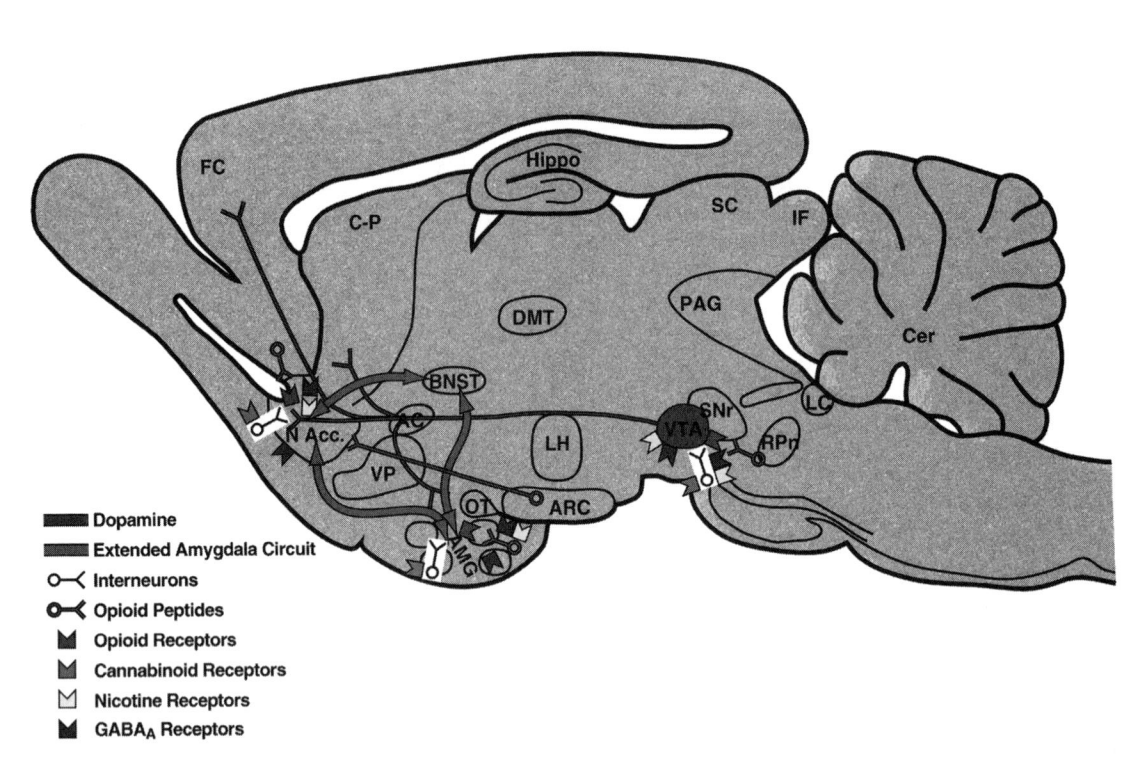

FIGURE 8.1.1. Sagittal section through a representative rodent brain illustrating the pathways and receptor systems implicated in the acute reinforcing actions of drugs of abuse. Cocaine and amphetamines activate the release of dopamine in the nucleus accumbens and amygdala via direct actions on dopamine terminals. Opioids activate opioid receptors in the ventral tegmental area, nucleus accumbens, and amygdala via direct actions on interneurons. Opioids facilitate the release of dopamine in the nucleus accumbens via an action either in the ventral tegmental area or the nucleus accumbens but also are hypothesized to activate elements independent of the dopamine system. Alcohol activates γ-aminobutyric acid-A (GABA$_A$) receptors in the ventral tegmental area, nucleus accumbens, and amygdala either by direct actions at the GABA$_A$ receptor or through indirect release of GABA. Alcohol is hypothesized to facilitate the release of opioid peptides in the ventral tegmental area, nucleus accumbens, and central nucleus of the amygdala. Alcohol facilitates the release of dopamine in the nucleus accumbens via an action either in the ventral tegmental area or the nucleus accumbens. Nicotine activates nicotinic acetylcholine receptors in the ventral tegmental area, nucleus accumbens, and amygdala, either directly or indirectly, via actions on interneurons. Nicotine also may activate opioid peptide release in the nucleus accumbens or amygdala independent of the dopamine system. Cannabinoids activate cannabinoid CB$_1$ receptors in the ventral tegmental area, nucleus accumbens, and amygdala via direct actions on interneurons. Cannabinoids facilitate the release of dopamine in the nucleus accumbens via an action either in the ventral tegmental area or the nucleus accumbens, but also are hypothesized to activate elements independent of the dopamine system. Endogenous cannabinoids may interact with postsynaptic elements in the nucleus accumbens involving dopamine and/or opioid peptide systems. The thick arrows connecting the nucleus accumbens, bed nucleus of the stria terminalis and amygdala represent the interactions within the extended amygdala hypothesized to have a key role in drug reinforcement. AC, anterior commissure; AMG, amygdala; ARC, arcuate nucleus; BNST, bed nucleus of the stria terminalis; Cer, cerebellum; C-P, caudate-putamen; DMT, dorsomedial thalamus; FC, frontal cortex; Hippo, hippocampus; IF, inferior colliculus; LC, locus coeruleus; LH, lateral hypothalamus; N Acc., nucleus accumbens; OT, olfactory tract; PAG, periaqueductal gray; RPn, reticular pontine nucleus; SC, superior colliculus; SNr, substantia nigra pars reticulata; VP, ventral pallidum; VTA, ventral tegmental area. Source: Reprinted with permission from [187]. (See Color Plate 8.1.1.)

compelling data suggesting that dopamine is released during drug self-administration. Intravenous cocaine, heroin, methamphetamine, and nicotine self-administration and oral ethanol self-administration increase extracellular dopamine during limited-access self-administration.[64,65] Binge cocaine self-administration also has profound effects on cocaine self-administration, with dramatic increases in extracellular dopamine followed by decreases during withdrawal (Fig. 8.1.2). The increase in dopamine in the nucleus accumbens produced by self-administration of different drugs of abuse varies by drug and may reflect the relative importance of the dopamine system in drug reward. For example, intravenous cocaine self-administration produces a 200% increase in extracellular dopamine[64] compared with ethanol (which produces a 20% increase

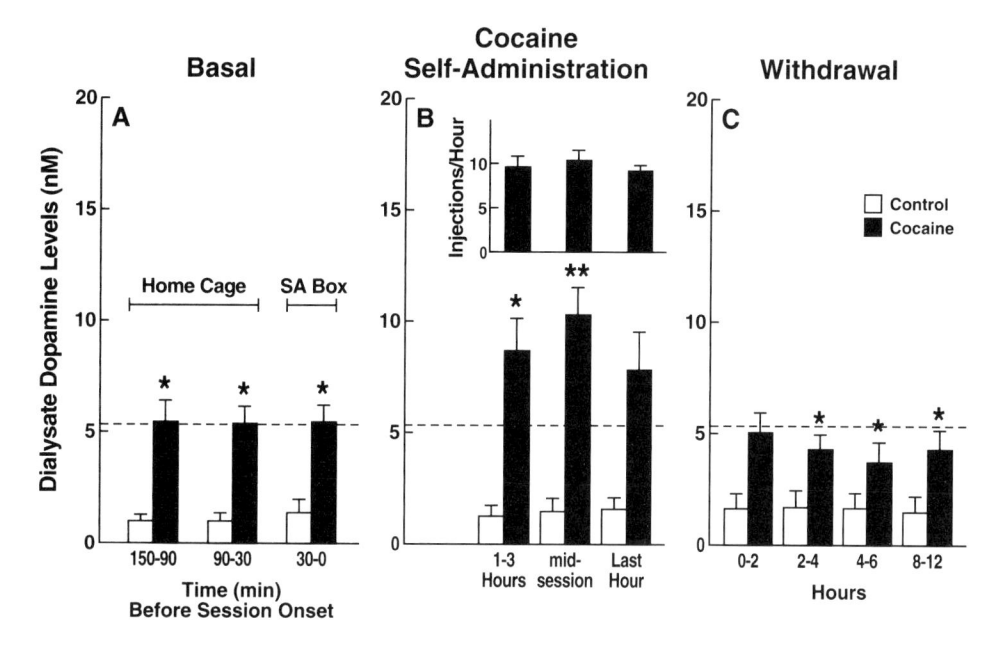

FIGURE 8.1.2. Mean (+ SEM) dopamine levels in microdialysate fractions collected from the nucleus accumbens of rats ($n = 5$) during unlimited-access cocaine self-administration (0.75 mg/kg/injection) and cocaine withdrawal. Control rats ($n = 3$) were drug-naive animals placed into the self-administration chambers for 30 min without access to cocaine. (A) Basal dopamine levels during two l-hr periods in the home cage and 30 min in the self-administration chamber (SA box) prior to cocaine access pre-cocaine basal dopamine levels in trained, self-administering rats were significantly higher than in drug-naive control rats (A): *P<0.02, significantly different from control. (B) Response rates for cocaine (inset) and dopamine levels during cocaine self-administration averaged over the first 3 hr, mid-session (total self-administration time minus the first 3 hr and last 1 hr) and the final 60 min of self-administration. (C) Dialysate dopamine concentrations during cocaine withdrawal. Dopamine release was significantly suppressed below basal levels between 2 and 6 hr after cocaine administration. Although dopamine levels tended to increase between 8 and 12 hr after the onset of the withdrawal period, dopamine overflow remained significantly below pre-session basal values. The dotted line represents mean presession basal dopamine levels for cocaine self-administering rats. *$p < 0.05$; **$p < 0.01$, significantly different from pre-session basal levels (Newman-Keuls post hoc tests). Control data in (B) and (C) are arranged with reference to the mean duration of approximately 14 hr in cocaine-self-administering rats. *Source*: Reprinted with permission from [64].

in extracellular dopamine in the nucleus accumbens[66]) and heroin (which does not increase extracellular dopamine in the nucleus accumbens) (Table 8.1.1). Such a relationship changes with the development of dependence. Ethanol-dependent animals show a much greater increase in extracellular dopamine in the N.Acc during ethanol self-administration during withdrawal.[67]

However, later work established that the nucleus accumbens is not a homogeneous structure, the *shell* part (medial and ventral) may be part of an extended amygdala system (see below) and the *core* resembles more the corpus striatum.[68,69] Most, if not all, drugs of abuse, when injected acutely into rats, stimulate dopamine transmission in the shell of the nucleus accumbens,[70–74] similar to nondrug rewards (e.g., Fonzies and chocolate).[75–78] Much less activation was observed in the core of the nucleus accumbens with both drug and nondrug rewards. The activation of dopamine release in the nucleus accumbens shell showed habituation with repeated administration of nondrug (food) rewards[79–81] but resistance to habituation with drug rewards. Based on these results, Di Chiara hypothesized that dopamine responsiveness in the nucleus accumbens shell may have a role in associative stimulus–reward

TABLE 8.1.1. *Effects of Intravenous Self-Administration of d-Amphetamine, Cocaine, and Heroin and Oral Self-Administration of Alcohol on Extracellular Dopamine Levels in the Nucleus Accumbens Using in Vivo Microdialysis*

Drug	% Increase Over Baseline	Reference
d-Amphetamine	700%	188
Cocaine	200%–500%	188, 189
Alcohol	25%–50%	64, 67
Heroin	<20%	190

learning.[75] However, this contrasts with neuropharmacological studies showing that the shell of the nucleus accumbens is more likely involved in the psychostimulant component of drug effects and that the core of the nucleus accumbens is more critical for imparting conditioned reinforcing properties to previously neutral stimuli.[82]

Pharmacological Studies

Dopamine D1, D2, and D3 receptor antagonists administered systemically block psychostimulant reward measured by conditioned place preference and self-administration.[83–85] Results obtained at the brain sites for pharmacological blockade of intravenous cocaine self-administration parallel those obtained with lesion studies. Delineation of the specific components of the dopamine projections of the mesocorticolimbic dopamine systems has pointed to the nucleus accumbens. Functional studies support the hypothesis that an extension of the central nucleus of the amygdala, described as the *extended amygdala*, may further delineate the neurobiological substrates of psychostimulant reinforcement. The extended amygdala has been conceptualized to be composed of several basal forebrain structures[68]: bed nucleus of the stria terminalis, central nucleus of the amygdala, and a transition area in the shell of the nucleus accumbens. Microinjections of a D1 receptor antagonist into the central nucleus of the amygdala, bed nucleus of the stria terminalis, and shell of the nucleus accumbens blocked cocaine self-administration, reflected in a decreased interinjection interval at all three sites, with the greatest effects in the shell of the nucleus accumbens.[86,87] Similar results have been obtained with microinjections of dopamine antagonists into the nucleus accumbens for intravenous self-administration of ethanol.[88] However, microinjection of D1 receptor antagonists into the nucleus accumbens did not affect heroin intake.[89] Additionally, whereas local and systemic administration of D1 antagonists produces what appears to be a competitive interaction with psychostimulant reinforcement (i.e., a decrease in interinjection interval and a shift to the right of the dose–response function), only a decrease in responding is observed with other drugs, raising the issue of the effects of dopamine blockade on voluntary movement produced by appetitive responding in general.[34] Microinjections of dopamine antagonists into the nucleus accumbens block the reinstatement of both cocaine and heroin seeking after extinction, suggesting a role for dopamine in the mesolimbic system in the motivational properties of relapse-like behavior.[90,91]

ROLE OF DOPAMINE IN THE MOTIVATIONAL DYSREGULATION OF DEPENDENCE

Pharmacological Studies

Motivational withdrawal as defined above involves two processes: decreases in the function of neurotransmitter systems involved in reward and motivation and recruitment of antireward systems such as the brain stress systems.[4] Decreases in dopamine system function form a key element of within-system neuroadaptations to chronic drug exposure that contribute to the negative emotional state of motivational withdrawal. In a within-system adaptation, repeated drug administration elicits an opposing reaction within the same system in which the drug elicits its primary reinforcing actions. For example, if the synaptic availability of the neurotransmitter dopamine is responsible for the acute reinforcing actions of cocaine, then the within-system opponent process neuroadaptation would be a decrease in synaptic availability of dopamine. One prominent hypothesis is that dopamine systems are compromised in crucial phases of the addiction cycle, such as withdrawal, which leads to decreased motivation for nondrug-related stimuli and increased sensitivity to the abused drug.[92]

Psychostimulant withdrawal in humans is associated with fatigue, depressed mood, and psychomotor retardation. In animals, psychostimulant withdrawal is associated with decreased motivation to work for natural rewards[93] and decreased locomotor activity,[94] behavioral effects that may involve decreased dopaminergic function. Animals during amphetamine withdrawal show decreased responding on a progressive-ratio schedule for a sweet solution.

Given the critical role of dopamine in the acute reinforcing effects of psychostimulant drugs, its contributory role to other drugs of abuse, and its dysregulation during withdrawal, a reasonable hypothesis is that a dopamine partial agonist may have efficacy in different components of the addiction cycle. A dopamine partial agonist has antagonist properties in situations of high intrinsic activity and agonist properties in situations of low intrinsic activity. Partial agonists also have fewer side effects than full agonists or antagonists.[95] Because of intermediate efficacy, a dopamine partial agonist acts as an agonist in the absence of dopamine and can act as an antagonist in the presence of dopamine.[96–98] The decreased responding on a progressive-ratio schedule for a sweet solution was reversed by the dopamine partial agonist terguride, suggesting that low dopamine tone contributes to the motivational deficits associated with psychostimulant withdrawal.[99]

In a series of studies, dopamine partial agonists have not only been shown to reverse psychostimulant withdrawal but also to block the increase in psychostimulant self-administration associated with extended access. Dopamine partial agonists decrease the reinforcing effects of psychostimulant drugs in nondependent, limited-access paradigms.[100,101] However, animals with extended access to intravenous methamphetamine self-administration show an increased sensitivity to a dopamine partial agonist.[102] Long-access rats that escalate their intravenous methamphetamine intake to the point of dependence, when administered the D2 partial agonist aripiprazole, showed a shift to the left of the dose–response function, similar to results observed with dopamine antagonists.[103] Another notable effect of aripiprazole on methamphetamine self-administration was a reduction in the maximum responding for methamphetamine in long-access rats under both progressive-ratio and fixed-ratio schedules, effects that were not apparent in short-access rats, again suggesting an increased sensitivity to the effects of the dopamine partial agonists in dependent rats. Dopamine partial agonists also decrease alcohol self-administration.[104] These results, combined with the observation that dopamine partial agonists can reverse psychostimulant withdrawal, suggest that dysregulation of dopamine tone may contribute to the motivational effects of drug withdrawal. A dopamine partial agonist with the appropriate neuropharmacological and pharmacokinetic profile may be effective in treating aspects of psychostimulant dependence.

Microdialysis Studies

Decreases in the activity of the mesolimbic dopamine system and decreases in serotonergic neurotransmission in the nucleus accumbens occur during drug withdrawal in animal studies[64,67,105] (Figs. 8.1.2, 8.1.3). Decreases in the firing of dopamine neurons in the ventral tegmental area have been observed during withdrawal from opioids, nicotine, and ethanol.[92] Imaging studies in drug-addicted humans have consistently shown long-lasting decreases in the number of dopamine D2 receptors in drug abusers compared with controls.[106] Additionally, cocaine abusers have reduced dopamine release in response to a pharmacological challenge with a stimulant drug.[107,108] The decrease in the number of dopamine D2 receptors, coupled with the decrease in dopaminergic activity in cocaine, nicotine, and alcohol abusers, results in decreased sensitivity of reward circuits to stimulation by natural reinforcers.[109,110] These findings suggest an overall reduction in the sensitivity of the dopamine component of reward circuitry to natural reinforcers and other drugs in drug-addicted individuals.

Advances in molecular biology have led to the ability to inactivate systematically the genes that control the expression of proteins that make up receptors or neurotransmitters/neuromodulators in the central nervous system using the gene knockout or knockin approach. Notable positive results with gene knockout studies in mice have focused on the different dopamine receptor subtypes and the dopamine transporter. Dopamine D1 receptor knockout mice show no response to D1 agonists or antagonists and show a blunted response to the locomotor-activating and rewarding effects of cocaine and amphetamine, confirming a key role for the dopamine system in psychostimulant reward.[111–114]

ROLE OF DOPAMINE IN REINSTATEMENT OF DRUG SEEKING

The mesolimbic dopamine system projections to the nucleus accumbens also have a key role in reinstatement of drug-seeking behavior induced by drug priming and cues (animal models of the preoccupation/anticipation stage). Dopamine D1 and D2 receptor antagonists block drug priming–induced reinstatement in the shell of the nucleus accumbens and prefrontal cortex,[115] and D3 antagonists localized specifically in the shell of the nucleus accumbens block cue-induced reinstatement.[84] For recent reviews on the role of dopamine in drug- and cue-induced reinstatement, see Zhai et al.[116] and Anderson and Pierce.[117]

ROLE OF DOPAMINE IN VULNERABILITY TO STIMULANT ADDICTION

Vulnerability to Drug Use and Individual Differences in Dopamine Utilization

Vulnerability is a construct used in all fields of medicine, particularly in psychiatry. A large proportion of the population takes one or several drugs at least once during their lifetime. Some individuals can even maintain prolonged recreational use, and comparatively few will develop the syndrome of addiction.[118,119] In addition to the differential intrinsic potential for abuse and addiction of a given drug,[120] many factors contribute to individual differences in the potential to develop addiction. The origins of such vulnerability are numerous and probably interactive and include genetic and environmental factors, aversive life events, age, early drug exposure, and gender. More generally,

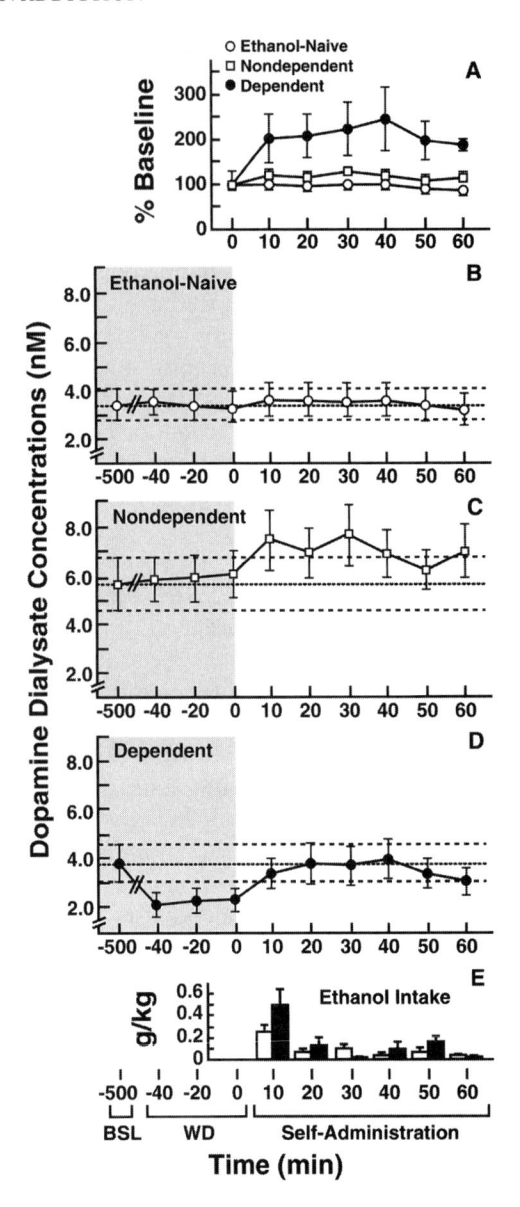

FIGURE 8.1.3. Effects of operant alcohol self-administration in nondependent and dependent rats undergoing ethanol withdrawal on dopamine efflux in the nucleus accumbens. Dialysate neurotransmitter levels are compared with those in ethanol-naive rats trained to self-administer water. Average water intake in this group was negligible (< 0.8 ml) and is not shown. (A) Changes in neurotransmitter output from levels recorded during the last hour of withdrawal. Data are expressed as a percentage of baseline values calculated as the average of three 20-min samples collected during hour 8 of withdrawal shown in (B)–(D). The corresponding dialysate neurotransmitter concentrations are shown in (B) (Ethanol-Naive), (C) (Nondependent), and (D) (Dependent). To illustrate the changes in neurotransmitter efflux over the various experimental phases, (B)–(D) also show prewithdrawal baseline (BSL) and withdrawal (WD) dialysate concentrations of dopamine during hour 8 of withdrawal. Dashed lines represent mean ± SEM prewithdrawal dialysate dopamine concentrations. (E) Amounts of self-administered ethanol (10% w/v) during 10-min intervals for the dependent (solid bars) and nondependent (open bars) groups. Ethanol self-administration in dependent rats restored dopamine levels to prewithdrawal values. *Source*: Reprinted with permission from [67].

epidemiological studies have shown that individuals who have been or will be diagnosed with an addiction disorder exhibit prior to the onset of addictive disorders one or more other observable manifestations of biopsychological pathologies, such as symptoms of another psychiatric disorder or dysfunctional behavior patterns.[121–124]

Emerging data from animal experiments have suggested vulnerable phenotypes that have the propensity to use drugs impulsively and are predisposed to administer drugs in large quantities.[125,126] Dopamine utilization and transmission have been shown to be involved in this process. Individual differences in the subjective and reinforcing effects of stimulants have been well documented in humans in large cohort studies.[127] For example, large individual differences exist in the reinforcing and subjective effects of amphetamine compared with placebo, with increased ratings of euphoria and positive mood compared with anxiety and depression when placebo is chosen.[128] Imaging techniques have provided information about the neural correlates of these subjective differences in healthy subjects. Volkow et al.[129] showed that the intensity of the methylphenidate "high" correlates significantly with the release of dopamine. Subjects who had the greatest increase perceived the most intense reinforcing effects. Importantly, the decrease in D2 receptor availability coincided with more intense reinforcing effects of the psychostimulant. Another study demonstrated that the differential decrease in extracellular dopamine induced after amphetamine administration among subjects correlated positively with self-reports of desire for the drug and, more interestingly, with the personality trait of novelty seeking.[130] These data suggest that individual differences exist for the rate of dopamine release and D2 receptor availability and that they correlate positively with the propensity to respond for psychostimulants.

A biological basis for the prediction of these individual differences is an important outcome of these studies. In drug-naive rats, differential locomotor reactivity in a novel environment and differential impulsivity and novelty seeking were predictive of individual vulnerability to self-administer a very low dose of amphetamine (20 μg/nosepoke) in an acquisition paradigm.[131] However, these individual responses to cocaine were not dependent on the dose per se. Vertical shifts in self-administration dose–response functions were predictive of a drug-vulnerable phenotype that is predisposed to drug use.[132]

Various environments influence psychostimulant self-administration, especially when moderate doses are presented. An environment associated with drug taking can alter both the intake of and motivation for the drug. Novelty facilitates acquisition, produces a shift to the left in the dose–effect function, and increases motivation in a progressive-ratio schedule.[133] This higher reactivity to the psychostimulant is predicted by dopamine utilization in the nucleus accumbens and prefrontal cortex. Vulnerable high-reactive rats displayed a specific pattern of dopamine utilization: a reduction in the prefrontal cortex and an increase in the nucleus accumbens.[134] A history of a relatively brief and common ecological stressor, such as a period of moderate food shortage, can reverse or abolish mouse strain differences in behavioral responses to a psychostimulant.[135] These data demonstrate the need for an integrated approach when considering the interaction between environmental and genetic factors.

Dopamine and Stress Glucocorticoid Interactions and Vulnerability to Psychostimulants

Stress activates the hypothalamic-pituitary-adrenal axis and elevates corticotropin-releasing factor (CRF) and glucocorticoid levels. Clear interactions exist between stress, glucocorticoids, and mesocorticolimbic dopaminergic neurons and between dopaminergic neurons and vulnerability to drugs of abuse (Fig. 8.1.4). Glucocorticoid receptors are localized in brain monoaminergic neurons, particularly in the ventral tegmental area,[136] although direct cellular interactions between stress hormones and dopamine neurons have been difficult to document. In normal situations, glucocorticoids state-dependently increase dopaminergic function, especially in mesolimbic regions, during various consummatory behaviors exhibited in the rodent's active period of the light/dark cycle and also in animals that self-administer psychostimulants.[137] However, glucocorticoid receptors have pivotal regulatory roles in many regions of the brain,[138–140] and glucocorticoids can interact with dopamine reward circuitry in the basal forebrain, which may be independent of direct glucocorticoid–dopamine interactions. More specifically, glucocorticoids modulate the transmission of the neuropeptides dynorphin, enkephalin, tachykinin, CRF, and neurotensin, especially in the basal ganglia and nucleus accumbens (for review, see [141,142]). Increased corticosterone secretion or higher sensitivity to the central effects of the hormone, either genetically present in certain individuals or induced by stress, increases the vulnerability to develop drug intake and may have a role in the transition to dependence and relapse via an enhancement of the activity of mesocorticolimbic dopaminergic neurons.

The enhancing effects of stress on amphetamine and cocaine self-administration have been documented for decades, and many of these effects are related to glucocorticoid release. These effects are observed for different doses of drugs during the acquisition phase and reinstatement and in motivational measures such as progressive-ratio schedules of reinforcement. Stress, through activation of the hypothalamic-pituitary-adrenal axis and the release of glucocorticoids, influences various regions of the brain, including dopamine neurons[125,143,144] that express corticosteroid receptors.[77] The interaction of stress, via the action of glucocorticoids,

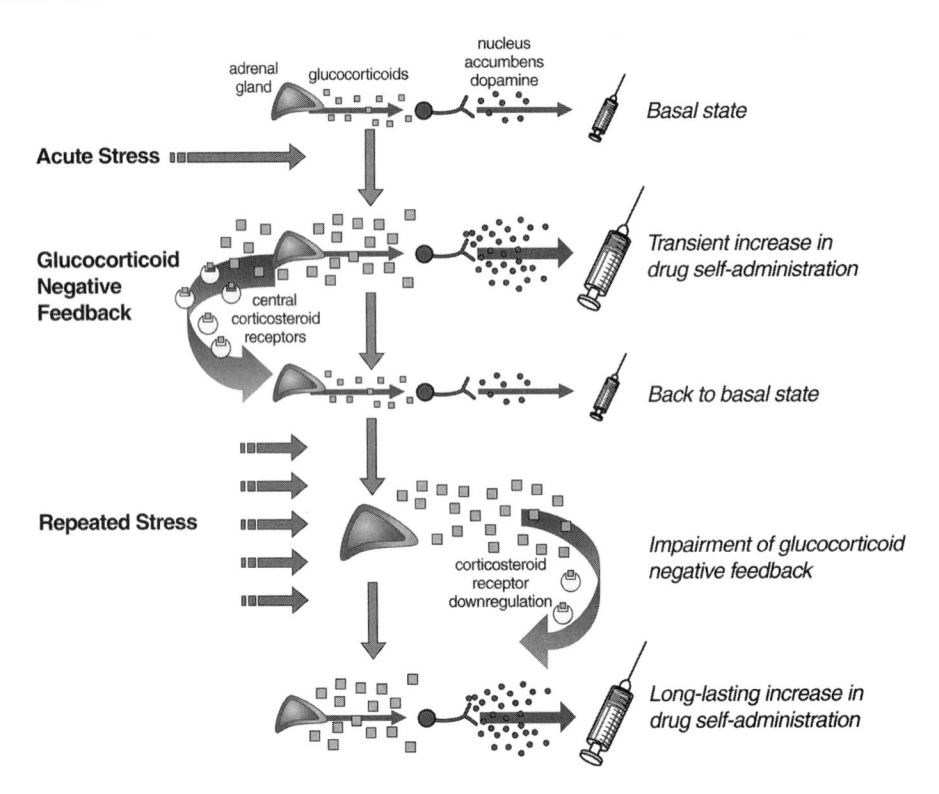

FIGURE 8.1.4. Possible pathophysiological mechanisms for the increase in drug self-administration induced by acute and repeated stress hypothesized by Piazza et al.[144] The interactions of two biological systems are schematically represented as the following: (1) the secretion of glucocorticoids (small square) from the adrenal gland, one of the principal hormonal responses to stress; (2) the release of dopamine (small circles) from the mesoaccumbens dopaminergic projection, one of the principal neurobiological substrates of the rewarding properties of drugs of abuse. These two systems interact in basal conditions and during stress. The concentrations of glucocorticoids determine the level of dopamine release in the nucleus accumbens. In basal conditions (basal state), glucocorticoid secretion and dopamine release are low, similar to sensitivity to drugs of abuse. An acute stress determines an increase in glucocorticoid secretion, which, by enhancing the release of dopamine, results in increased sensitivity to the reinforcing effects of drugs of abuse, which can result in an increase in self-administration. However, activation by glucocorticoids of the negative feedback that controls the secretion of these hormones returns the system to basal levels within 2 hr. Binding of glucocorticoids to hippocampal corticosteroid receptors is a key step in the activation of this negative feedback. The repeated increase in the concentrations of glucocorticoids induced by repeated exposure to stress will progressively impair glucocorticoid negative feedback by decreasing the number of central corticosteroid receptors in the hippocampus. The impairment of glucocorticoid negative feedback will result in a long-lasting increase in the secretion of these hormones and in the release of dopamine in the nucleus accumbens. These changes will, in turn, determine a long-lasting increase in the sensitivity to the reinforcing effects of drugs of abuse. The transient increase in glucocorticoids and dopamine observed after acute stress may explain why, in this context, an increase in drug self-administration is observed only if the exposure to drug closely follows the stressor. The long-lasting increase in the activity of these two biological factors could explain why, after repeated stress, an increase in the sensitivity to drugs is found even weeks after the end of the stressor. *Source*: Reprinted with permission from [144].

with the mesolimbic dopamine system may have a significant impact on vulnerability to self-administer psychostimulant drugs. Rats with initial high reactivity in a novel environment have high initial corticosterone responses in the hypothalamic-pituitary-adrenal axis and are more likely to self-administer psychostimulant drugs.[125,143] Additionally, rats receiving repeated injections of corticosterone acquire cocaine self-administration at lower doses than rats that receive vehicle.[145] Corticosterone administration causes rats that would not normally self-administer amphetamine at low doses (low-reactive rats) to self-administer amphetamine.[146] Conversely, adrenalectomy tends to suppress cocaine self-administration in rats.[137] Glucocorticoid hormones and stimulants interact at some of the same cellular levels, particularly the shell of the nucleus accumbens.[77] Animals, especially those that react more to stimulants, self-administer glucocorticoids similarly to cocaine and amphetamine.[147] These results suggest that glucocorticoids may be one of the biological factors determining vulnerability to substance use.[148]

Systematic studies from different models, including responses to novelty and responses to stressors, have led to the demonstration of increased drug intake across the full dose–effect function in high-reactive rats.[132,148–150] Levels of corticosterone measured 120 min after exposure to a stressor are positively correlated with the amount of drug consumed when the drug is presented for the first time (i.e., acquisition) to a high-reactive subject. Moreover, genetic inactivation of the glucocorticoid receptor gene reduces the excessive drug response not only in high-reactive animals but also in animals after long-term exposure to cocaine.[151] Indeed, the levels of the stress hormone before drug administration correlate with the extent of self-administration.[146,152] High-reactive rats also have a lower number of dopamine D1 and D2 receptors in the nucleus accumbens.[153] These changes in D2 receptors are similar to those reported in human drug addicts after the development of addiction.[129]

Progressive changes in the hypothalamic-pituitary-adrenal axis are observed during the transition from acute to chronic administration of drugs of abuse. Acute administration of most drugs of abuse in animals activates the hypothalamic-pituitary-adrenal axis, but with cocaine, these acute changes are blunted with repeated administration.[154,155] During withdrawal, increases in the activity of the hypothalamic-pituitary-adrenal axis occur for most drugs of abuse. In a rat model of the transition from moderate to excessive drug intake that is characterized by a change in the hedonic set point,[8] surgical adrenalectomy with corticosterone replacement slows the escalation of cocaine self-administration from day to day and prevents the augmentation of cocaine-induced reinstatement.[156] Corticosterone is hypothesized to be involved in the induction, but not expression, of addiction-related neuroplasticity and structural changes within the mesocorticolimbic system.

A key driver of the hypothalamic-pituitary-adrenal axis is CRF via the paraventricular nucleus of the hypothalamus. However, CRF also has a key role in behavioral responses to stressors, the anxiogenic-like and aversive effects of drug withdrawal, and stress-induced reinstatement via actions at extrahypothalamic sites.[157] CRF antagonists, when administered directly into the ventral tegmental area and bed nucleus of the stria terminalis, blocked footshock-induced reinstatement.[158–160] CRF receptors and the CRF binding protein have been implicated in these actions of CRF.[159] These results suggest that CRF via the ventral tegmental area and basal forebrain extrahypothalamic sites[158,160] may have a role in stress-induced reinstatement.

Dopamine Effects on Drug Use Vulnerability in Early Drug and Stress Exposure

At the origin of individual vulnerabilities, environmental events during critical periods of development produce enduring changes that influence drug reinforcement responsivity and the propensity to develop drug use. Although the development of an organism presumably has a strong genetic component, the organism's early environmental experience also has long-lasting influence. Both components shape psychobiological temperaments and are at the origin of these individual differences.

Prenatal stress has been found to have long-term effects on the activity of the dopamine system and on dopamine-related behaviors.[161,162] Psychostimulant self-administration has been studied in the offspring of mothers submitted to a restraint stress procedure during the last week of pregnancy.[163,164] These animals were also tested for locomotor reactivity to novelty and to psychostimulants; stressed animals were found to have increased and more rapid locomotor reactivity to amphetamine, particularly during the first hour of testing. The prenatal stress animals also had a propensity to develop rapid amphetamine self-administration. Although control and stressed animals did not differ during the first day of testing, animals in the prenatal stress group exhibited a higher intake of amphetamine on subsequent days. Repeated neonatal maternal separations (isolation stress experience) increased intravenous cocaine self-administration in rats in adulthood.[165] This reactivity was accompanied by structural differences in the mesolimbic dopamine system,[166] and the separations led to an increase in both stress-induced sensitization to amphetamine and acute mesolimbic dopamine release following cocaine administration.[167] Similar to maternal separation, isolation rearing of infant rats and isolation in general lead to enhanced responding for psychostimulants[168] and increased self-administration of almost all drugs of abuse, again with the general pattern of reduced dopamine turnover in the frontal cortex and increased turnover in limbic striatal regions.[134,169,170] Conversely, animals exposed to enriched environments have less dopamine and less dopamine transporter binding in the striatum,[171] and thus may be hypothesized to have protective effects against the misuse and abuse of stimulants.

Adolescence also is a critical period for drug effects. Acute stimulant exposure in adolescent rodents elicits behavioral responses different from those observed in adults. A single high dose of cocaine during adolescence can produce more robust long-term behavioral sensitization than in adulthood, suggesting that adolescents may be more vulnerable to neuroadaptations induced by drugs. Moreover, large individual differences are observed, and an initial ambulatory response to novelty correlated with ambulatory stimulant sensitization observed later.[172] Another study found that adolescent animals that approach a novel object faster and show higher novelty-induced impulsivity (high-reactive animals) than adults had an increased dopaminergic response to an acute cocaine challenge.[173] Fourteen days of repeated low-dose methylphenidate exposure in adolescent rats decreased dopamine neuronal impulse activity and increased cocaine use liability.[174,175]

At the receptor level, neonatal isolation not only enhances nucleus accumbens dopamine responses to cocaine that endure into adulthood, but also renders rats more sensitive to a D2 antagonist, reflecting decreased levels of this receptor.[176] Additionally, prenatal exposure to cocaine causes long-lasting neuroadaptive responses in the subcellular distribution of D1 receptors that affect cell signaling via these receptors, reduced receptor coupling to proteins, and hyperphosphorylation of the receptor.[177]

Prenatal stress also induces lasting effects. Cocaine-naive, stressed rats exhibited increased reactivity to novel environments and to a noncontingent stimulant injections, whereas stressed rats with a history of cocaine self-administration exhibited increased resistance to extinction and a greater vulnerability to reinstatement. The cocaine-naive animals exhibited increased basal mesolimbic dopamine with enhanced dopamine utilization after a drug challenge, whereas the cocaine-experienced animals exhibited increased mesofrontal dopamine and enhanced dopamine utilization in the mesolimbic and mesocortical systems after a cocaine challenge.[178]

Prenatal and postnatal life events and environments modify the activity of the hypothalamic-pituitary-adrenal axis,[179,180] and maternal glucocorticoids have a major role in the development of endocrine function in offspring. High levels of maternal glucocorticoids during prenatal stress have marked long-term repercussions on the efficiency of the offspring's negative feedback mechanisms in the hypothalamic-pituitary-adrenal axis. Thus, a modification of corticosterone secretion via changes in hypothalamic-pituitary-adrenal axis activity could be a biological substrate for the long-term behavioral effects of prenatal and postnatal events that could contribute to individual differences in vulnerability to drugs through long-lasting changes in dopamine utilization and receptor modifications within the mesocorticolimbic system.[181]

Summary: Interactions between Stress, Dopamine, and Vulnerability to Drug Use

Mild stressful situations and drugs of abuse have similar effects on dopamine neurotransmission (Fig. 8.1.4). Many functions have been attributed to dopamine neurons. Dopamine neurons are a part of the reticular formation and the limbic midbrain area. They receive ascending information from various regions from the lower brainstem and descending modulation from forebrain and cortical structures,[61,62] with a role of modulating functions integrated in 20 to 25 brain regions.[36] A parsimonious approach is to consider these neurons as participating in homeostatic and evolutionarily relevant integrated responses and for energy-related efforts to enable behavioral organization and activation.[34,36,37,182] Mild stressful situations increase dopamine transmission as an adaptive response to help the individual cope with mild stress and reduce the aversive effects of mild stress.[183] Increases in dopamine favor drug use–associated behaviors,[183] and stress-induced increases in dopamine utilization represent a complex mechanism by which the individual becomes more vulnerable to the effects of drugs, making the individual more susceptible to acquire stimulant self-administration. A positive relationship exists between the secretion of stress hormones, dopaminergic activity, and the subjective response to stimulants.[183,184] Stress-induced increases in reward-related behaviors induce not only changes in dopamine transmission. Stress can also bypass the dopamine terminal and act postsynaptically by increasing dopamine D1 receptor-mediated effects.[185] Stress also modifies, in a manner that could favor the development of drug use, the way in which dopamine neurons react to drugs. Analysis of dopamine overflow with microdialysis confirms that stress exacerbates the dopamine response to drugs. Stress-induced increases in cocaine-induced dopamine overflow are prevented by inhibiting stress-induced secretion of stress hormones. Moreover, blocking stress-induced increases in dopamine also blocks stress-induced increases in the response to the drug.[150] Stress-induced potentiation of the dopaminergic response to drugs is at least partially responsible for stress-induced potentiation of the behavioral effects of drugs. Conversely, stressed animals show not only greater dopaminergic reactivity to drugs but also greater dopaminergic reactivity to stress.[183] This reasoning suggests that drug-experienced individuals exhibit greater dopaminergic reactivity to stressful events, which could be responsible for the well-documented increases in drug responding and relapse produced by stress. Stress induces a parallel increase in mesolimbic dopamine and drug-seeking behavior, effects reversed by dopamine receptor antagonists.[186] However, when stress becomes excessive, dopaminergic activity decreases, which can offset the individual's homeostatic state and lead to decreased dopaminergic activity, a depressive-like state, and decreased reactivity to drugs.[183]

The role of dopamine in vulnerability to stimulants has been reviewed in terms of the developmental-environmental perspectives that account for individual differences in the propensity to take drugs and enter into the positive reinforcement process. However, these perspectives do not account for the transition to addiction and the passage from impulsive to compulsive intake driven by a negative reinforcement process.[4,182] Antireward mechanisms also are recruited in which extrahypothalamic CRF may play a central role. How dopamine interacts with the CRF system and exactly how CRF drives the dopamine system remain challenges for future work.

CONCLUSIONS

Research over the past 50 years has revealed that the mesocorticolimbic dopamine system has an essential role in the acute reinforcing effects of psychostimulant drugs and a contributory role in the acute reinforcing effects of nonstimulant drugs of abuse. Mesocorticolimbic dopamine systems contribute to motivational withdrawal and relapse with all drugs of abuse, and dopamine, by interacting with key elements of brain hormonal stress systems, also has a prominent role in individual differences for the vulnerability to initiate aspects of stimulant addiction that may extend to other drugs of abuse.

ACKNOWLEDGMENTS

This is Publication Number 19989 from The Scripps Research Institute. Research was supported by the Pearson Center for Alcoholism and Addiction Research and by National Institutes of Health grants AA06420 and AA08459 from the National Institute on Alcohol Abuse and Alcoholism, DA04043 and DA04398 from the National Institute on Drug Abuse, and DK26741 from the National Institute of Diabetes and Digestive and Kidney Diseases. The authors would like to thank Michael Arends for his assistance with manuscript preparation.

REFERENCES

1. American Psychiatric Association. *Diagnostic and Statistical Manual of Mental Disorders*. 4th ed. Washington, DC: American Psychiatric Press; 1994.
2. Koob GF, Le Moal M. Drug abuse: hedonic homeostatic dysregulation. *Science*. 1997;278:52–58.
3. Koob GF, Le Moal M. Plasticity of reward neurocircuitry and the "dark side" of drug addiction. *Nat Neurosci*. 2005;8:1442–1444.
4. Koob GF, Le Moal M. Addiction and the brain antireward system. *Annu Rev Psychol*. 2008;59:29–53.
5. Solomon RL. The opponent-process theory of acquired motivation: the costs of pleasure and the benefits of pain. *Am Psychol*. 1980;35:691–712.
6. Koob GF, Bloom FE. Cellular and molecular mechanisms of drug dependence. *Science*. 1988;242:715–723.
7. Koob GF. Allostatic view of motivation: implications for psychopathology. In: Bevins RA, Bardo MT, eds. *Motivational Factors in the Etiology of Drug Abuse*. Nebraska Symposium on Motivation). Vol. 50. Lincoln: University of Nebraska Press; 2004:1–18.
8. Ahmed SH, Koob GF. Transition from moderate to excessive drug intake: change in hedonic set point. *Science*. 1998;282:298–300.
9. Ahmed SH, Koob GF. Long-lasting increase in the set point for cocaine self-administration after escalation in rats. *Psychopharmacology*. 1999;146:303–312.
10. Deroche-Gamonet V, Belin D, Piazza PV. Evidence for addiction-like behavior in the rat. *Science*. 2004;305:1014–1017.
11. Mantsch JR, Yuferov V, Mathieu-Kia AM, et al. Effects of extended access to high versus low cocaine doses on self-administration, cocaine-induced reinstatement and brain mRNA levels in rats. *Psychopharmacology*. 2004;175:26–36.
12. Ahmed SH, Kenny PJ, Koob GF, Markou A. Neurobiological evidence for hedonic allostasis associated with escalating cocaine use. *Nat Neurosci*. 2002;5:625–626.
13. Kenny PJ, Chen SA, Kitamura O, et al. Conditioned withdrawal drives heroin consumption and decreases reward sensitivity. *J Neurosci*. 2006;26:5894–5900.
14. Ahmed SH, Walker JR, Koob GF. Persistent increase in the motivation to take heroin in rats with a history of drug escalation. *Neuropsychopharmacology*. 2000;22:413–421.
15. Kitamura O, Wee S, Specio SE, et al. Escalation of methamphetamine self-administration in rats: a dose-effect function. *Psychopharmacology*. 2006;186:48–53.
16. O'Dell LE, Roberts AJ, Smith RT, Koob GF. Enhanced alcohol self-administration after intermittent versus continuous alcohol vapor exposure. *Alcohol Clin Exp Res*. 2004;28:1676–1682.
17. George O, Ghozland S, Azar MR, et al. CRF-CRF$_1$ system activation mediates withdrawal-induced increases in nicotine self-administration in nicotine-dependent rats. *Proc Natl Acad Sci USA*. 2007;104:17198–17203.
18. Roberts DCS, Richardson NR. Self-administration of psychomotor stimulants using progressive ratio schedules of reinforcement. In: Boulton AA, Baker GB, W PH, eds. *Animal Models of Drug Addiction*. Neuromethods. Vol 24. Human Press, Totowa NJ: Humana Press; 1992:233–269.
19. Paterson NE, Markou A. Increased motivation for self-administered cocaine after escalated cocaine intake. *Neuroreport*. 2003;14:2229–2232.
20. Wee S, Mandyam CD, Lekic DM, Koob GF. α$_1$-Noradrenergic system role in increased motivation for cocaine intake in rats

21. Walker BM, Koob GF. The γ-aminobutyric acid-B receptor agonist baclofen attenuates responding for ethanol in ethanol-dependent rats. *Alcohol Clin Exp Res.* 2007;31:11–18.
22. Vanderschuren LJ, Everitt BJ. Drug seeking becomes compulsive after prolonged cocaine self-administration. *Science.* 2004;305:1017–1019.
23. Olds J, Milner P. Positive reinforcement produced by electrical stimulation of septal area and other regions of rat brain. *J Comp Physiol Psychol.* 1954;47:419–427.
24. Koob GF, Winger GD, Meyerhoff JL, Annau Z. Effects of D-amphetamine on concurrent self-stimulation of forebrain and brain stem loci. *Brain Res.* 1977;137:109–126.
25. Simon H, Le Moal M, Galey D, Cardo B. Silver impregnation of dopaminergic systems after radiofrequency and 6-OHDA lesions of the rat ventral. *Brain Res.* 1976;115:215–231.
26. Simon H, Le Moal M, Calas A. Efferents and afferents of the ventral tegmental-A10 region studied after local injection of [^{3}H]leucine and horseradish peroxidase. *Brain Res.* 1979;178:17–40.
27. Kornetsky C, Esposito RU. Euphorigenic drugs: effects on the reward pathways of the brain. *Fed Proc.* 1979;38:2473–2476.
28. Hernandez G, Hamdani S, Rajabi H, et al. Prolonged rewarding stimulation of the rat medial forebrain bundle: neurochemical and behavioral consequences. *Behav Neurosci.* 2006;120:888–904.
29. Simon H, Stinus L, Tassin JP, et al. Is the dopaminergic mesocorticolimbic system necessary for intracranial self-stimulation? Biochemical and behavioral studies from A10 cell bodies and terminals. *Behav Neural Biol.* 1979;27:125–145.
30. Garris PA, Kilpatrick M, Bunin MA, et al. Dissociation of dopamine release in the nucleus accumbens from intracranial self-stimulation. *Nature.* 1999;398:67–69.
31. Miliaressis E, Emond C, Merali Z. Re-evaluation of the role of dopamine in intracranial self-stimulation using in vivo microdialysis. *Behav Brain Res.* 1991;46:43–48.
32. Miliaressis E, Le Moal M. Stimulation of the medial forebrain bundle: behavioral dissociation of its rewarding and activating effects. *Neurosci Lett.* 1976;2:295–300.
33. Robinson S, Sandstrom SM, Denenberg VH, Palmiter RD. Distinguishing whether dopamine regulates liking, wanting, and/or learning about rewards. *Behav Neurosci.* 2005;119:5–15.
34. Salamone JD, Correa M, Farrar A, Mingote SM. Effort-related functions of nucleus accumbens dopamine and associated forebrain circuits. *Psychopharmacology.* 2007;191:461–482.
35. Yin HH, Ostlund SB, Balleine BW. Reward-guided learning beyond dopamine in the nucleus accumbens: the integrative functions of cortico-basal ganglia networks. *Eur J Neurosci.* 2008;28:1437–1448.
36. Le Moal M, Simon H. Mesocorticolimbic dopaminergic network: functional and regulatory roles. *Physiol Rev.* 1991;71:155–234.
37. Robbins TW, Everitt BJ. A role for mesencephalic dopamine in activation: commentary on Berridge (2006). *Psychopharmacology.* 2007;191:433–437.
38. Koob GF. Drugs of abuse: anatomy, pharmacology, and function of reward pathways. *Trends Pharmacol Sci.* 1992;13:177–184.
39. Di Chiara G, North RA. Neurobiology of opiate abuse. *Trends Pharmacol Sci.* 1992;13:185–193.
40. Nestler EJ. Is there a common molecular pathway for addiction? *Nat Neurosci.* 2005;8:1445–1449.
41. Hnasko TS, Sotak BN, Palmiter RD. Morphine reward in dopamine-deficient mice. *Nature.* 2005;438:854–857.
42. Pettit HO, Ettenberg A, Bloom FE, Koob GF. Destruction of dopamine in the nucleus accumbens selectively attenuates cocaine but not heroin self-administration in rats. *Psychopharmacology.* 1984;84:167–173.
43. Roberts DCS, Corcoran ME, Fibiger HC. On the role of ascending catecholaminergic systems in intravenous self-administration of cocaine. *Pharmacol Biochem Behav.* 1977;6:615–620.
44. Roberts DCS, Koob GF, Klonoff P, Fibiger HC. Extinction and recovery of cocaine self-administration following 6-hydroxydopamine lesions of the nucleus accumbens. *Pharmacol Biochem Behav.* 1980;12:781–787.
45. Koob GF, Le HT, Creese I. The D_1 dopamine receptor antagonist SCH 23390 increases cocaine self-administration in the rat. *Neurosci Lett.* 1987;79:315–320.
46. Lyness WH, Friedle NM, Moore KE. Destruction of dopaminergic nerve terminals in nucleus accumbens: effect on d-amphetamine self-administration. *Pharmacol Biochem Behav.* 1979;11:553–556.
47. McGregor A, Roberts DC. Dopaminergic antagonism within the nucleus accumbens or the amygdala produces differential effects on intravenous cocaine self-administration under fixed and progressive ratio schedules of reinforcement. *Brain Res.* 1993;624:245–252.
48. Deminiere JM, Taghzouti K, Tassin JP, et al. Increased sensitivity to amphetamine and facilitation of amphetamine self-administration after 6-hydroxydopamine lesions of the amygdala. *Psychopharmacology.* 1988;94:232–236.
49. Koob GF, Stinus L, Le Moal M. Hyperactivity and hypoactivity produced by lesions to the mesolimbic dopamine system. *Behav Brain Res.* 1981;3:341–359.
50. Tassin JP, Simon H, Herve D, et al. Non-dopaminergic fibres may regulate dopamine-sensitive adenylate cyclase in the prefrontal cortex and nucleus accumbens. *Nature.* 1982;295:696–698.
51. Dworkin SI, Guerin GF, Co C, et al. Lack of an effect of 6-hydroxydopamine lesions of the nucleus accumbens on intravenous morphine self-administration. *Pharmacol Biochem Behav.* 1988;30:1051–1057.
52. Rassnick S, Stinus L, Koob GF. The effects of 6-hydroxydopamine lesions of the nucleus accumbens and the mesolimbic dopamine system on oral self-administration of ethanol in the rat. *Brain Res.* 1993;623:16–24.
53. Lyness WH, Smith FL. Influence of dopaminergic and serotonergic neurons on intravenous ethanol self-administration in the rat. *Pharmacol Biochem Behav.* 1992;42:187–192.
54. Myers RD. Anatomical "circuitry" in the brain mediating alcohol drinking revealed by THP-reactive sites in the limbic system. *Alcohol.* 1990;7:449–459.
55. Stinus L, Koob GF, Ling N, et al. Locomotor activation induced by infusion of endorphins into the ventral tegmental area: evidence for opiate–dopamine interactions. *Proc Natl Acad Sci USA.* 1980;77:2323–2327.
56. Hyytia P, Koob GF. $GABA_A$ receptor antagonism in the extended amygdala decreases ethanol self-administration in rats. *Eur J Pharmacol.* 1995;283:151–159.
57. Rocha BA, Fumagalli F, Gainetdinov RR, et al. Cocaine self-administration in dopamine-transporter knockout mice. *Nat Neurosci.* 1998;1:132–137.

58. Chen R, Tilley MR, Wei H, et al. Abolished cocaine reward in mice with a cocaine-insensitive dopamine transporter. *Proc Natl Acad Sci USA*, 2006;103:9333–9338.
59. Lanteri C, Salomon L, Torrens Y, et al. Drugs of abuse specifically sensitize noradrenergic and serotonergic neurons via a non-dopaminergic mechanism. *Neuropsychopharmacology*. 2008;33:1724–1734.
60. Salomon L, Lanteri C, Glowinski J, Tassin JP. Behavioral sensitization to amphetamine results from an uncoupling between noradrenergic and serotonergic neurons. *Proc Natl Acad Sci USA*. 2006;103:7476–7481.
61. Nauta JH, Haymaker W. Hypothalamic nuclei and fiber connections. In: Haymaker W, Anderson E, Nauta WJH, eds. *The Hypothalamus*. Springfield, IL: Charles C. Thomas; 1969: 136–209.
62. Brodal A. *Neurological Anatomy in Relation to Clinical Medicine*. New York, NY: Oxford University Press; 1969.
63. Dahlstrom A, Fuxe K. Evidence for the existence of monoamine-containing neurons in the central nervous system: I. Demonstration of monoamines in the cell bodies of brain stem neurons. *Acta Physiol Scand Suppl*. 1964;232:1–55.
64. Weiss F, Markou A, Lorang MT, Koob GF. Basal extracellular dopamine levels in the nucleus accumbens are decreased during cocaine withdrawal after unlimited-access self-administration. *Brain Res*. 1992;593:314–318.
65. Weiss F, Lorang MT, Bloom FE, Koob GF. Oral alcohol self-administration stimulates dopamine release in the rat nucleus accumbens: genetic and motivational determinants *J Pharmacol Exp Ther*. 1993;267:250–258.
66. Doyon WM, York JL, Diaz LM, et al. Dopamine activity in the nucleus accumbens during consummatory phases of oral ethanol self-administration. *Alcohol Clin Exp Res*. 2003;27:1573–1582.
67. Weiss F, Parsons LH, Schulteis G, et al. Ethanol self-administration restores withdrawal-associated deficiencies in accumbal dopamine and 5-hydroxytryptamine release in dependent rats. *J Neurosci*. 1996;16:3474–3485.
68. Alheid GF, Heimer L. New perspectives in basal forebrain organization of special relevance for neuropsychiatric disorders: the striatopallidal, amygdaloid, and corticopetal components of substantia innominata. *Neuroscience*. 1988;27:1–39.
69. Groenewegen HJ, Berendse HW, Wolters JG, Lohman AH. The anatomical relationship of the prefrontal cortex with the striatopallidal system, the thalamus and the amygdala: evidence for a parallel organization. In: Uylings HBM, van Eden CG, de Bruin JPC, et al., eds. *The Prefrontal Cortex: Its Structure, Function, and Pathology*. Progress in Brain Research. Vol. 85. New York, NY: Elsevier; 1990:95–116.
70. Pontieri FE, Tanda G, Di Chiara G. Intravenous cocaine, morphine, and amphetamine preferentially increase extracellular dopamine in the "shell" as compared with the "core" of the rat nucleus accumbens. *Proc Natl Acad Sci USA*. 1995;92:12304–12308.
71. Pontieri FE, Tanda G, Orzi F, Di Chiara G. Effects of nicotine on the nucleus accumbens and similarity to those of addictive drugs. *Nature*. 1996;382:255–257.
72. Tanda G, Pontieri FE, Di Chiara G. Cannabinoid and heroin activation of mesolimbic dopamine transmission by a common μ$_1$ opioid receptor mechanism. *Science*. 1997;276: 2048–2050.
73. Tanda G, Di Chiara G. A dopamine-μ$_1$ opioid link in the rat ventral tegmentum shared by palatable food (Fonzies) and non-psychostimulant drugs of abuse. *Eur J Neurosci*. 1998;10: 1179–1187.
74. Cadoni C, Di Chiara G. Reciprocal changes in dopamine responsiveness in the nucleus accumbens shell and core and in the dorsal caudate-putamen in rats sensitized to morphine. *Neuroscience*. 1999;90:447–455.
75. Di Chiara G. Drug addiction as dopamine-dependent associative learning disorder. *Eur J Pharmacol*. 1999;375:13–30.
76. Barrot M, Marinelli M, Abrous DN, et al. Functional heterogeneity in dopamine release and in the expression of Fos-like proteins within the rat striatal complex. *Eur J Neurosci*. 1999;11:1155–1166.
77. Barrot M, Marinelli M, Abrous DN, et al. The dopaminergic hyper-responsiveness of the shell of the nucleus accumbens is hormone-dependent. *Eur J Neurosci*. 2000;12:973–979.
78. Bassareo V, De Luca MA, Di Chiara G. Differential expression of motivational stimulus properties by dopamine in nucleus accumbens shell versus core and prefrontal cortex. *J Neurosci*. 2002;22:4709–4719.
79. Bassareo V, Di Chiara G. Differential influence of associative and nonassociative learning mechanisms on the responsiveness of prefrontal and accumbal dopamine transmission to food stimuli in rats fed ad libitum. *J Neurosci*. 1997;17:851–861.
80. Bassareo V, Di Chiara G. Modulation of feeding-induced activation of mesolimbic dopamine transmission by appetitive stimuli and its relation to motivational state. *Eur J Neurosci*. 1999;11:4389–4397.
81. Bassareo V, Di Chiara G. Differential responsiveness of dopamine transmission to food-stimuli in nucleus accumbens shell/core compartments. *Neuroscience*. 1999;89:637–641.
82. Ito R, Robbins TW, Everitt BJ. Differential control over cocaine-seeking behavior by nucleus accumbens core and shell. *Nat Neurosci*. 2004;7:389–397.
83. Wise RA, Bozarth MA. A psychomotor stimulant theory of addiction. *Psychol Rev*. 1987;94:469–492.
84. Heidbreder CA, Gardner EL, Xi ZX, et al. The role of central dopamine D$_3$ receptors in drug addiction: a review of pharmacological evidence. *Brain Res Rev*. 2005;49:77–105.
85. Koob GF, Parsons LH, Caine SB, et al. Dopamine receptor subtype profiles in cocaine reward. In: Beninger RJ, Palomo T, Archer T, eds. *Dopamine Disease States*. Madrid, Spain: Editorial CYM; 1996:433–445.
86. Caine SB, Heinrichs SC, Coffin VL, Koob GF. Effects of the dopamine D-1 antagonist SCH 23390 microinjected into the accumbens, amygdala or striatum on cocaine self-administration in the rat. *Brain Res*. 1995;692:47–56.
87. Epping-Jordan MP, Markou A, Koob GF. The dopamine D-1 receptor antagonist SCH 23390 injected into the dorsolateral bed nucleus of the stria terminalis decreased cocaine reinforcement in the rat. *Brain Res*. 1998;784:105–115.
88. Hodge CW, Samson HH, Chappelle AM. Alcohol self-administration: further examination of the role of dopamine receptors in the nucleus accumbens. *Alcohol Clin Exp Res*. 1997;21: 1083–1091.
89. Gerrits MA, Ramsey NF, Wolterink G, van Ree JM. Lack of evidence for an involvement of nucleus accumbens dopamine D$_1$ receptors in the initiation of heroin self-administration in the rat. *Psychopharmacology*. 1994;114:486–494.
90. Bossert JM, Poles GC, Wihbey KA, et al. Differential effects of blockade of dopamine D$_1$-family receptors in nucleus accumbens core or shell on reinstatement of heroin seeking induced by contextual and discrete cues. *J Neurosci*. 2007;27: 12655–12663.
91. Schmidt HD, Pierce RC. Cooperative activation of D1-like and D2-like dopamine receptors in the nucleus accumbens shell is

required for the reinstatement of cocaine-seeking behavior in the rat. *Neuroscience.* 2006;142:451–461.
92. Melis M, Spiga S, Diana M. The dopamine hypothesis of drug addiction: hypodopaminergic state. *Int Rev Neurobiol.* 2005;63:101–154.
93. Barr AM, Phillips AG. Withdrawal following repeated exposure to *d*-amphetamine decreases responding for a sucrose solution as measured by a progressive ratio schedule of reinforcement. *Psychopharmacology.* 1999;141:99–106.
94. Pulvirenti L, Koob GF. Lisuride reduces psychomotor retardation during withdrawal from chronic intravenous amphetamine self-administration in rats. *Neuropsychopharmacology.* 1993; 8:213–218.
95. Pulvirenti L, Koob GF. Being partial to psychostimulant addiction therapy. *Trends Pharmacol Sci.* 2002;23: 151–153.
96. Clark D, Furmidge LJ, Petry N, et al. Behavioural profile of partial D2 dopamine receptor agonists: 1. Atypical inhibition of *d*-amphetamine-induced locomotor hyperactivity and stereotypy. *Psychopharmacology.* 1991;105:381–392.
97. Pulvirenti L, Koob GF. Dopamine receptor agonists, partial agonists and psychostimulant addiction. *Trends Pharmacol Sci.* 1994;15:374–379.
98. Svensson K, Ekman A, Piercey MF, et al. Effects of the partial dopamine receptor agonists SDZ 208-911, SDZ 208-912 and terguride on central monoamine receptors: a behavioral, biochemical and electrophysiological study. *Naunyn Schmiedebergs Arch Pharmacol.* 1991;344:263–274.
99. Orsini C, Koob GF, Pulvirenti L. Dopamine partial agonist reverses amphetamine withdrawal in rats. *Neuropsychopharmacology.* 2001;25:789–792.
100. Izzo E, Orsini C, Koob GF, Pulvirenti L. A dopamine partial agonist and antagonist block amphetamine self-administration in a progressive ratio schedule. *Pharmacol Biochem Behav.* 2001;68:701–708.
101. Pulvirenti L, Balducci C, Piercy M, Koob GF. Characterization of the effects of the partial dopamine agonist terguride on cocaine self-administration in the rat. *J Pharmacol Exp Ther.* 1998;286:1231–1238.
102. Wee S, Wang Z, Woolverton WL, et al. Effect of aripiprazole, a partial D$_2$ receptor agonist, on increased rate of methamphetamine self-administration in rats with prolonged access. *Neuropsychopharmacology.* 2007;32:2238–2247.
103. Ahmed SH, Koob GF. Changes in response to a dopamine antagonist in rats with escalating cocaine intake. *Psychopharmacology.* 2004;172:450–454.
104. Bono G, Balducci C, Richelmi P, et al. Dopamine partial receptor agonists reduce ethanol intake in the rat. *Eur J Pharmacol.* 1996;296:233–238.
105. Rossetti ZL, Hmaidan Y, Gessa GL. Marked inhibition of mesolimbic dopamine release: a common feature of ethanol, morphine, cocaine and amphetamine abstinence in rats. *Eur J Pharmacol.* 1992;221:227–234.
106. Volkow ND, Fowler JS, Wang GJ. Role of dopamine in drug reinforcement and addiction in humans: results from imaging studies. *Behav Pharmacol.* 2002;13:355–366.
107. Volkow ND, Wang GJ, Fischman MW, et al. Relationship between subjective effects of cocaine and dopamine transporter occupancy. *Nature.* 1997;386:827–830.
108. Martinez D, Narendran R, Foltin RW, et al. Amphetamine-induced dopamine release: markedly blunted in cocaine dependence and predictive of the choice to self-administer cocaine. *Am J Psychiatry.* 2007;164:622–629.
109. Martin-Solch C, Magyar S, Kunig G, et al. Changes in brain activation associated with reward processing in smokers and nonsmokers: a positron emission tomography study. *Exp Brain Res.* 2001;139:278–286.
110. Volkow ND, Fowler JS. Addiction, a disease of compulsion and drive: involvement of the orbitofrontal cortex. *Cereb Cortex.* 2000;10:318–325.
111. Xu M, Hu XT, Cooper DC, et al. Elimination of cocaine-induced hyperactivity and dopamine-mediated neurophysiological effects in dopamine D1 receptor mutant mice. *Cell.* 1994;79:945–955.
112. Xu M, Guo Y, Vorhees CV, Zhang J. Behavioral responses to cocaine and amphetamine administration in mice lacking the dopamine D1 receptor. *Brain Res.* 2000;852:198–207.
113. Smith DR, Striplin CD, Geller AM, et al. Behavioural assessment of mice lacking D$_{1A}$ dopamine receptors. *Neuroscience.* 1998;86:135–146.
114. Caine SB, Thomsen M, Gabriel KI, et al. Lack of self-administration of cocaine in dopamine D$_1$ receptor knock-out mice. *J Neurosci.* 2007;27:13140–13150.
115. Kalivas PW, McFarland K. Brain circuitry and the reinstatement of cocaine-seeking behavior. *Psychopharmacology.* 2003; 168:44–56.
116. Zhai H, Li Y, Wang X, Lu L. Drug-induced alterations in the extracellular signal-regulated kinase (ERK) signalling pathway: implications for reinforcement and reinstatement. *Cell Mol Neurobiol.* 2008;28:157–172.
117. Anderson SM, Pierce RC. Cocaine-induced alterations in dopamine receptor signaling: implications for reinforcement and reinstatement. *Pharmacol Ther.* 2005;106:389–403.
118. O'Brien CP, Ehrman RN, Ternes JM. Classical conditioning in human opioid dependence. In: Goldberg SR, Stolerman IP, eds. *Behavioral Analysis of Drug Dependence.* Orlando, FL: Academic Press; 1986:329–356.
119. Di Franza JR. Hooked from the first cigarette. *Sci Am.* 2008;298:82–87.
120. Substance Abuse and Mental Health Services Administration. *Results from the 2002 National Survey on Drug Use and Health: National Findings.* NHSDA Series H-22, DHHS Publication No. SMA 03-3836. Rockville, MD: Office of Applied Statistics; 2003.
121. Grant BF, Stinson FS, Dawson DA, et al. Prevalence and co-occurrence of substance use disorders and independent mood and anxiety disorders: results from the National Epidemiologic Survey on Alcohol and Related Conditions. *Arch Gen Psychiatry.* 2004;61:807–816.
122. Goodman A. Neurobiology of addiction: an integrative review. *Biochem Pharmacol.* 2008;75:266–322.
123. Shaffer HJ, LaPlante DA, LaBrie RA, et al. Toward a syndrome model of addiction: multiple expressions, common etiology. *Harvard Rev Psychiatry.* 2004;12:367–374.
124. Kessler RC, Berglund P, Demler O, et al. Lifetime prevalence and age-of-onset distributions of DSM-IV disorders in the National Comorbidity Survey Replication. *Arch Gen Psychiatry.* 2005;62:593-602 [erratum: 62:768].
125. Piazza PV, Le Moal M. Pathophysiological basis of vulnerability to drug abuse: role of an interaction between stress, glucocorticoids, and dopaminergic neurons. *Annu Rev Pharmacol Toxico.l* 1996;36:359–378.
126. Piazza PV, Le Moal M. The role of stress in drug self-administration. *Trends Pharmacol Sci.* 1998;19:67–74.
127. Abi-Dargham A, Kegeles LS, Martinez D, et al. Dopamine mediation of positive reinforcing effects of amphetamine in stimulant

128. de Wit H, Uhlenhuth EH, Johanson CE. Individual differences in the reinforcing and subjective effects of amphetamine and diazepam. *Drug Alcohol Depend.* 1986;16:341–360.
129. Volkow ND, Wang GJ, Fowler JS, et al. Reinforcing effects of psychostimulants in humans are associated with increases in brain dopamine and occupancy of D_2 receptors. *J Pharmacol Exp Ther.* 1999;291:409–415.
130. Leyton M, Boileau I, Benkelfat C, et al. Amphetamine-induced increases in extracellular dopamine, drug wanting, and novelty seeking: a PET/[^{11}C]raclopride study in healthy men. *Neuropsychopharmacology.* 2002;27:1027–1035.
131. Piazza PV, Deminiere JM, Le Moal M, Simon H. Factors that predict individual vulnerability to amphetamine self-administration. *Science.* 1989;245:1511–1513.
132. Piazza PV, Deroche-Gamonent V, Rouge-Pont F, Le Moal M. Vertical shifts in self-administration dose-response functions predict a drug-vulnerable phenotype predisposed to addiction. *J Neurosci.* 2000;20:4226–4232.
133. Caprioli D, Paolone G, Celentano M, et al. Environmental modulation of cocaine self-administration in the rat. *Psychopharmacology.* 2007;192:397–406.
134. Piazza PV, Rouge-Pont F, Deminiere JM, et al. Dopaminergic activity is reduced in the prefrontal cortex and increased in the nucleus accumbens of rats predisposed to develop amphetamine self-administration. *Brain Res.* 1991;567:169–174.
135. Cabib S, Orsini C, Le Moal M, Piazza PV. Abolition and reversal of strain differences in behavioral responses to drugs of abuse after a brief experience. *Science.* 2000;289:463–465.
136. Harfstrand A, Fuxe K, Cintra A, et al. Glucocorticoid receptor immunoreactivity in monoaminergic neurons of rat brain. *Proc Natl Acad Sci USA.* 1986;83:9779–9783.
137. Piazza PV, Rouge-Pont F, Deroche V, et al. Glucocorticoids have state-dependent stimulant effects on the mesencephalic dopaminergic transmission. *Proc Natl Acad Sci USA.* 1996;93:8716–8720.
138. de Kloet ER. Brain corticosteroid receptor balance and homeostatic control. *Front Neuroendocrinol.* 1991;12:95–164.
139. Joels M, de Kloet ER. Control of neuronal excitability by corticosteroid hormones. *Trends Neurosci.* 1992;15:25–30.
140. Joels M, de Kloet ER. Mineralocorticoid and glucocorticoid receptors in the brain: implications for ion permeability and transmitter systems. *Prog Neurobiol.* 1994;43:1–36.
141. Angulo JA, McEwen BS. Molecular aspects of neuropeptide regulation and function in the corpus striatum and nucleus accumbens. *Brain Res Rev.* 1994;19:1–28.
142. Schoffelmeer AN, Voorn P, Jonker AJ, et al. Morphine-induced increase in D-1 receptor regulated signal transduction in rat striatal neurons and its facilitation by glucocorticoid receptor activation: possible role in behavioral sensitization. *Neurochem Res.* 1996;21:1417–1423.
143. Piazza PV, Le Moal M. Glucocorticoids as a biological substrate of reward: physiological and pathophysiological implications. *Brain Res Rev.* 1997;25:359–372.
144. Piazza PV, Deroche V, Rouge-Pont F, Le Moal M. Behavioral and biological factors associated with individual vulnerability to psychostimulant abuse. In: Wetherington CL, Falk JL, eds. *Laboratory Behavioral Studies of Vulnerability to Drug Abuse.* NIDA Research Monograph. Vol 169. Rockville, MD: National Institute on Drug Abuse; 1998: 105–133.
145. Mantsch JR, Saphier D, Goeders NE. Corticosterone facilitates the acquisition of cocaine self-administration in rats: opposite effects of the type II glucocorticoid receptor agonist dexamethasone. *J Pharmacol Exp Ther.* 1998;287:72–80.
146. Piazza PV, Maccari S, Deminiere JM, et al. Corticosterone levels determine individual vulnerability to amphetamine self-administration. *Proc Natl Acad Sci USA.* 1991;88:2088–2092.
147. Piazza PV, Deroche V, Deminiere JM, et al. Corticosterone in the range of stress-induced levels possesses reinforcing properties: implications for sensation-seeking behaviors. *Proc Natl Acad Sci USA.* 1993;90:11738–11742.
148. Deroche V, Marinelli M, Le Moal M, Piazza PV. Glucocorticoids and behavioral effects of psychostimulants: II. Cocaine intravenous self-administration and reinstatement depend on glucocorticoid levels. *J Pharmacol Exp Ther.* 1997;281:1401–1407.
149. Rouge-Pont F, Piazza PV, Kharouby M, et al. Higher and longer stress-induced increase in dopamine concentrations in the nucleus accumbens of animals predisposed to amphetamine self-administration: a microdialysis study. *Brain Res.* 1993;602:169–174.
150. Rouge-Pont F, Marinelli M, Le Moal M, et al. Stress-induced sensitization and glucocorticoids: II. Sensitization of the increase in extracellular dopamine induced by cocaine depends on stress-induced corticosterone secretion. *J Neurosci.* 1995;15:7189–7195.
151. Deroche-Gamonet V, Sillaber I, Aouizerate B, et al. The glucocorticoid receptor as a potential target to reduce cocaine abuse. *J Neurosci.* 2003;23:4785–4790.
152. Goeders NE, Guerin GF. Non-contingent electric footshock facilitates the acquisition of intravenous cocaine self-administration in rats. *Psychopharmacology.* 1994;114:63–70.
153. Hooks MS, Juncos JL, Justice JB Jr, et al. Individual locomotor response to novelty predicts selective alterations in D_1 and D_2 receptors and mRNAs. *J Neurosci.* 1994;14:6144–6152.
154. Zhou Y, Spangler R, LaForge KS, et al. Corticotropin-releasing factor and type 1 corticotropin-releasing factor receptor messenger RNAs in rat brain and pituitary during "binge"-pattern cocaine administration and chronic withdrawal. *J Pharmacol Exp Ther.* 1996;279:351–358.
155. Zhou Y, Spangler R, Schlussman SD, et al. Alterations in hypothalamic-pituitary-adrenal axis activity and in levels of proopiomelanocortin and corticotropin-releasing hormone-receptor 1 mRNAs in the pituitary and hypothalamus of the rat during chronic "binge" cocaine and withdrawal. *Brain Res.* 2003;964:187–199.
156. Mantsch JR, Baker DA, Serge JP, et al. Surgical adrenalectomy with diurnal corticosterone replacement slows escalation and prevents the augmentation of cocaine-induced reinstatement in rats self-administering cocaine under long-access conditions. *Neuropsychopharmacology.* 2008;33:814–826.
157. Koob GF. A role for brain stress systems in addiction. *Neuron.* 2008;59:11–34.
158. Wang B, Shaham Y, Zitzman D, et al. Cocaine experience establishes control of midbrain glutamate and dopamine by corticotropin-releasing factor: a role in stress-induced relapse to drug seeking. *J Neurosci.* 2005;25:5389–5396.
159. Wang B, You ZB, Rice KC, Wise RA. Stress-induced relapse to cocaine seeking: roles for the CRF_2 receptor and CRF-binding protein in the ventral tegmental area of the rat. *Psychopharmacology.* 2007;193:283–294.
160. Shaham Y, Shalev U, Lu L, et al. The reinstatement model of drug relapse: history, methodology and major findings. *Psychopharmacology.* 2003;168:3–20.

161. Moyer JA, Herrenkohl LR, Jacobowitz DM. Stress during pregnancy: effect on catecholamines in discrete brain regions of offspring as adults. *Brain Res.* 1978;144:173–178.

162. Fride E, Weinstock M. Alterations in behavioral and striatal dopamine asymmetries induced by prenatal stress. *Pharmacol Biochem Behav.* 1989;32:425–430.

163. Maccari S, Piazza PV, Deminiere JM, et al. Life events–induced decrease of corticosteroid type I receptors is associated with reduced corticosterone feedback and enhanced vulnerability to amphetamine self-administration. *Brain Res.* 1991;547:7–12.

164. Deminiere JM, Piazza PV, Guegan G, et al. Increased locomotor response to novelty and propensity to intravenous amphetamine self-administration in adult offspring of stressed mothers. *Brain Res.* 1992;586:135–139.

165. Hall FS. Social deprivation of neonatal, adolescent, and adult rats has distinct neurochemical and behavioral consequences. *Crit Rev Neurobiol.* 1998;12:129–162.

166. Kosten TA, Miserendino MJ, Kehoe P. Enhanced acquisition of cocaine self-administration in adult rats with neonatal isolation stress experience. *Brain Res.* 2000;875:44–50.

167. Brake WG, Zhang TY, Diorio J, et al. Influence of early postnatal rearing conditions on mesocorticolimbic dopamine and behavioural responses to psychostimulants and stressors in adult rats. *Eur J Neurosci.* 2004;19:1863–1874.

168. Smith JK, Neill JC, Costall B. Post-weaning housing conditions influence the behavioural effects of cocaine and d-amphetamine. *Psychopharmacology.* 1997;131:23–33.

169. Blanc G, Herve D, Simon, H, et al. Response to stress of meso-cortico-frontal dopaminergic neurones in rats after long-term isolation. *Nature.* 1980;284:265–267.

170. Heidbreder CA, Weiss IC, Domeney AM, et al. Behavioral, neurochemical and endocrinological characterization of the early social isolation syndrome. *Neuroscience.* 2000;100:749–768.

171. Bezard E, Dovero S, Belin D, et al. Enriched environment confers resistance to 1-methyl-4-phenyl-1,2,3,6-tetrahydropyridine and cocaine: involvement of dopamine transporter and trophic factors. *J Neurosci.* 2003;23:10999–11007.

172. Caster JM, Walker QD, Kuhn CM. A single high dose of cocaine induces differential sensitization to specific behaviors across adolescence. *Psychopharmacology.* 2007;193:247–260.

173. Stansfield KH, Kirstein CL. Neurochemical effects of cocaine in adolescence compared to adulthood. *Dev Brain Res.* 2005;159:119–125.

174. Brandon CL, Marinelli M, Baker LK, White FJ. Enhanced reactivity and vulnerability to cocaine following methylphenidate treatment in adolescent rats. *Neuropsychopharmacology.* 2001;25:651–661.

175. Brandon CL, Marinelli M, White FJ. Adolescent exposure to methylphenidate alters the activity of rat midbrain dopamine neurons. *Biol Psychiatry.* 2003;54:1338–1344.

176. Kosten TA, Zhang XY, Kehoe P. Neurochemical and behavioral responses to cocaine in adult male rats with neonatal isolation experience. *J Pharmacol Exp Ther.* 2005;314:661–667.

177. Stanwood GD, Levitt P. Prenatal exposure to cocaine produces unique developmental and long-term adaptive changes in dopamine D_1 receptor activity and subcellular distribution. *J Neurosci.* 2007;27:152–157.

178. Kippin TE, Szumlinski KK, Kapasova Z, et al. Prenatal stress enhances responsiveness to cocaine. *Neuropsychopharmacology.* 2008; 33:769–782.

179. Caldji C, Tannenbaum B, Sharma S, et al. Maternal care during infancy regulates the development of neural systems mediating the expression of fearfulness in the rat. *Proc Natl Acad Sci USA.* 1998;95:5335–5340.

180. Ladd CO, Huot RL, Thrivikraman KV, et al. Long-term behavioral and neuroendocrine adaptations to adverse early experience. In: Mayer EA, Saper CB, eds. *The Biological Basis for Mind–Body Interactions.* Progress in Brain Research. Vol. 122. New York, NY: Elsevier; 2000:81–103.

181. Henry C, Guegant G, Cador M, et al. Prenatal stress in rats facilitates amphetamine-induced sensitization and induces long-lasting changes in dopamine receptors in the nucleus accumbens. *Brain Res.* 1995;685:179–186.

182. Koob GF, Le Moal M. *Neurobiology of Addiction.* London, England: Academic Press; 2006.

183. Marinelli M. Dopaminegic reward pathways and effects of stress. In: al'Absi M, ed. *Stress and Addiction: Biological and Psychological Mechanisms.* Amsterdam: Academic Press; 2007:41–83.

184. Oswald LM, Wong DF, McCaul M, et al. Relationships among ventral striatal dopamine release, cortisol secretion, and subjective responses to amphetamine. *Neuropsychopharmacology.* 2005;30:821–832.

185. Carr KD, Kim GY, Cabeza de Vaca S. Rewarding and locomotor-activating effects of direct dopamine receptor agonists are augmented by chronic food restriction in rats. *Psychopharmacology.* 2001;154:420–428.

186. Shaham Y, Stewart, J. Effects of opioid and dopamine receptor antagonists on relapse induced by stress and re-exposure to heroin in rats. *Psychopharmacology.* 1996;125:385–391.

187. Koob GF. The neurocircuitry of addiction: implications for treatment. *Clin Neurosci Res.* 2005;5:89–101.

188. Di Ciano, P, Coury A, Depoortere RY, et al. Comparison of changes in extracellular dopamine concentrations in the nucleus accumbens during intravenous self-administration of cocaine or d-amphetamine. *Behav Pharmacol.* 1995;6:311–322.

189. Weiss F, Hurd YL, Ungerstedt U, et al. Neurochemical correlates of cocaine and ethanol self-administration. In: Kalivas PW, Samson HH, eds. *The Neurobiology of Drug and Alcohol Addiction.* Annals of the New York Academy of Sciences. Vol. 654. New York, NY: New York Academy of Sciences; 1992:220–241.

190. Hemby SE, Martin TJ, Co C, et al. The effects of intravenous heroin administration on extracellular nucleus accumbens dopamine concentrations as determined by *in vivo* microdialysis. *J Pharmacol Exp Ther.* 1995;273:591–598.

8.2 Dopaminergic Mechanisms in Drug-Seeking Habits and the Vulnerability to Drug Addiction

BARRY J. EVERITT, DAVID BELIN, JEFFREY W. DALLEY, AND TREVOR W. ROBBINS

INTRODUCTION

Dopamine (DA) has been the focus of research on the neural mechanisms of addiction for four decades or more. This interest emerged in large part from the discovery that rats would electrically self-stimulate their brains (intracranial self-stimulation: ICSS)[1] and that the most effective sites at which electrodes would support ICSS lie on the projections of dopaminergic neurons from the ventral tegmental area (VTA) and substantia nigra pars compacta (SNc) to the ventral (nucleus accumbens, Acb, olfactory tubercle) and dorsal (caudate-putamen) striatum, respectively, as well as to limbic cortical structures including the amygdala, orbitofrontal and medial prefrontal cortices.[2] Not only is ICSS reduced after Acb DA depletion, but psychostimulant drugs such as amphetamine also enhance ICSS by shifting the response rate-current intensity function to the left and reducing the electrical threshold for ICSS to be sustained.[3] Thus, a link between addictive drug action, DA systems, and notions of *reward* was established. Much later, the rate at which rats learned to respond for ICSS was shown to be correlated with the degree of potentiation of synapses made by cortical afferents onto striatal neurons in a way that requires DA receptors.[4] This observation illustrates the link between reward, learning and DA that has become an important theoretical focus for current notions of neural plasticity and addiction.[5,6]

Initially it was suggested that DA release in the Acb mediated the pleasurable, or hedonic, aspects of reward, whether natural, such as food or sex, or drug rewards.[7] The hypothesis that drugs, especially psychostimulants such as cocaine and amphetamine, exerted their hedonic effects via an increase in Acb DA transmission had a powerful impact on drug addiction research, generating abundant experimental tests of a very difficult hypothesis to refute, not least because subjective states of *pleasure* or *liking* (as it is now so often referred to) cannot easily be measured in animals. Although Acb DA is not apparently involved in the presumed hedonic reactions to the taste of food reinforcers, whereas opiate mechanisms in the Acb and globus pallidus are,[8,9] there are no related data in animals concerning hedonic or liking responses to addictive drugs. However, it is clear that hedonic responses to addictive drugs must inevitably reflect the subjective perception of their neurochemical effects, including DA release in the Acb and elsewhere,[10] and these may be correlated with activity in other systems that are involved more directly with hedonic responses to natural and drug reinforcers. It also seems unlikely that pleasure can be mediated by neurochemical mechanisms occurring solely in subcortical structures such as the Acb or globus pallidus, and that activity in striatopallidal circuitry and in other sites following drug self-administration, or consumption of natural rewards, is subject to further processing in cortical, perhaps especially insular[11] and other prefrontal cortical areas, before attribution and accompanying subjective commentary — or *feelings* — can occur.[12,13]

DA AND REINFORCEMENT

There is, however, more general acceptance of the notion that DA transmission provides a neurochemical mechanism of reinforcement in the brain. Increased extracellular DA in the Acb is consistently seen in response to appetitive reinforcers, including and perhaps especially addictive drugs[14–17]; intra-Acb infusions of direct and indirect DA receptor agonists are reinforcing[18–20]; natural- and drug-reinforced responding depends on Acb DA.[21–24] The reinforcing effects of addictive drugs are multidimensional.[13] They can act as *instrumental reinforcers*, increasing the likelihood of responses that produce them and thereby resulting in drug self-administration or drug taking. Drugs produce subjective or *discriminative* effects, which include the sensing of autonomic activity (feelings) or distortions in sensory processing. Environmental stimuli that are closely associated with the effects of self-administered drugs gain incentive salience through the process of Pavlovian conditioning and may then act as conditioned reinforcers. Stimulant drugs such as cocaine and amphetamine, but other addictive drugs as well, can exaggerate the perceptual impact, or *incentive salience*, of environmental stimuli, especially conditioned stimuli (CSs) that already predict important environmental events. We have postulated that any combination of these effects may constitute the rewarding effect of a drug, that is, the subjective effects produced by attributions made about the CSs.[13] In particular, we have suggested that it is the sense of expectancy, or of control over such interoceptive and exteroceptive states—including the overall level of arousal accompanying them—acquired through instrumental action-outcome learning that constitutes instrumental drug reinforcement.[13]

MOLECULAR MECHANISMS OF ACTION OF ADDICTIVE DRUGS: DA AND PLASTICITY

One of the many successes arising out of the last 20 years of investigation of the neural mechanisms of action of addictive drugs has been the definition of their primary molecular targets in the brain and the cloning of the genes that encode these proteins (for reviews see[25–27]). They include the primary molecular targets of cocaine (DA transporter), amphetamine (synaptic vesicle amine transporter), the opioids heroin and morphine (μ-opiate receptor), nicotine (the nicotinic cholinergic receptor), cannabis (cannabinoid CB1 receptor), and alcohol (*N*-methyl-D-aspartate [NMDA] and GABA receptors, among other targets). Importantly, however, although addictive drugs of different classes have their own specific and discrete molecular targets and thereby produce different and discriminable subjective and other effects (including subjective "pleasure," "highs," and "rushes"), they also have in common the ability to increase DA transmission and thereby influence what is widely regarded to be a common reinforcement mechanism in the brain. Through these effects on DA transmission in the striatum, as well as in cortical sites also innervated by VTA and SNc DA neurons, addictive drugs can influence both Pavlovian and instrumental learning. This has become a central issue for contemporary theories of drug addiction—namely, that drugs not only have reinforcing effects mediated by alterations in DA transmission, but that they thereby also impact upon the learning and memory, or *plasticity*, mechanisms that underlie the development of addictive behavior.[13,27,28]

This view has been markedly strengthened by two sets of observations: (1) in vivo electrophysiological recordings of DA neuron activity indicating that DA neurons, firing in phasic mode, encode a reward prediction error[29]; (2) DA and, more especially, addictive drug-induced increases in DA, greatly influence the induction of long-term potentiation (LTP) and long-term depression (LTD) in these in vitro cellular models of plasticity.[30,31] Thus, in a series of experiments in monkeys, Wolfram Schultz[29,32,33] has shown that unexpected ingestive rewards result in a phasic increase in midbrain DA neuronal firing. But as the monkey learns through Pavlovian association that a previously neutral stimulus reliably predicts a reward of a particular magnitude, DA neurons no longer fire to the primary reward as it becomes expected, but instead fire to the earliest reliable predictor of the reward. Additionally, if an expected reward is omitted, the tonic firing of DA neurons is suppressed at the time that it was expected. These observations have led Schultz and others to propose that DA neuron firing is readily accommodated in reinforcement-learning models; that tonic firing of DA neurons reflects that positively reinforcing outcomes are as expected; that phasic burst firing signals that a reward is unexpected or better than expected (a positive reward prediction error); and that cessation of DA neuron firing signals a negative prediction error, that is, that the outcome is worse than expected.[34,35] Similar correlates of reward prediction errors have also been demonstrated in human functional imaging studies (e.g.,[36,37]).

PAVLOVIAN CONDITIONING, DA, AND ADDICTIVE BEHAVIOR

Contemporary theories of drug addiction have particularly emphasized Pavlovian mechanisms and the potential for powerful addictive drug-induced influences on DA transmission in the associative processes by which environmental stimuli have such an impact on drug seeking and relapse.[6,38–40] The possible relevance of these data for the potent reinforcing effects of addictive drugs and drug addiction is clear. Although the prediction error mechanism regulates the strength of conditioning to the level that is appropriate for the magnitude of a natural reinforcer, drug reinforcers, acting directly or indirectly as DA receptor agonists, could disrupt this regulation,[41,42] with the result that drug-associated CSs become 'supernormal' drug cues that are capable of establishing or influencing compulsive drug seeking habits.[41]

Despite the lack of direct experimental evidence, it has been argued persuasively by reinforcement-learning theorists[41,43] that drugs such as cocaine will, through their effects on central DA transmission, always generate a positive reward prediction error signaling that the drug reward is always 'better than expected'. The ability of cocaine to influence phasic DA release is consistent with this view.[44] Thus, these models predict that drug-associated CSs will be 'overlearned' and that actions directed at acquiring and taking drugs (see below) will be overlearned, thus setting up a scenario for craving and compulsive drug taking.[27]

Conditioned stimuli that predict natural reinforcers can have several effects on behavior, and it is not unreasonable to assume that this is also the case for CSs that predict the effects of addictive drugs, such as cocaine and heroin (for reviews, see[45,46]). Thus, through Pavlovian association with natural rewards, CSs can elicit Pavlovian (automatic or reflexive) *approach* (also called *sign tracking* and consummatory behavior (*goal tracking*).[45,47] Conditioned stimuli can have motivational effects, thereby increasing rates of responding for food when the CS is presented unexpectedly (called *Pavlovian–instrumental transfer* [PIT])[48,49] and also the ingestion of food.[50] These motivational effects of CSs reflect a Pavlovian arousal mechanism that serves to energize or activate responding, whether in terms of increasing rates of locomotor activity, instrumental behavior, or consummatory behavior.[45] Pavlovian CSs can also serve as goals of behavior, acting as *conditioned reinforcers* that can support long sequences of instrumental seeking responses and mediate the long delays that are often experienced by animals and humans when seeking to obtain primary goals.[51] We have reviewed in detail the neural systems basis of these Pavlovian influences on appetitive behavior, which involve dissociable contributions of subdivisions of the amygdala, the orbital prefrontal and anterior cingulate cortices, the shell and core of the Acb, and the mesolimbic DA system.[46,52]

The impact of dopaminergic modulation of limbic corticostriatal circuitry in these processes is supported by several key lines of neurobiological evidence, summarized as follows. Unexpected presentations of food- or drug-associated CSs increase extracellular DA in the Acb.[16,17] Consistent with these data, selective lesions of the AcbC[53] or infusions of NMDA or DA receptor antagonists into the AcbC during training[54] greatly retard the acquisition of a Pavlovian approach response, whereas infusions of NMDA or DA D1 receptor antagonists into this region after a conditioning trial disrupt the consolidation of this response into memory.[55] Lesions of the AcbC abolish PIT[56]; lesions of the AcbS apparently disrupt specific, rather than general, PIT— that is, when the CS specifically potentiates responding for the goal, or unconditioned stimulus (US), with which it is associated.[57] It is, however, quite difficult to reconcile these latter data with the effect of manipulations of the basolateral amygdala (BLA) on PIT, since the BLA projects predominantly to the AcbC, not the AcbS, yet BLA lesions disrupt the specific, but not general, form of PIT, as might be expected given the role of the BLA in sensory-specific associations of reinforcers with environmental stimuli.[46,58] Systemic treatment with a DA receptor antagonist decreases,[59] while increasing DA release in the AcbS potentiates, PIT.[60]

The neural basis of conditioned reinforcement has been investigated using the "acquisition of a new response" procedure, in which a new instrumental response is acquired and maintained solely by contingent presentation of a CS that has previously been associated with a reinforcer.[61] The BLA, orbital prefrontal cortex (OFC), and AcbC are important for the ability to respond for conditioned reinforcement.[53,62–64] Increasing DA transmission in the AcbS, by the infusion of the psychostimulant amphetamine, greatly potentiates the control over behavior by a food-associated conditioned reinforcer.[53,65] Conditioned reinforcement, then, depends on three major influences: the BLA, via its projections to the AcbC, underlies the conditioned reinforcement process[62,66]; the OFC is critical for what is termed the *outcome-specific* form of conditioned reinforcement, that is, the association of the CS with the specific properties of a reward.[64] Additionally, the mesolimbic DA projection, especially to the AcbS, mediates the response rate–increasing effects of psychomotor stimulants such as amphetamine and cocaine, hypothetically by simulating the behaviorally activating effects of Pavlovian arousal and affecting the incentive salience of the conditioned reinforcer.[13,45]

Data showing that DA may transiently impact upon the association between environmental stimuli and natural reinforcers are especially clear in the case of Pavlovian approach or sign tracking.[55] But conditioned reinforcement is acquired when DA is depleted from, or DA receptor antagonists are infused into, the Acb.[67,68] Dopaminergic activity in sites other than the Acb, such as the amygdala and OFC, might, however, be important recipients of a *teaching signal* important for stimulus–reward learning.[27] But it is clearly the case that increased DA transmission in the Acb, especially following stimulant drug exposure, greatly enhances the impact of CSs on sign tracking, PIT, and conditioned reinforcement (reviewed in[46]), and these effects may be a critical component of the reinforcing effects of cocaine, amphetamine, nicotine, and perhaps other drugs as well. Therefore, it is logical to suggest that Pavlovian approach is perhaps involved in maladaptively attracting humans toward sources of addictive drug reinforcers.[13] A sign-tracking response to a drug-associated CS has been demonstrated,[69] but it is far from clear that this response is aberrantly stronger, or that it is more difficult to extinguish, than the approach response to a food-associated CS as a consequence of the stronger conditioning that might have occurred as a result of the cocaine-enhanced increase in DA release.

It has also been argued that PIT by drug-associated CSs greatly invigorates instrumental drug seeking; indeed, the claim that PIT represents the very essence of wanting drugs is the cornerstone of the incentive salience theory of addiction.[70] However, the enhancement of responding for intravenously self-administered drugs by the unexpected presentation of a drug-associated CS has not easily been shown in experimental studies of drug seeking or relapse; indeed, Pavlovian drug CS presentations decrease, rather than increase, cocaine seeking.[71] Similarly, the reinstatement of drug seeking in extinction-relapse, or abstinence-relapse, procedures generally depends upon contingent CS presentations (i.e., the CSs are a consequence of responding, as in the case of conditioned reinforcement) rather than PIT.[72-74] Yet, PIT is readily demonstrated in animals responding for ingestive reinforcers. Thus, we must consider either that the experimental conditions for demonstrating drug-associated, CS-induced PIT have not been optimized or that the behavioral influences of CSs associated with drugs and natural reinforcers differ.[13,46] There is little direct evidence to suggest that drug-associated CSs that are "stamped in" by increasing DA transmission exert an undue motivational influence compared to, say, a food-associated CS, but this notion warrants focused experimental attention.

The conditioned reinforcing properties of drug-associated CSs are, however, critically important for extended periods of instrumental drug seeking[51] as well as for relapse.[75] The integrity of the BLA and AcbC is necessary for the acquisition of cocaine seeking[76,77] (see below). The self-administration of cocaine potentiates cocaine seeking under a second-order schedule of reinforcement[78] (Fig. 8.2.1), and this depends upon cocaine-induced increases in DA and the AcbS.[77] However, again, it is not clear that drug-associated conditioned reinforcers are aberrantly stronger in this regard than those associated with food or sex.

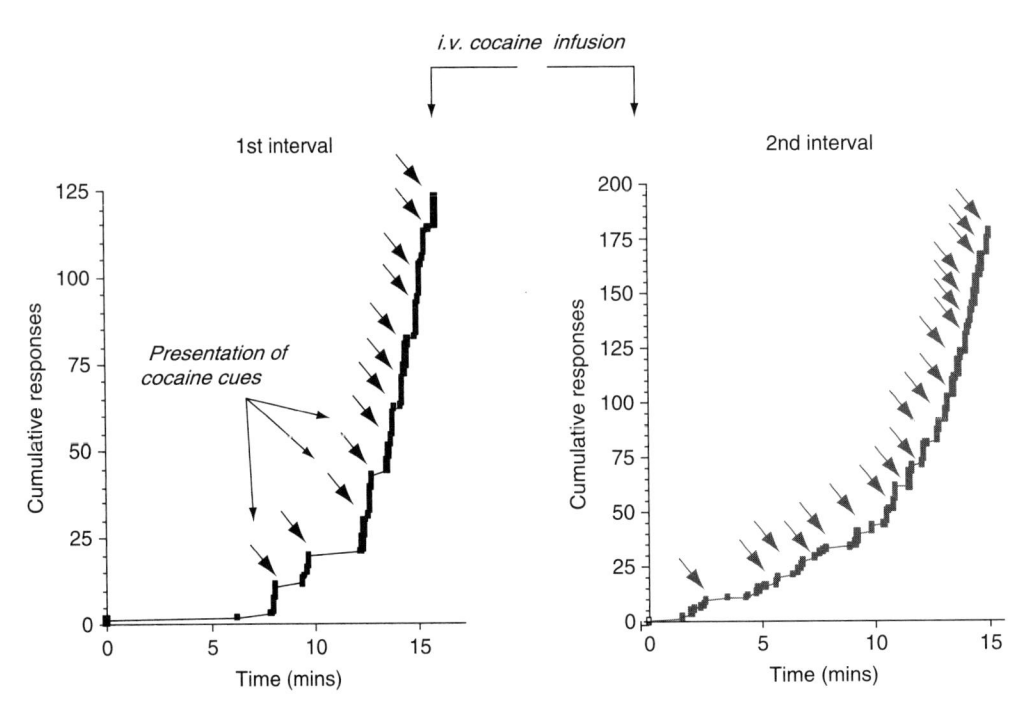

FIGURE 8.2.1. Cumulative response record of rats responding under a second-order schedule of cocaine reinforcement. The first, drug-free interval is shown in the left panel: the record shows a typical fixed-interval scalloped pattern of responding in which rats begin to respond at a low rate but accelerate as the interval proceeds. Each 10th response is reinforced by presentation of a cocaine-associated CS acting as a conditioned reinforcer. Once cocaine is self-administered intravenously, responding is greatly increased (right panel). This reflects the potentiation of conditioned reinforcement by cocaine-induced increases in DA, primarily in the shell of the Acb.[53,124] Source: Data taken from[78].

Although omission of a CS associated with cocaine has a greater disruptive effect on instrumental responding under a second-order schedule of reinforcement than omission of a CS associated with food,[79] sucrose- and cocaine-associated conditioned reinforcers appear equipotent in supporting the acquisition, as well as persistence, of a new instrumental response with conditioned reinforcement,[80] and both "incubate" with time.[81] Therefore, even though it is clear that drug-associated CSs exert marked effects on drug-seeking behavior[51,79] and relapse after extinction and abstinence,[74,82] it is not clear that they do so more powerfully than do CSs associated with natural reinforcers.

The involvement of DA in neuronal models of plasticity and learning (see below) has, however, encouraged the view that they do. Although these Pavlovian influences on addictive behavior are undoubtedly important, it should also be appreciated that addiction involves the repeated *self*-administration of drugs, that is, it depends upon *instrumental* behavior whereby individuals learn to *seek* and *take* drugs. Dopaminergic mechanisms, especially in the striatum, may have an even more important role in the learning of instrumental actions and habits in addition to determining the qualitative, and especially quantitative, impact of drug-associated CSs on drug seeking and taking behavior.

ADDICTIVE DRUGS AND CELLULAR MODELS OF LEARNING AND PLASTICITY

Perhaps the greatest impetus for viewing the development of addiction as a consequence of adaptive neuroplasticity has come from a plethora of studies on LTP and LTD in midbrain DA neurons and the targets of their projections, particularly the striatum, but increasingly in limbic and cortical targets as well (e.g., the hippocampus, amygdala, and prefrontal cortex). At the core of these studies is the view that the synaptic plasticity measured as LTP and LTD underpins alterations in neural circuitry induced by acute or chronic exposure to addictive drugs and thereby altered reward, Pavlovian, and instrumental learning processes.[30,31,83,84] Of course, as in the most studied form of LTP in the hippocampus, the link between LTP measured in vitro and the cellular mechanisms of learning and memory in behaving animals has not proved easy to forge[27]; this is also the case for the impact of addictive drugs on these processes.

There are abundant data showing that exposure to addictive drugs elicits LTP at excitatory synapses on VTA neurons. Measured 24 hr after a single in vivo exposure to cocaine, LTP was shown to be no longer induced at synapses onto DA neurons, suggesting that they were already potentiated.[85] In addition, the AMPA/NMDA ratio (the ratio between AMPA receptor–mediated and NMDA receptor–mediated excitatory postsynaptic currents [EPSCs]) was increased twofold in VTA slices. This effect was blocked by an NMDA receptor antagonist, indicating that NMDA receptor activation was necessary for cocaine to instantiate LTP.[85] The effect appears to be selective to DA neurons in the VTA, as there was no such effect of cocaine on LTP or AMPA/NMDA ratios on GABAergic neurons.[86] Other addictive drugs, including amphetamine, morphine, nicotine, and alcohol, have all been shown to increase AMPA/NMDA ratios 24 hr after treatment in vivo despite their very different primary mechanisms of action.[87] The LTP induced by cocaine is now known to involve the insertion of Glu-R2-lacking AMPA receptors into the neuronal membrane.[88] The cocaine-evoked plasticity persists for about 5 days but is not evident after 10 days,[85] although how and why it disappears is uncertain.

An important issue is whether this addictive drug–induced LTP in the VTA has a functional effect. It has been shown to correlate with behavioral sensitization[86]—the long-lasting increased locomotor response to stimulant drugs that is also observed after a single preexposure to cocaine—which is known to be blocked by glutamate receptor antagonists infused into the VTA.[89] Since overexpression of the GluR1 AMPA receptor subunit in the VTA has been reported to enhance sensitization and other motivational effects of drugs of abuse,[90,91] it has been hypothesized that LTP induced in VTA DA neurons by addictive drugs may have an important, albeit transient, effect that increases their rewarding effects.[27] Nevertheless, it is not clear how such acute and transient effects of addictive drugs on neuronal plasticity contribute to the development of some of the behavioral characteristics of addiction, such as escalation of drug intake, compulsive drug seeking, and an enhanced propensity to relapse during withdrawal (but see the review by [92]).

However, it has been shown that while the passive administration of cocaine results in the transient appearance of LTP in VTA neurons, cocaine *self-administration* results in a much more persistent potentiation of VTA excitatory synapses that is still detectable 3 months following the last cocaine exposure.[93] Moreover, food or sucrose ingestion results only in the transient form of LTP in VTA neurons. These studies are important both because they take into account that, in addiction, drugs are self-administered and not passively received and also because they suggest that plasticity induced by self-administered cocaine is different, being more persistent than that following exposure to voluntarily ingested natural rewards. These findings may indeed be more readily related to phenomena such as cued reinstatement or relapse to drug seeking after abstinence.[92,93]

Plasticity in the primary striatal target of mesolimbic DA neurons has also been demonstrated. But unlike VTA synaptic potentiation, the predominant response in Acb neurons is LTD-like rather than LTP-like, since AMPA/NMDA ratios are depressed, and this LTD is seen only after repeated (5 days) cocaine treatment, not after a single injection.[94,95] The picture is, though, somewhat more complicated, since after 1–2 weeks of withdrawal, there is an increase in AMPA/NMDA ratio and synaptic potentiation, not depression.[96] Moreover, if cocaine is self-administered rather than given noncontingently, there is in addition a marked increase in the AMPA receptor subunit GluR1 in Acb neurons and a reduction in GluR2 subunits[97]; thus, these neurons become calcium permeable and more excitable.[98] Again studying animals that had self-administered cocaine, Bonci and collaborators have shown that after 1 day of withdrawal, LTD is actually depressed in the Acb core (AcbC) and shell (AcbS), but after 3 weeks of withdrawal from cocaine, LTD was abolished specifically in the AcbC.[99] While cocaine-dosing procedures and other methodological differences do not allow a simple functional interpretation of these data, they do show that cocaine is able to induce long-lasting changes in the AcbC that may be related to drug-seeking behavior and relapse. As Martin et al. speculate, a failure to elicit LTD in the AcbC of rats having self-administered cocaine might be related to the consolidation of the instrumental drug-taking response, or to the readiness with which drug-associated CSs induce relapse, or other behavioral processes that depend upon the AcbC[99] (see below). Such speculations clearly justify specific experimental investigation. Given the role of the dorsal striatum in instrumental habit learning[100,101] and the enhancement of habit learning by stimulant drugs,[102] investigation of synaptic LTD in the dorsal striatum following the self-administration of cocaine and other drugs under conditions that are associated with the development of drug-seeking habits[103,104] would also provide a valuable means for linking neuronal plasticity and addictive behavior.[98]

INSTRUMENTAL LEARNING, DA, AND ADDICTIVE BEHAVIOR

The notion that DA in the striatum stamps in, or consolidates, the learning of stimulus–response (S-R) associations has considerable support, not least because DA has both acute effects to modulate corticostriatal transmission and lasting effects. This background has encouraged the view that DA neuron firing may be a teaching signal used for learning about actions that lead to reward.[32] It is not clear which targets of DA neuronal projections learn from the DA teaching signal, but they likely include the dorsal striatum (DS), amygdala, and prefrontal cortex, as well as the Acb. Thus, blockade of NMDA glutamate receptors in the AcbC has been shown to retard instrumental learning for food.[105] Concurrent blockade of NMDA and DA D1 receptors in the AcbC synergistically prevents learning of instrumental responding under a Variable Ratio (VR)-2 schedule of food reinforcement.[106] Once the response has been learned, subsequent performance on this schedule is not impaired by NMDA receptor blockade within the AcbC.[105] Furthermore, infusion of a protein kinase (PKA) inhibitor[107] or a protein synthesis inhibitor[108] into the AcbC *after* instrumental training sessions impairs subsequent performance, implying that PKA activity and protein synthesis in the AcbC contribute to the consolidation in memory of instrumental behavior.

However, it is apparent that the Acb is not *required* for simple instrumental conditioning—but, as discussed above, is implicated instead in providing extra motivation for behavior, for example when triggered by Pavlovian CSs as measured in PIT procedures (*Pavlovian arousal*) or when reinforcers are delayed.[109] Reinforcers that require substantial effort to obtain are especially affected by DA depletion or receptor antagonist infusions into the Acb.[110] But rats with Acb or AcbC lesions acquire lever-press responses on sequences of random ratio schedules at normal or slightly reduced levels[111,112] and remain sensitive to changes in the action–outcome contingency.[111,113]

In contrast, the dorsomedial striatum[101] and medial prefrontal cortical areas that project to this area[114,115] have been shown to be important for action-outcome (A-O) learning. Following lesioning of the dorsomedial striatum, instrumental responding is unaffected by reinforcer devaluation (for example, by feeding to satiety); that is, it shows the characteristics of habitual behavior arising from S-R associative mechanisms (see below). Whether DA transmission in the dorsomedial striatum is involved in the consolidation of A-O learning is unclear.

An operational definition of a habit is that the behavior continues even after the controlling influence of the goal is reduced by devaluation procedures (such as satiation or postingestive malaise induced by lithium chloride injection in the case of a food reinforcer), as well as by degrading the contingency between response and outcome.[116] The extent to which instrumental behavior is maintained under these conditions reveals the degree of control by S-R mechanisms. Balleine and colleagues have shown in a series of studies that S-R (habit) learning assessed in this way depends critically upon the dorsolateral striatum, since lesions or inactivation of this region result in instrumental behavior that remains sensitive to reinforcer devaluation, and under A-O control.[100,117] Dopaminergic mechanisms in the DS have also been shown to be involved in habit learning; 6-hydroxydopamine-induced DA depletions of the DS render instrumental responding sensitive to reinforcer devaluation.[118] Amphetamine treatment resulting in locomotor sensitization through an up-regulation of DA transmission enhances the development of habitual responding,[119] and although this effect has not been localized to the DS, similar treatments result in a marked propensity for stereotyped behavior[120] that reflects enhanced DA transmission in the DS rather than the Acb.[121]

FROM VOLUNTARY TO HABITUAL DRUG SEEKING: THE SHIFT FROM VENTRAL TO DORSAL STRIATUM

A key hypothesis guiding our research is that the development of drug addiction reflects interactions between Pavlovian and instrumental learning mechanisms that, as indicated in the discussion above, results from the neuroplasticity in both cortical and striatal structures that is induced by chronic self-administration of addictive drugs. Drug addiction can, we suggest, be seen as the endpoint of a series of transitions, or a progression, from initial drug use when a drug is self-administered because it has reinforcing effects, through establishment of an S-R *incentive habit*[122] when drug seeking takes on an automatic quality in the presence of drug-associated environmental stimuli, ultimately emerging as a compulsive habit as the addicted individual loses control over this behavior[13] (Fig. 8.2.2).

Although the picture is far from clear for all addictive drugs, with stimulant drugs such as cocaine, there is wide agreement that the dopaminergic innervation of the AcbS and olfactory tubercle underlies its primary reinforcing effects,[20,28,123] as measured in drug self-administration procedures. It may be that this form of drug taking, in which there is a highly stable and predictable relationship between actions and outcomes (e.g., each lever press results in cocaine self-administration), never shifts from A-O to S-R control.[116] It can be distinguished from drug-seeking behavior in which the relationship between drug-seeking responses and outcome is much weaker, must often be maintained over long and unpredictable periods of time, and is profoundly influenced by the conditioned reinforcing properties of environmental stimuli associated with self-administered drugs.[6] We have utilized a model of drug-seeking using a second-order schedule of cocaine reinforcement in which the behavior is sensitive both to the contingency between instrumental responses and an addictive drug reinforcer and to the presence of a conditioned reinforcer associated with the drug.[51] The initial *acquisition* of cocaine-seeking behavior depends upon the integrity of the AcbC and its afferents from the BLA, since selective lesions of the BLA or the AcbC prevent it[76,124] (Fig. 8.2.3), as predicted on the basis of their involvement in conditioned reinforcement.[45,62,66] In contrast, drug *taking* is unimpaired by BLA or AcbC lesions.[76,124] Further evidence suggesting that the BLA and AcbC function as elements of limbic cortical-ventral striatopallidal circuitry underlying the acquisition of drug seeking comes from our observation that disconnecting these structures by unilateral pharmacological blockade of DA and AMPA receptors in the BLA and AcbC, respectively, on opposite sides of the brain also greatly impairs cocaine seeking.[125] At this early stage, drug seeking is under the control of instrumental A-O contingencies, that is, the animals are responding in a goal-directed way for intravenous cocaine infusions. This is further indicated from studies using another model of drug seeking, namely, a *seeking-taking* chained schedule whereby it was shown that cocaine seeking was sensitive to devaluation of the drug-taking link soon after acquisition.[126]

There is now considerable evidence that the DS eventually dominates this Acb-mediated control over drug seeking and mediates the performance of well-established drug-seeking habits.[122] Our initial evidence for this derived from in vivo microdialysis measurement of extracellular DA in rats that had attained stable cocaine seeking under a second-order schedule over many weeks (Fig. 8.2.4). While self-administered cocaine increased DA release in the AcbS, AcbC, and caudate-putamen, extracellular DA was increased selectively in the AcbC in response to unexpected (i.e., non-response-contingent) presentations of a cocaine-associated stimulus. However, during a prolonged period of cocaine seeking maintained by contingent presentations of the same cocaine-associated CS, DA release was increased only in the DS, not in the AcbC or AcbS.[17,127] Furthermore, DA in the DS was subsequently shown to be causally important for the maintenance of drug seeking, since it was impaired by DA receptor blockade in the dorsolateral DS but not in the AcbC.[103] This is consistent with the habit hypothesis not only because of data implicating the dorsolateral DS in habit learning

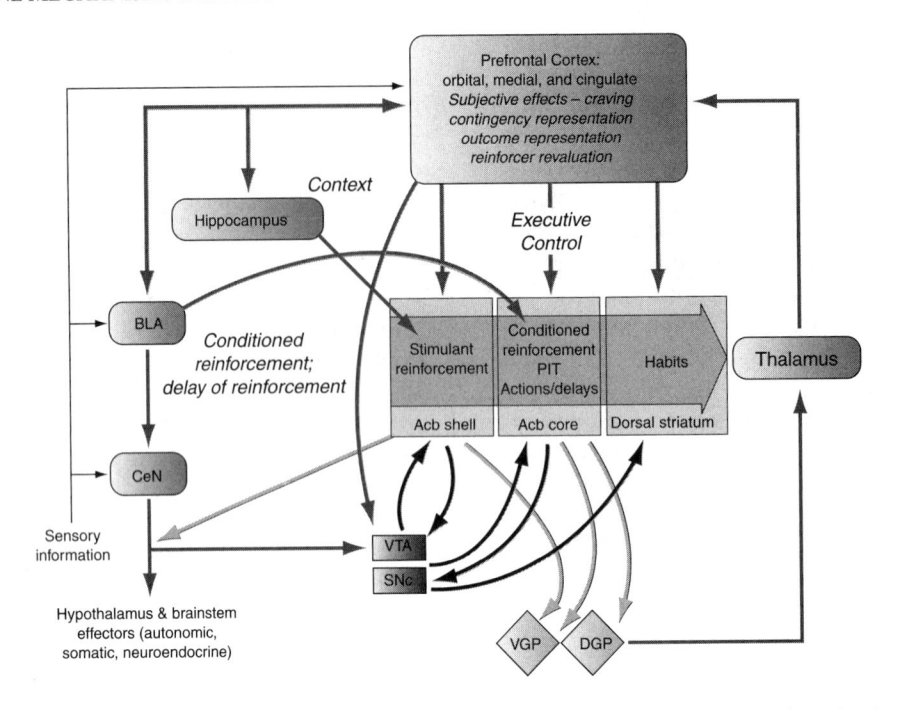

FIGURE 8.2.2. Drug addiction as a failure in top-down executive control over drug-oriented incentive habits. Basal ganglia circuitry is fundamentally involved in the mechanisms underlying the development and persistence of drug addiction. The reinforcing, and possibly the hedonic (**H**), effects of psychostimulants depend upon the shell of the nucleus accumbens (NAcS), the olfactory tubercle, and the ventral pallidum (GPe-GPi), whereas the motivational balance between natural and drug rewards (**NR/DR**) may depend upon the subthalamic nucleus. (STN). Exposure to addictive drugs triggers neurobiological and functional modifications, in neural networks involved in implicit sub-cortical, and declarative cortical, mechanisms. At the subcortical level, addictive drugs alter both Pavlovian and instrumental learning mechanisms, e.g.: (1) enhancing Pavlovian incentive influences from the basolateral amygdala (BLA) to the nucleus accumbens core (NAcC) and the BLA and the orbitofrontal cortex (OFc), thereby leading to increased incentive salience of drugs and environmental stimuli associated with them. 2) Addictive drugs facilitate the instantiation of habitual responding, whereby drug seeking behaviour is largely controlled by drug-associated stimuli in the environment. The development of habitual drug seeking and drug taking behaviour may be related to a ventral to dorsal striatal shift in the locus of control over behaviour ( ), which depends at least in part upon the ascending dopamine-dependent circuitry linking the ventral to the dorsolateral striatum (DLS) via recurrent connections with the dopaminergic neurons in the ventral tegmental area (VTA) and the substantia nigra pars compacta (SNc) in the ventral midbrain ( ). Thus, maladaptive, drug focused, Pavlovian incentive processes that control 'drug-oriented incentive impulses' in the NAcC eventually influence the dorsal striatum-dependent stimulus-response, or 'habit' system, thereby giving rise to 'incentive habits'. However, incentive habits alone cannot account for the development of compulsive drug seeking and taking behaviour which, instead, may arise from the interaction between sub-cortical mechanisms that tend to drive the addict towards drugs and drug-associated stimuli and declarative cortical mechanisms. Indeed, exposure to addictive drugs triggers a change in the balance of cortical neuroadaptations from ventromedial to dorsolateral prefrontal cortex ( ), which might be associated with drug-induced deficits in top-down executive control over instrumental behaviour. Drug addicts and drug-exposed animals display cognitive inflexibility, impaired decision making processes, and high rates of impulsivity, suggesting impairment of prefrontal cortical function. Thus, once incentive habits develop and are progressively less under the control of prefrontal executive function, drug use is no longer under the control of the individual and can be described as compulsive.

(see above), but also because the overall fixed-interval second-order schedule used in these studies is known to result in a more rapid development of S-R habits through the weaker relationship between action and outcome that is progressively established compared with ratio schedules.[128,129] Moreover, our studies with orally self-administered cocaine and alcohol that directly probed the associative structure underlying drug seeking by reinforcer devaluation have clearly shown the more rapid development of habitual drug seeking compared with the seeking of a natural sweet reward.[130,131]

Given the increasing importance of the DS in habitual drug seeking, the question arises as to how a shift in the locus of control from ventral striatum (VS) to DS might occur. The serial connectivity between Acb and DS and midbrain DA neurons[132,133] provides a possible mechanism. Thus, ventral domains of the striatum regulate the dopaminergic innervation of more dorsal domains through so-called spiraling connections with the midbrain. Thus, the AcbS projects to DA neurons in the VTA that innervate not only the AcbS, but also the more dorsally situated AcbC. Neurons in the AcbC DA neurons in the VTA that in turn project to the AcbC and substantia nigra DA neurons projecting to the immediately dorsal regions of the dorsomedial caudate-putamen, and so on, in a serially cascading pattern, ultimately to encompass more lateral parts of the DS—the site at which DA release is increased during habitual drug seeking and where DA receptor antagonist infusions reduce it.

We tested the hypothesis that the serial cascade of striato-nigro-striatal connectivity underlies progressively

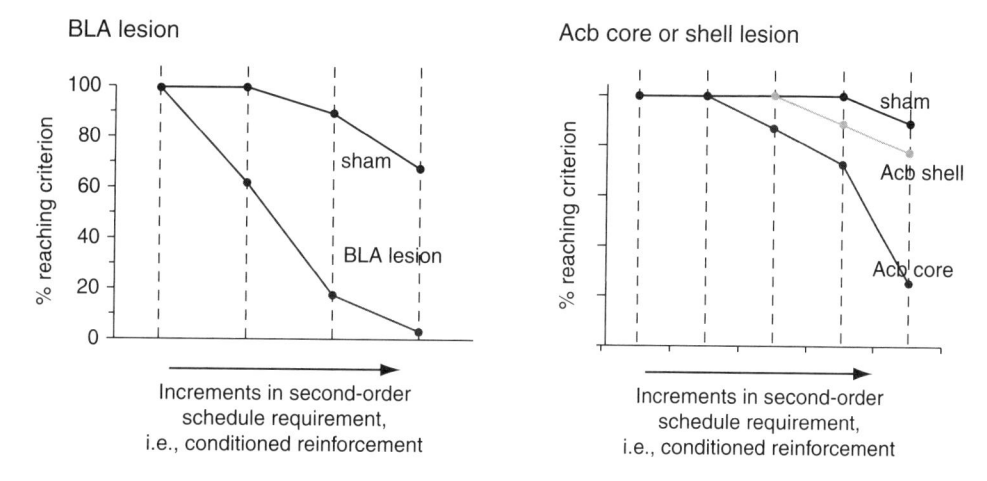

FIGURE 8.2.3. The effects of excitotoxic lesions of the basolateral amygdala (BLA) and nucleus accumbens (Acb) core or shell on the acquisition of cocaine seeking under a second-order schedule of reinforcement. While all groups of rats acquired cocaine self-administration and Acb shell lesions had no effect on the acquisition of cocaine seeking, rats with either BLA or AcbC lesions failed to reach criterion. *Source*: Data from[76,124].

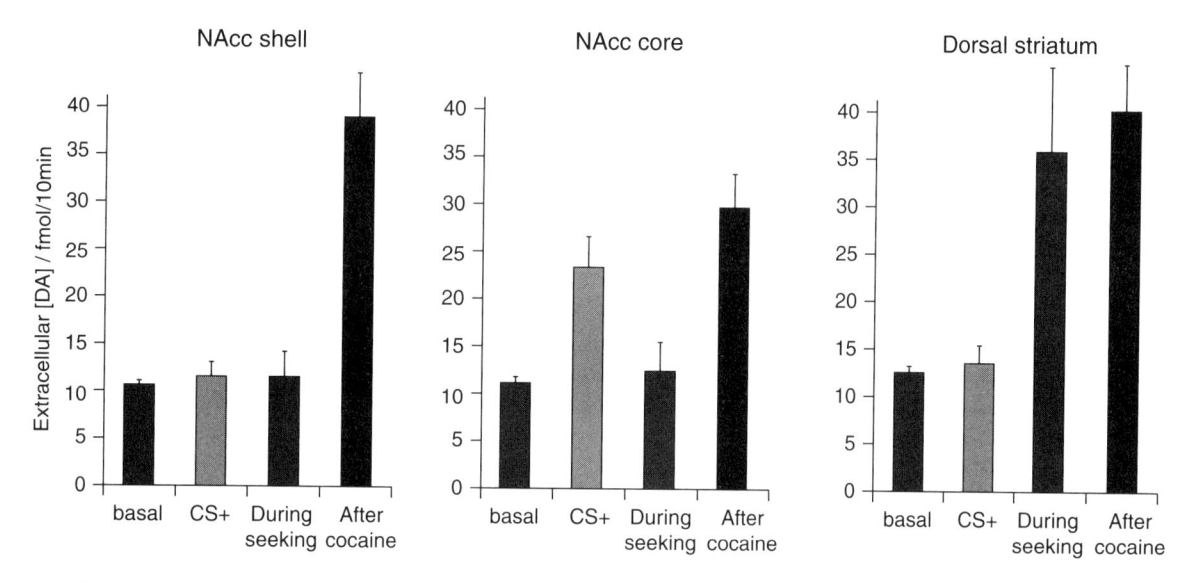

FIGURE 8.2.4. In vivo microdialysis study of rats responding for 60 min under a second-order schedule of cocaine reinforcement after an approximately 8-week training history. Unexpected presentations of the cocaine-associated CS resulted in increased extracellular DA selectively in the AcbC but not in the AcbS or DS. The prolonged period of drug seeking, in which each 10th response resulted in the presentation of the cocaine-associated CS, was associated selectively with increased extracellular DA in the DS but not in the AcbC or AcbS. When cocaine was self-administered, extracellular DA concentrations were increased in all three striatal domains. *Source*: Data from[17,127].

greater control over habitual cocaine seeking by the dorsolateral striatum by disconnecting the AcbC from the DS. The AcbC was selectively lesioned on one side of the brain and combined with DA receptor blockade in the contralateral dorsolateral striatum, thereby functionally disconnecting serial interactions between these VS and DS domains bilaterally.[104] The disconnection selectively decreased cocaine seeking in rats tested some weeks after stable responding for cocaine had been attained under a second-order schedule of reinforcement (Fig. 8.2.5). It is important to note that animals trained to perform a novel instrumental response for sucrose under a fixed ratio 1

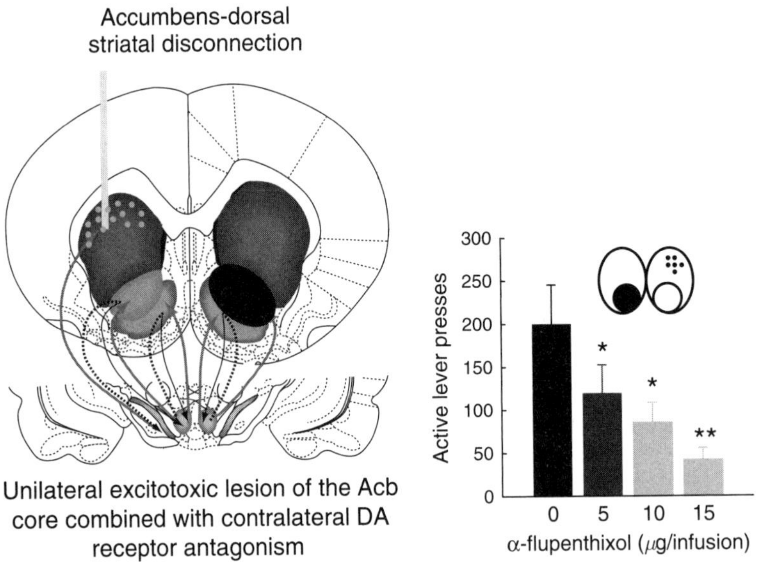

FIGURE 8.2.5. To disconnect the link between the Acb core and the dorsolateral striatum via the midbrain DA neurons, an excitotoxic lesion was made in the AcbC on one side of the brain (black shading) and a cannula was implanted into the contralateral DS through which the DA receptor antagonist, α-flupenthixol could be infused (left panel). In this way, AcbC neurons were unable to "recruit" DA neurons projecting to the DS on one side of the brain, whereas on the contralateral side, DA neurons could be recruited by the AcbC, but DA transmission was blocked by the antagonist infusion. In the right panel, it can be seen that in unilaterally AcbC-lesioned rats, α-flupenthixol dose-dependently reduced cocaine seeking under a second-order schedule of reinforcement. A unilateral AcbC lesion alone, or a unilateral infusion of α-flupenthixol, had no effect on cocaine seeking. Source: Data from[104].

schedule of reinforcement were completely unaffected, either by the AcbC-DS disconnection or by bilateral DS DA receptor antagonist infusions, immediately after acquisition when the behavior was under A-O control.[104] Moreover, bilateral DS DA receptor blockade or AcbC-DS DA-dependent disconnection had no effect on cocaine seeking when tested at a much earlier stage of acquisition of cocaine seeking when responding was under ratio, rather than interval, and therefore A-O control. (D. Belin and B.J. Everitt, 2008, unpublished observations).

Taken together, these data indicate the progressive shift from Acb to DS control over drug seeking. Other data also support the notion of this shift, including its mediation by dopaminergic mechanisms. Thus, neuroadaptations in DA D2/3 receptors (as well as other neurochemical or metabolic markers) are predominant in the Acb after a short period of cocaine self-administration by monkeys, but progressively spread to encompass the DS following a chronic cocaine history.[134,135] The DS has also been shown to be involved in relapse to a cocaine-seeking habit at a time when manipulations of the Acb had no effect.[136,137] Intriguingly, the presentation of drug cues to human cocaine addicts induced both drug craving and activation of the DS[138,139] in addition to the well-established activation of the amygdala and limbic prefrontal cortical areas.[138,140–142] These observations therefore strongly indicate a link between limbic cortical mechanisms and engagement of the DS in long-term drug abusers exposed to drug cues, whereas the results of AcB-DS disconnection suggest that this recruitment is mediated by antecedent limbic cortex–dependent activity in the AcbC and its regulation of DS dopaminergic projections, thereby underpinning drug seeking as an incentive habit.[122]

It seems likely that the VS to DS shift is not specific to drug seeking, but would apply equally to the control over instrumental responding for natural reinforcers under appropriate conditions. Indeed, lesioning or inactivation of the AcbC, dorsomedial, or dorsolateral striatum in rats responding for ingestive reinforcers does not globally impair instrumental behavior, but instead has major effects that depend upon the A-O or S-R associative structure underlying the behavior. Lesions or NMDA receptor blockade of the AcbC[105,111] or DM striatum[100,101] impair instrumental behavior under A-O control but actually enhance the development of S-R habits in which responding persists after reinforcer devaluation.[100] In contrast, dorsolateral striatal lesions, inactivation, or DA denervation return previously habitual responding to A-O control.[117] These observations emphasize that A-O and S-R learning mechanisms are likely engaged not serially, but in parallel, with dorsolateral striatum–dependent S-R mechanisms eventually dominating the control over behavior.

What remains is the attractive possibility that the shift from drug-seeking actions to habits, and from VS to DS, occurs more rapidly or that the instrumental S-R association is more firmly stamped in, or consolidated, in the DS because addictive drugs such as cocaine directly influence the neuronal plasticity mechanisms involved via their potent effects on DA transmission. There is some evidence that drug-induced increases in DA transmission potentiate S-R habit formation.[102] While there is not abundant evidence for the 'super-consolidation' of Pavlovian CS–drug associations, the demonstration of an amphetamine-enhanced shift in the balance of Pavlovian associative encoding from VS to DS[143] both suggests that this warrants further study and provides an experimental explanation for the observation of drug-associated, CS-induced activation of the DS in human cocaine abusers,[138,139] since it links craving and limbic cortical activation[141] with the DS and attendant drug seeking. The notion of the progressively more dominant control over drug seeking and taking by the DS, mediated by the effects of DA-potentiating addictive drugs on the control by the Acb over the dopaminergic innervation of the DS, could be investigated in cellular models of plasticity. Luscher and Bellone[98] have speculated that plasticity phenomena such as LTP in the VTA might determine plasticity in the Acb, which is intriguing given that the former is seen after very acute, and the latter after more chronic, cocaine exposure. Similarly, LTD-like plasticity in DS neurons might depend upon antecedent plasticity in Acb neurons mediated by the DA-dependent intermediary effects of circuitries involving the VTA and SNc that would result in the more effective consolidation of drug seeking as a S-R habit.[6] The influence of the Acb on DA-dependent functioning of the DS, is also revealed in mice with deletion of the Pitx-3 gene that lack a nigrostriatal pathway.[144]

DOPAMINERGIC MECHANISMS IN THE VULNERABILITY TO DRUG ADDICTION

As we have discussed previously, all animals responding for drugs or natural reinforcers will develop stimulus-bound incentive habits under appropriate reinforcement contingencies, because it is adaptive to do so.[145] The possibly enhanced consolidation of S-R habits induced by the effects of addictive drugs on plasticity mechanisms may be a key stage in the transition to addiction, but this does not capture the *compulsive* drug seeking that characterizes addiction. Nor does it capture the fact of individual *vulnerability to addiction*: some individuals more than others progressively lose control over their drug intake and over the drug-seeking habit. This may reflect individual differences in sensitivity to the reinforcing effects of drugs, or in the impact of these drugs on plasticity mechanisms, or in their toxic effects on corticostriatal function, especially drug-induced impairments in prefrontal cortical function.[13,145] These are but some of the possibilities, and they are not mutually exclusive. Full consideration of the evidence for impairments in top-down, executive control mechanisms as key factors underlying the development of compulsive drug seeking is beyond the scope of this chapter and has been reviewed extensively elsewhere.[145–149] The focus in the remainder of this discussion will be on the relationship between individual differences in DA transmission in the striatum that are linked to the behavioral characteristic of impulsivity and that predict the propensity to lose control over cocaine self-administration and to develop compulsive drug seeking, thereby capturing the essence of this symptom in DSM-IV.[150]

Impulsive behavior and sensation seeking have long been associated with drug addiction in humans, whether as a causal mechanism or a consequence of repeated episodes of drug taking. Behavioral impulsivity is a spontaneously occurring behavioral characteristic in rats,[151,152] and this discovery provided an opportunity to investigate its neural correlates and the predictive value of impulsivity for cocaine seeking and taking. About 10% of the outbred Lister-hooded strain of rats are impulsive, measured as having premature responses in the five-choice serial reaction-time task (5CSRTT).[153] They show high levels of anticipatory responses before the presentation of a food-predictive, brief light stimulus, especially when the stimulus presentation is delayed after trial onset.[153] In a positron emission tomography (PET) imaging study, these impulsive rats showed significantly reduced binding potential of the DA D2/3 receptor antagonist [18F]fallypride, specifically in the VS (including the Acb), but not in the caudate-putamen. In addition, DA D2/3 receptor availability in the Acb was correlated with impulsivity on the 5CSRTT: the greatest levels of impulsive responding were seen in individuals with the lowest amounts of fallypride binding.[153] Impulsive rats showed a marked escalation in cocaine intake when given the opportunity to self-administer cocaine[153] or nicotine[152] when compared with nonimpulsive controls. This distinguishes them from rats showing high locomotor responses to novelty (a "sensation-seeking" phenotype), which are more sensitive to cocaine and self-administer cocaine at low doses that do not sustain self-administration in those rats that show low locomotor responses to novelty.[154] Highly impulsive rats allowed to self-administer cocaine over an extended period of long-access sessions[155] also showed a greatly increased propensity to relapse after a period of abstinence.[156] Thus, the DA receptor–linked characteristic of impulsivity predicts two important further characteristics of addictive behavior: the tendency to escalate and lose control over drug intake and the increased likelihood of relapse after an extended period of withdrawal.

Abstinent cocaine and methamphetamine addicts, as well as alcoholics, show reduced DA D2 receptor binding in the DS,[10] and this is correlated with hypoactivity in the prefrontal cortex measured using PET.[157,158] This reduction in DA D2 receptor binding could easily be viewed as a consequence of chronic drug taking, not least because of the observations of Porrino and colleagues in monkeys chronically self-administering cocaine.[135,159] But in a series of elegant studies in drug-naive monkeys, low levels of striatal DA D2 receptors have been shown to predict subsequent self-administration of cocaine.[160,161] In humans, DA transporter occupancy or DA D2 receptors in the striatum predict the subjective response to methylphenidate,[162,163] a stimulant with effects somewhat similar to those of cocaine. Thus, low DA D2 receptor levels may also be a *causal* factor in the propensity to self-administer drugs in humans, as well as being a consequence of chronic drug effects. This view is supported by the observation that individuals at high familial risk for alcoholism also show a relationship between (low) striatal DA D2 receptors and prefrontal cortical metabolism.[164] Related observations further suggest a link between suboptimal functioning of DA systems in other compulsive disorders that may share features with compulsive drug taking. For example, there is a similar relationship to that seen in drug-addicted populations between striatal DA D2 receptors and prefrontal cortical metabolism in obese subjects,[165,166] while a blunted DS response to food is seen in individuals with the *Taq*IA A1 allele,[167] which is associated with DA D2 receptor binding in the striatum.[168] Thus, suboptimal DA signaling in individuals with the *Taq*IA A1 allele may lead to overeating to compensate for the attenuated striatal DA function.[167] Pathological gambling has also been linked to reduced activation of the mesolimbic DA system, since reduced ventral striatal and ventromedial prefrontal activation was seen to correlate with gambling severity in pathological gamblers engaged in a guessing game.[169]

The impact of preexisting low levels of Acb DA D2/3 receptors related to impulsivity, or of preexisting as well as drug-induced reductions in striatal DA D2/3 receptors, on the plasticity underlying instrumental learning and performance or Pavlovian drug cue-induced craving, are unclear. They may be linked to the putative sequential mechanisms within the VS and DS mediated by its dopaminergic innervation via the serial interconnectivity descried above. For example, the early vulnerability to escalate cocaine intake seen in impulsive rats that is predicted by low DA D2/3 receptor levels in the VS but not the DS[153] may lead to more rapid neuroadaptations, including down-regulation of DA D2/3 receptors, in the DS and may be mediated in part by aberrant engagement of the spiralling striato-nigro-striatal circuitry.[145] This may then lead to the more rapid consolidation of drug-seeking habits evoked and maintained by drug-associated stimuli that progressively more easily activate the DS.[139,143] However, a full explanation requires more detailed understanding of the relationship between pre- and postsynaptic dopaminergic mechanisms in the striatum in drug-naive, as well as acutely and chronically drug-exposed, individuals.

These studies still do not capture the key nature of the compulsive drug seeking that is core to substance dependence, or addiction, in DSM-IV. We have modeled *compulsive* drug seeking by its persistence despite negative or aversive outcomes and have shown that it only emerges following an extended, or chronic, history of cocaine taking.[170–172] Although different methodologies and different strains of rats were used, these two studies showed that after an extended, but not brief, history of cocaine self-administration, 17%–20% of rats continued to seek and take drugs despite the ongoing punishment of drug-seeking or -taking responses. This proportion is remarkably similar to that of the addiction-vulnerable subgroup of human subjects, which is also estimated to be about 20% of the population that initially use drugs.[173] We have subsequently shown that highly impulsive, but not novelty-seeking, rats were those that developed compulsive cocaine taking after protracted exposure to the drug, that is, they persisted in responding for cocaine despite punishment (Fig. 8.2.6). They also showed higher breakpoints for cocaine under a progressive ratio schedule of reinforcement and persistent responding even when cocaine unavailability was signaled.[174]

These data almost bring the story of DA and addiction full circle. Dopamine transmission has been central since its demonstrated importance in mediating the reinforcing effects of drugs was established through manipulations of the Acb dopaminergic innervation.

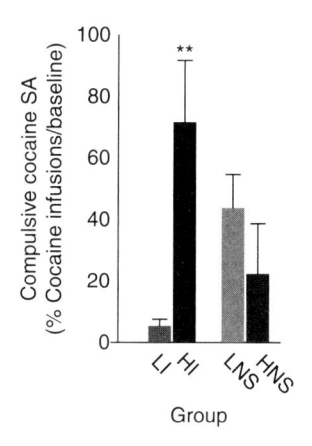

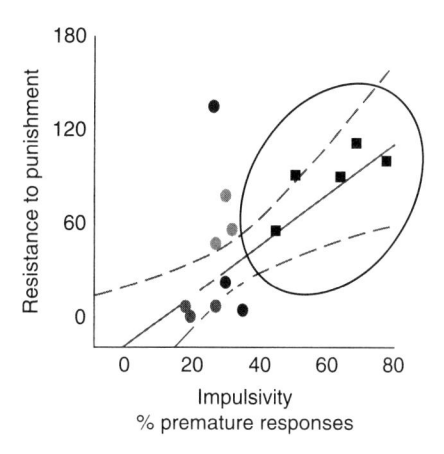

FIGURE 8.2.6. Rats screened for high impulsivity on the 5CSRTT (HI), but not low-impulsive rats or rats with high (HNS) or low (LNS) locomotor responses to novelty, showed persistent responding for intravenous cocaine despite the outcome of punishment by mild footshock (left panel). Rats showing high impulsivity (square symbols in right panel) prior to any experience with cocaine self-administration were those that developed this compulsive form of cocaine self-administration, which persisted despite the aversive consequence of punishment and thereby modeled one of the key diagnostic criteria of addiction in DSM-IV. Source: Data from[174].

Subsequently, as reviewed briefly above, the engagement of Pavlovian and instrumental learning mechanisms in addiction has become widely accepted. With it, there has been a focus on plasticity mechanisms in the Acb, DS, and limbic cortical sites and their modulation by DA, especially addictive drug-enhanced increases in DA. But the central involvement of DA transmission does not end there. It now seems very likely that individual differences in DA transmission may be linked to impulsivity and perhaps other endophenotypes that predict not only the response to self-administered drugs, but also the emergence of drug seeking as a compulsive incentive habit.

Many questions remain to be answered, not least the neural basis of compulsion, which is perhaps the most important one. Dopaminergic mechanisms have been implicated here too. For example, stimulant-induced sensitization of mesolimbic DA transmission has been argued to underlie pathological incentive motivational mechanisms characterized as the excessive 'wanting' of drugs.[175] Reductions in Acb DA transmission along with other, perhaps even more important neuroadaptations in stress circuitry, including up-regulation of CRF transmission in the extended amygdala, which encompasses the AcbS, have equally convincingly been argued to underlie compulsive drug taking as the self-medication—negative reinforcement—of anhedonic states.[26,176] There is also growing evidence, in cocaine and methamphetamine addicts as well as alcoholics, of reduced activity and impairments in prefrontal, particularly orbitofrontal, function and the loss of inhibitory control over maladaptive drug-seeking and -taking habits (reviewed in[145]).

Stimulant drugs may actually cause these reductions in prefrontal cortical function. In monkeys chronically self-administering cocaine, metabolic function is progressively altered first in ventromedial and then dorsolateral territories of the prefrontal cortex, the latter being unaffected after acute access to cocaine.[135] Remarkably, rats having self-administered cocaine or having been treated over much shorter periods with amphetamine show impaired reversal learning similar to that seen following orbital prefrontal cortex lesions.[177,178] A major challenge is to link such drug-induced alterations in prefrontal cortical function with notions of vulnerability that account for the relatively small proportion of chronic drug users that become addicted.

ACKNOWLEDGMENTS

The research summarized here was supported by grants from the Medical Research Council (G9536855 [BJE] and G0401068 [JWD]) and was conducted within the Behavioural and Clinical Neuroscience Institute, which is funded by a joint award from the MRC and the Wellcome Trust.

REFERENCES

1. Olds J, Milner P. Positive reinforcement produced by electrical stimulation of the septal area and other regions of the rat brain. *J Comp Physiol Psychol.* 1954;47:419–427.

2. Murray B, Shizgal P. Evidence implicating both slow- and fast-conducting fibers in the rewarding effect of medial forebrain bundle stimulation. *Behav Brain Res.* 1994;63:47–60.
3. Fibiger HC, LePiane FG, Jakubobic A, Phillips AG. The role of dopamine in intracranial self-stimulation of the ventral tegmental area. *J Neurosci.* 1987;7:3888–3896.
4. Reynolds JN, Hyland BI, Wickens JR. A cellular mechanism of reward-related learning. *Nature.* 2001;413(67-70).
5. Berke JD, Hyman SE. Addiction, dopamine and the molecular mechanisms of memory. *Neuron.* 2000;25:515–532.
6. Everitt BJ, Dickinson A, Robbins TW. The neuropsychological basis of addictive behaviour. *Brain Res Rev.* 2001;36(2-3):129–138.
7. Wise RA. Neuroleptics and operant behavior: the anhedonia hypothesis. *Behav Brain Sci.* 1982;5:39–87.
8. Berridge KC, Robinson TE. What is the role of dopamine in reward: hedonic impact, reward learning, or incentive salience? *Brain Res Rev.* 1998;28(3):309–369.
9. Kelley AE, Berridge KC. The neuroscience of natural rewards: relevance to addictive drugs. *J Neurosci.* 2002;22(9):3306–3311.
10. Volkow ND, Fowler JS, Wang GJ. Imaging studies on the role of dopamine in cocaine reinforcement and addiction in humans. *J Psychopharmacol.* 1999;13(4):337–345.
11. Critchley HD, Wiens S, Rotshtein P, Ohman A, Dolan RJ. Neural systems supporting interoceptive awareness. *Nat Neurosci.* 2004;7(2):189–195.
12. Altman J, Everitt BJ, Glautier S, et al. The biological, social and clinical bases of drug addiction: Commentary and debate. *Psychopharmacology.* 1996;125:285–345.
13. Everitt BJ, Robbins TW. Neural systems of reinforcement for drug addiction: from actions to habits to compulsion. *Nat Neurosci.* 2005;8(11):1481–1489.
14. Di Chiara G, Imperato A. Drugs abused by humans preferentially increase synaptic dopamine concentrations in the mesolimbic system of freely moving rats. *Proc Natl Acad Sci USA.* 1988;85:5274–5278.
15. Di Chiara G. Drug addiction and dysadaptive responsiveness of N. accumbens shell dopamine. *International J Neuropsychopharmacol.* 2004;7:S24-S24.
16. Phillips AG, Pfaus JG, Blaha CD. Dopamine and motivated behavior: insights provided by in vivo analyses. In: Willner P, Scheel-Kruger J, eds. *The Mesolimbic Dopamine System: From Motivation to Action.* New York, NY: Wiley; 1991:473–495.
17. Ito R, Dalley JW, Howes SR, Robbins TW, Everitt BJ. Dissociation in conditioned dopamine release in the nucleus accumbens core and shell in response to cocaine cues and during cocaine-seeking behavior in rats. *J Neurosci.* 2000;20(19):7489–7495.
18. Phillips GD, Robbins TW, Everitt BJ. Bilateral intra-accumbens self-administration of d-amphetamine: antagonism with intra-accumbens SCH-23390 and sulpiride. *Psychopharmacology.* 1994; 114:477–485.
19. Ikemoto S, Glazier BS, Murphy JM, McBride WJ. Role of dopamine D-1 and D-2 receptors in the nucleus accumbens in mediating reward. *J Neurosci.* 1997;17(21):8580–8587.
20. Ikemoto S, Qin M, Liu ZH. The functional divide for primary reinforcement of D-amphetamine lies between the medial and lateral ventral striatum: is the division of the accumbens core, shell, and olfactory tubercle valid? *J Neurosci.* 2005;25(20):5061–5065.
21. Pettit HO, Ettenberg A, Bloom FE, Koob GF. Destruction of dopamine in the nucleus accumbens selectively attenuates cocaine but not heroin self-administration in rats. *Psychopharmacology.* 1984;84:167–173.
22. Robbins TW, Everitt BJ. Functions of dopamine in the dorsal and ventral striatum. *Semin Neurosci.* 1992;4(2):119–128.
23. Salamone JD, Correa M, Mingote S, Weber SM. Nucleus accumbens dopamine and the regulation of effort in food-seeking behavior: Implications for studies of natural motivation, psychiatry, and drug abuse. *J Pharmacol Exp Ther.* 2003; 305(1):1–8.
24. Ikemoto S, Panksepp J. The role of nucleus accumbens dopamine in motivated behavior: a unifying interpretation with special reference to reward-seeking. *Brain Res Rev.* 1999;31(1):6–41.
25. Nestler EJ. Molecular mechanisms of drug addiction in the mesolimbic dopamine pathway. *Semin Neurosci.* 1993;5:369–376.
26. Koob GF, Le Moal M. *Neurobiology of Addiction.* San Diego, CA: Academic Press; 2005.
27. Hyman SE, Malenka RC, Nestler EJ. Neural mechanisms of addiction: the role of reward-related learning and memory. *Annu Rev Neurosci.* 2006;29:565–598.
28. Wise RA. Dopamine, learning and motivation. *Nat Rev Neurosci.* 2004;5(6):483–494.
29. Schultz W, Dickinson A. Neuronal coding of prediction errors. *Annu Rev Neurosci.* 2000;23:473–500.
30. Saal D, Malenka RC. The role of synaptic plasticity in addiction. *Clin Neurosci Res.* 2005;5(2-4):141–146.
31. Kauer JA, Malenka RC. Synaptic plasticity and addiction. *Nat Rev Neurosci.* 2007;8(11):844–858.
32. Schultz W, Dayan P, Montague PR. A neural substrate of prediction and reward. *Science.* 1997;275(5306):1593–1599.
33. Schultz W. Behavioral dopamine signals. *Trends Neurosci.* 2007;30(5):203–210.
34. Waelti P, Dickinson A, Schultz W. Dopamine responses comply with basic assumptions of formal learning theory. *Nature.* 2001;412(6842):43–48.
35. Montague PR, Hyman SE, Cohen JD. Computational roles for dopamine in behavioural control. *Nature.* 2004;431(7010):760–767.
36. O'Doherty JP, Dayan P, Friston K, Critchley H, Dolan RJ. Temporal difference models and reward-related learning in the human brain. *Neuron.* 2003;38(2):329–337.
37. Tobler PN, O'Doherty JP, Dolan RJ, Schultz W. Human neural learning depends on reward prediction errors in the blocking paradigm. *J Neurophysiol.* 2006;95(1):301–310.
38. O'Brien CP, Childress AR, McLellan AT, Ehrman R. A learning model of addiction. *Res Pub Assn Res Nerv Ment Dis.* 1992;70:157–177.
39. Robinson TE, Berridge KC. The neural basis of drug craving: an incentive-sensitization theory of addiction. *Brain Res Rev.* 1993;18:247–291.
40. Caggiula AR, Donny EC, White AR, et al. Cue dependency of nicotine self-administration and smoking. *Pharmacol Biochem Behav.* 2001;70(4):515–530.
41. Redish AD. Addiction as a computational process gone awry. *Science.* 2004;306(5703):1944–1947.
42. Schultz W. Getting formal with dopamine and reward. *Neuron.* 2002;36(2):241–263.
43. Chiu PH, Lohrenz TM, Montague PR. Smokers' brains compute, but ignore, a fictive error signal in a sequential investment task. *Nat Neurosci.* 2008;11(4):514–520.
44. Aragona BJ, Cleaveland NA, Stuber GD, Day JJ, Carelli RM, Wightman RM. Preferential enhancement of dopamine transmission within the nucleus accumbens shell by cocaine is

attributable to a direct increase in phasic dopamine release events. *J Neurosci.* 2008;28(35):8821–8831.
45. Cardinal RN, Parkinson JA, Hall J, Everitt BJ. Emotion and motivation: the role of the amygdala, ventral striatum, and prefrontal cortex. *Neurosci Biobehav Rev.* 2002;26(3):321–352.
46. Cardinal RN, Everitt BJ. Neural and psychological mechanisms underlying appetitive learning: links to drug addiction. *Curr Opin Neurobiol.* 2004;14(2):156–162.
47. Tomie A, Brooks W, Zito B. Sign-tracking: the search for reward. In: Klein SB, Mowrer RR, eds. *Contemporary Learning Theories—Pavlovian Conditioning and the Status of Traditional Learning Theory.* Hillsdale, NJ: Erlbaum; 1989: 191–226.
48. Estes WK. Discriminative conditioning. I. A discriminative property of conditioned anticipation. *J Exp Psychol.* 1943;32: 150–155.
49. Lovibond PF. Facilitation of instrumental behavior by a Pavlovian appetitive conditioned stimulus. *J Exp Psychol: Anim Behav Processes.* 1983;9:225–247.
50. Holland PC, Petrovich GD, Gallagher M. The effects of amygdala lesions on conditioned stimulus-potentiated eating in rats. *Physiol Behav.* 2002;76(1):117–129.
51. Everitt BJ, Robbins TW. Second-order schedules of drug reinforcement in rats and monkeys: measurement of reinforcing efficacy and drug-seeking behaviour. *Psychopharmacology.* 2000;153: 17–30.
52. Everitt BJ, Cardinal RN, Parkinson JA, Robbins TW. Appetitive behavior—impact of amygdala-dependent mechanisms of emotional learning. *Ann NY Acad Sci.* 2003;985:233–250.
53. Parkinson JA, Olmstead MC, Burns LH, Robbins TW, Everitt BJ. Dissociation in effects of lesions of the nucleus accumbens core and shell on appetitive Pavlovian approach behavior and the potentiation of conditioned reinforcement and locomotor activity by D-amphetamine. *J Neurosci.* 1999;19(6):2401-2411.
54. Di Ciano P, Cardinal RN, Cowell RA, Little SJ, Everitt BJ. Differential involvement of NMDA, AMPA/kainate, and dopamine receptors in the nucleus accumbens core in the acquisition and performance of Pavlovian approach behavior. *J Neurosci.* 2001;21(23):9471–9477.
55. Dalley JW, Laane K, Theobald DEH, et al. Time-limited modulation of appetitive Pavlovian memory by D1 and NMDA receptors in the nucleus accumbens. *Proc Natl Acad Sci USA.* 26 2005;102(17):6189–6194.
56. Hall J, Parkinson JA, Connor TM, Dickinson A, Everitt BJ. Involvement of the central nucleus of the amygdala and nucleus accumbens core in mediating Pavlovian influences on instrumental behaviour. *Eur J Neurosci.* 2001;13(10):1984–1992.
57. Corbit LH, Janak PH, Balleine BW. General and outcome-specific forms of Pavlovian-instrumental transfer: the effect of shifts in motivational state and inactivation of the ventral tegmental area. *Eur Jf Neurosci.* 2007;26(11):3141–3149.
58. Balleine BW, Killcross S. Parallel incentive processing: an integrated view of amygdala function. *Trends Neurosci.* 2006; 29(5):272–279.
59. Smith JW, Dickinson A. The dopamine antagonist, pimozide, abolishes Pavlovian-instrumental transfer. *J Psychopharmacol.* 1998;12(suppl A):A6.
60. Wyvell CL, Berridge KC. Intra-accumbens amphetamine increases the conditioned incentive salience of sucrose reward: enhancement of reward "wanting" without enhanced "liking" or response reinforcement. *J Neurosci.* 2000;20:8122–8130.
61. Mackintosh NJ. *Conditioning and Associative Learning.* Oxford, England: Oxford University Press; 1983.
62. Cador M, Robbins TW, Everitt BJ. Involvement of the amygdala in stimulus-reward associations: interaction with the ventral striatum. *Neuroscience.* 1989;30:77–86.
63. Burns LH, Annett L, Kelley AE, Everitt BJ, Robbins TW. Effects of lesions to amygdala, ventral subiculum, medial prefrontal cortex, and nucleus accumbens on the reaction to novelty: Implication for limbic-striatal interactions. *Behavioral Neuroscience.* 1996; 110:60–73.
64. Burke KA, Franz TM, Miller DN, Schoenbaum G. The role of the orbitofrontal cortex in the pursuit of happiness and more specific rewards. *Nature.* 2008;454(7202):340–345.
65. Taylor JR, Robbins TW. Enhanced behavioural control by conditioned reinforcers following microinjections of d-amphetamine into the nucleus accumbens. *Psychopharmacology.* 1984;84:405–412.
66. Robbins TW, Cador M, Taylor JR, Everitt BJ. Limbic-striatal interactions in reward-related processes. *Neurosci Biobehav Rev.* 1989;13(2-3):155–162.
67. Taylor JR, Robbins TW. 6-Hydroxydopamine lesions of the nucleus accumbens, but not of the caudate nucleus, attenuate enhanced responding with reward-related stimuli produced by intra-accumbens D-amphetamine. *Psychopharmacology.* 1986; 90:390–397.
68. Wolterink G, Phillips G, Cador M, Donselaar-Wolterink I, Robbins TW, Everitt BJ. Relative roles of ventral striatal D1 and D2 dopamine receptors in responding with conditioned reinforcement. *Psychopharmacology.* 1993;110:355–364.
69. Uslaner JM, Acerbo MJ, Jones SA, Robinson TE. The attribution of incentive salience to a stimulus that signals an intravenous injection of cocaine. *Behav Brain Res.* 2006;169(2):320–324.
70. Robinson TE, Berridge KC. Addiction. *Annu Rev Psychol.* 2003;54:25–53.
71. Di Ciano P, Everitt BJ. Differential control over drug-seeking behavior by drug-associated conditioned reinforcers and discriminative stimuli predictive of drug availability. *Behav Neurosci.* 2003;117(5):952–960.
72. Grimm JW, Kruzich PJ, See RE. Contingent access to stimuli associated with cocaine self-administration is required for reinstatement of drug-seeking behavior. *Psychobiology.* 2000; 28(3):383–386.
73. Kruzich PJ, Congleton KM, See RE. Conditioned reinstatement of drug-seeking behavior with a discrete compound stimulus classically conditioned with intravenous cocaine. *Behavl Neurosci.* 2001;115(5):1086–1092.
74. Shaham Y, Shalev U, Lu L, de Wit H, Stewart J. The reinstatement model of drug relapse: history, methodology and major findings. *Psychopharmacology.* 2003;168(1-2):3–20.
75. Lu L, Grimm JW, Hope BT, Shaham Y. Incubation of cocaine craving after withdrawal: a review of preclinical data. *Neuropharmacology.* 2004;47:214–226.
76. Whitelaw RB, Markou A, Robbins TW, Everitt BJ. Excitotoxic lesions of the basolateral amygdala impair the acquisition of cocaine-seeking behaviour under a second-order schedule of reinforcement. *Psychopharmacology.* 1996;127:213–224.
77. Ito R, Robbins TW, Everitt BJ. The nucleus accumbens and cocaine self-administration: dissociating the role of core and shell sub-regions. *Eur Neuropsychopharmacol.* 2004; 14:S26-S26.
78. Arroyo M, Markou A, Robbins TW, Everitt BJ. Acquisition, maintenance and reinstatement of intravenous cocaine self-administration under a second-order schedule of reinforcement in rats: effects of conditioned cues and continuous access to cocaine. *Psychopharmacology.* 1998;140(3): 331–344.

79. Goldberg SR, Kelleher RT, Goldberg DM. Fixed-ratio responding under second-order schedules of food presentation or cocaine injection. *J Pharmacol Exp Ther.* 1981;218: 271–281.
80. Di Ciano P, Everitt BJ. Conditioned reinforcing properties of stimuli paired with self-administered cocaine, heroin or sucrose: implications for the persistence of addictive behaviour. *Neuropharmacology.* 2004;47(suppl 1):202–213.
81. Grimm JW, Hope BT, Wise RA, Shaham Y. Neuroadaptation - Incubation of cocaine craving after withdrawal. *Nature.* 2001;412(6843):141–142.
82. Kruzich PJ, Grimm JW, Rustay NR, Parks CD, See RE. Predicting relapse to cocaine-seeking behavior: a multiple regression approach. *Behav Pharmacol.* 1999;10(5):513–521.
83. Jones S, Bonci A. Synaptic plasticity and drug addiction. *Curr Opin Pharmacol.* 2005;5(1):20–25.
84. Hyman SE, Malenka RC. Addiction and the brain: the neurobiology of compulsion and its persistence. *Nat Rev Neurosci.* 2001;2:695–703.
85. Ungless MA, Whistler JL, Malenka RC, Bonci A. Single cocaine exposure in vivo induces long-term potentiation in dopamine neurons. *Nature.* 2001;411(6837):583–587.
86. Kauer JA. Learning mechanisms in addiction: synaptic plasticity in the ventral tegmental area as a result of exposure to drugs of abuse. *Annu Rev Physiol.* 2004;66:447–475.
87. Saal D, Dong Y, Bonci A, Malenka RC. Drugs of abuse and stress trigger a common synaptic adaptation in dopamine neurons. *Neuron.* 2003;37(4):577–582.
88. Bellone C, Luscher C. Cocaine triggered AMPA receptor redistribution is reversed in vivo by mGluR-dependent long-term depression. *Nat Neurosci.* 2006;9(5):636–641.
89. Vanderschuren L, Kalivas PW. Alterations in dopaminergic and glutamatergic transmission in the induction and expression of behavioral sensitization: a critical review of preclinical studies. *Psychopharmacology.* 2000;151(2-3):99–120.
90. Carlezon WA, Boundy VA, Haile CN, et al. Sensitization to morphine induced by viral-mediated gene transfer. *Science.* 1997;277(5327):812–814.
91. Carlezon WA, Nestler EJ. Elevated levels of GluR1 in the midbrain: a trigger for sensitization to drugs of abuse? *Trends Neurosci.* 2002;25(12):610–615.
92. Thomas MJ, Kalivas PW, Shaham Y. Neuroplasticity in the mesolimbic dopamine system and cocaine addiction. *Br J Pharmacol.* 2008;154(2):327–342.
93. Chen BT, Bowers MS, Martin M, et al. Cocaine but not natural reward self-administration nor passive cocaine infusion produces persistent LTP in the VTA. *Neuron.* 2008;59(2):288–297.
94. Thomas MJ, Beurrier C, Bonci A, Malenka RC. Long-term depression in the nucleus accumbens: a neural correlate of behavioral sensitization to cocaine. *Nat Neurosci.* 2001;4(12):1217–1223.
95. Kourrich S, Rothwell PE, Klug JR, Thomas MJ. Cocaine experience controls bidirectional synaptic plasticity in the nucleus accumbens. *J Neurosci.* 2007;27(30):7921–7928.
96. Boudreau AC, Wolf ME. Behavioral sensitization to cocaine is associated with increased AMPA receptor surface expression in the nucleus accumbens. *J Neurosci.* 2005;25(40): 9144–9151.
97. Conrad KL, Tseng KY, Uejima JL, et al. Formation of accumbens GluR2-lacking AMPA receptors mediates incubation of cocaine craving. *Nature.* 2008;454(7200):118–119.
98. Luscher C, Bellone C. Cocaine-evoked synaptic plasticity: a key to addiction? *Nat Neurosci.* 2008;11(7):737–738.
99. Martin MT, Chen B, Hopf FW, Bowers MS, Bonci A. Cocaine self-administration selectively abolishes LTD in the core of the nucleus accumbens. *Nat Neurosci.* 2006;9(7):868–869.
100. Yin HH, Knowlton BJ, Balleine BW. Lesions of dorsolateral striatum preserve outcome expectancy but disrupt habit formation in instrumental learning. *Eur J Neurosci.* 2004;19(1): 181–189.
101. Yin HH, Ostlund SB, Knowlton BJ, Balleine BW. The role of the dorsomedial striatum in instrumental conditioning. *Eur J Neurosci.* 2005;22(2):513–523.
102. Nelson A, Killcross S. Amphetamine exposire enhances habit formation. *J Neurosci.* 2006;26:3805–3812.
103. Vanderschuren L, Di Ciano P, Everitt BJ. Involvement of the dorsal striatum in cue-controlled cocaine seeking. *J Neurosci.* 2005;25(38):8665–8670.
104. Belin D, Everitt BJ. Cocaine seeking habits depend upon dopamine-dependent serial connectivity linking the ventral with the dorsal striatum. *Neuron.* 2008;57:432–441.
105. Kelley AE, Smith Roe SL, Holahan MR. Response-reinforcement learning is dependent on N-methyl-D-aspartate receptor activation in the nucleus accumbens core. *Proc Natl Acad Sci USA.* 1997;94(22):12174–12179.
106. Smith-Roe SL, Kelley AE. Coincident activation of NMDA and dopamine D-1 receptors within the nucleus accumbens core is required for appetitive instrumental learning. *J Neurosci.* 2000;20:7737–7742.
107. Baldwin AE, Sadeghian K, Holahan MR, Kelley AE. Appetitive instrumental learning is impaired by inhibition of cAMP-dependent protein kinase within the nucleus accumbens. *Neurobiol Learn Memory.* 2002;77(1):44–62.
108. Hernandez PJ, Sadeghian K, Kelley AE. Early consolidation of instrumental learning requires protein synthesis in the nucleus accumbens. *Nat Neurosci.* 2002;5(12):1327–1331.
109. Cardinal RN, Pennicott DR, Sugathapala CL, Robbins TW, Everitt BJ. Impulsive choice induced in rats by lesions of the nucleus accumbens core. *Science.* 2001;292(5526):2499–2501.
110. Salamone JD, Correa M, Farrar A, Mingote SM. Effort-related functions of nucleus accumbens dopamine and associated forebrain circuits. *Psychopharmacology.* 2007;191(3):461–482.
111. Corbit LH, Muir JL, Balleine BW. The role of the nucleus accumbens in instrumental conditioning: evidence of a functional dissociation between accumbens core and shell. *J Neurosci.* 2001;21:3251–3260.
112. de Borchgrave R, Rawlins JNP, Dickinson A, Balleine BW. Effects of cytotoxic nucleus accumbens lesions on instrumental conditioning in rats. *Exp Brain Res.* 2002; 144(1):50–68.
113. Balleine B, Killcross S. Effects of ibotenic acid lesions of the nucleus accumbens on instrumental action. *Behav Brain Res.* 1994;65:181–193.
114. Killcross S, Coutureau E. Coordination of actions and habits in the medial prefrontal cortex of rats. *Cereb Cortex.* 2003; 13(4):400–408.
115. Ostlund SB, Balleine BW. Lesions of medial prefrontal cortex disrupt the acquisition but not the expression of goal-directed learning. *J Neurosci.* 2005;25:7763–7770.
116. Adams CD, Dickinson A. Instrumental responding following reinforcer devaluation. *Q J Exp Psychol B-Comp Physiol Psychol.* 1981;33:109–121.
117. Yin HH, Knowlton BJ, Balleine BW. Inactivation of dorsolateral striatum enhances sensitivity to changes in the action-outcome contingency in instrumental conditioning. *Behav Brain Res.* 2006;166(2):189–196.

118. Faure A, Haberland U, Conde F, El Massioui N. Lesion to the nigrostriatal dopamine system disrupts stimulus-response habit formation. *J Neurosci.* 2005;25(11):2771–2780.
119. Nelson A, Killcross S. Amphetamine exposure enhances habit formation. *J Neurosci.* 2006;26(14):3805–3812.
120. Ferrario CR, Gorny G, Crombag HS, Li YL, Kolb B, Robinson TE. Neural and behavioral plasticity associated with the transition from controlled to escalated cocaine use. *Biol Psychiatry.* 2005;58(9):751–759.
121. Kelly PH, Seviour PW, Iversen SD. Amphetamine and apomorphine responses in the rat following 6-OHDA lesions of the nucleus accumbens septi and corpus striatum. *Brain Res.* 1975;94:507–522.
122. Belin D, Jonkman S, Dickinson A, Robbins TW, Everitt BJ. Parallel and interactive learning processes within the basal ganglia: relevance for the understanding of addiction. *Behav Brain Res.* 2008;199:89–102.
123. Di Chiara G, Bassareo V, Fenu S, et al. Dopamine and drug addiction: the nucleus accumbens shell connection. *Neuropharmacology.* 2004;47:227–241.
124. Ito R, Robbins TW, Everitt BJ. Differential control over cocaine-seeking behavior by nucleus accumbens core and shell. *Nat Neurosci.* 2004;7(4):389–397.
125. Di Ciano P, Everitt BJ. Direct interactions between the basolateral amygdala and nucleus accumbens core underlie cocaine-seeking behavior by rats. *J Neurosci.* 2004;24(32):7167–7173.
126. Olmstead MC, Lafond MV, Everitt BJ, Dickinson A. Cocaine seeking by rats is a goal-directed action. *Behav Neurosci.* 2001;115(2):394–402.
127. Ito R, Dalley JW, Robbins TW, Everitt BJ. Dopamine release in the dorsal striatum during cocaine-seeking behavior under the control of a drug-associated cue. *J Neurosci.* 2002;22(14):6247–6253.
128. Dickinson A, Balleine B. Motivational control of goal-directed action. *Animal Learn Behav.* 1994;22(1):1–18.
129. Dickinson A. Actions and habits: the development of behavioural autonomy. *Philos Trans R Soc Lond B.* 1985;308:67–78.
130. Dickinson A, Wood N, Smith JW. Alcohol seeking by rats: action or habit? *Q J Exp Psychol B-Comp Physiol Psychol.* 2002;55(4):331–348.
131. Miles FJ, Everitt BJ, Dickinson A. Oral cocaine seeking by rats: action or habit? *Behav Neurosci.* 2003;117(5):927–938.
132. Haber SN, Fudge JL, McFarland NR. Striatonigral pathways in primates form an ascending spiral from the shell to the dorsolateral striatum. *J Neurosci.* 2000;20:2369–2382.
133. Ikemoto S. Dopamine reward circuitry: two projection systems from the ventral midbrain to the nucleus accumbens-olfactory tubercle complex. *Brain Res Rev.* 2007;56:27–78.
134. Letchworth SR, Nader MA, Smith HR, Friedman DP, Porrino LJ. Progression of changes in dopamine transporter binding site density as a result of cocaine self-administration in rhesus monkeys. *J Neurosci.* 2001;21(8):2799–2807.
135. Porrino LJ, Daunais JB, Smith HR, Nader MA. The expanding effects of cocaine: studies in a nonhuman primate model of cocaine self-administration. *Neurosci Biobehav Rev.* 2004;27(8):813–820.
136. Fuchs RA, Branham RK, See RE. Different neural substrates mediate cocaine seeking after abstinence versus extinction training: a critical role for the dorsolateral caudate-putamen. *J Neurosci.* 2006;26(13):3584–3588.
137. See RE, Elliott JC, Feltenstein MW. The role of dorsal vs. ventral striatal pathways in cocaine-seeking behavior following prolonged abstinence in rats. *Psychopharmacology.* 2007;195:321–331.
138. Garavan H, Pankiewicz J, Bloom A, et al. Cue-induced cocaine craving: neuroanatomical specificity for drug users and drug stimuli. *Am J Psychiatry.* 2000;157(11):1789–1798.
139. Volkow ND, Wang GJ, Telang F, et al. Cocaine cues and dopamine in dorsal striatum: mechanism of craving in cocaine addiction. *J Neurosci.* 2006;26(24):6583–6588.
140. Grant S, London ED, Newlin DB, et al. Activation of memory circuits during cue-elicited cocaine craving. *Proc Natl Acad Sci USA.* 1996;93:12040–12045.
141. Childress AR, Mozley PD, McElgin W, Fitzgerald J, Reivich M, O'Brien CP. Limbic activation during cue-induced cocaine craving. *Am J Psychiatry.* 1999;156(1):11–18.
142. Volkow ND, Fowler JS, Wang GJ, Goldstein RZ. Role of dopamine, the frontal cortex and memory circuits in drug addiction: insight from imaging studies. *Neurobiol Learn Memory.* 2002;78(3):610–624.
143. Takahashi Y, Roesch MR, Stanlaker TA, Schoenbaum G. Cocaine shifts the balance of cue-evoked firing from ventral to dorsal striatum. 2007;1:11
144. Beeler JA, Cao ZFH, Kheirbek MA, Zhuang XX. Loss of cocaine locomotor response in Pitx3-deficient mice lacking a nigrostriatal pathway. *Neuropsychopharmacology.* 2009;34:1149–1161.
145. Everitt BJ, Belin D, Economidou D, Pelloux Y, Dalley JW, Robbins TW. Neural mechanisms underlying the vulnerability to develop compulsive drug-seeking habits and addiction. *Philosophical Trans R Soc B-Biol Sci.* 2008;363(1507):3125–3135.
146. Jentsch JD, Taylor JR. Impulsivity resulting from frontostriatal dysfunction in drug abuse: implications for the control of behavior by reward-related stimuli. *Psychopharmacology.* 1999; 146(4):373–390.
147. Rogers RD, Robbins TW. Investigating the neurocognitive deficits associated with chronic drug misuse. *Curr Opin Neurobiol.* 2001;11(2):250–257.
148. Olausson P, Jentsch JD, Krueger DD, Tronson NC, Nairn AC, Taylor JR. Orbitofrontal cortex and cognitive-motivational impairments in psychostimulant addiction. *Linking Affect Action: Criti Contrib Orbitofront Cortex.* 2007;1121:610–638.
149. Garavan H, Kaufman JN, Hester R. Acute effects of cocaine on the neurobiology of cognitive control. *Philos Trans R Soc B-Biol Sci.* 2008;363(1507):3267–3276.
150. *Diagnostic and Statistical Manual of Mental Disorders.* 4th ed. Washington, DC: American Psychiatric Association; 1994.
151. Dalley JW, Theobald DE, Eagle DM, Passetti F, Robbins TW. Deficits in impulse control associated with tonically-elevated function in rat serotonergic prefrontal cortex. *Neuropsychopharmacology.* 2002;26(6):716–728.
152. Diergaarde L, Pattij T, Poortvliet I, et al. Impulsive choice and impulsive action predict vulnerability to distinct stages of nicotine seeking in rats. *Biol Psychiatry.* 2008;63(3):301–308.
153. Dalley JW, Fryer TD, Brichard L, et al. Nucleus accumbens D2/3 receptors predict trait impulsivity and cocaine reinforcement. *Science.* 2007;315(5816):1267–1270.
154. Piazza PV, Deminière JM, Le Moal M, Simon H. Factors that predict individual vulnerability to amphetamine self-administration. *Science.* 1989;245:1511–1513.
155. Ahmed SH, Koob GF. Long-lasting increase in the set point for cocaine self-administration after escalation in rats. *Psychopharmacology.* 1999;146(3):303–312.

156. Economidou D, Pelloux Y, Robbins TW, Dalley JW, Everitt BJ. High impulsivity predicts relapse to cocaine-seeking after punishment-induced abstinence. *Biol Psychiatry*. 2009;65:851–856.
157. Volkow ND, Fowler JS, Wang GJ, et al. Decreased dopamine-D(2) receptor availability is associated with reduced frontal metabolism in cocaine abusers. *Synapse*. 1993;14(2):169–177.
158. Volkow ND, Chang L, Wang GJ, et al. Low level of brain dopamine D-2 receptors in methamphetamine abusers: association with metabolism in the orbitofrontal cortex. *Am J Psychiatry*. 2001;158(12):2015–2021.
159. Nader MA, Daunais JB, Moore T, et al. Effects of cocaine self-administration on striatal dopamine systems in rhesus monkeys: initial and chronic exposure. *Neuropsychopharmacology*. 2002;27(1):35–46.
160. Nader MA, Czoty PW, Gould RW, Riddick NV. Positron emission tomography imaging studies of dopamine receptors in primate models of addiction. *Philos Trans R Soc B-Biol Sci*. 2008;363(1507):3223–3232.
161. Czoty PW, Gage HD, Nader SH, Reboussin BA, Bounds M, Nader MA. PET imaging of dopamine D2 receptor and transporter availability during acquisition of cocaine self-administration in rhesus monkeys. *J Addict Med*. 2007;1(1):33–39.
162. Volkow ND, Wang GJ, Fischman MW, et al. Relationship between subjective effects of cocaine and dopamine transporter occupancy. *Nature*. 1997;386(6627):827–830.
163. Volkow ND, Fowler JS, Wang GJ. Role of dopamine in drug reinforcement and addiction in humans: results from imaging studies. *Behav Pharmacol*. 2002;13(5-6):355–366.
164. Volkow ND, Wang GJ, Begleiter H, et al. High levels of dopamine D-2 receptors in unaffected members of alcoholic families—possible protective factors. *Arch Gen Psychiatry*. 2006;63(9):999–1008.
165. Volkow ND, Wang GJ, Telang F, et al. Low dopamine striatal D2 receptors are associated with prefrontal metabolism in obese subjects: possible contributing factors. *Neuroimage*. 2008;42(4):1537–1543.
166. Volkow ND, Wang GJ, Fowler JS, Telang F. Overlapping neuronal circuits in addiction and obesity: evidence of systems pathology. *Philos Trans R Soc B-Biol Sci*. 2008;363(1507):3191–3200.
167. Stice E, Spoor S, Bohon C, Small DM. Relation between obesity and blunted striatal response to food is moderated by TaqIA A1 allele. *Science*. 2008;322(5900):449–452.
168. Pohjalainen T, Rinne JO, Någren K, et al. The A1 allele of the human D2 dopamine receptor gene predicts low D2 receptor availability in healthy volunteers. *Mol Psychiatry*. 1998;3:256–260.
169. Reuter J, Raedler T, Rose M, Hand I, Glascher J, Buchel C. Pathological gambling is linked to reduced activation of the mesolimbic reward system. *Nat Neurosci*. 2005;8(2):147–148.
170. Vanderschuren LJMJ, Everitt BJ. Drug seeking becomes compulsive after prolonged cocaine self-administration. *Science*. 2004;305(5686):1017–1019.
171. Deroche-Gamonet V, Belin D, Piazza PV. Evidence for addiction-like behavior in the rat. *Science*. 2004;305(5686):1014–1017.
172. Pelloux Y, Everitt BJ, Dickinson A. Compulsive drug seeking by rats under punishment: effects of drug taking history. *Psychopharmacology*. 2007;194(1):127–137.
173. Anthony JC, Warner LA, Kessler RC. Comparative epidemiology of dependence on tobacco, alcohol, controlled substances, and inhalants. *Exp Clin Psychopharmacol*. 1994;2:244.
174. Belin D, Mar AC, Dalley JW, Robbins TW, Everitt BJ. High impulsivity predicts the switch to compulsive cocaine-taking. *Science*. 2008;320(5881):1352–1355.
175. Robinson TE, Berridge KC. Incentive-sensitization and addiction. *Addiction*. 2001;96(1):103–114.
176. Koob GF, Le Moal M. Plasticity of reward neurocircuitry and the "dark side" of drug addiction. *Nat Neurosci*. 2005;8(11):1442–1444.
177. Stalnaker TA, Roesch MR, Calu DJ, Burke KA, Singh T, Schoenbaum G. Neural correlates of inflexible behavior in the orbitofrontal—amygdalar circuit after cocaine exposure. *Link Affect Action: Crit Contrib Orbitofront Cortex*. 2007;1121:598–609.
178. Calu DJ, Stalnaker TA, Franz TM, Singh T, Shaham Y, Schoenbaum G. Withdrawal from cocaine self-administration produces long-lasting deficits in orbitofrontal-dependent reversal learning in rats. *Learn Memory*. 2007;14(5):325–328.

8.3 Imaging Dopamine's Role in Drug Abuse and Addiction

NORA D. VOLKOW, JOANNA S. FOWLER, GENE-JACK WANG, FRANK TELANG, AND RUBEN BALER

THE RELATIONSHIP BETWEEN ACUTE DOPAMINE INCREASES IN THE HUMAN BRAIN AND DRUG REINFORCEMENT

Multiple lines of evidence indicate that one of the major roles of dopamine (DA) is to optimize memory, learning, and attentional processes along the mesocorticolimbic axis (see also Chapter 5.1 in this volume). Because addictions are associated with profound disruptions in all three cognitive domains, it was not surprising to discover that some form of DA dysregulation can usually be found at the heart of most substance use disorders. Indeed, the vast majority of addictive drugs have been found to display an uncanny ability to acutely and dramatically increase extracellular DA levels in key regions of the limbic system.[1,2] Such DA surges resemble but greatly surpass the physiological increases triggered by the phasic DA cell firing that conveys information about saliency,[3,4] reward,[5,6] and reward expectation.[7] In addition, human brain imaging studies have largely corroborated that drug-induced increases in DA in the dorsal and ventral striatum (location of the nucleus accumbens, NAc) are closely linked to the subjective experience of reward or euphoria.[8,9]

However, as drug use continues, the repeated firing of DA cells begins to upset the balanced neurochemistry required to support plastic changes within associative learning circuits, facilitating the consolidation of maladaptive memory traces that are connected to the drug. These, in turn, will trigger DA cells firing upon exposure to any number of contextual stimuli that are merely associated with the drug (in expectation of the reward).[10] And because of DA's role in motivation, the DA increases associated with drug cues or the drug itself are also likely to modulate the drive to secure the reward.[11]

Our improved understanding of DA's multiple roles in the reinforcement process has led to a much more coherent model of drug addiction; that is, drugs are reinforcing not only because they are pleasurable but because, by increasing DA, they are being processed as salient stimuli that will inherently motivate the procurement of more drug (regardless of whether the drug is consciously perceived as pleasurable or not). This model continues to evolve, largely thanks to the increasing use of sophisticated brain imaging techniques that allow us to (1) measure neurochemical and metabolic processes in the living human brain,[12] (2) investigate the nature of the changes in DA induced by drugs of abuse and their behavioral impact, and (3) study the long-term plastic changes in brain DA activity and its functional consequences in drug-addicted subjects.

The use of positron emission tomography (PET) with D2 DA receptor radioligands (e.g., [^{11}C]raclopride, [^{18}F]N-methylspiroperidol, [^{11}C]-(+)-4-propyl-9-hydroxynaphthoxazine[13]) has had a particularly profound impact in the addiction field. The technique has proven invaluable for studying the relationships between the ability of many drugs to modulate DA and their reinforcing (i.e., euphorigenic, high-inducing, drug-liking) effects in the human brain. On the one hand, and consistent with the relatively smaller DA responses to opiates observed in rats,[2] it has been predictably more challenging to establish unequivocally a robust connection between the "high" from opioid administration and DA surges in dependent humans.[14] On the other hand, the [^{11}C]raclopride imaging approach has helped clarify the effects of stimulant drugs like methylphenidate, amphetamine, and cocaine on the DA system, as well as those of nicotine[15–18] and alcohol.[19] We now know, for example, that both the intravenous administration of methylphenidate (0.5 mg/kg), which, like cocaine, increases DA by blocking DA transporters (DATs), as well as that of amphetamine (0.3 mg/kg), which, like methamphetamine, increases DA by releasing it from the terminal via DATs, can increase the extracellular DA concentration in the striatum and that such increases are associated with self-reports of highs and euphoria.[20,21] In contrast, orally administered methylphenidate (0.75–1 mg/kg), which can also increase DA,[22] is not typically perceived as reinforcing.[23,24] It is also known that intravenous administration of methylphenidate leads to DA changes that are much faster than those observed after oral administration. Thus, the failure of oral methylphenidate—or amphetamine[25]—to induce a high is likely the reflection of slower pharmacokinetics.[26] The fact is that the speed with which drugs of abuse enter the brain is a key parameter that affects their reinforcing effects.[27–29] This is also likely to explain why the DA increases in ventral striatum induced by tobacco smoke,[30] which also has a very fast rate of brain uptake, are also associated with its reinforcing effects.[18]

The close correlation between the fast uptake of a drug into the brain, the rapid changes in extracellular DA in the striatum, and its reinforcing properties suggests the involvement of phasic DA firing. Phasic release at frequencies of >30 Hz cause abrupt fluctuations in DA levels that contribute to highlighting the saliency of a stimulus.[31] In contrast, tonic DA cell firing, with frequencies of ~5 Hz, serves to maintain the baseline steady-state DA levels that set the threshold of the DA system's responsiveness. Therefore, several authors have proposed that drugs of abuse manage to induce changes in DA concentration that mimic, but greatly exceed, those produced by physiological phasic DA cell firing. We must keep in mind, however, that the outcome of such DA increases—even when supraphysiological—will be contingent upon other factors, such as the expectation of a particular outcome and the context of administration.[7,32]

In contrast, oral administration of stimulant drugs, which is the therapeutic route, is more likely to induce slow DA changes, more akin to those provoked by tonic DA cell firing.[33] However, since stimulant drugs block DATs, which are the main mechanism for DA removal,[34] they have the potential to increase the reinforcing value of other reinforcers (natural or drug rewards) even when administered orally.[24] Similarly, nicotine, which facilitates DA cell firing,[35,36] can also enhance the reinforcing value of stimuli with which it is paired.[37] Thus, the combination of nicotine with the natural reward becomes inextricably linked to its reinforcing effects.

LONG-TERM EFFECTS OF DRUGS OF ABUSE ON DA IN THE HUMAN BRAIN: INVOLVEMENT IN ADDICTION

It is important to underscore the fact that even though drug-induced surges in synaptic DA occur in both addicted and nonaddicted individuals,[1,2] only a minority of exposed subjects—the actual fraction being a function of the type of drug used—ever develops a compulsive drive to continue taking the drug.[38] Clearly, DA increases alone are insufficient to explain the onset of an addiction trajectory. The fact that chronic drug administration is a *sine qua non* condition for the development of drug addiction suggests that addictions hinge—in vulnerable individuals—on the *repeated* perturbation of the DA system, which can, over time, induce neuroadaptations in reward/saliency, motivation/drive, inhibitory control/executive function, and memory/conditioning circuits, all of which are known to be modulated by dopaminergic pathways.[39,40]

In fact, there is growing evidence that supports this notion. Chronic exposure to stimulants, nicotine, or opiates can produce persistent adaptive changes in the structure of dendrites and dendritic spines on neurons in key brain circuits with roles in motivation, reward, judgment, and inhibitory control of behavior.[41] This observation becomes particularly significant when we consider that the induction of long-term potentiation (LTP) is often associated with measurable increases in the size of dendritic spines and their associated structures.[42,43]

Drug-induced DA perturbations are likely to have both direct and indirect effects on the maladaptive rewiring of neural circuits. Dopamine (but also glutamate,[44] GABA,[45] and other neurotransmitter systems[46]) is a versatile modulator of synaptic plasticity in its own right[47,48] but, in addition, chronic adaptations in DA receptor signaling may trigger, for example, compensatory glutamate receptor responses with the potential to affect synaptic plasticity.[49] It is relevant to point out in this context that, while DA receptors can be found throughout the neuron, there is growing evidence of their increased concentration in dendritic spines, which also feature the highest density of glutamatergic synapses.[50] Thus, the various combinations of postsynaptic DA receptor types are strategically located to influence the synaptic properties of spines via the accurate decodification of tonic and phasic trains of DA signals.

These observations draw a direct path connecting the effects of drugs of abuse with the adaptive alterations, not only in reward centers but also in many other circuits, through the strengthening, formation, and elimination of synapses.

Effects on Reward and Motivation Circuits

The availability of several radiotracers has allowed researchers to monitor both transient and persistent neurochemical changes in the DA network of the human brain (Table 8.3.1). It has been shown, using [^{18}F]N-methylspiroperidol or [^{11}C]raclopride,[51–53]

TABLE 8.3.1. *Summary of PET Findings Comparing Various Targets Involved in DA Neurotransmission between Substance Abusers and Control Subjects for Which Statistically Significant Differences between the Groups Were Identified*

Target Investigated	Drug Used	Finding	Ref.
D2 DA receptors	Cocaine	↓ Acute withdrawal	116
		↓ Detoxified	116, 117
	Alcohol	↓ 1- to 68-week abstinence	118
		↓ Detoxified	119
	Methamphetamine	↓ Detoxified	65, 120
	Heroin	↓ Active user	121
	Nicotine	↓ Active user	122
	Cannabis	0 Detoxified	123
		0 Early remission	123
DA transporters	Cocaine	↑ 4-week abstinence	124
		0 Detoxified	117
	Alcohol	↓ Acute withdrawal	125
		0 Detoxified	126
	Methamphetamine	↓ Detoxified	127
	Cigarettes	↓ Active user	128
Vesicular monoamine transporters-2	Methamphetamine	↓ Detoxified	127
Metabolism (monoamine oxidase A and B)	Cigarettes	↓ Active user	129
Synthesis (dopa decarboxylase)	Cocaine	↓ Detoxified	130
	Alcohol	0 Detoxified	131
DA release	Cocaine	↑ Active user	132
		↓ Detoxified	57
	Alcohol	↓ Detoxified	58

Source: Modified and updated from[64] with permission.

that subjects addicted to a wide range of drugs (cocaine, heroin, alcohol, and methamphetamine) exhibit significant reductions in D2 DA receptor availability in the striatum (including ventral striatum) that persist for months after protracted detoxification.[54,55] Similar findings were also recently reported for nicotine-dependent subjects.[56]

It has also been observed that the striatal increases in DA levels induced by intravenous (i.v.) methylphenidate or i.v. amphetamine (and assessed with [^{11}C]raclopride) in cocaine abusers and alcoholics are at least 50% lower than in control subjects.[53,57,58] Since DA increases induced by methylphenidate are dependent on DA release—a function of DA cell firing—it is reasonable to hypothesize that the difference likely reflects decreased dopaminergic cell activity in these drug-abusing populations.

While evaluating the results of PET studies based on the competition of [^{11}C]raclopride by endogenous DA, it is critical to remember that the results merely reflect the fraction of D2 DA receptors that is vacant and thus capable of binding the tracer. As a consequence, any reduction in D2 DA receptor availability measured with this technique could reflect either *decreases* in levels of D2 DA receptors and/or *increases* in DA release (competing for binding with [^{11}C]raclopride for the D2 DA receptors) in striatum (including NAc). However, the fact that cocaine abusers showed blunted reductions in specific binding (indicative of decreased DA release) when administered i.v. methylphenidate indicates that these individuals had both a reduction in the levels of D2 DA receptors and a decrease in DA release in striatum. Each deficiency would contribute to the overall decreased sensitivity in addicted subjects to natural reinforcers.[59]

On the other hand, drugs are much more potent in stimulating DA-regulated reward circuits[2] and in triggering persistent circuit changes[60] than natural reinforcers. Therefore, drugs would still have an advantage in individuals attempting to activate their depressed reward circuits. The decreased sensitivity, on the other hand, would result in a reduced interest in environmental stimuli, possibly predisposing subjects to seek drug stimulation as a means of temporarily activating an underresponsive reward network. As time progresses, the chronic nature of this behavior may sustain the transition from taking drugs in order to feel high to taking them just to feel normal.

Executive Function and Inhibitory Control

Predictably, there will be profound metabolic and functional consequences to such long-term drug-induced perturbations in the dopaminergic balance. Researchers have used the PET radiotracer [^{18}F]fluoro-deoxyglucose (FDG), which measures regional brain glucose metabolism, to document decreased activity in orbitofrontal cortex (OFC), cingulate gyrus (CG), and dorsolateral prefrontal cortex (DLPFC) in addicted subjects (alcoholics, cocaine abusers, marijuana abusers, methamphetamine abusers).[54,61–63] Moreover, significant correlations have been observed between reduced metabolic activity in OFC, CG, and DLPFC and decreased D2 DA receptor availability in the striatum of cocaine-[64] and methamphetamine-[65] addicted subjects and of alcoholics[58] (see Fig. 8.3.1 for cocaine and methamphetamine results). Since the OFC, CG, and DLPFC play critical roles in inhibitory control[66] and emotional processing,[67] it has been postulated that their abnormal regulation by DA, characteristic of addiction, could underlie the subjects' loss of control over drug intake and their poor emotional self-regulation. Indeed, in alcoholics, reductions in D2 DA receptor availability in ventral striatum have been shown to be associated with alcohol craving severity and with greater cue-induced activation of the medial prefrontal cortex and anterior CG, as assessed with functional magnetic resonance imaging (fMRI)[68]. In addition, because damage to the OFC results in perseverative behaviors[69]—and because, in humans, impairments in OFC and CG are associated with obsessive compulsive behaviors[70]—it has also been postulated that DA impairment of these regions could underlie the compulsive drug intake that characterizes addiction.[71]

The involvement of inhibitory control areas in the circuit abnormalities that underlie addiction disorders should give us pause, because weakened prefrontal regions could increase the drive to engage in risky behaviors in general, which could *secondarily* put individuals at risk for drug abuse. Alternatively, low D2 DA receptor levels during fetal development may also disrupt prefrontal activity in adulthood, resulting in impulsivity and the associated increases in risk for substance abuse.[72]

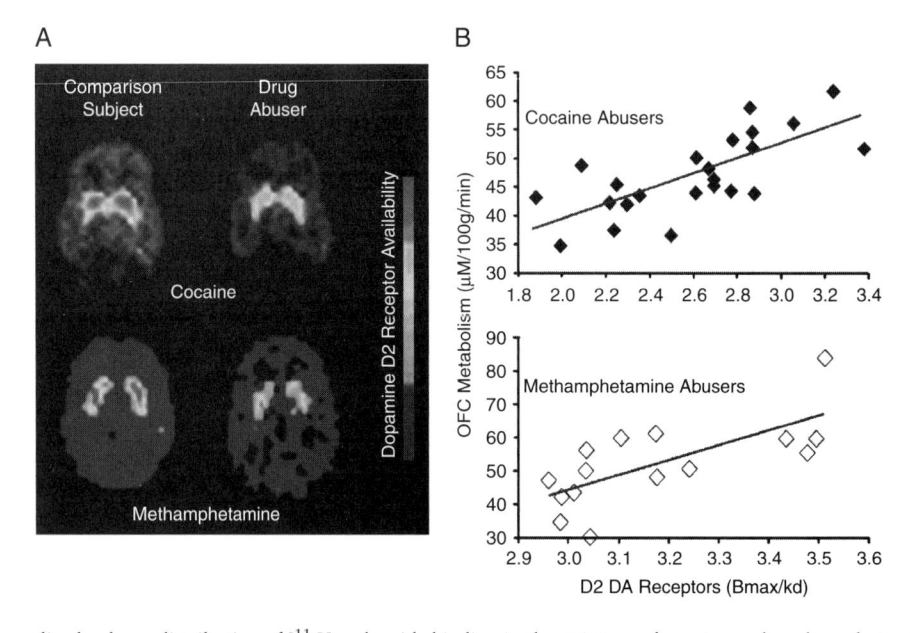

FIGURE 8.3.1. (A) Normalized volume distribution of [^{11}C] raclopride binding in the striatum of cocaine and methamphetamine abusers and non-drug-abusing control subjects. (B) Correlation of DA receptor availability (B_{max}/K_d) in the striatum with a measure of metabolic activity in the orbitofrontal cortex (OFC) in cocaine (closed diamonds) and methamphetamine (open diamonds) abusers. *Source*: Modified from[65,66] with permission. (See Color Plate 8.3.1.)

Effects on Conditioning Circuits

The hippocampus, the amygdala, the NAc, and the dorsal striatum are regions that play critical roles in learning and memory. Adaptations in these areas have been well documented in preclinical models of drug abuse[73] and have led to increasing recognition of the relevance and likely involvement of memory and learning mechanisms at different stages of an addiction trajectory.[74] For example, within the brain's reward center, the pathways that project from the ventral tegmental area (VTA) into the NAc and dorsal striatum are the primary targets where drugs like cocaine and amphetamine up-regulate neurotransmitter signaling,[75] while the secondary neuroplastic changes that occur in the NAc and dorsal striatum—and that become consolidated during periods of abstinence[76]—underlie the gradual transformation of a maladaptive habit into the enduring behavioral alterations that characterize addicted individuals[41] and animals that have been subjected to a conditioned reinforcement paradigm.[77] These observations are consistent with the fact that DA (interdependently with 5HT[78]) can modulate the activity of, and affect adaptive changes in, the circuits that support learning/memory, conditioning, and habit formation.[79] Thus, the effects of drugs of abuse on memory systems suggest a likely mechanism—conditioned-incentive learning—through which neutral stimuli can acquire reinforcing properties and motivational salience.[80]

The central, albeit multifaceted, question we need to answer through relapse research is as follows: why do drug-addicted subjects experience such an intense desire for the drug when exposed to places, people, or things associated with drug-taking behaviors? A better understanding of the mechanisms involved could have profound clinical implications, since exposure to conditioned cues (stimuli that had become strongly linked to the drug experience) is a key contributor (trigger) to relapse. Since DA is involved in the prediction of reward,[81] or more precisely perhaps with reward prediction error,[82] DA has been predicted to underlie the conditioned responses that trigger craving. Preclinical studies support this hypothesis: when neutral stimuli are paired with a drug, animals will—with repeated associations—acquire the ability to increase DA in NAc and dorsal striatum when exposed to the now conditioned cue. Predictably, these neurochemical responses have been found to be associated with drug-seeking behaviors.[74]

In humans, PET studies with [^{11}C]raclopride recently confirmed this hypothesis by showing that in cocaine abusers, drug cues (video recordings of subjects taking cocaine) significantly increased DA in dorsal striatum, and that these increases were also associated with cocaine craving[83,84] in a cue-dependent fashion.[85] Because the dorsal striatum is implicated in habit learning, this association is likely to reflect the strengthening of habits as chronicity of addiction develops. This suggests that the DA-triggered conditioned responses that form first habits and then compulsive drug consumption may reflect a fundamental neurobiological perturbation in addiction. In addition, it is likely that these conditioned responses involve adaptations in corticostriatal glutamatergic pathways that regulate DA release.[74]

To assess if cue-induced DA increases reflect a primary or a secondary response to the cue, a recent imaging study in cocaine-addicted individuals evaluated the effects of increasing DA (achieved by oral administration of methylphenidate), with and without the cue, in an attempt to determine whether DA increases by themselves could induce craving. The results of the study revealed a clear dissociation between oral methylphenidate–induced DA increases and cue-associated cravings,[85] suggesting that cue-induced DA increases are not the primary effectors but rather reflect downstream stimulation of DA cells (corticostriatal glutamatergic pathways that regulate DA release[86]). This observation further illuminates the subtle effects of DA firing rate upon addiction circuitry, for the failure of methylphenidate-induced DA increases to induce craving in this paradigm could be explained by the slow nature of the DA increases. On the other hand, fast DA changes triggered by phasic DA cell firing—as a secondary response to the activation of descending pathways—may underlie the successful induction of cravings during exposure to a cue. It is worth highlighting, that Martinez et al. reported a negative correlation between the DA increases induced by i.v. amphetamine in cocaine abusers and their choice of cocaine over money when tested on a separate paradigm.[53] That is, the subjects who showed the lower DA increases when given amphetamine were the ones more likely to select cocaine over a monetary reinforcer. Because in their studies Martinez et al. also reported reduced DA increases in cocaine abusers compared with controls, this could indicate that cocaine abusers with the most severe decreases in brain dopaminergic activity are the ones more likely to choose cocaine over other reinforcers.

DA AND VULNERABILITY TO DRUG ABUSE

Understanding why some individuals are more vulnerable to becoming addicted to drugs than others remains one of the most challenging questions in drug abuse research. The fact that only a largely unpredictable minority of drug abusers progresses to drug addiction hints at the complex interplay between genetic and environmental risk factors. Twin data, for example, suggest that about 50% of addiction vulnerability is heritable.[87] Not surprisingly, imaging studies can play a central role in solving this puzzle. The following examples should help to illustrate this point.

Recent studies found that D2 DA polymorphisms contribute to significantly higher scores in novelty seeking among methamphetamine-addicted patients[88] and among children who were reared in a punitive environment.[89] There is also preliminary evidence suggesting that specific DA receptor gene variants (particularly in the D2 and D4 subtypes) modulate smoking progression (initiation and continuation) in adolescence.[90] Finally, there is strong evidence that the availability of D2 DA receptors in the striatum can modulate the subjective responses of healthy non-drug-abusing controls to the stimulant drug methylphenidate.[91,92] In that experiment, subjects describing the experience as pleasant displayed significantly lower levels of receptors than those describing it as unpleasant (Fig. 8.3.2). This suggests that the relationship between DA levels and reinforcing responses follows an inverted U-shaped curve: too little is suboptimal for reinforcement, while too much may become aversive.

This last example is particularly intriguing, since, according to one possible interpretation, it is consistent with the notion that high D2 DA receptor levels may be protective against drug self-administration. Interestingly, there is a substantial amount of evidence that supports this hypothesis. On the preclinical front, higher levels of D2 DA receptors in NAc significantly reduced alcohol intake in animals previously trained to self-administer alcohol[93] or cocaine in animals trained to self-administer cocaine[94]; and switching cynomolgus macaques from individual to group-housing conditions exposes a robust (inverse) correlation between individual changes in striatal DA D2 receptor levels and the tendency to self-administer cocaine.[95] Evidence in favor of this relationship has also emerged from human studies. First, there is evidence of depressed DA activity in specific brain regions of adults with ADHD compared to controls.[96,97] Deficiencies were seen at the level of both D2 DA receptors and DA release in the caudate and in the ventral striatum. Importantly, and consistent with this model, the depressed DA phenotype was associated with higher scores on self-reports of methylphenidate liking.[96] It is not surprising, then, that, if left untreated, individuals with attention deficit hyperactivity disorder (ADHD) have a high risk of developing substance abuse disorders.[98] Second, it has been observed that subjects who, despite having a strong family history of alcoholism, were not alcoholics had significantly higher D2 DA

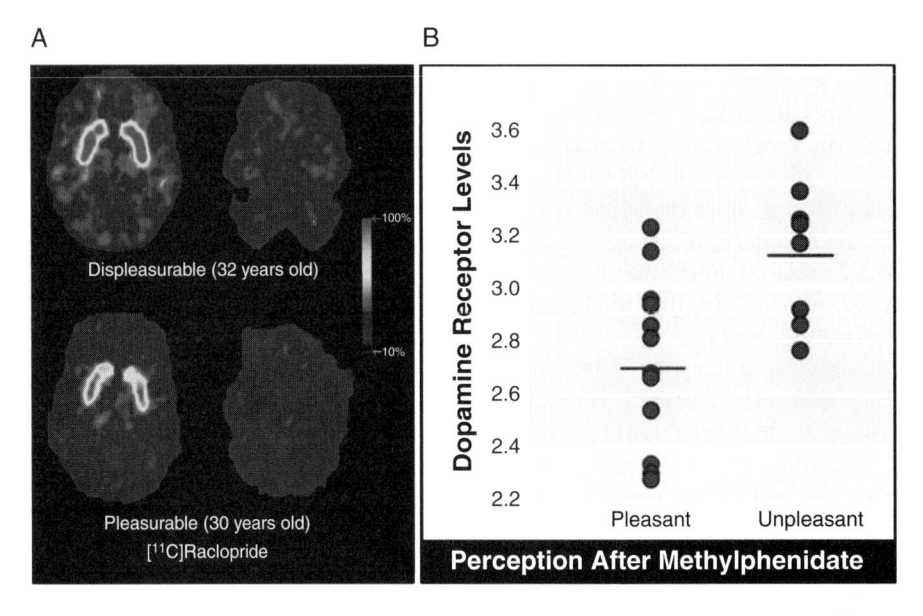

FIGURE 8.3.2. Striatal D2 DA is predictive of methylphenidate liking in humans (A) Distribution volume images of [11C]raclopride at the levels of the striatum (left) and cerebellum (right) in a healthy male subject who reported the effects of methylphenidate as pleasant and in a healthy male subject who reported them as unpleasant. (B) D2 DA receptor levels (bmax/kd) in 21 healthy male subjects who reported the effects of methylphenidate as pleasant or unpleasant. *Source:* Modified from[91] with permission. (See Color Plate 8.3.2.)

receptors in striatum than individuals without such family histories.[99] Interestingly, the higher the level of D2 DA receptors in these subjects, the higher their metabolic activity in OFC and CG. Thus, it can be postulated that high levels of D2 DA receptors may protect against alcoholism by modulating frontal circuits involved in salience attribution and inhibitory control. In this respect, it is worth noting that in the rodent model, low levels of D2 DA receptors are associated with impulsive behaviors,[100] which in turn predict compulsive self-administration of cocaine.[101] Inasmuch as the prefrontal cortex is involved in modulating impulsivity, this may be another mechanism by which low D2 DA levels may make an individual vulnerable to drug abuse and addictions and/or by which high D2 DA receptor levels may protect against drug abuse.

But there are many other variables that are likely or known to significantly modulate the risk of abuse, addiction, and/or relapse and to which we need to pay close attention. For example, sexual dimorphisms have been observed repeatedly in addictive disorders[102–104] and recently have been proposed to be strongly mediated by epigenetic mechanisms.[105] It would be reasonable to apply the power of brain imaging techniques to better understand the current preclinical evidence suggesting that such differences may be due in part to striatal DA system differences and/or differences in the activity of prefrontal regions.[106] Indeed, recent studies have documented sexually dimorphic patterns of amphetamine-induced striatal DA release[107,108] that could impact substance abuse vulnerability differently in men and women, although at this point, the data do not permit a clear-cut conclusion as to whether men or women display greater DA responses. It is also likely that the patterns will be sensitive to experimental conditions, such as context, age, and stage of the menstrual cycle.

It is also critically important that we continue to expand our research focused on the multiple connections that exist between the stress response and addiction vulnerabilities.[109] For, in addition to drug-related cues, stress is a major contributing factor to the increased risk of relapse in an addictive disorder. Indeed, there are substantial overlaps between the circuits in charge of processing stress signals and drug cues and those responsible for processing reward information.[110] Since chronic stress is often accompanied by some degree of sleep disturbances or full-fledged sleep deprivation (SD), it is pertinent to mention, in this context, the recent finding that a single night of SD was associated with a significant reduction in specific binding of [^{11}C]raclopride in the striatum, which was interpreted as a reflection of DA increases.[111] Thus, DA increases with SD may be one of the mechanisms linking sleep deprivation and relapse to drug taking.[112,113]

These and other observations combined provide critical insights into the contribution of the striatal DA system to addiction vulnerability, to the observed sexually dimorphic patterns of substance abuse, and to the emergence of frequent psychiatric comorbid conditions.

TREATMENT IMPLICATIONS

Imaging studies have corroborated the role of DA in the reinforcing effects of drugs of abuse in humans and have dramatically extended the traditional views of DA involvement in drug addiction. These findings suggest multipronged strategies for the treatment of drug addiction designed to (1) decrease the reward value of the drug of choice and increase the reward value of nondrug reinforcers; (2) weaken conditioned drug behaviors and the motivational drive to take the drug; and (3) strengthen frontal inhibitory and executive control. This review does not discuss at length the involvement of circuits that regulate emotions and the response to stress[114] or those responsible for interoceptive perception of needs and desires,[115] which are also potential targets for therapeutic interventions.

REFERENCES

1. Koob GF, Bloom FE. Cellular and molecular mechanisms of drug dependence. *Science*. 1988;242:715–23.
2. Di Chiara G, Imperato A. Drugs abused by humans preferentially increase synaptic dopamine concentrations in the mesolimbic system of freely moving rats. *Proc Natl Acad Sci USA*. 1988;85:5274–5278.
3. Zink CF, Pagnoni G, Martin ME, Dhamala M, Berns GS. Human striatal response to salient nonrewarding stimuli. *J Neurosci*. 2003;23:8092–8097.
4. Horvitz JC. (2000) Mesolimbocortical and nigrostriatal dopamine responses to salient non-reward events. *Neuroscience*. 2000;96:651–656.
5. Tobler PN, O'Doherty JP, Dolan RJ, Schultz W. Reward value coding distinct from risk attitude-related uncertainty coding in human reward systems. *J Neurophysiol*. 2007;97:1621–1632.
6. Schultz W, Tremblay L, Hollerman JR. Reward processing in primate orbitofrontal cortex and basal ganglia. *Cereb Cortex*. 2000;10:272–284.
7. Volkow ND, Wang GJ, Ma Y, et al. Expectation enhances the regional brain metabolic and the reinforcing effects of stimulants in cocaine abusers, *J Neurosci*. 2003;23: 11461–11468.
8. Drevets WC, Gautier C, Price JC, et al. (2001) Amphetamine-induced dopamine release in human ventral striatum correlates with euphoria. *Biol Psychiatry*. 2001;49:81–96.

9. Volkow ND, Wang GJ, Fowler JS, et al. Relationship between psychostimulant-induced "high" and dopamine transporter occupancy. *Proc Natl Acad Sci USA.* 1996;93:10388–103892.
10. Waelti P, Dickinson A, Schultz W. Dopamine responses comply with basic assumptions of formal learning theory. *Nature.* 2001;412:43–48.
11. McClure SM, Daw ND, Montague PR. A computational substrate for incentive salience. *Trends Neurosci.* 2003;26:423–428.
12. Volkow ND, Rosen B, Farde L. Imaging the living human brain: magnetic resonance imaging and positron emission tomography. *Proc Natl Acad Sci USA.* 1997;94:2787–2788.
13. Willeit M, Ginovart N, Graff A, et al. First human evidence of d-amphetamine induced displacement of a D2/3 agonist radioligand: A [11C]-(+)-PHNO positron emission tomography study. *Neuropsychopharmacology.* 2008;33:279–289.
14. Daglish MR, Williams TM, Wilson SJ, et al. Brain dopamine response in human opioid addiction. *Br J Psychiatry.* 2008;193:65–72.
15. Montgomery AJ, Lingford-Hughes AR, Egerton A, Nutt DJ, Grasby PM. The effect of nicotine on striatal dopamine release in man: a [11C]raclopride PET study. *Synapse.* 2007;61:637–645.
16. Takahashi H, Fujimura Y, Hayashi M, et al. Enhanced dopamine release by nicotine in cigarette smokers: a double-blind, randomized, placebo-controlled pilot study. *Int J Neuropsychopharmacol.* 2008;11:413–417.
17. Barrett SP, Boileau I, Okker J, Pihl RO, Dagher A. The hedonic response to cigarette smoking is proportional to dopamine release in the human striatum as measured by positron emission tomography and [11C]raclopride. *Synapse.* 2004;54:65–71.
18. Brody AL, Olmstead RE, London ED, et al. Smoking-induced ventral striatum dopamine release. *Am J Psychiatry.* 2004;161:1211–1218.
19. Boileau I, Assaad JM, Pihl RO, et al. Alcohol promotes dopamine release in the human nucleus accumbens. *Synapse.* 2003;49:226–231.
20. Villemagne VL, Wong DF, Yokoi F, et al. GBR12909 attenuates amphetamine-induced striatal dopamine release as measured by [(11)C]raclopride continuous infusion PET scans. *Synapse.* 1999;33:268–273.
21. Hemby SE. *Neurobiological Basis of Drug Reinforcement.* Philadelphia, PA: Lippincott-Raven; 1997.
22. Volkow ND, Wang GJ, Fowler JS, et al. Dopamine transporter occupancies in the human brain induced by therapeutic doses of oral methylphenidate. *Am J Psychiatry.* 1998;155:1325–1331.
23. Chait LD. Reinforcing and subjective effects of methylphenidate in humans. *Behav Pharmacol.* 1994;5:281–288.
24. Volkow ND, Wang G, Fowler JS, et al. Therapeutic doses of oral methylphenidate significantly increase extracellular dopamine in the human brain. *J Neurosci.* 2001;21:RC121.
25. Stoops WW, Vansickel AR, Lile JA, Rush CR. Acute d-amphetamine pretreatment does not alter stimulant self-administration in humans. *Pharmacol Biochem Behav.* 2007;87:20–29.
26. Parasrampuria DA, Schoedel KA, Schulle R, et al. Assessment of pharmacokinetics and pharmacodynamic effects related to abuse potential of a unique oral osmotic-controlled extended-release methylphenidate formulation in humans. *J Clin Pharmacol.* 2007;47:1476–1488.
27. Balster RL, Schuster CR. Fixed-interval schedule of cocaine reinforcement: effect of dose and infusion duration. *J Exp Anal Behav.* 1973;20:119–129.
28. Volkow ND, Wang J, Fischman MW, et al. Effects of route of administration on cocaine induced dopamine transporter blockade in the human brain. *Life Sci.* 2000;67:1507–1515.
29. Volkow ND, Ding YS, Fowler JS, et al. Is methylphenidate like cocaine? Studies on their pharmacokinetics and distribution in the human brain. *Arch Gen Psychiatry.* 1995;52:456–463.
30. Scott DJ, Domino EF, Heitzeg MM, et al. Smoking modulation of mu-opioid and dopamine D2 receptor-mediated neurotransmission in humans. *Neuropsychopharmacology.* 2007;32:450–457.
31. Grace AA. The tonic/phasic model of dopamine system regulation and its implications for understanding alcohol and psychostimulant craving. *Addiction.* 2000;95(suppl 2):S119-S128.
32. Kufahl P, Li Z, Risinger R, et al. Expectation modulates human brain responses to acute cocaine: a functional magnetic resonance imaging study. *Biol Psychiatry.* 2008;63:222–230.
33. Volkow ND, Swanson JM. Variables that affect the clinical use and abuse of methylphenidate in the treatment of ADHD. *Am J Psychiatry.* 2003;160:1909–1918.
34. Williams JM, Galli A. The dopamine transporter: a vigilant border control for psychostimulant action. *Handb Exp Pharmacol.* 2006;175:215–232.
35. Meyer EL, Yoshikami D, McIntosh JM. The neuronal nicotinic acetylcholine receptors alpha 4* and alpha 6* differentially modulate dopamine release in mouse striatal slices. *J Neurochem.* 2008;105:1761–1769.
36. Cao YJ, Surowy CS, Puttfarcken PS. Different nicotinic acetylcholine receptor subtypes mediating striatal and prefrontal cortical [3H]dopamine release. *Neuropharmacology.* 2005;48:72–79.
37. Donny EC, Chaudhri N, Caggiula AR, et al. Operant responding for a visual reinforcer in rats is enhanced by noncontingent nicotine: implications for nicotine self-administration and reinforcement. *Psychopharmacology (Berl).* 2003;169:68–76.
38. Schuh LM, Schuh KJ, Henningfield JE. Pharmacologic determinants of tobacco dependence. *Am J Ther.* 1996;3:335–341.
39. Volkow ND, Fowler JS, Wang GJ. The addicted human brain: insights from imaging studies. *J Clin Invest.* 2003;111:1444–1451.
40. Jones S, Bonci A. Synaptic plasticity and drug addiction. *Curr Opin Pharmacol.* 2005;5:20–25.
41. Robinson TE, Kolb B. Structural plasticity associated with exposure to drugs of abuse. *Neuropharmacology.* 2004;47(suppl 1):33–46.
42. Yuste R, Bonhoeffer T. Morphological changes in dendritic spines associated with long-term synaptic plasticity. *Annu Rev Neurosci.* 2001;24:1071–1089.
43. Matsuzaki M, Honkura N, Ellis-Davies GC, Kasai H. Structural basis of long-term potentiation in single dendritic spines. *Nature.* 2004;429:761–766.
44. Newpher TM, Ehlers MD. Glutamate receptor dynamics in dendritic microdomains. *Neuron.* 2008;58:472–497.
45. Nugent FS, Kauer JA. LTP of GABAergic synapses in the ventral tegmental area and beyond. *J Physiol.* 2008;586:1487–1493.
46. Spedding M, Gressens P. Neurotrophins and cytokines in neuronal plasticity. *Novartis Found Symp.* 2008;289:222-233; discussion 233–240.
47. Wolf ME, Sun X, Mangiavacchi S, Chao SZ. Psychomotor stimulants and neuronal plasticity. *Neuropharmacology.* 2004;47(suppl 1):61–79.
48. Liu QS, Pu L, Poo MM. Repeated cocaine exposure in vivo facilitates LTP induction in midbrain dopamine neurons. *Nature.* 2005;437:1027–1031.

49. Wolf ME, Mangiavacchi S, Sun X. Mechanisms by which dopamine receptors may influence synaptic plasticity. *Ann NY Acad Sci.* 2003;1003:241–249.
50. Yao WD, Spealman RD, Zhang J. Dopaminergic signaling in dendritic spines. *Biochem Pharmacol.* 2008;75:2055–2069.
51. Martinez D, Broft A, Foltin RW, et al. Cocaine dependence and D2 receptor availability in the functional subdivisions of the striatum: relationship with cocaine-seeking behavior. *Neuropsychopharmacology.* 2004;29:1190–1202.
52. Martinez D, Gil R, Slifstein M, et al. (2005) Alcohol dependence is associated with blunted dopamine transmission in the ventral striatum. *Biol Psychiatry.* 2005;58:779–786.
53. Martinez D, Narendran R, Foltin RW, et al. Amphetamine-induced dopamine release: markedly blunted in cocaine dependence and predictive of the choice to self-administer cocaine. *Am J Psychiatry.* 2007;164:622–629.
54. Volkow ND, Fowler JS, Wang GJ, Swanson JM, Telang F. Dopamine in drug abuse and addiction: results of imaging studies and treatment implications. *Arch Neurol.* 2007;64:1575–1579.
55. Johanson CE, Frey KA, Lundahl LH, et al. Cognitive function and nigrostriatal markers in abstinent methamphetamine abusers. *Psychopharmacology (Berl).* 2006;185:327–338.
56. Fehr C, Yakushev I, Hohmann N, et al. Association of low striatal dopamine D2 receptor availability with nicotine dependence similar to that seen with other drugs of abuse. *Am J Psychiatry.* 2008;165:507–514.
57. Volkow ND, Wang GJ, Fowler JS, et al. Decreased striatal dopaminergic responsiveness in detoxified cocaine-dependent subjects. *Nature.* 1997;386:830–833.
58. Volkow ND, Wang GJ, Telang F, et al. Profound decreases in dopamine release in striatum in detoxified alcoholics: possible orbitofrontal involvement. *J Neurosci.* 2007;7:12700–12706.
59. Volkow ND, Fowler JS, Wang, GJ. Role of dopamine in drug reinforcement and addiction in humans: results from imaging studies. *Behav Pharmacol.* 2002;13:355–366.
60. Chen BT, Bowers MS, Martin M, et al. Cocaine but not natural reward self-administration nor passive cocaine infusion produces persistent LTP in the VTA. *Neuron.* 2008;59:288–297.
61. London ED, Cascella NG, Wong DF, et al. Cocaine-induced reduction of glucose utilization in human brain. A study using positron emission tomography and [fluorine 18]-fluorodeoxyglucose. *Arch Gen Psychiatry.* 1990;47:567–574.
62. Galynker II, Watras-Ganz S, Miner C, et al. Cerebral metabolism in opiate-dependent subjects: effects of methadone maintenance. *Mt Sinai J Med.* 2000;67:381–387.
63. Ersche KD, Fletcher PC, Roiser JP, et al. Differences in orbitofrontal activation during decision-making between methadone-maintained opiate users, heroin users and healthy volunteers. *Psychopharmacology (Berl).* 2006;188:364–373.
64. Volkow ND, Fowler JS. Addiction, a disease of compulsion and drive: involvement of the orbitofrontal cortex. *Cereb Cortex.* 2000;10:318–325.
65. Volkow ND, Chang L, Wang GJ, et al. Low level of brain dopamine D2 receptors in methamphetamine abusers: association with metabolism in the orbitofrontal cortex. *Am J Psychiatry.* 2001;158:2015–2021.
66. Goldstein RZ, Volkow ND. (2002) Drug addiction and its underlying neurobiological basis: neuroimaging evidence for the involvement of the frontal cortex. *Am J Psychiatry.* 2002;159:1642–1652.
67. Phan KL, Wager T, Taylor SF, Liberzon I. Functional neuroanatomy of emotion: a meta-analysis of emotion activation studies in PET and fMRI. *Neuroimage.* 2002;16:331–348.
68. Heinz A, Siessmeier T, Wrase J, et al. Correlation between dopamine D(2) receptors in the ventral striatum and central processing of alcohol cues and craving. *Am J Psychiatry.* 2004;161:1783–1789.
69. Rolls ET. The orbitofrontal cortex and reward. *Cereb Cortex.* 2000;10:284–294.
70. Saxena S, Brody AL, Ho ML, et al. Differential cerebral metabolic changes with paroxetine treatment of obsessive-compulsive disorder vs major depression. *Arch Gen Psychiatry.* 2002;59:250–261.
71. Volkow ND, Wang GJ, Ma Y, et al. Activation of orbital and medial prefrontal cortex by methylphenidate in cocaine-addicted subjects but not in controls: relevance to addiction. *J Neurosci.* 2005;25:3932–3939.
72. Kellendonk C, Simpson EH, Polan HJ, et al. Transient and selective overexpression of dopamine D2 receptors in the striatum causes persistent abnormalities in prefrontal cortex functioning. *Neuron.* 2006;49:603–615.
73. Kauer JA, Malenka RC. Synaptic plasticity and addiction. *Nat Rev Neurosci.* 2007;8:844–858.
74. Vanderschuren LJ, Everitt BJ. Behavioral and neural mechanisms of compulsive drug seeking. *Eur J Pharmacol.* 2005;526:77–88.
75. Vezina P. Sensitization of midbrain dopamine neuron reactivity and the self-administration of psychomotor stimulant drugs. *Neurosci Biobehav Rev.* 2004;27:827–839.
76. Hyman SE, Malenka RC. Addiction and the brain: the neurobiology of compulsion and its persistence. *Nat Rev Neurosci.* 2001;2:695–703.
77. Belin D, Jonkman S, Dickinson A, Robbins TW, Everitt BJ. Parallel and interactive learning processes within the basal ganglia: relevance for the understanding of addiction. *Behav Brain Res.* 2009;99(1):89–102.
78. Gonzalez-Burgos I, Feria-Velasco A. Serotonin/dopamine interaction in memory formation. *Prog Brain Res.* 2008;172C:603–623.
79. Volkow ND, Fowler JS, Wang GJ, Goldstein RZ. Role of dopamine, the frontal cortex and memory circuits in drug addiction: insight from imaging studies. *Neurobiol Learn Mem.* 2002;78:610–624.
80. Wolfling K, Flor H, Grusser SM. Psychophysiological responses to drug-associated stimuli in chronic heavy cannabis use. *Eur J Neurosci.* 2008;27: 976–983.
81. Fiorillo CD, Newsome WT, Schultz W. The temporal precision of reward prediction in dopamine neurons. *Nat Neurosci.* 2008;11:966–973.
82. Bayer HM, Glimcher PW. Midbrain dopamine neurons encode a quantitative reward prediction error signal. *Neuron.* 2005;47:129–141.
83. Volkow ND, Wang GJ, Telang F, et al. (2006) Cocaine cues and dopamine in dorsal striatum: mechanism of craving in cocaine addiction. *J Neurosci.* 2006;26:6583–6588.
84. Wong DF, Kuwabara H, Schretlen DJ, et al. Increased occupancy of dopamine receptors in human striatum during cue-elicited cocaine craving. *Neuropsychopharmacology.* 2006;31:2716–2727.
85. Volkow ND, Wang GJ, Telang F, et al. Dopamine increases in striatum do not elicit craving in cocaine abusers unless they are coupled with cocaine cues. *Neuroimage.* 2008;39:1266–1273.

86. Kalivas PW, Volkow ND. The neural basis of addiction: a pathology of motivation and choice. *Am J Psychiatry.* 2005;162:1403–1413.
87. Uhl GR. Molecular genetics of addiction vulnerability. *Neuroreport.* 2006;3:295–301.
88. Han DH, Yoon SJ, Sung YH, et al. A preliminary study: novelty seeking, frontal executive function, and dopamine receptor (D2) TaqI A gene polymorphism in patients with methamphetamine dependence. *Compr Psychiatry.* 2008;49:387–392.
89. Keltikangas-Jarvinen L, Pulkki-Raback L, Elovainio M, et al. DRD2 C32806T modifies the effect of child-rearing environment on adulthood novelty seeking. *Am J Med Genet B Neuropsychiatr Genet.* 2009;150B(3):389–394.
90. Laucht M, Becker K, Frank J, et al. Genetic variation in dopamine pathways differentially associated with smoking progression in adolescence. *J Am Acad Child Adolesc Psychiatry.* 2008;47:673–681.
91. Volkow ND, Wang GJ, Fowler JS, et al. Prediction of reinforcing responses to psychostimulants in humans by brain dopamine D2 receptor levels. *Am J Psychiatry.* 1999;156:1440–1443.
92. Volkow ND, Wang GJ, Fowler JS, et al. Brain DA D2 receptors predict reinforcing effects of stimulants in humans: replication study. *Synapse.* 2002;46:79–82.
93. Thanos PK, Volkow ND, Freimuth P, et al. Overexpression of dopamine D2 receptors reduces alcohol self-administration. *J Neurochem.* 2001;78:1094–1103.
94. Thanos PK, Michaelides M, Umegaki H, Volkow ND. D2R DNA transfer into the nucleus accumbens attenuates cocaine self-administration in rats. *Synapse.* 2008;62:481–486.
95. Morgan D, Grant KA, Gage HD, et al. Social dominance in monkeys: dopamine D2 receptors and cocaine self-administration. *Nat Neurosci.* 2002;5:169–174.
96. Volkow ND, Wang GJ, Newcorn J, et al. Depressed dopamine activity in caudate and preliminary evidence of limbic involvement in adults with attention-deficit/hyperactivity disorder. *Arch Gen Psychiatry.* 2007;64:932–940.
97. Volkow ND, Wang GJ, Newcorn J, et al. Brain dopamine transporter levels in treatment and drug naive adults with ADHD. *Neuroimage.* 2007;34:1182–1190.
98. Elkins IJ, McGue M, Iacono WG. Prospective effects of attention-deficit/hyperactivity disorder, conduct disorder, and sex on adolescent substance use and abuse. *Arch Gen Psychiatry.* 2007;64:1145–1152.
99. Volkow ND, Wang GJ, Begleiter H, et al. High levels of dopamine D2 receptors in unaffected members of alcoholic families: possible protective factors. *Arch Gen Psychiatry.* 2006;63:999–1008.
100. Dalley JW, Fryer TD, Brichard L, et al. Nucleus accumbens D2/3 receptors predict trait impulsivity and cocaine reinforcement. *Science.* 2007;315:1267–1270.
101. Belin D, Mar AC, Dalley JW, Robbins TW, Everitt BJ. High impulsivity predicts the switch to compulsive cocaine-taking. *Science.* 2008;320:1352–1355.
102. Becker JB, Hu M. Sex differences in drug abuse. *Front Neuroendocrinol.* 2008;29:36–47.
103. Dahan A, Kest B, Waxman AR, Sarton E. Sex-specific responses to opiates: animal and human studies. *Anesth Analg.* 2008;107:83–95.
104. Quinones-Jenab V. Why are women from Venus and men from Mars when they abuse cocaine? *Brain Res.* 2006;1126:200–203.
105. Kaminsky Z, Wang SC, Petronis A. Complex disease, gender and epigenetics. *Ann Med.* 2006;38:530–544.
106. Koch K, Pauly K, Kellermann T, et al. Gender differences in the cognitive control of emotion: an fMRI study. *Neuropsychologia.* 2007;45:2744–2754.
107. Munro CA, McCaul ME, Wong DF, et al. Sex differences in striatal dopamine release in healthy adults. *Biol Psychiatry.* 2006;59:966–974.
108. Riccardi P, Zald D, Li R, et al. Sex differences in amphetamine-induced displacement of [(18)F]fallypride in striatal and extrastriatal regions: a PET study. *Am J Psychiatry.* 2006;163:1639–1641.
109. Koob GF, Le Moal M. Addiction and the brain antireward system. *Annu Rev Psychol.* 2008;59:29–53.
110. Sinha R, Li CS. Imaging stress- and cue-induced drug and alcohol craving: association with relapse and clinical implications. *Drug Alcohol Rev.* 2007;26:25–31.
111. Volkow ND, Wang GJ, Telang F, et al. Sleep deprivation decreases binding of [11C]raclopride to dopamine D2/D3 receptors in the human brain. *J Neurosci.* 2008;28:8454–8461.
112. Drummond SP, Gillin JC, Smith TL, DeModena A. The sleep of abstinent pure primary alcoholic patients: natural course and relationship to relapse. *Alcohol Clin Exp Res.* 1998;22:1796–1802.
113. Colrain IM, Trinder J, Swan GE. The impact of smoking cessation on objective and subjective markers of sleep: review, synthesis, and recommendations. *Nicotine Tob Res.* 2004;6:913–925.
114. Koob GF, Le Moal M. Drug abuse: hedonic homeostatic dysregulation. *Science.* 1997;278:52–58.
115. Gray MA, Critchley HD. Interoceptive basis to craving. *Neuron.* 2007;54:183–186.
116. Volkow ND, Fowler JS, Wang GJ, et al. Decreased dopamine D2 receptor availability is associated with reduced frontal metabolism in cocaine abusers. *Synapse.* 1993;14:169–177.
117. Volkow ND, Wang GJ, Fowler JS, et al. Cocaine uptake is decreased in the brain of detoxified cocaine abusers. *Neuropsychopharmacology.* 1996;14:159–168.
118. Hietala J, West C, Syvalahti E, et al. Striatal D2 dopamine receptor binding characteristics in vivo in patients with alcohol dependence. *Psychopharmacology (Berl).* 1994;116:285–290.
119. Volkow ND, Wang GJ, Maynard L, et al. Effects of alcohol detoxification on dopamine D2 receptors in alcoholics: a preliminary study. *Psychiatry Res.* 2002;116:163–172.
120. McCann UD, Kuwabara H, Kumar A, et al. Persistent cognitive and dopamine transporter deficits in abstinent methamphetamine users. *Synapse.* 2008;62:91–100.
121. Wang GJ, Volkow ND, Fowler JS, et al. Dopamine D2 receptor availability in opiate-dependent subjects before and after naloxone-precipitated withdrawal. *Neuropsychopharmacology.* 1997;16:174–182.
122. Fehr C, Yakushev I, Hohmann N, et al. Association of low striatal dopamine D2 receptor availability with nicotine dependence similar to that seen with other drugs of abuse *Am J Psychiatry.* 2008;165(4):507–514.
123. Sevy S, Smith GS, Ma Y, et al. Cerebral glucose metabolism and D2/D3 receptor availability in young adults with cannabis dependence measured with positron emission tomography. *Psychopharmacology (Berl).* 2008; 197:549–556.
124. Malison RT, Best SE, van Dyck CH, et al. Elevated striatal dopamine transporters during acute cocaine abstinence as measured by [123I] beta-CIT SPECT. *Am J Psychiatry.* 1998;155:832–834.
125. Laine TP, Ahonen A, Torniainen P, et al. Dopamine transporters increase in human brain after alcohol withdrawal. *Mol Psychiatry.* 1999;4:189–191, 104–105.

126. Volkow ND, Wang GJ, Fowler JS, et al. Decreases in dopamine receptors but not in dopamine transporters in alcoholics. *Alcohol Clin Exp Res.* 1996;20:1594–1598.
127. Chang L, Alicata D, Ernst T, Volkow N. Structural and metabolic brain changes in the striatum associated with methamphetamine abuse. *Addiction.* 2007;102(suppl 1):16–32.
128. Yang YK, Yao WJ, Yeh TL, et al. Decreased dopamine transporter availability in male smokers–a dual isotope SPECT study. *Prog Neuropsychopharmacol Biol Psychiatry.* 2008;32: 274–279.
129. Fowler JS, Logan J, Wang GJ, Volkow ND. Monoamine oxidase and cigarette smoking. *Neurotoxicology.* 2003;24: 75–82.
130. Wu JC, Bell K, Najafi A, et al. Decreasing striatal 6-FDOPA uptake with increasing duration of cocaine withdrawal. *Neuropsychopharmacology.* 1997;17:402–409.
131. Heinz A, Siessmeier T, Wrase J, et al. Correlation of alcohol craving with striatal dopamine synthesis capacity and D2/3 receptor availability: a combined [18F]DOPA and [18F]DMFP PET study in detoxified alcoholic patients. *Am J Psychiatry.* 2005;162:1515–1520.
132. Schlaepfer TE, Pearlson GD, Wong DF, Marenco S, Dannals RF. PET study of competition between intravenous cocaine and [11C]raclopride at dopamine receptors in human subjects. *Am J Psychiatry.* 1997;154:1209–1213.

9 | Parkinson's disease

9.1 Exploring the Myths about Parkinson's Disease

YVES AGID AND ANDREAS HARTMANN

MYTH 1: 'PARKINSON'S DISEASE': IN FACT, THERE ISN'T JUST ONE BUT SEVERAL PARKINSON'S DISEASES

After having eliminated secondary causes (Fig. 9.1.1), parkinsonian syndromes can be divided into two subgroups. (1) True Parkinson's disease (PD) displays exclusively or predominantly central dopaminergic lesions and responds to replacement therapy with L-DOPA. These lesions most commonly occur sporadically but hereditary forms are not uncommon, representing about 10%–15% of cases (Table 9.1.1). (2) The *Parkinson plus* syndromes are characterized by severe nondopaminergic lesions in addition to degeneration of the dopaminergic nigrostriatal pathway, with either an incomplete or absent response to L-DOPA. Since there is more than one PD, it therefore seems prudent to refer to a *parkinsonian syndrome* at the first clinical presentation, at which point the enlightened clinician can proceed to identify various subgroups on the basis of the different neural pathologies and enlist the patient in a treatment program appropriate for that individual. The symptomatic treatment of each PD subgroup will thus be optimized on a solid neuropathological basis. It is therefore vital to understand the complexity of parkinsonian syndromes: for the clinician, who must beware of hasty diagnosis and who must recognize the different forms of the disease and their different prognoses; and for the scientist, who must recognize not one but a whole array of histopathological indices with multiple causes, and who must further take into account a large number of interacting genetic and environmental causes as predisposing factors.

MYTH 2: 'PARKINSON'S DISEASE IS A MOVEMENT DISORDER CHARACTERIZED BY THE CLASSIC TRIAD OF AKINESIA, RIGIDITY, AND TREMOR': IN FACT, PARKINSONIAN MOTOR DISTURBANCES ARE MORE COMPLEX THAN WAS PREVIOUSLY THOUGHT

In its early stages, PD often manifests as hand tremor or small handwriting (micrographia). However, the feet are often the first affected extremities, reflected by a low-amplitude tremor (for example, under the stressful conditions of a consultation) or by a slight asymmetry of gait. This is also the reason why the feet show the first L-DOPA-induced dyskinesias.[1] These observations are not surprising since the initial dopaminergic terminal loss is seen in the dorsal striatum—a region that corresponds somatotopically to the cortical projection of the feet (Fig. 9.1.2).

The classic triad (akinesia, rigidity, and tremor) comprises great symptomological complexity because each of these three broad categories is underpinned by different mechanisms, each with its own neural pathologies. For this reason, it is essential to carry out a careful examination and, above all, to listen to the patient's own description of the symptoms. "My hand tremble at rest...my handwriting has shrunk...people tell me my arms swing less than before...I have difficulty getting coins out of my pocket," and so on. The most efficient means of confirming the existence of one or more of these symptoms is to ask the patient to walk and to write. The two major tools of the PD specialist are therefore the pen and the carpet!

Akinesia, often interpreted as "slowness," comprises several elements. (1) In true akinesia, or delayed movement initiation (measured by reaction time), the patient appears frozen. He remains immobile while describing his condition, and essential gestures are made economically. In some cases—for example, following a severe "off" episode—the patient is completely frozen, statue-like, often incapable of initiating any voluntary movement. (2) Bradykinesia, or slowing of movement, is easily observed during fine movements of the extremities or covering the whole body at more advanced disease stages. Speech is slow and monotone. Mobility is reduced, as the feet tend to drag on the ground. During speech, the face appears expressionless and immobile. (3) Hypokinesia occurs when movements are never terminated correctly, for example during alternating wrist movements (diadochokinesia). (4) There is difficulty in performing sequential or concurrent movements, as seen in the classical "beer drinker's" movement, which

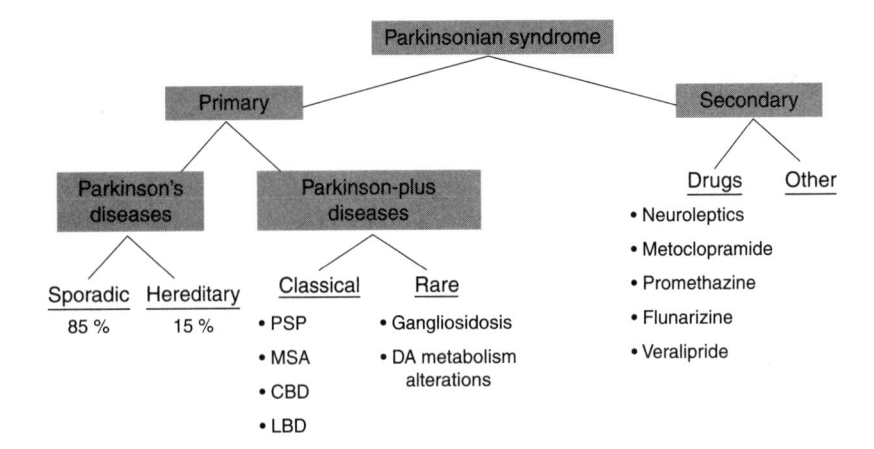

FIGURE 9.1.1. Differential diagnosis of Parkinsons's disease. CBD, corticobasal degeneration; LBD, lewy body disease; MSA, multiple system atrophy; PSP, progressive supranuclear palsy.

TABLE 9.1.1. *Hereditary Forms of Parkinson's Disease*

Locus/Gene	Inheritance	Onset	Pathology	Map Position	Gene
PARK1	Dominant	40s	Nigral degeneration with Lewy bodies	4q21	Alpha-synuclein
PARK2	Recessive	20s–40s	Nigral degeneration without Lewy bodies	6q25	Parkin
PARK3	Dominant	60s	Nigral degeneration with Lewy bodies, plaques, tangles	2p13	?
PARK4	Dominant	30s	Nigral degeneration with Lewy bodies	4q21	Alpha-synuclein duplications and triplications
PARK5	Dominant	50s	No pathology reported	4p14	Ubiquitin C-terminal hydrolase L1
PARK6	Recessive	30s–40s	No pathology reported	1p35-37	PINK1
PARK7	Recessive	30s–40s	No pathology reported	1p38	DJ-1
PARK8	Dominant	60s	Variable alpha-synuclein and tau pathology	12 cen	LRRK2
PARK9	Recessive	20s–40s	No pathology reported	1p36	ATP13A2
PARK10	Dominant (?)	50s–60s	No pathology reported	1p32	?
PARK11	Dominant (?)	Late	No pathology reported	2q34	?
PARK12	X-linked	Late	No pathology reported	Zq21	?
PARK13	Dominant (?)	Late	No pathology reported	2p12	HRTA2

combines grasping with the hand and flexion of the elbow to bring the glass to the mouth. (5) Finally, just as a lack of motivation and/or mood is accompanied by slowness in normal subjects, so are apathy and depression, common in PD patients, which contribute to bradykinesia. Consequently, these different aspects of akinesia are not simply and primarily of sensorimotor origin, but also comprise a cognitive and psychological component.

Rigidity is not a symptom but a sign. It is detectable in distal joints (for instance, the wrist). Parkinsonian rigidity is plastic, often giving way in a series of small jerks (*cogwheel rigidity*). If subtle, rigidity can be increased following the Froment maneuver (active mobilization of the contralateral limb). Parkinsonian (*lead pipe*) rigidity must be distinguished from pyramidal (*clasp knife*) rigidity and oppositional rigidity (*Gegenhalten*), the latter being provoked or increased by movements and due to diffuse brain lesions.

Resting tremor (4–6 Hz) is not easy to demonstrate because it often cannot be observed at rest, that is, during complete relaxation. It is best seen when the patient is walking with the arms held slightly rigid and in flexion. Usually intermittent and increased by stress, tremor may be absent in

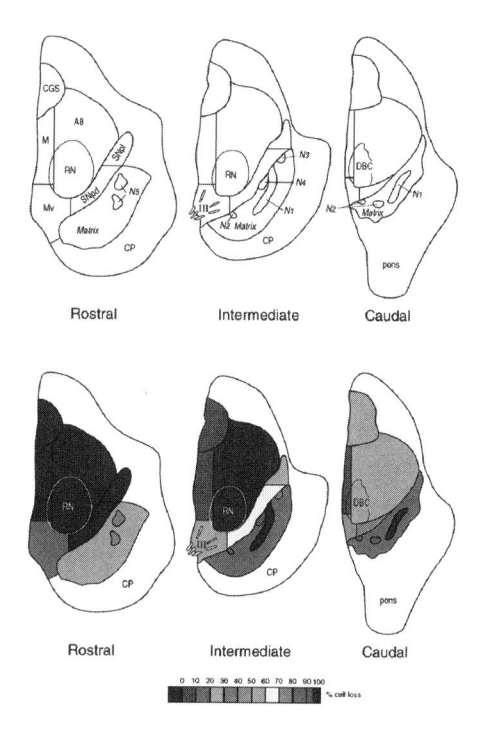

FIGURE 9.1.2. Regional and intranigral loss of dopamine-containing neurons in PD. A8, dopaminergic cell group A8; CGS, central gray substance; CP, cerebral peduncle; DBC, decussation of brachium conjunctivum; M, medial group; Mv, medioventral group; N, nigrosome; RN, red nucleus; SNpd, substantia nigra pars dorsalis; SNpl, substantia nigra pars lateralis; III, exiting fibers of the third cranial nerve. The colorimetric scale indicates the estimated amount of cell loss (least = blue; most = red). Across the mesencephalon, dopaminergic cell loss was weak in the central gray substance and the red nucleus and intermediate in the dopaminergic cell group A8 and the medioventral group of the ventral tegmental area. Within the substantia nigra pars compacta, dopamine-containing neurons in the calbindin-rich regions (*matrix*) and in five calbindin-poor pockets (*nigrosomes*) were identified. The spatiotemporal progression of neuronal loss in the substantia nigra pars compacta is as follows: depletion begins in the main pocket (nigrosome 1) and then spreads to other nigrosomes and the matrix along rostral, medial, and dorsal axes of progression. This pattern is reflected clinically by corresponding somatotopic progression of symptoms. *Source*: Adapted from [30]. (See Color Plate 9.1.2.)

certain forms of the illness or, by contrast, may be predominant in some patients. Treatment is considered difficult but, in true PD, tremor is usually improved by dopaminergic treatment if given in sufficiently high doses. In the absence of a response to adequate doses and prolonged L-DOPA treatment, the clinician should turn to other possible differential diagnoses such as essential tremor or, more rarely, rhythmic myoclonus, as seen in multisystem atrophy.

These three major symptoms respond to L-DOPA therapy because they result mainly from degeneration of the nigrostriatal dopamine pathway. However, there are differences in these signs and symptoms, which vary from one patient to another, implying either a widespread heterogeneous dopaminergic denervation varying among the different target brain structures or a nondopaminergic dysfunction. This is notably the case for the different component features of akinesia, each due to dysfunction within the different cortico-striato-pallido-thalamo-frontal pathways. The sensorimotor loops in particular are modified, but also the associative and limbic loops in certain patients (Fig. 9.1.3).[2,3] The plastic rigidity results primarily from a dopaminergic dysregulation of the cortico-subcortical sensorimotor loop. However, dysfunction of the descending striatal output to the spinal cord is also implicated to varying degrees, as shown by experimental and clinical studies, although the full details remain to be elucidated.[4] As for parkinsonian tremor, there is clearly a striatal component resulting from the dopaminergic denervation but cerebellar modifications are also present, perhaps of a secondary nature, as indicated by the presence of tremor frequencies characteristic of the basal ganglia (striato-pallido-thalamic pathway) but also of the cerebellum and thalamus via the olivo-dento-rubro-thalamic pathway.[5]

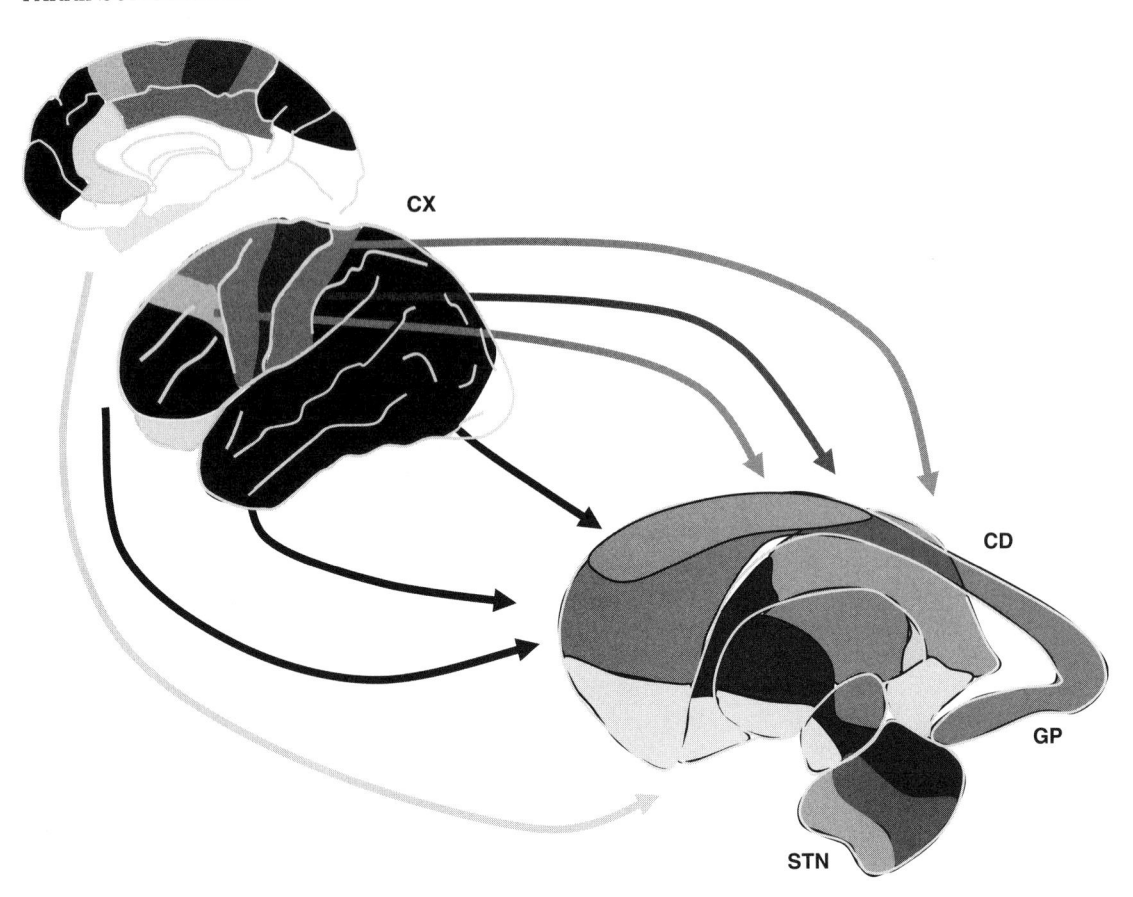

FIGURE 9.1.3. Schematic illustration of the convergence of projections from the cerebral cortex (CX), caudate nucleus (CD), and globus pallidus (GP) on the subthalamic nucleus (STN). These projections can be divided into three functional territories—sensorimotor (green), associative (mauve), and limbic (yellow)—and converge in the STN. *Source*: Adapted from [2]. (See Color Plate 9.1.3.)

MYTH 3: PARKINSON'S DISEASE IS A "MOVEMENT DISORDER": IN FACT, PD IS NOT JUST A MOTOR DISORDER, IT IS MULTISYMPTOMATIC AND HENCE A MULTISYSTEM DISORDER

Cognitive disorders, which do not necessarily imply dementia, are of two types. First, one must consider the dysexecutive syndrome,[6] which is hardly perceptible in early disease stages and worsens slowly over time. Its origin is subcortical and acts in two ways: either directly by dysregulation of the prefrontal cortex as a result of the progressive destruction of subcortical projection neurons (dopaminergic, noradrenergic, serotonergic, and cholinergic, arising in the ventral tegmental area, locus coeruleus, raphé, and the basal nucleus of Meynert [Fig. 9.1.4]); or indirectly by dysregulation of the striatum, a major relay for associative cortical input, principally prefrontal. Second, dementia, observed in advanced disease stages, involves additional extensive frontal and temporal neuronal loss. The nature of the histopathological signs associated with the cortical cell loss is not of major clinical relevance for the patient. We may encounter Alzheimer-type neurofibrillary tangles or distinct Lewy body deposits, these two histological signs being frequently found in association. However, we must be able to recognize true Lewy body disease, which comprises a predominance of Lewy bodies in the cerebral cortex, with hallucinatory dementia in its early stages and a rapid progression.[7]

Psychiatric problems in PD, largely ignored until recently, are of three types: depression, hypomania, and visual hallucinations. Depression, almost always associated with anxiety, classically precedes the first motor symptoms in 50% of cases.[8] However, we must distinguish true clinical depression with mood change from apathy-type depression.[9] These two components do not necessarily share the same pathological substrate and thus should not be treated the same way. In

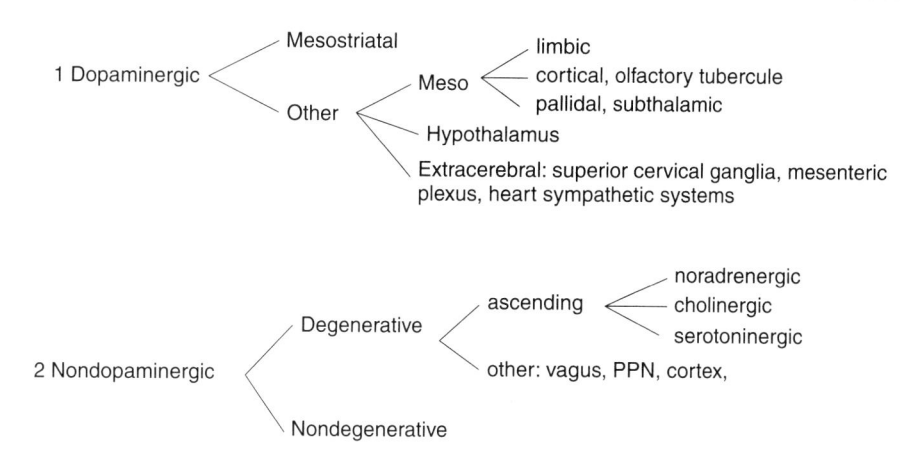

FIGURE 9.1.4. Schematic overview of the dopaminergic and nondopaminergic systems affected in pedunculopontine nucleus (PPN).

practice, depression in PD—which can potentially destroy the patient's' life—can be very effectively treated. An appropriate course of psychological treatment enables the debilitating mood changes to be controlled, but it should be combined with symptomatic replacement therapy for the organic basis of the disease. Depression in PD is partly caused by dopamine deficiency (hence the need for a sufficiently high dosage of dopamine replacement) but also by degeneration of noradrenergic and serotonergic neurons,[10] which stresses the need to prescribe the appropriate antidepressants: noradrenalin and serotonin reuptake inhibitors.

Hypomania is essentially characterized by disinhibition and hypersexuality. Usually, this is provoked in severely affected patients by high drug doses, which pushes the patient into an escalation of medication with all the characteristics of true addiction. This *dopamine dysregulation syndrome* is often associated with impulse control disorders[11] such as gambling (irrespective of financial means) and the appearance of repetitive, pointless behaviors or *punding*[12] (Table 9.1.2). Whereas the dopamine dysregulation syndrome and punding are more frequently observed under L-DOPA treatment, impulse control disorders are more frequently triggered by dopamine agonist intake. These observations raise the intriguing possibility of specific dopaminergic stimulation of the limbic circuits (and certain serotonergic neuronal systems) and the existence of an addiction-predisposing substrate.

Visual hallucinations are rarely terrifying and therefore not always reported spontaneously by the patient. Frequently, there is the impression that an animal is running across the floor or that someone is standing

TABLE 9.1.2. *Mental Symptoms Induced by Dopaminergic Replacement Therapy in PD Patients*

- Compulsive use of levodopa (addiction)
- Behavioral phenomena
  - Punding
  - Euphoria/hypomania
  - Food cravings
  - Hypersexuality
  - Pathological gambling and shopping
  - Heighted aggression
  - Psychosis

behind the patient. In advanced disease stages, hallucinations may become severe and associated with delusions. They are more easily triggered by dopamine agonists than by L-DOPA and may become associated with the full array of hypomanic symptoms, which may sometimes suggest the onset of cortical dementia. Moreover, patients may display symptoms of daydreaming, with similarities to the classic brainstem hallucinations, probably due to selective local lesions of the locus subcoeruleus.[13] This is a kind of genuine narcolepsy in a dreaming patient but one who can nevertheless move, reminding us of the classic cat experiments of Michel Jouvet.[14] In these experiments, cats could produce full physical movements while being clearly in a dream-like state, with the electroencephalogram displaying the characteristics of rapid eye movement (REM) sleep. These narcoleptic hallucinations in parkinsonism are not rare: the patient is not mentally ill and should not be treated with psychotropic drugs; rather, one should try to return him to a waking state. Whether one should distinguish between cortical (the principal type) and subcortical (waking dreaming) hallucinations remains to be seen, since the two are frequently associated.

Consequently, PD is not simply a motor disorder, but a genuine neuropsychiatric illness.[15] Its treatment is always difficult. Depression is difficult to treat because it requires strict adherence to a course of psychiatric treatment. Hypomania is also difficult to treat because it is triggered or increased by dopaminergic treatment; therefore, treatment reduction is tempting but may result in reemergence of motor symptoms. Finally, behavioral disorders are difficult to treat because the classical neuroleptics, which block dopamine transmission, are contraindicated. In these cases, atypical neuroleptics such as clozapine are particularly attractive, although these drugs require very close medical supervision.

Insomnia among PD patients is well known, with broken and reduced sleep patterns related for the most part to reduced nocturnal mobility. However, less well known but common (found to occur in approximately half of PD patients after careful history taking) are behavioral sleep disorders characterized by agitation and sometimes nocturnal violence. These are called *REM sleep behavior disorders*.[16] Daytime somnolence is another common problem and is apparently independent of nocturnal insomnia. This daytime somnolence is worsened by dopaminergic medication (dopamine agonists more so than L-DOPA) but can also be observed in one-third of drug-naive patients.[17] The fatigue is often so pronounced that patients have "sleep attacks" or return to bed after taking the first morning medication. Often ignored, these sleep disorders are frequent and can have serious consequences for the patient, causing car accidents and family conflicts due to insomnia and nocturnal agitation.[18]

Dysautonomia is manifest as orthostatic hypotension (rarely symptomatic and requiring some care in the administration of dopaminergic medication), constipation, and, most importantly, sexual and urinary problems.[19] The occurrence of impotence or frigidity is clearly underreported; when reported, it is too easily attributed to aging. However, it is a critical subject because relationships, as pillars of the quality of life, often depend on it. Furthermore, these difficulties may bias the clinician toward other diagnoses, such as multisystem atrophy. In these cases, a diagnosis of PD can be confirmed by cystomanometry.[20]

Hyposmia—probably due to dopaminergic denervation of the olfactory lobe—is present in almost all PD patients even at very early disease stages. These disorders are only mildly disabling, but they can serve as predictive signs for future neuroprotective therapies.[21]

Sympathetic cardiac denervation (as measured by heart muscle radioactivity levels after administration of a sympathetic radioisotope) may be more frequent than was previously thought; however, its clinical consequences remain to be elucidated.[22]

These symptoms, and many others, can be thought of as being the hidden part of the iceberg described in Langston's *Parkinson's complex*.[23] To simply treat the motor symptoms satisfies the doctor at first but is not sufficient to ensure the patient's comfort ("The doctor is happy but not the patient.") For the clinician, this leads to the therapeutic challenge of finding a cocktail of nondopaminergic drugs for the amelioration of the cognitive, psychiatric, and other symptoms while avoiding dangerous and expensive overmedication. Finally, we must keep in mind that at the diagnostic level, brain dysfunction is best reflected by the parkinsonian symptoms. At the physiological level, this multiplicity of symptoms implies a dysfunction that goes well beyond the degeneration of the nigrostriatal dopamine pathway (Table 9.1.3).

TABLE 9.1.3. *Parkinsonian Symptoms Beyond the Motor Triad*

Neuropsychiatric symptoms	Gastrointestinal symptoms
Depression, anxiety	Drooling
Apathy	Dysphagia
Anhedonia	Nausea ((often drug-induced)
Attention deficit	Constipation
Hallucinations, delusions, illusions (often drug-induced)	
Dementia	
Obsessive and repetitive behaviors (often drug-induced	
Confusion (often drug-induced)	
Sleep disorders	*Sensory symptoms*
Restless legs and periodic limb movements	Pain
REM sleep behavior disorder	Paresthesia
Excessive daytime sleepiness	Olfactory disturbance (hyposmia, anosmia)
Sleep attacks	
Insomnia	
Autonomic symptoms	*Other symptoms*
Bladder disturbances	Fatigue
Sweating	Diplopia
Orthostatic hypotension	Blurred vision
Erectile dysfunction	Seborrhea
Hypersexuality (often drug-induced)	

MYTH 4: FIRST SYMPTOMS APPEAR IN THE PATIENT'S 60S: IN FACT, PD IS NOT JUST AN ILLNESS OF OLD AGE

Parkinson's disease begins on average in the patient's 60s, but 10% of PD cases start before the age of 45 and precocious forms are not exceptional, beginning even before the age of 20. The latter cases are often hereditary forms of the disease, either recessive autosomal or dominant (Table 9.1.1). The consequences are far from negligible in physiological terms since if hereditary factors are predominant, the illness is causally unrelated to age. Nevertheless, neural aging, especially of dopamine neurons, necessarily renders them more fragile and therefore vulnerable to unknown pathological processes. However, if aging does not play a major role in causing the disorder, such is not the case for some motor and cognitive symptoms, which are quite different, depending on whether they appear in younger subjects (who respond well to replacement therapy because they are L-DOPA-responsive) or older subjects (who often show only partial improvement with dopamine substitution therapy because of additional nondopaminergic lesions—see above).

MYTH 5: 'CLINICAL DIAGNOSIS IS SIMPLE': IN FACT, DIAGNOSIS IS RATHER DIFFICULT AND MISTAKES ARE FREQUENTLY MADE

It is often difficult to separate motor parkinsonian symptoms from those of other neurodegenerative diseases such as Alzheimer's disease. It is even more difficult to be certain of a diagnosis of primary degenerative PD, and yet more so to distinguish the different forms of the disease within the PD complex. Even for the most competent experts, there will be at least a 10% error rate.[24] This is particularly understandable when other diseases such as Parkinson plus syndromes or Alzheimer's disease are present. Misdiagnosis is rarer when dealing with Creutzfeld-Jacob disease, amyotrophic lateral sclerosis, or alcoholic encephalopathy.

In practice, we must distinguish two situations in which the patient is seen either early or late in the disease course. In the early stages, there is usually little or no discussion concerning diagnosis of a dystonic tremor of the extremities or skeletal myoclonus due to brainstem lesions. The real problems arise with the possible diagnosis of asymmetrical essential tremor or, in the case of parkinsonian tremor, when it is dominant during the upper extremities posture. Slow development and the presence of a family history are characteristics of essential tremor but are not the definitive disease markers. In these cases, it is helpful to carry out tremor recordings with rigorous frequency analysis. Such studies demonstrate that although an association between essential tremor and PD possibly exists, it is extremely rare.[25] If there is doubt, a dopamine transporter (DAT) scan, using a radioactive dopamine uptake ligand, can be used to detect an asymmetrical striatal dopaminergic denervation.

Later in the disease course, the problem of the Parkinson plus syndromes appears. Foremost is multisystem atrophy, which may present as true PD for a long period of time, showing a good initial response to L-DOPA therapy. The dysautonomia characteristic of multisystem atrophy may sometimes be delayed; it may also be expressed at a moderate level in true PD patients. Progressive supranuclear palsy should not cause too much confusion except in the few cases where supranuclear ophthalmoplegia cannot be clearly assessed. Corticobasal degeneration is unlikely to be a cause of misdiagnosis for long because of the clear lack of an L-DOPA response in these patients. Furthermore, dyspraxia in corticobasal degeneration can be easily identified when the patient is asked to manipulate a pen or carry out other fine finger movements. The patient typically describes his disability as his hand "not obeying his wishes." As for Lewy body disease, with its rapidly appearing dementia and often spectacular visual hallucinations, the diagnosis is often made too rapidly, without clear postmortem confirmation.

A much more serious medical situation arises when the diagnosis of PD is either not made or is delayed to the point where the patient has suffered a disability that could have been substantially improved earlier. This is mainly the case at the two extremes of the age spectrum. Parkinson's disease runs the risk of being identified as a psychogenic disorder in the young adult or adolescent, a plight that is all the more infuriating because the psychotherapy that ensues will obviously have no effect, whereas dopaminergic treatment is spectacularly effective in these readily L-DOPA-responsive young patients. In the very old patient, the classic symptoms of gradual motor slowing and walking difficulties are all too easily attributed to normal aging. The diagnosis is often difficult, but nothing is lost by testing with small doses of L-DOPA.

MYTH 6: 'THERE IS LITTLE OR NO NEED FOR NEUROIMAGING': IN FACT, IT CAN HELP

In Western countries at least, it is rare for patients not to have received a computed tomography (CT) or magnetic resonance imaging (MRI) brain scan at least once in their lives. Brain imaging is often carried out immediately when PD is diagnosed, often for psychological reasons, because the patient finds it difficult to believe that the diagnosis of such a serious illness can possibly be made without "a photo of the inside" of the brain. Later, during disease progression, brain imaging is often prescribed by the specialist when he is not sure what to do and has the impression that the patient is beginning to be disappointed. Yet, even if the diagnosis of PD is based on a good response to dopamine replacement therapy, there are situations where brain imaging is justified, whether it is T1- or T2-weighted MRI or a DAT scan. Brain MRI may be of interest early in the disease, when the symptoms are still incomplete (e.g., isolated unilateral tremor for several years), or are atypical (e.g., predominantly postural tremor), or are accompanied by other symptoms (e.g., apathy unexplained by a simple depressive state). Most importantly, one must eliminate curable conditions such as frontal tumors (with classic postural tremor) or hydrocephalus, whether communicating or not.

Brain imaging remains useful in later disease stages, when the response to dopaminergic treatment becomes weak or absent. Then the scan has two essential goals: (1) It is used to investigate various brain structures for signs of atrophy, such as that observed in Parkinson plus syndromes, particularly multisystem atrophy and progressive supranuclear palsy.[26] Third ventricle dilation (indicating atrophy of the basal ganglia) and midbrain atrophy (with enlargement of the sylvian aqueduct) suggest progressive supranuclear palsy. Particularly evocative of multisystem atrophy is the presence of a hyperdense putaminal border visualized in T2-weighted images, together with atrophy of the pons (sometimes with the classic cross sign appearing later) and signs of cerebellar atrophy. (2) The second goal is to eliminate nondegenerative lesions, which may obscure the clinical picture, in particular a vascular leukoencephalopathy with hypersignals spread throughout the white matter in T2-weighted images. In some cases, this may explain the poor therapeutic response to L-DOPA, and cardiovascular investigations should then be initiated.

Single photon emission computed tomography (SPECT) imaging, using a radioactive dopamine uptake inhibitor, allows the visualization of dopaminergic terminals in the striatum. The application of this technique is, however, limited to three situations. (1) parkinsonian syndromes arising from prolonged administration of neuroleptic dopamine receptor blockers; this usually occurs when small neuroleptic doses are administered as tranquilizers or antinausea agents early in the disease course (2) certain forms of atypical essential tremor, asymmetrical and recently identified, particularly in older patients; (3) rare psychogenic parkinsonian syndromes (the doctor being careful to distinguish between conversion/hysteria and malingering). In all of these cases, a normal DAT scan allows the doctor to eliminate a parkinsonian syndrome and points to other diagnoses.

MYTH 7: "A CLEAR PROGNOSIS IS ALWAYS DIFFICULT TO MAKE": IN FACT, IT IS VITAL TO GIVE A PROGNOSIS; THIS IS ALWAYS POSSIBLE EVEN IF IT IS SOMETIMES ONLY APPROXIMATE

The patient and his friends and family are naturally concerned about the future but rarely dare ask any questions. The medical position at the announcement of the diagnosis cannot always be uniform and must take into account the patient's pyschological profile. A patient who demands the truth may break down when he realizes the significance of the diagnosis; another is immediately overwhelmed but finally accepts the news; still another does not want to hear a word about a truth that he cannot accept. Whatever the case, as for any chronic illness, the specialist must state a prognosis so that the patient can receive the best possible course of treatment.

How does one make this prognosis? (1) First, the doctor must consider the rate of symptom progression, judged by estimating the relationship between the severity of the motor symptoms and the time since they first appeared. This subjective evaluation of the primary symptoms or the totality of the motor syndrome can also take into account the progression of the signs of the disease from one limb to another (Fig. 9.1.5). (2) The evaluation of parkinsonian rigidity provides a good prognostic measure. We can compare its severity in the extremities to that seen in the upper arms and in the neck, where, if it were to occur, would provide a poor prognosis. We can also distinguish a low level of rigidity with cogwheeling, which gives a good prognosis, from severe lead pipe rigidity, which we know to be extremely disabling if it spreads to the rest of the body. (3) The presence of axial signs (Fig. 9.1.6) is worrying because it may be an indication of further afflictions,[27] ranging from cognitive disorders (Lewy body disease) to postural instability with falls (progressive supranuclear palsy), through bladder and erectile dysfunction (multisystem atrophy). (4) A weak or absent L-DOPA response after prolonged and adequate dosage must be noted. If the motor disability is not improved by adequate dopamine replacement therapy, then the motor problems are probably due to

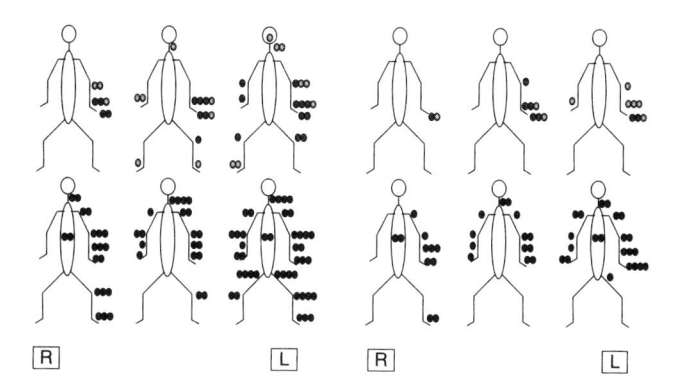

FIGURE 9.1.5. Illustration of the topographical progression of rest tremor (upper row) and rigidity (lower row) in two de novo PD patients, from M0 (left) to M12 (right) (Schüpbach et al., in press). Symptom severity is indicated by the number of dots (medium gray/light gray for tremor; dark gray for rigidity) according to the 6-point rating scale. Medium gray dots stand for rest tremor under relaxed conditions; orange dots indicate the worsening of rest tremor during a mental task. R, right; L, left. (See Color Plate 9.1.5.)

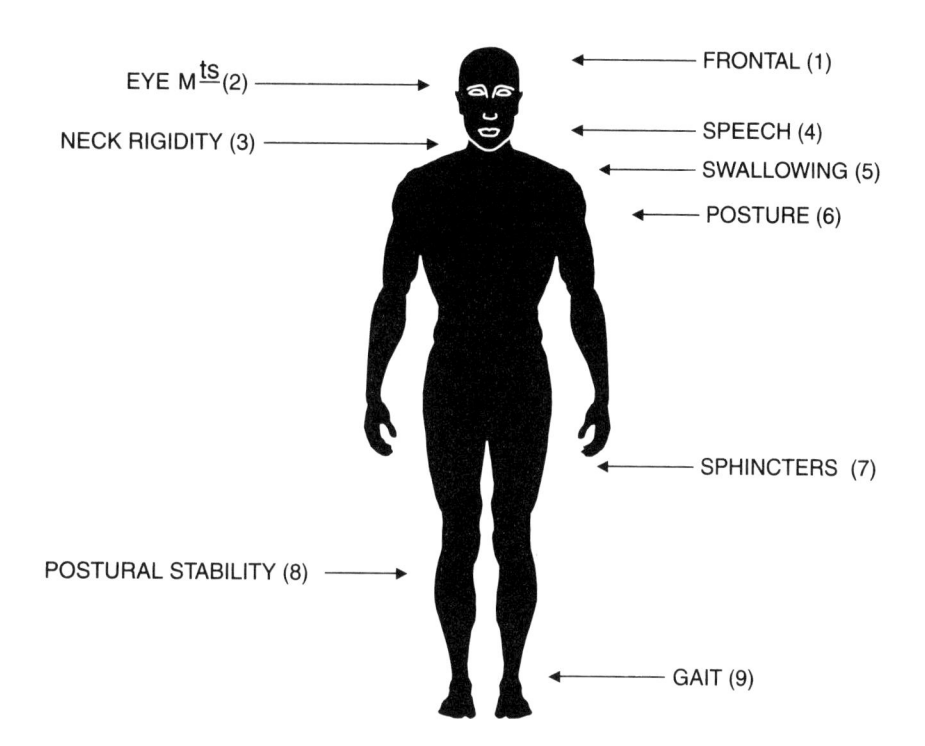

FIGURE 9.1.6. The nine axial signs to be examined in a parkinsonian patient.

nondopaminergic lesions[27] (Fig. 9.1.4). In contrast, a spectacular "honeymoon" response confirms a dopaminergic dysfunction but could also herald the appearance of severe motor complications (fluctuations and dyskinesia). The early appearance of dystonia of the feet and toes, especially if painful, is worrying because it probably indicates that there are even more troublesome dystonias to come. Conversely, the delayed appearance of motor fluctuations during the progression of the illness in a good L-DOPA responder, or the fact that the patient awakes in the morning with no serious motor symptoms (the *sleep beneficial effect*), indicates a good prognosis, suggesting that dopamine levels are, at least partially, replenished during the night. (5) In case of doubt, an MRI brain scan can be useful to reveal unfavorable signs such as associated vascular leukoencephalopathy, as well as signs of atrophy of the midbrain (progressive supranuclear palsy), cerebellum, and brainstem (multiple system atrophy) and of the parietal cortex (corticobasal degeneration) (see above).

MYTH 8: 'TREATMENT IS GENERALLY STRAIGHTFORWARD AND IS CONSIDERABLY AIDED BY THE USE OF MOTOR SCALES': IN FACT, THE GOOD DOCTOR LISTENS AND DOESN'T REQUIRE A SCORING SYSTEM

In a typical outpatient setting, motor scales are a waste of time and are too simplified to accurately evaluate the patient's quality of life. These scales, however, are indispensable for clinical research in order to evaluate the effectiveness of drug treatments and the natural history of the disease. In practice, the evaluation of the state of the motor system is necessary but not sufficient, because to ensure the patient's well-being and comfort, we must take into account his family, professional, and social circumstances. Our understanding of the severity of the motor disability is necessarily imperfect because of the motor fluctuations resulting from the brief time period in which dopaminergic medications, in particular L-DOPA, are effective. Every neurologist knows that a patient with normal mobility can, the next minute, be "statufied." Some practitioners use motor scales such as the UPDRS III, but apart from the fact that they fail to take into account the whole range of motor symptomatology, they are laborious to use in practice and do not help the patient–doctor dialogue—rather the opposite. In order to preserve this vital dialogue with the patient and his family, one needs to discuss the following subjects, not necessarily in this order: "How well did you sleep? How did you feel when you woke up? At what time did you take your first drug dose? Did you feel an improvement? After how much time? When you get better, exactly how long does it last? Do you feel any unpleasant effects during this improvement period, such as drowsiness, difficulties with concentration, or pain?"

Listening to the patient therefore allows the doctor to reconstitute the motor state over a 24-hr period, but it also has the advantage of allowing the patient to talk about himself. It also gives the doctor an opportunity to listen and to discuss the professional and social context of a patient who has a tendency to be rather introverted. Here again, scales appear derisory due to their simplification and the excessive weight given to the motor items. Questionnaires evaluating social maladaptation exist, but they are very difficult to handle in clinical routine situations.

MYTH 9: 'PARKINSON'S DISEASE IS CHARACTERIZED BY THE SELECTIVE LOSS OF DOPAMINERGIC NEURONS OF THE NIGROSTRIATAL PATHWAY. PARKINSONIAN SYMPTOMS APPEAR WHEN THE DOPAMINERGIC CELL LOSS EXCEEDS 50%': IN FACT, COMPLETE NEURONAL LOSS WITHIN THE NIGROSTRIATAL PATHWAY NEVER OCCURS; RATHER, IT IS HETEROGENEOUS, INVOLVING SEVERAL OTHER DOPAMINERGIC AND NONDOPAMINERGIC NEURONAL SYSTEMS

Parkinson's disease is defined anatomo-clinically as an akinetic-rigid syndrome—usually, but not always, associated with resting tremor—due to nigrostriatal dopamine loss.[28] This definition allows the doctor to exclude akinetic-rigid syndromes due to frontal or diffuse basal ganglia lesions with no associated nigrostriatal dopamine loss. If this definition is adopted, and it is probably the least bad, then the restoration of dopamine transmission within the nigrostriatal pathway should result in the disappearance or improvement of the cardinal signs of PD, namely, akinesia, rigidity, and resting tremor.

Experience shows, however, that this is not always the case. Well-conducted dopamine replacement therapy in PD patients leads to three types of clinical response.[27] (1) In approximately 15% of cases, the response is spectacular. When asked "What do you estimate the percentage improvement in your symptoms to be?", the answer is generally 75%–100%. The forms of the disease that fall into this category are generally those seen in young patients showing severe akinesia. The reason for this high success rate of clinical improvement is that the dopaminergic lesions are severe and confined to the nigrostriatal pathway (Fig. 9.1.7). They are of the same type as the 6-hydroxydopamine (6-OHDA) or MPTP-induced experimental lesions in animals. (2) In another 15% of cases, the L-DOPA response is limited (<30%) or nonexistent. This is the response seen in the Parkinson plus syndromes, which additionally include basal ganglia output lesions affecting the striato-pallido-thalamo-cortical circuits (see above)—lesions that can be described as in series (Fig. 9.1.7). In these cases, striatal dopamine transmission reestablishment, even with high doses of L-DOPA, is inefficient because the neuronal messages destined for the cortex are blocked downstream of the striatum. (3) In the remaining 70% of patients, the L-DOPA response is intermediate, usually good at the start but becoming progressively weaker over time. The restoration of striatal dopamine transmission remains efficient, particularly concerning akinesia, rigidity, and rest tremor. However, other symptoms reveal themselves as a result of progressive nondopaminergic neuron loss, which can

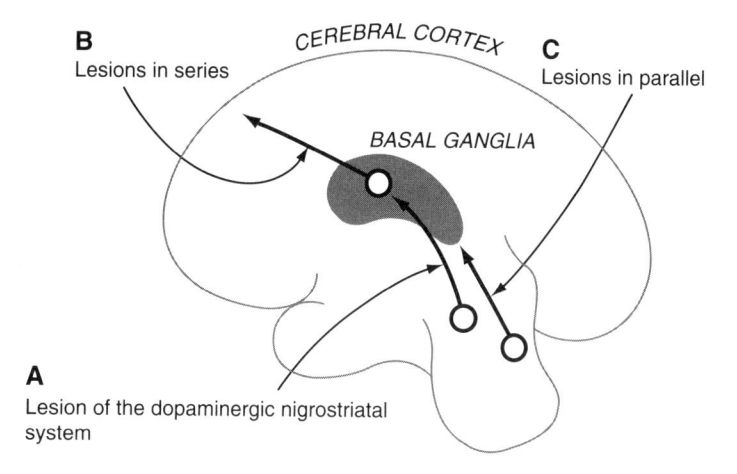

FIGURE 9.1.7. Brain lesions in patients with parkinsonism. The response to L-DOPA treatment depends on the presence of dopaminergic and nondopaminergic lesions. When lesions affect only the nigrostriatal dopaminergic pathway (A), patients display an excellent response to L-DOPA (15% of patients). When lesions occur postsynaptically of the nigrostriatal dopaminergic pathway (B), the L-DOPA response is poor or absent (15% of patients). When lesions occur in parallel to the nigrostriatal dopaminergic pathway (C), the L-DOPA response is intermediate (70% of patients).

be described as in parallel (Fig. 9.1.7). This classic long-term treatment problem is not due to a reduction in the efficiency of the dopaminergic medication but rather to the appearance of nondopaminergic lesions, which are unresponsive to L-DOPA therapy. Therefore, the neuronal degeneration seen in PD can be usefully divided into two categories due to dopaminergic and nondopaminergic lesions.

The degeneration of the dopaminergic nigrostriatal "interrupt" circuit is the touchstone of PD, but lesions of other brain dopaminergic neurons can be implicated. We are in the habit of hearing that the first parkinsonian symptoms occur when dopamine terminal loss reaches 70% in the striatum or 50% in the cells of origin in the substantia nigra. However, this is not the case, as two observations, one experimental and one pathological, suggest. In the rat rendered parkinsonian, symptoms occur only when 90% dopamine cell loss is achieved.[29] In autopsy material from PD patients, the initial neuronal loss is highly limited, beginning in the posterolateral part of the substantia nigra and progressively spreading anterolaterally, following a well-established gradient pattern[30] (Fig. 9.1.2). This and the somatotopic organization of the substantia nigra explains why the symptoms are first seen unilaterally in the extremities in the majority of patients, slowly gaining ground to become bilateral and more severe. Moreover, all dopaminergic neurons in the nervous system can degenerate (Fig. 9.1.4), to different degrees in different patients and at different points in the disease progression, but sometimes very early.[28]

The meso-cortico-limbic neurons, arising in the ventromedial midbrain (ventral tegmental area) and projecting to cortical (especially frontal) and limbic structures (amygdala, nucleus accumbens, hippocampus, etc.), play an important role in the cognitive and emotional symptoms of the disease. Hypothalamic neurons receive a small midbrain dopamine projection but also possess an intrinsic dopamine system (arising in the arcuate nucleus and projecting to the supraoptic nucleus), with the endocrine consequences that ensue. Dopaminergic denervation of the two parts of the pallidum and the subthalamic nucleus (of nigral origin) probably also plays a role in the dysfunction of the striatal output pathways. Other dopaminergic neurons are affected, such as those in the spinal cord originating in the A11 region; these are suspected to play a role in restless legs syndrome. Similarly, degeneration of olfactory lobe neurons leads to a decrease in the sense of smell. Apart from damage to the brain's dopaminergic systems, there exist extracerebral dopamine lesions. Examples are the superior cervical ganglion (dysautonomia), the mesenteric plexus (constipation), and the sympathetic cardiac systems (primarily in multisystem atrophy).

The degeneration of multiple nondopaminergic systems may become quantitavely greater than the dopaminergic cell loss (Fig. 9.1.4). The damage best documented is that sustained by the major ascending neural pathways to the cortex. Pharmacologically, these are the noradrenergic, serotonergic, and cholinergic fibers arising, respectively, from the locus coeruleus, the raphé, and the basal nucleus of Meynert. Neuronal damage averaging 50% is usually observed in these systems, contributing to significant cognitive and psychiatric disorders.[15] Among other neuronal systems frequently damaged are the catecholamine system (e.g., vagal and bulbar neurons) and other brain structures, such as the pendunculopontine nucleus, which receives substantial input from the basal ganglia and probably plays an important role in the gait and posture abnormalities observed in PD patients. Also implicated in dementia are other brain structures such the centromedian and parafasicular nuclei of the thalamus and the cerebral cortex.

CONCLUSION

The symptoms of PD are often hidden in plain sight because we have become accustomed to think of this disease in certain slightly calcified ways. However, pathophysiological and therapeutic progress constantly challenges our understanding of PD and, ultimately, our approach to patient diagnosis and care. Although PD remains the paradigmatic dopaminergic disease, we now appreciate that it is a multisystem brain disorder. More importantly, we wish to emphasize that understanding PD—and other chronic neurodegenerative disorders—depends heavily on a precise semiologic analysis of each individual patient. Semiology, then, is our key to understanding brain function and dysfunction. Once these principles have been applied, rational therapy can be initiated or, when lacking, developed. Thus is our hope for the next 50 years after the discovery of dopamine.

REFERENCES

1. Vidailhet M, Bonnet AM, Marconi R, Gouider-Khouja N, Agid Y. Do parkinsonian symptoms and levodopa induced dyskinesias start in the foot? *Neurology.* 1994;44:1613–1616.
2. Mallet L., Schüpbach M, N'diaye K, Remy P, Bardinet E, Czernecki V, Welter ML, Pelissolo A, Ruberg M, Agid Y, Yelnik J. Stimulation of subterritories of the subthalamic nucleus reveals its role in the integration of the emotional and motor aspects of behavior. Proc Natl Acad Sci USA. 2007;104(25): 10661–10666.
3. Marsden CD, Obeso JA. The functions of the basal ganglia and the paradox of stereotaxic surgery in Parkinson's disease. *Brain.* 1994;117:877–897.
4. Hallett M. Parkinson revisited: pathophysiology of motor signs. *Adv Neurol.* 2003;91:19–28.
5. Bergman H, Deuschl G. Pathophysiology of Parkinson's disease: from clinical neurology to basic neuroscience and back. *Mov Disord.* 2002;17(suppl 3):S28–S40.
6. Pillon B, Dubois B, Agid Y. Testing cognition may contribute to the diagnosis of movement disorders. *Neurology.* 1996;46:329–334 .
7. Geser F, Wenning GK, Poewe W, McKeith I. How to diagnose dementia with Lewy bodies: state of the art. *Mov Disord.* 2005;20(suppl 12):S11–S20.
8. Burn DJ. Beyond the iron mask: towards better recognition and treatment of depression associated with Parkinson's disease. *Mov Disord.* 2002;17:445–454.
9. Czernecki V, Schüpbach M, Yaici S, Lévy R, Bardinet E, Yelnik J, Dubois B, Agid Y. Apathy following subthalamic stimulation in Parkinson disease: a dopamine responsive symptom. *Mov Disord.* 2008;15(23):964–969.
10. Remy P, Doder M, Lees A, Turjanski N, Brooks D. Depression in Parkinson's disease: loss of dopamine and noradrenaline innervation in the limbic system. *Brain.* 2005;128:1314–1322.
11. Potenza MN, Voon V, Weintraub D. Drug Insight: impulse control disorders and dopamine therapies in Parkinson's disease. *Nat Clin Pract Neurol.* 2007;3:664–672.
12. Evans AH, Katzenschlager R, Paviour D, O'Sullivan JD, Appel S, Lawrence AD, Lees AJ. Punding in Parkinson's disease: its relation to the dopamine dysregulation syndrome. *Mov Disord.* 2004;19:397–405.
13. Arnulf I, Konofal E, Merino-Andreu M, Houeto JL, Mesnage V, Welter ML, Lacomblez L, Golmard JL, Derenne JP, Agid Y. Parkinson's disease and sleepiness: an integral part of PD. *Neurology.* 2002;58:1019–1024.
14. Sastre JP, Jouvet M. [Oneiric behavior in cats]. *Physiol Behav.* 1979;22:979–989.
15. Agid Y, Arnulf I, Bejjani P, Bloch F, Bonnet AM, Damier P, Dubois B, François C, Houeto JL, Iacono D, Karachi C, Mesnage V, Messouak O, Vidailhet M, Welter ML, Yelnik J. Parkinson's disease is a neuropsychiatric disorder. *Adv Neurol.* 2003;91:365–370.
16. Gagnon JF, Postuma RB, Mazza S, Doyon J, Montplaisir J. Rapid-eye-movement sleep behaviour disorder and neurodegenerative diseases. *Lancet Neurol.* 2006;5:424–432.
17. Schifitto G, Friedman JH, Oakes D, Shulman L, Comella CL, Marek K, Fahn S; Parkinson Study Group ELLDOPA Investigators. Fatigue in levodopa-naive subjects with Parkinson disease. *Neurology.* 2008;71:481–485.
18. Arnulf I. Excessive daytime sleepiness in parkinsonism. *Sleep Med Rev.* 2005;9:185–200.
19. Chaudhuri KR, Healy DG, Schapira AH; National Institute for Clinical Excellence. Non-motor symptoms of Parkinson's disease: diagnosis and management. *Lancet Neurol.* 2006;5:235–245.
20. Winge K, Fowler CJ. Bladder dysfunction in Parkinsonism: mechanisms, prevalence, symptoms, and management. *Mov Disord.* 2006;21:737–745.
21. Berendse HW, Ponsen MM. Detection of preclinical Parkinson's disease along the olfactory tract. *J Neural Transm Suppl.* 2006;70:321–325.
22. Goldstein DS. Dysautonomia in Parkinson's disease: neurocardiological abnormalities. *Lancet Neurol.* 2003;2: 669–676.
23. Langston JW. The Parkinson's complex: parkinsonism is just the tip of the iceberg. *Ann Neurol.* 2006;59:591–596.
24. Hughes AJ, Daniel SE, Ben-Shlomo Y, Lees AJ. The accuracy of diagnosis of parkinsonian syndromes in a specialist movement disorder service. *Brain.* 2002;125:861–870.

25. Louis ED, Frucht SJ. Prevalence of essential tremor in patients with Parkinson's disease vs. Parkinson-plus syndromes. *Mov Disord*. 2007;22:1402–1407.
26. Seppi K, Schocke MF. An update on conventional and advanced magnetic resonance imaging techniques in the differential diagnosis of neurodegenerative parkinsonism. *Curr Opin Neurol*. 2005;18:370–375.
27. Agid Y. Parkinson's disease: pathophysiology. *Lancet*. 1991;337:1321–1324.
28. Agid Y, Ruberg M, Dubois B, Javoy-Agid F. Biochemical substrates of mental disturbances in Parkinson's disease. *Adv Neurol*. 1984;40:211–218.
29. Mendez JS, Finn BW. Use of 6-hydroxydopamine to create lesions in catecholamine neurons in rats. *J Neurosurg*. 1975;42:166–173.
30. Damier P, Hirsch EC, Agid Y, Graybiel AM. The substantia nigra of the human brain. II. Patterns of loss of dopamine-containing neurons in Parkinson's disease. *Brain*. 1999;122:1437–1448.

9.2 | Pathophysiology of L-DOPA-Induced Dyskinesia in Parkinson's Disease

M. ANGELA CENCI

MOTOR COMPLICATIONS OF L-DOPA PHARMACOTHERAPY: A CLINICAL PERSPECTIVE

The scientific breakthroughs celebrated in this book had an immediate impact on the treatment of Parkinson's disease (PD). Soon after Carlsson et al. reported that dopamine (DA) depletion produces a parkinsonian-like syndrome in rabbits,[1] Hornykiewicz and Birkmayer discovered DA deficiency in the brains of PD patients and proposed L-DOPA as a treatment[2,3]. The proposal turned into an effective clinical treatment thanks to the efforts of many investigators, among whom Cotzias and collaborators played a particularly important role[4]. Because of its impressive efficacy in treating all parkinsonian motor features (resting tremor, rigidity, akinesia, and postural instability), L-DOPA revolutionized the management of PD. Even today, this amino acid is recognized as the most efficacious drug to alleviate the typical signs and symptoms of the disease, and it is also the least expensive treatment[5–7]. Unfortunately, however, the response to L-DOPA changes during the progression of PD. As the disease becomes more severe, the need for symptomatic medications becomes greater, while the threshold dose of L-DOPA inducing unwanted movement becomes smaller[8]. At this point, it becomes increasingly difficult to define a dosing regimen for L-DOPA to relieve parkinsonism without inducing abnormal involuntary movements (dyskinesia) and motor fluctuations (Fig. 9.2.1). Meta-analyses of published studies indicate that these motor complications affect approximately 40% of PD patients after 4–6 years of L-DOPA therapy[9] and up to 90% of patients by 10 years of treatment[10]. Although the incidence of motor complications is lower when DA receptor agonists are used instead of L-DOPA, these agents achieve poorer symptomatic control and have important side effects[11]. The vast majority of PD patients will therefore continue to require treatment with L-DOPA at some point during the course of the disease. Nonpharmacological methods of DA replacement are being evaluated to avoid or alleviate motor complications (reviewed in[12]). Some of these methods, however, can induce dyskinesias of their own (reviewed in[13]). Dyskinesias and motor fluctuations thus remain major clinical therapeutic problems in PD, and their treatment is recognized as an unmet medical need[6].

The most common pattern of L-DOPA-induced dyskinesia (LID) consists of choreiform movements that are most severe at the time when the drug is producing the maximal relief of parkinsonian motor symptoms, hence the term *peak-of-dose* or *on* dyskinesia. In some patients, involuntary movements are most prominent at the beginning and at the end of the L-DOPA dosing cycle, a pattern referred to as *diphasic dyskinesia*[14]. Motor fluctuations are rapid transitions between periods of good motor function (on phase) and periods of severe parkinsonian immobility (off phase)[15,16]. The earliest and most common type of motor fluctuation consists of a decreased duration of the effect of single L-DOPA doses, termed the *wearing-off phenomenon* or *end-of-dose deterioration*. This usually calls for an adjustment in the treatment regimen whereby the daily L-DOPA dosage becomes divided into a larger number of doses per day. With increased dosage fractionation, however, the response to L-DOPA becomes more erratic[17], and fluctuations between on and off time become unpredictable[18]. Risk factors common to all motor complications are young age at PD onset, duration and severity of PD, and cumulative exposure to L-DOPA (i.e., treatment duration and dosage) (reviewed in[10]).

ANIMAL MODELS OF TREATMENT-INDUCED MOTOR COMPLICATIONS

The motor complications of L-DOPA pharmacotherapy are attracting growing attention on the part of basic investigators, who use animal models to address the underlying mechanisms. Different behavioral paradigms have been established to model specific aspects of the L-DOPA motor complication syndrome in rodents or nonhuman primate models of PD. In all experimental models, animals are subjected to nigrostriatal DA lesioning, followed

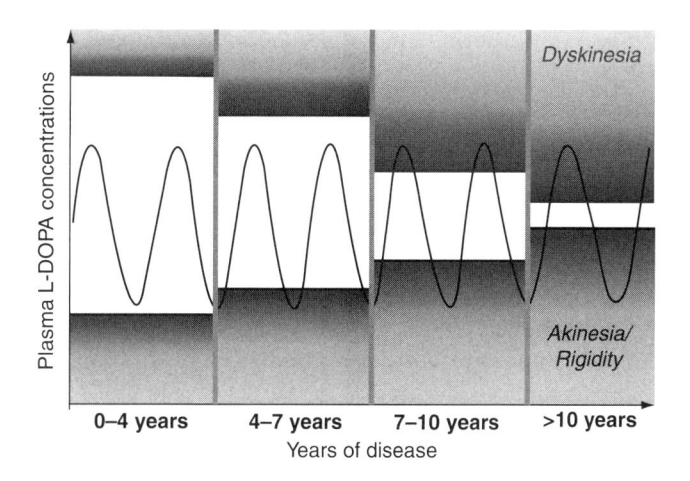

FIGURE 9.2.1. Theoretical model illustrating how the therapeutic window of L-DOPA changes during the progression of PD. The upper black line indicates the threshold L-DOPA concentration in plasma above which patients exhibit dyskinesia; the lower black line indicates the threshold concentration required to reverse PD motor features. The empty area indicates the range of L-DOPA concentrations at which the patient exhibits relief of PD motor symptoms without dyskinesia. In early disease stages, standard regimens of L-DOPA pharmacotherapy achieve good, stable control of the clinical status. As the disease advances, the dose of L-DOPA required to provide symptomatic benefit becomes larger, while the dyskinesia-threshold dose becomes smaller. As the therapeutic window of L-DOPA narrows, the daily medication-induced swings in plasma L-DOPA levels (sinuous line) cause pronounced fluctuations between on time with dyskinesia and off time with severe parkinsonism.

by a course of L-DOPA administration (single or twice-daily doses) for a time sufficient to induce a particular behavior, which lends itself to quantification. The nigrostriatal lesion is most commonly produced using 1-methyl-4-phenyl-1,2,3,6-tetrahydropyridine (MPTP) in nonhuman primates or 6-hydroxydopamine (6-OHDA) in rats and mice.

In rats with unilateral 6-OHDA lesions, the induction of contralateral turning by L-DOPA has been variably used to model either a reversal of akinesia or the occurrence of dyskinesia (reviewed in[19–21]). Some groups use the sensitization of turning as a behavioral correlate of dyskinesia[22,23], although this behavior also is induced by DA agonists with low dyskinetic potential[24–26], The gradual shortening of L-DOPA-induced contralateral turning during 3 weeks of treatment has been considered as a model of wearing-off fluctuations[27,28]. Some specific types of motor fluctuations, such as beginning-of-dose and rebound worsening of parkinsonism, have been recently modeled in MPTP-intoxicated marmosets[29]. Peak-of-dose dyskinesia is, however, the type of motor complication that has been most extensively studied in nonhuman primate models of PD[30,31]. More recently, rating scales for L-DOPA-induced abnormal involuntary movements (AIMs) also have been developed for rats and mice (reviewed in[19,32]). The phenomenology of L-DOPA-induced AIMs includes both dystonic and hyperkinetic components, and the movement patterns are largely species-specific[19,21,33]. There is a growing interest in using genetic models of PD for dyskinesia research, and encouraging results have been recently obtained in the aphakia mouse[34]. As new genetic models of PD are developed[35], this area of research will certainly expand in the future. All the basic research on LID reviewed in this chapter stems, however, from conventional neurotoxic models of PD, that is, rats with unilateral 6-OHDA lesions of the nigrostriatal pathway or nonhuman primates intoxicated with MPTP.

THE MULTILAYERED PATHOPHYSIOLOGY OF L-DOPA-INDUCED DYSKINESIA

According to a consolidated hypothesis[36–38], LID stems from two main interacting factors: a severe nigrostriatal lesion and intermittent surges in brain DA levels concomitant with standard L-DOPA medication. The ensuing nonphysiological stimulation of brain DA receptors is posited to cause abnormal plastic responses in dopaminoceptive neurons. In line with this overall model, the most recent studies on the subject have crystallized three main layers of alterations at the basis of LID, namely: (1) presynaptic abnormalities in the handling of exogenous L-DOPA, which are intimately linked to an abnormal release and clearance of extracellular DA; (2) maladaptive molecular and synaptic plasticity in striatal neurons; and (3) pathological oscillatory activities and altered firing patterns in the "deep" basal ganglia nuclei, including the subthalamic nucleus (STN), the internal segment of the globus pallidus (GPi), and the substantia nigra pars reticulata (SNr) (Fig. 9.2.2). Mechanistic explanations within and between layers remain, however, quite incomplete[39]. Moreover, as new approaches are applied to investigating LID, additional previously unsuspected alterations are revealed and new links are uncovered. This is exemplified by the recent discovery of prominent microvascular changes induced by L-DOPA treatment in the basal ganglia, in which dyskinetic subject exhibit angiogenesis[40] and abnormal hemodynamic responses[41,42]. This discovery has provided a new framework to try and interpret the effects of antidyskinetic treatments with unknown mechanisms of action[42,43].

The following review will focus on some main categories of alterations that have been documented by several independent studies. Particular attention will be paid to the most recent literature on the subject, and unresolved issues will be highlighted in order to stimulate further research.

PRESYNAPTIC CHANGES IN DA RELEASE AND CLEARANCE

Studies in 6-OHDA-lesioned rats have revealed a close temporal relationship between the expression of AIMs and a rise in striatal levels of L-DOPA and DA following a peripheral drug injection[44–46]. Moreover, higher striatal levels of these compounds have been measured in L-DOPA-treated dyskinetic rats compared to nondyskinetic cases[44,46]. The experimental findings are in keeping with the results of positron emission tomography (PET) studies in human patients. A study using [^{11}C] raclopride-PET found that PD patients with peak-dose dyskinesia exhibited greater changes in putaminal DA levels than did stable L-DOPA responders 1 hr after a standard L-DOPA dose[47]. Using a similar approach, Pavese et al.[48] found a highly significant positive correlation between putaminal changes in raclopride binding and the patients' dyskinesia scores. The critical role played by striatal DA levels in dyskinesia was demonstrated in 6-OHDA-lesioned rats using a reverse microdialysis approach. Intrastriatal infusion of L-DOPA[44,49] promptly elicited AIMs, which were blocked by

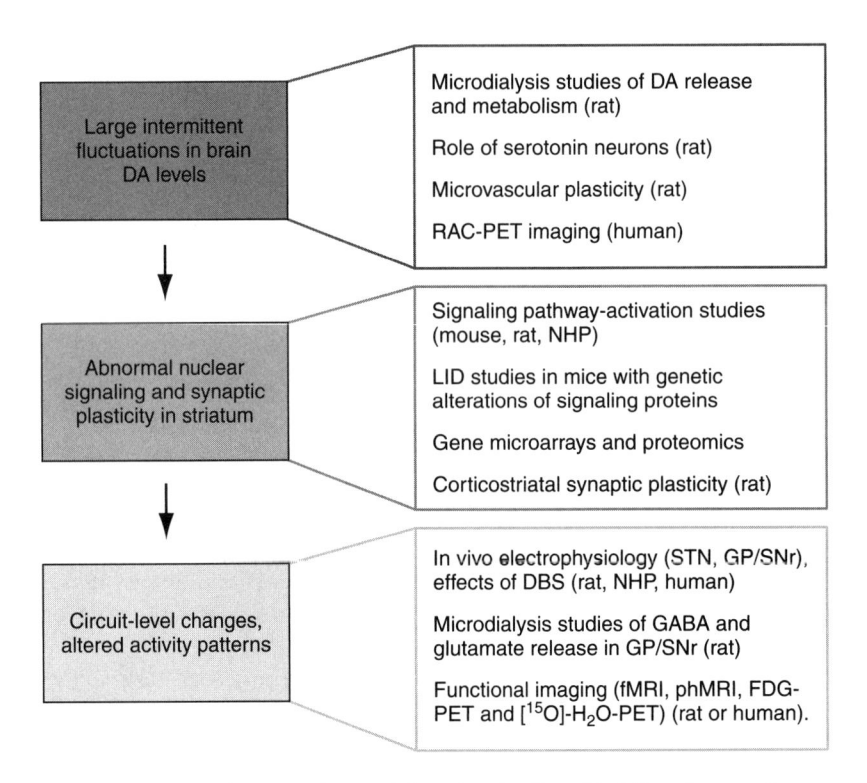

FIGURE 9.2.2. This schematic drawing illustrates three main layers of alterations implicated in LID and summarizes recent approaches with which such alterations have been addressed. With the standard medication regimens, presynaptic abnormalities in the handling of exogenous L-DOPA cause large intermittent surges of brain DA levels (upper box). Large fluctuations in brain DA levels are posited to cause exuberant activation of nuclear signaling pathways and maladaptive synaptic plasticity in striatal neurons (mid box). The altered activity in striatal efferent pathways is believed to contribute to system-level changes affecting cortico-basal ganglionic-thalamocortical loops (lower box). DA, dopamine; DBS, deep brain stimulation; FDG, [^{18}F]-fluorodeoxyglucose; fMRI, functional magnetic resonance imaging; GP, globus pallidus; LID, L-DOPA-induced dyskinesia; NHP, nonhuman primate; PET, positron emission tomography; phMRI, pharmacological MRI; RAC, [^{11}C]-raclopride; SNr, substantia nigra pars reticulata; STN, subthalamic nucleus.

intrastriatal inhibition of aromatic amino acid decarboxylase[49]. Taken together, the findings above indicate that a rise in striatal DA levels post-L-DOPA administration is the prime trigger of peak-dose dyskinesia. Moreover, the human PET studies suggest that the varying susceptibility to dyskinesia among patients may be related to individual differences in the extent of putaminal DA release post-L-DOPA administration. It is therefore important to identify the factors contributing to these differences. Seminal studies by Abercrombie and colleagues demonstrated that the striatal increase in extracellular DA after L-DOPA administration is greater in 6-OHDA-lesioned rats compared to intact rats, the difference being due to a loss of high-affinity DA uptake capacity[50]. More recent studies in animal models of LID, however, indicate that the extent of the nigrostriatal DA lesion is not the only determinant of large increases in striatal DA levels post-L-DOPA administration. Indeed, the magnitude of the surge in central DA levels may vary among subjects who show very similar degrees and patterns of nigrostriatal DA degeneration[44,51,52]. A growing literature in rats indicates that, when nigrostriatal DA neurons are damaged, serotonin neurons become the main site of decarboxylation of exogenous L-DOPA to DA[53,54]. Serotonin neurons can decarboxylate L-DOPA, store DA in synaptic vesicles, and release it together with serotonin[53,55,56], but they lack both DA autoreceptors and the DA transporter. Their handling of exogenous L-DOPA results in unregulated DA efflux and defective DA clearance (reviewed in[37]). The causal role of brain serotonergic systems in LID has been demonstrated by a recent study in 6-OHDA-lesioned rats[57], in which the severity of dyskinesia was dramatically reduced following either serotonin-specific lesioning or combined treatment with agonists at 5-HT1A and 5-HT1B autoreceptors (which reduce transmitter release from serotonergic neurons). This effect was specific for LID, as the 5-HT-autoreceptor agonists did not reduce apomorphine-induced AIMs. The data thus pointed to a presynaptic site of action for the antidyskinetic effects of 5-HT1A and 5-HT1B-receptor agonists[57]. Taken together, the findings above suggest that the integrity of forebrain serotonergic projections (which are variably affected in PD[58]) may condition the susceptibility to LID by exaggerating the increase in striatal DA levels post L-DOPA administration. Preliminary data from 6-OHDA-lesioned rats support this suggestion[44,59]. The extent to which serotonin neurons are involved in human LID is, however, unknown. In a recent postmortem study of caudate and putamen samples from PD patients, both serotonin and serotonin transporter (SERT) levels were found to be reduced to a similar extent in dyskinetic and non-dyskinetic cases[58]. This observation remains, however, tentative because of the small number of patients examined and because of difficulties in accurately establishing the presence or absence of dyskinesia in a retrospective examination of clinical notes. The question of whether a preserved raphe-striatal serotonin innervation is (or not) a major susceptibility factor for human LID should therefore be addressed in prospective brain imaging studies, which are presently being undertaken in several centers. Further investigation also is required to establish whether the angiogenic effects of L-DOPA observed in the rat[40] also occur in people with PD and whether they contribute to dysregulated swings in extracellular DA levels post-L-DOPA administration.

IMBALANCE IN THE ACTIVITY OF STRIATAL EFFERENT PATHWAYS

The striatum is the brain structure richest in DA receptors and the main anatomical site through which the motor effects of L-DOPA are produced (reviewed in[37,39]). Approximately 95% of all neurons in the striatum are medium spiny neurons (MSN)[60], which belong to either of two classes. The *direct-pathway* MSN send their main axonal projection to the output stations of the basal ganglia (i.e., the GPi and the SNr), they are rich in D1 DA receptors, and they express neuropeptide genes coding for prodynorphin (preproenkephalin-B [PPE-B]) and preprotachykinin (the substance P precursor). The *indirect-pathway* MSN project to the external segment of the globus pallidus (GPe), they are rich in D2 receptors, and they express preproenkephalin-A (PPE-A; reviewed in[61]). These two main neuronal populations have opposing roles in the control of movement. Direct pathway neurons are referred to as the *Go* pathway because their activation favors the selection/execution of specific cortically driven motor commands[62–64]. By contrast, indirect pathway neurons provide *NoGo* signals to the cortex. Their activity is believed to prevent particular cortical actions from being facilitated, thus suppressing competing, unwanted motor responses[63]. The dynamic interplay between these pathways, enabling a fluid execution of cortically initiated movements, is critically dependent on physiological levels of DA receptor stimulation. By activating D1 receptors on direct-pathway MSN, DA amplifies the excitatory effects of glutamate. By contrast, D2 receptor activation on indirect-pathway MSN opposes the excitatory effects of glutamate[65].

High striatal levels of DA, such as those documented to occur in LID, are bound to have opposite effects on the activity of the *direct* and *indirect* output pathways. The strong simultaneous stimulation of both D1 and D2 receptors would be expected to produce excess activation of the *Go* pathway and excess inhibition of the *NoGo* pathway, with the net result of greatly facilitating multiple motor representations in the cortex. Accordingly, the emerging behavioral output will be characterized by excessive (hyperkinetic or dystonic) movement. This view is in keeping with both traditional anatomo-functional models[62,66] and recent neurocomputational models of the basal ganglia[63]. More importantly, it can accommodate a large number of experimental findings. For example, several independent studies performed in 6-OHDA-lesioned rats have shown that L-DOPA causes up-regulation of transcription factors and plasticity genes specifically in the direct-pathway MSN[67–70]. Moreover, both the development of L-DOPA-induced AIMs and the associated plastic changes are dose-dependently blocked by D1 receptor antagonists, whereas D2 antagonists have no effect[71,72]. In L-DOPA-treated, dyskinetic macaques, the efficiency of G protein-mediated signal transduction is abnormally elevated at striatal D1 but not D2 receptors[73], pointing to an enhanced responsiveness to L-DOPA in the direct-pathway MSN. Furthermore, in both rat[74] and nonhuman primate models of LID[75] the severity of dyskinesia is positively correlated with up-regulation of prodynorphin mRNA levels in the striatum. This effect is paralled by decreased radioligand binding densities to kappa opioid receptors in the basal ganglia output nuclei[75,76], which is indicative of an increased receptor occupancy by dynorphins, released from the direct pathway axons. All these experimental findings point to a hyperactivity of the direct pathway in LID. Such hyperactivity, however, has not yet been demonstrated with electrophysiological techniques, and it is apparently at odds with the results of a recent study. Using urethane-anesthetized, 6-OHDA-lesioned rats, Gonon and collaborators have examined the spike response to cortical stimulation of direct-pathway and indirect-pathway MSN, which were identified by the presence and absence (respectively) of an antidromic response to SNr stimulation[77,78]. As expected, the 6-OHDA lesion caused a decreased sensitivity to cortical stimulation in direct-pathway MSN, and the opposite effect in the indirect-pathway MSN[77]. Unexpectedly, however, treatment with L-DOPA, or D1 receptor agonists either did not correct or further worsened the depressed response of direct-pathway MSN to cortical stimulation[78]. Because of technical reasons (use of a urethane-anesthetized preparation, simple stimulation parameters), these results may however not reflect the dynamic responses of direct-pathway MSN in a freely moving, dyskinetic rat. It is hoped that future investigations will succeed in elucidating the electrophysiological responsiveness of direct-pathway MSN in awake animals during the actual expression of LID.

Alterations of indirect-pathway MSN that are specifically linked with LID have been difficult to pin down, and most of the available data pertain to the regulation of PPE-A mRNA, a marker of transcriptional activity in these neurons. Several studies have described an association between LID and high levels of PPE-A mRNA in the striatum[74,79–81], but the interpretation of these findings is far from clear. Indeed, activation of D2 receptors by L-DOPA would be expected to cause down-regulation of PPE-A mRNA[82]. Accordingly, increased striatal levels of PPE-A are promptly induced by D2 receptor antagonists or DA-denervating lesions (reviewed in[83]), this being a compensatory response that limits the hyperactivity of striatopallidal MSN[83,84]. Most studies of opioid mRNA expression in animals treated with L-DOPA have been carried out after short periods of treatment washout. The only study where dyskinetic animals were killed 1 hr post L-DOPA administration reported a significant reduction in PPE-A mRNA levels in the striatum[75]. Overall, the available data thus indicate that PPE-A gene transcription decreases during the on phase of the L-DOPA action cycle but increases upon cessation of L-DOPA treatment, the latter effect being greater in dyskinetic subjects. The pathophysiological significance of these changes is presently unknown, and the specific contribution of D2/PPE-A-positive MSN to LID remains an unresolved issue. Novel approaches and/or outcome measures that can reveal specific dysfunctions of the indirect-pathway MSN in LID would greatly increase our pathophysiological understanding of this movement disorder.

ALTERED PLASTICITY OF CORTICOSTRIATAL SYNAPSES

Clinical and preclinical observations leave little doubt that a disorder of brain plasticity is implicated in LID. The severity of dyskinesia increases gradually during a course of L-DOPA treatment, and severe LID is promptly induced by challenge drug doses even after long periods of treatment discontinuation (reviewed in[37]). Once fully established, LID can be triggered by drugs with low dyskinesiogenic potential (reviewed in[83]). Moreover, both the incidence and the severity of LID are particularly pronounced in young PD patients, which has been attributed to a higher capacity for neuroplasticity in the young brain[86].

Some investigators have equated LID with a form of maladaptive neuroplasticity that involves corticostriatal synapses[85-87]. The first pioneering study on this subject[88] was performed in the rat model of LID, which conveniently allows researchers to distinguish between animals that develop AIMs and animals that exhibit a normal behavioral response to the same L-DOPA treatment. In this study, corticostriatal synaptic plasticity was examined in acute brain slices from L-DOPA-treated dyskinetic and nondyskinetic animals. Long-term potentiation (LTP) was induced by high-frequency stimulation (HFS) of cortical afferents, and recordings were performed from striatal MSN. While the inducibility of LTP did not differ between dyskinetic and nondyskinetic rats, the former group lacked the ability to reverse LTP following low-frequency stimulation of the cortical afferent pathway[88]. This study provided the first demonstration that corticostriatal synaptic plasticity is indeed impaired in LID and prompted the suggestion that abnormal information storage in corticostriatal synapses is key to the movement disorder[88]. The incidence of both AIMs and loss of depotentiation increases when rats are administered higher doses of L-DOPA[89], pointing to a close link between the synaptic abnormality and the dyskinetic behavior. Although the mechanisms underlying the loss of synaptic depotentiation are not fully understood, this alteration was attributed to an overactive signaling downstream of D1 receptors, leading to persistent blockade of intracellular phosphatases by DA- and cAMP-regulated phosphoprotein of 32 kDa (DARPP-32)[88]. Further investigations are, however, required to clarify which specific systems of intracellular kinases and phosphatases become unbalanced in the dyskinetic state and which phosphorylated substrates impede the reversal of LTP. Increased striatal phosphorylation of extracellular signal-regulated kinases 1 and 2 (ERK1/2) has been associated with LID in both rat[71] and mouse models of PD[90], but a role for this pathway in the loss of synaptic depotentiation has not yet been explored. Another unresolved issue is the extent to which a lack of depotentiation affects the indirect pathway versus the direct-pathway MSN. In line with recent concepts[91], intermittent increases in extracellular DA levels, such as those associated with LID, would be expected to produce opposite changes in synaptic strength in the two populations of MSN, that is, the strength of corticostriatal synaptic transmission would increase in the Go pathway and decrease in the NoGo pathway. This prediction needs to be tested by recording synaptically evoked responses separately in D1/prodynorphin neurons and D2/PPE-A neurons in future experiments.

ALTERED ACTIVITY IN PEPTIDERGIC AND GABAERGIC PATHWAYS TO THE BASAL GANGLIA OUTPUT NUCLEI

Maladaptive plasticity of corticostriatal synapses is not the only culprit in LID. Exuberant activation of nuclear signaling pathways and persistent changes in gene and protein expression are seen in the striatum in DA-denervated animals treated with L-DOPA (recently reviewed in[37,83,90]). The pattern of striatal mRNA expression was recently compared in L-DOPA-treated dyskinetic and nondyskinetic rats using gene microarray technology[93]. A salient feature of the gene expression profile associated with dyskinesia was the up-regulation of many genes involved in GABA transmission, such as glutamic acid decarboxylase isoform 67 (GAD67), several GABA-receptor subunits, and the vesicular GABA transporter[93]. These findings are in keeping with in situ hybridization studies performed in 6-OHDA-lesioned rats chronically treated with L-DOPA, showing that GAD67 mRNA is up-regulated in direct-pathway MSN[94] and that the striatal levels of GAD67 mRNA are positively correlated with the L-DOPA-induced AIM scores[74]. Moreover, in both rodents and nonhuman primates, L-DOPA-induced AIM scores are highly correlated with the striatal levels of prodynorphin mRNA[52,74,75,95], which is induced by L-DOPA via ΔFosB-like transcription factors[68,96]. Prodynorphin and ΔFosB show cellular colocalization[68] and are persistently up-regulated in the striatum both during and following chronic courses of L-DOPA treatment[97,98]. Taken together, these findings indicate that the development of LID goes hand in hand with increased transcriptional activity in striatal direct-pathway MSN. The transcriptional changes include a pronounced up-regulation of mRNAs coding for opioid precursors and GABA biosynthetic enzymes. The consequences of such changes on the physiology of striatal neurons remain to be unraveled. On the other hand, several lines of evidence indicate that changes in gene expression in direct-pathway MSN are paralleled by increased GABAergic and opioidergic transmission in the basal ganglia output stations, that is, the GPi (or its rat equivalent, the entopeduncular nucleus) and the SNr. In particular, increased expression and/or radioligand binding activity of GABA-A receptors has been seen in the GPi/SNr in both rodent[94,99] and nonhuman primate models of LID[100], and also in dyskinetic PD patients in a postmortem analysis[101]. Moreover, a large elevation of GABA release in the SNr was found to occur specifically in dyskinetic rats following a peripheral injection of L-DOPA, and this elevation was blunted by treatments that reduced dyskinesia[102]. Studies in rats have shown that GABA and dynorphin are coreleased in the SNr upon stimulation of striatal D1 receptors[103]. These two neurotransmitters have additive inhibitory effects on neuronal activity in the SNR[104], paralleled by disinhibitory effects on movement[105].

All these findings indicate that persistent molecular and neurochemical changes in striatofugal GABAergic/dynorphinergic pathways are a crucial component of the maladaptive plasticity at the basis of LID. These changes augment inhibitory neurotransmission in the basal ganglia output stations, and most likely contribute to generating slow oscillatory activities and reduced firing rates in these structures[39,92]. In turn, these electrophysiological alterations provide a neural code for the emergence of dyskinetic movements, as described in the next section.

SYSTEM-LEVEL CHANGES IN CORTICO-BASAL GANGLIONIC CIRCUITS

To this day, our knowledge of system-level alterations in LID relies on a limited number of studies that have utilized either in vivo electrophysiological recordings or ex vivo metabolic mapping techniques[106]. Metabolic patterns associated with LID have been studied using 2-deoxyglucose (2-DG) autoradiography in the MPTP-lesioned macaque[107,108]. In a seminal study, DA agonist-induced dyskinesia was found to be associated with increased 2-DG uptake in the GPi but not the GPe[108]. Moreover, the regional brain uptake of 2-DG was significantly reduced in the ventral anterior (VA) and ventrolateral (VL) thalamic nuclei, which receive input from the GPi. Because 2-DG uptake mainly reflects metabolic changes in axon terminals, these results indicated that LID is linked to underactivity of the basal ganglia output nuclei (GPi and SNr), a suggestion later confirmed by single-unit electrophysiological recordings[109]. The finding of reduced basal ganglia output in LID was in complete agreement with pathophysiological models proposed in the late 1980s and early 1990s[62,66]. In these classical models, akinesia and dyskinesia were attributed to opposite changes in overall neuronal activity in the GPi/SNr, where a reduced activity rate was proposed to release movement through disinhibition of thalamic and brainstem targets. Despite the great merits of these models, ascribing dyskinesia to reduced output from the GPi/SNr appeared simplistic, being at odds with the pronounced antidyskinetic effects of GPi lesions (pallidotomy)[110]. This paradox has been overcome by the realization that altered patterns of ensemble activity are key to movement disorders of basal ganglia origin. Electrophysiological recordings from PD patients undergoing deep brain stimulation (DBS) have now established that both untreated parkinsonism and LID imply an exaggerated synchronization of neuronal activities within the STN and GPi. This results in oscillations of the local field potential (LFP) at characteristic frequencies. Thus, untreated PD patients with bradykinesia-rigidity show increased beta (10–30 Hz) oscillations of the LFP, whereas increased gamma (30–60 Hz) oscillations are seen after dopaminergic treatment that relieve PD motor symptoms (recently reviewed in[111]). Interestingly, LID is specifically associated with increased oscillatory activity in the theta band[112], similar to that found in patients affected by primary dystonia[113-114]. Oscillations of the LFP in this low-frequency band also have been recorded from the SNr in L-DOPA-treated dyskinetic rats, showing associations with altered firing patterns on single-unit recordings[46]. Whether these slow oscillatory activities and altered firing patterns are causal in LID remains to be proven, but it is noteworthy that low-frequency stimulation of the STN at 5 Hz induces choreiform movements of the contralateral upper limb[115]. Taken together, the findings of these and other studies indicate that electrical stimulation of the STN or GPi at therapeutically high frequencies (above 100 Hz) inhibits involuntary movements by desynchronizing low-frequency oscillations in the basal ganglia output nuclei. This principle applies to many forms of dyskinetic/dystonic movements, regardless of their primary origin[116-117]. To explain the beneficial effects of pallidotomy, one could then posit that dyskinesia stems from pathological *patterns* of activity in the GPi/SNr, not from overall changes in firing *rates*.

Given the importance of oscillatory neural activities to both PD and LID, it is hoped that future investigations will uncover the underlying electrophysiological, neurochemical, and molecular mechanisms. Moreover, it seems important to establish whether and how slow LFP oscillations in the STN and GPi/SNr are transmitted across basal ganglia–thalamocortical circuits, interfering with the selection and control of cortically driven movements.

CONCLUDING REMARKS

Research on LID has recently attracted the interest of many investigators.

Studies in animal models have crystallized a backbone of pathophysiological events, linking altered DA release to perturbations of intracellular signaling cascades and gene expression in striatal neurons, and ultimately leading to abnormal output from the basal ganglia nuclei. Similar presynaptic alterations, metabolic changes, and pathological oscillatory activities in the basal ganglia have been found in PD patients affected by peak-dose LID and in rats and nonhuman primate models of the movement disorder. While the *molecular signature* of human LID is still unknown, some common molecular alterations have been identified in rodent and nonhuman primate models. Many crucial questions, however, require further investigation. What factors contribute to altered DA efflux in the dyskinetic brain? What mechanisms render striatal neurons supersensitive to DA? What neurochemical and electrophysiological imbalances cause pathological oscillatory activities in the deep basal ganglia nuclei? How does altered activity in the basal ganglia output stations reverberate across thalamocortical networks to generate dyskinetic movements? What are the overall patterns of brain activity associated with LID?

Thanks to recent methodological advances, it has become possible to address the above questions conclusively, and unraveling the causal links between different levels of alterations will be an exciting task for the future. This research will reveal the impact of dysregulated DA transmission on the cells and circuits of the basal ganglia and will, in addition, inform the development of novel treatments for PD.

REFERENCES

1. Carlsson A, Lindqvist M, Magnusson T. 3,4-Dihydroxyphenylalanine and 5-hydroxytryptophan as reserpine antagonists. *Nature*. 1957;180:1200.
2. Hornykiewicz O. Die Topische Lokalisation und das Verhalten von Noradrenalin und Dopamin in der Substantia Nigra des normalen und Parkinsonkranken Menschen. *Wien Klin Wochenschr*. 1963;75:309–312.
3. Birkmayer W, Hornykiewicz O. Der L-Dioxyphenylalanin (=DOPA)-Effekt bei der Parkinson-akinase. *Wien Klin Wochenschr*. 196173:787–788.
4. Cotzias G, Van Woert M, Schiffer L. Aromatic amino acids and modification of parkinsonism. *N Engl J Med*. 1967;276:374–379.
5. Mercuri NB, Bernardi G. The "magic" of L-dopa: why is it the gold standard Parkinson's disease therapy? *Trends Pharmacol Sci*. 2005;26:341–344.
6. Olanow CW, Agid Y, Mizuno Y, Albanese A, Bonuccelli U, Damier P, De Yebenes J, Gershanik O, Guttman M, Grandas F, Hallett M, Hornykiewicz O, Jenner P, Katzenschlager R, Langston WJ, LeWitt P, Melamed E, Mena MA, Michel PP, Mytilineou C, Obeso JA, Poewe W, Quinn N, Raisman-Vozari R, Rajput AH, Rascol O, Sampaio C, Stocchi F. Levodopa in the treatment of Parkinson's disease: current controversies. *Mov Disord*. 2004;19:997–1005.
7. Rascol O, Payoux P, Ory F, Ferreira JJ, Brefel-Courbon C, Montastruc JL. Limitations of current Parkinson's disease therapy. *Ann Neurol*. 2003;53(suppl 3):S3–S12.
8. Mouradian MM, Heuser IJ, Baronti F, Fabbrini G, Juncos JL, Chase TN. Pathogenesis of dyskinesias in Parkinson's disease. *Ann Neurol*. 1989;25:523–526.
9. Ahlskog JE, Muenter MD. Frequency of levodopa-related dyskinesias and motor fluctuations as estimated from the cumulative literature. *Mov Disord*. 2001;16:448–458.
10. Manson A, Schrag A. Levodopa-induced dyskinesias, the clinical problem: clinical features, incidence, risk factors, management and impact on quality of life. In: Bezard E, ed. *Recent Breakthroughs in Basal Ganglia Research*. New York, NY: Nova Science; 2006:369–380.
11. Stowe RL, Ives NJ, Clarke C, van Hilten J, Ferreira J, Hawker RJ, Shah L, Wheatley K, Gray R. Dopamine agonist therapy in early Parkinson's disease. *Cochrane Database Syst Rev*. 2008;CD006564.
12. Cenci MA, Odin P. Dopamine replacement therapy in Parkinson's disease: past, present and future. In: Tseng K-Y, ed. *Cortico-Subcortical Dynamics in Parkinson's Disease*. New York, NY: Humana Press, c/o Springer Science + Business Media, LLC; 2009:309–334.
13. Cenci MA, Hagell P. Dyskinesia and neural grafting in Parkinson's disease. In: Brundin P, Olanow W, eds. *Movement Disorders*. New York, NY: Springer Science + Business Media, LLC; 2006:184–224.
14. Luquin MR, Scipioni O, Vaamonde J, Gershanik O, Obeso JA. Levodopa-induced dyskinesias in Parkinson's disease: clinical and pharmacological classification. *Mov Disord*. 1992;7:117–124.
15. Marsden CD, Parkes JD, Quinn N. Fluctuations in disability in Parkinson's disease—clinical aspects. In: Marsden CD, Fahn S, eds. *Movement Disorders*. London: Butterworths; 1981: 96–122.
16. Quinn NP. Classification of fluctuations in patients with Parkinson's disease. *Neurology*. 1998;51:S25–S29.
17. Nyholm D, Lennernas H, Gomes-Trolin C, Aquilonius SM. Levodopa pharmacokinetics and motor performance during activities of daily living in patients with Parkinson's disease on individual drug combinations. *Clin Neuropharmacol*. 2002;25:89–96.
18. Nutt JG, Holford NH. The response to levodopa in Parkinson's disease: imposing pharmacological law and order. *Ann Neurol*. 1996;39:561–573.
19. Cenci MA, Whishaw IQ, Schallert T. Animal models of neurological deficits: how relevant is the rat? *Nat Rev Neurosc.i* 2002;3:574–579.
20. Lane EL, Cheetham SC, Jenner P. Does contraversive circling in the 6-OHDA-lesioned rat indicate an ability to induce motor complications as well as therapeutic effects in Parkinson's disease? *Exp Neurol*. 2006;197:284–290.
21. Marin C, Rodriguez-Oroz MC, Obeso JA. Motor complications in Parkinson's disease and the clinical significance of rotational behavior in the rat: have we wasted our time? *Exp Neurol*. 2006;197:269–274.
22. Carta AR, Pinna A, Morelli M. How reliable is the behavioural evaluation of dyskinesia in animal models of Parkinson's disease? *Behav Pharmacol*. 2006;17:393–402.
23. Henry B, Crossman AR, Brotchie JM. Characterization of enhanced behavioral responses to L-DOPA following repeated administration in the 6-hydroxydopamine-lesioned rat model of Parkinson's disease. *Exp Neurol*. 1998;151:334–342.
24. Lindgren HS, Rylander D, Ohlin KE, Lundblad M, Cenci MA. The "motor complication syndrome" in rats with 6-OHDA lesions treated chronically with L-DOPA: relation to dose and route of administration. *Behav Brain Res*. 2007;177:150–159.

25. Lundblad M, Andersson M, Winkler C, Kirik D, Wierup N, Cenci MA. Pharmacological validation of behavioural measures of akinesia and dyskinesia in a rat model of Parkinson's disease. *Eur J Neurosci.* 2002;15:120–132.
26. Ravenscroft P, Chalon S, Brotchie JM, Crossman AR. Ropinirole versus L-DOPA effects on striatal opioid peptide precursors in a rodent model of Parkinson's disease: implications for dyskinesia. *Exp Neurol.* 2004;185:36–46.
27. Marin C, Aguilar E, Bonastre M. Effect of locus coeruleus denervation on levodopa-induced motor fluctuations in hemiparkinsonian rats. *J Neural Transm.* 2008;115:1133–1139.
28. Papa SM, Engber TM, Kask AM, Chase TN. Motor fluctuations in levodopa treated parkinsonian rats: relation to lesion extent and treatment duration. *Brain Res.* 1994;662:69–74.
29. Kuoppamaki M, Al-Barghouthy G, Jackson M, Smith L, Zeng BY, Quinn N, Jenner P. Beginning-of-dose and rebound worsening in MPTP-treated common marmosets treated with levodopa. *Mov Disord.* 2002;17:1312–1317.
30. Jenner P. The MPTP-treated primate as a model of motor complications in PD: primate model of motor complications. *Neurology.* 2003;61:S4-S11.
31. Petzinger GM, Quik M, Ivashina E, Jakowec MW, Jakubiak M, Di Monte D, Langston JW. Reliability and validity of a new global dyskinesia rating scale in the MPTP-lesioned non-human primate. *Mov Disord.* 2001;16:202–207.
32. Cenci MA, Lundblad M. Ratings of L-DOPA-induced dyskinesia in the unilateral 6-OHDA lesion model of Parkinson's disease in rats and mice. *Curr Protocol Neurosci* Chapter 9: Unit 9 25, 2007:9.25.1–9.25.23.
33. Fox SH, Lang AE, Brotchie JM. Translation of nondopaminergic treatments for levodopa-induced dyskinesia from MPTP-lesioned nonhuman primates to phase IIa clinical studies: keys to success and roads to failure. *Mov Disord.* 2006;21: 1578–1594.
34. Ding Y, Restrepo J, Won L, Hwang DY, Kim KS, Kang UJ. Chronic 3,4-dihydroxyphenylalanine treatment induces dyskinesia in aphakia mice, a novel genetic model of Parkinson's disease. *Neurobiol Dis.* 2007;27:11–23.
35. Litvan I, Chesselet MF, Gasser T, Di Monte DA, Parker D, Jr., Hagg T, Hardy J, Jenner P, Myers RH, Price D, Hallett M, Langston WJ, Lang AE, Halliday G, Rocca W, Duyckaerts C, Dickson DW, Ben-Shlomo Y, Goetz CG, Melamed E. The etiopathogenesis of Parkinson disease and suggestions for future research. Part II. *J Neuropathol Exp Neurol.* 2007;66:329–336.
36. Bezard E, Brotchie JM, Gross CE. Pathophysiology of levodopa-induced dyskinesia: potential for new therapies. *Nat Rev Neurosci.* 2001;2:577–588.
37. Cenci MA, Lundblad M. Post- versus presynaptic plasticity in L-DOPA-induced dyskinesia. *J Neurochem.* 2006;99:381–392.
38. Chase TN. Levodopa therapy: consequences of the nonphysiologic replacement of dopamine. *Neurology.* 1998;50:S17–S25.
39. Cenci MA. Dopamine dysregulation of movement control in L-DOPA-induced dyskinesia. *Trends Neurosci.* 2007;30: 236–243.
40. Westin JE, Lindgren HS, Gardi J, Nyengaard JR, Brundin P, Mohapel P, Cenci MA. Endothelial proliferation and increased blood-brain barrier permeability in the basal ganglia in a rat model of 3,4-dihydroxyphenyl-L-alanine-induced dyskinesia. *J Neurosci.* 2006;26:9448–9461.
41. Delfino M, Kalisch R, Czisch M, Larramendy C, Ricatti J, Taravini IR, Trenkwalder C, Murer MG, Auer DP, Gershanik OS. Mapping the effects of three dopamine agonists with different dyskinetogenic potential and receptor selectivity using pharmacological functional magnetic resonance imaging. *Neuropsychopharmacology.* 2007;32:1911–1921.
42. Hirano S, Asanuma K, Ma Y, Tang C, Feigin A, Dhawan V, Carbon M, Eidelberg D. Dissociation of metabolic and neurovascular responses to levodopa in the treatment of Parkinson's disease. *J Neurosci.* 2008;28:4201–4209.
43. Cenci MA. L-DOPA-induced dyskinesia: cellular mechanisms and approaches to therapy. *Parkinsonism Relat Disord.* 2007;13/S3:S263-S267.
44. Carta M, Lindgren HS, Lundblad M, Stancampiano R, Fadda F, Cenci MA. Role of striatal L-DOPA in the production of dyskinesia in 6-hydroxydopamine lesioned rats. *J Neurochem.* 2006;96:1718–1727.
45. Lee J, Zhu WM, Stanic D, Finkelstein DI, Horne MH, Henderson J, Lawrence AJ, O'Connor L, Tomas D, Drago J, Horne MK. Sprouting of dopamine terminals and altered dopamine release and uptake in Parkinsonian dyskinaesia. *Brain.* 2008;131:1574–1587.
46. Meissner W, Ravenscroft P, Reese R, Harnack D, Morgenstern R, Kupsch A, Klitgaard H, Bioulac B, Gross CE, Bezard E, Boraud T. Increased slow oscillatory activity in substantia nigra pars reticulata triggers abnormal involuntary movements in the 6-OHDA-lesioned rat in the presence of excessive extracellular striatal dopamine. *Neurobiol Dis.* 2006;22:586–598.
47. de la Fuente-Fernandez R, Sossi V, Huang Z, Furtado S, Lu JQ, Calne DB, Ruth TJ, Stoessl AJ. Levodopa-induced changes in synaptic dopamine levels increase with progression of Parkinson's disease: implications for dyskinesias. *Brain.* 2004;127:2747–2754.
48. Pavese N, Evans AH, Tai YF, Hotton G, Brooks DJ, Lees AJ, Piccini P. Clinical correlates of levodopa-induced dopamine release in Parkinson disease: a PET study. *Neurology.* 2006;67:1612–1617.
49. Buck K, Ferger B. Intrastriatal inhibition of aromatic amino acid decarboxylase prevents l-DOPA-induced dyskinesia: a bilateral reverse in vivo microdialysis study in 6-hydroxydopamine lesioned rats. *Neurobiol Dis.* 2008;29:210–220.
50. Abercrombie ED, Bonatz AE, Zigmond MJ. Effects of L-dopa on extracellular dopamine in striatum of normal and 6-hydroxydopamine-treated rats. *Brain Res.* 1990;525:36–44.
51. Guigoni C, Dovero S, Aubert I, Li Q, Bioulac BH, Bloch B, Gurevich EV, Gross CE, Bezard E. Levodopa-induced dyskinesia in MPTP-treated macaques is not dependent on the extent and pattern of nigrostrial lesioning. *Eur J Neurosci.* 2005;22: 283–287.
52. Winkler C, Kirik D, Bjorklund A, Cenci MA. L-DOPA-induced dyskinesia in the intrastriatal 6-hydroxydopamine model of parkinson's disease: relation to motor and cellular parameters of nigrostriatal function. *Neurobiol Dis.* 2002;10:165–186.
53. Arai R, Karasawa N, Geffard M, Nagatsu I. L-DOPA is converted to dopamine in serotonergic fibers of the striatum of the rat: a double-labeling immunofluorescence study. *Neurosci Lett.* 1995;195:195–198.
54. Tanaka H, Kannari K, Maeda T, Tomiyama M, Suda T, Matsunaga M. Role of serotonergic neurons in L-DOPA-derived extracellular dopamine in the striatum of 6-OHDA-lesioned rats. *Neuroreport.* 1999;10: 631–634.
55. Arai R, Karasawa N, Geffard M, Nagatsu T, Nagatsu I. Immunohistochemical evidence that central serotonin neurons produce dopamine from exogenous L-DOPA in the rat, with reference to the involvement of aromatic L-amino acid decarboxylase. *Brain Res.* 1994;667:295–299.
56. Kitahama K, Geffard M, Araneda S, Arai R, Ogawa K, Nagatsu I, Pequignot JM. Localization of L-DOPA uptake and decarboxylating neuronal structures in the cat brain using dopamine immunohistochemistry. *Brain Res.* 2007;1167:56–70.

57. Carta M, Carlsson T, Kirik D, Bjorklund A. Dopamine released from 5-HT terminals is the cause of L-DOPA-induced dyskinesia in Parkinsonian rats. *Brain.* 2007;130:1819–1833.
58. Kish SJ, Tong J, Hornykiewicz O, Rajput A, Chang LJ, Guttman M, Furukawa Y. Preferential loss of serotonin markers in caudate versus putamen in Parkinson's disease. *Brain.* 2008;131:120–131.
59. Rylander MD, Strome E, Mela F, Mercanti G, Cenci MA. The severity of L-DOPA-induced dyskinesia in the rat is positively correlated with the density of striatal serotonin afferents *Parkinsonism Relat Disord.* 2007;13(suppl 2):S90.
60. Reiner A, Anderson KD. The patterns of neurotransmitter and neuropeptide co-occurrence among striatal projection neurons: conclusions based on recent findings. *Brain Res Brain Res Rev.* 1990;15:251–265.
61. Gerfen CR. The neostriatal mosaic: multiple levels of compartmental organization in the basal ganglia. *Annu Rev Neurosc.* 1992;15:285–320.
62. Albin RL, Young AB, and Penney JB. The functional anatomy of basal ganglia disorders. *Trends Neurosci.* 1989;12:366–375.
63. Cohen MX, and Frank MJ. Neurocomputational models of basal ganglia function in learning, memory and choice. *Behav Brain Res.* In press: 2009, 199:141–156.
64. Mink JW. The basal ganglia: focused selection and inhibition of competing motor programs. *Prog Neurobiol.* 1996;50:381–425.
65. Nicola SM, Surmeier J, Malenka RC. Dopaminergic modulation of neuronal excitability in the striatum and nucleus accumbens. *Annu Rev Neurosci.* 2000;23:185–215.
66. DeLong MR. Primate models of movement disorders of basal ganglia origin. *Trends Neurosci.* 1990;13:281–285.
67. Carta AR, Tronci E, Pinna A, Morelli M. Different responsiveness of striatonigral and striatopallidal neurons to L-DOPA after a subchronic intermittent L-DOPA treatment. *Eur J Neurosci.* 2005;21:1196–1204.
68. Andersson M, Hilbertson A, Cenci MA. Striatal fosB expression is causally linked with l-DOPA-induced abnormal involuntary movements and the associated upregulation of striatal prodynorphin mRNA in a rat model of Parkinson's disease. *Neurobiol Dis.* 1999;6:461–474.
69. Sgambato-Faure V, Buggia V, Gilbert F, Levesque D, Benabid AL, Berger F. Coordinated and spatial upregulation of arc in striatonigral neurons correlates with L-dopa-induced behavioral sensitization in dyskinetic rats. *J Neuropathol Exp Neurol.* 2005;64:936–947.
70. St-Hilaire M, Landry E, Levesque D, Rouillard C. Denervation and repeated L-DOPA induce complex regulatory changes in neurochemical phenotypes of striatal neurons: implication of a dopamine D1-dependent mechanism. *Neurobiol Dis.* 2005; 20:450–460.
71. Westin JE, Vercammen L, Strome EM, Konradi C, Cenci MA. Spatiotemporal pattern of striatal ERK1/2 phosphorylation in a rat model of L-DOPA-induced dyskinesia and the role of dopamine D1 receptors. *Biol Psychiatry.* 2007;62:800–810.
72. Lindgren HS, Ohlin KE, and Cenci MA. Differential Involvement of D1 and D2 Dopamine Receptors in L-DOPA-Induced Angiogenic Activity in a Rat Model of Parkinson's Disease. Neuropsychopharmacology 2009; ISSN 1470-634X (Electronic).
73. Aubert I, Guigoni C, Hakansson K, Li Q, Dovero S, Barthe N, Bioulac BH, Gross CE, Fisone G, Bloch B, Bezard E. Increased D1 dopamine receptor signaling in levodopa-induced dyskinesia. *Ann Neurol.* 2005;57:17–26.
74. Cenci MA, Lee CS, Bjorklund A. L-DOPA-induced dyskinesia in the rat is associated with striatal overexpression of prodynorphin- and glutamic acid decarboxylase mRNA. *Eur J Neurosci.* 1998;10:2694–2706.
75. Aubert I, Guigoni C, Li Q, Dovero S, Bioulac BH, Gross CE, Crossman AR, Bloch B, Bezard E. Enhanced preproenkephalin-B-derived opioid transmission in striatum and subthalamic nucleus converges upon globus pallidus internalis in L-3,4-dihydroxyphenylalanine-induced dyskinesia. *Biol Psychiatry.* 2007; 61:836–844.
76. Johansson PA, Andersson M, Andersson KE, Cenci MA. Alterations in cortical and basal ganglia levels of opioid receptor binding in a rat model of l-DOPA-induced dyskinesia. *Neurobiol Dis.* 2001;8:220–239.
77. Mallet N, Ballion B, Le Moine C, and Gonon F. Cortical inputs and GABA interneurons imbalance projection neurons in the striatum of parkinsonian rats. *J Neurosci* 26: 3875–3884, 2006.
78. Ballion B, Frenois F, Zold CL, Chetrit J, Murer MG, and Gonon F. D2 receptor stimulation, but not D1, restores striatal equilibrium in a rat model of Parkinsonism. *Neurobiol Dis*: 2009; ISSN 1095–953X(Electronic).
79. Herrero MT, Augood SJ, Hirsch EC, Javoy-Agid F, Luquin MR, Agid Y, Obeso JA, Emson PC. Effects of L-DOPA on preproenkephalin and preprotachykinin gene expression in the MPTP-treated monkey striatum. *Neuroscience.* 1995;68:1189–1198.
80. Tel BC, Zeng BY, Cannizzaro C, Pearce RK, Rose S, Jenner P. Alterations in striatal neuropeptide mRNA produced by repeated administration of L-DOPA, ropinirole or bromocriptine correlate with dyskinesia induction in MPTP-treated common marmosets. *Neuroscience.* 2002;115:1047–1058.
81. Zeng BY, Pearce RK, MacKenzie GM, Jenner P. Alterations in preproenkephalin and adenosine-2a receptor mRNA, but not preprotachykinin mRNA correlate with occurrence of dyskinesia in normal monkeys chronically treated with L-DOPA. *Eur J Neurosci.* 2000;12:1096–1104.
82. Chen JF, Aloyo VJ, Weiss B. Continuous treatment with the D2 dopamine receptor agonist quinpirole decreases D2 dopamine receptors, D2 dopamine receptor messenger RNA and proenkephalin messenger RNA, and increases mu opioid receptors in mouse striatum. *Neuroscience.* 1993;54:669–680.
83. Steiner H, Gerfen CR. Role of dynorphin and enkephalin in the regulation of striatal output pathways and behavior. *Exp Brain Res.* 1998;123:60–76.
84. Steiner H, Gerfen CR. Enkephalin regulates acute D2 dopamine receptor antagonist-induced immediate-early gene expression in striatal neurons. *Neuroscience.* 1999;88:795–810.
85. Jenner P. Molecular mechanisms of L-DOPA-induced dyskinesia. *Nat Rev Neurosci.* 2008;9:665–677.
86. Linazasoro G. New ideas on the origin of L-dopa-induced dyskinesias: age, genes and neural plasticity. *Trends Pharmacol Sci.* 2005;26:391–397.
87. Calabresi P, Giacomini P, Centonze D, Bernardi G. Levodopa-induced dyskinesia: a pathological form of striatal synaptic plasticity? *Ann Neurol.* 2000;47:S60-S68; discussion S68-S69.
88. Picconi B, Centonze D, Hakansson K, Bernardi G, Greengard P, Fisone G, Cenci MA, Calabresi P. Loss of bidirectional striatal synaptic plasticity in L-DOPA-induced dyskinesia. *Nat Neurosci.* 2003;6:501–506.
89. Picconi B, Paille V, Ghiglieri V, Bagetta V, Barone I, Lindgren HS, Bernardi G, Angela Cenci M, Calabresi P. l-DOPA dosage is critically involved in dyskinesia via loss of synaptic depotentiation. *Neurobiol Dis.* 2008;29:327–335.
90. Santini E, Valjent E, Usiello A, Carta M, Borgkvist A, Girault JA, Herve D, Greengard P, Fisone G. Critical involvement of cAMP/DARPP-32 and extracellular signal-regulated protein kinase

signaling in L-DOPA-induced dyskinesia. *J Neurosci.* 2007;27:6995–7005.
91. Shen W, Flajolet M, Greengard P, Surmeier DJ. Dichotomous dopaminergic control of striatal synaptic plasticity. *Science.* 2008;321:848–851.
92. Cenci MA, Lindgren HS. Advances in understanding L-DOPA-induced dyskinesia. *Curr Opin Neurobiol.* 2007;17:665–671.
93. Konradi C, Westin JE, Carta M, Eaton ME, Kuter K, Dekundy A, Lundblad M, Cenci MA. Transcriptome analysis in a rat model of L-DOPA-induced dyskinesia. *Neurobiol Dis.* 2004;17:219–236.
94. Nielsen KM, Soghomonian JJ. Normalization of glutamate decarboxylase gene expression in the entopeduncular nucleus of rats with a unilateral 6-hydroxydopamine lesion correlates with increased GABAergic input following intermittent but not continuous levodopa. *Neuroscience.* 2004;123:31–42.
95. Lundblad M, Picconi B, Lindgren H, Cenci MA. A model of L-DOPA-induced dyskinesia in 6-hydroxydopamine lesioned mice: relation to motor and cellular parameters of nigrostriatal function. *Neurobiol Dis.* 2004;16:110–123.
96. Andersson M, Konradi C, Cenci MA. cAMP response element-binding protein is required for dopamine-dependent gene expression in the intact but not the dopamine-denervated striatum. *J Neurosci.* 2001;21:9930–9943.
97. Andersson M, Westin JE, Cenci MA. Time course of striatal DeltaFosB-like immunoreactivity and prodynorphin mRNA levels after discontinuation of chronic dopaminomimetic treatment. *Eur J Neurosci.* 2003;17:661–666.
98. Westin JE, Andersson M, Lundblad M, Cenci MA. Persistent changes in striatal gene expression induced by long-term L-DOPA treatment in a rat model of Parkinson's disease. *Eur J Neurosci.* 2001;14:1171–1176.
99. Katz J, Nielsen KM, Soghomonian JJ. Comparative effects of acute or chronic administration of levodopa to 6-hydroxydopamine-lesioned rats on the expression of glutamic acid decarboxylase in the neostriatum and GABAA receptors subunits in the substantia nigra, pars reticulata. *Neuroscience.* 2005;132:833–842.
100. Calon F, Goulet M, Blanchet PJ, Martel JC, Piercey MF, Bedard PJ, Di Paolo T. Levodopa or D2 agonist induced dyskinesia in MPTP monkeys: correlation with changes in dopamine and GABAA receptors in the striatopallidal complex. *Brain Re.s* 1995;680:43–52.
101. Calon F, Morissette M, Rajput AH, Hornykiewicz O, Bedard PJ, Di Paolo T. Changes of GABA receptors and dopamine turnover in the postmortem brains of parkinsonians with levodopa-induced motor complications. *Mov Disord.* 2003;18:241–253.
102. Mela F, Marti M, Dekundy A, Danysz W, Morari M, Cenci MA. Antagonism of metabotropic glutamate receptor type 5 attenuates l-DOPA-induced dyskinesia and its molecular and neurochemical correlates in a rat model of Parkinson's disease. *J Neurochem.* 2007;101:483–497.
103. You ZB, Herrera-Marschitz M, Nylander I, Goiny M, O'Connor WT, Ungerstedt U, Terenius L. The striatonigral dynorphin pathway of the rat studied with in vivo microdialysis–II. Effects of dopamine D1 and D2 receptor agonists. *Neuroscience.* 1994;63:427–434.
104. Robertson BC, Hommer DW, Skirboll LR. Electrophysiological evidence for a non-opioid interaction between dynorphin and GABA in the substantia nigra of the rat. *Neuroscience.* 1987;23:483–490.
105. Herrera-Marschitz M, Christensson-Nylander I, Sharp T, Staines W, Reid M, Hokfelt T, Terenius L, Ungerstedt U. Striato-nigral dynorphin and substance P pathways in the rat. II. Functional analysis. *Exp Brain Res.* 1986;64: 193–207.
106. Chakraborty S, Guigoni C, Gross CE, Bezard E, Boraud T. The pathophysiology of dyskinesia: electrophysiological recordings and system-level changes in cortico-basal ganglia circuit. In: Bezard E, ed. *Recent Breakthroughs in Basal Ganglia Research.* New York, NY: Nova Science; 2006: 361–367.
107. Guigoni C, Li Q, Aubert I, Dovero S, Bioulac BH, Bloch B, Crossman AR, Gross CE, Bezard E. Involvement of sensorimotor, limbic, and associative basal ganglia domains in L-3,4-dihydroxyphenylalanine-induced dyskinesia. *J Neurosci.* 2005;25:2102–2107.
108. Mitchell IJ, Boyce S, Sambrook MA, Crossman AR. A 2-deoxyglucose study of the effects of dopamine agonists on the parkinsonian primate brain. Implications for the neural mechanisms that mediate dopamine agonist-induced dyskinesia. *Brain.* 1992;115(pt 3):809–824.
109. Papa SM, Desimone R, Fiorani M, Oldfield EH. Internal globus pallidus discharge is nearly suppressed during levodopa-induced dyskinesias. *Ann Neurol.* 1999;46:732–738.
110. Marsden CD, Obeso JA. The functions of the basal ganglia and the paradox of stereotaxic surgery in Parkinson's disease. *Brain.* 1994;117(pt 4):877–897.
111. Brown P. Abnormal oscillatory synchronisation in the motor system leads to impaired movement. *Curr Opin Neurobiol.* 2007;17:656–664.
112. Alonso-Frech F, Zamarbide I, Alegre M, Rodriguez-Oroz MC, Guridi J, Manrique M, Valencia M, Artieda J, and Obeso JA. Slow oscillatory activity and levodopa-induced dyskinesias in Parkinson's disease. *Brain* 2006;129: 1748–1757.
113. Liu X, Griffin IC, Parkin SG, Miall RC, Rowe JG, Gregory RP, Scott RB, Aziz TZ, Stein JF. Involvement of the medial pallidum in focal myoclonic dystonia: a clinical and neurophysiological case study. *Mov Disord.* 2002;17:346–353.
114. Silberstein P, Kuhn AA, Kupsch A, Trottenberg T, Krauss JK, Wohrle JC, Mazzone P, Insola A, Di Lazzaro V, Oliviero A, Aziz T, Brown P. Patterning of globus pallidus local field potentials differs between Parkinson's disease and dystonia. *Brain.* 2003;126:2597–2608.
115. Liu X, Ford-Dunn HL, Hayward GN, Nandi D, Miall RC, Aziz TZ, Stein JF. The oscillatory activity in the Parkinsonian subthalamic nucleus investigated using the macro-electrodes for deep brain stimulation. *Clin Neurophysiol.* 2002;113: 1667–1672.
116. Bittar RG, Yianni J, Wang S, Liu X, Nandi D, Joint C, Scott R, Bain PG, Gregory R, Stein J, Aziz TZ. Deep brain stimulation for generalised dystonia and spasmodic torticollis. *J Clin Neurosci.* 2005;12:12–16.
117. Herzog J, Pogarell O, Pinsker MO, Kupsch A, Oertel WH, Lindvall O, Deuschl G, Volkmann J. Deep brain stimulation in Parkinson's disease following fetal nigral transplantation. *Mov Disord.* 2008;23:1293–1296.

9.3 Progression of Parkinson's Disease Revealed by Imaging Studies

DAVID J. BROOKS

INTRODUCTION

Parkinson's disease (PD) is characterized pathologically by loss of dopamine (DA) neurons in the substantia nigra pars compacta (SNc) in association with intracellular Lewy body inclusions.[1] The neuronal loss is characteristically asymmetric and, when it reaches around 50%, patients manifest combinations of resting tremor, rigidity, and bradykinesia in their limbs. Later, additional degeneration of nondopaminergic pathways can result in complications such as postural instability, bulbar dysfunction, impaired autonomic function, dementia, and depression. Braak staging suggests that Lewy body pathology first arises in the medulla and then spreads in an ascending fashion to involve the pons and midbrain followed by limbic areas and the association cortices.[2] Despite this, function of the dopamine system seems most susceptible to the presence of intraneuronal Lewy body inclusions. The factors underlying the pathogenesis of PD remain uncertain, though several genetic mutations that predispose to the condition have now been identified.[3] The greatest risk factors for idiopathic PD are age, a positive family history, late-onset idiopathic impaired sense of smell (hyposmia), and rapid eye movement (REM) sleep behavior disorder (RBD).[4]

Structural neuroimaging with magnetic resonance imaging (MRI) allows changes in regional brain volumes, water T1 and T2 relaxation times, water apparent diffusion coefficients (ADC), and magnetic susceptibility to be detected. Transcranial sonography (TCS), an ultrasound-based modality, can reveal hyperechogenicity of the substantia nigra when cell loss occurs.[5] Positron emission tomography (PET) and single photon computed emission tomography (SPECT) are both radiotracer-based imaging modalities that can be used as in vivo biomarkers of the functional integrity of the dopaminergic and other systems in PD. Markers of presynaptic DA terminal function include dopamine transporter (DAT) and vesicular monoamine transporter 2 (VMAT2) binding and the activity of dopa decarboxylase (DDC). Postsynaptic DA D1 and D2 receptor availability can also be monitored, as can the function of the cholinergic system and the glial reaction to neurodegeneration.

Parkinson's disease patients are six times more likely than healthy age-matched controls to develop dementia, while the prevalence of dementia averages out at 40% across series.[6] Dementia in PD patients differs from that in Alzheimer's disease (AD) in that it is characterized by a dysexecutive syndrome along with impairment of visuospatial capacities, attentional control, and short-term memory, while verbal skills are relatively preserved. These cognitive deficits are associated with loss of mesolimbic and mesocortical dopaminergic projections, but also with direct involvement of the cortex by Lewy body pathology, cholinergic cell loss in the nucleus basalis of Meynert, and in many cases incidental AD or vascular pathology.

IMAGING NIGRAL STRUCTURE WITH TCS

Ninety percent of patients with clinically established PD have been reported to show increased midbrain echogenicity with TCS.[5] However, this is also true of 10% of elderly normal individuals, 15% of essential tremor patients,[7] and 40% of depressed patients,[8] raising questions about the specificity of this finding. Transcranial sonography has been reported to have positive and negative predictive values of 86% and 83%, respectively, for clinically probable PD compared with healthy controls.[9] While hyperechogenicity is most noticeable contralateral to the more clinically affected limbs in PD, the intensity does not correlate significantly with disability scores. Over 5 years of follow-up of PD patients showed no significant change in TCS findings, although their clinical disability progressed – Figure 9.3.1.[10] It has been suggested that the presence of midbrain hyperechogenicity reflects the presence of perivascular iron deposition rather than loss of dopaminergic function.[11] As such, it may represent a trait rather than a state marker for susceptibility to parkinsonism. In support

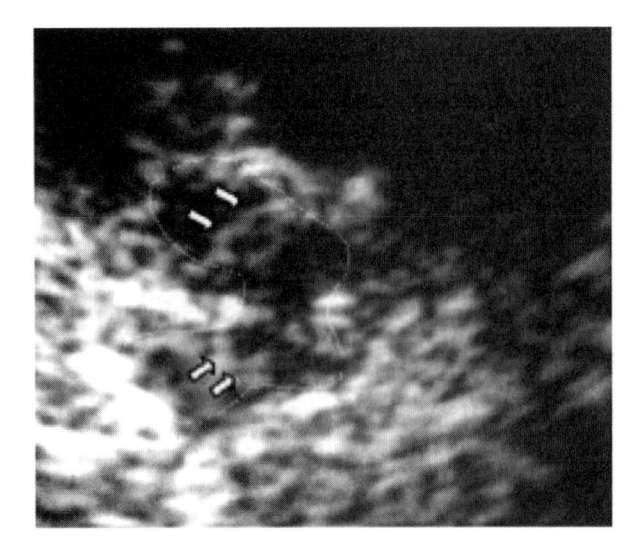

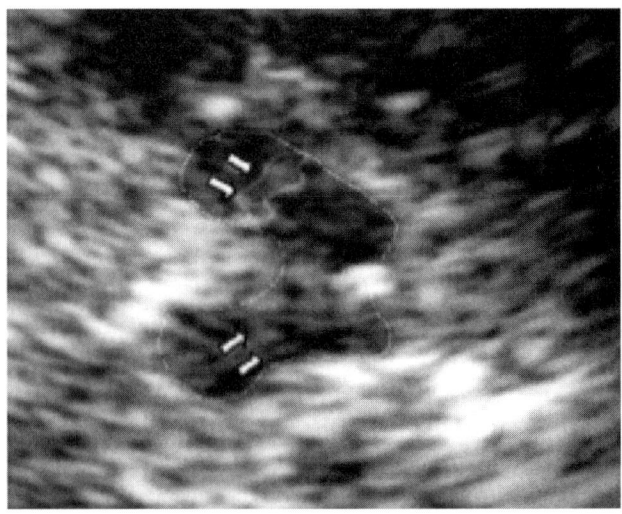

FIGURE 9.3.1. Transcranial sonography of the midbrain in a PD patient at baseline and 5 years later. No significant change in echogenicity has occurred despite deterioration of the patient. *Source*: Pictures courtesy of D. Berg from[10].

of this viewpoint, a raised TCS signal can be seen in carriers of alpha-synuclein, lysine rich repeat kinase 2 (LRRK2), parkin, and DJ1 gene mutations, who are all at risk for PD.[12,13]

MRI VOLUMETRIC STUDIES OF PD PROGRESSION

To date, MRI has not been able to demonstrate atrophic changes in PD substantia nigra, partly because of the difficulty in defining its borders. However, MRI with voxel-based morphometry (VBM) can localize significant regional brain volume reductions in both nondemented PD patients and those who developed later dementia (PDD).[14] The structural correlates of dementia in PDD are hippocampal, thalamic, and anterior cingulate atrophy. Subclinical volume loss in these areas can be detected with VBM in nondemented PD. When PDD patients are followed serially, further volume loss can be detected in neocortical areas over 2 years.[15] In contrast, nondemented PD patients tend to show further volume loss primarily in limbic and temporal association areas. Alzheimer disease patients have been reported to lose brain volume at a rate of 2% per annum, 10 times the rate seen in healthy normal controls.[16] Burton and colleagues compared whole brain volume reductions over 1 year in PD and PDD cases. Loss of brain volume occurred at a normal rate in PD (0.31% per annum), whereas it was raised in PDD (1.12% per annum)—around 50% of the rate reported for AD.[17]

GLUCOSE METABOLISM AND PD

Function of the DA system can also be indirectly monitored with ^{18}F-fluorodeoxyglucose (FDG) PET, a marker of hexokinase activity that, in turn, reflects resting regional cerebral glucose metabolism (rCMRGlc). While absolute levels of striatal rCMRGlc are generally normal in PD, covariance analysis reveals an abnormal pattern of rCMRGlc, the lentiform nucleus showing relatively raised and frontal, temporal and parietal cortex lowered activity. This abnormal pattern of regional metabolic covariation can be quantified as a PD-related profile (PDRP) in individual patients and correlates with their severity of disability when withdrawn from medication, normalizing after successful L-DOPA therapy.[18,19] In contrast, atypical parkinsonian syndromes, such as multiple system atrophy and progressive supranuclear palsy, show reduced lentiform nucleus glucose metabolism and can be discriminated from PD, where this is relatively raised.[20] In principle, FDG PET can be used to follow the progression of PD, as evidenced by increasing expression of the PDRP – see Figure 9.3.3; however, as this expression is also influenced by dopaminergic medication, changes in treatment could result in a potential confound.

Dementia with Lewy bodies (DLB) is characterized by dementia, parkinsonism, visual hallucinations, psychosis, and fluctuating confusion. The dementia is present at disease onset or within the first year of parkinsonism development. ^{18}F-fluorodeoxyglucose PET reveals a consistent pattern of reduced rCMRGlc in posterior cingulate, parietal, and temporal association regions, later spreading to prefrontal cortex in PD patients who develop dementia.[21,22] This pattern of reduced rCMRGlc is reminiscent of that seen in AD,[23] though there may be more severe occipital cortex involvement in DLB.[24]

In a recent PET study, Yong and colleagues compared patterns of glucose metabolism in PD patients with (PDD) and without later dementia and patients fulfilling consensus criteria for DLB.[25] Compared to normal controls, both PDD and DLB patients showed significant metabolic decreases in the parietal lobe, occipital lobe, temporal lobe, frontal lobe, and anterior cingulate. When DLB patients and PDD patients were compared with PD patients without dementia, both dementia groups showed relative reductions of glucose metabolism in inferior and medial frontal lobes bilaterally and in the right parietal lobe. These metabolic deficits were greater in DLB patients. A direct comparison between DLB and PDD patients showed a relative metabolic decrease in the anterior cingulate in patients with DLB. These findings support the concept that PDD and DLB have a similar underlying pattern of cortical dysfunction reminiscent of AD, though the anterior cingulate and occipital lobe may be more involved in patients with DLB.

One-third of nondemented PD patients with established disease also show temporoparietal hypometabolism.[26] This may reflect the presence of occult primary cortical pathology or be secondary to a loss of cholinergic or monoaminergic input. It remains to be determined whether the observed glucose hypometabolism in these patients is a predictive factor for later onset of dementia.

IMAGING THE DA SYSTEM

Presynaptic DA reuptake through the DAT is the primary mechanism of DA removal from the striatal synaptic cleft. Availability of striatal DATs can be assessed using tropane-based SPECT radiotracers such as ^{123}I-β-CIT (carboxymethoxy-3 beta-(4h-iodophenyl) tropane), ^{123}I-FP-CIT, ^{123}I-altropane, and ^{99m}Tc-TRODAT (2-[[2-[[[3-(4-chlorophenyl)-8-methyl-8-azabicyclo [3.2.1] oct-2-yl]methyl](2-mercaptoethyl)amino]ethyl] amino]ethanethiolato(3-)-oxo-[1R-(exo-exo)]-technecium or PET tracers such as ^{18}F-CFT (18F)-2β-carbomethoxy-3β-(4-fluorophenyl)tropane and ^{18}F-FP-CIT.[27] ^{123}I-β-CIT is only slowly taken up by the striatum, so DAT imaging is performed 24 hr after radioligand injection. Brainstem ^{123}I-β-CIT uptake and washout are more rapid and reflect serotonin transporter (SERT) rather than DAT availability.[28] ^{123}I-FP-CIT, ^{123}I-altropane, and ^{99m}Tc-TRODAT allow striatal DAT binding to be assessed 2–4 hr after intravenous tracer administration; however, their nonspecific background signals are significantly higher than that of ^{123}I-ß-CIT. ^{99m}Tc-TRODAT has the advantage that it can be produced from a kit without a cyclotron; however, it provides the lowest specific-to-nonspecific signal ratios.

The integrity of dopaminergic terminals can also be assessed with ^{18}F-DOPA PET. The uptake over 90 minutes of ^{18}F-DOPA primarily reflects DDC activity and terminal density, while its subsequent washout over 3–4 hr is a marker of ^{18}F-DA metabolism to homovanillic acid (HVA) and dehydroxyphenyl acetic acid (DOPAC). Vesicular monoamine transporters function to store presynaptic DA in synaptic vesicles for subsequent release, protecting DA from catabolism. VMAT2 binding in striatal dopamine terminals can be imaged with ^{11}C-dihydrotetrabenazine (DHTBZ) PET.

PRESYNAPTIC DOPAMINERGIC FUNCTION IN PD

Hoehn and Yahr stage 1 hemiparkinsonian patients show bilaterally reduced putamen dopaminergic terminal function, activity being more depressed in the putamen contralateral to the affected limbs. Head of caudate and ventral striatal function are initially preserved but decrease later. It has been estimated that clinical parkinsonism occurs when PD patients have lost around 50% of their posterior putamen DA terminal function.[29,30] Figure 9.3.2 shows progressive loss of putamen 18F-dopa uptake over 5 years in an initial asymptomatic identical twin of a symptomatic PD patient.

Several studies have reported that ^{18}F-DOPA PET, beta-CIT and FP-CIT SPECT can all detect progressive loss of DA terminal function in PD.[31–33] This loss ranged from 4% to 12% per annum of the baseline putamen levels in patients treated with L-DOPA; however, findings varied greatly among subjects, and it is possible that uptake of PET and SPECT markers may be directly influenced by the type of dopaminergic medication used during the course of the disease.[34] Dopa decarboxylase activity is known to fall when rats are exposed to high doses of bromocriptine and to rise when neuroleptic medications are administered.[35] Dopamine transporter binding falls when rats are DA-depleted with reserpine.[36] Administration of L-DOPA and DA agonists, therefore, could potentially influence ^{18}F-DOPA PET and beta-CIT or FP-CIT SPECT studies on rates of PD progression. Against this viewpoint are the findings of the InSPECT study, where 12 weeks of treatment with clinical doses of L-DOPA or

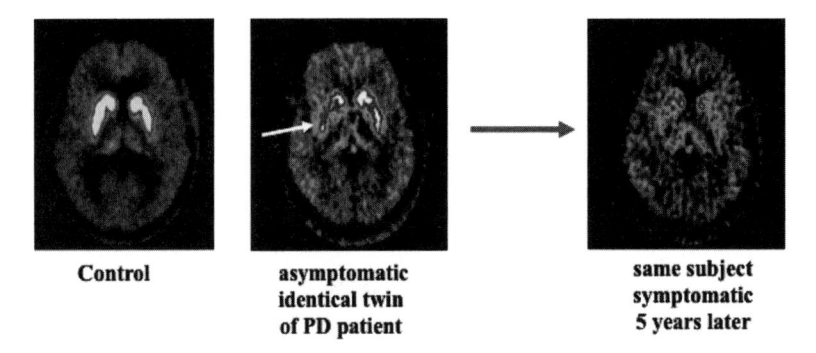

FIGURE 9.3.2. ^{18}F-DOPA PET images of a healthy control and a monozygotic twin of a PD patient when asymptomatic at baseline and 5 years later when parkinsonian. The twin shows subclinical reduction of left putamen ^{18}F-DOPA uptake when asymptomatic and further bilateral reductions 5 years later. *Source*: Picture courtesy of P. Piccini. (See Color Plate 9.3.2.)

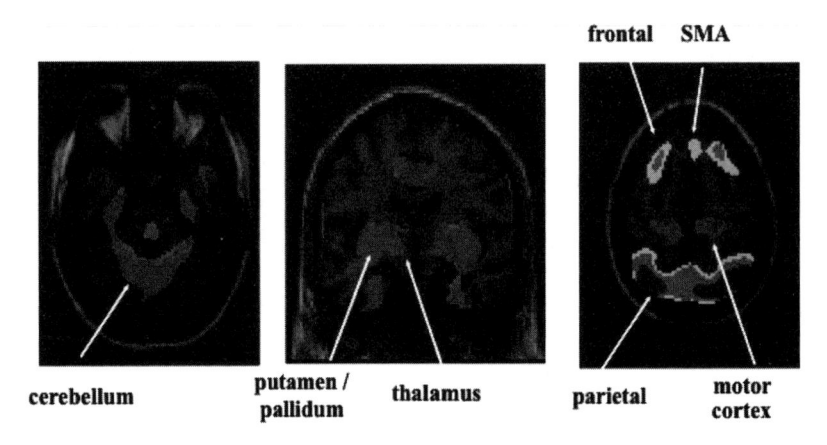

FIGURE 9.3.3. The PDRP of abnormal FDG uptake with relatively raised basal ganglia and reduced frontal and parietal glucose metabolism. *Source*: Picture courtesy of D. Eidelberg. (See Color Plate 9.3.3.)

pramipexole failed to significantly alter striatal beta-CIT binding.[37] However, as exposure to medications is usually for months or years rather than weeks, short-term washout or washin study designs will only partially exclude possible treatment confounds on imaging findings. Despite these potential difficulties, several clinical trials of putative neuroprotective agents have imaged the function of the nigrostriatal dopaminergic system to evaluate its response to putative neuroprotective treatments and to correlate imaging findings with clinical disease progression.

NEUROPROTECTION STUDIES IN PD

It has been argued that L-DOPA may be neurotoxic to DA neurons due to its oxidative metabolism, which potentially leads to increased generation of hydrogen peroxide, formation of toxic hydroxyl radicals, and enhanced oxidative stress.[38] Use of DA agonists avoids this oxidative pathway. In addition, some DA agonists have antioxidant and mitochondrial membrane potential–stabilizing properties and so may be neuroprotective in their own right.[39] The REAL PET trial was a 2-year double-blind, randomized, controlled multinational trial in PD patients that used ^{18}F-DOPA PET to compare loss of putamen DA storage capacity and clinical outcomes in de novo patients assigned to receive either L-DOPA or ropinirole therapy.[40] At the end of 2 years, the results showed significantly less reduction in putamen Ki in patients treated with ropinirole compared to those treated with L-DOPA (-13.4% vs. -20.3%, respectively; $p = 0.022$). Clinical evaluations showed that significantly fewer ropinirole-treated patients developed dyskinesias compared to L-DOPA-treated patients and that there was a significant difference in the time to the development of dyskinesias in favor of ropinirole. However, motor Unified

Parkinson's Disease Rating Scale (UPDRS) scores of functional impairment rated while patients were receiving treatment showed better control of motor symptoms in the L-DOPA-treated patients.

The CALM-PD trial was a multicenter, double-blind, 24-month randomized trial that compared outcomes in patients with early PD after initial treatment with the DA agonist pramipexole or with L-DOPA.[41] As in the REAL PET trial, this study found that initial treatment with an agonist reduced the prevalence of dopaminergic complications compared with initial treatment with L-DOPA but that L-DOPA was more effective in ameliorating symptoms of PD. A subgroup of patients underwent sequential β-CIT SPECT imaging to compare the rate of loss of DAT between the groups initially treated with pramipexole and those initially treated with L-DOPA. Over 4 years, there was a mean annual decline in striatal DAT binding of 5.2% per year; however, this decline was one-third slower in the pramipexole treatment arm (4.0% vs. 6.4% loss per annum; $p = 0.01$).

While the results of these two trials suggest that, relative to L-DOPA, DA agonists were associated with a slower loss of DA terminal function, the studies could not distinguish between a neuroprotective effect of ropinirole and a toxic effect of L-DOPA, as no placebo control arms were present. Additionally, potential confounds were differential direct effects of L-DOPA and DA agonist exposure on PET and SPECT imaging over 2–4 years of exposure and masking of faster clinical disease progression in the L-DOPA group by its superior symptomatic efficacy.

The ELLDOPA trial was designed to address the issue of whether L-DOPA was neurotoxic or protective to PD patients.[42] This randomized, placebo-controlled study used β-CIT SPECT to compare rates of loss of striatal DAT binding in de novo PD cohorts treated for 9 months with either placebo or doses of L-DOPA varying from 150 to 600 mg a day. Fourteen percent of the enrolled PD patients were found to have no dopaminergic deficit on SPECT. After their exclusion, there was a significantly greater decrease in ^{123}I β-CIT uptake among those patients receiving L-DOPA than among those receiving placebo ($p = 0.036$). The imaging data, therefore, were compatible with a toxic effect of L-DOPA on DA neuronal function. In contrast to these imaging results, L-DOPA significantly reduced the progression of PD symptoms, rated with the UPDRS, in a dose-dependent manner ($p < 0.001$). A 2-week washout only partially reversed the clinical improvements in the treated arms, which could be interpreted as L-DOPA having a neuroprotective effect in patients with PD. This apparent conflict between imaging and clinical findings could again be artifactual as it cannot be ruled out that the observed reductions in striatal β-CIT binding may in be attributable in part to direct down-regulation of DAT activity by L-DOPA.

The conflicting interpretations across studies investigating the effects of either DA agonists or L-DOPA on disease progression highlight the limitations of using currently available imaging methods to analyze DA neuronal degeneration. The CALM-PD, REAL-PET, and ELLDOPA studies were also complicated by the fact that 4%, 11%, and 14% of patients (respectively) had normal baseline PET or SPECT scans against the presence of a DA-deficient parkinsonian syndrome. Two neuroprotection trials have been reported in which the clinical and imaging findings were concordant—both of these were negative. In the first study, riluzole failed to slow the clinical progression of early PD or alter the rate of decline of putamen ^{18}F-dopa uptake (N. Pavese and D.J. Brooks, unpublished observations). In the second study, the PRECEPT trial, the mixed-lineage kinase inhibitor CEP1347 had no effect on the clinical progression of early PD patients but decreased their striatal beta-CIT uptake relative to placebo.[43]

RESTORATIVE THERAPIES IN PD

To date, restorative therapies in PD have aimed at correcting the biochemical and functional defects of the disease by increasing the production of DA in the striatum. These procedures include cell transplantation, intraparenchymal injection of growth factors, and, more recently, gene therapy.

Implantation of human fetal midbrain neurons into the striatum has been the most widely investigated strategy so far. Several small open-label uncontrolled studies reported significant clinical improvements following striatal transplants. ^{18}F-DOPA PET was used as a biomarker of graft survival in many of these studies and showed increased striatal DA storage capacity after surgery.[44–49] While the DAT marker ^{123}I-IPT SPECT was used to demonstrate graft survival in two transplanted PD patients over an 8-year follow-up period,[50] Remy and coworkers were unable to detect increased DAT binding after transplantation in PD patients who demonstrated increased ^{18}F-DOPA uptake.[44] It is possible that fetal DA cells do not always express DATs.

Using combined ^{18}F-DOPA and ^{11}C-raclopride PET, it has been possible to demonstrate in vivo that implanted fetal DA cells survive up to 10 years in striatum and are able to release DA following a methamphetamine challenge.[49,51] A significant correlation between ^{18}F-DOPA uptake and methamphetamine-induced reductions in ^{11}C-raclopride binding was also found in the putamen containing the graft, suggesting that pharmacologically induced levels of DA release by grafts relate to their DA storage capacity.[51]

The effects of fetal grafts on movement-related activation of frontal cortical areas have also been investigated in four PD patients with H$_2$^{15}O PET while they performed a joystick task.[48] A significant increase in striatal ^{18}F-DOPA uptake in these patients was detectable 6 months after transplantation but was associated with only a modest clinical improvement on the UPDRS. The impaired mesial premotor and dorsal prefrontal activation seen preoperatively during performance of freely chosen, paced joystick movements was unchanged at this time point. By 18 months after surgery, there was a significant clinical improvement in the absence of any additional increase in striatal ^{18}F-DOPA uptake. Rostral supplementary motor area (SMA) and dorsal prefrontal cortical activation during performance of joystick movements, however, had now significantly improved. These findings suggest that initially the graft acts purely as a DA resevoir but that later it is able to form connections in the host brain, restoring the activation of motor cortical areas.

Based on the encouraging results of these open studies, two prospective, randomized, double-blind, controlled trials were performed.[52,53] Despite both postmortem and ^{18}F-DOPA evidence of some graft function, there was no significant improvement in the primary outcome measures in either of these trials. Younger transplanted patients, however, showed a significant improvement in measures of motor severity at 1 year, and all patients were improved after 3 years in the Freed et al. study. In the Olanow et al. study there was a significant clinical improvement at 6 months, but this was subsequently lost—possibly coinciding with withdrawal of immunosuppression. Troublesome off-period dyskinesias occurred in 15% of cases in the Freed et al. series and 56% of the implanted patients in the Olanow et al. series.

It has been suggested that graft-induced dyskinesias may be associated with greater ^{18}F-DOPA uptake in the ventral putamen[54] or may be due to clumping of transplant cells. When statistical parametric mapping is used to interrogate ^{18}F-DOPA uptake images of individual PD patients before and after neural transplantation, the patients with the best functional outcome appear to show no dopaminergic denervation in areas outside the grafted areas postoperatively.[55] In contrast, patients with no or only modest clinical benefit show reduction of ^{18}F-DOPA in ventral striatum prior to or following transplantation. These findings indicate that a poor outcome after transplantation is associated with progressive dopaminergic denervation in areas outside those grafted.

^{18}F-DOPA PET has been employed to assess the effects of intrastriatal implantation of carotid body (CB) glomus cells in PD.[56] Carotid body cells are dopaminergic and also express glial cell line–derived neurotrophic factor (GDNF). A mild clinical improvement, which was maximal at 6–12 months after transplantation (5%–74%), was seen in this study. ^{18}F-DOPA PET showed a non-significant 5% increase in mean putaminal uptake.

Putaminal infusion of GDNF via an indwelling catheter has been trialed as a restorative approach for PD. GDNF stimulates embryonic stem cells to differentiate into DA cells and protects DA neurons against nigral toxins such as MPTP and 6-OHDA in rodent and primate models of PD. In an early open, uncontrolled trial, five PD patients received continuous unilateral (one) or bilateral (four) putaminal infusions of GDNF at a dose of 14 µg per day.[57] A significant improvement in the UPDRS score was seen after 12 months of treatment in all patients. Positron emission tomography studies revealed a postoperative 28% increase in putaminal ^{18}F-DOPA uptake. The patient who received a unilateral GDNF infusion later died from an unrelated cause; the postmortem revealed DA terminal sprouting in the ipsilateral putamen.[58] A subsequent randomized, placebo-controlled study in 34 PD patients confirmed the local increase in ^{18}F-DOPA uptake following putaminal infusion of GDNF but failed to show any consistent clinical benefit of this procedure.[59] It seems probable that, while GDNF induces DA terminal sprouting in PD striatum, this may not necessarily communicate with postsynaptic receptors in an effective manner. At present, the future of GDNF infusion as a possible restorative treatment for PD is uncertain.

CHOLINERGIC FUNCTION IN PD

The SPECT tracer ^{123}I-iodobenzovesamicol (^{123}I-BVM), an acetylcholine vesicle transporter marker, has been employed to assess the association of cholinergic deficiency in PD patients with dementia. Parkinson's disease patients without dementia showed selectively reduced binding of ^{123}I-BVM in parietal and occipital cortex, whereas PDD and AD patients had more globally reduced cortical binding.[60]

More recently, cortical acetylcholinesterase (AChE) activity in PD with and without dementia has been investigated with the PET ligands ^{11}C-MP4A

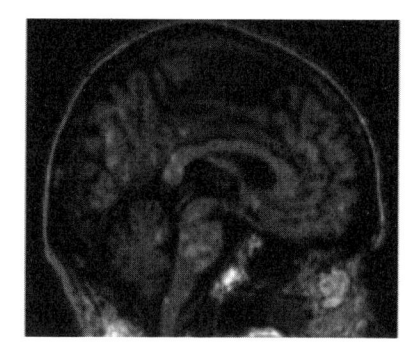

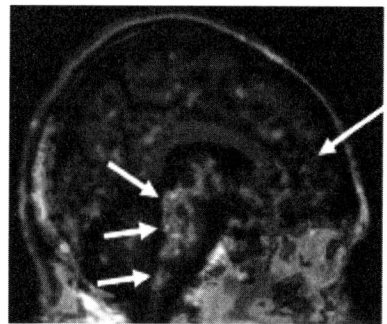

FIGURE 9.3.4. ^{11}C-PK11195 PET images of an elderly normal subject (left) and a PD patient (right). The PD patient shows extensive microglial activation of the entire brainstem, striatum, and frontal cortex in line with the distribution of Lewy body pathology described by Braak et al.[2] Source: Picture courtesy of A. Gerhard. (See Color Plate 9.3.4.)

(N-11C-methyl-4-piperidyl acetate) and ^{11}C-PMP N-11C-methylpiperidin-4-yl propionate, acetylcholine analogues that serve as selective substrates for AChE hydrolysis. Global cortical ^{11}C-MP4A binding was reduced by 30% in PDD but only 11% in PD.[61] The PDD group had significantly lower parietal ^{11}C-MP4A uptake than the PD patients, and loss of frontal and temporoparietal ^{11}C-MP4A binding correlated with striatal reduction of ^{18}F-DOPA uptake. The authors concluded that as PD progresses, there is a parallel reduction in both dopaminergic and cholinergic function, and this is most severe when dementia is present. Interestingly, while AChE deficiency correlated with performance on tests of working memory and attention, it did not correlate with motor symptoms.[62]

MICROGLIAL ACTIVATION IN PD

Microglia constitute 10%–20% of white cells in the brain and are normally in a resting state, but local injury causes them to activate and swell, expressing human leukocyte antigens (HLAs) on the cell surface and to release cytokines such as tumor necrosis factor-α (TNFα) and interleukins. The mitochondria of activated but not resting microglia express peripheral benzodiazepine (BDZ) sites – now known as translocator protein – that can be visualized with ^{11}C-PK11195 PET.

Loss of substantia nigra neurons in PD has been shown to be associated with microglial activation. More recently, histochemical studies have shown that microglial activation can also be seen in other basal ganglia, the cingulate, the hippocampus, and cortical areas.[63] ^{11}C-PK11195 PET has been used to study microglial activation in PD, and an increased midbrain signal has been reported to correlate inversely with levels of posterior putamen DAT binding measured with ^{11}C-CFT PET.[64] Gerhard and coworkers subsequently reported additional microglial activation in the brainstem, striatum, pallidum, and frontal cortex in line with the distribution of Lewy body pathology reported by Braak and colleagues in advanced PD – Figure 9.3.4.[2] Interestingly, little change in the level of microglial activation was seen over a 2-year follow-up period, although the patients all deteriorated clinically. This could imply that microglial activation is merely an epiphenomenon in PD; however, postmortem studies have shown that these cells continue to express cytokine mRNA suggesting that they could be driving disease progression.

CONCLUSIONS

Imaging dopaminergic function with PET and SPECT or changes in the expression of a PDRP with FDG PET currently remain the best biomarkers for monitoring disease progression. These measurements correlate significantly with clinical disability in PD and are able to detect preclinical dysfunction; however, the modalities cannot be regarded as surrogate markers, as they do not correlate well with clinical outcome in practice and may well be directly influenced by medication changes. While structural changes in PD substantia nigra can be detected with TCS, the associated hyperechogenicity does not appear to alter as patients clinically deteriorate. Volumetric MRI is valuable for detecting progressive brain atrophy in PDD but currently is unable to detect nigral volume changes. In the future, it is likely that PET will be increasingly used to reveal the pharmacological changes underlying many of the nonmotor complications of PD including depression, sleep disturbance, and dysautonomia.

REFERENCES

1. Fearnley JM, Lees AJ. Ageing and Parkinson's disease: substantia nigra regional selectivity. *Brain.* 1991;114:2283–2301.
2. Braak H, Ghebremedhin E, Rub U, et al. Stages in the development of Parkinson's disease-related pathology. *Cell Tissue Res.* 2004;318:121–134.
3. Gasser T. Update on the genetics of Parkinson's disease. *Mov Disord.* 2007;22(suppl 17:S343–S350.
4. Siderowf A, Stern MB. Preclinical diagnosis of Parkinson's disease: are we there yet? *Curr Neurol Neurosci Rep.* 2006;6: 295–301.
5. Berg D, Siefker C, Becker G. Echogenicity of the substantia nigra in Parkinson's disease and its relation to clinical findings. *J Neurol.* 2001;248:684–689.
6. Emre M. Dementia associated with Parkinson's disease. *Lancet Neurol.* 2003;2:229–237.
7. Stockner H, Sojer M, K KS, Mueller, J. Wenning, G. K. Schmidauer, C. Poewe, W. Midbrain sonography in patients with essential tremor. *Mov Disord.* 2007;22:414–417.
8. Walter U, Hoeppner J, Prudente-Morrissey L. Horowski S, Herpertz SC, Benecke R. Parkinson's disease-like midbrain sonography abnormalities are frequent in depressive disorders. *Brain.* 2007;130:1799–1807.
9. Prestel J, Schweitzer KJ, Hofer A, et al. Predictive value of transcranial sonography in the diagnosis of Parkinson's disease. *Mov Disord.* 2006;21:1763–1765.
10. Berg D, Merz B, Reiners K, et al. Five-year follow-up study of hyperechogenicity of the substantia nigra in Parkinson's disease. *Mov Disord.* 2005;20:383–385.
11. Berg D, Roggendorf W, Schroder U, et al. Echogenicity of the substantia nigra: association with increased iron content and marker for susceptibility to nigrostriatal injury. *Arch Neurol.* 2002;59:999–1005.
12. Schweitzer KJ, Brussel T, Leitner P, et al. Transcranial ultrasound in different monogenetic subtypes of Parkinson's disease. *J Neurol.* 2007;254:613–616.
13. Walter U, Klein C, Hilker R, et al. Brain parenchyma sonography detects preclinical parkinsonism. *Mov Disord.* 2004;19: 1445–1449.
14. Summerfield C, Junque C, Tolosa E, et al. Structural brain changes in Parkinson disease with dementia: a voxel-based morphometry study. *Arch Neurol.* 2005;62:281–285.
15. Ramirez-Ruiz B, Marti MJ, Tolosa E, et al. Longitudinal evaluation of cerebral morphological changes in Parkinson's disease with and without dementia. *J Neurol.* 2005;252:1345–1352.
16. Fox NC, Crum WR, Scahill RI, et al. Imaging of onset and progression of Alzheimer's disease with voxel-compression mapping of serial magnetic resonance images. *Lancet.* 2001;358: 201–205.
17. Burton EJ, McKeith IG, Burn DJ, O'Brien JT. Brain atrophy rates in Parkinson's disease with and without dementia using serial magnetic resonance imaging. *Mov Disord.* 2005;20:1571–1576.
18. Eidelberg D, Moeller JR, Dhawan V, et al. The metabolic topography of parkinsonism. *J Cereb Blood Flow Metab.* 1994;14:783–801.
19. Feigin A, Fukuda M, Dhawan V, et al. Metabolic correlates of levodopa response in Parkinson's disease. *Neurology.* 2001;57:2083–2088.
20. Eckert T, Barnes A, Dhawan V, et al. FDG PET in the differential diagnosis of parkinsonian disorders. *Neuroimage.* 2005;26: 912–921.
21. Peppard RF, Martin WRW, Carr GD, et al. Cerebral glucose metabolism in Parkinson's disease with and without dementia. *Arch Neurol.* 1992;49:1262–1268.
22. Vander-Borght T, Minoshima S, Giordani B, et al. Cerebral metabolic differences in Parkinson's and Alzheimer's disease matched for dementia severity. *J Nucl Med.* 1997;38:797–802.
23. Foster NL, Chase TN, Fedio P, et al. Alzheimer's disease: focal cortical changes shown by positron emission tomography. *Neurology.* 1983;33:961–965.
24. Albin RL, Minoshima S, D'Amato CJ, et al. Fluoro-deoxyglucose positron emission tomography in diffuse Lewy body disease. *Neurology.* 1996;47:462–466.
25. Yong SW, Yoon JK, An YS, Lee PH. A comparison of cerebral glucose metabolism in Parkinson's disease, Parkinson's disease dementia and dementia with Lewy bodies. *Eur J Neurol.* 2007;14:1357–1362.
26. Hu MTM, Taylor-Robinson SD, Chaudhuri KR, et al. Cortical dysfunction in non-demented Parkinson's disease patients: a combined 31phosphorus MRS and 18FDG PET study. *Brain.* 2000;123:340–352.
27. Brooks DJ. Neuroimaging in Parkinson's disease. *Neuroreport.* 2004;1:243–254.
28. Laruelle M, Baldwin RM, Malison RT, et al. SPECT imaging of dopamine and serotonin transporters with [123I]beta-CIT: pharmacological characterization of brain uptake in nonhuman primates. *Synapse.* 1993;13:295–309.
29. Morrish PK, Sawle GV, Brooks DJ. Clinical and [18F]dopa PET findings in early Parkinson's disease. *J Neurol Neurosurg Psychiatry.* 1995;59:597–600.
30. Marek K, Seibyl JP, Zoghbi SS, et al. [I-123] beta-CIT SPECT imaging demonstrates bilateral loss of dopamine transporters in hemiparkinsons disease. *Neurology.* 1996;46:231–237.
31. Morrish PK, Rakshi JS, Sawle GV, Brooks DJ. Measuring the rate of progression and estimating the preclinical period of Parkinson's disease with [18F]dopa PET. *J Neurol Neurosurg Psychiatry.* 1998;64:314–319.
32. Marek K, Innis R, van Dyck C, et al. [123I]beta-CIT SPECT imaging assessment of the rate of Parkinson's disease progression. *Neurology.* 2001;57:2089–2094.
33. Winogrodzka A, Bergmans P, Booij J, et al. [123I]FP-CIT SPECT is a useful method to monitor the rate of dopaminergic degeneration in early-stage Parkinson's disease. *J Neural Transm.* 2001;108:1011–1019.
34. Ahlskog JE. Slowing Parkinson's disease progression: recent dopamine agonist trials. *Neurology.* 2003;60:381–389.
35. Hadjiconstantinou M, Wemlinger TA, Sylvia CP, et al. Aromatic L-amino acid decarboxylase activity of mouse striatum is modulated via dopamine receptors. *J Neurochem.* 1993;60:2175–2180.
36. Kim SE, Scheffel U, Boja JW, Kuhar MJ. Effect of reserpine on binding of H-3 WIN-35,428 to dopamine uptake sites. *J Nucl Med.* 1994;35:199P.
37. Marek K, Jennings D, Tabamo R, Seibyl J. InSPECT: An investigation of the effect of short-term treatment with pramipexole or levodopa on [123I] B-CIT and SPECT imaging in early Parkinson disease. *Neurology.* 2006;66:A112.
38. Jenner P, Olanow CW. Understanding cell death in Parkinson's disease. *Ann. Neurol.* 1998;44:S72–S84.
39. Schapira AH, Olanow CW. Neuroprotection in Parkinson disease: mysteries, myths, and misconceptions. *JAMA.* 2004;291:358–364.
40. Whone AL, Watts RL, Stoessl J, et al. Slower progression of PD with ropinirol versus L-dopa: the REAL-PET study. *Ann Neurol.* 2003;54:93–101.

41. Parkinson Study Group. Dopamine transporter brain imaging to assess the effects of pramipexole vs levodopa Parkinson disease progression. *JAMA.* 2002;287:1653–1661.
42. Fahn S, Oakes D, Shoulson I, et al. Levodopa and the progression of Parkinson's disease. *N Engl J Med.* 2004;351:2498–2508.
43. Parkinson Study Group. Mixed lineage kinase inhibitor CEP-1347 fails to delay disability in early Parkinson disease. *Neurology.* 2007;69:1480–1490.
44. Remy P, Samson Y, Hantraye P, et al. Clinical correlates of [18F]fluorodopa uptake in five grafted parkinsonian patients. *Ann Neurol.* 1995;38:580–588.
45. Wenning GK, Odin P, Morrish PK, et al. Short- and long-term survival and function of unilateral intrastriatal dopaminergic grafts in Parkinson's disase. *Ann Neurol.* 1997;42:95–107.
46. Brundin P, Pogarell O, Hagell P, et al. Bilateral caudate and putamen grafts of embryonic mesencephalic tissue treated with lazaroids in Parkinson's disease. *Brain.* 2000; 123:1380–1390.
47. Hauser RA, Freeman TB, Snow BJ, et al. Long-term evaluation of bilateral fetal nigral transplantation in Parkinson disease. *Arch Neurol.* 1999;56:179–187.
48. Piccini P, Lindvall O, Bjorklund A, et al. Delayed recovery of movement-related cortical function in Parkinson's disease after striatal dopaminergic grafts. *Ann Neurol.* 2000; 48:689–695.
49. Piccini P, Brooks DJ, Bjorklund A, et al. Dopamine release from nigral transplants visualised in vivo in a Parkinson's patient. *Nat Neurosci.* 1999;2:1137–1140.
50. Pogarell O, Koch W, Gildehaus FJ, et al. Long-term assessment of striatal dopamine transporters in Parkinsonian patients with intrastriatal embryonic mesencephalic grafts. *Eur J Nucl Med Mol Imaging.* 2006;33:407–411.
51. Pavese N, Evans AH, Tai YF, et al. Clinical correlates of levodopa-induced dopamine release in Parkinson disease: a PET study. *Neurology.* 2006;67:1612–1617.
52. Freed CR, Greene PE, Breeze RE, et al. Transplantation of embryonic dopamine neurons for severe Parkinson's disease. *N Engl J Med.* 2001;344:710–719.
53. Olanow CW, Goetz CG, Kordower JH, et al. A double-blind controlled trial of bilateral fetal nigral transplantation in Parkinson's disease. *Ann Neurol.* 2003;54:403–414.
54. Ma Y, Feigin A, Dhawan V, et al. Dyskinesia after fetal cell transplantation for parkinsonism: a PET study. *Ann Neurol.* 2002;52:628–634.
55. Piccini P, Pavese N, Hagell P, et al. Factors affecting the clinical outcome after neural transplantation in Parkinson's disease. *Brain.* 2005;128:2977–2986.
56. Minguez-Castellanos A, Escamilla-Sevilla F, Hotton GR, et al. Carotid body autotransplantation in Parkinson disease: a clinical and positron emission tomography study. *J Neurol Neurosurg Psychiatry.* 2007;78:825–831.
57. Gill SS, Patel NK, Hotton GR, et al. Direct brain infusion of glial cell line–derived neurotrophic factor in Parkinson disease. *Nat Med.* 2003;9:589–595.
58. Love S, Plaha P, Patel NK, et al. Glial cell line–derived neurotrophic factor induces neuronal sprouting in human brain. *Nat Med.* 2005;11:703–704.
59. Lang AE, Gill S, Patel NK, et al. Randomized controlled trial of intraputamenal glial cell line-derived neurotrophic factor infusion in Parkinson disease. *Ann Neurol.* 2006;59:459–466.
60. Kuhl DE, Minoshima S, Fessler JA, et al. In vivo mapping of cholinergic terminals in normal aging, Alzheimer's disease, and Parkinson's disease. *Ann Neurol.* 1996;40:399–410.
61. Hilker R, Thomas AV, Klein JC, et al. Dementia in Parkinson disease: functional imaging of cholinergic and dopaminergic pathways. *Neurology.* 2005;65:1716–1722.
62. Bohnen NI, Kaufer DI, Hendrickson R, et al. Cognitive correlates of cortical cholinergic denervation in Parkinson's disease and parkinsonian dementia. *J Neurol.* 2005;380: 127–132
63. Imamura K, Hishikawa N, Sawada M, et al. Distribution of major histocompatibility complex class II-positive microglia and cytokine profile of Parkinson's disease brains. *Acta Neuropathol (Berl).* 2003;106:518–526.
64. Ouchi Y, Yoshikawa E, Sekine Y, et al. Microglial activation and dopamine terminal loss in early Parkinson's disease. *Ann Neurol.* 2005;57:168–175.
65. Gerhard A, Pavese N, Hotton G, et al. In vivo imaging of microglial activation with [(11)C](R)-PK11195 PET in idiopathic Parkinson's disease. *Neurobiol Dis.* 2006;21: 404–412.

9.4 | Transplantation of Dopamine Neurons: Extent and Mechanisms of Functional Recovery in Rodent Models of Parkinson's Disease

STEPHEN B. DUNNETT AND ANDERS BJÖRKLUND

Over the last three decades, transplantation of tissues rich in dopaminergic (DA) neurons has been the most widely studied model system within the field of neural transplantation. Transplantation of DA neurons has provided a powerful model system for understanding the basic biology and methods for achieving viable cell transplantation in the brain; it has contributed major insights of the mechanisms for structural repair and functional recovery; and it has paved the way for the first clinical trials of cell therapies in neurological disease (see Chapter 9.5, this volume).

TRANSPLANTATION METHODS

The first studies demonstrating the feasibility of dopamine (DA) cell transplantation in the adult mammalian brain showed the survival of small pieces of embryonic ventral mesencephalon (VM), rich in developing DA neurons, implanted into the anterior eye chamber[1] or into ventricular spaces in the choroidal fissure adjacent to the hippocampus.[2,3] However, the power of DA tissue grafts for the study of functional cell transplantation really attracted widespread attention with the commencement of functional studies in animals that had sustained selective dopaminergic denervation of the forebrain by nigrostriatal lesions.[4,5]

The 6-OHDA Lesion Model

Injection of the catecholamine neurotoxin 6-hydroxydopamine (6-OHDA) into appropriate sites in the fore- or midbrain allows selective destruction of the ascending nigrostriatal DA fibers in the vicinity.[6,7] The 6-OHDA lesion model has several distinct advantages for studies of regeneration and functional repair: the system is well characterized anatomically, biochemically, and pharmacologically[8]; unilateral 6-OHDA lesions are associated with a well-characterized syndrome of lateralized motor impairments[9–11]; and the resulting syndrome reflects the neuropathology and symptoms of Parkinson's disease (PD), providing potential clinical relevance for transplantation studies.[4,12]

In recent years, the 6-OHDA lesion model has been refined considerably, in particular using injections into terminal areas to produce partial lesions that are more slowly progressive and more suitable for evaluating drugs or cell therapies targeted at neuroprotection and endogenous regeneration.[13] By contrast, cell replacement and repair strategies remain better evaluated using acute bundle lesions, since these are less prone to spontaneous recovery processes[14] that can confound the attribution of recovery specifically to the implanted cells.

Solid and Cell Suspension Grafts

The two main criteria for effective transplantation of neurons into the adult central nervous system (CNS) are harvesting the donor tissue close to the time of cell birth, which for most CNS tissues implies harvesting the relevant cells from donor embryos at specific time windows of gestational development,[15,16] and selecting an implantation site that provides adequate energy and nutritional supply to sustain the newly transplanted cells prior to their incorporation into the microvasculature network of the host brain.[3] The first studies to achieve effective functional transplantation of DA neurons in fragments of tissue dissected from embryonic VM of E16- to E18-day embryos for implantation into either the lateral ventricle[4] or an artificial cortical cavity,[5] immediately adjacent to the 6-OHDA-lesioned striatum. Both studies demonstrated effective survival of catecholamine-fluorescent (presumed DA) cells within the grafts, outgrowth of fluorescent fibers into the DA-denervated host striatum, and alleviation of rotational asymmetries induced by the dopaminergic drugs apomorphine and amphetamine, respectively.

Early studies were constrained by the need to use natural ventricular spaces or to create additional artificial cavities to accommodate solid pieces of graft tissue. This constraint was largely overcome by the introduction of cell suspension methods, in which cell culture-derived enzymatic digestion and dissociation protocols are used to prepare embryonic tissues as cell suspensions for stereotaxic delivery into deep brain sites[17,18] (Fig. 9.4.1). Suspension grafts avoid the need to create additional cavities, allow systematic selection and manipulation of cells prior to transplantation, enable effective placement and good survival throughout the CNS neuropil, permit combination of graft cell types or graft placements at will, and provide a standardization of methods that facilitates well-controlled experimental design.[19] Again, the first studies using this method demonstrated effective survival of DA cells implanted directly into the striatum, reinnervation of the denervated host brain, and recovery of the amphetamine-induced rotation response.[17] Dozens of subsequent studies have amply confirmed the ability of embryonic VM grafts to alleviate deficits in rotation, and it has been determined that better yields of surviving cells and a more rapid recovery in rotation are achieved with somewhat younger donor tissues than those originally used. Most studies today target approximately E14 in rats or mice as the optimal donor age,[20,21] with some reports suggesting that rat donors as young as E12 can be highly effective,[22] although accurate dissection of VM from such young, small embryos presents its own challenges.

ANATOMICAL AND NEUROCHEMICAL RECONSTRUCTION

Fetal mesencephalic DA neuroblasts, taken from fetuses at the stage of cell-cycle exit, have a striking capacity to substitute for the lost DA innervation by extensive axonal growth into the denervated target. Tyrosine hydroxylase (TH)-positive fiber outgrowth from a single deposit of DA neurons in the denervated striatum will cover a distance of about 1–2 mm from the graft core,[23,24] with an average DA terminal density within the reinnervated area of around 40% of the intact striatal innervation.[24,25] Single intrastriatal grafts can thus provide reinnervation of a limited subregion of the striatal complex. In order to achieve more complete reinnervation of the entire striatal target, it is necessary to use multiple graft deposits[23,26,27] (Fig. 9.4.2).

The ability to innervate the striatum efficiently appears to be specific for the midbrain DA neuron phenotypes. Dopamine neurons obtained from other fetal brain parts, that is, neurons that do not normally have any axonal connections with the striatum, survive but

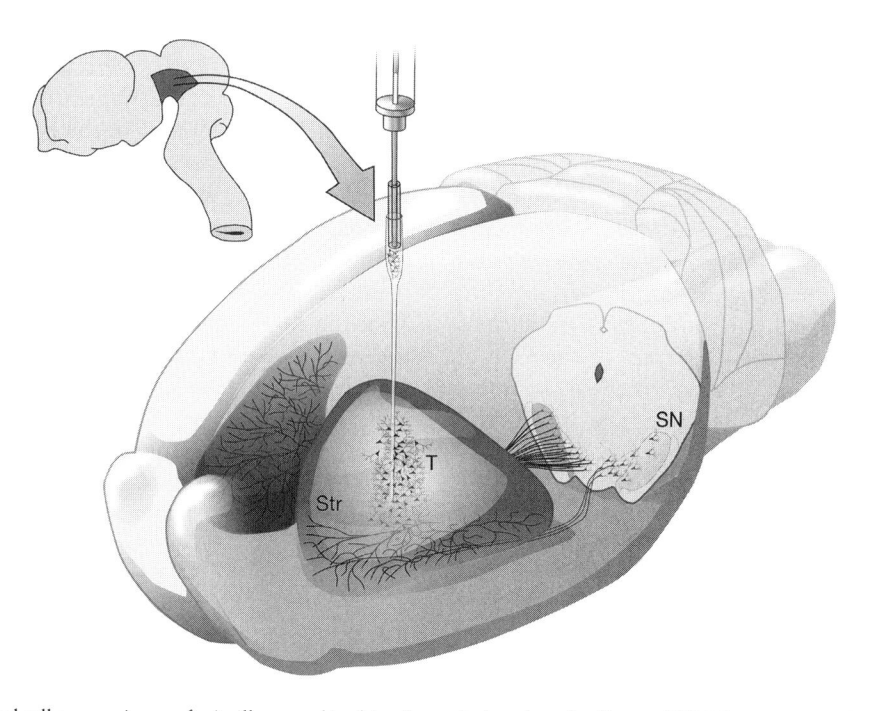

FIGURE 9.4.1. Intrastriatal cell suspension graft. As illustrated in this schematic drawing, the dissected VM tissue is injected as a cell suspension into the head of the caudate-putamen, in one or several deposits, using a stainless steel cannula or glass capillary attached to the tip of a Hamilton syringe. In most cases, the host nigrostriatal DA projection is removed by injection of 6-OHDA into the medial forebrain bundle or the striatum. SN, substantia nigra; Str, striatum; T, transplant.

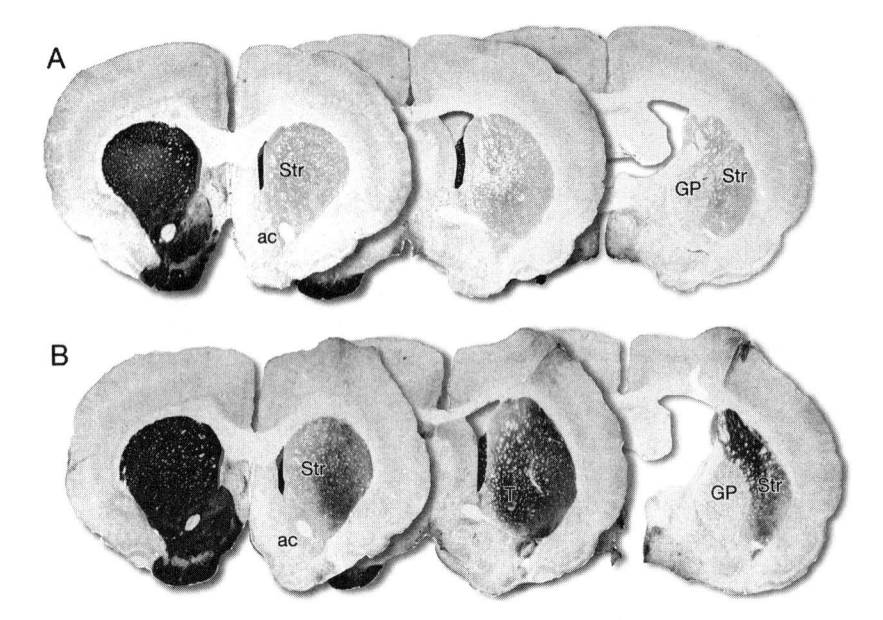

FIGURE 9.4.2. Graft-derived innervation of the host striatum as revealed by TH immunostaining. (A) Control rat, no transplant. The recipient received a 6-OHDA injection into the right medial forebrain bundle, which resulted in a complete denervation of both striatal and limbic areas in the forebrain. (B) Grafted rat. Ventral mesencephalic tissue from 14-day-old rat embryos were dissociated into a single cell suspension and injected in five sites in the right (6-OHDA-lesioned) striatum. The total number of cells injected was 450,000. These grafts result in a widespread reinnervation of the host caudate-putamen, while the limbic areas (including the nucleus accumbens and olfactory tubercle) remained completely denervated. ac, nucleus accumbens; GP, globus pallidus; Str, striatum; T, transplant. Data from Breysse et al.[102]

do not innervate the host striatum.[28,29] This suggests that cells used for DA neuron replacement in PD may have to be of the correct midbrain phenotype. The DA neuroblasts contained in the developing VM, however, are not a homogeneous population but comprise two major distinctive subtypes: (1) the neurons of the substantia nigra pars compacta (SNc; the A9 neurons, according to the nomenclature of Dahlström and Fuxe[30]) that give rise to the nigrostriatal pathway and innervate the major dorsal part of the striatum in rodents (the putamen and part of the caudate nucleus in primates) and (2) the A10 neurons of the ventral tegmental area (VTA) that give rise to the mesolimbic and mesocortical pathways that innervate the ventral striatum and parts of the limbic system and the neocortex (see [8] for a recent review). Early studies suggested that these two DA neuron phenotypes may possess different growth characteristics and that only neurons of the nigral (A9) subtype are able to reinnervate the denervated striatum after transplantation.[31,32] In a recent study, Thompson and colleagues[33] made use of a transgenic mouse expressing green fluorescent protein (GFP) under the TH promoter,[34] allowing visualization of the grafted DA neurons and their axonal projections by means of their expression of the GFP protein (Fig. 9.4.3). The two major A9 and A10 DA neuron subtypes could be identified on the basis of their expression of Girk2 and calbindin, respectively. The A9 cells, which expressed Girk2, were large, angular, and typically elongated in shape, with an average mean diameter of about 19 μm, located in the periphery of the grafts. The calbindin-expressing A10 cells were small and rounded overall, with an average diameter of about 13 μm. Most of them were located in the central core of the graft. These characteristics match well with the two principal TH +ve cell types in adult mouse and rat VM: the small calbindin-positive cells in the VTA and the medial aspect of the SNc and the larger, angular Girk2-positive cells in the SNc. These distinctive features (morphology, location, and Girk2/calbindin expression) thus seem to be retained after transplantation, and can be used to distinguish the SNc and VTA subtypes in intrastriatal VM grafts. By retrograde axonal tracing, it was found that the dopaminergic innervation of the striatum is derived almost exclusively from the SNc cells within the graft, while the VTA neurons project to the frontal cortex and probably also other forebrain areas.[33]

The ability to establish functional synaptic connections with the denervated striatal target may thus be a specific property of the DA neurons of the SNc. These data, moreover, suggest the presence of axon guidance

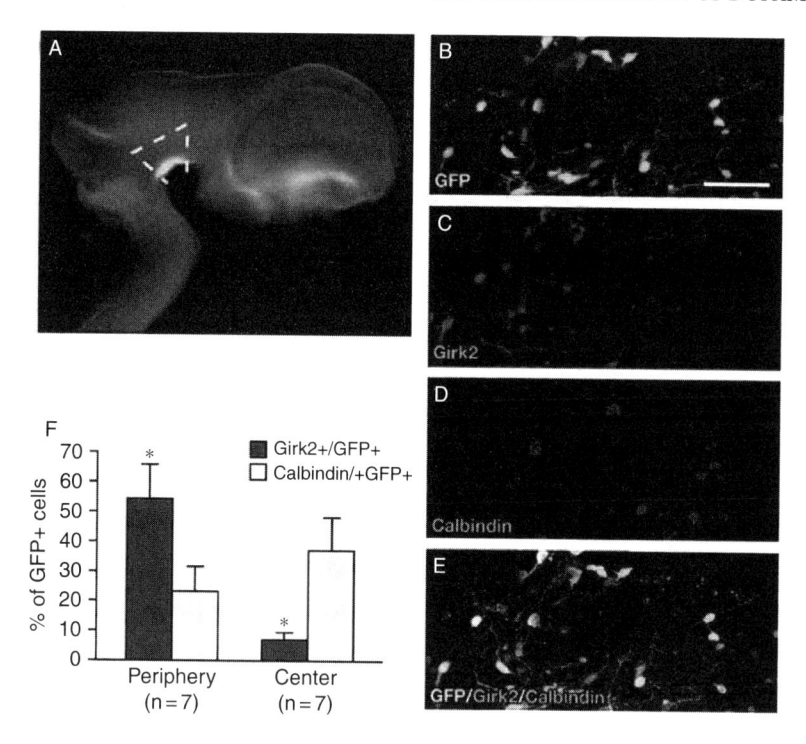

FIGURE 9.4.3. Distribution of nigral and VTA cell types in fetal VM transplants. Nigral and VTA neuron subtypes can be distinguished using antibodies to Girk2 (expressed primarily in the neurons of the SNc) and calbindin (expressed mostly in the neurons in the VTA). The intrastriatal fetal VM grafts contain both subtypes, in about equal numbers, but they are differentially localized: the Girk2+, SN-type neurons mostly in the periphery of the transplants (C, F) and the calbindin+, VTA-type neurons in the core (D, F). The VM tissue were dissected from a TH-GFP-expressing transgenic mouse, as illustrated in (A), which allowed the TH-expressing cells to be detected by staining for the GFP reporter (B, E). Data from Thompson et al.[33] (See Color Plate 9.4.3.)

and target recognition mechanisms in the DA-denervated forebrain that allow guidance of the growing axons to their appropriate targets. This growth-regulating mechanism may involve specific recognition molecules present on the appropriate target cells or growth-stimulating factors acting over some distance from the target. This possibility is further supported by early studies showing that midbrain DA neurons placed at the border of the denervated striatum extend their axons into the denervated striatum and not into the adjacent (non-DA-innervated) parietal cortex.[23] By contrast, midbrain DA neurons placed in non-DA-innervated areas extend axons abundantly within the graft itself, and around the margin of the implant, but do not extend into the host tissue.[23]

Interactions between the grafted cells and the denervated target are thus likely to play a crucial role in the establishment of a new functional innervation from the grafted DA neurons. The placement of the grafts ectopically in the striatum, rather than in the substantia nigra (SN), where they normally reside, is likely to impose some important limitations on the functionality of the grafted cells. In their normal location, nigral DA neurons are known to receive afferents from a number of brain regions, including striatum, pallidum, subthalamic nucleus, neocortex, and brainstem. Dopamine neurons placed within the striatum, by contrast, are likely to lack many of these regulatory afferent inputs. Nevertheless, neuroanatomical and electrophysiological studies have shown that intrastriatal VM transplants do receive some afferent connections from host cortex, striatum, and brainstem raphé nuclei,[35–37] suggesting that intrastriatal DA neurons may become integrated partially, but incompletely, into the host basal ganglia circuitry. For example, Fisher and colleagues[37] observed that burst firing, which is a characteristic feature of mature nigral DA neurons in situ, was present in the intrastriatal DA neuron grafts but developed very slowly and retained immature features.

These observations lend support to the view that ectopic DA neuron transplants may be efficient in restoring baseline tonic DA release within the reinnervated part of the striatum, but that the phasic aspects of DA neuron function are incompletely restored by such grafts. There is plenty of experimental data to show that

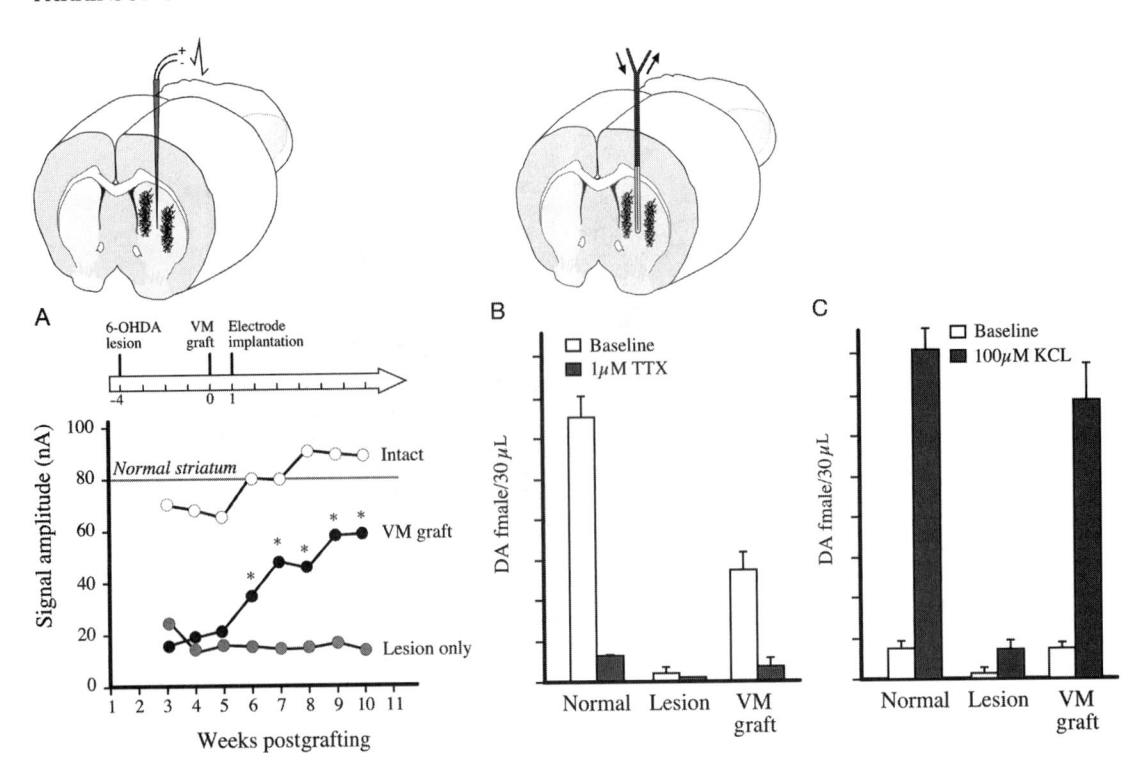

FIGURE 9.4.4. In vivo DA release as monitored by voltammetry (A) and microdialysis (B, C). In the experiments illustrated here, *in vivo* release of DA from intrastriatal VM grafts was assessed by two complementary techniques. As illustrated in the insets, the probes were implanted into the host striatum, into the area innervated by the grafted DA neurons. Panel A shows the recovery in DA release over time after transplantation in a group of grafted rats (filled circles), compared to the intact side (open circles) and nongrafted lesioned rats (gray circles), measured by chronically implanted voltammetry electrodes (data from Forni et al.[172]). Panels B and C show the recovery of the extracellular level of DA, as monitored by microdialysis, in a group of grafted animals compared to lesion-only controls, as well as the effect of blockade of action potentials by tetrodotoxin (TTX) (in B) and the effect of addition of KCl to the perfusion medium (in C). Data from Cenci et al.[233]

grafted DA neurons are spontaneously active and secrete DA at near-normal rates (Fig. 9.4.4). This supports the view that the functional effects induced by intrastriatal DA grafts are mostly due to restoration of tonic transmitter release and that this may take place, at least in part, at ultrastructurally normal synaptic sites.[38,38] Neurochemical studies have shown that multiple intrastriatal VM grafts can restore total striatal DA levels to about 30% of normal[26] and DA release in the reinnervated area (as measured by microdialysis) varying from 40% to 100% of normal.[40,41] This is further supported by studies using Fos immunohistochemistry, DA receptor binding, and in situ hybridization to monitor postsynaptic changes in DA receptor function in the striatal projection neurons. Changes seen in these cellular markers in the DA-denervated striatum (D2 receptor binding, D1 receptor–related changes in Fos expression, and changes in glutamic acid decarboxylase [GAD] mRNA and proenkephalin mRNA) are all partly or completely normalized in the grafted animals.[42–48]

TRANSPLANTATION INTO THE SN

In the studies so far reviewed, the DA neuron grafts were placed either within or very close to the denervated striatal target. In an ideal scenario, however, the cells should be implanted into the SN to allow reconstruction of the entire nigrostriatal DA pathway. Attempts to reestablish a functional nigrostriatal connection from grafts placed in the SN, however, have so far met with only limited success. In rodent experiments, intranigral DA neuron grafts have shown no or very limited growth of axons along the nigrostriatal pathway toward the striatum.[49–51] Previous studies in rats have suggested that the growth of the grafted cells may depend on the age of the recipient, and that the ability of the grafted DA neuroblasts to extend axons along the nigrostriatal pathway may be reduced or lost during the early postnatal period.[49,52] Based on these observations, it has been suggested that the properties of the axonal growth territory changes during postnatal development to become nonpermissive for the outgrowing axons—

for example, by down-regulation of growth-promoting factors and/or expression of growth-inhibiting molecules along the nigrostriatal trajectory. However, studies of fetal neuron transplants in other areas of the adult rodent CNS, such as striatum,[53–56] cortex,[56,57] hippocampus,[58] and spinal cord,[59,60] have shown that immature developing neuroblasts, or young postmitotic neurons, in many cases retain the capacity to extend axons in a target-specific manner over large distances even in adult recipients. An intriguing finding has been that cells grafted across the species barrier, that is, fetal human or porcine neurons grafted to the rat brain, can grow axons over much greater distances, along the entire length of the nigrostriatal pathway, and reinnervate the striatum from afar[53,61] (see below).

These observations raise the possibility that the failure in previous studies to detect any significant long-distance axon growth from intranigral transplants of fetal DA neuroblasts may be due, at least in part, to insufficient sensitivity of the TH immunohistochemistry used to trace the graft-derived axonal projections. As described above, the GFP transgenic mouse in which the marker gene is expressed under the control of the TH promoter[34] allows unequivocal identification of the transplanted DA neurons and their axonal projections in their entirety, and with exquisite sensitivity, within the host brain. With this tool, we have been able to show that fetal DA neuroblasts implanted into the SN in 6-OHDA-lesioned adult mice are indeed capable of extending their axons along the nigrostriatal pathway and reestablishing a terminal network with a distribution in striatal and limbic forebrain areas that closely matches that seen in the intact animal[62] (Fig. 9.4.5). The extent of striatal innervation for the intranigral grafts was further enhanced by overexpression of glial cell line–derived neurotrophic factor (GDNF), by injection of an adeno-associated virus-GDNF vector at the time of transplantation, suggesting that GDNF can act as a diffusible attractant to promote the regrowth of graft-derived axons over larger distances. In these animals, the more extensive striatal reinnervation was accompanied by a near-complete reversal of motor asymmetry in the amphetamine rotation test.[62]

The success of earlier studies using xenogeneic transplants, that is, from human or pig cells grafted to the nigra in adult rats (see above), raised the possibility that axonal growth inhibitory factors may operate poorly between species, that is, that cells derived from human or pig donors may not recognize the growth inhibitory molecules present along the growth trajectory in adult hosts. The results obtained in our recent study[62] show that this is not the case. This raises the question of why this long-distance axonal projection failed to be detected in many previous studies. *First,* the earlier allografting studies were performed in rats, that is, in a larger brain where the increased distance between nigra and striatum may provide a greater challenge for the regrowing axons. *Second,* in these earlier studies, the investigators had to rely on TH immunohistochemistry to visualize the graft-derived axons. This made it necessary to perform the graft experiments in rats with complete lesions of the nigrostriatal pathway (by injection of 6-OHDA into the medial forebrain bundle). Although there are as yet no data, it may be that axonal growth along the nigrostriatal pathway is facilitated by the presence of spared DA axons. If so, animals with complete lesions may not provide the right conditions for regrowth to occur. The use of donor cells from the TH-GFP mouse made it possible for Thompson and colleagues[62] to perform their experiment in mice with subtotal lesions of the midbrain DA projection. The GFP reporter allowed visualization of all fibers and their fine-beaded terminal branches with high sensitivity, even in the presence of spared intrinsic TH-positive fibers.

BEHAVIORAL RESPONSES

Behavioral Testing in Rats

Rotation

Rats with unilateral 6-OHDA lesions exhibit postural biases toward the side of the lesion soon after surgery. This motor asymmetry can be markedly augmented in magnitude if the animals are activated, for example by a stressor or a stimulant drug. Thus, after intraperitoneal injection of amphetamine, the animals exhibit a strong turning response, known as *rotation*, which is easily quantified in automated *rotometer* bowls.[9] Rotation is believed to be the consequence of a dual process: an asymmetry between the hemispheres in DA activation in the dorsal striatum to impose the side bias, combined with a net locomotor activation via DA stimulation in the ventral striatum.[62]

Rats with unilateral 6-OHDA lesions exhibit ipsilateral rotation (i.e., toward the side of the lesion) in response to presynaptic stimulant drugs such as amphetamine.[63] The demonstration of recovery of amphetamine-induced rotation in VM-grafted animals suggests functional DA release from graft-derived nerve terminals in the host brain,[5,64,65] which has been confirmed by biochemical measurement of DA turnover in the graft-reinnervated striatum, measured both postmortem[66] and by in vivo microdialysis.[41,67] Moreover, removal of the grafts, whether by aspiration, subsequent

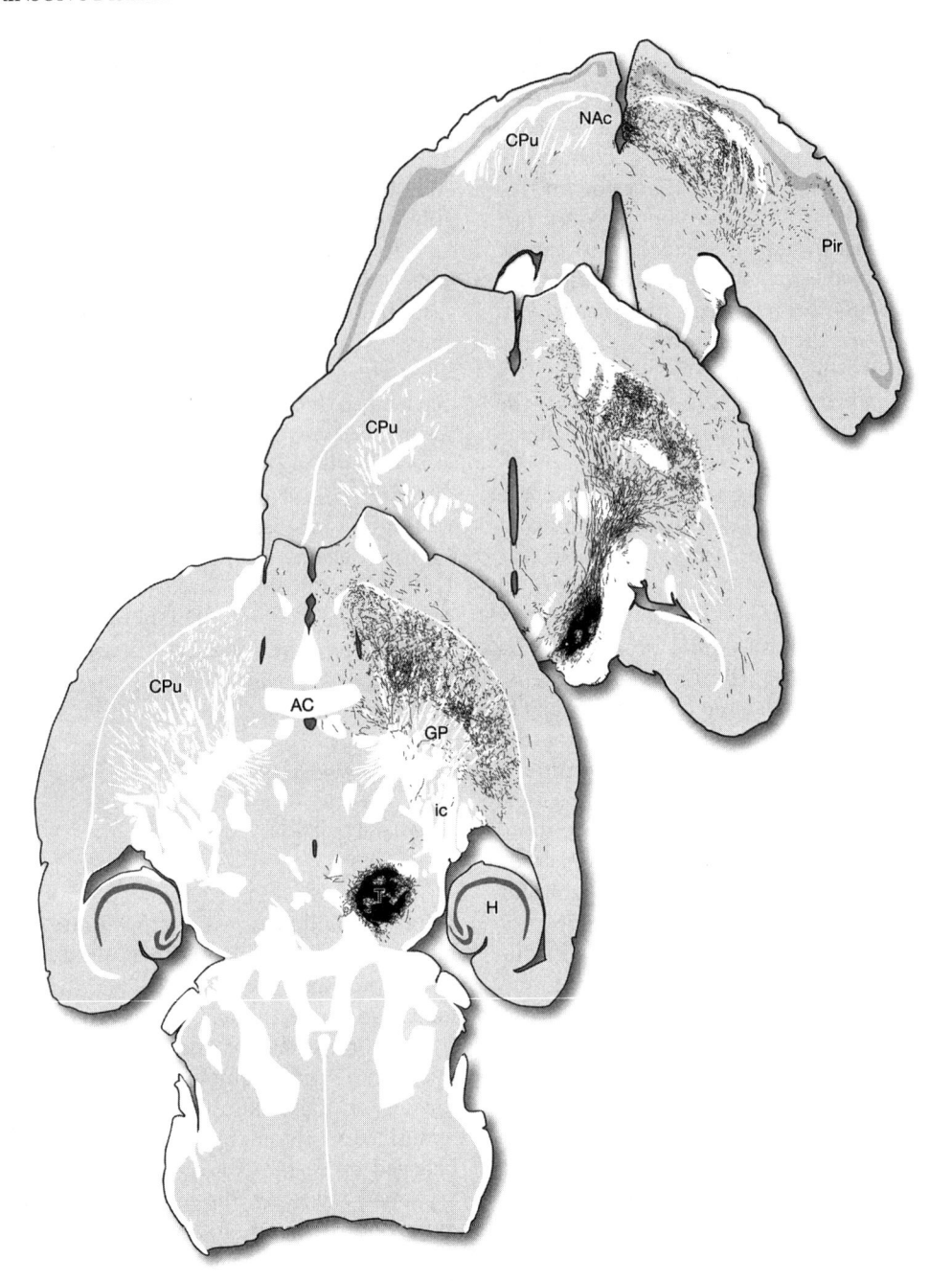

FIGURE 9.4.5. Reinnervation of the host striatum from VM grafts placed in the SN. In this experiment (Thompson et al.[62]), fetal VM tissue was taken from a transgenic mouse expressing GFP under the TH enhancer and injected as a single-cell suspension into the SN in adult 6-OHDA-lesioned mice. The exquisite sensitivity of the GFP reporter made it possible to trace the axons from the graft deposit (T) along the trajectory of the nigrostriatal pathway, in the internal capsule (ic), toward striatal and limbic forebrain areas. The distribution of the GFP-expressing fibers matched well that of the intrinsic DA projection system. AC, anterior commissure; CPu, caudate-putamen; GP, globus pallidus; H, hippocampus; NAc, nucleus accumbens, Pir, piriform cortex. Drawing made from material reported in Thompson et al.[62]

6-OHDA lesioning, or immunological rejection in each case, immediately restored rotational asymmetry,[65,67,68] confirming that the observed recovery is indeed dependent upon the continued survival of dopaminergic neurons within the grafts and their integration into the host brain.

In contrast to rats administered presynaptic stimulants, unilaterally lesioned rats rotate in the opposite, contralateral direction in response to DA receptor agonists such as apomorphine. This is attributed to the compensatory development of receptor supersensitivity of the receptors on the postsynaptic striatal neurons on the lesioned side.[69] Again, VM grafts alleviate apomorphine rotation, although typically not to as great a degree as that seen on amphetamine tests, suggesting that diffuse release of DA from grafted cells can normalize receptor sensitivity.[4,64] Endogenous DA release correlating with recovery has been confirmed in receptor binding studies.[46,70] It is noteworthy that grafts of encapsulated PC12 cells that excrete DA diffusely into the striatum, chronic infusion of DA into the striatum, implantation of viral vectors that can enable host striatal neurons to secrete DA locally, and carotid body or adrenal grafts can all produce recovery on the apomorphine rotation test, while having little or no effect on amphetamine rotation, in animals with complete nigrostriatal depletion.[71–74] This indicates that apparent functional recovery can be achieved through a variety of mechanisms to which different tests are differentially sensitive[72] (see below).

Rotation has been the most widely used test of functional recovery, not just for the very practical reasons of sensitivity, objectivity, and ease of use. More importantly, it provides a reliable and accurate index of the integrity of the underlying DA system, yielding close correlations between a simple, noninvasive behavioral measure in vivo and postmortem biochemical measurement of DA loss and restitution following experimental lesioning and transplantation.[65,66,75] It has consequently provided an effective behavioral screen in studies designed to compare the functional efficacy of alternative graft preparation paradigms,[20,76] transplantation methods,[77–79] graft placements,[64,80] the viability of alternative tissues,[72,77,81] or probing mechanisms of neuroprotection and functional recovery.[72,73,79,82,83]

Nevertheless, rotation used in isolation as an index of functional recovery needs to be treated with caution. Its very sensitivity can mean that functional effects are seen even with low levels of repair and reconstruction[83,84]—as few as 400 surviving neurons may be sufficient to sustain recovery[85]—which would not extend to other equally valid, but less sensitive, behavioral measures. Moreover, animals can show apparent recovery of apomorphine rotation following reduction of striatal activation not only as a consequence of receptor normalization but also, less specifically, simply by eliminating striatal overactivation through a lesion of the target neurons themselves.[86] Consequently, claims of functional efficacy based on recovery of apomorphine-induced rotation alone need to be complemented both with amphetamine rotation and preferably with other tests of motor behavior not dependent upon additional pharmacological activation.

Other simple motor functions

Over the last 25 years, a large range of simple motor behaviors have been used to assess functional recovery in VM-grafted rats, seeking indices that have better face validity representative of parkinsonian symptoms, and/or that are not dependent upon the use of pharmacological stimuli to drive the effects artificially (see Table 9.4.1). The first such tests evolved from similarities noted between nigrostriatal and lateral hypothalamic lesions, and involved neglect of motivational stimuli in contralateral space after unilateral lesions,[87] and generalized akinesia and regulatory failure in feeding and drinking after bilateral lesioning.[88,89] Some, but not all, of these voluntary behaviors exhibit enhanced recovery in transplanted animals, in particular in various tests of contralateral neglect, catalepsy, and akinesia.[64,90–92] An early principle to emerge was that graft placement is critical: the forebrain striatal complex is heterogeneously organized, at least in part reflecting the separation of parallel but functionally distinct corticostriatal circuits, so that grafts are needed to reinnervate the areas critical for each class of functional

TABLE 9.4.1. *Profiles of Behavioral Recovery Following DA Transplantation in Animals with Unilateral or Bilateral* 6-OHDA Lesions*

Recovered	Not Recovered	Worsened
Amphetamine rotation	Disengage test	±L-DOPA-primed dyskinesia
Apomorphine rotation	Hoarding	
Spontaneous rotation	Skilled paw reaching	Graft-induced dyskinesia
Contralateral neglect		
± Staircase test	± Staircase test	Tumors: focal neurological signs of raised intracranial pressure
± Cylinder test	± Cylinder test	
Stepping test	Aphagia*	
Placing test	Adipsia*	
Corridor test		
Emergent dyskinesia		
Intra-cranial self stimulation		
Choice reaction time		
Rotarod		
Beam balance		
Akinesia*		
Catalepsy*		

* Tested after bilateral 6-OHDA lesioning; ±, conflicting results in different studies. For detailed reviews, see[105,208].

impairment. As a consequence, a series of single and double dissociations between graft placement and functional recovery have been mapped in the striatum, reflecting known topography of the system, and additive patterns of recovery are readily achieved using multiple graft placements.[64,92,93]

There remain, nevertheless, a subset of behaviors that appear to be resistant to recovery, even with multiple graft deposits providing comprehensive striatal reinnervation (see Table 9.4.1). Thus, for example, unilateral 6-OHDA lesions induce consistent impairments of contralateral limb use in tests of skilled paw use that require the rats to reach, grasp, and retrieve food pellets. In early studies, paw reaching consistently failed to show recovery after VM grafting in animals where the impairment was caused by nigrostriatal bundle lesions, regardless of where in the striatal terminal fields the graft was placed.[94–96] One interpretation of this failure has been that some functions, such as skilled reaching, are dependent upon the integrity of the nigrostriatal pathway for signaling (e.g., somesthetic feedback relating to successful grasping).[94] This led to the introduction of a variety of strategies seeking to improve recovery by providing a more complete circuit reconstruction, either by the combined placement of grafts into striatum and SN[97] or by combining intranigral VM graft placements with a bridge graft that would allow the implanted DA neurons to extend long-distance axons to the distal striatum (see below).[83,98,99] These strategies achieved, at best, only limited success. Attention therefore turned to the contribution of extrastriatal contributions to dopaminergic denervation, since recovery on several measures, such as stepping paw use and the corridor test, is more readily achieved in animals with partial lesions restricted to the striatum than in bundle-lesioned animals, even though the initial deficit may seem similar[100,101] and recovery in terminal lesioned rats is abolished by subsequent nigrostriatal bundle lesions.[102] Moreover, recovery may be promoted by additional nigral placement of either VM grafts[103–105] or nondopaminergic grafts rich in GABA neurons.[27]

More recently, attention has focused on developing a range of improved tests of motor function. These include tests of voluntary and reflexive use of individual paws—for example, in the cylinder, stepping, and corner placing tests—and tests that involved more comprehensive motor coordination and balance—for example, beam balance or rotarod tests.[105–108] In addition, the corridor test has recently provided a simple method of recording recovery in contralateral neglect using the objective measure of food pellets collected from the two sides of the body,[101] in contrast to the rating scores used in earlier sensorimotor batteries. On each of these tests, again, incomplete profiles of recovery are reported in VM-grafted animals that nevertheless exhibit complete recovery in rotation, again related to a variety of factors including the number of cells surviving, graft placement, extent of reinnervation, and its relation to both the magnitude and the spatial extent of the initial lesion-induced denervation.[100,101,104,105,109–113]

Motor learning and cognition

Cognitive symptoms are an established feature of PD, and marginal changes have been reported in cognitive and neuropsychological function following VM transplantation in several studies in patients.[114,115] However, although clear-cut cognitive deficits of the frontal type can be recorded in rats with bilateral striatal lesions, it has proved difficult to study similar effects following nigrostriatal 6-OHDA lesioning, due to the profound regulatory and motivational changes that result from bilateral nigrostriatal denervation in animals.[89,116] Bilateral 6-OHDA lesions in discrete striatal terminal zones can reveal specific cognitive impairments in the immediate postoperative period.[117,118] However, any such lesion that is sufficiently mild to avoid the attendant problems in feeding and drinking will also exhibit a marked capacity for spontaneous recovery of function by a variety of compensatory cellular and subcellular mechanisms,[14,119] no longer providing stable, long-term impairment against which graft function can be assessed. As a consequence, there have been no studies of VM graft effects on frontal-type cognitive functions in experimental rodents.

However, one aspect of cognition that has been amenable to analysis in unilaterally lesioned animals concerns the aspect of motor learning relating to the establishments of stimulus-response (S-R) associations underlying habits—that is, motor skills acquired through repetition and practice. *Habit formation* is a specific aspect of procedural learning that has been particularly associated with the striatum, as distinct from the *episodic* (factual or knowledge-based) learning established via more posterior hippocampal and cortical systems.[120–122] Recent physiological evidence has highlighted the involvement of the nigrostriatal dopaminergic projection to the striatum in providing the signals of *reward* necessary to establish and modify associations between imperative stimuli and the animal's responses, depending upon the outcome of the response.[123,124] Rats with unilateral nigrostriatal lesions exhibit selective impairments in the speed and accuracy of making contralateral responses in a choice reaction time task,[125] a deficit that is significantly alleviated by VM grafts.[126] In particular, not only did grafted animals exhibit recovery on a range of parameters in task performance, but the profile of responding and recovery over multiple trials clearly suggested an underlying learning rather than a simple motor impairment, most plausibly in terms of loss and restitution of the motivational or reward-related signals necessary to maintain the learned S-R habit.[126] Interestingly, with hindsight, this result was presaged in the early graft literature by the demonstration that implanted DA neurons could provide an effective substrate for signaling reward in a self-stimulation paradigm.[127]

Behavioral Testing in Mice

Until recently, most anatomical and behavioral analysis was undertaken in rats, but increasingly, mice are emerging as an additional important experimental species because of the opportunities provided by the new range of genetic tools for transgenesis of marker genes (see the anatomical studies using GFP transgenic donors[33] above), and for genetic manipulation to provide improved models of disease, or as a tool to manipulate the cells and their host environment.

The standard unilateral 6-OHDA lesion used in rats can be transferred to mice, and there are a number of reports of successful alleviation of rotational impairments following intrastriatal transplantation of embryonic nigral cells[128–130] in such mice, similar to previous observations in rats. Nevertheless, such surgery is technically more difficult in mice due to the smaller brain size and the consequent difficulty of avoiding a significant death rate under surgery due to diffusion of the toxin to the contralateral side of the brain.

An alternative lesion method for mice is the peripheral administration of the toxin 1-methyl-4-phenyl-tetrahydropiridine (MPTP), which causes extensive bilateral dopaminergic depletions in mice and induces a parkinsonian syndrome in monkey and man. Nevertheless, because the lesions are partial and largely spare the DA innervation of the ventral striatum, the lesioned mice survive and have been used as recipients for embryonic nigral grafts,[131] adrenal medulla tissue grafts,[132,133] and, more recently, as a platform for exploring alternative sources of mouse and human neural stem cells.[134–136] This model system was important not only for demonstrating the functional viability of adrenal grafts, but also for highlighting the fact that functional recovery was not primarily due to the replacement of lost dopaminergic cells, but to trophic influences of the grafted tissues on sprouting of spared host DA neurons, enhancing their capacity to reinnervate areas of lesion-induced denervation.[132,133] An additional advantage of the MPTP model is that the behavioral consequences in mice are relatively well characterized.[137] Nevertheless, the utility of the model in mice is complicated by the fact that smaller lesions are associated with significant spontaneous recovery over the time course required for graft studies, whereas larger lesions can result in many fatalities, in particular in female mice.[137]

The first studies of DA-rich grafts in genetic mutant animals were undertaken in a series of experiments by Triarhou and colleagues in the mutant weaver ($^{wv/wv}$) strain of mice.[138] The weaver strain involves a single base mutation in the *Girk2* gene, and the homozygous mice are characterized by a marked degeneration of nigrostriatal DA neurons, resulting in 60% cell loss by 3 months of age and up to 85% loss over the life span, as well as additional pathology in the cerebellum and hippocampus. As a consequence of the combined pathology, the mice exhibit progressive impairments in motor functions, including impaired locomotion, instability of gait, poor limb coordination, and tremor. Ventral mesencephalic grafts survive well in the weaver striatum, restore synaptic connectivity, and alleviate functional impairments on a broad battery of behavioral tests including beam balance, locomotor coordination, and locomotor activity.[138–141] These studies provide the first clear evidence of nigral graft survival in a progressive neurodegenerative model relevant to most human neurodegenerative diseases, in comparison to the acute lesions that are typically utilized prior to grafting in a stable environment of chronic denervation.

It will now be of interest to explore the survival and integration of grafted DA neurons and the functional viability of VM transplants in transgenic models of genetic forms of PD. Where the new transgenic technologies have already proved of value is in manipulating the specific molecular phenotypes of implanted neurons: to provide explicit markers of the fates of dopaminergic neurons[33,142] or astrocytes[143] of donor origin by using different promoters to a GFP transgene; to explore the effects of inhibition of oxidative stress on graft cell survival using transgenic embryos that overexpress Cu/Zn superoxide dismutase as VM graft donors[144]; and, most recently, by using mice that overexpress GFP under the control of the Pitx3 gene as VM tissue donors, along with fluorescent-activated cell sorting, in order to select a purified subpopulation of cells destined to develop a dopaminergic fate at an earlier age than their explicit expression of dopaminergic phenotypes.[145]

Dyskinesia

Dyskinesia is a common side effect of L-DOPA therapy in PD patients. As discussed in detail in Chapter 9.2 in this volume, these abnormal movements are caused by the chronic, intermittent L-DOPA medication, which leads to nonphysiological activation of DA receptors and the development of abnormal postsynaptic responses in the dopaminoceptive neurons. Since dyskinesia develops only in patients with severe lesions of the nigrostriatal DA system, one would predict that functionally effective DA neuron grafts would have a beneficial effect on this unwanted side effect. Experiments performed in 6-OHDA-lesioned rats have shown that intrastriatal DA neuron grafts are indeed effective in reducing (by 60%–80%) established dyskinesias induced by daily injections of low-dose L-DOPA[147–149] (Fig. 9.4.6). Such grafts are also effective in preventing the development of dyskinesia in previously nondyskinetic animals.[150] In open-label clinical studies, reductions in dyskinesias (percentage of time "on" with dyskinesias) were reported in some PD patients, while the remaining patients either showed no change or became worse.[151–154] The reason for this variable outcome is not clear. Differences in graft placement and DA fiber outgrowth are one possibility. In their 2006 study, Carlsson and colleagues[146] showed that the effect of single VM graft deposits is markedly different when the grafts are placed in the rostral or caudal part of the striatum: animals with caudal grafts showed a 60% reduction, compared to about 20% in the rostral graft group.

Graft-induced dyskinesia was first observed as a troublesome side effect in the two National Institutes of Health-sponsored double-blind studies published in 2001 and 2003.[155,156] In the first study, 4 out of 33 grafted patients developed severe involuntary movements over time in the absence of L-DOPA; in the second study, off-state dyskinesia was observed in 13 out of 23 grafted patients. These dyskinesias that persisted after L-DOPA withdrawal were observed as repetitive, stereotypic movements in the lower extremities, with residual parkinsonism in other body regions. In a recent retrospective analysis, Olanow and colleagues[157] suggested that these graft-induced involuntary movements may represent a prolonged form of diphasic dyskinesia.

Further investigation of the dyskinetic patients in the first trial using [^{18}F]-DOPA positron emission tomography (PET) showed asymmetric, localized increases in the PET signal in the ventral putamen that were not present in the grafted nondyskinetic patients,[158] suggesting that unbalanced local increases in striatal DA function caused by patchy DA innervation may contribute to the development of these dyskinesias. This effect may be particularly pronounced if the grafts are placed in the posterior putamen, that is, close to the areas that showed localized increases in the [^{18}F]-DOPA PET scans in the dyskinetic patients[158] (see Chapter 9.5 in this volume for further discussion).

These unexpected clinical findings have stimulated a new wave of experimental studies aimed at elucidating the mechanisms underlying graft-induced dyskinesia. It is an interesting fact that none of the early preclinical studies performed in rodent or primate models of PD before the onset of the clinical studies had observed any signs of graft-induced involuntary movements. Studies in animals made dyskinetic by chronic L-DOPA treatment prior to transplantation were clearly warranted. Those performed so far have focused on the role of focal, patchy or unevenly distributed innervation,[146,147,159] dysregulated DA release (induced by an amphetamine pulse),[146,147,159] an induced inflammatory response,[160] or inclusion of serotonin neurons in the transplanted cell preparation.[150,161] As in nondyskinetic animals, off-state dyskinesia (i.e., dyskinesia in the absence of L-DOPA) was not observed in grafted, dyskinetic animals in any of the animal studies performed so far. Prominent dyskinesia, however, was seen when the grafted and chronically L-DOPA-treated animals were given a single dose of the DA-releasing compound amphetamine.[146,147] This effect was not seen in the nongrafted controls and was correlated with the extent of graft-derived DA innervation in the caudal part of the striatum.[146] Although it is unclear whether this form of graft-induced dyskinesia is relevant to the understanding of clinical dyskinesias, it is the only experimental model of graft-induced dyskinesia currently available.

During the last few years, the Lund laboratory has been particularly interested in the role of serotonin neurons in the induction of L-DOPA- and graft-induced dyskinesias. The serotonergic neurons not only innervate the striatum, but are also capable of decarboxylating L-DOPA to DA, and store and release DA in the DA-denervated striatum. This may be particularly important in advanced stages of PD when a major part of the nigral DA system has degenerated and the remaining DA neurons are in a compromised functional state. As the striatum loses its dopaminergic innervation, it is likely that the spared serotonin innervation comes to play an increasing role in the handling of systemically administered L-DOPA and provides an additional site for synthesis, storage, and release of DA formed from L-DOPA. In a recent study, Carta and colleagues[162] showed that the striatal serotoninergic

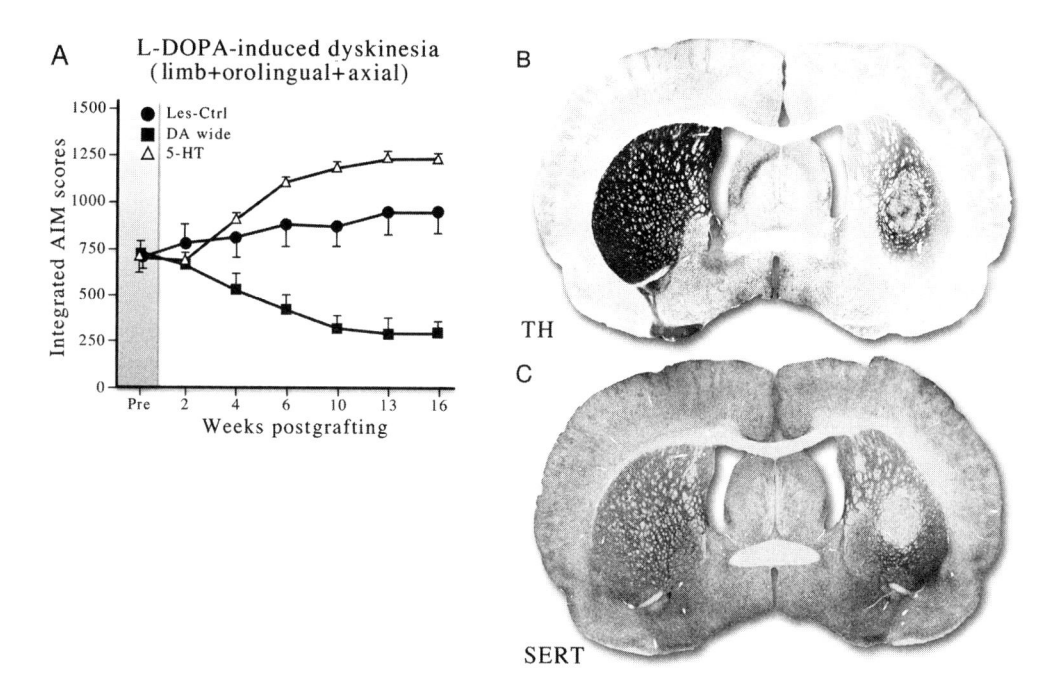

FIGURE 9.4.6. Effect of grafts rich in DA or 5-HT neurons on L-DOPA-induced dyskinesia. In this experiment 6-OHDA-lesioned rats were made dyskinetic by daily injections of L-DOPA (6 mg/kg + benserazide) and grafted with either fetal VM tissue (rich in DA neurons; see the TH-stained section in panel B) or fetal raphé tissue (rich in serotonin neurons; see the 5-HT-stained section in panel C). The two types of grafts had opposite effects: a marked reduction in dyskinesia over time in the animals that received DA neuron–rich grafts (filled squares) and a marked exacerbation over time in the animals that received serotonin-rich grafts (open triangles). Data from Carlsson et al.[150]

afferents play a key role in the induction and maintenance of L-DOPA-induced dyskinesia in 6-OHDA-lesioned rats. In animals with either partial or complete lesions of the nigrostriatal DA system, dyskinesia induced by daily L-DOPA treatment was almost completely eliminated when the serotonin afferents were removed. Dampening of the serotonin neuron activity by $5-HT_{1A}$ and $5-HT_{1B}$ autoreceptor agonists provided a near-complete blockade of L-DOPA-induced dyskinesia in L-DOPA-primed animals.

These results indicate that dyskinetic movements induced by repetitive low doses of L-DOPA are triggered by DA released from serotonin terminals in the DA-denervated striatum. Although the serotonin terminals are capable of synthesizing and storing L-DOPA-derived DA, DA released from serotonin terminals is not regulated in a normal way. In dopaminergic synapses, extracellular DA concentrations are kept within a narrow physiological range through a combination of autoreceptor-mediated feedback and reuptake via the DA transporter. Since DA released from serotonin terminals is not subjected to any autoregulatory feedback control, the extracellular levels of DA released from the serotonin afferents would be expected to show excessive swings in response to a systemic L-DOPA injection. Such dysregulated release of L-DOPA-derived DA is likely to be the main trigger of dyskinesia in L-DOPA-primed animals.

These results suggest the possibility that serotonin neurons included in the grafted VM tissue could play an important role as a source of excessive, dysregulated release of DA and serve as a trigger for the induction of L-DOPA-induced, and possibly also graft-induced, dyskinesias. The result of two recent studies[150,161] show that grafts containing serotonin neurons indeed have a detrimental effect on L-DOPA-induced dyskinesia. In these 6-OHDA-lesioned rats, which showed only low-level dyskinesia at the time of transplantation, serotonin grafts induced a worsening in the severity of dyskinesia that developed during continued L-DOPA treatment, while grafts rich in DA neurons had the opposite, dampening effect. The detrimental effect seen in animals with serotonin neuron grafts was dramatically increased when the residual DA innervation in the striatum was removed by a second 6-OHDA lesion. FosB expression in the striatal projection neurons, which is closely linked to dyskinesia, was normalized by DA neuron grafts but not by serotonin neuron grafts. The results, moreover, suggested that the increased serotonin innervation generated by the grafted serotonin neurons had a limited effect on the development or severity of L-DOPA-induced dyskinesias as long as a sufficient portion, some 10%–20%, of the DA innervation remained. At more advanced stages of the disease, when the DA innervation of the putamen is reduced below this critical threshold, grafted serotonin neurons are likely to aggravate L-DOPA-induced dyskinesia, but only in those cases where the DA reinnervation derived from the grafted neurons is insufficient in magnitude to restore the striatal DA innervation above this threshold.

The conclusion of these studies is that it is not the absolute number of serotonin neurons in the transplants, but the relative proportion of DA and serotonin neurons, that is the main factor determining the beneficial or detrimental effects of VM tissue grafts on L-DOPA-induced dyskinesia in grafted PD patients. Whether the serotonin neurons play any role in the development of the off-state graft-induced dyskinesia in patients remains unclear. In the Carlsson et al study,[161] dyskinesia induced by an amphetamine pulse (see above) was observed in all animals with DA neuron grafts, independent of their content of serotonin neurons, but not in the rats with serotonin-only transplants. These results support the idea that dysregulated release of DA from the graft-derived DA innervation is the primary cause of off-drug dyskinesia.

MECHANISMS OF RECOVERY

With the demonstration of functional recovery following transplantation of embryonic dopaminergic neurons that not only survived but restored connections with the host brain, it was natural for early studies to conclude that this new technique offered a strategy for surgical repair based on replacement of lost neurons and reconstruction of damaged circuits in the brain. However, as already suggested when comparing the effects of different graft types on amphetamine- and apomorphine-induced rotation (above), subsequent analyses indicated that grafts might influence host function via a variety of different more or less specific mechanisms.[163–166] A more refined theoretical analysis is therefore required of the alternative mechanisms (see Table 9.4.2) that apply to distinct classes of structural damage and functional impairment if we are to implement the most effective and efficient strategies for treatment, whether symptomatic, neuroprotective, or truly reparative.

Nonspecific (Surgical) Effects

The past decade has seen a dramatic rise in surgical therapies for advanced PD, which increases the

TABLE 9.4.2. *Mechanisms Influencing Functional Outcome After VM Cell Transplantation*

Mechanism	Description	Example(s)	Reference(s)
Trauma	Adverse effects as a consequence of surgical trauma or damage	Abdominal surgery for adrenal tissues; tumor formation from grafted tissues	209, 215
Nonspecific	Surgical lesion produces restorative balance in output systems	Deep brain stimulation or lesions of subthalamic r pallidal nuclei	216, 217
Trophic—protective	Grafts release trophic molecules that protect neurons against disease progression	In vivo and ex vivo gene therapy	218, 219
Trophic—restorative	Grafts release trophic molecules that stimulate endogenous plasticity, sprouting, and reorganization	Adrenal grafts (?); in vivo and ex vivo gene transfer of trophic factors	133, 193, 220
Pharmacological	Diffuse release of DA into host neuropil	DA-secreting polymers or minipumps; in vivo "tricistronic" gene transfer	73, 176
Pathway repair	Grafts provide substrate to stimulate and direct axon growth to remote targets	Bridge grafts of peripheral glial cells to allow nigrostriatal reconstruction	83, 98, 204
Neuronal replacement and innervation	Grafted neurons reinnervate host brain and restore locally regulated transmitter release at physiological levels at synaptic sites	Ectopic nigral grafts in the striatum	164
Full circuit reconstruction	Full replacement of lost DA neurons in nigra, receiving appropriate inputs from the host brain and restitution of signaling via a reconstructed nigrostriatal pathway	Combination intranigral and bridge grafts (but not reliably achieved in practice)	3, 98, 204

expectation of safety and efficacy demanded of an alternative transplantation strategy. Surgical approaches initially involved lesions in basal ganglia output nuclei, including ventrolateral nuclei of the thalamus, globus pallidus, and subthalamic nucleus.[167] These have largely been replaced more recently by deep brain stimulation via implanted electrodes, which offers a significant safety advantage due to its reversibility (i.e., simply by switching off the stimulator) and its scope for patient-specific titration to optimize the therapeutic response.[167,168] The rationale for both lesion and stimulation surgeries has emerged from models of basal ganglia function, in which blocking activity of specific structures in the output pathways can restore a degree of balance disrupted by striatal disinhibition resulting from the primary DA denervation of the disease state.[169] These studies highlight the fact that targeted surgical lesions can by themselves yield significant recovery in PD—and indeed, striatal lesions were themselves suggested as an early surgical target.[170] Consequently, the first issue for cell transplantation is to determine that any functional effects are not simply attributable to nonspecific damage associated with the implantation surgery. Experimentally, this issue may be addressed by using a variety of control procedures including implantation of nondopaminergic tissues,[65,171] showing that the functional changes develop progressively,[68,172,173] along with graft integration and growth, and correlate specifically with profiles of DA replacement and reinnervation,[66,68,127] as well as demonstration of relapse when the grafts are removed.[65,67,68]

Pharmacological Effects

A more plausible mechanism for the functional efficacy of nigral grafts is that the implanted cells secrete neuroactive molecules, in particular DA, into the host neuropil, replacing the endogenous DA lost through a lesion or disease. As such, the grafts would restore dopaminergic activation at receptors in the denervated striatum via a process of diffuse, chronic release and restore function by a mechanism similar to the DA receptor activation provided by L-DOPA or DA agonists. Indeed, the functional benefit of grafts may be greater than that of drugs, as they offer the possibility of providing chronic, stable delivery of DA, at physiological levels, to selected sites in the brain, circumventing the pharmacodynamic and peripheral side effects of conventional pharmaceuticals.

The hypothesis that VM grafts exert their effects through a pharmacological mechanism of action has provided the stimulus for the search for alternative implantable DA delivery systems, including DA-secreting polymers or minipumps,[73,174] neuroendocrine cells or cell lines that secrete catecholamines including DA,[71,175] or direct engineering of exogenous or host cells to synthesize DA.[176] It is difficult to determine the extent to which diffuse release of DA from VM grafts contributes to their functional effects, but we have argued that—since the profile of functional recovery is both broader in extent and greater in magnitude in VM-grafted animals than in animals with any of the alternative DA-secreting implants—the observed recovery provided by the VM grafts is not simply pharmacological, but rather is dependent upon the re-formation of synaptic innervation of the host brain from implanted neurons that is subject to local phasic regulation.[164,166] As we have seen (above), even in ectopic sites, the substrates for such local regulation are observed at anatomical, physiological, and biochemical levels of analysis.

Trophic Grafts

In addition to replacing lost neurons, grafted embryonic cells and cell lines can provide a rich source of trophic factors that can promote the survival and plasticity of endogenous neurons. Early studies noted unidentified trophic responses in the host brain, such as the induction of anatomical sprouting, associated with grafted tissues[133,177]; subsequently, interest focused on designing cells, for example by genetic engineering ex vivo, for their capacity to secrete specific trophic molecules.[178,179] Trophic processes can involve several distinct components: neuroprotection against prodegenerative processes associated with trauma and disease or with the graft preparation process itself; the provision of alternative targets to protect against retrograde degeneration following loss of a neuron's normal targets; or the induction of biochemical and/or anatomical plasticity to compensate for neurodegeneration that has already been sustained.[180,181]

DA cell survival in VM grafts

A key issue in VM transplantation is that only small numbers of implanted neurons survive the first week following transplantation.[182] Analysis of DA cell death in the immediate posttransplantation period led to the introduction of a variety of strategies designed to enhance DA cell survival. These include the exogenous application of trophic molecules (such as FGF or GDNF) or other neuroprotective agents (such as lazaroids, antioxidants, etc.) into the host graft environment,[182–187] engineering the host striatal neurons to express increased levels of the relevant survival factors,[188] or engineering the grafted cells themselves to express trophic or antiapoptotic factors.[144,189–191] Although in most cases safety considerations have required that they remain as research tools, some of these methods have been found to transfer effectively to the development of improved preparation protocols for clinical application.[153]

Protection of endogenous DA neurons

Although not relating specifically to DA grafts themselves, the same considerations have resulted in similar strategies being applied for neuroprotection of host DA neurons whose survival is compromised by the endogenous disease process. This has been widely studied in the context of application of neurotrophic factors, in particular GDNF, to retard or reduce the retrograde degeneration associated with intrastriatal 6-OHDA lesions. As with neuroprotection of grafts, early studies focused on exogenous delivery of the trophic factors into the forebrain by direct injection or chronic infusion, whereas recent studies have focused on use of viral vectors (in particular based on lentivirus and adeno-associated virus) to transfect striatal or nigral neurons with the relevant neuroprotective genes.[192,193]

Pilot open label studies have suggested promising results of intrastriatal GDNF infusion in PD, following patients for up to 2 years, demonstrating modest alleviation of symptoms and slowing of disease progression.[194] However, a similar benefit did not translate into a first full randomized control trial, which may be attributable to important technical differences.[195] Further trials using a gene transfer strategy instead of direct infusions are believed to be in progress.

Enhanced DA axon growth

A third method for trophic stimulation of DA graft integration is the use of trophic/tropic molecules or cells to stimulate and direct DA axon growth, whether this be developmental, from grafted DA neurons, or regenerative, from axotomized host nigral neurons. Cografting of VM tissues with embryonic striatal tissue can both increase the magnitude of DA neurite outgrowth from VM grafts and direct fiber regeneration toward the trophic stimulus,[196,197] although this may be detrimental to the extent that the developing embryonic axons exhibit a preference for the embryonic striatum rather than for adult host striatal targets.[98,198] A specific aspect of the challenge of promoting and targeting axon growth is the search for strategies to promote pathway repair, in particular across long-distance tracts such as the nigrostriatal projection (as described above).

Bridge Grafts

Ventral mesencephalic grafts placed into the striatum are unlikely to have the capacity to fully restore all aspects of normal dopaminergic function, not least because their ectopic location precludes the possibility of their providing a relay of information afferent to the dopaminergic neurons that would normally reside in the SNc. Conversely, we now know, using new, more-sensitive methods of visualization (as described above), that VM grafts implanted into the SN can provide the source of at least some long-distance axon growth back to the striatum and recovery on some tests of motor asymmetry, such as DA agonist-induced rotation.[103,199,200] Nevertheless, the extent of such growth and any associated functional recovery in most studies is extremely limited.[64,97,103] Thus, if we wish to achieve effective reconstruction of the nigrostriatal circuitry following transplantation of replacement DA neurons into the homotopic area from which endogenous cells are lost, new strategies need to be found to promote the long-distance growth of connections to their appropriate distant targets. This challenge led to studies using alternative tissues, such as peripheral nerve, that are known to provide an effective substrate for CNS axon regeneration[201,202] as bridges for nigrostriatal regeneration. The feasibility of this strategy was first achieved by the demonstration of effective long-distance growth of DA axons from a solid VM graft implanted onto the dorsal tectal surface, cografted with a segment of peripheral nerve overlying the cortex and exiting the distal end back into the striatum.[203,204] This basic strategy was then repeated using a more practical stereotaxic placement of alternative bridge cells via an oblique intracerebral track directly between a VM graft placed into the SN and striatum.[98,205] Using a variety of different tissues, in each case the bridges have been shown to enable limited numbers of DA axons from VM grafts positioned in the midbrain to grow back the full distance to innervate the host striatum.[83,98,99,205] However, the extent of growth remains extremely limited with all present bridge graft protocols, and efficient nigrostriatal reconstruction at a level likely to be necessary for an effective profile of functional benefit cannot yet be achieved.

ALTERNATIVE DA TISSUE SOURCES

Although embryonic VM grafts can alleviate a broad range of motor deficits in 6-OHDA-lesioned rats, and there is now clear proof of the principle that human VM grafts can provide significant clinical benefit to some PD

TABLE 9.4.3. *Alternative Sources of DA Cells for Transplantation*

Type	Source	Graft Survival	Recovery in 6-OHDA Model	Significant Issues	Example(s)
Embryonic VM	E12-E16 (rat) or human embryos	Excellent	Extensive	Tissue availability and quality for clinical use	4,221
Xeno embryos	Porcine, rodent	Moderate	Moderate	Immunological rejection; zoonoses	222,223
Adrenal medulla	Autografts	Poor	Poor	Poor morbidity; limited efficacy in clinical trials	175,224
Carotid gland	Autografts	Moderate	Poor	Limited evidence for neuronal differentiation or efficacy	71,225
PC12 cells	Banked cell line	Moderate	Poor	Rejection; poor differentiation; limited function	71
Engineered cells				Poor differentiation; limited function	212,226
Fetal neural progenitors	Rat, mouse, or human embryos	Moderate	Poor	Poor survival, limited differentiation, migration	227,228
ES stem cells	Banked mouse or human lines	Good	Good	Tumor formation; Effective differentiation prior to, and de-differentiation following, transplantation	229,230
Adult neural stem cells	Fresh or banked cells	Poor	Poor	Limited evidence for specific differentiation or efficacy	231
Adult peripheral cells	Fresh or banked cells (e.g., cord blood)	Poor	Poor	Limited evidence for specific differentiation or efficacy	232

patients (for a review, see Chapter 9.5 in this volume), a significant issue remains: the choice of appropriate tissues for clinical application. Widespread clinical development of cell transplantation is currently constrained largely by the ethical sensitivity, the limited availability, and the difficulty of maintaining an appropriate level of quality control and standardization that attend the clinical use of human fetal donor tissues derived from elective abortion. As a consequence, there is a long-standing experimental search for alternative sources of tissue of suitable quality, specificity, and standardization that could be as effective as primary fetal tissues for clinical transplantation, whether in PD or in other neurodegenerative diseases[206] (Table 9.4.3).

The first alternatives to be explored were the use of peripheral neuroendocrine tissues containing dopaminergic cells, such as adrenal medulla or carotid body. Experimental studies indicated functional benefit in the 6-OHDA-lesioned rat on simple rotation measures,[71,175,207] but peripheral tissue grafts never achieved the broad profile of recovery exhibited by fetal VM cells.[208] Moreover, clinical trials using adrenal tissue autografts proved disappointing because of their short duration of benefit, limited efficacy, and significant morbidity.[209,210] A second approach has been to use immortalized and engineered cells and cell lines that secrete DA,[71,211–214] but these have also proved disappointing because of their inability to provide a sustained functional benefit.

More promisingly, recent attention has turned to the prospect of using stem cells as a source for directing differentiation of dopaminergic neurons with functional capacity comparable to that of embryonic-derived neurons. Interesting progress has been made over the last few years, particularly in the development of protocols that allow the generation of DA neurons, or DA neuron precursors, in large numbers from embryonic stem cells. As discussed in some detail in Chapter 9.5 of this volume, this line of research shows great promise, but important issues remain to be resolved, relating both to the reliability of specific cell differentiation and to safety, before these stem cell–derived DA neuron preparations can be used effectively in experimental research.

SUMMARY

In this chapter, we have reviewed the transplantation of DA neurons as a powerful model for understanding the basic neurobiology and methods for achieving viable cell transplantation in the brain. Analysis of the mechanisms involved in structural repair and functional recovery indicates that there are particular requirements for the implanted cells to differentiate into specific brainstem phenotypes for effective integration into the host brain and broad functionally efficacy. Cell implantation into DA-denervated rats and mice has provided effective animal models for the preclinical analyses required for translating novel cell therapies into applications in human neurodegenerative disease and for resolving specific issues, such as potential dyskinetic side effects, that have been raised in the course of the pilot clinical trials. Although most studies have used primary fetal DA neurons, attention is increasingly turning to understanding the developmental, molecular, and genetic principles that will allow selection and differentiation of alternative cell sources, such as stem cells, into a dopaminergic phenotype suitable for future clinical development.

ACKNOWLEDGMENTS

We acknowledge the United Kingdom and Swedish Medical Research Councils, the Parkinson's Disease Society of Great Britain, and the Michael J. Fox Foundation for their long-term financial support of our respective laboratories and studies in this field. We thank Bengt Mattsson for preparation of the illustrations.

REFERENCES

1. Olson L, Seiger Å. Brain tissue transplanted to the anterior chamber of the eye. 1. Fluorescence histochemistry of immature catecholamine and 5-hydroxytryptamine neurons innervating the iris. *Z Zellforsch*. 1972;195:175–194.
2. Björklund A, Stenevi U, Svendgaard N-A. Growth of transplanted monoaminergic neurones into the adult hippocampus along the perforant path. *Nature*. 1976;262:787–790.
3. Stenevi U, Björklund A, Svendgaard N-A. Transplantation of central and peripheral monoamine neurons to the adult rat brain: techniques and conditions for survival. *Brain Re.s* 1976;114:1–20.
4. Perlow MJ, Freed WJ, Hoffer BJ, Seiger Å, Olson L, Wyatt RJ. Brain grafts reduce motor abnormalities produced by destruction of nigrostriatal dopamine system. *Science*. 1979;204:643–647.
5. Björklund A, Stenevi U. Reconstruction of the nigrostriatal dopamine pathway by intracerebral transplants. *Brain Res*. 1979;177:555–560.
6. Ungerstedt U. 6-Hydroxy-dopamine induced degeneration of central monoamine neurons. *Eur J Pharmacol*. 1968;5:107–110.
7. Schwarting RKW, Huston JP. Unilateral 6-hydroxydopamine lesions of meso-striatal dopamine neurons and their physiological sequelae. *Prog Neurobiol*. 1996;49:215–266.
8. Björklund A, Dunnett SB. Dopamine neuron systems in the brain: an update. *Trends Neurosci*. 2007;30:194–202.
9. Ungerstedt U, Arbuthnott GW. Quantitative recording of rotational behaviour in rats after 6-hydroxydopamine lesions of the nigrostriatal dopamine system. *Brain Res*. 1970;24:485–493.
10. Schwarting RKW, Huston JP. The unilateral 6-hydroxydopamine lesion model in behavioral brain research. Analysis of functional deficits, recovery and treatments. *Prog Neurobiol*. 1996;50:275–331.
11. Dunnett SB. Motor functions of the nigrostriatal dopamine system: studies of lesions and behaviour. In: Dunnett SB, Bentivoglio M, Björklund A, Hökfelt T, eds. *Handbook of Chemical Neuroanatomy. Vol. 21. Dopamine*. Amsterdam: Elsevier; 2005;235–299.
12. Dunnett SB, Björklund A. Parkinson's disease: prospects for novel restorative and neuroprotective treatments. *Nature Suppl*. 1999;399:32–39.
13. Kirik D, Rosenblad C, Björklund A. Characterization of behavioral and neurodegenerative changes following partial lesions of the nigrostriatal dopamine system induced by intrastriatal 6-hydroxydopamine in the rat. *Exp Neurol*. 1998;152:259–277.
14. Zigmond MJ, Abercrombie ED, Berger TW, Grace AA, Stricker EM. Compensations after lesions of central dopaminergic neurons: some clinical and basic implications. *Trends Neurosci*. 1990;13:290–296.
15. Olson L, Seiger Å, Strömberg I. Intraocular transplantation in rodents: a detailed account of the procedure and examples of its use in neurobiology with special reference to brain tissue grafting. *Adv Cell Neurobiol*. 1983;4:407–442.
16. Sinclair SR, Fawcett JW, Dunnett SB. Dopamine cells in nigral grafts differentiate prior to implantation. *Eur J Neurosci*. 1999;11:4341–4348.
17. Björklund A, Schmidt RH, Stenevi U. Functional reinnervation of the neostriatum in the adult rat by use of intraparenchymal grafting of dissociated cell suspensions from the substantia nigra. *Cell Tissue Res*. 1980;212:39–45.
18. Dunnett SB, Torres EM, Gates MA, Fricker-Gates RA. Neural transplantation. In: Tatlisumak T, Fisher M, eds. *Handbook of Experimental Neurology: Methods and Techniques in Animal Research*. Cambridge, England: Cambridge University Press; 2006;269–307.
19. Schmidt RH, Björklund A, Stenevi U. Intracerebral grafting of dissociated CNS tissue suspensions: a new approach for neuronal transplantation to deep brain sites. *Brain Res*. 1981;218:347–356.
20. Brundin P, Barbin G, Strecker RE, Isacson O, Prochiantz A, Björklund A. Survival and function of dissociated rat dopamine neurons grafted at different developmental stages or after being cultured in vitro. *Dev Brain Res*. 1988;467:233–243.
21. Torres EM, Monville C, Lowenstein PR, Castro MG, Dunnett SB. Delivery of the sonic hedgehog or glial derived neurotrophic factor to dopamine-rich grafts in a rat model of Parkinson's disease using adenoviral vectors. Increased yield of dopamine cells is dependent on embryonic donor age. *Brain Res Bull*. 2005;68:31–41.
22. Torres EM, Monville C, Gates MA, Bagga V, Dunnett SB. Improved survival of young donor age dopamine grafts in a rat model of Parkinson's disease. *Neuroscience*. 2007;146:1606–1617.
23. Björklund A, Stenevi U, Schmidt RH, Dunnett SB, Gage FH. Intracerebral grafting of neuronal cell suspensions. II. Survival and growth of nigral cell suspensions implanted in different brain sites. *Acta Physiol Scand Suppl*. 1983;522:9–18.
24. Doucet G, Brundin P, Descarries L, Björklund A. Effect of prior dopamine denervation on survival and fiber outgrowth from

intrastriatal fetal mesencephalic grafts. *Eur J Neurosci.* 1990;2:279–290.

25. Manier M, Abrous DN, Feuerstein C, Le Moal M, Herman JP. Increase of striatal methionin enkephalin content following lesion of the nigrostriatal dopaminergic pathway in adult rats and reversal following the implantation of embryonic dopaminergic neuron—a quantitative immunohistochemical analysis. *Neuroscience.* 1991;42:427–439.

26. Nikkhah G, Cunningham MG, Jödicke A, Knappe U, Björklund A. Improved graft survival and striatal reinnervation by microtransplantation of fetal nigral cell suspensions in the rat parkinson model. *Brain Res.* 1994;633:133–143.

27. Winkler C, Bentlage C, Nikkhah G, Samii M, Björklund A. Intranigral transplants of GABA-rich striatal tissue induce behavioral recovery in the rat Parkinson model and promote the effects obtained by intrastriatal dopaminergic transplants. *Exp Neurol.* 1999;155:165–186.

28. Hudson JL, Bickford P, Johansson M, Hoffer BJ, Strömberg I. Target and neurotransmitter specificity of fetal central nervous system transplants: importance for functional reinnervation. *J Neurosci.* 1994;14:283–290.

29. Zuddas A, Corsini GU, Barker JL, Kopin IJ, Di Porzio U. Specific reinnervation of lesioned mouse striatum by grafted mesencephalic dopaminergic neurons. *Eur J Neurosci.* 1991;3:72–85.

30. Dahlström A, Fuxe K. Evidence for the existence of monoamine-containing neurons in the central nervous system. I. Demonstration of monoamines in the cell bodies of brain stem neurons. *Acta Physiol Scand Suppl.* 1964;232:1–55.

31. Haque NSK, Borghesani P, Isacson O. Differential dissection of the rat E16 ventral mesencephalon and survival and reinnervation of the 6-OHDA lesioned striatum by a subset of aldehyde dehydrogenase-positive TH neurons. *Cell Transplant.* 1997;6:239–248.

32. Schultzberg M, Dunnett SB, Björklund A, et al. Dopamine and cholecystokinin immunoreactive neurons in mesencephalic grafts reinnervating the neostriatum: evidence for selective growth regulation. *Neuroscience.* 1984;12:17–32.

33. Thompson L, Barraud P, Andersson E, Kirik D, Björklund A. Identification of dopaminergic neurons of nigral and ventral tegmental area subtypes in grafts of fetal ventral mesencephalon based on cell morphology, protein expression, and efferent projections. *J Neurosci.* 2005;25:6467–6477.

34. Sawamoto K, Nakao N, Kakishita K, et al. Generation of dopaminergic neurons in the adult brain from mesencephalic precursor cells labeled with a *nestin*-GFP transgene. *J Neurosci.* 2001;21:3895–3903.

35. Arbuthnott GW, Dunnett SB, MacLeod N. The electrophysiological properties of single units in mesencephalic transplants in rat brain. *Neurosci Lett.* 1985;57:205–210.

36. Doucet G, Murata Y, Brundin P, et al. Host afferents into intrastriatal transplants of fetal ventral mesencephalon. *Exp Neurol.* 1989;106:1–9.

37. Fisher LJ, Young SJ, Tepper JM, Groves PM, Gage FH. Electrophysiological characteristics of cells within mesencephalon suspension grafts. *Neuroscience.* 1991;40:109–122.

38. Freund TF, Bolam JP, Björklund A, et al. Efferent synaptic connections of grafted dopaminergic-neurons reinnervating the host neostriatum: a tyrosine hydroxylase immunocytochemical study. *J Neurosci.* 1985;5:603–616.

39. Mahalik TJ, Finger TE, Strömberg I, Olson L. Substantia nigra transplants into denervated striatum of the rat: ultrastructure of graft and host interconnections. *J Comp Neurol.* 1985;240:60–70.

40. Rioux L, Gaudin DP, Bui LK, Grégoire L, DiPaolo T, Bédard PJ. Correlation of functional recovery after a 6-hydroxydopamine lesion with survival of grafted fetal neurons and release of dopamine in the striatum of the rat. *Neuroscience.* 1991;40:123–131.

41. Strecker RE, Sharp T, Brundin P, Zetterström T, Ungerstedt U, Björklund A. Auto-regulation of dopamine release and metabolism by intrastriatal nigral grafts as revealed by intracerebral dialysis. *Neuroscience.* 1987;22:169–178.

42. Cenci MA, Kalén P, Mandel RJ, Björklund A. Regional differences in the regulation of dopamine and noradrenaline release in medial frontal cortex, nucleus accumbens and caudate-putamen: a microdialysis study in the rat. *Brain Res.* 1992;581:217–228.

43. Cenci MA, Kalén P, Mandel RJ, Wictorin K, Björklund A. Dopaminergic transplants normalize amphetamine- and apomorphine-induced fos expression in the 6-hydroxydopamine-lesioned striatum. *Neuroscience.* 1992;46:943–957.

44. Cenci MA, Campbell K, Björklund A. Glutamic acid decarboxylase gene expression in the dopamine-denervated striatum: effects of intrastriatal fetal nigral transplants or chronic apomorphine treatment. *Mol Brain Res.* 1997;48:149–155.

45. Chritin M, Savasta M, Mennicken F, et al. Intrastriatal dopamine-rich implants reverse the increase of dopamine D2 receptor mRNA levels caused by lesion of the nigrostriatal pathway: a quantitative in situ hybridization study. *Eur J Neurosci.* 1992;4:663–672.

46. Dawson TM, Dawson VL, Gage FH, Fisher LJ, Hunt MA, Wamsley JK. Functional recovery of supersensitive dopamine receptors after intrastriatal grafts of fetal substantia nigra. *Exp Neurol.* 1991;111:282–292.

47. Savasta M, Mennicken F, Chritin M, et al. Intrastriatal dopamine-rich implants reverse the changes in dopamine D2 receptor densities caused by 6-hydroxydopamine lesion of the nigrostriatal pathway in rats: an autoradiographic study. *Neuroscience.* 1992;46:729–738.

48. Sirinathsinghji DJS, Dunnett SB. Increased proenkephalin mRNA levels in the rat neostriatum following lesion of the ipsilateral nigrostriatal dopamine pathway with 1-methyl-4-phenylpyridinium ion (MPP+): reversal by embryonic nigral dopamine grafts. *Mol Brain Res.* 1991;9:263–269.

49. Bentlage C, Nikkhah G, Cunningham MG, Björklund A. Reformation of the nigrostriatal pathway by fetal dopaminergic micrografts into the substantia nigra is critically dependent on the age of the host. *Exp Neurol.* 1999;159:177–190.

50. Mendez I, Sadi D, Hong M. Reconstruction of the nigrostriatal pathway by simultaneous intrastriatal and intranigral dopaminergic transplants. *J Neurosci.* 1996;16:7216–7227.

51. Mukhida K, Baker KA, Sadi D, Mendez I. Enhancement of sensorimotor behavioral recovery in hemiparkinsonian rats with intrastriatal, intranigral, and intrasubthalamic nucleus dopaminergic transplants. *J Neurosci.* 2001;21: 3521–3530.

52. Nikkhah G, Cunningham MG, Cenci MA, McKay RD, Björklund A. Dopaminergic microtransplants into the substantia nigra of neonatal rats with bilateral 6-OHDA lesions. I. Evidence for anatomical reconstruction of the nigrostriatal pathway. *J Neurosci.* 1995;15:3548–3561.

53. Isacson O, Deacon TW, Pakzaban P, Galpern WR, Dinsmore J, Burns LH. Transplanted xenogeneic neural cells in neurodegenerative disease models exhibit remarkable axonal target specificity and distinct growth patterns of glial and axonal fibres. *Nat Med.* 1995;1:1189–1194.

54. Wictorin K, Brundin P, Gustavii B, Lindvall O, Björklund A. Reformation of long axon pathways in adult rat central nervous system by human forebrain neuroblasts. *Nature*. 1990;347:556–558.
55. Wictorin K, Lagenaur CF, Lund RD, Björklund A. Efferent projections to the host brain from intrastriatal striatal mouse-to-rat grafts: time course and tissue-type specificity as revealed by a mouse specific neuronal marker. *Eur J Neurosci*. 1991;3:86–101.
56. Isacson O, Deacon TW. Specific axon guidance factors persist in the adult brain as demonstrated by pig neuroblasts transplanted to the rat. *Neuroscience*. 1996;75:827–837.
57. Gaillard A, Prestoz L, Dumartin B, et al. Reestablishment of damaged adult motor pathways by grafted embryonic cortical neurons. *Nat Neurosci*. 2007;10:1294–1299.
58. Davies SJA, Field PM, Raisman G. Long interfascicular axon growth from embryonic neurons transplanted into adult myelinated tracts. *J Neurosci*. 1994;14:1596–1612.
59. Björklund A, Nornes H, Gage FH. Cell suspension grafts of noradrenergic locus coeruleus neurons in rat hippocampus and spinal cord—reinnervation and transmitter turnover. *Neuroscience*. 1986;18:685–698.
60. Li Y, Raisman G. Long axon growth from embryonic neurons transplanted into myelinated tracts of the adult rat spinal cord. *Brain Res*. 1993;629:115–127.
61. Wictorin K, Brundin P, Sauer H, Lindvall O, Björklund A. Long distance directed axonal growth from human dopaminergic mesencephalic neuroblasts implanted along the nigrostriatal pathway in 6-hydroxydopamine lesioned adult rats. *J Comp Neurol*. 1992;323:475–494.
62. Thompson LH, Grealish S, Kirik D, Björklund A. Reconstruction of the nigro-striatal dopamine pathway in the adult mouse brain. *J Neurosci*. 2009; in press.
63. Ungerstedt U. Striatal dopamine release after amphetamine or nerve degeneration revealed by rotational behaviour. *Acta Physiol Scand Supp*. 1971;367:49–68.
64. Dunnett SB, Björklund A, Schmidt RH, Stenevi U, Iversen SD. Intracerebral grafting of neuronal cell suspensions. IV. Behavioral recovery in rats with unilateral 6-OHDA lesions following implantation of nigral cell suspensions in different forebrain sites. *Acta Physiol Scand Suppl*. 1983;522:29–37.
65. Dunnett SB, Hernandez TD, Summerfield A, Jones GH, Arbuthnott GW. Graft-derived recovery from 6-OHDA lesions: specificity of ventral mesencephalic graft tissues. *Exp Brain Res*. 1988;71:411–424.
66. Schmidt RH, Björklund A, Stenevi U, Dunnett SB, Gage FH. Intracerebral grafting of neuronal cell suspensions. III. Activity of intrastriatal nigral suspension implants as assessed by measurements of dopamine synthesis and metabolism. *Acta Physiol Scand Suppl*. 1983;522:19–28.
67. Brundin P, Widner H, Nilsson OG, Strecker RE, Björklund A. Intracerebral xenografts of dopamine neurons: the role of immunosuppression and the blood-brain barrier. *Exp Brain Res*. 1989;75:195–207.
68. Björklund A, Dunnett SB, Stenevi U, Lewis ME, Iversen SD. Reinnervation of the denervated striatum by substantia nigra transplants: functional consequences as revealed by pharmacological and sensorimotor testing. *Brain Res*. 1980;199:307–333.
69. Ungerstedt U. Postsynaptic supersensitivity after 6-hydroxydopamine-induced degeneration of the nigro-striatal dopamine system. *Acta Physiol Scand Suppl*. 1971;367:69–93.
70. Freed WJ, Ko GN, Niehoff DL, et al. Normalization of spiroperidol binding in the denervated rat striatum by homologous grafts of substantia nigra. *Science*. 1983;222:937–939.
71. Bing G, Notter MFD, Hansen JT, Gash DM. Comparison of adrenal medullary, carotid body and PC12 cell grafts in 6-OHDA lesioned rats. *Brain Res Bull*. 1988;20:399–406.
72. Brown VJ, Dunnett SB. Comparison of adrenal and fetal nigral grafts on drug-induced rotation in rats with 6-OHDA lesions. *Exp Brain Res*. 1989;78:214–218.
73. Hargraves RW, Freed WJ. Chronic intrastriatal dopamine infusions in rats with unilateral lesions of the substantia nigra. *Life Sci*. 1987;40:959–966.
74. Kirik D, Georgievska B, Burger C, et al. Reversal of motor impairments in parkinsonian rats by continuous intrastriatal delivery of L-dopa using rAAV-mediated gene transfer. *Proc Natl Acad Sci USA*. 2002;99:4708–4713.
75. Hefti F, Melamed E, Sahakian BJ, Wurtman RJ. Circling behavior in rats with partial, unilateral nigro-striatal lesions: effects of amphetamine, apomorphine, and DOPA. *Pharmacol Biochem Behav*. 1980;12:185–188.
76. Sauer H, Brundin P. Effects of cool storage on survival and function of intrastriatal ventral mesencephalic grafts. *Rest Neurol Neurosci*. 1991;2:123–135.
77. Freeman TB, Sanberg PR, Nauert GM, et al. The influence of donor age on the survival of solid and suspension intraparenchymal human embryonic nigral grafts. *Cell Transplant*. 1995;4:141–154.
78. Heim RC, Willingham G, Freed WJ. A comparison of solid intraventricular and dissociated intraparenchymal fetal substantia nigra grafts in a rat model of Parkinson's disease: impaired graft survival is associated with high baseline rotational behavior. *Exp Neurol*. 1993;122:5–15.
79. Leigh K, Elisevich K, Rogers KA. Vascularization and microvascular permeability in solid versus cell-suspension embryonic neural grafts. *J Neurosurg*. 1994;81:272–283.
80. Nikkhah G, Rosenthal C, Hedrich HJ, Samii M. Differences in acquisition and full performance in skilled forelimb use as measured by the "staircase test" in five rat strains. *Behav Brain Res*. 1998;92:85–95.
81. Galpern WR, Burns LH, Deacon TW, Dinsmore J, Isacson O. Xenotransplantation of porcine fetal ventral mesencephalon in a rat model of Parkinson's disease: functional recovery and graft morphology. *Exp Neurol*. 1996;140:1–13.
82. Rosenblad C, Kirik D, Devaux B, Moffat B, Phillips HS, Björklund A. Protection and regeneration of nigral dopaminergic neurons by neurturin or GDNF in a partial lesion model of Parkinson's disease after administration into the striatum or the lateral ventricle. *Eur J Neurosci*. 1999;11:1554–1566.
83. Wilby M, Sinclair SR, Muir EM, et al. A GDNF-secreting clone of the Schwann cell line SCTM41 enhances survival and fibre outgrowth from embryonic nigral neurones grafted to the striatum and the lesioned substantia nigra. *J Neurosci*. 1999;19:2301–2312.
84. Castilho RF, Hansson O, Brundin P. Improving the survival of grafted embryonic dopamine neurons in rodent models of Parkinson's disease. *Prog Brain Res*. 2000;127:203–231.
85. Nakao N, Frodl EM, Duan WM, Widner H, Brundin P. Lazaroids improve the survival of grafted rat embryonic dopamine neurons. *Proc Natl Acad Sci USA*. 1994;91:12408–12412.
86. Barker RA, Dunnett SB. Ibotenic acid lesions of the striatum reduce drug-induced rotation in the 6-hydroxydopamine-lesioned rat. *Exp Brain Res*. 1994;101:365–374.

87. Marshall JF, Levitan D, Stricker EM. Activation-induced restoration of sensorimotor functions in rats with dopamine-depleting brain lesions. *J Comp Physiol Psychol*. 1976;90: 536–546.
88. Ungerstedt U. Adipsia and aphagia after 6-hydroxydopamine-induced degeneration of the nigro-striatal dopamine system. *Acta Physiol Scand Suppl*. 1971;367:96–122.
89. Zigmond MJ, Stricker EM. Deficits in feeding behavior after intraventricular injection of 6-hydroxydopamine in rats. *Science*. 1972;177:1211–1214.
90. Dunnett SB, Björklund A, Stenevi U, Iversen SD. Behavioral recovery following transplantation of substantia nigra in rats subjected to 6-OHDA lesions of the nigrostriatal pathway. 2. Bilateral lesions. *Brain Res*. 1981;229:457–470.
91. Dunnett SB, Björklund A, Stenevi U, Iversen SD. Grafts of embryonic substantia nigra reinnervating the ventrolateral striatum ameliorate sensorimotor impairments and akinesia in rats with 6-OHDA lesions of the nigrostriatal pathway. *Brain Res*. 1981;229:209–217.
92. Dunnett SB, Björklund A, Schmidt RH, Stenevi U, Iversen SD. Intracerebral grafting of neuronal cell suspensions. V. Behavioral recovery in rats with bilateral 6-OHDA lesions following implantation of nigral cell suspensions. *Acta Physiol Scand Suppl*. 1983;522:39–47.
93. Brundin P, Strecker RE, Londos E, Björklund A. Dopamine neurons grafted unilaterally to the nucleus accumbens affect drug-induced circling and locomotion. *Exp Brain Res*. 1987;69:183–194.
94. Dunnett SB, Whishaw IQ, Rogers DC, Jones GH. Dopamine-rich grafts ameliorate whole body motor asymmetry and sensory neglect but not independent limb use in rats with 6-hydroxydopamine lesions. *Brain Res*. 1987;415:63–78.
95. Montoya CP, Astell S, Dunnett SB. Effects of nigral and striatal grafts on skilled forelimb use in the rat. *Prog Brain Res*. 1990;82:459–466.
96. Abrous DN, Torres EM, Dunnett SB. Dopaminergic grafts implanted into the neonatal or adult striatum: comparative effects on rotation and paw reaching deficits induced by subsequent unilateral nigrostriatal lesions in adulthood. *Neuroscience*. 1993;54:657–668.
97. Nikkhah G, Cunningham MG, McKay R, Björklund A. Dopaminergic microtransplants into the substantia nigra of neonatal rats with bilateral 6-OHDA lesions. II. Transplant-induced behavioral recovery. *J Neurosci*. 1995;15:3562–3570.
98. Dunnett SB, Rogers DC, Richards SJ. Nigrostriatal reconstruction after 6-OHDA lesions in rats: combination of dopamine-rich nigral grafts and nigrostriatal bridge grafts. *Exp Brain Res*. 1989;75:523–535.
99. Brecknell JE, Haque NS, Du JS, et al. Functional and anatomical reconstruction of the 6-hydroxydopamine lesioned nigrostriatal system of the adult rat. *Neuroscience*. 1996;71: 913–925.
100. Kirik D, Winkler C, Björklund A. Growth and functional efficacy of intrastriatal nigral transplants depends on the extent of nigrostriatal degeneration. *J Neurosci*. 2001;21: 2889–2896.
101. Dowd E, Monville C, Torres EM, Dunnett SB. The corridor task: a simple test of lateralised response selection and neglect sensitive to unilateral dopamine deafferentation and graft-derived replacement in the striatum. *Brain Res Bull*. 2005;68:24–30.
102. Breysse N, Carlsson T, Winkler C, Björklund A, Kirik D. The functional impact of the intrastriatal dopamine neuron grafts in parkinsonian rats is reduced with advancing disease. *J Neurosci*. 2007;27:5849–5856.
103. Nikkhah G, Bentlage C, Cunningham MG, Björklund A. Intranigral fetal dopamine grafts induce behavioral compensation in the rat Parkinson model. *J Neurosci*. 1994;14:3449–3461.
104. Baker KA, Sadi D, Hong M, Mendez I. Simultaneous intrastriatal and intranigral dopaminergic grafts in the Parkinsonian rat model: role of the intranigral graft. *J Comp Neurol*. 2000;426:106–116.
105. Winkler C, Kirik D, Björklund A, Dunnett SB. Transplantation in the rat model of Parkinson's disease: ectopic versus homotopic graft placement. *Prog Brain Res*. 2000;127:233–265.
106. Rozas G, Guerra MJ, Labandeira-García JL. An automated rotarod method for quantitative drug-free evaluation of overall motor deficits in rat models of parkinsonism. *Brain Res Prot*. 1997;2:75–84.
107. Schallert T, Tillerson JL. Intervention strategies for degeneration of dopamine neurons in Parkinsonism: optimizing behavioral assessment of outcome. In: Emerich DF, Dean RL, Sanberg PR, eds. *Central Nervous System Diseases: Innovative Animal Models from Lab to Clinic*. Totowa, NJ: Humana Press; 2000;131–151.
108. Cenci MA, Lundblad M. Utility of 6-hydroxydopamine lesioned rats in the preclinical screening of novel treatments for Parkinson disease. In: LeDoux M, ed. *Animal Models of Movement Disorders*. Amsterdam, the Netherlands: Elsevier; 2005;193–209.
109. Olsson M, Nikkhah G, Bentlage C, Björklund A. Forelimb akinesia in the rat Parkinson model: differential effects of dopamine agonists and nigral transplants as assessed by a new stepping test. *J Neurosci*. 1995;15:3863–3875.
110. Rozas G, Labandeira-García JL. Drug-free evaluation of rat models of parkinsonism and nigral grafts using a new automated rotarod test. *Brain Res*. 1997;749:188–199.
111. Nikkhah G, Rosenthal C, Falkenstein G, Samii M. Dopaminergic graft-induced long-term recovery of complex sensorimotor behaviors in a rat model of Parkinson's disease. *Zentralbl Neurochir*. 1998;59:97–103.
112. Roedter A, Winkler C, Samii M, Nikkhah G. Complex sensorimotor behavioral changes after terminal striatal 6-OHDA lesion and transplantation of dopaminergic embryonic micrografts. *Cell Transplant*. 2001;9:197–214.
113. Torres EM, Dowd E, Dunnett SB. Recovery of functional deficits following early donor age (E12) ventral mesencephalic grafts in a rat model of Parkinson's disease. *Neuroscience*. 2008;154: 631–640.
114. Sass KJ, Buchanan CP, Westerveld M, et al. General cognitive ability following unilateral and bilateral fetal ventral mesencephalic tissue transplantation for treatment of Parkinson's disease. *Arch Neurol*. 1995;52:680–686.
115. Leroy A, Michelet D, Mahieux F, et al. [Neuropsychological testing of 5 patients with Parkinson's disease before and after neuron graft]. *Rev Neurol (Paris)*. 1996;152:158–164.
116. Marshall JF, Richardson JS, Teitelbaum P. Nigrostriatal bundle damage and the lateral hypothalamic syndrome. *J Comp Physiol Psychol*. 1974;87:808–830.
117. Simon H. Dopaminergic A10 neurons and frontal system. *J Physiol (Paris)*. 1981;77:81–95.
118. Dunnett SB, Iversen SD. Neurotoxic lesions of ventrolateral but not anteromedial neostriatum in rats impair differential reinforcement of low rates (DRL) performance. *Behav Brain Res*. 1982;6:213–226.

119. Zigmond MJ, Stricker EM. Recovery of feeding and drinking by rats after intraventricular 6-hydroxydopamine or lateral hypothalamic lesions. *Science*. 1973;182:717–720.
120. Mishkin M, Malamut B, Bachevalier J. Memories and habits: two neural systems. In: Lynch G, McGaugh JL, Weinberger NM, eds. *Neurobiology of Learning and Memory*. New York, NY: Guilford Press; 1984:65–77.
121. Gaffan D. Memory, action and the corpus striatum—current developments in the memory–habit distinction. *Semin Neurosci*. 1996;8:33–38.
122. White NM. Mnemonic functions of the basal ganglia. *Curr Opin Neurobiol*. 1997;7:164–169.
123. Schultz W. Getting formal with dopamine and reward. *Neuron*. 2002;36:241–263.
124. Calabresi P, Picconi B, Tozzi A, Di Filippo M. Dopamine-mediated regulation of corticostriatal synaptic plasticity. *Trends Neurosci*. 2007;30:211–219.
125. Carli M, Evenden JL, Robbins TW. Depletion of unilateral striatal dopamine impairs initiation of contralateral actions and not sensory attention. *Nature*. 1985;313:679–682.
126. Dowd E, Dunnett SB. Deficits in a lateralised associative learning task in dopamine-depleted rats, with selective recovery by dopamine-rich transplants. *Eur J Neurosci*. 2004;20:1953–1959.
127. Fray PJ, Dunnett SB, Iversen SD, Björklund A, Stenevi U. Nigral transplants reinnervating the dopamine-depleted neostriatum can sustain intracranial self-stimulation. *Science*. 1983;219:416–419.
128. Shimizu K, Tsuda N, Okamoto Y, et al. Transplant-induced recovery from 6-OHDA lesions of the nigrostriatal dopamine neurones in mice. *Acta Neurochir Suppl*. 1988;43:149–153.
129. Brundin P, Isacson O, Gage FH, Prochiantz A, Björklund A. The rotating 6-hydroxydopamine lesioned mouse as a model for assessing functional effects of neuronal grafting. *Brain Res*. 1986;366:346–349.
130. Witt TC, Triarhou LC. Transplantation of mesencephalic cell suspensions from wild-type and heterozygous weaver mice into the denervated striatum—assessing the role of graft-derived dopaminergic dendrites in the recovery of function. *Cell Transplant*. 1995;4:323–333.
131. Di Porzio U, Zuddas A. Embryonic dopaminergic neuron transplants in MPTP lesioned mouse striatum. *Neurochem Int*. 1992;20(suppl):309S–320S.
132. Bohn MC, Kanuicki M. Bilateral recovery of striatal dopamine after unilateral adrenal grafting into the striatum of the 1-methyl-4-(2'methylphenyl)-1,2,3,6-tetrahydropyridine (2'CH3-MPTP)-treated mice. *J Neurosci Res*. 1990;25:281–286.
133. Bohn MC, Cupit L, Marciano F, Gash DM. Adrenal grafts enhance recovery of striatal dopaminergic fibers. *Science*. 1987;237:913–916.
134. Kong XY, Cai Z, Pan L, et al. Transplantation of human amniotic cells exerts neuroprotection in MPTP-induced Parkinson disease mice. *Brain Res*. 2008;1205:108–115.
135. Liker MA, Petzinger GA, Nixon K, McNeill T, Jakowec MW. Human neural stem cell transplantation in the MPTP-lesioned mouse. *Brain Res*. 2003;971:168–177.
136. Ourednik J, Ourednik V, Lynch WP, Schachner M, Snyder EY. Neural stem cells display an inherent mechanism for rescuing dysfunctional neurons. *Nat Biotechnol*. 2002;20:1103–1110.
137. Sedelis M, Schwarting RK, Huston JP. Behavioral phenotyping of the MPTP mouse model of Parkinson's disease. *Behav Brain Res*. 2001;125:109–125.
138. Triarhou LC. *Dopaminergic Neuron Transplantation in the Weaver Mouse Model of Parkinson's Disease*. New York, NY: Kluwer Academic; 2002.
139. Triarhou LC, Low WC, Ghetti B. Transplantation of ventral mesencephalic anlagen to hosts with genetic nigrostriatal dopamine deficiency. *Proc Natl Acad Sci USA*. 1986;83:8789–8793.
140. Triarhou LC, Low WC, Norton J, Ghetti B. Reinstatement of synaptic connectivity in the striatum of weaver mutant mice following transplantation of ventral mesencephalic anlagen. *J Neurocytol*. 1988;17:233–243.
141. Triarhou LC, Norton J, Hingtgen JN. Amelioration of the behavioral phenotype in weaver mutant mice through bilateral intrastriatal grafting of fetal dopamine cells. *Exp Brain Res*. 1995;104:191–198.
142. Sorensen AT, Thompson L, Kirik D, Björklund A, Lindvall O, Kokaia M. Functional properties and synaptic integration of genetically labelled dopaminergic neurons in intrastriatal grafts. *Eur J Neurosci*. 2005;21:2793–2799.
143. Quintana JG, Lopez-Colberg I, Cunningham LA. Use of GFAP-lacZ transgenic mice to determine astrocyte fate in grafts of embryonic ventral midbrain. *Dev Brain Res*. 1998;105:147–151.
144. Nakao N, Frodl EM, Widner H, et al. Overexpressing Cu/Zn superoxide dismutase enhances survival of transplanted neurons in a rat model of Parkinson's disease. *Nat Med*. 1995;1:226–231.
145. Hedlund E, Pruszak J, Lardaro T, et al. Embryonic stem cell-derived Pitx3-enhanced green fluorescent protein midbrain dopamine neurons survive enrichment by fluorescence-activated cell sorting and function in an animal model of Parkinson's disease. *Stem Cells*. 2008;26:1526–1536.
146. Carlsson T, Winkler C, Lundblad M, Cenci MA, Björklund A, Kirik D. Graft placement and uneven pattern of reinnervation in the striatum is important for development of graft-induced dyskinesia. *Neurobiol Dis*. 2006;21:657–668.
147. Lane EL, Winkler C, Brundin P, Cenci MA. The impact of graft size on the development of dyskinesia following intrastriatal grafting of embryonic dopamine neurons in the rat. *Neurobiol Dis*. 2006;22:334–346.
148. Lee CS, Cenci MA, Schulzer M, Björklund A. Embryonic ventral mesencephalic grafts improve levodopa-induced dyskinesia in a rat model of Parkinson's disease. *Brain*. 2000;123:1365–1379.
149. Steece-Collier K, Collier TJ, Danielson PD, Kurlan R, Yurek DM, Sladek JR. Embryonic mesencephalic grafts increase levodopa-induced forelimb hyperkinesia in Parkinsonian rats. *Mov Disord*. 2003;18:1442–1454.
150. Carlsson T, Carta M, Munoz A, et al. Impact of grafted serotonin and dopamine neurons on development of L-DOPA-induced dyskinesias in parkinsonian rats is determined by the extent of dopamine neuron degeneration. *Brain* 2008;132:319–335.
151. Defer GL, Gény C, Ricolfi F, et al. Long-term outcome of unilaterally transplanted Parkinsonian patients. 1. Clinical approach. *Brain*. 1996;119:41–50.
152. Hauser RA, Freeman TB, Snow BJ, et al. Long-term evaluation of bilateral fetal nigral transplantation in Parkinson disease. *Arch Neurol*. 1999;56:179–187.
153. Brundin P, Pogarell O, Hagell P, et al. Bilateral caudate and putamen grafts of embryonic mesencephalic tissue treated with lazaroids in Parkinson's disease. *Brain*. 2000;123:1380–1390.
154. Hagell P, Schrag A, Piccini P, et al. Sequential bilateral transplantation in Parkinson's disease—effects of the second graft. *Brain*. 1999;122:1121–1132.

155. Freed CR, Greene PE, Breeze RE, et al. Transplantation of embryonic dopamine neurons for severe Parkinson's disease. *N Engl J Med*. 2001;344:710–719.
156. Olanow CW, Goetz CG, Kordower JH, et al. A double-blind controlled trial of bilateral fetal nigral transplantation in Parkinson's disease. *Ann Neurol*. 2003;54:403–414.
157. Olanow CW, Gracies JM, Goetz CG, et al. Clinical pattern and risk factors for dyskinesias following fetal nigral transplantation in Parkinson's disease: a double blind video-based analysis. *Mov Disord*. 2009; 24:335–343.
158. Ma YL, Feigin A, Dhawan V, et al. Dyskinesia after fetal cell transplantation for parkinsonism: a PET study. *Ann Neurol*. 2002;52:628–634.
159. Maries E, Kordower JH, Chu Y, et al. Focal not widespread grafts induce novel dyskinetic behavior in parkinsonian rats. *Neurobiol Dis*. 2006;21:165–180.
160. Lane EL, Dunnett SB. Animal models of Parkinson's disease and l-dopa induced dyskinesia. How close are we to the clinic? *Psychopharmacology (Berl)*. 2008;199:303–312.
161. Carlsson T, Carta M, Winkler C, Björklund A, Kirik D. Serotonin neuron transplants exacerbate L-DOPA-induced dyskinesias in a rat model of Parkinson's disease. *J Neurosci*. 2007;27:8011–8022.
162. Carta M, Carlsson T, Kirik D, Björklund A. Dopamine released from 5-HT terminals is the cause of L-DOPA-induced dyskinesia in parkinsonian rats. *Brain*. 2007;130:1819–1833.
163. Freed WJ. Repairing neuronal circuits with brain grafts: where can brain grafts be used as a therapy? *Neurobiol Aging*. 1985;6:153–156.
164. Björklund A, Lindvall O, Isacson O, et al. Mechanisms of action of intracerebral neural implants—studies on nigral and striatal grafts to the lesioned striatum. *Trends Neurosci*. 1987;10:509–516.
165. Gage FH, Buzsaki G. CNS grafting: potential mechanisms of action. In: Seil FJ, ed. *Neural Regeneration and Transplantation*. New York, NY: Alan R. Liss; 1989;211–226.
166. Dunnett SB. Cell replacement therapy: mechanisms of functional recovery. In: Squire LR, ed. *New Encyclopedia of Neuroscience*. Amsterdam, the Netherlands: Elsevier; 2009; pp 643–647.
167. Krauss JK, Jankovic J, Grossman RG. *Surgery for Parkinson's Disease and Movement Disorders*. Philadelphia, PA: Lippincott, Williams & Wilkins; 2001.
168. Perlmutter JS, Mink JW. Deep brain stimulation. *Annu Rev Neurosci*. 2006;29:229–257.
169. Albin RL, Young AB, Penney JB. The functional anatomy of basal ganglia disorders. *Trends Neurosci*. 1989;12:366–375.
170. Meyers R. Surgical experiments in therapy of certain "extrapyramidal" diseases: current evaluation. *Acta Psychiatr Neurol Suppl* 1951;67:1–42.
171. Johnston RE, Becker JB. Behavioral changes associated with grafts of embryonic ventral mesencephalon tissue into the striatum and/or substantia nigra in a rat model of Parkinson's disease. *Behav Brain Res*. 1999;104:179–187.
172. Forni C, Brundin P, Strecker RE, Elganouni S, Björklund A, Nieoullon A. Time-course of recovery of dopamine neuron activity during reinnervation of the denervated striatum by fetal mesencephalic grafts as assessed by in vivo voltammetry. *Exp Brain Res*. 1989;76:75–87.
173. Strömberg I, Adams C, Bygdeman M, Hoffer B, Boyson S, Humpel C. Long-term effects of human-to-rat mesencephalic xenografts on rotational behavior, striatal dopamine receptor binding, and mRNA levels. *Brain Res Bull*. 1995;38: 221–233.
174. McRae-Degueurce A, Hjorth S, Dillon DL, Mason DW, Tice TR. Implantable microencapsulated dopamine (DA): a new approach for slow release DA delivery into brain tissue. *Neurosci Lett*. 1988;92:303–309.
175. Freed WJ, Morihisa JM, Spoor E, et al. Transplanted adrenal chromaffin cells in rat brain reduce lesion-induced rotational behavior. *Nature*. 1981;292:351–352.
176. Azzouz M, Martin-Rendon E, Barber RD, et al. Multicistronic lentiviral vector-mediated striatal gene transfer of aromatic L-amino acid decarboxylase, tyrosine hydroxylase, and GTP cyclohydrolase I induces sustained transgene expression, dopamine production, and functional improvement in a rat model of Parkinson's disease. *J Neurosci*. 2002;22:10302–10312.
177. Gage FH, Björklund A. Trophic and growth-regulating mechanisms in the central nervous system monitored by intracerebral neural transplants. *Ciba Found Symp*. 1987;126: 143–159.
178. Strömberg I, Wetmore CJ, Ebendal T, Ernfors P, Persson H, Olson L. Rescue of basal forebrain cholinergic neurons after implantation of genetically modified cells producing recombinant NGF. *J Neurosci Res*. 1990;25:405–411.
179. Raymon HK, Thode S, Gage FH. Application of *ex vivo* gene therapy in the treatment of Parkinson's disease. *Exp Neurol*. 1997;144:82–91.
180. Olson L, Ayer-LeLievre C, Ebendal T, et al. Grafts, growth factors and grafts that make growth factors. *Prog Brain Res*. 1990;82:55–66.
181. Dunnett SB, Mayer E. Neural grafts, growth factors and trophic mechanisms of recovery. In: Hunter AJ, Clarke M, eds. *Neurodegeneration*. New York, NY: Academic Press; 1992:183–217.
182. Brundin P, Karlsson J, Emgård M, et al. Improving the survival of grafted dopaminergic neurons: a review over current approaches. *Cell Transplant*. 2000;9:179–195.
183. Mayer E, Fawcett JW, Dunnett SB. Basic fibroblast growth factor promotes the survival of embryonic ventral mesencephalic dopaminergic neurons. II. Effects on neural transplants in vivo. *Neuroscience*. 1993;56:389–398.
184. Strömberg I, Björklund L, Johansson M, et al. Glial cell line–derived neurotrophic factor is expressed in the developing but not adult striatum and stimulates developing dopamine neurons *in vivo*. *Exp Neurol*. 1993;124:401–412.
185. Rosenblad C, Martinez-Serrano A, Björklund A. Glial cell line–derived neurotrophic factor increases survival, growth and function of intrastriatal fetal nigral dopaminergic grafts. *Neuroscience*. 1996;75:979–985.
186. Sinclair SR, Svendsen CN, Torres EM, Martin D, Fawcett JW, Dunnett SB. GDNF enhances dopaminergic cell survival and fibre outgrowth in embryonic nigral grafts. *Neuroreport*. 1996;7:2547–2552.
187. Hansson O, Castilho RF, Schierle GSK, et al. Additive effects of caspase inhibitor and lazaroid on the survival of transplanted rat and human embryonic dopamine neurons. *Exp Neurol*. 2000;164:102–111.
188. Sánchez-Capelo A, Corti O, Mallet J. Adenovirus-mediated over-expression of TGFb1 in the striatum decreases dopaminergic cell survival in embryonic nigral grafts. *Neuroreport*. 1999;10:2169–2173.
189. Takayama H, Ray J, Raymon HK, et al. Basic fibroblast growth factor increases dopaminergic graft survival and function in a rat model of Parkinson's disease. *Nat Med*. 1995;1: 53–58.

190. Barkats M, Nakao N, Grasbon-Frodl EM, et al. Intrastriatal grafts of embryonic mesencephalic rat neurons genetically modified using an adenovirus encoding human Cu/Zn superoxide dismutase. *Neuroscience*. 1997;78:703–713.
191. Ostenfeld T, Tai YT, Martin P, Déglon N, Aebischer P, Svendsen CN. Neurospheres modified to produce glial cell line-derived neurotrophic factor increase the survival of transplanted dopamine neurons. *J Neurosci Res*. 2002;69:955–965.
192. Connor B, Kozlowski DA, Unnerstall JR, et al. Glial cell line–derived neurotrophic factor (GDNF) gene delivery protects dopaminergic terminals from degeneration. *Exp Neurol*. 2001;169:83–95.
193. Georgievska B, Kirik D, Rosenblad C, Lundberg C, Björklund A. Neuroprotection in the rat Parkinson model by intrastriatal GDNF gene transfer using a lentiviral vector. *Neuroreport*. 2002;13:75–82.
194. Patel NK, Bunnage M, Plaha P, Svendsen CN, Gill SS. Intraputamenal infusion of glial cell line–derived neurotrophic factor in PD: a two-year outcome study. *Ann Neurol*. 2005;57:298–302.
195. Patel NK, Gill SS. GDNF delivery for Parkinson's disease. *Acta Neurochir Suppl*. 2007;97:135–154.
196. Brundin P, Isacson O, Gage FH, Björklund A. Intrastriatal grafting of dopamine-containing neuronal cell suspensions: effects of mixing with target or non-target cells. *Dev Brain Res*. 1986;24:77–84.
197. Jaeger CB. Axon terminal clustering in nigrostriatal double grafts. *Dev Brain Res*. 1986;24:309–314.
198. de Beaurepaire R, Freed WJ. Embryonic substantia nigra grafts innervate embryonic striatal co-grafts in preference to mature host striatum. *Exp Neurol*. 1987;95:448–454.
199. Yurek DM. Intranigral transplants of fetal ventral mesencephalic tissue attenuate D1-agonist-induced rotational behavior. *Exp Neurol*. 1997;143:1–9.
200. Palmer MR, Granholm AC, van Horne CG, et al. Intranigral transplantation of solid tissue ventral mesencephalon or striatal grafts induces behavioral recovery in 6-OHDA-lesioned rats. *Brain Res*. 2001;890:86–99.
201. David S, Aguayo AJ. Axonal elongation into peripheral nervous system "bridges" after central nervous system injury in adult rats. *Science*. 1981;214:931–933.
202. Olson L. Regeneration in the adult central nervous system: experimental repair strategies. *Nat Med*. 1997;3:1329–1335.
203. Aguayo AJ, Björklund A, Stenevi U, Carlstedt T. Fetal mesencephalic neurons survive and extend long axons across peripheral nervous system grafts inserted into the adult rat striatum. *Neurosci Lett*. 1984;45:53–58.
204. Gage FH, Stenevi U, Carlstedt T, Foster G, Björklund A, Aguayo AJ. Anatomical and functional consequences of grafting mesencephalic neurons into a peripheral nerve bridge connected to the denervated striatum. *Exp Brain Res*. 1985;60:584–589.
205. Zhou FC, Chiang YH, Wang Y. Constructing a new nigrostriatal pathway in the Parkinsonian model with bridged neural transplantation in substantia nigra. *J Neurosci*. 1996;16:6965–6974.
206. Christophersen NS, Correia AS, Roybon L, Li JY, Brundin P. Developing novel stem cell sources for transplantation in Parkinson's disease. In: Davis Sanberg C, Sanberg PR, eds. *Cell Therapy, Stem Cells, and Brain Repair*. Totowa, NJ: Humana Press; 2006:31–59.
207. Espejo EF, Montoro RJ, Armengol JA, López-Barneo J. Cellular and functional recovery of Parkinsonian rats after intrastriatal transplantation of carotid body cell aggregates. *Neuron*. 1998;20:197–206.
208. Brundin P, Duan WM, Sauer H. Functional effects of mesencephalic dopamine neurons and adrenal chromaffin cells grafted to the rodent striatum. In: Dunnett SB, Björklund A, eds. *Functional Neural Transplantation*. New York, NY: Raven Press; 1994:9–46.
209. Quinn NP. The clinical application of cell grafting techniques in patients with Parkinson's disease. *Prog Brain Res*. 1990;82:619–625.
210. Goetz CG, Stebbins GT, Klawans HL, et al. United Parkinson Foundation neurotransplantation registry on adrenal medullary transplants: presurgical, and 1-year and 2-year follow-up. *Neurology*. 1991;41:1719–1722.
211. Adams FS, La Rosa FG, Kumar S, et al. Characterization and transplantation of two neuronal cell lines with dopaminergic properties. *Neurochem Res*. 1996;21:619–627.
212. Horellou P, Brundin P, Kalén P, Mallet J, Björklund A. In vivo release of DOPA and dopamine from genetically engineered cells grafted to the denervated rat striatum. *Neuron*. 1990;5:393–402.
213. Uchida K, Takamatsu K, Kaneda N, et al. Synthesis of L-3,4-dihydroxyphenylalanine by tryosine hydroxylase cDNA-transfected C6 cells: application for intracerebral grafting. *J Neurochem*. 1989;53:728–732.
214. Jaeger CB. Immunocytochemical study of PC12 cells grafted to the brain of immature rats. *Exp Brain Res*. 1985;59:615–624.
215. Folkerth RD, Durso R. Survival and proliferation of non-neural tissues, with obstruction of cerebral ventricles, in a parkinsonian patient treated with fetal allografts. *Neurology*. 1996;46:1219–1225.
216. Bergman H, Wichmann T, DeLong MR. Reversal of experimental parkinsonism by lesions of the subthalamic nucleus. *Science*. 1990;249:1436–1438.
217. Limousin P, Pollak P, Benazzouz A, et al. Effect on parkinsonian signs and symptoms of bilateral subthalamic nucleus stimulation. *Lancet*. 1995;345:91–95.
218. Isacson O, Frim DM, Galpern WR, Tatter SB, Breakefield XO, Schumacher JM. Cell-mediated delivery of neurotrophic factors and neuroprotection in the neostriatum and substantia nigra. *Rest Neurol Neurosci*. 1995;8:59–61.
219. Åkerud P, Canals JM, Snyder EY, Arenas E. Neuroprotection through delivery of glial cell line–derived neurotrophic factor by neural stem cells in a mouse model of Parkinson's disease. *J Neurosci*. 2001;21:8108–8118.
220. Bankiewicz KS, Mandel RJ, Sofroniew MV. Trophism, transplantation, and animal models of Parkinson's disease. *Exp Neurol*. 1993;124:140–149.
221. Lindvall O, Brundin P, Widner H, et al. Grafts of fetal dopamine neurons survive and improve motor function in Parkinson's disease. *Science*. 1990;247:574–577.
222. Barker RA, Ratcliffe E, Richards A, Dunnett SB. Fetal porcine dopaminergic cell survival *in vitro* and its relationship to embryonic age. *Cell Transplant*. 1999;8:593–599.
223. Schumacher JM, Ellias SA, Palmer EP, et al. Transplantation of embryonic porcine mesencephalic tissue in patients with PD. *Neurology*. 2000;54:1042–1050.
224. Backlund EO, Granberg PO, Hamberger B, et al. Transplantation of adrenal medullary tissue to striatum in parkinsonism. First clinical trials. *J Neurosurg*. 1985;62:169–173.
225. Arjona V, Mínguez-Castellanos A, Montoro RJ, et al. Autotransplantation of human carotid body cell aggregates for treatment of Parkinson's disease. *Neurosurgery*. 2003;53:321–328.

226. Wolff JA, Fisher LJ, Xu L, et al. Grafting fibroblasts genetically modified to produce l-DOPA in a rat model of Parkinson disease. *Proc Natl Acad Sci USA.* 1989;86:9011–9014.
227. Svendsen CN, Clarke DJ, Rosser AE, Dunnett SB. Survival and differentiation of rat and human EGF responsive precursor cells following grafting into the lesioned adult CNS. *Exp Neurol.* 1996;137:376–388.
228. Studer L, Tabar V, McKay RDG. Transplantation of expanded mesencephalic precursors leads to recovery in parkinsonian rats. *Nat Neurosci.* 1998;1:290–295.
229. Björklund L, Sánchez-Pernaute R, Chung S, et al. Embryonic stem cells develop into functional dopaminergic neurons after transplantation in a Parkinson rat model. *Proc Natl Acad Sci USA.* 2002;99:2344–2349.
230. Kim JH, Auerbach JM, Rodriguez-Gomez JA, et al. Dopamine neurons derived from embryonic stem cells function in an animal model of Parkinson's disease. *Nature.* 2002;418:50–56.
231. Dziewczapolski G, Lie DC, Ray J, Gage FH, Shults CW. Survival and differentiation of adult rat-derived neural progenitor cells transplanted to the striatum of hemiparkinsonian rats. *Exp Neurol.* 2003;183:653–664.
232. Ende N, Chen R. Parkinson's disease mice and human umbilical cord blood. *J Med.* 2002;33:173–180.
233. Cenci MA, Kalén P, Duan WM, Björklund A. Transmitter release from transplants of fetal ventral mesencephalon and locus coeruleus in the rat frontal cortex and nucleus accumbens: effects of pharmacological and behavioural activating stimuli. *Brain Res.* 1994;641:225–248.

9.5 | Clinical Experiences with Dopamine Neuron Replacement in Parkinson's Disease: What Is the Future?

OLLE LINDVALL

INTRODUCTION

Parkinson's disease (PD) is a chronic neurodegenerative disorder characterized by tremor, rigidity, and hypokinesia. Although L-DOPA is effective early in the disease, there is a need for new therapeutic approaches in advanced stages of PD. Patients with PD seem to be ideal for testing whether function in the human brain, affected by a neurodegenerative disorder, can be restored by replacing dead neurons with new, healthy neurons through transplantation. The main pathology underlying motor symptoms in PD is a rather selective degeneration of mesencephalic dopamine (DA) neurons. There is also a solid experimental basis showing the functional efficacy of transplantation of embryonic mesencephalic tissue to the striatum in animal models of PD and a biological mechanism underlying the observed improvement, that is, restoration of striatal DA transmission (see Chapter 9.4, this volume).

When the clinical trials with transplantation of human embryonic mesencephalic tissue, rich in DA neuroblasts, started in PD patients about two decades ago, it was unknown whether cell replacement can work in the human brain. Therefore, the first phase of transplantation research in PD aimed at addressing the following questions: Can the grafted DA neurons survive and form connections? Can the patient's brain integrate and use the grafted DA neurons? Can the grafts induce a measurable clinical improvement in PD patients? So far, 300–400 patients with PD have been grafted with human embryonic mesencephalic tissue. The results have provided proof of the principle that cell replacement can work in the human PD brain.

Cell therapy research in PD now has entered its second phase, and the main objective is to develop this approach into a clinically competitive treatment. It should be emphasized, though, that during the more than 20 years since the clinical cell therapy trials started, several new therapeutic options for the PD patient have been added. Most importantly, deep-brain stimulation (DBS), in most cases in the subthalamic nucleus, has been developed and shown to substantially improve motor deficits in advanced PD.[1] Therefore, in order to become clinically useful, cell replacement has to give rise to long-lasting, major improvement in mobility, suppression of dyskinesias, and amelioration of symptoms resistant to other treatments or to counteract disease progression. In this chapter, I will describe what has been learned from the clinical trials with transplantation of human embryonic mesencephalic tissue in patients with PD, the major scientific and clinical problems to be solved, and how far stem cells have reached toward the clinical application.

CAN THE GRAFTS SURVIVE AND BECOME FUNCTIONALLY INTEGRATED IN THE PATIENT'S BRAIN?

It has been convincingly shown that mesencephalic DA neuroblasts, obtained from 6- to 9-week-old human embryos, can survive transplantation into the brain of PD patients. Significant increases in ^{18}F-DOPA uptake in the grafted striatum measured with positron emisson tomography (PET) have been reported in more than 40 PD patients.[2–16] Histopathological studies have confirmed the long-term survival of the dopaminergic grafts and demonstrated persistent, extensive reinnervation of the striatum in several PD patients who died after transplantation.[16–22] The dopaminergic innervation was distributed in a patch-matrix pattern, and there were synaptic connections between graft and host.

The grafts can restore DA synthesis and release in the denervated striatum. Concomitantly with major clinical improvement, one patient with unilateral putaminal grafts showed a gradual increase in ^{18}F-DOPA uptake in the implanted putamen, which reached a normal level after 3 years and remained stable thereafter up to 10 years[23] (Fig. 9.5.1A, B). In contrast, there was a continuous loss of ^{18}F-DOPA uptake in the nongrafted putamen, indicating ongoing degeneration of the patient's

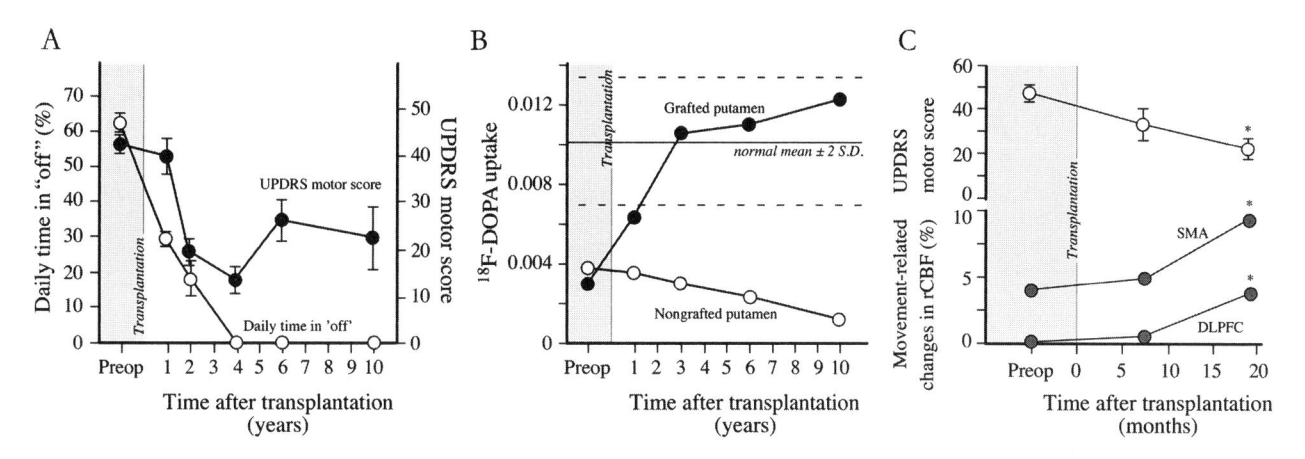

FIGURE 9.5.1. Grafted embryonic mesencephalic dopaminergic neurons can become functionally integrated in PD patient's brain, restore striatal DA synthesis and release to normal levels, and give rise to major long-lasting improvements in some patients. (A) Percentage of the day spent in the off-phase and the UPDRS motor score in the off-phase preoperatively and at various time points after transplantation to the right putamen in a PD patient. Mean ±95% confidence interval. (B) ^{18}F-DOPA uptake in grafted and nongrafted putamen of the same patient with comparative values from a group of 16 healthy volunteers. (C) UPDRS motor score and percentage of movement-related levels of regional cerebral blood flow in comparison to rest in the supplementary motor area (SMA) and dorsolateral prefrontal cortex (DLPFC) before surgery and at 6.5 and 18.3 months after bilateral implantation of embryonic mesencephalic tissue in the caudate and putamen of four PD patients. The grafts restored the activation of frontal cortical areas associated with movements. *, $p < 0.001$ compared with the preoperative value, Student's t-test. *Source:* Modified from Piccini et al.[23,25].

own DA neurons concomitantly with regeneration by the grafted neurons. In this patient, the basal and drug-induced DA release, as assessed using ^{11}C-raclopride binding, was normalized in the grafted putamen but severely impaired in the nongrafted putamen 10 years after transplantation.[23]

In further support of normal regulation of DA release from the grafted neurons, Piccini et al.[24] showed a positive correlation between ^{18}F-DOPA uptake and the percentage reduction of ^{11}C-raclopride binding potential induced by methamphetamine in the grafted putamen in eight patients (Fig. 9.5.2A). These measures of DA storage and drug-induced DA release were compared with those in a similar analysis using data from the nongrafted putamen in another group of PD patients. Interstingly, there was a clear trend: for a

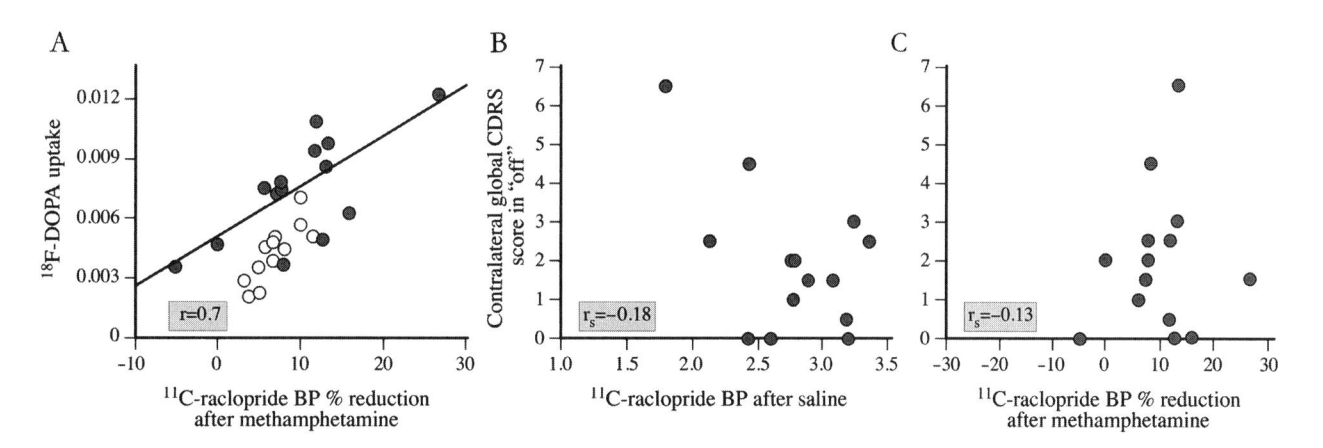

FIGURE 9.5.2. Transmitter release from grafted embryonic dopaminergic neurons resembles that of intrinsic neurons, and off-phase dyskinesias are not due to excessive DA release. (A) Correlation between ^{18}F-DOPA uptake and percentage reduction of ^{11}C-raclopride binding potential after methamphetamine in the grafted putamen (filled circles and line; data from the two sides are pooled). For comparison, values from the left and right putamen in a group of nongrafted PD patients are given (open circles). (B, C) Correlation between ^{11}C-raclopride binding potential in the putamen after saline administration (B) or its percentage reduction after methamphetamine administration (C) and the global dyskinesia (clinical dyskinesia rating scale [CDRS]) score in the off-phase on the contralateral side of the body. Data from six PD patients with bilateral grafts and two patients with unilateral grafts. *Source:* Modified from Piccini et al.[24]

similar level of ^{18}F-DOPA uptake in grafted and non-grafted patients, the percentage change in ^{11}C-raclopride binding potential induced by methamphetamine was less pronounced in the transplanted patients (Fig. 9.5.2A). These findings argue against the possibility of abnormal, excessive DA release from the terminals of grafted neurons.

The grafts can also become functionally integrated into the neural circuitries in the PD patient's brain.[25] The supplementary motor area (SMA) and the dorsolateral prefrontal cortex (DLPFC) are influenced by the basal ganglia-thalamo-cortical neural circuitries, and their impaired activation is believed to underlie parkinsonian akinesia. In four patients grafted bilaterally in the caudate and putamen, preoperative regional cerebral blood flow measurements showed only a small activation of SMA and no activation of DLPFC (Fig. 9.5.1C). No significant differences in activation were observed in these patients at 6.5 months after grafting, while at 18.3 months there was increased activation of both SMA and DLPFC compared to their preoperative status. This time course paralleled that of the clinical improvement (Fig. 9.5.1C). Taken together, these findings indicate that successful grafts in patients with PD, by improving striatal dopaminergic neurotransmission, restore movement-related cortical activation, which probably is necessary to induce substantial clinical improvement.

CAN THE GRAFTS GIVE RISE TO CLINICALLY DETECTABLE SYMPTOMATIC IMPROVEMENT?

Open-label trials have reported clear clinical benefit associated with survival of the human embryonic mesencephalic grafts.[2-11,16,26-28] In the most successful cases, patients have withdrawn from L-DOPA treatment and have exhibited major clinical improvement several years after transplantation.[8,9,11,23] Table 9.5.1 summarizes the magnitude of the clinical benefit at 10–24 months and 3 years postoperatively in four open-label trials. All patients received bilateral grafts of tissue from two to five donors in each putamen. In some cases, tissue was also implanted in the caudate nucleus and, in one patient, into substantia nigra. The Unified Parkinson's Disease Rating Scale (UPDRS) motor score during the practically defined off-phase (i.e., in the morning, at least 12 hr after the last dose of antiparkinsonian medication) revealed 30% to 50% symptomatic relief. The daily time spent in the off-phase decreased by 43% to 59% and the mean daily L-DOPA requirements by 16% to 45%. All patients showed significant increases in ^{18}F-DOPA uptake in the operated putamen, indicating graft survival. However, in three of these studies, uptake after transplantation was still only about 50% of the normal mean. This probably explains, at least to some extent, the incomplete functional recovery and indicates that there is room for improvement. Some support for this idea is the more pronounced reduction of the UPDRS motor score in the patients of Mendez et al.,[16] in whom 72% of the normal ^{18}F-DOPA uptake was restored (Table 9.5.1).

The first double-blind, sham surgery-controlled study[12] demonstrated only a modest clinical response, with an 18% reduction of the UPDRS motor scores in the off-phase at 12 months after bilateral putaminal grafts, but no improvement was seen in the sham-operated group. In patients younger than 60 years, the UPDRS improvement was 34%. No immunosuppression was given. These data are important because they

TABLE 9.5.1. *Magnitude of Postoperative Changes in Symptomatology and Putaminal ^{18}F-DOPA Uptake in Open-Label Trials with Embryonic Mesencephalic Tissue Transplantation Compared to Subthalamic Deep-Brain Stimulation in Patients with PD*

	Hauser et al.[10] (n = 6)	Hagell et al.[9] (n = 4)	Brundin et al.[11] (n = 5)	Mendez et al.[16] (n = 2)	DBS-STN (22 studies)
UPDRS motor score in off-phase (Δ)	−30%	−30%	−40%	−51%	−52%
Daily time in off-phase (Δ)	−43%	−59%	−43%	−50%	−68%
Daily L-DOPA dose (Δ)	−16%	−37%	−45%	±0/−30%	−56%
^{18}F-DOPA uptake (putamen; % of normal mean)					
Preop	34%	31%	31%	28%	
Postop	55%	52%	48%	72%	

Notes: n, number of patients; DBS-STN, deep-brain stimulation in subthalamic nucleus; UPDRS, Unified Parkinson's Disease Rating Scale.
Source: DBS-STN data are from Kleiner-Fisman et al.[1]

provide some evidence of a specific graft-induced symptomatic improvement, distinguishable from a placebo effect.

In the second, sham surgery-controlled clinical trial,[13] human embryonic mesencephalic tissue from one or four donors was implanted in each postcommissural putamen. Immunosuppressive treatment with cyclosporine was given for 6 months after surgery, and patients were followed for 2 years. The trial failed to meet its primary outcome, that is, a group difference in the change in UPDRS motor scores at 24 months compared to baseline. However, similar to the time course of improvement in the open-label trials, the patients grafted with tissue from four donors showed progressive symptomatic relief up to 6 to 9 months after surgery (but deteriorated thereafter). Putaminal ^{18}F-DOPA uptake was significantly increased in grafted patients at 12 months, compared to controls and nongrafted striatal areas, and remained largely stable at 2 years after transplantation.

The most troublesome complication caused by transplantation of embryonic mesencephalic tissue in PD patients has been the occurrence of dyskinesias in the off-phase.[29] In the study of Freed et al.,[12] 15% of the grafted patients developed severe dyskinesias. Hagell et al.[30] found that among 14 grafted PD patients, 8 displayed mild off-phase dyskinesias. The remaining six patients exhibited dyskinesias of moderate severity, which in one patient constituted a clinical therapeutic problem. Olanow et al.[13] reported that 56.5% of the grafted patients developed off-phase dyskinesias. Dyskinesia severity appeared to be generally mild, but was disabling and required surgery in three cases. The off-phase dyskinesias in grafted patients have been effectively treated with DBS of the globus pallidus internus.[31,32]

Off-phase dyskinesias are most likely not caused by dopaminergic overgrowth or excessive DA release from the grafts. No correlation has been found between the magnitude of dyskinesias and that of the antiparkinsonian response.[13,30] Dyskinesias and functional improvements have shown different time courses following transplantation.[12,13,30] Moreover, off-phase dyskinesias have not been associated with high postoperative striatal ^{18}F-DOPA uptake with the most pronounced graft-induced increases in striatal ^{18}F-DOPA uptake.[13,30] Ma et al.[33] found evidence of an imbalance between the dopaminergic innervation (regional putaminal ^{18}F-DOPA uptake) in the ventral and dorsal putamen in dyskinetic grafted patients. However, no differences were observed in either regional or global levels of striatal F-DOPA uptake between patients with and without off-phase dyskinesias by Olanow et al.[13]

Piccini et al.[24] found no correlation between ^{11}C-raclopride binding (as a measure of DA release) in the putamen and dyskinesia severity scores (Fig. 9.5.2B, C). Finally, off-phase dyskinesias have resembled biphasic dyskinesias,[13,29,33] suggesting intermediate (not excessive) DA levels.

Whether off-phase dyskinesias only develop in patients with already established L-DOPA-induced dyskinesias is unclear. Three main hypotheses regarding the mechanisms underlying off-phase dyskinesias can be proposed: First, they may be due to small grafts giving rise to islands of reinnervation, surrounded by supersensitive denervated striatal areas. Consistent with this idea, Maries et al.[34] found that grafts forming "hot spots" in the rat striatum gave rise to dyskinetic behavior, which was not observed when the grafts were more evenly distributed over the structure. Second, the underlying mechanism could be an unfavorable cellular composition in the graft. Particular attention is now focused on the serotonergic component. Carta et al.[35] have shown that L-DOPA-induced dyskinesias in the rat PD model can be suppressed by lesions of the serotonergic neuron system or pharmacological blockade of serotonergic neuron firing and transmitter release. Thus, L-DOPA-derived DA, released as a "false transmitter" from serotonergic terminals, could be the main trigger of dyskinesias. In accordance, Carlsson et al.[36] reported a detrimental role of serotonin neurons in grafts with few dopaminergic neurons leading to worsening of L-DOPA-induced dyskinesias. In contrast, large numbers of DA neurons in the grafts counteracted the swings of DA released from serotonergic terminals. Third, off-phase dyskinesias may be dependent on chronic inflammatory and immune responses around the graft. Off-phase dyskinesias did not develop until after withdrawal of immunosuppression at 6 months in the study of Olanow et al.[13] When autopsies were performed at later time points, an inflammatory reaction with activated microglia was observed around the graft. Piccini et al.[24] found that withdrawal of immunosuppression at 29 months after transplantation caused no reduction of striatal ^{18}F-DOPA uptake or worsening of other PD motor symptoms, but it was accompanied by increased dyskinesia severity scores (Fig. 9.5.3). Interstingly, Soderstrom et al.[37] recently reported that dyskinesia-like behavior correlated with increases in aberrant synaptic features of grafted dopaminergic neurons in rats with an elevated immune response. In contrast, Lane et al.[38] found that inflammation caused by interleukin 2 infusion did not worsen or induce dyskinesia-like behavior in grafted rats.

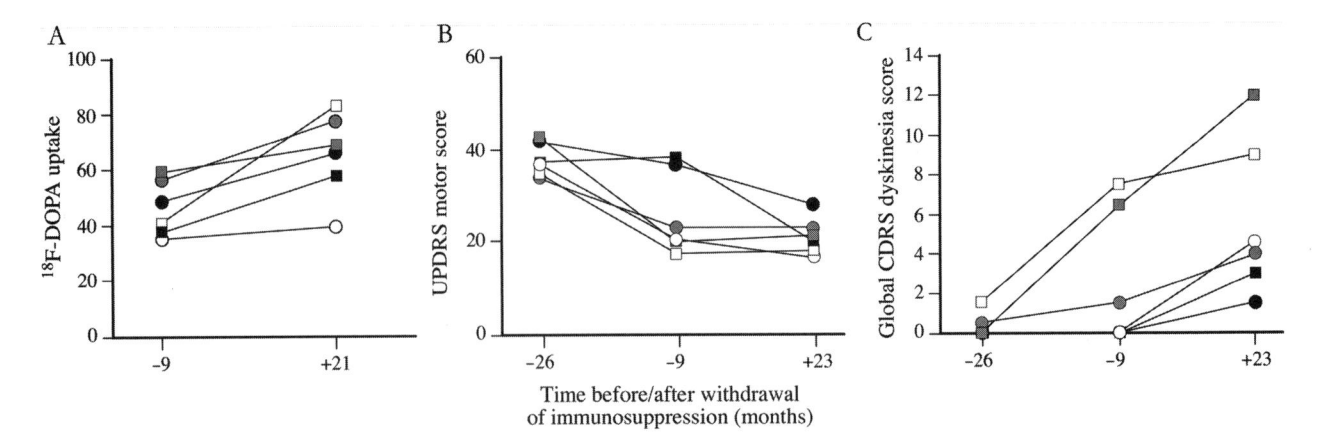

FIGURE 9.5.3. Withdrawal of immunosuppression late after transplantation of embryonic mesencephalic tissue does not compromise survival or antiparkinsonian action of the grafts but may contribute to worsening of off-phase dyskinesias. (A–C) Effects of withdrawal of immunosuppression in six PD patients on (A) ^{18}F-DOPA uptake in grafted putamen (expressed as a percentage of the normal mean) and (B) the UPDRS motor score and (C) the global dyskinesia (CDRS) score in the off-phase. Immunosuppression was completely withdrawn at a mean time of 29 months after the last transplantation, and the time points given on the x-axis depict when in relation to withdrawal respective data were collected. *Source*: Modified from Piccini et al.[24]

WHY WAS THE OUTCOME NEGATIVE IN THE TWO SHAM SURGERY-CONTROLLED CLINICAL TRIALS?

The functional improvement in the sham surgery-controlled clinical trials was only modest and clearly less than that in the open-label trials. Four main explanations to this lack of efficacy can be proposed:

1. *Clinical benefits in open-label trials have been placebo effects.* Arguing against such an interpretation, improvements in motor function after unilateral grafting have been predominantly on the contralateral side of the body. In several patients, their "worse" side switched after transplantation. Functional recovery developed gradually, from about 3 months up to 1 to 2 years after grafting. Objective neurophysiological methods measuring arm and hand movements confirmed the improvements. Some patients recovered to the extent that they were able to return to work, and L-DOPA treatment was withdrawn for many years. Finally, changes in motor function corresponded broadly to the degree of graft survival and restoration of movement-related cortical activation.
2. *The number of surviving grafted DA neurons has been too low.* In the first sham surgery-controlled study,[12] less tissue was implanted compared to the open-label trials and the tissue was stored in cell culture for up to 4 weeks. Two patients who died after grafting had only 7,000 to 40,000 grafted dopaminergic neurons in each putamen,[12] which was much lower than the number found in two patients in one of the open-label trials (80,000–135,000).[17–19]
3. *Dopaminergic denervation in the patient's brain was too widespread at the time of transplantation.* The patients of Olanow et al.[13] were more severely disabled and required higher doses of antiparkinsonian medication than, for example, the Lund patients. When Olanow and coworkers[13] analyzed the outcome in their less severely disabled patients, they found a significant difference from sham-operated patients at 2 years. It is conceivable that the extent of degeneration of dopaminergic and nondopaminergic neurons in the patient's brain prior to transplantation will influence the magnitude of functional recovery induced by a dopaminergic graft. As described in more detail below, Piccini et al.[24] reported that patients with widespread denervation in several brain areas and a dopaminergic graft in the putamen will exhibit only modest clinical recovery. In contrast, patients with dopaminergic denervation largely restricted to the putamen are more likely to benefit from an intraputaminal graft.
4. *Immune reactions have compromised the survival and function of the grafts.* The poor outcome in the two sham surgery-controlled clinical trials in which either no[12] or short-term, low-dose immunosuppression[13] was given has raised the possibility that such treatment, at least during the first year, is necessary for major and persistent

symptomatic relief. Several open-label trials reporting a clear clinical benefit have used strictly controlled immunosuppressive regimens for 1 to 2 years after transplantation. The low numbers of surviving DA neurons in the study of Freed et al.[12] suggest that the lack of immunosuppressive treatment had compromised graft survival. Although the grafts survived in the study of Olanow et al.,[13] their function may have been impaired due to an immune reaction. The improvement compared to that seen in the sham surgery-controlled study up to 6 to 9 months, and the deterioration thereafter, in Olanow and collaborators' patients are consistent with an immune reaction after withdrawal of immunosuppression after 6 months. In two patients who came to autopsy, the grafts were surrounded by activated microglia, suggesting a delayed immune response.[13] In contrast, in two patients who had been subjected to 6 months of immunosuppressive treatment in the study of Mendez et al.[16] and showed clinical improvement (Table 9.5.1), only a few macrophages and activated microglia were found in grafted regions 3 to 4 years after surgery.

ARE THE GRAFTS AFFECTED BY THE DISEASE PROCESS?

One important question in the clinical trials, which could not be addressed in the experimental studies, is whether the grafted neurons would be affected by the disease process. Two recent reports on patients transplanted with human embryonic mesencephalic tissue provide evidence that PD pathology may propagate from the host to the graft.[20,21] In three patients who died 11 to 16 years after surgery, a fraction (1%–4%; J.-Y. Li et al., unpublished observations) of the grafted DA neurons contained α-synuclein-rich Lewy bodies (LBs) and Lewy neurites (LNs). These characteristic pathological features are observed in affected brain regions, including the substantia nigra, of patients with PD. One study reported no LBs in a patient 14 years after transplantation.[22] Interestingly, four patients who had survived for 4 or 9 years following transplantation showed no α-synuclein pathology in the grafts,[20,22] suggesting that at least one decade is required for the development of LBs. The pathological changes are probably progressive because LBs were more frequent in the grafts implanted 16 years prior to death than in those transplanted in the contralateral hemisphere 4 years later in one PD patient (J.-Y. Li et al., unpublished observations).

The new observations suggest that certain pathogenetic events in PD are non-cell-autonomous and that affected neurons or glia may transfer the disease process to healthy neurons. Potential mechanisms that may underlie the pathological changes include inflammation, oxidative stress, excitotoxicity, reduced levels of trophic factors, or a prion-like mechanism.[39] Obviously, these findings raise several important issues regarding the cause of PD pathology, but what are the consequences for the DA cell replacement strategy? Imaging studies have shown that embryonic mesencephalic grafts can synthesize and release normal DA levels 10 years after transplantation associated with major clinical improvement.[23] This indicates that the majority of the grafted cells are not functionally impaired after one decade. Thus, cell replacement is still a viable therapeutic option because (1) the process is slow, (2) the majority of grafted neurons are unaffected after a decade, and (3) the patients experience long-term symptomatic relief. It is important to point out, though, that in one patient, who died 14 years after transplantation, the immunostaining for the DA transporter (DAT) was very light or absent in the graft despite robust staining for tyrosine hydroxylase (TH). This patient was reported to experience progressive worsening of PD beginning 11 years after surgery, and the reduced DAT staining combined with TH and vesicular monoamine transporter (VMAT) staining may indicate an early compensatory response to graft failure.[20]

HOW CAN WE MAKE CELL THERAPY WORK IN PD PATIENTS?

If DA cell replacement becomes a clinically competitive therapy in PD, it has to provide advantages over currently available, rather effective treatments for alleviation of motor symptoms in PD patients. So far, the improvements after intrastriatal transplantation of DA neurons in patients[12,13] (Table 9.5.1) have not exceeded those found with DBS in the subthalamic nucleus,[1] and there is no convincing evidence that drug-resistant symptoms are reversed by these grafts.[40] For the development of a clinically competitive DA cell replacement therapy in PD, four major scientific advancements will be necessary:

1. *Generation of large numbers of DA neurons in standardized, quality-controlled preparations.* Human embryonic mesencephalic grafts will probably continue to be the gold standard in cell therapy research for PD. However, it is unlikely that transplantation of human embryonic

mesencephalic tissue will become routine treatment for PD due to problems with tissue availability and standardization of the grafts, leading to too much variation in functional outcome. Small groups of patients could still be operated on, using this type of tissue to explore specific scientific issues and thereby pave the way for stem cell–based approaches. The main interest is now focused on the production of DA neuroblasts for transplantation from stem cells in culture. It should be emphasized, though, that after maturation, these neurons have to work at least as well as those in the embryonic mesencephalic grafts. Conceivably, the stem cell–derived cells have to fulfill the following requirements in order to induce marked clinical improvement after transplantation: (a) release DA in a regulated manner and exhibit the molecular, morphological, and electrophysiological properties of substantia nigra neurons[16,41]; (b) reverse the motor deficits in animal models resembling the symptoms in patients; (c) allow for 100,000 or more grafted DA neurons to survive long-term in each human putamen[42]; (d) reestablish a dense terminal network throughout the striatum; and (e) become functionally integrated into host neural circuitries.[25]

2. *Improved patient selection.* Better criteria for selection of the most suitable patients for transplantation with respect to stage and type of disease have to be defined, and the preoperative degeneration pattern has to be determined. Dopaminergic cell therapy will most likely be successful only in those patients who can exhibit a marked symptomatic benefit in response to L-DOPA and in whom the main pathology is a loss of DA neurons. Debilitating symptoms in PD and related disorders are also caused by pathological changes in nondopaminergic systems. Until it is known how to repair these systems, patients with such symptoms should not undergo cell transplantation.

3. *Improved functional efficacy of grafts.* The transplantation procedure needs to be tailor-made with respect to dose and location of grafted cells based on preoperative imaging so that the repair of the DA system in striatum and extrastriatal areas in each patient's brain is as complete as possible. Piccini et al.[24] have explored the possibility that the extent of dopaminergic denervation in areas not reached by putaminal grafts influences the functional outcome after transplantation. Statistical parametric mapping (SPM) of ^{18}F-DOPA uptake, comparing each patient with an appropriate control across the whole brain, was used to show the pattern of dopaminergic denervations outside the grafted areas and also to explore whether these changes were present prior to transplantation or developed during the postoperative assessment period. Out of eight patients, all having surviving grafts bilaterally in the putamen, three showed no reduction in ^{18}F-DOPA uptake outside the grafted areas either before or after transplantation, indicating that the dopaminergic denervation remained confined to the caudate-putamen throughout the period of assessment. Three patients showed denervation outside the areas to be grafted prior to transplantation. Two patients developed such denervation during the first 2 years after surgery. Remarkably, the three patients with denervation confined to grafted areas exhibited major improvements, whereas those who had widespread denervation prior to transplantation showed no overall benefit or even deteriorated. The two patients who developed denervation outside the grafted areas postoperatively showed modest overall benefit. These findings indicate that the occurrence of dopaminergic denervation in nongrafted areas before or after transplantation exerts a marked influence on the overall outcome. The results also provide evidence that patients with denervations that remain restricted to the caudate-putamen are likely to have major long-term benefits from grafts placed in these areas. In contrast, a long-lasting successful outcome in patients with more widespread denervations probably will require that grafts also be placed in areas outside the caudate-putamen.

4. *Strategies to avoid adverse effects.* The risk of off-phase dyskinesias following cell transplantation has to be minimized. Available data indicate that this could be achieved by excluding serotonergic neurons from the graft material, by giving carefully monitored immunosuppression for 6–12 months, and by using a surgical procedure that gives rise to an optimum distribution of tissue over the putamen and complete, even reinnervation without hot spots. The risk of tumor formation from pluripotent stem cells, and the consequences of the introduction of new genes in stem cell–derived neurons, should be carefully evaluated after transplantation in animal models prior to clinical application. To improve safety, it may be necessary to engineer stem cells with regulatable suicide genes or to use cell sorting to eliminate those cells that could give rise to tumors.

CAN DA NEURONS FOR CLINICAL APPLICATION BE GENERATED FROM STEM CELLS?

In a clinical setting, the stem cell–derived DA neuroblasts used for transplantation most likely have to be of human origin. Cells exhibiting at least some characteristics of mesencephalic DA neurons have been produced from stem cells from different sources: embryonic stem cells, embryonic brain, adult brain, and other tissues, obtained from rodents, nonhuman primates, and humans. Table 9.5.2 summarizes the reported data when stem cell-derived neurons were tested in animal models of PD, with a focus on properties of particular importance for deciding whether the cells are suitable for clinical application. In most cases, it has not been demonstrated that the stem cell–derived cells can substantially reinnervate the striatum, restore DA release, and markedly improve deficits resembling the PD patient's symptoms (Table 9.5.2). Thus, much experimental work remains to be done before any stem cell–derived DA neuroblast can be selected as a clinical candidate cell for transplantation in a PD patient.

One of the most exciting recent developments is the demonstration that somatic cells such as skin cells can be reprogrammed to a pluripotent state and become indistinguishable from embryonic stem cells. Wernig et al.[43] recently reported that DA neurons can be generated from induced pluripotent stem cells derived from mouse fibroblasts and can ameliorate behavioral deficits after transplantation in a rodent PD model. The major potential advantage of this approach is that patient-specific DA neurons suitable for transplantation, avoiding immune reactions, can be produced without the use of human embryonic stem cells. However, several problems have to be solved before induced pluripotent stem cells can even be considered in a clinical setting. First, the risk of tumor formation, which resembles that of embryonic stem cells, has to be eliminated. The development of small molecules for reprogramming and cell sorting to separate the tumorigenic cells could be the solution to this problem. Much work is also needed to improve the efficacy of induced pluripotent stem cell generation and to determine the functionality of the generated dopamine neurons. A specific problem could be envisioned if the DA neurons are generated from the patient's own skin cells. It seems possible that this process could be associated with increased susceptibility to the degenerative process, making the neurons more susceptible to the disease.

Another possible way to avoid immune reactions is by therapeutic cloning. Genetically identical embryonic stem cell–derived DA neurons, generated by transfer of autologous nuclei from fibroblasts, ameliorate functional deficits without producing an immune reaction in parkinsonian mice.[44] However, to produce cells using therapeutic cloning for clinical application would be a logistical challenge. It has to be shown that therapeutic cloning leading to DA neurons also works with human cells and that substantial recovery can be obtained. Tumor formation has to be eliminated. The patient may exhibit a gene profile that would make the cells particularly susceptible to pathological changes. Finally, it can be questioned whether all the efforts to produce patient-specific DA neurons for PD are justified. Immune reactions to brain allografts are moderate, and survival can be obtained even without immunosuppression, although most investigators favor immunosuppression for 6 to 12 months after transplantation.

TABLE 9.5.2. *Clinically Important Properties of Dopaminergic Stem/Precursor Cell Grafts in Animal Models of PD*

Cell Source	Striatal Reinnervation	In Vivo Dopamine Release	Improvement of Parkinson-Like Symptoms
Mouse ES cells	Partial	Significant	Partial
Mouse fibroblasts(iPS cells)	Fibers	N.D.	N.D.
Mouse ES cells (therapeutically cloned)	N.D	N.D	Partial
Monkey ES cells	Partial	N.D.	Partial
Human ES cells	N.D.	N.D.	Partial
Rat embryonic VM-derived NSCs	Partial	N.D.	Partial
Human embryonic VM-derived NSCs	Fibers	N.D.	N.D.
Rat adult SVZ-derived NSCs	N.D.	N.D.	N.D.
Rat bone marrow stem cells	Fibers	N.D.	Partial

Note: ES cells, embryonic stem cells; iPS cells, induced pluripotent stem cells; N.D., not demonstrated; NSCs, neural stem cells; SVZ, subventricular zone; VM, ventral mesencephalon.
Source: Based on data from [43,44,47–60]

WILL CELL REPLACEMENT EVER BECOME A CLINICALLY COMPETITIVE THERAPY FOR PD PATIENTS?

Parkinson's disease is progressive and affects areas outside the putamen, where most grafts have been placed, as well as nondopaminergic systems, which are not replaced by embryonic mesencephalic tissue or stem cell–derived dopaminergic neurons. Moreover, it is not yet possible to reconstruct the nigrostriatal pathway. Therefore, the dopaminergic grafts have in virtually all cases been placed in an ectopic location, namely, the striatum. Several arguments support the belief that cell replacement research, despite these problems, should continue, with the aim of developing a clinically useful transplantation treatment for PD patients. Dopaminergic cell therapy leads to replacement specifically of those neurons that have died because of the disease process, and thereby targets the impaired biological mechanism underlying a substantial part of the patient's symptoms. In successful cases, dopaminergic cell therapy has induced major, long-lasting clinical improvements and allowed PD patients to stop taking medication for several years. Moreover, imaging techniques, in particular PET, have improved to the extent that it is now possible, with high resolution, to monitor the extent and pattern of innervation as well as the function of different neural systems (e.g., the nigrostriatal DA system). Finally, in the future, it may also be possible to implant stem cell–derived neurons with other phenotypes, as well as to reconstruct the nigrostriatal pathway by suppressing axonal growth inhibitory mechanisms. However, for long-lasting symptomatic relief in PD, cell replacement therapy probably has to be combined with strategies to hinder disease progression. Possible approaches to prevent the death of existing neurons could include transplanting human stem cells engineered to express neuroprotective molecules such as glial cell line–derived neurotrophic factor (GDNF)[15] or using direct gene delivery of a trophic factor, such as neurturin.[46]

ACKNOWLEDGMENTS

Our own research was supported by by the Swedish Research Council and the Söderberg, Crafoord, and Kock Foundations. The Lund Stem Cell Center is supported by a Center of Excellence grant in Life Sciences from the Swedish Foundation for Strategic Research.

REFERENCES

1. Kleiner-Fisman G, Herzog J, Fisman DN, et al. Subthalamic deep brain stimulation: summary and meta-analysis of outcomes. *Mov Disord.* 2006;21(suppl 14): S290–S304.
2. Lindvall O, Brundin P, Widner H, et al. Grafts of fetal dopamine neurons survive and improve motor function in Parkinson's disease. *Science.* 1990;247:574–577.
3. Lindvall O, Sawle G, Widner H, et al. Evidence for long-term survival and function of dopaminergic grafts in progressive Parkinson's disease. *Ann Neurol.* 1994;35:172–180.
4. Sawle GV, Bloomfield PM, Björklund A, et al. Transplantation of fetal dopamine neurons in Parkinson's disease: PET [^{18}F]6-L-fluorodopa studies in two patients with putaminal implants. *Ann Neurol.* 1992;31:166–173.
5. Peschanski M, Defer G, N'Guyen JP, et al. Bilateral motor improvement and alteration of L-dopa effect in two patients with Parkinson's disease following intrastriatal transplantation of foetal ventral mesencephalon. *Brain.* 1994;117: 487–499.
6. Freeman TB, Olanow CW, Hauser RA, et al. Bilateral fetal nigral transplantation into the postcommissural putamen in Parkinson's disease. *Ann Neurol.* 199538:379–388.
7. Remy P, Samson Y, Hantraye P, et al. Clinical correlates of [^{18}F]fluorodopa uptake in five grafted parkinsonian patients. *Ann Neurol.* 1995;38:580–588.
8. Wenning GK, Odin P, Morrish P, et al. Short- and long-term survival and function of unilateral intrastriatal dopaminergic grafts in Parkinson's disease. *Ann Neurol.* 1997;42:95–107.
9. Hagell P, Schrag A, Piccini P, et al. Sequential bilateral transplantation in Parkinson's disease: effects of the second graft. *Brain.* 1999;122:1121–1132.
10. Hauser RA, Freeman TB, Snow BJ, et al. Long-term evaluation of bilateral fetal nigral transplantation in Parkinson disease. *Arch Neurol.* 1999;56:179–187.
11. Brundin P, Pogarell O, Hagell P, et al. Bilateral caudate and putamen grafts of embryonic mesencephalic tissue treated with lazaroids in Parkinson's disease. *Brain.* 2000;123: 1380–1390.
12. Freed CR, Greene PE, Breeze RE, et al. Transplantation of embryonic dopamine neurons for severe Parkinson's disease. *N Engl J Med.* 2001;344:710–719.
13. Olanow CW, Goetz CG, Kordower JH, et al. A double-blind controlled trial of bilateral fetal nigral transplantation in Parkinson's disease. *Ann Neurol.* 2003;54:403–414.
14. Cochen V, Ribeiro MJ, Nguyen JP, et al. Transplantation in Parkinson's disease: PET changes correlate with the amount of grafted tissue. *Mov Disord.* 2003;18:928–932.
15. Mendez I, Dagher A, Hong M, et al. Simultaneous intrastriatal and intranigral fetal dopaminergic grafts in patients with Parkinson disease: a pilot study. Report of three cases. *J Neurosurg.* 2002;96:589–596.
16. Mendez I, Sanchez-Pernaute R, Cooper O, et al. Cell type analysis of functional fetal dopamine cell suspension transplants in the striatum and substantia nigra of patients with Parkinson's disease. *Brain.* 2005;128:1498–1510.
17. Kordower JH, Freeman TB, Snow BJ, et al. Neuropathological evidence of graft survival and striatal reinnervation after the transplantation of fetal mesencephalic tissue in a patient with Parkinson's disease. *N Engl J Med.* 1995;332:1118–1124.
18. Kordower JH, Rosenstein JM, Collier TJ, et al. Functional fetal nigral grafts in a patient with Parkinson's disease:

chemoanatomic, ultrastructural, and metabolic studies. *J Comp Neurol.* 1996;370:203–230.
19. Kordower JH, Freeman TB, Chen EY, et al. Fetal nigral grafts survive and mediate clinical benefit in a patient with Parkinson's disease. *Mov Disord.* 1998;13:383–393.
20. Kordower JH, Chu Y, Hauser RA, et al. Lewy body–like pathology in long-term embryonic nigral transplants in Parkinson's disease. *Nat Med.* 2008;14:504–506.
21. Li J-Y, Englund E, Holton JL, et al. Lewy bodies in grafted neurons in subjects with Parkinson's disease suggest host-to-graft disease propagation. *Nat Med.* 2008;14:501–503.
22. Mendez I, Vinuela A, Astradsson A, et al. Dopamine neurons implanted into people with Parkinson's disease survive without pathology for 14 years. *Nat Med.* 2008;14:507–509.
23. Piccini P, Brooks DJ, Björklund A, et al. Dopamine release from nigral transplants visualized *in vivo* in a Parkinson's patient. *Nat Neurosci.* 1999;2:1137–1140.
24. Piccini P, Pavese N, Hagell P, et al. Factors affecting the clinical outcome after neural transplantation in Parkinson's disease. *Brain.* 2005;128:2977–2986.
25. Piccini P, Lindvall O, Björklund A, et al. Delayed recovery of movement-related cortical function in Parkinson's disease after striatal dopaminergic grafts. *Ann Neurol.* 2000;48: 689–695.
26. Lindvall O, Widner H, Rehncrona S, et al. Transplantation of fetal dopamine neurons in Parkinson's disease: 1-year clinical and neurophysiological observations in two patients with putaminal implants. *Ann Neurol.* 1992;31:155–165.
27. Defer GL, Geny C, Ricolfi F, et al. Long-term outcome of unilaterally transplanted parkinsonian patients: I. Clinical approach. *Brain.* 1996;119:41–50.
28. Mendez I, Dagher A, Hong M, et al. Enhancement of survival of stored dopaminergic cells and promotion of graft survival by exposure of human fetal nigral tissue to glial cell line–derived neurotrophic factor in patients with Parkinson's disease. *J Neurosurg.* 2000;92:863–869.
29. Hagell P, Cenci MA. Dyskinesias and dopamine cell replacement in Parkinson's disease: a clinical perspective. *Brain Res Bull.* 2005;68:4–15.
30. Hagell P, Piccini P, Björklund A, et al. Dyskinesias following neural transplantation in Parkinson's disease. *Nat Neurosci.* 2002;5:627–628.
31. Graff-Radford J, Foote KD, Rodriguez RL, et al. Deep brain stimulation of the internal segment of the globus pallidus in delayed runaway dyskinesia. *Arch Neurol.* 2006;63:1181–1184.
32. Herzog J, Pogarell O, Pinsker MO, et al. Deep brain stimulation in Parkinson's disease following fetal nigral transplantation. *Mov Disord.* 2008;23:1293–1296.
33. Ma Y, Feigin A, Dhawan V, et al. Dyskinesia after fetal cell transplantation for parkinsonism: a PET study. *Ann Neurol.* 2002;52:628–634.
34. Maries E, Kordower JH, Chu Y, et al. Focal, not widespread, grafts induce novel dyskinetic behavior in parkinsonian rats. *Neurobiol Dis.* 2006;21:165–180.
35. Carta M, Carlsson T, Kirik D, Bjorklund A. Dopamine released from 5-HT terminals is the cause of L-DOPA-induced dyskinesia in parkinsonian rats. *Brain.* 2007;130:1819–1833.
36. Carlsson T, Carta M, Winkler C, et al. Serotonin neuron transplants exacerbate L-DOPA-induced dyskinesias in a rat model of Parkinson's disease. *J Neurosci.* 2007;27:8011–8022.
37. Soderstrom KE, Meredith G, Freeman TB, et al. The synaptic impact of the host immune response in a parkinsonian allograft rat model: influence on graft-derived aberrant behaviors. *Neurobiol Dis.* 2008;32:229–242.
38. Lane EL, Soulet D, Vercammen L, et al. Neuroinflammation in the generation of post-transplantation dyskinesia in Parkinson's disease. *Neurobiol Dis.* 2008;32:220–228.
39. Brundin P, Li J-Y, Holton JL, et al. Research in motion: the enigma of Parkinson's disease pathology spread. *Nat Rev Neurosci.* 2008;9:741–745.
40. Lindvall O, Hagell P. Clinical observations after neural transplantation in Parkinson's disease. *Prog Brain Res.* 2000;127: 299–320.
41. Isacson O, Bjorklund LM, Schumacher JM. Toward full restoration of synaptic and terminal function of the dopaminergic system in Parkinson's disease by stem cells. *Ann Neurol.* 2003;53(suppl 3):S135–S146.
42. Hagell P, Brundin P. Cell survival and clinical outcome following intrastriatal transplantation in Parkinson's disease. *J Neuropathol Exp Neurol.* 2001;60:741–752.
43. Wernig M, Zhao J-P, Pruszak J, et al. Neurons derived from reprogrammed fibroblasts functionally integrate into the fetal brain and improve symptoms of rats with Parkinson's disease. *Proc Natl Acad Sci USA.* 2008;105:5856–5861.
44. Tabar V, Tomishima M, Panagiotakos G, et al. Therapeutic cloning in individual parkinsonian mice. *Nat Med.* 2008;14: 379–381.
45. Behrstock S, Ebert A, McHugh J, et al. Human neural progenitors deliver glial cell line–derived neurotrophic factor to parkinsonian rodents and aged primates. *Gene Ther.* 2006;13: 379–388.
46. Marks WJ Jr, Ostrem JL, Verhagen L, et al. Safety and tolerability of intraputaminal delivery of CERE-120 (adeno-associated virus serotype 2-neurturin) to patients with idiopathic Parkinson's disease: an open-label, phase I trial. *Lancet Neurol.* 2008;7: 400–408.
47. Björklund LM. Sánchez-Pernaute R, Chung S, et al. Embryonic cells develop into functional dopaminergic neurons after transplantation in a Parkinson rat model. *Proc Natl Acad Sci USA.* 2002;99:2344–2349.
48. Cho MS, Lee Y-E, Kim JY, et al. Highly efficient and large-scale generation of functional dopamine neurons from human embryonic stem cells. *Proc Natl Acad Sci USA.* 2008;105: 3392–3397.
49. Dezawa M, Kanno H, Hoshino M, et al. Specific induction of neuronal cells from bone marrow stromal cells and application for autologous transplantation. *J Clin Invest.* 2004;113: 1701–1710.
50. Kawasaki H, Mizuseki K, Nishikawa S, et al. Induction of midbrain dopaminergic neurons from ES cells by stromal cell-derived inducing activity. *Neuron.* 2000;28:31–40.
51. Kawasaki H, Suemori H, Mizuseki K, et al. Generation of dopaminergic neurons and pigmented epithelia from primate ES cells by stromal cell–derived inducing activity. *Proc Natl Acad Sci USA.* 2002;99:1580–1585.
52. Kim JH, Auerbach JM, Rodriguez-Gomez JA, et al. Dopamine neurons derived from embryonic stem cells function in an animal model of Parkinson's disease. *Nature.* 2002;418:50–56.
53. O'Keeffe FE, Scott SA, Tyers P, et al. Induction of A9 dopaminergic neurons from neural stem cells improves motor function in an animal model of Parkinson's disease. *Brain.* 2008;131: 630–641.
54. Parish CL, Castelo-Branco G, Rawal N, et al. Wnt5a-treated midbrain neural stem cells improve dopamine cell replacement therapy in parkinsonian mice. *J Clin Invest.* 2008;118: 149–160.

55. Rodriguez-Gomez JA, Lu J-Q, Velasco I, et al. Persistent dopamine functions of neurons derived from embryonic stem cells in a rodent model of Parkinson disease. *Stem Cells.* 2007;25:918–928.
56. Roy NS, Cleren C, Singh SK, et al. Functional engraftment of human ES cell–derived dopaminergic neurons enriched by coculture with telomerase-immortalized midbrain astrocytes. *Nat Med.* 2006;12:1259–1268.
57. Sanchez-Pernaute R, Studer L, Bankiewicz KS, et al. In vitro generation and transplantation of precursor-derived human dopamine neurons. *J Neurosci Res.* 2001;65:284–288.
58. Shim J-W, Park C-H, Bae Y-C, et al. Generation of functional dopamine neurons from neural precursor cells isolated from the subventricular zone and white matter of the adult rat brain using Nurr1 overexpression. *Stem Cells.* 2007;25:1252–1262.
59. Studer L, Tabar V, McKay RDG. Transplantation of expanded mesencephalic precursors leads to recovery in parkinsonian rats. *Nat Neurosci.* 1998;1:290–295.
60. Takagi Y, Takahashi J, Saiki H, et al. Dopaminergic neurons generated from monkey embryonic stem cells function in a Parkinson primate model. *J Clin Invest.* 2005;115:102–109.

9.6 | Novel Gene-Based Therapeutics Targeting the Dopaminergic System in Parkinson's Disease

DENIZ KIRIK, TOMAS BJÖRKLUND, SHILPA RAMASWAMY, AND JEFFREY H. KORDOWER

INTRODUCTION

Novel therapeutic intervention based on gene therapy has moved the field of Parkinson's disease (PD) research forward during the last decade. The process of supplementing cells with genes that promote normal, healthy function promises to be an efficient way of treating diseases like PD, above and beyond what has been possible to achieve with traditional pharmacotherapy or deep brain stimulation. Studies examining gene therapy for PD usually have one of two goals: (1) to replace dopamine (DA) that is depleted in the striatum or (2) to administer factors that would prevent the degeneration of dopaminergic neurons in the substantia nigra (SN), as this disease is known to lead to a dramatic reduction in levels of DA in the striatum due to the loss and dysfunction of nigral neurons. Several techniques to target the dopaminergic system in the brain have entered into the clinical testing phase using these currently experimental procedures, and others are expected to be tested in the near future. This chapter will discuss the status of these therapeutic interventions in both animal models and patients.

The limited permeability of the blood-brain barrier, combined with limited targeting capabilities of the current generation of viral vectors, restricts the systemic delivery of most gene therapies, thus requiring intracerebral injections. Ex vivo gene therapy involves genetically engineering cells to express a gene with therapeutic value and then transplanting the cells into the patient's brain, where the gene product can improve the function in the region of interest. Several different cell lines have been engineered for grafting in animal models, including human neural stem cells and autologous fibroblasts. Although this has been a successful method of gene therapy under some experimental circumstances, its clinical utility is limited due to rapid transgene downregulation and impoverished distribution of the transgene. By contrast, in vivo gene therapy is capable of transducing endogenous cells of the brain and can result in widespread transgene expression for at least 10 years (K. Bankiewicz, personal communication). As a result, in vivo gene therapy has replaced the use of ex vivo procedures in the treatment of PD.

PRINCIPLES OF IN VIVO GENE TRANSFER AND AVAILABLE VECTORS

Introduction of novel genes into a cell can be achieved by a number of different techniques. The earliest methods utilized electroporation to deliver naked DNA plasmids into cells in vitro. In a small fraction of these cells, the plasmid will integrate in the host genome and give rise to a clonal population that expresses the transgene. By integration of toxin resistance genes, the cells with the integrated transgene can be enriched. For successful transcription to an mRNA and subsequent expression of any protein, the introduced plasmid needs to contain, besides the gene of interest, a promoter sequence, typically from another gene, that would be expressed in the targeted cell type. This promoter can be either ubiquitously active (e.g., the beta-actin promoter) or display a cell type–specific pattern (e.g., the synapsin promoter) that is only expressed in cells of neuronal origin.

Introduction of novel genes by naked DNA has very low efficacy and is not well suited for in vivo applications. For this purpose, the highly efficient machinery of viruses has been harnessed. By removing essentially all genes needed for viral replication and capsid/envelope production from wild-type virus, recombinant viral vectors have been developed that can infect a single target cell only once. Recombinant viral vectors have been created from numerous viruses, and each has specific characteristic advantages and limitations. In vivo applications for transduction of the nondividing neurons in the central nervous system (CNS) have utilized four main vector types: adenoviruses (Ad), adeno-associated viruses (AAV),

HIV-1-derived lentiviruses (LV), and herpes simplex virus (HSV). Of these, AAV and LV vectors have been the preferred choices for experimental studies in the CNS due to their high efficiency and safety (for more information, see the review by Mandel et al.[1]).

Although infections with wild-type AAV are common, no known pathologies have been associated with the virus.[2,3] Furthermore, 96% of the viral genome of the wild-type AAV is removed during the generation of the recombinant vectors, which further reduces the risk of an immune reaction at least for the first administration.[4,5] This fact, together with the very low integration frequency of recombinant AAV, gives this vector a strong safety profile.[6] Although the majority of the delivered genes remain as episomal plasmids, their introduction into nondividing cells like neurons results in a long-term, perhaps lifelong, stable expression. Lentivirus vectors also provide long-term expression in neurons and have the advantage of a larger packaging capacity than AAV. When larger genes are of interest or if multiple genes need to be delivered simultaneously, LV might be the vector of choice. However, the drawback is that the LV vectors cause random integration of the transgene sequence into the host genome.[7] Theoretically, this might cause unpredictable transgene expression levels and, in the worst case, give rise to insertional mutagenesis in tumor suppressor genes. However, no such result has been observed in any of the experimental animal studies thus far.

DA REPLACEMENT BY GENE THERAPY

Since the discovery of DA as a neurotransmitter and its involvement in PD in 1957, the focus of pharmacological therapies has been on restoring the DAergic tone in the brain. The reconstitution of striatal DA via peripheral 3,4-dihydroxyphenylalanine (L-DOPA) administration, combined with peripheral decarboxylase inhibitors, proved to be a very successful therapy and became the gold standard for treatment of PD patients in the 1970s. In the early stages of the disease, L-DOPA medication provides excellent symptomatic relief and can greatly improve the patient's quality of life. However, long-term treatment with L-DOPA is not without limitations and adverse events, which inevitably emerge in more than 80% of all PD patients within the first 10 years of disease onset.[8] The majority of patients eventually develop involuntary movements, so-called dyskinesias.[9] Other adverse conditions include hypotension, sexual dysfunction, and psychiatric side effects.[10]

Because large fluctuations were observed in the serum levels of L-DOPA after oral administration, it was hypothesized that the development of dyskinesias might be a result of the large variations in DA concentrations at the synaptic sites in the denervated striatum.[11] This hypothesis is supported by clinical data showing that continuous infusion of L-DOPA, delivered either intravenously or via duodenal pump, or continuous infusion of DA agonists, can significantly reduce the occurrence and magnitude of dyskinesias and decrease the daily off-phase time. There are, however, several limitations to these approaches. First, systemically delivered DAergic drugs reach the whole brain at high concentration. This is clearly not the best approach since not all brain regions suffer from DAergic degeneration to the same extent. For example, the requirement of additional DAergic tone might be substantially less in the limbic and cortical areas than in the severely affected striatum. Thus, in this mode of treatment, these regions might be constantly overstimulated with high DA tone. Second, the chronic implantation of an infusion catheter creates the opportunity for opportunistic infections. And finally, the continuous infusion of L-DOPA in the duodenum is susceptible to variations in the uptake capacity of the gut as a result of food intake.

Therefore, other treatment approaches that can locally enhance the DA concentrations in the striatum could prove to be more beneficial and limit the occurrence and severity of side effects to levels not achievable with currently available treatment modalities. As will be detailed below, three major gene therapy strategies have been developed to synthesize DA locally in the brain. The main differentiating factor among these approaches is the interpretation of which enzymes are necessary and sufficient to express ectopically in the target area of the brain in order to reconstitute the DA synthesis capacity. It is widely accepted that the tyrosine hydroxylase (TH) enzyme is significantly reduced in the parkinsonian striatum, severely compromising the rate of synthesis of DOPA from tyrosine. Thus, it is clear that striatal TH enzyme activity must be restored. Whether the amount of aromatic acid decarboxylase (AADC), the enzyme responsible for the conversion of DOPA into DA, is available in the diseased brain for synthesis of therapeutic levels of DA in the appropriate target regions, however, is a matter of debate.

Pro-Drug Approach for Enhanced DA Synthesis

The AADC enzyme is present in the striatum, not only in DAergic axons but also in serotonergic terminals,[12,13] but it has been shown that its levels are decreased in the striatum of PD patients. The reported level of residual

AADC activity is variable among patients and also among studies. It may be as low as 5% in the most severely affected patients, and usually larger decreases are found in the putamen than in the caudate nucleus.[14,15] If the level of AADC enzyme were increased or even restored to normal selectively in the striatum with gene therapy, then a larger fraction of the total systemic L-DOPA would be converted to DA in this part of the brain. As a result, the dose of oral L-DOPA could be decreased, resulting in reduced side effects but still with maintained efficacy, whereas the effects due to extrastriatal DA synthesis could be minimized. The pro-drug approach is based on this strategy.

The first proof-of-principle for this therapy was demonstrated in primates with a unilateral MPTP (1-methyl 4-phenyl 1,2,3,6-tetrahydropyridine) lesion, which later received injection of AAV2 vectors coding for the human AADC gene. These animals showed increased conversion efficacy of peripheral L-DOPA to DA, as seen by biochemical analysis of tissue punches from the transduced striatum.[16] In a follow-up study, Bankiewicz and colleagues achieved long-term behavioral improvement and in vivo AADC enzyme activity for up to 6 years after AAV-AADC transduction of MPTP-treated monkeys.[17] The animals displayed a 50% improvement on the clinical rating scale after a single injection of L-DOPA at a dose that is not sufficient to induce a significant improvement in control vector–injected animals. However, the duration of action for peripheral L-DOPA in these AAV-AADC-transduced primates was not reported. Nevertheless, the behavioral effects were coupled with a normalized striatal [^{18}F]-MT (fluoro-L-m-tyrosine) uptake that was stable for the full duration of the study.[17] Some of these animals have been kept alive and studied for over 10 years with maintained AADC expression (K. Bankiewicz, personal communication).

One potential concern about this approach is the increase in the fluctuations of DA supply that might lead to aggravation of dyskinesias. In fact, it has been shown in primates that AADC overexpression can potentiate L-DOPA-induced dyskinesias if the transduction is heterogeneous.[18] If this were to happen in a clinical setting, the daily L-DOPA dose would have to be decreased or even discontinued. However, as oral L-DOPA is the main pharmacotherapy for patients, this might leave them worse off than before the intervention. On the other hand, a phase I clinical safety trial utilizing AAV-mediated gene delivery of AADC as a therapy for PD has recently reported data from five patients injected bilaterally into the postcommisural putamen with an AAV2-AADC vector. The treatment was well tolerated and induced a robust increase in [^{18}F]-MT uptake 6 months post-injection. The patients, however, did not display any improvement in the Unified Parkinson's Disease Rating Scale (UPDRS) on-phase score (i.e., on oral L-DOPA medication) but appeared to have improved off medication.[19] The interpretation of these results regarding preliminary efficacy information remains thus unclear.

Restoration of DA Neurotransmission by Triple-Gene Transfer

One alternative to the pro-drug approach is to reconstitute some or all of the enzymes required for DA synthesis in the parkinsonian striatum. In the normal brain, the DOPA substrate used by the AADC enzyme to synthesize DA is generated from dietary tyrosine by the TH enzyme. This enzymatic conversion is efficient only in the presence of the cofactor tetrahydrobiopterin (BH4), which, in turn, is synthesized in a three-step reaction where the guanosine triphosphate (GTP) cyclohydrolase 1 (GCH1) acts as the rate-limiting enzyme.[20,21] The TH and GCH1 enzymes can then be combined with AADC to provide a three-enzyme replacement strategy for ectopic synthesis of DA in any transduced cell in the brain. This can be done either by using a mixture of three vectors coding for TH, GCH1, or AADC genes or by designing vectors that can deliver multiple genes in one viral particle. Ectopic DA production in striatal cells, however, raises concerns due to the fact that the DA synthesis is localized to cells that have no vesicular storage and release mechanism for this neurotransmitter. The first problem that needs to be resolved in this scenario is the strong negative feedback of free cytosolic DA on the TH enzyme. It is known that the enzymatic properties remain and are even slightly enhanced after digestion of the first 158 amino acids of the TH enzyme.[22,23] Thus, the truncated form of the TH (tTH) enzyme that lack the regulatory N-terminal fragment becomes constitutively active regardless of cytosolic DA.[24]

The multiple AAV vectors coding for single transgenes have been shown to induce ectopic DA synthesis in the DA-denervated rodent striatum, which can reduce apomorphine-induced rotation by up to 80% for at least 12 months.[25] The same mixture of AAV vectors was later applied to parkinsonian cynomolgus monkeys; the animals were reported to improve by up to 64% on the Primate Parkinsonian Rating Scale (PPRS) at 2 weeks post transduction and remained stable throughout the study up to 10 months.[26] In another approach, the equine infectious anemia virus (EIAV) was used as a vector platform to carry a tricistronic construct encoding for the tTH, GCH1, and AADC genes from a single vector.[27] Injection of this multicistronic vector into the striatum of hemiparkinsonian rats resulted in a partial decrease in apomorphine-induced rotation but did not result in any detectable increase in striatal DA levels. These results were considered a sufficient basis for continued development of this vector as a product for clinical testing (ProSavin, Oxford Biomedica, UK). In follow-up studies performed by Oxford Biomedica, the ProSavin vector was injected into MPTP-lesioned primates. The company reported in a press release that, at repeated time points up to 15 months after the vector injection, the motor performance of the monkeys improved significantly.[28] With these results as a basis, a phase I/II clinical trial has now been initiated in patients with PD. To date, three patients have been injected with the vector and followed for 6 months. At this time point, the patients displayed average improvement in the UPDRS motor off-phase score of 30%.[28] In our opinion this result should be viewed with caution, as the changes seen in these patients are within the range that can be expected due to placebo in an open-label trial. Three additional patients have received a higher dose of the vector, and their evaluation is pending. The outcome of these patients might give further indications of the efficacy of the therapy.

Continuous DOPA Delivery Strategy Using Viral Vector–Mediated Gene Transfer

As we described above, the cotransduction of TH and GCH1 genes is sufficient to sustain high levels of DOPA synthesis in various cell types both in vitro and in vivo (see also [29] for a detailed review of this topic). Continuous DOPA delivery relies on endogenous AADC activity for synthesis of DA locally in the brain. Two major sources of AADC in the striatum are the DA and serotonin (5HT) terminals. Thus, in the parkinsonian brain, the remaining DA axons and the serotonergic terminals are the two most likely places where conversion to (and release of) DA takes place. As the disease progresses, it is anticipated that fewer and fewer DA terminals will remain. Nevertheless, the serotonergic denervation of the striatum is significantly less than the DAergic one in PD patients, so it may remain as a reliable long-term source in the majority of patients.[30]

In the rat model of PD, the efficiency of the combined AAV2-TH and rAAV2-GCH1 strategy was explored by utilizing a new generation of AAV2 vectors with which therapeutic levels of DOPA synthesis could be reached. The treated animals not only recovered in drug-induced rotation tests but also showed improvements on a spontaneous motor test.[31] Moreover, Carlsson and colleagues showed that AAV5-mediated DOPA delivery could reverse previously manifest L-DOPA-induced dyskinesias in rats.[32] These encouraging results were recently complemented by proof that DA synthesized after this gene therapy approach resulted in normalization of the DA concentrations at the receptor sites on the striatal neurons and that this was correlated with behavioral recovery.[33] Taken together, these data show that viral vector–mediated, continuous DOPA delivery is an attractive strategy for enzyme replacement in PD and should be pursued, with the demonstration of efficacy in the MPTP primate model prior to first clinical trials.

GENE THERAPY USING NEUROTROPHIC FACTORS

The studies mentioned above are of therapeutic value primarily to patients who are in advanced stages of PD and have lost innervation from the nigra to the striatum. Alternative therapies may be targeted to patients in earlier stages of the disease to halt nigral cell death and prevent disease progression as soon as a definitive diagnosis has been reached. The most powerful method of doing so involves the use of neurotrophic factors that support the damaged but surviving DA neurons in the SN and sustained dopaminergic innervation of the striatum at a normal or supranormal level. The trophic factors most commonly used for this purpose are members of the glial cell line–derived neurotrophic factor (GDNF) family of ligands (GFLs).

GDNF

Interest in GDNF for PD began with the initial demonstration by Lin et al. that GDNF supported the viability of dopaminergic cells in vitro.[34] GDNF binds to GDNFR-alpha, and the ligand receptor complex transmits its action via the RET oncogene receptor.[35–38] Subsequent in vivo studies demonstrated that infusion of GDNF protected nigral neurons from various types of experimental insults (see [39] or [40] for a review of the topic). Ex vivo delivery of GDNF in PD models employed several different cell types as gene delivery systems. Rat fibroblasts genetically engineered to secrete GDNF and packaged into microcapsules were transplanted into the striatum of 6-hydroxydopamine (6-OHDA)-lesioned rats. In the first study, the BHK-GDNF cell line was transplanted in capsules into the striatum of unilaterally lesioned rats. The neuroprotective effect of the GDNF was strictly dependant on the timing of the delivery. When transplanted before or within 2 hr of 6-OHDA lesioning, these cells provided significant functional improvement. However, at later time points of administration, functional recovery was not statistically significant.[41] In a second study, immortalized fibroblasts were engineered to secrete GDNF and placed within a capsule that could be easily implanted and explanted.[42] This study used 6-OHDA bilaterally lesioned rats and implanted them with these encapsulated cells 1 week after lesioning. Rats that received GDNF cell transplants showed an almost immediate recovery on motor tasks like the swim test and movement initiation tests. This behavioral recovery was associated with a reinnervation of striatal dopaminergic fibers. To determine whether these motor improvements persisted even after GDNF treatment was halted, capsules were explanted 8 weeks after initial implantation. Motor performance on these tasks persisted even after GDNF-secreting cells were removed. In a similar study, unilaterally lesioned rats displayed significant functional recovery from capsule implantation immediately following transplantation and for up to 24 weeks afterward.[43] Additionally, this process of gene delivery was viable for up to 6 months, with no detectable immune response.

GDNF-secreting cell lines have also been used to promote the survival of embryonic stem cell transplants. A Schwann cell line expressing a cDNA for GDNF when cocultured with embryonic dopaminergic cells promotes the survival of the latter.[44] These GDNF-expressing Schwann cells enhance the survival of ventral mesencephalic cells transplanted into the striatum and increase neuritic outgrowth into host cells. However, motor benefits are not enhanced further than those seen in animals receiving only ventral mesencephalic transplants. When cocultures are transplanted into the nigra, there is enhanced survival of nigral neurons and an outgrowth of axons along the striatonigral pathway, a phenomenon not seen in mesencephalic implants alone.

A recent study by Emborg et al. reported on the use of human neural progenitor cells that secrete GDNF.[45] MPTP-lesioned cynomolgus monkeys received nigral transplants of GDNF-secreting cells 1 week following lesioning. They were monitored for 3 months following transplantation surgeries. Limited preservation of host nigral neurons was observed, and very modest GDNF production was seen localized to the injected nigra. Additionally, there was an increase in TH- and vesicular monoamine transporter 2 (VMAT2)-positive fibers in two of the three animals, indicating sprouting of host fibers. These two animals showed an improvement in Clinical Rating Scale (CRS) scores.

While ex vivo gene delivery methods are efficient at transducing cells, they have several limitations. As mentioned above, the challenges of long-term gene expression and limited transgene distribution needs to be addressed. Additional drawbacks include a limited availability of cell lines, potential immune reactions, the prospect of having to employ potentially toxic immunosuppression therapies, and aberrant migration and sprouting of implanted cells. To counteract many of these problems, in vivo gene therapy using viral vectors have been explored. In vivo gene therapy is an effective and safe way of delivering trophic factors for very long periods of time and over long distances. The first viral vector used for GDNF delivery was the Ad vector. Bohn and colleagues made a single injection of Ad-GDNF supranigrally to sustain trophic factor expression for 7 weeks.[46] While this method significantly protected TH-positive neurons in the SN from 6-OHDA-induced toxicity, it did not change the expression of TH-positive fibers in the striatum, suggesting for the first time that treatment of the nigra alone may not be sufficient to protect fibers in the striatum. This study was repeated, administering Ad-GDNF to the striatum, the site of DA fiber loss.[47] There was 40% protection in the SN but no preservation of TH in fibers of the striatum. An improvement in motor performance in treated rats indicated that there might have been a modest increase in TH levels that was undetected by the methods used. The Ad-GDNF vector yielded variable results in a separate study where both nigral cells and striatal fibers were protected,[48] and behavioral deficits were reduced in the amphetamine-induced rotational test. The use of adenovirus was abandoned due to its highly immunogenic properties and alternate vehicles of gene delivery were subsequently utilized, although more modern studies employing the so-called gutless adenovirus may provide efficient transduction with less immunogenicity.

In subsequent studies, AAV vectors were utilized and AAV-GDNF has been shown to be neuroprotective in 6-OHDA-lesioned rats when administered to the nigra 3 weeks before partial lesioning.[49] Nigral neurons were almost completely protected, but there was no detectable functional recovery. These disappointing results may have been due to the site of AAV-GDNF administration. Administration of AAV-GDNF to either the striatum or the nigra can efficiently protect nigral cell bodies. However, only striatal delivery of AAV-GDNF protects TH-positive striatal fibers, a protection that is sustained for prolonged periods of time (4–5 months).[50] Indeed, only in rats in which striatal DA innervation was preserved was functional recovery achieved. In a nonhuman primate model of PD, AAV-GDNF was administered 4 weeks prior to 6-OHDA lesioning in marmoset monkeys.[51,52] These studies showed protection of up to 84% of the cells in the SN and preservation of TH-immunoreactive fibers in the striatum of some monkeys. In those monkeys that showed striatal preservation, there was amelioration of behavioral deficits in amphetamine- and apomorphine-induced rotations, improvements noted on a clinical rating scale as well as in other behavioral tests.

Recombinant LV vectors expressing GDNF have been tested extensively in rodent models of PD. Two areas of vector administration are potentially therapeutic: the striatum and the substantia nigra. Studies have compared both of these delivery sites by either infusion of recombinant GDNF protein[53] or LV-GDNF injection in the 6-OHDA rat model.[54] Both striatal and nigral administration of GDNF significantly protected dopaminergic neurons in the SN regardless of the vehicle used. However, only striatal delivery of recombinant protein in the striatum protected TH-positive fiber innervation, coupled with an improvement in drug-induced rotation tests. When LV-GDNF was administered to the striatum, there was partial protection of nigrostriatal fibers, as seen by preservation of TH fibers in the globus pallidus. This partial protection was accompanied by improvements in amphetamine-induced rotational behavior, indicating an increase in DA function on the GDNF-treated side. These studies indicated that a striatal route of delivery might be preferred to nigral administration. In LV-GDNF delivery there was also a dose-dependent response, with higher doses of GDNF producing stronger neuroprotection. Surprisingly, there was no increase in fiber protection from long-term treatment with LV-GDNF (9 months).[55] As an alternative vector to LV, a recombinant equine infectious anemia virus (EIAV) vector was utilized to deliver GDNF in the rat PD model as well. This study showed an amelioration of deficits on several motor tasks in lesioned rats, indicating that further exploration of this method of GDNF delivery may be beneficial.[56]

Lentivirus-GDNF was the first vector system examined in two nonhuman primate models of PD, aged monkeys and MPTP-treated monkeys.[57] In our hands, aged monkeys rarely respond to levodopa, and thus the focus of this experiment was on neuroanatomical findings. Aged monkeys showed a clear, robust increase in dopaminergic function in the striatum, as demonstrated by TH-optical density measurements and quantification of DA and metabolites from striatal punches. Lentivirus-GDNF-treated monkeys also displayed an 85% increase in TH-immunoreactive neurons within the SN. This increase is thought to be due not to neurogenesis but rather to up-regulation of TH in viable cells that had previously undergone age-related dopaminergic down-regulation. Additionally, these neurons had a 35% increase in volume, supporting the belief that in addition to frank neural protection, gene delivery of GDNF had restorative properties as well.

This is a critical issue, because if GDNF gene delivery has only neuroprotective properties and not restorative ones, the number of patients needed to demonstrate this neuroprotection clinically would be prohibitive. This study also demonstrated structural and functional neuroprotection in MPTP-lesioned monkeys. Neuroanatomically, LV-GDNF completely protected nigral neurons from degeneration and completely prevented the loss of striatal insufficiency that occurs following the injection of the dopaminergic toxin. GDNF-treated monkeys also improved on a clinical rating scale and a limb use task for up to 3 months, corresponding to an increase in TH activity in the striatum.

Most of GDNF's neuroprotective potential has been demonstrated in toxin-induced models of PD, where the effect is robust. However, in a genetic model of the disease, where viral overexpression of alpha-synuclein in the SN causes progressive degeneration, LV-GDNF delivery did not result in protection from the insult.[58] This finding indicates the need for further development of more clinically relevant models of PD, and the implications for clinical efficacy of GDNF need to be thoroughly assessed.

Neurturin

Neurturin (NTN) was the second family member of the GFLs to be discovered in 1996[59] and has 40% homology with GDNF. Neurturin signals in a manner similar to that of GDNF by binding to the GFRα-2 receptor and also signals through a RET receptor.[60] In the adult brain, the GFRα-2 receptor is expressed at readily detectable levels in the SN but is low or undetectable in the striatum.[61] However, if NTN is expressed at high enough levels, it can bind to the GFRα-1 receptor.[62] As the GFRα-1 receptor is abundantly expressed in the striatum, striatal delivery of NTN has been tested in animal models of PD.

Similar to GDNF, NTN has been delivered by both ex vivo and in vivo gene therapy approaches. When polymer-encapsulated fibroblasts releasing NTN were implanted near the SN 1 week before a unilateral medial forebrain bundle axotomy,[63] sparing of TH neurons in the nigra was observed but this was not linked to behavioral recovery. This was likely due to the failure to preserve striatal innervation. One of the first published studies using an LV-NTN gene expression cassette reported several problems with the expression and secretion of wild-type NTN.[64] By replacing the pro-NTN sequence with the signal peptide from the mouse immunoglobulin heavy chain gene, the authors were able to circumvent problems. From this point on, most of the gene therapy studies using NTN have been conducted by CEREGENE Inc. This company has developed an AAV2-NTN gene delivery system (commercially known as CERE-120) that efficiently delivers NTN to striatal neurons and contains a pre-pro region of the nerve growth factor in the vector that also promotes secretion of NTN.

In the 6-OHDA rodent model, AAV2-NTN delivered to the striatum completely protects nigral TH-positive neurons in a manner similar to that seen with AAV2-GDNF[65] and results in stable expression for up to at least 1 year, with no visible toxicity.[66] In aged rhesus monkeys receiving unilateral injections of AAV2-NTN into the striatum,[67] there was effective NTN expression for at least 8 months and an increase in fluorodopa uptake in the treated striatum. Furthermore, there was an increase in TH-fiber expression in the striatum, efficient retrograde transport of NTN to the SN, and up-regulation of phophorylated Erk (a downstream signaling marker for NTN trophic activity) in the nigra. In the MPTP nonhuman primate model of PD, AAV2-NTN, when administered 4 days after a lesion to both the striatum and the SN, caused both behavioral and neuroanatomical improvements.[68] Animals that received AAV2-NTN showed improvements in motor performance starting 4 months after treatment, which lasted until the end of the study at 10 months. This motor benefit likely resulted from the observed increase in striatal levels of TH and significant protection of nigral neurons.

These positive results prompted CEREGENE Inc. to initiate clinical trials using AAV2-NTN. An initial phase I trial recruited 12 advanced PD patients who received either a low dose or a high dose into the putamen.[69] Observed for 1 year, these patients showed no adverse reactions and a statistically significant improvement in the UPDRS score in the off-state. This safety study prompted a phase II double-blinded trial including 58 patients; two-thirds received CERE-120 and one-third received a placebo. Unfortunately, this study failed to reach its primary endpoint.[70] However, this failure might be attributed to technical issues regarding gene delivery, as well as the status of the PD striatum with regard to the number of available dopaminergic fibers to bind NTN and retrogradely transport the NTN to the nigral perikarya.[71] Future studies addressing these issues may yield better results.

CONCLUSION

The clinical trials described above have had varying degrees of success regarding the efficacy of the chosen strategy. Nevertheless, they all have one thing in common: gene delivery with AAV in the human CNS has been shown to be both safe and efficient, at least regarding transgene expression. No major adverse events have been reported from any of the trials so far. This shows that gene delivery by viral vectors is not only a valuable research tool, as described in this chapter, but also an important clinical vehicle. With the ongoing clinical trials and a number additional trials in the pipeline, the future for clinical gene therapy is bright, and we are optimistic that gene therapy–based medications will form an important addition to the neurologist's arsenal of treatments for patients with PD.

References

1. Mandel RJ, Burger C, Snyder RO. Viral vectors for in vivo gene transfer in Parkinson's disease: properties and clinical grade production. *Exp Neurol.* 2008;209(1):58–71.
2. Schnepp BC, Jensen RL, Chen CL, Johnson PR, Clark KR. Characterization of adeno-associated virus genomes isolated from human tissues. *J Virol.* 2005;79(23):14793–14803.
3. Chen CL, Jensen RL, Schnepp BC, et al. Molecular characterization of adeno-associated viruses infecting children. *J Virol.* 2005;79(23):14781–14792.
4. Mastakov MY, Baer K, Symes CW, Leichtlein CB, Kotin RM, During MJ. Immunological aspects of recombinant adeno-associated virus delivery to the mammalian brain. *J Virol.* 2002;76(16):8446–8454.
5. Peden CS, Burger C, Muzyczka N, Mandel RJ. Circulating anti-wild-type adeno-associated virus type 2 (AAV2) antibodies inhibit recombinant AAV2 (rAAV2)-mediated, but not rAAV5-mediated, gene transfer in the brain. *J Virol.* 2004;78(12):6344–6359.
6. Kay MA, Nakai H. Looking into the safety of AAV vectors. *Nature.* 2003;424(6946):251.
7. Mitchell RS, Beitzel BF, Schroder AR, et al. Retroviral DNA integration: ASLV, HIV, and MLV show distinct target site preferences. *PLoS Biol.* 2004;2(8):1127–1137.
8. Ahlskog JE, Muenter MD. Frequency of levodopa-related dyskinesias and motor fluctuations as estimated from the cumulative literature. *Mov Disord.* 2001;16(3):448–458.
9. Obeso JA, Olanow CW, Nutt JG. Levodopa motor complications in Parkinson's disease. *Trends Neurosci.* 2000; 23(10 suppl):S2–S7.
10. Moskovitz C, Moses H, Klawans HL. Levodopa-induced psychosis: a kindling phenomenon. *Am J Psychiatry.* 1978;135(6):669–675.
11. Olanow CW, Obeso JA, Stocchi F. Continuous dopamine-receptor treatment of Parkinson's disease: scientific rationale and clinical implications. *Lancet Neurol.* 2006;5(8):677–687.
12. Arai R, Karasawa N, Nagatsu I. Aromatic L-amino acid decarboxylase is present in serotonergic fibers of the striatum of the rat. A double-labeling immunofluorescence study. *Brain Res.* 1996;706(1):177–179.
13. Arai R, Karasawa N, Geffard M, Nagatsu T, Nagatsu I. Immunohistochemical evidence that central serotonin neurons produce dopamine from exogenous L-DOPA in the rat, with reference to the involvement of aromatic L-amino acid decarboxylase. *Brain Res.* 1994;667(2):295–299.

14. Lloyd K, Hornykiewicz O. Parkinson's disease: activity of L-dopa decarboxylase in discrete brain regions. *Science.* 1970;170(963):1212–1213.
15. Nagatsu T, Yamamoto T, Kato T. A new and highly sensitive voltammetric assay for aromatic L-amino acid decarboxylase activity by high-performance liquid chromatography. *Anal Biochem.* 1979;100(1):160–165.
16. Bankiewicz KS, Eberling JL, Kohutnicka M, et al. Convection-enhanced delivery of AAV vector in parkinsonian monkeys; in vivo detection of gene expression and restoration of dopaminergic function using pro-drug approach. *Exp Neurol.* 2000;164(1):2–14.
17. Bankiewicz KS, Forsayeth J, Eberling JL, et al. Long-term clinical improvement in MPTP-lesioned primates after gene therapy with AAV-hAADC. *Mol Ther.* 2006;14(4):564–570.
18. Bankiewicz KS, Daadi M, Pivirotto P, et al. Focal striatal dopamine may potentiate dyskinesias in parkinsonian monkeys. *Exp Neurol.* 2006;197(2):363–372.
19. Eberling JL, Jagust WJ, Christine CW, et al. Results from a phase I safety trial of hAADC gene therapy for Parkinson disease. *Neurology.* 2008;70(21):1980–1983.
20. Levine RA, Miller LP, Lovenberg W. Tetrahydrobiopterin in striatum: localization in dopamine nerve terminals and role in catecholamine synthesis. *Science.* 1981;214(4523):919–921.
21. Nagatsu T. Biopterin cofactor and monoamine-synthesizing monooxygenases. *Neurochem Int.* 1983;5(1):27–38.
22. Abate C, Joh TH. Limited proteolysis of rat brain tyrosine hydroxylase defines an N-terminal region required for regulation of cofactor binding and directing substrate specificity. *J Mol Neurosci.* 1991;2(4):203–215.
23. Abate C, Smith JA, Joh TH. Characterization of the catalytic domain of bovine adrenal tyrosine hydroxylase. *Biochem Biophys Res Commun.* 1988;151(3):1446–1453.
24. Moffat M, Harmon S, Haycock J, O'Malley KL. L-Dopa and dopamine-producing gene cassettes for gene therapy approaches to Parkinson's disease. *Exp Neurol.* 1997;144(1):69–73.
25. Shen Y, Muramatsu SI, Ikeguchi K, et al. Triple transduction with adeno-associated virus vectors expressing tyrosine hydroxylase, aromatic-L-amino-acid decarboxylase, and GTP cyclohydrolase I for gene therapy of Parkinson's disease. *Hum Gene Ther.* 2000;11(11):1509–1519.
26. Muramatsu S, Fujimoto K, Ikeguchi K, et al. Behavioral recovery in a primate model of Parkinson's disease by triple transduction of striatal cells with adeno-associated viral vectors expressing dopamine-synthesizing enzymes. *Hum Gene Ther.* 2002;13(3):345–354.
27. Azzouz M, Martin-Rendon E, Barber RD, et al. Multicistronic lentiviral vector–mediated striatal gene transfer of aromatic L-amino acid decarboxylase, tyrosine hydroxylase, and GTP cyclohydrolase I induces sustained transgene expression, dopamine production, and functional improvement in a rat model of Parkinson's disease. *J Neurosci.* 2002;22(23):10302–10312.
28. Oxford Biomedica. Oxford Biomedica announces six-month efficacy results with low dose of ProSavin® in phase I/II trial in Parkinson's disease. Nov. 19 2008.
29. Carlsson T, Björklund T, Kirik D. Restoration of the striatal dopamine synthesis for Parkinson's disease: viral vector-mediated enzyme replacement strategy. *Curr Gene Ther.* 2007;7(2):109–120.
30. Kish SJ, Tong J, Hornykiewicz O, et al. Preferential loss of serotonin markers in caudate versus putamen in Parkinson's disease. *Brain.* 2008;131(pt 1):120–131.
31. Kirik D, Georgievska B, Burger C, et al. Reversal of motor impairments in parkinsonian rats by continuous intrastriatal delivery of L-dopa using rAAV-mediated gene transfer. *Proc Natl Acad Sci USA.* 2002;99(7):4708–4713.
32. Carlsson T, Winkler C, Burger C, et al. Reversal of dyskinesias in an animal model of Parkinson's disease by continuous L-DOPA delivery using rAAV vectors. *Brain.* 2005;128(Pt 3):559–569.
33. Leriche L, Björklund T, Breysse N, et al. Positron emission tomography imaging demonstrates correlation between behavioral recovery and correction of dopamine neurotransmission after gene therapy. *J Neurosci.* 2009;29(5):1544–1553.
34. Lin LF, Doherty DH, Lile JD, Bektesh S, Collins F. GDNF: a glial cell line–derived neurotrophic factor for midbrain dopaminergic neurons. *Science.* 1993;260(5111):1130–1132.
35. Treanor JJ, Goodman L, de Sauvage F, et al. Characterization of a multicomponent receptor for GDNF. *Nature.* 1996;382(6586):80–83.
36. Jing S, Wen D, Yu Y, et al. GDNF-induced activation of the ret protein tyrosine kinase is mediated by GDNFR-alpha, a novel receptor for GDNF. *Cell.* 1996;85(7):1113–1124.
37. Durbec P, Marcos-Gutierrez CV, Kilkenny C, et al. GDNF signalling through the Ret receptor tyrosine kinase. *Nature.* 1996;381(6585):789–793.
38. Trupp M, Arenas E, Fainzilber M, et al. Functional receptor for GDNF encoded by the c-ret proto-oncogene. *Nature.* 1996;381(6585):785–789.
39. Kordower JH. In vivo gene delivery of glial cell line–derived neurotrophic factor for Parkinson's disease. *Ann Neurol.* 2003;53(suppl 3):S120-S132; discussion S132-S134.
40. Kirik D, Georgievska B, Bjorklund A. Localized striatal delivery of GDNF as a treatment for Parkinson disease. *Nat Neurosci.* 2004;7(2):105–110.
41. Yasuhara T, Shingo T, Muraoka K, et al. Early transplantation of an encapsulated glial cell line–derived neurotrophic factor-producing cell demonstrating strong neuroprotective effects in a rat model of Parkinson disease. *J Neurosurg.* 2005; 102(1):80–89.
42. Sajadi A, Bensadoun JC, Schneider BL, Lo Bianco C, Aebischer P. Transient striatal delivery of GDNF via encapsulated cells leads to sustained behavioral improvement in a bilateral model of Parkinson disease. *Neurobiol Dis.* 2006;22(1):119–129.
43. Grandoso L, Ponce S, Manuel I, et al. Long-term survival of encapsulated GDNF secreting cells implanted within the striatum of parkinsonized rats. *Int J Pharm.* 2007;343(1–2):69–78.
44. Wilby MJ, Sinclair SR, Muir EM, et al. A glial cell line–derived neurotrophic factor–secreting clone of the Schwann cell line SCTM41 enhances survival and fiber outgrowth from embryonic nigral neurons grafted to the striatum and to the lesioned substantia nigra. *J Neurosci.* 1999;19(6):2301–2312.
45. Emborg ME, Ebert AD, Moirano J, et al. GDNF-secreting human neural progenitor cells increase tyrosine hydroxylase and VMAT2 expression in MPTP-treated cynomolgus monkeys. *Cell Transplant.* 2008;17(4):383–395.
46. Choi-Lundberg DL, Lin Q, Chang YN, et al. Dopaminergic neurons protected from degeneration by GDNF gene therapy. *Science.* 1997;275(5301):838–841.
47. Choi-Lundberg DL, Lin Q, Schallert T, et al. Behavioral and cellular protection of rat dopaminergic neurons by an adenoviral vector encoding glial cell line–derived neurotrophic factor. *Exp Neurol.* 1998;154(2):261–275.
48. Bilang-Bleuel A, Revah F, Colin P, et al. Intrastriatal injection of an adenoviral vector expressing glial-cell-line-derived neurotrophic factor prevents dopaminergic neuron degeneration and

behavioral impairment in a rat model of Parkinson disease. *Proc Natl Acad Sci USA.* 1997;94(16):8818–8823.
49. Mandel RJ, Spratt SK, Snyder RO, Leff SE. Midbrain injection of recombinant adeno-associated virus encoding rat glial cell line–derived neurotrophic factor protects nigral neurons in a progressive 6-hydroxydopamine-induced degeneration model of Parkinson's disease in rats. *Proc Natl Acad Sci USA.* 1997;94(25):14083–14088.
50. Kirik D, Rosenblad C, Bjorklund A, Mandel RJ. Long-term rAAV-mediated gene transfer of GDNF in the rat Parkinson's model: intrastriatal but not intranigral transduction promotes functional regeneration in the lesioned nigrostriatal system. *J Neurosci.* 2000;20(12):4686–4700.
51. Eslamboli A, Cummings RM, Ridley RM, et al. Recombinant adeno-associated viral vector (rAAV) delivery of GDNF provides protection against 6-OHDA lesion in the common marmoset monkey (*Callithrix jacchus*). *Exp Neurol.* 2003;184(1):536–548.
52. Eslamboli A, Georgievska B, Ridley RM, et al. Continuous low-level glial cell line–derived neurotrophic factor delivery using recombinant adeno-associated viral vectors provides neuroprotection and induces behavioral recovery in a primate model of Parkinson's disease. *J Neurosci.* 2005;25(4):769–777.
53. Kirik D, Rosenblad C, Björklund A. Preservation of a functional nigrostriatal dopamine pathway by GDNF in the intrastriatal 6-OHDA lesion model depends on the site of administration of the trophic factor. *Eur J Neurosci.* 2000;12(11):3871–3882.
54. Georgievska B, Kirik D, Rosenblad C, Lundberg C, Bjorklund A. Neuroprotection in the rat Parkinson model by intrastriatal GDNF gene transfer using a lentiviral vector. *Neuroreport.* 2002;13(1):75–82.
55. Georgievska B, Kirik D, Bjorklund A. Aberrant sprouting and downregulation of tyrosine hydroxylase in lesioned nigrostriatal dopamine neurons induced by long-lasting overexpression of glial cell line derived neurotrophic factor in the striatum by lentiviral gene transfer. *Exp Neurol.* 2002;177(2):461–474.
56. Dowd E, Monville C, Torres EM, et al. Lentivector-mediated delivery of GDNF protects complex motor functions relevant to human Parkinsonism in a rat lesion model. *Eur J Neurosci.* 2005;22(10):2587–2595.
57. Kordower JH, Emborg ME, Bloch J, et al. Neurodegeneration prevented by lentiviral vector delivery of GDNF in primate models of Parkinson's disease. *Science.* 2000;290(5492):767–773.
58. Lo Bianco C, Deglon N, Pralong W, Aebischer P. Lentiviral nigral delivery of GDNF does not prevent neurodegeneration in a genetic rat model of Parkinson's disease. *Neurobiol Dis.* 2004;17(2):283–289.
59. Kotzbauer PT, Lampe PA, Heuckeroth RO, et al. Neurturin, a relative of glial-cell-line-derived neurotrophic factor. *Nature.* 1996;384(6608):467–470.
60. Creedon DJ, Tansey MG, Baloh RH, et al. Neurturin shares receptors and signal transduction pathways with glial cell line–derived neurotrophic factor in sympathetic neurons. *Proc Natl Acad Sci USA.* 1997;94(13):7018–7023.
61. Burazin TC, Gundlach AL. Localization of GDNF/neurturin receptor (c-ret, GFRalpha-1 and alpha-2) mRNAs in postnatal rat brain: differential regional and temporal expression in hippocampus, cortex and cerebellum. *Brain Res Mol Brain Res.* 1999;73(1–2):151–171.
62. Wang LC, Shih A, Hongo J, Devaux B, Hynes M. Broad specificity of GDNF family receptors GFRalpha1 and GFRalpha2 for GDNF and NTN in neurons and transfected cells. *J Neurosci Res.* 2000;61(1):1–9.
63. Tseng JL, Bruhn SL, Zurn AD, Aebischer P. Neurturin protects dopaminergic neurons following medial forebrain bundle axotomy. *Neuroreport.* 1998;9(8):1817–1822.
64. Fjord-Larsen L, Johansen JL, Kusk P, et al. Efficient in vivo protection of nigral dopaminergic neurons by lentiviral gene transfer of a modified neurturin construct. *Exp Neurol.* 2005;195(1):49–60.
65. Gasmi M, Herzog CD, Brandon EP, et al. Striatal delivery of neurturin by CERE-120, an AAV2 vector for the treatment of dopaminergic neuron degeneration in Parkinson's disease. *Mol Ther.* 2007;15(1):62–68.
66. Gasmi M, Brandon EP, Herzog CD, et al. AAV2-mediated delivery of human neurturin to the rat nigrostriatal system: long-term efficacy and tolerability of CERE-120 for Parkinson's disease. *Neurobiol Dis.* 2007;27(1):67–76.
67. Herzog CD, Dass B, Holden JE, et al. Striatal delivery of CERE-120, an AAV2 vector encoding human neurturin, enhances activity of the dopaminergic nigrostriatal system in aged monkeys. *Mov Disord.* 2007;22(8):1124–1132.
68. Kordower JH, Herzog CD, Dass B, et al. Delivery of neurturin by AAV2 (CERE-120)-mediated gene transfer provides structural and functional neuroprotection and neurorestoration in MPTP-treated monkeys. *Ann Neurol.* 2006;60(6):706–715.
69. Marks WJ, Ostrem JL, Verhagen L, et al. Safety and tolerability of intraputaminal delivery of CERE-120 (adeno-associated virus serotype 2-neurturin) to patients with idiopathic Parkinson's disease: an open-label, phase I trial. *Lancet Neurol.* 2008;7(5):400–408.
70. Ceregene Presents Additional Clinical Data from Phase 2 Trial of CERE-120 for Parkinson's Disease. 2009. http://www.ceregene.com/press_052709.asp.

9.7 Neuroprotective Strategies in Parkinson's Disease

C. WARREN OLANOW

INTRODUCTION

Parkinson's disease (PD) is characterized by degeneration of dopamine neurons in the substantia nigra pars compacta (SNc) and a reduction in striatal dopamine. The current therapy for PD is based primarily on a dopamine replacement strategy using the dopamine precursor levodopa.[1] Levodopa improves the principal motor features of the disease including tremor, rigidity, and bradykinesia. Almost all patients exhibit a good response to levodopa and demonstrate increased independence, a better quality of life, and even prolonged survival. Since its introduction in the late 1960s, levodopa has provided benefit for millions of PD patients throughout the world. Levodopa is routinely administered in combination with a peripheral decarboxylase inhibitor in order to prevent the peripheral accumulation of dopamine and the consequent nausea and vomiting that occur due to stimulation of dopamine receptors in the nausea and vomiting center of the brain (area postrema) that are not protected by the blood-brain barrier. Shortly after its introduction, it was appreciated that chronic levodopa treatment is associated with motor complications in the majority of patients.[2] These generally take the form of a fluctuating motor response and involuntary movements or dyskinesia. Over the past several decades, several different classes of antiparkinsonian agents that act primarily on the dopamine system have become available. These provide incremental benefits with respect to motor complications but are not superior in efficacy to levodopa[3] (Table 9.7.1). Surgical therapies such as pallidotomy or deep brain stimulation targeting the subthalamic nucleus (STN) or the globus pallidus pars interna (GPi) provide effective treatment for established motor complications,[4] but again do not provide antiparkinsonian benefits exceeding those that can be achieved with levodopa. Forty years after its introduction, levodopa remains the most effective symptomatic therapy for PD and the gold standard against which other treatments must be compared.

It is now appreciated that pathology in PD is widespread and extends to involve multiple areas of the nervous system beyond the nigrostriatal dopamine system. These include the olfactory system, the cerebral hemisphere, and particularly the hippocampus, upper and lower brainstem, spinal cord, and peripheral autonomic nervous system.[5,6] This nondopaminergic pathology is thought to underlie the development of nondopaminergic clinical features such as freezing, postural instability, falling, autonomic dysfunction, mood disorders, sensory alterations, sleep abnormalities, and dementia, which are not well controlled with dopaminergic therapies. Indeed, these nondopaminergic features represent the major source of disability and the primary reason for nursing home placement for patients with advanced PD.[7] Thus, despite the many benefits of levodopa and other currently available therapies, the majority of PD patients eventually develop unacceptable levels of disability. A neuroprotective, or disease-modifying, therapy that can slow or stop disease progression and prevent the emergence of nondopaminergic features is urgently required and is the major unmet medical need in PD today.

CLINICAL TRIALS OF PUTATIVE NEUROPROTECTIVE AGENTS IN PD

A number of putative neuroprotective agents have been identified based on laboratory studies, and several of these have been tested in clinical trials in PD patients. A list of these agents, along with their proposed mechanism of action and the primary endpoint that was employed, are presented in Tables 9.7.2a and 9.7.2b. To date, no agent has been determined to have a neuroprotective effect in PD despite the fact that several of these clinical trials have been positive. This is because confounding symptomatic or pharmacological effects cannot be excluded, and it cannot be ascertained with certainty that positive results in the clinical trial were in fact due to a protective effect.[8]

TABLE 9.7.1. *Drugs Used in Treatment of PD and their Mechanism of Action*

Agent	Mechanism of Action
Dopamine agonists	Activate postsynaptic dopamine receptors
Monoamine oxidase-B (MAO-B) inhibitors	Block dopamine metabolism and increase synaptic dopamine levels
Catechol-O-methyltransferase (COMT) inhibitors	Block peripheral metabolism of levodopa and increase brain dopamine levels

TABLE 9.7.2a. *Negative Trials*

	Proposed Mechanism	Endpoint
Vitamin E[9]	Antioxidant	Time to need for levodopa
TCH346[10]	Antiapoptotic	Time to need for levodopa
CEP346[11]	Antiapoptotic	Time to need for levodopa
Immunophilin[12]	Trophic effect	Time to need for levodopa
Glial cell line–derived neurotrophic factor (GDNF)[13]	Trophic factor	Δ Unified Parkinson's Disease Rating Scale (UPDRS) in off phase between baseline and final visit
Neurturin[14]	Trophic factor	Δ UPDRS in off phase between baseline and final visit

TABLE 9.7.2b. *Positive Trials*

	Proposed Mechanism	Endpoint
Selegiline[9]	Antiapoptotic	Time to need for levodopa
Selegiline[15]	Antiapoptotic	Time to need for levodopa
Coenzyme Q10[16]	Bioenergetic	Change in UPDRS score
Ropinirole[17]	Antiapoptotic	Δ from baseline in striatal fluorodopa FD uptake on positron emission tomography (PET)
Pramipexole[18]	Antiapoptotic	Δ from baseline in striatal β-CIT uptake on SPECT
Minocycline[19]	Anti-inflammatory	Change in UPDRS score
Creatine[19]	Bioenergetic	Change in UPDRS score

Positive Clinical Trials Using Clinical Endpoints

The first major double-blind, controlled trial to test the possibility of achieving neuroprotection in PD was the DATATOP study.[9] The monoamine oxidase-B (MAO-B) inhibitor selegiline was evaluated based on its capacity to prevent 1-methyl-4-phenyl-1,2,3,6-tetrahydropyridine (MPTP) parkinsonism in animal models and to inhibit the formation of free radicals generated by the oxidative metabolism of dopamine. Subsequent studies demonstrated that the drug also has antiapoptotic effects, probably related to a propargyl ring incorporated within its molecular structure that binds to reduced glyceraldehyde-phosphate dehydrogenase (GAPDH) and prevent its translocation to the nucleus and the inhibition of transcriptional up-regulation of protective molecules such as BCL-2, BCL-X_L, catalase, and superoxide dismutase (SOD).[20] The DATATOP study used the time to development of a milestone of PD progression (i.e., time to the development of disability necessitating the need for levodopa therapy in an untreated patient) as the primary outcome measure. Selegiline significantly delayed the time to the need for levodopa compared to placebo, but the drug was shown to have previously unappreciated symptomatic effects, and it could not be determined if the benefit observed was due to the drug's having a neuroprotective effect that slowed disease progression or to a symptomatic effect that merely masked underlying neurodegeneration. The Sinemet-Deprenyl-Parlodel (SINDEPAR) study attempted to resolve this confusion by randomizing untreated PD patients to receive selegiline or placebo in combination with dopaminergic therapy for 12 months. The final visit was performed after selegiline had been withdrawn for 2 months. The primary endpoint was the change between untreated baseline and untreated final visit (14 months).[15] Deterioration from baseline in the Unified Parkinson's Disease Rating Scale (UPDRS) score was significantly less in the selegiline group than in the placebo group. However, here too, it could not be determined for certain that this benefit was due to a protective effect, as the possibility of a long-acting symptomatic effect that persisted despite 2 months of drug washout could not be excluded. Two subsequent long-term studies indicated that PD patients receiving levodopa had better UPDRS scores and fewer motor complications if they received selegiline rather than placebo at the start of treatment,[21,22] so it remains possible that the drug does have a neuroprotective property that we have not yet been able to delineate.

Coenzyme Q10 is an antioxidant and a bioenergetic agent that is thought to enhance mitochondrial function and has been shown to protect against MPTP toxicity in animal models.[23] The QE2 study compared 3 doses of coenyme Q10 to placebo in untreated PD patients using the change in UPDRS score between baseline and final visit as the primary endpoint.[16] Patients receiving coenzyme Q10 had a trend toward less deterioration from baseline in the UPDRS score compared to those receiving placebo, with the highest dose almost reaching significance in this underpowered pilot trial. However, there were short-term improvements in the activities of daily living (ADL) score, and here too a symptomatic effect could not be completely excluded. A large-scale National Institutes of Health (NIH)–sponsored trial is currently underway. Creatine is another bioenergetic agent that showed positive results in a short-term futility study that also measured change from baseline in the UPDRS score,[19] and here too the benefits could easily have been confounded by the drug's having a symptomatic effect. It is now being tested in a long-term simple study (see below).

Positive Clinical Trials Using Surrogate Neuroimaging Biomarkers as the Primary Endpoint

Dopamine agonists have been shown to have antiapoptotic effects in laboratory models of PD. Recent studies have shown that the dopamine agonist ropinirole induces protection through activation of the phosphoinositide-3 (PI-3) kinase/Akt pathway with phosphorylation and inhibition of glycogen synthase kinase (GSK)-3β.[24,25] Other studies have shown that pramipexole protects dopamine neurons through a receptor-independent mechanism that is presently not defined.[26] When clinical trials for dopamine agonists were being considered, it was appreciated that they have relatively powerful antiparkinsonian effects that would confound any neuroprotective study that relied on the clinical endpoints used in previous trials. For this reason, surrogate imaging biomarkers of nigrostriatal dopamine function were used as the primary endpoint in two clinical trials of dopamine agonists seeking a putative disease-modifying effect. The first compared ropinirole to levodopa using fluorodopa-positron emission tomography (FD-PET)[17] and the second compared pramipexole to levodopa with β-CIT single photon emission computed tomography (SPECT).[18] Both studies showed that the rate of decline in striatal uptake of the imaging biomarker was significantly less in patients receiving the dopamine agonist than in those treated with levodopa. As there was no placebo control, these results were compatible with the agonists having a protective effect or with levodopa having a toxic effect. However, it has also been suggested that dopamine agonists and levodopa might have differential pharmacological effects on the neuroimaging biomarker that could confound interpretation of this study and therefore do not permit a final conclusion to be reached.[27] The INSPECT study tried to resolve this issue by performing repeat SPECT studies before and 12 weeks after administration of levodopa or pramipexole, and showed no evidence of any short-duration regulatory effect.[28] However, this does not exclude the possibility that such an effect might occur over a longer period of time. So, at present, dopamine agonists cannot be definitely determined to have neuroprotective properties in PD despite positive study results in the laboratory and in clinical trials.

THE SEARCH FOR A NEUROPROTECTIVE AGENT

The previous section illustrates that despite the many candidate agents, and positive study results in both the laboratory and the clinic, we have not yet been able to define a neuroprotective agent in PD. The search for a neuroprotective agent has been hampered by several obstacles whose resolution might considerably facilitate progress. These include the need for (1) insight into the etiology and pathogenesis of cell death in PD, (2) an animal model that reflects the etiopathogenesis, nondopaminergic pathology, and chronic progressive course of

PD, (3) a methodology for determining the optimal dose to employ in a clinical trial, and (4) a clinical endpoint that reflects the underlying disease state that is not readily confounded and can provide an accurate measure of disease progression.

Etiopathogenesis

The rational development of a neuroprotective agent in PD would be greatly enhanced by better understanding the cause and mechanism responsible for the cell death process. Several factors have been implicated in the pathogenesis of PD.[29,30] These include oxidative stress, mitochondrial dysfunction, excitotoxicity, inflammation, protein accumulation, and signal-mediated apoptosis (Fig. 9.7.1). However, efforts to manipulate these pathways have not yet led to the development of a neuroprotective therapy. This reflects in part our uncertainty as to which one, if any, of these factors is the primary driver of cell death and which ones are secondary, although still possibly contributing, factors. Indeed, it is possible that cell death involves multiple pathogenic factors that are incorporated into a complex network where the initiating factor may be different in different individuals. If that is the case, a cocktail of agents acting against multiple pathogenic pathways may be required to achieve neuroprotection. It is also possible that the factors that have been implicated to date are merely epiphenomena and develop coincident to a still unidentified alternate pathogenic mechanism. Among the more intriguing new possibilities are agents that block the 1.3L-type calcium channel. This is based on recent evidence demonstrating that dopamine cells switch from sodium ion channels to 1.3 L-type calcium channels to maintain pacemaker activity, and that blocking these channels causes dopamine neurons to revert back to using sodium channels and protects them from a variety of toxins including MPTP, rotenone, and 6-hydroxydopamine (6-OHDA).[31]

Genetic and environmental targets have also attracted attention in attempting to define a neuroprotective treatment for PD. Epidemiological studies suggest that environmental factors are likely to be important in sporadic forms of PD,[32] but no specific environmental cause of PD has been identified. Toxins such as MPTP and rotenone, which can cause dopaminergic lesions in the laboratory, have attracted considerable interest but have not been demonstrated to cause PD. Familial forms of PD have been described in association with mutations in a variety of genes including alpha-synuclein,[33] UCH-L1,[34] Parkin,[35] DJ-1,[36] PINK1,[37] LRRK2,[38,39] and, more recently, OMI/HTRA2[40] and ATP13A2.[41] These are potentially more interesting because they offer the possibility of understanding the mechanism leading to cell death in a model that is directly involved in the etiopathogenesis of at least one form of PD. Several of the mutations that have been discovered to date provide support for the possibility that proteolytic stress and/or mitochondrial dysfunction are key factors in the etiopathogenesis of PD and suggest novel targets for candidate neuroprotective agents.

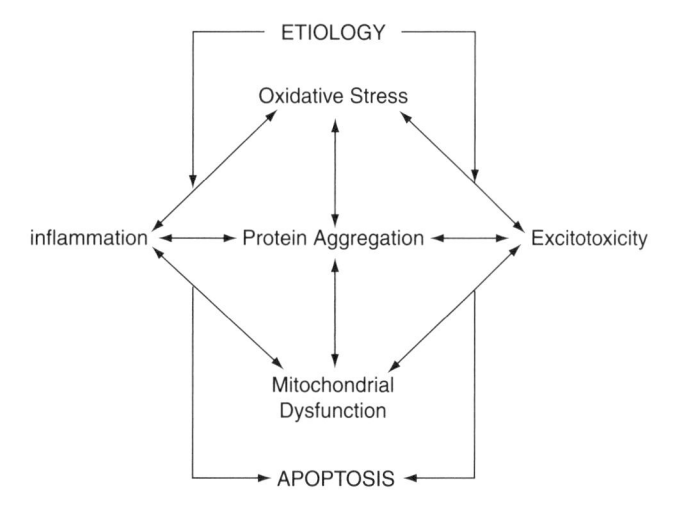

FIGURE 9.7.1. Schematic representation showing how factors thought to be involved in the pathogenesis of PD might interact with one another, eventually leading to cell death. This hypothesis suggests that PD might be caused by multiple etiological agents inducing a neurodegenerative cascade through activation of pathogenic factors that might be different in different individuals. It also suggests that blocking a single pathogenic factor might not be able to prevent the network of events leading to the cell death process. *Source:* Adapted from[29].

Proteolytic stress could result from increased production or impaired clearance of abnormal and misfolded proteins. That protein accumulation and aggregation is an important factor in the cause of cell death in PD is a natural concern, as the disease is characterized by the presence of Lewy bodies and Lewy neurites, which are comprised of protein aggregates in affected cell bodies and nerve terminals. These inclusions stain positively for alpha-synuclein, a natively unfolded protein that is prone to misfold and assume a beta sheet conformation when it accumulates. Mutations in alpha-synuclein as found in familial cases make the protein even more prone to misfold[33] and suggest that increased formation of unwanted proteins can drive the cell death process. Even more intriguing are the cases of familial PD that are associated with duplication or triplication of wild-type alpha-synuclein,[42,43] which indicate that increased production of even the normal protein can lead to the development of PD. This concept is supported by studies showing that gene delivery of alpha-synuclein to the region of the SNc can replicate many of the behavioral and pathological features of PD in rodents and primates.[44,45] There is also evidence suggesting that impairment in the capacity to clear unwanted proteins can lead to PD. Pathological studies in patients with sporadic PD show evidence of structural and functional impairment in the 20S proteasome.[46] Several gene mutations associated with familial PD also support this concept. Parkin is a ubiquitin ligase that attaches ubiquitin to misfolded proteins to signal for their proteasomal transport and clearance.[35] Mutations associated with PD impair this process. Interestingly, parkin protects against alpha-synuclein toxicity, even though alpha-synuclein is not a substrate protein.[47-49] UCH-l1 is a de-ubiquitinating enzyme necessary for cleaving ubiquitin from protein conjugates to permit it to enter the proteasome and to free up ubiquitin monomers to facilitate clearance of additional unwanted proteins.[34] Mutations in these two proteins could thus interfere with normal function of the ubiquitn proteasome system (UPS). Regardless of the cause of alpha-synuclein accumulation, aggregation of the protein can damage UPS and lysosomal functions, thereby further impairing the clearance of both itself and other proteins.[50,51] Thus, a vicious cycle can be envisioned to occur whereby increased production of mutant or even wild-type proteins could interfere with clearance mechanisms, while impaired clearance could result in further protein accumulation. Continued protein accumulation resulting from this vicious cycle could eventually exceed the capacity of the cell to degrade unwanted proteins, thereby leading to a state of proteolytic stress with protein accumulation, oligomer formation, aggregation, and cell death. There is some evidence suggesting that oligomers rather than polymers may be the toxic form of alpha-synuclein,[52] and that Lewy bodies may be protective.[53]

The possibility that proteolytic stress may be a key factor in the cell death process in PD provides several novel candidate targets for a putative neuroprotective therapy. Such therapies could be designed to prevent the production of misfolded proteins, facilitate their refolding to a normal state, enhance proteasomal or lysosomal degradation, and promote the dissolution or prevent the formation of toxic oligomers or polymers. Preliminary studies in animal models have demonstrated the potential value of approaches such as heat shock proteins or geldanamycin, a drug that blocks the suppression of HSp70 expression.[54-56] Further, vaccination-induced production of immunoantibodies directed against oligomeric alpha-synuclein have been demonstrated to decrease aggregate formation and PD pathology in mice models.[57] The antibiotic rifampicin has also been shown to inhibit oligomerization, disaggregate alpha synuclein and thereby protect dopaminergic cells.[58,59] The sirtuin family of proteins (SIRTs) that are involved in histone deacetylation and autophagy have also begun to attract attention as candidate targets for a neuroprotective therapy. SIRT2 has been postulated to adversely affect protein clearance, and inhibition of SIRT2 has been shown to diminish alpha-synuclein-mediated toxicity in cell culture and in drosophila.[60,61] In contrast, it has also been reported that SIRT1 protects against alpha-synuclein toxicity.[62] Interestingly, SIRT2 inhibitors appear to act by promoting inclusion body formation,[63] leading to the hypothesis that SIRT2 promotes alpha-synuclein oligomer formation while SIRT1 promotes their disassembly.[64] The SIRT family thus represents another series of potential targets for a neuroprotective therapy. While these various approaches are interesting, none has yet been tested in PD patients, and safety issues remain to be more fully evaluated before clinical trials can be considered.

There has been great excitement about the recent finding that embryonic dopamine neurons implanted into the striatum of advanced PD patients develop alpha-synuclein-positive inclusions that are identical to Lewy bodies.[65-67] While the mechanism responsible for these pathological changes is not known, the fact that genetically independent embryonic dopamine neurons can develop PD pathology after such a short latency raises the possibility that alpha-synuclein can act like a prion and can be transmitted from host to implanted dopamine neurons.[67a] Indeed, there is evidence that neurons can release and take up alpha-synuclein by exocytosis and endocytosis,[68,69] and one could envision that this form of transmission could account for the pattern of evolution of alpha-synuclein pathology described by Braak and colleagues.[6] Studies are currently underway to determine if inoculates derived from alpha-synuclein aggregates in PD patients can transmit the disease to other species. This concept provides yet another source of targets for therapeutic agents that could interfere with a prion-like process in PD.

Mitochondria have also been implicated as an important target for a possible neuroprotective therapy in PD. A defect in mitochondrial complex 1 staining and activity has been detected in the SNc of patients with sporadic PD.[70,71] In addition, mutations have been identified in the mitochondrial DNA polymerase gamma (POLG) in patients with both sporadic and hereditary forms of PD, further implicating the mitochondria in PD.[72,73] Further, several nuclear mutations have been identified in proteins that are linked to mitochondria. Phosphatase and tensin homologue (PTEN)-induced putative kinase (PINK1) is a serine/threonine protein kinase that has a mitochondrial targeting sequence.[37] PINK1 is involved in sensing mitochondrial stress and protecting against apoptosis, possibly through interactions with its binding partners TRAP1 and HtrA2.[74,75] It is noteworthy that mutations in HtrA2 have also been associated with familial PD.[76] Removal of the PINK1 homologue in drosophila causes mitochondrial dysfunction with enlargement and fragmentation of christae.[77,78] Defects in the parkin gene also lead to alterations in mitochondrial morphology and enhance the degree of mitochondrial damage seen with PINK1 mutations. Overexpression of wild-type parkin restores mitochondrial morphology in PINK1 mutant drosophila or following RNAi-induced reduction of PINK1, but PINK1 does not reverse the mitochondrial changes due to the parkin mutation.[79,80] These observations suggest that PINK1 and parkin may act in a common pathway that is critical for normal mitochondrial function, with parkin being downstream. DJ-1 mutations are also associated with familial PD[36] and provide additional potential targets for neuroprotective therapies. DJ-1 acts as an antioxidant or sensor of oxidative stress in mitochondria, and this function is lost when the protein is in a mutant form.[81] DJ-1 has also been shown to maintain protein levels of the antioxidant transcriptional master regulator NF-E2-related factor 2 (Nrf2), which helps ensure an adequate stress response.[82] Overexpression of DJ-1 protects dopamine neurons from oxidative stress in rats,[83] while knockdown of the protein increases susceptibility to oxidative stress, endoplasmic reticulum stress, and proteasomal inhibition.[84] DJ-1 has also been shown to interact with Daxx and to inhibit apoptotic cell signaling.[85] Interestingly, wild-type DJ-1 inhibits aggregation of alpha-synuclein, while this effect is lost when the protein is in its mutant form.[86] For all of these reasons, DJ-1 and its signaling partners are potential targets for putative neuroprotective therapies. Collectively, these studies suggest that the mitochondrion and related proteins might also be appropriate targets for a neuroprotective therapy in PD. Bioenergetics such as creatine and coenzyme Q10 are currently being investigated in PD (see above).

It is noteworthy that proteasomal and mitochondria functions are interdependent and that damage to the one can lead to dysfunction in the other.[87] Adenosine triphosphate (ATP) generated by mitochondria is essential for normal proteasomal function, and mitochondrial toxins lead to proteasomal impairment, while proteasome inhibitors result in mitochondrial dysfunction.[88,89] Further, many toxic models such as alpha-synuclein overexpression induce both mitochondrial and proteasomal dysfunction.[90,91]

Leucine-rich repeat kinase 2 (LRRK2) has received particular attention as a possible target for a neuroprotective therapy because mutations have been described in patients with apparently sporadic PD[92] and because this mutation accounts for as many as 40% of PD cases in Ashkenazy Jews and some North African populations.[93,94] LRRK2 is linked to the outer mitochondrial membrane. The precise manner by which LRRK2 causes neurodegeneration in PD is not known, but recent studies indicate that LRRK2 has kinase[95] and guanosine triphosphatase (GTPase)[96] activities. Mutations found in PD are associated with reduced guanosine triphosphate (GTP) hydrolysis and altered kinase activity, and alterations in the LRRK2 protein that reduce kinase activity in the mutant LRRK2 are associated with a reduction in neuronal toxicity.[97] These observations suggest that cell death may relate to altered phosphorylation of target proteins, possibly inducing the accumulation of misfolded substrate proteins. These findings suggest that LRRK2 may be a target for novel neuroprotective drugs, and drugs that alter or inhibit its kinase activity are currently being actively explored.

While the large majority of PD cases occur sporadically and are of unknown etiology, the identification of mutations associated with some forms of PD permits us to elicit the precise mechanism and signaling pathways that are associated with the cell death process in one form of PD and to identify candidate targets for novel neuroprotective agents. While the causes of genetic forms of PD may differ from each other and from sporadic cases, it is not unreasonable to anticipate that they might share a common pathogenic pathway, and that interventions that are protective against one of these forms might also be applicable to others.

This brief sample of experimental studies illustrates the many possible candidate targets that are currently being pursued, and it is likely that many more will be identified as new gene mutations associated with PD continue to be identified and explored.

Animal Models of PD

Another major obstacle to the development of a neuroprotective drug for PD is a reliable animal model that replicates the etiopathogenesis, the pathology, and the chronic progressive course of the disease in which to test putative new agents. Current preclinical studies primarily utilize the MPTP monkey and the 6-OHDA-lesioned rat for testing putative neuroprotective agents.[98] While these toxins adequately model the dopaminergic lesions of PD, they do not replicate the non-dopaminergic features of the disease and they likely do not reflect its etiopathogenesis. Thus, there is no assurance that positive results in these preclinical studies will translate into positive results in clinical trials or, alternatively, that negative studies in the models necessarily mean that the drugs won't be protective in PD. Clearly, a better model of PD is required. With the identification of several gene and protein mutations that are associated with PD, it had been hoped that this would provide the basis for developing transgenic models that bear directly on the disease process. Unfortunately, this has proven difficult to accomplish, and none of the transgenic models developed to date completely reflects the pattern and distribution of dopaminergic and nondopaminergic pathology that is found in PD. This may reflect the fact that a protein that accumulates and causes pathology in humans may be handled differently and may not be toxic in different species.

An alternate approach that is currently being used takes advantage of the fact that famillial forms of PD are found with mutations, as well as duplication and triplication of the wild-type alpha-synuclein gene. This can be modeled by using viral vectors to deliver wild-type or mutant alpha-synuclein to rodents or primates by direct injection into the supranigral region.[37,38] This strategy has been shown to induce dopamine neuronal cell death with inclusion body formation and behavioral changes replicating the findings in PD. This model provides an opportunity to test multiple agents that interfere with alpha-synuclein aggregation or that facilitate clearance of the protein in species such as worms or drosophila and are most promising in primates. It remains to be seen if this approach will prove more useful in defining therapies for the clinic.

A validated model of PD would be of enormous value in facilitating preclinical testing of promising candidate neuroprotective agents and increasing the likelihood that positive results in the laboratory would translate into positive results in the clinic[98a].

Dosing

Another major problem in trying to bring a putative neuroprotective drug from the laboratory to the clinic is the difficulty of determining the optimal dose to employ. Determining the correct dose is difficult with a putative neuroprotective drug because there is no biomarker against which to titrate the compound, as there generally is for symptomatic agents. Thus, selecting the dose to use in a clinical trial is often based solely on attempts to replicate concentrations that provide positive results in tissue culture studies. Translation of the concentration that protects cells in culture into a dose for testing in humans is difficult to estimate. For examples, propargylamines, which have protective effects in laboratory models, often show their benefit with extremely low concentrations (approximately 10^{-10} M) and lose this benefit with lower or higher concentrations of the drug. It is hard to determine the dose to use in humans that would give the desired concentration at the cell level given the variability in absorption and protein binding, the multiple factors that can influence CNS transport, and variables within the brain that might influence the local tissue concentration. For example, the propargylamine TCH346 showed powerful protective effects in the laboratory but failed to show benefits in a clinical trial.[99] It remains uncertain why the drug does not have protective effects in PD, as it did in the laboratory and calls into question the reliability of the preclinical studies and models, or if we simply chose the wrong doses to study in the clinical trial.

A resolution of this problem would be of great value in ensuring that potentially effective drugs are not studied in ineffective doses that cannot achieve the desired goal.

Clinical Endpoints

One of the most serious obstacles to developing a neuroprotective therapy for PD is the lack of an outcome measure that accurately and reliably reflects the underlying disease state. As discussed above, the clinical endpoints that have been used in clinical trials to date, such as measures of change in the parkinsonian score (UPDRS), time to a milestone of disease progression (e.g., time to the need for levodopa), and washout studies, are all readily confounded by potential symptomatic effects of the study intervention. The use of biomarkers has also led to uncertain results because of the potential of study interventions to directly influence the biomarker without necessarily affecting the disease process. For these reasons, there has been a search for a more reliable endpoint or study design that could be used in a clinical trial.

The delayed start study was proposed by Paul Leber to help resolve this dilemma.[100,101] This type of study is done in two phases. In phase I, subjects are randomized to active study drug or placebo. Any differences between the groups at the end of this phase could be due to symptomatic and/or neuroprotective effects. In phase II, subjects in both study groups are placed on the same active intervention. If, at the end of phase II, benefits in the early treatment group are no different than those seen in the delayed start group, then it is likely that the benefit observed at the end of phase I was symptomatic in origin. However, a symptomatic benefit cannot readily explain a difference between the groups if it is still present at the end of phase II, when patients in both study groups are taking the same medication. In this scenario, the difference in parkinsonian scores at the end of the study must somehow relate to the early treatment and is consistent with the possibility that the drug has a neuroprotective or disease modifying effect. This study design was employed in the ADAGIO trial,[102] which demonstrated that early treatment with rasagiline, 1 mg per day, provided benefits that could not be achieved with later introduction of the same agent.[103] While there are alternate explanations for a positive result, such as preserving a beneficial compensatory response or avoiding a detrimental maladaptive response, this is the closest we have been able to come to date in trying to demonstrate that a study intervention has a disease-modifying property. This design is now being used in evaluating the putative disease-modifying effects of other compounds such as the dopamine agonist pramipexole (the PROUD study).

There is also a question of whether or not it is even possible to identify a neuroprotective effect in a clinical trial, as neuroprotection is in reality is a laboratory concept. Rather, interest is now beginning to focus on the use of outcome measures that provide an assessment of cumulative disability, with the idea that benefits with respect to this type of endpoint are important and clinically relevant regardless of their mechanism. Toward this end, the NIH has initiated the NET-PD study. In this clinical trial, a number of possible neuroprotective agents are examined in short-term futility studies designed to determine if any of these agents can be rejected as being futile.[104] The most promising agent is then evaluated in what is known as a *long-term simple study* where patients are randomized to the active agent or placebo and then followed for a prolonged period of time (5 years) during which the treating investigator may employ any therapy deemed appropriate for patients in either study group. The outcome measure for this trial is a composite endpoint that includes UPDRS scores, a quality-of-life measure, and a battery of tests assessing nondopaminergic functions such as cognition and gait.

CONCLUSIONS

Parkinson's disease patients inevitably develop disability despite currently available medical and surgical therapies. Accordingly, a neuroprotective therapy that slows or stops disease progression is an urgent requirement. While there are many promising candidate agents based on laboratory studies, the translation of a novel study intervention into a viable disease-modifying therapy has proven to be extremely difficult to achieve; to date, no agent has been determined to be neuroprotective by either regulatory authorities or physicians. Among the limiting factors are uncertainty as to the etiology and pathogenesis of cell death in PD and what precisely to target, a reliable animal model in which to test putative neuroprotective therapies, a method for accurately determining the optimal dose range to employ in clinical trials, and a clinical outcome measure that accurately reflects the status of the underlying disease state. There is some optimism that we are beginning to be able to overcome some of these obstacles. While we don't yet know the precise cause of PD, genes associated with familial cases of the disease have implicated mitochondrial defects and/or proteolytic stress, with increased production or impaired clearance of misfolded proteins being at the heart of the disease. These genes have led to discovery of a host of novel candidate targets for putative neuroprotective agents. While no satisfactory animal model currently exists, it is not unreasonable to consider that gene mutations that have been or will be discovered will eventually replicate PD in an animal model. Determining the dose levels to study in a clinical trial of a putative neuroprotective agent is currently a problem, but it should be solvable with a concerted effort. Finally, there remains the issue of how to measure the impact of a protective agent on disease progression. The study designs used to date are too readily compromised by potential symptomatic effects of the study intervention and therefore cannot determine with certainty that positive results with a study agent imply that the intervention has a disease-modifying effect. There is an intensive search for a biomarker of disease progression that could be used as a surrogate, but at present, none has been delineated. The delayed start design offers the opportunity to determine that an agent provides benefits that cannot be defined by purely symptomatic effects, but even here it is not possible to discriminate neuroprotective effects from those that act on compensatory mechanisms. In the long run, it may not be that important to ascertain the underlying mechanism responsible for a drug effect. Rather, there is an attempt from the clinical perspective to begin to focus on cumulative disability, with the thought that an intervention that slows the development of disability resulting from features such as gait impairment or dementia is worthwhile regardless of the responsible mechanism. Such trials, however, tend to be longer and more expensive, and may not be feasible for the pharmaceutical industry.

There have been major advances in our understanding and treatment of PD. It must now be determined if some of the strategies described above can be translated into effective treatment strategies that will improve the quality of life for PD patients.

REFERENCES

1. Olanow CW. The scientific basis for the current treatment of Parkinson's disease. *Ann Rev Med.* 2004;55:41–60.
2. Ahlskog JE, Muenter MD. Frequency of levodopa-related dyskinesias and motor fluctuations as estimated from the cumulative literature. *Mov Disord.* 2001;16:448–458.
3. Schapira AHV, Olanow CW. *Principles of Treatment for Parkinson's Disease.* Philadelphia, PA: Butterworth Heinemann, Elsevier; 2005.
4. The Deep Brain Stimulation for PD Study Group. Deep brain stimulation of the subthalamic nucleus or globus pallidus pars interna in Parkinson's disease. *N Engl J Med.* 2001;345:956–963.
5. Forno LS. Neuropathology of Parkinson's disease. *J Neuropathol Exp Neurol.* 1996;55:259–272.
6. Braak H, Del Tredici K, Rub U, et al. Staging of brain pathology related to sporadic Parkinson's disease. *Neurobiol Aging.* 2003;24:197–211.
7. Hely MA, Morris JG, Reid WG, Trafficante R. Sydney Multicenter Study of Parkinson's disease: non-L-dopa-responsive problems dominate at 15 years. *Mov Disord.* 2005;20:190–199.
8. Schapira AHV, Olanow CW. Neuroprotection in Parkinson's disease: myths, mysteries, and misconceptions. *JAMA.* 2004;291:358–364.
9. Parkinson's Study Group, Olanow CW, Steering Committee. Effects of tocopherol and deprenyl on the progression of disability in early Parkinson's disease. *N Engl J Med.* 1993;328:176–183.
10. Olanow CW, Schapira AHV, Lewitt PA, Kieburtz K, Sauer D, Olivieri G, Pohlmann H, Hubble J. TCH346 as a neuroprotective drug in Parkinson's disease: a double-blind, randomised, controlled trial. *Lancet Neurol.* 2006;5:1013–1020.
11. Parkinson Study Group PRECEPT Investigators. Mixed lineage kinase inhibitor CEP-1347 fails to delay disability in early Parkinson disease. *Neurology.* 2007;69:1480–1490.
12. Gold BG, Nutt JG. Neuroimmunophilin ligands in the treatment of Parkinson's disease. *Curr Opin Pharmacol.* 2002;2:82–86.
13. Lang AE, Gill S, Patel NK, et al. Randomized controlled trial of intraputamenal glial cell line–derived neurotrophic factor infusion in Parkinson disease. *Ann Neurol.* 2006;59:459–466.
14. C.W. Olanow, personal observation.
15. Olanow CW, Hauser RA, Gauger L, et al. The effect of deprenyl and levodopa on the progression of signs and symptoms in Parkinson's disease. *Ann Neurol.* 1995;38:771–777.
16. Shults CW, Oakes D, Kieburtz K, et al. Effects of coenzyme Q10 in early Parkinson disease: evidence of slowing of the functional decline. *Arch Neurol.* 2002;59:1541–1550.
17. Whone A, Watts R, Stoessl J, et al. Slower progression of Parkinson's disease with ropinirole versus levodopa: the REAL-PET Study. *Ann Neurol.* 2003;54:93–101.
18. Parkinson Study Group. Dopamine transporter brain imaging to assess the effects of pramipexole vs levodopa on Parkinson disease progression. *JAMA.* 2002;287:1653–1661.
19. NINDS NET-PD Investigators. A randomized, double-blind, futility clinical trial of creatine and minocycline in early Parkinson disease. *Neurology.* 2006;66:664–671.
20. Carlile GW, Chalmers-Redman RM, Tatton NA, Pong A, Borden KE, Tatton WG. Reduced apoptosis after nerve growth factor and serum withdrawal: conversion of tetrameric glyceraldehyde-3-phosphate dehydrogenase to a dimer. *Mol Pharmacol.* 2000;57:2–12.
21. Larsen JP, Boas J, Erdal JE. Does selegiline modify the progression of early Parkinson's disease? Results from a five-year study. The Norwegian-Danish Study Group. *Eur J Neurol.* 1999;6:539–547.
22. Pålhagen S, Heinonen E, Hägglund J, Kaugesaar T, Mäki-Ikola O, Palm R; Swedish Parkinson Study Group. Selegiline slows the progression of the symptoms of Parkinson disease. *Neurology.* 2006;66:1200–1206.
23. Beal MF. Coenzyme Q10 as a possible treatment for neurodegenerative diseases. *Free Radic Res.* 2002;36:455–460.
24. Nair VD, Olanow CW, Sealfon SC. Activation of phosphoinositide 3-kinase by D2 receptor prevents apoptosis in dopaminergic cell lines. *Biochem J.* 2003;373:25–32.
25. Nair VD, Olanow CW. Differential modulation of Akt/GSK-3β pathway regulates apoptotic and cytoprotective signaling responses. *J Biol Chem.* 2008;283:15469–15478.
26. Gu M, Iravani MM, Cooper JM, King D, Jenner P, Schapira AH. Pramipexole protects against apoptotic cell death by non-dopaminergic mechanisms. *J Neurochem.* 2004;91:1075–1081.
27. Ahlskog JE. Slowing Parkinson's disease progression: recent dopamine agonist trials. *Neurology.* 2003;60:381–389.
28. Jennings DL, Tabamo R, Seibyl JP, Marek K. InSPECT: investigating the effect of short-term treatment with pramipexole or levodopa on [123I]β-CIT and SPECT imaging. *Mov Disord.* 2007;22(abstract):A143.
29. Olanow CW. Pathogenesis of cell death in Parkinson's disease—2007. *Mov Disord.* 2007;22:S335–342.
30. Gupta A, Dawson VL, Dawson TM. What causes cell death in Parkinson's disease? *Ann Neurol.* 2008;64(suppl 2):S3–S15.
31. Chan CS, Guzman JN, Ilijic E, et al. "Rejuvenation" protects neurons in mouse models of Parkinson's disease. *Nature.* 2007;447:1081–1086.
32. Tanner CM, Ottman R, Goldman SM, et al. Parkinson disease in twins: an etiologic study. *JAMA.* 1999;281:341–346.

33. Polymeropoulos MH, Lavedan C, Leroy E, et al. Mutation in the alpha-synuclein gene identified in families with Parkinson's disease. *Science*. 1997;276:2045–2047.
34. Leroy E, Boyer R, Auburger G, et al. The ubiquitin pathway in Parkinson's disease. *Nature*. 1998;395:451–452.
35. Kitada T, Asakawa S, Hattori N, et al. Mutations in the parkin gene cause autosomal recessive juvenile parkinsonism. *Nature*. 1998;392:605–608.
36. Bonifati V, Rizzu P, van Baren MJ, et al. Mutations in the DJ-1 gene associated with autosomal recessive early-onset parkinsonism. *Science*. 2003;299:256–259.
37. Valente EM, Abou-Sleiman PM, Caputo V, et al. Hereditary early-onset Parkinson's disease caused by mutations in PINK1. *Science*. 2004;304:1158–1160.
38. Paisán-Ruíz C, Jain S, Evans EW, et al. Cloning of the gene containing mutations that cause PARK8-linked Parkinson's disease. *Neuron*. 2004;44:595–600.
39. Zimprich A, Biskup S, Leitner P, et al. Mutations in LRRK2 cause autosomal-dominant parkinsonism with pleomorphic pathology. *Neuron*. 2004;44:601–607.
40. Strauss KM, Martins LM, Plun-Favreau H, et al. Loss of function mutations in the gene encoding Omi/HtrA2 in Parkinson's disease. *Hum Mol Genet*. 2005;14:2099–2111.
41. Ramirez A, Heimbach A, Gründemann J, et al. Hereditary parkinsonism with dementia is caused by mutations in ATP13A2, encoding a lysosomal type 5 P-type ATPase. *Nat Genet*. 2006;38:1184–1191.
42. Chartier-Harlin MC, Kachergus J, Roumier C, et al. Alpha-synuclein locus duplication as a cause of familial Parkinson's disease. *Lancet*. 2004;364:1167–1169.
43. Singleton AB, Farrer M, Johnson J. Alpha-synuclein locus triplication causes Parkinson's disease. *Science*. 2003;302:841.
44. St Martin JL, Klucken J, Outeiro TF, Nguyen P, Keller-McGandy C, Cantuti-Castelvetri I, Grammatopoulos TN, Standaert DG, Hyman BT, McLean PJ. Dopaminergic neuron loss and up-regulation of chaperone protein mRNA induced by targeted over-expression of alpha-synuclein in mouse substantia nigra. *J Neurochem*. 2007;100:1449–1457.
45. Kirik D, Annett LE, Burger C, Muzyczka N, Mandel RJ, Björklund A. Nigrostriatal alpha-synucleinopathy induced by viral vector–mediated overexpression of human alpha-synuclein: a new primate model of Parkinson's disease. *Proc Natl Acad Sci USA*. 2003;100:2884–2889.
46. McNaught K St. P, Olanow CW, Halliwell, Isacson O, Jenner P. Failure of the ubiquitin-proteasome system in Parkinson's disease. *Nat Rev Neurosci*. 2001;2: 589–594.
47. Petrucelli L, O'Farrell C, Lockhart PJ, et al. Parkin protects against the toxicity associated with mutant alpha-synuclein: proteasome dysfunction selectively affects catecholaminergic neurons. *Neuron*. 2002;36:1007–1019.
48. Yang Y, Gehrke S, Imai Y, Huang Z, Ouyang Y, Wang JW, Yang L, Beal MF, Vogel H, Haywood AF, Staveley BE. Mutant alpha-synuclein-induced degeneration is reduced by parkin in a fly model of Parkinson's disease. *Genome*. 2006;49:505–510.
49. Lo Bianco C, Schneider BL, Bauer M, Sajadi A, Brice A, Iwatsubo T, Aebischer P. Lentiviral vector delivery of parkin prevents dopaminergic degeneration in an alpha-synuclein rat model of Parkinson's disease. *Proc Natl Acad Sci USA*. 2004;101:17510–17515.
50. Snyder H, Mensah K, Theisler C, Lee JM, Matouschek A, Wolozin B. Aggregated and monomeric alpha-synuclein bind to the S6' proteasomal protein and inhibit proteasome function. *J Biol Chem*. 2003;278:11753–11759.
51. Martinez-Vicente M, Talloczy Z, Kaushik S, Massey AC, Mazzulli J, Mosharov EV, Hodara R, Fredenburg R, Wu DC, Follenzi A, Dauer W, Przedborski S, Ischiropoulos H, Lansbury PT, Sulzer D, Cuervo AM. Dopamine-modified alpha-synuclein blocks chaperone-mediated autophagy. *J Clin Invest*. 2008;118:777–788.
52. Cookson MR, van der Brug M. Cell systems and the toxic mechanism(s) of alpha-synuclein. *Exp Neurol*. 2008; 209:5–11.
53. Olanow CW, Perl DP, DeMartino GN, McNaught K. Lewy-body formation is an aggresome-related process: a hypothesis. *Lancet Neurol*. 2004;3:496–503.
54. Klucken J, Shin Y, Masliah E, Hyman BT, McLean PJ. Hsp70 reduces alpha-synuclein aggregation and toxicity. *J Biol Chem*. 2004;279:25497–25502.
55. Auluck PK, Chan E, Trojanowski JQ, Lee V, Bonini NM. Chaperone suppression of alpha-synuclein toxicity in a drosophila model of Parkinson's disease. *Science*. 2002;295:865–868.
56. McLean PJ, Klucken J, Shin Y, Hyman BT. Geldanamycin induces Hsp70 and prevents alpha-synuclein aggregation and toxicity in vitro. *Biochem Biophys Res Commun*. 2004;321:665–669.
57. Masliah E, Rockenstein E, Adame A, et al. Effects of alpha-synuclein immunization in a mouse model of Parkinson's disease. *Neuron*. 2005;46:857–868.
58. Li J, Zhu M, Rajamani S, et al. Rifampicin inhibits alpha-synuclein fibrillation and disaggregates fibrils. *Chem Biol*. 2004;11:1513–1521.
59. Xu J, Wei C, Xu C, et al. Rifampicin protects PC12 cells against MPP+-induced apoptosis and inhibits the expression of an alpha-synuclein multimer. *Brain Res*. 2007;1139:220–225.
60. Inoue T, Hiratsuka M, Osaki M, Oshimura M. The molecular biology of mammalian SIRT proteins: SIRT2 in cell cycle regulation. *Cell Cycle*. 2007;6:1011–1018.
61. Outeiro TF, Kontopoulos E, Altmann SM, et al. Sirtuin 2 inhibitors rescue alpha-synuclein-mediated toxicity in models of Parkinson's disease. *Science*. 2007;317:516–519.
62. Okawara M, Katsuki H, Kurimoto E, et al. Resveratrol protects dopaminergic neurons in midbrain slice culture from multiple insults. *Biochem Pharmacol*. 2007;73:550–560.
63. Bodner RA, Outeiro TF, Altmann S, et al. Pharmacological promotion of inclusion formation: a therapeutic approach for Huntington's and Parkinson's diseases. *Proc Natl Acad Sci USA*. 2006;103:4246–4251.
64. Dillin A, Kelly JW. Medicine. The yin-yang of sirtuins. *Science*. 2007;317:461–462.
65. Kordower JH, Chu Y, Hauser RA, Freeman TB, Olanow CW. Parkinson's disease pathology in long-term embryonic nigral transplants in Parkinson's disease. *Nat Med*. 2008;14:504–506.
66. Kordower JH, Chu Y, Hauser RA, Olanow CW, Freeman TB. Transplanted dopaminergic neurons develop PD pathologic changes: a second case report. *Mov Disord*. 2008;23:2303–2306.
67. Li JY, Englund E, Holton JL, Soulet D, Hagell P, Lees AJ, Lashley T, Quinn NP, Rehncrona S, Björklund A, Widner H, Revesz T, Lindvall O, Brundin P. Lewy bodies in grafted neurons in subjects with Parkinson's disease suggest host-to-graft disease propagation. *Nat Med*. 2008;14:501–503.
67a. Olanow CW, Prusiner SB. Is Parkinson's disease a prion disorder. *Proc Natl Acad Sci*. 2009 (in press).
68. Lee H-J, Patel S, Lee S-J. Intravesicular localization and exocytosis of alpha-synuclein and its aggregates. *J Neurosci*. 2005;25:6016–6024.
69. Lee H-J, Suk JE, Bae EJ, Lee JH, Paik SR, Lee S-J Assembly-dependent endocytosis and clearance of extracellular alpha-synuclein. *Int J Biochem Cell Biol*. 2008;40:1835–1849.

70. Schapira AHV, Cooper JM, Dexter D, et al. Mitochondrial complex I deficiency in Parkinson's disease. *J Neurochem.* 1990;54:823–827.
71. Mizuno Y, Ohta S, Tanaka M, et al. Deficiencies in complex I subunits of the respiratory chain in Parkinson's disease. *Biochem Biophys Res Commun.* 1989;163:1450–1455.
72. Davidzon G, Greene P, Mancuso M, et al. Early-onset familial parkinsonism due to POLG mutations. *Ann Neurol.* 2006;59:859–862.
73. Luoma PT, Eerola J, Ahola S, et al. Mitochondrial DNA polymerase gamma variants in idiopathic sporadic Parkinson disease. *Neurology.* 2007;69:1152–1159.
74. Pridgeon JW, Olzmann JA, Chin LS, Li L. PINK1 Protects against oxidative stress by phosphorylating mitochondrial chaperone TRAP1. *PLoS Biol.* 2007;5:e172.
75. Alnemri ES. HtrA2 and Parkinson's disease: think PINK? *Nat Cell Biol.* 2007;9:1227–1229.
76. Strauss KM, Martins LM, Plun-Favreau H, et al. Loss of function mutations in the gene encoding Omi/HtrA2 in Parkinson's disease. *Hum Mol Genet.* 2005;14:2099–2111.
77. Clark IE, Dodson MW, Jiang C, Cao JH, Huh JR, Seol JH, Yoo SJ, Hay BA, Guo M. Drosophila pink1 is required for mitochondrial function and interacts genetically with parkin. *Nature.* 2006;441:1162–1116.
78. Park J, Lee SB, Lee S, Kim Y, Song S, Kim S, Bae E, Kim J, Shong M, Kim JM, Chung J. Mitochondrial dysfunction in *Drosophila* PINK1 mutants is complemented by parkin. *Nature.* 2006;441:1157–1161.
79. Yang Y, Gehrke S, Imai Y, Huang Z, Ouyang Y, Wang JW, Yang L, Beal MF, Vogel H, Lu B. Mitochondrial pathology and muscle and dopaminergic neuron degeneration caused by inactivation of *Drosophila* Pink1 is rescued by Parkin. *Proc Natl Acad Sci USA.* 2006;103:10793–10798.
80. Exner N, Treske B, Paquet D, et al. Loss-of-function of human PINK1 results in mitochondrial pathology and can be rescued by parkin. *J Neurosci.* 2007;27:12413–12418.
81. Yang Y, Gehrke S, Haque ME, Imai Y, Kosek J, Yang L, Beal MF, Nishimura I, Wakamatsu K, Ito S, Takahashi R, Lu B. Inactivation of *Drosophila* DJ-1 leads to impairments of oxidative stress response and phosphatidylinositol 3-kinase/Akt signaling. *Proc Natl Acad Sci USA.* 2005;102:13670–13675.
82. Clements CM, McNally RS, Conti BJ, et al. DJ-1, a cancer- and Parkinson's disease-associated protein, stabilizes the antioxidant transcriptional master regulator Nrf2. *Proc Natl Acad Sci USA.* 2006;103:15091–15096.
83. Inden M, Taira T, Kitamura Y, Yanagida T, Tsuchiya D, Takata K, Yanagisawa D, Nishimura K, Taniguchi T, Kiso Y, Yoshimoto K, Agatsuma T, Koide-Yoshida S, Iguchi-Ariga SM, Shimohama S, Ariga H. PARK7 DJ-1 protects against degeneration of nigral dopaminergic neurons in Parkinson's disease rat model. *Neurobiol Dis.* 2006;24:144–158.
84. Yokota T, Sugawara K, Ito K, Takahashi R, Ariga H, Mizusawa H. Down regulation of DJ-1 enhances cell death by oxidative stress, ER stress, and proteasome inhibition. *Biochem Biophys Res Commun.* 2003;312:1342–1348.
85. Junn E, Taniguchi H, Jeong BS, et al. Interaction of DJ-1 with Daxx inhibits apoptosis signal-regulating kinase 1 activity and cell death. *Proc Natl Acad Sci USA.* 2005;102:9691–9696.
86. Shendelman S, Jonason A, Martinat C, Leete T, Abeliovich A. DJ-1 is a redox-dependent molecular chaperone that inhibits alpha-synuclein aggregate formation. *PLoS Biol.* 2004;2(11):e362.
87. Hoglinger GU, Carrard G, Michel PP, et al. Dysfunction of mitochondrial complex I and the proteasome: interactions between two biochemical deficits in a cellular model of Parkinson's disease. *J Neurochem.* 2003;86:1297–1307.
88. Shamoto-Nagai M, Maruyama W, Kato Y, et al. An inhibitor of mitochondrial complex I, rotenone, inactivates proteasome by oxidative modification and induces aggregation of oxidized proteins in SH-SY5Y cells. *J Neurosci Res.* 2003;74:589–597.
89. Sullivan PG, Dragicevic NB, Deng JH, et al. Proteasome inhibition alters neural mitochondrial homeostasis and mitochondria turnover. *J Biol Chem.* 2004;279:20699–20707.
90. Chen L, Thiruchelvam MJ, Madura K, Richfield EK. Proteasome dysfunction in aged human alpha-synuclein transgenic mice. *Neurobiol Dis.* 2006;23:120–126.
91. Martin LJ, Pan Y, Price AC, Sterling W, Copeland NG, Jenkins NA, Price DL, Lee MK. Parkinson's disease alpha-synuclein transgenic mice develop neuronal mitochondrial degeneration and cell death. *J Neurosci.* 2006;26:41–50.
92. Gilks WP, Abou-Sleiman PM, Gandhi S, et al. A common LRRK2 mutation in idiopathic Parkinson's disease. *Lancet.* 2005;365:415–416.
93. Ozelius LJ, Senthil G, Saunders-Pullman R, et al. LRRK2 G2019S as a cause of Parkinson's disease in Ashkenazi Jews. *N Engl J Med.* 2006;354:424–425.
94. Lesage S, Dürr A, Tazir M, et al. LRRK2 G2019S as a cause of Parkinson's disease in North African Arabs. *N Engl J Med.* 2006;354:422–423.
95. West AB, Moore DJ, Biskup S, et al. Parkinson's disease–associated mutations in leucine-rich repeat kinase 2 augment kinase activity. *Proc Natl Acad Sci USA.* 2005;102:16842–16847.
96. Li X, Tan Y, Poulose S, Olanow CW, Huang X-Y, Yue Z. Leucine-rich repeat kinase 2 /PARK8 possesses GTPase activity that is altered in familial Parkinson's disease R1441C/G mutant. *J Neurochem.* 2007;103:238–247.
97. Smith WW, Pei Z, Jiang H, Dawson VL, Dawson TM, Ross CA. Kinase activity of mutant LRRK2 mediates neuronal toxicity. *Nat Neurosci.* 2006;9:1231–1233.
98. Jenner P. Functional models of Parkinson's disease. *Ann Neurol.* 2008;64(suppl 2):S16–S29.
98a. Olanow CW, Kordower J. Modeling Parkinson's disease. *Ann Neurol.* (in press).
99. Olanow CW, Schapira AHV, Lewitt PA, Kieburtz K, Sauer D, Olivieri G, Pohlmann H, Hubble J. TCH346 as a neuroprotective drug in Parkinson's disease: a double-blind, randomised, controlled trial. *Lancet Neurol.* 2006;5:1013–1020.
100. Leber P. Observations and suggestions on anti-dementia drug development. *Alzheimer's Dis Assoc Disord.* 1996;10(suppl 1):S31–S35.
101. Leber P. Slowing the progression of Alzheimer disease: methodologic issues. *Alzheimer's Dis Assoc Disord.* 1997;11(suppl 5):S10–S21.
102. Olanow CW, Hauser R, Jankovic J, Langston W, Lang A, Poewe W, Tolosa E, Stocchi F, Melamed E, Eyal E, Rascol O. A Randomized, double-blind, placebo-controlled, delayed start study to assess rasagiline as a disease modifying therapy in Parkinson's disease (the ADAGIO study): rationale, design, and baseline characteristics. *Mov Disord.* 2008;23:2194–2201.
103. Olanow CW, Rascol O. Early rasagiline treatment slows UPDRS decline in the ADAGIO delayed start study. *Ann Neurol.* 2008;64(suppl 12):S68.
104. Tilley BC, Palesch YY, Kieburtz K, et al. Optimizing the ongoing search for new treatments for Parkinson disease: using futility designs. *Neurology.* 2006;66:628–633.

10 | Schizophrenia and other psychiatric illnesses

10.1 Dopamine Dysfunction in Schizophrenia

ANISSA ABI-DARGHAM, MARK SLIFSTEIN, LARRY KEGELES, AND MARC LARUELLE

INTRODUCTION

Schizophrenia presents with multiple clinical features, ranging from positive symptoms (hallucinations, delusions, and thought disorder) to negative symptoms (social withdrawal, poverty of speech and thought, flattening of affect, and lack of motivation) and disturbances in cognitive processes (attention, working memory, verbal fluency and learning, social cognition, and executive function). Dopamine (DA) dysregulation was thought to play a role within each of these clinical dimensions.[1] In the last decade, imaging methodology has allowed testing and confirmation of these initial hypotheses, yielding evidence that striatal DA is increased, and cortical DA transmission is altered. Furthermore the studies indicated a direct relationship between striatal DA excess and the positive symptoms of the illness as well as the magnitude and speed of their response to antipsychotic treatment, while cognitive and negative symptoms were related to cortical DA dysfunction. New evidence from both animal studies and studies in prodromal patients suggests that both sets of symptoms may emerge in relation to the striatal dopaminergic excess, the mechanisms of which are not well understood. We will first describe the evidence derived from imaging studies using measures of cortical and subcortical dopaminergic parameters and then speculate on the cellular significance of the imaging findings. We will then describe the information gained from animal models regarding regulation of DA function by other transmitters and the circuits that may be involved, possibly leading to the dopaminergic phenotype. Finally, we will emphasize the need for translational studies to be able to understand the effects of early dopaminergic dysregulation on the function of the relevant circuits and how these effects may mediate the various symptom domains.

PREFRONTAL CORTICAL D1 RECEPTORS

Preclinical studies have documented the importance of DA function in the prefrontal cortex (PFC) for cognitive processes (for review, see[2,3]). This important role has been recently confirmed in humans by the repeated observation that carriers of the high-activity allele of catechol-O-methyltransferase (COMT), an enzyme involved in DA metabolism, display lower performance in various cognitive tasks compared to carriers of the allele associated with higher concentrations of DA in PFC (for review, see[4] and Chapter 4.4 in this volume). Furthermore, it has been suggested that DA is decreased in the dorsolateral prefrontal cortex (DLPFC) in schizophrenia,[5] and one postmortem study has found a decrease in tyrosine hydroxylase immunolabeling in DLPFC of patients with schizophrenia.[6] The D1 receptor is the main mediator of DA effects in PFC and is present at levels that allow quantification with imaging. Three published positron emission tomography (PET) studies of prefrontal D1 receptor availability in patients with schizophrenia have yielded discrepant results. Two studies were performed with the D1 radiotracer [^{11}C]SCH 23390. The first reported decreased [^{11}C]SCH 23390 binding potential in the PFC,[7] and the other reported no change.[8] One study was performed with [^{11}C]NNC 112.[9] It reported increased [^{11}C]NNC 112 binding potential in the DLPFC and no change in other regions of the PFC, such as the medial prefrontal cortex (MPFC) or the orbitofrontal cortex. In patients with schizophrenia, increased [^{11}C]NNC 112 binding in the DLPFC was predictive of poor performance on a working memory task.[9] A parallel relationship was found between the decrease in PFC [^{11}C]SCH 23390 binding potential in one study[7] and the severity of the cognitive deficit, suggesting that although the two radiotracers detect different types of alterations, both sets of alterations relate to cognitive impairment and may reflect a common underlying deficit in DA.

In an effort to understand the reasons for these discrepant results, and assuming that the described D1 alterations may be a consequence of an underlying deficit in DA transmission, as suggested by the decrease in tyrosine hydroxylase immunolabeling, we assessed the impact of acute and subchronic DA depletion on the in vivo binding of [¹¹C]SCH 23390 and [¹¹C]NNC 112 in rats by administering a combined regimen of reserpine and alpha-methyl-para-tyrosine.[10] Acute DA depletion in rats did not affect the in vivo binding of [¹¹C]NNC 112 but resulted in decreased in vivo binding of [³H]SCH 23390, a paradoxical response that might be related to DA depletion–induced translocation of D1 receptors from the cytoplasm to the cell surface compartment.[11–13] In contrast, chronic DA depletion achieved with daily administration of reserpine for 14 days is associated with increased in vivo [¹¹C]NNC 112 binding, presumably reflecting a compensatory up-regulation of D1 receptors. Interestingly, chronic DA depletion did not result in enhanced in vivo binding of [³H]SCH 23390, possibly as a result of opposite effects of receptor up-regulation and externalization on the binding of this tracer.

Consistent with the interpretation that up-regulation of D1 receptors measured with [¹¹C]NNC 112 reflects a dopaminergic deficit, we have observed a similar increase in human volunteers who abuse N-methyl-D-aspartate (NMDA) antagonists.[14] Preclinical studies suggest that chronic hypofunction of NMDA receptors is associated with dysregulated prefrontal DA function (for review, see[15,16]). Furthermore, in nonhuman primates, chronic intermittent exposure to the NMDA antagonist MK-801 resulted in decreased performance on prefrontal tasks, decreased extracellular levels of DA in the DLPFC, and an increase in [¹¹C]NNC 112 BP in the DLPFC.[17,18] Together, these preclinical data suggest that NMDA dysfunction might lead to decreased prefrontal DA activity and increased D1 receptor availability, all three dysregulations contributing to deficits in working memory.

Finally, we recently compared [¹¹C]NNC 112 binding in healthy volunteers homozygous for the val[158] allele compared to met[158] carriers of COMT.[19] Subjects were otherwise matched for parameters known to affect [¹¹C]NNC 112 binding. Subjects with val/val alleles had significantly higher cortical [¹¹C]NNC 112 binding compared to met carriers but did not differ in striatal binding. These results confirm the prominent role of COMT in regulating DA transmission in cortex but not striatum and the reliability of [¹¹C]NNC 112 as a marker for low DA tone, as previously suggested by studies in patients with schizophrenia.

Summary and Functional Implications

Based on the differential effects of DA depletion on the different D1 radiotracers, one is justified in examining the results obtained in different conditions with one tracer. Doing so with [¹¹C]NNC 112, we detect a relatively consistent pattern of up-regulation, modest in magnitude but statistically significant throughout conditions, showing increased cortical D1 in schizophrenia in a DA depletion rat model and in the human ketamine user model of NMDA dysfunction. The functional meaning of this up-regulation is subject to speculation and has different therapeutic implications. A most likely interpretation, supported by the rat model of DA depletion, is that D1 increases represent a compensatory up-regulation in response to a chronic deficiency in DA and a chronic hypostimulation of D1 receptors. This explanation is congruent with all the clinical data reviewed above. The hypothesis of D1 hypostimulation suggests that the relationship described between D1 up-regulation and poor performance on working memory tasks, also present for severity of negative symptoms (A. Abi-Dargham, unpublished observations), implies that the cognitive deficit and negative symptoms are mediated to some extent by a lack of cortical D1 stimulation. This can be tested with the use of D1 agonists in conjunction with antipsychotic treatment as a therapeutic enhancement strategy. Many challenges exist in terms of implementing this strategy in the treatment of cognitive deficits in schizophrenia, as the availability of D1 agonists is limited, and the appropriate level and mode of D1 stimulation, as well as the corresponding D1 occupancy, are all unclear at this point. Recently, an investigational drug, DAR100, with poor bioavailability, has been tested in patients with schizophrenia by subcutaneous administration. Initial studies showed its safety and the absence of negative side effects.[20] Further testing is currently underway to test its efficacy against cognitive impairment. Some preclinical studies suggested long-lasting effects after a single administration of DAR100 in antipsychotic-induced cognitive impairment in nonhuman primates.[21,22] An additional benefit of D1 stimulation is the potential to enhance NMDA transmission in schizophrenia, as these two systems exhibit many synergies.[12,23,24]

As an alternative to the hypothesis of D1 hypostimulation in schizophrenia, the data may suggest an inverted U-shaped curve in schizophrenia, with D1 stimulation oscillating between low levels at baseline and superstimulation under conditions of stress-evoked DA release, because of the increased expression of the receptors suggested by at least some of the imaging studies. Direct measurement of DLPFC DA release, which is now feasible using high-affinity D2 tracers such as [¹¹C]FLB 457,[25] is needed to address more definitively the issue of cortical levels of DA in schizophrenia (Fig. 10.1.1).

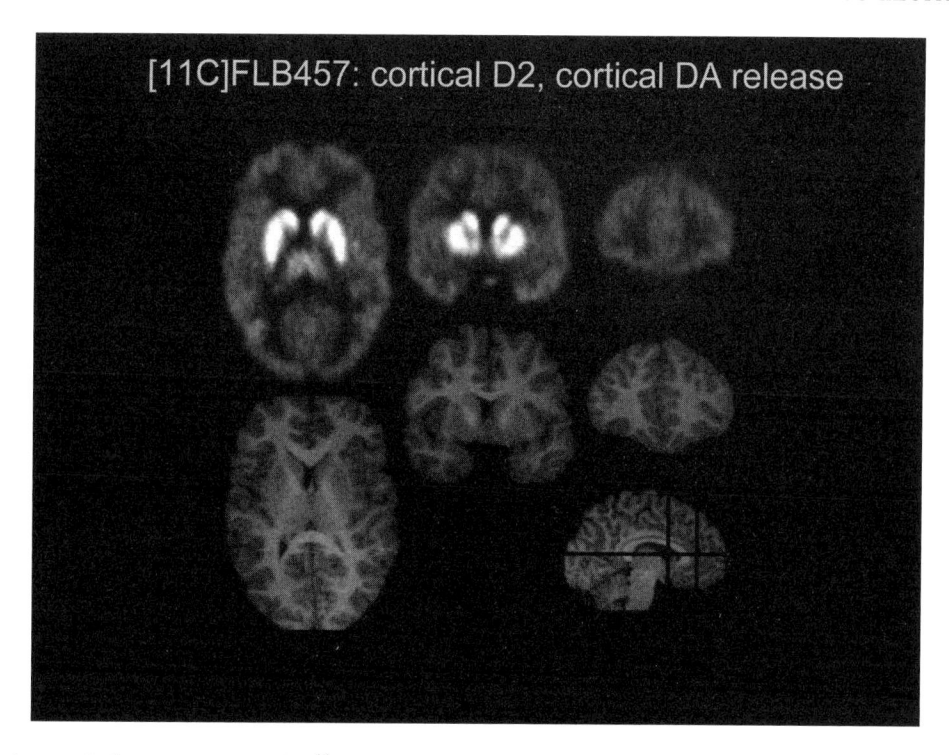

FIGURE 10.1.1. Imaging cortical D2 receptors with [^{11}C] FLB 457 scan: here, a scan under baseline conditions with coregistered (SPGR, Spoiled gradient recalled; MRI, Magnetic resonance imaging). Shown are a transverse slice (left) at the level of the striatum and thalamus and coronal slices at the level of the striatum (center) and frontal cortex (right). The sagittal slice (bottom right, not to scale) has lines through the three slice levels. Note the pituitary gland in the center coronal slice. (See Color Plate 10.1.1.)

INCREASED STRIATAL DA AND POSITIVE SYMPTOMS

Positron emission tomography imaging studies provide reliable and convergent evidence for increased striatal DA transmission in schizophrenia, particularly at D2 receptors, and evidence that this phenotype is related to the positive symptoms and to their response to antipsychotics. Furthermore, these studies have recently demonstrated that the dorsal caudate, within the associative striatum, is the area most affected.[26,27]

The evidence for excessive D2 stimulation by DA in the striatum, and in particular the associative striatum, derives from four lines of investigation:

1. Striatal D2 receptors: A meta-analysis of 17 imaging studies comparing D2 receptor parameters in patients with schizophrenia and controls[28] revealed a small (12%) but significant elevation of striatal D2 receptors in untreated patients with schizophrenia. No alterations in striatal D1 receptors were reported.[7–9]
2. Striatal DOPA decarboxylase activity: Studies of striatal DA synthesis estimated via the activity of the enzymatic step involving DOPA decarboxylase (amino acid decarboxylase, AADC) in patients with schizophrenia compared to controls using [^{18}F]DOPA or [^{11}C]DOPA contributed importantly to the understanding of the DA alteration in the illness. Six studies reported increased accumulation of DOPA in the striatum of patients with schizophrenia,[29–36] one reported no change,[29] and one reported reduced uptake.[30] Poor prefrontal activation was related to elevated [^{18}F]DOPA accumulation in the striatum, adding evidence to the link between cortical and subcortical dysfunction in schizophrenia.[35] Grunder et al. reported a decrease in [^{18}F]DOPA uptake following subchronic treatment with haloperidol[37] in patients with schizophrenia, suggesting that chronic neuroleptic administration tends to decrease AADC activity and hence DA synthesis. Finally, a very recent study showed that increased [^{18}F]DOPA uptake precedes the onset of schizophrenia, is located in the associative striatum in a magnitude similar to that in patients with schizophrenia, and relates to positive like symptoms as well as verbal fluency deficits.[27]

3. Striatal amphetamine-induced DA release is increased: Three studies[38–40] showed that the amphetamine-induced decrease in [^{11}C]raclopride or [^{123}I]IBZM binding, a validated measure of DA release, is larger in untreated patients with schizophrenia compared to well-matched controls. These studies showed that the increase in DA response to amphetamine is observed in never-treated patients; is related to the transient induction or worsening of positive symptoms by amphetamine; is larger in patients experiencing an episode of illness exacerbation; and is unrelated to stress.[41] Furthermore, it is present in patients with schizotypal personality disorders,[42] albeit to a smaller extent, suggesting that at least in part, it is mediated by genetic factors, and it is absent in nonpsychotic subjects with unipolar depression,[43] showing specificity to psychosis and schizophrenia spectrum disorders.

4. Baseline occupancy of striatal D2 receptors by DA: This can be measured using a DA depletion paradigm. A D2/3 scan is obtained before and 48 hr after administration of alpha-methyl-para-tyrosine (α–MPT). α–MPT inhibits tyrosine hydroxylase, the rate-limiting step in DA synthesis. After 48 hr substantial DA depletion is obtained, and the postdepletion scan shows higher binding potential (BP) as a reflection of a certain proportion of D2/3 receptors that were occupied by DA in the first scan. Two studies reported higher occupancy of D2 receptors by DA in patients with schizophrenia, that is, a larger α–MPT effect,[44,45] and a better response of positive symptoms to antipsychotics at 6 weeks in patients with the highest D2 occupancy by DA. The data are consistent with higher synaptic DA levels in patients with schizophrenia if one assumes normal affinity of D2 receptors for DA. Recently, D2/3 agonist radiotracers became available that bind only to the receptors that are in the high-affinity state, not those in the low-affinity state for DA. By using these tracers, one can compare the proportion of receptors that are in the high-affinity state. A recent study showed no differences in the proportion of D2 receptors in the high-affinity state between patients with schizophrenia and matched controls, as measured by the binding of a D2/3 agonist tracer [^{11}C]PHNO,[46] supporting the assumption of no changes in the affinity of D2 to DA in schizophrenia. Furthermore, these data shows that treatment response is driven by DA dysregulation, an observation consistent with the fact that all antipsychotics lower D2 stimulation.

The most recent study suggests that in patients with schizophrenia, striatal dopaminergic hyperfunction is predominant in the precommissural dorsal caudate (preDCA) within the anterior striatum.[45] The first study measuring baseline DA activity with a depletion paradigm used [^{123}I]IBZM and single photon emission computed tomography (SPECT)[44] did not allow exploration of regional differences across the striatum due to the low resolution of the SPECT scanner. The second study used [^{11}C]raclopride and PET, and a similar depletion paradigm, to measure the in vivo occupancy of D2 receptors by DA in subregions of the striatum in 18 untreated patients with schizophrenia and 18 matched controls by comparing D2 receptor availability before and during acute DA depletion. Based on the cortical inputs to the striatum, according to Haber et al.'s neuroanatomical findings,[47–49] we examined D2 receptor occupancy in the limbic, associative, and sensorimotor striatum. Acute DA depletion resulted in a larger increase in D2 receptor availability in associative rather than limbic regions of the striatum, and the anterior dorsal caudate was most affected.[45] This result is in agreement with the finding in patients with prodromal schizophrenia.[27] As the preDCA receives a prominent input from the DLPFC, these convergent observations further suggest that elevated subcortical DA function may adversely affect DLPFC function and cognitive functions such as working memory in schizophrenia.

In summary, studies converge to demonstrate an increase in presynaptic DA synthesis, storage, and transmission in the striatum in patients with schizophrenia, in particular in the dorsal caudate within the anterior striatum. This is accompanied by an increase in D2 receptor density, leading to excessive D2 stimulation. The presence of this excessive DA function explains the good therapeutic response of positive symptoms to antipsychotics.

EXTRASTRIATAL D2 RECEPTORS

The recent availability of high-affinity D2 radiotracers allowed the study of D2 receptors in low-density regions such as the substantia nigra, thalamus, and temporal cortex in patients with schizophrenia compared to controls. Lower D2 receptor density has been described in untreated schizophrenia in the thalamus,[50–54] as well as in the midbrain,[55] temporal cortex,[50] and cingulate cortex.[54,56] One study showed an increase in D2 in the substantia nigra.[57] A very recent large study using similar methodology did not confirm any of these alterations in extrastriatal D2.[58] Additional studies are needed to resolve the discrepancies and to expand beyond measuring levels of D2 receptors to assess alterations in levels of the transmitter itself in extra-striatal areas. As in striatum, alterations in levels of DA may mask potential differences in receptor density between patients and controls.

CELLULAR IMPLICATIONS

As reviewed above, there is substantial evidence for DA dysregulation in schizophrenia. Dopamine synthesis or storage capacity is increased (F-DOPA studies) and DA release is increased (amphetamine studies), leading to higher occupancy of the D2 receptors (α–MPT studies), which are also up-regulated (D2 studies). This increase is highest in the dorsal caudate, an area of projection of the DLPFC and of the orbitofrontal cortex (OFC)[59] (Fig. 10.1.2); thus, the cortical area is crucial not only for cognitive processing but also for integration across emotional and cognitive domains. Furthermore the increase is present in spectrum disorder patients and precedes onset of the illness.

Physiological Meaning of the Imaging Measures

Midbrain DA neurons can fire in two modes: tonically and in bursts. Tonic firing contributes to basal DA tone, while burst firing produces *phasic* release, which reaches much higher levels. Because the dopamine transporter (DAT) is activated mostly by high levels of DA, basal tonically released DA diffuses out of the synapse and can be measured with microdialysis, while phasically released DA affects mostly intrasynaptic levels and can be detected with microdialysis only in the presence of a DAT inhibitor.[60] Since the magnitude of the DA increases measured with microdialysis or by the magnitude of displacement of D2 radiotracers in response to

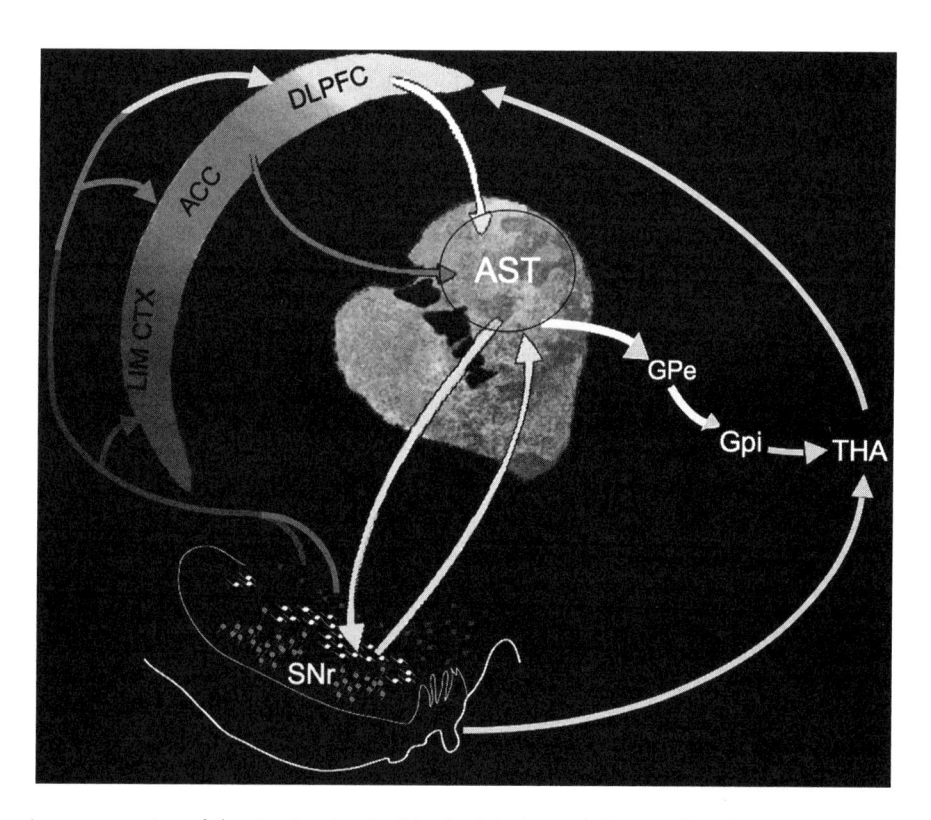

FIGURE 10.1.2. Schematic representation of the circuitry involved in the DA dysregulation in schizophrenia. Cortical DA dysfunction and DA hyperactivity at D2 receptors in striatum (left), more specifically preDCA in the associative striatum (AST) (circle), may be linked. The anatomical substrates and circuits that may mediate this relationship are illustrated in this figure. ACC, anterior cingulate prefrontal cortex; DLPFC, dorsolateral prefrontal cortex; GPe, globus pallidum external; GPi, globus pallidum internal; LIM, LIMBIC; CTX; SNR, substantia nigra pars reticulta; THA thalamus. *Source*: Adapted from[59]. (See Color Plate 10.1.2.)

different challenge drugs does not correlate, it is assumed that the imaging measures of striatal DA transmission are largely measures of intrasynaptic phasically released DA (for review, see[13]). Drugs that block the DAT, such as amphetamine and GBR12909, increase microdialysis-measured DA more than those that do not block the DAT, such as ketanserin and nicotine, yet the range of D2 radiotracer displacement is similar for all, suggesting that it is affected only by changes in intrasynaptic DA. Thus, the imaging measures of DA in the stimulated (obtained with the amphetamine challenge) or baseline condition (obtained with the α-MPT depletion paradigm) are both measures of intrasynaptic or phasic release of DA. No information on the basal release of DA is available, which relates to extrasynaptic DA levels in the extracellular milieu, an area not accessible to imaging. In support of this analysis is our finding of a significant correlation between the amphetamine-induced release of DA and the baseline intrasynaptic occupancy of striatal D2 receptors in drug-naive patients with schizophrenia but not in controls. This suggests that findings with these different paradigms reflect the same alteration in presynaptic DA transmission. This alteration, probably due to an increase in the activity of midbrain DA neurons, can lead to increased baseline occupancy of D2/3 receptors, revealed with the α-MPT paradigm, as well as a higher DA synthesis and storage capacity, revealed with the amphetamine paradigm.[61]

Animal Models

Animal models can inform us mechanistically about potential pathogenic contributions to the DA phenotype observed in imaging studies. Most accepted animal models in the field of schizophrenia research show dysregulated subcortical DA transmission. Models involving hippocampal lesions[62]; a model of disruption of cortical development by a methylating agent administered prenatally, methylazoxymethanol acetate (MAM),[63,64] which leads to ventral hippocampal overdrive; and models of immunological interventions[65] all lead to alterations in subcortical DA. Furthermore, with the use of PET imaging in small animals, these models can be used to test the presence of alterations in DA imaging parameters similar to those described in patients, thus clarifying the correspondence between the imaging measures and the cellular alterations producing them. This type of translational research is an important development that can expand in the future to define the effects of specific genetic interventions on DA transmission and on the imaging measurements of DA function. By understanding the pathogenic mechanisms, we can better understand the pathophysiology and develop more focused and even preventive therapeutic approaches.

Corticostriatal Circuits

Cortical and striatal dopaminergic dysfunctions are likely to be linked. It has been shown that selective lesions of DA neurons in the PFC are associated with increased striatal DA release in rats[66–69] and primates,[70] while augmentation of DA or GABA activity in the PFC inhibits striatal DA release[71,72] (for a review, see[73]). More recently, the opposite effect was shown in a mouse model of striatal D2 overexpression associated with cognitive impairments in tasks of working memory and behavioral flexibility, as well as altered PFC DA levels, rates of DA turnover, and activation of prefrontal D1 receptors.[74] This set of observations suggests that striatal and cortical DA alterations in schizophrenia are very likely linked to each other, regardless of whether the primary dopaminergic problem is cortical (top-down) or striatal (bottom-up) dysfunction.

The mechanisms by which this opposite modulation may occur are unknown and may involve various factors: (1) Within the striatum: striatal D2 opposes the glutamate-mediated cortical flow of information along the pyramidal corticostriatal projections.[75,76] (2) Prefrontal cortical DA exerts negative feedback on ventral tegmentum (VTA) activity. It has been shown that selective lesions of DA neurons in the PFC are associated with increased striatal DA release in rats[66–69] and primates.[70] On the other hand, augmentation of DA or GABA activity in the PFC inhibits striatal DA release.[71,72] Regardless of the exact mechanisms and pathways, the converging evidence suggests that the flow of information in cortico-striato-thalamo-cortical loops can be impaired in schizophrenia at the cortical or at the striatal level, with resulting effects on the overall function of the integrated circuit (Fig. 10.1.2).

SUMMARY AND FUTURE DIRECTIONS

Studies of cortical D1 receptors have suggested alterations related to poor cognition and negative symptoms. Negative symptoms are heterogeneous and can be thought of as related to a cortical cognitive dysfunction (thought and language deficits) versus a limbic reward-related dysfunction (anhedonia, deficits in motivation, affective flattening). The role for DA release in the striatum in mediating some aspects of reward functions, and the alterations of reward processes in schizophrenia,[77] suggest that a subset of negative symptoms may relate to alterations in ventrostriatal DA transmission. Future studies aimed at characterizing multidimensional factors within the negative symptoms domain as they relate to DA transmission in cortical versus striatal substructures are needed to test this conceptualization. Furthermore, a more direct characterization of cortical DA and of extrastriatal DA transmission in schizophrenia is needed, as the evidence for the cortical deficit exists largely by inference. These studies are now feasible with the development of high-affinity benzamide radiotracers that allow visualization of D2 receptors in low-density areas of the brain and are sensitive to acute changes in DA tone.[25]

The dopaminergic phenotype in the striatum can help to explain and to bring together many of the different elements of pathology that have been documented in schizophrenia, as the striatum performs an essential integrative function, by receiving input from the cortex and the hippocampus, two areas of pathology in schizophrenia,[78–80] and by modulating DA midbrain neurons, projecting indirectly to the cortex, thus affecting the input to the cortex. Testing for associations in patients between these three areas of pathology—cortical, striatal, and hippocampal—using multimodal imaging may be a first step in understanding their relatedness. Testing for their occurrence in prodromal patients can address the issue of which area of pathology may be primary or secondary. It is interesting to note that the initial studies in prodromal patients suggest that dyregulation of striatal dopamine function is already present and is similar in localization and magnitude to that observed in patients with schizophrenia.[27] A better understanding of the pathogenesis and mechanisms of the DA dysfunction in schizophrenia will lead to better treatment development and potentially preventive strategies.

REFERENCES

1. Weinberger DR. Implications of the normal brain development for the pathogenesis of schizophrenia. *Arch Gen Psychiatry.* 1987;44:660–669.
2. Goldman-Rakic P. Working memory dysfunction in schizophrenia. *J Neuropsychiatry Clin Neurosci.* 1994;6(4):348–357.
3. Goldman-Rakic PS, Muly EC 3rd, Williams GV. D(1) receptors in prefrontal cells and circuits. *Brain Res Brain Res Rev.* 2000;31(2–3):295–301.
4. Goldberg TE, Weinberger DR. Genes and the parsing of cognitive processes. *Trends Cogn Sci.* 2004;8(7):325–335.
5. Weinberger DR, Berman KF, Chase TN. Mesocortical dopaminergic function and human cognition. *Ann NY Acad Sci.* 1988;537:330–338.
6. Akil M, Edgar CL, Pierri JN, Casali S, Lewis DA. Decreased density of tyrosine hydroxylase–immunoreactive axons in the entorhinal cortex of schizophrenic subjects. *Biol Psychiatry.* 2000;47(5):361–370.
7. Okubo Y, Suhara T, Suzuki K, et al. Decreased prefrontal dopamine D1 receptors in schizophrenia revealed by PET. *Nature.* 1997;385(6617):634–636.
8. Karlsson P, Farde L, Halldin C, Sedvall G. PET study of D(1) dopamine receptor binding in neuroleptic-naive patients with schizophrenia. *Am J Psychiatry.* 2002;159(5):761–767.
9. Abi-Dargham A, Mawlawi O, Lombardo I, et al. Prefrontal dopamine D1 receptors and working memory in schizophrenia. *J Neurosci.* 2002;22(9):3708–3719.
10. Guo, N. N., D. R. Hwang, et al. (2003). "Dopamine depletion and in vivo binding of PET D-1 receptor radioligands: Implications for imaging studies in schizophrenia." Neuropsychopharmacology 28(9):1703–1711.
11. Dumartin B, Jaber M, Gonon F, Caron MG, Giros B, Bloch B. Dopamine tone regulates D1 receptor trafficking and delivery in striatal neurons in dopamine transporter–deficient mice. *Proc Natl Acad Sci USA.* 2000;97(4):1879–1884.
12. Scott L, Kruse MS, Forssberg H, Brismar H, Greengard P, Aperia A. Selective up-regulation of dopamine D1 receptors in dendritic spines by NMDA receptor activation. *Proc Natl Acad Sci USA.* 2002;99(3):1661–1664.
13. Laruelle M. Imaging synaptic neurotransmission with in vivo binding competition techniques: a critical review. *J Cereb Blood Flow Metab.* 2000;20(3):423–451.
14. Narendran R, Frankle WG, Keefe R, et al. Altered prefrontal dopaminergic function in chronic recreational ketamine users. *Am J Psychiatry.* 2005;162(12):2352–2359.
15. Jentsch JD, Roth RH. The neuropsychopharmacology of phencyclidine: from NMDA receptor hypofunction to the dopamine hypothesis of schizophrenia. *Neuropsychopharmacology.* 1999;20(3):201–225.
16. Jentsch JD, Redmond DE Jr, Elsworth JD, Taylor JR, Youngren KD, Roth RH. Enduring cognitive deficits and cortical dopamine dysfunction in monkeys after long-term administration of phencyclidine. *Science.* 1997;277(5328):953–955.
17. Kakiuchi T, Nishiyama S, Sato K, Ohba H, Nakanishi S, Tsukada H. Effect of MK801 on dopamine parameters in the monkey brain. *Neuroimage.* 2001;16:110P.
18. Tsukada H, Nishiyama S, Fukumoto D, Sato K, Kakiuchi T, Domino EF. Chronic NMDA antagonism impairs working memory, decreases extracellular dopamine, and increases D1 receptor binding in prefrontal cortex of conscious monkeys. *Neuropsychopharmacology.* 2005;30(10):1861–1869.
19. Slifstein M, Kolachana B, Simpson EH, et al. COMT genotype predicts cortical-limbic D1 receptor availability measured with [(11)C]NNC112 and PET. *Mol Psychiatry.* 2008;13(8):821–827.
20. George MS, Molnar CE, Grenesko EL, et al. A single 20 mg dose of dihydrexidine (DAR-0100), a full dopamine D1 agonist, is safe and tolerated in patients with schizophrenia. *Schizophr Res.* 2007;93(1–3):42–50.
21. Castner SA, Williams GV, Goldman-Rakic PS. Reversal of antipsychotic-induced working memory deficits by short-term dopamine D1 receptor stimulation. *Science.* 2000;287(5460):2020–2022.
22. Castner SA, Goldman-Rakic PS. Enhancement of working memory in aged monkeys by a sensitizing regimen of dopamine D1 receptor stimulation. *J Neurosci.* 2004;24(6):1446–1450.
23. Flores-Hernandez J, Cepeda C, Hernandez-Echeagaray E, et al. Dopamine enhancement of NMDA currents in dissociated medium-sized striatal neurons: role of D1 receptors and DARPP-32. *J Neurophysiol.* 2002;88(6):3010–3020.

24. Dunah AW, Standaert DG. Dopamine D1 receptor–dependent trafficking of striatal NMDA glutamate receptors to the postsynaptic membrane. *J Neurosci.* 2001;21(15):5546–5558.
25. Narendran, R., W. G. Frankle, et al. "Positron Emission Tomography Imaging of Amphetamine-Induced Dopamine Release in the Human Cortex: A Comparative Evaluation of the High Affinity Dopamine D-2/3 Radiotracers [C-11]FLB 457 and [C-11]Fallypride." *Synapse* 2009 63(6):447–461.
26. Kegeles L, Frankle W, Gil R, et al. Schizophrenia is associated with increased synaptic dopamine in associative rather than limbic regions of the striatum: implications for mechanisms of action of antipsychotic drugs. *J Nucl Med.* 2006(47):139P.
27. Howes OD, Montgomery AJ, Asselin MC, et al. Elevated striatal dopamine function linked to prodromal signs of schizophrenia. *Arch Gen Psychiatry.* 2009;66(1):13–20.
28. Weinberger D, Laruelle M. Neurochemical and neuropharmacological imaging in schizophrenia. In: Davis KL, Charney DS, Coyle J, Nemeroff C, eds. *Neuropsychopharmacology—The Fifth Generation of Progress*: Lippincott Williams and Wilkins; 2001:Philadelphia, PA, 833–855.
29. Dao-Castellana MH, Paillere-Martinot ML, Hantraye P, et al. Presynaptic dopaminergic function in the striatum of schizophrenic patients. *Schizophr Res.* 1997;23(2):167–174.
30. Elkashef AM, Doudet D, Bryant T, Cohen RM, Li SH, Wyatt RJ. 6-(18)F-DOPA PET study in patients with schizophrenia. Positron emission tomography. *Psychiatry Res.* 2000;100(1):1–11.
31. Hietala J, Syvalahti E, Vilkman H, et al. Depressive symptoms and presynaptic dopamine function in neuroleptic-naive schizophrenia. *Schizophr Res.* 1999;35(1):41–50.
32. Hietala J, Syvalahti E, Vuorio K, et al. Presynaptic dopamine function in striatum of neuroleptic-naive schizophrenic patients. *Lancet.* 1995;346(8983):1130–1131.
33. Lindstrom LH, Gefvert O, Hagberg G, et al. Increased dopamine synthesis rate in medial prefrontal cortex and striatum in schizophrenia indicated by L-(beta-11C) DOPA and PET. *Biol Psychiatry.* 1999;46(5):681–688.
34. McGowan S, Lawrence AD, Sales T, Quested D, Grasby P. Presynaptic dopaminergic dysfunction in schizophrenia: a positron emission tomographic [18F]fluorodopa study. *Arch Gen Psychiatry.* 2004;61(2):134–142.
35. Meyer-Lindenberg A, Miletich RS, Kohn PD, et al. Reduced prefrontal activity predicts exaggerated striatal dopaminergic function in schizophrenia. *Nat Neurosci.* 2002;5(3):267–271.
36. Reith J, Benkelfat C, Sherwin A, et al. Elevated dopa decarboxylase activity in living brain of patients with psychosis. *Proc Natl Acad Sci USA.* 1994;91(24):11651–11654.
37. Grunder G, Vernaleken I, Muller MJ, et al. Subchronic haloperidol downregulates dopamine synthesis capacity in the brain of schizophrenic patients in vivo. *Neuropsychopharmacology.* 2003;28(4):787–794.
38. Laruelle M, Abi-Dargham A, van Dyck CH, et al. Single photon emission computerized tomography imaging of amphetamine-induced dopamine release in drug-free schizophrenic subjects. *Proc Natl Acad Sci USA.* 1996;93(17):9235–9240.
39. Abi-Dargham A, Gil R, Krystal J, et al. Increased striatal dopamine transmission in schizophrenia: confirmation in a second cohort. *Am J Psychiatry.* 1998;155(6):761–767.
40. Breier A, Su TP, Saunders R, et al. Schizophrenia is associated with elevated amphetamine-induced synaptic dopamine concentrations: evidence from a novel positron emission tomography method. *Proc Natl Acad Sci USA.* 1997;94(6):2569–2574.
41. Laruelle M, Abi-Dargham A, Gil R, Kegeles L, Innis R. Increased dopamine transmission in schizophrenia: relationship to illness phases. *Biol Psychiatry.* 1999;46(1):56–72.
42. Abi-Dargham A, Kegeles L, Zea-Ponce Y, et al. Amphetamine-induced dopamine release in patients with schizotypal personality disorders studied by SPECT and [123I]IBZM. *Biological psychiatry* 2004 55(10): 1001–1006.
43. Parsey RV, Oquendo MA, Zea-Ponce Y, et al. Dopamine D(2) receptor availability and amphetamine-induced dopamine release in unipolar depression. *Biol Psychiatry.* 2001;50(5):313–322.
44. Abi-Dargham A, Rodenhiser J, Printz D, et al. Increased baseline occupancy of D2 receptors by dopamine in schizophrenia. *Proc Natl Acad Sci USA.* 2000;97(14):8104–8109.
45. Kegeles L, Frankle W, Gil R, et al. Schizophrenia is associated with increased synaptic dopamine in associative rather than limbic regions of the striatum: implications for mechanisms of action of antipsychotic drugs. *J Nucl Med.* 2006(47):139P.
46. Graff-Guerrero A, Mizrahi R, Agid O, et al. The dopamine D(2) receptors in high-affinity state and D(3) receptors in schizophrenia: a clinical [(11)C]-(+)-PHNO PET study. *Neuropsychopharmacology.* 2008; 34(4):1078–1086.
47. Haber SN, Fudge JL, McFarland NR. Striatonigrostriatal pathways in primates form an ascending spiral from the shell to the dorsolateral striatum. *J Neurosci.* 2000;20(6):2369–2382.
48. Haber SN, McFarland NR. The concept of the ventral striatum in nonhuman primates. *Ann NY Acad Sci.* 1999;877:33–48.
49. Haber SN, Ryoo H, Cox C, Lu W. Subsets of midbrain dopaminergic neurons in monkeys are distinguished by different levels of mRNA for the dopamine transporter: comparison with the mRNA for the D2 receptor, tyrosine hydroxylase and calbindin immunoreactivity. *J Comp Neurol.* 1995;362(3):400–410.
50. Tuppurainen H, Kuikka J, Viinamaki H, Husso-Saastamoinen M, Bergstrom K, Tiihonen J. Extrastriatal dopamine D2/3 receptor density and distribution in drug-naive schizophrenic patients. *Mol Psychiatry.* 2003;8(4):453–455.
51. Talvik M, Nordstrom AL, Olsson H, Halldin C, Farde L. Decreased thalamic D2/D3 receptor binding in drug-naive patients with schizophrenia: a PET study with schizophrenia: a PET study with [11C]FLB 457. *Int J Neuropsychopharmacol.* 2003;6(4):361–370.
52. Talvik M, Nordstrom AL, Okubo Y, et al. Dopamine D2 receptor binding in drug-naive patients with schizophrenia examined with raclopride-C11 and positron emission tomography. *Psychiatry Res.* 2006;148(2–3):165–173.
53. Yasuno F, Suhara T, Okubo Y, et al. Low dopamine D(2) receptor binding in subregions of the thalamus in schizophrenia. *Am J Psychiatry.* 2004;161(6):1016–1022.
54. Suhara T, Okubo Y, Yasuno F, et al. Decreased dopamine D2 receptor binding in the anterior cingulate cortex in schizophrenia. *Arch Gen Psychiatry.* 2002;59(1):25–30.
55. Tuppurainen H, Kuikka JT, Laakso MP, Viinamaki H, Husso M, Tiihonen J. Midbrain dopamine D2/3 receptor binding in schizophrenia. *Eur Arch Psychiatry Clin Neurosci.* 2006;256(6):382–387.
56. Glenthoj BY, Mackeprang T, Svarer C, et al. Frontal dopamine D(2/3) receptor binding in drug-naive first-episode schizophrenic patients correlates with positive psychotic symptoms and gender. *Biol Psychiatry.* 2006;60(6):621–629.
57. Kessler, R., N. Woodward, et al. Dopamine D2 Receptor Levels in Striatum, Thalamus, Substantia Nigra, Limbic Regions, and Cortex in Schizophrenic Subjects. *Biological psychiatry* 2009 65(12): 1024–1031.
58. Kegeles LS, Slifstein M, Xu X, Hackett E, Castrillon J, Bae S-A, Urban N, Laruelle M, Abi-Dargham A: [18F]Fallypride PET

assessment of D2/D3 receptor binding in schizophrenia. *J Nucl Med* 2008;49:36P.
59. Haber SN, Kim KS, Mailly P, Calzavara R. Reward-related cortical inputs define a large striatal region in primates that interface with associative cortical connections, providing a substrate for incentive-based learning. *J Neurosci.* 2006;26(32): 8368–8376.
60. Grace AA. Phasic versus tonic dopamine release and the modulation of dopamine system responsivity: a hypothesis for the etiology of schizophrenia. *Neuroscience.* 1991;41(1):1–24.
61. Abi-Dargham A, Van de Giessen E, Slifstein M, Kegeles L, Laruelle M. "Baseline and Amphetamine-Stimulated Dopamine Activity Are Related in Drug-Naïve Schizophrenic Subjects." *Biological psychiatry* 2009 65(12):1091–1093.
62. Chrapusta SJ, Egan MF, Wyatt RJ, Weinberger DR, Lipska BK. Neonatal ventral hippocampal damage modifies serum corticosterone and dopamine release responses to acute footshock in adult Sprague-Dawley rats. *Synapse.* 2003;47(4):270–277.
63. Lodge DJ, Grace AA. Aberrant hippocampal activity underlies the dopamine dysregulation in an animal model of schizophrenia. *J Neurosci.* 2007;27(42):11424–11430.
64. Moore H, Jentsch JD, Ghajarnia M, Geyer MA, Grace AA. A neurobehavioral systems analysis of adult rats exposed to methylazoxymethanol acetate on E17: implications for the neuropathology of schizophrenia. *Biol Psychiatry.* 2006;60(3): 253–264.
65. Ozawa K, Hashimoto K, Kishimoto T, Shimizu E, Ishikura H, Iyo M. Immune activation during pregnancy in mice leads to dopaminergic hyperfunction and cognitive impairment in the offspring: a neurodevelopmental animal model of schizophrenia. *Biol Psychiatry.* 2006;59(6):546–554.
66. Pycock CJ, Kerwin RW, Carter CJ. Effect of lesion of cortical dopamine terminals on subcortical dopamine receptors in rats. *Nature.* 1980;286:74–77.
67. Bubser M, Koch M. Prepulse inhibition of the acoustic startle response of rats is reduced by 6-hydroxydopamine lesions of the medial prefrontal cortex. *Psychopharmacology.* 1994;113 (3–4):487–492.
68. Deutch A, Clark WA, Roth RH. Prefrontal cortical dopamine depletion enhances the responsiveness of the mesolimbic dopamine neurons to stress. *Brain Res.* 1990;521:311–315.
69. Thompson TL, Moss RL. In vivo stimulated dopamine release in the nucleus accumbens: modulation by the prefrontal cortex. *Brain Res.* 1995;686(1):93–98.
70. Roberts AC, Desalvia MA, Wilkinson LS, et al. 6-Hydroxydopamine lesions of the prefrontal cortex in monkeys enhance performance on an analog of the Wisconsin card sort test: possible interactions with subcortical dopamine. *J. Neurosci.* 1994;14(5 pt 1):2531–2544.
71. Kolachana BS, Saunders R, Weinberger D. Augmentation of prefrontal cortical monoaminergic activity inhibits dopamine release in the caudate nucleus: an in vivo neurochemical assessment in the rhesus monkey. *Neuroscience.* 1995;69:859–868.
72. Karreman M, Moghaddam B. The prefrontal cortex regulates the basal release of dopamine in the limbic striatum: an effect mediated by ventral tegmental area. *J Neurochem.* 1996;66:589–598.
73. Tzschentke TM. Pharmacology and behavioral pharmacology of the mesocortical dopamine system. *Prog Neurobiol.* 2001;63(3):241–320.
74. Kellendonk C, Simpson EH, Polan HJ, et al. Transient and selective overexpression of dopamine D2 receptors in the striatum causes persistent abnormalities in prefrontal cortex functioning. *Neuron.* 2006;49(4):603–615.
75. Konradi C. The molecular basis of dopamine and glutamate interactions in the striatum. *Adv Pharmacol.* 1998;42:729–733.
76. Leveque JC, Macias W, Rajadhyaksha A, et al. Intracellular modulation of NMDA receptor function by antipsychotic drugs. *J Neurosci.* 2000;20(11):4011–4020.
77. Krystal JH, D'Souza DC, Gallinat J, et al. The vulnerability to alcohol and substance abuse in individuals diagnosed with schizophrenia. *Neurotox Res.* 2006;10(3–4):235–252.
78. Heckers S, Rauch SL, Goff D, et al. Impaired recruitment of the hippocampus during conscious recollection in schizophrenia. *Nat Neurosci.* 1998;1(4):318–323.
79. Medoff DR, Holcomb HH, Lahti AC, Tamminga CA. Probing the human hippocampus using rCBF: contrasts in schizophrenia. *Hippocampus.* 2001;11(5):543–550.
80. Meyer-Lindenberg AS, Olsen RK, Kohn PD, et al. Regionally specific disturbance of dorsolateral prefrontal-hippocampal functional connectivity in schizophrenia. *Arch Gen Psychiatry.* 2005;62(4):379–386.

10.2 | Neuropharmacological Profiles of Antipsychotic Drugs

BRYAN L. ROTH AND SARAH C. ROGAN

THE FIRST ANTIPSYCHOTICS

Before the 1950s, no effective pharmacological treatments for psychosis existed. Indeed, the mainstays of medical therapy were the insulin coma and electroconvulsive therapy (ECT). Then, in the 1950s, the discovery of psychotropic drugs revolutionized psychiatric medicine.

Henri Laborit synthesized chlorpromazine in an effort to develop a superior antihistamine for use as a calming agent before surgery. He and others then realized that chlorpromazine was effective at calming patients but did so without sedating them. Labhardt was the first to administer chlorpromazine to psychotic patients, and he found that it decreased their agitation and aggression.[1] Almost immediately, chlorpromazine, followed shortly by reserpine, radically altered the approach to psychiatric care, replacing ECT as the central tenet of medical therapy. These and other similar drugs, the first-generation, or typical, antipsychotics, enabled the treatment of the positive symptoms of schizophrenia, namely, hallucinations and delusions. However, typical antipsychotics are relatively ineffective at treating the negative symptoms (e.g., anhedonia, lack of motivation, poverty of speech) of schizophrenia, or do so indirectly, through the treatment of positive symptoms.[2] Additionally, typical antipsychotics do not address the cognitive dysfunction (e.g., working memory deficits) that is so pervasive in schizophrenia, and they are associated with significant extrapyramidal side effects (EPS) and serum prolactin elevation.

The search for chlorpromazine analogs led to the discovery of clozapine in the 1950s (see[3] for review), and the so-called second-generation, or atypical, antipsychotics, in the 1980s–1990s, issued in the next era in medical therapy for schizophrenia.[3,4] Atypical drugs, characterized by a different receptor profile than typical drugs (see below), cause minimal EPS and serum prolactin elevations. Clozapine is the current gold standard of antipsychotic drugs and demonstrates superior clinical efficacy, including the ability to improve treatment-resistant schizophrenia[5–7]; however its own severe and potentially life-threatening side effects (e.g., agranulocytosis, diabetes) prevent clozapine from being the ideal antipsychotic drug. Indeed, the ideal antipsychotic continues to elude researchers and clinicians, despite intensive efforts to develop safer clozapine-like compounds. While other atypical drugs do not cause agranulocytosis, as clozapine does, they are frequently, though not always,[8] associated with weight gain and metabolic disturbances.

Several challenges have prevented the development of better therapies for schizophrenia. The first is the inherent nature of schizophrenia as a clinical syndrome that may include several separate disease entities.[9] Schizophrenia is clearly a multifactorial, heterogeneous, and polygenetic disorder. Differences between individual patients and among patient populations preclude facile treatment with a small arsenal of drugs, and distinct treatments for the various symptom classes (positive, negative, cognitive) might ultimately be necessary. Secondly, preclinical models might not always be predictive of clinical efficacy.[4] While existing models can reflect atypicality, they do not inform on the overall efficacy, efficacy superior to that of conventional treatment, or side effect profiles of antipsychotic drugs (see[4] for review). Thirdly, despite decades of research, researchers and clinicians still do not understand the genetic, neurobiological, and molecular mechanisms of the disease. A better understanding of the contributions to disease of each of these components could enable segregation of schizophrenic patients into more homogeneous groups (e.g., those with a particular set of predisposing genetic risk factors) and improved target-based drug design that is specific to each group.

This chapter reviews the state of psychopharmacological therapy for schizophrenia, covering both Food and Drug Administration (FDA)–approved typical and atypical drugs and emerging molecular targets for new and developmental drugs.

TYPICAL ANTIPSYCHOTICS: THE D2 DOPAMINE RECEPTOR

The clinical success of chlorpromazine led researchers to synthesize thousands of structurally similar compounds.

By 1969 at least 28 phenothiazines were available, and this drug class dominated the market.[10] Most studies showed these drugs to be more effective than placebo; however, efficacy varied across studies, depending on the patient population.[10] In addition to the phenothiazines, reserpine, butyrophenones (e.g., haloperidol), benzoquinolizines (e.g., tetrabenazine), and thioxanthenes (e.g., chlorprothixene) were also available.

At the time of these drugs' initial discovery in the 1950s and 1960s, their mechanism of action and molecular targets were unknown. The development of radioligand binding assays in the 1970s allowed the determination of ligand binding affinity, and shortly thereafter followed evidence that all clinically effective antipsychotic drugs had affinity for dopamine receptors[11] and that dopamine receptor binding correlated with antipsychotic potency,[12] leading to the prediction that dopamine receptor affinity could serve as a screen for new antipsychotic compounds.[12] For the first time, schizophrenia had a molecular target, and the search for a "magic bullet"[3] began. These and other data also led Solomon Snyder and others to propose the *dopamine hypothesis* of schizophrenia,[13,14] namely, that psychosis is due to hyperactivity of the dopaminergic system. This hypothesis was based on the following four observations: (1) amphetamine and cocaine, which increase dopamine release, induce psychosis in normal patients; (2) amphetamine worsens symptoms in schizophrenic patients; (3) chlorpromazine and other typical antipsychotic drugs antagonize dopamine receptors; and (4) antipsychotic efficacy correlates with dopamine receptor affinity. More modern techniques, such as positron emission tomography (PET), have since provided further evidence of the importance of D2 receptor binding, showing, for example, that the antipsychotic activity of many drugs correlates with striatal D2 receptor occupancy.[15,16]

Not all antipsychotic drugs, however, preferentially bind striatal D2 receptors. Recent PET studies using more selective radioligands have suggested that extrastriatal D2 occupancy could also be critical for antipsychotic efficacy and that striatal D2 receptor occupancy might be the molecular correlate of EPS, although these hypotheses remain controversial. In studies with the radioligand [^{8}F]fallypride, the atypical antipsychotics clozapine and quetiapine preferentially occupied D2 receptors in the temporal cortex and demonstrated low striatal receptor occupancy at therapeutic doses,[17,18] while olanzapine spared D2 receptors in the substantia nigra and ventral tegmental area relative to haloperidol.[19] The former finding suggests that the superior clinical efficacy of clozapine could be due in part to its cortical activity, while the latter finding suggests that the high incidence of EPS observed with typical antipsychotic drugs could be due to excessive striatal D2 receptor blockade. Indeed, PET studies have revealed higher D2 occupancy in patients with EPS than in patients who do not experience EPS.[15] The relationship between striatal and extrastriatal D2 occupancy is still controversial; published data are contradictory as to whether antipsychotics preferentially occupy striatal or extrastriatal receptors (compare, for example,[20] and[21]), but some of these contradictory results might be due to improper methodology.[17,22–24] While the precise mechanism of antipsychotic drug action remains undetermined, all currently approved drugs have some affinity for the D2 dopamine receptor, and it remains a critical molecular target for antipsychotic drug development.

ATYPICAL ANTIPSYCHOTICS: D2 AND THE 5-HT$_{2A}$ SEROTONIN RECEPTOR

While first-generation antipsychotics were effective at reducing positive symptoms in most patients, these drugs were far from ideal. They did not ameliorate the negative or cognitive symptoms of the disease, and their side effects (elevations of serum prolactin, EPS, and tardive dyskinesia) were severe and occasionally debilitating. The introduction to the market of clozapine and the documentation of its superior efficacy[6,25] ushered in a new era of schizophrenia treatment. The second generation of drugs included the dibenzodiazepines (e.g., clozapine), the thienobenzodiazepines (e.g., olanzapine), the dibenzothiazepines (e.g., quetiapine), the benzisothiazolyl piperazines (e.g., ziprasidone), the benzamides (e.g., sulpride and amisulpride), and the benzisoxazoles (e.g., risperidone, 9-OH-risperidone). These drugs were deemed superior to first-generation drugs in that they did not induce EPS or serum prolactin elevations, but at least some of the drugs can cause severe, life-threatening metabolic disturbances including weight gain, hyperlipidemia, and diabetes. The severity of these side effects correlates with H$_1$ histamine receptor affinity.[8]

There have been proposals that atypical antipsychotic drugs as a class are characterized as having a "rapid dissociation" from D2 receptors compared to typical antipsychotic drugs that dissociate "slowly."[26] We have examined a large number of antipsychotic drugs, however, and have found no correlation between dissociation rates and degree of atypicality (Fig. 10.2.1)

The finding that clozapine was associated with less D2 receptor occupancy in vivo than typical antipsychotics[27] suggested that its efficacy was due to a different

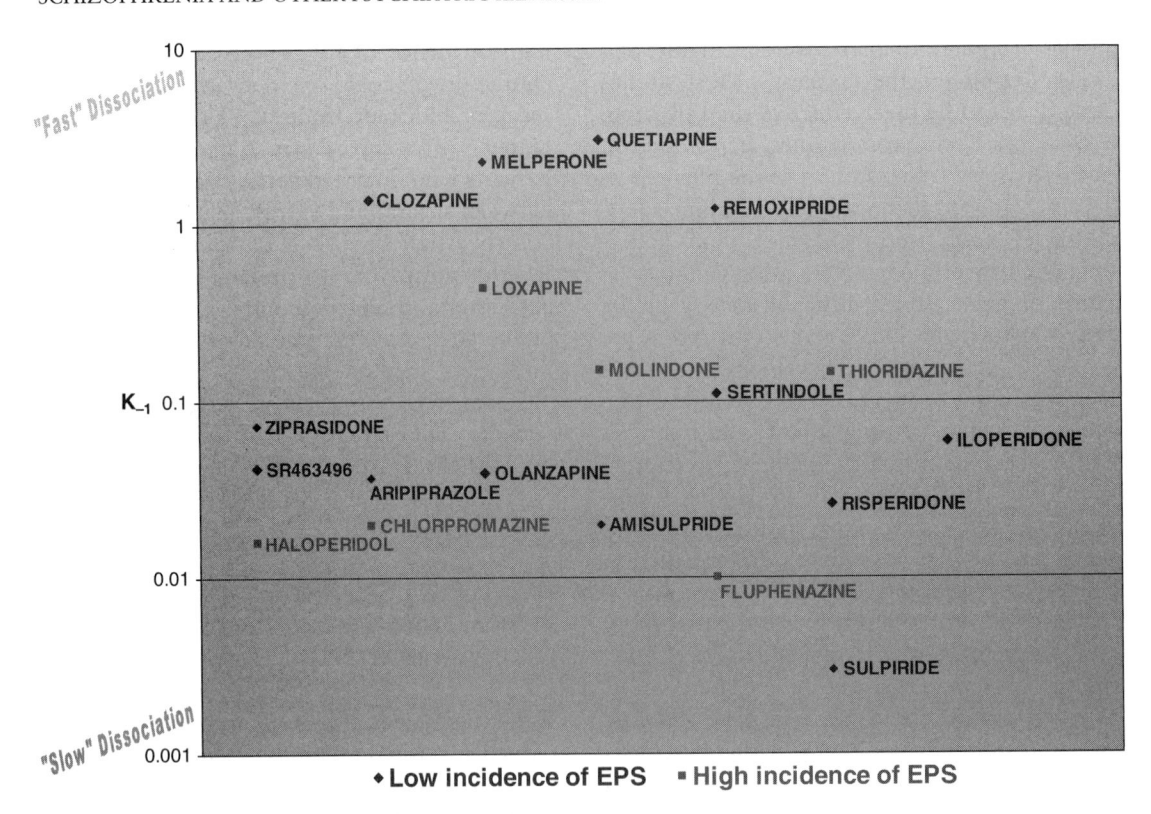

FIGURE 10.2.1 Drug dissociation rate from the D2 dopamine receptor does not predict atypicality. Dissociation rates (K$_{-1}$) of select antipsychotic drugs at the D2 dopamine receptor do not correlate with the extent to which these drugs induce EPS. Atypical drugs are considered those with a low incidence of EPS. K$_{-1}$ values determined by the National Institute of Mental Health Psychoactive Drug Screening Program (NIMH-PDSP; pdsp.med.unc.edu).

molecular target. A screen of drug affinity for the 5-HT$_2$, D1, and D2 receptors revealed that atypical drugs display 10-fold selectivity for the 5-HT$_2$ receptor over the D2 receptor, while typical drugs are more selective for the D2 receptor.[28] From this discovery, the *dopamine-serotonin hypothesis*[29] of schizophrenia emerged, which stated that schizophrenia might involve a disruption of the normal balance between serotonergic and dopaminergic signaling that is restored by clozapine and other antipsychotic drugs.

The emergence of the dopamine-serotonin hypothesis led to a search for 5-HT$_{2A}$-selective antagonists. Unfortunately, such compounds have not been as efficacious as was hoped. Both the 5-HT$_{2A}$-selective antagonist M-100907[30] and the 5-HT$_{2A/2C}$ antagonist SR46349B[31] were only marginally more effective than placebo and no more effective than the comparator in clinical trials. These findings were important because they demonstrated that 5-HT$_{2A}$ antagonists might have antipsychotic properties, but they also demonstrated the key feature of D2 receptor antagonism.

Since the advent of the dopamine-serotonin hypothesis in 1989, researchers have identified many additional molecular targets of atypical drugs (serotonergic, dopaminergic, muscarinic cholinergic, and histaminergic), yet they have found no "magic receptor." These data suggest that the role of 5-HT$_{2A}$ receptors may be to modulate dopaminergic tone[32,33] and that compounds with complex neuropharmacological profiles—"magic shotguns"—may be more effective than "magic bullets."[3] Indeed, we have discovered that typical and atypical antipsychotic drugs have an exceedingly rich pharmacology,[3] with many of them interacting with more than 50 G protein-coupled receptors (GPCRs) (Fig. 10.2.2). Additionally, studies using drugs such as aripiprazole have revealed novel mechanisms of drug action[34] (i.e., partial agonism and functional selectivity). The remainder of this chapter will focus on those other targets and the roles they may play in the pathogenesis and treatment of schizophrenia.

OTHER DOPAMINERGIC TARGETS

While D2 receptors remain the primary dopaminergic target of approved antipsychotic drugs, evidence

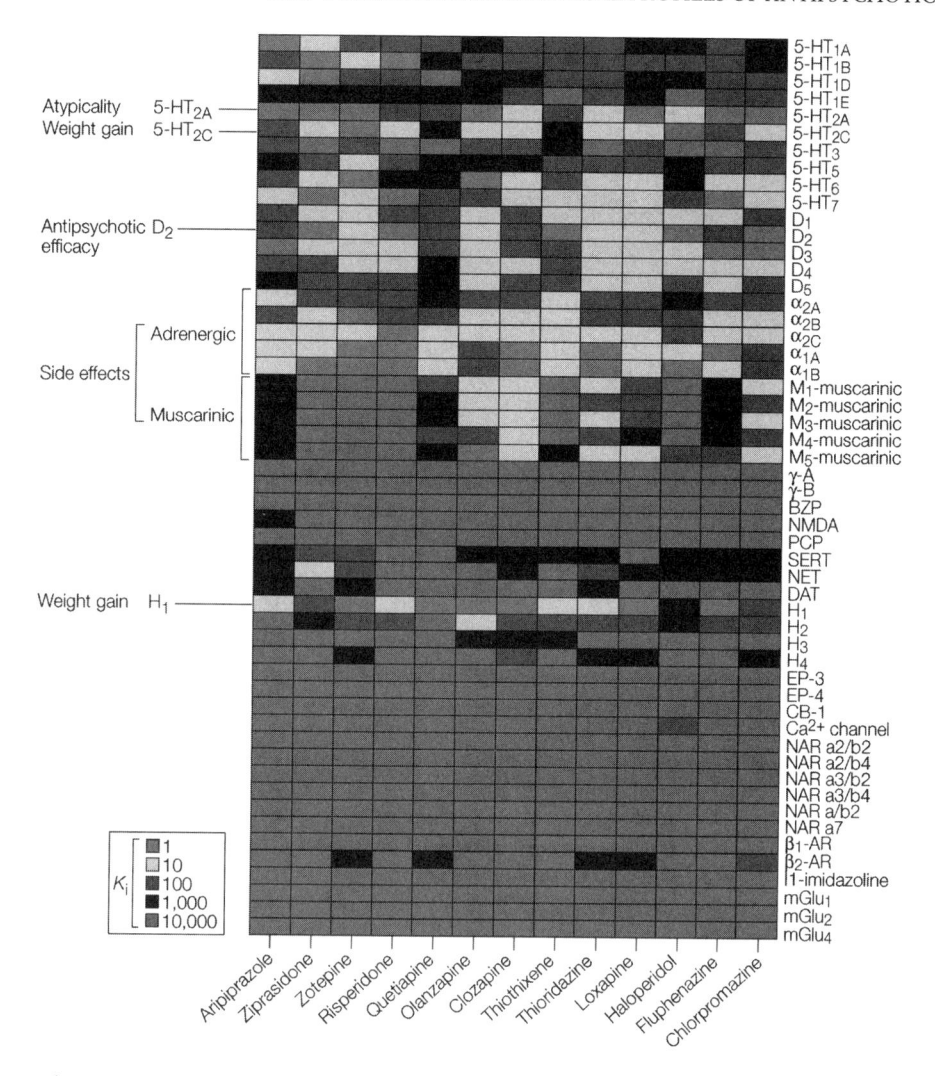

FIGURE 10.2.2. Screening the receptorome reveals multiple molecular targets implicated in antipsychotic drug actions. The affinity (K_i) values for clozapine and a large number of other biologically active compounds at various receptors can be found at the PDSP K_i Database (pdsp.med.unc.edu); the database is part of the National Institute of Mental Health Psychoactive Drug Screening Program and represents the largest database of its kind in the public domain. At present, the PDSP K_i database has >47,000 K_i values for more than 700 targets. *Source*: Reprinted from[3] with permission. (See Color Plate 10.2.2.)

suggests the D1, D3, and D4 dopamine receptors also represent therapeutic targets.

D1 Dopamine Receptors

Research implicates D1 receptor signaling in the prefrontal cortex (PFC) in working memory deficits and cognitive dysfunction in schizophrenia.[35,36] As working memory ability and cognitive function correlate with clinical outcome after antipsychotic treatment, including social reintegration and rehospitalization,[37–39] modulation of D1 signaling could be important in the treatment of schizophrenia. In nonhuman primates, short-term treatment with the D1 agonist ABT 431 reversed working memory deficits that were induced by chronic D2 receptor blockade,[40] while intermittent treatment provided long-lasting (>1 year) improvements in working memory.[41] In other studies, the D1 antagonist SCH23390 impaired memory in monkeys, while the D1 receptor full agonist dihydrexidine improved memory performance; SCH23390 blocked this agonist-induced enhancement of cognitive function.[42–44] Similar studies in both rodents and nonhuman primates with the D1 agonists A77636[45,46] and SKF81297[45] and with the D1 antagonist SCH39166[44] have replicated these findings.

While the aforementioned data may suggest that the role of D1 dopamine receptors in the pathophysiology and treatment of cognitive dysfunction in schizophrenia is relatively straightforward, three issues significantly complicate the picture. First, D1 agonism follows an inverted U-shaped dose–response curve with regard to efficacy.[47] Low levels of D1 activation improve cognitive performance, while higher levels are deleterious.[42,45,48,49] Second, chronic D1 agonism may stimulate receptor down-regulation and restoration of basal tone. Third, spinal D1 receptors contribute to hypotension[50,51]; therefore, chronic D1 agonism could elicit hypotensive crises. In fact, the selective D1 agonist fenoldopam is a potent vasodilator and a remedy for malignant hypertension, though it acts at peripheral receptors. Thus, while D1 receptors represent promising pharmacological targets for the treatment of memory deficits and cognitive dysfunction, these potential pitfalls may require that drugs be partial, rather than full, agonists or that they be administered using an intermittent regimen.

D3 Dopamine Receptors

The discovery and cloning in 1990 of the D3 dopamine receptor[52] revealed a novel target of existing antipsychotic drugs. Due to the structural similarity between D2 and D3 receptors, Sokoloff et al.[52] found that widely used antipsychotic drugs have similar affinity for the two receptors, and for the first time suggested that the different profiles of atypical antipsychotic drugs may be due in part to activity at the D3 receptor. Postmortem analysis of D3 receptor expression levels in human brain tissue revealed elevations in untreated schizophrenic patients compared to control patients; treatment with antipsychotic drugs near the time of death normalized those levels.[53,54] Thus, D3 receptor blockade may restore balance to a hyperfunctioning mesolimbic dopamine system.[55] As D3 expression is highest in the limbic system, D3 blockade does not cause EPS and in fact may have pro-motor effects.[56] Additionally, research implicates D3 receptors in cognitive dysfunction[56] and suggests that D3 antagonists may enhance cognition. Indeed, SB-773812 is currently in phase I and II clinical testing and S33138, a preferential D3 antagonist,[57,58] is in phase II testing. This drug has antipsychotic and pro-cognitive properties in rodent models of psychosis and is not associated with catalepsy.[59] Thus, D3 antagonism is an attractive option for new antipsychotic treatments due to its potential pro-cognitive and pro-motor effects. However, its ability to lessen positive symptoms and the optimal balance between D2 and D3 affinity are still undefined.

D4 Dopamine Receptors

The D4 dopamine receptor, cloned shortly after the characterization of the D3 receptor,[60] identified yet another target of existing drugs. In particular, clozapine's affinity for D4 was an order of magnitude higher than for the D2 and D3 receptors, implicating D4 in contributing to the superior efficacy of clozapine and suggesting the D4 receptor as a potential target for novel therapeutic approaches. Additionally, D4 expression may be elevated in the brains of schizophrenic patients,[61] although data are conflicting and inconclusive.[62] However, D4-selective agents have not shown clinical success, and D4 affinity does not distinguish between typical and atypical antipsychotic drugs.[63] The D4 antagonists L-745,870[64,65] and sonepiprazole,[66] as well as the D4/5-HT$_{2A}$ coantagonist fananserin,[67] have shown no antipsychotic efficacy in clinical trials. Despite these negative data, D4 drugs may still be of some benefit. Surprisingly, various antagonists (PNU-101387G,[68] NGD94-1[69]), as well as agonists (A-412997[70]), show cognitive benefits in animals. D4 receptors modulate cortical glutamatergic excitatory activity[71] and inhibit GABA$_A$ channel activity[72]; thus, perhaps some optimal level of D4 activity enhances cognition.

Catechol-O-Methyltransferase

The enzyme catechol-O-methyltransferase (COMT) catalyzes the deactivation of monoamines in the synaptic cleft and is particularly important for the breakdown of dopamine in the PFC.[73,74] In this region, COMT might be important for cognitive function. Catechol-O-methyltransferase knockout mice show perturbations suggestive of a role for the enzyme in schizophrenia psychopathology. These mice have selective alterations in dopamine levels and enhanced working memory abilities.[75] The COMT inhibitor and anti-Parkinson drug tolcapone can improve working memory in rodents[76] and executive function in humans.[77] Concerns about liver toxicity from tolcapone initially hampered use of this drug, but recent studies show the risk to be minor.[78,79] Further clinical study of COMT inhibitors for schizophrenia is ongoing.

Genetic studies lend further support to a role for COMT in the pathogenesis of schizophrenia. A particular single-nucleotide polymorphism (SNP) in the COMT gene, encoding a Val158Met mutation,[80] results in an enzyme with less thermostability and, consequently, with 40% less enzymatic activity than the Val variant. As the alleles are codominant, heterozygotes express an intermediate phenotype.[73] Individuals who are homozygous for the Val allele have impaired cognitive processing and lower physiological efficiency in the PFC than individuals who are homozygous for the Met allele.[81] Moreover, a family-based association showed that the Val allele is preferentially transmitted to schizophrenic offspring,[81] and other analyses also support a genetic link between this COMT SNP and schizophrenia.[73] The Val allele could contribute not only to an individual's risk of developing schizophrenia, as seen in this study, but also to the response to drug treatment. Schizophrenic patients who are homozygous for the Val allele showed less improvement in cognitive tasks after 6 months of clozapine therapy than patients who were heterozygous or homozygous for the Met allele.[82] Thus, COMT is a promising target for drug and gene-based approaches to schizophrenia therapy.

OTHER SEROTONERGIC TARGETS

The affinity of clozapine and the other atypicals for serotonin receptors spawned the aforementioned search for selective 5-HT$_{2A}$ antagonists. While that quest has not yielded effective drugs, several other serotonin receptors continue to be targets of antipsychotic drug development.

5-HT$_{1A}$ Serotonin Receptors

Clozapine, aripiprazole, and other atypical drugs are agonists at 5-HT$_{1A}$ receptors and may owe some of their efficacy to actions at those receptors.[83–85] Similar to the proposed role of 5-HT$_{2A}$ receptors, 5-HT$_{1A}$ receptors may modulate dopaminergic tone. 5-HT$_{1A}$ receptor expression is high in the PFC.[86] Research suggests that in this area, D2 and 5-HT$_{2A}$ antagonism result in 5-HT$_{1A}$ activation, which in turn enhances dopamine release.[87,88] Thus, 5-HT$_{1A}$ agonists have potential as pro-cognitive agents. It is unclear, however, to what extent the efficacy and atypicality of some antipsychotics are due to 5-HT$_{1A}$ agonism.

5-HT$_{2C}$ Serotonin Receptors

In 1992 we discovered that a variety of typical and atypical antipsychotic drugs have high affinities for 5-HT$_{2C}$ receptors,[89] and this finding implied that 5-HT$_{2C}$ receptors might represent a potential target for antipsychotic drug development. 5-HT$_{2C}$ receptor expression is primarily in the ventral tegmental area and the substantia nigra, where the receptors inhibit dopamine release.[33] Thus, 5-HT$_{2C}$ agonists, rather than inverse agonists,[90] have potential as antipsychotic agents. For example, the selective 5-HT$_{2C}$ agonist Ro 60-0175 effectively decreases cortical dopamine levels in rats.[91] Other 5-HT$_{2C}$ agonists, including WAY-163909[92] and CP-809,101,[93] showed antipsychotic efficacy without inducing EPS in rodent models of psychosis. Additionally, 5-HT$_{2C}$ agonists have anorexic properties. As a major side effect of atypical antipsychotics is metabolic disturbances (some of which may be due to 5-HT$_{2C}$ antagonism), 5-HT$_{2C}$ agonists could be an effective alternative. Because 5-HT$_{2A}$ agonism could exacerbate psychosis and 5-HT$_{2B}$ agonism could result in valvular heart disease,[94] a high degree of 5-HT$_{2C}$ selectivity is an important characteristic of any antipsychotic agent targeting the 5-HT$_{2C}$ receptor.

5-HT$_4$ Serotonin Receptors

5-HT$_4$ receptors are intriguing targets for antipsychotic drug development due to their potentially pro-cognitive characteristics. Their levels are decreased in the brains of Alzheimer's disease patients,[95] and 5-HT$_4$ agonists inhibit β-amyloid secretion and enhance neuronal survival in vitro.[96] Moreover, 5-HT$_4$ agonism enhances cholinergic neurotransmission and thus may be effective at improving cognitive function.[97] For these reasons, 5-HT$_4$ agonists are promising treatments for Alzheimer's disease and possibly for the cognitive dysfunction that is so prevalent in schizophrenia, although none have advanced to clinical testing.

5-HT$_6$ and 5-HT$_7$ Serotonin Receptors

Many antipsychotics, both typical and atypical, have low nanomolar affinity for the 5-HT$_6$ and 5-HT$_7$ receptors.[98] Genetic polymorphisms of the 5-HT$_6$ receptor may confer susceptibility to schizophrenia, and levels of the 5-HT$_6$ receptor,[99] as well as both 5-HT$_6$ and 5-HT$_7$ mRNA,[100] are decreased in certain brain regions of schizophrenic individuals. As both receptors could have roles in learning and memory, these receptors are potential molecular targets for treating cognitive dysfunction in schizophrenia. Antagonism of the 5-HT$_6$ receptor reverses the amnesic effects of anticholinergic drugs, improves performance in diverse memory tasks, and enhances memory consolidation,[99] while 5-HT$_7$ receptor activation increases hippocampal neuron excitability.[101] Furthermore, 5-HT$_7$ knockout mice have deficits in contextual fear conditioning,[102] a behavioral trait that characterizes psychosis. Together, these data suggest that drugs that act at 5-HT$_6$ and 5-HT$_7$ receptors may have antipsychotic or pro-cognitive properties.

CHOLINERGIC TARGETS

In addition to acetylcholine's well-documented role in motor function, it contributes to sensory processing, attention, learning and memory, and other cognitive processes.[103] Accordingly, dysfunction of the cholinergic system contributes not only to Alzheimer's disease pathology, for which it is most well known, but also to other neuropsychiatric diseases, including schizophrenia.[104] Postmortem analysis of brains from schizophrenic and control patients has demonstrated a reduction in cholinergic interneuron number in schizophrenia[105,106] and alterations in nicotinic and muscarinic receptor expression (see below). Studies have also revealed a correlation between low cortical levels of choline acetyltransferase (ChAT; the enzyme that catalyzes the last step in acetylcholine synthesis) and poor cognitive functioning.[107] Measurement of the effects of antipsychotic drugs on cortical acetylcholine release provides further evidence of a link between the cholinergic system and schizophrenia: atypical antipsychotics induce acetylcholine release[108,109] by a largely serotonin receptor–independent mechanism.[110] Enhancing cortical cholinergic signaling might therefore have beneficial effects on cognitive dysfunction in schizophrenia.

Cholinesterase inhibitors, which inhibit acetylcholine breakdown and thus prolong its action, are an effective therapy for Alzheimer's disease[111] and theoretically might be beneficial in schizophrenia too. Chronic treatment with the cholinesterase inhibitor donepezil and an atypical antipsychotic provides superior functional normalization of brain activity in a verbal fluency task compared to atypical antipsychotic treatment alone.[112] Yet, clinical trials of donepezil as an adjunct to typical or atypical antipsychotic treatment have shown no or minimal efficacy.[113–115] Clinical trials with donepezil, as well as other cholinesterase inhibitors, have also yielded inconsistent, incomplete, and confounded data.[104,116] A cholinergic approach to antipsychotic drug development might be more effective if it directly targets cholinergic receptors rather than nonspecifically raising acetylcholine levels.

M₁ Muscarinic Receptors

While various data have implicated the M₂–M₄ muscarinic receptors in cognition and schizophrenia, the M₁ receptor has been the focus of most of the research on the role of the muscarinic cholinergic system in schizophrenia.[104,117] In vivo radioligand binding studies in schizophrenic patients have revealed reductions in M₁ receptor binding sites in several brain regions.[117–120] While treatment with clozapine[121,122] and olanzapine[123,124] both decrease available muscarinic binding sites, indicating receptor occupancy by these drugs, unmedicated schizophrenics also have lower levels of M₁[119]; antipsychotic treatment does not explain the alterations in muscarinic receptor number.

Muscarinic agonists may normalize cholinergic signaling and improve schizophrenia symptoms. Patients who chew betel nuts, which contain the muscarinic agonist arecoline, have less severe positive and negative symptoms.[125] Xanomeline, an arecoline derivative, is an M₁ and M₄ agonist that might also be effective in schizophrenia.[126] Xanomeline has antipsychotic and antidopaminergic properties in animal models of psychosis[127–129]; it selectively inhibits firing of mesolimbic dopaminergic neurons without striatal effects, suggesting that it could be free of EPS.[126] In humans, xanomeline improves symptoms of schizophrenia, including cognitive function,[130] but is poorly tolerated.[126] Clozapine, the gold standard antipsychotic drug, has both muscarinic agonist and antagonist properties.[117] Its metabolite, N-desmethylclozapine (NDMC; ACP-104), is an agonist at the M₁ receptor, and this action could contribute to clozapine's superior efficacy.[131,132] In addition, NDMC has affinity for 5-HT$_{2A/2C}$ receptors and D2/3 receptors, which suggests that it might have antipsychotic potential beyond its muscarinic activity.[133] Moreover, a high NDMC-to-clozapine ratio correlates with greater improvement in cognition and quality of life than that associated with either compound alone.[132] This finding suggests that clozapine's in vivo effects are indeed partially mediated through its metabolite. One cautionary remark must be inserted here: xanomeline and NMDC have robust "off-target" pharmacology (see Table 10.2.1). Both drugs are potent 5-HT receptor antagonists and are agonists at all cloned muscarinic receptor sybtypes. More recent studies with M₁- and M₄-selective compounds[134,135] suggest, however, that drugs selectively targeting these receptors may be antipsychotic.

Nicotinic Receptors

Rates of cigarette smoking are twofold to fourfold higher among individuals with schizophrenia than in the general population, and schizophrenics have a higher nicotine intake than other smokers.[136] Nicotine use in this patient population may represent self-medication.[136,137] Research has documented nicotine-induced improvements in sensory gating[138] and working memory and attention[139] in schizophrenic patients, as well as alleviation of haloperidol-induced EPS[140] and cognitive dysfunction[141]—data that support

TABLE 10.2.1. *Xanomeline, N-desmethyl-Clozapine, and Clozapine: Nonselective Muscarinic Agents*

Receptor	Xanomeline	N-Desmethyl-Clozapine	Clozapine
M_1	776 (Full Agonist)	60 (Full Agonist)	23 (Partial Agonist)
M_2	501 (Full Agonist)	339 (Full Agonist)	589 (Partial Agonist)
M_3	234 (Partial Agonist)	324 (Partial Agonist)	36 (Antagonist)
M_4	35 (Full Agonist)	135 (Full Agonist)	45 (Partial Agonist)
M_5	257 (Full Agonist)	23 (Partial Agonist)	11 (Antagonist)
5-HT$_{2A}$	125 (Antagonist)	11 (Antagonist)	6.5 (Antagonist)
5-HT$_{2C}$	40 (Antagonist)	11 (Antagonist)	9 (Antagonist)
5-HT$_6$	1258 (Antagonist)	9 (Antagonist)	23 (Antagonist)
5-HT$_7$	126 (Antagonist)	15 (Antagonist)	13 (Antagonist)
D_2	1000 (Antagonist)	89 (Partial Agonist)	343 (Antagonist)
D_3	398 (Antagonist)	153 (Unknown)	319 (Antagonist)

Notes: The affinity (Ki, nM) values for xanomeline, N-desmethyl-clozapine, and clozapine at a variety of CNS targets demonstrate the robust pharmacology of these drugs.
Source: Compiled from [31,132,268], NIMH-PDSP K$_i$ database at pdsp.med.unc.edu, Acadia Pharmaceuticals, and GlaxoSmithKline.

the self-medication hypothesis. An intriguing study by Smith et al. found improvements in negative symptoms due to acute cigarette smoking but found denicotinized cigarettes to have a similar effect,[142] suggesting that other components of cigarettes may contribute to their therapeutic effects.

Both genetic and postmortem studies lend further support to a role for nicotinic receptors in schizophrenia psychopathology. Linkage analysis[143] and cloning studies[144] have identified alterations in schizophrenia of the α$_7$ nicotinic receptor gene on chromosome 15, while measurement of receptor expression reveals less nicotinic receptor expression in the hippocampus of schizophrenics than in controls that is not attributable to drug treatment, smoking history, or generalized cell loss.[145,146] Along those lines, α$_7$ receptor antagonists induce sensory gating deficits similar to those of schizophrenia,[147] and the α$_7$ receptor mediates clozapine-induced improvements in sensory processing.[148]

These data suggest a therapeutic potential of nicotinic agonists, both of the low-affinity α$_7$ receptor and of the high-affinity α$_4$β$_2$ receptor, the two most prevalent nicotinic receptors.[149,150] DMXB-A, a selective α$_7$ receptor agonist, improved sensory gating symptoms,[151] and SIB-1553A, a selective β$_4$ subunit drug, improved attention and working memory[152] in rodents. DMXB-A was also effective at improving cognitive symptoms in a small proof-of-concept clinical trial.[153] Unfortunately, trials of DMXB-A in schizophrenia showed no apparent efficacy for cognition enhancement.[154] A potential caveat of nicotinic receptor agonism is desensitization of receptors.[149] This phenomenon could explain the lack of efficacy of cholinesterase inhibitors too. Daily nicotine injections in rodents result in desensitization of dopamine release in the nucleus accumbens and of the locomotion-enhancing effects of nicotine.[155] Partial agonists (e.g., GTS-21[56]) or allosteric potentiators (e.g., galantamine) might be beneficial in activating these receptors without inducing desensitization.[36,157] However, recent clinical trials of galantamine have yielded mixed results.[158–163] Thus, the value of nicotinic receptors as molecular targets for antipsychotic drugs is still unclear.

GLUTAMATERGIC TARGETS

Since the 1960s, NMDA antagonism by the *dissociative anesthetics* phencyclidine (PCP) and ketamine has served as a pharmacological model of psychosis.[164–168] These drugs produce psychotomimetic effects, including positive, negative, and cognitive symptoms. By inference, NMDA hypofunction, or glutamatergic hypoactivity in general, could contribute to the pathophysiology of schizophrenia. Although alternative hypotheses implicate glutamatergic *hyper*activity or NMDA-induced apoptotic change,[166] most data suggest that up-regulation of glutamatergic signaling via a variety of receptors and mechanisms could have beneficial therapeutic effects in schizophrenia.

NMDA Glutamate Receptors

N-methyl-D-aspartate (NMDA) receptors are ligand-gated ion channels with both a primary glutamate binding site and a secondary allosteric glycine binding site,[166] providing dual sites for drug targeting. Unfortunately, direct agonism of the glutamate binding site could induce excitotoxicity and seizures; thus, this site is an unsafe target. The glycine binding site, however, has therapeutic potential, and drugs targeting this site are not hindered by excitotoxicity.[169] Chronic glycine administration to rodents does not induce excitotoxicity or neuronal pathology.[170,171] Evidence demonstrating low plasma levels of glycine and D-serine in schizophrenic patients implicates a hypofunctioning NMDA receptor in the pathology of schizophrenia. Data from small clinical trials with glycine,[172,173] D-serine, and other amino acids with affinity for the glycine allosteric site, such as D-alanine and D-cycloserine, are promising.[169] These compounds, in conjunction with a typical or atypical antipsychotic, improve negative symptoms and cognitive dysfunction.[169] D-Cycloserine is a partial agonist at the NMDA receptor and acts as an antagonist at high doses[174]; thus, it has less efficacy than the other amino acids and even worsens symptoms at high doses.[169] Achieving an optimal level of receptor occupancy will be a challenging but crucial aspect of the proper dosing regimen. Additionally, glycine[175] and D-serine[176] lose their efficacy when coadministered with clozapine, while D-cycloserine worsens symptoms,[177] suggesting that clozapine itself modulates glutamatergic neurotransmission and that direct targeting of the NMDA receptor provides no further benefit.

These glycine-based approaches to increasing glutamatergic signaling have three potential pitfalls. First, the glycine site is already half-saturated under physiological conditions; thus, there is little room for stimulating signaling further. Second, high (gram-level) doses of the amino acids are required to sufficiently elevate central nervous system (CNS) levels. Finally, the molecular target size for the glycine site is small enough to prohibit the identification of high-affinity agonists.[169] These pitfalls could interfere with the development of glycine-based approaches to schizophrenia treatment.

Another approach to targeting the glycine binding site of the NMDA receptor is to antagonize the glycine transporter with glycine transporter inhibitors (GTIs) and thus raise synaptic levels of glycine and promote saturation of the glycine binding site. The GTI SSR504734 was effective in a variety of rodent models of schizophrenia.[178] Sarcosine (N-methylglycine), another GTI and the only one to undergo clinical testing thus far, improved all symptom domains when added to a patient's existing antipsychotic regimen and was more effective than D-serine as an adjuvant to risperidone therapy,[179] but like NMDA allosteric modulators, it was not effective when a patient was also on clozapine.[180] Phase II clinical trials with sarcosine are ongoing and could better delineate the potential benefits of its use. Development of other GTIs, including Org 24598[181] and N-[3-(4'-fluorophenyl)-3-(4'-phenylphenoxy)propyl]sarcosine (NFPS),[182] is in progress.

AMPA Glutamate Receptors

Although AMPA receptor antagonists are not psychotomimetic, agonism of this receptor might improve cognitive dysfunction in schizophrenia. AMPA receptors, along with kainate glutamate receptors, mediate fast glutamatergic signaling, and signaling through AMPA receptors is closely tied to NMDA receptor activation.[166] Like NMDA receptors, orthosteric agonists are not useful; in the case of AMPA receptors, they cause rapid desensitization. Thus, allosteric modulators, termed *ampakines*, are the chief approach to activating AMPA receptor signaling. Ampakines mediate receptor desensitization kinetics[183] and can enhance glutamatergic neurotransmission with effects on synaptic plasticity and learning and memory. These drugs have pro-cognitive potential.[184] The ampakine CX-516, when coadministered with clozapine to schizophrenic patients, improved cognitive parameters.[185] However, a small subsequent study using CX-516 as monotherapy[186] and a recent study of CX-516 as an adjuvant to clozapine, olanzapine, or risperidone[187] found no beneficial effect of the drug. While ampakines induce less desensitization than direct AMPA receptor ligands, down-regulation of receptors could still be extensive and might limit the efficacy of ampakine treatment.[188] Further characterization of ampakines and their effects is necessary to determine their efficacy in treating schizophrenia.

In addition to evidence suggesting that ampakines might improve cognition through enhancement of AMPA receptor signaling, preclinical data conversely suggest that AMPA antagonists might have therapeutic potential. For example, the AMPA antagonist LY32 6325 suppressed the conditioned avoidance response (CAR)[189] and apomorphine- and amphetamine-induced stereotypy[190] in rodents. The joint AMPA and kainate receptor antagonists CNQX and LY293558 attenuated ketamine-induced dopamine release in the PFC,[191] while CNQX normalized PCP-induced changes in neuronal firing rates.[192] Other AMPA antagonists have demonstrated varied efficacy in rodent models of psychosis.[190] Further elucidation of AMPA receptor signaling pathways could clarify the dual and contrasting effects of AMPA ligands and enable the development of effective glutamatergic drugs for the treatment of schizophrenia.

Metabotropic Glutamate Receptors

In addition to the three types of ionotropic glutamate receptors (NMDA, AMPA, kainate), three classes of metabotropic glutamate receptors (mGluRs) are also present in the CNS, and drugs targeting these receptors are in preclinical development.[193] Group I mGluRs (mGluR$_1$, mGluR$_5$) increase presynaptic glutamate release and potentiate NMDA receptor–mediated neurotransmission. Thus, group I mGluR agonism represents another approach to stimulating NMDA signaling. Such drugs have effectively normalized PCP- and amphetamine-induced disruptions of pre-pulse inhibition (PPI) in rodents[194,195]; their efficacy in humans is still undetermined.

Conversely, group II mGluRs (mGluR$_2$, mGluR$_3$) inhibit presynaptic glutamate release, yet agonists of this receptor class are also under investigation for antipsychotic efficacy.[166] One hypothesis regarding this seemingly contradictory finding is that Group II mGluR agonists normalize excess glutamate release induced by NMDA antagonism, which would otherwise overactivate non-NMDA glutamate receptors and trigger cognitive impairment.[193] The mGluR$_{2/3}$ agonists LY354740 and LY379268 normalize both increases in synaptic glutamate levels and behavioral changes following NMDA antagonist treatment.[166,196–199] Highly selective positive allosteric modulators of mGluR$_{2/3}$ signaling[200,201] might also prove effective if they induce less receptor desensitization than direct agonists.[202] Biphenyl-indanone A (BINA), an mGluR$_2$ positive allosteric modulator, attenuated psychotic behavior in mice,[203] for example. A recent clinical trial of an mGluR$_{2/3}$ pro-drug showed efficacy,[204] indicating the clear potential utility of this approach.

OTHER TARGETS

α_2-Adrenergic Receptors

The α_2 adrenergic receptor signals in the central noradrenergic system, which projects from the locus ceruleus to the prefrontal cortex. Activation of α_2 receptors modulates dopamine release,[88] strengthens working memory, and enhances cognitive function.[205,206] Clozapine and other atypical antipsychotics have high affinity for α_2 receptors.[207] In schizophrenic patients, the α_2 agonists guanfacine[208] and clonidine[209] improve performance on memory tasks. However, when the α_2 *antagonist* idazoxan is coadministered with the typical antipsychotic fluphenazine to patients with treatment-resistant schizophrenia, the patients show significant improvement compared to those given fluphenazine treatment alone and similar to that seen with clozapine.[210] Thus, a proper balance of receptor activity may be necessary to moderate both the antipsychotic activity and pro-cognitive effects of α_2 signaling.

Cannabinoid Receptors

Numerous studies implicate the endocannabinoid system, which consists of the CB$_1$ and CB$_2$ receptors, in schizophrenia. First, schizophrenic patients have decreased levels of both CB$_1$ mRNA and protein in their prefrontal cortices.[211] Second, studies consistently associate cannabis use with an increased risk of psychosis, with greater risk among more frequent users.[212–214] Whether cannabis use precipitates psychosis, or whether vulnerable individuals are more likely to use cannabis for self-medication or other purposes, is unclear.[213]

Cannabis contains multiple cannabinoids with opposing actions: while Δ9-tetrahydrocannabinol (Δ9-THC) is psychotomimetic,[215] cannabidiol has antipsychotic efficacy.[216,217] A recent study examining cannabinoid levels in hair samples found a greater incidence of positive schizophrenia-like symptoms in patients whose hair samples contained only Δ9-THC and lower levels of anhedonia in patients whose hair contained both Δ9-THC and cannabidiol, compared to controls.[218] SR141716 (rimonabant), a selective CB$_1$ antagonist, demonstrated antipsychotic properties in signaling studies[219] and in a rodent model of psychosis.[220] SR141716 and the CB$_1$ antagonist AM251 both had antipsychotic efficacy in the PCP-induced disruption of PPI rodent model of psychosis.[221] Unfortunately however, the effect of SR141716 was indistinguishable from that of placebo in a recent clinical trial.[31] Another potential advantage of cannabinoid receptor antagonists is their anorexic qualities. Many clinical trials have investigated or are currently testing the efficacy of SR141716 as a weight loss drug, and it remains to be seen whether CB$_1$ antagonists could present an alternative to atypical antipsychotics in patients at increased risk for metabolic disturbances. More study is needed to determine whether or not cannabinoid receptors represent an effective molecular target for antipsychotic drug development.

Neurokinin Receptors

The neurokinin system is involved in pain modulation, emesis, depression, drug abuse, Parkinson's disease, and, potentially, schizophrenia. The endogenous peptide ligands substance P, neurokinin A, and neurokinin B preferentially bind the NK$_1$, NK$_2$, and NK$_3$

neurokinin receptors, respectively, and modulate the mesolimbic dopamine pathway.[222] In schizophrenics, reports have demonstrated increases in brain levels of substance P[223] and NK_1,[224] as well as neurokinin receptors; however, the results are inconsistent.[222] These data suggest that neurokinin antagonists might have antipsychotic properties. A small exploratory clinical trial found the NK_1 antagonist aprepitant (MK869) ineffective,[225] but data from studies of NK_3 antagonists are more promising. NK_3 activation induces dopamine, 5-HT, and norepinephrine release; thus, blocking these receptors could hypothetically reduce excitatory activation of these systems.[226] The NK_1 antagonist talnetant (SB-223412) can modulate mesolimbic and mesocortical dopamine levels and has antipsychotic properties in guinea pigs[227]; clinical trials with talnetant have recently been completed, but the findings were reported to be negative. Osanetant (SR142801) effected significant symptomatic improvement compared to placebo and similar to that of haloperidol.[31] What is unknown is the extent to which neurokinin receptor antagonism is additive with regard to antipsychotic potential. As all three receptors are potential neuropsychiatric drug targets, a nonselective antagonist could possibly achieve greater efficacy, for example by attenuating both the positive symptoms and affective disturbances.[226]

Neurotensin Receptors

Neurotensin is another neuropeptide with suggested ties to neuropsychiatric diseases, including schizophrenia, drug abuse, and Parkinson's disease, and might even act as an endogenous antipsychotic.[228] Neurotensin antagonizes dopaminergic signaling and favors D1 over D2 activation through a variety of mechanisms.[228] Numerous studies have examined neurotensin concentrations in the cerebrospinal fluid (CSF) and neurotensin receptor expression in postmortem brain specimens from schizophrenic patients (see[29]). While the latter studies present no consistent abnormalities, the former studies reveal a correlation between low neurotensin concentrations in CSF and the severity of symptoms (positive, negative, and cognitive) in schizophrenics. This finding is specific to schizophrenia (as opposed to other psychiatric diseases), and neurotensin levels normalize as patients show clinical improvement following antipsychotic drug treatment. These data suggest that neurotensin agonists could have antipsychotic properties, and in vivo evidence from rodent experiments supports that hypothesis (see[28,229]). For example, the neurotensin peptide analog PD149163 attenuates amphetamine-induced inhibition of PPI,[230] and the analog NT69L blocks amphetamine- and cocaine-induced hyperlocomotion[231] in rats. Despite significant and abundant preclinical evidence suggesting that neurotensin agonists could be effective antipsychotic agents, no clinical trials have taken place.

Paradoxically, neurotensin antagonists may also be antipsychotic[228,232]. Chronic administration of the potent antagonist SR48692[233] both lowers by 50% extracellular dopamine in the nucleus accumbens shell[234] and blocks cocaine-induced hyperlocomotion,[235] demonstrating an ability of the drug to modulate mesolimbic dopamine circuitry. However, a recent clinical trial found SR48692 devoid of clinical efficacy in treating schizophrenia.[31] Clearly, the complex circuitry of the neurotensin system still requires exploration and evaluation to determine the exact mechanisms of the antipsychotic effects of both neurotensin agonists and antagonists.

Sigma Receptors

In contrast to all of the aforementioned cell membrane-bound receptor targets, σ_1 receptors are intracellular, endoplasmic reticulum-bound proteins. Martin et al. incorrectly designated σ receptors as a class of opioid receptors but suggested a dopaminergic mechanism for their activity.[236] That finding, along with evidence that haloperidol and other antipsychotics have affinity for σ receptors[237] and that σ receptor polymorphisms confer increased susceptibility to schizophrenia[238], suggested σ ligands as antipsychotic agents.[239] The antagonist NE-100 improves PCP-induced psychosis in animal models.[240] Some σ antagonists attenuate stimulant-induced locomotion, while other, more selective antagonists do not.[241] In humans, the antagonist SL 82.0715 (eliprodil) might improve negative symptoms,[242] and EMD-57445 (panamesine) demonstrates modest efficacy for positive and negative symptoms,[243,244] but BW 23FU (rimcazole)[245] and BMY 14802[246] are ineffective. Several in vivo animal studies have shown that agonists of the σ receptor improve memory and cognition, suggesting that they might work as pro-cognitive agents in schizophrenia via a mechanism involving a complex interplay with neurosteroids[36,241] (see below). One potential pitfall of σ ligands as antipsychotics is that they might contribute to EPS.[241] Clinical trials of σ-selective compounds will require careful monitoring and evaluation to determine the antipsychotic benefits of such ligands.

Additional Approaches

Other molecular targets implicated in schizophrenia include nitric oxide synthase, neurosteroids, secretin, cyclooxygenase-2 (COX2), neurotrophic factors, and phosphodiesterase 10A (PDE10A). Several rodent studies have validated the ability of nitric oxide synthase inhibitors to normalize behavior in NMDA antagonist models of psychosis,[247–250] and one clinical study using the nitric oxide synthase inhibitor methylene blue found clinical benefit for management of treatment-resistant schizophrenia.[251] Neurosteroids, such as dehydroepiandrosterone (DHEA) and pregnenolone, interact with σ_1 receptors[252,253] (see above) and might be effective drugs, particularly for improving negative symptoms in women.[254] A phase III clinical trial for management of autism with secretin, a gastrointestinal peptide, has recently been completed. Secretin can dose-dependently attenuate PCP-induced disruption of PPI in rodents[255] and has some antipsychotic efficacy in patients with schizophrenia.[256,257] Its effects in these studies were transient, however, and secretin administration via an intravenous route, as in these studies, is not feasible for long-term treatments. Cyclooxygenase-2 inhibitors, such as celecoxib, modulate the immune system and reduce inflammation. As neuroinflammation interferes with cognitive processes, COX2 inhibitors could act as pro-cognitive agents in schizophrenia,[258,259] and a recent study showed benefit of celecoxib as an adjunct to risperidone therapy.[260] Both neurodevelopmental[261] and neurodegenerative[262] hypotheses of schizophrenia implicate neurotrophic factors, either in the proper development of the nervous system or in its maintenance, respectively. Although the factors themselves

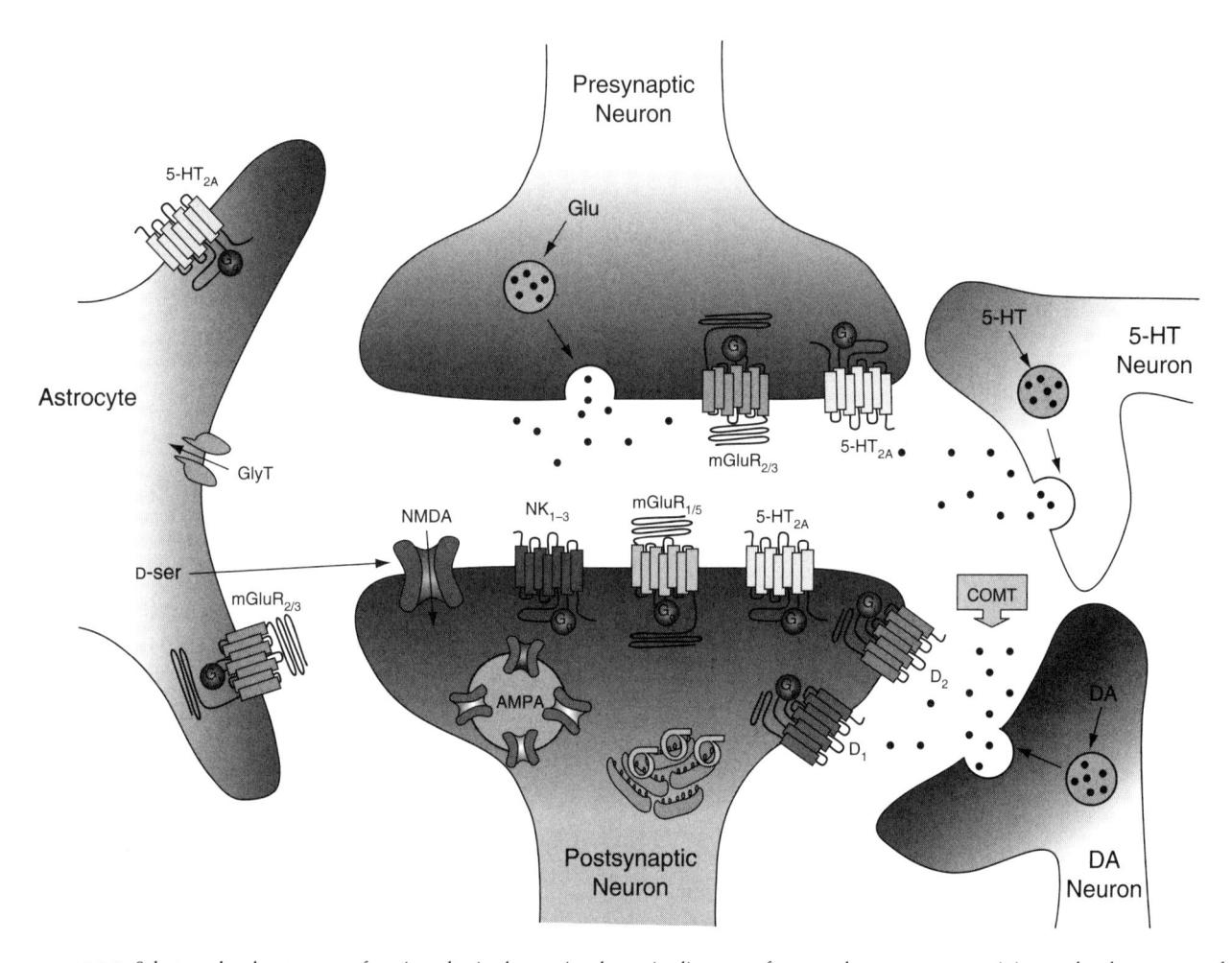

FIGURE 10.2.3. Select molecular targets of antipsychotic drugs. A schematic diagram of a complex synapse containing molecular targets of antipsychotic drugs is depicted. These targets include D1 and D2 dopamine receptors (D1, D2), the 5-HT$_{2A}$ serotonin receptor (5-HT$_{2A}$), metabotropic glutamate receptors 1/5 and 2/3 (mGluR$_{1/5 \text{ or } 2/3}$), neurokinin receptors 1–3 (NK$_{1–3}$), the N-methyl-D-aspartate glutamate receptor (NMDA), the alpha-amino-3-hydroxy-5-methyl-4-isoxazolepropionic acid glutamate receptor (AMPA), the sigma receptor (σ), the glycine transporter (GlyT), and catechol-O-methyltransferase (COMT). DA, dopamine; D-ser, D-serine; Glu, glutamate; G$_q$/G$_s$/G$_i$, G-protein; 5-HT, serotonin. Source: Adapted from[157].

cannot cross the blood-brain barrier and thus are not feasible drugs, compounds that modulate their expression and/or signaling might be beneficial.[263] Finally, evidence demonstrates that PDE 10A, a phosphodiesterase with high striatal expression, has potential as a target of antipsychotic drugs. PDE10A-selective inhibitors such as papaverine show antipsychotic efficacy in rodent models of psychosis,[263,264] but no data in humans are available.

CONCLUSIONS AND FUTURE DIRECTIONS

Despite decades of research, the state of schizophrenia therapy is much the same today as it was 20 years ago, when clozapine returned to the market. Clozapine remains the gold standard drug, and all therapeutically effective treatments act at the D2 dopamine receptor—a target first identified in the 1970s. Current approaches are largely modeled on the *signal transduction hypothesis* of schizophrenia[157,265] (Figure 10.2.3). Such methods might yet have potential if we develop "selectively nonselective" drugs[3] with binding affinity profiles similar to that of clozapine or if we employ polypharmacy to treat the distinct symptom domains of schizophrenia. Employing functionally selective ligands[266] and modulating noncanonical GPCR signaling (i.e., β-arrestin[67]) also represent new opportunities for drug development within the signal transduction model of therapy. Clearly, however, we need to develop a new paradigm for antipsychotic therapy.

A *molecular-genetics* approach[157,265] would identify specific susceptibility genes, and then treatment could be individualized to target a patient's altered genes or gene products. The *neural network hypothesis* of schizophrenia represents yet another paradigm for antipsychotic drug development.[157,265] This theory postulates that schizophrenia results from abnormal neurodevelopment and suggests that treatments should target neuronal migration, pruning, and synapse formation—a therapeutic approach that will require early identification of at-risk individuals at the presymptomatic or prodromal stage. This method perhaps seems the least promising, at least for the time being, as we currently have no reliable method of predicting who will develop schizophrenia. But as our understanding of the molecular genetics improves, we might accomplish that feat. In the meantime, development of superior preclinical models, validation of existing molecular targets, and differentiation of toxic versus therapeutic mediators should be priorities of academic research and are likely to yield more effective, safer medications.

REFERENCES

1. Labhardt F. Largactil therapy in schizophrenia and other psychotic conditions. *Schweiz Arch Neurol Psychiatr*. 1954;73:309–338.
2. Murphy BP, Chung YC, Park TW, et al. Pharmacological treatment of primary negative symptoms in schizophrenia: a systematic review. *Schizophr Res*. 2006;88:5–25.
3. Roth BL, Sheffler DJ, Kroeze WK. Magic shotguns versus magic bullets: selectively non-selective drugs for mood disorders and schizophrenia. *Nat Rev Drug Discov*. 2004;3:353–359.
4. Conn PJ, Roth BL. Opportunities and challenges of psychiatric drug discovery: roles for scientists in academic, industry, and government settings. *Neuropsychopharmacology*. 2008;33:2048–2060.
5. McEvoy JP, Lieberman JA, Stroup TS, et al. Effectiveness of clozapine versus olanzapine, quetiapine, and risperidone in patients with chronic schizophrenia who did not respond to prior atypical antipsychotic treatment. *Am J Psychiatry*. 2006;163:600–610.
6. Kane J, Honigfeld G, Singer J, et al. Clozapine for the treatment-resistant schizophrenic. A double-blind comparison with chlorpromazine. *Arch Gen Psychiatry*. 1988;45:789–796.
7. Wahlbeck K, Cheine M, Essali MA. Clozapine versus typical neuroleptic medication for schizophrenia. *Cochrane Database Syst Rev*. 2000;CD000059.
8. Kroeze WK, Hufeisen SJ, Popadak BA, et al. H1-histamine receptor affinity predicts short-term weight gain for typical and atypical antipsychotic drugs. *Neuropsychopharmacology*. 2003;28:519–526.
9. Carpenter WT Jr. The schizophrenia paradigm: a hundred-year challenge. *J Nerv Ment Dis*. 2006;194:639–643.
10. Klein DF, Davis JM. *Diagnosis and Drug Treatment of Psychiatric Disorders*. Baltimore, MD: Williams & Wilkins; 1969.
11. Seeman P, Lee T, Chau-Wong M, et al. Antipsychotic drug doses and neuroleptic/dopamine receptors. *Nature*. 1976;261:717–719.
12. Creese I, Burt DR, Snyder SH. Dopamine receptor binding predicts clinical and pharmacological potencies of antischizophrenic drugs. *Science*. 1976;192:481–483.
13. Snyder SH, Banerjee SP, Yamamura HI, et al. Drugs, neurotransmitters, and schizophrenia. *Science*. 1974;184:1243–1253.
14. Snyder SH. The dopamine hypothesis of schizophrenia: focus on the dopamine receptor. *Am J Psychiatry*. 1976;133:197–202.
15. Nordstrom AL, Farde L, Wiesel FA, et al. Central D2-dopamine receptor occupancy in relation to antipsychotic drug effects: a double-blind PET study of schizophrenic patients. *Biol Psychiatry*. 1993;33:227–235.
16. Peroutka SJ, Synder SH. Relationship of neuroleptic drug effects at brain dopamine, serotonin, alpha-adrenergic, and histamine receptors to clinical potency. *Am J Psychiatry*. 1980;137:1518–1522.
17. Kessler RM, Ansari MS, Riccardi P, et al. Occupancy of striatal and extrastriatal dopamine D2 receptors by clozapine and quetiapine. *Neuropsychopharmacology*. 2006;31:1991–2001.
18. Grunder G, Landvogt C, Vernaleken I, et al. The striatal and extrastriatal D2/D3 receptor-binding profile of clozapine in patients with schizophrenia. *Neuropsychopharmacology*. 2006;31:1027–1035.
19. Kessler RM, Ansari MS, Riccardi P, et al. Occupancy of striatal and extrastriatal dopamine D2/D3 receptors by olanzapine and haloperidol. *Neuropsychopharmacology*. 2005;30:2283–2289.

20. Pilowsky LS, Mulligan RS, Acton PD, et al. Limbic selectivity of clozapine. *Lancet*. 1997;350:490–491.
21. Talvik M, Nordstrom AL, Nyberg S, et al. No support for regional selectivity in clozapine-treated patients: a PET study with [(11)C]raclopride and [(11)C]FLB 457. *Am J Psychiatry*. 2001;158:926–930.
22. Kessler RM, Meltzer HY. Regional selectivity in clozapine treatment? *Am J Psychiatry*. 2002;159:1064–1065.
23. Olsson H, Farde L. Potentials and pitfalls using high affinity radioligands in PET and SPET determinations on regional drug induced D2 receptor occupancy–a simulation study based on experimental data. *Neuroimage*. 2001;14: 936–945.
24. Erlandsson K, Bressan RA, Mulligan RS, et al. Analysis of D2 dopamine receptor occupancy with quantitative SPET using the high-affinity ligand [123I]epidepride: resolving conflicting findings. *Neuroimage*. 2003;19:1205–1214.
25. Claghorn J, Honigfeld G, Abuzzahab FS Sr, et al. The risks and benefits of clozapine versus chlorpromazine. *J Clin Psychopharmacol*. 1987;7:377–384.
26. Kapur S, Seeman P. Does fast dissociation from the dopamine D(2) receptor explain the action of atypical antipsychotics?: A new hypothesis. *Am J Psychiatry*. 2001;158:360–369.
27. Farde L, Wiesel FA, Nordstrom AL, et al. D1- and D2-dopamine receptor occupancy during treatment with conventional and atypical neuroleptics. *Psychopharmacology (Berl)*. 1989;99(suppl):S28–S31.
28. Meltzer HY, Matsubara S, Lee JC. Classification of typical and atypical antipsychotic drugs on the basis of dopamine D-1, D-2 and serotonin2 pKi values. *J Pharmacol Exp Ther*. 1989;251:238–246.
29. Meltzer HY. Clinical studies on the mechanism of action of clozapine: the dopamine-serotonin hypothesis of schizophrenia. *Psychopharmacology (Berl)*. 1989;99(suppl):S18–S27.
30. de Paulis T. M-100907 (Aventis). *Curr Opin Investig Drugs*. 2001;2:123–132.
31. Meltzer HY, Arvanitis L, Bauer D, et al. Placebo-controlled evaluation of four novel compounds for the treatment of schizophrenia and schizoaffective disorder. *Am J Psychiatry*. 2004;161:975–984.
32. Pehek EA, Nocjar C, Roth BL, et al. Evidence for the preferential involvement of 5-HT2A serotonin receptors in stress- and drug-induced dopamine release in the rat medial prefrontal cortex. *Neuropsychopharmacology*. 2006;31:265–277.
33. Alex KD, Pehek EA. Pharmacologic mechanisms of serotonergic regulation of dopamine neurotransmission. *Pharmacol Ther*. 2007;113:296–320.
34. Shapiro DA, Renock S, Arrington E, et al. Aripiprazole, a novel atypical antipsychotic drug with a unique and robust pharmacology. *Neuropsychopharmacology*. 2003;28:1400–1411.
35. Goldman-Rakic PS, Castner SA, Svensson TH, et al. Targeting the dopamine D1 receptor in schizophrenia: insights for cognitive dysfunction. *Psychopharmacology (Berl)*. 2004;174:3–16.
36. Gray JA, Roth BL. Molecular targets for treating cognitive dysfunction in schizophrenia. *Schizophr Bull*. 2007;33:1100–1119.
37. Meltzer HY, Thompson PA, Lee MA, et al. Neuropsychologic deficits in schizophrenia: relation to social function and effect of antipsychotic drug treatment. *Neuropsychopharmacology*. 1996;14:27S–33S.
38. Keks NA. Impact of newer antipsychotics on outcomes in schizophrenia. *Clin Ther*. 1997;19:148-158; discussion 126–147.
39. Lysaker PH, Bell MD, Bioty S, et al. Performance on the Wisconsin Card Sorting Test as a predictor of rehospitalization in schizophrenia. *J Nerv Ment Dis*. 1996;184:319–321.
40. Castner SA, Williams GV, Goldman-Rakic PS. Reversal of antipsychotic-induced working memory deficits by short-term dopamine D1 receptor stimulation. *Science*. 2000;287:2020–2022.
41. Castner SA, Goldman-Rakic PS. Enhancement of working memory in aged monkeys by a sensitizing regimen of dopamine D1 receptor stimulation. *J Neurosci*. 2004;24:1446–1450.
42. Arnsten AF, Cai JX, Murphy BL, et al. Dopamine D1 receptor mechanisms in the cognitive performance of young adult and aged monkeys. *Psychopharmacology (Berl)*. 1994;116:143–151.
43. Schneider JS, Sun ZQ, Roeltgen DP. Effects of dihydrexidine, a full dopamine D-1 receptor agonist, on delayed response performance in chronic low dose MPTP-treated monkeys. *Brain Res*. 1994;663:140–144.
44. Sawaguchi T, Goldman-Rakic PS. D1 dopamine receptors in prefrontal cortex: involvement in working memory. *Science*. 1991;251:947–950.
45. Cai JX, Arnsten AF. Dose-dependent effects of the dopamine D1 receptor agonists A77636 or SKF81297 on spatial working memory in aged monkeys. *J Pharmacol Exp Ther*. 1997;283:183–189.
46. Stuchlik A, Vales K. Effect of dopamine D1 receptor antagonist SCH23390 and D1 agonist A77636 on active allothetic place avoidance, a spatial cognition task. *Behav Brain Res*. 2006;172:250–255.
47. Williams GV, Castner SA. Under the curve: critical issues for elucidating D1 receptor function in working memory. *Neuroscience*. 2006;139:263–276.
48. Goldman-Rakic PS, Muly EC 3rd, Williams GV. D(1) receptors in prefrontal cells and circuits. *Brain Res Brain Res Rev*. 2000;31:295–301.
49. Williams GV, Goldman-Rakic PS. Modulation of memory fields by dopamine D1 receptors in prefrontal cortex. *Nature*. 1995;376:572–575.
50. Pellissier G, Demenge P. Hypotensive and bradycardic effects elicited by spinal dopamine receptor stimulation: effects of D1 and D2 receptor agonists and antagonists. *J Cardiovasc Pharmacol*. 1991;18:548–555.
51. Lahlou S. Enhanced hypotensive response to intravenous apomorphine in chronic spinalized, conscious rats: role of spinal dopamine D(1) and D(2) receptors. *Neurosci Lett*. 2003;349:115–119.
52. Sokoloff P, Giros B, Martres MP, et al. Molecular cloning and characterization of a novel dopamine receptor (D3) as a target for neuroleptics. *Nature*. 1990;347:146–151.
53. Gurevich EV, Bordelon Y, Shapiro RM, et al. Mesolimbic dopamine D3 receptors and use of antipsychotics in patients with schizophrenia. A postmortem study. *Arch Gen Psychiatry*. 1997;54:225–232.
54. Joyce JN, Gurevich EV. D3 receptors and the actions of neuroleptics in the ventral striatopallidal system of schizophrenics. *Ann NY Acad Sci*. 1999;877:595–613.
55. Joyce JN. Dopamine D3 receptor as a therapeutic target for antipsychotic and antiparkinsonian drugs. *Pharmacol Ther*. 2001;90:231–259.
56. Joyce JN, Millan MJ. Dopamine D3 receptor antagonists as therapeutic agents. *Drug Discov Today*. 2005;10:917–925.
57. Millan MJ, Svenningsson P, Ashby CR Jr, et al. S33138 [N-[4-[2-[(3aS,9bR)-8-cyano-1,3a,4,9b-tetrahydro[1]-benzopyrano[3,4-c]pyrr ol-2(3H)-yl]-ethyl]phenylacetamide], a preferential dopamine D3 versus D2 receptor antagonist and potential antipsychotic agent. II. A neurochemical, electrophysiological and behavioral characterization in vivo. *J Pharmacol Exp Ther*. 2008;324:600–611.

58. Millan MJ, Mannoury la Cour C, Novi F, et al. S33138 [N-[4-[2-[(3aS,9bR)-8-cyano-1,3a,4,9b-tetrahydro[1]-benzopyrano[3,4-c]pyrr ol-2(3H)-yl]-ethyl]phenylacetamide], a preferential dopamine D3 versus D2 receptor antagonist and potential antipsychotic agent: I. Receptor-binding profile and functional actions at G-protein-coupled receptors. *J Pharmacol Exp Ther.* 2008;324:587–599.
59. Millan MJ, Loiseau F, Dekeyne A, et al. S33138 (N-[4-[2-[(3aS,9bR)-8-cyano-1,3a,4,9b-tetrahydro[1] benzopyrano[3,4-c]pyrrol-2(3H)-yl)-ethyl]phenyl-acetamide), a preferential dopamine D3 versus D2 receptor antagonist and potential antipsychotic agent: III. Actions in models of therapeutic activity and induction of side effects. *J Pharmacol Exp Ther.* 2008;324:1212–1226.
60. Van Tol HH, Bunzow JR, Guan HC, et al. Cloning of the gene for a human dopamine D4 receptor with high affinity for the antipsychotic clozapine. *Nature.* 1991;350:610–614.
61. Seeman P, Guan HC, Van Tol HH. Dopamine D4 receptors elevated in schizophrenia. *Nature.* 1993;365:441–445.
62. Tarazi FI, Zhang K, Baldessarini RJ. Dopamine D4 receptors: beyond schizophrenia. *J Recept Signal Transduct Res.* 2004;24:131–147.
63. Roth BL, Tandra S, Burgess LH, et al. D4 dopamine receptor binding affinity does not distinguish between typical and atypical antipsychotic drugs. *Psychopharmacology (Berl).* 1995;120:365–368.
64. Kramer MS, Last B, Getson A, et al. The effects of a selective D4 dopamine receptor antagonist (L-745,870) in acutely psychotic inpatients with schizophrenia. D4 Dopamine Antagonist Group. *Arch Gen Psychiatry.* 1997;54:567–572.
65. Bristow LJ, Kramer MS, Kulagowski J, et al. Schizophrenia and L-745,870, a novel dopamine D4 receptor antagonist. *Trends Pharmacol Sci.* 1997;18:186–188.
66. Corrigan MH, Gallen CC, Bonura ML, et al. Effectiveness of the selective D4 antagonist sonepiprazole in schizophrenia: a placebo-controlled trial. *Biol Psychiatry.* 2004;55:445–451.
67. Truffinet P, Tamminga CA, Fabre LF, et al. Placebo-controlled study of the D4/5-HT2A antagonist fananserin in the treatment of schizophrenia. *Am J Psychiatry.* 1999;156:419–425.
68. Arnsten AF, Murphy B, Merchant K. The selective dopamine D4 receptor antagonist, PNU-101387G, prevents stress-induced cognitive deficits in monkeys. *Neuropsychopharmacology.* 2000;23:405–410.
69. Jentsch JD, Taylor JR, Redmond DE Jr, et al. Dopamine D4 receptor antagonist reversal of subchronic phencyclidine-induced object retrieval/detour deficits in monkeys. *Psychopharmacology (Berl).* 1999;142:78–84.
70. Browman KE, Curzon P, Pan JB, et al. A-412997, a selective dopamine D4 agonist, improves cognitive performance in rats. *Pharmacol Biochem Behav.* 2005;82:148–155.
71. Rubinstein M, Cepeda C, Hurst RS, et al. Dopamine D4 receptor-deficient mice display cortical hyperexcitability. *J Neurosci.* 2001;21:3756–3763.
72. Wang X, Zhong P, Yan Z. Dopamine D4 receptors modulate GABAergic signaling in pyramidal neurons of prefrontal cortex. *J Neurosci.* 2002;22:9185–9193.
73. Tunbridge EM, Harrison PJ, Weinberger DR. Catechol-O-methyltransferase, cognition, and psychosis: Val158Met and beyond. *Biol Psychiatry.* 2006;60:141–151.
74. Karoum F, Chrapusta SJ, Egan MF. 3-Methoxytyramine is the major metabolite of released dopamine in the rat frontal cortex: reassessment of the effects of antipsychotics on the dynamics of dopamine release and metabolism in the frontal cortex, nucleus accumbens, and striatum by a simple two pool model. *J Neurochem.* 1994;63:972–979.
75. Gogos JA, Morgan M, Luine V, et al. Catechol-O-methyltransferase-deficient mice exhibit sexually dimorphic changes in catecholamine levels and behavior. *Proc Natl Acad Sci USA.* 1998;95:9991–9996.
76. Liljequist R, Haapalinna A, Ahlander M, et al. Catechol-O-methyltransferase inhibitor tolcapone has minor influence on performance in experimental memory models in rats. *Behav Brain Res.* 1997;82:195–202.
77. Apud JA, Mattay V, Chen J, et al. Tolcapone improves cognition and cortical information processing in normal human subjects. *Neuropsychopharmacology.* 2007;32:1011–1020.
78. Olanow CW, Watkins PB. Tolcapone: an efficacy and safety review (2007). *Clin Neuropharmacol.* 2007;30:287–294.
79. Lew MF, Kricorian G. Results from a 2-year centralized tolcapone liver enzyme monitoring program. *Clin Neuropharmacol.* 2007;30:281–286.
80. Lachman HM, Papolos DF, Saito T, et al. Human catechol-O-methyltransferase pharmacogenetics: description of a functional polymorphism and its potential application to neuropsychiatric disorders. *Pharmacogenetics.* 1996;6:243–250.
81. Egan MF, Goldberg TE, Kolachana BS, et al. Effect of COMT Val108/158 Met genotype on frontal lobe function and risk for schizophrenia. *Proc Natl Acad Sci USA.* 2001;98:6917–6922.
82. Woodward ND, Jayathilake K, Meltzer HY. COMT val108/158met genotype, cognitive function, and cognitive improvement with clozapine in schizophrenia. *Schizophr Res.* 2007;90:86–96.
83. Jordan S, Koprivica V, Chen R, et al. The antipsychotic aripiprazole is a potent, partial agonist at the human 5-HT1A receptor. *Eur J Pharmacol.* 2002;441:137–140.
84. Meltzer HY, Li Z, Kaneda Y, et al. Serotonin receptors: their key role in drugs to treat schizophrenia. *Prog Neuropsychopharmacol Biol Psychiatry.* 2003;27:1159–1172.
85. Millan MJ. Improving the treatment of schizophrenia: focus on serotonin (5-HT)(1A) receptors. *J Pharmacol Exp Ther.* 2000;295:853–861.
86. Azmitia EC, Gannon PJ, Kheck NM, et al. Cellular localization of the 5-HT1A receptor in primate brain neurons and glial cells. *Neuropsychopharmacology.* 1996;14:35–46.
87. Ichikawa J, Ishii H, Bonaccorso S, et al. 5-HT(2A) and D(2) receptor blockade increases cortical DA release via 5-HT(1A) receptor activation: a possible mechanism of atypical antipsychotic-induced cortical dopamine release. *J Neurochem.* 2001;76:1521–1531.
88. Gobert A, Rivet JM, Audinot V, et al. Simultaneous quantification of serotonin, dopamine and noradrenaline levels in single frontal cortex dialysates of freely-moving rats reveals a complex pattern of reciprocal auto- and heteroreceptor-mediated control of release. *Neuroscience.* 1998;84:413–429.
89. Roth BL, Ciaranello RD, Meltzer HY. Binding of typical and atypical antipsychotic agents to transiently expressed 5-HT1C receptors. *J Pharmacol Exp Ther.* 1992;260:1361–1365.
90. Rauser L, Savage JE, Meltzer HY, et al. Inverse agonist actions of typical and atypical antipsychotic drugs at the human 5-hydroxytryptamine(2C) receptor. *J Pharmacol Exp Ther.* 2001;299:83–89.
91. Millan MJ, Dekeyne A, Gobert A. Serotonin (5-HT)2C receptors tonically inhibit dopamine (DA) and noradrenaline (NA), but not 5-HT, release in the frontal cortex in vivo. *Neuropharmacology.* 1998;37:953–955.

92. Marquis KL, Sabb AL, Logue SF, et al. WAY-163909 [(7bR,10aR)-1,2,3,4,8,9,10,10a-octahydro-7bH-cyclopenta-[b][1,4]diazepino[6,7,1hi]indole]: a novel 5-hydroxytryptamine 2C receptor-selective agonist with preclinical antipsychotic-like activity. *J Pharmacol Exp Ther.* 2007;320:486–496.
93. Siuciak JA, Chapin DS, McCarthy SA, et al. CP-809,101, a selective 5-HT2C agonist, shows activity in animal models of antipsychotic activity. *Neuropharmacology.* 2007;52:279–290.
94. Roth BL. Drugs and valvular heart disease. *N Engl J Med.* 2007;356:6–9.
95. Reynolds GP, Mason SL, Meldrum A, et al. 5-Hydroxytryptamine (5-HT)4 receptors in post mortem human brain tissue: distribution, pharmacology and effects of neurodegenerative diseases. *Br J Pharmacol.* 1995;114:993–998.
96. Cho S, Hu Y. Activation of 5-HT4 receptors inhibits secretion of beta-amyloid peptides and increases neuronal survival. *Exp Neurol.* 2007;203:274–278.
97. Meneses A, Hong E. Effects of 5-HT4 receptor agonists and antagonists in learning. *Pharmacol Biochem Behav.* 1997;56:347–351.
98. Roth BL, Craigo SC, Choudhary MS, et al. Binding of typical and atypical antipsychotic agents to 5-hydroxytryptamine-6 and 5-hydroxytryptamine-7 receptors. *J Pharmacol Exp Ther.* 1994.268:1403–1410.
99. Mitchell ES, Neumaier JF. 5-HT6 receptors: a novel target for cognitive enhancement. *Pharmacol Ther.* 2005;108:320–333.
100. East SZ, Burnet PW, Kerwin RW, et al. An RT-PCR study of 5-HT(6) and 5-HT(7) receptor mRNAs in the hippocampal formation and prefrontal cortex in schizophrenia. *Schizophr Res.* 2002;57:15–26.
101. Thomas DR, Hagan JJ. 5-HT7 receptors. *Curr Drug Targets CNS Neurol Disord.* 2004;3:81–90.
102. Roberts AJ, Krucker T, Levy CL, et al. Mice lacking 5-HT receptors show specific impairments in contextual learning. *Eur J Neurosci.* 2004;19:1913–1922.
103. Sarter M, Bruno JP. Cognitive functions of cortical acetylcholine: toward a unifying hypothesis. *Brain Res Brain Res Rev.* 1997;23:28–46.
104. Friedman JI. Cholinergic targets for cognitive enhancement in schizophrenia: focus on cholinesterase inhibitors and muscarinic agonists. *Psychopharmacology (Berl).* 2004;174:45–53.
105. Holt DJ, Bachus SE, Hyde TM, et al. Reduced density of cholinergic interneurons in the ventral striatum in schizophrenia: an in situ hybridization study. *Biol Psychiatry.* 2005;58: 408–416.
106. Holt DJ, Herman MM, Hyde TM, et al. Evidence for a deficit in cholinergic interneurons in the striatum in schizophrenia. *Neuroscience.* 1999;94:21–31.
107. Powchik P, Davidson M, Haroutunian V, et al. Postmortem studies in schizophrenia. *Schizophr Bull.* 1998;24:325–341.
108. Ichikawa J, Dai J, O'Laughlin IA, et al. Atypical, but not typical, antipsychotic drugs increase cortical acetylcholine release without an effect in the nucleus accumbens or striatum. *Neuropsychopharmacology.* 2002;26:325–339.
109. Ichikawa J, Li Z, Dai J, et al. Atypical antipsychotic drugs, quetiapine, iloperidone, and melperone, preferentially increase dopamine and acetylcholine release in rat medial prefrontal cortex: role of 5-HT1A receptor agonism. *Brain Res.* 2002;956:349–357.
110. Ichikawa J, Dai J, Meltzer HY. 5-HT(1A) and 5-HT(2A) receptors minimally contribute to clozapine-induced acetylcholine release in rat medial prefrontal cortex. *Brain Res.* 2002;939:34–42.
111. Birks J. Cholinesterase inhibitors for Alzheimer's disease. *Cochrane Database Syst Rev.* 2006;CD005593.
112. Nahas Z, George MS, Horner MD, et al. Augmenting atypical antipsychotics with a cognitive enhancer (donepezil) improves regional brain activity in schizophrenia patients: a pilot double-blind placebo controlled BOLD fMRI study. *Neurocase.* 2003;9:274–282.
113. Buchanan RW, Summerfelt A, Tek C, et al. An open-labeled trial of adjunctive donepezil for cognitive impairments in patients with schizophrenia. *Schizophr Res.* 2003;59:29–33.
114. Friedman JI, Adler DN, Howanitz E, et al. A double-blind placebo-controlled trial of donepezil adjunctive treatment to risperidone for the cognitive impairment of schizophrenia. *Biol Psychiatry.* 2002;51:349–357.
115. Stryjer R, Strous RD, Bar F, et al. Beneficial effect of donepezil augmentation for the management of comorbid schizophrenia and dementia. *Clin Neuropharmacol.* 2003;26:12–17.
116. Ferreri F, Agbokou C, Gauthier S. Cognitive dysfunctions in schizophrenia: potential benefits of cholinesterase inhibitor adjunctive therapy. *J Psychiatry Neurosci.* 2006;31:369–376.
117. Raedler TJ, Bymaster FP, Tandon R, et al. Towards a muscarinic hypothesis of schizophrenia. *Mol Psychiatry.* 2007;12:232–246.
118. Dean B, Crook JM, Opeskin K, et al. The density of muscarinic M1 receptors is decreased in the caudate-putamen of subjects with schizophrenia. *Mol Psychiatry.* 1996;1:54–58.
119. Crook JM, Tomaskovic-Crook E, Copolov DL, et al. Low muscarinic receptor binding in prefrontal cortex from subjects with schizophrenia: a study of Brodmann's areas 8, 9, 10, and 46 and the effects of neuroleptic drug treatment. *Am J Psychiatry.* 2001;158:918–925.
120. Dean B, McLeod M, Keriakous D, et al. Decreased muscarinic1 receptors in the dorsolateral prefrontal cortex of subjects with schizophrenia. *Mol Psychiatry.* 2002;7:1083–1091.
121. Raedler TJ. Comparison of the in-vivo muscarinic cholinergic receptor availability in patients treated with clozapine and olanzapine. *Int J Neuropsychopharmacol.* 2007;10:275–280.
122. Raedler TJ, Knable MB, Jones DW, et al. Central muscarinic acetylcholine receptor availability in patients treated with clozapine. *Neuropsychopharmacology.* 2003;28:1531–1537.
123. Raedler TJ, Knable MB, Jones DW, et al. In vivo olanzapine occupancy of muscarinic acetylcholine receptors in patients with schizophrenia. *Neuropsychopharmacology.* 2000;23:56–68.
124. Lavalaye J, Booij J, Linszen DH, et al. Higher occupancy of muscarinic receptors by olanzapine than risperidone in patients with schizophrenia. A[123I]-IDEX SPECT study. *Psychopharmacology (Berl).* 2001;156:53–57.
125. Sullivan RJ, Allen JS, Otto C, et al. Effects of chewing betel nut (*Areca catechu*) on the symptoms of people with schizophrenia in Palau, Micronesia. *Br J Psychiatry.* 2000;177:174–178.
126. Mirza NR, Peters D, Sparks RG. Xanomeline and the antipsychotic potential of muscarinic receptor subtype selective agonists. *CNS Drug Rev.* 2003;9:159–186.
127. Shannon HE, Hart JC, Bymaster FP, et al. Muscarinic receptor agonists, like dopamine receptor antagonist antipsychotics, inhibit conditioned avoidance response in rats. *J Pharmacol Exp Ther.* 1999;290:901–907.
128. Andersen MB, Fink-Jensen A, Peacock L, et al. The muscarinic M1/M4 receptor agonist xanomeline exhibits antipsychotic-like activity in *Cebus apella* monkeys. *Neuropsychopharmacology.* 2003;28:1168–1175.
129. Stanhope KJ, Mirza NR, Bickerdike MJ, et al. The muscarinic receptor agonist xanomeline has an antipsychotic-like profile in the rat. *J Pharmacol Exp Ther.* 2001;299:782–792.

130. Shekhar A, Potter WZ, Lightfoot J, et al. Selective muscarinic receptor agonist xanomeline as a novel treatment approach for schizophrenia. *Am J Psychiatry.* 2008;165:1033–1039.
131. Davies MA, Compton-Toth BA, Hufeisen SJ, et al. The highly efficacious actions of N-desmethylclozapine at muscarinic receptors are unique and not a common property of either typical or atypical antipsychotic drugs: is M1 agonism a pre-requisite for mimicking clozapine's actions? *Psychopharmacology (Berl).* 2005;178:451–460.
132. Weiner DM, Meltzer HY, Veinbergs I, et al. The role of M1 muscarinic receptor agonism of N-desmethylclozapine in the unique clinical effects of clozapine. *Psychopharmacology (Berl).* 2004;177:207–216.
133. Burstein ES, Ma J, Wong S, et al. Intrinsic efficacy of antipsychotics at human D2, D3, and D4 dopamine receptors: identification of the clozapine metabolite N-desmethylclozapine as a D2/D3 partial agonist. *J Pharmacol Exp Ther.* 2005;315:1278–1287.
134. Chan WY, McKinzie DL, Bose S, et al. Allosteric modulation of the muscarinic M4 receptor as an approach to treating schizophrenia. *Proc Natl Acad Sci USA.* 2008;105:10978–10983.
135. Shirey J, Xiang Z, Orton D, et al. An allosteric potentiator of M4 mAChR modulates hippocampal synaptic transmission. *Nat Chem Biol.* 2008;4:42–50.
136. Kumari V, Postma P. Nicotine use in schizophrenia: the self medication hypotheses. *Neurosci Biobehav Rev.* 2005; 29:1021–1034.
137. Dalack GW, Healy DJ, Meador-Woodruff JH. Nicotine dependence in schizophrenia: clinical phenomena and laboratory findings. *Am J Psychiatry.* 1998;155:1490–1501.
138. Adler LE, Hoffer LD, Wiser A, et al. Normalization of auditory physiology by cigarette smoking in schizophrenic patients. *Am J Psychiatry.* 1993;150:1856–1861.
139. Sacco KA, Termine A, Seyal A, et al. Effects of cigarette smoking on spatial working memory and attentional deficits in schizophrenia: involvement of nicotinic receptor mechanisms. *Arch Gen Psychiatry.* 2005;62:649–659.
140. Yang YK, Nelson L, Kamaraju L, et al. Nicotine decreases bradykinesia-rigidity in haloperidol-treated patients with schizophrenia. *Neuropsychopharmacology.* 2002;27:684–686.
141. Levin ED, Wilson W, Rose JE, et al. Nicotine–haloperidol interactions and cognitive performance in schizophrenics. *Neuropsychopharmacology.* 1996;15:429–436.
142. Smith RC, Singh A, Infante M, et al. Effects of cigarette smoking and nicotine nasal spray on psychiatric symptoms and cognition in schizophrenia. *Neuropsychopharmacology.* 2002;27: 479–497.
143. Freedman R, Coon H, Myles-Worsley M, et al. Linkage of a neurophysiological deficit in schizophrenia to a chromosome 15 locus. *Proc Natl Acad Sci USA.* 1997;94:587–592.
144. Leonard S, Breese C, Adams C, et al. Smoking and schizophrenia: abnormal nicotinic receptor expression. *Eur J Pharmacol.* 2000;393:237–242.
145. Freedman R, Hall M, Adler LE, et al. Evidence in postmortem brain tissue for decreased numbers of hippocampal nicotinic receptors in schizophrenia. *Biol Psychiatry.* 1995;38:22–33.
146. Breese CR, Lee MJ, Adams CE, et al. Abnormal regulation of high affinity nicotinic receptors in subjects with schizophrenia. *Neuropsychopharmacology.* 2000;23:351–364.
147. Luntz-Leybman V, Bickford PC, Freedman R. Cholinergic gating of response to auditory stimuli in rat hippocampus. *Brain Res.* 1992;587:130–136.
148. Simosky JK, Stevens KE, Adler LE, et al. Clozapine improves deficient inhibitory auditory processing in DBA/2 mice, via a nicotinic cholinergic mechanism. *Psychopharmacology (Berl).* 2003;165:386–396.
149. Simosky JK, Stevens KE, Freedman R. Nicotinic agonists and psychosis. *Curr Drug Targets CNS Neurol Disord.* 2002;1: 149–162.
150. Levin ED, McClernon FJ, Rezvani AH. Nicotinic effects on cognitive function: behavioral characterization, pharmacological specification, and anatomic localization. *Psychopharmacology (Berl).* 2006;184:523–539.
151. Simosky JK, Stevens KE, Kem WR, et al. Intragastric DMXB-A, an alpha7 nicotinic agonist, improves deficient sensory inhibition in DBA/2 mice. *Biol Psychiatry.* 2001;50:493–500.
152. Bontempi B, Whelan KT, Risbrough VB, et al. SIB-1553A, (+/−)-4-[[2-(1-methyl-2-pyrrolidinyl)ethyl]thio]phenol hydrochloride, a subtype-selective ligand for nicotinic acetylcholine receptors with putative cognitive-enhancing properties: effects on working and reference memory performances in aged rodents and nonhuman primates. *J Pharmacol Exp Ther.* 2001;299:297–306.
153. Olincy A, Harris JG, Johnson LL, et al. Proof-of-concept trial of an alpha7 nicotinic agonist in schizophrenia. *Arch Gen Psychiatry.* 2006;63:630–638.
154. Freedman R, Olincy A, Buchanan RW, et al. Initial phase 2 trial of a nicotinic agonist in schizophrenia. *Am J Psychiatry.* 2008;165:1040–1047.
155. Benwell ME, Balfour DJ, Birrell CE. Desensitization of the nicotine-induced mesolimbic dopamine responses during constant infusion with nicotine. *Br J Pharmacol.* 1995;114:454–460.
156. Briggs CA, Anderson DJ, Brioni JD, et al. Functional characterization of the novel neuronal nicotinic acetylcholine receptor ligand GTS-21 in vitro and in vivo. *Pharmacol Biochem Behav.* 1997;57:231–241.
157. Gray JA, Roth BL. The pipeline and future of drug development in schizophrenia. *Mol Psychiatry.* 2007;12:904–922.
158. Kelly DL, McMahon RP, Weiner E, et al. Lack of beneficial galantamine effect for smoking behavior: a double-blind randomized trial in people with schizophrenia. *Schizophr Res.* 2008;103:161–168.
159. Dyer MA, Freudenreich O, Culhane MA, et al. High-dose galantamine augmentation inferior to placebo on attention, inhibitory control and working memory performance in nonsmokers with schizophrenia. *Schizophr Res.* 2008;102:88–95.
160. Allen TB, McEvoy JP. Galantamine for treatment-resistant schizophrenia. *Am J Psychiatry.* 2002;159:1244–1245.
161. Buchanan RW, Conley RR, Dickinson D, et al. Galantamine for the treatment of cognitive impairments in people with schizophrenia. *Am J Psychiatry.* 2008;165:82–89.
162. Lee SW, Lee JG, Lee BJ, et al. A 12-week, double-blind, placebo-controlled trial of galantamine adjunctive treatment to conventional antipsychotics for the cognitive impairments in chronic schizophrenia. *Int Clin Psychopharmacol.* 2007; 22:63–68.
163. Sacco KA, Creeden C, Reutenauer EL, et al. Effects of galantamine on cognitive deficits in smokers and non-smokers with schizophrenia. *Schizophr Res.* 2008;103:326–327.
164. Domino EF, Chodoff P, Corssen G. Pharmacologic effects of Ci-581, a new dissociative anesthetic, in man. *Clin Pharmacol Ther.* 1965;6:279–291.
165. Javitt DC. Negative schizophrenic symptomatology and the PCP (phencyclidine) model of schizophrenia. *Hillside J Clin Psychiatr.* 1987;9:12–35.
166. Javitt DC. Glutamate as a therapeutic target in psychiatric disorders. *Mol Psychiatry.* 2004;9:984–997, 979.

167. Olney JW, Newcomer JW, Farber NB. NMDA receptor hypofunction model of schizophrenia. *J Psychiatr Res.* 1999;33:523–533.
168. Newcomer JW, Farber NB, Jevtovic-Todorovic V, et al. Ketamine-induced NMDA receptor hypofunction as a model of memory impairment and psychosis. *Neuropsychopharmacology.* 1999;20:106–118.
169. Javitt DC. Is the glycine site half saturated or half unsaturated? Effects of glutamatergic drugs in schizophrenia patients. *Curr Opin Psychiatry.* 2006;19:151–157.
170. Shoham S, Javitt DC, Heresco-Levy U. High dose glycine nutrition affects glial cell morphology in rat hippocampus and cerebellum. *Int J Neuropsychopharmacol.* 1999;2:35–40.
171. Shoham S, Javitt DC, Heresco-Levy U. Chronic high-dose glycine nutrition: effects on rat brain cell morphology. *Biol Psychiatry.* 2001;49:876–885.
172. Heresco-Levy U, Javitt DC, Ermilov M, et al. Double-blind, placebo-controlled, crossover trial of glycine adjuvant therapy for treatment-resistant schizophrenia. *Br J Psychiatry.* 1996;169:610–617.
173. Heresco-Levy U, Silipo G, Javitt DC. Glycinergic augmentation of NMDA receptor-mediated neurotransmission in the treatment of schizophrenia. *Psychopharmacol Bull.* 1996;32:731–740.
174. Hood WF, Compton RP, Monahan JB. D-Cycloserine: a ligand for the N-methyl-D-aspartate coupled glycine receptor has partial agonist characteristics. *Neurosci Lett.* 1989;98:91–95.
175. Evins AE, Fitzgerald SM, Wine L, et al. Placebo-controlled trial of glycine added to clozapine in schizophrenia. *Am J Psychiatry.* 2000;157:826–828.
176. Tsai GE, Yang P, Chung LC, et al. D-Serine added to clozapine for the treatment of schizophrenia. *Am J Psychiatry.* 1999;156:1822–1825.
177. Goff DC, Tsai G, Manoach DS, et al. D-Cycloserine added to clozapine for patients with schizophrenia. *Am J Psychiatry.* 1996;153:1628–1630.
178. Depoortere R, Dargazanli G, Estenne-Bouhtou G, et al. Neurochemical, electrophysiological and pharmacological profiles of the selective inhibitor of the glycine transporter-1 SSR504734, a potential new type of antipsychotic. *Neuropsychopharmacology.* 2005;30:1963–1985.
179. Lane HY, Chang YC, Liu YC, et al. Sarcosine or D-serine add-on treatment for acute exacerbation of schizophrenia: a randomized, double-blind, placebo-controlled study. *Arch Gen Psychiatry.* 2005;62:1196–1204.
180. Lane HY, Huang C, Wu PL, et al. Glycine transporter I inhibitor, N-methylglycine (sarcosine), added to clozapine for the treatment of schizophrenia. *Biol Psychiatry.* 2006;60:645–649.
181. Brown A, Carlyle I, Clark J, et al. Discovery and SAR of org 24598—a selective glycine uptake inhibitor. *Bioorg Med Chem Let.* 2001;11:2007–2009.
182. Aubrey KR, Vandenberg RJ. N[3-(4'-fluorophenyl)-3-(4'-phenylphenoxy)propyl]sarcosine (NFPS) is a selective persistent inhibitor of glycine transport. *Br J Pharmacol.* 2001;134:1429–1436.
183. Suppiramaniam V, Bahr BA, Sinnarajah S, et al. Member of the ampakine class of memory enhancers prolongs the single channel open time of reconstituted AMPA receptors. *Synapse.* 2001;40:154–158.
184. Black MD. Therapeutic potential of positive AMPA modulators and their relationship to AMPA receptor subunits. A review of preclinical data. *Psychopharmacology (Berl).* 2005;179:154–163.
185. Goff DC, Leahy L, Berman I, et al. A placebo-controlled pilot study of the ampakine CX516 added to clozapine in schizophrenia. *J Clin Psychopharmacol.* 2001;21:484–487.
186. Marenco S, Egan MF, Goldberg TE, et al. Preliminary experience with an ampakine (CX516) as a single agent for the treatment of schizophrenia: a case series. *Schizophr Res.* 2002;57:221–226.
187. Goff DC, Lamberti JS, Leon AC, et al. A placebo-controlled add-on trial of the ampakine, CX516, for cognitive deficits in schizophrenia. *Neuropsychopharmacology.* 2008;33:465–472.
188. Lauterborn JC, Truong GS, Baudry M, et al. Chronic elevation of brain-derived neurotrophic factor by ampakines. *J Pharmacol Exp Ther.* 2003;307:297–305.
189. Mathe JM, Fagerquist MV, Svensson TH. Antipsychotic-like effect of the AMPA receptor antagonist LY326325 as indicated by suppression of conditioned avoidance response in the rat. *J Neural Transm.* 1999;106:1003–1009.
190. Vanover KE. Effects of AMPA receptor antagonists on dopamine-mediated behaviors in mice. *Psychopharmacology (Berl).* 1998;136:123–131.
191. Moghaddam B, Adams B, Verma A, et al. Activation of glutamatergic neurotransmission by ketamine: a novel step in the pathway from NMDA receptor blockade to dopaminergic and cognitive disruptions associated with the prefrontal cortex. *J Neurosci.* 1997;17:2921–2927.
192. Katayama T, Jodo E, Suzuki Y, et al. Activation of medial prefrontal cortex neurons by phencyclidine is mediated via AMPA/kainate glutamate receptors in anesthetized rats. *Neuroscience.* 2007;150:442–448.
193. Moghaddam B. Targeting metabotropic glutamate receptors for treatment of the cognitive symptoms of schizophrenia. *Psychopharmacology (Berl).* 2004;174:39–44.
194. Kinney GG, Burno M, Campbell UC, et al. Metabotropic glutamate subtype 5 receptors modulate locomotor activity and sensorimotor gating in rodents. *J Pharmacol Exp Ther.* 2003;306:116–123.
195. Maeda J, Suhara T, Okauchi T, et al. Different roles of group I and group II metabotropic glutamate receptors on phencyclidine-induced dopamine release in the rat prefrontal cortex. *Neurosci Lett.* 2003;336:171–174.
196. Moghaddam B, Adams BW. Reversal of phencyclidine effects by a group II metabotropic glutamate receptor agonist in rats. *Science.* 1998;281:1349–1352.
197. Lorrain DS, Baccei CS, Bristow LJ, et al. Effects of ketamine and N-methyl-D-aspartate on glutamate and dopamine release in the rat prefrontal cortex: modulation by a group II selective metabotropic glutamate receptor agonist LY379268. *Neuroscience.* 2003;117:697–706.
198. Schoepp DD, Marek GJ. Preclinical pharmacology of mGlu2/3 receptor agonists: novel agents for schizophrenia? *Curr Drug Targets CNS Neurol Disord.* 2002;1:215–225.
199. Homayoun H, Jackson ME, Moghaddam B. Activation of metabotropic glutamate 2/3 receptors reverses the effects of NMDA receptor hypofunction on prefrontal cortex unit activity in awake rats. *J Neurophysiol.* 2005;93:1989–2001.
200. Johnson MP, Baez M, Jagdmann GE Jr, et al. Discovery of allosteric potentiators for the metabotropic glutamate 2 receptor: synthesis and subtype selectivity of N-(4-(2-methoxyphenoxy)phenyl)-N-(2,2,2- trifluoroethylsulfonyl)pyrid-3-ylmethylamine. *J Med Chem.* 2003;46:3189–3192.
201. Govek SP, Bonnefous C, Hutchinson JH, et al. Benzazoles as allosteric potentiators of metabotropic glutamate receptor 2

(mGluR2): efficacy in an animal model for schizophrenia. *Bioorg Med Chem Lett.* 2005;15:4068–4072.
202. Marino MJ, Conn PJ. Glutamate-based therapeutic approaches: allosteric modulators of metabotropic glutamate receptors. *Curr Opin Pharmacol.* 2006;6:98–102.
203. Galici R, Jones CK, Hemstapat K, et al. Biphenyl-indanone A, a positive allosteric modulator of the metabotropic glutamate receptor subtype 2, has antipsychotic- and anxiolytic-like effects in mice. *J Pharmacol Exp Ther.* 2006;318:173–185.
204. Patil ST, Zhang L, Martenyi F, et al. Activation of mGlu2/3 receptors as a new approach to treat schizophrenia: a randomized Phase 2 clinical trial. *Nat Med.* 2007;13:1102–1107.
205. Coull JT. Pharmacological manipulations of the alpha 2-noradrenergic system. Effects on cognition. *Drugs Aging.* 1994;5:116–126.
206. Arnsten AF. Adrenergic targets for the treatment of cognitive deficits in schizophrenia. *Psychopharmacology (Berl).* 2004;174:25–31.
207. Richelson E, Nelson A. Antagonism by neuroleptics of neurotransmitter receptors of normal human brain in vitro. *Eur J Pharmacol.* 1984;103:197–204.
208. Friedman JI, Adler DN, Temporini HD, et al. Guanfacine treatment of cognitive impairment in schizophrenia. *Neuropsychopharmacology.* 2001;25:402–409.
209. Fields RB, Van Kammen DP, Peters JL, et al. Clonidine improves memory function in schizophrenia independently from change in psychosis. Preliminary findings. *Schizophr Res.* 1988;1:417–423.
210. Litman RE, Su TP, Potter WZ, et al. Idazoxan and response to typical neuroleptics in treatment-resistant schizophrenia. Comparison with the atypical neuroleptic, clozapine. *Br J Psychiatry.* 1996;168:571–579.
211. Eggan SM, Hashimoto T, Lewis DA. Reduced cortical cannabinoid 1 receptor messenger RNA and protein expression in schizophrenia. *Arch Gen Psychiatry.* 2008;65:772–784.
212. Henquet C, Krabbendam L, Spauwen J, et al. Prospective cohort study of cannabis use, predisposition for psychosis, and psychotic symptoms in young people. *BMJ.* 2005;330:11–14.
213. Henquet C, Murray R, Linszen D, et al. The environment and schizophrenia: the role of cannabis use. *Schizophr Bull.* 2005;31:608–612.
214. Moore TH, Zammit S, Lingford-Hughes A, et al. Cannabis use and risk of psychotic or affective mental health outcomes: a systematic review. *Lancet.* 2007;370:319–328.
215. D'Souza DC, Perry E, MacDougall L, et al. The psychotomimetic effects of intravenous delta-9-tetrahydrocannabinol in healthy individuals: implications for psychosis. *Neuropsychopharmacology.* 2004;29:1558–1572.
216. Zuardi AW, Crippa JA, Hallak JE, et al. Cannabidiol, a *Cannabis sativa* constituent, as an antipsychotic drug. *Braz J Med Biol Res.* 2006;39:421–429.
217. Leweke FM, Schneider U, Radwan M, et al. Different effects of nabilone and cannabidiol on binocular depth inversion in Man. *Pharmacol Biochem Behav.* 2000;66:175–181.
218. Morgan CJ, Curran HV. Effects of cannabidiol on schizophrenia-like symptoms in people who use cannabis. *Br J Psychiatry.* 2008;192:306–307.
219. Alonso R, Voutsinos B, Fournier M, et al. Blockade of cannabinoid receptors by SR141716 selectively increases Fos expression in rat mesocorticolimbic areas via reduced dopamine D2 function. *Neuroscience.* 1999;91:607–620.
220. Poncelet M, Barnouin MC, Breliere JC, et al. Blockade of cannabinoid (CB) receptors by 141716 selectively antagonizes drug-induced reinstatement of exploratory behaviour in gerbils. *Psychopharmacology (Berl).* 1999;144:144–150.
221. Ballmaier M, Bortolato M, Rizzetti C, et al. Cannabinoid receptor antagonists counteract sensorimotor gating deficits in the phencyclidine model of psychosis. *Neuropsychopharmacology.* 2007;32:2098–2107.
222. Chahl LA. Tachykinins and neuropsychiatric disorders. *Curr Drug Targets.* 2006;7:993–1003.
223. Roberts GW, Ferrier IN, Lee Y, et al. Peptides, the limbic lobe and schizophrenia. *Brain Res.* 1983;288:199–211.
224. Tooney PA, Crawter VC, Chahl LA. Increased tachykinin NK(1) receptor immunoreactivity in the prefrontal cortex in schizophrenia. *Biol Psychiatry.* 2001;49:523–527.
225. Rupniak NM, Kramer MS. Discovery of the antidepressant and anti-emetic efficacy of substance P receptor (NK1) antagonists. *Trends Pharmacol Sci.* 1999;20:485–490.
226. Spooren W, Riemer C, Meltzer H. Opinion: NK3 receptor antagonists: the next generation of antipsychotics? *Nat Rev Drug Discov.* 2005;4:967–975.
227. Dawson LA, Cato KJ, Scott C, et al. In vitro and in vivo characterization of the non-peptide NK3 receptor antagonist SB-223412 (talnetant): potential therapeutic utility in the treatment of schizophrenia. *Neuropsychopharmacology.* 2008;33:1642–1652.
228. Caceda R, Kinkead B, Nemeroff CB. Neurotensin: role in psychiatric and neurological diseases. *Peptides.* 2006;27:2385–2404.
229. Binder EB, Kinkead B, Owens MJ, et al. The role of neurotensin in the pathophysiology of schizophrenia and the mechanism of action of antipsychotic drugs. *Biol Psychiatry.* 2001;50:856–872.
230. Feifel D, Reza TL, Wustrow DJ, et al. Novel antipsychotic-like effects on prepulse inhibition of startle produced by a neurotensin agonist. *J Pharmacol Exp Ther.* 1999;288:710–713.
231. Boules M, Warrington L, Fauq A, et al. A novel neurotensin analog blocks cocaine- and D-amphetamine-induced hyperactivity. *Eur J Pharmacol.* 2001;426: 73–76.
232. Scatton B, Sanger DJ. Pharmacological and molecular targets in the search for novel antipsychotics. *Behav Pharmacol.* 2000;11:243–256.
233. Azzi M, Gully D, Heaulme M, et al. Neurotensin receptor interaction with dopaminergic systems in the guinea-pig brain shown by neurotensin receptor antagonists. *Eur J Pharmacol.* 1994;255:167–174.
234. Azzi M, Betancur C, Sillaber I, et al. Repeated administration of the neurotensin receptor antagonist SR 48692 differentially regulates mesocortical and mesolimbic dopaminergic systems. *J Neurochem.* 1998;71:1158–1167.
235. Betancur C, Cabrera R, de Kloet ER, et al. Role of endogenous neurotensin in the behavioral and neuroendocrine effects of cocaine. *Neuropsychopharmacology.* 1998;19: 322–332.
236. Martin WR, Eades CG, Thompson JA, et al. The effects of morphine- and nalorphine-like drugs in the nondependent and morphine-dependent chronic spinal dog. *J Pharmacol Exp Ther.* 1976;197:517–532.
237. Tam SW, Cook L. Sigma opiates and certain antipsychotic drugs mutually inhibit (+)-[3H] SKF 10,047 and [3H]haloperidol binding in guinea pig brain membranes. *Proc Natl Acad Sci USA.* 1984;81:5618–5621.
238. Ishiguro H, Ohtsuki T, Toru M, et al. Association between polymorphisms in the type 1 sigma receptor gene and schizophrenia. *Neurosci Lett.* 1998;257:45–48.

239. Debonnel G, de Montigny C. Modulation of NMDA and dopaminergic neurotransmissions by sigma ligands: possible implications for the treatment of psychiatric disorders. *Life Sci.* 1996;58:721–734.
240. Okuyama S, Imagawa Y, Sakagawa T, et al. NE-100, a novel sigma receptor ligand: effect on phencyclidine-induced behaviors in rats, dogs and monkeys. *Life Sci.* 1994;55:PL133-PL138.
241. Hayashi T, Su TP. Sigma-1 receptor ligands: potential in the treatment of neuropsychiatric disorders. *CNS Drugs.* 2004;18:269–284.
242. Modell S, Naber D, Holzbach R. Efficacy and safety of an opiate sigma-receptor antagonist (SL 82.0715) in schizophrenic patients with negative symptoms: an open dose-range study. *Pharmacopsychiatry.* 1996;29:63–66.
243. Frieboes RM, Murck H, Wiedemann K, et al. Open clinical trial on the sigma ligand panamesine in patients with schizophrenia. *Psychopharmacology (Berl).* 1997;132:82–88.
244. Muller MJ, Grunder G, Wetzel H, et al. Antipsychotic effects and tolerability of the sigma ligand EMD 57445 (panamesine) and its metabolites in acute schizophrenia: an open clinical trial. *Psychiatry Res.* 1999;89:275–280.
245. Borison RL, Diamond BI, Dren AT. Does sigma receptor antagonism predict clinical antipsychotic efficacy? *Psychopharmacol Bull.* 1991;27:103–106.
246. Gewirtz GR, Gorman JM, Volavka J, et al. BMY 14802, a sigma receptor ligand for the treatment of schizophrenia. *Neuropsychopharmacology.* 1994;10:37–40.
247. Klamer D, Engel JA, Svensson L. Phencyclidine-induced behaviour in mice prevented by methylene blue. *Basic Clin Pharmacol Toxicol.* 2004;94: 65–72.
248. Deutsch SI, Rosse RB, Paul SM, et al. 7-Nitroindazole and methylene blue, inhibitors of neuronal nitric oxide synthase and NO-stimulated guanylate cyclase, block MK-801-elicited behaviors in mice. *Neuropsychopharmacology.* 1996;15:37–43.
249. Johansson C, Jackson DM, Svensson L. Nitric oxide synthase inhibition blocks phencyclidine-induced behavioural effects on prepulse inhibition and locomotor activity in the rat. *Psychopharmacology (Berl).* 1997;131:167–173.
250. Wiley JL. Nitric oxide synthase inhibitors attenuate phencyclidine-induced disruption of prepulse inhibition. *Neuropsychopharmacology.* 1998;19:86–94.
251. Deutsch SI, Rosse RB, Schwartz BL, et al. Methylene blue adjuvant therapy of schizophrenia. *Clin Neuropharmacol.* 1997;20:357–363.
252. Bergeron R, de Montigny C, Debonnel G. Potentiation of neuronal NMDA response induced by dehydroepiandrosterone and its suppression by progesterone: effects mediated via sigma receptors. *J Neurosci.* 1996;16:1193–1202.
253. Debonnel G, Bergeron R, de Montigny C. Potentiation by dehydroepiandrosterone of the neuronal response to N-methyl-D-aspartate in the CA3 region of the rat dorsal hippocampus: an effect mediated via sigma receptors. *J Endocrinol.* 1996;150(suppl):S33–S42.
254. Strous RD, Maayan R, Lapidus R, et al. Dehydroepiandrosterone augmentation in the management of negative, depressive, and anxiety symptoms in schizophrenia. *Arch Gen Psychiatry.* 2003;60:133–141.
255. Myers KM, Goulet M, Rusche J, et al. Partial reversal of phencyclidine-induced impairment of prepulse inhibition by secretin. *Biol Psychiatry.* 2005;58:67–73.
256. Alamy SS, Jarskog LF, Sheitman BB, et al. Secretin in a patient with treatment-resistant schizophrenia and prominent autistic features. *Schizophr Res.* 2004;66:183–186.
257. Sheitman BB, Knable MB, Jarskog LF, et al. Secretin for refractory schizophrenia. *Schizophr Res.* 2004;66:177–181.
258. Riedel M, Strassnig M, Schwarz MJ, et al. COX-2 inhibitors as adjunctive therapy in schizophrenia: rationale for use and evidence to date. *CNS Drugs.* 2005;19:805–819.
259. Muller N, Riedel M, Schwarz MJ, et al. Clinical effects of COX-2 inhibitors on cognition in schizophrenia. *Eur Arch Psychiatry Clin Neurosci.* 2005;255:149–151.
260. Muller N, Ulmschneider M, Scheppach C, et al. COX-2 inhibition as a treatment approach in schizophrenia: immunological considerations and clinical effects of celecoxib add-on therapy. *Eur Arch Psychiatry Clin Neurosci.* 2004;254:14–22.
261. Thome J, Foley P, Riederer P. Neurotrophic factors and the maldevelopmental hypothesis of schizophrenic psychoses. Review article. *J Neural Transm.* 1998;105:85–100.
262. Lieberman JA. Is schizophrenia a neurodegenerative disorder? A clinical and neurobiological perspective. *Biol Psychiatry.* 1999;46:729–739.
263. Miyamoto S, Duncan GE, Marx CE, et al. Treatments for schizophrenia: a critical review of pharmacology and mechanisms of action of antipsychotic drugs. *Mol Psychiatry.* 2005;10:79–104.
264. Menniti FS, Chappie TA, Humphrey JM, et al. Phosphodiesterase 10A inhibitors: a novel approach to the treatment of the symptoms of schizophrenia. *Curr Opin Investig Drugs.* 2007;8:54–59.
265. Roth BL. Contributions of molecular biology to antipsychotic drug discovery: promises fulfilled or unfulfilled? *Dialogues Clin Neurosci.* 2006;8:303–309.
266. Urban JD, Clarke WP, von Zastrow M, et al. Functional selectivity and classical concepts of quantitative pharmacology. *J Pharmacol Exp Ther.* 2007;320:1–13.
267. Masri B, Salahpour A, Didriksen M, et al. Antagonism of dopamine D2 receptor/beta-arrestin 2 interaction is a common property of clinically effective antipsychotics. *Proc Natl Acad Sci USA.* 2008;105:13656–13661.
268. Watson J, Brough S, Coldwell MC, et al. Functional effects of the muscarinic receptor agonist, xanomeline, at 5-HT1 and 5-HT2 receptors. *Br J Pharmacol.* 1998;125:1413–1420.

10.3 | How Antipsychotics Work: Linking Receptors to Response

NATHALIE GINOVART AND SHITIJ KAPUR

SCHIZOPHRENIA

Schizophrenia is a chronic and disabling disease that typically begins during adolescence or early adult life and severely impacts psychosocial functioning. Schizophrenia is characterized by a myriad of symptoms that are generally divided into three categories: positive, negative, and cognitive.[1,2] The positive symptoms are typically regarded as manifestations of psychosis and include hallucinations, delusions, and disorganized thoughts. The negative symptoms consist of severe disturbances in social interaction and include blunted affect, poverty of speech, anhedonia, loss of drive (avolition), and social withdrawal. Cognitive deficits affect attention, working and verbal memory, social cognition, and executive function.

There is no known single cause of schizophrenia. It is hypothesized that genetic factors and early neurodevelopmental abnormalities (including apoptosis, disruption of neuronal migration, or alteration of synaptogenesis) may confer a constitutional vulnerability to the disease.[3] Subsequent environmental factors (including obstetric complications, exposure to viral infection in utero, or exposure to psychosocial stress during childhood) may then trigger the behavioral expression of this vulnerability, perhaps via subtle alterations of brain development.[4] Within this framework, dysregulations of the dopamine (DA) and glutamate neurotransmitter systems have been most intimately associated with the physiopathology of schizophrenia. It is this aspect of the illness that is the focus of this chapter, with special attention given to the DA receptors.

The DA Hypothesis

The DA hypothesis, which postulates that schizophrenia is caused by overactivity of DA neurotransmission, was first proposed in the 1960s.[5] The basis for this hypothesis was that the most widely used drugs for the treatment of schizophrenia, the antipsychotic drugs, were suspected to act through the blockade of central DA receptors,[6] and there was a tight correlation between the clinical potency and D2 receptor binding affinity of antipsychotic drugs.[7] Accordingly, the D2 receptor has been a major focus of schizophrenia research for several decades. Initially, postmortem and in vivo imaging studies reported increased densities of D2 receptors in the brain of schizophrenic patients.[8-11] However, a lack of consistent replication and a potential confounding effect of prior exposure to antipsychotic medication made this finding controversial.[12-22] Studies of D1 receptors showed no elevation in striatum[23,24] but remained inconsistent with regard to the density of D1 receptors in frontal cortex, with studies showing a decreased,[25] unchanged,[23] or increased[24] D1 receptor level in this brain region. Thus far, only one study has investigated D3 receptors in schizophrenia and found elevated D3 receptor levels in striatum,[26] a finding that still need to be replicated in other studies. As for D4, an elevated striatal density of D4 receptors has been reported in schizophrenic patients,[27-29] but this finding was not replicated in other studies.[30-33] Thus, there is still no clear evidence about the contributions of postsynaptic DA receptors to the pathophysiology of schizophrenia.

More recently, in vivo investigations of humans in both clinical and experimental settings have yielded evidence that disturbances of DA function in schizophrenia involve dysregulation of presynaptic rather than postsynaptic DA function. Indeed, it has been shown, using positron emission tomography (PET) and the DA precursor analog radioligand [^{18}F]fluorodopa, that the synthesis of DA is increased in the striatum of drug-naive schizophrenic patients compared to healthy controls.[34-37] Moreover, further in vivo imaging studies have provided evidence for an exaggerated release of DA in the striatum of schizophrenic patients both under basal conditions[38] and following an amphetamine challenge.[39-42] Interestingly, the amphetamine-induced elevation of DA release was found to correlate with the induction of positive psychotic symptoms. Taken together, these data suggest that an increased presynaptic capacity of DA synthesis and release may constitute part of the dysfunctional neural connectivity underlying schizophrenia and may be the concurring proximate causes of psychoses. In contrast, DA function may be decreased in the neocortex in schizophrenia.[43] The reconceptualized DA hypothesis of schizophrenia thus posits that the positive and negative symptoms of schizophrenia arise from a cortical/subcortical imbalance of DA tone in the brain.[44] The positive symptoms would result from an excess of DA in the subcortical mesolimbic DA projections, whereas the negative symptoms and cognitive deficits would result from a concomitant deficit of DA in the mesocortical DA projections to the frontal cortex.

The Glutamate Hypothesis

While most emphasis has centered on DA, a dysfunction of glutamate transmission has also been implicated in schizophrenia. It has long been known that antagonists of the glutamatergic N-methyl-D-aspartic acid (NMDA) receptor complex, such as phencyclidine (PCP) and ketamine, can produce psychotic symptoms and cognitive deficits in normal subjects that closely mimic those of schizophrenia[45-48] and can precipitate psychoses in schizophrenic patients.[45,49-51] In addition, postmortem studies have identified abnormalities in NMDA receptor subunit composition and signal transduction pathways in limbic structures implicated in schizophrenia, including the frontal cortex, temporal cortex, cingulate cortex, hippocampus, and thalamus.[52-54] Further, abnormal concentrations of glutamate in the hippocampus and prefrontal cortex (PFC)[55-57] have recently been revealed in schizophrenic patients. Taken together, these findings suggest that excitatory glutamatergic transmission, specifically that mediated by NMDA receptors, may be dysregulated in schizophrenia.[58,59] Pathophysiologically, the glutamate hypothesis of schizophrenia holds that hypofunction of the NMDA receptor results in a disinhibition of gamma-aminobutyric acid (GABA)–mediated processes and a compensatory excessive release of glutamate in cerebral cortex, which, in turn, overactivates corticolimbic afferents (via the non-NMDA receptors such as the metabotropic glutamate receptors[60,61]) and leads to the clinical symptoms of schizophrenia.[62]

Integration of the DA and Glutamate Hypotheses

The DA hypothesis of schizophrenia is not inconsistent with a proposed dysfunction of the glutamatergic neurotransmission system, since reciprocal anatomical and functional interactions between forebrain DA and glutamate systems have been well described.[63] These interactions control the functioning of the mesocorticolimbic loop, and drug treatments or illnesses that alter one neurotransmitter system are expected to alter the other. For instance, acute disruption of the glutamatergic neurotransmission with PCP or ketamine increases the firing rate of midbrain DA neurons[64-67] and the release of DA in the PFC and subcortical structures.[67-70] Interestingly, subchronic administration of NMDA receptor antagonists in animals differentially affects DA release in terminal regions of midbrain DA neurons: DA release is decreased in the PFC and increased in the ventral striatum.[71,72] The mechanism by which a decrease in PFC excitatory glutamatergic output leads to subcortical DA hyperfunction may involve a direct or indirect (via GABA interneurons) disinhibition of mesolimbic DA neurons.[73] Because the positive and negative symptoms of schizophrenia have been associated with a mesolimbic/mesocortical imbalance of DA function, the effects produced by NMDA receptor hypofunction on DA neurotransmission are most consistent with the DA hypothesis of schizophrenia. It remains unclear, though, whether NMDA receptor hypofunction represents a primary abnormality that interferes with normal DA function and contributes to the symptoms of schizophrenia, or if it occurs secondarily to structural brain alteration known to exist in the illness.[74,75]

ANTIPSYCHOTIC TREATMENT

Effective drug treatment of schizophrenia has been available since the early 1950s with the introduction of chlorpromazine.[76] Originally developed as a preanesthetic agent,[77] it was quickly found to reduce the psychotic symptoms of schizophrenia.[76] Other drugs followed in the late 1950s, including haloperidol, thioridazine, trifluoperazine, and loxapine. These first-generation drugs, also referred to as *typical antipsychotics*, are effective against the positive symptoms but are associated with side effects such as sedation, hyperprolactinemia, and acute extrapyramidal side effects (EPS), which manifest as dystonic reactions, parkinsonian symptoms, and akathisia.[78] On long-term use, typical antipsychotics are also associated with the risk of developing tardive dyskinesia (TD), a serious and potentially irreversible side effect characterized by repetitive and involuntary muscle contractions that is very stigmatizing to the patient. Besides producing these motor side effects, typical antipsychotics have limited efficacy in ameliorating the negative and cognitive symptoms.[79] Another disadvantage of these drugs is their failure to control positive symptoms in 15%–30% of schizophrenic patients,[80] who have the so-called refractory form of schizophrenia. The relative inability of typical antipsychotics to treat some symptoms of schizophrenia, together with their association with significant EPS, is thought to contribute to treatment nonadherence,[81,82] which often leads to relapse,[83] and thus motivated the search for better treatment options.

The development of second-generation antipsychotics, also referred to as *atypical antipsychotics*, was viewed as a major advance in the treatment of schizophrenia, primarily because they confer less risk of EPS and TD than typical antipsychotics.[84–86] Clozapine, the prototype of atypical antipsychotic drugs, was introduced in Europe in the mid-1970s and proved to be effective in treating patients with refractory schizophrenia.[87–89] It was followed by a myriad of other atypical drugs, including olanzapine, risperidone, sertindole, amisulpride and quetiapine, and, more recently, ziprasidone and aripiprazole, which share clozapine's lower liability for motor side effects but seem to be less beneficial for refractory schizophrenia.[90,91] Besides their superior motor side effect profile and their minimal effect on prolactinemia,[92,93] it has been claimed that atypical antipsychotics have greater effects against the negative and cognitive symptoms of schizophrenia than typical drugs.[87,88,93–101] However, several studies have failed to demonstrate superior efficacy of atypical antipsychotics in these domains.[102–107] Atypical antipsychotics may thus offer only modest efficacy advantages over typical antipsychotics. In contrast, because of their low incidence of EPS, they offer notable benefits in terms of tolerability and compliance with treatment[107,108] and may thus achieve a better overall prognosis. This must be balanced, though, against the higher risk of metabolic side effects associated with atypical drugs, such as diabetes, hypercholesterolemia, and weight gain.[109,110]

DA RECEPTORS INVOLVED IN ANTIPSYCHOTIC DRUG ACTION

In the 1960s, the pioneering work of Carlsson and Lindqvist[6] led to the prevailing view that the main mechanism by which antipsychotics exert their benefits was their blockage of DA transmission in the brain. This concept was substantiated by subsequent work by Seeman and Lee[7] and Snyder et al.,[111] who showed that antipsychotics act as DA receptor blockers. Since then, five different subtypes of DA receptors (D1–D5) have been identified in brain, which were classified as D1-like (D1, D5) and D2-like receptor subtypes (D2, D3, D4) based on their similar linkage to adenylate cyclase and their similar pharmacological properties.[112] D1-like receptors are positively coupled to adenylate cyclase, whereas D2-like receptors are negatively coupled to the enzyme. D1 receptors are highly expressed in terminal regions of the ventral tegmental area (VTA) and substantia nigra (SN) such as the dorsal striatum (caudate-putamen), the nucleus accumbens (NAcc), the cortex, and, to a lesser extent, the amygdala, globus pallidus, and hippocampus.[113] Compared to D1 receptors, D5 receptors are much less widely expressed in the human brain.[114] D2 receptors are widely expressed in the brain, with high densities found in DA-rich regions such as the dorsal striatum, NAcc, SN, and VTA.[113] They can function either as postsynaptic receptors or as presynaptic autoreceptors. In the latter capacity, the presynaptic D2 autoreceptors located on nerve terminals provide feedback modulation of DA synthesis and release, whereas those located in somatodendritic regions of midbrain DAergic neurons modulate neuronal firing.[115–117] D3 receptors are expressed with high density in the NAcc and the islands of Calleja and, to a lesser extent, in the dorsal striatum, SN, and thalamus.[118] Like D2 receptors, D3 receptors can be postsynaptic or function as presynaptic autoreceptors to negatively modulate DA cell activity.[119,120] D4 receptors are located mainly in cortex and hippocampus, with low density levels being found in the caudate-putamen.[121] It thus appears that while the anatomical distribution of DA receptors largely overlaps in the brain, their quantitative ratios differ among brain structures, supporting the view that different subtypes subserve different functions.

Role of D2 Receptor Antagonism

The importance of the DA D2 receptor in the mechanism of antipsychotic action was first highlighted by the demonstration that the clinical potency of antipsychotic drugs correlates closely with their ability to block D2 receptors.[7,122] Such a correlation is specific to the D2 receptors; it has not been found for any other DA receptor subtypes.[123] Antagonism at the D2 receptors is thus a central mechanism in the treatment of schizophrenia and, to date, no drugs with a proven antipsychotic activity lacks D2 receptor antagonistic properties. The primary mechanism by which antipsychotics achieve their therapeutic action is thus to reduce DA overactivity in the mesolimbic pathway (which underlies the positive symptoms) through blockade of the D2 receptors. By contrast, the concurrent blockade of D2 receptors in the mesocortical pathway, where DA activity is already reduced in schizophrenia, produces few benefits and may even exacerbate the negative symptoms and cognitive deficits of the disease. Action at the D2 receptors is also primarily involved in the induction of EPS, as evidenced by the direct relationship between the blockade of nigrostriatal D2 receptors and the induction of catalepsy in rodents, a model used to predict the motor side effect liability of antipsychotic drugs,[124] while excessive action on the tuberoinfundibular tract produces hyperprolactinemia.

While converging evidence from both in vitro and animal experiments has pointed to the central involvement of D2 receptor blockade in antipsychotic action, neuroimaging studies in human have provided additional insights into the therapeutic efficacy and the risk of EPS associated with these drugs. With a few exceptions, most antipsychotics are effective in a therapeutic window in which 60%–80% of D2 receptors are blocked.[125] Occupancies below 60% are usually ineffective against the positive symptoms of schizophrenia, and occupancies above 80% lead to EPS.[126] Clozapine and quetiapine do not conform to that profile; both drugs achieve therapeutic efficacy at D2 receptor occupancies that are clearly lower, in the 20%–68% range.[127–129]

Role of D2 Receptor Partial Agonism

The development of D2 partial agonists has recently emerged as an alternative approach for the treatment of schizophrenia. Because of their lower intrinsic activity than the natural full agonist receptor ligand DA, these drugs elicit a submaximal response in the absence of DA and block the maximal response to excessive concentrations of DA. D2 partial agonists would thus stabilize DA activity in the schizophrenic brain by dampening excessive D2 stimulation in the mesolimbic system and by restoring deficient D2 stimulation in the mesocortical system.[130] On the other hand, low intrinsic activity at the D2 receptors would prevent a complete D2 receptor blockade in the nigrostriatal system and would thus confer a low propensity to cause EPS and prolactin elevation.

Several D2 partial agonists, often referred as to *DA stabilizers*, have been evaluated in schizophrenia, including Preclamol, also known as (−)-3-(3-hydroxyphenyl)-N-n-propylpiperidine or (3PPP), talipexole (B-HT 920), roxindole (EMD 49980), terguride, and SDZ-HDC 912. These compounds, though, either failed to show antipsychotic efficacy[131–133] or were associated with significant motor side effects[134] or tolerance[135] upon treatment, which limited their usefulness for the treatment of schizophrenia.

Aripiprazole is the first D2 partial agonist to have been successfully introduced into clinical practice. Pharmacologically, it is a D2 receptor partial agonist with partial agonist activity at the serotonin $5HT_{1A}$ receptors and antagonist activity at the $5HT_{2A}$ receptors (see[136,137] for a review). Clinically, aripiprazole has efficacy comparable to that of existing typical and atypical antipsychotics in treating the positive, negative, and cognitive symptoms of schizophrenia and shows a low incidence of EPS and prolactin elevation.[138–145] At therapeutic doses, aripiprazole occupies 85% to 95% of the striatal D2 receptors without causing the EPS and prolactin elevation commonly seen at such high levels of D2 occupancy with D2 antagonists.[146–148] The relatively low intrinsic activity of aripiprazole[149,150] when compared to other clinically unsuccessful D2 partial agonists such as 3PPP or terguride is probably the reason for aripiprazole's favorable clinical profile. Indeed, aripiprazole's intrinsic activity is approximately 30% of the full effect of DA,[151] which is ideally suited for treating the positive and negative symptoms of schizophrenia without causing undesirable motor effects when about 90% of D2 receptors are occupied.[147] Taken together, studies on the pharmacological action of D2 antagonists and D2 partial agonists converge to underline the importance of fine tuning of D2 receptor blockade for achieving an optimal antipsychotic benefit and thus further emphasize the central role of this receptor subtype in antipsychotic action.

Role of D1 Receptor Antagonism and Agonism

Based on clozapine's greater affinity for D1 than for D2 receptors in vitro, preferential antagonism at the D1 receptor has been regarded as one potential mechanism by which atypical drugs could mediate antipsychotic action.[152] In vivo, whereas therapeutic doses of clozapine produce similar occupancies of the D1 and D2 receptors, others atypical drugs such as quetiapine and risperidone occupy only low levels of D1 receptors, indicating that D1 antagonism is a poor predictor of atypicality.[153,154] Moreover, clinical trials with two selective D1 antagonists, SCH39166[155,156] and NNC01-0687,[157] have failed to demonstrate antipsychotic efficacy, indicating that D1 antagonism by itself does not confer antipsychotic activity.

The D1 receptor is, however, considered as a most promising therapeutic target for improving the cognitive deficits and negative symptoms in schizophrenia. Indeed, activation of the D1 receptor in the PFC is strongly involved in the control of higher cognitive functions such as working memory.[158–160] For instance, local injection of a D1 antagonist into the PFC disrupts working memory in primates,[158,161] while systemic administration of a D1 agonist improves working memory performance, the latter effect being blocked by a D1 antagonist.[162] There is also evidence of abnormalities in the density of D1 receptors in the PFC of drug-naive schizophrenic patients[24,25] (but also see[156]), thus supporting the view that altered DA transmission at D1 receptors in PFC might be involved in the working memory deficits observed in schizophrenia.[163,164] In addition, the chronic D2 receptor blockade associated with antipsychotic treatment has been shown to down-regulate D1 receptors in PFC in experimental animals[165,166] and possibly in patients with schizophrenia,[167] resulting in cognitive deficits. In support of this finding, working memory deficits induced in monkeys by chronic D2 antagonist treatment can be reversed by a brief cotreatment with a D1 full agonist.[168] Therefore, treatments that increase DA in the PFC or that directly stimulate D1 receptors could be useful in restoring cognitive function in schizophrenia.[169,170]

Role of D3 Receptor Blockade

Antipsychotic drugs generally display limited selectivity between D2 and D3 receptors.[171] While the involvement of the D2 subtype has unequivocally been associated with antipsychotic action, the role of the D3 subtype is less clearly defined, though it has been suggested that D3 receptors may represent an important target for antipsychotic drugs.[172,173] A fundamental basis for this assertion is the great abundance and preferential localization of D3 receptors in target regions (e.g., NAcc, ventral putamen) of the mesolimbic DA system. Support for this hypothesis also comes from a postmortem study showing elevated levels of D3 receptors in the limbic striatum of drug-free schizophrenic patients and a potential D3 receptor normalization by antipsychotic treatment.[26] Elevated D3 receptor levels may thus account for the hyperactivity of the DA mesolimbic system postulated in schizophrenia, and selective blockade of these receptors may thus be beneficial for the resolution of positive symptoms. Preclinical studies in rodents also suggest that selective D3 blockade, in contrast to D2 blockade, may enhance motor function and have beneficial effects on cognition.[173,174] Given these data, it has been suggested that "optimized" levels of D3 versus D2 antagonists may permit enhanced effectiveness against cognitive dysfunction and may reduce motor side effects compared to currently available D2-preferring antipsychotics. Such preferential D3 versus D2 receptor antagonists are currently in experimental phases,[175,176] and clinical trials with such compounds are still needed to better understand the significance of D3 receptor blockade in the treatment of schizophrenia.

Role of D4 Receptor Blockade

Unlike most antipsychotics, clozapine has a higher affinity for D4 than for D2 receptors, a finding that was thought to explain the unique efficacy of clozapine in the treatment of schizophrenia.[177] Postmortem studies showing elevated densities of D4 receptors in the striatum of schizophrenic patients[27-29] have further sparked interest in this receptor subtype, although this finding has not been consistently confirmed.[30-33] However, the initial hope that D4 receptors may be an important target of antipsychotic action has been dashed by clinical trials that failed to prove any antipsychotic effect of D4 antagonists.[178-180]

PROPOSED MECHANISMS OF ATYPICALITY

There have been a number of hypotheses concerning the mechanisms underlying antipsychotic atypicality. Most of the hypotheses postulate that the different side effect profiles of typical and atypical drugs result mainly from differences in their receptor binding profile. Indeed, while typical drugs are usually selective for D2-like receptors, atypical antipsychotics usually bind to a larger spectrum of receptor systems. For instance, clozapine exerts its action through D2 receptors but also through D1, serotonin, muscarinic, adenosine, and adrenergic receptors; this multireceptor action has been proposed to be the main determinant of atypicality. This has given rise to a number of ideas explaining atypicality.

Combined Blockade of D2 and Serotonin 5HT$_{2A}$ Receptors

In addition to D2 mechanisms, additional pharmacological mechanisms are thought to contribute to atypicality; among them is antagonism to the serotonin 5-HT$_{2A}$ receptor subtype. Serotonin 5-HT$_{2A}$ blockade per se does not confer efficacy against positive symptoms, though there is some suggestion of a direct effect on negative symptoms.[181] As expected, for atypical antipsychotics such as risperidone and olanzapine, which show high affinity for that receptor subtype, there is no correlation between 5HT$_{2A}$ occupancy and clinically effective doses.[182] In fact, despite nearly full saturation of 5-HT$_{2A}$ receptors, risperidone and olanzapine become effective only at doses that exceed the conventional 65% level of D2 occupancy.[182]

Although combined 5-HT$_{2A}$/D2 antagonism is not essential for antipsychotic action, Meltzer and colleagues[183] have proposed that a high 5-HT$_{2A}$/D2 affinity ratio is the pharmacological feature that best distinguishes typical from atypical antipsychotics. Serotonin 5HT$_{2A}$ antagonism has been proposed to modulate the DA system and, through this modulation, to reduce the motor effects associated with D2 blockade and improve the negative and cognitive symptoms of schizophrenia.[184,185] Serotonin 5HT$_{2A}$ receptor blockade, in combination with D2 receptor antagonism, facilitates DA release in PFC[186,187] but not in NAcc,[187] suggesting that concomitant blockade of 5-HT$_{2A}$ and D2 may stimulate the mesocortical DA pathway relative to the nigrostriatal and mesolimbic pathways. Given the central role of prefrontal DA in cognitive function, the increase in PFC DA release may underlie the effects of atypical antipsychotic drugs on the negative and cognitive symptoms by normalizing a putative cortical hypodopaminergic transmission. The mechanism by which 5HT$_{2A}$ blockade may alleviate EPS remains unclear, as 5-HT$_{2A}$ blockade, when combined with D2 blockade, has been found to reduce DA release in the striatum[188,189] and to have no effect[190,191] or even to potentiate the catalepsy induced by D2 blockade.[192] However, atypical neuroleptics also antagonize 5-HT$_{2C}$ receptors,[193,194] and striatal 5-HT$_{2A}$ and 5-HT$_{2C}$ blockade has the opposite influence on striatal DA release. Indeed, it has been shown that 5-HT$_{2C}$ blockade increases striatal DA release and prevents the catalepsy induced by D2 blockade, suggesting that 5-HT$_{2C}$ rather than 5-HT$_{2A}$ antagonism may be the mechanism by which atypical neuroleptics reduce EPS.[188] However, the fact that atypicality is possible without significant 5-HT$_2$ binding, as observed with remoxipride[195] and amisulpiride,[196,197] and that, in contrast, typicality is possible with high levels of 5-HT$_2$ binding, as observed with chlorpromazine and loxapine,[197,198] suggests that 5-HT$_2$ blockade is not a central determinant of EPS liability. However, it is quite possible that 5-HT$_2$ blockade may have supplementary advantages with regard to negative symptoms, affect, and sleep—though this has yet to be proven in the context of antipsychotic action.

Preferential Limbic D2 Receptor Blockade

One widely accepted hypothesis to explain the separation of antipsychotic activity from EPS liability is the regional specificity (often called *limbic selectivity*) of antipsychotic action. According to this hypothesis, action at D2 receptors is sufficient to explain atypicality through preferential effects on limbic and cortical brain structures.[199] Such limbic selectivity has been widely demonstrated in preclinical studies, based on the regional effects of antipsychotics on DA metabolism, DA cell activity, and DA-mediated c-Fos expression.

For instance, while an acute dose of haloperidol increases DA neuronal activity in both the SN and VTA and preferentially increases DA release in the dorsolateral striatum, atypical drugs such as clozapine, which is not associated with EPS in humans, affect DA neuronal activity in the VTA only and preferentially increase DA output in the NAcc.[200-203] Such regional specificity is also seen with repeated treatment, as typical drugs such as haloperidol cause a depolarization inactivation of DA neurons in both the SN and VTA, while atypical drugs such as clozapine only cause inactivation of VTA DA neurons, sparing the SN.[200-202,204-206] Thus, the therapeutic action and EPS liability associated with antipsychotic drugs correlate with the regional selectivity of these drugs in inducing depolarization blockade of the mesolimbic and nigrostriatal DA neurons, respectively. A further index of atypical drug limbic selectivity comes from studies on the effect of antipsychotics on immediate early gene expression, such as c-*fos* expression. While haloperidol increases c-*fos* expression both in the limbic-associated accumbens shell region and in the motor-related dorsolateral striatum and accumbens core regions, atypical drugs only increase c-*fos* expression in the accumbens shell.[207-211] Exactly how drugs may lead to limbically selective effects is unclear. In principle, two mechanisms can be hypothesized: differential occupancy (i.e., atypical drugs occupy more receptors in the limbic regions) or differential sensitivity (i.e., the limbic regions and the motor striatum show differential sensitivity to similar levels of blockade).

In clinical settings, the concept of *differential occupancy* leading to limbic selectivity predicts that atypical drugs produce higher levels of D2 receptor blockade in limbic regions than in the dorsolateral striatum. In vivo imaging studies in the field have provided inconsistent results, though, with some studies showing that therapeutic doses of atypical drugs produce preferential D2 blockade in limbic cortical regions than in striatum[212-216] and others finding similar levels of D2 blockade in striatum and limbic cortical areas.[217-221] These inconsistencies thus called into question the concept of limbic selectivity as a possible mechanism of atypicality. Moreover, neuroimaging data recently challenged the concept that blockade of limbic D2 receptors is solely associated with antipsychotic efficacy, while blockade of the striatal D2 receptors is solely associated with EPS. Indeed, two independent imaging studies found that, in schizophrenic patients treated with atypical antipsychotics, improvement in positive (but not negative) symptoms correlated with striatal but not extrastriatal (i.e., frontal, temporal, thalamic) D2 receptor blockade.[148,220] Thus, contrary to common belief, it is possible that in addition to inducing EPS, striatal D2 receptors are also likely to be an important site of antipsychotic action. Clearly, more clinical investigations are needed to determine the exact role of limbic D2 receptors in the treatment of schizophrenia.

Fast Dissociation of D2 Receptors

In contrast to the multireceptor hypothesis, Kapur and collaborators[125,222,223] introduced a new concept to account for atypicality. They proposed that the different liability of typical versus atypical drugs to induce EPS is due largely to the speed with which these drugs dissociate from D2 receptors. Indeed, the pharmacological property that best distinguishes atypical from typical drugs is their low affinity for D2 receptors. Affinity is by definition the ratio between the rate at which a drug dissociates from receptors (k_{off}) and the rate at which it associates with receptors (k_{on}). The lower affinity of atypical versus typical drugs has been shown to be completely independent of the drug association rate and to be entirely accounted for by rapid dissociation of the drug from the receptors.[223] Atypical drugs thus bind more loosely and dissociate faster from the receptors than typical antipsychotics.[224] As a consequence, atypical drugs would produce appropriate levels of D2 receptor blockade for an antipsychotic effect but still would dissociate fast enough from the receptors to preserve some level of cellular DA tone. Rather than suppressing it, atypical antipsychotics would thus only briefly silence DA neurotransmission, thereby allowing antipsychotic action while diminishing the risk of D2-related EPS and hyperprolactinemia.

Transient versus Continuous D2 Receptor Blockade

In addition to reaching a threshold of D2 receptor blockade, the within-day pattern of D2 occupancy kinetics achieved during antipsychotic treatment is an important determinant of the treatment outcome. Accumulating preclinical evidence indicates that, even for a given drug such as haloperidol, transient D2 blockade achieved by pulsatory drug delivery has different effects than continuous D2 blockade achieved by continuous infusion of the drug. For instance, in animal models used to test haloperidol antipsychotic-like efficacy, within-day transient D2 blockade was found to be more effective than continuous D2 blockade.[225,226] Importantly, while continuously high levels (>70%–80% for 24 hours per day) of D2 receptor blockade lead to the development of D2 receptor supersensitivity and functional tolerance, and to an increased risk for the development of vacuous chewing movements (i.e., an animal model for TD), transiently high levels (>80% for a few hours per day) of D2 receptor blockade do not produce any of these effects.[225–228] Transient D2 blockade may thus be sufficient to induce an antipsychotic response and may even improve therapeutic efficacy while minimizing the risks of EPS, and may thus be a key component of atypicality. In vivo imaging data complement this view. Indeed, it appears that atypical drugs such as clozapine and quetiapine give rise to only transient D2 blockade, with no incidence of EPS and prolactin elevation.[129,229] It is thus possible that transient D2 blockade is all that is needed to achieve therapeutic efficacy, and that continuous blockade is unnecessary and may even be detrimental. The question remains as to how long transient blockade should last to achieve a therapeutic effect while minimizing the risk of EPS.

OTHER RECEPTOR MECHANISMS INVOLVED IN ANTIPSYCHOTIC ACTION

Although action at the D2 receptor appears to be the most relevant marker for understanding antipsychotic activity, especially for controlling hallucinations, delusions, and thought disorder, current antipsychotics still do not meet the clinical need for improvement of the negative and cognitive symptoms. Research to develop novel agents with greater efficacy in these domains has focused on compounds targeting nondopaminergic receptors but acting through neurotensin 1 (NTS_1), neurokinin 3 (NK3), cannabinoid CB_1 receptors, or glutamatergic mechanisms. Thus far, drugs acting at the NTS_1, NK_3, or CB_1 receptors, such as SR48692, SR142801, and SR141716, respectively, have demonstrated only modest, if any, antipsychotic efficacy.[181] Drugs acting on the glutamatergic system include agonists of the glycine recognition site of the NMDA receptor (e.g., glycine, D-serine, D-cycloserine), glycine reuptake inhibitors, glutamate release inhibitors (e.g., LY-354740 and lamotrigine), AMPA agonists and antagonists (e.g., LY-293558 and GYKI 52466), ampakines (e.g., CX-516), and mGlu receptor agonists (for review, see[230]). The efficacy of these compounds appears somewhat limited, with some studies suggesting some beneficial effect on the negative and cognitive symptoms and others being equivocal.[231–235] However, a recent breakthrough indicates the clinical efficacy of a drug stimulating the metabotropic glutamate receptor 2/3 (mGlu2/3), LY2140023, which has improved efficacy for the negative and cognitive symptoms of schizophrenia.[236] Studies of larger patient samples are required to consolidate these data (see the review in[237]) but, if replicated, mGlu2/3 agonists, might be the first class of antipsychotic not acting through the D2 receptor. If this is sustained, it would be of tremendous interest to examine whether these new mGlu2/3 agonists exert their effect directly or do so by modulating the DA system.

CONSIDERATIONS CRITICAL FOR UNDERSTANDING RECEPTOR INVOLVEMENT IN ANTIPSYCHOTIC ACTION

Speed of Onset and Implications for Mechanism

One of the major challenges to receptor accounts of antipsychotic action has been the idea of a *delayed onset* of antipsychotic action, which proposes that the clinical effects of these drugs are significantly delayed (2–3 weeks) after the onset of administration.[238] Given that receptor occupancy is established within hours and days of starting antipsychotic medication, this delay between occupancy and onset of action seriously calls into question the direct role of receptors. According to this account, while binding of receptors may initiate the cascade, it is the secondary and tertiary downstream effects (e.g., depolarization blockade) that are critical for patient improvement.[238] Recent clinical studies have questioned this conventional wisdom. Using data from nearly 7000 patients in over 100 clinical comparisons, Agid et al.[239] have now demonstrated that the onset of antipsychotic activity is almost immediate. It is evident in clinical trials at the earliest measurement (usually the first week), and this early improvement is specific, not just sedation, and is clearly differentiable from the result of placebo. Further, it has now been demonstrated that there is robust and specific antipsychotic improvement within the very first day[240] and that early improvement is strongly predictive of eventual improvement.[241,242] Thus, the recent clinical findings remove the final hurdle in linking receptors directly to response, and a recent imaging study shows that receptor occupancy within the first 48 hr predicts the clinical response that is achieved at the 2-week mark, providing further clinical evidence to substantiate and link receptors to response.[243]

Relapse on Withdrawal and Supersensitivity

Two of the long-term consequences of antipsychotic treatment is the development of a behavioral supersensitivity to direct or indirect DA agonists in experimental animals[244–246] and the emergence of rebound psychosis (also called *supersensitivity psychosis*) in some schizophrenic patients upon withdrawal of antipsychotic medication.[247–251] These effects are seen with both typical and atypical drugs and are thought to reflect a DA supersensitized state due to an increase in postsynaptic DA receptor–mediated processes. Support for this comes from studies showing that chronic treatment with typical and atypical antipychotics, such as haloperidol and remoxipride, elevates D2 receptor density in the striatum[252–258] and increases the proportion of D2 receptors with functional high affinity for DA, the $D2^{High}$.[226,259,260] D2 up-regulation has been proposed to mediate the DA supersensitivity leading to TD and supersensitivity psychosis upon withdrawal,[261–264] although other factors, including GABA insufficiency or excitotoxic neuronal damage, have also been involved in TD.[265–267] However, atypical antipsychotics such as clozapine and quetiapine are also associated with supersensitivity psychosis[248–250] and with behavioral DA supersensitivity[244,246,268] but do not induce D2 receptor elevation,[258,269,270], though both of these drugs do induce an increase in the proportion of $D2^{High}$.[259] Similarly, aripiprazole was recently found to elevate $D2^{High}$ despite the absence of any elevation in the total number of D2 receptors.[271] Such a shift toward a greater population in the $D2^{High}$ state is expected to result in increased D2 signaling and DA supersensitivity despite the lack of D2 up-regulation. This lack of striatal D2 receptor up-regulation has been proposed to underlie the low propensity of clozapine to induce TD in humans.[272] Thus, there are two types of D2 receptor–mediated supersensitivity: one associated with increased $D2^{High}$ and potentially leading to supersensitivity psychosis and one associated with increased D2 density and potentially leading to TD. It should be pointed out that hypotheses regarding $D2^{High}$ and its implications are derived mainly from animal studies and in vitro binding. With the advent of $[^{11}C]$-(+)-PHNO and the ability to image the $D2^{High}$ state in patients, it is now possible to test these ideas in humans.[273,274]

LINKING THE BIOLOGY, PHARMACOLOGY, AND PSYCHOLOGY OF ANTIPSYCHOTICS

While the preceding sections highlight the biological disturbance in schizophrenia (likely due to increased presynaptic DA function) and the pharmacological action of antipsychotics (likely due to postsynaptic D2 receptor blockade), they leave unanswered the central question: how does this relate to the symptoms experienced by the patient? In other words, how does one link the biology and pharmacology to the psychological expression of the disease? To do this, one requires a framework that links DA to symptom expression and its blockade to symptom resolution.

Dopamine is recognized to play a central role in reward-seeking[275,276] and in reward-based learning.[277] Accumulating evidence from nonhuman primate studies demonstrates that DA neurons of the mesolimbic system are activated not when a reward occurs or is consumed, but when the reward is predicted or when a predicted reward is not received.[278,279] Rather than mediating the hedonic impact of natural reward, as originally thought,[280] DA neurons of the mesolimbic system thus encode reward prediction errors and serve as an important teaching signal by which animals can learn environmental cues associated with rewards.[281,282] While DA neurons fire with subsecond timing, microdialysis studies suggest that the DA released by sustained firing or pharmacological means remains elevated for longer periods (minutes and even an hour). This released DA is thought to attribute incentive salience to stimuli—such that in future interactions those stimuli, whether they are aversive or pleasurable, motivate goal-directed behaviors.[283,284] Thus, under normal circumstances, the DA system is involved in detecting new rewards, in learning cues predicting those rewards, and in motivating behaviors to salient stimuli so as to maximize rewards.

How can an abnormality in such a system make a patient paranoid about the police? It is hypothesized that chaotic fluctuations of DA release in schizophrenia disrupt the normal stimulus–reward association learning and misattribute motivational salience to otherwise irrelevant stimuli.[285,286] This aberrant attribution of salience directs overwhelming attention to neutral stimuli, and it is proposed that the patient then provides a culturally and personally consistent cognitive scheme (the delusion) to account for these aberrantly salient experiences. Such a model is supported by observations of an abnormally high ventral striatal response to neutral stimuli recently evidenced in schizophrenic patients during reward learning.[287] In this context, by blocking DA transmission, antipsychotics would work by dampening or attenuating the motivational salience attributed to environmental cues or events.[286] Antipsychotic treatment would not obliterate positive symptoms but rather dampen the motivational salience accorded to them—explaining why stopping treatment leads to a resurgence of the muted symptoms. Moreover, while the salience of delusional ideation and hallucinatory experiences may be dampened rather quickly by antipsychotics, the full resolution of symptoms requires not only the change in DA tone (and the immediately dampened salience) but also the change in the cognitive schema that subserves the delusion and hallucinations—a process that takes longer, thus explaining both the immediate onset of antipsychotic effects and the longer time to the complete resolution of these symptoms. However, the antipsychotics cannot selectively distinguish between aberrant motivational drives and normative ones, and by indistinguishably dampening the motivational salience of both irrelevant and relevant stimuli, antipsychotics may give rise to dysphoric symptoms of depression and anxiety, leading to noncompliance and relapse. This cycle—in which genetic and environmental influences lead to abnormal DA release, the chaotic DA release leads to aberrant salience that leads to delusions and from delusions to aberrant behavior, which brings the patients to treatment, and treatment that dampens symptoms and normal motivational drives until the next relapse—is represented in Figure 10.3.1.

CONCLUSION AND FUTURE DIRECTIONS

Thus, the last 50 years of research have provided unquestionable evidence supporting a key role of DA receptor blockade in the mechanism of action of antipsychotics. The optimum modulation of DA function during antipsychotic treatment depends on adequate levels of D2 receptor blockade, the daily duration of D2 blockade, and the regional distribution of this effect. While action at the D2 receptor remains indispensable for controlling the positive symptoms of schizophrenia, activity at other receptors is probably required for amelioration of the negative and cognitive symptoms. As outlined above, the development of improved antipsychotics is currently expanding along several lines, with promising reports of glutamatergic interventions, and efforts targeted to DA D1, nicotinic, and/or muscarinic acetylcholine receptors. A major conceptual challenge for the field is whether to look for "silver bullets" that will have an optimal combination of action at several receptors or to move toward rational pharmacotherapy, in which a combination of agents is used to treat the different domains of the illness in a given patient. While in theory the specific targeted approach seems more attractive, in practice efforts to find subgroups of schizophrenia or to find non-D2 axes that provide specific efficacy have so far been unsuccessful. While the history of failed efforts to discover new antipsychotics cautions one against making predictions, the success of the 50 years of dopaminergic treatments is certainly a triumph of modern medicine against the scourge of schizophrenia.

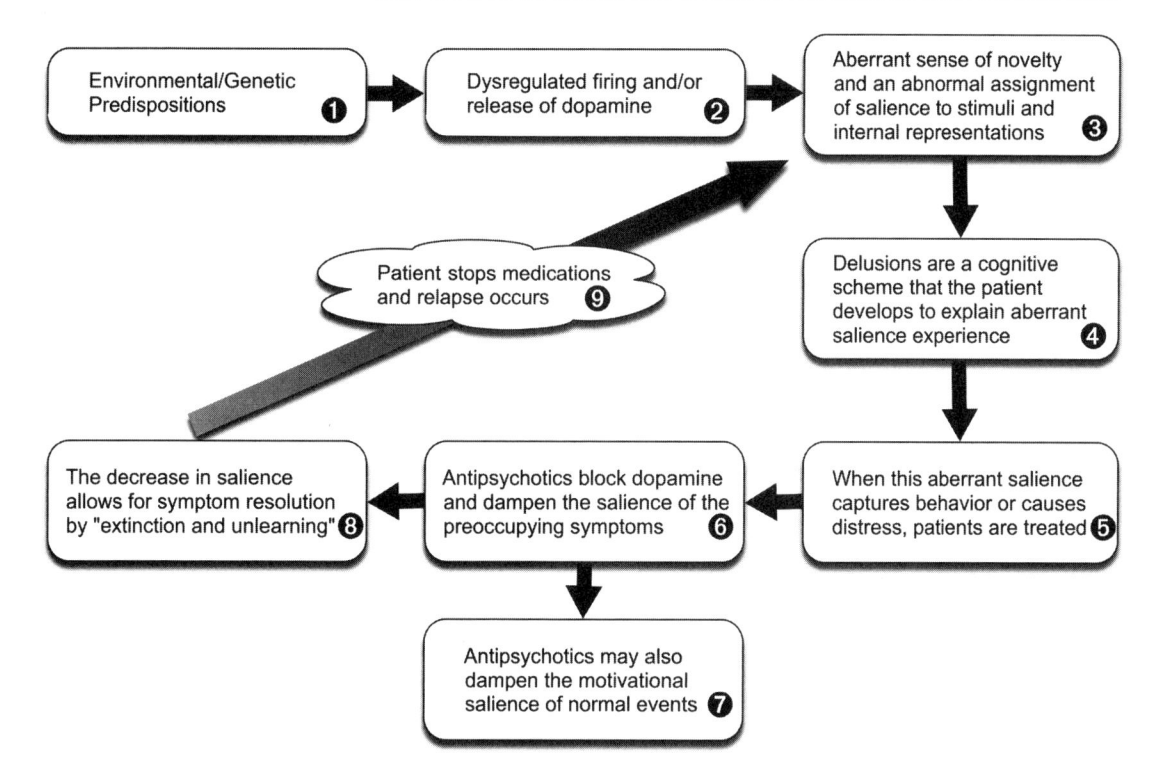

FIGURE 10.3.1. The hypothesis linking DA to psychosis and antipsychotics. The diagram shows a scheme for the chronological evolution of symptoms as a consequence of alterations in DA transmission and the effects of antipsychotics on these symptoms via blocking the effects of DA. The number in each box indicates the order of the event in the sequence. Boxes 1–5 show the etiology and pathophysiology of symptoms and how aberrant DA transmission, via aberrant salience, leads to psychosis; boxes 6–8 show the therapeutic effects and side effects of antipsychotic treatment as related to their actions on the DA system; and box 9 depicts the common consequence of stopping antipsychotics and how the resulting relapse leads to a reentry into the cycle of events. *Source*: Reprinted from *Trends in Pharmacological Sciences*, Vol. 25, S. Kapur, How antipsychotics become anti-"psychotic"—from dopamine to salience to psychosis, 402–406. Copyright 2004, with permission from Elsevier.

REFERENCES

1. Kapur S, Mamo D. Half a century of antipsychotics and still a central role for dopamine D2 receptors. *Prog Neuropsychopharmacol Biol Psychiatry*. 2003;27(7):1081–1090.
2. Lieberman JA, Stroup TS, McEvoy JP, et al. Effectiveness of antipsychotic drugs in patients with chronic schizophrenia. *N Engl J Med*. 2005;353(12):1209–1223.
3. Walker E, Kestler L, Bollini A, Hochman KM. Schizophrenia: etiology and course. *Annu Rev Psychol*. 2004;55:401–430.
4. Tsuang M. Schizophrenia: genes and environment. *Biol Psychiatry*. 2000;47(3):210–220.
5. van Rossum JM. The significance of dopamine-receptor blockade for the mechanism of action of neuroleptic drugs. *Arch Int Pharmacodyn Ther*. 1966;160(2):492–494.
6. Carlsson A, Lindqvist M. Effect of chlorpromazine or haloperidol on formation of 3-methoxytyramine and normetanephrine in mouse brain. *Acta Pharmacol Toxicol*. 1963;20:140–144.
7. Seeman P, Lee T. Antipsychotic drugs: direct correlation between clinical potency and presynaptic action on dopamine neurons. *Science*. 1975;188(4194):1217–1219.
8. Lee T, Seeman P, Tourtellotte WW, Farley IJ, Hornykeiwicz O. Binding of 3H-neuroleptics and 3H-apomorphine in schizophrenic brains. *Nature*. 1978;274(5674):897–900.
9. Seeman P, Ulpian C, Bergeron C, et al. Bimodal distribution of dopamine receptor densities in brains of schizophrenics. *Science*. 1984;225(4663):728–731.
10. Wong DF, Wagner HN Jr, Tune LE, et al. Positron emission tomography reveals elevated D2 dopamine receptors in drug-naive schizophrenics. *Science*. 1986;234(4783):1558–1563.
11. Tune LE, Wong DF, Pearlson G, et al. Dopamine D2 receptor density estimates in schizophrenia: a positron emission tomography study with 11C-N-methylspiperone. *Psychiatry Res*. 1993;49(3):219–237.
12. Reynolds GP, Riederer P, Jellinger K, Gabriel E. Dopamine receptors and schizophrenia: the neuroleptic drug problem. *Neuropharmacology*. 1981;20(12B):1319–1320.
13. Mackay AV, Bird ED, Spokes EG, et al. Dopamine receptors and schizophrenia: drug effect or illness? *Lancet*. 1980;2(8200):915–916.
14. Mackay AV, Iversen LL, Rossor M, et al. Increased brain dopamine and dopamine receptors in schizophrenia. *Arch Gen Psychiatry*. 1982;39(9):991–997.
15. Farde L, Wiesel FA, Stone-Elander S, et al. D2 dopamine receptors in neuroleptic-naive schizophrenic patients. A positron emission tomography study with [11C]raclopride. *Arch Gen Psychiatry*. 1990;47(3):213–219.
16. Hietala J, Syvalahti E, Vuorio K, et al. Striatal D2 dopamine receptor characteristics in neuroleptic-naive schizophrenic

patients studied with positron emission tomography. *Arch Gen Psychiatry.* 1994;51(2):116–123.
17. Nordstrom AL, Farde L, Eriksson L, Halldin C. No elevated D2 dopamine receptors in neuroleptic-naive schizophrenic patients revealed by positron emission tomography and [11C]N-methylspiperone. *Psychiatry Res.* 1995;61(2):67–83.
18. Lomena F, Catafau AM, Parellada E, et al. Striatal dopamine D2 receptor density in neuroleptic-naive and in neuroleptic-free schizophrenic patients: an 123I-IBZM-SPECT study. *Psychopharmacology (Berl).* 2004;172(2):165–169.
19. Yang YK, Yu L, Yeh TL, Chiu NT, Chen PS, Lee IH. Associated alterations of striatal dopamine D2/D3 receptor and transporter binding in drug-naive patients with schizophrenia: a dual-isotope SPECT study. *Am J Psychiatry.* 2004;161(8):1496–1498.
20. Glenthoj BY, Mackeprang T, Svarer C, et al. Frontal dopamine D(2/3) receptor binding in drug-naive first-episode schizophrenic patients correlates with positive psychotic symptoms and gender. *Biol Psychiatry.* 2006;60(6):621–629.
21. Buchsbaum MS, Christian BT, Lehrer DS, et al. D2/D3 dopamine receptor binding with [F-18]fallypride in thalamus and cortex of patients with schizophrenia. *Schizophr Res.* 2006;85(1–3):232–244.
22. Talvik M, Nordstrom AL, Okubo Y, et al. Dopamine D2 receptor binding in drug-naive patients with schizophrenia examined with raclopride-C11 and positron emission tomography. *Psychiatry Res.* 2006;148(2–3):165–173.
23. Karlsson P, Farde L, Halldin C, Sedvall G. PET study of D(1) dopamine receptor binding in neuroleptic-naive patients with schizophrenia. *Am J Psychiatry.* 2002;159(5):761–767.
24. Abi-Dargham A, Mawlawi O, Lombardo I, et al. Prefrontal dopamine D1 receptors and working memory in schizophrenia. *J Neurosci.* 2002;22(9):3708–3719.
25. Okubo Y, Suhara T, Suzuki K, et al. Decreased prefrontal dopamine D1 receptors in schizophrenia revealed by PET. *Nature.* 1997;385(6617):634–636.
26. Gurevich EV, Bordelon Y, Shapiro RM, Arnold SE, Gur RE, Joyce JN. Mesolimbic dopamine D3 receptors and use of antipsychotics in patients with schizophrenia. A postmortem study. *Arch Gen Psychiatry.* 1997;54(3):225–232.
27. Seeman P, Guan H, Van Tol HHM. Dopamine D4 receptors elevated in schizophrenia,. *Nature.* 1995;365:441–445.
28. Murray AM, Hyde TM, Knable MB, et al. Distribution of putative D4 dopamine receptors in postmortem striatum from patients with schizophrenia. *J Neurosci.* 1995;15(3 Pt 2):2186–2191.
29. Sumiyoshi T, Stockmeier CA, Overholser JC, Thompson PA, Meltzer HY. Dopamine D4 receptors and effects of guanine nucleotides on [3H]raclopride binding in postmortem caudate nucleus of subjects with schizophrenia or major depression. *Brain Res.* 1995;681(1–2):109–116.
30. Reynolds GP, Mason SL. Are striatal dopamine D4 receptors increased in schizophrenia? *J Neurochem.* 1994;63(4):1576–1577.
31. Reynolds GP, Mason SL. Absence of detectable striatal dopamine D4 receptors in drug-treated schizophrenia. *Eur J Pharmacol.* 1995;281(2):R5–6.
32. Lahti RA, Roberts RC, Conley RR, Cochrane EV, Mutin A, Tamminga CA. D2-type dopamine receptors in postmortem human brain sections from normal and schizophrenic subjects. *Neuroreport.* 1996;7(12):1945–1948.
33. Lahti RA, Roberts RC, Cochrane EV, et al. Direct determination of dopamine D4 receptors in normal and schizophrenic postmortem brain tissue: a [3H]NGD-94-1 study. *Mol Psychiatry.* 1998;3(6):528–533.

34. Hietala J, Syvalahti E, Vilkman H, et al. Depressive symptoms and presynaptic dopamine function in neuroleptic-naive schizophrenia. *Schizophr Res.* 1999;35(1):41–50.
35. Hietala J, Syvalahti E, Vuorio K, et al. Presynaptic dopamine function in striatum of neuroleptic-naive schizophrenic patients. *Lancet.* 1995;346(8983):1130–1131.
36. McGowan S, Lawrence AD, Sales T, Quested D, Grasby P. Presynaptic dopaminergic dysfunction in schizophrenia: a positron emission tomographic [18F]fluorodopa study. *Arch Gen Psychiatry.* 2004;61(2):134–142.
37. Bose SK, Turkheimer FE, Howes OD, et al. Classification of schizophrenic patients and healthy controls using [18F] fluorodopa PET imaging. *Schizophr Res.* 2008;106(2–3):148–155.
38. Abi-Dargham A, Rodenhiser J, Printz D, et al. Increased baseline occupancy of D2 receptors by dopamine in schizophrenia. *Proc Natl Acad Sci USA.* 2000;97(14):8104–8109.
39. Abi-Dargham A, Gil R, Krystal J, et al. Increased striatal dopamine transmission in schizophrenia: confirmation in a second cohort. *Am J Psychiatry.* 1998;155(6):761–767.
40. Bertolino A, Breier A, Callicott JH, et al. The relationship between dorsolateral prefrontal neuronal N-acetylaspartate and evoked release of striatal dopamine in schizophrenia. *Neuropsychopharmacology.* 2000;22(2):125–132.
41. Breier A, Su TP, Saunders R, et al. Schizophrenia is associated with elevated amphetamine-induced synaptic dopamine concentrations: evidence from a novel positron emission tomography method. *Proc Natl Acad Sci USA.* 1997;94(6):2569–2574.
42. Laruelle M, Abi-Dargham A, van Dyck CH, et al. Single photon emission computerized tomography imaging of amphetamine-induced dopamine release in drug-free schizophrenic subjects. *Proc Natl Acad Sci USA.* 1996;93(17):9235–9240.
43. Grace AA. Phasic versus tonic dopamine release and the modulation of dopamine system responsivity: a hypothesis for the etiology of schizophrenia. *Neuroscience.* 1991;41(1):1–24.
44. Davis KL, Kahn RS, Ko G, Davidson M. Dopamine in schizophrenia: a review and reconceptualization. *Am J Psychiatry.* 1991;148(11):1474–1486.
45. Luby ED, Cohen BD, Rosenbaum G, Gottlieb JS, Kelley R. Study of a new schizophrenomimetic drug; sernyl. *AMA Arch Neurol Psychiatry.* 1959;81(3):363–369.
46. Johnstone M, Evans V, Baigel S. Sernyl (CI-395) in clinical anaesthesia. *Br J Anaesth.* 1959;31:433–439.
47. Krystal JH, Karper LP, Seibyl JP, et al. Subanesthetic effects of the noncompetitive NMDA antagonist, ketamine, in humans. Psychotomimetic, perceptual, cognitive, and neuroendocrine responses. *Arch Gen Psychiatry.* 1994;51(3):199–214.
48. Malhotra AK, Pinals DA, Weingartner H, et al. NMDA receptor function and human cognition: the effects of ketamine in healthy volunteers. *Neuropsychopharmacology.* 1996;14(5):301–307.
49. Itil T, Keskiner A, Kiremitci N, Holden JM. Effect of phencyclidine in chronic schizophrenics. *Can Psychiatr Assoc J.* 1967;12(2):209–212.
50. Lahti AC, Koffel B, LaPorte D, Tamminga CA. Subanesthetic doses of ketamine stimulate psychosis in schizophrenia. *Neuropsychopharmacology.* 1995;13(1):9–19.
51. Malhotra AK, Pinals DA, Adler CM, et al. Ketamine-induced exacerbation of psychotic symptoms and cognitive impairment in neuroleptic-free schizophrenics. *Neuropsychopharmacology.* 1997;17(3):141–150.
52. Clinton SM, Meador-Woodruff JH. Abnormalities of the NMDA receptor and associated intracellular molecules in the thalamus in schizophrenia and bipolar disorder. *Neuropsychopharmacology.* 2004;29(7):1353–1362.

53. Kristiansen LV, Huerta I, Beneyto M, Meador-Woodruff JH. NMDA receptors and schizophrenia. *Curr Opin Pharmacol.* 2007;7(1):48–55.
54. Beneyto M, Meador-Woodruff JH. Lamina-specific abnormalities of NMDA receptor-associated postsynaptic protein transcripts in the prefrontal cortex in schizophrenia and bipolar disorder. *Neuropsychopharmacology.* 2008;33(9):2175–2186.
55. Olbrich HM, Valerius G, Rusch N, et al. Frontolimbic glutamate alterations in first episode schizophrenia: evidence from a magnetic resonance spectroscopy study. *World J Biol Psychiatry.* 2008;9(1):59–63.
56. van Elst LT, Valerius G, Buchert M, et al. Increased prefrontal and hippocampal glutamate concentration in schizophrenia: evidence from a magnetic resonance spectroscopy study. *Biol Psychiatry.* 2005;58(9):724–730.
57. Theberge J, Jensen JE, Rowland LM. Regarding "Increased prefrontal and hippocampal glutamate concentration in schizophrenia: evidence from a magnetic resonance spectroscopy study." *Biol Psychiatry.* 2007;61(10):1218–1219; author reply 1219–1220.
58. Javitt DC, Zukin SR. Recent advances in the phencyclidine model of schizophrenia. *Am J Psychiatry.* 1991;148(10):1301–1308.
59. Olney JW, Farber NB. Glutamate receptor dysfunction and schizophrenia. *Arch Gen Psychiatry.* 1995;52(12):998–1007.
60. Moghaddam B, Adams BW. Reversal of phencyclidine effects by a group II metabotropic glutamate receptor agonist in rats. *Science.* 1998;281(5381):1349–1352.
61. Duncan GE, Miyamoto S, Leipzig JN, Lieberman JA. Comparison of the effects of clozapine, risperidone, and olanzapine on ketamine-induced alterations in regional brain metabolism. *J Pharmacol Exp Ther.* 2000;293(1):8–14.
62. Farber NB. The NMDA receptor hypofunction model of psychosis. *Ann NY Acad Sci.* 2003;1003:119–130.
63. Sesack SR, Carr DB, Omelchenko N, Pinto A. Anatomical substrates for glutamate–dopamine interactions: evidence for specificity of connections and extrasynaptic actions. *Ann NY Acad Sci.* 2003;1003:36–52.
64. French ED. Effects of phencyclidine on ventral tegmental A10 dopamine neurons in the rat. *Neuropharmacology.* 1986;25(3):241–248.
65. French ED, Ceci A. Non-competitive N-methyl-D-aspartate antagonists are potent activators of ventral tegmental A10 dopamine neurons. *Neurosci Lett.* 1990;119(2):159–162.
66. Murase S, Mathe JM, Grenhoff J, Svensson TH. Effects of dizocilpine (MK-801) on rat midbrain dopamine cell activity: differential actions on firing pattern related to anatomical localization. *J Neural Transm Gen Sect.* 1993;91(1):13–25.
67. Schmidt CJ, Fadayel GM. Regional effects of MK-801 on dopamine release: effects of competitive NMDA or 5-HT2A receptor blockade. *J Pharmacol Exp Ther.* 1996;277(3):1541–1549.
68. Takahata R, Moghaddam B. Activation of glutamate neurotransmission in the prefrontal cortex sustains the motoric and dopaminergic effects of phencyclidine. *Neuropsychopharmacology.* 2003;28(6):1117–1124.
69. Lorrain DS, Baccei CS, Bristow LJ, Anderson JJ, Varney MA. Effects of ketamine and N-methyl-D-aspartate on glutamate and dopamine release in the rat prefrontal cortex: modulation by a group II selective metabotropic glutamate receptor agonist LY379268. *Neuroscience.* 2003;117(3):697–706.
70. Bristow LJ, Hutson PH, Thorn L, Tricklebank MD. The glycine/NMDA receptor antagonist, R-(+)-HA-966, blocks activation of the mesolimbic dopaminergic system induced by phencyclidine and dizocilpine (MK-801) in rodents. *Br J Pharmacol.* 1993;108(4):1156–1163.
71. Svensson TH. Dysfunctional brain dopamine systems induced by psychotomimetic NMDA-receptor antagonists and the effects of antipsychotic drugs. *Brain Res Brain Res Rev.* 2000;31(2–3):320–329.
72. Jentsch JD, Roth RH. The neuropsychopharmacology of phencyclidine: from NMDA receptor hypofunction to the dopamine hypothesis of schizophrenia. *Neuropsychopharmacology.* 1999;20(3):201–225.
73. Carlsson A, Waters N, Holm-Waters S, Tedroff J, Nilsson M, Carlsson ML. Interactions between monoamines, glutamate, and GABA in schizophrenia: new evidence. *Annu Rev Pharmacol Toxicol.* 2001;41:237–260.
74. Keshavan MS, Prasad KM, Pearlson G. Are brain structural abnormalities useful as endophenotypes in schizophrenia? *Int Rev Psychiatry.* 2007;19(4):397–406.
75. Meisenzahl EM, Koutsouleris N, Bottlender R, et al. Structural brain alterations at different stages of schizophrenia: A voxel-based morphometric study. *Schizophr Res.* 2008;104(1–3):44–60.
76. Delay J, Deniker P, Harl JM. Utilisation en thérapeutique psychiatrique d'une phénothiazine d'action centrale élective (4560RP). *Ann Méd Psychol.* 1952(110):112–117.
77. Laborit H, Huguenard P, Alluaume R. A new vegetative stabilizer; 4560 R.P. *Presse Med.* 1952;60(10):206–208.
78. Levinson DF, Simpson GM, Singh H, et al. Fluphenazine dose, clinical response, and extrapyramidal symptoms during acute treatment. *Arch Gen Psychiatry.* 1990;47(8):761–768.
79. Meltzer HY, Sommers AA, Luchins DJ. The effect of neuroleptics and other psychotropic drugs on negative symptoms in schizophrenia. *J Clin Psychopharmacol.* 1986;6(6):329–338.
80. Meltzer HY, Lee MA, Ranjan R. Recent advances in the pharmacotherapy of schizophrenia. *Acta Psychiatr Scand Suppl.* 1994;384:95–101.
81. Naber D, Karow A. Good tolerability equals good results: the patient's perspective. *Eur Neuropsychopharmacol.* 2001;11(suppl 4):S391–S396.
82. Casey DE. Implications of the CATIE trial on treatment: extrapyramidal symptoms. *CNS Spectr.* 2006;11(7 suppl 7):25–31.
83. Dossenbach M, Arango-Davila C, Silva Ibarra H, et al. Response and relapse in patients with schizophrenia treated with olanzapine, risperidone, quetiapine, or haloperidol: 12-month follow-up of the Intercontinental Schizophrenia Outpatient Health Outcomes (IC-SOHO) study. *J Clin Psychiatry.* 2005;66(8):1021–1030.
84. Glick ID, Marder SR. Long-term maintenance therapy with quetiapine versus haloperidol decanoate in patients with schizophrenia or schizoaffective disorder. *J Clin Psychiatry.* 2005;66(5):638–641.
85. Haro JM, Salvador-Carulla L. The SOHO (Schizophrenia Outpatient Health Outcome) study: implications for the treatment of schizophrenia. *CNS Drugs.* 2006;20(4):293–301.
86. Correll CU, Schenk EM. Tardive dyskinesia and new antipsychotics. *Curr Opin Psychiatry.* 2008;21(2):151–156.
87. Kane J, Honigfeld G, Singer J, Meltzer H. Clozapine for the treatment-resistant schizophrenic. A double-blind comparison with chlorpromazine. *Arch Gen Psychiatry.* 1988;45(9):789–796.
88. Breier A, Buchanan RW, Kirkpatrick B, et al. Effects of clozapine on positive and negative symptoms in outpatients with schizophrenia. *Am J Psychiatry.* 1994;151(1):20–26.
89. Rosenheck R, Cramer J, Xu W, et al. A comparison of clozapine and haloperidol in hospitalized patients with refractory

schizophrenia. Department of Veterans Affairs Cooperative Study Group on Clozapine in Refractory Schizophrenia. *N Engl J Med*. 1997;337(12):809–815.
90. Taylor DM, Duncan-McConnell D. Refractory schizophrenia and atypical antipsychotics. *J Psychopharmacol*. 2000;14(4): 409–418.
91. Chakos M, Lieberman J, Hoffman E, Bradford D, Sheitman B. Effectiveness of second-generation antipsychotics in patients with treatment-resistant schizophrenia: a review and meta-analysis of randomized trials. *Am J Psychiatry*. 2001;158(4): 518–526.
92. Beasley CM Jr, Tollefson G, Tran P, Satterlee W, Sanger T, Hamilton S. Olanzapine versus placebo and haloperidol: acute phase results of the North American double-blind olanzapine trial. *Neuropsychopharmacology*. 1996;14(2):111–123.
93. Zimbroff DL, Kane JM, Tamminga CA, et al. Controlled, dose-response study of sertindole and haloperidol in the treatment of schizophrenia. Sertindole Study Group. *Am J Psychiatry*. 1997;154(6):782–791.
94. Petit M, Raniwalla J, Tweed J, Leutenegger E, Dollfus S, Kelly F. A comparison of an atypical and typical antipsychotic, zotepine versus haloperidol, in patients with acute exacerbation of schizophrenia: a parallel-group double-blind trial. *Psychopharmacol Bull*. 1996;32(1):81–87.
95. Claus A, Bollen J, De Cuyper H, et al. Risperidone versus haloperidol in the treatment of chronic schizophrenic inpatients: a multicentre double-blind comparative study. *Acta Psychiatr Scand*. 1992;85(4):295–305.
96. Chouinard G, Jones B, Remington G, et al. A Canadian multicenter placebo-controlled study of fixed doses of risperidone and haloperidol in the treatment of chronic schizophrenic patients. *J Clin Psychopharmacol*. 1993;13(1):25–40.
97. Carman J, Peuskens J, Vangeneugden A. Risperidone in the treatment of negative symptoms of schizophrenia: a meta-analysis. *Int Clin Psychopharmacol*. 1995;10(4):207–213.
98. Buchanan RW. Clozapine: efficacy and safety. *Schizophr Bull*. 1995;21(4):579–591.
99. Tollefson GD, Beasley CM Jr, Tran PV, et al. Olanzapine versus haloperidol in the treatment of schizophrenia and schizoaffective and schizophreniform disorders: results of an international collaborative trial. *Am J Psychiatry*. 1997;154(4):457–465.
100. Marder SR, Davis JM, Chouinard G. The effects of risperidone on the five dimensions of schizophrenia derived by factor analysis: combined results of the North American trials. *J Clin Psychiatry*. 1997;58(12):538–546.
101. Beasley CM Jr, Tollefson GD, Tran PV. Efficacy of olanzapine: an overview of pivotal clinical trials. *J Clin Psychiatry*. 1997;58(suppl 10):7–12.
102. Arvanitis LA, Miller BG. Multiple fixed doses of "Seroquel" (quetiapine) in patients with acute exacerbation of schizophrenia: a comparison with haloperidol and placebo. The Seroquel Trial 13 Study Group. *Biol Psychiatry*. 1997;42(4): 233–246.
103. Peuskens J, Link CG. A comparison of quetiapine and chlorpromazine in the treatment of schizophrenia. *Acta Psychiatr Scand*. 1997;96(4):265–273.
104. Leucht S, Pitschel-Walz G, Abraham D, Kissling W. Efficacy and extrapyramidal side-effects of the new antipsychotics olanzapine, quetiapine, risperidone, and sertindole compared to conventional antipsychotics and placebo. A meta-analysis of randomized controlled trials. *Schizophr Res*. 1999;35(1):51–68.
105. Buchanan RW, Breier A, Kirkpatrick B, Ball P, Carpenter WT Jr. Positive and negative symptom response to clozapine in schizophrenic patients with and without the deficit syndrome. *Am J Psychiatry*. 1998;155(6):751–760.
106. Copolov DL, Link CG, Kowalcyk B. A multicentre, double-blind, randomized comparison of quetiapine (ICI 204,636, 'Seroquel') and haloperidol in schizophrenia. *Psychol Med*. 2000;30(1):95–105.
107. Moller HJ, Riedel M, Jager M, et al. Short-term treatment with risperidone or haloperidol in first-episode schizophrenia: 8-week results of a randomized controlled trial within the German Research Network on Schizophrenia. *Int J Neuropsychopharmacol*. 2008;11(7):985–997.
108. Haro JM, Novick D, Suarez D, Roca M. Antipsychotic treatment discontinuation in previously untreated patients with schizophrenia: 36-month results from the SOHO study. *J Psychiatr Res*. 2009;43(3):265–273.
109. Newcomer JW. Second-generation (atypical) antipsychotics and metabolic effects: a comprehensive literature review. *CNS Drugs*. 2005;19(suppl 1):1–93.
110. Luft B, Taylor D. A review of atypical antipsychotic drugs versus conventional medication in schizophrenia. *Expert Opin Pharmacother*. 2006;7(13):1739–1748.
111. Snyder SH, Creese I, Burt DR. The brain's dopamine receptor: labeling with (3H) dopamine and (3H) haloperidol. *Psychopharmacol Commun*. 1975;1(6):663–673.
112. Gingrich JA, Caron MG. Recent advances in the molecular biology of dopamine receptors. *Annu Rev Neurosci*. 1993;16:299–321.
113. Palacios JM, Camps M, Cortes R, Probst A. Mapping dopamine receptors in the human brain. *J Neural Transm Suppl*. 1988;27:227–235.
114. Khan ZU, Gutierrez A, Martin R, Penafiel A, Rivera A, de la Calle A. Dopamine D5 receptors of rat and human brain. *Neuroscience*. 2000;100(4):689–699.
115. Onali P, Olianas MC. Involvement of adenylate cyclase inhibition in dopamine autoreceptor regulation of tyrosine hydroxylase in rat nucleus accumbens. *Neurosci Lett*. 1989;102(1):91–96.
116. Limberger N, Trout SJ, Kruk ZL, Starke K. "Real time" measurement of endogenous dopamine release during short trains of pulses in slices of rat neostriatum and nucleus accumbens: role of autoinhibition. *Naunyn Schmiedebergs Arch Pharmacol*. 1991;344(6):623–629.
117. Mercuri NB, Calabresi P, Bernardi G. The electrophysiological actions of dopamine and dopaminergic drugs on neurons of the substantia nigra pars compacta and ventral tegmental area. *Life Sci*. 1992;51(10):711–718.
118. Gurevich EV, Joyce JN. Distribution of dopamine D3 receptor expressing neurons in the human forebrain: comparison with D2 receptor expressing neurons. *Neuropsychopharmacology*. 1999;20(1):60–80.
119. Tepper JM, Sun BC, Martin LP, Creese I. Functional roles of dopamine D2 and D3 autoreceptors on nigrostriatal neurons analyzed by antisense knockdown in vivo. *J Neurosci*. 1997;17(7):2519–2530.
120. Zapata A, Witkin JM, Shippenberg TS. Selective D3 receptor agonist effects of (+)-PD 128907 on dialysate dopamine at low doses. *Neuropharmacology*. 2001;41(3):351–359.
121. Lahti RA, Primus RJ, Gallager DW, Roberts R, Tamminga CA. Distribution of dopamine D4 receptor in human postmortem brain sections: autoradiographic studies with [3H]NGD-94-I. *Schizophr Res*. 1996;18:173.
122. Creese I, Burt DR, Snyder SH. Dopamine receptor binding predicts clinical and pharmacological potencies of antischizophrenic drugs. *Science*. 1976;192(4238):481–483.

123. Seeman P. Dopamine receptors and the dopamine hypothesis of schizophrenia. *Synapse.* 1987;1(2):133–152.
124. Wadenberg ML, Kapur S, Soliman A, Jones C, Vaccarino F. Dopamine D2 receptor occupancy predicts catalepsy and the suppression of conditioned avoidance response behavior in rats. *Psychopharmacology (Berl).* 2000;150(4):422–429.
125. Kapur S, Seeman P. Does fast dissociation from the dopamine D(2) receptor explain the action of atypical antipsychotics?: A new hypothesis. *Am J Psychiatry.* 2001;158(3):360–369.
126. Kapur S, Zipursky R, Jones C, Remington G, Houle S. Relationship between dopamine D(2) occupancy, clinical response, and side effects: a double-blind PET study of first-episode schizophrenia. *Am J Psychiatry.* 2000;157(4):514–520.
127. Farde L, Nordstrom AL, Wiesel FA, Pauli S, Halldin C, Sedvall G. Positron emission tomographic analysis of central D1 and D2 dopamine receptor occupancy in patients treated with classical neuroleptics and clozapine. Relation to extrapyramidal side effects. *Arch Gen Psychiatry.* 1992;49(7):538–544.
128. Farde L, Wiesel FA, Nordstrom AL, Sedvall G. D1- and D2-dopamine receptor occupancy during treatment with conventional and atypical neuroleptics. *Psychopharmacology (Berl).* 1989;99(suppl):S28–S31.
129. Kapur S, Zipursky R, Jones C, Shammi CS, Remington G, Seeman P. A positron emission tomography study of quetiapine in schizophrenia: a preliminary finding of an antipsychotic effect with only transiently high dopamine D2 receptor occupancy. *Arch Gen Psychiatry.* 2000;57(6):553–559.
130. Tamminga CA. Partial dopamine agonists in the treatment of psychosis. *J Neural Transm.* 2002;109(3):411–420.
131. Wetzel H, Hillert A, Grunder G, Benkert O. Roxindole, a dopamine autoreceptor agonist, in the treatment of positive and negative schizophrenic symptoms. *Am J Psychiatry.* 1994;151(10):1499–1502.
132. Ohmori T, Koyama T, Inoue T, Matsubara S, Yamashita I. B-HT 920, a dopamine D2 agonist, in the treatment of negative symptoms of chronic schizophrenia. *Biol Psychiatry.* 1993;33(10):687–693.
133. Olbrich R, Schanz H. An evaluation of the partial dopamine agonist terguride regarding positive symptoms reduction in schizophrenics. *J Neural Transm Gen Sect.* 1991;84(3):233–236.
134. Naber D, Gaussares C, Moeglen JM, Tremmel L, Bailey PE, Group tSHCS. Efficacy and tolerability of SDZ HDC 912, a partial dopamine D-2 agonist, in the treatment of schizophrenia. In: Meltzer HY, ed. *Novel Antipsychotic Drugs.* New York, NY: Raven Press; 1992:99–107.
135. Lahti AC, Weiler MA, Corey PK, Lahti RA, Carlsson A, Tamminga CA. Antipsychotic properties of the partial dopamine agonist (-)-3-(3-hydroxyphenyl)-N-n-propylpiperidine(preclamol) in schizophrenia. *Biol Psychiatry.* 1998;43(1):2–11.
136. DeLeon A, Patel NC, Crismon ML. Aripiprazole: a comprehensive review of its pharmacology, clinical efficacy, and tolerability. *Clin Ther.* 2004;26(5):649–666.
137. Lieberman JA. Aripiprazole. In: Schatzberg AF, Nemeroff CB, eds. *Texbook of Psychopharmacology.* 3rd. ed. Washington, DC: American Psychiatric Press; 2004:487–494.
138. Kane JM, Carson WH, Saha AR, et al. Efficacy and safety of aripiprazole and haloperidol versus placebo in patients with schizophrenia and schizoaffective disorder. *J Clin Psychiatry.* 2002;63(9):763–771.
139. Kasper S, Lerman MN, McQuade RD, et al. Efficacy and safety of aripiprazole vs. haloperidol for long-term maintenance treatment following acute relapse of schizophrenia. *Int J Neuropsychopharmacol.* 2003;6(4):325–337.
140. Kern RS, Green MF, Cornblatt BA, et al. The neurocognitive effects of aripiprazole: an open-label comparison with olanzapine. *Psychopharmacology (Berl).* 2006;187(3):312–320.
141. Potkin SG, Saha AR, Kujawa MJ, et al. Aripiprazole, an antipsychotic with a novel mechanism of action, and risperidone vs placebo in patients with schizophrenia and schizoaffective disorder. *Arch Gen Psychiatry.* 2003;60(7):681–690.
142. Chrzanowski WK, Marcus RN, Torbeyns A, Nyilas M, McQuade RD. Effectiveness of long-term aripiprazole therapy in patients with acutely relapsing or chronic, stable schizophrenia: a 52-week, open-label comparison with olanzapine. *Psychopharmacology (Berl).* 2006;189(2):259–266.
143. Zimbroff D, Warrington L, Loebel A, Yang R, Siu C. Comparison of ziprasidone and aripiprazole in acutely ill patients with schizophrenia or schizoaffective disorder: a randomized, double-blind, 4-week study. *Int Clin Psychopharmacol.* 2007;22(6):363–370.
144. Bhattacharjee J, El-Sayeh HG. Aripiprazole versus typical antipsychotic drugs for schizophrenia. *Cochrane Database Syst Rev.* 2008(3):CD006617.
145. McEvoy JP, Daniel DG, Carson WH Jr, McQuade RD, Marcus RN. A randomized, double-blind, placebo-controlled, study of the efficacy and safety of aripiprazole 10, 15 or 20 mg/day for the treatment of patients with acute exacerbations of schizophrenia. *J Psychiatr Res.* 2007;41(11):895–905.
146. Grunder G, Fellows C, Janouschek H, et al. Brain and plasma pharmacokinetics of aripiprazole in patients with schizophrenia: an [18F]fallypride PET study. *Am J Psychiatry.* 2008; 165(8):988–995.
147. Mamo D, Graff A, Mizrahi R, Shammi CM, Romeyer F, Kapur S. Differential effects of aripiprazole on D(2), 5-HT(2), and 5-HT(1A) receptor occupancy in patients with schizophrenia: a triple tracer PET study. *Am J Psychiatry.* 2007;164(9): 1411–1417.
148. Kegeles LS, Slifstein M, Frankle WG, et al. Dose-occupancy study of striatal and extrastriatal dopamine D2 receptors by aripiprazole in schizophrenia with PET and [18F]fallypride. *Neuropsychopharmacology.* 2008;33(13):3111–3125.
149. Tadori Y, Miwa T, Tottori K, et al. Aripiprazole's low intrinsic activities at human dopamine D2L and D2S receptors render it a unique antipsychotic. *Eur J Pharmacol.* 2005;515(1–3): 10–19.
150. Jordan S, Johnson JL, Regardie K, et al. Dopamine D2 receptor partial agonists display differential or contrasting characteristics in membrane and cell-based assays of dopamine D2 receptor signaling. *Prog Neuropsychopharmacol Biol Psychiatry.* 2007; 31(2):348–356.
151. Weiden PJ, Preskorn SH, Fahnestock PA, Carpenter D, Ross R, Docherty JP. Translating the psychopharmacology of antipsychotics to individualized treatment for severe mental illness: a roadmap. *J Clin Psychiatry.* 2007;68(suppl 7):1–48.
152. Coward DM, Imperato A, Urwyler S, White TG. Biochemical and behavioural properties of clozapine. *Psychopharmacology (Berl).* 1989;99(suppl):S6–S12.
153. Tauscher J, Hussain T, Agid O, et al. Equivalent occupancy of dopamine D1 and D2 receptors with clozapine: differentiation from other atypical antipsychotics. *Am J Psychiatry.* 2004;161(9):1620–1625.
154. Reimold M, Solbach C, Noda S, et al. Occupancy of dopamine D(1), D (2) and serotonin (2A) receptors in schizophrenic patients treated with flupentixol in comparison with risperidone and haloperidol. *Psychopharmacology (Berl).* 2007;190(2): 241–249.

155. de Beaurepaire R, Labelle A, Naber D, Jones BD, Barnes TR. An open trial of the D1 antagonist SCH 39166 in six cases of acute psychotic states. *Psychopharmacology (Berl)*. 1995;121(3): 323–327.
156. Karlsson P, Smith L, Farde L, Harnryd C, Sedvall G, Wiesel FA. Lack of apparent antipsychotic effect of the D1-dopamine receptor antagonist SCH39166 in acutely ill schizophrenic patients. *Psychopharmacology (Berl)*. 1995;121(3):309–316.
157. Karle J, Clemmesen L, Hansen L, et al. NNC 01-0687, a selective dopamine D1 receptor antagonist, in the treatment of schizophrenia. *Psychopharmacology (Berl)*. 1995;121(3):328–329.
158. Sawaguchi T, Goldman-Rakic PS. D1 dopamine receptors in prefrontal cortex: involvement in working memory. *Science*. 1991;251(4996):947–950.
159. Muller U, von Cramon DY, Pollmann S. D1- versus D2-receptor modulation of visuospatial working memory in humans. *J Neurosci*. 1998;18(7):2720–2728.
160. Williams GV, Goldman-Rakic PS. Modulation of memory fields by dopamine D1 receptors in prefrontal cortex. *Nature*. 1995;376(6541):572–575.
161. Sawaguchi T, Goldman-Rakic PS. The role of D1-dopamine receptor in working memory: local injections of dopamine antagonists into the prefrontal cortex of rhesus monkeys performing an oculomotor delayed-response task. *J Neurophysiol*. 1994;71(2):515–528.
162. Arnsten AF, Cai JX, Murphy BL, Goldman-Rakic PS. Dopamine D1 receptor mechanisms in the cognitive performance of young adult and aged monkeys. *Psychopharmacology (Berl)*. 1994;116(2):143–151.
163. Goldman-Rakic PS. Working memory dysfunction in schizophrenia. *J Neuropsychiatry Clin Neurosci*. 1994;6(4):348–357.
164. Goldman-Rakic PS, Muly EC 3rd, Williams GV. D(1) receptors in prefrontal cells and circuits. *Brain Res Brain Res Rev*. 2000;31(2–3):295–301.
165. Lidow MS, Goldman-Rakic PS. A common action of clozapine, haloperidol, and remoxipride on D1- and D2-dopaminergic receptors in the primate cerebral cortex. *Proc Natl Acad Sci USA*. 1994;91(10):4353–4356.
166. Lidow MS, Elsworth JD, Goldman-Rakic PS. Down-regulation of the D1 and D5 dopamine receptors in the primate prefrontal cortex by chronic treatment with antipsychotic drugs. *J Pharmacol Exp Ther*. 1997;281(1):597–603.
167. Hirvonen J, van Erp TG, Huttunen J, et al. Brain dopamine d1 receptors in twins discordant for schizophrenia. *Am J Psychiatry*. 2006;163(10):1747–1753.
168. Castner SA, Williams GV, Goldman-Rakic PS. Reversal of antipsychotic-induced working memory deficits by short-term dopamine D1 receptor stimulation. *Science*. 2000;287(5460): 2020–2022.
169. Friedman JI, Temporini H, Davis KL. Pharmacologic strategies for augmenting cognitive performance in schizophrenia. *Biol Psychiatry*. 1999;45(1):1–16.
170. Goldman-Rakic PS, Castner SA, Svensson TH, Siever LJ, Williams GV. Targeting the dopamine D1 receptor in schizophrenia: insights for cognitive dysfunction. *Psychopharmacology (Berl)*. 2004;174(1):3–16.
171. Burstein ES, Ma J, Wong S, et al. Intrinsic efficacy of antipsychotics at human D2, D3, and D4 dopamine receptors: identification of the clozapine metabolite N-desmethylclozapine as a D2/D3 partial agonist. *J Pharmacol Exp Ther*. 2005;315(3):1278–1287.
172. Schwartz JC, Diaz J, Pilon C, Sokoloff P. Possible implications of the dopamine D(3) receptor in schizophrenia and in antipsychotic drug actions. *Brain Res Brain Res Rev*. 2000; 31(2–3):277–287.
173. Joyce JN, Millan MJ. Dopamine D3 receptor antagonists as therapeutic agents. *Drug Discov Today*. 2005;10(13):917–925.
174. Millan MJ, Brocco M. Cognitive impairment in schizophrenia: a review of developmental and genetic models, and pro-cognitive profile of the optimised D(3) > D(2) antagonist, S33138. *Therapie*. 2008;63(3):187–229.
175. Millan MJ, Loiseau F, Dekeyne A, et al. S33138 (N-[4-[2-[(3aS,9bR)-8-cyano-1,3a,4,9b-tetrahydro[1] benzopyrano[3,4-c]pyrrol-2(3H)-yl]-ethyl]phenyl-acetamide), a preferential dopamine D3 versus D2 receptor antagonist and potential antipsychotic agent: III. Actions in models of therapeutic activity and induction of side effects. *J Pharmacol Exp Ther*. 2008;324(3): 1212–1226.
176. Thomasson-Perret N, Penelaud PF, Theron D, Gouttefangeas S, Mocaer E. Markers of D(2) and D(3) receptor activity in vivo: PET scan and prolactin. *Therapie*. 2008;63(3):237–242.
177. Sanyal S, Van Tol HH. Review the role of dopamine D4 receptors in schizophrenia and antipsychotic action. *J Psychiatr Res*. 1997;31(2):219–232.
178. Kramer MS, Last B, Getson A, Reines SA. The effects of a selective D4 dopamine receptor antagonist (L-745,870) in acutely psychotic inpatients with schizophrenia. D4 Dopamine Antagonist Group. *Arch Gen Psychiatry*. 1997;54(6):567–572.
179. Bristow LJ, Collinson N, Cook GP, et al. L-745,870, a subtype selective dopamine D4 receptor antagonist, does not exhibit a neuroleptic-like profile in rodent behavioral tests. *J Pharmacol Exp Ther*. 1997;283(3):1256–1263.
180. Corrigan MH, Gallen CC, Bonura ML, Merchant KM. Effectiveness of the selective D4 antagonist sonepiprazole in schizophrenia: a placebo-controlled trial. *Biol Psychiatry*. 2004;55(5):445–451.
181. Meltzer HY, Arvanitis L, Bauer D, Rein W. Placebo-controlled evaluation of four novel compounds for the treatment of schizophrenia and schizoaffective disorder. *Am J Psychiatry*. 2004;161(6):975–984.
182. Kapur S, Zipursky RB, Remington G. Clinical and theoretical implications of 5-HT2 and D2 receptor occupancy of clozapine, risperidone, and olanzapine in schizophrenia. *Am J Psychiatry*. 1999;156(2):286–293.
183. Meltzer HY, Matsubara S, Lee JC. The ratios of serotonin2 and dopamine2 affinities differentiate atypical and typical antipsychotic drugs. *Psychopharmacol Bull*. 1989;25(3):390–392.
184. Meltzer HY, Li Z, Kaneda Y, Ichikawa J. Serotonin receptors: their key role in drugs to treat schizophrenia. *Prog Neuropsychopharmacol Biol Psychiatry*. 2003;27(7):1159–1172.
185. Meltzer HY. What's atypical about atypical antipsychotic drugs? *Curr Opin Pharmacol*. 2004;4(1):53–57.
186. Ichikawa J, Ishii H, Bonaccorso S, Fowler WL, O'Laughlin IA, Meltzer HY. 5-HT(2A) and D(2) receptor blockade increases cortical DA release via 5-HT(1A) receptor activation: a possible mechanism of atypical antipsychotic-induced cortical dopamine release. *J Neurochem*. 2001;76(5):1521–1531.
187. Liegeois JF, Ichikawa J, Meltzer HY. 5-HT(2A) receptor antagonism potentiates haloperidol-induced dopamine release in rat medial prefrontal cortex and inhibits that in the nucleus accumbens in a dose-dependent manner. *Brain Res*. 2002;947(2): 157–165.
188. Lucas G, De Deurwaerdere P, Caccia S, Umberto S. The effect of serotonergic agents on haloperidol-induced striatal dopamine release in vivo: opposite role of 5-HT(2A) and 5-HT(2C)

189. Peinado J, Hameg A, Garay RP, Bayle F, Nuss P, Dib M. Reduction of extracellular dopamine and metabolite concentrations in rat striatum by low doses of acute cyamemazine. *Naunyn Schmiedebergs Arch Pharmacol.* 2003;367(2):134–139.
190. Kalkman HO, Neumann V, Nozulak J, Tricklebank MD. Cataleptogenic effect of subtype selective 5-HT receptor antagonists in the rat. *Eur J Pharmacol.* 1998;343(2–3):201–207.
191. Reavill C, Kettle A, Holland V, Riley G, Blackburn TP. Attenuation of haloperidol-induced catalepsy by a 5-HT2C receptor antagonist. *Br J Pharmacol.* 1999;126(3):572–574.
192. Wadenberg MG, Browning JL, Young KA, Hicks PB. Antagonism at 5-HT(2A) receptors potentiates the effect of haloperidol in a conditioned avoidance response task in rats. *Pharmacol Biochem Behav.* 2001;68(3):363–370.
193. Herrick-Davis K, Grinde E, Teitler M. Inverse agonist activity of atypical antipsychotic drugs at human 5-hydroxytryptamine2C receptors. *J Pharmacol Exp Ther.* 2000;295(1):226–232.
194. Di Matteo V, Cacchio M, Di Giulio C, Di Giovanni G, Esposito E. Biochemical evidence that the atypical antipsychotic drugs clozapine and risperidone block 5-HT(2C) receptors in vivo. *Pharmacol Biochem Behav.* 2002;71(4):607–613.
195. Seeman P. Antipsychotic drugs, dopamine receptors, and schizophrenia. *Clin Neurosci Res.* 2001;1:53–60.
196. Schoemaker H, Claustre Y, Fage D, et al. Neurochemical characteristics of amisulpride, an atypical dopamine D2/D3 receptor antagonist with both presynaptic and limbic selectivity. *J Pharmacol Exp Ther.* 1997;280(1):83–97.
197. Trichard C, Paillere-Martinot ML, Attar-Levy D, Recassens C, Monnet F, Martinot JL. Binding of antipsychotic drugs to cortical 5-HT2A receptors: a PET study of chlorpromazine, clozapine, and amisulpride in schizophrenic patients. *Am J Psychiatry.* 1998;155(4):505–508.
198. Kapur S, Zipursky R, Remington G, Jones C, McKay G, Houle S. PET evidence that loxapine is an equipotent blocker of 5-HT2 and D2 receptors: implications for the therapeutics of schizophrenia. *Am J Psychiatry.* 1997;154(11):1525–1529.
199. Strange PG. Antipsychotic drugs: importance of dopamine receptors for mechanisms of therapeutic actions and side effects. *Pharmacol Rev.* 2001;53(1):119–133.
200. White FJ, Wang RY. Differential effects of classical and atypical antipsychotic drugs on A9 and A10 dopamine neurons. *Science.* 1983;221(4615):1054–1057.
201. Goldstein JM, Litwin LC, Sutton EB, Malick JB. Seroquel: electrophysiological profile of a potential atypical antipsychotic. *Psychopharmacology (Berl).* 1993;112(2–3):293–298.
202. Stockton ME, Rasmussen K. Electrophysiological effects of olanzapine, a novel atypical antipsychotic, on A9 and A10 dopamine neurons. *Neuropsychopharmacology.* 1996;14(2):97–105.
203. Hertel P. Comparing sertindole to other new generation antipsychotics on preferential dopamine output in limbic versus striatal projection regions: mechanism of action. *Synapse.* 2006;60(7):543–552.
204. Skarsfeldt T. Differential effects after repeated treatment with haloperidol, clozapine, thioridazine and tefludazine on SNC and VTA dopamine neurones in rats. *Life Sci.* 1988;42(10):1037–1044.
205. Skarsfeldt T. Electrophysiological profile of the new atypical neuroleptic, sertindole, on midbrain dopamine neurones in rats: acute and repeated treatment. *Synapse.* 1992;10(1):25–33.
206. Chiodo LA, Bunney BS. Typical and atypical neuroleptics: differential effects of chronic administration on the activity of A9 and A10 midbrain dopaminergic neurons. *J Neurosci.* 1983;3(8):1607–1619.
207. Deutch AY, Lee MC, Iadarola MJ. Regionally specific effects of atypical antipsychotic drugs on striatal Fos expression: the nucleus accumbens shell as a locus of antipsychotic action. *Mol Cell Neuroscience.* 1992;3:332–341.
208. MacGibbon GA, Lawlor PA, Bravo R, Dragunow M. Clozapine and haloperidol produce a differential pattern of immediate early gene expression in rat caudate-putamen, nucleus accumbens, lateral septum and islands of Calleja. *Brain Res Mol Brain Res.* 1994;23(1–2):21–32.
209. Robertson GS, Fibiger HC. Neuroleptics increase c-*fos* expression in the forebrain: contrasting effects of haloperidol and clozapine. *Neuroscience.* 1992;46(2):315–328.
210. Robertson GS, Matsumura H, Fibiger HC. Induction patterns of Fos-like immunoreactivity in the forebrain as predictors of atypical antipsychotic activity. *J Pharmacol Exp Ther.* 1994;271(2):1058–1066.
211. Wan W, Ennulat DJ, Cohen BM. Acute administration of typical and atypical antipsychotic drugs induces distinctive patterns of Fos expression in the rat forebrain. *Brain Res.* 1995;688(1–2):95–104.
212. Pilowsky LS, Mulligan RS, Acton PD, Ell PJ, Costa DC, Kerwin RW. Limbic selectivity of clozapine. *Lancet.* 1997;350(9076):490–491.
213. Kessler RM, Ansari MS, Riccardi P, et al. Occupancy of striatal and extrastriatal dopamine D2 receptors by clozapine and quetiapine. *Neuropsychopharmacology.* 2006;31(9):1991–2001.
214. Bressan RA, Erlandsson K, Jones HM, Mulligan RS, Ell PJ, Pilowsky LS. Optimizing limbic selective D2/D3 receptor occupancy by risperidone: a [123I]-epidepride SPET study. *J Clin Psychopharmacol.* 2003;23(1):5–14.
215. Bigliani V, Mulligan RS, Acton PD, et al. Striatal and temporal cortical D2/D3 receptor occupancy by olanzapine and sertindole in vivo: a [123I]epidepride single photon emission tomography (SPET) study. *Psychopharmacology (Berl).* 2000;150(2):132–140.
216. Stephenson CM, Bigliani V, Jones HM, et al. Striatal and extrastriatal D(2)/D(3) dopamine receptor occupancy by quetiapine in vivo. [(123)I]-epidepride single photon emission tomography(SPET) study. *Br J Psychiatry.* 2000;177:408–415.
217. Nyberg S, Olsson H, Nilsson U, Maehlum E, Halldin C, Farde L. Low striatal and extra-striatal D2 receptor occupancy during treatment with the atypical antipsychotic sertindole. *Psychopharmacology (Berl).* 2002;162(1):37–41.
218. Kessler RM, Ansari MS, Riccardi P, et al. Occupancy of striatal and extrastriatal dopamine D2/D3 receptors by olanzapine and haloperidol. *Neuropsychopharmacology.* 2005;30(12):2283–2289.
219. Talvik M, Nordstrom AL, Nyberg S, Olsson H, Halldin C, Farde L. No support for regional selectivity in clozapine-treated patients: a PET study with [(11)C]raclopride and [(11)C]FLB 457. *Am J Psychiatry.* 2001;158(6):926–930.
220. Agid O, Mamo D, Ginovart N, et al. Striatal vs extrastriatal dopamine D2 receptors in antipsychotic response—a double-blind PET study in schizophrenia. *Neuropsychopharmacology.* 2007;32(6):1209–1215.
221. Ito H, Arakawa R, Takahashi H, et al. No regional difference in dopamine D2 receptor occupancy by the second-generation antipsychotic drug risperidone in humans: a positron emission

222. Kapur S, Remington G. Dopamine D(2) receptors and their role in atypical antipsychotic action: still necessary and may even be sufficient. *Biol Psychiatry*. 2001;50(11):873–883.
223. Kapur S, Seeman P. Antipsychotic agents differ in how fast they come off the dopamine D2 receptors. Implications for atypical antipsychotic action. *J Psychiatry Neurosci*. 2000;25(2):161–166.
224. Seeman P. Atypical antipsychotics: mechanism of action. *Focus*. 2004;2:48–58
225. Samaha AN, Reckless GE, Seeman P, Diwan M, Nobrega JN, Kapur S. Less is more: antipsychotic drug effects are greater with transient rather than continuous delivery. *Biol Psychiatry*. 2008;64(2):145–152.
226. Samaha AN, Seeman P, Stewart J, Rajabi H, Kapur S. "Breakthrough" dopamine supersensitivity during ongoing antipsychotic treatment leads to treatment failure over time. *J Neurosci*. 2007;27(11):2979–2986.
227. Ginovart N, Wilson A, Hussey D, Houle S, Kapur S. D2-receptor upregulation is dependent upon temporal course of D2-occupancy: a longitudinal [11C]-raclopride PET study in cats. *Neuropsychopharmacology*. 2009;34(3):662–671.
228. Turrone P, Remington G, Kapur S, Nobrega JN. Differential effects of within-day continuous vs. transient dopamine D2 receptor occupancy in the development of vacuous chewing movements (VCMs) in rats. *Neuropsychopharmacology*. 2003;28(8):1433–1439.
229. Gefvert O, Bergstrom M, Langstrom B, Lundberg T, Lindstrom L, Yates R. Time course of central nervous dopamine-D2 and 5-HT2 receptor blockade and plasma drug concentrations after discontinuation of quetiapine (Seroquel) in patients with schizophrenia. *Psychopharmacology (Berl)*. 1998;135(2):119–126.
230. Abi-Saab WM, D'Souza DC, Madonick SH, Krystal JH. Targeting the glutamate system. In: Breier A, Tran PV, Herrera JM, Ollefson GD, Bymaster FP, eds. *Current Issues in the Psychopharmacology of Schizophrenia*. Philadelphia, PA: Lippincott, Williams & Wilkins Healthcare; 2001:304–332.
231. Goff DC, Keefe R, Citrome L, et al. Lamotrigine as add-on therapy in schizophrenia: results of 2 placebo-controlled trials. *J Clin Psychopharmacol*. 2007;27(6):582–589.
232. Zoccali R, Muscatello MR, Bruno A, et al. The effect of lamotrigine augmentation of clozapine in a sample of treatment-resistant schizophrenic patients: a double-blind, placebo-controlled study. *Schizophr Res*. 2007;93(1–3):109–116.
233. Premkumar TS, Pick J. Lamotrigine for schizophrenia. *Cochrane Database Syst Rev*. 2006(4):CD005962.
234. Duncan EJ, Szilagyi S, Schwartz MP, et al. Effects of D-cycloserine on negative symptoms in schizophrenia. *Schizophr Res*. 2004;71:239–248.
235. Carpenter WT, Buchanam RW, Javitt DC, et al. Is glutamatergic therapy efficacious in schizophrenia? *Neuropsychopharmacology*. 2004;29:S110.
236. Patil ST, Zhang L, Martenyi F, et al. Activation of mGlu2/3 receptors as a new approach to treat schizophrenia: a randomized Phase 2 clinical trial. *Nat Med*. 2007;13(9):1102–1107.
237. Sodhi M, Wood KH, Meador-Woodruff J. Role of glutamate in schizophrenia: integrating excitatory avenues of research. *Expert Rev Neurother*. 2008;8(9):1389–1406.
238. Grace AA, Bunney BS, Moore H, Todd CL. Dopamine-cell depolarization block as a model for the therapeutic actions of antipsychotic drugs. *Trends Neurosci*. 1997;20(1):31–37.
239. Agid O, Kapur S, Arenovich T, Zipursky RB. Delayed-onset hypothesis of antipsychotic action: a hypothesis tested and rejected. *Arch Gen Psychiatry*. 2003;60(12):1228–1235.
240. Kapur S, Mizrahi R, Li M. From dopamine to salience to psychosis–linking biology, pharmacology and phenomenology of psychosis. *Schizophr Res*. 2005;79(1):59–68.
241. Agid O, Kapur S, Warrington L, Loebel A, Siu C. Early onset of antipsychotic response in the treatment of acutely agitated patients with psychotic disorders. *Schizophr Res*. 2008;102(1–3):241–248.
242. Kinon BJ, Chen L, Ascher-Svanum H, et al. Predicting response to atypical antipsychotics based on early response in the treatment of schizophrenia. *Schizophr Res*. 2008;102(1–3):230–240.
243. Catafau AM, Corripio I, Perez V, et al. Dopamine D2 receptor occupancy by risperidone: implications for the timing and magnitude of clinical response. *Psychiatry Res*. 2006;148(2–3):175–183.
244. Seeger TF, Thal L, Gardner EL. Behavioral and biochemical aspects of neuroleptic-induced dopaminergic supersensitivity: studies with chronic clozapine and haloperidol. *Psychopharmacology (Berl)*. 1982;76(2):182–187.
245. Smith RC, Davis JM. Behavioral supersensitivity to apomorphine and amphetamine after chronic high dose haloperidol treatment. *Psychopharmacol Commun*. 1975;1(3):285–293.
246. Smith RC, Davis JM. Behavioral evidence for supersensitivity after chronic administration of haloperidol, clozapine, and thioridazine. *Life Sci*. 1976;19(5):725–731.
247. Kahne GJ. Rebound psychoses following the discontinuation of a high potency neuroleptic. *Can J Psychiatry*. 1989;34(3):227–229.
248. Margolese HC, Chouinard G, Beauclair L, Belanger MC. Therapeutic tolerance and rebound psychosis during quetiapine maintenance monotherapy in patients with schizophrenia and schizoaffective disorder. *J Clin Psychopharmacol*. 2002;22(4):347–352.
249. Meltzer HY, Lee MA, Ranjan R, Mason EA, Cola PA. Relapse following clozapine withdrawal: effect of neuroleptic drugs and cyproheptadine. *Psychopharmacology (Berl)*. 1996;124(1–2):176–187.
250. Ekblom B, Eriksson K, Lindstrom LH. Supersensitivity psychosis in schizophrenic patients after sudden clozapine withdrawal. *Psychopharmacology (Berl)*. 1984;83(3):293–294.
251. Llorca PM, Vaiva G, Lancon C. Supersensitivity psychosis in patients with schizophrenia after sudden olanzapine withdrawal. *Can J Psychiatry*. 2001;46(1):87–88.
252. Liskowsky DR, Potter LT. Dopamine D2 receptors in the striatum and frontal cortex following chronic administration of haloperidol. *Neuropsychopharmacology*. 1987;26(5):481–483.
253. Srivastava LK, Morency MA, Bajwa SB, Mishra RK. Effect of haloperidol on expression of dopamine D2 receptor mRNAs in rat brain. *J Mol Neurosci*. 1990;2(3):155–161.
254. Young KA, Zavodny R, Hicks PB. Subchronic buspirone, mesulergine, and ICS 205–930 lack effects on D1 and D2 dopamine binding in the rat striatum during chronic haloperidol treatment. *J Neural Transm Gen Sect*. 1991;86(3):223–228.
255. Sakai K, Gao XM, Hashimoto T, Tamminga CA. Traditional and new antipsychotic drugs differentially alter neurotransmission markers in basal ganglia-thalamocortical neural pathways. *Synapse*. 2001;39(2):152–160.
256. Silvestri S, Seeman MV, Negrete JC, et al. Increased dopamine D2 receptor binding after long-term treatment with antipsychotics in humans: a clinical PET study. *Psychopharmacology (Berl)*. 2000;152(2):174–180.

257. Dean B, Hussain T, Scarr E, Pavey G, Copolov DL. Extended treatment with typical and atypical antipsychotic drugs differential effects on the densities of dopamine D2-like and GABAA receptors in rat striatum. *Life Sci.* 2001;69(11):1257–1268.
258. Tarazi FI, Zhang K, Baldessarini RJ. Long-term effects of olanzapine, risperidone, and quetiapine on dopamine receptor types in regions of rat brain: implications for antipsychotic drug treatment. *J Pharmacol Exp Ther.* 2001;297(2):711–717.
259. Seeman P, Weinshenker D, Quirion R, et al. Dopamine supersensitivity correlates with D2^{High} states, implying many paths to psychosis. *Proc Natl Acad Sci USA.* 2005;102(9):3513–3518.
260. Hall H, Sallemark M. Effects of chronic neuroleptic treatment on agonist affinity states of the dopamine-D2 receptor in the rat brain. *Pharmacol Toxicol.* 1987;60(5):359–363.
261. Klawans HL Jr, Rubovits R. An experimental model of tardive dyskinesia. *J Neural Transm.* 1972;33(3):235–246.
262. Tarsy D, Baldessarini RJ. The pathophysiologic basis of tardive dyskinesia. *Biol Psychiatry.* 1977;12(3):431–450.
263. Chouinard G, Jones BD. Neuroleptic-induced supersensitivity psychosis: clinical and pharmacologic characteristics. *Am J Psychiatry.* 1980;137(1):16–21.
264. Creese I, Snyder S. Chronic neuroleptic treatment and dopamine receptor regulation. In: Cattebeni F, Racagani G, Spano P, Coata E, eds. *Long-Term Effects of Neuroleptics.* Advances in Biochemical Psychopharmacology. Vol 24. New York, NY: Raven Press; 1980:89–94.
265. Tsai G, Goff DC, Chang RW, Flood J, Baer L, Coyle JT. Markers of glutamatergic neurotransmission and oxidative stress associated with tardive dyskinesia. *Am J Psychiatry.* 1998;155(9):1207–1213.
266. Casey DE. Tardive dyskinesia: pathophysiology and animal models. *J Clin Psychiatry.* 2000;61 (suppl 4):5–9.
267. Lohr JB, Kuczenski R, Niculescu AB. Oxidative mechanisms and tardive dyskinesia. *CNS Drugs.* 2003;17(1):47–62.
268. Halperin R, Guerin JJ Jr, Davis KL. Regional differences in the induction of behavioral supersensitivity by prolonged treatment with atypical neuroleptics. *Psychopharmacology (Berl).* 1989;98(3):386–391.
269. Rupniak NM, Hall MD, Mann S, et al. Chronic treatment with clozapine, unlike haloperidol, does not induce changes in striatal D-2 receptor function in the rat. *Biochem Pharmacol.* 1985;34(15):2755–2763.
270. Florijn WJ, Tarazi FI, Creese I. Dopamine receptor subtypes: differential regulation after 8 months' treatment with antipsychotic drugs. *J Pharmacol Exp Ther.* 1997;280(2):561–569.
271. Seeman P. Dopamine D2^{High} receptors moderately elevated by Bifeprunox and aripiprazole. *Synapse.* 2008;62:902–908.
272. Baldessarini RJ, Frankenburg FR. Clozapine. A novel antipsychotic agent. *N Engl J Med.* 1991;324(11):746–754.
273. Graff-Guerrero A, Mizrahi R, Agid O, et al. The dopamine D(2) receptors in high-affinity state and D(3) receptors in schizophrenia: a clinical [(11)C]-(+)-PHNO PET study. *Neuropsychopharmacology.* 2009;34(4):1078–1086.
274. Ginovart N, Willeit M, Rusjan P, et al. Positron emission tomography quantification of [11C]-(+)-PHNO binding in the human brain. *J Cereb Blood Flow Metab.* 2007;27(4):857–871.
275. Ikemoto S, Panksepp J. The role of nucleus accumbens dopamine in motivated behavior: a unifying interpretation with special reference to reward-seeking. *Brain Res Brain Res Rev.* 1999;31(1):6–41.
276. Wise RA. Dopamine, learning and motivation. *Nat Rev Neurosci.* 2004;5(6):483–494.
277. Robbins TW, Everitt BJ. Neurobehavioural mechanisms of reward and motivation. *Curr Opin Neurobiol.* 1996;6(2):228–236.
278. Schultz W, Dayan P, Montague PR. A neural substrate of prediction and reward. *Science.* 1997;275(5306):1593–1599.
279. Schultz W, Dickinson A. Neuronal coding of prediction errors. *Annu Rev Neurosci.* 2000;23:473–500.
280. Wise RA. Neuroleptics and operant behavior: the anhedonia hypothesis. *Behav Brain Sci.* 1982;5:39–87.
281. Schultz W. Getting formal with dopamine and reward. *Neuron.* 2002;36(2):241–263.
282. Schultz W. Behavioral theories and the neurophysiology of reward. *Annu Rev Psychol.* 2006;57:87–115.
283. Berridge KC, Robinson TE. What is the role of dopamine in reward: hedonic impact, reward learning, or incentive salience? *Brain Res Brain Res Rev.* 1998;28(3):309–369.
284. Berridge KC. The debate over dopamine's role in reward: the case for incentive salience. *Psychopharmacology (Berl).* 2007;191(3):391–431.
285. Kapur S. Psychosis as a state of aberrant salience: a framework linking biology, phenomenology, and pharmacology in schizophrenia. *Am J Psychiatry.* 2003;160(1):13–23.
286. Kapur S. How antipsychotics become anti-"psychotic"–from dopamine to salience to psychosis. *Trends Pharmacol Sci.* 2004;25(8):402–406.
287. Jensen J, Willeit M, Zipursky RB, et al. The formation of abnormal associations in schizophrenia: neural and behavioral evidence. *Neuropsychopharmacology.* 2008;33(3):473–479.

10.4 | Dopamine Dysfunction in Schizophrenia: From Genetic Susceptibility to Cognitive Impairment

HEIKE TOST, SHABNAM HAKIMI AND ANDREAS MEYER-LINDENBERG

INTRODUCTION

Mental illness is a common phenomenon in our society and places an enormous burden on the affected individuals and their families. Schizophrenia stands out as one of the most severe and disabling psychiatric conditions, affecting about 1% of the population.[1] Contemporary disease models posit that the path to psychopathology is laid by the adverse interaction of susceptibility genes and environmental factors. At the level of neural systems, the dominant pathophysiological hypotheses suggest that psychotic symptoms are the result of neurotransmitter dysregulation in the brain, a concept that is closely related to the discovery of both dopaminergic neurotransmission and the first effective antipsychotic drugs in the early 1950s. Over the years, new empirical evidence has directed the continual refinement and integration of the dopamine hypothesis. The development of elaborate brain network models has been promoted substantially by significant advances in biomedical research techniques.

Among the medical disciplines, psychiatric research faces the exceptional challenge of developing pathophysiological models that span multiple levels of description, from elementary biological processes to complex behavioral and cognitive phenomena. In the neuroimaging field, magnetic resonance imaging (MRI) and positron emission tomography (PET) have emerged as pivotal tools capable of bridging the gap between psychopathology and brain dysfunction. At the molecular level, the sequencing of the complete human genome in 2003 was a critical accomplishment that had a profound impact on the way biomedical research is conducted today. The availability of such research techniques as linkage analysis and positional cloning has resulted in particular in the identification of a number of gene loci associated with risk and protection for schizophrenia. Further methodological advances have facilitated the successful integration of psychiatric neuroimaging and molecular genetics (*imaging genetics*), a powerful research strategy that allows for a detailed characterization of schizophrenia risk gene effects at the brain systems level.[2] These developments have significantly advanced our understanding of the neural mechanisms that mediate the link between genetic susceptibility leading to dopamine dysfunction and the phenotypic markers of schizophrenia, cognitive deficits, and psychopathology. This chapter attempts to provide an overview of the causes and effects of dopamine dysfunction in schizophrenia. In doing so, it summarizes historical perspectives and our current scientific knowledge about the susceptibility genes, neural system anomalies, and cognitive symptoms that link the disorder to disturbances in dopamine neurotransmission (see also Chapter 4.4 in this volume).

HISTORICAL ROOTS

The dopamine hypothesis is the oldest and best-established pharmacological theory of schizophrenia. The roots of the theory can be traced back to a fortuitous scientific discovery that ultimately transformed the way that mental disease is conceptualized today. In 1949, the French physician Henri Laborit experimented with different antihistamine compounds in an attempt to develop a new pharmacotherapy for surgery-related shock symptoms. The unexpected outcome of these trials turned out to be remarkable. One of the substances, chlorpromazine (CZP), induced a significant state of mental indifference in Laborit's patients, a condition he later described as "sedation without narcosis."[3] Based on his clinical observations, Laborit convinced Pierre Hamon, a psychiatrist at the Val de Grâce Military Hospital in Paris, to apply the substance to one of his patients suffering from severe agitation and mania.[4] As a result of the patient's considerable improvement, the substance was soon applied to psychiatric patients throughout the country.

Although serendipitous in nature, the discovery of CZP provided the first effective chemotherapy for schizophrenia and later became the prototype of the phenothiazine class, a group of first-generation neuroleptics with a similar pharmacological profile. Ultimately, this development revolutionized psychiatric healthcare by replacing older invasive and ineffective procedures and even enabled the reintegration of long-term hospitalized patients into the community. The popularity of the substance also evoked substantial scientific interest in the pathophysiological basis of the disease. In 1966, Van Rossum was the first scientist to speculate that "overstimulation of dopamine receptors could be part of the aetiology."[5] This assumption was based on several independent observations. First, the most prominent side effects of the phenothiazines involved extrapyramidal symptoms (EPS) similar to those of Parkinson's disease, a disorder that was linked to a central dopamine deficit.[6,7] It was also observed that the application of the dopamine precursor L-DOPA not only alleviated neuroleptic-induced EPS, but also triggered psychosis in some cases.[8] In the mid-1970s, a systematic search for the biochemical mode of action revealed that the clinical efficacy of neuroleptics relates to their potency to block dopamine D2 receptors.[9,10] The theory of a dopamine imbalance in schizophrenia has continued to receive considerable scientific attention and is an integral part of contemporary pathophysiological disease models (see Chapters 10.1–10.3 in this volume for details).

CLINIC, COGNITION, AND FRONTAL LOBE FUNCTION

Clinical Picture

Schizophrenic patients suffer from a wide array of clinical symptoms that impact significantly on their ability to function in society. The onset of psychopathology is frequently preceded by an extended prodromal phase in which impairments in psychosocial adaptation and nonspecific cognitive or affective symptoms prevail. Most patients experience their first clinical signs during early adolescence, followed by an often chronic episodic course. The hallmark feature of the disorder is psychosis. Psychotic manifestations include *positive* symptoms such as hearing voices in the absence of auditory input (auditory hallucinations), the development of false personal beliefs in spite of invalidating evidence (delusions), and disorganized, bizarre, or catatonic behaviors. The *negative* symptom cluster, in contrast, is typically less accessible to the observer and the impairments are more persistent. The manifestations stand out by a diminution or loss of normal functions, such as affective flattening, anhedonia, or avolition. It is of note that some characteristics of the disease course are of limited predictive clinical value. For instance, the abrupt onset of symptoms in the context of psychosocial stressors is suggestive of a positive outcome. In contrast, the constellation of a positive family history of psychosis, gradual onset, predominance of negative symptoms, and neurostructural anomalies is often indicative of a poor prognosis. Several pathophysiological models have explained the development of positive and negative symptoms in the context of a central dopamine dysfunction. One influential neurodevelopmental hypothesis[11] proposes an intrauterine disturbance during neural network formation that triggers an imbalance in dopamine transmission during adulthood. The resulting disturbed frontal-temporal neural interaction is thought to lead to a functional impairment of prefrontal cortex (PFC) function that manifests clinically as negative symptoms and cognitive deficits. At the same time, dysregulation of control of the prefrontal lobe over lower brain areas is thought to be at the root of disinhibited subcortical dopamine release, a dynamic that promotes the development of positive symptoms, most likely through the functional destabilization of cortical neural assemblies.[12]

Cognitive Deficits

Cognitive dysfunction is a core feature of schizophrenia. Disease-related deficits are broad and impact on a wide array of higher-order intellectual functions such as memory, learning, and attention.[13] In recent decades, cognitive neuroscience has accumulated a wealth of empirical evidence for cognitive dysfunctions in schizophrenia at both the psychological and neural systems levels.[14] Although not predominantly impaired in schizophrenia, executive functional domains that are known to be dependent on the efficiency of the PFC have generated much research interest, such as working memory, cognitive flexibility, attention, and interference control. Among these functions, working memory deficits have been examined most extensively in schizophrenia research. Unlike short-term memory, working memory performance requires the maintenance and active manipulation of memorized items. A standard experiment to challenge working memory functions is the so-called n-back task. In this experiment, participants are asked to monitor a series of stimuli and react to items that match the stimulus presented *n* stimuli before. The popularity of the paradigm is explained by the fact that the task load can be manipulated by parametrically increasing the factor *n* (i.e., to 2-back, 3-back, etc.), while the stimulus and response conditions are kept constant. During the performance of the n-back task, schizophrenic patients typically exhibit characteristic significant capacity constraints of the working memory buffer, as indicated by a significantly enhanced rate of omissions and false-positive responses.[15] The Wisconsin Card Sorting Test (WCST) is another popular measure in cognitive neuroscience that, in addition to its working memory requirements, challenges abstract reasoning and cognitive flexibility. Several studies have shown that patients with schizophrenia perform poorly on the WCST,[12,16] notably because of their proneness to frequent perseverative errors, a diagnostic marker for deficits in cognitive flexibility. A large body of evidence suggests that the performance deficits in both the n-back task and the WCST relate to anomalies in dopamine signaling and functional impairments of the PFC[17,18] (Fig. 10.4.1). Moreover, the severity of the deficits has some predictive value with regard to the clinical course and the degree of the developing social and occupational impairment.[19–21] It has become increasingly obvious in the past decade that first-degree relatives of schizophrenic patients exhibit similar, albeit less extensive, cognitive impairments. Disease-related evidence for heritability can also be derived for neural systems supporting working memory, especially PFC activation and signal to noise.[22,23] From a pathophysiological standpoint, these observations strongly suggest that the cognitive deficits in schizophrenia relate to the genetic susceptibility to the disease.

Neural Network Dysfunction

The flexible adaptation of behavioral patterns to changing environmental demands is a core function of the frontal lobes. Among others, the predominance of executive dysfunctions in schizophrenia suggests involvement of the PFC in the pathogenesis of the disorder.[24] As such, the neural mechanisms of the assumed prefrontal impairment have been the subject of intense scientific interest. Single-cell recording studies in animal models have demonstrated that mesocortical dopamine signaling, especially the binding potential of dopamine D1 receptors in the PFC, is a crucial modulator of cognitive function.[25–27] Previous evidence has indicated that the relationship between PFC activation and D1 receptor stimulation follows an inverted U-shaped curve,[25] with an optimal "tuning" of prefrontal networks at intermediate receptor occupancy levels[28] (Fig. 10.4.2; see also Chapter 5.3 in this volume). Thus, it appears very likely that the cognitive impairments in schizophrenia relate to a dysregulation in PFC dopamine signaling. Consequently, a substantial number of neuroimaging studies in schizophrenia have been done to characterize the neural basis of working memory dysfunction. While earlier work from the

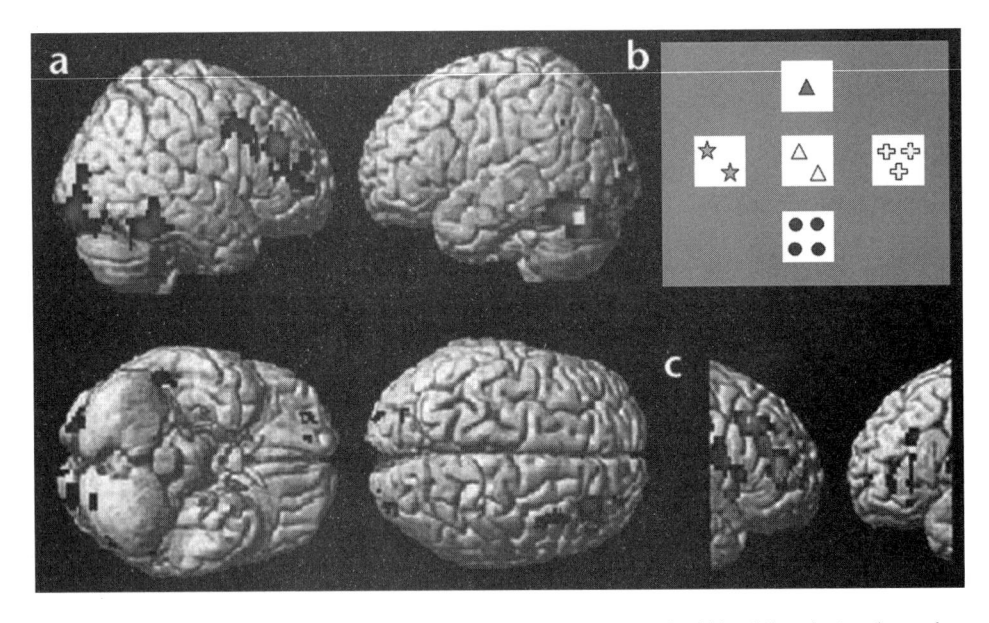

FIGURE 10.4.1. Neural mechanisms of cognitive dysfunction. Statistical maps of regional cerebral blood flow during the performance of the WCST in schizophrenia patients and healthy controls. (a) Conjunction analysis showing voxels with significantly ($p < 0.01$, voxel level) higher regional cerebral blood flow (rCBF) during WCST performance than the control task. (b) Computer screen showing the WCST stimuli. (c) Voxels showing significantly ($p < 0.05$) higher rCBF in the task-minus-control contrast in the frontal lobes of controls compared to patients. *Source*: Reprinted with permission from[12]. (See Color Plate 10.4.1.)

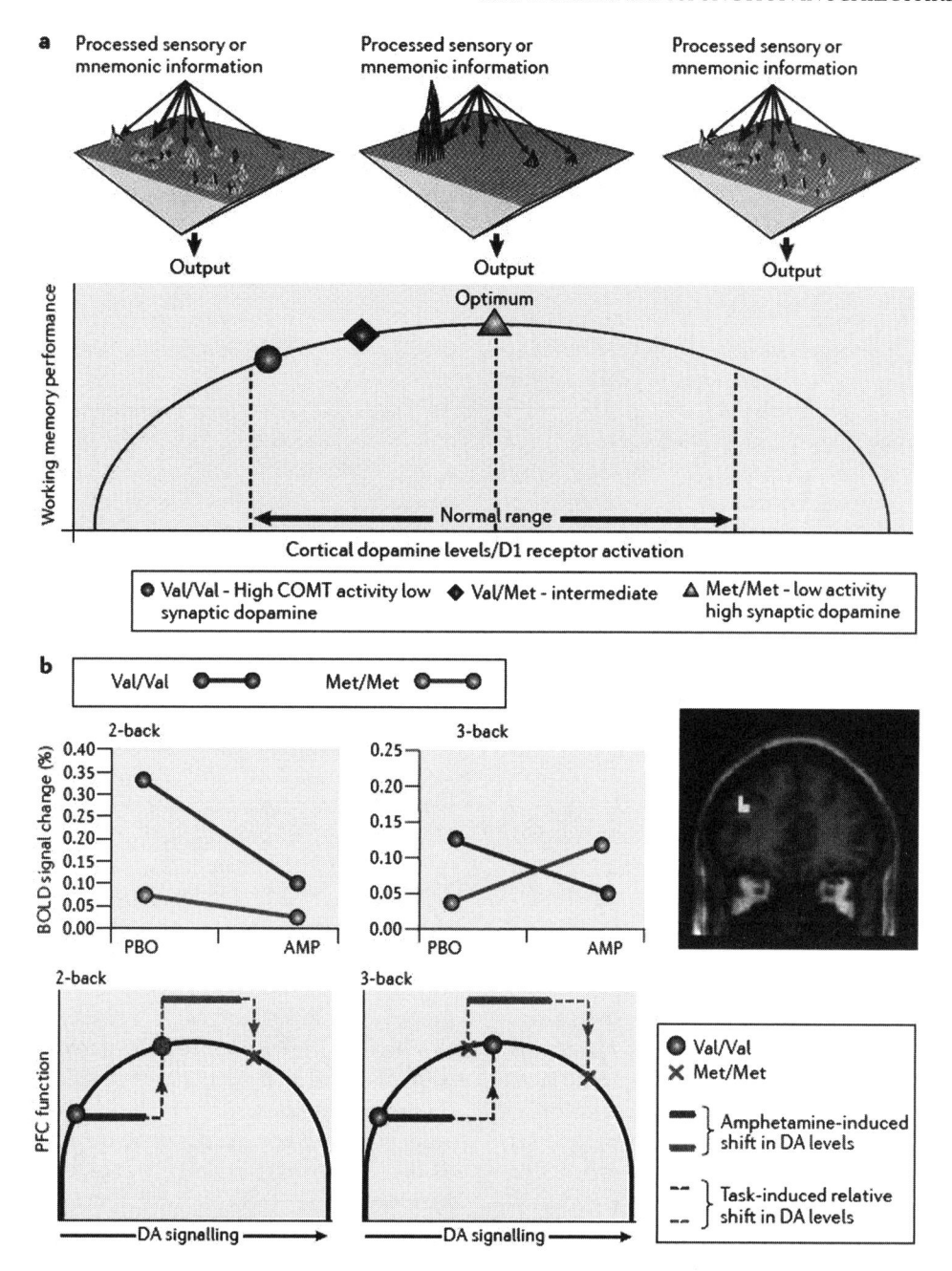

FIGURE 10.4.2. Dopamine, genetic variation in COMT, and prefrontal cortex function. (a) An inverted U-shaped curve links extracellular dopamine to prefrontal signal-to-noise ratio (top) and working memory performance (bottom): homozygotes for the Val[158]-encoding allele are positioned at the left (high COMT efficacy, low dopamine), while Met[158] homozygotes are near the optimum of the curve (low COMT efficacy, high dopamine). Heterozygotes are intermediate. (b) Increasing synaptic dopamine by the administration of amphetamine dissociates the functional states of Val[158] and Met[158] homozygotes. At a medium working memory load level (2-back task), Val[158] homozygotes profit from increased dopamine, whereas Met[158] homozygotes, near the optimum, show little change (left). At a high load level (3-back task), dopamine increase by drug intervention pushes Met[158] homozygotes into the suboptimally high range of dopamine stimulation, leading to reduced prefrontal efficiency (right; localization of activity in the prefrontal cortex is shown on the far right). *Source*: Reprinted with permission from[2].

1990s predominantly suggested hypoactivation of the dorsolateral PFC in schizophrenia (DLPFC, Brodmann areas 46, 9),[29–31] later studies yielded divergent findings; for example, PFC hypoactivation,[29,30,32,33] increased activation,[34] normal activation,[35] and combinations of hyper- and hypoactive states[36] have been reported. These findings demonstrate that any theory suggesting a simple hypo- or hyperactivation of prefrontal neural resources in schizophrenia seems to underestimate the real complexity of the disease.[18,36]

Not surprisingly, the exact nature of the prefrontal deficit in schizophrenia is still a matter of debate. Recent neuroimaging evidence has demonstrated anomalies in the functional coupling of the DLPFC and hippocampus[37,38] (Fig. 10.4.1), as well as the coupling of the DLPFC to hierarchically lower areas in the ventrolateral PFC (VLPFC; Brodmann areas 44, 45, 47).[39] Based on these observations, it has been hypothesized that the functional efficacy of the DLPFC is compromised in schizophrenia, resulting in a compensatory recruitment of subordinate and less-specialized neural areas.[40] This is just one example of the successful refinement of pathophysiological models using empirical data, explaining both clinic and cognition in the context of dynamic prefrontal neural networks. One particularly influential theory is the dual-state model of PFC function framed by Jeremy Seamans and Daniel Durstewitz[41,42] (see also Chapter 5.5 in this volume). Based on neurocomputational simulations of dopamine-induced currents in the PFC, the model predicts that extrasynaptic (i.e., tonic) D1 receptor stimulation, provided it is in the optimal range, promotes the formation of sustained and noise-resistant neural network states. On the cognitive level, this dynamic is thought to resemble the neural mechanism by which the PFC actively maintains memorized items during n-back performance, as it seems to "lock working memory buffers into a single mode of action, such that one or a few representations completely guide action at the expense of response flexibility"[42]). In contrast, high levels of intrasynaptic (i.e., phasic) dopamine seem to promote the formation of more transient, D2-dominated network states with a comparatively lower signal-to-noise ratio. Under physiological circumstances, these network states enable the access of new information to PFC networks, a dynamic that facilitates less perseverative cognitive functions (e.g., cognitive flexibility, set shifting). In the healthy brain, the overall PFC network dynamic is thought to be in equilibrium. In the case of a dopamine imbalance, however, the PFC network dynamic might become skewed toward one of the extremes, thereby promoting the development of cognitive deficits and psychotic symptoms. Several hallmark features of psychosis are consistent with this idea. On the one hand, a predominantly D2-dominated network state might make the PFC less resilient to interfering neural noise, disrupt the more enduring cognitive representations (e.g., working memory functions, selective attention), and give rise to hallucinations and delusions. On the other hand, a predominantly D1-related DA signaling state might introduce a disproportionately high energy barrier to PFC networks. Several pathophysiological features of schizophrenia are also consistent with this idea, especially negative symptoms like anhedonia, avolition, and deficits in cognitive flexibility.[43] As both pathophysiological extremes usually coexist, further research must face the challenge of integrating the temporal development and potential superposition of both network dynamics into one comprehensive model.

DOPAMINE RECEPTOR DYSREGULATION

Although the roots of the dopamine hypothesis can be traced back to the 1950s, the molecular basis of the assumed dopamine dysfunction in schizophrenia was largely inaccessible until the 1980s. Technical advances in the fields of molecular biology and nuclear neuroimaging subsequently enabled a detailed examination of the DA receptor status in vivo and in postmortem brain tissue samples.[44] The dopamine D2 receptor family, which includes the dopamine receptor subtypes D2, D3, and D4, is expressed at highest concentrations in the striatum.[45,46] Although less abundant, the D2 receptors are also distributed ubiquitously in the cerebral cortex and may impact on a diverse range of neural systems. From the beginning, the close relationship between the clinical efficacy and D2 receptor affinity of antipsychotics prompted interest in whether or not an increase in D2 receptor density is evident in schizophrenia. In the 1980s, several postmortem studies used the D2 receptor ligand [^{3}H]spiperone and reported an increase in the number of D2 receptors in the caudate nucleus of schizophrenics.[47–49] A subsequent study with the ligand [^{3}H]raclopride failed to detect significant differences in striatal D2 receptor densities and raised the issue of whether or not previously reported findings were influenced by drug treatment effects.[50] Later work identified several regional D2 receptor anomalies in schizophrenia, such as differences in the laminar arrangements of D2 receptors in the temporal lobe,[51] elevated D3 receptor levels in the striatum,[52] and decreased D3 but normal D1 and D2 mRNA expression levels in the PFC.[53]

In 1986, the dopamine hypothesis of schizophrenia was revitalized by nuclear neuroimaging; one of the first PET experiments in the field described an increase in striatal D2 receptor density in schizophrenic patients naive to neuroleptics.[54] Not all subsequent studies were able to replicate this finding, however, and a series of discussions examined this disparity in detail.[55–60] Nonetheless, recent PET studies have provided convincing evidence for an increase in the phasic activity of dopaminergic neurons in schizophrenia, as indicated by excessive striatal dopamine release and increased availability of D2 receptors[60,61] (see also Chapter 10.1 in this volume). These studies demonstrated that the striatal D2 receptor binding profile can provide valuable information about the clinical outcome, particularly with respect to positive symptom severity and treatment response. A PET study from our own group examined the relationship between exaggerated striatal dopamine synthesis and PFC dysfunction in unmedicated schizophrenic patients.[12] As hypothesized, PFC hypoactivation, which was measured during the WCST, predicted striatal dopamine transmission anomalies in patients but not in healthy controls. This finding supports the hypothesis that the subcortical disinhibition of dopamine transmission is secondary to a top-down control deficit of the PFC in schizophrenia. Furthermore, the results of a recent [^{11}C]raclopride PET study complemented this model by showing that an increase in caudate D2 receptor density was evident not only in schizophrenic patients but also in their unaffected monozygotic cotwins.[62] This finding strongly supports the hypothesis that dopamine receptor dysregulation in schizophrenia is a neural mechanism that is related to the genetic risk for the disorder.

The dopamine D1 receptor family, which includes the dopamine receptor subtypes D1 and D5, is widely distributed in the brain and is highly expressed in the amygdala, putamen, caudate nucleus, hippocampus, and PFC. For instance, alterations in the DLPFC D1 receptor status have been postulated because of the characteristic working memory deficits in schizophrenia and their links to inappropriate mesocortical stimulation of these D1 receptors.[24] In addition, several postmortem studies have reported D1 receptor anomalies in schizophrenia. In a series of [^{3}H]SCH23390 autoradiography studies, Knable and colleagues observed an increase in D1 receptor binding in the PFC[63] but not the striatum[50] of patients who had undergone chronic neuroleptic treatment. Although the etiology of these findings remains somewhat ambiguous, other postmortem evidence has suggested that such changes in D1 receptor functioning are intrinsic to the pathophysiology of schizophrenia and are not treatment artifacts. Supporting this claim, Domyo and colleagues found that [^{3}H]SCH23390 binding in the temporal and parietal cortex was significantly increased in patients who had been drug-free for more than 40 days at the time of death.[64] It should be noted, however, that most nuclear imaging studies in schizophrenia have, until recently, been confined to the examination of D2 receptor status, as high-affinity D1 receptor radioligands were not available. In 2002, the first [^{11}C]NNC-112 PET study in drug-free patients observed an increase in prefrontal D1 receptor binding that strongly predicted the degree of working memory dysfunction.[65] In line with current pathophysiological models of the disorder, both the cognitive and neuroimaging data support the idea that a sustained deficiency in mesocortical dopamine signaling is present in schizophrenia.

NEURAL MECHANISMS OF DOPAMINERGIC RISK GENES

A complex set of interactions underlies the transition from genetic predisposition to psychopathology. Multiple gene variants can interact with one another and with the environment, working at various levels and to varying degrees to shape the neural circuits that produce behavior. While a great deal of attention has been paid to identifying the basis of genetic susceptibility to schizophrenia, the mechanisms by which putative susceptibility genes affect the neural systems that are dysfunctional in psychopathology have long remained elusive. There are several reasons that characterizing these systems poses peculiar challenges. First, the small effect size of psychiatric risk genes requires that very large samples be used to characterize the associated neural correlates. Second, diagnostic criteria are wholly descriptive, making it difficult to pinpoint the set of gene variants or environmental factors that characterize the majority of individuals within a given diagnostic label. Recent technological advances in multimodal neuroimaging and genetic mapping, however, have proven uniquely valuable in the effort to overcome these obstacles.[2] Imaging genetics employs an intermediate phenotype approach, which takes advantage of the fact that many genetic variants linked to psychopathology are commonly expressed in the normal population, to examine the neurobiology underlying the phenotypes associated with these genes. The intermediate phenotype approach assumes that gene penetrance is higher at the neurobiological level than at the level of complex behavior. The effects of genetic variation can, therefore, be seen at the neural systems level even when the associated cognitive-behavioral phenotype itself is not expressed.[66] Using this method, the effects of risk genes can be studied in the absence of confounding effects of treatment or duration of illness. Given these considerations, this research strategy has significant advantages for the study of gene effects in psychiatric illness.

Catechol-O-Methyltransferase (COMT)

One of the most promising candidate genes in psychiatric research is the gene encoding catechol-O-methyl transferase (COMT), an enzyme that catabolizes the catecholamines dopamine, norepinephrine, and epinephrine through 3-O-methylation of the benzene ring. Naturally, the *COMT* gene is of particular interest to schizophrenia researchers. One reason is that this gene is located on 22q11.2, a chromosomal region that has been associated with schizophrenia in linkage studies.[67,68] Additionally, the *COMT* gene is susceptible to the microdeletion syndrome VCFS, or velo-cardio-facial syndrome, which is linked to high psychosis rates.[69] Finally, COMT plays a crucial role in regulating dopamine flux in PFC,[70] where there is low availability of dopamine transporters.[71] Of particular interest is the Val[158]Met single-nucleotide polymorphism (SNP) in the *COMT* gene, a common DNA sequence variation that affects the thermostability of the encoded protein. Met[158] allele carriers display a three- to fourfold reduction in COMT activity,[72] which translates into alterations in prefronto-cortical cognitive functionality.[73,74] Because of the relative increase in dopamine catabolism and the resulting impairment in PFC function in Val[158] allele carriers, it is thought that this variant may increase the risk for schizophrenia.[72,75] Despite extensive evidence linking the *COMT* Val[158]Met SNP with measures of prefrontal efficacy, association studies on this polymorphism in schizophrenia have provided inconsistent results.[75–77] A possible explanation for this is that it may be important consider other sources of genetic variability in *COMT*; recent work has suggested that the analysis of *COMT* haplotypes, which take into account specific interactions between various closely linked risk alleles, may provide more robust information than single markers about the association between genes and schizophrenia.[78,79] Nicodemus and colleagues,[80] Tan et al.,[81] and Straub et al.[82] have posited that epistatic interactions between *COMT* and other risk genes, including those involved in the regulation of glutamatergic and GABAergic signaling, are central to determining risk for schizophrenia.

These studies on *COMT* have laid a strong foundation for the use of the intermediate phenotype approach in psychosis. The Val[(108/158)]Met SNP has been shown to impact PFC function during working memory[75] and other cognitive tests.[73] Specifically, it is thought that the cognitive deficits and positive symptoms present in carriers of the Val[158] allele are mediated by a decreased signal-to-noise ratio in prefronto-cortical networks.[24,42] Furthermore, homozygotes for this *COMT* polymorphism were shown to exhibit differential prefrontal efficiency while performing a working memory task during an amphetamine challenge.[76] Because of the acute increase in dopaminergic tone, the performance of Val[158]/Val[158] homozygotes improved while that of Met[158]/Met[158] homozygotes declined, despite the latter having superior baseline function (Fig. 10.4.2). These findings are consistent with previous results proposing that dopamine signaling in the PFC can be modeled by an inverted-U functional response curve.[25] Thus, individuals can be placed along this curve according to their *COMT* genotype: Val[158]/Val[158] homozygotes fall to the left of Met[158] allele carriers due to decreased PFC efficiency (high COMT efficacy, low dopamine), while Met[158] allele carriers are located at the peak of the curve, the supposed functional optimum (low COMT efficacy, high dopamine). Additional work from our group identified molecular processes underlying this genotype effect using PET.[83] We found that activation in PFC is inversely related to markers of midbrain dopamine synthesis, which are dependent on the Val[(108/158)]Met genotype status. Carriers of the Val allele had higher midbrain dopamine synthesis as well as reduced cerebral blood flow in PFC. It is clear from these data that pharamacogenomics can offer unique insights into the link between the neural basis of cognition and the risk for psychosis.

Protein Kinase B (AKT1)

Two classes of G protein–coupled receptor (GPCR) subtypes mediate dopamine function. One of these subtypes, represented by the D1 receptor family, leads to increased intracellular production of cyclic adenosine monophosphate (cAMP), while the other, representing the D2 receptor family, reduces the intracellular production of protein kinase A (PKA). The D2 receptors also engage in signaling activity outside the cAMP/PKA pathway through an AKT1/glycogen synthase kinase (GSK3) pathway mediated by β-arrestin 2; this signaling cascade also serves to modulate the expression on dopaminergic activity.[84,85]

Consequently, *AKT1* has been investigated as a susceptibility gene for schizophrenia. Association studies have linked the *AKT1* gene, which is located at 14q32.32, to schizophrenia[86,87]; other work has suggested that this gene may produce a risk for schizophrenia by its interaction with significant environmental risk factors such as obstetric complications.[88] Some of the most convincing work linking *AKT1* to schizophrenia was done by Emamian and coworkers. In examining the postmortem brains and peripheral lymphocytes of schizophrenic patients, they found significantly reduced levels of AKT1 and GSK3β phosphorylation.[89] They also reported that the administration of amphetamine to AKT1-deficient mice resulted in disrupted prepulse inhibition; these data mirror the sensorimotor gating deficits characteristic of psychosis. Further, they discovered that one of the molecular mechanisms of the drug haloperidol is to increase the phosphorylation activity of AKT1 and GSK3β, thereby compensating for deficient AKT1.

Imaging genetics methods have also been applied to the study of *AKT1*. Tan and colleagues[81] examined the neural impact of the interactions between multiple risk gene variants involved in the integration of dopaminergic signals, taking advantage of the fact that the cAMP/PKA pathways act in tandem with other biochemical pathways. They demonstrated a relationship between genetic variation in *AKT1* and changes in prefronto-striatal structure and function, as well as the risk for psychosis. The authors hypothesized that these effects are grounded in alterations in dopaminergic signaling, which result from dysfunction in the AKT1/GSK3 signaling cascade. Finally, Tan and coworkers identified significant genetic epistasis with the *COMT* Val[108/158]Met SNP, reaffirming the relationship between *AKT1* and dopamine signaling.[81]

Dopamine and cAMP-Regulated Phosphoprotein 32 (DARPP-32)

Dopamine and cAMP-regulated phosoprotein of molecular weight 32 kDa (DARPP-32) is encoded by the gene *PPP1R1B* (located at 17q12) and also plays an important role in dopaminergic neurotransmission, regulating and integrating neural signals. DARPP-32 is expressed in the efferent pathways of the striatum, particularly on medium spiny neurons. When dopamine D1 receptors are stimulated, DARPP-32 is phosphorylated via cAMP and PKA, facilitating its conversion to the physiologically active protein phosphatase 1 (PP-1) inhibitor.[90] The activity of this phosphatase to inhibit PP-1 serves to modulate other downstream effectors, such as receptors, ion channels, and transcription factors.[91] Therefore, the activity of DARPP-32 performs the critical task of integrating local dopaminergic signals with other converging neural signals, including glutamate, serotonin, neuropeptides, and steroid hormones[92] (see also Chapter 3.3 in this volume).

Because of the importance of DARPP-32 in modulating neural signals, it has become the focus of recent studies on the pathophysiology of psychosis. Knockout mice for *PPP1R1B* exhibit a decreased response to amphetamines; similarly, mice with point mutations in regions coding for DARPP-32 phosphorylation sites show reduced responses to phencyclidine administration.[93] Moreover, postmortem studies in schizophrenia have observed attenuated levels of DARPP-32 in dorsolateral PFC,[94,95] and linkage studies have identified the *PPP1R1B* gene as a schizophrenia risk gene.[96,97] Despite the mounting evidence highlighting the fundamental role of DARPP-32 in neuronal signaling, however, studies examining variation in *PPP1R1B* in humans have been few. A recent report from our own group employed a translational approach to characterize the impact of complex genetic variation in *PPP1R1B* on human neural structure and function.[98] Using multimodal techniques, we showed that a frequent *PPP1R1B* haplotype impacts striatal volume, activation, and prefronto-striatal functional connectivity (Fig. 10.4.3) while also predicting cognitive function and mRNA expression in postmortem human brains. Moreover, this same haplotype and its variants were associated with the risk of developing psychosis and schizophrenia. Collectively, these data provide strong initial evidence that the *PPP1R1B* gene and its variants impact cognitive function and the integrity of network communication between striatum and frontal cortex through the modulation of DARPP-32 expression.

Proline Dehydrogenase (PRODH)

Another potential candidate gene is *PRODH*, which encodes the mitochondrial enzyme proline oxidase (POX). Proline oxidase catabolizes the amino acid proline into its metabolites, one of which is the neurotransmitter glutamate. Proline's role in glutamatergic neurotransmission is further underscored by the observation that a subset of excitatory neurons express high-affinity proline transporters in their axon terminals and synapses.[99,100] Several lines of evidence support a potential role for variation in *PRODH* in schizophrenia. Like *COMT*, the *PRODH* gene is located at 22q11.2, the same chromosomal region implicated in VCFS. Concordantly, hyperprolinemia has been associated with increased susceptibility to psychosis,[101–104] and behaviors analogous to those seen in schizophrenia have been observed in mice with mutations in *PRODH*.[105,106] Additional work has also found links between schizophrenia and variation at the *PRODH* locus,[107–109] although conflicting evidence has also been found (e.g., see [110,111]). Recent work in mice with *PRODH* mutations has shown that COMT is the most dysregulated gene product in these animals.[106] These data establish a potential mechanistic link between *PRODH* dysfunction and dopaminergic neurotransmission. A recent imaging genetics study has also supported this hypothesis by demonstrating a consistent effect of *PRODH* variation on interactions between prefrontal and subcortical regions.[109] This study demonstrated that risk and protective variants of PRODH are associated with separable effects on the enzymatic activity of POX as well as fronto-striatal connectivity on both the structural and functional levels. The risk haplotype investigated demonstrated reduced gray matter volume

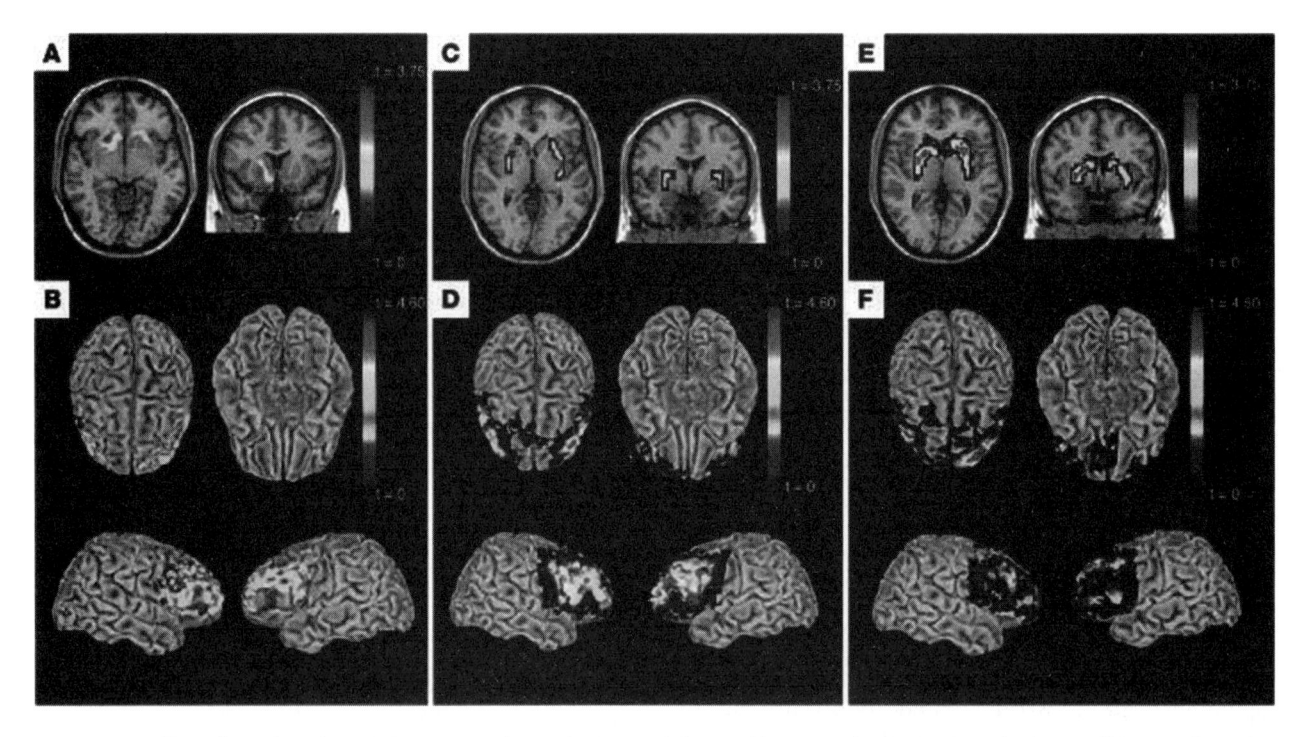

FIGURE 10.4.3. Effects of genetic variation in *PPP1R1B* on human brain morphology and function. The top row shows haplotype effects on volume (A) or activation (C and E) in the striatum; the bottom row shows haplotype effects on structural (B) and functional (D and F) connectivity of the striatum to the PFC. Voxel-based morphometry: (A) significantly reduced volume in the striatum ($P < 0.05$) for carriers of the frequent (CGCACTC) haplotype; (B) greater structural connectivity between PFC and striatum for homozygotes for the frequent (CGCACTC) haplotype. Functional MRI, n-back task: (C) significantly reduced reactivity in putamen ($P < 0.05$) for carriers of the frequent (CGCACTC) haplotype; (D) greater functional connectivity between PFC and striatum for homozygotes for the frequent (CGCACTC) haplotype. Functional MRI, face-matching task; (E) significantly reduced reactivity in striatum ($P < 0.05$) for carriers of the frequent (CGCACTC) haplotype; (F) greater functional connectivity between PFC and striatum for homozygotes for the frequent (CGCACTC) haplotype. *Source*: Reprinted with permission from [98]. (See Color Plate 10.4.3.)

in the striatum and increased functional connectivity between this region and frontal areas; the protective haplotype, on the other hand, was associated with opposite effects.[109] Together, these findings bring attention to the contributions of genetic variance in *PRODH* to protection and susceptibility factors for schizophrenia and to the central role of the frontostriatal circuitry in shaping the pathophysiology of psychosis.

DO GENETIC VARIANTS ENCODE FOR COGNITIVE DYSFUNCTION?

Schizophrenia is a highly heritable disorder with cognitive dysfunction. Previous evidence suggests that schizophrenia risk genes, especially the *COMT* Val[108/158]Met coding variant, predict performance in tests that challenge attention, working memory, and executive function.[112] The nature of the relationship between genes and cognition, however, is far from clear. Do schizophrenia risk genes code for impairments in cognitive task performance? If this were the case, one would expect genetic effects to be penetrant at the level of behavior at least as much as, or even more so than, measures that are more proximate to the underlying neurobiology, such as neuroimaging measures. Alternatively, one might argue that susceptibility genes facilitate anomalies at the neural systems level, and that these neural dysfunctions in turn trigger the development of psychopathological phenomena. In this case, cognitive dysfunction would be just another intermediate phenotype of genetic susceptibility, indexing a state of cortical pathophysiology that is primary. The intermediate phenotype concept favors the second interpretation. According to this idea, multiple risk gene variants impact, through interactions with one other and with the environment, on multiple neural systems that mediate the cognitive, emotional, and behavioral impairments observable in schizophrenia.[2] This suggests that the more "remote" or behavioral a given phenotype is from the biological cascade that mediates the genetic effects, the less directly it will be predicted by genotypic variation. In line with this notion, the effect size for most genes is higher for imaging genetics studies, that is, larger sample sizes are needed to show the same genetic effect at the cognitive-behavioral level.[113]

If cognitive dysfunction in schizophrenia reflects the impact of susceptibility genes at the neural systems level, then insights into their downstream molecular effects should stimulate the discovery of new therapeutic targets for the treatment of cognitive deficits. Recent studies on the effects of tolcapone, a brain-penetrant COMT inhibitor, support this notion. Tolcapone improves cognition and PFC function in both humans and rodents, an effect that is probably related to the increased bioavailability of dopamine in the PFC.[115,116] In agreement with previous findings,[74,75] a significant COMT genotype-by-drug interaction on neuropsychological performance has been demonstrated (Fig. 10.4.2b). While individuals with the Val^{158}/Val^{158} genotype benefited from the drug, the cognitive performance of subjects with the Met^{158}/Met^{158} genotype worsened.[114]

Tolcapone is the prototype for a novel pharmacological treatment strategy in psychiatry. Due to its focused action in the PFC, the substance lacks the characteristic neurological side effects of less specific psychostimulants (e.g., potential for abuse, EPS). The application of tolcapone as a cognitive enhancer also lacks the touch of "scientific serendipity" associated with previous psychopharmacological inventions (e.g., chlorpromazine). Instead, its use is rooted in the thorough understanding of the neurobiological mechanisms of a schizophrenia risk gene variant that translate into cognitive dysfunction.[117] Tolcapone itself is expected be of limited use in clinical practice because of its inherent hepatotoxicity. The exemplified approach, however, represents an advance that will likely open a new chapter in the history of psychopharmacology, an era characterized by the pursuit of individualized, regionally selective, and genotype-based treatment approaches.[118]

PERSPECTIVES

Half a century after the first pharmacological theory of schizophrenia was formulated, substantial insights into the underlying pathological mechanisms have been achieved. Current evidence suggests that the interaction of multiple risk gene variants and environmental factors paves the way to psychopathology through dopamine dysregulation. At the neural systems level, core pathophysiological processes such as subcortical dopamine disinhibition, PFC inefficiency, and cognitive deficits have been identified. Schizophrenic patients have benefited from this progress because it has encouraged the demystification and destigmatization of the public's perception of their illness. Yet, there is no place for complacency. Outside the abstract reality of our laboratories, MRI scanners, and testing environments, schizophrenic patients and their families still battle with a devastating disorder. The psychopathological symptoms are highly distressing for the patients themselves and for their social environment. Typical disease management involves recurrent hospitalizations, a fact that limits the chances of a patient's successfully reintegrating into society at multiple levels, despite a chain of professional support programs for this patient group. As therapists, we face the considerable problem that our current treatment options are usually effective but still entirely palliative, while bearing the risk of substantial side effects and noncompliance.

The primary goal of medical research is the transformation of scientific insights into practical solutions that change the fate of our patients for the better. In light of exploding health care costs and the immense personal suffering that is caused by chronic debilitating diseases, the success of basic research is increasingly evaluated by its capacity to motivate successful translational applications. Schizophrenia research in particular faces the challenge of bridging the gap between bench and bedside, fostering innovative approaches, and properly allocating available resources to scientific questions with the potential to lead to effective treatments. Research funding must reflect this reality if physician-researchers are to prioritize these approaches over more predictable, higher-impact basic research. As part of this translational agenda, imaging genetics has proved to be a valuable tool in identifying the genetic risk factors and system-level dysfunctions that manifest in psychotic symptoms. The future goal of this research is to further dissect the converging molecular pathways and their potential neural system targets that lead to the development of schizophrenia—and, ultimately, to its therapy.

ACKNOWLEDGMENT

This work was supported by the Intramural Research Program of the National Institute of Mental Health, NIH, and the German Research Foundation, DFG.

REFERENCES

1. Sullivan PF, Neale MC, Kendler KS. Genetic epidemiology of major depression: review and meta-analysis. *Am J Psychiatry*. 2000; 157(10):1552–1562.
2. Meyer-Lindenberg A, Weinberger DR. Intermediate phenotypes and genetic mechanisms of psychiatric disorders. *Nat Rev Neurosci*. 2006;7(10):818–827.

3. Seeman P. Dopamine receptors and the dopamine hypothesis of schizophrenia. *Synapse*. 1987;1(2):133–152.
4. Hamon J, Paraire J, Velluz J. Remarques sur l'action du 4560 R. P. sur l'agitation maniaque. *Ann Med Psychol (Paris)*. 1952;110(1:3):331–335.
5. Van Rossum JM. The significance of dopamine-receptor blockade for the mechanism of action of neuroleptic drugs. *Arch Int Pharmacodyn Ther*. 1966;160:492–494.
6. Ehringer H, Hornykiewicz O. Verteilung von Noradrenalin und Dopamin (3-Hydroxytyramin) im Gehirn des Menschen bei Erkrankungen des extrapyramidalen Systems. *Klin Wochenschr*. 1960;38:1236–1239.
7. Carlsson A. Evidence for a role of dopamine in extrapyramidal functions. *Acta Neuroveg (Wien)*. 1964;26:484–493.
8. Yaryura-Tobias JA, Wolpert A, Dana L, Merlis S. Action of L-Dopa in drug induced extrapyramidalism. *Dis Nerv Syst*. 1970;31(1):60–63.
9. Creese I, Burt DR, Snyder SH. Dopamine receptor binding predicts clinical and pharmacological potencies of antischizophrenic drugs. *Science*. 1976;192(4238):481–483.
10. Seeman P, Chau-Wong M, Tedesco J, Wong K. Brain receptors for antipsychotic drugs and dopamine: direct binding assays. *Proc Natl Acad Sci USA*. 1975;72(11):4376–4380.
11. Weinberger DR. Implications of normal brain development for the pathogenesis of schizophrenia. *Arch Gen Psychiatry*. 1987;44(7):660–669.
12. Meyer-Lindenberg A, Miletich RS, Kohn PD, et al. Reduced prefrontal activity predicts exaggerated striatal dopaminergic function in schizophrenia. *Nat Neurosci*. 2002;5(3):267–271.
13. Joyce EM, Roiser JP. Cognitive heterogeneity in schizophrenia. *Curr Opin Psychiatry*. 2007;20(3):268–272.
14. Barch DM, Smith E. The cognitive neuroscience of working memory: relevance to CNTRICS and schizophrenia. *Biol Psychiatry*. 2008;64(1):11–17.
15. Krieger S, Lis S, Janik H, Cetin T, Gallhofer B, Meyer-Lindenberg A. Executive function and cognitive subprocesses in first-episode, drug-naive schizophrenia: an analysis of n-back performance. *Am J Psychiatry*. 2005;162(6):1206–1208.
16. Prentice KJ, Gold JM, Buchanan RW. The Wisconsin Card Sorting impairment in schizophrenia is evident in the first four trials. *Schizophr Res*. 2008;106(1):81–87.
17. Callicott JH, Mattay VS, Bertolino A, et al. Physiological characteristics of capacity constraints in working memory as revealed by functional MRI. *Cereb Cortex*. 1999;9(1):20–26.
18. Manoach DS. Prefrontal cortex dysfunction during working memory performance in schizophrenia: reconciling discrepant findings. *Schizophr Res*. 2003;60(2–3):285–298.
19. Cervellione KL, Burdick KE, Cottone JG, Rhinewine JP, Kumra S. Neurocognitive deficits in adolescents with schizophrenia: longitudinal stability and predictive utility for short-term functional outcome. *J Am Acad Child Adolesc Psychiatry*. 2007;46(7):867–878.
20. Heinrichs RW, Goldberg JO, Miles AA, McDermid Vaz S. Predictors of medication competence in schizophrenia patients. *Psychiatry Res*. 2008;157(1–3):47–52.
21. Niendam TA, Bearden CE, Rosso IM, et al. A prospective study of childhood neurocognitive functioning in schizophrenic patients and their siblings. *Am J Psychiatry*. 2003;160(11):2060–2062.
22. Callicott JH, Egan MF, Mattay VS, et al. Abnormal fMRI response of the dorsolateral prefrontal cortex in cognitively intact siblings of patients with schizophrenia. *Am J Psychiatry*. 2003;160(4):709–719.
23. Winterer G, Coppola R, Goldberg TE, et al. Prefrontal broadband noise, working memory, and genetic risk for schizophrenia. *Am J Psychiatry*. 2004;161(3):490–500.
24. Goldman-Rakic PS. Working memory dysfunction in schizophrenia. *J Neuropsychiatry Clin Neurosci*. 1994;6(4):348–357.
25. Williams GV, Goldman-Rakic PS. Modulation of memory fields by dopamine D1 receptors in prefrontal cortex. *Nature*. 1995;376(6541):572–575.
26. Goldman-Rakic PS. Cellular basis of working memory. *Neuron*. 1995;14(3):477–485.
27. Fuster JM. Prefrontal cortex and the bridging of temporal gaps in the perception-action cycle. *Ann NY Acad Sci*. 1990;608:318–329; discussion 330–316.
28. Durstewitz D, Seamans JK, Sejnowski TJ. Dopamine-mediated stabilization of delay-period activity in a network model of prefrontal cortex. *J Neurophysiol*. 2000;83(3):1733–1750.
29. Paulman RG, Devous MD Sr, Gregory RR, et al. Hypofrontality and cognitive impairment in schizophrenia: dynamic single-photon tomography and neuropsychological assessment of schizophrenic brain function. *Biol Psychiatry*. 1990;27(4):377–399.
30. Andreasen NC, O'Leary DS, Flaum M, et al. Hypofrontality in schizophrenia: distributed dysfunctional circuits in neuroleptic-naive patients. *Lancet*. 1997;349(9067):1730–1734.
31. Volz H, Gaser C, Hager F, et al. Decreased frontal activation in schizophrenics during stimulation with the continuous performance test—a functional magnetic resonance imaging study. *Eur Psychiatry*. 1999;14(1):17–24.
32. Barch DM, Carter CS, Braver TS, et al. Selective deficits in prefrontal cortex function in medication-naive patients with schizophrenia. *Arch Gen Psychiatry*. 2001;58(3):280–288.
33. Weinberger DR, Berman KF, Zec RF. Physiologic dysfunction of dorsolateral prefrontal cortex in schizophrenia. I. Regional cerebral blood flow evidence. *Arch Gen Psychiatry*. 1986;43(2):114–124.
34. Callicott JH, Bertolino A, Mattay VS, et al. Physiological dysfunction of the dorsolateral prefrontal cortex in schizophrenia revisited. *Cereb Cortex*. 2000;10(11):1078–1092.
35. Manoach DS, Press DZ, Thangaraj V, et al. Schizophrenic subjects activate dorsolateral prefrontal cortex during a working memory task, as measured by fMRI. *Biol Psychiatry*. 1999;45(9):1128–1137.
36. Callicott JH, Mattay VS, Verchinski BA, Marenco S, Egan MF, Weinberger DR. Complexity of prefrontal cortical dysfunction in schizophrenia: more than up or down. *Am J Psychiatry*. 2003;160(12):2209–2215.
37. Meyer-Lindenberg AS, Olsen RK, Kohn PD, et al. Regionally specific disturbance of dorsolateral prefrontal-hippocampal functional connectivity in schizophrenia. *Arch Gen Psychiatry*. 2005;62(4):379–386.
38. Meyer-Lindenberg A, Poline JB, Kohn PD, et al. Evidence for abnormal cortical functional connectivity during working memory in schizophrenia. *Am J Psychiatry*. 2001;158(11):1809–1817.
39. Tan HY, Choo WC, Fones CS, Chee MW. fMRI study of maintenance and manipulation processes within working memory in first-episode schizophrenia. *Am J Psychiatry*. 2005;162(10):1849–1858.
40. Tan HY, Sust S, Buckholtz JW, et al. Dysfunctional prefrontal regional specialization and compensation in schizophrenia. *Am J Psychiatry*. 2006;163(11):1969–1977.
41. Durstewitz D, Seamans JK. The dual-state theory of prefrontal cortex dopamine function with relevance to catechol-O-methyltransferase genotypes and schizophrenia. *Biol Psychiatry*. 2008;64(9):739–749.

42. Seamans JK, Yang CR. The principal features and mechanisms of dopamine modulation in the prefrontal cortex. *Prog Neurobiol.* 2004;74(1):1–58.

43. Tost H, Meyer-Lindenberg A, Klein S, et al. D2 antidopaminergic modulation of frontal lobe function in healthy human subjects. *Biol Psychiatry.* 2006;60(11):1196–1205.

44. Harrison PJ. Advances in postmortem molecular neurochemistry and neuropathology: examples from schizophrenia research. *Br Med Bull.* 1996;52(3):527–538.

45. Farde L, Pauli S, Hall H, et al. Stereoselective binding of 11C-raclopride in living human brain—a search for extrastriatal central D2-dopamine receptors by PET. *Psychopharmacology (Berl).* 1988;94(4):471–478.

46. Lidow MS, Goldman-Rakic PS, Rakic P, Innis RB. Dopamine D2 receptors in the cerebral cortex: distribution and pharmacological characterization with [3H]raclopride. *Proc Natl Acad Sci USA.* 1989;86(16):6412–6416.

47. Hess EJ, Bracha HS, Kleinman JE, Creese I. Dopamine receptor subtype imbalance in schizophrenia. *Life Sci.* 1987;40(15):1487–1497.

48. Mjorndal T, Winblad B. Alteration of dopamine receptors in the caudate nucleus and the putamen in schizophrenic brain. *Med Biol.* 1986;64(6):351–354.

49. Mita T, Hanada S, Nishino N, et al. Decreased serotonin S2 and increased dopamine D2 receptors in chronic schizophrenics. *Biol Psychiatry.* 1986;21(14):1407–1414.

50. Knable MB, Hyde TM, Herman MM, Carter JM, Bigelow L, Kleinman JE. Quantitative autoradiography of dopamine-D1 receptors, D2 receptors, and dopamine uptake sites in postmortem striatal specimens from schizophrenic patients. *Biol Psychiatry.* 1994;36(12):827–835.

51. Goldsmith SK, Shapiro RM, Joyce JN. Disrupted pattern of D2 dopamine receptors in the temporal lobe in schizophrenia. A postmortem study. *Arch Gen Psychiatry.* 1997;54(7):649–658.

52. Gurevich EV, Bordelon Y, Shapiro RM, Arnold SE, Gur RE, Joyce JN. Mesolimbic dopamine D3 receptors and use of antipsychotics in patients with schizophrenia. A postmortem study. *Arch Gen Psychiatry.* 1997;54(3):225–232.

53. Meador-Woodruff JH, Haroutunian V, Powchik P, Davidson M, Davis KL, Watson SJ. Dopamine receptor transcript expression in striatum and prefrontal and occipital cortex. Focal abnormalities in orbitofrontal cortex in schizophrenia. *Arch Gen Psychiatry.* 1997;54(12):1089–1095.

54. Wong DF, Wagner HN Jr, Tune LE, et al. Positron emission tomography reveals elevated D2 dopamine receptors in drug-naive schizophrenics. *Science.* 1986;234(4783):1558–1563.

55. Farde L, Wiesel FA, Hall H, Halldin C, Stone-Elander S, Sedvall G. No D2 receptor increase in PET study of schizophrenia. *Arch Gen Psychiatry.* 1987;44(7):671–672.

56. Farde L, Wiesel FA, Stone-Elander S, et al. D2 dopamine receptors in neuroleptic-naive schizophrenic patients. A positron emission tomography study with [11C]raclopride. *Arch Gen Psychiatry.* 1990;47(3):213–219.

57. Young LT, Wong DF, Goldman S, et al. Effects of endogenous dopamine on kinetics of [3H]N-methylspiperone and [3H]raclopride binding in the rat brain. *Synapse.* 1991; 9(3):188–194.

58. Chugani DC, Ackermann RF, Phelps ME. In vivo [3H]spiperone binding: evidence for accumulation in corpus striatum by agonist-mediated receptor internalization. *J Cereb Blood Flow Metab.* 1988;8(3):291–303.

59. Zawarynski P, Tallerico T, Seeman P, Lee SP, O'Dowd BF, George SR. Dopamine D2 receptor dimers in human and rat brain. *FEBS Lett.* 1998;441(3):383–386.

60. Abi-Dargham A, Rodenhiser J, Printz D, et al. Increased baseline occupancy of D2 receptors by dopamine in schizophrenia. *Proc Natl Acad Sci USA.* 2000;97(14):8104–8109.

61. Laruelle M, Abi-Dargham A, van Dyck CH, et al. Single photon emission computerized tomography imaging of amphetamine-induced dopamine release in drug-free schizophrenic subjects. *Proc Natl Acad Sci USA.* 1996;93(17):9235–9240.

62. Hirvonen J, van Erp TG, Huttunen J, et al. Increased caudate dopamine D2 receptor availability as a genetic marker for schizophrenia. *Arch Gen Psychiatry.* 2005;62(4):371–378.

63. Knable MB, Hyde TM, Murray AM, Herman MM, Kleinman JE. A postmortem study of frontal cortical dopamine D1 receptors in schizophrenics, psychiatric controls, and normal controls. *Biol Psychiatry.* 1996;40(12):1191–1199.

64. Domyo T, Kurumaji A, Toru M. An increase in [3H]SCH23390 binding in the cerebral cortex of postmortem brains of chronic schizophrenics. *J Neural Transm.* 2001;108(12):1475–1484.

65. Abi-Dargham A, Mawlawi O, Lombardo I, et al. Prefrontal dopamine D1 receptors and working memory in schizophrenia. *J Neurosci.* 2002;22(9):3708–3719.

66. Goldberg TE, Ragland JD, Torrey EF, Gold JM, Bigelow LB, Weinberger DR. Neuropsychological assessment of monozygotic twins discordant for schizophrenia. *Arch Gen Psychiatry.* 1990;47(11):1066–1072.

67. Owen MJ, Williams NM, O'Donovan MC. The molecular genetics of schizophrenia: new findings promise new insights. *Mol Psychiatry.* 2004;9(1):14–27.

68. Stefansson H, Rujescu D, Cichon S, et al. Large recurrent microdeletions associated with schizophrenia. *Nature.* 2008;455(7210):232–236.

69. Murphy KC. Schizophrenia and velo-cardio-facial syndrome. *Lancet.* 2002;359(9304):426–430.

70. Tunbridge EM, Harrison PJ, Weinberger DR. Catechol-O-methyltransferase, cognition, and psychosis: Val(158)Met and beyond. *Biol Psychiatry.* 2006;60(2):141–151.

71. Lewis DA, Melchitzky DS, Sesack SR, Whitehead RE, Auh S, Sampson A. Dopamine transporter immunoreactivity in monkey cerebral cortex: regional, laminar, and ultrastructural localization. *J Comp Neurol.* 2001;432(1):119–136.

72. Chen J, Lipska BK, Halim N, et al. Functional analysis of genetic variation in catechol-O-methyltransferase (COMT): effects on mRNA, protein, and enzyme activity in postmortem human brain. *Am J Hum Genet.* 2004;75(5):807–821.

73. Goldberg TE, Egan MF, Gscheidle T, et al. Executive subprocesses in working memory: relationship to catechol-O-methyltransferase Val158Met genotype and schizophrenia. *Arch Gen Psychiatry.* 2003;60(9):889–896.

74. Mattay VS, Callicott JH, Bertolino A, et al. Effects of dextroamphetamine on cognitive performance and cortical activation. *NeuroImage.* 2000;12(3):268–275.

75. Egan MF, Goldberg TE, Kolachana BS, et al. Effect of COMT Val$^{108/158}$ Met genotype on frontal lobe function and risk for schizophrenia. *Proc Natl Acad Sci USA.* 2001;98(12):6917–6922.

76. Mattay VS, Goldberg TE, Fera F, et al. Catechol-O-methyltransferase val158-met genotype and individual variation in the brain response to amphetamine. *PNAS.* 2003;100(10):6186–6191.

77. Fan JB, Zhang CS, Gu NF, et al. Catechol-O-methyltransferase gene Val/Met functional polymorphism and risk of schizophrenia: a large-scale association study plus meta-analysis. *Biol Psychiatry.* 2005;57(2):139–144.

78. Shifman S, Bronstein M, Sternfeld M, et al. A highly significant association between a COMT haplotype and schizophrenia. *Am J Hum Genet.* 2002;71(6):1296–1302.

79. Meyer-Lindenberg A, Nichols T, Callicott JH, et al. Impact of complex genetic variation in COMT on human brain function. *Mol Psychiatry*. 2006;11(9):867–877, 797.
80. Nicodemus KK, Kolachana BS, Vakkalanka R, et al. Evidence for statistical epistasis between catechol-O-methyltransferase (COMT) and polymorphisms in RGS4, G72 (DAOA), GRM3, and DISC1: influence on risk of schizophrenia. *Hum Genet*. 2007;120(6):889–906.
81. Tan HY, Chen Q, Sust S, et al. Epistasis between catechol-O-methyltransferase and type II metabotropic glutamate receptor 3 genes on working memory brain function. *Proc Natl Acad Sci USA*. 2007;104(30):12536–12541.
82. Straub RE, Lipska BK, Egan MF, et al. Allelic variation in GAD1 (GAD67) is associated with schizophrenia and influences cortical function and gene expression. *Mol Psychiatry*. 2007;12(9):854–869.
83. Meyer-Lindenberg A, Kohn PD, Kolachana B, et al. Midbrain dopamine and prefrontal function in humans: interaction and modulation by COMT genotype. *Nat Neurosci*. 2005;8(5):594–596.
84. Beaulieu JM, Sotnikova TD, Marion S, Lefkowitz RJ, Gainetdinov RR, Caron MG. An Akt/beta-arrestin 2/PP2A signaling complex mediates dopaminergic neurotransmission and behavior. *Cell*. 2005;122(2):261–273.
85. Beaulieu JM, Sotnikova TD, Yao WD, et al. Lithium antagonizes dopamine-dependent behaviors mediated by an AKT/glycogen synthase kinase 3 signaling cascade. *Proc Natl Acad Sci USA*. 2004;101(14):5099–5104.
86. Schwab SG, Hoefgen B, Hanses C, et al. Further evidence for association of variants in the *AKT1* gene with schizophrenia in a sample of European sib-pair families. *Biol Psychiatry*. 2005;58(6):446–450.
87. Tan HY, Nicodemus KK, Chen Q, et al. Genetic variation in *AKT1* is linked to dopamine-associated prefrontal cortical structure and function in humans. *J Clin Invest*. 2008;118(6):2200–2208.
88. Nicodemus KK, Marenco S, Batten AJ, et al. Serious obstetric complications interact with hypoxia-regulated/vascular-expression genes to influence schizophrenia risk. *Mol Psychiatry*. 2008;13(9):873–877.
89. Emamian ES, Hall D, Birnbaum MJ, Karayiorgou M, Gogos JA. Convergent evidence for impaired AKT1-GSK3beta signaling in schizophrenia. *Nat Genet*. 2004;36(2):131–137.
90. Fernandez E, Schiappa R, Girault JA, Le Novere N. DARPP-32 is a robust integrator of dopamine and glutamate signals. *PLoS Comput Biol*. 2006;2(12):e176.
91. Svenningsson P, Nishi A, Fisone G, Girault JA, Nairn AC, Greengard P. DARPP-32: an integrator of neurotransmission. *Annu Rev Pharmacol Toxicol*. 2004;44:269–296.
92. Svenningsson P, Tzavara ET, Liu F, Fienberg AA, Nomikos GG, Greengard P. DARPP-32 mediates serotonergic neurotransmission in the forebrain. *Proc Natl Acad Sci USA*. 2002;99(5):3188–3193.
93. Svenningsson P, Tzavara ET, Carruthers R, et al. Diverse psychotomimetics act through a common signaling pathway. *Science*. 2003;302(5649):1412–1415.
94. Albert KA, Hemmings HC Jr, Adamo AI, et al. Evidence for decreased DARPP-32 in the prefrontal cortex of patients with schizophrenia. *Arch Gen Psychiatry*. 2002;59(8):705–712.
95. Ishikawa M, Mizukami K, Iwakiri M, Asada T. Immunohistochemical and immunoblot analysis of dopamine and cyclic AMP-regulated phosphoprotein, relative molecular mass 32,000 (DARPP-32) in the prefrontal cortex of subjects with schizophrenia and bipolar disorder. *Prog Neuropsychopharmacol Biol Psychiatry*. 2007;31(6):1177–1181.
96. Lewis CM, Levinson DF, Wise LH, et al. Genome scan meta-analysis of schizophrenia and bipolar disorder, part II: schizophrenia. *Am J Hum Genet*. 2003;73(1):34–48.
97. Cardno AG, Holmans PA, Rees MI, et al. A genomewide linkage study of age at onset in schizophrenia. *Am J Med Genet*. 2001;105(5):439–445.
98. Meyer-Lindenberg A, Straub RE, Lipska BK, et al. Genetic evidence implicating DARPP-32 in human frontostriatal structure, function, and cognition. *J Clin Invest*. 2007;117(3):672–682.
99. Parra LA, Baust TB, El Mestikawy S, et al. The orphan transporter Rxt1/NTT4 (SLC6A17) functions as a synaptic vesicle amino acid vesicular transporter selective for proline, glycine, leucine, and alanine. *Mol Pharmacol*. 2008;74(6):1521–1532.
100. Shafqat S, Velaz-Faircloth M, Henzi VA, et al. Human brain-specific L-proline transporter: molecular cloning, functional expression, and chromosomal localization of the gene in human and mouse genomes. *Mol Pharmacol*. 1995;48(2):219–229.
101. Raux G, Bumsel E, Hecketsweiler B, et al. Involvement of hyperprolinemia in cognitive and psychiatric features of the 22q11 deletion syndrome. *Hum Mol Genet*. 2007;16(1):83–91.
102. Jacquet H, Demily C, Houy E, et al. Hyperprolinemia is a risk factor for schizoaffective disorder. *Mol Psychiatry*. 2005;10(5):479–485.
103. Jacquet H, Raux G, Thibaut F, et al. PRODH mutations and hyperprolinemia in a subset of schizophrenic patients. *Hum Mol Genet*. 2002;11(19):2243–2249.
104. Bender HU, Almashanu S, Steel G, et al. Functional consequences of PRODH missense mutations. *Am J Hum Genet*. 2005;76(3):409–420.
105. Gogos JA, Santha M, Takacs Z, et al. The gene encoding proline dehydrogenase modulates sensorimotor gating in mice. *Nat Genet*. 1999;21(4):434–439.
106. Paterlini M, Zakharenko SS, Lai WS, et al. Transcriptional and behavioral interaction between 22q11.2 orthologs modulates schizophrenia-related phenotypes in mice. *Nat Neurosci*. 2005;8(11):1586–1594.
107. Liu H, Heath SC, Sobin C, et al. Genetic variation at the 22q11 PRODH2/DGCR6 locus presents an unusual pattern and increases susceptibility to schizophrenia. *Proc Natl Acad Sci USA*. 2002;99(6):3717–3722.
108. Li T, Ma X, Sham PC, et al. Evidence for association between novel polymorphisms in the *PRODH* gene and schizophrenia in a Chinese population. *Am J Med Genet B Neuropsychiatr Genet*. 2004;129B(1):13–15.
109. Kempf L, Nicodemus KK, Kolachana B, et al. Functional polymorphisms in PRODH are associated with risk and protection for schizophrenia and fronto-striatal structure and function. *PLoS Genet*. 2008;4(11):e1000252.
110. Williams HJ, Williams N, Spurlock G, et al. Detailed analysis of PRODH and PsPRODH reveals no association with schizophrenia. *Am J Med Genet B Neuropsychiatr Genet*. 2003;120B(1):42–46.
111. Glaser B, Moskvina V, Kirov G, et al. Analysis of ProDH, COMT and ZDHHC8 risk variants does not support individual or interactive effects on schizophrenia susceptibility. *Schizophr Res*. 2006;87(1–3):21–27.
112. Diaz-Asper CM, Goldberg TE, Kolachana BS, Straub RE, Egan MF, Weinberger DR. Genetic variation in catechol-O-methyltransferase: effects on working memory in schizophrenic patients, their siblings, and healthy controls. *Biol Psychiatry*. 2008;63(1):72–79.

113. Munafo MR, Brown SM, Hariri AR. Serotonin transporter (5-HTTLPR) genotype and amygdala activation: a meta-analysis. *Biol Psychiatry.* 2008;63(9):852–857.
114. Apud JA, Mattay V, Chen J, et al. Tolcapone improves cognition and cortical information processing in normal human subjects. *Neuropsychopharmacology.* 2007;32(5):1011–1020.
115. Tunbridge EM, Bannerman DM, Sharp T, Harrison PJ. Catechol-O-methyltransferase inhibition improves set-shifting performance and elevates stimulated dopamine release in the rat prefrontal cortex. *J Neurosci.* 2004;24(23):5331–5335.
116. Lapish CC, Ahn S, Evangelista LM, So K, Seamans JK, Phillips AG. Tolcapone enhances food-evoked dopamine efflux and executive memory processes mediated by the rat prefrontal cortex. *Psychopharmacology (Berl).* 2008;202(1–3): 521–530.
117. Diaz-Asper CM, Weinberger DR, Goldberg TE. Catechol-O-methyltransferase polymorphisms and some implications for cognitive therapeutics. *Neuroreport.* 2006;3(1):97–105.
118. Apud JA, Weinberger DR. Treatment of cognitive deficits associated with schizophrenia: potential role of catechol-O-methyltransferase inhibitors. *CNS Drugs.* 2007;21(7):535–557.

10.5 | The Role of Dopamine in the Pathophysiology and Treatment of Major Depressive Disorder

BOADIE W. DUNLOP AND CHARLES B. NEMEROFF*

INTRODUCTION

It is difficult to overstate the public health importance of majpr depressive disorder (MDD). The lifetime prevalence of MDD is 16%, and the 12-month prevalence is 6.6%.[1] The lifetime risk for developing the illness in women is approximately double the risk in men. A widely cited study, the Global Burden of Disease, a collaborative effort of the World Bank, the World Health Organization, and the Harvard School of Public Health, predicts that by the year 2020, MDD will be the second leading cause of disability worldwide, trailing only cardiovascular disease.[2] Major depressive disorder is also a leading cause of premature death due to suicide. Depressive symptoms contribute to the risk of several other important diseases, including coronary artery disease and stroke.[3,4] Major depressive disorder follows a chronic course in about 20% of those affected, and of those who remit, approximately 85% will experience another episode of depression within 15 years.[5] Finally, the economic burden of MDD is enormous, with conservatively estimated annual direct costs of $2.1 billion and indirect costs of $4.2 billion per year in the United States alone.[6]

Major depressive disorder, also known as *unipolar depression*, to distinguish it from depression occurring in bipolar disorder (manic-depressive illness), is a multidimensional disorder. Only one major depressive episode (see Table 10.5.1) is required for the diagnosis of MDD, though major depressive episodes can also occur in patients with other psychiatric disorders. The primary clinical characteristics that distinguish these disorders from MDD are presented in Table 10.5.2. The clinical diagnosis of a major depressive episode refers to a syndrome in which there is a significant change in (1) mood state: either prominent feelings of sadness and/or anhedonia, along with the presence of several other symptoms. These other symptoms can be grouped into additional categories: (2) neurovegetative systems: disturbances in sleep and appetite and reduction in energy; (3) cognitive functions: excessive thoughts of guilt or worthlessness, poor concentration or indecisiveness, and thoughts of suicide; and (4) altered psychomotor performance: either slowed (*retarded*) or agitated. The symptoms in each of these categories are believed to have their own specific neurobiological basis. Because the diagnosis of a major depressive episode can be made when all four categories of symptoms are present or

---

* *Sources of Support:* Dr. Nemeroff is supported by the following NIH grants: MH-39415, MH-42088, MH-58299, and MH-69056. Dr. Dunlop is supported (in part) by a K12 grant from NIH National Center for Research Resources, K12 RR 017643 and 1KL2RR025009

Financial disclosures, current and past 12 months:

Dr. Dunlop: Consultant: BMS, Imedex
      Honoraria: BMS, Wyeth
      Research Support: AstraZeneca, Novartis, Ono Pharmaceuticals, Takeda.

Dr. Nemeroff:

Dr. Nemeroff currently serves on the scientific advisory boards of the American Foundation for Suicide Prevention (AFSP); AstraZeneca; National Alliance for Research on Schizophrenia and Depression (NARSAD); Quintiles; Janssen/Ortho-McNeil, and PharmaNeuroboost. He holds stock/equity in Corcept; Revaax; NovaDel Pharma; CeNeRx, and PharmaNeuroboost. He is on the board of directors of the AFSP; George West Mental Health Foundation; NovaDel Pharma; and Mt. Cook Pharma, Inc. Dr. Nemeroff holds a patent on the method and devices for transdermal delivery of lithium (US 6,375,990 B1) and the method to estimate serotonin and norepinephrine transporter occupancy after drug treatment using patient or animal serum (provisional filing April, 2001). In the past year, he also served on the Scientific Advisory Board for Forest Laboratories; received grant support from the National Institute of Mental Health (NIMH), NARSAD, and American Foundation for Suicide Prevention (AFSP); and served on the board of directors of American Psychiatric Institute for Research and Education (APIRE).

TABLE 10.5.1. *Diagnostic Criteria for a Major Depressive Episode*

Symptom Category	Symptom
Mood Change	1. Excessive sadness
	2. Anhedonia/loss of interest
Neurovegetative	3. Insomnia or hypersomnia
	4. Weight or appetite change
	5. Diminished energy
Cognitive	6. Poor concentration or indecisiveness
	7. Excessive guilt or feeling of worthlessness
	8. Thoughts of own death or suicide
Psychomotor Speed	9. Psychomotor agitation or retardation

Notes: To diagnose a major depressive episode, at least five of the above symptoms must be present for most of the day, nearly every day for the past 2 weeks, and the symptoms must cause some level of impairment. At least one of the symptoms must be either excessive sadness or anhedonia. One major depressive episode justifies the diagnosis of MDD as long as criteria for other disorders higher in the diagnostic hierarchy are not met.

TABLE 10.5.2. *Other DSM-IV Diagnoses with Prominent Depression without Psychotic Symptoms*

Diagnosis	Primary Characteristic Distinguishing Diagnosis from MDD
Bipolar Disorder	If a major depressive episode is present, the patient has also experienced at least one episode of elevated, irritable, or expansive mood.
Dysthymia	Chronically ($\geq$2 years) depressed mood of lower intensity, along with fewer associated symptoms, than a major depressive episode.
Posttraumatic Stress Disorder (PTSD)	In addition to some symptoms of depression, the patient also re-experiences a past traumatic life event (e.g., through nightmares, flashbacks, or intrusive memories). The patient may have MDD and PTSD concurrently.
Substance-Induced Mood Disorder	Depressed mood stemming directly from a state of intoxication or withdrawal from substance use (e.g., alcohol or cocaine).
Mood Disorder Due to a Medical Condition	Depressed mood derived directly from the pathophysiological processes of a medical disorder (e.g., hypothyroidism).

when as few as two categories are present, great heterogeneity among equivalently diagnosed patients exists, both phenomenologically and biologically.

Motivation, psychomotor speed, concentration, and the ability to experience pleasure are all regulated in part by dopamine (DA)-containing circuits in the central nervous system (CNS). To a lesser extent, the neurovegetative symptoms of sleep and appetite are also regulated in part by DA. Despite the influence of DA on these multiple aspects of depression, research on the role of DA in depression has been largely overshadowed by research on norepinephrine (NE)- and serotonin (5HT)-containing circuits. Recent findings clearly warrant scrutiny of the role of DA in the pathophysiology of depression and, moreover, whether there exists a "dopaminergic dysfunction" subtype, characterized by a poor response to antidepressants that act primarily on 5HT or NE neurons.[7,8] There is now an emerging consensus that the majority of depressed patients treated with selective serotonin reuptake inhibitors (SSRIs) and selective serotonin/norepinephrine reuptake inhibitors (SNRIs) do not attain remission.[9] The relatively limited effects of SSRIs and SNRIs on DA neurons may contribute to this unsatisfactory success rate.

The original monoamine hypothesis of depression emerged largely from the observed effects on mood of reserpine, which depletes vesicular monoamine stores and reduces mood; of amphetamine, which briefly increases synaptic concentrations of monoamines and raises mood; and of monoamine oxidase inhibitors (MAOIs), which increase the CNS concentrations of monoamines and are, of course, effective antidepressants.[10,11] Although these agents all affect DA similarly to NE and 5HT, it wasn't until the mid-1970s that a role for DA in depression was postulated.[12] The primary reason for the limited focus on DA was the finding that the efficacy of tricyclic antidepressants (TCAs) stemmed from their ability to inhibit the reuptake of NE and/or 5HT. However, a long-standing conundrum associated with the original monoamine hypothesis is that the reuptake-inhibiting effects of TCAs (and SSRIs and SNRIs) occur within hours of drug ingestion, but their antidepressant effects take longer to occur. This temporal discrepancy implies that other mechanisms must be involved in recovery from a depressive episode.

As detailed in this chapter, many of the studies exploring DA function in depression have produced inconsistent findings. Contributors to this inconsistency include the diagnostic heterogeneity of MDD; failure to control for age, bipolar disorder, and comorbid diagnoses; and variation in patient medication treatment status at the time of the study. Despite this variability, there is now a convergence of data from animal models, genetics, neuroimaging, and human clinical trials that strengthens the case for DA dysfunction in the pathophysiology of major depression, at least in a significant subgroup of patients. This chapter comprehensively reviews the current evidence, with subsequent recommendations for future studies of dopaminergic signaling in depression and its treatment.

ANIMAL MODELS OF DA FUNCTION IN DEPRESSION

Rodent models of depression demonstrate altered mesolimbic DA system function; moreover, certain antidepressants act to enhance DA transmission.[13] Whether these effects stem from induction of subsensitivity of DA autoreceptors or heightened responsivity of postsynaptic receptors, or both, is unclear, though the weight of the evidence most supports increased postsynaptic sensitivity, as first proposed by Spyraki and Fibiger.[14] This heightened sensitivity seems to be limited to the ventral striatum (nucleus accumbens), because the dose—response curve for DA agonist—induced stereotypies (stemming from dorsal striatal DA receptor binding) is not shifted to the left by chronic antidepressant treatment. Evidence supporting this theory includes the findings that chronic treatment with electroconvulsive therapy (ECT), sleep deprivation, and virtually all antidepressants increase the motor stimulant effects of DA receptor agonists.[15] Chronic treatment with antidepressants (TCAs, SSRIs, or MAOIs) for 21 days or 10 days of ECT results in increased D3 receptor mRNA expression in the nucleus accumbens.[16] A potential contributor to altered DA receptor sensitivity is the prostate apoptosis response (Par-4) protein, a leucine zipper—containing protein that regulates the activity of the D2 receptor in neurons. Mutant mice lacking the component of Par-4 that interacts with the D2 receptor demonstrate depressive behaviors.[17]

Impaired DA release is also proposed to contribute to the pathophysiology of depression. In *effort expenditure* rodent models of depression, reduced DA concentrations in the nucleus accumbens correlate with reduced efforts by rodents to work for specific rewards.[18,19] Additionally, administration of TCAs or fluoxetine increases DA concentrations in the nucleus accumbens and functionally up-regulates D2 and D3 receptors in the striatum.[20,21] Transcranial magnetic stimulation, a new Food and Drug Administration (FDA)—approved treatment for depression, applied to the rat frontal cortex increases extracellular DA concentrations in the striatum.[22]

The chronic mild stress model has been suggested to have the best face validity of any animal model of depression, in that repeated mild stresses over time gradually induce a state of decreased responsiveness to rewards and reduced sexual and aggressive behaviors.[23] Rodents exposed to this model demonstrate decreased D2/D3 receptor binding in the nucleus accumbens, which is reversed by chronic antidepressant treatment (TCAs, SSRIs or mianserin).[24] When these "recovered" rodents are exposed to D2/D3 antagonists, decreased reward responding reemerges.[25,26] Rodents exposed to chronic mild stress also show reduced responsiveness to the stimulatory effects on locomotion and reward of the D2/D3 agonist quinpirole.[26]

Two other animal models, *learned helplessness* and the *forced swim test*, both use a reduction in locomotor activity under stress as proxies for depression.[27] Animals experiencing learned helplessness exhibit DA depletion in the caudate nucleus and nucleus accumbens, which can be prevented by pretreatment with DA agonists.[28,29] In the forced swim test, immobility in rodents is reversed by D2/D3 agonists, nomifensine (a DA/NE reuptake inhibitor), and TCAs, and the effect of antidepressants can be inhibited by D2/D3 antagonists.[30,31]

HUMAN GENETIC AND NEUROCHEMICAL STUDIES

The heritability of major depression is estimated to be 31%–42% and is likely higher for individuals with recurrent major depressive episodes.[32] Although major depression is almost certainly a polygenetic illness, certain genes may influence the subtype of depression expressed, and the presence of more than one vulnerability gene may significantly increase the likelihood of developing this disorder.[32] As mentioned in the Introduction, the large degree of heterogeneity subsumed under the syndrome of major depression creates challenges in identifying shared genetic underpinnings among depressed patients as a whole.

At least three genes related to dopamine signaling are known to possess functional polymorphisms: the DA D4 receptor, the dopamine transporter (DAT) and catechol-O-methyltransferase (COMT). The D4 receptor exhibits the most polymorphisms of the DA receptors, possessing a 48 base pair variable number tandem repeat (VNTR) polymorphism in exon 3 of the gene, with alleles in humans encoding for 2 to 10 repeats.[33] To date, studies investigating associations between dopaminergic candidate genes and MDD either have found no relationship or have not been replicated. A large study of Jewish patients with major depression found no difference in the allelic distributions of the D4, DAT, or COMT polymorphisms.[34] Among Japanese patients with depression, a polymorphism in the COMT gene was associated with depression.[35] In another study, a poor response to 6 weeks of treatment with the mixed noradrenergic-serotonergic antidepressant mirtazapine was found in depressed patients homozygous for methionine at the COMT[158] (val-met) polymorphism.[36] A meta-analysis of 2071 subjects in 12 studies identified the 2-repeat allele of the D4 receptor as a vulnerability allele for depression.[37] Others have identified a possible association between the Bal I polymorphism of the D3 receptor and unipolar and bipolar depression.[38,39] Consistent associations between D2 receptor or DAT polymorphisms and major depression have not been identified. Mutations in the gene for dopamine ß-hydroxylase (DBH), the enzyme that converts DA to NE, can lead to elevations in the DA/NE ratio, potentially increasing the risk for psychotic symptoms in depression.[40]

Negative findings for associations between genotype and illness in complex disorders such as major depression may arise due to the failure to consider the effects of the environment on gene expression, so-called gene-by-environment (GxE) interactions. Over the past several years, a compelling example of a GxE interaction in depression has emerged for polymorphisms of the serotonin transporter. In this case, carriers of the less efficient "short" allele of the gene for this transporter are at higher risk for developing depression if exposed to significant early life stressors, such as child abuse, than individuals homozygous for the more efficient long-arm variant.[41] Recently, a study of juvenile detainees who reported high levels of maternal rejection during development found that they were at significantly greater risk for current depression if they possessed the TT genotype of the rs40184 polymorphism for the DAT1 gene (located in intron 14 of the gene).[42] Although this finding must be considered tentative until it is replicated in other samples, it suggests that continued exploration of GxE interactions may further elucidate the contribution of variations in DA-related genes to the depression phenotype.

NEUROTRANSMISSION

Studies comparing measures of DA neurotransmission between depressed and control groups require careful age matching, because there is a functionally significant and progressive loss of DA activity with advancing age, largely due to a loss of DA neurons.[43] The majority of studies examining the concentration of DA metabolites in cerebrospinal fluid (CSF), primarily homovanillic acid (HVA), found lower concentrations in depressed patients compared to controls, particularly in patients with psychomotor retardation.[12,44-51] Some discrepant results have also appeared.[52,53] Low pretreatment CSF HVA concentrations have failed to consistently predict the response to TCA treatment,[54] though individual studies have found an inverse association between CSF HVA concentrations and the magnitude of the clinical response to L-DOPA,[50] piribedil,[55] and nomifensine[56]. Of note, however, is one study of 40 unipolar or bipolar depressed inpatients with psychomotor retardation in which the rank order of effectiveness of three antidepressants and placebo correlated with their pro-dopaminergic effects.[57]

In a unique study employing internal jugular venous sampling, medication-free, treatment-resistant, unipolar depressed patients were found to exhibit reduced concentrations of both NE and its metabolites, and of HVA but not 5-HIAA, compared to healthy controls.[58] Estimates of brain DA turnover were inversely correlated with the severity of depressive illness, as measured by the Hamilton Depression Rating Scale (HDRS). Others have reported that the lymphocytes of depressed patients have significantly lower D4 receptor mRNA expression compared to controls, with normalization after 8 weeks of paroxetine treatment.[59] In contrast to the above findings, psychotically depressed patients demonstrate elevated concentrations of plasma DA and HVA, lower serum DBH activity, and increased CSF concentrations of HVA.[60]

Apomorphine, a DA agonist, has been used as a probe to assess DA receptor responsiveness in depression. Acting on DA receptors in the arcuate nucleus of the hypothalamus, apomorphine stimulates the release of growth hormone-releasing hormone (GHRH), which acts to increase peripheral growth hormone (GH) concentrations. The majority of studies have found no difference in the GH response to apomorphine between depressed and healthy control subjects.[61] However, a Belgian group has repeatedly reported a blunted GH response to apomorphine administration in suicidal but not nonsuicidal depressed patients.[62,63] Similar mixed findings exist for the effect of apomorphine on peripheral prolactin concentrations.[61,64] The extent to which DA modulation of an endocrine response reflects DA functioning in the mesocortical, mesolimbic, and nigrostriatal pathways is unknown.

An additional impetus to seek DA involvement in depression is the unduly high frequency of depression among patients with Parkinson's disease. The incidence of major depression in community samples of Parkinson's disease patients is 5%–10%, with an additional 10%–30% experiencing subsyndromal depressive symptoms.[65] In addition, high-frequency deep-brain stimulation of the left substantia nigra led to dramatic and severe transient depression in one subject with Parkinson's disease.[66]

STRESS AND DA FUNCTION

In animal models of MDD, stress potently activates ventral tegmental area (VTA) DA neurons and DA release in the nucleus accumbens.[67] Stress-induced activation of the VTA (along with the hippocampus, prefrontal cortex, and amygdala) may reflect a positive short-term coping mechanism by enhancing the motivation to deal with the stressor. Sustained exposure to stress, however, can produce long-term changes in the VTA-accumbens pathway similar to those seen after chronic exposure to drugs of abuse.[68]

Increased DA signaling may contribute to the development of depression under situations of stress, recently demonstrated using a social defeat stress model, in which mice were subjected to social defeat through exposure to aggressor mice daily for 10 days.[69] This model produces marked social withdrawal behavior, a common observation in patients with MDD. In mice exposed to the model, the VTA neurons that project to the nucleus accumbens exhibit dramatic elevations in brain-derived neurotrophic factor (BDNF). BDNF potentiates DA release in the nucleus accumbens, and this signal likely encodes the motivational salience of the experience of these social interactions.[70] The accumbens may play a role in the assessment of social status and appraisal of threats stemming from the social environment, and socially aversive stimuli are associated with chronic alterations in DA function.[71,72] The social withdrawal induced after social defeat is reversed by local deletion of the BDNF gene, and chronic 4-week treatment with fluoxetine reverses most of the changes in gene expression induced by the exposure.[69] These data indicate that BDNF signaling in VTA neurons, and its effects on DA signaling, may contribute to the learning of social defeat and subsequent social aversion.

Reduction of subjective distress is part of the frequently observed benefit of placebo in the treatment of MDD and may be mediated, at least in part, by DA transmission. One component of the placebo effect is the patients' expectation of improvement, which may be mediated by a form of reward expectation processing.[73] In Parkinson's disease, placebo administration in the setting of expected administration of a dopaminergic agent induced DA release in the nucleus accumbens.[74] In a pain challenge paradigm, placebo administration when subjects expected an effective analgesic produced a significant reduction of ^{11}C-raclopride D2/D3 receptor binding potential in the nucleus accumbens, suggesting increased DA release. Subjectively reported analgesia correlated positively with a change in D2/D3 receptor binding and μ-opiod receptor binding in the nucleus accumbens.[75] Furthermore, subjects who reported a heightened pain experience after receiving placebo (the *nocebo* effect) experienced increased ^{11}C-raclopride binding, suggesting diminished DA release.

DISTURBED REWARD SYSTEM FUNCTION IN DEPRESSION

Anhedonia, the absolute or relative inability to experience pleasure, is one of two symptoms required for the diagnosis of major depression. Of the putative endophenotypes of major depression, the anhedonic form is one of the most well supported.[76] Dopamine neurons have long been known to be critical to a wide variety of pleasurable experiences and reward. The severity of MDD has been found to correlate highly with the magnitude of reward experienced after administration of oral d-amphetamine, which increases DA availability by a variety of mechanisms.[77] In particular, medication-free, severely depressed subjects experienced greater reward than controls, while those with milder forms of depression did not differ from the control group. One explanation for these findings is that in severe depression, there is a reduction in DA release, resulting in compensatory mechanisms, such as up-regulation of postsynaptic DA receptors and decreased DAT density, which, taken together, would increase DA signal transduction resulting from amphetamine-induced DA release into the synapse. These findings have now been confirmed and extended in a recent study employing functional magnetic resonance imaging (fMRI) to assess the activity of brain reward systems after d-amphetamine challenge in 12 drug-free depressed patients and 12 matched controls. The depressed subjects had a markedly greater behavioral response to the rewarding effects of the psychostimulant, as well as altered brain activation of the ventrolateral prefrontal cortex, orbitofrontal cortex, caudate, and putamen.[78] These findings further implicate DA circuit dysfunction in major depression.

The finding of increased reward with psychostimulant administration in severely depressed patients may be related to the finding that glucocorticoids may selectively facilitate DA transmission in the nucleus accumbens.[79] In healthy control subjects, cortisol levels are positively associated with d-amphetamine-induced DA release in the ventral striatum and dorsal putamen. Subjects with higher plasma cortisol concentrations report greater positive drug effects.[80] This work is supported by the finding that, when exposed to a psychosocial stressor, ventral striatal DA concentrations are increased among subjects who report poor early life maternal care compared to those who do not, and the DA increase is correlated with the increase in salivary cortisol concentrations.[81] The high incidence of hypercortisolemia in depression, particularly in severe depression, raises the speculation that elevated cortisol concentrations alter dopaminergic reward systems, thereby altering hedonic responsiveness. One proposed model posits that over time, frequent bouts of stress associated with intermittent increased exposure to glucocorticoids sensitize the mesolimbic DA system.[80] In a test of this model, dexamethasone added to the drinking water of maternal rats both pre- and postpartum resulted in a 50% greater survival rate of midbrain dopaminergic neurons in the adult offspring.[82] Such a model also provides a potential explanatory framework for the high comorbidity between major depression and substance abuse.

The activity of cyclic adenosine monophosphate response element binding protein (CREB) may be another important mediator of the altered reward responsiveness in MDD. In the nucleus accumbens, CREB function is regulated by both glutamatergic and dopaminergic inputs.[83] It has been argued that CREB thereby regulates the set point of accumbens neurons in gating behavioral responses to stimuli.[84] Sustained elevations of CREB activity in the accumbens may cause a nonspecific numbing to emotional stimuli, producing an anhedonia-like state. In transgenic mice, overexpression of CREB produces a phenotype of depression and reduces reward responsiveness to cocaine administration.[85,86] Reductions in CREB activity in the accumbens produce antidepressant-like effects. Sustained pathological elevations in CREB activity in the accumbens may produce diminished emotional reactivity, reminiscent of anhedonia, whereas sustained reductions in CREB activity may produce excessive emotional reactivity, perhaps associated with anxiety.[84] It should be noted that CREB activity in the accumbens differs from its role in the hippocampus, where enhanced CREB function may mediate the effects of antidepressants.[87]

RESPONSE TO STARTLE

The startle reflex is a set of involuntary responses to a sudden strong stimulus, such as a loud noise, that can be measured in humans via the amplitude of the electromyographic response of the eye blink. The acoustic startle response is mediated by a simple subcortical circuit that is modulated by inputs from several areas, including the nucleus accumbens, striatum, and prefrontal cortex. Preclinical and clinical studies indicate that enhancement of dopaminergic neurotransmission increases startle responding.[88,89] Administration of a D1 receptor antagonist in rats significantly reduces the startle-enhancing effects of intracerebroventricularly administered corticotropin-releasing factor.[90] Although serotonin also modulates the startle response, serotonin's effects in animal models are less generalizable to human responses than those of DA.[88]

Despite the extensive demonstration of startle abnormalities in diseases thought to be characterized by disrupted DA functioning, including attention deficit hyperactivity disorder (ADHD) and schizophrenia, little work has been done to examine the startle response in MDD. The findings to date have been somewhat mixed, likely due to a failure to control for confounding factors, including antidepressant treatment and common comorbid disorders that heighten startle responding, such as anxiety disorders.[91–93] The most consistent finding is that more severely depressed patients demonstrate a lower startle magnitude. Intriguingly, in the only prospective study that employed startle to assess the response to an adequate trial of antidepressant treatment, there was a positive correlation between baseline startle magnitude and improvement of depressive symptoms with sertraline (an SSRI) or reboxetine (a specific noradrenergic reuptake inhibitor).[94] This study did not control for comorbid anxiety disorders, but these findings suggest that subjects with a low baseline startle magnitude (perhaps associated with diminished DA function) may carry a lower likelihood of benefit from SSRI treatment. Another study using an affective startle paradigm (in which emotional film clips are used in an attempt to modulate the subject's mood) found that the most anhedonic patients displayed lower startle magnitude across all mood conditions.[92] Finally, a small study of patients in remission from MDD who demonstrated attenuated startle responding were more likely to have relapsed at follow-up 2 years later than patients with higher startle responsiveness.[95]

POSTMORTEM FINDINGS

Postmortem studies of the DA system in depressed patients are relatively few and, not surprisingly, have provided conflicting results, due at least in part to variability in the age of the subjects, agonal states, the presence of psychotropic medications, and the inclusion in some studies of victims of suicide, which may have its own unique pathobiology.[96] Brain concentrations of DA in suicide victims are unchanged compared to controls.[97–100] Homovanillic acid concentrations have been found to be elevated[100,101] or unaltered[102] in the frontal cortex and unaltered in the basal ganglia[99,100] of suicide victims. Cerebrospinal fluid HVA concentrations have been found to be lower in suicide attempters than in controls,[103] but not different between patients with a high- versus low-lethality attempt.[104] Concentrations of dihydroxyphenyl acetic acid (DOPAC) in the caudate, putamen, and nucleus accumbens were reduced in antidepressant-free depressed patients who died by suicide compared to controls.[105]

In one elegant postmortem study using immunohistochemical and autoradiographic methods with high anatomical resolution, depressed subjects, most of whom died by suicide, demonstrated reduced DAT density and elevated D2/3 receptor binding in the central and basal nuclei of the amygdala compared with psychiatrically normal controls.[106] A second study using different methods found no difference in D2 receptor number or affinity.[105] Neither of these studies reported a difference in D1 receptor binding between depressed subjects who died by suicide and controls.[105,106] A third study found no difference in D2 receptor binding among subjects with nonspecific "depression" who died by suicide compared with controls, but it did find increased D2 binding among the depressed subjects who met the full criteria for MDD.[107]

NEUROIMAGING FINDINGS

Relatively few studies have examined DA system alterations in depression with neuroimaging methods. Published studies have focused largely on D2 receptor or DAT occupancy. Interpreting the results of earlier studies employing ^{123}I-2ß-carboxymethoxy-3ß-(4-iodophenyl) tropane (^{123}I-ß-CIT) to image the DAT are problematic in that the binding profile for this ligand is not specific for this monoamine transporter, though when focused on the striatum, the vast majority of binding is indeed to the DAT.[108] Few studies of DAT binding or uptake have been performed with more specific ligands.

Results of neuroimaging studies of D2 receptor binding in MDD have been inconsistent[109–116] (Table 10.5.3). Early studies found elevated striatal D2 binding levels in depressed inpatients, either in whole group samples[109,116] or when limited to a psychomotor retarded group.[110] Elevated D2 receptor binding may reflect increased numbers of D2 receptors in depression, an increase in affinity of the receptor for the ligand, or a decrease in availability of synaptic DA (which competes with the radiolabeled ligand, albeit weakly, for D2 binding). The subjects in the studies finding no difference between depressed and control subjects included patients who were less ill than those in the previous studies, with little psychomotor retardation, or used an unhealthy control group.[111,112,115] A major confound across the studies was the medication status of the subjects, because most were either currently treated with antidepressants or had only a 7-day washout prior to the imaging procedure. Variability in the level of anxiety may also confound the results, as anxiety has been associated with reduced D2 receptor expression.[117]

Conflicting results have also been found in other types of imaging studies. In studies comparing D2 binding before and after antidepressant treatment for depression, clinical improvement was noted with either an increase or a decrease in D2 receptor binding, perhaps due to the differing mechanism of action of the drugs employed[110,112,113,118,119] (Table 10.5.4). Studies of DAT expression have also found conflicting results, though the most comprehensive positron emission tomography (PET) study observed reduced DAT binding in depression.[120] In a PET study assessing DA neuronal function by measuring [^{18}F]-fluorodopa uptake in the striatum, depressed patients with psychomotor retardation exhibited reduced striatal uptake of the radioligand compared to anxious, depressed inpatients and healthy volunteers[120–124] (Table 10.5.5).

CLINICAL THERAPEUTICS

Of the antidepressants either currently or previously available, those that are likely to enhance DA neurotransmission include nomifensine and amineptine,

TABLE 10.5.3. *Summary of Published Studies of Striatal D2 Receptor Binding in Patients with MDD versus Controls Prior to Treatment Intervention*

Reference	Method	N (P/C)	Sample Characteristics	Comment
Studies reporting statistically significant findings on primary outcome				
D'Haenen and Bossuyt[109]	SPECT [123]IZBM	21/11	Unselected MDD patients, ≥1 week AD free	11% greater striatal/cerebellum D2 binding ratio in MDD patients.
Shah et al.[116]	SPECT [123]IZBM	15/15	8 MDD patients on AD, 7 currently not on AD	4% greater binding ratio in right striatum in MDD patients. Binding ratios correlated with reaction time and verbal fluency.
Meyer et al.[114]	PET ^{11}C raclopride	21/21	≥26 weeks AD free.	6%–8% greater striatal binding potential among all MDD patients. In subgroup analysis, PMR MDD patients, but not non-PMR MDD patients had greater striatal binding versus controls.
Studies reporting no significant difference on primary outcome				
Ebert et al.[110]	SPECT [123]IZBM	20/10	2 MDD groups: 1. 10 AD free ≥ 6 months 2. 10 on AMI for 2 weeks	PMR MDD patients had 6% increase in striatal D2 binding ratio compared to all others.
Klimke et al.[112]*	SPECT [123]IZBM	15/17	≥1 week AD free	No difference in D2 binding at baseline between eventual responders and nonresponders to AD treatment.
Parsey et al.[115]	SPECT [123]IZBM	9/10	≥2 weeks AD free	No difference in striatal D2 binding after amphetamine administration.
Kuroda et al.[113]	PET ^{11}C raclopride	9/14	8/9 MDD patients on fluvoxamine	No subgroup analyses performed.
Hirvonen et al.[111]	PET ^{11}C raclopride	25/19	24/25 MDD patients AD naive	PMR MDD subgroup also showed no difference versus controls.

* Klimke et al.[112] report updated results of their group's original publication (Larisch et al.[202]). AD, antidepressant; AMI, amitriptyline; D2, DA type 2 receptor; [123]IBZM, ^{123}I-iodobenzamide; MDD, major depressive disorder; P/C, patients/controls; PET, positron emission tomography; PMR, psychomotor retardation; SPECT, single photon emission computed tomography; SSRI, selective serotonin reuptake inhibitor.

potent DA reuptake inhibitors[125] (both withdrawn from the market due to adverse events); sertraline, an SSRI that also blocks DA reuptake at high doses; and MAOIs, which prevent degradation of DA, NE, and 5HT. Moreover, the absence of DAT in the prefrontal cortex and the role of the norepinephrine transporter (NET) in inactivating the DA signal in this critical brain region, taken together, have revealed an effect of NE reuptake inhibitors in increasing DA availability in this area.[126] Antidepressants and ECT share the effect of increasing binding at D2-family receptors, though it is unknown whether this increase stems from greater expression of D2 receptors or a change in the state of existing receptors.[127] The greater efficacy of MAOIs over TCAs in atypical depression and anergic bipolar depression suggests that alterations in DA metabolism may be particularly important in these conditions.[128]

Although bupropion is often considered to produce its antidepressant effects via DAT blockade, at clinically significant doses the drug occupies less than 22%–26% of DAT binding sites.[129,130] In contrast, SSRIs typically inhibit 80% or more of serotonin transporter binding sites at minimally effective doses.[131] Microdialysis experiments demonstrate that bupropion does raise extracellular DA concentrations in the nucleus accumbens, though the mechanism that drives this change is uncertain, and the concentrations used in these preclinical studies are not attained with customary clinical doses.[132,133] This action may also contribute to the effectiveness of bupropion in tobacco smoking cessation treatment.

In addition, several drugs acting on the DA system have been evaluated for their efficacy in major depression. The first agents employed to treat depression that directly altered dopaminergic signaling were the psychostimulants, acting through increases in DA release and blockade of the DAT, though these agents also act upon 5HT and NE neurons. In double-blind,

TABLE 10.5.4. *Summary of Studies Examining the Effects of Treatment on Striatal D2 Binding in Depressed Patients*

Reference	Method	N	Treatment	Primary Findings and Comment
Ebert et al.[110]	SPECT [123]IZBM	10 inpatients	150 mg AMI/day for 3 weeks	Nonresponders to 3 weeks of AMI showed increased or no change in striatal D2 binding. Responders to AMI significantly decreased D2 binding.
Klimke et al.[112]	SPECT [123]IZBM	15 inpatients (3 with BP)	Fluoxetine or paroxetine for 6 weeks	After 3–7 weeks of SSRI treatment: Nonresponders had decreased D2 binding in striatum. Responders had increased D2 binding in striatum, and increased D2 binding correlated with reduction in HDRS score ($r = 0.54$, $p < 0.04$).
Pogarell et al.[119]	SPECT [123]IZBM	5 outpatients	Single rTMS bolus challenge	9.6% reduction in striatal D2 binding; 4/5 patients completed 15 sessions of rTMS treatment; no change in resting state D2 binding from pre- to posttreatment found.
Kuroda et al.[113]	PET ^{11}C raclopride	9 outpatients	10 daily sessions of rTMS	No change in striatal D2 binding; 8/9 patients were on fluvoxamine throughout the study.
Montgomery et al.[118]	PET ^{11}C raclopride	8 AD-treated outpatients (HDRS ≤ 10)	SSRI, duration not reported	Lower D2 binding in dorsal but not ventral striatum versus controls ($n = 8$).

AD, antidepressant; AMI, amitriptyline; D2, DA type 2 receptor; BP, bipolar disorder; HDRS, Hamilton Depression Rating Scale; [123]IBZM, ^{123}I-iodobenzamide; PET, positron emission tomography; rTMS, rapid transcranial magnetic stimulation; SPECT, single photon emission computed tomography; SSRI, selective serotonin reuptake inhibitor.

TABLE 10.5.5. *Studies of Presynaptic DA Turnover and DAT Density in Patients with MDD*

Reference	Method	N (P/C)	Sample Characteristics	Primary Findings and Comment
Paillere-Martinot et al.[123]	PET ^{18}F DOPA	12/10	6 inpatients with PMR and AF; 6 inpatients with impulsivity/anxiety; 3 in each group on antidepressant	Patients with PMR and AF had lower K_i values for ^{18}F DOPA uptake in left caudate (−12%) than impulsive/anxious depressed patients or controls.
Meyer et al.[120]	PET [11C]RTI-32	9/23	≥12 weeks AD washout	Striatal DAT levels lower bilaterally in patients than in controls.
Brunswick et al.[122]	SPECT [99mTc]-TRODAT-1	15/46	≥1 week AD washout	DAT levels higher in bilateral putamen and left caudate (+12%–36%) in patients versus controls.
Argyelan et al.[121]	SPECT [99mTc]-TRODAT-1	16/12	≥2 weeks AD washout	No statistically significant differences between patients and controls in striatal-occipital ratio.
Sarchiapone et al.[124]	SPECT DATSCAN	11/9	≥4 weeks AD washout; selected for anhedonic patients	Striatal DAT levels lower in bilateral striatum (−17% to −23%) in patients versus controls.

AD, antidepressant; AF, affective flattening; AMI, amitriptyline; [11C]RTI-32, [11C]methyl (1R-2-exo-3-exo)-8-methyl-3-(4-methylphenyl)-8-azabicyclo[3.2.1]octane-2-carboxylate; DAT, DA transporter; DATSCAN, [123I]N-fluoropropyl-carbomethoxy-3β-(4-iodophenyl)tropane; [18F]DOPA, [18F]fluorodopa; PET, positron emission tomography; P/C, patients/controls; PMR, psychomotor retardation; SPECT, single photon emission computed tomography; [99mTc]TRODAT-1, [99mTc][2][2-]{[[3-94-chlorophenyl)-8-methyl-8-azabicyclo[3.2.1]oct-2-yl]-methyl}(2-mercaptoentyl)amino]ethyl]amino]ethane-thiolato (3-)-N2,N2',S2,S2']oxo-[1R-(exo-exo).

placebo-controlled studies of unselected depressed patients, psychostimulants were inferior to TCAs and MAOIs.[134] Studies employing methylphenidate or dextroamphetamine as a predictor of response to TCAs found inconsistent results, though design limitations likely contributed to these results.[135]

Bromocriptine, a D2 agonist, was found to be as efficacious as TCAs in depression in three small double-blind studies, though the absence of a placebo confounds interpretation of these findings.[136] Open-label studies suggest that bromocriptine may provide antidepressant benefit in treatment-resistant depression and tachyphylaxis-associated relapses.[137,138] In a small double-blind trial, the DA agonist piribedil was efficacious in depression, with low pretreatment CSF HVA concentrations predictive of a response.[55] Pergolide, a DA agonist used for Parkinson's disease, suggested efficacy in two open-label augmentation trials for major depression,[139,140] but a placebo-controlled augmentation study did not demonstrate benefit.[141]

Pramipexole, a nonergot DA agonist used in the treatment of Parkinson's disease and restless legs syndrome, exhibits marked selectivity for D2-like receptors, particularly the D3 receptor. Several case series and reports suggested antidepressant efficacy for pramipexole in refractory bipolar depression[142,143] or as an augmentation agent with SSRIs, TCAs, or psychotherapy.[144–147] In a study of baboons, pramipexole reduced cerebral blood flow in the orbitofrontal cortex, subgenual anterior cingulate cortex, and insula, all regions thought to contribute significantly to mood regulation.[148] Although acute administration of pramipexole inhibits neuronal DA firing, with sustained treatment the firing rate normalizes.[149,150] In rats, 14-day treatment with pramipexole increased 5HT neuron firing rates and induced desensitization of the D2/D3, 5HT1a, and α2 cell body autoreceptors.[149]

Three double-blind, placebo-controlled trials have explored the use of pramipexole for the treatment of major depressive episodes. In unipolar depression, pramipexole (5 mg/day) was superior to placebo and equivalent to fluoxetine (20 mg/day) among completers of an 8-week trial.[151] Two studies of patients with bipolar depression on mood stabilizer therapy found significantly greater response rates in pramipexole- versus placebo-treated patients.[152,153] Open-label treatment with ropinirole was also reported to be effective in 4 of 10 patients with treatment-resistant MDD or Bipolar II disorder.[154]

Atypical antipsychotics have convincingly been demonstrated to be effective when added to an SSRI or SNRI in converting partially responsive and nonresponsive depressed patients to responders, and aripiprazole has an FDA indication for such treatment in MDD (i.e., not antidepressant in itself, but demonstrating antidepressant efficacy when combined with a proven antidepressant).[155,156] The mechanism of action in this regard is uncertain.[157] Interestingly, addition of an atypical antipsychotic typically induces improvement in depressive symptoms within the first few weeks of treatment, achieved more rapidly than the response to the antidepressant itself. All atypical antipsychotics are relatively potent inhibitors of the 5HT2A receptor, which increases NE release.[158,159] These medications also exert some antagonism at the 5HT2C receptor, which can increase DA signaling. Although the evidence base is smaller, typical antipsychotics (which are believed to exert their antipsychotic effect through D2 antagonism) have also demonstrated efficacy when used adjunctively at low doses with an antidepressant.[156] Unlike other antipsychotics, aripiprazole is a partial agonist at D2 receptors, so its antidepressant effect may arise through direct dopaminergic activation.[160] Quetiapine is an atypical antipsychotic that has demonstrated efficacy as monotherapy for both MDD and bipolar depression. Quetiapine has a relatively weak affinity for DA receptors and the 5HT2C receptor, but its primary active metabolite, norquetiapine, is a potent inhibitor of NE reuptake.[161] Thus, although many antipsychotics can enhance antidepressant responsiveness, they may do so through different mechanisms.

In contrast to the depressive relapse induced by dietary depletion of tryptophan in SSRI responders or tyrosine depletion in TCA responders, dietary depletion of the DA precursors phenylalanine and tyrosine does not induce a recurrence in remitted depressed patients.[162,163] However, availability of these amino acid precursors to DA, unlike 5HT, is not rate-limiting in DA synthesis. Administration of α-methylparatyrosine, an inhibitor of tyrosine hydroxylase, rapidly reduces levels of catecholamine metabolites and induces a robust increase in depressive symptoms, particularly anhedonia, poor concentration, and loss of energy in patients treated with NE reuptake inhibitors.[164]

Several pharmaceutical companies are now evaluating the efficacy and safety of compounds that act either via combined serotonin transporter (SERT)/DAT reuptake inhibition (www.rexahn.com) or via combined triple reuptake inhibition of the SERT, DAT, and NET (www.DOVpharm.com; www.neurosearch.com; www.sepracor.com). The results of these studies should help clarify whether enhancing DA transmission improves the speed of or the overall response to treatment in MDD.

BRAIN STIMULATION

Repetitive transcranial magnetic stimulation (rTMS) and deep brain stimulation (DBS) are two brain stimulation approaches to depression treatment that may induce changes in dopaminergic function. In macaque monkeys, stimulation of the primary motor cortex induces DA release in the ventral striatum, suggesting that rTMS may activate the mesolimbic DA pathway.[165] In healthy human controls, rTMS applied over either the prefrontal cortex or the motor cortex induces DA release in the caudate and putamen, respectively.[166–168] Studies of the efficacy of rTMS in MDD have generally found antidepressant benefit, but the treatment does not have an FDA indication for depression.[169] Two studies examining the effects of acute rTMS on striatal D2 binding in depressed patients found no change in resting state D2 binding after 10–15 sessions of rTMS treatment.[113,119] Both studies had significant limitations, however (small sample size, concomitant antidepressant treatment), so the question of whether rTMS treatment can produce sustained changes in DA signaling in MDD patients remains unanswered.

Deep brain stimulation is a novel treatment approach for depression that is being explored in patients demonstrating inadequate benefit from standard treatments. The first double-blind study using DBS for MDD targeted the subgenual cingulate cortex.[170] Recently, the preliminary results of double-blind stimulation of the nucleus accumbens in three patients with MDD reported partial improvement in depressive symptoms, with subjective reports of increased motivation and reduced anhedonia.[171]

DA INTERACTIONS WITH 5HT AND NE

Determining the interrelationships among DA, NE, and 5HT activity faces several challenges. The factors to be considered in assessing the impact of one monoamine system on another include separating acute from chronic effects, distinguishing high versus low levels of stimulation, different forms of cell firing patterns and frequency, differential response to modulation by subpopulations of neurons within a group of monoamine neurons, and feedback and control from other brain regions. Thus, lesion, stimulation, and local infusion techniques may be used to study interactions between monoamine nuclei in the brainstem, whereas systemic or intracerebroventricular administration of agonists and antagonists informs more global effects on monoamine firing rates. In all such laboratory animal experiments, the question of pharmacological dose equivalence to clinical studies is of paramount importance. Often the doses used in preclinical studies far exceed those used clinically, and therefore the translation of the preclinical findings to the clinical setting may be problematic.

An important unresolved question is how SSRIs and SNRIs alter, or fail to alter, DA systems. It is now clear that treatment with these antidepressants, though clearly superior to placebo treatment, frequently fails to render patients symptom free, that is, the majority do not achieve remission.[9] Such a partial response may result from a failure of increased serotonergic or noradrenergic neurotransmission to induce similar alterations in DA signaling. Supporting this hypothesis is the finding that SSRI responders, but not nonresponders, exhibited increased DA binding to D2 receptors in the striatum, and that the degree of increase in D2 binding correlated with an improvement in the Hamilton Depression Scale (HAMD) score.[110]

There are substantial and complex interactions between the CNS serotonergic and dopaminergic systems, with the DA cell bodies in the VTA and substantia nigra pars compacta being targets for the serotonergic cells of the midbrain raphe.[172] The effects of 5HT signaling on the VTA are mixed, though overall it is likely that the net effect of 5HT is to inhibit DA neuronal activity. In the brainstem raphe cells, firing of serotonergic neurons reduces spontaneous activity of DA neurons in the VTA but not in the substantia nigra pars compacta, and inhibits DA-related behaviors such as locomotor and exploratory behavior. Enhanced serotonergic signaling produced by administration of an SSRI reduces VTA firing rates.[173,174] 5-HT1a agonists, which decrease 5HT neuronal firing in the dorsal raphe (DR), increase VTA firing rates.[175,176] However, in the medial prefrontal cortex, activation of 5HT1a or 5HT2a receptors enhances the activity of VTA DA neurons.[177,178] Atypical antipsychotics increase prefrontal cortex extracellular DA concentrations via a 5HT1a-dependent mechanism.[179]

The mesocortical and mesolimbic DA projections from the VTA are tonically inhibited by the action of GABA interneurons, which are stimulated via 5HT2C receptor activation.[180–182] Desensitization of 5HT2C receptors occurs after chronic treatment with the antidepressants fluoxetine, paroxetine, and clomipramine, which should reduce the inhibition over DA cell firing.[183,184] Agomelatine, a novel antidepressant, is thought to act primarily via 5HT2C antagonism.[185]

An increase in extracellular concentrations of 5HT in the striatum, whether by exogenous application or through use of an SSRI, results in uptake of 5HT into dopaminergic terminals via the DAT.[186] This 5HT is then coreleased with DA from the terminal vesicles when the dopaminergic cell fires. Whether this effect contributes to the antidepressant action of SSRIs is unknown.

Noradrenergic signaling may affect DA transmission through pathways projecting from the frontal cortex, or via interactions between the locus ceruleus (LC) and VTA. Separating the contributions of NE and DA to the pathophysiology of MDD is of value in determining whether treatment approaches should incorporate DA-specific targets, or whether altering NE systems is sufficient to induce corrections in DA signaling. Stimulation of $\alpha 1$ receptors exerts a direct excitatory effect on DA VTA neurons, but also acts to inhibit those neurons via activation of inhibitory GABA interneurons.[187,188] Lesions of the locus ceruleus decrease DA turnover in the dorsal and ventral striatum, and lesion of fibers projecting from the LC to the VTA reduce DA turnover in the prefrontal cortex.[189,190] Systematic administration of low doses of clonidine, an agonist at somatodendritic $\alpha 2$ receptors, decreases DA release in the striatum by reducing burst firing of VTA neurons, but higher doses increase the firing rate of these neurons.[191] Agonism of $\alpha 2$ heteroreceptors located on the DA terminals in the striatum acts to reduce DA release in this region.[192] In contrast, micro-iontophoretic administration of an $\alpha 2$ antagonist to VTA cell bodies attenuates DA neuronal activity.[193]

These contradictory findings of the effects of adrenergic receptors on VTA activity may arise from the effects of long-loop feedback mechanisms from the frontal cortex after systemic administration of adrenergic agents. Depletion of medial prefrontal cortex NE greatly reduces the amount of DA release in the accumbens after d-amphetamine administration.[194] This finding suggests that intact medial prefrontal cortex NE transmission is necessary for d-amphetamine-stimulated release of DA in the accumbens. Cortical NE may contribute to DA release in the accumbens via excitatory prefrontal cortex-VTA projections.[195] Alternatively, glutamatergic projections from the prefrontal cortex to VTA nerve terminals in the accumbens may stimulate DA release, or prefrontal cortex NE function may alter tonic inhibitory GABAergic control over DA cells.[196]

The VTA exerts a tonic excitatory effect on the DR, probably through activation of D2 receptors located on 5HT neurons.[197,198] The effects of VTA activity on the LC are less clear, with both increases and decreases in NE neuronal activity reported.[197,199,200] A further complication is that DA modulation of glutamatergic afferents to the LC may increase NE activation.[201] Clearly, the interaction between monoamine systems is one of the knottier problems to solve in determining the role of DA transmission in depressive pathology and treatment.

CONCLUSION

The question of what role DA circuit dysfunction plays in the pathophysiology of depression remains largely unanswered. The importance of DA in processing signals related to motivation, reward, sleep, appetite, and psychomotor speed suggests that its modulation is of fundamental importance to the biology of MDD. A crucial unanswered question is how existing treatments do or do not rectify disturbances in DA function in depressed patients. Identifying a subtype of depression that is not responsive to serotonergic modulation approaches would be of enormous benefit to the field of clinical psychiatry. The thoughtful combination of neuroimaging, genomic, pharmacological, and psychophysiological challenges with brain stimulation techniques may provide the means to better delineate DA dysfunction in depression and its response to treatment. The most fruitful investigations may involve patients who fail to respond to existing treatments, including those with bipolar depression. Pre- and posttreatment studies could identify state versus trait disturbances in DA signaling. Clinical trials with pure-DA acting compounds as monotherapy and for augmentation in SSRI/SNRI nonresponders would also be valuable. Further elucidation of the role of DA dysfunction is clearly warranted.

REFERENCES

1. Kessler RC, Berglund, P, Demler O, et al. The epidemiology of major depressive disorder: results from the National Comorbidity Survey Replication (NCS-R). *JAMA*. 2003;289:3095–3105.
2. Murray CJL, Lopez AD. *The Global Burden of Disease: A Comprehensive Assessment of Mortality and Disability from Diseases, Injuries, and Risk Factors in 1990 and Projected to 2020*. Cambridge, MA: Harvard University Press; 1996.
3. Anda R, Williamson D, Jones D, et al. Depressed affect, hopelessness, and the risk of ischemic heart disease in a cohort of U.S. adults. *Epidemiology*. 1993;4:285–294.
4. Jonas BS, Mussolino ME. Symptoms of depression as a prospective risk factor for stroke. *Psychosom Med*. 2000;62:463–471.
5. Mueller IM, Leon AC, Keller MB, et al. Recurrence after recovery from major depressive disorder during 15 years of observational follow-up. *Am J Psychiatry*. 1999;156:1000–1006.

6. Jones ME, Cockrum PC. A critical review of published economic modeling studies in depression. *Pharmacoeconomics.* 2000;17:555–583.
7. Dunlop BW, Nemeroff CB. The role of dopamine in the pathophysiology of depression. *Arch Gen Psychiatry.* 2007;64:327–337.
8. Kapur S, Mann JJ. Role of the dopaminergic system in depression. *Biol Psychiatry.* 1992;32:1–17.
9. Association AP. *Practice Guideline for the Treatment of Patients with Major Depression.* 2nd ed. Washington, DC: American Psychiatric Association; 2000.
10. Coppen A. The biochemistry of affective disorders. *Br J Psychiatry.* 1967;113:1237–1264.
11. Schildkraut JJ. The catecholamine hypothesis of affective disorders: a review of supporting evidence. *Am J Psychiatry.* 1965;122:509–522.
12. Randrup A, Munkvad I, Fog R, et al. Mania, depression and brain dopamine. In: Essman WB, Valzelli S, eds. *Current Developments in Psychopharmacology.* New York, NY: Spectrum Publications; 1975:206–248.
13. Willner P. The mesolimbic dopamine system as a target for rapid antidepressant action. *Int Clin Psychopharmacol.* 1997;12(suppl 3):S7–SS14.
14. Spyraki C, Fibiger HC. Behavioural evidence for supersensitivity of postsynaptic dopamine receptors in the mesolimbic system after chronic administration of desipramine. *Eur J Pharmacol.* 1981;74:195–206.
15. D'Aquila PS, Collu M, Gessa GL, et al. The role of dopamine in the mechanism of action of antidepressant drugs. *Eur J Pharmacol.* 2000;405:365–373.
16. Lammers C-H, Diaz J, Schwartz J-C, et al. Selective increase of dopamine D3 receptor gene expression as a common effect of chronic antidepressant treatments. *Mol Psychiatry.* 2000;5:378–388.
17. Park SK, Nguyen MD, Fischer A, et al. Par-4 links dopamine signaling and depression. *Cell.* 2005;122:275–287.
18. Salamone JD, Aberman JE, Sokolowski JD, et al. Nucleus accumbens dopamine and rate of responding: neurochemical and behavioral studies. *Psychobiology.* 1999;27:236–247.
19. Neill DB, Fenton H, Justice JB. Increase in accumbal dopaminergic transmission correlates with response cost, not reward, of hypothalamic stimulation. *Behav Brain Res.* 2002;137:129–138.
20. Dziedzicka-Wasylewska M, Rogoz Z, Skuza G, et al. Effect of repeated treatment with tianeptine and fluoxetine on central dopamine D(2)/D(3) receptors. *Behav Pharmacol.* 2002;13:127–138.
21. Ichikawa J, Meltzer HY. Effect of antidepressants on striatal and accumbens extracellular dopamine levels. *Eur J Pharmacol.* 1995;281:255–261.
22. Zangen A, Nakash R, Overstreet DH, et al. Association between depressive behavior and absence of serotonin–dopamine interaction in the nucleus accumbens. *Psychopharmacology.* 2001;155:434–439.
23. D'Aquila PS, Collu M, Pani L, et al. Antidepressant-like effect of selective D1 receptor agonists in the behavioural despair animal model of depression. *Eur J Pharmacology.* 1994;262:107–111.
24. Papp M, Klimek V, Willner P. Parallel changes in dopamine D2 receptor binding in limbic forebrain associated with chronic mild stress-induced anhedonia and its reversal by imipramine. *Psychopharmacology.* 1994;115:441–446.
25. Cheeta S, Broekkamp C, Willner P. Stereospecific reversal of stress-induced anhedonia by mianserin and its (+)-enantiomer. *Psychopharmacology.* 1994;116:523–528.
26. Willner P, Muscat R, Papp M. Chronic mild stress-induced anhedonia: a realistic animal model of depression. *Neurosci Biobehav Rev.* 1992;16:525–534.
27. Sherman AD, Sacquitine JL, Petty F. Specificity of the learned helplessness model of depression. *Pharmacol Biochem Behav.* 1982;16:449–454.
28. Anisman H, Irwin J, Sklar LS. Deficits of escape performance following catecholamine depletion: implications for behavioral deficits induced by uncontrollable stress. *Psychopharmacology.* 1979;64:163–170.
29. Anisman H, Remington G, Sklar LS. Effect of inescapable shock on subsequent escape performance: catecholamine and cholinergic mediation of response initiation and maintenance. *Psychopharmacology.* 1979;61:107–124.
30. Basso AM, Gallagher KB, Bratcher NA, et al. Antidepressant-like effect of D2/3 receptor- but not D4 receptor-activation in the rat forced swim test. *Neuropsychopharmacology.* 2005;30:1257–1268.
31. Borsini F, Meli A. The forced swimming test: its contribution to the understanding of the mechanisms of action of antidepressants. In: Gessa GL, Serra G, eds. *Dopamine and Mental Depression.* Oxford, England: Pergamon Press, 1990;63–76.
32. Sullivan PF, Neale MC, Kendler KS. Genetic epidemiology of major depression: review and meta-analysis. *Am J Psychiatry.* 2000;157:1552–1662.
33. Cravchik A, Goldman D. Neurochemical individuality. Genetic diversity among human dopamine and serotonin receptors and transporters. *Arch Gen Psychiatry.* 2000;57:1105–1114.
34. Frisch A, Postilnick D, Rockah R, et al. Association of unipolar major depressive disorder with genes of the serotonergic and dopaminergic pathways. *Mol Psychiatry.* 1999;4:389–392.
35. Ohara K, Nagai M, Suzaki Y, et al. Low activity allele of catechol-O-methyltransferase gene and Japanese unipolar depression. *Neuroreport.* 1998;9:1305–1308.
36. Szegedi A, Rujescu D, Tadic A, et al. The catechol-O-methyltransferase Val[108/158]Met polymorphism affects short-term treatment response to mirtazapine, but not to paroxetine in major depression. *Pharmacogenomics J.* 2005;5:49–53.
37. Leon SL, Croes EA, Sayed-Tabatabaei FA, et al. The dopamine D4 receptor gene 48-base-pair-repeat polymorphism and mood disorders: a meta-analysis. *Biol Psychiatry.* 2005;57:999–1003.
38. Dikeos DG, Papadimitriou GN, Avramopoulos D, et al. Association between the dopamine D3 receptor gene locus (DRD3) and unipolar affective disorder. *Psychiatr Genet.* 1999;9:189–195.
39. Chiaroni P, Azorin JM, Dassa D, et al. Possible involvement of the dopamine D3 receptor locus in subtypes of bipolar affective disorder. *Psychiatr Genet.* 2000;10:43–49.
40. Cubells JF, Zabetian CP. Human genetics of plasma dopamine beta-hydroxylase activity: applications to research in psychiatry and neurology. *Psychopharmacology.* 2004;174:463–476.
41. Zammit S, Owen MJ. Stressful life events, 5-HTT genotype and risk of depression. *Br J Psychiatry.* 2006;188:199–201.
42. Haeffel GJ, Getchell M, Koposov RA, et al. Association between polymorphisms in the dopamine transporter gene and depression: evidence for a gene–environment interaction in a sample of juvenile detainees. *Psychol Sci.* 2008;19:62–69.
43. Volkow ND, Logan J, Fowler JS, et al. Association between age-related decline in dopamine activity and impairment in frontal and cingulate metabolism. *Am J Psychiatry.* 2000;157:75–80.

44. Banki CM. Correlation between cerebrospinal fluid amine metabolites and psychomotor activity in affective disorders. *J Neurochem.* 1977;28:255–257.
45. Bowers MB Jr, Heninger GR, Gerbode F. Cerebrospinal fluid 5-hydroxyindoleacetic acid and homovanillic acid in psychiatric patients. *Int J Neuropharmacol.* 1969;8:255–262.
46. Mendels J, Frazer A, Fitzgerald RG, et al. Biogenic amine metabolites in cerebrospinal fluid of depressed and manic patients. *Science.* 1972;175:1380–1382.
47. van Praag HM, Korf J. Retarded depression and the dopamine metabolism. *Psychopharmacologia.* 1971;19:199–203.
48. Nordin C, Siwers B, Bertilsson L. Bromocriptine treatment of depressive disorders: clinical and biochemical effects. *Acta Psychiatr Scand.* 1981;64:25–33.
49. Roy A, de Jong J, Linnoila M. Cerebrospinal fluid monoamine metabolites and suicidal behavior in depressed patients: a 5-year follow-up study. *Arch Gen Psychiatry.* 1989;46:609–612.
50. van Praag HM, Korf J, Lakke JPWF, et al. Dopamine metabolism in depressions, psychoses and Parkinson's disease: the problem of the specificity of biological variables in behavior disorders. *Psychol Med.* 1975;5:138–146.
51. Goodwin FK, Post RM, Dunner DL, et al. Cerebrospinal fluid amine metabolites in affective illness: the probenecid technique. *Am J Psychiatry.* 1973;130:73–79.
52. Jimerson DC. Role of dopamine mechanisms in the affective disorders. In: Meltzer HY, ed. *Psychopharmacology: The Third Generation of Progress.* 1987, New York, NY: Raven Press; 1987:505–511.
53. Vestergaard P, Sorensen T, Hoppe E, et al. Biogenic amine metabolites in cerebrospinal fluid of patients with affective disorders. *Acta Psychiatr Scand.* 1978;58:88–96.
54. Maas JW, Koslow SH, Katz MM, et al. Pretreatment neurotransmitter metabolite levels and response to tricyclic antidepressant drugs. *Am J Psychiatry.* 1984;141:1159–1171.
55. Post RM, Gerner RH, Carman JS, et al. Effects of a dopamine agonist, piribedil, in depressed patients. *Arch Gen Psychiatry.* 1978;35:609–615.
56. van Scheyen JD, van Praag HM, Korf J. Controlled study comparing nomifensine and clomipramine in unipolar depression, using the probenecid technique. *Br J Clin Pharmacol.* 1977;4:179S–184S.
57. Rampello L, Nicoletti G, Raffaele R. Dopaminergic hypothesis for retarded depression: a symptom profile for predicting therapeutical responses. *Acta Psychiatr Scand.* 1991;84:552–554.
58. Lambert G, Johansson M, Agren H, et al. Reduced brain norepinephrine and dopamine release in treatment-refractory depressive illness. *Arch Gen Psychiatry.* 2000;57:787–793.
59. Rocca P, De Leo C, Eva C, et al. Decrease of D4 dopamine receptor messenger RNA expression in lymphocytes from patients with major depression. *Prog Neuropsychopharmacol Biol Psychiatry.* 2002;26:1155–1160.
60. Cubells JF, Price LH, Meyers BS, et al. Genotype-controlled analysis of plasma dopamine beta-hydroxylase activity in psychotic unipolar major depression. *Biol Psychiatry.* 2002;51:358–364.
61. McPherson H, Walsh A, Silverstone T. Growth hormone and prolactin response to apomorphine in bipolar and unipolar depression. *J Affect Disord.* 2003;76:121–125.
62. Ansseau M, von Frenckell R, Cerfontaine JL, et al. Blunted response of growth hormone to clonidine and apomorphine in endogenous depression. *Br J Psychiatry.* 1988; 53:65–71.
63. Scantamburlo G, Hansenne M, Fuchs S, et al. AVP- and OT-neurophysins response to apomorphine and clonidine in major depression. *Psychoneuroendocrinology.* 2005;30:839–845.
64. Monreal J, Duval F, Mokrani M-C, et al. Exploration de la fonction dopaminergique dans les depressions bipolaires et unipolares. *Ann Med Psychol.* 2005;163:399–404.
65. Tandberg E, Larsen JP, Aarsland D, et al. The occurrence of depression in Parkinson's disease. *Arch Neurol.* 1996;53:175–179.
66. Bejjani BP, Damier P, Arnulf I, et al. Transient acute depression induced by high-frequency deep-brain stimulation. *N Engl J Med.* 1999;340:1476–1480.
67. Nieoullon A, Coquerel A. Dopamine: a key regulator to adapt action, emotion, motivation and cognition. *Curr Opin Neurol.* 2003;16(suppl 2):S3–S9.
68. Saal D, Dong Y, Bonci A, et al. Drugs of abuse and stress trigger a common synaptic adaptation in dopamine neurons [erratum appears in *Neuron.* 2003; 24;38(2):359]. *Neuron.* 2003;37:577–582.
69. Berton O, McClung CA, Dileone RJ, et al. Essential role of BDNF in the mesolimbic dopamine pathway in social defeat stress. *Science.* 2006;311:864–868.
70. Singer T, Kiebel SJ, Winston JS, et al. Brain responses to the acquired moral status of faces. *Neuron.* 2004;41:653–662.
71. Insel TR, Fernald RD. How the brain processes social information: searching for the social brain. *Annu Rev Neurosci.* 2004;27:697–722.
72. Isovich E, Engelmann M, Landgraf R, et al. Social isolation after a single defeat reduces striatal dopamine transporter binding in rats. *Eur J Neurosci.i* 2001;13:1254–1256.
73. de la Fuente-Fernandez R, Schulzer M, Stoessl AJ. Placebo mechanisms and reward circuitry: clues from Parkinson's disease. *Biol Psychiatry.* 2004;56:67–71.
74. de la Fuente-Fernandez R, Ruth TJ, Sossi V, et al. Expectation and dopamine release: mechanism of the placebo effect in Parkinson's disease. *Science.* 2001;293:1164–1166.
75. Scott DJ, Stohler CS, Egnatu CM, et al. Placebo and nocebo effects are defined by opposite opioid and dopaminergic responses. *Arch Gen Psychiatry.* 2008;65:220–231.
76. Hasler G, Drevets WC, Charney DS. Discovering endophenotypes for major depression. *Neuropsychopharmacology.* 2004;29:1765–1781.
77. Tremblay LK, Naranjo CA, Cardenas L, et al. Probing brain reward system function in major depressive disorder: altered response to dextroamphetamine. *Arch Gen Psychiatry.* 2002;59:409–416.
78. Tremblay LK, Naranjo CA, Graham SJ, et al. Functional neuroanatomical substrates of altered reward processing in major depressive disorder revealed by a dopaminergic probe. *Arch Gen Psychiatry.* 2005;62:1228–1236.
79. Marinelli M, Piazza PV. Interaction between glucocorticoid hormones, stress and psychostimulant drugs. *Eur J Neurosci* 2002;16:387–394.
80. Oswald LM, Wong DF, McCaul M, et al. Relationships among ventral striatal dopamine release, cortisol secretion, and subjective responses to amphetamine. *Neuropsychopharmacology.* 2005;30:821–832.
81. Pruessner JC, Champagne F, Meaney MJ, et al. Dopamine release in response to a psychological stress in humans and its relationship to early life maternal care: a positron emission tomography study using [11C] raclopride. *J Neurosci.* 2004;24:2825–2831.

82. McArthur S, McHale E, Dalley JW, et al. Altered mesencephalic dopaminergic populations in adulthood as a consequence of brief perinatal glucocorticoid exposure. *J Neuroendocrinol.* 2005;17:475–482.
83. Dudman JT, Eaton ME, Rajadhyaksha A, et al. Dopamine D1 receptors mediate CREB phosphorylation via phosphorylation of the NMDA receptor at Ser897-NR1. *J Neurochem.* 2003;87:922–934.
84. Nestler EJ, Carlezon WA Jr. The mesolimbic dopamine reward circuit in depression. *Biol Psychiatry.* 2006;59:1151–1159.
85. McClung CA, Nestler EJ. Regulation of gene expression and cocaine reward by CREB and DeltaFosB. *Nat Neurosc.* 2003;6:1208–1215.
86. Newton SS, Thome J, Wallace TL, et al. Inhibition of cAMP response element-binding protein or dynorphin in the nucleus accumbens produces an antidepressant-like effect. *J Neurosci.* 2002;22:10883–10890.
87. Duman RS, Heninger GR, Nestler EJ. A molecular and cellular theory of depression. *Arch Gen Psychiatry.* 1997;54:597–606.
88. Braff DL, Geyer MA, Swerdlow NR. Human studies of prepulse inhibition of startle: normal subjects, patient groups, and pharmacological studies. *Psychopharmacology.* 2001;156:234–258.
89. Swerdlow NR, Stephany N, Wasserman LC, et al. Amphetamine effects on prepulse inhibition across species: replication and parametric extension. *Neuropsychopharmacology.* 2003;28:640–650.
90. Meloni EG, Gerety LP, Knoll AT, et al. Behavioral and anatomical interactions between dopamine and corticotropin-releasing factor in the rat. *J Neurosci.* 2006;26:3855–3863.
91. Allen NB, Trinder J, Brennen C. Affective startle modulation in clinical depression: preliminary findings. *Biol Psychiatry.* 1999;46:542–550.
92. Kaviani H, Gray JA, Checkley SA, et al. Affective modulation of the startle response in depression: influence of the severity of depression, anhedonia, and anxiety. *J Affect Disord.* 2004;83:21–31.
93. Perry W, Minassian A, Feifel D. Prepulse inhibition in patients with non-psychotic major depressive disorder. *J Affect Disord.* 2004;81:179–184.
94. Quednow BB, Kuhn KU, Stelzenmueller R, et al. Effects of serotonergic and noradrenergic antidepressants on auditory startle response in patients with major depression. *Psychopharmacology.* 2004;175:399–406.
95. O'Brien-Simpson L, Di Parsia P, Simmons JG, et al. Recurrence of major depressive disorder is predicted by inhibited startle magnitude while recovered. *J Affect Disord.* 2008;112:243–249.
96. Oquendo MA, Mann JJ. The biology of impulsivity and suicidality. *Psychiatr Clin North Am.* 2000;23:11–25.
97. Pare CMB, Yeung DPH, Price K, et al. 5-Hydroxytryptamine, noradrenaline and dopamine in brainstem, hypothalamus and caudate nucleus of controls and patients committing suicide by coal gas poisoning. *Lancet II.* 1969;294:133–135.
98. Moses SG, Robins E. Regional distribution of norepinephrine and dopamine in brains of depressive suicides and alcoholic suicides. *Psychopharmacol Commun.* 1975;1:327–337.
99. Bowden C, Cheetham SC, Lowther S, et al. Reduced dopamine turnover in the basal ganglia of depressed suicides. *Brain Res.* 1997;769:135–140.
100. Beskow J, Gotffries CG, Winblad B. Determination of monoamine and monoamine metabolites in the human brain: postmortem studies in a group of suicides and in a control group. *Acta Psychiatr Scand.* 1976;53:7–20.
101. Ohmori T, Arora RC, Meltzer HY. Serotonergic measures in suicide brain: the concentration of 5-HIAA, HVA and tryptophan in frontal cortex of suicide victims. *Biol Psychiatry.* 1992;32:57–71.
102. Crow TJ, Cross AJ, Cooper SJ, et al. Neurotransmitter receptors and monoamine metabolites in the brains of patients with Alzheimer-type dementia and depression, and suicides. *Neuropharmacology.* 1984;23:1561–1569.
103. Engstrom G, Alling C, Blennow K, et al. Reduced cerebrospinal HVA concentration and HVA/5-HIAA ratios in suicide attempters. *Eur Neuropsychopharmacol.* 1999;9:399–405.
104. Mann JJ, Malone KM. Cerebrospinal fluid amines and higher-lethality suicide attempts in depressed inpatients. *Biol Psychiatry.* 1997;41:162–171.
105. Bowden C, Theodorou AE, Cheetham SC, et al. Dopamine D1 and D2 receptor binding sites in brain samples from depressed suicides and controls. *Brain Res.* 1997;752:227–233.
106. Klimek V, Schenck JE, Han H, et al. Dopaminergic abnormalities in amygdaloid nucleus in major depression: a postmortem study. *Biol Psychiatry.* 2002;52:740–748.
107. Allard P, Norlen M. Caudate nucleus dopamine D2 receptors in depressed suicide victims. *Neuropsychobiology.* 2001;44:70–73.
108. Malison RT, Price LH, Berman R, et al. Reduced brain serotonin transporter availability in major depression as measured by 123I-2Beta-carbomethoxy-3Beta-(4-iodophenyl)tropane and single photon emission computed tomography. *Biol Psychiatry.* 1998;44:1090–1098.
109. D'Haenen HA, Bossuyt A. Dopamine D2 receptors in depression measured with single photon emission computed tomography. *Biol Psychiatry.* 1994;35:128–132.
110. Ebert D, Feistel H, Loew T, et al. Dopamine and depression—striatal dopamine D2 receptor SPECT before and after antidepressant therapy. *Psychopharmacology.* 1996;126:91–94.
111. Hirvonen J, Karlsson H, Kajander J, et al. Striatal dopamine D2 receptors in medication-naive patients with major depressive disorder as assessed with [11C]raclopride PET. *Psychopharmacology.* 2008;197:581–590.
112. Klimke A, Larisch R, Janz A, et al. Dopamine D2 receptor binding before and after treatment of major depression measured by. *Psychiatry Res.* 1999;90:91–101.
113. Kuroda Y, Motohashi N, Ito H, et al. Effects of repetitive transcranial magnetic stimulation on [^{11}C] raclopride binding and cognitive function in patients with depression. *J Affect Disord.* 2006;95:35–42.
114. Meyer JH, McNeely HE, Sagrati S, et al. Elevated putamen D(2) receptor binding potential in major depression with motor retardation: an [11C]raclopride positron emission tomography study. *Am J Psychiatry.* 2006;163:1594–1602.
115. Parsey RV, Oquendo MA, Zea-Ponce Y, et al. Dopamine D2 receptor availability and amphetamine-induced dopamine release in unipolar depression. *Biol Psychiatry.* 2001;50:313–322.
116. Shah PJ, Ogilvie AD, Goodwin GM, et al. Clinical and psychometric correlates of dopamine D2 binding in depression. *Psychol Med.* 1997;27:1247–1256.
117. Schneier FR, Liebowitz MR, Abi-Dargham A, et al. Low dopamine D(2) receptor binding potential in social phobia. *Am J Psychiatry.* 2000;157:457–459.
118. Montgomery AJ, Stokes P, Kitamura Y, et al. Extrastriatal D2 and striatal D2 receptors in depressive illness: pilot PET studies using [11C]FLB 457 and [11C]raclopride. *J Affect Disord.* 2007;101:113–122.

119. Pogarell O, Koch W, Popperl G, et al. Striatal dopamine release after prefrontal repetitive transcranial magnetic stimulation in major depression: preliminary results of a dynamic [123I] IBZM SPECT study. *J Psychiatr Res.* 2006;40:307–314.
120. Meyer JH, Kruger S, Wilson AA, et al. Lower dopamine transporter binding potential in striatum during depression. *Neuroreport.* 2001;12:4121–4125.
121. Argyelan M, Szabo Z, Kanyo B, et al. Dopamine transporter availability in medication free and in bupropion treated depression: a 99mTc-TRODAT-1 SPECT study. *J Affect Disord.* 2005;89:115–123.
122. Brunswick DJ, Amsterdam JD, Mozley PD, et al. Greater availability of brain dopamine transporters in major depression shown by [^{99m}Tc]Trodat-1 SPECT imaging. *Am J Psychiatry.* 2003;160:1836–1841.
123. Paillere-Martinot M, Bragulat V, Artiges E, et al. Decreased presynaptic dopamine function in the left caudate of depressed patients with affective flattening and psychomotor retardation. *Am J Psychiatry.* 2001;158:314–316.
124. Sarchiapone M, Carli V, Camardese G, et al. Dopamine transporter binding in depressed patients with anhedonia. *Psychiatry Res.* 2006;147:243–248.
125. Gillings D, Grizzle J, Koch G, et al. Pooling 12 nomifensine studies for efficacy generalizability. *J Clin Psychiatry.* 1984;45:78–84.
126. Sesack SR, Hawrylak VA, Matus C, et al. Dopamine axon varicosities in the prelimbic division of the rat prefrontal cortex exhibit sparse immunoreactivity for the dopamine transporter. *J Neurosci.* 1998;18:2697–2708.
127. Gershon AA, Vishne T, Grunhaus L. Dopamine D2-like receptors and the antidepressant response. *Biol Psychiatry.* 2007;61:145–153.
128. Thase ME, Trivedi MH, Rush AJ. MAOIs in the contemporary treatment of depression. *Neuropsychopharmacology.* 1995;12:185–219.
129. Learned-Coughlin SM, Bergstrom M, Savitcheva I, et al. In vivo activity of bupropion at the human dopamine transporter as measured by positron emission tomography. *Biol Psychiatry.* 2003;54:800–805.
130. Meyer JH, Goulding VS, Wilson AA, et al. Bupropion occupancy of the dopamine transporter is low during clinical treatment. *Psychopharmacology.* 2002;163:102–105.
131. Meyer JH, Wilson AA, Sagrati S, et al. Serotonin transporter occupancy of five selective serotonin reuptake inhibitors at different doses: an [11C]DASB positron emission tomography study. *Am J Psychiatry.* 2004;161:826–835.
132. Nomikos GG, Damsma G, Wenkstern D, et al. Acute effects of bupropion on extracellular dopamine concentrations in rat striatum and nucleus accumbens studied by in vivo microdialysis. *Neuropsychopharmacology.* 1989;2:273–279.
133. Nomikos GG, Damsma G, Wenkstern D, et al. Effects of chronic bupropion on interstitial concentrations of dopamine in rat nucleus accumbens and striatum. *Neuropsychopharmacology.* 1992;7:7–14.
134. Satel SL, Nelson JC. Stimulants in the treatment of depression: a critical review. *J Clin Psychiatry.* 1989;50:242–249.
135. Little KY. Amphetamine, but not methylphenidate, predicts antidepressant efficacy. *J Clin Psychopharmacol.* 1988;8:177–183.
136. Sitland-Marken PA, Wells BG, Froemming JH, et al. Psychiatric applications of bromocriptine therapy. *J Clin Psychiatry.* 1990;51:68–82.
137. McGrath PJ, Quitkin FM, Klein DF. Bromocriptine treatment of relapses seen during selective serotonin re-uptake inhibitor treatment of depression. *J Clin Psychopharmacol.* 1995;15:289–291.
138. Inoue T, Tsuchiya K, Miura J, et al. Bromocriptine treatment of tricyclic and heterocyclic antidepressant–resistant depression. *Biol Psychiatry.* 1996;40:151–153.
139. Izumi T, Inoue T, Kitagawa N, et al. Open pergolide treatment of tricyclic and heterocyclic antidepressant–resistant depression. *J Affect Disord.* 2000;61:127–132.
140. Bouckoms A, Mangini L. Pergolide: an antidepressant adjuvant for mood disorders? *Psychopharmacol Bull.* 1993;29:207–211.
141. Mattes JA. Pergolide to augment the effectiveness of antidepressants: clinical experience and a small double-blind study. *Ann Clin Psychiatry.* 1997;9:87–88.
142. Goldberg JF, Frye MA, Dunn RT. Pramipexole in refractory bipolar depression. *Am J Psychiatry.* 1999;156:798.
143. Perugi G, Toni C, Ruffolo G, et al. Adjunctive dopamine agonists in treatment-resistant bipolar II depression: an open case series. *Pharmacopsychiatry.* 2001;34:137–141.
144. Sporn J, Ghaemi S, Sambur MR, et al. Pramipexole augmentation in the treatment of unipolar and bipolar depression: a retrospective chart review. *Ann Clin Psychiatry.* 2000;12:137–140.
145. Lattanzi L, Dell'Osso L, Cassano P, et al. Pramipexole in treatment-resistant depression: a 16-week naturalistic study. *Bipolar Disord.* 2002;4:307–314.
146. DeBattista C, Solvason H, Heilig Breen JA, et al. Pramipexole augmentation of a selective serotonin reuptake inhibitor in the treatment of depression. *J Clin Psychopharmacol.* 2000;20:274–275.
147. Ostow M. Pramipexole for depression. *Am J Psychiatry.* 2002;159:320–321.
148. Black KJ, Hershey T, Koller JM, et al. A possible substrate for dopamine-related changes in mood and behavior: prefrontal and limbic effects of a D3-preferring dopamine agonist. *Proc Natl Acad Sci USA.* 2002;99:17113–17118.
149. Chernoloz O, El Mansari M, Blier, P. Sustained administration of pramipexole modifies the spontaneous firing of dopamine, norepinephrine, and serotonin neurons in the rat brain. *Neuropsychopharmacology.* 2008;34:651–661.
150. Piercey MF, Hoffman WE, Smith MW, et al. Inhibition of dopamine neuron firing by pramipexole, a dopamine D3 receptor–preferring agonist: comparison to other dopamine receptor agonists. *Eur J Pharmacol.* 1996;312:35–44.
151. Corrigan MH, Denahan AQ, Wright E, et al. Comparison of pramipexole, fluoxetine, and placebo in patients with major depression. *Depress Anxiety.* 2000;11:58–65.
152. Zarate CA Jr, Payne JL, Singh J, et al. Pramipexole for bipolar II depression: a placebo-controlled proof of concept study. *Biol Psychiatry.* 2004;56:54–60.
153. Goldberg JF, Burdick KE, Endick CJ. Preliminary randomized, double-blind, placebo-controlled trial of pramipexole added to mood stabilizers for treatment-resistant bipolar depression. *Am J Psychiatry.* 2004;161:564–566.
154. Cassano P, Lattanzi L, Fava M, et al. Ropinirole in treatment-resistant depression: a 16-week pilot study. *Can J Psychiatry.* 2005;50:357–360.
155. Berman RM, Marcus RN, Swanink R, et al. The efficacy and safety of aripiprazole as adjunctive therapy in major depressive disorder: a multicenter, randomized, double-blind, placebo-controlled study. *J Clin Psychiatry.* 2007;68:843–853.
156. Montgomery SA. The under-recognized role of dopamine in the treatment of major depressive disorder. *Int Clin Psychopharmacol.* 2008;23:63–69.

157. Blier P, Szabo ST. Potential mechanisms of action of atypical antipsychotic medications in treatment-resistant depression and anxiety. *J Clin Psychiatry*. 2005;66(suppl 8):30–40.
158. Richtand NM, Welge JA, Logue AD, et al. Dopamine and serotonin receptor binding and antipsychotic efficacy. *Neuropsychopharmacology*. 2007;32:1715–1726.
159. Szabo ST, Blier P. Effects of serotonin (5-hydroxytryptamine, 5-HT) reuptake inhibition plus 5-HT(2A) receptor antagonism on the firing activity of norepinephrine neurons. *J Pharmacol Exp Ther*. 2002;302:983–991.
160. Burris KD, Molski TF, Xu C, et al. Aripiprazole, a novel antipsychotic, is a high-affinity partial agonist at human dopamine D2 receptors. *J Pharmacol Exp Ther*. 2002;302:381–389.
161. McIntyre RS, Soczynska JK, Woldeyohannes HO, et al. A preclinical and clinical rationale for quetiapine in mood syndromes. *Curr Med Res Opin*. 2007;8:1211–1219.
162. McTavish SFB, Mannie ZN, Cowen PJ. Tyrosine depletion does not cause depressive relapse in antidepressant-treated patients. *Psychopharmacology*. 2004;175:124–126.
163. Roiser JP, McLean A, Ogilvie AD, et al. The subjective and cognitive effects of acute phenylalanine and tyrosine depletion in patients recovered from depression. *Neuropsychopharmacology*. 2005;30:755–785.
164. Miller HL, Delgado PL, Salomon RM, et al. Clinical and biochemical effects of catecholamine depletion on antidepressant-induced remission of depression. *Arch Gen Psychiatry*. 1996;53:117–128.
165. Ohnishi T, Hayashi T, Okabe S, et al. Endogenous dopamine release induced by repetitive transcranial magnetic stimulation over the primary motor cortex: an [11C]raclopride positron emission tomography study in anesthetized macaque monkeys. *Biol Psychiatry*. 2004;55:484–489.
166. Strafella AP, Paus T, Barrett J, et al. Repetitive transcranial magnetic stimulation of the human prefrontal cortex induces dopamine release in the caudate nucleus. *J Neurosci*. 2001;21:RC157 (1–4).
167. Strafella AP, Paus T, Fraraccio M, et al. Striatal dopamine release induced by repetitive transcranial magnetic stimulation of the human motor cortex. *Brain*. 2003;126:2609–2615.
168. Zangen A, Hyodo K. Transcranial magnetic stimulation induces increases in extracellular levels of dopamine and glutamate in the nucleus accumbens. *Neuroreport*. 2002;13:2401–2405.
169. O'Reardon JP, Solvason HB, Janicak PG, et al. Efficacy and safety of transcranial magnetic stimulation in the acute treatment of major depression: a multisite randomized controlled trial. *Biol Psychiatry*. 2007;62:1208–1216.
170. Mayberg HS, Lozano AM, Voon V, et al. Deep brain stimulation for treatment-resistant depression. *Neuron*. 2005;45:651–60.
171. Schlaepfer TE, Cohen MX, Frick C, et al. Deep brain stimulation to reward circuitry alleviates anhedonia in refractory major depression. *Neuropsychopharmacology*. 2008; 33:368–377.
172. Steinbush HWM. Serotonin-immunoreactive neurons and their projections in the CNS. Classical transmitters and transmitter receptors in the CNS. Part II. In: Bjorklun A, Hokfelt T, Kuhar MJ, eds. *Handbook of Chemical Neuroanatomy*. Amsterdam, the Netherlands: Elsevier; 1984:68–125.
173. Prisco S, Esposito E. Differential effects of acute and chronic fluoxetine administration on the spontaneous activity of dopaminergic neurones in the ventral tegmental area. *Br J Pharmacol*. 1995;116:1923–1931.
174. Di Mascio M, Di Giovanni G, Di Matteo V, et al. Selective serotonin reuptake inhibitors reduce the spontaneous activity of dopaminergic neurons in the ventral tegmental area. *Brain Res Bull*. 1998;46:547–554.
175. Lejeune F, Millan MJ. Induction of burst firing in ventral tegmental area dopaminergic neurons by activation of serotonin (5-HT)1A receptors: WAY 100,635-reversible actions of the highly selective ligands, flesinoxan and S 15535. *Synapse*. 1998;30:172–180.
176. Lejeune F, Millan MJ. Pindolol excites dopaminergic and adrenergic neurons, and inhibits serotonergic neurons, by activation of 5-HT1A receptors. *Eur J Neurosci*. 2000;12:3265–3275.
177. Bortolozzi A, Diaz-Mataix L, Scorza MC, et al. The activation of 5-HT receptors in prefrontal cortex enhances dopaminergic activity. *J Neurochem*. 2005;95:1597–1607.
178. Diaz-Mataix L, Scorza MC, Bortolozzi A, et al. Involvement of 5-HT1A receptors in prefrontal cortex in the modulation of dopaminergic activity: role in atypical antipsychotic action. *J Neurosci*. 2005;25:10831–10843.
179. Ichikawa J, Ishii H, Bonaccorso S, et al. 5-HT(2A) and D(2) receptor blockade increases cortical DA release via 5-HT(1A) receptor activation: a possible mechanism of atypical antipsychotic-induced cortical dopamine release. *J Neurochem*. 2001;76:1521–1531.
180. Dremencov E, Newman ME, Kinor N, et al. Hyperfunctionality of serotonin-2C receptor-mediated inhibition of accumbal dopamine release in an animal model of depression is reversed by antidepressant treatment. *Neuropharmacology*. 2005;48:34–42.
181. Di Matteo V, De Blasi A, Di Giulio CD, et al. Role of 5-HT2C receptors in control of central dopamine function. *Trends Pharmacol Sci*. 2001;22:229–232.
182. Ichikawa J, Meltzer HY. Relationship between dopaminergic and serotonergic neuronal activity in the frontal cortex and the action of typical and atypical antipsychotic drugs. *Eur Arch Psychiatry Clin Neurosci*. 1999;249(suppl. 4):90–98.
183. Kennett GA, Lightowler S, de Biasi V, et al. Effect of chronic administration of selective 5-hydroxytryptamine and noradrenaline uptake inhibitors on a putative index of 5-HT2C/2B receptor function. *Neuropharmacology*. 1994;33:1581–1588.
184. Ni YG, Miledi R. Blockage of 5HT2C serotonin receptors by fluoxetine (Prozac). *Proc Natl Acad Sci USA*. 1997;94:2036–2040.
185. Millan MJ, Gobert A, Lejeune F, et al. The novel melatonin agonist agomelatine (S20098) is an antagonist at 5-hydroxytryptamine2C receptors, blockade of which enhances the activity of frontocortical dopaminergic and adrenergic pathways. *J Pharmacol Exp Ther*. 2003;306:954–964.
186. Zhou F-M, Liang Y, Salas R, et al. Corelease of dopamine and serotonin from striatal dopamine terminals. *Neuron*. 2005;46:65–74.
187. Grenhoff J, North RA, Johnson SW. Alpha 1-adrenergic effects on dopamine neurons recorded intracellularly in the rat midbrain slice. *Eur J Neurosci*. 1995;7:1707–1713.
188. Steffensen SC, Svingos AL, Pickel VM, et al. Electrophysiological characterization of GABAergic neurons in the ventral tegmental area. *J Neurosci*. 1998;18:8003–8015.
189. Herve D, Blanc G, Glowinski J, et al. Reduction of dopamine utilization in the prefrontal cortex but not in the nucleus accumbens after selective destruction of noradrenergic fibers innervating the ventral tegmental area in the rat. *Brain Res*. 1982;237:510–516.

190. Lategan AJ, Marien MR, Colpaert FC. Suppression of nigrostriatal and mesolimbic dopamine release in vivo following noradrenaline depletion by DSP-4: a microdialysis study. *Life Sci.* 1992;50:995–999.
191. Gobbi G, Muntoni AL, Gessa GL, et al. Clonidine fails to modify dopaminergic neuronal activity during morphine withdrawal. *Psychopharmacology.* 2001;158:1–6.
192. Yavich L, Lappalainen R, Sirvio J, et al. Alpha2-adrenergic control of dopamine overflow and metabolism in mouse striatum. *Eur J Pharmacol.* 1997;339:113–119.
193. Guiard BP, El Mansari M, Blier P. Cross-talk between dopaminergic and noradrenergic systems in the rat ventral tegmental area, locus ceruleus, and dorsal hippocampus. *Mol Pharmacol.* 2008;74:1–13.
194. Ventura R, Cabib S, Alcaro A, et al. Norepinephrine in the prefrontal cortex is critical for amphetamine-induced reward and mesoaccumbens dopamine release. *J Neurosc.* 2003;23:1879–1885.
195. Shi W-X, Pun CL, Zhang XX, et al. Dual effects of d-amphetamine on dopamine neurons mediated by dopamine and nondopamine receptors. *J Neurosci.* 2000;20:3504–3511.
196. Darracq L, Drouin C, Blanc G, et al. Stimulation of metabotropic but not ionotropic glutamatergic receptors in the nucleus accumbens is required for the D-amphetamine-induced release of functional dopamine. *Neuroscience.* 2001;103:395–403.
197. Guiard BP, El Mansari M, Merali Z, et al. Functional interactions between dopamine, serotonin and norepinephrine neurons: an in-vivo electrophysiological study in rats with monoaminergic lesions. *Int J Neuropsychopharmacol.* 2008;11:625–639.
198. Haj-Dahmane S: D2-like dopamine receptor activation excites rat dorsal raphe 5-HT neurons in vitro. *Eur J Neurosci.* 2001;14:125–134.
199. Deutch AY, Goldstein M, Roth RH. Activation of the locus coeruleus induced by selective stimulation of the ventral tegmental area. *Brain Res.* 1986;363:307–314.
200. Elam M, Clark D, Svensson TH. Electrophysiological effects of the enantiomers of 3-PPP on neurons in the locus coeruleus of the rat. *Neuropharmacology.* 1986;25:1003–1008.
201. Nilsson LK, Schwieler L, Engberg G, et al. Activation of noradrenergic locus coeruleus neurons by clozapine and haloperidol: involvement of glutamatergic mechanisms. *Int J Neuropsychopharmacol.* 2005;8: 329–339.
202. Larisch R, Klimke A, Vosberg H, et al. In vivo evidence for the involvement of dopamine-D2 receptors in striatum and anterior cingulate gyrus in major depression. *Neuroimage.* 1997; 5:251–260.

10.6 | Dopamine Modulation of Forebrain Pathways and the Pathophysiology of Psychiatric Disorders

ANTHONY A. GRACE

The neurotransmitter dopamine (DA) has received substantial attention due to its involvement in a wide array of neurological and psychiatric disorders, ranging from Parkinson's disease to affective disorders and schizophrenia. As a result, this system has been studied extensively at many levels of analysis. This is an exciting time for research into psychiatric disorders and the DA system, as evidenced by the convergence of basic neuroscience and clinical research studies on common pathophysiological targets. Dopamine itself has been described as involved in reward and addiction, in attention and compulsions, in cognition and affect. However, recent studies suggest that the DA system may act to coordinate integration of information via selective potentiation of circuits or pathways. This suggests that DA is acting as a "glue" that holds together plastic relationships among diverse brain structures. This chapter focuses on the system physiology of the DA system in intact animals, how the DA system is regulated, and how dysregulation of this system may contribute to the pathophysiology of major psychiatric disorders.

THE DA NEURON: IDENTIFICATION AND PHYSIOLOGICAL PROPERTIES

Dopamine neurons are medium-sized neurons located in the midbrain. The midbrain DA neuron population is generally divided into three classes based on their projection sites: (1) the nigrostriatal DA system, which is involved in movement; loss of DA neurons in this region underlies Parkinson's disease[1]; (2) the mesolimbic DA system, which projects to limbic structures including the nucleus accumbens, amygdala, hippocampus, and other regions; this system is involved in reward and emotion[2,3]; and (3) the mesocortical system, which projects to frontal cortical regions and has a role in cognitive processes such as executive function.[4] The neurons that comprise the DA system are generally similar in physiology, although there are differences related to the presence of autoreceptors[5,6] and action potential duration.[7] Nonetheless, these neurons exhibit a characteristic action potential shape that allows them to be readily distinguished during electrophysiological recordings. Thus, owing to the large calcium component, the pacemaker potential that drives spike firing, and the dendritic origin of the initial segment,[8,9] the neurons exhibit unique action potential spike trains. This is reflected in the characteristic long-duration action potentials exhibiting a variable shape even when recording from a single neuron, presumably due to variability in the level of pacemaker-current-induced membrane depolarization from which the spike is triggered.[10,11] It is important to note that identification using classical criteria[10,12] depends on open filter settings; improper truncation of action potentials from overfiltering can lead to misidentification of the neurons.[13]

Dopamine neurons exhibit distinct activity states that appear to have functional relevance. First, studies indicate that not all DA neurons are firing spontaneously; at least 50% of DA neurons are nonfiring but can be activated by distinct stimuli.[14] The fact that the neurons depend on pacemaker conductance to drive activity is consistent with the major input to these neurons being inhibitory in nature. In particular, the GABAergic input from the ventral pallidum provides an important source of regulation. The ventral pallidum consists of rapidly firing primarily GABAergic neurons[15]; as a result, it provides a strong inhibitory clamp on its postsynaptic structures, including the DA neurons of the ventral tegmental area (VTA). Indeed, in vivo intracellular recordings from DA neurons show that they are under constant bombardment with high-amplitude GABAergic inhibitory postsynaptic potentials (IPSPs)[16]. This ventral pallidal input appears to be important for controlling the proportion of DA neurons firing spontaneously. Thus, inactivation of the ventral pallidum will cause an increase in the number of DA neurons firing (assessed by passing an electrode through the region in a reproducible pattern and counting the number of active neurons[17]). The propensity of the ventral pallidum to regulate DA neuron activity appears to depend on a multisynaptic circuit originating in the ventral subiculum of the hippocampus[18,19] (Fig. 10.6.1). Thus, the ventral subiculum, via glutamatergic drive of the nucleus accumbens and subsequent accumbens inhibition of the ventral pallidum, controls the population activity of the DA neurons, defined as the number of DA neurons firing spontaneously.[19]

Spontaneously firing DA neurons are driven primarily via a pacemaker conductance, which brings the neurons from their resting membrane potential to their atypically high spike threshold.[11,14,20] When the DA neuron is deprived of its inputs, it will exhibit a very regular pacemaker firing, as observed in vitro[20] or in vivo after blocking DA receptors and administering a GABA-B agonist.[21] Pacemaker firing in DA neurons in vivo is rarely observed. In contrast, the baseline firing condition of DA neurons is one of an irregular single-spiking pattern[14] due to the interaction of the pacemaker drive with bombardment by IPSPs originating from long-loop and local interneurons.[16,22] This baseline activity state has been termed the *tonic* firing state.[19,23]

In addition to single-spike firing, when a spontaneously firing DA neuron is activated via glutamatergic inputs, it will exhibit burst firing.[24] Burst firing appears to be the functionally relevant rapid change in state that signifies a behaviorally relevant event.[25] As such, DA neuron burst firing has been termed the *phasic* DA response.[19,23] Burst firing is not observed in vitro in the basal state, but will be readily driven in vivo by direct application of glutamate to DA neurons[24] or by stimulation of glutamatergic afferents.[19,26] The pedunculopontine tegmentum, in particular, is uniquely effective in driving burst firing in spontaneously firing VTA DA neurons[19] (Fig. 10.6.1), whereas subthalamic nucleus glutamatergic input is particularly effective in driving burst firing in the nigrostriatal DA neuron population.[27] Because burst firing depends on activation of *N*-methyl-D-aspartate (NMDA) receptors,[28,29] the DA neuron must be in a depolarized, spontaneously discharging state for it to exhibit glutamate-mediated

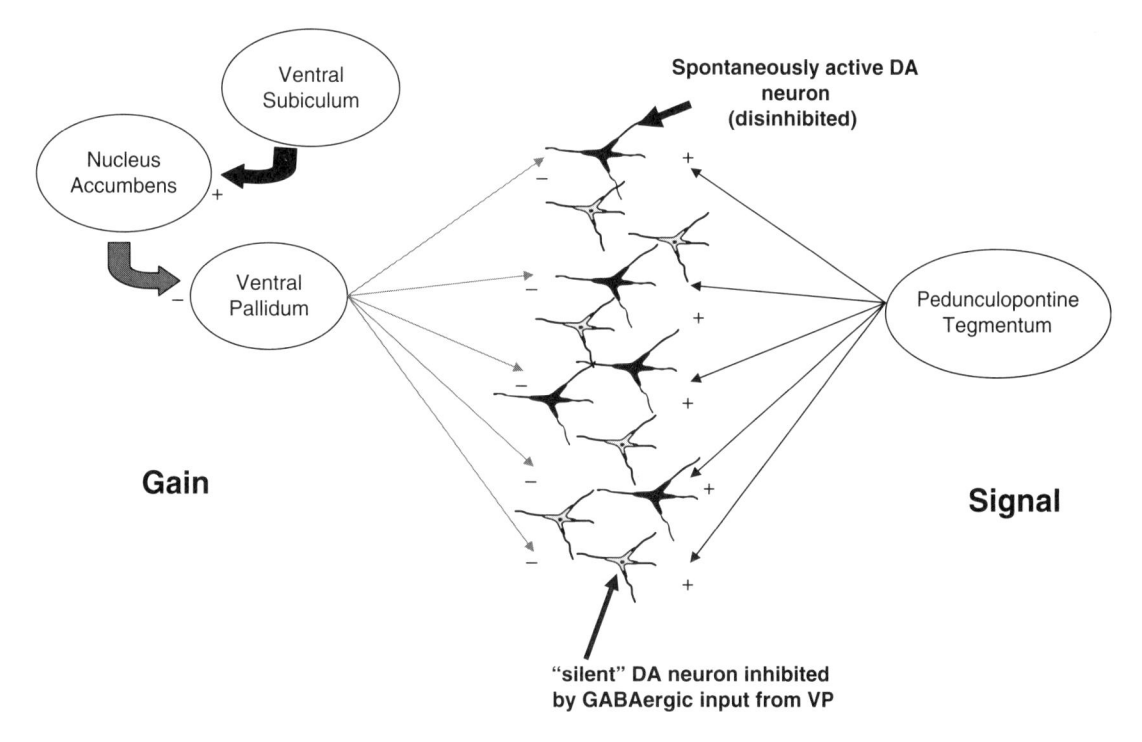

FIGURE 10.6.1. Circuit diagram illustrating the control of tonic and phasic DA neuron activity states. The ventral pallidum holds a proportion of VTA DA neurons in a hyperpolarized, nonfiring state via a potent GABAergic inhibition. However, when the ventral subiculum is activated, it drives the nucleus accumbens, which, in turn, inhibits the ventral pallidum and releases DA neurons from inhibition, causing them to begin to fire spontaneously. In contrast, the pedunculopontine tegmentum provides a phasic glutamatergic input to the DA neurons that causes them to fire in bursts, with burst firing believed to be the behaviorally salient signal coming from the DA neurons. However, only spontaneously firing DA neurons can be made to burst fire by the pedunculopontine tegmentum. Therefore, the ventral subiculum by controlling the number of DA neurons firing spontaneously, determines the proportion of DA neurons that can be made to burst fire by the pedunculopontine tegmentum, thereby setting the *amplification factor* (gain) of the burst firing signal. With more DA neurons firing, glutamate input from the pedunculopontine would cause more DA neurons to burst fire, thereby increasing the amplitude of the phasic DA signal. VP, ventral pallidum.

burst firing[19]; otherwise, the magnesium blockade present in hyperpolarized DA neurons prevents this action.[30] In the VTA, a contribution of another brainstem nucleus, the lateral dorsal tegmentum, is apparently required for burst firing. Thus, inactivation of the lateral dorsal tegmentum will prevent DA neuron burst firing elicited by either direct glutamate application or activation of pedunculopontine glutamate afferents.[31] Thus, the lateral dorsal tegmentum provides a permissive "gate" over burst discharge.

The fact that burst firing can only be driven in spontaneously firing DA neurons reveals a unique role for the ventral subiculum–controlled population activity versus the pedunculopontine glutamate–driven burst firing. By determining the number of DA neurons that are firing, the ventral subiculum is positioned to provide the "gain" in the behaviorally relevant burst firing "signal"[26] (Fig. 10.6.1). Thus, activation of the ventral subiculum increases DA neuron population activity, enabling the pedunculopontine to activate burst firing in a greater number of DA neurons. This is also consistent with the role of the pedunculopontine region, which is reported to control conditioned responses in DA neurons.[32]

ROLE OF THE VENTRAL SUBICULUM IN REGULATING DA NEURON ACTIVITY

The ventral subiculum is the limbic output of the hippocampus, with projections to the nucleus accumbens, prefrontal cortex, and other limbic regions.[33,34] Studies suggest that the ventral subiculum is involved in context-related processes. This context-related gating is believed to underlie the bistable "up" states in the nucleus accumbens that are driven by the ventral subiculum.[35] Context-dependent gating is essential in guiding the correct response to stimuli. Thus, the response that a DA neuron would emit would be substantially different in a contextual environment that is highly rewarding, highly threatening, or benign. By controlling the amplitude of the DA neuron phasic response, the ventral subiculum is positioned to determine the level of responsivity that can be assigned to a given condition. Indeed, the subiculum receives a potent excitatory drive from the amygdala and the locus coeruleus,[36] regions that are known to be involved in modulating responses to stress and fear.

Of course, it would not be behaviorally effective to maintain an activated state; instead, the responsivity of the DA system should be modulated in a manner that is consistent with a given context. Thus, the ventral subiculum should be positioned to provide bidirectional modulation of the DA system. In a context in which stimuli are highly salient for optimal responding, such as in a highly rewarding or highly threatening environment, the ventral subiculum should be activated in order to increase the responsivity of the DA system, thereby assigning high motivational salience to all stimuli.[37,38] However, if the organism is in a benign environment, responsivity should be lower so that resources are not utilized to provide inappropriately high levels of activation for each nonsalient stimulus that arises.

THE HIPPOCAMPUS AND THE PATHOPHYSIOLOGY OF SCHIZOPHRENIA

Although, as mentioned above, the ventral subiculum should be capable of adjusting the DA system based on the context rather than via continuous activation, this overactive state appears to be precisely the condition that may arise in disorders in which context is a variable. Thus, schizophrenia patients are known to exhibit deficits in context-dependent processing. They respond to stimuli in a manner that is inappropriate or ineffective for the given context, as well as showing an inability to respond selectively to highly salient versus nonsalient stimuli. Moreover, schizophrenia patients exhibit an abnormally heightened response to DA-releasing drugs such as amphetamine. Thus, amphetamine can mimic psychosis in normal individuals[39] and will exacerbate the psychotic symptoms of schizophrenia. Positron emission tomography (PET) imaging studies show that schizophrenia subjects exhibit greater raclopride displacement (indicative of greater DA release) that is proportional to the exacerbation of the psychosis by this drug.[40] Therefore, the DA system appears to be hyperresponsive in the schizophrenia patient. Studies show that psychosis in schizophrenia patients may be correlated with hippocampal activity,[41] and numerous recent studies have shown that there is hyperactivity in the anterior hippocampus of schizophrenia patients,[42–47] which is functionally analogous to the ventral subiculum in rats. Finally, anatomical studies have demonstrated a loss of parvalbumin interneuron staining in the hippocampus of schizophrenia patients,[48] which could potentially be a factor in the hyperactivity in this region.

In fact, a similar condition was observed in a rat developmental model of schizophrenia. We have shown in a model developed in our laboratory,[49,50] which has been replicated by others,[51–53] that rats given a mitotoxin, methylazoxymethanol acetate (MAM), during gestational day 17 exhibit deficits consistent with schizophrenia in humans, including thinning of homologous limbic cortical structures without substantial neuronal loss, deficits in prepulse inhibition of startle, deficits in social and executive function, increased responsivity to phencyclidine, and increased locomotor response to amphetamine postpubertally but not prepubertally. Electrophysiological analyses showed that MAM-treated rats examined as adults exhibit a dramatic increase in the number of DA neurons firing.[54] Moreover, this increase in DA population activity occurs in concert with increases in ventral subicular neuronal firing. Furthermore, inactivation of the ventral subiculum restores DA neuron population activity to control levels and eliminates the behavioral hyperresponsivity to amphetamine[54] (Fig. 10.6.2).

Histological examination revealed that the MAM-treated rats exhibit decreased parvalbumin interneuron staining in both the hippocampus subiculum and the medial prefrontal cortex.[55] Interestingly, this corresponds to the same regions that exhibit a deficit in conditioned stimulus–evoked gamma oscillations,[55] a phenomenon that is also characteristic of schizophrenia[56–59] and that is believed to be dependent on parvalbumin interneurons.[60–63] Taken together, these studies suggest that in both the MAM-treated rat model of schizophrenia and the human schizophrenia patient, a deficit in parvalbumin interneurons results in hyperactivity within the limbic region of the hippocampus, leading to increased DA neuron responsivity and psychosis. Therefore, the "gain" of the DA system is always set to maximum, causing all stimuli to be judged as highly salient and causing a continuous overdrive of the attentional system. It also suggests that a better therapeutic method for treating schizophrenia would not be to block what otherwise appears to be a normal DA system. Instead, a more effective approach would be restoration of ventral subicular function by increasing the functionality of the parvalbumin interneuron class, thereby reversing the origin of the deficit proposed to underlie this disorder.

SCHIZOPHRENIA RISK FACTORS EXHIBIT COMMON PATHOPHYSIOLOGICAL INDICES

Several factors are known to increase the risk for schizophrenia, as well as cause exacerbation of psychosis. Interestingly, two of these factors—stress and drug abuse—are also context-dependent phenomena that activate the DA system. Furthermore, both repeated administration of amphetamine[64,65] and exposure to

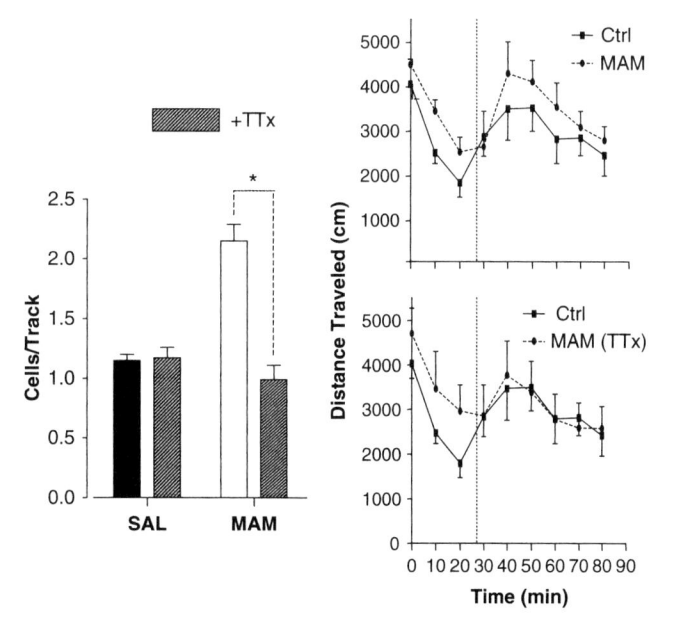

FIGURE 10.6.2. Hyperactivity in the ventral subiculum of the hippocampus drives the hyperdopaminergic state in an animal model of schizophrenia. Methylaxozymethanol (MAM)-treated rats exhibit a greater number of DA neurons firing spontaneously (left) as well as a greater locomotor response to amphetamine administration (right, top). This is believed to be due to an overdrive from the ventral subiculum, causing an abnormally large number of DA neurons to fire spontaneously and thereby increasing the amplitude of the phasic signal. If the ventral subiculum is inactivated by injection of tetrodotoxin (TTx), the number of DA neurons firing returns to control levels (left), and the heightened behavioral response to amphetamine is brought back to control levels (right, bottom). Therefore, inactivation of the hyperactive ventral subiculum reverses the hyperdopaminergic state present in the MAM rat. Saline (SAL). *Source:* Adapted from [54].

stressful stimuli[66–68] are known to lead to hyperresponsivity of the DA system in terms of exacerbating the locomotor response to acute amphetamine challenge similar to that observed in schizophrenia. Our studies show that this may be due to common actions within the context-dependent circuitry of the ventral subiculum. Thus, we found that repeated cocaine administration will induce long-term potentiation within the ventral subiculum–nucleus accumbens circuitry.[69] Moreover, repeated amphetamine also activates this pathway, but in this case does so by increasing ventral subicular neuronal activity.[70] This is consistent with recent data showing that cocaine sensitization occurs with changes in AMPA receptor surface expression consistent with long-term potentiation (LTP), but amphetamine sensitization does not.[71,72] Thus, both cocaine sensitization and amphetamine sensitization appear to lead to increased ventral subicular–nucleus accumbens drive. Moreover, this sensitization is known to be context-dependent, in that the sensitized response is greatest when tested in the same environment in which the drug was administered.[73–75]

How does the increased ventral subiculum–nucleus accumbens drive affect the DA system? Our studies show that following 1 week of amphetamine administration and 1 week of withdrawal, there is an increase in the number of DA neurons firing.[70] Moreover, inactivation of the ventral subiculum restores DA neuron population activity to control levels while eliminating the sensitized response to amphetamine. Therefore, stimulants appear to cause hyperresponsivity in the DA system in a manner analogous to that observed in schizophrenia; however, since the disruption is one of activity levels rather than interneuron-dependent modulation of rhythmicity, the effects seem to be more limited in scope.

Stress is also a known risk factor in both the onset and the exacerbation of psychosis.[76,77] Stress, also a context-dependent phenomenon, cross-sensitizes with amphetamine.[68] We have found that several stressors, including maintained footshock as well as restraint stress, will cause an increase in the number of DA neurons firing spontaneously. Moreover, inactivation of the ventral subiculum reverses this sensitized state.[78] This finding is consistent with studies identifying the ventral subiculum as playing an essential role in mediating stress responsivity.[79] Indeed, the potent activation of the ventral subiculum by the amygdala and the locus coeruleus,[36] both stress-related structures, positions the ventral subiculum to mediate both normal and abnormal modulation of DA system responsivity.

POSTSYNAPTIC ACTIONS OF DA AND ITS RELEVANCE TO PSYCHIATRIC DISORDERS

Therefore, in schizophrenia, drug abuse, and stress, increased responsivity of the DA system in terms of the number of neurons that can be activated into a burst-firing mode by the presentation of a stimulus is abnormally high. But how does this translate into alterations in goal-directed behavior? This depends strongly on how DA acts on its postsynaptic targets. Within the prefrontal cortex, data indicate that DA is essential in working memory and cognition, factors that are discussed extensively in other chapters in this volume.

Dopamine also plays a substantial role in gating information flow within limbic circuits. It has been proposed that DA actions within the hippocampus, and the feedback loop from the hippocampus to the DA neuron, are involved in gating the incorporation of information into memory processes.[80] Within the nucleus accumbens, the DA system has a complex role in modulating goal-directed behavior. The nucleus accumbens is a crossroads for information flow within the limbic system; it receives glutamatergic inputs from the prefrontal cortex, the hippocampus, and the amygdala. Moreover, each of these afferent systems is involved in different aspects of behavioral regulation. Thus, the prefrontal cortex is involved in behavioral flexibility, with disruptions in this area causing perseverative behavior.[81,82] In contrast, the hippocampus subicular input is involved in context-dependent processing and therefore has the opposite function: keeping the subject on task. Moreover, the DA system affects each of these afferents in an opposite manner, with phasic DA release causing D1-mediated potentiation of the hippocampal subicular input and tonic DA causing D2-mediated attenuation of the prefrontal cortical input.[82] It is proposed that these systems act in balance, with the ventral subiculum keeping the subject focused on a task and the prefrontal cortex allowing the subject to break out from the task and seek out another strategy when the task is ineffective.[23,83] This is further modulated by the DA system. Therefore, if a task is effective at achieving a goal, the reinforcing properties of the outcome would activate the DA system, leading to potentiation of the hippocampal subicular input (which would further reinforce keeping on task) and attenuation of the prefrontal cortical input (which would prevent switching strategies). However, if there is a mismatch between expectation and outcome, indicative of a wrong choice, studies show that there is an interruption in the DA signal.[19,83] As a consequence, the prefrontal cortical input would be disinhibited and the hippocampal subicular input attenuated, thus favoring

Conditioning

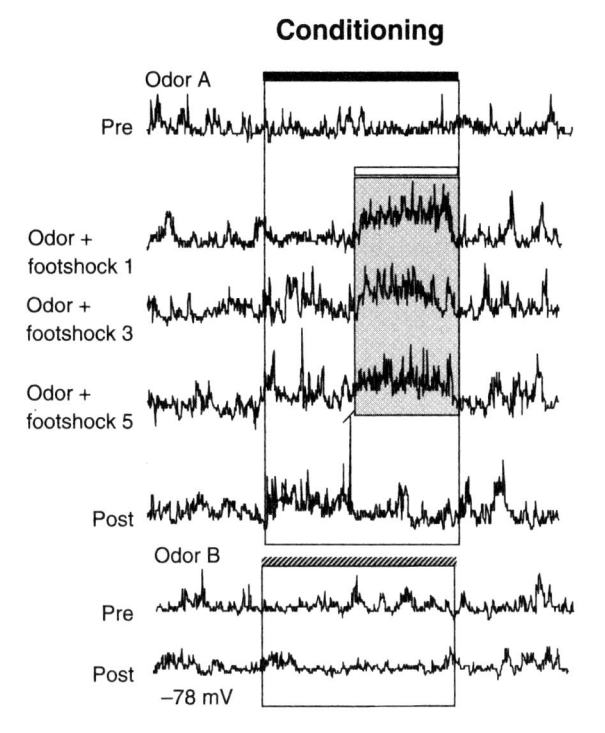

[Figure showing intracellular recording traces labeled: Odor A (Pre, Odor + footshock 1, Odor + footshock 3, Odor + footshock 5, Post), Odor B (Pre, Post), with −78 mV reference.]

FIGURE 10.6.3. Conditioning in the basolateral amygdala. During in vivo intracellular recordings, exposure of a rat to an odor (Odor A) does not alter baseline activity substantially. In contrast, a footshock (open bar) will activate amygala neuron firing. When the odor is paired with the footshock (traces 2–4), the activation is shifted to the onset of the odor. After five pairings, presentation of the conditioned odor alone (Post) causes activation of the basolateral amygdala neuron. In contrast, presentation of a different odor (Odor B) before and after conditioning to Odor A fails to alter the activity of the basolateral amygdala neuron. *Source*: Adapted from [85].

switching strategies. On the other hand, if the DA system is overdriven, the subject would not be capable of showing adequate behavioral flexibility, and instead would continue to respond in a manner that is ineffective at achieving the goal. Such a condition would lead to perseveration and disruption of executive function.

Another site at which the DA system is positioned to regulate behavior is within the amygdala. The amygdala is a region involved in the expression of emotion.[84] It also plays a major role in emotional learning. Thus, when a neutral stimulus is paired with a noxious event, the basolateral amygdala complex is involved in establishing an association between these inputs. Indeed, this learning is expressed even at the level of individual basolateral amygdala neurons. Thus, exposing a rat to a neutral stimulus such as an odor causes a small activation of basolateral amygdala neurons. Repeated exposure to the same stimulus eventually causes the response to habituate, with odor-evoked activity gradually decreasing in amplitude. On the other hand, if the odor is paired with a highly activating stimulus such as a footshock, which by itself causes profound activation of the amygdala, after several pairings the odor presented by itself will cause a powerful activation of the basolateral amygdala neurons[85] (Fig. 10.6.3). This association is dependent on an input from the DA system; if the DA system is blocked, this emotional learning fails to occur. In contrast, if the DA system is overstimulated (e.g., with amphetamine), the amygdala begins to respond to stimuli other than those associated with the noxious event (J.A. Rosenkranz and A.A. Grace, in preparation). Emotional responses that occur to what otherwise should be benign events could be an example of paranoid responses and may play into the subjective emotional experience of schizophrenia patients.

SUMMARY

Dopamine clearly has multifaceted actions throughout the central nervous system, playing a central role in enabling associations between stimuli to occur and to gate behavioral output based on external feedback. Therefore, it may be incorrect to suggest that the DA system "does" anything—reward, executive function, decision making, goal-directed behaviors, and so on. Instead, the DA system appears positioned to alter the relationships among structures involved in performing a task. Furthermore, these modulatory actions are based on the multifaceted way in which the DA system can respond, including DA neuron population activity and burst firing,[23] tonic-phasic DA regulation within postsynaptic structures,[86,87] and the impact of DA on synaptic plasticity within the limbic circuits.[69,85] By understanding the factors that modulate the DA neuron, and how dysfunctions within the circuitry can impact DA neuron activity, we may be better prepared to treat disorders more effectively, not by directly changing DA transmission, but instead by restoring activity within the structures that modulate DA neuron function.

REFERENCES

1. Hornykiewicz O. [Dopamine (3-hydroxytyramine) in the central nervous system and its relation to the Parkinson syndrome in man.]. *D Med Wochenschr.* 1946/1962;87:1807–1810.
2. Mogenson GJ, Brudzynski SM, Wu M, Yang CR, Yim CCY. (1993) From motivation to action: a review of dopaminergic regulation of limbic -> nucleus accumbens -> ventral pallidum -> pedunculopontine nucleus circuitries involved in limbic-motor integration. In: Kalivas PW, Barnes CW, eds. *Limbic Motor Circuits and Neuropsychiatry.* Boca Raton, FL: CRC Press; 1993:193–236.
3. Mogenson GJ, Jones DL, Yim CY. From motivation to action: functional interface between the limbic system and the motor system. *Prog Neurobiol.* 1980;14:69–97.
4. Goldman-Rakic PS. The cortical dopamine system: role in memory and cognition. *Adv Pharmacol.* 1998;42:707–711.
5. Bannon MJ, Michaud RL, Roth RH. Mesocortical dopamine neurons. Lack of autoreceptors modulating dopamine synthesis. *Mol Pharmacol.* 1981;19:270–275.
6. Chiodo LA, Bannon MJ, Grace AA, Roth RH, Bunney BS. Evidence for the absence of impulse-regulating somatodendritic and synthesis-modulating nerve terminal autoreceptors on subpopulations of mesocortical dopamine neurons. *Neuroscience.* 1984;12:1–16.
7. Lammel S, Hetzel A, Hackel O, Jones I, Liss B, Roeper J. Unique properties of mesoprefrontal neurons within a dual mesocorticolimbic dopamine system. *Neuron.* 2008;57:760–773.
8. Grace AA. Evidence for the functional compartmentalization of spike generating regions of rat midbrain dopamine neurons recorded in vitro. *Brain Res.* 1990;524:31–41.
9. Grace AA. Regulation of spontaneous activity and oscillatory spike firing in rat midbrain dopamine neurons recorded in vitro. *Synapse.* 1991;7:221–234.
10. Grace AA, Bunney BS. Intracellular and extracellular electrophysiology of nigral dopaminergic neurons—1. Identification and characterization. *Neuroscience.* 1983;10:301–315.
11. Grace AA, Bunney BS. Intracellular and extracellular electrophysiology of nigral dopaminergic neurons—2. Action potential generating mechanisms and morphological correlates. *Neuroscience.* 1983;10:317–331.
12. Grace AA, Bunney BS. Nigral dopamine neurons: intracellular recording and identification with L-dopa injection and histofluorescence. *Science.* 1980;210:654–656.
13. Ungless MA, Magill PJ, Bolam JP. Uniform inhibition of dopamine neurons in the ventral tegmental area by aversive stimuli. *Science.* 2004;303:2040–2042.
14. Grace AA, Bunney BS. The control of firing pattern in nigral dopamine neurons: single spike firing. *J Neurosci.* 1984;4:2866–2876.
15. Tsai CT, Mogenson GJ, Wu M, Yang CR. A comparison of the effects of electrical stimulation of the amygdala and hippocampus on subpallidal output neurons to the pedunculopontine nucleus. *Brain Res.* 1989;494:22–29.
16. Grace AA, Bunney BS. Opposing effects of striatonigral feedback pathways on midbrain dopamine cell activity. *Brain Res.* 1985;333:271–284.
17. Bunney BS, Grace AA. Acute and chronic haloperidol treatment: comparison of effects on nigral dopaminergic cell activity. *Life Sci.* 1978;23:1715–1727.
18. Floresco SB, Todd CL, Grace AA. Glutamatergic afferents from the hippocampus to the nucleus accumbens regulate activity of ventral tegmental area dopamine neurons. *J Neurosci.* 2001;21:4915–4922.
19. Floresco SB, West AR, Ash B, Moore H, Grace AA. Afferent modulation of dopamine neuron firing differentially regulates tonic and phasic dopamine transmission. *Nat Neurosci.* 2003;6:968–973.
20. Grace AA, Onn SP. Morphology and electrophysiological properties of immunocytochemically identified rat dopamine neurons recorded in vitro. *J Neurosci.* 1989;9:3463–3481.
21. Grace AA, Bunney BS. Effects of baclofen on nigral dopaminergic cell activity following acute and chronic haloperidol treattment. *Brain Res Bull.* 1980;5:537–543.
22. Grace AA, Bunney BS. Paradoxical GABA excitation of nigral dopaminergic cells: indirect mediation through reticulata inhibitory neurons. *Eur J Pharmacol.* 1979; 59:211–218.
23. Grace AA, Floresco SB, Goto Y, Lodge DJ. Regulation of firing of dopaminergic neurons and control of goal-directed behaviors. *Trends Neurosci.* 2007;30:220–227.
24. Grace AA, Bunney BS. The control of firing pattern in nigral dopamine neurons: burst firing. *J Neurosci.* 1984;4:2877–2890.
25. Schultz W. The phasic reward signal of primate dopamine neurons. *Adv Pharmacol.* 1998;42:686–690.
26. Lodge DJ, Grace AA. The hippocampus modulates dopamine neuron responsivity by regulating the intensity of phasic neuron activation. *Neuopsychopharmacology.* 2006; 31:1356–1361.
27. Smith ID, Grace AA. Role of the subthalamic nucleus in the regulation of nigral dopamine neuron activity. *Synapse.* 1992;12:287–303.
28. Chergui K, Charlety PJ, Akaoka H, Saunier CF, Brunet JL, Buda M, Svensson TH, Chouvet G. Tonic activation of NMDA receptors causes spontaneous burst discharge of rat midbrain dopamine neurons in vivo. *Eur J Neurosci.* 1993;5: 137–144.
29. Overton P, Clark D. Iontophoretically administered drugs acting at the N-methyl-D-aspartate receptor modulate burst firing in A9 dopamine neurons in the rat. *Synapse.* 1992;10:131–140.
30. Mayer ML, Westbrook GL, Guthrie PB. Voltage-dependent block by Mg^{2+} of NMDA responses in spinal cord neurones. *Nature.* 1984;309:261–263.
31. Lodge DJ, Grace AA. The laterodorsal tegmentum is essential for burst firing of ventral tegmental area dopamine neurons. *Proc Natl Acad Sci USA.* 2006;103:5167–5172.
32. Pan WX, Hyland B. Pedunculopontine tegmental nucleus controls conditioned responses of midbrain dopamine neurons in behaving rats. *J Neurosci.* (online) 2005;25:4725–4732.

33. Witter MP, Groenewegen HJ. The subiculum: cytoarchitectonically a simple structure, but hodologically complex. *Prog Brain Res.* 1990;83:47–58.
34. O'Mara S. The subiculum: what it does, what it might do, and what neuroanatomy has yet to tell us. *J Anat.* 2005;207:271–282.
35. O'Donnell P, Grace AA. Synaptic interactions among excitatory afferents to nucleus accumbens neurons: hippocampal gating of prefrontal cortical input. *J Neurosci.* 1995;15:3622–3639.
36. Lipski WJ, Grace AA. Neurons in the ventral subiculum are activated by noxious stimuli and are modulated by noradrenergic afferents. *Program No 1951, 2008 Neuroscience Meeting Planner.* Washington, DC: Society for Neuroscience; 2008 Online.
37. Kapur S. Psychosis as a state of aberrant salience: a framework linking biology, phenomenology, and pharmacology in schizophrenia. *Am J Psychiatry.* 2003;160:13–23.
38. Berridge KC, Robinson TE. What is the role of dopamine in reward: hedonic impact, reward learning, or incentive salience? *Brain Res Brain Res Rev.* 1998;28:309–369.
39. Angrist B, Sathananthan G, Wilk S, Gershon S. Amphetamine psychosis: behavioral and biochemical aspects. *J Psychiatr Res.* 1974;11:13–23.
40. Laruelle M, Abi-Dargham A, Van Dyck CH, Gil R, D'Souza CD, Erdos J, McCance E, Rosenblatt W, Fingado C, Zoghbi SS, Baldwin RM, Seibyl JP, Krystal JH, Charney DS, Innis RB. Single photon emission computerize d tomography imaging of amphetamine-induced dopamine release in drug-free schizophrenic subjects. *Proc Natl Acad Sci USA.* 1996;93:9235–9240.
41. Silbersweig DA, Stern E, Frith C, Cahill C, Holmes A, Grootoonk S, Seaward J, McKenna P, Chua SE, Schnorr L, et al. A functional neuroanatomy of hallucinations in schizophrenia. *Nature.* 1995;378:176–179.
42. Heckers S, Rauch SL, Goff D, Savage CR, Schacter DL, Fischman AJ, Alpert NM. Impaired recruitment of the hippocampus during conscious recollection in schizophrenia. *Nat Neurosci.* 1998;1:318–323.
43. Lahti AC, Weiler MA, Holcomb HH, Tamminga CA, Carpenter WT, McMahon R. Correlations between rCBF and symptoms in two independent cohorts of drug-free patients with schizophrenia. *Neuropsychopharmacology.* 2006;31:221–230.
44. Medoff DR, Holcomb HH, Lahti AC, Tamminga CA. Probing the human hippocampus using rCBF: contrasts in schizophrenia. *Hippocampus.* 2001;11:543–550.
45. Meyer-Lindenberg A, Poline JB, Kohn PD, Holt JL, Egan MF, Weinberger DR, Berman KF. Evidence for abnormal cortical functional connectivity during working memory in schizophrenia. *Am J Psychiatry.* 2001;158:1809–1817.
46. Weiss AP, Goff D, Schacter DL, Ditman T, Freudenreich O, Henderson D, Heckers S. Fronto-hippocampal function during temporal context monitoring in schizophrenia. *Biol Psychiatry.* 2006;60:1268–1277.
47. Malaspina D, Storer S, Furman V, Esser P, Printz D, Berman A, Lignelli A, Gorman J, Van Heertum R. SPECT study of visual fixation in schizophrenia and comparison subjects. *Biol Psychiatry.* 1999;46:89–93.
48. Lewis DA, Hashimoto T, Volk DW. Cortical inhibitory neurons and schizophrenia. *Nat Rev Neurosci.* 2005;6:312–324.
49. Grace AA, Moore H. Regulation of information flow in the nucleus accumbens: a model for the pathophysiology of schizophrenia. In: Lenzenweger MF, Dworkin RH, eds. *Origins and Development of Schizophrenia: Advances in Experimental Psychopathology.* Washington, DC: American Psychological Association Press; 1998:123–157.
50. Moore H, Jentsch JD, Ghajarnia M, Geyer MA, Grace AA. A neurobehavioral systems analysis of adult rats exposed to methylazoxymethanol acetate on E17: implications for the neuropathology of schizophrenia. *Biol Psychiatry.* 2006;60:253–264.
51. Gourevitch R, Rocher C, Le Pen G, Krebs MO, Jay TM. Working memory deficits in adult rats after prenatal disruption of neurogenesis. *Behav Pharmacol.* 2004;15:287–292.
52. Le Pen G, Gourevitch R, Hazane F, Hoareau C, Jay TM, Krebs MO. Peri-pubertal maturation after developmental disturbance: a model for psychosis onset in the rat. *Neuroscience.* 2006;143:395–405.
53. Flagstad P, Mork A, Glenthoj BY, van Beek J, Michael-Titus AT, Didriksen M. Disruption of neurogenesis on gestational day 17 in the rat causes behavioral changes relevant to positive and negative schizophrenia symptoms and alters amphetamine-induced dopamine release in nucleus accumbens. *Neuropsychopharmacology.* 2004;29:2052–2064.
54. Lodge DJ, Grace AA. Aberrant hippocampal activity underlies the dopamine dysregulation in an animal model of schizophrenia. *J Neurosci.* 2007;27:11424–11430.
55. Lodge DJ, Behrens MM, Grace AA. A loss of parvalbumin-containing interneurons is associated with diminished oscillatory activity in an animal model of schizophrenia. *J Neurosci.* 2009;29:2344–2354.
56. Gonzalez-Hernandez JA, Cedeno I, Pita-Alcorta C, Galan L, Aubert E, Figueredo-Rodriguez P. Induced oscillations and the distributed cortical sources during the Wisconsin Card Sorting Test performance in schizophrenic patients: new clues to neural connectivity. *Int J Psychophysiol.* 2003;48:11–24.
57. Schmiedt C, Brand A, Hildebrandt H, Basar-Eroglu C. Event-related theta oscillations during working memory tasks in patients with schizophrenia and healthy controls. *Brain Res Cogn Brain Res.* 2005;25:936–947.
58. Cho RY, Konecky RO, Carter CS. Impairments in frontal cortical gamma synchrony and cognitive control in schizophrenia. *Proc Natl Acad Sci USA.* 2006;103:19878–19883.
59. Basar-Eroglu C, Brand A, Hildebrandt H, Karolina Kedzior K, Mathes B, Schmiedt C. Working memory related gamma oscillations in schizophrenia patients. *Int J Psychophysiol.* 2007;64:39–45.
60. Vida I, Bartos M, Jonas P. Shunting inhibition improves robustness of gamma oscillations in hippocampal interneuron networks by homogenizing firing rates. *Neuron.* 2006;49:107–117.
61. Bartos M, Vida I, Jonas P. Synaptic mechanisms of synchronized gamma oscillations in inhibitory interneuron networks. *Nat Rev Neurosci.* 2007;8:45–56.
62. Fuchs EC, Zivkovic AR, Cunningham MO, Middleton S, Lebeau FE, Bannerman DM, Rozov A, Whittington MA, Traub RD, Rawlins JN, Monyer H. Recruitment of parvalbumin-positive interneurons determines hippocampal function and associated behavior. *Neuron.* 2007;53:591–604.
63. Tukker JJ, Fuentealba P, Hartwich K, Somogyi P, Klausberger T. Cell type–specific tuning of hippocampal interneuron firing during gamma oscillations in vivo. *J Neurosci.* 2007;27:8184–8189.
64. Segal DS, Mandell AJ. Long-term administration of d-amphetamine: progressive augmentation of motor activity and stereotypy. *Pharmacol Biochem Behav.* 1974;2:249–255.
65. Post RM, Rose H. Increasing effects of repetitive cocaine administration in the rat. *Nature.* 1976;260:731–732.

66. Piazza PV, Le Moal M. The role of stress in drug self-administration. *Trends Pharmacol Sci.* 1998;19:67–74.
67. Antelman SM. Stressor-induced sensitization to subsequent stress: implications for the development and treatment of clinical disorders. In: Kalivas PW, Barnes CD, eds. *Sensitization in the Nervous System.* Caldwell, NJ: Telford Press; 1988:227–254.
68. Antelman SM, Eichler AJ, Black CA, Kocan D. Interchangeability of stress and amphetamine in sensitization. *Science.* 1980;207:329–331.
69. Goto Y, Grace AA. Dopamine-dependent interactions between limbic and prefrontal cortical plasticity in the nucleus accumbens: disruption by cocaine sensitization. *Neuron.* 2005;47:255–266.
70. Lodge DJ, Grace AA. Amphetamine activation of hippocampal drive of mesolimbic dopamine neurons: a mechanism of behavioral sensitization. *J Neurosci.* 2008;28:7876–7882.
71. Nelson CL, Milovanovic M, Wetter JB, Ford KA. Wolf ME Behavioral sensitization to amphetamine is not accompanied by changes in glutamate receptor surface expression in the rat nucleus accumbens. *J Neurochem.* 2009;109:35–51.
72. Tucker DC, Campioni M, Bubula N, Suto N, McGehee DS, Vezina P. Previous exposure to amphetamine functionally up-regulates nucleus accumbens glutamate transmission without changing basal AMPA receptor surface expression. *Program No 3588, 2008 Neuroscience Meeting Planner.* Washington, DC: Society for Neuroscience; 2008 Online.
73. Badiani A, Browman KE, Robinson TE. Influence of novel versus home environments on sensitization to the psychomotor stimulant effects of cocaine and amphetamine. *Brain Res.* 1995;674:291–298.
74. Crombag HS, Badiani A, Maren S, Robinson TE. The role of contextual versus discrete drug-associated cues in promoting the induction of psychomotor sensitization to intravenous amphetamine. *Behav Brain Res.* 2000;116:1–22.
75. Vezina P, Giovino AA, Wise RA, Stewart J. Environment-specific cross-sensitization between the locomotor activating effects of morphine and amphetamine. *Pharmacol Biochem Behav.* 1989;32:581–584.
76. Thompson JL, Pogue-Geile MF, Grace AA. The interactions among developmental pathology, dopamine, and stress as a model for the age of onset of schizophrenia symptomatology. *Schizophr Bull.* 2004;30:875–900.
77. Norman RMG, Malla AK. Stressful life events and schizophrenia I: a review of research. *Br J Psychiatry.* 1993;162:161–166.
78. Valenti O, Grace AA. Acute and repeated stress induce a pronounced and sustained activation of VTA DA neuron population activity. *Program No 47911, 2008 Neuroscience Meeting Planner.* Washington, DC: Society for Neuroscience; 2008; Online.
79. Herman JP, Mueller NK. Role of the ventral subiculum in stress integration. *Behav Brain Res.* 2006;174:215–224.
80. Lisman JE, Grace AA. The hippocampal-VTA loop: controlling the entry of information into long-term memory. *Neuron.* 2005;46:703–713.
81. Ragozzino ME. The contribution of the medial prefrontal cortex, orbitofrontal cortex, and dorsomedial striatum to behavioral flexibility. *Ann NY Acad Sci.* 2007;1121:355–375.
82. Goto Y, Grace AA. Dopaminergic modulation of limbic and cortical drive of nucleus accumbens in goal-directed behavior. *Nat Neurosci.* 2005;8:805–812.
83. Goto Y, Grace AA. Limbic and cortical information processing in the nucleus accumbens. *Trends Neurosci.* 2008;31:552–558.
84. LeDoux JE. Emotion circuits in the brain. *Annu Rev Neurosci.* 2000;23:155–184.
85. Rosenkranz JA, Grace AA. Dopamine-mediated modulation of odour-evoked amygdala potentials during pavlovian conditioning. *Nature.* 2002;417:282–287.
86. Grace AA. Phasic versus tonic dopamine release and the modulation of dopamine system responsivity: a hypothesis for the etiology of schizophrenia. *Neuroscience.* 1991;41:1–24.
87. Grace AA. The tonic/phasic model of dopamine system regulation: its relevance for understanding how stimulant abuse can alter basal ganglia function. *Drug Alcohol Depend.* 1995; 37:111–129.

Index

Note: Page numbers with '*f*' and '*t*' in the index denote figures and tables in the text.

AADC (aromatic acid decarboxylase), 490–491
α-Adrenergic receptors, as atypical antipsychotic drug targets, 529
AAV vectors, for GDNF-based gene therapy, 493–494
Abnormal involuntary movements (AIM), 435, 436–437, 438, 439
Acb. *See* Nucleus accumbens
Acb/dorsal striatum distinction, 303
ACC (anterior cingulate frontal cortex), 39, 42
Acetylcholine, 358
Action-outcome learning, 394
Action potentials
  backpropagating, 351
  definition of, 118
  dopamine neuron, 118
  midbrain dopamine neurons
    in vitro, 124
    in vivo, 120, 122–123
Activation. *See also specific types of activation*
  behavioral (*See* Behavior, activation)
  DA neuronal, in basolateral amygdala, 595, 595*f*
  definition of, 203
  gain-amplificatory mode, 203
  vs. arousal, 203
  Yerkes-Dodson principle and, 203–204, 204*f*
Addiction
  a-process or euphoria, 371
  b-process or opponent process, 371
  cycle, 371–372
    binge/intoxication, 371–372
    preoccupation/anticipation, 371–372
    withdrawal/negative affect, 371–372
  diagnostic criteria, DSM-IV, 371
  drug exposure (*See* Drug exposure)
  impulsivity and, 399
  instrumental learning, DA and, 394–395
  long-term effects of drugs of abuse on dopamine, 408–411
  motivational view of, 371–372
  natural reinforcers, 391
  negative reinforcement and, 371
  Pavlovian conditioning of, DA and, 390–393, 391*f*
  positive reinforcement and, 371
  reinforcement, DA and, 389–390
  relapse, stress and, 413
  reward, 389
  sensation seeking and, 399–400
  vulnerability, 382–383, 411–413, 412*f*
    dopaminergic mechanisms in, 399–401, 401*f*
    in early drug and stress exposure, 381–382
    individual differences in DA utilization, 377–379
    stress, DA and glucosteroid interactions, 377–381, 380*f*
Adenosine, interaction with DA, 292–293
Adenosine A_1 antagonist DPCPX, 292
Adenosine A_{A2} receptors, 292
Adenylyl cyclase regulation, 102
ADHD. *See* Attention deficit hyperactivity disorder
Adolescence
  addiction vulnerability and, 381–382
  prefrontal cortex dopamine activity
    GABA interneurons and, 180–181, 180*f*, 181*f*
    in pyramidal neurons, 179–180, 179*f*
Afterhyperpolarization (AHP), 123
AIM (abnormal involuntary movements), 435, 436–437, 438, 439
A-kinase anchoring protein (AKAP), 103
Akinesia
  DA replacement therapy for, 279
  in Parkinson's disease, 421
  single forelimb, 280–281
AKT1, epistatic interaction with COMT, 194–195, 196*f*
AKT1 gene, in schizophrenia, 564–565
AKT/GSK-3 cascade, D2R signaling via, 109
Akt/protein kinase B, 170
Alcohol self-administration, 374–375, 375*t*
Alpha-methyl-tyrosine, 514
Alpha-synuclein, 281, 502
AM251, 357, 529
AMPA, modulation of pyramidal cell excitability, 178–180, 179*f*
Ampakines, 528
AMPA/NMDA ratio, 393–394
AMPA receptor (AMPAR)
  blockade, 305
  GluR1-type, 339–340
  in schizophrenia, 528
  signaling, 339–340
    GluR1-type, 344
    in VTA, 341–342
Amphetamines
  activation of early response genes, 336
  addiction vulnerability, 377–381, 380*f*
  DA concentrations and, 407–408
  DAT and, 92–93, 516
  dopamine hypothesis of schizophrenia and, 521
  dose, behavioral effects of, 205, 205*f*
  IEG induction, 17–18
  induction of synaptic plasticity in NAcb, 342–344
  intra-Acb drug infusion studies, 308–310, 309*f*
  mediation, NAcb DA and, 288
  microinfusion mapping study, 303–305, 304*f*
  rate-increasing effects, 205–207, 207*f*
  reward effects, 371
  self-administration, 374–375, 375*t*
  VMAT2 and, 89
Amylin, 305
Anhedonia, 576
Animal models. *See* Primate models; Rodent models
Anterior cingulate frontal cortex (ACC), 39, 42
Anticipation-invigoration mechanism, 287–288
Antiparkinson agents, 498, 499*t*
Antipsychotic drugs. *See also specific antipsychotic drugs*
  D2DAR blocking, 70–71, 71*t*
  first-generation or typical, 520–521
    disadvantages of, 541–542
    for schizophrenia, 541–542
  future research directions, 548
  mechanism of action, 540–549
    DA receptors and, 542–544

599

Antipsychotic drugs (*Continued*)
    DA transmission alterations, 548, 549*f*
    delayed onset, 547
    implications for, 547
    other receptors in, 546–547
    speed of onset and, 547
    rebound psychosis or supersensitivity psychosis, 547
    for schizophrenia, 541–542, 547–548
    second-generation (*See* Atypical antipsychotic drugs)
    targets, 523*f*, 530–532, 531*f*
        α-adrenergic receptors, 529
        cannabinoid receptors, 529
        catechol-O-methyltransferase, 524–5525
        cholinergic, 526–527, 527*t*
        D1 receptor, 523–524
        D2 receptor, 521–522
        D3 receptor, 524
        D4 receptor, 524
        glutamatergic, 527–529
        neurokinin receptors, 529–530
        neurotensin receptors, 530
        serotonergic, 525
        sigma receptors, 530
    withdrawal, relapse on, 547
Apomorphine
    D2DAR agonism, 71*t*
    growth hormone response in depression, 575–576
Apoptosis, in SNpc during postnatal development, 160, 162*f*
Aripiprazole
    mechanism of action, 71, 71*t*
    therapeutic/experimental use, 71*t*
Aromatic acid decarboxylase (AADC), 490–491
Artificial intelligence, 268–269
Artificial learning, 269
Ascending dopaminergic pathways, 301
AT$_1$R signaling, 79
Attention, dopamine neurons and, 326–328, 327*f*, 328*f*
Attention deficit hyperactivity disorder (ADHD)
    DAT gene polymorphisms, 93–94
    DA transmission in PFC and, 242–243
    D1R actions in PFC and, 241
    startle reflex in, 577
    substance abuse risk and, 412–413
    treatment, 88
Attractor states, 265*f*
Atypical antipsychotic drugs. *See also specific antipsychotic drugs*
    D4DAR and, 76
    drug dissociation rates, 521, 522*f*
    mechanisms of action, 544–546
        combined blockade of D2 and serotonin 5HT$_{2A}$ receptors, 544–545
        preferential limbic D2R blockade, 545–536
    for schizophrenia, 542

targets, 523*f*, 530–532, 531*f*
    α-adrenergic receptors, 529
    cannabinoid receptors, 529
    catechol-O-methyltransferase, 524–525
    cholinergic, 526–527, 527*t*
    D1 receptor, 523–524
    D2 receptor, 521–522
    D3 receptor, 524
    D4 receptor, 524
    glutamatergic, 527–529
    neurokinin receptors, 529–530
    neurotensin receptors, 530
    serotonergic, 525
    sigma receptors, 530
withdrawal, rebound or supersensitivity psychosis, 547
Autonomic symptoms, in Parkinson's disease, 426, 426*t*
Autoradiography, receptor localization, 23–24
Axonal tracing studies, 11, 12

Backpropagating action potentials (bAPs), 351
Basal ganglia
    dopamine cells, 38
    dopamine neurons, 49, 50*f*
    dopaminergic synapses, 49–50
    feedforward system, 49
    functional projections through, 39–40
    functions, 49
    inputs, 49, 50*f*
    parallel processing, 38, 39–40
    pathways
        in Parkinson's disease, 15–17, 16*f*
Basins of attraction, 265*f*
Basolateral amygdala (BLA)
    DA neuronal activation, 595, 595*f*
    drug-seeking behavior and, 395–396, 396*f*, 397*f*
    inactivation, 293
    PIT and, 391
BDNF. *See* Brain-derived neurotrophic factor
Behavior
    activation
        anticipation-invigoration mechanism, 287–288
        DA-adenosine interaction and, 292–293
        NAcb DA and, 287–288
    addictive (*See* Addiction)
    dopamine and, 209–212, 211*f*
    dopamine neurons and, 316–328
    dose-relationship, Yerkes-Dodson principle and, 205, 205*f*
    drug-seeking (*See* Drug-seeking behavior)
    effort-related choice, NAcb DA and, 289–292, 291*t*
    flexibility, 221–226
    food-motivated instrumental, NAcb DA in, 288–289
    response-reinforcer contingencies, 307

responses to dopamine neuron transplantation, 459–466
reward (*See* Reward)
striatal regions, functional heterogeneity of, 307
striosome predominance and, 336
Behavioral economics, 288
Behavioral response processing, in NAcb, 344–345
Behavioral testing, in mice, 463
BINA (biphenyl-indanone A), 529
Biophysical model, 263*f*, 264, 267
Biphenyl-indanone A (BINA), 529
Bipolar disorder, diagnostic characteristics, 573*f*
Bistability, cellular, 262, 263*f*
BLA. *See* Basolateral amygdala
Bradykinesia, in Parkinson's disease, 421
Brain-derived neurotrophic factor (BDNF)
    DA response to stress and, 576
    regulation of NCD in mdDA neurons, 165, 169–170
Brain-stimulation reward (BSR), 308
Bridge grafts, in dopamine neuron transplantation, 468
Bromocriptine
    D2DAR agonism, 71*t*
    regional specificity, 252, 253*f*, 254–257
BSR (brain-stimulation reward), 308
Burst activity, of midbrain dopamine neurons, 129
Butaclamol, D2DAR antagonism, 71*t*
BW 23FU (rimcazole), 530

Cabergoline, D2DAR agonism, 71*t*
Caffeine, 292
Ca^{2+} ion channels. *See* Calcium ion channels
Calbindin, 12, 42, 305
Calcium ion channels (Ca^{2+} ion channels)
    Cav1.3, 354, 358, 360
    Cav2, 354, 358
    K-ATP-type, 132–133
    L-type, 132–133
        blocking, 353
        Cav1.3, 354
        differential vulnerability of dopamine midbrain neurons and, 133
        NMDA receptor currents and, 354
        opening after PKA phosphorylation, 354
    voltage-gated, 307, 307*f*
    voltage-dependent, D1 receptors control of, 105–106
    voltage-gated, NMDA and, 271
CalDAG-GEF1, 336
CalDAG-GEF2, 336
Calmodulin, 307
CALM-PD trial, 449
CaMKII-ItTA, 167
cAMP, 353
CAMP/PKA/DARPP-32 cascade, 107
cAMP-regulated phosphoprotein. *See* DAARP-32

cAMP response element binding protein (CREB)
  activation, 67, 105
  in neural plasticity, 306–307, 307f
  reward responsiveness in depression, 577
Candidate gene studies, 251–252
Cannabinoid receptors, as atypical antipsychotic drug targets, 529
Cannabis, 529
CANTAB (Cambridge Neuropsychological Test Automated Battery), 209
CAR (conditioned avoidance response), 528
Carlson, Arvid, 301
Carotid body glomus cell implantation, for Parkinson's disease, 450
Catechol-O-methyltransferase (COMT). See also COMT gene
  atypical antipsychotic drug target, 524–5525
  dopamine in PFC and, 252
  epistatic interaction with AKT1, 194–195, 196f
  genetic variation, in behavioral syndromes, 189–190
  in prefrontal cortex, 189–192, 191f, 193f
β-Catenin, 147
Ca^{2+}-transients, bAP-evoked, 351, 352f–353f, 360–361
Caudate nucleus projections, convergence, 423, 424f
CBLG-DT, 167
CDK5 (cyclin-dependent kinase 5), 94
Cerebral cortex projections, convergence, 423, 424f
Cerebrospinal fluid (CSF), HVA concentration, 575, 578
CGS21680, 292–294
ChAT (choline acetyltransferase), 52, 526
Chlorpromazine
  analogs, 520–521
  development, 520
  dopamine hypothesis of schizophrenia and, 521
Choline acetyltransferase (ChAT), 52, 526
Cholinergic function, in Parkinson's disease, 450–451
Cholinergic interneurons, 357–358
Cholinergic targets, of atypical antipsychotic drugs, 526–526, 527t
Cholinesterase inhibitors, 526
Cingulate cortex lesions, effort-related processes and, 293
Clozapine
  α-adrenergic receptor affinity, 529
  with ampakines, 528
  D2DAR antagonism, 71t
  D2 receptor occupancy, 521–522
  efficacy, 521
  M$_1$ muscarinic receptors and, 526, 527t
  side effects, 520
CMKII control, of D2R-Gαi/o coupling, 109
CNQX, 528

Cocaine
  abuse/addiction
    D2 dopamine receptors and, 400
    executive function and, 410, 410f
    pattern of, 372
    vulnerability, DA-stress-glucocorticoid interactions and, 377–381, 380f
  activation of ERK and Elk-1, 106f, 107
  cue-induced DA increase in dorsal striatum, 411
  DAT interaction, 92–93, 94
  dopamine hypothesis of schizophrenia and, 521
  drug-seeking behavior, acquisition of, 400
  exposure
    induction of synaptic plasticity in NAcb, 342–344
    synaptic plasticity in VTA, 341–342
  IEG induction, 17–18
  induced reinstatement, 344
  LTP-induced by, 393–3934
  mechanisms of action, 390
  mediation, NAcb DA and, 288
  pleasure or high, 390
  priming dose, 344
  reexposure during abstinence, 343
  reinforcement, 392–393, 392f
  second-order schedule, 400
  self-administration, 374–375, 375t, 376f, 393–394
Coenzyme Q10 neuroprotectivity, in Parkinson's disease, 499t, 500
Cognition
  dopamine and, 209–212, 211f
  flexible control, 249–257
    dopaminergic drug effects, 250–252, 251f, 253f, 254–257
    paradox of flexible mind, 249–250
  mesocorticolimbic dopamine and, 215–226
  in 6-OHDA lesion model, after DA neuron transplantation, 462–463
  paradox of flexible mind, 249–250
  prefrontal cortical, 230–244
Compulsivity, definition of, 371
Computed tomography, Parkinson's disease, 427–428
COMT. See Catechol-O-methyltransferase
COMT gene
  neural mechanisms of, 563–564
  polymorphisms, 575
  prefrontal DA function and, 209–210
  in schizophrenia, 561f, 563–564
Conditioned avoidance response (CAR), 528
Conditioned place preference (CPP), 93
Conditioned reinforcement (CR), 304, 391
Conditioned response, 306
Conditioned stimulus (CS)
  DA-based reinforcement learning and, 268
  drug-associated, 392–393, 392f
Connectionist models, 263–264

Cortical terminals, 54
Corticobasal ganglia pathways
  dual parallel and integrative processing, 45, 45f
  parallel processing in, 39–40, 40f
  system-level changes in L-DOPA-induced dyskinesia, 440
Cortico-stiato-midbrain pathway, 42
Corticostriatal circuits, in schizophrenia, 516
Corticostriatal projects, 14–15
Cortico-striato-pallido-thalamo-frontal pathway, 423
Cost/benefit analysis, effort-related choice behavior and, 289–292, 291t
CPP (conditioned place preference), 93
CR (conditioned reinforcement), 304, 391
[^{11}C]raclopride, 407, 409, 410, 410f
^{13}C-raclopride D2/D3 receptor, 576
Creatine neuroprotectivity, in Parkinson's disease, 499t, 500
CREB. See cAMP response element binding protein
CRF (corticotropin releasing factor), 379, 381
Cross-talk, between cAMP/PKA and DARPP-32, 106–107, 106f
CS. See Conditioned stimulus
CSF (cerebrospinal fluid), HVA concentration, 575, 578
5CSRTT (five-choice serial reaction-time task), 399–400, 401f
CX-516, 528
Cyclic AMP
  cross-talk with PKA and DARPP-32, 106–107, 106f
  D1R control of voltage-dependent Ca^{2+} channels, 105–106
  regulation by intracellular dopamine signaling, 102
Cyclin D1, 148–149
Cyclin-dependent kinase 5 (CDK5), 94
Cyclooxygenase-2 inhibitors, 531
D-Cycloserine, 528

DAARP-32, 301
dACC, 39, 41, 43
DAG (diacylglycerol), 354
D1 agonists
  D1Rs and, 523
  modulation of pyramidal cell excitability, 179–180, 179f
D2 agonists, 124
DALA, intra-Acb drug infusion studies, 308–309, 309f
DAMGO, intra-Acb infusions, 309, 310
d-Amphetamine. See Amphetamines
D1 and D2 receptor agonists, 17, 18
D1 antagonists, 523
DAR100, 512

DARPP-32
    cross-talk with cAMP and PKA, 106–107, 106f
    DA and, 565
    ERK1/2 activation, 18–19
    intracellular dopamine signaling and, 103–104, 104f
    intrinsic excitability and, 353
DAT. See Dopamine transporter
Daytime somnolence, dopaminergic medication and, 426
D1b (D1β). See D5 dopamine receptor
DBS (deep brain stimulation), for major depressive disorder, 582
D1DAR. See D1 dopamine receptor
D2DAR. See D2 dopamine receptor
D2DAR agonists, 70–71, 71f
D2DAR antagonists, 70–71, 71f
D1 dopamine receptor (D1DAR; D1R)
    AC5 and, 102
    activation
        coincident with NMDAR activation, 306–307, 307f
        LTP and LTD, 269
    antipsychotic action and, 543–544
    bidirectional STDP and, 356–357, 357f
    in biophysical network model, 270
    blockage, 307, 394
    in direct and indirect striatal projects, 15–17, 16f
    distribution, 100, 334, 542
    function, 542
    functional domains, 64–65
    inhibitory actions, in mechanistic primate studies, 237–238
    interactions
        forming higher order structures, 65
        with G$_S$ and G$_{OLF}$ proteins, 101–102
    intracellular dopamine signaling, 65–67, 354
        future research, 110–111
        G protein coupling, 65–67, 101–102
        mechanisms, 101–102, 110, 110f
        nonneuronal, 69
        in PFC, working memory and, 216–221, 217f
        through cAMP and PKA, 67
    inverted-U physiological actions, 235–236
    L-DOPA-induced dyskinesia in Parkinson's disease, 17
    ligands, 65, 66t (See also D1DAR agonists; D1DAR antagonists)
    localization
        in brain, 65
        peripheral, 69
    long-term depression and, 101
    long-term potentiation and, 101
    mediation of gene regulation, 18–19, 18f
    modulation
        of cellular bistability, 262, 263f
        of single cell input/output, 261–263, 262f, 263f
        of working memory, 264, 265f, 266–267, 266f

    PFC neuronal physiology and, 235
    pharmacology, 65
        as antipsychotic target, 523–524
        therapeutic potential, 65
    posttranslational modifications, 63–64
    in prefrontal cortex
        excessive stimulation during stress, 242
        optimal actions, 240–242, 241f
        in schizophrenia, 511–512, 513f
        stimulation of, 252
    in primate prefrontal cortex, 232–234, 233f
    regulation, 67–69
    in schizophrenia, 540
    spatial working memory and, 235, 236f
    specificity of dopaminergic drugs and, 257
    in striatonigral MSNs, 351–353
    structure, 63, 64f
    synaptic plasticity and, 270
    up-states, 204
    voltage-gated ion channels and, 270, 271
    Yerkes-Dodson principle and, 203–204, 204f
D2 dopamine receptor (D2DAR; D2R)
    activation, 268, 355
    addiction and, 400
    antipsychotic drugs and, 542–543, 547
    bidirectional STDP and, 356–357, 356f
    binding
        in major depressive disorder, 578, 579t, 580t
        typical antipsychotics and, 521
    blockade
        combined with serotonin 5HT$_{2A}$ receptor blockade, 544–545
        preferential limbic, 545–536
    in depression, 578
    distribution, 25–30, 100, 334
        forebrain, 26–27
        midbrain, 25–26
    drug self-administration and, 412
    functions, 22, 25–30
        glutamate transmission modulation, 22, 23f
        in striatal projections, 15–17
        structural plasticity, 29
    gating, 204
    G protein interactions, 107–109
    inhibition of indirect pathway NAcb neurons, 344
    intracellular dopamine signaling, 110–111, 111f, 354
        control of cAMP/PKA/DARPP-32 cascade, 107
        future research, 110–111
        mechanisms, 72–73
        NMDR regulation, 108
        in PFC, working memory and, 216–221, 217f
        regulation of voltage-dependent K$^+$ channels, 108
        via AKT/GSK-3 cascade, 109
    in working memory, 189

    isoforms, 72
    D2$_L$, 73
    postsynaptic, 110
    presynaptic, 110
    ligands, 407 (See also [^{11}C]raclopride; D2DAR agonists; D2DAR antagonists)
    localization, 22–31, 73
        cellular studies, 27–28
        on DA neurons, 25–26
        in forebrain, 29–30
        on Glut nerve terminals, 29–30
        on Glut neurons, 26–27
        in midbrain, 30
        on postsynaptic targets of DA and Glut inputs, 27–29
        on secondary neurons, 27
        subcellular studies, 28
    long-term depression and, 101
    long-term potentiation and, 101
    modulation of PFC attractor dynamics/function, 267–268
    NMDR regulation and, 108
    partial agonism, antipsychotic action and, 542–543
    pharmacology, 70–72
        atypical antipsychotics and, 521–522, 522f
        therapeutic potential, 65
        vs. D4DAR, 76
    prefrontal cortex primate studies, 238–239, 240f
    in primate prefrontal cortex, 232, 233f
    regulation, 73
    in schizophrenia, 513–515, 540, 562–563
    specificity of dopaminergic drugs and, 257
    stimulation
        LTP conversion to LTD, 271
        in PFC, 252
    striatal, baseline occupancy of, 514
    structure, 69–70, 70f
    in substance abusers, 410, 410f
    in substance abusers, PET findings, 409–410, 409t
D3 dopamine receptor (D3DAR; D3R)
    antipsychotic action and, 544
    glutamate and, 30–31
    localization, 75
    pharmacology, 73–74, 524
    regulation, 75
    in schizophrenia, 540
    signaling mechanisms, 74–75
    splice variants, 74
    structure, 73
D4 dopamine receptor (D4DAR; D4R)
    antipsychotic action and, 544
    discovery of, 75
    gene polymorphisms, 575
    glutamate and, 31
    ligands, 76
    localization, 77
    molecular cloning, 75
    pharmacology, 76, 524

prefrontal cortex primate studies, 239
primate prefrontal cortex, 234
regulation
　of expression, 77
　by protein-protein interactions, 77
signaling mechanisms, 77
structure, 75
　genomic, 75–76, 76f
　molecular, 75
D5 dopamine receptor (D5DAR; D5R)
localization, 79–80
pharmacology, 77–78
primate prefrontal cortex, 232–234, 233f
signaling mechanisms, 78–79
structure, 77–78
D1 dopamine receptors (D1DAR; D1R)
activation of cAMP signaling, 105–106
control of voltage-dependent Ca^{2+} channels, 105–106
ERKs signaling and, 106–107, 106f
NMDA transmission and, 104–105
PSD-95 binding, 105
regulation
　of AMPA, 103
　of GABA$_A$, 103
　of NA+ channels, 105
Deep brain stimulation (DBS), for major depressive disorder, 582
Dehydroepiandrosterone (DHEA), 531
Delayed response tasks, 268
Dementia, 424
Dementia with Lewy bodies (DLB), 446–447
Dendrite length, total, 349–350
Depolarization of ion channels, in midbrain dopamine neurons, 125–127, 126f
Depression. See also Major depressive disorder
animal models of dopamine function, 574
dopaminergic dysfunction subtype, 573
genetic studies, 574–575
neurochemical studies, 574–575
in Parkinson's disease, 424–425, 426, 576
reward system dysfunction, 576–577
unipolar (See Major depressive disorder)
Desensitization, D1DAR regulation, 67–68
N-Desmethyl-clozapine, 526, 527t
Developmental cell death, mesencephalic dopamine neurons, 160–163, 161f, 162f
DHEA (dehydroepiandrosterone), 531
Diacylglycerol (DAG), 354
Diagnostic and Statistical Manual IV (DSM-IV)
diagnoses with depression without psychotic symptoms, 573f
diagnostic criteria for addiction, 371
Direct pathway. See Striatonigral pathway
DISC1, 183
Dissociative anesthetics, NMDA antagonism by, 527
DJ-1 mutations, 503
DLB (dementia with Lewy bodies), 446–447

DLPFC (dorsolateral prefrontal cortex), dysfunction in schizophrenia, 511–512, 513f, 559–562, 561f
DMXB-A, 527
DOPAC (dihydroxyphenyl acetic acid), 578
DOPA decarboxylase
striatal, 513
in substance abusers, PET findings, 409–410, 409t
DOPA delivery, continuous using viral vector, 492
Dopamine. See also specific aspects of
discovery of, 11
early research, 3–4
functions (See also specific dopamine functions)
　in behavior, 209–212, 211f
　in cognition, 209–212, 211f
　in Parkinson's disease motor function, 279–283
　single cell input/output, 261–263, 262f, 263g
　stress and, 576
　in striatum, 101
intracellular signaling (See Intracellular dopamine signaling)
neural network dynamics, 263–264, 263f, 265f, 266–268
neurotransmission control, 88
Dopamine agonists, 448
Dopamine antagonists
fixed ratio 5 lever pressing and, 288–289
food-reinforced lever pressing and, 288
with SCH-23390, 308–309, 309f
Dopamine axons, 11
Dopamine cell replacement
as alternative tissue source for transplantation, 468–469, 469t
clinical competition with DA neuron transplantation, 486
as PD treatment, 483–484
Dopamine clearance, presynaptic changes in, 436–437
Dopamine D2 antagonists, 292–293
Dopamine depletion
Cav1.3 subunit of L-type Ca^{2+} ion channels, 108
parkinsonian-like syndromes and, 434
Dopamine dysregulation syndrome, 425
Dopamine hypothesis
of reward, problems with, 286–287
of schizophrenia (See Schizophrenia, dopamine hypothesis of)
Dopamine islands, in striatal development, 334, 334f
Dopamine neurons
action potential, 118, 120, 122–123
activity states
　phasic, 591–592, 591f
　tonic, 591–592, 591f
　ventral subiculum regulation of, 592
attention functions, 326–328, 327f, 328f
in basal ganglia, 49–50, 50f
behavioral functions, 316–328, 316–329

in reward learning, 317–319, 317f, 318f–324f, 322–324
in reward processing, 316–317, 317f
burst activity mode, 121–122, 121f, 123
for clinical applications, generation of, 485, 485t
in conditioning experiments, 333–334
dorsal tier pars compacta, 12
economic value functions, 324–326, 325f, 326f
electrophysiologic analysis, 119–120
　dynamic clamp configuration, 120
　Hodgkin-Huxley-style computer models, 120
　membrane potential recordings, 119–120
endogenous, protection during DA neuron grafts, 468
forebrain input, 42
functions, 38
glutamate colocalization in, 25–26
historical perspective, 3–6
homeostasis, 88–95
　DAT and, 88
　neurotransmission, 90–94, 91f
　storage, 89–90
identification, 590–592, 591f
late development stages, 160–172
loss
　in major depressive disorder, 575
　in Parkinson's disease, 421, 423f
medium spiny (See Medium spiny neurons)
mesencephalic, naturally occurring cell death, 160–163, 161f, 162f
　regulation by extrinsic molecules, 166–171
　regulation by intrinsic molecules, 171–172, 171f
midbrain (See Midbrain dopamine neurons)
motor functions, 316, 317f
nigrostriatal depletion, 6-OHDA-induced (See 6-OHDA lesion model)
nigrostriatal, 437
novelty functions, 326–328, 327f
phasic signaling or burst activity, 118
physiologic properties, 590–592, 591f
plasticity and, 390
pleasure and, 576
postsynaptic targets, 5–6
replacement therapy (See Dopamine neuron transplantation)
reward and, 576
reward prediction error, 42
signal-to-noise ratio, 262
size/number of, 160
striatal, 49–50
subsets, 12, 13f
substantia nigra pars compacta
　degeneration, in Parkinson's disease, 279
tonic activity mode, midbrain, 120–121, 121f, 123
tonic signaling, 118
transplantation, 454–470
ventral tier, 12

Dopamine neuron transplantation
  alternative tissue sources, 468–469, 469t
  anatomical reconstruction, 455–458, 456f–458f
  behavioral recovery
    cognition, 462–463
      motor learning, 462–463
      other motor functions, 461–462, 461t
      rotation, 459–461
  behavioral responses, 459–466
  bridge grafts, 468
  enhanced axonal growth, 468
  methods
    in 6-OHDA model, 454
    solid and cell suspension grafts, 454–455, 455f
  neurochemical reconstruction, 455–458, 456f–458f
  for Parkinson's disease, 478–486, 502–503
    clinical responses, 430–432, 431f
    disease process, grafts and, 483
    functional integration, 478–480, 479f
    generation of DA neurons from stem cells, 485, 485t
    graft survival, 478–480, 479f
    sham surgery controlled trials, 482–483
    symptom improvement, 480–481, 480t, 482f
  protection of endogenous DA neurons, 468
  rodent models, 454–470
    methods, 454–455, 455
  into substantia nigra, 458–459, 460f
Dopamine partial agonists, reinforcing effects of psychostimulants, 376
Dopamine pathways. *See specific dopamine pathways*
Dopamine projections, in primate prefrontal cortex, 232
Dopamine receptors
  antipsychotic mechanism of action and, 542–544
  in dendrites, 349
  in depression, 576–577
  distribution of, 63
  dopamine neuronal action potential and, 118
  excessive signaling, 94
  inhibition in NAcb, reward-predicting cue and, 344
  localization methods, 23–25
    autoradiography, 23–24
    immunocytochemistry, 24–25
    in situ hybridization, 24
  molecular pharmacology (*See under specific dopamine receptor subtypes*)
  in schizophrenia, 562–563
  sensitivity, prostate apoptosis response protein and, 574
  signaling, 339, 408
  stimulation, baseline working memory capacity, 250–251, 251f
  stimulation of, 63
  subtypes, 63, 101, 542
    D2 (*See* D2 dopamine receptor)
    D3 (*See* D3 dopamine receptor)
    D4 (*See* D4 dopamine receptor)
    D5 (*See* D5 dopamine receptor)
    D1 or D1A (*See* D1 dopamine receptor)
Dopamine release
  DAT deletion/absence and, 91, 91f
  presynaptic changes in, 436–437
  in schizophrenia, 513
  in substance abusers, PET findings, 409–410, 409t
Dopamine replacement therapy
  cellular (*See* Dopamine cell replacement)
  by gene therapy, 490–492
  mental symptoms induced by, 425, 425t
  neuronal (*See* Dopamiine neuron transplantation)
Dopaminergic drugs
  baseline dependency, 250–251, 251f
  daytime somnolence and, 426
  functional specificity, 251–252
  receptor specificity, 257
  regional specificity, 252, 253f, 254–257
Dopaminergic neurotransmission regulation, by plasma membrane DAT, 90–94, 91f
Dopamine-serotonin hypothesis, 522
Dopamine signaling, tonic, 118
Dopamine synthesis
  enhanced, pro-drug approach for, 490–491
  restoration by triple-gene transfer, 491–492
Dopamine systems
  activation, 203
  dual midbrain, 129–132, 131f
  imaging, in Parkinson's disease, 447
  mesocortical, 207–209, 208f
  mesolimbic, 205–207, 207f
  mesostriatal (*See* Mesostriatal dopamine system)
  midbrain (*See* Midbrain dopamine neurons; Substantia nigra pars compacta; Ventral tegmental area)
Dopamine transporter (DAT)
  activation, 515–516
  blockade, 516
  in DA neuronal homeostasis, 88–95
    neurotransmission, 90–94, 91f
    storage, 89–90
  deletion/absence of, 91, 91f
  in depression, 578
  expression, 90
  extracellular dopamine and, 92
  function, 88
  gene polymorphisms, 575
  knockout mice, 90–92, 91f
  localization/distribution, 90
  protein interactions, 90
  psychostimulants and, 90–94
  structure, 90
  in substance abusers, PET findings, 409–410, 409t
Dopamine transporter reuptake blockers, 92–93. *See also* Amphetamines; Cocaine; Methylphenidate
Dorolateral prefrontal cortex (DLPFC), 192, 193f
Dorsal prefrontal cortex (DPFC), 39, 41, 42
Dorsal tier pars compacta dopamine neurons, 12
Dorsolateral prefrontal cortex (DLPFC), dysfunction in schizophrenia, 511–512, 513f, 559–562, 561f
DPFC. *See* Dorsal prefrontal cortex
D1R. *See* D1 dopamine receptor
D2R. *See* D2 dopamine receptor
Drd1a-agonist. *See* L-DOPA
D1 receptor agonists
  induction of IEGs, 17, 18
  SKF-81297, 208, 218–219
D1 receptor antagonist. *See* SCH-23390
Drive, 287
Drug abuse. *See also* Addiction
  DA transmission in PFC and, 243
  risk, in attention deficit hyperactivity disorder, 412–413
  as schizophrenia risk factors, 593–594
  vulnerability, dopamine and, 411–413, 412f
Drug dependence. *See also* Addiction
  motivational dysregulation, DA role in
    microdialysis studies, 377, 378f
    pharmacological studies, 376–377
Drug-seeking behavior, 395–399, 396f–398f
  acquisition of, 400
  compulsive
    rat models of, 400–401, 401f
  nucleus accumbens core and, 395–398, 396f–398f
  reinstatement, DA role in, 377
  vulnerability to addiction and (*See* Addiction, vulnerability)
Drugs of abuse, 407–408. *See also specific drugs of abuse*
  addictive (*See also* Addiction)cellular models of learning, 393–394
  molecular mechanisms, 390
  patterns, 372
  plasticity and, 393–394
  cue-induced DA increase in dorsal striatum, 411
  discriminative effects, 389–390
  experience, altered glutamatergic signaling in NAcb and, 344
  exposure
    induction of synaptic plasticity in NAcb, 342–344
    repeated, NAcb glutamatergic signaling and, 343
    rodent models, 341
    synaptic plasticity after, 340–341
    synaptic plasticity in VTA and, 341–342
  exposure and abstinence, repeated, 343–344

INDEX

long-term effects on dopamine, 408–411
  executive function, 410, 410f
  inhibitory control, 410, 410f
  motivation and, 408–410
  reward and, 408–410, 409t
pleasure/liking, 389
reward effects, DA role in, 372–376
  lesion studies, 370–371, 372–373, 372f, 374f
  microdialysis studies, 371–376, 375f, 375t
  pharmacological studies, 376
self-administration
  D2DARs and, 400
  S-R incentive habit, 400
DSM-IV. *See Diagnostic and Statistical Manual IV*
Dysautonomia, in Parkinson's disease, 426
Dysexecutive syndrome, 424
Dyskinesia
  graft-induced or off-phase, 450, 464–466, 465f, 481, 482f, 484
  L-DOPA-induced (*See* L-DOPA-induced dyskinesia)
  reduction after intrastriatal DA neuron grafts, 464–466, 465f
Dysthymia, diagnostic characteristics, 573f
Dystonia, DOPA-responsive, 336

Ecopipam, effort-related choice behavior and, 290–291, 291t
Effort-related processes, NAcb DA and, 293–294, 294f
EGFP (enhanced green fluorescent protein), 349
EIAV (equine infectious anemia virus), 491
Electrophysiological studies
  of dopamine neurons, 328–329
  of prefrontal cortex
    dopamine activity, 177–178, 178f
    working memory and, 219
Eligibility trace (synaptic tagging), 270
Eliprodil (SL 82.0715), 530
Elk-1, cocaine activation of, 106f, 107
ELLDOPA trial, 449
EMD-57445 (panamesine), 530
Emotion control, 337
En1/2, mdDA neuron maintenance, 154
Endocannabinoid production, 354
Endocytosis, D1DAR, 69
Enhanced green fluorescent protein (EGFP), 349
Enkephalin, 335
Episodic learning, 462
EPSP (excitatory postsynaptic potential), 269
Equine infectious anemia virus (EIAV), 491
ERKs. *See* Extracellular signal-regulated kinases
Ethanol withdrawal, DA efflux in NAcb and, 377, 378f

Eticlopride, effort-related choice behavior and, 290–291, 291t
Excitatory postsynaptic potential (EPSP), 269
Executive function
  definition of, 39
  mesocorticolimbic DA and, 215–226
  substance abuse long-term effects and, 410, 410f
Exocytosis, calcium-mediated, 118
External pathway. *See* Striatonigral pathway
Extracellular signal-regulated kinases (ERKs)
  activation, 105
    in dopamine-depleted striatum, 18
    by D1R, 17
    D2R isoform inhibition of, 110
    via cocaine, 106f, 107
  in neural plasticity, 307, 307f
  phosphorylation, 110
  signaling, D1Rs and, 106–107, 106f
Extrapyramidal motor movement disorders, 301. *See also* Parkinson's disease
Extrastriosomal matrix, 334–336, 337f
Extrasynaptic receptor function, dopamine and, 333

Fast-spiking interneurons, 52
Feedforward organization, 45, 45f
Feeding behavior, microinfusion mapping study, 303–305, 304f
Fetal graft therapy, for Parkinson's disease, 449–450
^{18}F-fluorodeoxyglucose PET, in Parkinson's disease, 446–447
Fgfr genes, 146–147
Five-choice serial reaction-time task (5CSRTT), 399–400, 401f
Fluorescent catecholamine histochemical methods, 11
Fluoxetine exposure, synaptic plasticity in VTA, 341–342
cis-Flupenthixol, effort-related choice behavior and, 290–291, 291t
Fluphenazine, D2DAR antagonism, 71t
Food-associated motivational arousal, intra-Acb drug infusion studies, 308–310, 309f
Food-reinforced lever-press response, 305
Forebrain
  circuitry, in effort-related processes, 293–294, 294f
  localization, of D2 receptors, 29–30
Foxa1/2, 148
Foxa2 (Hnf3β), 148, 153–154
Frontal cortex. *See also specific frontal cortex regions*
  divisions, 38
  functional organization, 39
  lesions, effort-related processes and, 293
Functional magnetic resonance imaging, 410

GABA$_A$-currents, 266, 271
GABA$_A$R
  D5DAR and, 78
  NMDA responses and, 178
  reinforcing effects of alcohol and, 373
GABA$_A$-receptor antagonists, 124
GABAergic neurons, 293, 590
GABA-ergic pathway alterations, in L-DOPA-induced dyskinesia, 439–440
Gastrointestinal symptoms, in Parkinson's disease, 426, 426t
GBR 12909, 516
GCH1, 491
GDNF. *See* Glial cell line-derived neurotrophic factor
GDP (guanosine diphosphate), 102
GEF (guanine nucleotide exchange factor), 336
Geldanamycin, 502
Gene-based therapy, 489–495
  DA replacement by, 490–492
  delivery, 489
  with neurturin, 494–495
  in Parkinson's disease, goals of, 489
  principles of in vivo gene transfer, 489–490
  using neurotrophic factors, 492–495
  vector availability, 489–490
Gene-by-environment interactions (GxE), 575
Gene mutations. *See also specific genes*
  in familial Parkinson's disease, 501–502
Genetic mouse models of parkinsonism, sensorimotor tests, 281–283
  challenging beam transversal, 282
  dot test of somatosensory neglect, 282
  nest building, 282–283
  pole test and inverted grid, 282
  spontaneous movement, 282
Genetic studies, dopaminergic mechanisms in PFC, 189–192, 191f
Gene transfer, mediated, 492
GIRK (G-protein-activated inwardly rectifying potassium channels), 124
GIRK channels, D3 signaling and, 74
Glial cell line-derived neurotrophic factor (GDNF)
  gene therapy with, 492–494
  intrastriatal infusion, in Parkinson's disease, 468
  overexpression, 459
  putaminal infusion, 450
  regulation of NCD in mdDA neurons, 166–168
Globus pallidus projections, convergence, 423, 424f
Glucocorticoids, addiction vulnerability and, 377–381, 380f
Glucose metabolism, in Parkinson's disease, 446–447, 448f
GluR1, 339, 343, 394
GluR2, 339, 343
GluR3, 339

Glutamate
    afferents, 25
    colocalization in DA neurons, 25–26
    convergence with DA, functional
        implications, 28–29
    D2R and, 23
    D3R and, 30–31
    D4R and, 31
Glutamate cells, D2 receptors on, 26–27
Glutamate-dopamine signaling, 308
Glutamate hypothesis of schizophrenia, 541
Glutamate neurons (Glut neurons)
    D2 dopamine receptor localization, 22–31
    D2 receptors, 26–27
Glutamate receptors
    AMPA, 528
    compensatory responses, 408
    gating, D1 receptor stimulation of, 353
    metabotropic, 529
    NMDA, 527–528
Glutamatergic targets, of atypical
        antipsychotic drugs, 527–529
    AMPA glutamate receptors, 528–529
    NMDA glutamate receptors, 527–528
Glutamatergic transmission modulation,
    54–55
Glutamate signaling
    convergence with dopaminergic signals, 55,
        55f
    in Parkinson's disease, DA modulation of,
        358, 359f–360f, 360–362
Glycine transporter inhibitors (GTIs), 528
Goal tracking, 391
GPCR-kinase-mediated desensitization
    of D1Rs, 102–103
    of fragile X mental retardation protein,
        102–103
G-protein-activated inwardly rectifying
    potassium channels (GIRK), 124
G protein-coupled receptors (GPCRs).
    See also Dopamine receptors
    atypical antipsychotics and, 522, 523f
    DA activation of, 353
    intracellular signaling and, 101
    posttranslational modifications, 63–64
    subtypes, 564
G proteins
    activation
        by D1DAR, 65–67
        by D2DAR, 72–73
    definition of, 101
    G_αi/o
        D2R coupling, control of, 109
        interaction with D2Rs, 107–108, 354
    $G\alpha$olf, 101–102
    $G\beta\gamma$, 107–108
    $G_{i/o}$, 107
    NMDA transmission control by D1
        receptors and, 104–105
Greengard, Paul, 301
GRK6 control, of D2R-Gαi/o coupling,
    109
GTIs (glycine transporter inhibitors), 528

GTP binding proteins. See G proteins
Guanine nucleotide exchange factor (GEF),
    336
Guanosine diphosphate (GDP), 102
Guanosine triphosphate binding proteins. See
    G proteins
GxE (gene-by-environment interactions),
    575

Habit
    formation, 462
    operational definition of, 394–395
Haloperidol
    D2DAR antagonism, 71t
    effort-related choice behavior and,
        290–291, 291t
    food-reinforced lever pressing and, 288
    motor function in rats and, 279
    T-maze choice task and, 292
Hamilton Depression Rating Scale (HDRS),
    575
HCN channels
    D1R and, 236
    D2R and, 239–240
    gating, 128
    inhibition, 131
HDRS (Hamilton Depression Rating Scale),
    575
Heat shock proteins, 502
Hebb-like correlation rule, 269
Hedonia, 287
Heroin
    mechanisms of action, 390
    pleasure or high, 390
    self-administration, 374–375, 375t
High-frequency stimulation (HFS), of LTD,
    339, 354–355
Hippocampus, in schizophrenia
    pathophysiology, 592–593, 593f
Hnf3α/β, 148
Hnf3β (Foxa2), 148, 153–154
Homer1a, 308
Homovanillic acid, in CSF, 575, 578
5HT, dopamine interactions, 582–583
Huntington's disease, 94, 336
6-Hydroxydopamine model. See 6-OHDA
    lesion model
Hypercortisolemia, in depression, 577
Hyperphagia, 305, 309–310
Hypertension onset, D5DAR and, 79
Hypokinesia, in Parkinson's disease,
    421–422
Hypomania, in Parkinson's disease,
    424, 425, 426
Hyposmia, in Parkinson's disease, 426

Idazoxan, 529
IEGs (immediate early genes),
    17, 307
Imaging. See Neuroimaging
Imaging genetics, 558

Immediate early genes (IEGs), 17, 307
Immunocytochemistry, receptor localization,
    24–25
Immunohistochemistry methods, 11
Impulse control disorder, 425
Impulsivity
    addiction and, 399–400
    definition of, 371
Incentive, 287
Incentive salience, 203, 206, 390
Indirect pathway. See Striatopallidal
    pathway
Inhibitory control, substance abuse long-term
    effects, 410, 410f
In situ hybridization, receptor localization, 24
Insomnia, in Parkinson's disease, 426
Instrumental learning, DA and, 3934–395
Instrumental reinforcers, 389
Integrative dopamine pathways, 40–46
    convergence, 41
    corticostriatal projections, 41–42, 41f
    functional considerations, 45–46, 45f
    midbrain dopamine system and, 42–45,
        42f, 43f
Internal pathway. See Striatopallidal pathway
Interspike interval (ISI), 118, 127–129
Intra-Acb drug infusion studies, 308–309,
    309f
Intracellular dopamine signaling, 100–111
    A-kinase anchoring protein and, 103
    behavioral effects, 333
    DARPP-32 and, 103–104, 104f
    dopamine receptor classification and, 101
    future research, 110–111
    GPCR-kinase-mediated desensitization
        of D1Rs, 102–103
    of fragile X mental retardation protein,
        102–103
    medium spiny neurons and, 100–101
    protein kinase A and, 103
    regulation
        of AC, 102
        of adenylyl cyclase, 102
        of cAMP, 102
    via D1R, 101–102, 110, 110f
        AMP regulation, 103
        control of voltage-dependent Ca^{2+}
            channels, 105–106
        $GABA_A$ regulation, 103
        GPCR-kinase-mediated desensitization,
            102–103
        regulation of NA+ channels, 105
    via D2R, 107–111, 110–111, 111f
        control of cAMP/PKA/DARPP-32
            cascade, 108
        G_αi/o protein interaction, 107–108
        $G\beta\gamma$ and, 107–108
        isoforms and, 110
        NMDR regulation, 108–109
        regulation of voltage-dependent Ca^{2+}
            channels, 108
        regulation of voltage-dependent K^+
            channels, 108

Intradimensional shift test, 209, 211f
Intrastriatal cell suspension graft
  anatomical reconstruction, 455–458, 456f, 457f
  method, 454–455, 455f
  neurochemical reconstruction, 455–458, 456f, 457f
  into substantia nigra, 458–459, 460f
Intrinsic excitability modulation
  by dopamine, 351, 353–354
  by glutamatergic signaling by D2Rs, 354
Inverted-U dose-response curve, 230, 235–238
Ion channels
  Ca^{2+} (See Calcium ion channels)
  Ca^{2+}-dependent, 261
  D1DAR-PKA regulation of, 67
  in dopamine neurons, 118
  K+ (See Potassium ion channels)
  KCNQ, 123–124
  ligand-gated, 118
  in midbrain dopamine neurons, 122, 123–124
    depolarization, 125–127, 126f
    functional diversity and, 129–132, 131f
    interspike interval, 127–129
    repolarization, 127
    in vitro burst activity and, 129
  regulation
    by D2 dopamine receptors, 108–109
    upstream, 118–119
  repolarization, in midbrain dopamine neurons, 127
  sodium or Na+
    tetrodotoxin-sensitive voltage gated, 126–127, 126f
    voltage-dependent, 353–354
  voltage-gated, 118
ISI (interspike interval), 118, 127–129
Isthemic organizer (IsO), 146

Kandel, Eric, 301
KCNQ channels, 123–124
Ketamine, 527
KF17837, 292
Kir3, 124
Knockdown mice, VMAT2, 90
Knockout mice
  DAT, 90–94, 91f
  D1DAR, 63
  Gαolf, 101–102
  VMAT2, 89
KW6002, 292

Law of initial value, 250
L-DOPA
  activation of ERK 1/2 in dopamine-depleted striatum, 18–19, 19f
  adverse effects, 490
  dyskinesia induced by
    (See L-DOPA-induced dyskinesia)
  for Parkinson's disease, 423, 490
    clinical response, 430–432, 431f, 575
    COMT phenotype modulation of, 210–211
    neuroprotective studies, 448–449
    risk factors for motor complications, 434
    side effects, 434
    sleep beneficial effect, 429
    therapeutic rationale, 434
    weak/absent response, 428–429
  in rodent parkinsonism model, 336
L-DOPA-induced dyskinesia (LID)
  altered plasticity in corticostriatal synapses and, 438–439
  animal models, 434–435, 435f
  corticobasal ganglionic circuits, system-level changes in, 440
  diphasic, 434
  dopamine neuron grafts and, 464–466, 465f
  end-of-dose deterioration, 434
  GABA-ergic pathway alterations, 439–440
  multilayered pathophysiology of, 435–436, 436f
  in Parkinson's disease, 434–441
  pathophysiology, 490
  peak-of-dose, 434, 435
  peptidergic pathway alterations, 439–440
  striatal dopamine, high, 437–438
  striatal efferent pathway imbalance in, 437–438
  wearing-off phenomenon, 434, 435
Learning
  action-outcome, 394
  appetitively motivated
    future research directions, 310–311
    strial role in, 306–308, 307f
  artificial or machine, 268–269
  cellular models, addictive drugs and, 393–394
  episodic, 462
  instrumental, DA and, 3934–395
  motor, in 6-OHDA lesion model after DA neuron transplantation, 462–463
  reinforcement (See Reinforcement learning)
  reward-based (See Reward-based learning)
  stimulus-reward, 391
  temporal-difference-error, 269
Leucine-rice repeat kinase 2 (LRRK2), 503
Lewy bodies, 281, 502
Lewy body disease, 424, 427
LFP (local field potential), 440
LID. See L-DOPA-induced dyskinesia
Limbic corticostriatal circuitry, DA modulation of, 391
Limbic midbrain area, 373
Limbic selectivity, of atypical antipsychotic drugs, 545–546
Limb-use asymmetry, in unilateral 6-OHDA lesion model, 280–281
Lisuride, D2DAR agonism, 71t
Lmx1a, 148–149
Lmxb1, mdDA precursors and, 151

Local field potential (LFP), 440
Long-term depression (LTD)
  AMPAR levels and, 342–343
  behavioral sensitization and, 344
  DA modulation of, 354–358, 356f, 357f
  DA regulation of, 268
  dopamine receptors and, 101
  D2R-dependent, 362
  D1R effect on synaptic plasticity and, 270
  endocannabinoid-mediated, 342
  Hebbian, induction of, 356
  high-frequency stimulation of, 339, 354–355
  induction, 339, 354–355
    DA and, 390
    loss of D2R stimulation and, 361
    in NAcb, 340
    in VTA DA neurons, 340
  in midbrain DA neurons, addictive drugs and, 393–394
  production sequence, 6
  spike-timing dependent plasticity and, 269
  in VTA DA neurons, 25
Long-term potentiation (LTP)
  $A2_a$-dependent, 362
  blockage by D1 antagonists, 306
  cocaine-dependent, 342, 594
  conversion to LTD, 270, 271
  D1 agonists and, 178
  dopamine receptors and, 101
  drug-induced, 341–342
  Hebbian, induction of, 356
  induction, 355, 408, 439
    by cocaine, 341
    DA and, 390
    in midbrain DA neurons, addictive drugs and, 393–394
  NMDA-dependent, 306
  NMDA receptor-dependent, 339
  production sequence, 6
  regulation
    by dopamine, 268
    by in NAcb DA receptors, 340
  spike-timing dependent plasticity and, 269
  in VTA DA neurons, 25, 340
LRRK2 (leucine-rice repeat kinase 2), 503
LTD. See Long-term depression
LTP. See Long-term potentiation
LY293558, 528
LY354750, 529
LY379268, 529

Machine learning, 268–269
Magnetic resonance imaging (MRI), in Parkinson's disease, 427–428, 446
Major depressive disorder (MDD)
  costs of, 572
  DA, stress and, 576
  deaths related to, 572
  diagnostic criteria, 572–573, 573t
  inheritance, 574

Major depressive disorder (MDD) (*Continued*)
  pathophysiology, 572–583
    dopamine neurotransmission and, 575–576
    neuroimaging findings, 578, 579t
    postmortem findings, 578
    startle reflex, 577–578
  placebo effect, 576
  prevalence, lifetime, 572
  risk for developing, 572
  severity, 576
  treatment, 573–574, 578–579, 581
    DA interactions with 5HT and NE, 582–583
    deep brain stimulation, 582
    repetitive transcranial magnetic stimulation, 582
MAM rat model
  cortical interneuron deficits, 183
  hippocampus in schizophrenia pathophysiology, 516, 592–593, 593f
MAO (monoamine oxidase), 91
MAO-A (monoamine oxidase A), 409–410, 409t
MAO-B (monoamine oxidase B), 409–410, 409t
MAP (mitogen-activated protein), 307
MAPKs (mitogen-activated protein kinases), 74–75, 105
Mash1, 148
Matrisomes, 51
MDD. See Major depressive disorder
md DA neurons. See Meso-diencephalic DA neurons
MDMA, 93
Medium spiny neurons (MSNs)
  as coincidence detectors, 302
  cultured, 351
  D1DAR expressing, 349–351, 350f
    DA depletion and, 358, 359f–360f, 360
    dendritic anatomy, 351, 350f
  D2DAR expressing, 349–351, 350f
    bAP-evoked Ca^{2+}-transients, 351, 352f–353f, 361
    DA depletion and, 358, 359f–360f, 369
    dendritic anatomy, 351, 350f
  direct-pathway, 437–438
  D2R-expressing, 270
  excitation, 354
  GABAergic, 302
  indirect-pathway, 437–438
  inhibition, 354
  NAcb, cocaine effects on LTD, 343
  peptide markers in, 302
  single cell input/output function, 262–263
  striatal, 51, 100–101, 349, 437
  striatonigral, 362
    activation, 100
    bidirectional STDP, 356–357, 356f, 357f
    DA depletion, 358, 359f–360f, 360
    D1Rs, 353
    LTD induction in, 355
  striatopallidal, 362–363
    activation, 100
    DA depletion, 358, 359f–360f, 360
    thalamic innervation of, 355, 362
    thalmostriatal neurons and, 52f, 52–53
Membrane potential recordings, in dopamine neurons, 119–120
Memory. See Working memory
Mesencephalic dopamine neurons, naturally occurring cell death, 160–163, 161f, 162f
  regulation by extrinsic molecules, 166–171
  regulation by intrinsic molecules, 171–172, 171f
Mesocortical dopamine system, 207–209, 208f
Mesocorticolimbic dopamine
  behavioral flexibility and, 221–226
    primate studies, 221–222
    rodent studies, 222–224
  cognition and, 215–226
  executive function and, 215–226
  working memory
    primate studies, 215–216
    rodent studies, 216–221, 217f
Meso-diencephalic DA neurons (mdDA neurons)
  development
    early differentiation, 143, 143f, 144f
    induction, 143, 143f, 144f
    induction progenitor domain, 144–148
    initiation, 147–148
    late differentiation, 143, 143f, 144f
    maintenance, 143, 143f, 144f
    spatial origin, 141–142, 142f
    specification, 143, 143f, 144f
    time course, 142–144, 143f, 144f
  maintenance, 153–154
  MHB position and, 145–147
  progenitors/precursors, 148–152
    mdDA-specific characteristics of, 150–152
    neuronal properties of, 148–150
    terminal or late differentiation, 152–153
  size of population, MHB position and, 145–147
Mesolimbic dopamine system
  functions, 205–207, 207f
  as limbic-motor interface, 339
  reward and, 286–287
  synaptic plasticity, 339–340
  ventral striatum (See Nucleus accumbens)
  ventral tegmental area (See Ventral tegmental area)
Mesostriatal dopamine system
  activational effects, 204–205, 205f
  cell groups, 11
  functions, 203–212
  operant stereotyped behavior, 205, 206f
  overactivation, 205, 205f
  projections, 11
  Yerkes-Dodson principle and, 205, 205f
  behavioral flexibility and, 224–226

Metabotropic glutamate receptors, 529
Methamphetamine (METH)
  abuse/addiction
    D2 DA polymorphisms and, 412, 412f
    D2 dopamine receptors and, 400
    executive function and, 410, 410f
    motor activity effects, 250
  neurotoxicity, 94
  VMAT2 and, 89
Methylazoxymethanol acetate (MAM). See MAM rat model
Methylphenidate
  DAT interaction, 92–93
  reinforcement, DA and, 407–408
  vulnerability to abuse, 411–413, 412f
mGluR$_{2/3}$ agonists, 529
MHB position, location/size of mdDA population and, 145–147
MHO (mid-/hindbrain organizer), 146
Microglial activation, in Parkinson's disease, 451, 451f
Midbrain
  D2 receptors, 30
  D3 receptors, 30
  spiraling connections with, 396, 396f
Midbrain dopamine neurons
  afferent projections, 43, 43f
  anesthetized animals, in vivo activity patterns in, 122–124
  awake animals, in vivo activity patterns in, 120–122, 121f
  burst activity mode, 121–125, 121f, 125f
  classes of, 590
  D2 autoreceptors, 25
  degeneration, differential vulnerability, 132–133
  efferent projections, 43–44, 43f
  firing modes, 515–516
  functional considerations, 45–46, 45f
  functional diversity, 129–132
    electrophysiologic analysis, 129–130
    ion channels and, 131–132, 131f
  ion channels, interspike interval, 127–129
  malfunctions, 22
  meso-cortico-limbic, cognitive/emotional symptoms in PD and, 431
  organization, 42–43, 42f
  pacemaker or tonic activity mode, 120–121, 121f, 124–125, 125f, 130–132, 131f
  striato-nigro-striatal projections, 44–45, 45f
  in vitro activity patterns, 124–129, 125f, 126f
Mid-/hindbrain organizer (MHO), 146
Minocycline neuroprotectivity, in Parkinson's disease, 499t
Mitogen-activated protein (MAP), 307
Mitogen-activated protein kinases (MAPKs), 74–75, 105
MK869, 530
M1-like receptors, 358
M2-like receptors, 358

M1 muscarinic cholinergic receptors, striatal distribution, 335
Molecular-genetics approach, 532
Monoamine hypothesis of depression, 573–574
Monoamine oxidase (MAO), 91
Monoamine oxidase A (MAO-A), 409–410, 409t
Monoamine oxidase B (MAO-B), 409–410, 409t
Monoamines, extracellular concentrations, 88
Mood control, 337
Mood disorders
  depression (See Depression; Major depressive disorder)
  due to medical disorder, diagnostic characteristics, 573f
  substance-induced, diagnostic characteristics, 573f
  substance-induced, diagnostic characteristics of, 573f
Morphine
  BSR threshold lowering, 308
  conditioned place preference to, 342
  exposure, synaptic plasticity in VTA and, 341–342
  mechanisms of action, 390
  pleasure or high, 390
Motivation
  activational aspects, 287, 288
  control, 337
  definition of, 287
  directional aspects, 287
  dopamine and, 407
  drive and, 287
  incentive and, 287
  substance abuse, long-term effects on DA, 408–410, 409t
Motivational withdrawal, definition of, 376
Motor learning, in 6-OHDA lesion model after DA neuron transplantation, 462–463
Motor scales, for Parkinson's disease, 430
Mouse models
  behavioral testing in, 463
  knockout
    DAT, 90–94, 91f
    D1DAR, 63
    Gαolf, 101–102
    VMAT2, 89
  of Parkinson's disease
    MPTP lesion model (See MPTP lesion model, of PD)
  of Parkinson's disease
    motor function, 279–283
    sensorimotor tests, 281–283
    transgenic, 6, 18
    VMAT2 knockdown, 90
    Weaver, 133
Movement disorders. See Parkinson's disease
MPP+ toxicity, 89–90
MPTP lesion model, of PD
  behavioral testing in, 463

corticobasal ganglionic circuit alterations, 435
differential vulnerability of dopamine midbrain neurons, 133–134
GDNF-based gene therapy, 493
methodology, 89, 94
motor complications, 435
pro-drug approach for enhanced DA synthesis, 490–491
MRI (magnetic resonance imaging), in Parkinson's disease, 427–428, 446
MSNs. See Medium spiny neurons
Msx1, 148–149
MSX-3, 292
Mu opioid receptors, 335
Muscarinic receptors, M_1, as atypical antipsychotic targets, 526, 527t
Muscimol, 293–294
Myr-Akt, 171–172, 171f

NAcb. See Nucleus accumbens
Na+ ion channels (sodium)
  tetrodotoxin-sensitive voltage gated, 126–127, 126f
  voltage-dependent, 353–354
Naturally occurring cell death, in mesencephalic dopamine neurons, 160–163, 161f, 162f
  apoptotic, 160–163, 162f
  molecular regulation of, 166–172, 171f
  systems regulation of, 163–165
  time course, 160–161, 162f
Natural reinforcers, of additive behavior, 391
Negative reinforcement, 371
Neonatal ventral hippocampal lesion (NVHL), abnormal periadolescent maturation of DA actions, 181–183, 182f–184f
Nestin-Cre/+ gene, 148
NET inhibitors, 93
Neural network dynamics
  biophysical modeling, 263f, 264
  D1R modulation of working memory, 264, 265f, 266–267, 266f
Neural network hypothesis of schizophrenia, 532
Neurocomputational analysis, of dopamine function, 261–272
  neural network dynamics, 263–264, 265f, 266–268, 266f
  reinforcement learning, 268–270, 270f
  signal cell input/output, 261–263, 262f, 263f
Neurog2 (Ngn2), 149
Neuroimaging
  dopamine system
    in major depressive disorder, 578, 579t, 580t
    in prefrontal cortex, 189–192, 191f
    long-term effects of substance abuse on dopamine

conditioning circuits, 411
executive function, 410, 410f
inhibitory control, 410, 410f
of Parkinson's disease, 427–428
microglial activation and, 451, 451f
progression, 445–451
Neurokinin receptors, as atypical antipsychotic drug targets, 529–530
Neurokinin system, 529–530
Neuroplasticity. See Plasticity
Neurotensin antagonists, 530
Neurotensin receptors, as atypical antipsychotic drug targets, 530
Neurotrophic factor regulation, of NCD in mdDA neurons, 169–171
Neurturin
  gene therapy with, 494–495
  regulation of NCD in mdDA neurons, 168
Ngn2 (Neurog2), 149
Nicotine
  chronic exposures, 408
  exposure, synaptic plasticity in VTA and, 341–342
Nicotinic agonists, 526–527
Nicotinic receptors, as atypical antipsychotic targets, 526–527
Nigrostriatal dopamine system (pathway)
  degeneration in Parkinson's disease, 423, 430, 431, 431f
  developmental milestones, 160, 161f
  organization, 12, 13f
Nisoxetine, 93
NK$_1$ antagonists, 530
NMDA (N-methyl-D-aspartate)
  conductance in biophysical model, 263f, 264
  D1 receptor transmission control, 104–105, 178
  modulation of pyramidal cell excitability, 179–180, 179f
  voltage-gated Ca^{+2} ion channels, 271
NMDA antagonist MK-801, 93
NMDA receptor antagonist (AP-5), 305
NMDA receptors (NMDARs)
  activation, 18, 261, 339
  coincident with D1R activation, 306–307, 307f
  in instrumental learning, 308
  blockade, 394
  D5DAR and, 78, 79f
  D1/PKA cascade and, 353
  D1R stimulation, 355
  hypofunction, in schizophrenia, 528
  long-term potentiation and, 339
  in schizophrenia, 541
  structure, 527–528
  up-states, 204
N-methyl-D-aspartate. See NMDA
Nomifensine, 574, 575
Norepinephrine-dopamine interaction, 582–583
NSD-1015 model, 92

Nucleus accumbens (NAcb; Acb)
  AMPAR signaling, 343–344
  behavioral responding, 344–345
  core, 305, 339
    amylin receptors, 305
    cocaine self-administration and, 394
    drug-seeking behavior and, 395–398, 396f–398f
    functional distinctions from shell, 305–306
    inputs, 305
    motor output, 305–306
    structure of, 375
  dopamine, 286–287
    behavioral activation functions, 287–288
    depletion, 288–289
    in effort-related choice behavior, 289–292, 291t
    in effort-related processes, 292–294, 294f
    in food-motivated instrumental behavior, 288–289
    transmission, vs. caudate transmission, 303
  drug infusion studies, 308–309, 309f
  glutamate, during drug exposure and reinstatement, 343
  glutamatergic inputs, 339
  as limbic-motor interface, 303
  opioid peptide infusions, 310
  reward processing, 344–345
  shell, 39, 305, 339
    cocaine self-administration and, 394
    DA responsiveness in, 375–376
    functional distinctions from core, 305–306
    GABA receptor stimulation, 305
    inhibition of cocaine-induced reinstatement, 344
    inputs, 305
    structure of, 375
    as viscero-endocrine region, 305
  synaptic plasticity
    DA and, 340
    induced by drugs of abuse, 342–344
  taste reactivity, 310
Nullclines, 265f
Nurr1 (Nr4a2)
  mdDA neuron maintenance, 154
  mdDA precursors and, 150–151, 153
NVHL (neonatal ventral hippocampal lesion), abnormal periadolescent maturation of DA actions, 181–183, 182f–184f

Oculomotor delayed response task (ODR), dopamine effects, 234–236
OFC (orbitofrontal cortex), 39, 41, 43, 410
6-OHDA lesion model
  alternative tissue sources for transplantation, 468–469, 469t
  anatomical reconstruction, 455–458, 456f
  behavioral testing
    in mice, 463
    in rats, 459–462, 461t
  BSR threshold lowering by morphine, 308
  DA neuron transplantation
    cognition and, 462–463
    dyskinesia and, 464–466, 465f
    method, 454
    motor learning and, 462–463
    into substantia nigra, 458–459, 460f
  GDNF-based gene therapy, 492–493
  L-DOPA-induced dyskinesia and, 430, 435–437
  mapping DA effects in striatal subregions, 302
  neurochemical reconstruction, 455–458, 456f
  neurturin-based gene therapy, 495
  nigrostriatal nerve cell death in, 164
  presynaptic changes in DA release/clearance, 436–437
  reward effect from drugs of abuse, DA role in, 371–373, 372f, 374f
  tests, 280
    catch-up steps, 281
    limb-use asymmetry, 280–281
    voluntary movement initiation, 280–281
  unilateral technique, 204
Olanzapine, 528
Operant behavior, stereotyped, 205, 206f
Operant choice task, effort-related choice behavior and, 290
Operant conditioning, 268
Opiates, chronic exposures, 408
Opioid-dopamine interactions, 308
Orbital prefrontal cortex (OPF), 391
Orbitofrontal cortex (OFC), 39, 41, 43, 410
Organizers, 145
Orthostatic hypotension, 426
Osanetant (SR142801), 530
Otx2, mdDA neuron development and, 143f, 144–145, 149–150
Overdosing effect, 211–212

Pacemaker (tonic activity), midbrain dopamine neurons, 124–125, 125f
Pallidotomy, for Parkinson's disease, 440
Pamipexol, D2DAR agonism, 71t
Panamesine (EMD-57445), 530
Par-4 (prostate apoptosis response protein), 574
Parallel processing circuits, 38, 39–40
Parkin, 502
Parkinsonian syndrome, 421
Parkinson plus syndromes, 421
Parkinson's complex, 426
Parkinson's disease (PD)
  age of onset, 422t, 427
  animal models
    dopamine neuronal transplantation, 454–470

L-DOPA therapy, 336
MPTP-lesion (See MPTP lesion model, of PD)
basal ganglia pathways in, 15–17, 16f
definition of, 430
with dementia, 445, 446–447
depression in, 576
diagnosis, 427–428
differential diagnosis, 421, 422f, 427
D1R action in PFC and, 241
dyskinesia, L-DOPA-induced, 17–19, 18f, 434–441
etiopathogenesis, 15, 501–504, 501f
  environmental, 501
  familial, 501–502
Gαolf expression, 102
glucose metabolism and, 446–447, 448f
glutamate signaling, DA modulation of, 358, 359f–360f, 360–363
hereditary forms, 421, 422t
imaging, 427–428
  of cholinergic function, 450–451
  of dopamine system, 447
  of later disease stages, 428
  of microglial activation, 451, 451f
  MRI, 427–428, 446
  neuroprotective studies, 448–449
  of nigral structures, 445–446, 446f
  of presynaptic dopaminergic function, 447–448, 448f
  of progression, 446
  of restorative therapies, 449–450
indirect pathway and, 15–17, 16f
microglial activation, 451, 451f
midbrain dopamine neurons, differential vulnerability to degeneration, 132
misdiagnosis, 427
motor impairments
  animal studies, 279–283
  dopamine neurons and, 316, 317f
  striatopallidal MSNs and, 363
as multisystem disorder, 424–426, 425t
neuroprotective strategies, 498–506
  animal models and, 504
  clinical endpoints, 504–505
  dosing, 504
  etiopathogenesis and, 501–504, 501f
  negative clinical trials, 499t
  positive clinical trials, 498–500, 499t
nigrostriatal dopamine in, 204
6-OHDA, urinary, 280
pathophysiology, 430
  degeneration of multiple nondopaminergic systems, 431–432
  degeneration of nigrostriatal pathway, 430–431, 431f
  loss of dopamine neurons, 421, 423f
  substantia nigra pars compacta degeneration, 38
postural instability, push-pull test for, 281
prognosis, 428–430, 429f

progression, 445–451
  MRI volumetric studies, 446
  neuroimaging of, 428
psychiatric problems in, 424–426, 425t
risk factors, 445
rodent models, 17, 89–90, 204–205
  genetic, development of, 279–283
  motor function, 279–283
  MPTP-induced in mice, 279, 430
  neuroprotective drug development and, 504
  6-OHDA-induced, 430
  6-OHDA-lesioned rats, 436–437
signs
  axial, prognosis and, 428–429, 429f
  rigidity (*See* Rigidity, in Parkinson's disease)
striatal glutamatergic signaling, 349–363
symptoms, 430
  akinesia, 15, 16, 421
  beyond motor triad, 426, 426t
  bradykinesia, 421
  classic triad, 421
  dysautonomia, 426
  dystonia of feet, 429
  in feet, 421
  hand tremor, 421
  hypokinesia, 421–422
  hyposmia, 426
  improvement from dopamine neuron transplantation, 480–481, 480t, 482f
  insomnia, 426
  Lewy body inclusions, 445
  resting tremor, 422–423
treatment, 88, 430
  DA cell replacement, 483–484
  DA neuron replacement (*See* Dopamine neuron transplantation)
  DA transmission in PFC and, 242
  D2-like agonists, 74
  gene-based therapy (*See* Gene-based therapy)
  L-DOPA, 17–19, 423
  motor scales and, 430
  nocebo effect, 576
  pallidotomy, 440
  pharmacologic, 498, 499t
  placebo effect, 576
  restorative therapies, 449–450
  surgical, 498
Pavlovian associations, 305–306
Pavlovian conditioning, of additive behavior, 390–393, 391f
Pavlovian-instrumental transfer (PIT), 391–392, 394
PD. *See* Parkinson's disease
PD149163, 530
Pedunculopontine nucleus (PPN), 425, 425f
Peptidergic pathway alterations, in L-DOPA-induced dyskinesia, 439–440
Pergolide, D2DAR agonism, 71t
PET. *See* Positron emission tomography
PFC. *See* Prefrontal cortex

Phencyclidine, 527
Phosphatidylinostiol-3 kinase (PI3K), 170
Phospholipase C (PLC), 358
Phospholipase D (PLD), 79
Phosphoproteins, 301
Phosphorylation, in D1DAR regulation, 68–69
Physiologic models, cognitive implications, 263–264, 263f, 265f, 266–268
PI3K (phosphatidylinostiol-3 kinase), 170
Piribedil, 575
PIT (Pavlovian-instrumental transfer), 391–392, 394
Pitx3, mdDA precursors and, 151–152
PKA. *See* Protein kinase A
PKC (protein kinase C), activation, 354
Plasma membrane regulation, of dopaminergic neurotransmission, 90–94, 91f
Plasticity. *See also* Long-term depression; Long-term potentiation
  addictive drugs and, 393–394
  alterations in corticostriatal synapses, L-DOPA-induced dyskinesia and, 438–439
  circuit-level, in striatal gene activation, 335–336
  dopamine neurons and, 390
  future research directions, 310–311
  in instrumental learning, 308
  in mesolimbic neurons, 394
  spike-timing dependent or STDP, 269–270, 270f, 355–356, 356f
  structural, D2 receptor regulation, 29
  synaptic, 269
    after drug exposure, 340–341
    dopamine and, 333
    intracellular signaling, 306
    in mesolimbic circuits, dopamine and, 339–340
    in mesolimbic system, 339–340
    in nucleus accumbens, 340
    striatal, forms of, 6
    in ventral tegmental area, 340
    in VTA DA neurons, 341–342
  systems-level, dopamine and, 333–334
  of thalamostriatal synapses, 53
PLC (phospholipase C), 358
PLD (phospholipase D), 79
PLS (posterior lateral striatum), 308
Policy, in machine learning, 269
Positive reinforcement, 371
Positron emission tomography (PET)
  with [^{11}C]raclopride, cocaine drug cues increase DA in dorsal striatum, 411
  with D2 dopamine receptor ligands, applications, 407
  D2R binding, in major depressive disorder, 578, 579t, 580t
Posterior lateral striatum (PLS), 308
Postmortem studies, of major depressive disorder, 578
Posttraumatic stress disorder (PTSD), 573f

Potassium ion channels (K+ ion channels)
  ATP-sensitive, 129
  A-type, 128
  Ca^{2+}-activated, 127
  G-protein-activated inwardly rectifying or GIRK, 124
  G-protein-coupled inwardly rectifiying (Girk; Kir3), 128
  KCNQ, 358
  Kir2, 126, 358
  Kv4, 358, 361
  leak, 126
  voltage-activated, 127
  voltage-gated, D2R regulation of, 108
PPN (pedunculopontine nucleus), 425, 425f
PPP1R1B genetic variation, 565, 566f
Pramipexole neuroprotectivity, in Parkinson's disease, 499t
Prefrontal cortex (PFC)
  attractor dynamics/function, D2R modulation of, 267–268
  DA activity, 177–184
    abnormal periadolescent maturation, 181–184, 182f–184f
    dopamine receptors and, 215
    electrophysiological analysis, 177–178, 178f
    GABA interneurons in adolescence, 180–181, 180f, 181f
    in pyramidal neurons in adolescence, 179–180, 179f
  DA depletion, 230, 251
  dopaminergic mechanisms
    genetic studies, 189–192, 191f
    imaging studies, 189–192, 191f
  dopamine transmission
    ADHD and, 242–243
    drug abuse and, 243
    future research directions, 244
    Parkinson's disease and, 242
    schizophrenia and, 243, 244f
  dorsal (*See* Dorsal prefrontal cortex)
  dorsolateral, dysfunction in schizophrenia, 511–512, 513f, 559–562, 561f
  D1R actions, optimal, 240–242, 241f
  D2 receptor on, 6
  D1 stimulation, 264
  dysfunction in schizophrenia, 511–512, 513f, 559–562, 561f
  functions, 230–232
    DA effects on, 234–239, 236f, 237f
    genetic links with DA-associated intracellular signaling models, 194–195, 196f
    working memory (*See* Working memory)
  inputs, 41
  network connectivity, 239–242
  neurocircuitry, 230–232
  neurons
    delay-related firing, 231–232
    preferred direction of, 230–231, 231f
    pyramidal, 230, 262–263

Prefrontal cortex (PFC) (*Continued*)
  response-related firing or saccadic activity, 238–239
  orbital, 230
  outputs, 215
  physiology, DA effects on, 230–232, 234–239, 236f, 237f
  primate studies
    D2 actions, 238–239, 240f
    D4 dopamine receptor, 239
    D1 dopamine receptors, 232
    D2 dopamine receptors, 232, 233f, 234
    D4 dopamine receptors, 234
    D5 dopamine receptors, 232–234, 233f
    mechanistic, 237–238
  pyramidal neurons, gating network inputs, 239
  signal-to-noise processing, 188–189
  ventral medial, 39, 43
  ventrolateral, 192, 193f
  working memory and, 188
Prefrontal cortical-striatal mechanisms
  epistatic gene, 192, 194, 194f
  of schizophrenia, 187–197
Prenatal stress, addiction vulnerability and, 381–382
Preproenkephalin, 302
Primate models
  mesocorticolimbic DA, behavioral flexibility and, 221–222
  ODR task, D1R inverted-U physiological actions, 235–236
  prefrontal cortex
    cognition, 230–244
    DA projections, 232
    D1 dopamine receptors, 232–234, 233f
    D2 dopamine receptors, 232, 233f, 234
    D4 dopamine receptors, 234
    D5 dopamine receptors, 232–234, 233f
    synapses in, 232
  of spatial working memory, ODR task, 230–231, 231f
  working memory, 215–216
PRODH (proline dehydrogenase gene), 565–566
Progressive supranuclear palsy, 429–430
Proline dehydrogenase gene *(PRODH)*, 565–566
Propargylamine THC346, 504
ProSavin vector, 492
Prostate apoptosis response protein (Par-4), 574
Protein kinase A (PKA), 306
  blockage of D1Rs in PFC, 178
  cross-talk with cAMP and DARPP-32, 106–107, 106f
  D1DAR activation of, 67, 353
  instrumental learning and, 394
  intracellular dopamine signaling and, 103
  phosphorylation, 353–354
  RIIβ isoform, 103
Protein kinase B gene, 564–565
Protein kinase C (PKC), activation, 354

Proteolytic stress, in cell death process in Parkinson's disease, 502
PSD-95 binding, to D1 receptors, 105
Psychiatric disorders. *See also specific psychiatric disorders*
  in Parkinson's disease, 424–426, 426t
  postsynaptic actions of DA and, 594–595, 595f
Psychostimulants. *See also* Amphetamines; Cocaine
  addiction vulnerability
    in early drug and stress exposure, 381–382
    individual differences in DA utilization, 377–379
    stress, DA and glucosteroid interactions, 377–381, 380f
  addiction vulnerability to, 382–383
  chronic exposures, 408
  DAT and, 90–94
  DAT interaction, 92–94
  IEG induction, 17–18
  incentive salience, 390
  oral, 408
  reinforcing effects, 376
  withdrawal, 376
PTSD (posttraumatic stress disorder), 573f
Punding, 425, 425t
Punishment, 327

Quinpirole, D2DAR agonism, 71t

Raclopride, effort-related choice behavior and, 290–291, 291t
Radial-arm maze task, DA neurotransmission in PFC, 216–221, 217f
Rat models
  of compulsive drug seeking, 400–401, 401f
  of depression, reward system dysfunction and, 577
  food-reinforced lever pressing, DA antagonism and, 288
  motor functions, dopamine requirements for, 279–280
  NAcb DA depletion, 288–289
  6-OHDA-lesioned (*See* 6-OHDA lesion model)
  of Parkinson's disease (*See* Parkinson's disease, rodent models)
  self-activated movement, 279
  stability-regaining reactions, 279
  voluntary movement, 279
REAL PET trial, 448
Real-time reverse transcriptase polymerase chain reaction (RT-PCR), receptor localization, 24
Rebound psychosis or supersensitivity psychosis, antipsychotic drugs and, 547
Reboxetine, 93, 577

Recombinant LV vectors, for GDNF-based gene therapy, 494
Regulators of G protein signaling (RGSs), 73, 102, 109
Reinforcement learning
  DA and, 389–390
  DA-based models, 268–272, 270f
  dopamine and, 407
  of drug addiction, treatment implications, 413
  synaptic weights and, 268
  ventral striatum and, 38
REM sleep behavior disorders, 425, 426
Repetitive transcranial magnetic stimulation (rTMS), for major depressive disorder, 582
Reserpine, 279
Resting tremor, in Parkinson's disease, 428, 429f
Retrorubral cells, 42
Reward
  activation, transfer to reward-predicting stimulus, 318, 319f, 320f
  definition of, 287
  dopamine hypothesis of
    conceptual problems with, 286–287
    empirical problems with, 286–287
  from drugs of abuse, DA role in, 372–376
    lesion studies, 372–373, 374f
    pharmacological studies, 376
  magnitude coding of dopamine neurons, 317–318, 318f
  prediction error coding, 318–319, 322–323, 323f
  prediction errors, 318–319, 321f, 322f
  processing, 462
  dopamine neurons and, 316–317, 317f
  in NAcb, 344–345
  risk, 325–326, 326f
  substance abuse, long-term effects on dopamine, 408–410, 409t
  ventral striatum and, 38
Reward-based learning
  blocking paradigm, 319, 322, 322f
  conditioned inhibitors, 323–324, 324f
  dopamine and, 548
  dopamine neurons and, 317–319, 317f, 318f–324f, 322–324
  plasticity in, 307, 307f
  temporal difference model, 318–319
Reward expectancy, 333–334
Reward/motor dichotomy, 302
Reward prediction error, 390
Reward-related behavior, modulation, 302
Reward-seeking behavior, dopamine and, 548
Reward system dysfunction, in depression, 576–577
RGSs (regulators of G protein signaling), 73, 102, 109
RhoA activation, by D2DAR, 72–73
Rifampicin, 502
Rigidity
  clasp knife, 421

oppositional or Gegenhalten, 421
in Parkinson's disease
cogwheel, 421
lead pipe, 421
prognosis and, 428, 429*f*
Rimcazole (BW 23FU), 530
Risperidone, 71*t*, 528
Rodent models. *See also* Mouse models; Rat models
of depression, 574
chronic mild stress, 574
effort expenditure, 574
forced swim test, 574
learned helplessness, 574
of drug exposure, 341
mesocorticolimbic DA, behavioral flexibility and, 222–224
of Parkinson's disease (*See* Parkinson's disease, rodent models)
of schizophrenia, 181–183, 182*f*, 183*f*, 516
T-maze procedure, 290
working memory, DA neurotransmission in PFC, 216–221, 217*f*
Ropinirole
D2DAR agonism, 71*t*
neuroprotectivity, in Parkinson's disease, 499*t*
Rotometer bowl quantitation, of rotation in 6-OHDA-lesioned rats, 459–461
rTMS (repetitive transcranial magnetic stimulation), for major depressive disorder, 582
RT-PCR (real-time reverse transcriptase polymerase chain reaction), receptor localization, 24

SCH-23390
blockage
of D1rs, 307–308, 356–357
of single cell input/output function, 262
DA neurotransmission in PFC and, 216
effort-related choice behavior and, 290–291, 291*t*
memory impairment, 523
mesocortical DA system and, 208
receptor affinity, 102
SCH39166, 523
Schedule-induced locomotor activity, NAcb DA and, 288
Schizophrenia
animal models, 181–183, 182*f*, 183*f*, 516
clinical presentation, 559–562, 560*f*, 561*f*
cognitive abnormalities, 188
cognitive dysfunction
genetic variants and, 566–567
DA transmission in PFC and, 243, 244*f*
dopamine dysregulation, 511–517
cellular implications, 515–516, 515*f*
in corticostriatal circuits, 516
extrastriatal D2 receptors, 514–515
future research directions, 516–517

increased striatal, 513–514
in prefrontal cortical D1R, 511–512, 513*f*
dopamine hypothesis of, 520, 540–541
historical perspective, 558–559
historical perspective of, 558–559
integration with glutamate hypothesis, 541
dopamine-mediated prefrontal-striatal mechanisms, 187–197
dopamine receptor dysregulation, 562–563
dopaminergic risk genes, 563–566
etiology, 540
genetics of prefrontal cognitive functions, 187–197
intermediate phenotype approach, 188
neural network dysfunction, 559–562, 561*f*
neurokinin system in, 529–530
neurotensin in, 530
pathophysiology
cortical DA hypofunction, 187
DA hypothesis, 540–541
DA transmission alterations, 548, 549*f*
glutamate hypothesis, 541
hippocampus and, 592–593, 593*f*
integration of DA and glutamate hypothesis, 541
neural network hypothesis of, 532
preadolescent maturation of PFC circuits, 181–184, 182*f*–184*f*
prefrontal cortex dopaminergic mechanisms, 189–192, 191*f*, 193*f*
refractory form, 542
research, future, 567
risk factors, 593–594
signal transduction hypothesis of, 531*f*, 532
startle reflex in, 577
symptoms
cognitive, 540
cognitive deficits, 559–560
negative, 540, 559
positive, 513–514, 540, 559
treatment, 88
antipsychotic drugs for, 541–542, 547–548
Second-messenger systems
plasticity-related, 306
signaling cascades, 301
Selective serotonin reuptake inhibitors (SSRIs), 577–578, 582
Selegiline neuroprotectivity, in Parkinson's disease, 499, 499*t*
Sensation seeking, addiction and, 399–400
Sensorimotor tests, in genetic mouse models of parkinsonism, 281–283
challenging beam transversal, 282
dot test of somatosensory neglect, 282
nest building, 282–283
pole test and inverted grid, 282
spontaneous movement, 282
Sensory symptoms, in Parkinson's disease, 426, 426*t*
Serotonin 5HT$_{2A}$ receptor blockade, with D2 dopamine receptor blockade, 544–545

Serotonin neurons, 437
Serotonin receptors, 525
Serotonin transporter (SERT), 437
Sertraline, 577
SIB-1553A, 527
Sigma receptors, as atypical antipsychotic drug targets, 530
Signal-to-noise enhancing effect, 204
Signal transduction hypothesis of schizophrenia, 531*f*, 532
Sign tracking, 391
Single photon emission computed tomography (SPECT)
D2R binding, in major depressive disorder, 578, 579*t*, 580*t*
Parkinson's disease, 428
Sirtuin family of proteins (SIRTs), 502
SKF38393, modulation of pyramidal cell excitability, 179–180, 179*f*
SKF81297, 18*f*
SKF83566, effort-related choice behavior and, 290–291, 291*t*
Sleep behavior disorders, in Parkinson's disease, 425, 426, 426*t*
SNc (substantia nigra pars compacta), 42, 49, 100
S/N ratio, 267
Sodium ion channels
tetrodotoxin-sensitive voltage gated, 126–127, 126*f*
voltage-dependent, 353–354
Sonic hedgehog *(Shh)*, 144, 147–148
SOP (spontaneous oscillatory potential), 125–126, 126*f*
SPECT. *See* Single photon emission computed tomography
Spike-timing dependent plasticity (STDP), 269–270, 270*f*, 355–357, 356*f*, 357*f*
Spiperone, D2DAR antagonism, 71*t*
Spontaneous oscillatory potential (SOP), 125–126, 126*f*
SR48692, 530
SR141716, 529
SSR504734, 528
SSRIs (selective serotonin reuptake inhibitors), 577–578, 582
Startle reflex, 577–578
Stem cell generation, of DA neurons for transplantation, 485, 485*t*
Stimulants. *See* Psychostimulants
Stimulus-reward learning, teaching signal, 391
Stress
addiction relapse and, 413
addiction vulnerability and, 377–381, 380*f*, 381–383
dopamine function and, 576
proteolytic, in cell death process in Parkinson's disease, 502
as schizophrenia risk factors, 593–594
Striatal neuron projections, dopamine receptors and, 15–17
Striatal patch-matrix components, 11–12, 13*f*, 14–15, 14*f*

Striatonigral pathway (direct; external)
activation, 100
drug-seeking behavior and, 396–397, 396f
dual parallel and integrative processing, 45, 45f
Go signals, 437–438
MSNs, 6, 351
Striatopallidal pathway (indirect; internal)
activation, 100
MSNs, 6
NoGo signals, 437–438
Striatum
in appetitively motivated learning, 306–308, 307f
ascending dopamine pathways, 301–302
circuitry, 302
clinical disorders of, 336
compartmental interactions, 335
cortical innervation, synaptic organization of, 50–52, 52f
development, dopamine islands and, 334, 334f
dopamine-depleted, ERK 1/2 activation, 18
dopamine functions
augmenting, 301
selective-strengthening, 301
dopamine-glutamate interactions, 53
dopamine innervation, 49–50
principles of, 53–56, 54f, 55f
quantitative analysis, 53–56, 54f, 55f
synaptic organization, 50, 51f, 51t, 52f
synaptic organization of, 50, 51
dopamine release, 306
dopaminergic-cholinergic balance, 334, 334f
efferent pathway imbalances, 437–438
extrastriosomal matrix vulnerability, 336
functional neuronanatomy of dopamine in, 11–19
functional specificity, neurochemical systems of, 308–310, 309f
glutamatergic signaling, in Parkinson's disease, 349–363
inputs, 302
NMDAR activation, 307
organization, 302
outputs, 302
striosome vulnerability, 336
subregions, mapping dopamine effects in, 302–305, 304f
thalamic innervation, synaptic organization of, 52–53
Striosomes
circuit-level plasticity in striatal gene activation, 335–336
definition of, 334
outputs, 337
predominance, behavior and, 336
structure, 334, 335f
value processing, 336–337, 337f
Substance abuse. *See* Drug abuse
Substance dependence. *See* Addiction

Substance-induced mood disorder, diagnostic characteristics, 573f
Substance P, 302, 305
Substantia nigra pars compacta (SNc), 42, 49, 100
Subthalamic nucleus projections, convergence, 423, 424f
Sulpiride, D2DAR antagonism, 71t
Synapses, dopaminergic
in basal ganglia, 49–50
dopaminergic axons and, 55, 56t
excitatory synapses and, 54, 54f
functions, DA control of, 333
glutamatergic terminals, 54
plasticity (*See* Plasticity, synaptic)
in primate prefrontal cortex, 232
in striatal dopamine neurons, 50, 51f, 51t, 52f
in striatum, 50, 51t
cortical, 50–52, 52f, 55, 56t
at cortical terminals, 50–52
thalamic, 55, 56t
thalamic innervation, 52–53
TH-positive terminals, 51, 51t, 52t
Synaptic pruning determinants, 360
Synaptic tagging (eligibility trace), 270

Talnetanat (SB 223412), 530
TANs (tonically active neurons), 334
TaqIA A1 allele, 400
Taste reactivity, Acb, 310
TCAs (tricyclic antidepressants), efficacy in depression, 574
TCS (transcranial sonography), of nigral structures in Parkinson's disease, 445–446, 446f
TDE-L (temporal-difference-error learning), 269
Temporal-difference-error learning (TDE-L), 269
Δ9-Tetrahydrocannabinol (Δ9-THC), 529
TEXAN (toxin-extruding antiporter gene family), 89
TFs. *See* Transcription factors
Tgfs. *See* Transforming growth factors
Thalamostriatal synapses, plasticity of, 54
TH enzyme, 491
Theophylline, 292
Th gene, 141
Thioridiazine, D2DAR antagonism, 71t
TH protein, DAT and, 91–92, 91f
T-maze choice task procedure, 290, 292, 293
Tolcapone, 567
Tonically active neurons (TANs), 334
Tourette's syndrome, 80
Toxin-extruding antiporter gene family (TEXAN), 89
Transcranial magnetic stimulation, 574
Transcranial sonography (TCS), of nigral structures in Parkinson's disease, 445–446, 446f
Transcription factors (TFs)

mdDA-specific characteristics of mdDA precursors and, 150–152
neuronal properties of mdDA precursors and, 148–150
Transforming growth factors (Tgfs)
differentiation of mdDA precursors, 153
regulation of NCD in mdDA neurons, 170–171
Transgenic mice, 6, 18
Tricyclic antidepressants (TCA), efficacy in depression, 574

Ubiquitin proteasome system (UPS), 502
Unconditioned behavior, Acb shell and, 305
Unified Parkinson's Disease Rating Scale (UPDRS), 480, 481t, 500
Unipolar depression. *See* Major depressive disorder
UPDRS (Unified Parkinson's Disease Rating Scale), 480, 481t, 500
UPS (ubiquitin proteasome system), 502

Velocardiofacial syndrome, 189
Ventral medial prefrontal cortex (vmPFC), 39, 43
Ventral pallidum
in effort-related processes, 293
GABA receptor stimulation, 293
Ventral striatum, 39
Ventral subiculum regulation, DA neuron activity, 592
Ventral tegmental area (VTA)
DA-depleting lesions of, 308
DA neurons, 25, 42, 49
drug-induced LTP, 393–394
glutamatergic inputs, 339
inactivation, 345
LTP in, 339–340
modulation of dopamine cells, 93
reward/reinforcement and, 38
stress-induced activation, 576
synaptic plasticity in, 340
Ventrolateral prefrontal cortex (VLPFC), 192, 193f
Ventrolateral striatum (VLS), d-amphetamine microinfusion mapping study, 303–305, 304f
Vesicular glutamate transporter type 2 (VGlut2), 26
Vesicular monoamine transporter 1 (VMAT1), 88–90
Vesicular monoamine transporter 2 (VMAT2)
function, 88, 94–95
control of neuronal DA storage/release, 89–90
prevention of MPP+ toxicity, 89–90
in substance abusers, PET findings, 409–410, 409t
VGlut2 (vesicular glutamate transporter type 2), 26

Visual hallucinations, in Parkinson's disease, 424, 425–426
VLPFC (ventrolateral prefrontal cortex), 192, 193f
VLS (ventrolateral striatum), d-amphetamine microinfusion mapping study, 303–305, 304f
VMAT1 (vesicular monoamine transporter 1), 88–90
VMAT2. *See* Vesicular monoamine transporter 2
VM cell transplantation, 462, 463
   bridge grafts, 468
   DA cell survival, 467–468
   mechanisms of recovery, 466–468, 466t
   nonspecific surgical effects, 466–467, 466t
   pharmacologic effects, 466t, 467
   trophic effects, 466t, 467
vm-PFC (ventral medial prefrontal cortex), 39, 43

Volume transmission, 4, 55
VTA. *See* Ventral tegmental area
VTA/SN neurons, 268

WCST (Wisconsin Card Sorting Test), 210, 560, 560f
Weaver mouse model, 133
Wisconsin Card Sort Test (WCST), 210, 560, 560f
Wnt1, 148–149
Wnt5a, 153
Wnt protein, mdDA neuron development and, 143f, 145, 147–148, 153
Working memory
   cerebral blood flow and, 209
   DA neurotransmission in PFC
     primate studies, 215–216
     rodent studies, 216–221, 217f
   D2 dopamine receptor and, 238–239

D1 dopamine receptor modulation of, 264, 265f, 266–267, 266f
   dopamine agonist and, 208, 208f
   in prefrontal cortex, signal-to-noise processing, 188–189
   prefrontal cortex and, 188, 230
   spatial, D1 dopamine receptor and, 235, 236f
   testing, ODR task for, 230–231, 231f

Xanomeline, 526, 527t
X-linked dystonia parkinsonism, 336

Yerkes-Dodson principle
   behavioral or cognitive output, DA and, 208–211, 211f
   description of, 203–204, 204f
   dose-behavior relationship, 205, 205f

Zif268, 308